IET TELECOMMUNICATIONS SERIES 96

Edge Caching for Mobile Networks

Other volumes in this series:

Volume 9 **Phase noise in signal sources** W.P. Robins
Volume 12 **Spread spectrum in communications** R. Skaug and J.F. Hjelmstad
Volume 13 **Advanced signal processing** D.J. Creasey (Editor)
Volume 19 **Telecommunications traffic, tariffs and costs** R.E. Farr
Volume 20 **An introduction to satellite communications** D.I. Dalgleish
Volume 26 **Common-channel signalling** R.J. Manterfield
Volume 28 **Very small aperture terminals (VSATs)** J.L. Everett (Editor)
Volume 29 **ATM: the broadband telecommunications solution** L.G. Cuthbert and J.C. Sapanel
Volume 31 **Data communications and networks, 3rd edition** R.L. Brewster (Editor)
Volume 32 **Analogue optical fibre communications** B. Wilson, Z. Ghassemlooy and I.Z. Darwazeh (Editors)
Volume 33 **Modern personal radio systems** R.C.V. Macario (Editor)
Volume 34 **Digital broadcasting** P. Dambacher
Volume 35 **Principles of performance engineering for telecommunication and information systems** M. Ghanbari, C.J. Hughes, M.C. Sinclair and J.P. Eade
Volume 36 **Telecommunication networks, 2nd edition** J.E. Flood (Editor)
Volume 37 **Optical communication receiver design** S.B. Alexander
Volume 38 **Satellite communication systems, 3rd edition** B.G. Evans (Editor)
Volume 40 **Spread spectrum in mobile communication** O. Berg, T. Berg, J.F. Hjelmstad, S. Haavik and R. Skaug
Volume 41 **World telecommunications economics** J.J. Wheatley
Volume 43 **Telecommunications signalling** R.J. Manterfield
Volume 44 **Digital signal filtering, analysis and restoration** J. Jan
Volume 45 **Radio spectrum management, 2nd edition** D.J. Withers
Volume 46 **Intelligent networks: principles and applications** J.R. Anderson
Volume 47 **Local access network technologies** P. France
Volume 48 **Telecommunications quality of service management** A.P. Oodan (Editor)
Volume 49 **Standard codecs: image compression to advanced video coding** M. Ghanbari
Volume 50 **Telecommunications regulation** J. Buckley
Volume 51 **Security for mobility** C. Mitchell (Editor)
Volume 52 **Understanding telecommunications networks** A. Valdar
Volume 53 **Video compression systems: from first principles to concatenated codecs** A. Bock
Volume 54 **Standard Codecs: image compression to advanced video coding, 3rd edition** M. Ghanbari
Volume 59 **Dynamic Ad Hoc Networks** H. Rashvand and H. Chao (Editors)
Volume 60 **Understanding Telecommunications Business** A Valdar and I Morfett
Volume 65 **Advances in Body-Centric Wireless Communication: Applications and State-of-the-art** Q. H. Abbasi, M. U. Rehman, K. Qaraqe and A. Alomainy (Editors)
Volume 67 **Managing the Internet of Things: Architectures, Theories and Applications** J. Huang and K. Hua (Editors)
Volume 68 **Advanced Relay Technologies in Next Generation Wireless Communications** I. Krikidis and G. Zheng
Volume 69 **5G Wireless Technologies** A. Alexiou (Editor)
Volume 70 **Cloud and Fog Computing in 5G Mobile Networks** E. Markakis, G. Mastorakis, C. X. Mavromoustakis and E. Pallis (Editors)
Volume 71 **Understanding Telecommunications Networks, 2nd edition** A. Valdar
Volume 72 **Introduction to Digital Wireless Communications** Hong-Chuan Yang
Volume 73 **Network as a service for next generation internet** Q. Duan and S. Wang (Editors)
Volume 74 **Access, Fronthaul and Backhaul Networks for 5G & Beyond** M. A. Imran, S. A. R. Zaidi and M. Z. Shakir (Editors)
Volume 76 **Trusted Communications with Physical Layer Security for 5G and Beyond** T. Q. Duong, X. Zhou and H. V. Poor (Editors)
Volume 77 **Network Design, Modelling and Performance Evaluation** Q. Vien
Volume 78 **Principles and Applications of Free Space Optical Communications** A. K. Majumdar, Z. Ghassemlooy, A. A. B. Raj (Editors)
Volume 79 **Satellite Communications in the 5G Era** S. K. Sharma, S. Chatzinotas and D. Arapoglou
Volume 80 **Transceiver and System Design for Digital Communications 5th Edition** Scott R. Bullock
Volume 81 **Applications of Machine Learning in Wireless Communications** R. He and Z. Ding (Editors)
Volume 83 **Microstrip and Printed Antenna Design 3rd Edition** R. Bancroft
Volume 84 **Low Electromagnetic Emission Wireless Network Technologies: 5G and beyond** M.A. Imran, F. Héliot and Y.A. Sambo (Editors)
Volume 86 **Advances in Communications Satellite Systems Proceedings of The 36th International Communications Satellite Systems Conference (ICSSC-2018)** I. Otung, T. Butash and P. Garland (Editors)
Volume 87 **Real Time Convex Optimisation for 5G Networks and Beyond** T.Q. Duong, L. D. Nguyen and H. D. Tuan
Volume 89 **Information and Communication Technologies for Humanitarian Services** M. N. Islam (Editor)
Volume 90 **Communication Technologies for Networked Smart Cities** S.K. Sharma, N. Jayakody, S. Chatzinotas and A. Anpalagan (Editors)
Volume 91 **Green Communications for Energy-Efficient Wireless Systems and Networks** Himal Asanga Suraweera, Jing Yang, Alessio Zappone and John S. Thompson (Editors)

Volume 92 **Flexible and Cognitive Radio Access Technologies for 5G and Beyond** H. Arslan and E. Başar (Editors)
Volume 93 **Antennas and Propagation for 5G and Beyond** Q. Abbasi, S.F. Jilani, A. Alomainy and M. A. Imran (Editors)
Volume 94 **Intelligent Wireless Communications** G. Mastorakis, C. X. Mavromoustakis, J. M. Batalla and E. Pallis (Editors)
Volume 95 **ISDN applications in education and training** R. Mason and P.D. Bacsich
Volume 98 **Artificial Intelligence Applied to Satellite-based Remote Sensing Data for Earth Observation** M. P. Del Rosso, A. Sebastianelli and S.L. Ullo (Editors)

Edge Caching for Mobile Networks

Edited by
Wei Chen and H. Vincent Poor

The Institution of Engineering and Technology

Published by The Institution of Engineering and Technology, London, United Kingdom

The Institution of Engineering and Technology is registered as a Charity in England & Wales (no. 211014) and Scotland (no. SC038698).

First published 2021

The Institution of Engineering and Technology
Michael Faraday House
Six Hills Way, Stevenage
Herts, SG1 2AY, United Kingdom

www.theiet.org

British Library Cataloguing in Publication Data
A catalogue record for this product is available from the British Library

ISBN 978-1-83953-122-4 (hardback)
ISBN 978-1-83953-123-1 (PDF)

Typeset in India by Exeter Premedia Services Private Limited
Printed in the UK by CPI Group (UK) Ltd, Croydon

Contents

About the Editors xxi

1 Introduction 1
Wei Chen and H. Vincent Poor

1.1 Fundamental limits of caching: An information-theoretic perspective 2
1.1.1 Coded caching for error-free channels 2
1.1.2 Caching over wireless channels 3
1.1.3 Data compression and secrecy preservation for caching 4
1.2 Caching in large mobile networks: Asymptotic analysis 4
1.2.1 Scaling laws for mobile networks without fixed infrastructure 5
1.2.2 Scaling laws for caching with hierarchical cooperation 6
1.2.3 Stochastic geometry approaches for caching 6
1.3 Resource allocation for caching: A game-theoretic perspective 7
1.3.1 Dynamic pricing for caching 7
1.3.2 Auction approaches for caching 8
1.3.3 Contract theory for caching 8
1.3.4 Mean-field game theory for caching 9
1.4 Caching with time-varying popularity and recommendation systems 10
1.4.1 Popularity dynamics in the time and spatial domain 10
1.4.2 Joint caching and recommendation 11
1.5 Some applications of caching in mobile networks 12
1.5.1 Caching for mobile social networks 12
1.5.2 Caching for CRANs, UAVs, and MEC 13
1.6 Organization of the book 13
References 14

2 Physical layer schemes for coded caching 17
Giuseppe Caire, Mari Kobayashi, and Sheng Yang

2.1 Introduction 17
2.2 Single-cell network 18
2.2.1 System model 18
2.2.2 Scalable content delivery with wireless multicasting 20

2.2.3 Extensions 25
2.3 Multi-cell network 28
2.3.1 System model 29
2.3.2 Performance analysis 31
2.3.3 Numerical results 38
References 40

3. Cache-aided content delivery over noisy broadcast channels **43**
Mohammad Mohammadi Amiri and Deniz Gündüz

3.1 Error-free BC 44
3.1.1 MAN scheme 44
3.1.2 Generalization of the MAN scheme 47
3.2 Erasure BC 49
3.2.1 Problem formulation 49
3.2.2 Preliminaries 51
3.2.3 Achievable scheme 52
3.2.4 Achievable rate-memory pair 55
3.2.5 Upper bound 58
3.3 Gaussian BC 59
3.3.1 Problem formulation 59
3.3.2 Preliminaries 60
3.3.3 Centralized cache-aided broadcasting 61
3.3.4 Decentralized cache-aided broadcasting 66
3.3.5 Lower bound 68
3.4 Numerical experiments 70
3.5 Summary 72
References 72

4 Multi-antenna cache-aided channels **75**
Eleftherios Lampiris and Petros Elia

4.1 Abstract 75
4.2 The subpacketization bottleneck of coded caching 76
4.2.1 Coded caching with unbounded file sizes 76
4.2.2 Exponential subpacketization of coded caching 76
4.2.3 Algorithmic improvements for reducing subpacketization 78
4.2.4 Subpacketization bottleneck of separable multiple antenna coded caching 79
4.3 Setting and notation 80
4.4 Using multiple antennas to dramatically alleviate the subpacketization bottleneck of coded caching 81
4.4.1 Impact and insight 81
4.4.2 Scheme description 82
4.4.3 Intuition 84
4.4.4 Extensions 84

4.5 The feedback bottleneck of multi-antenna systems 84
4.5.1 Feedback bottleneck of multi-antenna coded caching 85
4.6 Joint caching and multi-antenna delivery with low CSIT 85
4.6.1 Scheme description 88
4.6.2 Additional feedback-related open problems 92
4.7 Conclusions – Further impact of combining multiple transmitters with coded caching 93
4.7.1 From caching to distributed computing 94
References 94

5 Caching for interference networks: a separation architecture 99
Jad Hachem, Urs Niesen, and Suhas Diggavi

5.1 Introduction 99
5.2 Cache-aided interference networks 101
5.3 Performance limits 103
5.3.1 Scaling with normalized transmitter cache size $\mu_t \triangleq M_t/N$ 104
5.3.2 Scaling with normalized receiver cache size $\mu_r \triangleq M_r/N$ 105
5.3.3 Scaling with number of receivers K_r 105
5.3.4 Scaling with number of transmitter K_t 106
5.4 Separation architecture 106
5.4.1 Separating the physical and network layers 109
5.4.2 The network layer 110
5.4.3 The physical layer 110
5.4.4 Information-theoretic outer bounds 112
5.4.5 A note on the low-SNR regime 113
5.5 Related work and outlook 114
5.5.1 Related work 114
5.5.2 Outlook 115
References 115

6 Scalable delivery of correlated video content over cache-aided broadcast networks 119
Parisa Hassanzadeh, Antonia M. Tulino, Jaime Llorca, and Elza Erkip

6.1 Introduction 120
6.1.1 Edge caching 120
6.1.2 Turning memory into bandwidth 120
6.1.3 Exploiting content correlations 121
6.2 Delivering correlated content over cache-aided broadcast networks 122
6.2.1 Correlation-aware cache-aided coded multicasting 123
6.2.2 Distortion-memory trade-offs 126
6.3 Caching networks with correlated sources 128
6.3.1 Problem formulation 128
6.3.2 Peak and the average rate-memory functions 130

6.4 Caching networks with correlated sources – the static case 131
6.4.1 Gray-Wyner network 132
6.4.2 Gray-Wyner multiple-request CACM 133
6.4.3 Multiple-request CACM 135
6.4.4 GW-MR for two files and *K* receivers 138
6.5 Caching networks with correlated sources – the dynamic case 141
6.5.1 Dynamic correlation-aware scheme 142
6.5.2 Greedy group coloring 147
6.6 Illustration results and conclusions 147
References 149

7 Rate-distortion theory for caching 155
Shirin Saeedi Bidokhti, Roy Timo, and Michèle Wigger

7.1 Introduction 155
7.2 Rate-distortion functions 157
7.3 Description of the system model 158
7.4 Single-user systems 163
7.4.1 The RDM functions of the single-user system 163
7.4.2 Lower bounds, critical cache sizes, and common informations 169
7.5 Two-user systems 173
7.5.1 A genie-aided lower bound 174
7.5.2 Two users and one cache 175
7.5.3 Two users and two files 178
7.6 *K*-user systems with uncorrelated sources 179
7.6.1 Scalable source coding 180
7.6.2 The layered approach in caching 180
7.6.3 Optimal cache allocation and uncoded placement – linear delivery schemes 183
7.6.4 Cache allocation across users 183
7.6.5 Minimizing the average distortions 184
7.7 Summary and Conclusions 185
References 186

8 Secrecy of edge caching 189
Mohamed Nafea and Aylin Yener

8.1 Introduction 189
8.2 Coded caching with secure delivery 191
8.3 Private coded caching: secure caching 194
8.4 Combination network with security requirements 197
8.5 The caching broadcast channel with a wire and cache tapping adversary 199
8.5.1 System model 199
8.5.2 Main results 201

8.5.3 Proof of Theorem 9 203
8.5.4 Proof of Theorem 10 212
8.5.5 Proof of Theorem 11 215
8.6 Conclusion 217
References 217

9 Scaling laws for cache-aided device-to-device networks for wireless video 221
Ming-Chun Lee, Mingyue Ji, and Andreas F Molisch

9.1 Introduction 221
9.1.1 Motivation for D2D-based video distribution 221
9.1.2 D2D scaling laws without caching 223
9.2 System model 224
9.2.1 Popularity model 224
9.2.2 Network model 225
9.2.3 Target functions for scaling 225
9.3 Single-hop networks 227
9.3.1 Scaling law results 228
9.3.2 Numerical studies under realistic setups 232
9.4 Multihop networks 234
9.5 D2D coded caching 239
9.5.1 Single-hop-coded D2D caching 239
9.5.2 An improved coded D2D caching schemes 241
9.5.3 Extension to multihop coded D2D caching networks 242
9.6 Practical implementation considerations for D2D caching networks 245
9.6.1 D2D caching under realistic channel model 245
9.6.2 D2D caching in mobile networks 246
9.6.3 Complexity of D2D-coded caching 247
9.7 Conclusions and further readings 249
References 249

10 Caching increases the capacity of wireless networks 255
Li Qiu and Guohong Cao

10.1 Introduction 255
10.2 Related work 257
10.3 Preliminaries 258
10.3.1 Network model 258
10.3.2 Content access and Zipf distribution 258
10.3.3 Capacity 259
10.4 Capacity under uniform content popularity 260
10.4.1 Capacity upper bound 260
10.4.2 Capacity lower bound 261
10.4.3 Numerical results 262

10.5 Capacity under Zipf content popularity 263
10.5.1 Problem formulation 263
10.5.2 Capacity upper bound 265
10.5.3 Analytic results 269
10.5.4 Numerical results 272
10.6 Caching algorithms 275
10.6.1 Algorithm description 275
10.6.2 Evaluation results 276
10.7 Conclusions 278
References 278

11 Cache-induced hierarchical cooperation in wireless D2D networks 281
An Liu, Vincent Lau, and Giuseppe Caire

11.1 Introduction 281
11.1.1 Related work 282
11.1.2 Summary of the main results 285
11.1.3 Organization 287
11.2 System model 287
11.2.1 Wireless device-to-device caching networks 287
11.2.2 Wireless channel model 290
11.3 Preliminaries on hierarchical cooperation and classical multihop schemes 290
11.3.1 Throughput performance of hierarchical cooperation under uniform permutation traffic 291
11.3.2 Throughput performance of multihop scheme under uniform permutation traffic 294
11.3.3 Extension to per cluster uniform permutation traffic 295
11.4 Cache-induced hierarchical cooperation 296
11.4.1 Key components of the cache-induced hierarchical cooperation scheme 296
11.4.2 Hierarchical cache content placement 297
11.4.3 Capacitated graph for content delivery 298
11.5 Cache content placement optimization 302
11.5.1 Problem formulation 302
11.5.2 Low-complexity cache content placement algorithm 303
11.6 Throughput performance of cache-induced hierarchical cooperation 306
11.6.1 Throughput bounds under Zipf popularity distribution 307
11.6.2 Comparison with cache-assisted multihop scheme 308
11.7 Scaling laws in extended networks 310
11.7.1 System scaling regime 310
11.7.2 Throughput scaling laws of cache-induced hierarchical cooperation 311
11.7.3 Main converse results 314

11.7.4 Converse proof 314
11.8 Discussions on the scaling laws 317
11.8.1 Extension for $n \neq 4^M$ 317
11.8.2 Extension to coded-caching-enabled multicasting 318
11.8.3 Extension to more general PHY model 319
11.8.4 Limitation of the user request model 320
11.9 Conclusion 320
References 335

12 Wireless device-to-device caching networks with distributed MIMO and hierarchical cooperations 339
Jiajia Guo, Jinhong Yuan, and J Andrew Zhang

12.1 Introduction 339
12.2 Wireless caching networks 340
12.2.1 Caching in centralized networks 340
12.2.2 Caching in decentralized networks 341
12.2.3 Motivations 344
12.3 System model and problem formulation 344
12.3.1 Definition 1 (Caching placement strategy) 345
12.3.2 Definition 2 (Users' request) 345
12.3.3 Definition 3 (Transmission policy) 345
12.3.4 Definition 4 (Average aggregate throughput) 346
12.3.5 Definition 5 (Outage probability) 346
12.4 Achievable scaling law and throughput 347
12.5 An achievable scheme 349
12.5.1 Caching placement phase 349
12.5.2 Caching delivery phase 351
12.6 Performance of the proposed scheme 354
12.6.1 Aggregate throughput of the proposed scheme 354
12.6.2 Outage probability of the proposed scheme 357
12.6.3 Throughput scaling law of the proposed scheme 357
12.6.4 Optimized throughput under a target outage Probability 359
12.7 Numerical results 361
12.8 Conclusions 363
12.9 Appendix 364
12.9.1 Proof of Lemma 1 364
12.9.2 Proof of Theorem 3 364
References 367

13 Randomised geographic caching and its applications in wireless networks 371
Anastasios Giovanidis and Bartłomiej Błaszczyszyn

13.1 Introduction 371
13.1.1 The FemtoCaching content placement 372

13.1.2 The randomised approach 373
13.2 Random independent placement Policy 376
13.3 The typical user 380
13.3.1 Coverage number 380
13.3.2 Content catalog and popularity 383
13.4 Optimal randomised caching in cellular networks 383
13.4.1 Solution for [RandK1F2] 384
13.4.2 Problem solution [RandCache] 387
13.5 Treating unequal file sizes 394
13.5.1 Method: overflow constraint 396
13.5.2 [RandCache] with cache overflow constraint 396
13.6 Node mobility 397
13.7 Competition among content providers 399
13.8 Beyond independent caches 399
References 400

14 Pricing-aided proactive caching for mobile services **405**
John Tadrous, Atilla Eryilmaz, and Hesham El-Gamal

14.1 Summary 405
14.2 Introduction 406
14.3 Content provider network setup 408
14.3.1 Users' demand profiles: 409
14.3.2 Proactive caching decisions 410
14.3.3 Incurred cost 410
14.3.4 Revenue 411
14.4 The single-user scenario: decoupled users' demand 412
14.4.1 Single-user iterative pricing and proactive download 413
14.4.2 Performance analysis 414
14.4.3 Distributed implementation 416
14.4.4 Numerical results for the single-user scenario 416
14.5 The multi-user scenario: coupled demand 417
14.5.1 Multi-user iterative pricing and proactive download 418
14.5.2 Performance analysis 419
14.5.3 Distributed implementation 423
14.5.4 Numerical results for the multi-user scenario 424
14.6 Conclusion 427
14.7 Proofs of main results 428
14.7.1 Proof of Theorem 2 428
14.7.2 Proof of Theorem 3 429
14.7.3 Proof of Theorem 4 430
14.7.4 Proof of Theorem 5 431
14.7.5 Proof of Theorem 6 431
14.7.6 Proof of Theorem 7 435
14.7.7 Proof of Theorem 8 435
References 436

15 Mobile edge caching: an optimal auction approach 439
Xuanyu Cao, Junshan Zhang, and H. Vincent Poor

15.1 Introduction 440
15.2 System model and problem formulation 441
15.2.1 System model 442
15.2.2 Problem formulation for auction mechanism design 443
15.3 Optimal auction design 444
15.3.1 Optimal auction with fixed content delivery quality 445
15.3.2 Optimal determination of content delivery quality θ 452
15.4 Numerical results 453
15.5 Conclusion 458
References 460

16 Contract theory-based approach for caching 463
Tingting Liu, Jun Li, Feng Shu, and Zhu Han

16.1 Introduction 463
16.1.1 Related works 464
16.1.2 Small-cell caching issues 465
16.2 System model 465
16.2.1 Network model 466
16.2.2 Popularity and preferences 467
16.2.3 Caching procedures 468
16.3 Contract-based service model 469
16.3.1 NSP model 469
16.3.2 CP model 470
16.4 Optimal contract design and solution 471
16.4.1 Contract formulation 471
16.4.2 Constraints reduction 472
16.4.3 Solution to optimal contract 473
16.5 Discussions on other schemes 478
16.5.1 Uniform quality scheme 478
16.5.2 Stackelberg game scheme 478
16.5.3 Discussions on the Stackelberg game scheme and the contract scheme 480
16.5.4 Discussions on auction scheme 480
16.5.5 Discussions on a heterogeneous system 480
16.6 Numerical results 480
16.6.1 Discussions on R_v and $\hat{y}(v)$ 481
16.6.2 Impact of γ on optimal price and quality 481
16.6.3 Comparisons of NSPs' and CPs' profits achieved by the three different schemes 483
16.6.4 Performance of social welfare among three schemes 486
16.7 Conclusions 487
References 488

17 Mean-field game-theoretic edge caching **493**
Hyesung Kim, Jihong Park, Mehdi Bennis, Seong-Lyun Kim, and Mérouane Debbah

17.1 Introduction 493
17.2 Related works 494
17.3 System model 495
17.3.1 Network, channel, and caching model 495
17.3.2 Spatio-temporal dynamics of demand, caching, and interference 496
17.4 Game-theoretic formulation for edge caching 499
17.4.1 Cost functions 500
17.4.2 Interactions through content overlap and interference 500
17.4.3 Stochastic differential game for edge caching 502
17.4.4 Mean-field game for edge caching 503
17.5 Numerical results 508
17.5.1 Mean-field equilibrium achieved by the proposed MF caching algorithm 508
17.5.2 Performance evaluation in terms of long run average caching cost 509
17.6 Conclusion 514
References 514

18 Caching with time-varying popularity profiles **519**
B N Bharath, K G Nagananda, D Gündùz, and H V Poor

18.1 Introduction 519
18.2 User request model 522
18.3 Problem statement 523
18.4 Main results 525
18.5 Bernoulli and Poisson requests 529
18.5.1 Bernoulli request model 529
18.5.2 Poisson request model 530
18.6 Cache update algorithm and numerical results 531
18.7 Concluding remarks 533
References 534

19 Reinforcement learning for caching with space-time popularity dynamics **537**
Alireza Sadeghi, Georgios B Giannakis, Gang Wang, and Fatemeh Sheikholeslami

19.1 Introduction 537
19.2 Single-node caching: Modeling and problem statement 539
19.2.1 Costs and caching strategies 540
19.2.2 Markovian popularity profile evolution 541

19.2.3 Reinforcement learning formulation 543
19.2.4 Optimality conditions 543
19.2.5 Optimal caching via Q-learning 544
19.2.6 Scalable caching 546
19.2.7 Learning Λ 547
19.2.8 Numerical tests 549
19.3 Network caching with space-time popularity dynamics 551
19.3.1 Modeling and problem statement 552
19.3.2 Two-timescale problem formulation 553
19.3.3 Deep reinforcement learning-based caching 554
19.3.4 Numerical tests 557
19.4 Discussion 560
References 560

20 Push-based wireless converged networks with user caching 565
Zhiyong Chen, Kongtao Wang, and Hui Liu

20.1 Modeling of the converged network 566
20.1.1 System model 566
20.1.2 Characterization of a multimedia content 567
20.1.3 Traffic offloading effect in the converged network 568
20.2 Characterization of traffic offloading 568
20.2.1 Size of M′ 568
20.2.2 Traffic offloading effects 569
20.3 Performance analysis of the converged network 570
20.3.1 The capacity improvement of the converged network 570
20.3.2 The average downloading time $\mathbb{E}\left[T_w\right]$ 571
20.3.3 System energy consumption 572
20.3.4 Extreme case analysis for the converged network 573
20.4 Numerical results 574
20.4.1 Impact of the network resources 574
20.4.2 Impact of the content characteristics 575
20.4.3 Impact of the user request traffics 577
20.4.4 User request classification 577
20.5 Conclusions 579
References 579

21 On the joint optimization of content caching and recommendations 581
Livia Elena Chatzieleftheriou, Merkourios Karaliopoulos, and Iordanis Koutsopoulos

21.1 Introduction 582
21.2 System model 584
21.2.1 Caches, content, users 585
21.2.2 Content preferences of users 586

	21.2.3	The impact of recommendations on user content requests	587
	21.2.4	Engineering the user recommendations	588
21.3	The joint caching and recommendations problem		590
	21.3.1	Problem complexity and approximability properties	591
21.4	An algorithm for the joint caching and recommendation problem		592
	21.4.1	Description of the algorithm	592
	21.4.2	Complexity of the CawR algorithm	594
21.5	Properties and performance bounds of our algorithm		595
	21.5.1	Benchmark recommender schemes	595
	21.5.2	Sensitivity analysis to parameters of CawR	596
21.6	Experimental evaluation of CawR		597
	21.6.1	Datasets and default parameter settings	597
	21.6.2	Trace-Driven simulations	598
	21.6.3	Simulations with synthetic datasets	600
21.7	User associations as additional control parameter: extending the model		601
	21.7.1	Model extension	602
	21.7.2	The Joint Caching, Recommendations and Association problem	603
	21.7.3	Heuristic algorithms for the JCRA problem	604
21.8	Open directions		606
21.9	Related work		607
21.10	Conclusions		610
Funding			611
References			611

22 Caching with recommendation and reinforcement learning **615**
Dong Liu and Chenyang Yang

22.1	Introduction		615
22.2	System model		617
	22.2.1	Basic terminology and concepts in RL	618
	22.2.2	Learning-enabled network architecture	618
	22.2.3	Recommendation, pushing, and user requests	619
22.3	Recommendation and pushing by DRL		621
	22.3.1	Problem formulation	621
	22.3.2	RL framework for joint optimization	622
	22.3.3	Decomposed RL framework	624
	22.3.4	DRL with dueling DDQN	627
22.4	Simulation results		631
	22.4.1	Simulation environment setup	631
	22.4.2	Fine-tuned hyper-parameters of Algorithm 1	633
	22.4.3	Comparing with policy-based DRL algorithms	634

22.4.4 Comparing with baseline policies 635
22.4.5 Behavior of the pushing policy 637
22.5 Conclusion and future works 639
References 639

23 Social networking-driven caching **643**
Kuo Chun Tsai, Li Wang, Huaqing Wu, Lianming Xu, and Zhu Han

23.1 Introduction 643
23.1.1 Concepts of caching 643
23.1.2 Socially aware caching networks 645
23.1.3 Organization 646
23.2 Exploit socially enabled D2D communications in caching 646
23.2.1 Social characteristics consideration 646
23.2.2 Socially enabled D2D caching with hypergraph 648
23.2.3 Socially aware caching scheme 650
23.3 D2X-based caching with deep learning 655
23.3.1 System model 658
23.3.2 Problem formulation 659
23.3.3 Mobile social media natural language processing using deep learning 661
23.3.4 Proposed algorithm 664
23.3.5 Performance evaluation 667
23.4 Conclusion 670
References 671

24 Caching for secure social networks **675**
F. Richard Yu and Tiffany Ying He

24.1 Social networks with caching, computing, and communications 675
24.1.1 In-network caching 676
24.1.2 Mobile edge computing 676
24.1.3 D2D communications 677
24.1.4 Integration of in-network caching, MEC, and communications 677
24.2 Social trust scheme with uncertain reasoning 678
24.2.1 Trust evaluation from direct observations 679
24.2.2 Trust evaluation from indirect observations 681
24.3 System model with 3C framework 683
24.3.1 Network model 683
24.3.2 Communication model 683
24.3.3 Cache model 684
24.3.4 Computing model 685
24.4 Formulation as a reinforcement learning problem 686
24.4.1 Deep Q-learning 686
24.4.2 System state 688

24.4.3 System action 688
24.4.4 Reward function 689
24.5 Simulation results and discussions 691
24.5.1 Simulation settings 691
24.5.2 Simulation results 692
24.6 Chapter conclusion 694
References 695

25 Caching for CRAN and UAVs **699**
Mingzhe Chen, Zhaohui Yang, Walid Saad, Changchuan Yin, and Shuguang Cui

25.1 Introduction 699
25.2 Echo state networks for proactive caching 700
25.2.1 Basic problem 700
25.2.2 Echo state networks for content prediction and mobility 705
25.2.3 Sublinear algorithm for clustering and caching 708
25.2.4 Simulation results 711
25.3 Echo state networks for cache-enabled UAVs 713
25.3.1 Optimal location and content caching for UAVs 719
25.3.2 Simulation results 722
25.4 Summary 724
References 725

26 Dynamic computation offloading and resource allocation in cache-assisted mobile edge computing systems **729**
Samrat Nath and Jingxian Wu

26.1 Introduction 729
26.2 System model and problem formulation 732
26.2.1 Task model 733
26.2.2 Caching model 734
26.2.3 Communication model 735
26.2.4 Computation model 735
26.2.5 Problem formulation 737
26.3 DRL-based solution for dynamic caching, computation offloading, and resource allocation 738
26.3.1 RL framework 738
26.3.2 DRL-based solution with DDPG 740
26.4 Simulation results 743
26.5 Conclusion 748
References 748

Index **751**

About the Editors

Wei Chen is a tenured Full Professor and a university council member at Tsinghua University as well as a standing committee member of the All-China Youth Federation and the secretary-general of its education board. He has been supported by the national program for special support for eminent professionals and the Cheung Kong young scholar program. He received the IEEE Marconi Prize Paper Award, the IEEE Comsoc Asia Pacific Board Best Young Researcher Award, the National May 1st Labor Medal, and the China Youth May 4th Medal. He received his Ph.D. Degree with the highest honor from Tsinghua University, China.

H. Vincent Poor is the Michael Henry Strater University Professor of Electrical Engineering and former Dean of the School of Engineering and Applied Science at Princeton University, USA. His research interests are in the areas of information theory, machine learning and network science, and their applications in wireless communications, energy systems and related fields. A Fellow of the IEEE and the IET, Vince is also a member of the U. S. National Academy of Engineering and the U. S. National Academy of Sciences, a Foreign Member of the Royal Society and an International Fellow of the Royal Academy of Engineering. He received a Guggenheim Fellowship in 2002, and the IEEE Education Medal in 2005. Recent recognition of his work includes the 2017 IEEE Alexander Graham Bell Medal and the 2019 ASEE Benjamin Garver Lamme Award. He holds a Ph.D. Degree in EECS from Princeton University.

Chapter 1
Introduction

Wei Chen[1] and H. Vincent Poor[2]

Caching has attracted significant attention recently because it holds the promise of scaling the service capability of mobile networks. Modern mobile networks are capable of achieving data rates of Gbps, yet they may still fail to meet the predicted bandwidth requirements of future mobile applications. A recent report from Cisco [1] forecasts that mobile data traffic will grow to 77.49 EB per month in 2022. In theory, a human brain may process up to 100 T bits per second. Even though this does not necessarily reflect the ultimate bandwidth required by a human being, a huge gap between bandwidth demand and provisioning may still exist in the near future. Unfortunately, on-demand transmission which dominates current mobile network architectures has almost achieved its performance limits revealed by Shannon in 1948. Because of this and other limitations it is time to envision paradigm-shifting mobile network architectures for the sixth generation (6G) of mobile networks [2]. Cache-aided mobile networking is regarded as one of the potential solutions for beyond fifth generation (5G) systems.

Caching techniques were originally developed for computer systems in the 1960s. Web caching was conceived for the Internet due to the explosively increasing number of websites in the 2000s. In contrast to on-demand transmission, caching allows proactive content placement before being requested, which has motivated some novel infrastructures such as information-centric networks and content delivery networks. More recently, caching has been found to substantially benefit data transmissions over harsh wireless channels and meet growing demands with restrained radio resources. The key purpose of this book is to introduce the cutting-edge theory and techniques in the area of edge caching for mobile networks.

In this chapter, we shall present a brief overview of the technical content of this book. In particular, this book mainly covers five major topics belonging to edge caching for mobile networks. They are fundamental limits of caching, asymptotic analysis of caching in large mobile networks, resource allocation for edge caching, caching with time-varying popularity and recommendation systems, and some typical applications of edge caching.

[1]Tsinghua University, Beijing, China
[2]Princeton University, New Jersey, United States

1.1 Fundamental limits of caching: An information-theoretic perspective

Since 1948, information theory has been widely recognized as a powerful tool to reveal the fundamental limits of mobile networks. More specifically, for caching-aided mobile networks, it has been demonstrated that judiciously designed coding schemes may allow caching to attain substantial bandwidth gain, thereby making mobile networks scalable in terms of their service capabilities. The state-of-the-art information-theoretic analysis of caching mainly focuses on a simple but typical topology in which multiple users are connected to a common base-station through a shared link. Its core idea is that coding helps to attain the multicasting gain so as to satisfy the bandwidth requirement in peak-traffic hours. The investigation of the fundamental limits of caching over various channels has significantly broadened the area of modern information theory. In this section, we shall review the information-theoretic analysis of caching, which is discussed in much more detail in Chapters 2–8 of this book.

1.1.1 Coded caching for error-free channels

Coded caching enables multicasting in the delivery phase even when users ask for different files, thereby holding the potential of scaling a mobile network's service capability. Coded caching was first investigated in an ideal shared-link network, in which all users are connected to a common content server by the same error-free broadcast channel, as shown in Figure 1.1. Consider a coded caching scenario with only two users, U_1 and U_2, who are interested in either file A or B. We split each file into two sub-files of size 1/2, i.e., $\mathsf{A} = \{\mathsf{A}_1, \mathsf{A}_2\}$ and $\mathsf{B} = \{\mathsf{B}_1, \mathsf{B}_2\}$ and let U_1 cache $\{\mathsf{A}_1, \mathsf{B}_1\}$, and U_2 cache $\{\mathsf{A}_2, \mathsf{B}_2\}$ in the placement phase [3]. The files are placed and cached in the off-peak time, in which the bandwidth cost is low. In the delivery

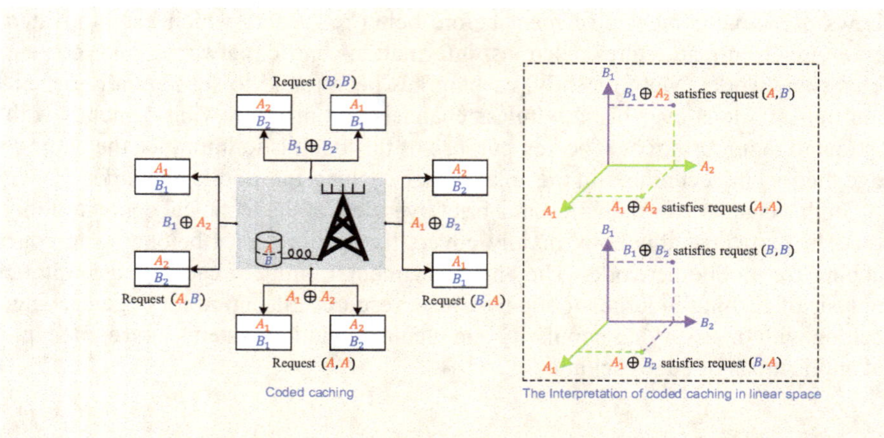

Figure 1.1 Coded caching

phase, there could be potentially four different request patterns. If both U_1 and U_2 ask for A, then the server transmits the bit-by-bit XOR output of A_1 and A_2. If both U_1 and U_2 ask for B, then the server transmits the bit-by-bit XOR output of B_1 and B_2. If U_1 asks for A while U_2 asks for B, then the server transmits the bit-by-bit XOR output of A_2 and B_1. If U_1 asks for B while U_2 asks for A, then the server transmits the bit-by-bit XOR output of A_1 and B_2. It is not difficult to see that both users' demands can be satisfied by sending only half of a file in the delivery phase, no matter what they ask for. By this means, the bandwidth demand in the peak-traffic hours can be substantially reduced. The effective throughput of coded caching with arbitrary user preference is presented in [4].

In error-free networks, coded caching can be characterized by the assignment of a vector space [5], which is also adopted to characterize linear network coding [6]. Most research in this area of the past decade has focused on the tradeoff between the caching gain and storage cost of error-free networks. However, the error-free channel model might be impractical due to the existence of noise, fading, and mutual interference in real-world mobile networks. As a result, Chapters 2–8 are devoted to the information-theoretic analysis of edge caching with physical-layer constraints, compressible sources, and secrecy demands.

1.1.2 Caching over wireless channels

The design of caching schemes for practical mobile networks becomes much more challenging than that for ideal shared-link networks. In particular, the performance limits along with the approaching scheme relies on the specified physical-layer model. In this book, four chapters are devoted to the information-theoretic analysis of caching over practical wireless channels. In particular, Chapters 2 and 3 focus on caching over noisy broadcast channels. The Maddah-Ali and Niesen (MAN) scheme proposed in [3] is introduced and then generalized to various broadcast channels. These are time-varying Gaussian broadcast channel and erasure broadcast channel, which are capable of characterizing a single-cell wireless access system. The MAN scheme for multi-cell scenarios is further discussed as an extension.

Since the fourth generation (4G) mobile communications was deployed, multi-antenna techniques have been extensively adopted in Radio Access Networks (RANs). As a result, there is a need to introduce the information-theoretic bounds for coded caching for multi-antenna wireless communication systems. Therefore, Chapter 4 is devoted to this topic. In particular, this chapter demonstrates that the joint design of caching and multi-antenna transmission may bring a powerful performance boost. The Degree of Freedom (DoF) analysis plays a key role in this chapter, which provides both theoretical and engineering insights. Finally, when multiple transceiver pairs share common spectrum simultaneously, mutual interference may become the bottleneck that limits the overall spectral efficiency of mobile networks. As a result, how to efficiently perform caching over interference networks is of practical importance and has attracted considerable recent attentions. Chapter 5 presents a separation architecture for cache-aided interference networks. The scaling of DoF gains of various caching schemes are derived for interference channels. The chapter also demonstrates that a separation architecture that layers coded caching on top of channel coding schemes is approximately optimal, thereby reducing the complexity in both algorithm design and potential hardware implementation.

1.1.3 Data compression and secrecy preservation for caching

In an information-theoretic framework, data compression or source coding has been extensively investigated due to its potential to reduce the bandwidth demand while assuring the quality or resolution of content files. A number of lossless and lossy compression schemes have been investigated for single source and correlated source scenarios in information theory. In cache-aided broadcast networks, there is a need to revisit the data compression theory and algorithms from an information-theoretic perspective. For instance, novel lossless and lossy edge data compression can further save the bandwidth requirement in mobile networks by exploiting user preferences [7, 8]. In this book, Chapters 6 and 7 are devoted to the discussion of typical lossless and lossy data compression for edge caching, respectively.

In Chapter 6, lossless data compression for caching is considered. In particular, the correlated content items distributed throughout a cache-aided network are jointly compressed to provide additional gains for bandwidth saving. The server is assumed to access uniformly popular content files generated by a discrete memoryless source. This chapter discusses both static and dynamic settings, in which the content library can be either fixed or updated during the placement or delivery phase. More particularly, the Gray-Wyner coding scheme is adopted to optimize the rate-memory trade off of caching. In Chapter 7, lossy data compression for caching is discussed. The rate-distortion functions of distortion-tolerant sources in caching-aided networks are investigated to characterize the minimum data rate needed to reproduce a content file while meeting a target distortion under a given distortion measure. More specifically, rate-distortion functions of single-user and two-user systems as well as K-user systems with uncorrelated sources are studied in Chapter 7.

In addition to the reliable transmission and data compression that are regarded as the most classical issues of information theory, security preserving mechanisms have attracted considerable attention from information theorists since information theory was established. Cache-aided mobile networks, no doubt, are vulnerable to various types of security threats. As a result, Chapter 8 is devoted to typical confidentiality concerns and adversarial attacks in cache-aided mobile networks. More particularly, rate-memory tradeoff functions for secure or private-coded caching in broadcast channels and two-hop layered network topologies are, respectively, conceived based on the assumption that the cache placement occurs over a secure channel. Cache-tapping is further introduced to the information-theoretic model of coded caching, in which the secrecy capacity is derived.

1.2 Caching in large mobile networks: Asymptotic analysis

Conventional network information theory focuses on the capacity region analysis of typical multi-terminal communication systems, including the broadcast channel, multi-access channel, relay channel, interference channel, etc. Hence it is also referred to as multi-terminal information theory. In practice, the size of mobile networks can be very large, while its topology can be extremely complicated. For large-scale mobile networks, accurate capacity regions can hardly be obtained by conventional information-theoretic analysis. To deal with this, Gupta and Kumar presented in their milestone work [9] an asymptotic analysis approach to study the scalability of mobile networks as the density or the number of users increases. On the other hand, a stochastic geometry approach was presented to reveal the performance limit of large-scale mobile networks in [10]. Both approaches have now been generalized to investigate caching gains in large-scale mobile networks. In this section, we shall briefly review whether the benefits of caching remain as the size of mobile networks grows, which is discussed in much more detail in Chapters 9–13 of this book.

1.2.1 Scaling laws for mobile networks without fixed infrastructure

Device-to-device (D2D) networks are important alternative to conventional cellular networks. In traditional cellular systems, all the mobile users or devices must transmit to their base stations. Even if two nearby devices located in the same cell want to talk to each other, their messages have to be relayed by a base station. By contrast, D2D communication allows a mobile device to talk to its peers directly, thereby holding the potential of improving the energy and/or spectral efficiency of mobile networks. In particular, D2D networks are similar to conventional mobile ad-hoc networks because their data packets are not relayed by a fixed infrastructure. Since scaling law analysis has played a key role in revealing the performance limits of various mobile ad-hoc networks, there is, no doubt, a need to investigate scaling laws of cache-aided D2D or ad-hoc networks.

The scaling analysis mainly targets the asymptotic performance of cache-aided D2D networks when the numbers of both devices and files grow without bound. The performance metrics to be considered in the scaling analysis, but are not limited to, the average per-device throughput, throughput-outage tradeoff, etc. Both uncoded and coded caching schemes are analyzed and compared to show that coding may bring substantial performance gains even in D2D caching networks with complicated and general topologies in Chapter 9. Moreover, the scaling laws of various caching schemes are thoroughly analyzed in both single-hop and multi-hop topologies. The conclusions demonstrate that caching holds significant potential not only in the simple star topology, e.g., those adopting metropolitan-area networking schemes, but also in more complicated and practical mobile network topologies.

The throughput of cache-aided ad-hoc networks relies heavily on the popularity distribution of content items and user preferences. As a result, a natural question is raised whether caching increases the capacity of large wireless networks with various popularity distributions. In addition to the widely adopted Zipf distribution, uniform content popularity is also discussed in Chapter 10. It is shown that caching enables wireless ad-hoc networks to enjoy scalable sum capacity or unchanged per-node capacity even as the number of devices increases. This is in contrast to conventional mobile ad-hoc networks without hierarchical cooperation, the service capability of which is not scalable. In Chapter 10, the impact of content popularity on scaling laws is extensively investigated. In particular, when skewness exists in content popularity, the per-node capacity is found to increase with the number of nodes monotonically.

1.2.2 Scaling laws for caching with hierarchical cooperation

Since the scaling analysis has been proposed by Gupta and Kumar [9], there has been considerable effort towards assuring the scalability of mobile ad-hoc networks [11, 12]. However, before hierarchical cooperation was proposed, it was long recognized that the per-node capacity of mobile ad-hoc networks approaches zero when the number of nodes grows to infinity. In the seminal work of Ozgur, Leveque, and Tse [13], it was found that hierarchical cooperation based on distributed multiple-input multiple-output (MIMO) techniques may achieve the optimal capacity scaling in mobile ad-hoc networks. As a result, the scaling analysis of cache-aided mobile ad-hoc networks with hierarchical cooperation is well motivated. Chapters 11 and 12 focus on the cache-aided hierarchical cooperation in wireless D2D networks.

Chapter 11 addresses a natural question as to whether hierarchical cooperations can still achieve much better scaling performance than conventional multi-hop schemes in mobile ad-hoc networks when caching is enabled. Furthermore, this chapter investigates how the cache size affects scaling laws, motivated by the fact that the size of local cache is rather limited compared with the library that users are interested in. It is shown that with hierarchical cooperation and uncoded random caching placement, the average aggregate throughput may scale almost linearly with the number of nodes in a mobile ad-hoc network, while the outage probability vanishes. In Chapter 12, how to achieve an order gain in the throughput scaling law is investigated. To this end, physical-layer caching, also referred to as cache-induced MIMO cooperation, and hierarchical cooperation are jointly designed. The proposed cache-induced hierarchical cooperation technique may not only achieve a better scaling law but also attain substantial throughput gain in practice. Based on the solution to an integer programming problem, a low-complexity cache placement algorithm is presented. The algorithm is shown to be order-optimal. The per-node capacity scaling law for cache-aided D2D networks under the physical-layer model and Zipf distribution of content popularity has been completely characterized.

1.2.3 Stochastic geometry approaches for caching

The stochastic geometry approach is adopted to characterize the performance of large-scale mobile networks, in which both devices and base stations are randomly placed. In the scaling analysis of mobile ad-hoc or D2D networks, Poisson Point Processes (PPPs) are widely assumed to generate the network topology. More recently, the stochastic geometry approach has been exploited to analyze the coverage and data rate in cellular networks with fixed infrastructures. In particular, Andrews, Baccelli, and Ganti presented a tractable approach based on stochastic geometry theory in [14]. As a result, a natural question is raised whether the stochastic geometry approach can be helpful in analyzing and optimizing cache-aided cellular networks with fixed but randomly placed base stations.

In Chapter 13, randomized geographic caching which greatly simplifies the cache-management problem is discussed based on the stochastic geometry approach. With fixed infrastructure, a user can be served by multiple base stations that are randomly distributed and equipped with finite cache memories. By using the stochastic geometry approach, the distribution of the number of covering base stations can be obtained. Furthermore, the probabilistic content placement over these cache memories allows the formulation of a continuous optimization problem, thereby reducing the complexity in optimizing the content placement over large-scale cellular networks. Chapter 13 demonstrates that stochastic geometry is a powerful tool with the potential to address the content placement issue in practical cellular systems.

1.3 Resource allocation for caching: A game-theoretic perspective

The information-theoretic analysis and asymptotic analysis reveal the performance limits of caching schemes over typical star topology and large-scale mobile networks with random node placement, respectively. In practice, each individual user has its own utility function to be maximized. As a result, fair and efficient resource allocation and scheduling protocols are of practical importance for cache-aided mobile networks. Moreover, centralized scheduling may result in substantial signaling overhead, especially in cache-aided networks with very huge resource and system state dimensions. When multiple virtual network operators share a common infrastructure, centralized resource allocation might be even prohibited, because each virtual operator would make its own decision to maximize its own revenue. Game theory, auction theory, and contract theory are considered as emerging and powerful tools to address fair and efficient decentralized resource allocation or scheduling problems. In this book, Chapters 14–17 are devoted to the economic-theoretic perspective study of caching.

1.3.1 Dynamic pricing for caching

Pricing wireless services has mattered since commercial cellular systems were first deployed. Conventionally, a user is charged for her or his access time or data volume received, because RANs provide on-demand services only. With caching, however, not all the received data is needed by a user. As no one is willing to pay for what he or she does not need, revolutionary pricing mechanisms are desired. One straightforward solution is to charge a user for what he or she requests rather than what he or she receives. In addition, pricing should ensure that a user pays a lower service fee, while the mobile network's operator has more profit; otherwise no one is willing to adopt caching. The two goals seem to be contradictory but can be achieved simultaneously due to caching gains. As a result, recent works [15, 16] focus on caching-friendly pricing mechanisms.

A dynamic pricing mechanism is introduced in Chapter 14 to motivate the use of caching. Proactive caching can offload the service demand during peak times to off-peak times by pushing content items in which a user might be interested, in idle time-slots before the user requests. By this means, the peak-time service outage probability or latency can be reduced, while the spectrum during off-peak times can be fully utilized. However, no one is willing to pay for uncertain demands in the future unless the bandwidth in off-peak times is much cheaper compared with that in peak times or even free. Intuitively, reducing the price in off-peak times may advocate proactive pushing and caching, as discussed in [15, 16]. An alternative way to motivate caching is to offer pricing discounts to render the users' demand profiles more predictable and enhance the certainty about future demand. Such a problem is tackled in Chapter 14, which conceives a pricing-aided proactive caching policy for mobile services.

1.3.2 Auction approaches for caching

Auction theory is an applied branch of economics that studies how bidders act in auction markets. It is recognized as an emerging and powerful tool to allocate wireless spectrum efficiently. For instance, a spectrum auction may increase the bandwidth price during peak times while reducing it during off-peak times. In other words, spectrum auctions give incentives to implement proactive caching during off-peak times. Recent work [17] presents an auction-based proactive caching policy in which virtual network operators bid for bandwidth to serve their users, either by proactive caching or on-demand transmission. If an operator fails in getting bandwidth when it has to launch on-demand transmissions, its users will suffer from poor Quality-of-Experience (QoE). An operator providing poor QoE frequently will lose its users as they choose to be served by other operators. Accordingly, an operator has to spend a lot of money to win the bandwidth auction if an on-demand transmission is necessary. In other words, the greater the hit ratio (i.e., the fraction of pushed content that is actually consumed by users) is, the more profit an operator has. Therefore, bandwidth auctions motivate the evolution of prediction and caching policies.

Chapter 15 focuses on auction-based caching at the network edge or in small-cell base stations. The users may enjoy better QoE because of the latency reduction and would pay the service provider for. An action mechanism is designed for the service provider to push content items and charge users. The design purpose is to maximize the service provider's expected profit while ensuring truthful reports and user participation. Based on the auction approach, both the space allocation in cache memories and choice of the content delivery quality are optimized. Chapter 15 demonstrates the potential of applying auction theory in cache-aided mobile networks.

1.3.3 Contract theory for caching

Contract theory is also an applied branch of economics that deals with how economic actors can construct contractual arrangements, when trading is modeled in the presence of information asymmetry. More specifically, in contract theory, one side has its private information that is unknown by the other side. In some cache-aided mobile networks, the utility functions of network service providers and content providers are unknown to each other. As a result, contract theory is expected to play a central role in fair and efficient decentralized resource allocation. In particular, various contract-theoretic incentive mechanisms have been designed recently.

Chapter 16 focuses on a small-cell caching system, in which the network facility monopolist intents to lease its small base stations to content providers to maximize their profits. Since the network service provider has limited knowledge about the content providers, contract theory is applied to conceive trading schemes and offer economic incentives to the network facility monopolist. By this means, the cache-aided mobile network is characterized by a monopoly market. More particularly, content providers can be classified according to their popularity among mobile users. Then the optimal contact problem is formulated and discussed in the presence of information asymmetry in the sense that the network facility monopolist only knows the distribution of content providers' popularity. In such typical cache-aided hierarchical cellular systems, contract theory is demonstrated to be helpful and efficient in profit maximization.

1.3.4 Mean-field game theory for caching

Mean-field is a branch of game theory that investigates the strategic decision-making in very large populations of small interacting agents. It is inspired by the mean-field theory, originally developed in modern physics, in which the effect of all the other agents in a system on any given agent is approximated by a simple averaged effect. By this means, a high-dimensional stochastic system can be approximated by a much simpler one-dimensional system. A mean-field game is usually equivalent to a partial differential equation referred to as the Hamilton-Jacobi-Bellman equation. Since the mean-field game theory takes into account a large number of agents, it fits naturally with an ultra-dense network (UDN). Consequently, there is a need to discuss the application of the mean-field game theory in cache-aided ultra-dense mobile networks.

Chapter 17 is devoted to mean-field game-theoretic edge caching in a downlink UDN, in which small base stations and mobile users are randomly placed obeying PPPs. Each small base station is equipped with a finite cache memory in order to reduce the peak-hour bandwidth demand in the backhaul. Each small base station intends to minimize its long-run average cost in a decentralized manner by solving a non-cooperative stochastic differential game. By applying the mean-field approximation, each small base station only interacts with an equivalent small base station, the behavior of which follows a mean-field distribution. By this means, the mean-field game-theoretic caching problem in cache-aided ultra-dense networks can be solved with relatively low complexity.

1.4 Caching with time-varying popularity and recommendation systems

Caching relies on the prediction of users' requests and content popularity. In practice, the content popularity and/or user preference are time-varying. Moreover, they can be heavily affected by various recommendation systems (RSs). For instance, users tend to generate more video requests during recreation times, resulting in a periodic fluctuation of request frequency in every single day. In addition, a lot of users may choose to watch video clips to kill time in the subway, but Internet surfing is strictly prohibited during car driving. There have been some models to characterize the time-varying content popularity and demand probability. A user-specific model of the time-varying content popularity is presented in [18], in which the probability distribution of a user's request time for a content item is referred to as the statistical request delay information (RDI). The RDI model relies on the observation in [19] that a content item is usually requested only once by a user. RDI provides more knowledge than demand probability and popularity because its integration over a time period gives a user's demand probability in this period. By summarizing the real-time demand probabilities of all users for a content item, we further obtain the content item's real-time popularity. The model enables the energy-efficient caching [20], storage-efficient time-domain memory sharing [21, 22], and joint pushing and caching [23, 24]. Nowadays a large portion of data services are essentially driven by recommendation systems. Not only content providers such as YouTube, Netflix, Tik-Tok, etc. but also news websites and even search engines "recommend" what may interest a user. In this book, Chapters 18–20 are devoted to caching with time-varying popularity. The joint design of caching and RS will be discussed in Chapters 21–22.

1.4.1 Popularity dynamics in the time and spatial domain

Driven by the ever-increasing demand for high-quality recommendation systems, popularity analysis and prediction have attracted much attention recently. The quantitative study of content popularity is enabled based on historical request data of popular video clips and movies, which is made available by content providers, analyzers, or operators of mobile social networks, such as Netflix, YouTube, Movielens, Facebook, etc. From these data, a conclusion can be drawn for typical content items that popularity dynamics exist in both the time and spatial domains. The popularity dynamics or evolution trends can be predicted, e.g., by temporal analysis such as time series analysis. As a result, how to design and optimize caching schemes for content items with popularity dynamics in the time and spatial domains needs careful investigation.

In Chapter 18, caching with time-varying popularity profiles is conceived for a heterogenous wireless network, in which both macro and small base stations along with users are randomly distributed according to independent PPPs. The design purpose is to minimize the offloading loss in backhaul links. Risk bounds on the offloading loss are presented under unknown non-stationary popularity profiles. Based on these risk bounds, a dynamic cache update policy is conceived, in which cached content items are updated only when the popularity profile is significantly changed.

Since the content popularity may fluctuate in both the temporal and spatial domains, Chapter 19 is devoted to caching with space-time popularity dynamics, which can be characterized by a Markov process. Based on this formulation, reinforcement learning, or more specifically, Q-learning is adopted to optimize caching policies for both single-node and multi-node scenarios. Aiming at further reducing the computational complexity, the chapter exploits linear function approximation and deep neural networks to overcome the curse of dimensionality. Finally, in Chapter 20, a lifetime-based model is formulated to characterize the content popularity dynamics. When a content item becomes unpopular, it runs out of its lifetime. The joint scheduling of multicast pushing and conventional unicast transmission is then investigated based on the simplified popularity model. In [25], a Markov decision process approach is exploited to jointly schedule the cache placement of content items with time-varying popularity and on-demand transmission over a common channel. It is demonstrated that lower latency can be attained by proactive caching.

1.4.2 Joint caching and recommendation

RSs, long recognized as an area of computer science, have been widely used by content providers. Both RSs and caching systems predict what a user is likely to read, as noted in [19]. An RS shows a user her or his favorite content items, while a cache-aided mobile network steps further by sending them to the user before being requested. Naturally, joint caching and recommendation has attracted recent attention. An intuitive idea is to push what an RS would recommend and let the user know. By this means, the hit ratio can be improved, as are caching gains. Meanwhile, the RS recommends essentially what it wants to recommend only and avoids showing a user what it is not interested in. When the cached files are not user's most favorite ones, cache-friendly recommendation was conceived to recommend alternative content items that are cached but not the most favorable, e.g., [26]. Moreover, there has been some research on the joint design of content recommendation with beamforming-assisted content caching [27] and coded caching [28]. In these cases, a cost to be paid is that users might be less satisfied with the RS if they frequently find unwanted or useless recommendations. Chapter 21 is devoted to the discussion of cache-friendly recommendation with quality of recommendation assurance.

When a user asks for a content item that has not been cached, the RS finds some relevant content items from the local cache, if there are any, and recommends them to the user. If the user accepts the recommendation, a soft hit is achieved [29]. Otherwise, on-demand transmission will be launched to satisfy the user. In addition, recommendation may enhance users' common interests, thereby grouping them to achieve coded-caching or multi-casting gain [28]. The "Show Me the Cache" approach allows mobile networks to affect users pro-actively, which may result in potential ethical problems in that users are manipulated. Personalized content recommendation can also reduce the uncertainty of user requests in mobile networks. Since both a user's request and response to a content recommendation are random, reinforcement learning is adopted to maximize the average accumulated profit of mobile network operators in Chapter 22. To overcome the curse of dimensionality, the joint caching and recommendation problem is decomposed into two sub problems by letting the recommendation agent and pushing agent work together with individual goals. By this means, the dimensions of state and action spaces of each agent are substantially reduced, which makes the problem solvable.

1.5 Some applications of caching in mobile networks

Edge caching is expected to revolutionize mobile applications and network infrastructures. After the discussion of information-theoretic analysis, scaling laws, resource allocation, and popularity dynamics of caching, we finally turn our attention to some of its typical applications in mobile networks. For instance, edge caching has been applied in multiple unmanned aerial vehicle (UAV)-enabled RANs, both the three-dimensional model and spectral efficiency optimization of which are discussed in [30]. At the end of this book, Chapters 23 and 24 are devoted to caching-aided mobile social networks, while the application of caching schemes in cloud radio access networks (CRANs), UAVs, and mobile edge computing (MEC) systems are reviewed in Chapters 25 and 26. These chapters demonstrate that the integration of computing, caching, and communication is drawing increasing attention from both academia and industry.

1.5.1 Caching for mobile social networks

Mobile social networks enable people having common interests to interact with each other using mobile phones, tablets, or laptops. By utilizing mobile messaging applications, mobile users may form virtual communities, which effectively engage users by providing smooth interaction among them. In particular, users may not only contribute to a virtual community by uploading or posting social media but also consume social media by viewing or experiencing content items. Most recently, mobile social networks have been rapidly developing based on emerging cellular and Internet technologies. In mobile social networks, most content items are cacheable because they are uploaded earlier than being viewed. As a result, proactive caching may be adopted to reduce duplicate transmissions, exploit off-peak spectrum, and attain multicasting gain in mobile social networks, by formulating a "Human-in-the-Loop" communication architecture [31]. On the other hand, social ties that are related to user interactions enable the prediction of user requests and assure mutual trust in D2D caching.

Chapter 23 is devoted to the discussion of social networking-driven caching policies. Various fundamental design principles for caching with D2D social communications are discussed based on hyper-graph models that take into account social ties, interest similarity, and physical links. Resource allocation for socially enabled D2D caching is investigated to enhance the spectral and energy efficiencies. Device-to-infrastructure- communication-based caching schemes are also designed for mobile social networks in this chapter. More specifically, a context-aware caching policy is obtained based on matching theory and deep learning. In Chapter 24, security issues in cache-aided mobile social networks are considered. The knowledge of social relationships or social trust is taken into account to enhance the security of mobile social networks. To this end, social trust among users is evaluated from both direct observation and indirect observation using Bayesian inference and the Dempster-Shafer theory, respectively.

1.5.2 Caching for CRANs, UAVs, and MEC

Motivated by the substantial gain brought by edge caching in mobile networks, emerging studies aim at combining edge caching with various novel network infrastructures, e.g., CRANs, UAV networks, and mobile edge computing systems. In CRANs, radio access functions are realized by a generic compute platform rather than a purpose-built hardware platform. More specifically, mobile users are served by remote radio heads (RRHs) or distributed antennas controlled by cloud-based baseband units using cloud-native principles. UAV-empowered communication networks bring another novel and promising infrastructure using flying access points or base stations. The high mobility of UAVs enables quick and flexible deployment of flying base stations to provide emergent data access to areas lacking telecommunication infrastructures. Furthermore, mobile edge computing, which provides cloud computing capabilities at the edge of RANs, holds the potential of running or processing applications closer to end devices of mobile users. By this means, MEC is capable of running new applications flexibly and rapidly, as well as, reducing the backhaul congestion. As a result, it is of practical importance to combine caching with CRANs, UAVs, and MEC.

In Chapter 25, caching policies for CRANs and UAVs are presented. More specifically, remote RRH caching, cloud caching, and RRH caching are extensively discussed. The quality-of-service performance in terms of delay violation probability is studied based on large deviation theory. A low-complexity algorithm for clustering and caching is presented with mobility prediction. Furthermore, in this chapter, cache-enabled UAVs are exploited to provide data access services to ground mobile users. Echo state networks are introduced to predict content request distributions and user mobility patterns, based on which user association, content caching, and UAV locations are jointly optimized. Chapter 26 is devoted to cache-aided MEC systems with time-varying wireless links with random channel coefficients. The dynamic caching, computation offloading, and resource allocation are jointly and judiciously optimized. In particular, a Markov decision process is formulated to minimize the average energy consumption and fetching cost given both the computation and storage constraints. A deep deterministic policy gradient algorithm is leveraged to efficiently fetch popular tasks from a remote cloud to an MEC server.

1.6 Organization of the book

Given the above, we summarize the organization of this book in Figure 1.2.

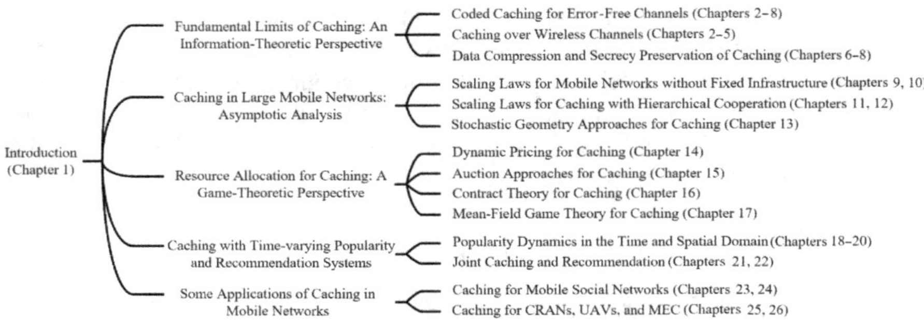

Figure 1.2 Organization of this book

References

[1] Cisco. Cisco Visual Networking Index: Global Mobile Data Traffic Forecast Update, 2017–2022, White Paper. [online]. 2019. Available from https://www.cisco.com/c/en/us/solutions/collateral/service-provider/visual-networking-index-vni/white-paper-c11-741490.html [Accessed 11 Nov 2019].

[2] Letaief K.B., Chen W., Shi Y., Zhang J., Zhang Y.-J.A. 'The roadmap to 6G: AI empowered wireless networks'. *IEEE Communications Magazine.* 2019;**57**(8):84–90.

[3] Maddah-Ali M.A., Niesen U. 'Fundamental limits of caching'. *IEEE Transactions on Information Theory*. 2014;**60**(5):2856–67.

[4] Lu Y., Li C., Chen W. 'On the effective throughput of coded caching with heterogeneous user preferences: A game theoretic perspective'. *IEEE Transactions on Communications*. 2020;**69**(3):1387–402.

[5] Lu Y., Chen W., Poor H.V. 'A unified framework for caching in arbitrary networks'. Proceedings of the 2018 IEEE 23rd International Conference on Digital Signal Processing; 2018. pp. 1–5.

[6] SYR L., Yeung R.W., Cai N. 'Linear network coding'. *IEEE Transactions on Information Theory*. 2003;**49**(2):371–81.

[7] Lu Y., Chen W., Poor H.V. 'User preference aware lossless data compression at the edge'. *IEEE Transactions on Communications*. 2020;**68**(6):3792–807.

[8] Lu Y., Chen W., Poor H.V. 'User preference aware lossy data compression for edge caching'. Proceedings of the IEEE Global Communications Conference; 2020. pp. 1–6.

[9] Gupta P., Kumar P.R. 'The capacity of wireless networks'. *IEEE Transactions on Information Theory*. 2000;**46**(2):388–404.

[10] Haenggi M., Andrews J.G., Baccelli F., Dousse O., Franceschetti M. 'Stochastic geometry and random graphs for the analysis and design of wireless networks'. *IEEE Journal on Selected Areas in Communications*. 2009;**27**(7):1029–46.

[11] Grossglauser M., Tse D.N.C. 'Mobility increases the capacity of ad-hoc wireless networks'. *IEEE/ACM Transactions on Networking*. 2002;**10**(4):477–86.

[12] Weber S., Andrews J.G., Jindal N. 'An overview of the transmission capacity of wireless networks'. *IEEE Transactions on Communications*. 2010;**58**(12):3593–604.

[13] Ozgur A., Leveque O., Tse D.N.C. 'Hierarchical cooperation achieves optimal capacity scaling in ad-hoc networks'. *IEEE Transactions on Information Theory*. 2007;**53**(10):3549–72.

[14] Andrews J.G., Baccelli F., Ganti R.K. 'Atractable approach to coverage and rate in cellular networks'. *IEEE Transactions on Communications*. 2011;**59**(11):3122–3134.

[15] Huang W., Chen W., Poor H.V. 'Request delay-based pricing for proactive caching: A stackelberg game approach'. *IEEE Transactions on Wireless Communications*. 2019;**18**(6):2903–2918.

[16] Lin Z., Huang W., Chen W. 'Bandwidth and storage efficient caching based on dynamic programming and reinforcement learning'. *IEEE Wireless Communications Letters*. 2020;**9**(2):206–209.

[17] Hui H., Chen W. 'A pricing-based joint scheduling of pushing and on-demand transmission over shared spectrum'. Proceedings of the IEEE Global Communications Conference; 2020. pp. 1–6.

[18] Chen W., Poor H.V. 'Content pushing with request delay information'. *IEEE Transactions on Communications*. 2017;**65**(3):1146–61.

[19] Paschos G., Bastug E., Land I., Caire G., Debbah M. 'Wireless caching: technical misconceptions and business barriers'. *IEEE Communications Magazine*. 2016;**54**(8):16–22.

[20] Huang W., Chen W., Poor H.V. 'Energy efficient pushing in AWGN channels based on content request delay information'. *IEEE Transactions on Communications*. 2018;**66**(8):3667–82.

[21] Chen W., Poor H.V. 'Caching with time domain buffer sharing'. *IEEE Transactions on Communications*. 2019;**67**(4):2730–45.

[22] Hui H., Chen W., Wang L. 'Caching with finite buffer and request delay information: a Markov decision process approach'. *IEEE Transactions on Wireless Communications*. 2020;**19**(8):5148–61.

[23] Lu Y., Chen W., Poor H.V. 'Coded joint pushing and caching with asynchronous user requests'. *IEEE Journal on Selected Areas in Communications*. 2018;**36**(8):1843–56.

[24] Lu Y., Chen W., Poor H.V. 'Multicast pushing with content request delay information'. *IEEE Transactions on Communications*. 2018;**66**(3):1078–92.

[25] Li C., Chen W. 'Energy efficient joint pushing and on-demand transmission over shared spectrum'. Proceedings of the IEEE Global Communications Conference; 2020. pp. 1–6.

[26] Chatzieleftheriou L.E., Karaliopoulos M., Koutsopoulos I. 'Jointly optimizing content caching and recommendations in small cell networks'. *IEEE Transactions on Mobile Computing*. 2019;**18**(1):125–38.

[27] Lin Z., Chen W. 'Content pushing over multiuser MISO downlinks with multicast beamforming and recommendation: A cross-layer approach'. *IEEE Transactions on Communications*. 2019;**67**(10):7263–76.

[28] Zhu B., Chen W. 'Coded caching with moderate recommendation: balancing delivery rate and quality of experience'. *IEEE Wireless Communications Letters*. 2019;**8**(5):1456–9.

[29] Sermpezis P., Giannakas T., Spyropoulos T., Vigneri L. 'Soft cache hits: Improving performance through recommendation and delivery of related content'. *IEEE Journal on Selected Areas in Communications*. 2018;**36**(6):1300–13.

[30] Zhou F., Wang N., Luo G., Fan L., Chen W. 'Edge caching in multi-UAV-enabled radio access networks: 3D modeling and spectral efficiency optimization'. *IEEE Transactions on Signal and Information Processing over Networks*. 2020;**6**:329–41.

[31] Yan Q., Chen W., Poor H.V. 'Big data driven wireless communications: a human-in-the-loop pushing technique for 5G systems'. *IEEE Wireless Communications*. 2018;**25**(1):64–9.

Chapter 2

Physical layer schemes for coded caching

Giuseppe Caire[1], Mari Kobayashi[2], and Sheng Yang[2]

2.1 Introduction

With the forthcoming explosion of data traffic generated by the wireless Internet, it is important to consider schemes that take advantage of the user data consumption patterns. In particular, multimedia content delivery (especially video) is by far the most bandwidth consuming and rapidly growing application. In LTE, the *evolved Multimedia Broadcast Multicast Service* (eMBMS) standard was designed to simultaneously multicast popular common content (e.g., live TV channels) to a large number of users in a wide area via the simultaneous transmission from multiple base stations (multipoint multicasting). On the other hand, the by-far favorite media user consumption pattern over Internet consists of on-demand content delivery (e.g., Netflix, iTunes, Hulu, Amazon Prime, Spotify). In this case, the strongly asynchronous nature of the users' requests prevents from taking advantage of simple multicasting. In order to exploit the inherent *asynchronous content reuse* of on-demand content delivery, wireless edge caching has been widely advocated in several contexts (e.g., see [1–5]).

In this chapter, we focus on the elegant and information-theoretic near-optimal *coded caching* scheme introduced by Maddah-Ali and Niesen in [6, 7], for brevity referred to in the following as MAN and reviewed in Section 2.2.1. The MAN scheme was originally proposed for an ideal shared link network, where all users receive the same error-free broadcast transmission from the content server.

When applying coded caching to practical wireless networks, affected by small-scale fading (e.g., Rayleigh fading) and large-scale (distance-dependent) pathloss, several options are possible, as evidenced in a significant amount of recent literature which would be impossible to review here (see also other chapters in this book). Nevertheless, two conceptually distinct approaches can be identified. On one hand, we have schemes based on the separation between caching and physical layer (PHY). Schemes following the separation paradigm aim at delivering to the users the same coded multicast messages generated by the MAN scheme (or by some of

[1]Faculty of Electrical Engineering and Computer Science, Technische Universität Berlin, Germany
[2]Laboratory of Signals and Systems, CentraleSupélec, Paris-Saclay University, France

its variants) for the single shared link, by exploiting features of the PHY such as multiple antennas (e.g., by multicast beamforming), routing through multiple base stations, and combination thereof. On the other hand, we find schemes that jointly design the coded caching and the PHY. By doing so, certain additional non-trivial advantages can be obtained, as for example, the dramatic reduction in the MAN subpacketization order obtained by exploiting multiantenna spatial multiplexing as in [8]. While joint coded caching and PHY design is attractive in many aspects, it is often incompatible with the current technology development, where the PHY is fixed by some standard and where intermediate nodes (e.g., access points, base stations) are agnostic of the end-to-end traffic. For example, in practice video streaming users place encrypted HTTP requests to video packets that can only be decrypted by the video server somewhere in the cloud, and not by the Edge Node (e.g., a base station) which may belong to a cellular operator and not to the streaming service provider [3]. Therefore, the separation paradigm is certainly attractive since it allows the coded caching system to be designed independently on the underlying physical network.

Motivated by this consideration, in this chapter, we follow this separation principle and focus on the efficient delivery of the coded multicast message generated by the MAN scheme over a wireless network. The central issue treated here is the *scalability* of the delivery rate as the number of users K grows large. In particular, we call *scalable* a scheme that achieves a non-vanishing per-user delivery rate as $K \to \infty$.

In Section 2.2, we consider a single-cell scenario where the broadcast transmission of the base station undergoes independent small-scale Rayleigh fading to the users, with limited time-frequency diversity. In this case, the worst-case user with respect to the small-scale fading represents the system bottleneck limiting the broadcast rate. Countermeasures to achieve scalability in this regime are presented, by using multiple antennas at the base station side.

In Section 2.3, we consider an extended network formed by K users and N_E edge nodes (e.g., base stations, access points), such that the ratio K/N_E is an arbitrary constant, and the area of the network grows as $A = O(N_E)$ when N_E becomes large. In this case, we focus on the scalability with respect to the network size and focus on the main limiting factor which is distance-dependent pathloss. For extended networks, $K \to \infty$ implies that the network coverage area A must grow linearly with K. In this case, a scalable scheme shall be referred to as *spatially scalable*, since the scheme can provide a fixed positive delivery rate for an arbitrarily large coverage area.

2.2 Single-cell network

2.2.1 System model

Content delivery model We consider a single-cell wireless network where a n_t-antenna base station serves K single-antenna users through a wireless downlink (DL) channel. The base station has access to a library of N files indicated as $\mathcal{F} = \{W_1, \ldots, W_N\}$, assumed to be equally popular and with equal size F bits for simplicity. Each user k is equipped with a cache of size MF bits, where $M \geq 1$ denotes the cache size measured in files. Building on the separation paradigm between caching and PHY transmission, we follow the original Maddah-Ali and Niesen coded caching scheme [6, 7], hereafter called MAN scheme, for the pre-fetching and formation of the coded multicast messages. For the sake of completeness, we briefly recall centralized coded caching [6] although any other scheme such as decentralized caching [7] or improved versions can be applied to our PHY transmission.

By assuming that $t = MK/N$ is an integer, we partition each file i into segments[a] $\left\{W_i^S : S \subseteq [K], [S = t]\right\}$. Each user k caches all the segments W_i^S for all the subsets S such that $S \ni k$ for each file i. In the delivery phase, given a set of user demands $(d_1, \ldots, d_K) \in [N]^K$, the base station computes the message X obtained from the concatenation of sub-messages X_S for all $S \subseteq [K], |S| = t + 1$ and sends it over the DL channel. Each message X_S intended to a subset S of users is given by

$$X_S = \bigoplus_{k \in S} W_{d_k}^{S \setminus \{k\}}, \tag{2.1}$$

where for any S of size $t + 1$ including k, $W_{d_k}^{S \setminus \{k\}}$ corresponds to the segments of the file desired by user k and stored by all users in S except k. Then, by combining X_S and the cached contents, user k is able to decode the desired file d_k. The number of multicast bits, or the length of the message X, necessary to satisfy K requests from K users is given by

$$T(\mu, K) := \begin{cases} (1 - \mu) \frac{1}{1/K + \mu} F, & \text{for centralized caching,} \\ (1 - \mu) \frac{1 - (1 - \mu)^K}{\mu} F, & \text{for decentralized caching,} \end{cases} \tag{2.2}$$

where $\mu := \frac{M}{N}$ is the normalized cache memory. The striking result is that the number of required transmissions converges to a constant $\frac{1-\mu}{\mu} F$ as the system dimension K grows, yielding a *scalable* scheme that achieves a non-vanishing per-user delivery rate.

Physical-layer channel We provide the PHY channel to deliver the message created by the MAN scheme on the application layer. The message of $T(\mu, K)$ information bits is encoded via a suitable PHY variable-rate encoder into a large number of codewords. For the sake of reduced decoding complexity, it is preferable to keep each codeword much shorter than the coherence time of the underlying channel. We consider L-parallel fading channels such that the channel remains unchanged during the transmission duration of each codeword, referred to as a coherence block, and changes from one block to another in an independently and identically distributed (IID) manner. The parallel channels correspond, e.g., to OFDM with L sub-carriers, so that each codeword spans L sub-channels. A special case of L = 1 is referred to

[a]For an integer n, we let $[n] = \{1, \ldots, n\}$.

the quasi-static fading channel. Within a given resource block, receiver k observes at time t and in sub-channel l the signal

$$y_{l,k}[t] = \boldsymbol{h}_{l,k}^{\mathsf{T}} \boldsymbol{x}_l[t] + z_{l,k}[t], \quad t = 1, 2, \ldots, n, \; l = 1, \ldots, \mathsf{L}, \tag{2.3}$$

where $\boldsymbol{x}_l[t] \in \mathbb{C}^{m_t \times 1}$ is the input vector at time t in sub-channel l satisfying the average power constraint $\frac{1}{n\mathsf{L}} \sum_{t=1}^{n} \sum_{l=1}^{\mathsf{L}} \|\boldsymbol{x}_l[t]\|^2 \leq P$; $z_{l,k}[t]$ is a spatially and temporally white Gaussian process of zero mean and unit variance, denoted by $z_{l,k}[t] \sim \mathcal{CN}(0, 1)$, IID across l, k, and t. For simplicity, we mainly focus on the case of symmetric Rayleigh fading channels, i.e., $\boldsymbol{h}_{l,k} \sim \mathcal{CN}(0, \mathbf{I}_{n_t})$ for any sub-channel l and user k. The extension to the asymmetric fading channels will be provided in Section 2.2.3.

In a practical wireless system, the channel state information (CSI) is not perfectly known at the transmitter, typically due to limited resource for uplink channel training in time division duplex (TDD) or limited channel feedback bandwidth in frequency division duplex (FDD). In the regime of a large number of users, it is reasonable to assume that the base station has only statistical channel knowledge. We assume that CSI is perfect at the receivers.

Equivalent content delivery rate and scalability Our performance metric is the *equivalent content delivery rate*, or simply content delivery rate, defined as the number of total demanded information bits, including those already in the cache, that can be delivered per unit of time in average. Let $\bar{R}_0$ be the average multicast rate of the underlying wireless channel in bits/second/Hz. To satisfy the demands of K users, we need to multicast $T(\mu, K)$ bits, which takes $T(\mu, K)/\bar{R}_0$ units of time. Therefore, the equivalent content delivery rate with coded caching is

$$\mathcal{R}_{\text{mul}} = \frac{KF}{T(\mu,K)} \bar{R}_0(K, P) \quad \text{bits/second/Hz.} \tag{2.4}$$

We say that the system is *scalable* with the number K of users if the equivalent content delivery rate scales at least linearly with an increasing K. Since the number of multicast bits $T(\mu, K)$ in (2.2) converges to a constant, we observe from (2.4) that scalability can be guaranteed if the average multicast rate $\bar{R}_0(K, P)$ is not vanishing with K. Indeed, the multicast rate, essentially the communication rate of the user with the *worst* channel condition, can be arbitrarily small in wireless channels. We shall show how to prevent a vanishing multicast rate.

2.2.2 *Scalable content delivery with wireless multicasting*

In this section, we study the content delivery that combines coded caching and wireless multicasting for the system model described in the previous section. We reveal how the system parameters, such as the number of users and the number of transmit antennas, impact the equivalent content delivery rate. The main results are summarized in the following theorem.

Theorem 1. Let us consider a content delivery system as described in the Section 2.2.1. With coded caching and wireless multicasting, the system is scalable in the following cases:

1. With high *diversity*: the product of the numbers of transmit antennas n_t and independent sub-channels L is large w.r.t. the number of users, namely, $\liminf_{K\to\infty} \frac{n_t \mathsf{L}}{\log K} > 0$;
2. With one-bit feedback from each user: the base station selects a large subset of users whose SNR is above a predetermined threshold.

In the rest of the section, we show the scalability of each case. For Case 1, we first consider $\mathsf{L} = 1$ and then extend the result to an arbitrary $\mathsf{L} \geq 1$.

Quasi-static fading channels ($\mathsf{L} = 1$) In order to multicast the message created by MAN scheme on the application layer over wireless channels, the message is encoded by a suitable coded modulation scheme into a large number of PHY codewords. For a given realization of the channel $\boldsymbol{H} = [\boldsymbol{h}_1 \cdots \boldsymbol{h}_K]^{\mathsf{T}}$[b], the achievable multicast rate is given by

$$R_0(\boldsymbol{H}) = \max_{\boldsymbol{Q}_0 : \mathrm{tr}(\boldsymbol{Q}_0) \leq P} \min_{k \in [K]} \log(1 + \boldsymbol{h}_k^{\mathsf{T}} \boldsymbol{Q}_0 \boldsymbol{h}_k^*), \tag{2.5}$$

where the minimum over users reflects that the codeword must be decoded by all K users. We also assume that a Gaussian codebook is used to encode the input vectors $\{\boldsymbol{x}[t]\}_{t=1}^n$ with input covariance matrix $\boldsymbol{Q}_0$ indicating the precoding and spatial power allocation strategy. The outer maximization means that the transmitter can choose a strategy that maximizes the multicast rate. Since we assume that the channel is not known at the transmitter and the channel is isotropic with IID Rayleigh fading, it is reasonable to use isotropic signaling, i.e., $\boldsymbol{Q}_0 = \frac{P}{n_t} \mathbf{I}_{n_t}$. Assuming that a large number of codewords experience IID fading blocks, the average multicast rate tends to the long-term average multicast rate, given by

$$\bar{R}_0 := \mathbb{E}[R_0] = \mathbb{E}\left[\log\left(1 + \min_{k \in [K]} \frac{P}{n_t} \|\boldsymbol{h}_k\|^2\right)\right]. \tag{2.6}$$

Proposition 1. The long-term average multicast rate is non-vanishing with K if and only if the number n_t of transmit antennas increases at least logarithmically with K, namely, $\liminf_{K\to\infty} \frac{n_t}{\log K} > 0$.

Proof. It is easy to verify that $\bar{R}_0$ is bounded from 0 if and only if we have $\liminf_{K\to\infty} \mathbb{P}\left(\min_{k\in[K]} \frac{P}{n_t} \|\boldsymbol{h}_k\|^2 \geq c\right) > 0$ for some constant $c > 0$. By noticing that $\frac{1}{n_t} \|\boldsymbol{h}_k\|^2$ is Gamma distributed with parameters $(n_t, 1/n_t)$, denoted by $G(n_t, 1/n_t)$, and using the independence of the channel coefficients across users, it follows that

$$\mathbb{P}\left(\min_{k \in [K]} \frac{P}{n_t} \|\boldsymbol{h}_k\|^2 \geq c\right) = \left(\bar{F}_{G(n_t, 1/n_t)}\left(\tfrac{c}{P}\right)\right)^K =: f_{n_t, K}\left(\tfrac{c}{P}\right), \tag{2.7}$$

where $\bar{F}_{G(n_t,1/n_t)}(x)$ is the complementary cumulative distribution function (CDF) of $G(n_t, 1/n_t)$ distributed random variable. Noticing that the CDF is given by

[b]With some abuse of notation, $\boldsymbol{h}_k$ denotes both a random variable and a particular realization.

$F_{G(n_t,1/n_t)}(x) = \alpha_x \left(xe^{1-x-\frac{\ln n_t}{2n_t}}\right)^{n_t}$ for some $\alpha_x \in [a, b]$ with $a, b > 0$, the probability $f_{n_t,K}(x)$ in (2.7) is non-vanishing with K if and only if $F_{G(n_t,1/n_t)}(x) \to 0$ and $KF_{G(n_t,1/n_t)}(x)$ must be upper bounded (must not go to infinity). We can then show that if $\liminf_{K\to\infty} \frac{n_t}{\log K} = 0$, $KF_{G(n_t,1/n_t)}(x) \to \infty$, for all $x > 0$. Otherwise, if $\liminf_{K\to\infty} \frac{n_t}{\log K} = \beta > 0$, then there exist $x_\beta > 0$ such that $KF_{G(n_t,1/n_t)}(x) \to 0$ for all $x \leq x_\beta$. Combining both arguments, we complete the proof of the proposition.

Intuitively, for a fixed number of transmit antennas, the channel quality of the worst user degrades with the total number of users. A remedy for this is to increase the transmit power with K, which is however not desirable (if not impossible) in many practical situations. Another solution is to increase the number of transmit antennas with K. The above result shows that if n_t increases at least logarithmically with K, a constant amount of transmit power suffices to maintain a non-vanishing multicast rate even without any CSI at the transmitter side. We refer to this behavior as the *channel hardening* effect, i.e., increasing transmit antennas number decreases the variance of the individual SNR so that the worst user can still have a constant rate.

L-parallel channels When a codeword can span $\mathsf{L} \geq 1$ independent fading blocks, e.g., coding across independent sub-carriers in OFDM, the instantaneous multicast rate becomes

$$R_0(\{\boldsymbol{h}_{k,l}\}_{k,l}) = \min_{k\in[K]} R_k(\{\boldsymbol{h}_{k,l}\}_l), \tag{2.8}$$

where $R_k(\{\boldsymbol{h}_{k,l}\}_l)$ denotes the instantaneous rate of user k defined as

$$R_k(\{\boldsymbol{h}_{k,l}\}_l) := \frac{1}{\mathsf{L}} \sum_{l=1}^{\mathsf{L}} \log(1 + \frac{P}{n_t} \|\boldsymbol{h}_{k,l}\|^2). \tag{2.9}$$

By taking a large number of channel realizations, the long-term average multicast rate is

$$\bar{R}_0 = \mathbb{E}[R_0(\{\boldsymbol{h}_{k,l}\}_{k,l})]. \tag{2.10}$$

It turns out that the channel hardens faster as compared to the single-block case, which is precisely stated in the following proposition.

Proposition 2. The long-term average multicast rate is not vanishing with K if and only if the product of the numbers of transmit antennas and independent sub-channels, $n_t\mathsf{L}$, increases at least logarithmically with K, namely, $\liminf_{K\to\infty} \frac{n_t\mathsf{L}}{\log K} > 0$.

Proof. The converse is straightforward from Jensen's inequality. For the average over $l \in [\mathsf{L}]$ can be put inside the logarithm such that the problem becomes a single-block channel with effectively $n_t\mathsf{L}$ antennas. Applying Proposition 1 proves the converse. To prove the direct part, we need to show that with the proper scaling of $n_t\mathsf{L}$, the instantaneous multicast rate $R_0(\{\boldsymbol{h}_{k,l}\}_{k,l})$ in (2.8) is larger than a constant c with non-zero probability when $K \to \infty$. Since the probability of the aforementioned event is $(1 - F_{n_t,L}(c))^K$, where $F_{n_t,\mathsf{L}}(x)$ the CDF of R_k for any k in (2.9), it is enough to show that $KF_{n_t,\mathsf{L}}(c)$ is upper bounded when $K \to \infty$. Note that we have the following upper bound

$$\begin{aligned} F_{n_t,\mathsf{L}}(x) &\le \binom{\mathsf{L}}{\mathsf{L}/2} F_{G(n_t,1/n_t)}((2^{2x}-1)/P)^{\mathsf{L}/2} \\ &\le (4F_{G(n_t,1/n_t)}((2^{2x}-1)/P))^{\mathsf{L}/2}, \end{aligned} \tag{2.11}$$

where the first inequality is from the union bound applying on the fact that the average is smaller than x implies that at least half of the terms should be smaller than $2x$; the second one is from $\binom{2n}{n} \le 4^n$ for all positive integer n. Following the proof of Proposition 1, where we use $F_{G(n_t,1/n_t)}(x) \approx \left(xe^{1-x-\frac{\ln n_t}{2n_t}}\right)^{n_t}$, we can show that if $\liminf_{K\to\infty} \frac{n_t\mathsf{L}}{\log K} = \beta > 0$, $K(4F_{G(n_t,1/n_t)}(x'))^{\mathsf{L}/2} \to 0$ for some $x' > 0$. The proof is complete.

The above result demonstrates an interesting trade off between the number of transmit antennas and the number of independent sub-channels for a scalable multicast rate. A large number of sub-channels can compensate for the limited number of transmit antennas and vice versa. Since n_t and L are respectively the spatial and temporal/frequency diversity per user, the product $n_t\mathsf{L}$ can be interpreted as the *total diversity* that can be exploited by each user. Proposition 2 says that the multicast rate is non-vanishing if and only if the *total diversity* scales logarithmically with K. As shown previously, in the quasi-static fading channel ($\mathsf{L} = 1$), a large array of antennas should be exploited to obtain scalability. The other extreme case is when the base station has a single antenna, i.e., $n_t = 1$ as assumed in [6, 7] and many follow-up works, but spans the multicast transmission over L independent sub-channels with L increasing logarithmically with K.

User selection Since the multicast transmission is limited by the channel quality of the worst user, the transmission rate can be improved if we only serve users with better quality at a given channel realization. In other words, we eliminate users with "*unacceptable*" channel qualities. For instance, if we transmit at the average (median) rate over the channel gain, then the number of users being able to decode the transmitted codeword is roughly $K/2$, and we can still guarantee a linear sum rate scaling with K. The trade off between the multicast rate and the number of served users should be balanced so as to maximize the sum rate.[c] In order to achieve linear scaling with the total number of users, a non-negligible fraction of the K users should be selected. In the following, we shall show that a simple threshold-based user selection scheme is enough. For simplicity, we focus on the case $n_t = \mathsf{L} = 1$ although the proposed scheme can be easily extended to the case $n_t\mathsf{L} > 1$.

We assume that the base station fixes a SNR threshold s that will be shared *a priori* among all users in the network. Upon the reception of the useful signals, each user k computes its instantaneous SNR given by $\mathrm{SNR}_k = P|h_k|^2$ and sends back a one-bit feedback indicating whether it is above the threshold. Let the random variable $\mathrm{K}^*(s)$ be the number of users with rate above the threshold, i.e., $\mathrm{K}^*(s) := |\{k : \mathrm{SNR}_k \ge s\}|$. The base station multicasts at rate s so that every selected user is able to successfully decode the transmitted codewords. Since the set of active users changes dynamically

[c]The sum rate maximization is appropriate for delay-tolerant applications under the symmetric channel assumption. When the fading statistics are asymmetric across users, the performance metric shall be modified to incorporate fairness among users.

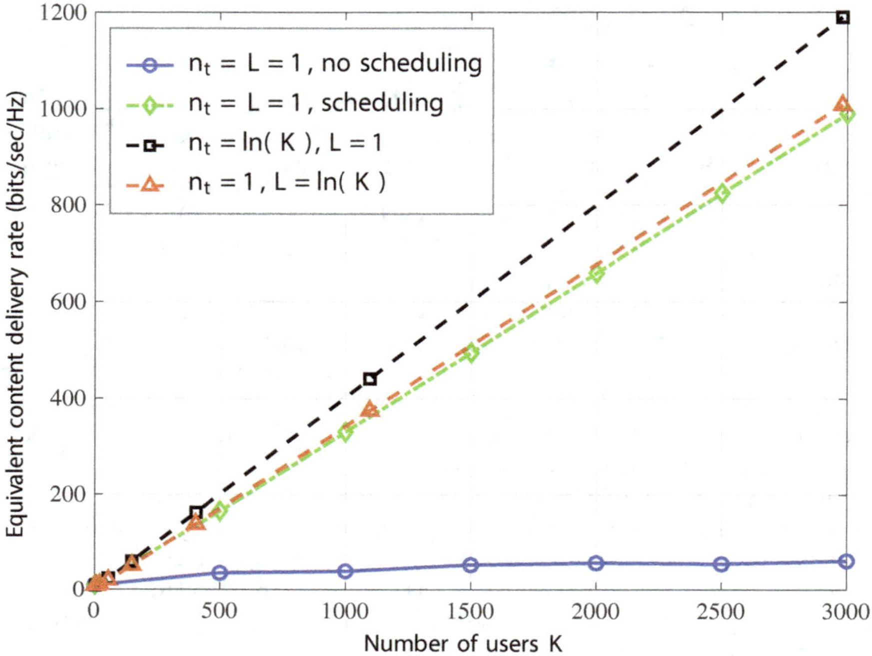

Figure 2.1 The equivalent content delivery rate versus the number of users K for $\mu = 5\%$ and $P = 30$ dB

in the presence of user selection, we focus on decentralized placement [7] and obtain the equivalent content delivery rate from (2.2) and (2.4) as

$$\mathcal{R}_{\text{mul}} = \frac{1}{\ln(2)} \mathbb{E}\left[\frac{\frac{\mu}{1-\mu}\text{K}^*(s)}{1-(1-\mu)^{\text{K}^*(s)}} \ln(1+s) \right]. \tag{2.12}$$

From the strong law of large numbers, we know that $\frac{\text{K}^*(s)}{K}$ converges almost surely to $\mathbb{P}(\text{SNR}_k \geq s) = e^{-s/P}$. Using the dominated convergence, it readily follows that the equivalent content delivery rate scales linearly with K, i.e.,

$$\frac{\mathcal{R}_{\text{mul}}}{K} \sim \frac{\mu}{1-\mu} \ln(1+s) e^{-s/P} \tag{2.13}$$

for *any non-zero thresholds*. One can further maximize the content delivery rate by choosing an optimal threshold s^* maximizing $\ln(1+s)e^{-s/P}$. This can be solved numerically. We can verify that

$$s^* = e^{W(P)} - 1 = \frac{P}{W(P)} - 1, \tag{2.14}$$

where $W(\cdot)$ is the Lambert-W function such that $W(x)e^{W(x)} = x$.

Numerical Results To validate Theorem 1, we calculate numerically the equivalent content delivery rate $\mathcal{R}_{\text{mul}}$ and observe its behavior when K increases. Figure 2.1 shows the equivalent content delivery rate achieved by different

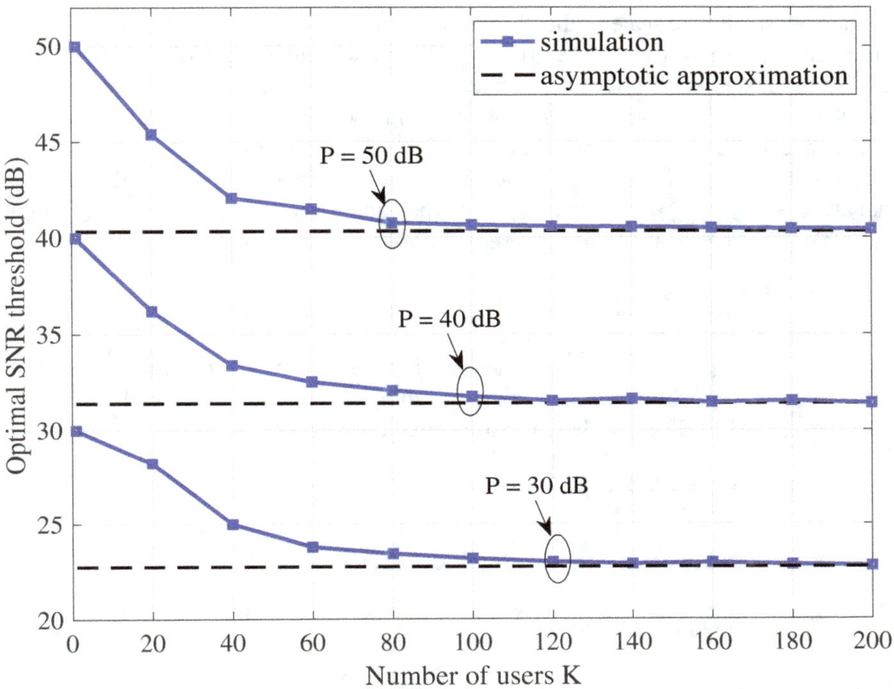

Figure 2.2 *The optimal rate threshold for user selection scheduling with* $\mu = 5\%$, $P = 30, 40, 50$ *dB and* $n_t = \mathsf{L} = 1$

multicasting schemes listed in Theorem 1 as a function of the number of users K for normalized cache size $\mu = 5\%$ and a fixed total power P. Specifically, we consider multicasting with (1) $n_t = \lfloor \ln(K) \rfloor$ antennas, (2) single antenna and transmission spanning $\mathsf{L} = \lfloor \ln(K) \rfloor$ sub-channels, and (3) single antenna and threshold-based user selection scheduling.

Figure 2.2 shows the asymptotically optimal SNR threshold s^* given in (2.13) for user selection and the exact optimal solution obtained by the simulation. We observe that the analytical solution converges to the exact optimal one when K is large. Since the analytical optimal SNR threshold (2.14) only depends on the total power, the base station can compute it beforehand.

2.2.3 *Extensions*

In this section, we briefly provide a few scenarios of interest that extend our previous results presented in Theorem 1.

Simultaneous multicasting and multiplexing Let us consider the case when the base station has some imperfect CSI. In this case, the content delivery rate can be further improved by simultaneous multicasting and multiplexing with $n_t \geq K$ antennas. We assume that each user k has a sequence of demands $d_k^{(1)}, d_k^{(2)}, \ldots$ for $k \in [K]$ over a relatively small time scale so that the cached contents remain unchanged. By focusing on the first two sequences, we describe the delivery scheme that exploits coded caching gain and spatial multiplexing gain. In order to deliver $d_1^{(1)}, \dots, d_K^{(1)}$, the base station forms a multicast codeword containing $T(\mu, K)$ bits and encodes it in the common signal $\boldsymbol{x}_0$. Meanwhile, it encodes the uncached fraction of $d_k^{(2)}$ containing $(1-\mu)F$ bits in the private signal $\boldsymbol{x}_k$ for each k. The server then transmits a multiplexed version of all the encoded signals $\boldsymbol{x} = \boldsymbol{x}_0 + \sum_{k=1}^{K} \boldsymbol{w}_k x_k$, where $\boldsymbol{w}_k$ is the precoding vector chosen according to the available CSI. The received signal at user k is

$$y_k = \boldsymbol{h}_k^{\mathsf{T}} \boldsymbol{x}_0 + g_k \boldsymbol{x}_k + \sum_{l \neq k} \tilde{g}_{k,l} x_l + z_k, \tag{2.15}$$

where $g_k := \boldsymbol{h}_k^{\mathsf{T}} \boldsymbol{w}_k$ and $\tilde{g}_{k,l} := \tilde{\boldsymbol{h}}_k^{\mathsf{T}} \boldsymbol{w}_l$. Each receiver decodes first the common message for coded caching while treating the private signals as interference and then the private message. The SINR of the common signal at receiver k is

$$\mathrm{SINR}_k^{(0)} := \frac{\frac{P_0}{n_t} \|\boldsymbol{h}_k\|^2}{1 + |g_k|^2 P_k + \sum\limits_{l \neq k} |\tilde{g}_{k,l}|^2 P_l}, \tag{2.16}$$

and the long-term average common (multicast) rate is

$$\bar{R}_0^{\mathrm{mix}} = \mathbb{E}\left[\log\left(1 + \min_{k \in [K]} \mathrm{SINR}_k^{(0)}\right)\right]. \tag{2.17}$$

Then, the private messages are decoded after removing the already decoded common signal. With uniform private power allocation $P_k = \frac{P - P_0}{K} = p, \ \forall k \in [K]$, we have the following average symmetric private rate $\bar{R}_{\mathrm{sym}}^{\mathrm{mix}}$

$$\bar{R}_{\mathrm{sym}}^{\mathrm{mix}} := \mathbb{E}\left[\log\left(1 + \frac{|g_k|^2}{p^{-1} + \sum_{l \neq k} |\tilde{g}_{k,l}|^2}\right)\right], \tag{2.18}$$

where we considered user k without loss of generality due to the symmetric fading process. Each user k can retrieve $d_k^{(1)}$ using the cached content and the multicast codeword decoded from the common signal, while it can also decode $d_k^{(2)}$ from the cached content and the message decoded from the private signal. Thus, each user can receive two independent flows, i.e., the first flow carrying the common message and the second flow carrying the private message. By noticing that each private message is delivered by sending missing $(1-\mu)F$ bits, the aggregated content delivery rate with the proposed mixed scheme is given by

$$\mathcal{R}_{\mathrm{mix}} = \frac{KF}{T(\mu, K)} \bar{R}_0^{\mathrm{mix}} + \frac{K}{1-\mu} \bar{R}_{\mathrm{sym}}^{\mathrm{mix}}, \tag{2.19}$$

which reduces to (2.4) for the case of the first flow only ($P_0 = P$). To maximize the above aggregated content delivery rate (2.18 and 2.19), we need to jointly optimize over the precoding vectors $\{\boldsymbol{w}_k\}_k$ and the power allocation P_0. Such an optimization problem is non-convex and is not trivial to solve even numerically due to the expectation.

Finally, it is worth noting that the interplay between coded caching gain and spatial multiplexing gain has been extensively studied under different assumptions and setups in the literature (see, e.g., [9–12] and references therein). These works mostly consider a finite dimension system such that $n_t \leq K$ contrary to our case where the system dimension is large, where rate splitting is also considered in [9, 10] by focusing on a high SNR regime.

Fair scheduling under asymmetric fading statistics We consider a more general single-cell network such that K users are located at different distances from the base station, while restricting to a single antenna at the base station and the quasi-static fading channel, i.e., $n_t = \mathsf{L} = 1$. We assume that $\mathrm{SNR}_k = P|h_k|^2$ of user k is exponentially distributed with mean $1/\gamma_k$ where γ_k captures the path loss (the symmetric fading case is $\gamma_k = P$ for all k). In order to deal with asymmetric fading statistics across users, we are interested in designing a fair scheduling policy $\pi \in \Pi$ that maximizes some utility function of the long-term rates over all admissible policies Π. By letting $\mathcal{J}^\pi(\boldsymbol{h}) \subseteq [K]$ denote the subset selected by a policy π for a given channel realization $\boldsymbol{h} = (h_1, \ldots, h_K)$, the long-term average rate of user k under a policy π is given by

$$U_k^\pi = \mathbb{E}\left[\frac{1\{k \in \mathcal{J}^\pi(\boldsymbol{h})\}}{T(\mu, |\mathcal{J}^\pi(\boldsymbol{h})|)/F} \log\left(1 + \min_{j \in \mathcal{J}^\pi(\boldsymbol{h})} \mathrm{SNR}_j(\boldsymbol{h})\right)\right]. \tag{2.20}$$

Our objective is to solve the fair scheduling problem

$$\pi^\star = \arg\max_{\pi \in \Pi} \left\{\frac{1}{K}\sum_{i=1}^{K} g_\alpha(U_i^\pi)\right\}, \tag{2.21}$$

where $g_\alpha(x)$ denotes the α-fair family of concave functions

$$g_\alpha(x) = \begin{cases} \frac{x^{1-\alpha}-1}{1-\alpha}, & \text{if } \alpha \neq 1, \\ \log(x), & \text{if } \alpha = 1. \end{cases} \tag{2.22}$$

Notice that $\alpha = 0$ corresponds to the sum rate maximization, $\alpha = 1$ corresponds to the proportional fair scheduling, and $\alpha \to \infty$ yields the max-min fairness such that $U_1 = \cdots = U_K$. Interestingly, it can be proved that a well-designed threshold policy $\pi_{s^\star}^{th}$ that selects users whose SNR is above a threshold $s^\star$, is asymptotically utility optimal in the sense that ([13], Theorem 3)

$$\frac{1}{K}\sum_{k=1}^{K} g_\alpha(U_k^{\pi_{s^\star}^{th}}) = \max_{\pi \in \Pi}\left\{\frac{1}{K}\sum_{k=1}^{K} g_\alpha(U_k^\pi)\right\} + o(1), K \to \infty. \tag{2.23}$$

For the case of practical interest $\alpha \geq 1$, the threshold $s^\star$ is given as a unique solution to

$$(1+s)\log(1+s) = \frac{\sum_{k=1}^{K} e^{-\frac{s(1-\alpha)}{\gamma_i}}}{\sum_{k=1}^{K} \frac{1}{\gamma_i} e^{-\frac{s(1-\alpha)}{\gamma_k}}}, \tag{2.24}$$

yielding a closed-form solution for $\alpha = 1$ similarly to (2.13) by replacing P with $K\left(\sum_k 1/\gamma_k\right)^{-1}$.

2.3 Multi-cell network

So far we have considered the case of a single cell with a large number of users K, where the main problem was to cope with the worst-user rate bottleneck of that limits the multicast rate as $K \to \infty$. This problem has been tackled in the regime of limited diversity and large number of antennas at the base station. In particular, the main focus is on quasi-static small-scale fading (Rayleigh fading), while the average receiver SNR for all users in the cell is the same, modeling the fact that in a single (small) cell, the pathloss is not the main performance limiting factor.

In this section, for the same MAN delivery model introduced before and following again the separation paradigm between caching PHY, we focus on a different aspect of scalability for *extended multi-cell networks*. We call *spatially scalable* a scheme that achieves a non-vanishing per-user delivery rate as the network coverage area A grows to infinity, for fixed user and base station densities. In an extended network, users are affected by both small-scale fading (e.g., Rayleigh fading as considered before) and by large-scale distance-dependent pathloss. A single base station multi-antenna approach as in [14, 15] will not be effective against the pathloss since the worst-case user rate can be made arbitrarily small with a network whose area grows as $A = O(K)$, such that the distance from the worst-case user and the base station grows as $O(\sqrt{K})$. In order to clearly separate the effect of small-scale fading and and large-scale pathloss, in this section we consider the case where the relevant PHY transmission rate is given by averaging with respect to the small-scale fading statistics. This is motivated by the fact that when there is enough time-frequency diversity in the transmission of a single PHY codeword, the so-called *ergodic rate* (averaging with respect to the small-scale fading) rather than the instantaneous rate per fading state becomes a much more relevant performance metric.

In the scheme presented in this section, the content server is connected through high-capacity backhaul links to N_{E} wireless *Edge Nodes* (ENs) (e.g., base stations, access points, or remote radio heads). In the presence of a given set of users' demands, the server computes the MAN coded message, splits it in subblocks, adds some additional subblocks of parity bits, and routes such subblocks to the ENs, which transmits them simultaneously and on the same frequency channel to the users. The additional parity blocks form a *Maximum Distance Separable* (MDS) code, such that the whole message can be retrieved from any L out of N_{E} MDS-coded blocks, and where $L \geq 1$ is an integer parameter referred to as *macro-diversity order* of the system. Each user in the system can choose to receive data from its best L ENs in a completely autonomous user-centric way. As long as a user is able to decode at least L EN transmissions, it will retrieve the whole MAN coded message and therefore its own requested file. In this sense, the system is reminiscent of single-frequency networks for content distribution (eMBMS) already used in 4G-LTE systems.

As extensively discussed in [16, 17], in this context it is meaningful to consider the per-user ergodic rate where averaging (ergodic behavior) is with respect to the small-scale fading and conditioning is with respect to the random realization of the pathloss, which in turn depends on the random placement of users and ENs on the plane. Since such conditional ergodic rate is a function of the random network geometry, it is a random variable itself. Using stochastic geometry, we are able to characterize the distribution of such random variable under different detection schemes at the users' receivers. In particular, the tail of such distribution yields the user outage probability, which can be interpreted as the fraction of users in the system that are not able to decode at least L ENs transmissions. Since this probability is a constant strictly less than 1 (in fact, it can be made sufficiently small for a good choice of the system parameters), this means that a linear fraction of users can be served at non-vanishing delivery rate. In turn, this shows that the proposed system is spatially scalable in the presence of both (ergodic) small-scale Rayleigh fading and distance dependent pathloss.

2.3.1 System model

We consider a wireless network with N_{E} single-antenna ENs and K users, each equipped with an n_r-antenna array.[d] Server is connected to the ENs via an error-free backhaul data network of very high capacity and has access to a content library formed by N files of F bits. As before, each user has a cache memory of capacity MF bits. Following the separation paradigm between caching and PHY, the pre-fetching and formation of the coded caching messages are according to the MAN scheme reviewed before. In the delivery phase, the multicast codeword X obtained by the concatenation of the blocks X_S, $\forall\, S \subseteq [K]$ of size $t+1$, is divided into L blocks of equal size, for some integer $L \leq N_{\text{E}}$, and MDS coding is applied to append $N_{\text{E}} - L$ parity blocks. Each MDS-coded block is treated as a symbol over a large finite field. The resulting N_{E} MDS-coded blocks are sent to the N_{E} ENs, such that each EN is associated to a distinct MDS-coded block. Eventually, the blocks are transmitted through an appropriate PHY-coded modulation scheme and sent in parallel, on the same time-frequency slot, from the different ENs. The users use their antenna array to detect and decode the PHY codewords of L ENs. Thanks to the MDS coding, whenever a user k is able to decode L ENs messages, then it can reconstruct the whole multicast codeword X.[e] At this point, user k can decode the desired file W_{d_k} from X and the cached content, as explained before and detailed in [6]. Figure 2.3 shows an example with $N_{\text{E}} = 7$ and $L = 2$, where all users that can decode 2 data streams from two distinct ENs can retrieve their requested file.

The delivery time, i.e., the time necessary to deliver all requests, is given by

[d]Typical hand-held devices operating at conventional cellular and WiFi frequency bands cannot host a large number of antenna elements. Nevertheless, for a relevant scenario we may think of a vehicular network where users are cars, which can host relatively large arrays, or a mmWave network, where a large number of antenna elements can be accommodated in a small form factor even on hand-held devices.

[e]The basic property of an MDS code is that it satisfies the Singleton bound with equality. In our case, a (L,N_E) code has minimum Hamming distance (in symbols) equal to $N_E - L + 1$. This implies that any $(N_E - L)\times(NE - L)$ submatrix of the party-check matrix has full rank. In turn, this means that the transmitted codeword can be uniquely identified if atleast L MDS-coded symbols are received.

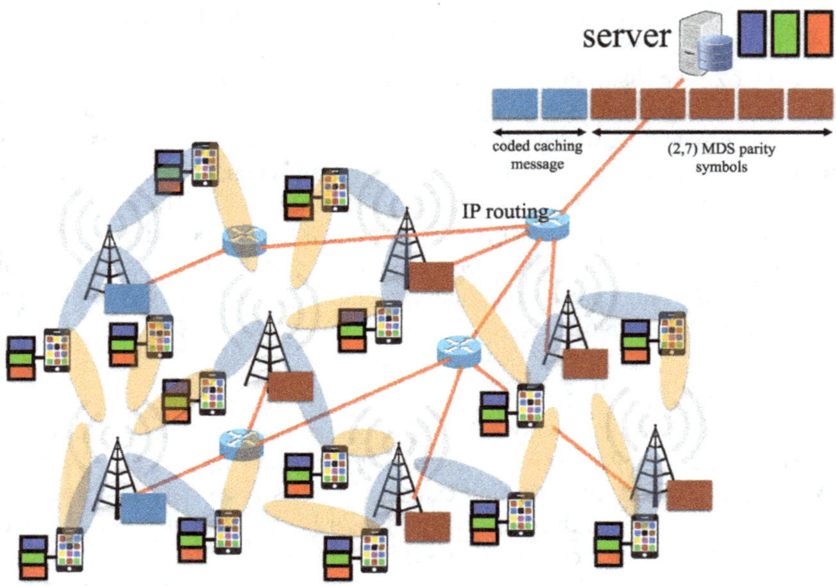

Figure 2.3 A conceptual example of the proposed system with $N_E = 7$ and $L = 2$. User nodes make use of their beamforming antenna array to decode two out of seven EN transmissions, by treating the others as noise.

$$D = \frac{1}{LR} \times \frac{T(\mu, K)}{w}, \tag{2.25}$$

where $T(\mu, K)$ is the already discussed MAN codeword length (bits), w is the bandwidth of the wireless channel (Hz), and R is the PHY coding rate (bit/s/Hz). Since $\frac{T(\mu,K)}{w}$ tends to a constant as K grows large, the main system design goal here consists of determining L such that the product LR is maximized (i.e., the delivery latency is minimized) for a given target per-user outage probability.

From the PHY viewpoint, the goal for each user is to successfully decode the L strongest ENs. Assuming that the PHY codewords span a sufficiently large number of small-scale fading states, we denote by $C_{k,\ell}(\Phi)$ the achievable ergodic rate that user k can successfully decode with high probability (w.h.p.) from EN ℓ for a given decoding strategy (to be detailed later) and a given geometry Φ, i.e., a given placement of the ENs relative to user k. Without loss of generality, we can sort these achievable rates such that $C_{k,1}(\Phi), \ldots, C_{k,L}(\Phi)$ are the largest rates over all the ENs. It follows that user k can obtain the requested file if

$$C_k^{(L)}(\Phi) = \min_{\ell \in [L]} C_{k,\ell}(\Phi) > R. \tag{2.26}$$

For a random geometry Φ, we can define the user outage probability as

$$P_{\text{out},k}(R) := \mathbb{P}(C_k^{(L)}(\Phi) \leq R), \tag{2.27}$$

given by the CDF of $C_k^{(L)}(\Phi)$.

2.3.2 Performance analysis

We assume that ENs are randomly located on the two-dimensional plane and are distributed according to a two-dimensional homogeneous Poisson point process (PPP) of density λ. The ensemble of the locations of the ENs is denoted by Φ. The propagation is characterized by a small-scale Rayleigh fading that remains constant on time-frequency coherence blocks and evolves according to a stationary ergodic process from block to block. Each receiving user acquires CSI from the surrounding ENs through periodically broadcasted DL pilot symbols.

Without loss of generality, we can consider the performance of the typical user indexed by k and located at the origin. We sort the ENs according to their distance from user k in non-decreasing order, and let $j \in [N_{\text{E}}]$ be the index of the j-th closest EN to user k. The space-time signal received by user k on a symbol interval is given by

$$\boldsymbol{y}_k = \sum_{j=1}^{N_{\text{E}}} \sqrt{\beta r_{k,j}^{-\eta}} \boldsymbol{h}_{k,j} x_j + \boldsymbol{z}_k, \tag{2.28}$$

where $\boldsymbol{h}_{k,j} \in \mathbb{C}^{n_r \times 1}$ is the channel vector containing the small-scale fading coefficients from EN j to the array of user k, x_j is the coded-modulation symbol sent by EN j, $r_{k,j}$ is the distance between EN j and user k, η is the pathloss exponent, and β is the path loss intercept. The ENs transmit at a constant average power $\mathbb{E}[|x_j|^2] = P$. The noise samples in $\boldsymbol{z}_k$ are IID $\sim \mathcal{CN}(0, N_0)$. According to the independent Rayleigh statistics, $\boldsymbol{h}_{k,j}$ are IID across users and ENs and have IID components $\sim \mathcal{CN}(0, 1)$.

Partial Zero-Forcing detection We introduce here a *Partial Zero-Forcing* (PZF) receiver strategy for the system at hand. For a subset $\mathcal{B} \subseteq [N_{\text{E}}]$, let $\boldsymbol{H}_{k,\mathcal{B}}$ denote the sub matrix formed by the columns (channel vectors) $\boldsymbol{h}_j$ with indices $j \in \mathcal{B}$. PZF consists of applying linear zero-forcing only with respect to the signals of the L nearest ENs, whereas the signals of the remaining ENs with indices $j > L$ are treated as noise. Assuming $n_r \geq L$, the receive filtering matrix of user k is defined as the pseudo-inverse of the $n_r \times L$ matrix $\boldsymbol{H}_{k,1:L}$. This is given by

$$\boldsymbol{H}_{k,1:L}^{\dagger} = \boldsymbol{H}_{k,1:L}(\boldsymbol{H}_{k,1:L}^{\mathsf{H}} \boldsymbol{H}_{k,1:L})^{-1}. \tag{2.29}$$

Denoting by $\boldsymbol{h}_{k,\ell}^{\dagger}$, $1 \leq \ell \leq L$, the ℓ-th column of $\boldsymbol{H}_{k,1:L}^{\dagger}$ and by $\boldsymbol{q}_{k,\ell} = \boldsymbol{h}_{k,\ell}^{\dagger}/\|\boldsymbol{h}_{k,\ell}^{\dagger}\|$ the normalized receive filtering vector corresponding to the ℓ-th stream, the filtered output $\underline{\tilde{y}}_{k,\ell} = \boldsymbol{q}_{k,\ell}^{\mathsf{H}} \boldsymbol{Y}_k$ is written as

$$\underline{\tilde{y}}_{k,\ell} = \frac{(\boldsymbol{h}_{k,\ell}^{\dagger})^{\mathsf{H}}}{\|\boldsymbol{h}_{k,\ell}^{\dagger}\|} \boldsymbol{h}_{k,\ell} \sqrt{\beta r_{k,\ell}^{-\eta}}\, \underline{\boldsymbol{x}}_{\ell} + \sum_{j=L+1}^{N_{\text{E}}} \frac{(\boldsymbol{h}_{k,\ell}^{\dagger})^{\mathsf{H}}}{\|\boldsymbol{h}_{k,\ell}^{\dagger}\|} \boldsymbol{h}_{k,j} \sqrt{\beta r_{k,\ell}^{-\eta}}\, \underline{\boldsymbol{x}}_j + \underline{\tilde{z}}_k, \tag{2.30}$$

where $\underline{\tilde{z}}_k$ is the filtered noise vector with IID components $\sim \mathcal{CN}(0, N_0)$. In the rest of this section, for the sake of analytical tractability, we focus on the case of an infinitely extended network (i.e., $N_E \to \infty$) and high-SNR interference limited performance (i.e., $N_0 \to 0$). As a result, the system becomes "scale-free," i.e., the results of our analysis are invariant to any non-zero scaling factor of the physical distances $r_{k,\ell}$ appearing in (2.30). The signal-to-interference ratio (SIR) at user k receiver is given by

$$\mathsf{SIR}_{k,\ell} = \frac{r_{k,\ell}^{-\eta}\|\boldsymbol{h}_{k,\ell}^{\dagger}\|^{-2}}{\sum_{j=L+1}^{\infty} r_{k,j}^{-\eta}|\tilde{h}_{k,j}|^2}. \tag{2.31}$$

Since $\boldsymbol{q}_{k,l}$ is a unit vector that is independent of $\{\boldsymbol{h}_{k,j} : j = L+1, L+2, \dots\}$, it follows that $\tilde{h}_{k,j} = \boldsymbol{q}_{k,l}^{\mathsf{H}}\boldsymbol{h}_{k,j} \sim \mathcal{CN}(0,1), \forall j \geq L+1$. Furthermore, a well-known property of the ZF linear detector is that in the case of Gaussian IID channel vectors the useful signal coefficient $\|\boldsymbol{h}_{k,l}^{\dagger}\|^{-2} = \mathcal{X}_{2(n_r-L+1)}$ is chi-squared distributed with $2(n_r - L + 1)$ degrees of freedom [18] and mean $\mathbb{E}[\mathcal{X}_{2(n_r-L+1)}] = n_r - L + 1$. The ergodic rate with Gaussian inputs and treating interference as noise for the transmission of the ℓ-th closest EN at user k is given by

$$C_{k,\ell}(\Phi) = \mathbb{E}\left[\log\left(1 + \frac{r_{k,\ell}^{-\eta}\mathcal{X}_{2(n_r-L+1)}}{\sum_{j=L+1}^{\infty} r_{k,j}^{-\eta}|\tilde{h}_{k,j}|^2}\right)\middle|\, \Phi\right], \tag{2.32}$$

where the random variables $\mathcal{X}_{2(n_r-L+1)}$ and $\{|\tilde{h}_{k,j}|^2 : j > L\}$ are mutually independent, and the conditioning with respect to the geometry Φ determines the distances $r_{k,\ell}$ and $\{r_{k,j} : j > L\}$. In order to obtain analytically tractable expressions, we consider the following lower bound.

Proposition 3. The ergodic rate in (2.32) is lower-bounded as

$$C_{k,\ell}(\Phi) \geq \mathbb{E}\left[\log\left(1 + \rho_{k,\ell}\mathcal{X}_{2(n_r-L+1)}\right)\middle|\, \Phi\right], \tag{2.33}$$

where we define the *local-average SIR*

$$\rho_{k,\ell} := \frac{r_{k,\ell}}{\sum_{j=L+1}^{\infty} r_{k,j}^{-\eta}}, \quad 1 \leq \ell \leq L. \tag{2.34}$$

Proof. From the convexity of the function $f(x) = \log(1 + a/x)$ for $a > 0$ and $x > 0$, and using $\mathbb{E}[|\tilde{h}_{k,j}|^2] = 1$, the result follows by applying Jensen's inequality with respect to the interference fading coefficients $|\tilde{h}_{k,j}|^2$, $j \geq L+1$ in (2.33).

Next, we would like to eliminate the dependency on the distances $r_{k,j}$ with $j \geq L+1$. For this purpose, we replace the conditioning with respect to Φ by the conditioning with respect to only $r_{k,\ell}$ and $r_{k,L}$, i.e., the ℓ-th and the L-th shortest distances from the origin. Furthermore, letting $I = \sum_{j=L+1}^{\infty} r_{k,j}^{-\eta}$ and noticing that $r_{k,j}$ for $j \geq L+1$ depends on PPP points outside the disk of radius $r_{k,L}$ around user k while $r_{k,\ell}$ depends on PPP points inside the disk, we have that [19], by applying this approximation and Jensen's inequality in (2.35), we obtain

$$C_{k,\ell}(\Phi) \gtrapprox \mathbb{E}\left[\log\left(1 + \frac{r_{k,\ell}^{-\eta}\mathcal{X}_{2(n_r-L+1)}}{\sum_{j=L+1}^{\infty} r_{k,j}^{-\eta}}\right)\middle|\, r_{k,L}, r_{k,\ell}\right] \tag{2.35a}$$

$$\geq \mathbb{E}\left[\log\left(1 + \frac{r_{k,\ell}^{-\eta}\mathcal{X}_{2(n_r-L+1)}}{\mathbb{E}\left[I \middle| r_{k,L}\right]}\right)\middle|\, r_{k,L}, r_{k,\ell}\right]. \tag{2.35b}$$

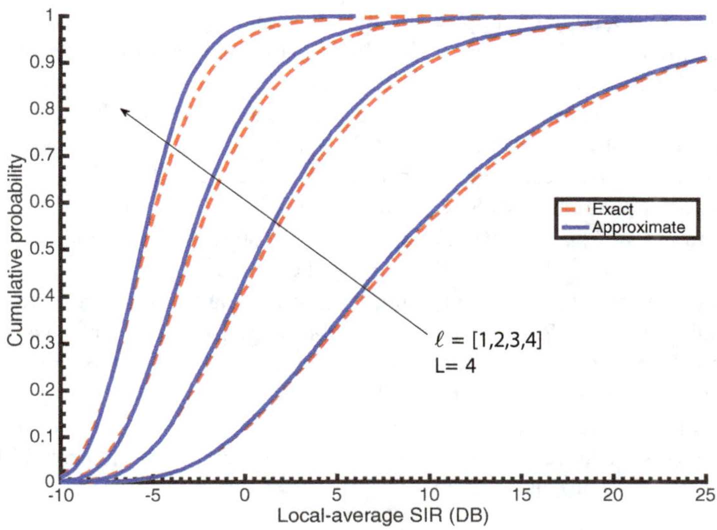

Figure 2.4 CDFs of the exact and approximate local-average SIRs $\rho_{k,\ell}$ *and* $\tilde{\rho}_{k,\ell}$ *with* $n_r = 8$

In order to calculate the conditional average interference term in (2.35b), we apply Campbell's theorem [19, theorem 4.1] and, for $\eta > 2$, we obtain

$$\mathbb{E}\left[I|r_{k,L}\right] = \frac{2\pi\lambda}{\eta-2} r_{k,L}^{2-\eta}. \tag{2.36}$$

Plugging (2.36) in (2.35b), we obtain the quasi-lower bound[f] on the ergodic rate.

Proposition 4. The quasi-lower bound on the ergodic rate is given by

$$C_{k,\ell}^{\text{qlb}}(\Phi) = \mathbb{E}\left[\log\left(1 + \tilde{\rho}_{k,\ell}\mathcal{X}_{2(n_r-L+1)}\right)\middle|\,\Phi\right], \tag{2.37}$$

where we define the approximated conditional local-average SIR as

$$\tilde{\rho}_{k,\ell} := \frac{r_{k,L}^{\eta-2}}{r_{k,\ell}^{\eta}} \frac{\eta-2}{2\pi\lambda}, \quad 1 \le \ell \le L. \tag{2.38}$$

Notice that $\tilde{\rho}_{k,\ell}$ is related to $\rho_{k,\ell}$ defined in (2.34) by replacing the denominator I of (2.34) with its conditional expectation given in (2.36). This approach yields analytically tractable expressions and is expected to provide an accurated approximation of the actual achievable ergodic rate for fixed geometry Φ because the interference term I contains the cumulative effect of many "far" ENs, such that it is expected that a sort of self-averaging effect kicks in Figure 2.4 compares the CDF of the local

[f]This is referred to as quasi-lower bound because of the approximation in (2.36a).

average SIR $\tilde{\rho}_{k,\ell}$ obtained via approximation in (2.38) against its exact form $\rho_{k,\ell}$ in (2.34). All the curves are obtained via Monte-Carlo simulations by setting the EN density $\lambda = 8$ ENs/km^2 (which means on average there is one EN in a circular cell of radius 200 m), the pathloss exponent $\eta = 3.75$. We assumed that the PZF is applied to cancel the $L = 4$ nearest interferers at the receiving user equipped with antenna array size $n_r = 8$. As it can be seen from Figure 2.4, the CDFs of $\tilde{\rho}_{k,\ell}$ and $\rho_{k,\ell}$ are very close for a wide range of SIR. Similar results have been observed for different system parameters L, n_r, and η. Notice also that the quasi-lower bound in (2.37) and the actual lower bound in (2.33) differ only by the replacement of $\rho_{k,\ell}$ with $\tilde{\rho}_{k,\ell}$. Since these terms appear inside the logarithm, their very small difference (as evidenced by the CDFs of Figure 2.4) plays an even more negligible role with respect to the ergodic rates. In order to proceed further, we make use of the following results on distance distribution in PPPs from [20, 21].

Lemma 1. Given a homogeneous PPP Φ of density of λ, let r_n denote the n-th shortest distance of points of Φ from the origin. Then, the probability density function (PDF) of r_n is given by

$$f_{r_n}(v) = \tfrac{2(\pi\lambda)^n}{(n-1)!} v^{2n-1} e^{-\pi\lambda v^2}, \quad v \geq 0, \tag{2.39}$$

and the joint PDF of r_ℓ and r_n with $1 \leq \ell < n$ is given by

$$\begin{aligned} f_{r_\ell, r_n}(u, v) &= \tfrac{4(\pi\lambda)^n e^{-\pi\lambda v^2}}{(n-\ell-1)!\,(\ell-1)!}(v^2 - u^2)^{n-\ell-1} v u^{2\ell-1}, \\ u, v &\geq 0. \end{aligned} \tag{2.40}$$

Using Lemma 1, we can obtain closed-form accurate approximations for the outage probability and the average delivery rate in terms of the macro-diversity order L, the receive antenna array size n_r, the pathloss exponent η, and the PHY coding rate R.

Outage probability derivation Due to the difficulty of obtaining an exact closed-form expression for the local-average SIR, we derive an approximate expression based on the approach introduced in [22, 23]. First, the Laplace transform of $1/\tilde{\rho}_{k,\ell}$ is derived. Then, this is numerically inverted via Euler series expansion. Finally, the CDF of $\tilde{\rho}_{k,\ell}$ can be obtained in the form of finite series yielding accurate results with only a few terms (see [22]).

Lemma 2. The sought approximated form for the CDF of $\tilde{\rho}_{k,\ell}$ is

$$F_{\tilde{\rho}_{k,\ell}}(\gamma) \approx 1 - \gamma \tfrac{e^{A/2}}{2^B} \sum_{b=0}^{B} \binom{B}{b} \sum_{g=0}^{G+b} \tfrac{(-1)^g}{D_g} \mathrm{Re}\left\{ \tfrac{\mathcal{L}_{1/\tilde{\rho}_{k,\ell}}(\tau)}{\tau} \right\}, \tag{2.41}$$

where

$$\tau := \tfrac{(A + i2\pi g)\gamma}{2}. \tag{2.42}$$

The values $D_0 = 2$ and $D_g = 1$ for $g > 0$ and $A = 9.21$, $B = 5$ and $G = 8$ provide a satisfactory accuracy [24].

The Laplace transform of $1/\tilde{\rho}_{k,\ell}$, for $s \in \mathcal{C}$, is given by the following compact expressions, proved in [25].

Proposition 5. The Laplace transform of $1/\tilde{\rho}_{k,\ell}$ takes on the following form:

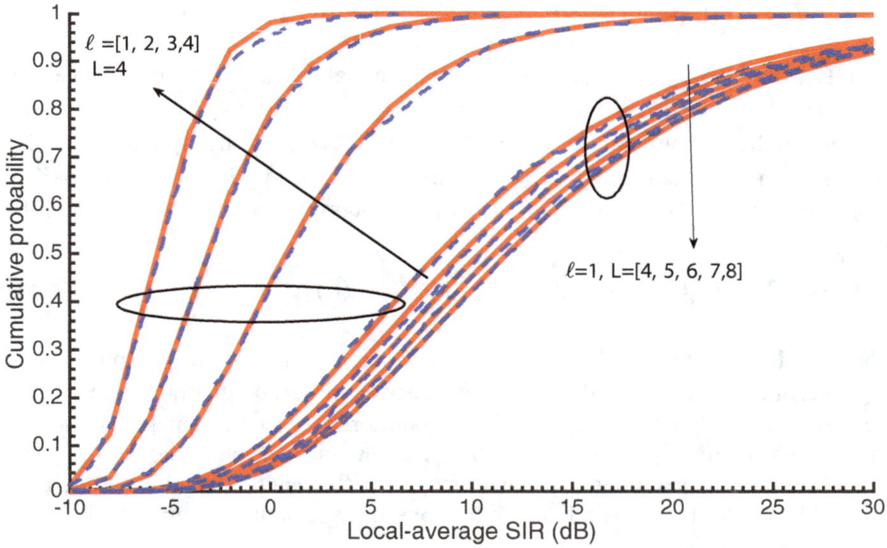

Figure 2.5 *CDF of the approximated local-average SIR ($\tilde{\rho}_{k,\ell}$) with $n_r = 8$. Solid curves correspond to the analysis in (2.40) while the dashed curves are obtained via simulation.*

Case 1: For $\ell < L$

$$\begin{aligned}\mathcal{L}_{1/\tilde{\rho}_{k,\ell}}(s) = \; & \sum_{n=0}^{L-\ell-1} \frac{(-1)^n \Gamma(L)}{(L-\ell-1)!\,(\ell-1)!\,(n+\ell)} \binom{L-\ell-1}{n} \\ & \times {}_2F_1(L, \eta'; \eta' + 1; \tfrac{-2s}{\eta-2}),\end{aligned} \tag{2.43}$$

where $\eta' := 2\frac{n+\ell}{\eta}$ and ${}_2F_1(a, b; c; z)$ is the Gaussian hypergeometric function.

Case 2: For $\ell = L$

$$\mathcal{L}_{1/\tilde{\rho}_{k,\ell}}(s) = \left(\tfrac{2s}{\eta-2} + 1\right)^{-L}. \tag{2.44}$$

Figure 2.5 illustrates the CDF of the approximated local-average SIR given in (2.43) with different sub-stream indices and different values of L with a received antenna array size $n_r = 8$, EN density $\lambda = 8$ ENs/km^2, and $\eta = 3.75$. Monte-Carlo simulation of the CDF is also reported for comparison and shows that our analytical approximation is accurate. The expectation with respect to $\mathcal{X}_{2(n_r-L+1)}$ in (2.37) can be calculated in closed form as

$$\begin{aligned}C^{\text{qlb}}_{k,\ell}(\tilde{\rho}_{k,\ell}) &= \mathbb{E}\left[\log\left(1 + \tilde{\rho}_{k,\ell}\mathcal{X}_{n_r-L+1}\right) \middle| \Phi\right] \\ &= \mathcal{I}_{(n_r-L+1)}(\tilde{\rho}_{k,\ell}) \log(e),\end{aligned} \tag{2.45}$$

where with a slight abuse of notation we use $C^{\text{qlb}}_{k,\ell}(\tilde{\rho}_{k,\ell})$ instead of $C^{\text{qlb}}_{k,\ell}(\Phi)$ since the dependency on the PPP points reduces to the single random variable $\tilde{\rho}_{k,\ell}$, and where we define

$$\mathcal{I}_M(\mu) := \Pi_M(-1/\mu)E_i(1, 1/\mu) + \sum_{m=1}^{M-1} \tfrac{1}{m}\Pi_m(1/\mu)\Pi_{M-m}(-1/\mu), \tag{2.46}$$

with $\Pi_n(x) := e^{-x}\sum_{i=0}^{n-1} \frac{x^i}{i!}$ and with the exponential integral function defined as $E_i(n,x) := \int_1^\infty t^{-n}e^{-xt}dt$.

Due to the monotonicity of the log function, $C_{k,\ell}^{\text{qlb}}(\tilde{\rho}_{k,\ell})$ is strictly monotonically increasing in $\tilde{\rho}_{k,\ell}$. Then, using $C_{k,L}^{\text{qlb}}(\tilde{\rho}_{k,\ell})$ *instead* of $C_k^{(L)}(\Phi)$ in the outage probability expression (2.27), we arrive at the desired quasi-upper bound to the outage probability in the form

$$P_k^{\text{out}}(R) \lessapprox \mathbb{P}\left(C_{k,L}^{\text{qlb}}(\tilde{\rho}_{k,L}) \le R\right) = F_{\tilde{\rho}_{k,L}}\left(\left(C_{k,L}^{\text{qlb}}\right)^{-1}(R)\right), \tag{2.47}$$

where the inverse function $\left(C_{k,L}^{\text{qlb}}\right)^{-1}(\cdot)$ is easily obtained numerically from (2.45).

Average delivery rate derivation In order to appreciate the trade-off between the macro-diversity order L and the PHY coding rate R, and to quickly see which value of L minimizes the delivery latency, we can consider the *average rate of the typical user* when averaging is also with respect to the random locations of the ENs (i.e., with respect to Φ). We indicate by "overline" quantities averaged also with respect to the geometry. Then, we have

$$rl\overline{C}_{k,\ell} = \mathbb{E}\left[\log\left(1 + \frac{r_{k,\ell}^{-\eta}\mathcal{X}_{2(n_r-L+1)}}{\sum_{j=L+1}^{\infty} r_{k,j}^{-\eta}}\right)\right] \tag{2.48}$$

$$= \mathbb{E}\left[\mathbb{E}\left[\log\left(1 + \frac{r_{k,\ell}^{-\eta}\mathcal{X}_{2(n_r-L+1)}}{\sum_{j=L+1}^{\infty} r_{k,j}^{-\eta}}\right)\middle| r_{k,L}, r_{k,\ell}\right]\right] \tag{2.49}$$

$$\ge \mathbb{E}\left[\mathbb{E}\left[\log\left(1 + \tilde{\rho}_{k,\ell}\mathcal{X}_{2(n_r-L+1)}\right)\middle| r_{k,L}, r_{k,\ell}\right]\right] \tag{2.50}$$

$$= \mathbb{E}\left[\log_2\left(1 + \tilde{\rho}_{k,\ell}\mathcal{X}_{2(n_r-L+1)}\right)\right]. \tag{2.51}$$

Remark 1. Comparing (2.51) with (2.37) and using the law of iterated expectation, for which $\mathbb{E}\left[\log\left(1 + \tilde{\rho}_{k,\ell}\mathcal{X}_{2(n_r-L+1)}\right)\right] = \mathbb{E}\left[\mathbb{E}\left[\log\left(1 + \tilde{\rho}_{k,\ell}\mathcal{X}_{2(n_r-L+1)}\right)\middle| \Phi\right]\right]$, we have that the expectation of our quasi-lower bound $C_{k,\ell}^{\text{qlb}}(\Phi)$ on the ergodic mutual information for a given geometry Φ yields a true lower bound $\overline{C}_{k,\ell}^{\text{lb}}$ on the average (w.r.t. the random geometry) ergodic mutual information.

At this point, it is convenient to define

$$\widetilde{\text{SIR}}_{k,\ell} := \tilde{\rho}_{k,\ell}\mathcal{X}_{2(n_r-L+1)}, \tag{2.52}$$

as the "SIR" quantity appearing in (2.51) and evaluate the expectation by following the approach of [16]. We can write

$$\overline{C}_{k,\ell}^{\text{lb}} = \int_0^\infty \log_2(1+\gamma)\,\mathrm{d}F_{\widetilde{\text{SIR}}_{k,\ell}}(\gamma) = \int_0^\infty \left(1 - F_{\widetilde{\text{SIR}}_{k,\ell}}(2^\gamma - 1)\right)\mathrm{d}\gamma. \tag{2.53}$$

Notice that, for fixed $\tilde{\rho}_{k,\ell}$, $\widetilde{\text{SIR}}_{k,\ell}$ is chi-squared distributed (up to a scaling factor). Hence, its conditional CDF is given by

$$F_{\widetilde{\text{SIR}}_{k,\ell}|\tilde{\rho}_{k,\ell}}(\gamma) = 1 - \frac{\Gamma\left(n_r-L+1, \gamma/\tilde{\rho}_{k,\ell}\right)}{\Gamma(n_r-L+1)}, \tag{2.54}$$

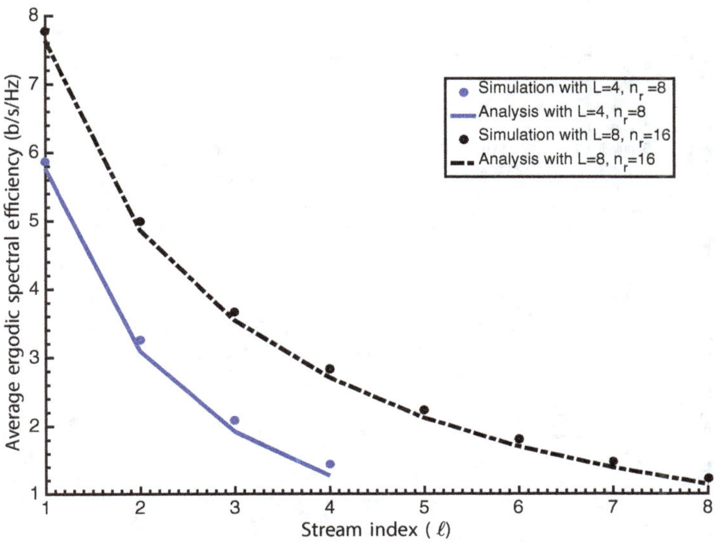

Figure 2.6 Average ergodic rate versus stream index

where $\Gamma(a,x) =: \int_x^{\infty} t^{a-1} e^{-t} dt$ is the upper incomplete gamma function. The unconditional CDF of $\widetilde{\mathsf{SIR}}_{k,\ell}$ is derived by expressing $\tilde{\rho}_{k,\ell}$ in terms of $r_{k,\ell}$ and $r_{k,L}$ (cf. (2.38)) and using their joint PDF to average these variables out, by treating the two cases $\ell < L$ and $\ell = L$ separately. The following result (see [25]) provides the sought lower bound of the average ergodic rate corresponding to the ℓ-th data stream:

Proposition 6. A lower bound of the average ergodic rate is given by

Case 1: For $\ell < L$

$$\overline{C}_{k,\ell}^{\text{lb}} = \sum_{m=0}^{n_r-L} \sum_{n=0}^{L-\ell-1} \frac{2(-1)^n}{m!\,(L-\ell-1)!\,(\ell-1)!\,\eta} \binom{L-\ell-1}{n} \cdot G_{3,3}^{2,3}\left(\begin{matrix} -(m+1), -(m+L), -m-\eta' \\ -(m+1), -1, -(m+1+\eta') \end{matrix} \middle| \frac{2}{\eta-2} \right) \cdot \left(\frac{2}{\eta-2}\right)^{m+1} \log_2(e), \tag{2.55}$$

where $G_{m,n}^{p,q}\left(\begin{matrix} a_1, \ldots, a_n, a_{n+1}, \ldots, a_p \\ b_1, \ldots, b_m, b_{m+1}, \ldots, b_q \end{matrix} \middle| z \right)$ is Meijer-G function.

Case 2: For $\ell = L$

$$\overline{C}_{k,L}^{\text{lb}} = \sum_{m=0}^{n_r-L} \frac{\log_2(e)}{m+L}\, {}_2F_1\left(1, L; m+L+1; 1-\frac{2}{\eta-2}\right). \tag{2.56}$$

The tightness of these lower bounds is illustrated in Figure 2.6, comparing $\overline{C}_{k,L}^{\text{lb}}$ and $\overline{C}_{k,L}$ for the settings: $\lambda = 8$ ENs/km^2 and $\eta = 3.75$ with $n_r = 8, L = 4$ and with $n_r = 16, L = 8$. Since eventually the delivery rate is defined by the worst-case data stream, we define the lower bound for minimum average ergodic rate as follows.

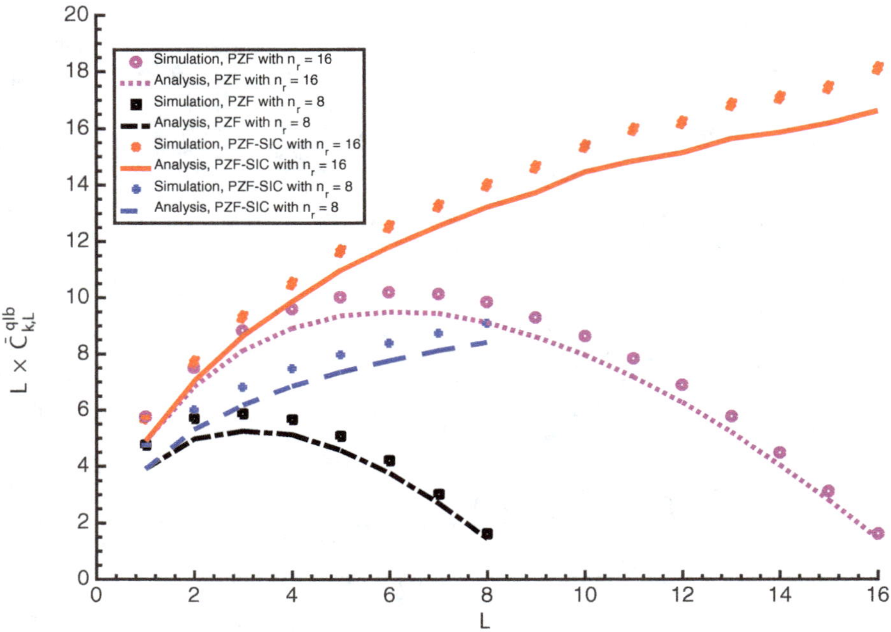

Figure 2.7 $L \times \overline{C}^{\text{lb}}_{k,\text{PZF}}$ *and* $L \times \overline{C}^{\text{qlb}}_{k,\text{PZF-SIC}}$ *versus* $L \leq n_r$ *for a typical user with* $n_r = 8, 16$

$$\overline{C}^{\text{lb}}_{k,\text{PZF}} := \overline{C}^{\text{lb}}_{k,L}. \tag{2.57}$$

It should be mentioned that beyond the simple linear PZF, each user can also apply successive interference cancellation. The resulting strategy, indicated as PZF-SIC, consists of decoding the strongest EN signal in sequence, applying PZF at each stage. Namely, the L strongest EN signals are sorted by their pathloss, and the strongest signal is decoded by zero-forcing the remaining $L - 1$ signals (and treating the remaining $N_{\text{E}} - L$ weakest signals as noise as before). Then, the decoded signal is subtracted from the received signal, and the procedure is repeated for the second strongest signal. This goes on until all L strongest EN signals are decoded. The analysis of PZF-SIC is more complicated and it is omitted here for the sake of space limitation, and can be found in [25].

2.3.3 Numerical results

In this section we provide some numerical results illustrating the behavior and the performance of the proposed content delivery wireless caching network. In order to generate the simulation results, the locations of ENs are distributed according to a homogeneous PPP restricted to a disk radius $R_{\text{area}} = 3$ km with density $\lambda = 8$ ENs/km^2. The number of ENs is a Poisson random variable with mean $\pi R^2_{\text{area}}\lambda$ and the EN positions are generated independently with uniform probability over the disk. We considered realistic values of the pathloss exponent $\eta = 3.75$ and the number of antennas $n_r = 8, 16$ at the user receivers.

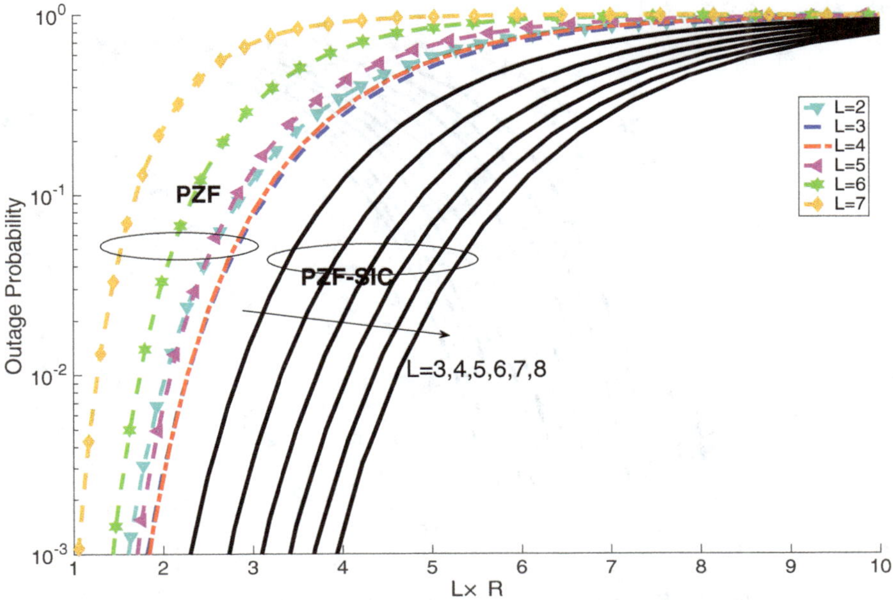

Figure 2.8 *Outage probability versus $L \times R$ with $n_r = 8$*

Figure 2.7 compares, as a function of n_r, $L \times \overline{C}^{\text{lb}}_{k,\text{PZF}}$ (cf. (2.57)) and $L \times \overline{C}^{\text{qlb}}_{k,\text{PZF-SIC}}$ (analogous rate achievable by PZF-SIC) versus L for PZF and PZF-SIC receiver, which serves for the system designer a quick way to choose the best macro-diversity order L in order to minimize the delivery latency. With PZF, the trade-off between PHY rate and macro-diversity order L is evident. In fact, the product $L \times \overline{C}^{\text{lb}}_{k,\text{PZF}}$ increases for small L, reaches a maximum at $L = 3$ for $n_r = 8$ and at $L = 6$ for $n_r = 16$, and then decreases, because the PHY rate at which the typical user can decode the L-th strongest EN decreases faster than increase in L. In contrast, with PZF-SIC, the product $L \times \overline{C}^{\text{qlb}}_{k,\text{PZF-SIC}}$ monotonically increases with $L \in \{1, \dots, n_r\}$ showing that, at least for these system parameters, the best macro-diversity order is $L = n_r$.

Figures 2.8 and 2.9 illustrate the outage probability versus $L \times R$ with PZF and PZF-SIC receiver, $n_r = 8$ and $n_r = 16$, respectively. For better exposition and simplicity of illustration, we only bring the plots for some macro-diversity orders whose outage probability is the lowest. The missing marco diversity orders in these two figures have higher outage probability than the illustrated plots, so they do not play a role to find the best trade-off between the latency and the macro-diversity order. The delivery latency varies with $L \times R$ in an inverse manner and therefore moving to the right of these curves yields shorter delivery latency at higher outage probability (i.e., less and less users will be able to retrieve their requests, but those who

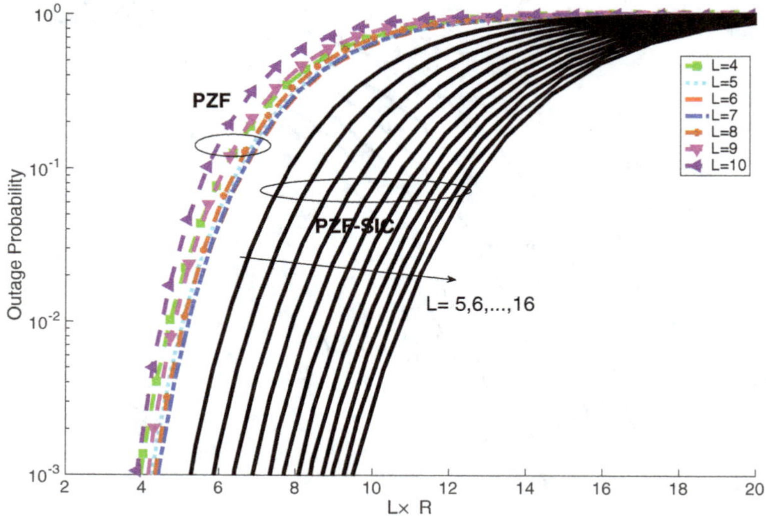

Figure 2.9 *Outage probability versus $L \times R$ with $n_r = 16$*

can, receive it faster). In Figure 2.8, by using PZF alone the macro-diversity order $L = 3, 4$, achieves better overall product $L \times R$ and therefore lower delivery latency. In PZF-SIC, by increasing L the curves move to right side, indicating that for given target outage probability the delivery latency decreases by increasing the macro-diversity order L, confirming the behavior already observed for the average ergodic rate in Figure 2.7. Similarly for $n_r = 16$, with PZF receiver macro-diversity $L = 6, 7$ yield uniformly the best trade-off between latency and the outage probability while for PZF-SIC the best trade-off is obtained by increasing L.

References

[1] Golrezaei N., Molisch A.F., Dimakis A.G., Caire G. 'Femtocaching and device-to-device collaboration: a new architecture for wireless video distribution'. *IEEE Communications Magazine*. 2013;**51**(4):142–9.

[2] Liu D., Chen B., Yang C., Molisch A.F. 'Caching at the wireless edge: design aspects, challenges, and future directions'. *IEEE Communications Magazine*. 2016;**54**(9):22–8.

[3] Paschos G., Bastug E., Land I., Caire G., Debbah M. 'Wireless caching: technical misconceptions and business barriers'. *IEEE Communications Magazine*. 2016;**54**(8):16–22.

[4] Sengupta A., Tandon R., Simeone O. 'Fog-aided wireless networks for content delivery: fundamental latency tradeoffs'. *IEEE Transactions on Information Theory*. 2017;**63**(10):6650–78.

[5] Poularakis K., Tassiulas L. 'On the complexity of optimal content placement in hierarchical caching networks'. *IEEE Transactions on Communications*. 2016;**64**(5):2092–103.

[6] Maddah-Ali M.A., Niesen U. 'Fundamental limits of caching'. *IEEE Transactions on Information Theory*. 2014;**60**(5):2856–67.

[7] Maddah-Ali M.A., Niesen U. 'Decentralized coded caching attains order-optimal memory-rate tradeoff'. *IEEE/ACM Transactions on Networking*. 2015;**23**(4):1029–40.

[8] Lampiris E., Elia P. 'Adding transmitters dramatically boosts coded-caching gains for finite file sizes'. *IEEE Journal on Selected Areas in Communications*. 2018;**36**(6):1176–88.

[9] Zhang J., Elia P. 'Fundamental limits of cache-aided wireless BC: interplay of coded-caching and CSIT feedback'. *IEEE Transactions on Information Theory*. 2017;**63**(5):1–3160.

[10] Piovano E., Joudeh H., Clerckx B. 'Generalized degrees of freedom of the symmetric cache-aided miso broadcast channel with partial CSIT'. *IEEE Transactions on Information Theory*. 2019;**65**(9):5799–815.

[11] Shariatpanahi S.P., Caire G., Khalaj B.H. 'Physical-layer schemes for wireless coded caching'. *IEEE Transactions on Information Theory*. 2018;**65**(5):2792–807.

[12] Tölli A., Shariatpanahi S.P., Kaleva J., Khalaj B.H. 'Multi-antenna interference management for coded caching'. *IEEE Transactions on Wireless Communications*. 2020;**19**(3):2091–106.

[13] Combes R., Ghorbel A., Kobayashi M., Yang S. 'Utility optimal scheduling for coded caching in general topologies'. *IEEE Journal on Selected Areas in Communications*. 2018;**8**:1692–705.

[14] Ngo K.-H., Yang S., Kobayashi M. 'Scalable content delivery with coded caching in multi-antenna fading channels'. *IEEE Transactions on Wireless Communications*. 2018;**17**(1):548–62.

[15] Ghorbel A., Ngo K., Combes R. 'Opportunistic content delivery in fading broadcast channels'. GLOBECOM 2017 - 2017 IEEE Global Communications Conference; 2017. pp. 1–6.

[16] Mungara R.K., Morales-Jimenez D., Lozano A. 'System-level performance of interference alignment'. *IEEE Transactions on Wireless Communications*. 2015;**14**(2):1060–70.

[17] George G., Mungara R.K., Lozano A., Haenggi M. 'Ergodic spectral efficiency in MIMO cellular networks'. *IEEE Transactions on Wireless Communications*. 2017;**16**(5):2835–49.

[18] Marzetta T.L., Larsson E.G., Yang H. *Fundamentals of massive MIMO*. Cambridge, UK: Cambridge Univ Press; 2016.

[19] Haenggi M. *Stochastic geometry for wireless networks*. Cambridge University Press; 2012.

[20] Haenggi M. 'On distances in uniformly random networks'. *IEEE Transactions on Information Theory*. 2005;**51**(10):3584–6.

[21] Martin-Vega F.J., Lopez-Martinez F.J., Gomez G. 'Multi-user coverage probability of uplink cellular systems: a stochastic geometry approach'. 2014

IEEE Global Communications Conference (GLOBECOM) IEEE; 2014. pp. 3989–94.

[22] Błaszczyszyn B., Karray M.K. 'Spatial distribution of the SINR in Poisson cellular networks with sector antennas'. *IEEE Transactions on Wireless Communications*. 2016;**15**(1):581–93.

[23] Abate J., Whitt W. 'Numerical inversion of Laplace transforms of probability distributions'. *ORSA Journal on Computing*. 1995;7(1):36–43.

[24] O'cinneide C.A. 'Euler summation for Fourier series and Laplace transform inversion'. *Communications in Statistics. Stochastic Models*. 1997;**13**(2):315–37.

[25] Bayat M., Mungara R.K., Caire G. 'Achieving spatial scalability for coded caching via coded multipoint multicasting'. *IEEE Transactions on Wireless Communications*. 2018;**18**(1):227–40.

Chapter 3
Cache-aided content delivery over noisy broadcast channels

Mohammad Mohammadi Amiri[1] *and Deniz Gündüz*[2]

Emerging new wireless applications along with the growing number and diversity of wireless devices with their ever-increasing capacity to process and consume information have brought new challenges on wireless network architectures. One of the fastest growing traffic types over current wireless networks is traffic due to video content, and this trend is expected to continue, and even accelerate further with the introduction of AR/VR technologies, and new gaming and teleconferencing technologies. To alleviate the load on the network due to the growing demand for video content during peak traffic periods, proactive caching and coded delivery techniques can be exploited at the physical layer [1–17].

The focus of this chapter is on the benefits of coded caching over wireless broadcast channels (BCs) taking into account the physical characteristics of the wireless medium. The seminal coded caching technique by Maddah-Ali and Niesen [1], referred to as the MAN scheme in this chapter, considers delivery over an error-free rate limited BC from the server to the users. However, in practice the delivery takes place over a noisy BC, and users may experience different quality channels. Hence, a more practical delivery technique must take the varying channel quality of the users into account. We will first consider a memoryless packet erasure BC from the server to the users, where the users have arbitrary erasure probabilities during the delivery phase. Equal-size caches are assigned to the users with the worse channel conditions in order to compensate for the weaker channel quality. The trade-off between the rate of the files in the library that can be delivered to the users reliably and the cache size available at the users is provided. Next, a Gaussian BC from the server to the users is considered during the delivery phase, while the placement phase is carried out without any channel state information (CSI). The goal is to characterize the trade-off between the transmit power at the server and the cache size available at the users.

[1]Department of Electrical and Computer Engineering, Princeton University, USA
[2]Department of Electrical and Electronic Engineering, Imperial College London, UK

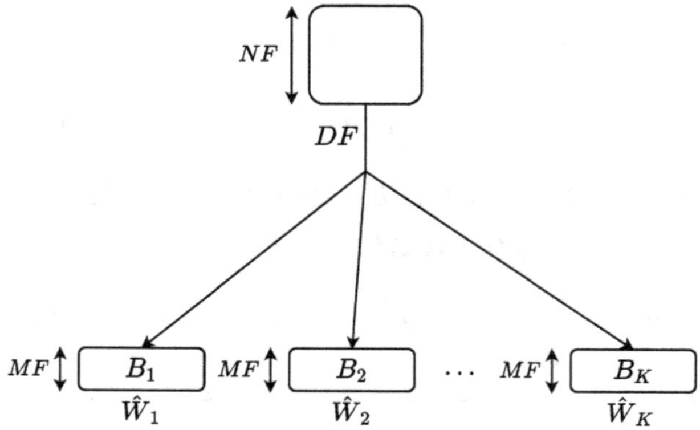

Figure 3.1 Illustration of the caching system model consisting of a server with a database of N *files, each of size* F *bits, serving* K *users, each equipped with a cache memory of size* MF *bits*

3.1 Error-free BC

To lay the foundations for the more advanced broadcasting schemes over noisy channels, we first present an overview of the caching and coded delivery technique over an error-free shared link from the server to the users. Here, the goal is to reduce the communication load at the server, which is equivalent to the communication delay, during the delivery phase.

3.1.1 MAN scheme

Consider a server hosting a library of N files, $\mathbf{W} \triangleq (W_1, \ldots, W_N)$, each with F bits, serving K users, each equipped with a cache of size MF bits, as depicted in Figure 3.1. Assume that file W_n is uniformly distributed over $[2^F]$. Given a demand vector $\mathbf{d} \triangleq (d_1, \ldots, d_K)$ during the delivery phase, a (D, M, F) *caching and delivery code* consists of

i. K caching functions:

$$\phi_k : \left[2^F\right]^N \to \left[2^{\lfloor FM \rfloor}\right], \quad k \in [K], \tag{3.1}$$

each of which maps the library to the cache content B_k of user k, i.e., $B_k = \phi_k(\mathbf{W})$;

ii. Encoding function:

$$\psi : \left[2^F\right]^N \times [N]^K \to \left[2^{\lfloor FD \rfloor}\right], \tag{3.2}$$

which maps the library and demand vector $\mathbf{d}$ to a common message X, i.e., $X = \psi(\mathbf{W}, \mathbf{d})$;

iii. K decoding functions:

$$\mu_k : \left[2^{\lfloor FM \rfloor}\right] \times \left[2^{\lfloor FD \rfloor}\right] \times [N]^K \to \left[2^F\right], \quad k \in [K], \tag{3.3}$$

each of which maps the cache content B_k at user k, common message X, and demand vector $\mathbf{d}$ to the reconstructed message $\hat{W}_k$ at user k, i.e., $\hat{W}_k = \mu_k\left(B_k, X, \mathbf{d}\right)$.

The error probability of the (D, M, F) caching and delivery code is defined as

$$P_e \triangleq \Pr\left(\bigcup_{\mathbf{d}\in[N]^K} \bigcup_{k=1}^{K} \left\{ \hat{W}_k \neq W_{d_k} \right\} \right). \tag{3.4}$$

The delivery load and cache size pair (D, M) is *achievable* if there exists a caching and delivery code (D, M, F) such that, for large enough F, $P_e < \varepsilon$, for any $\varepsilon > 0$. The trade-off between the delivery load and the cache size is defined as follows:

$$D^*\left(M\right) \triangleq \inf\left\{D : \left(D, M\right) \text{ is achievable}\right\}. \tag{3.5}$$

We first recall the MAN scheme for the *centralized caching* scenario, where the identity of the users participating in the delivery phase is known during the placement phase. This allows us to coordinate the cache contents of the users during the placement phase.

Placement phase. We denote the t-element subsets of set $[K]$ by $\mathcal{S}^t_{K,1}, \ldots, \mathcal{S}^t_{K,\binom{K}{t}}$. File W_i is split into $\binom{K}{t}$ equal-length subfiles $W_{i,1}, \ldots, W_{i,\mathcal{S}^t_{K,\binom{K}{t}}}$, each consisting of $F/\binom{K}{t}$ bits, for some $t \in [0 : K]$. During the placement phase, user k caches

$$B_k = \bigcup_{i\in[N]} \bigcup_{l\in\left[\binom{K}{t}\right]:k\in\mathcal{S}^t_{K,l}} W_{i,\mathcal{S}^t_{K,l}}, \quad k \in [K], \tag{3.6}$$

which requires a total normalized cache size of

$$M = \frac{\binom{K-1}{t-1}N}{\binom{K}{t}} = \frac{tN}{K}, \tag{3.7}$$

where t is referred to as the *caching factor*.

Delivery phase. Given a demand vector $\mathbf{d}$, the delivery phase follows, in a coded or uncoded manner, whichever results in a smaller delivery load. With the coded delivery procedure, the server broadcasts the following coded packet to the users in $\mathcal{S}^{t+1}_{K,l}$:

$$W_{\mathcal{S}^{t+1}_{K,l}} \triangleq \bigoplus_{k\in\mathcal{S}^{t+1}_{K,l}} W_{d_k,\mathcal{S}^{t+1}_{K,l}\backslash\{k\}}, \quad \forall l \in \left[\binom{K}{t+1}\right], \tag{3.8}$$

from which device $k \in \mathcal{S}^{t+1}_{K,l}$ can retrieve the missing subfile $W_{d_k,\mathcal{S}^{t+1}_{K,l}\backslash\{k\}}$ having cached all the subfiles $W_{d_k,\mathcal{S}^{t+1}_{K,l}\backslash\{k'\}}$, $\forall k' \in \mathcal{S}^{t+1}_{K,l}\backslash\{k\}$. It is easy to verify that, after delivering the coded packets given in (3.8), device k can retrieve all its missing subfiles $\bigcup_{l:k\notin\mathcal{S}^t_{K,l}} W_{d_k,\mathcal{S}^t_{K,l}}$, which, along with the cached subfiles of its requested file, $\bigcup_{l:k\in\mathcal{S}^t_{K,l}} W_{d_k,\mathcal{S}^t_{K,l}}$, enable user k to decode its desired file W_{d_k}. On the other hand, with the uncoded delivery procedure, the server broadcasts the missing subfiles of each file directly in an uncoded manner.

Delivery load. For the coded delivery procedure, users' demands can be satisfied after receiving $\binom{K}{t+1}$ coded packets, each of size $1/\binom{K}{t}$, for some $t \in [0:K]$. Thus, given a cache size $M = tN/K$ at each device, the delivery load

$$D = \frac{\binom{K}{t+1}}{\binom{K}{t}} = \frac{K-t}{t+1} = \frac{K(1-M/N)}{1+KM/N} \tag{3.9}$$

is achievable. On the other hand, for the uncoded delivery procedure, a total of $(1 - M/N) \cdot \min\{N, K\} \cdot F$ bits are delivered by the server directly to the users in an uncoded manner. As a result, the MAN scheme achieves the lower convex envelope of the delivery load-cache size trade-off

$$D(M) = (1 - M/N) \cdot \min\left\{\frac{K}{1+KM/N}, N\right\}, \; M = tN/K, \forall t \in [0:K], \tag{3.10}$$

which is achievable through memory-sharing.

Next, we consider the *decentralized caching* scenario, where the server knows neither the identity nor the number of users participating in the delivery phase during the placement phase.

Placement phase. Since it is not possible to coordinate the cache contents of users, each user caches a random MF/N bits of each file independently. Let $\mathcal{S} \subset [K]$. We denote the subfile of file W_i cached exclusively by users in $\mathcal{S}$ by $W_{i,\mathcal{S}}$. We note that, for large enough F, the size of subfile $W_{i,\mathcal{S}}$ is approximately $\left(\frac{M}{N}\right)^{|\mathcal{S}|}\left(1 - \frac{M}{N}\right)^{K-|\mathcal{S}|}F$ bits. Accordingly, the cache content at user k is given by

$$B_k = \bigcup_{i\in[N]} \bigcup_{\mathcal{S}\subset[K]:k\in\mathcal{S}} W_{i,\mathcal{S}}, \quad \text{for } k \in [K]. \tag{3.11}$$

Delivery phase. Similarly to the delivery phase of the centralized caching scenario, both coded and uncoded delivery procedures are considered, and the one resulting in a lower delivery load is performed. Consider a non-empty subset of users $\mathcal{S} \subset [K]$. With the coded delivery procedure, the server broadcasts coded packet

$$W_{\mathcal{S}} = \bigoplus\nolimits_{k\in\mathcal{S}} W_{d_k,\mathcal{S}\setminus\{k\}} \tag{3.12}$$

to the users in $\mathcal{S}$. After receiving $W_{\mathcal{S}}$, user $k \in \mathcal{S}$ can recover subfile $W_{d_k,\mathcal{S}\setminus\{k\}}$ since it has cached all the subfiles $W_{d_{k'},\mathcal{S}\setminus\{k'\}}$, $\forall k' \in \mathcal{S}\setminus\{k\}$. Thus, delivering $\bigcup_{\mathcal{S}\subset[K]:k\in\mathcal{S}} W_{\mathcal{S}}$ enables user k to recover all the subfiles $\bigcup_{\mathcal{S}\subset[K]:k\in\mathcal{S}} W_{\mathcal{S}\setminus\{k\}}$, and satisfies its demand together with its cached content given in (3.11). On the other hand, with the uncoded delivery procedure, the server broadcasts enough number of linear combinations of the requested files satisfying the demands of all the users.

Delivery load. The coded packet delivered in (3.12) for the coded delivery procedure is approximately of size $\left(\frac{M}{N}\right)^{|\mathcal{S}|-1}\left(1 - \frac{M}{N}\right)^{K-|\mathcal{S}|+1}F$ bits. Therefore, counting over all non-empty subsets of $[K]$, the delivery load for the coded delivery procedure is given by

$$D = \sum_{s=1}^{K} \binom{K}{s} (\tfrac{M}{N})^{s-1} (1 - \tfrac{M}{N})^{K-s+1} = \tfrac{N-M}{M} (1 - (1 - \tfrac{M}{N})^{K}). \tag{3.13}$$

On the other hand, it is easy to verify that the uncoded delivery procedure requires a delivery load of $(1 - M/N) \cdot \min\{N, K\}$. As a result, the delivery load for a cache size of M is given by

$$D(M) = \tfrac{N-M}{M} \cdot \min\left\{1 - (1 - M/N)^{K}, M\right\}. \tag{3.14}$$

3.1.2 Generalization of the MAN scheme

The MAN scheme is designed to minimize the delivery load for the worst-case demand combination, which maximizes the delivery load. Indeed, it is shown in [10] that the MAN scheme is optimal when constrained to uncoded cache placement with $N \geq K$. However, the coded delivery procedure of the MAN scheme does not take into account the potential repetitions among requested files, i.e., when a file is requested by more than one user. This happens for $N < K$ even if we consider only the worst-case demand combinations. Also, when we are interested in the average delivery load, averaged over the demand combinations, rather than focusing only on the worst-case delivery rate, accommodating for repetitions in the most efficient manner is essential [5–10].

Here we elaborate the coded delivery technique introduced in [10]. For the delivery phase, given a demand vector $\mathbf{d}$, let $N_{\mathbf{d}}$ denote the number of distinct demands. Also, let $\mathcal{U}_{\mathbf{d}}$ denote the set of $N_{\mathbf{d}}$ users with distinct demands, referred to as the *leaders*, where one user is selected at random among the users with the same demand. The achievability of a delivery load and cache size pair $(D(\mathbf{d}), M)$ with respect to a specific demand vector $\mathbf{d}$ at the delivery phase is defined in the following. Given a specific demand vector $\mathbf{d}$, the error probability is defined as

$$P_e(\mathbf{d}) \triangleq \Pr\left\{\bigcup_{k=1}^{K} \left\{\hat{W}_k \neq W_{d_k}\right\}\right\}, \tag{3.15}$$

and the delivery load and cache size pair $(D(\mathbf{d}), M)$ is *achievable* if there exists a caching and delivery code $(M, D(\mathbf{d}), F)$ such that, for large enough F, $P_e(\mathbf{d}) < \varepsilon$, for any $\varepsilon > 0$.

We first focus on the delivery phase of the centralized caching scenario, where the same placement phase as in the MAN scheme is performed. Given a demand vector $\mathbf{d}$, we show that the server does not need to broadcast the coded packet $W_{\mathcal{S}_{K,l}^{t+1}}$ when $\mathcal{S}_{K,l}^{t+1} \cap \mathcal{U}_{\mathbf{d}} = \emptyset$, $\forall l \in \left[\binom{K}{t+1}\right]$, and all the users can still retrieve the missing subfiles of their demands after receiving the remaining coded packets $W_{\mathcal{S}_{K,l}^{t+1}}$ for which $\mathcal{S}_{K,l}^{t+1} \cap \mathcal{U}_{\mathbf{d}} \neq \emptyset$, $\forall l \in \left[\binom{K}{t+1}\right]$. Following [10, Lemma 1], for a demand vector $\mathbf{d}$ with $N_{\mathbf{d}}$ distinct requests, let $\mathcal{U}_{\mathbf{d}} \subset \mathcal{C} \subset [K]$. Also, let $\mathcal{G}_{\mathcal{C}}$ be defined as the set consisting of all the subsets of $\mathcal{C}$ with cardinality $N_{\mathbf{d}}$ such that all $N_{\mathbf{d}}$ users in each subset request distinct files. For any $\mathcal{C}$, it is proved in [10, Lemma 1] that

$$\bigoplus_{\mathcal{G}\in\mathcal{G}_{\mathcal{C}}} W_{\mathcal{C}\backslash\mathcal{G}} = \mathbf{0}, \tag{3.16}$$

where $\mathbf{0}$ denotes the zero vector.

Consider a demand vector $\mathbf{d}$ with $N_{\mathbf{d}} < K$, and any set $\mathcal{S} \subset [K]\backslash\mathcal{U}_{\mathbf{d}}$ of users. By setting $\mathcal{C} = \mathcal{S} \cup \mathcal{U}_{\mathbf{d}}$, it follows from (3.16) that

$$\begin{aligned}\bigoplus_{\mathcal{G}\in\mathcal{G}_{\mathcal{C}}} W_{\mathcal{C}\backslash\mathcal{G}} &= \left(\bigoplus_{\mathcal{G}\in\mathcal{G}_{\mathcal{C}}\backslash\mathcal{U}_{\mathbf{d}}} W_{\mathcal{C}\backslash\mathcal{G}}\right) \oplus W_{\mathcal{C}\backslash\mathcal{U}_{\mathbf{d}}}\\ &= \left(\bigoplus_{\mathcal{G}\in\mathcal{G}_{\mathcal{C}}\backslash\mathcal{U}_{\mathbf{d}}} W_{\mathcal{C}\backslash\mathcal{G}}\right) \oplus W_{\mathcal{S}} = \mathbf{0},\end{aligned} \tag{3.17}$$

which is equivalent to

$$W_{\mathcal{S}} = \bigoplus_{\mathcal{G}\in\mathcal{G}_{\mathcal{C}}\backslash\mathcal{U}_{\mathbf{d}}} W_{\mathcal{C}\backslash\mathcal{G}}. \tag{3.18}$$

Therefore, having received all the coded packets $W_{\mathcal{C}\backslash\mathcal{G}}$, $\forall\mathcal{G} \in \mathcal{G}_{\mathcal{C}}\backslash\mathcal{U}_{\mathbf{d}}$, $W_{\mathcal{S}}$ can be recovered through (3.18). It is easy to verify that, for any $\mathcal{G} \in \mathcal{G}_{\mathcal{C}}\backslash\mathcal{U}_{\mathbf{d}}$, we have

$$|\mathcal{C}\backslash\mathcal{G}| = |\mathcal{S}|, \tag{3.19a}$$

$$(\mathcal{C}\backslash\mathcal{G}) \cap \mathcal{U}_{\mathbf{d}} \neq \emptyset; \tag{3.19b}$$

i.e., each coded packet on the right-hand side (RHS) of (3.18) is targeted for a set of $|\mathcal{S}|$ users, at least one of which is in set $\mathcal{U}_{\mathbf{d}}$. Therefore, by broadcasting only the coded packets targeted for at least one user in $\mathcal{U}_{\mathbf{d}}$ the coded packets targeted for users in $[K]\backslash\mathcal{U}_{\mathbf{d}}$ can be retrieved.

With this coded delivery technique for demand vector $\mathbf{d}$, the server does not broadcast coded packets $W_{\mathcal{S}_l^{t+1}}$ targeted exclusively for users in $[K]\backslash\mathcal{U}_{\mathbf{d}}$, where $|[K]\backslash\mathcal{U}_{\mathbf{d}}| = K - N_{\mathbf{d}}$. Therefore, given cache size $M = tN/K$, for some $t \in [0 : K]$, this coded delivery technique delivers $\binom{K}{t+1} - \binom{K-N_{\mathbf{d}}}{t+1}$ coded packets, each of size $F/\binom{K}{t}$ bits. As a result, the delivery load for $M = tN/K$ and demand vector $\mathbf{d}$ is given by

$$D(\mathbf{d}) = \frac{\binom{K}{t+1}-\binom{K-N_{\mathbf{d}}}{t+1}}{\binom{K}{t}}. \tag{3.20}$$

The worst-case demand combination maximizing $D(\mathbf{d})$ corresponds to $N_{\mathbf{d}} = \min\{K, N\}$, for which the lower convex envelope of

$$D(tN/K) = \frac{\binom{K}{t+1}-\binom{K-\min\{K,N\}}{t+1}}{\binom{K}{t}}, \quad \text{for } t \in [0 : K], \tag{3.21}$$

is achieved, and it is proved to be optimal under uncoded cache placement [10].

For the decentralized caching scenario, given a demand vector $\mathbf{d}$, for any non-empty subset of users $\mathcal{S} \subset [K]$, the coded packet $W_{\mathcal{S}}$, given in (3.12), is delivered only if $\mathcal{S} \cap \mathcal{U}_{\mathbf{d}} \neq \emptyset$. Consequently, considering all the s-element subsets of users, the server delivers $\binom{K}{s} - \binom{K-N_{\mathbf{d}}}{s}$ coded packets, each of size $\left(\frac{M}{N}\right)^{s-1}\left(1-\frac{M}{N}\right)^{K-s+1} F$ bits, where $\binom{K-N_{\mathbf{d}}}{s}$ coded packets targeted for users in $\mathcal{S}\backslash\mathcal{U}_{\mathbf{d}}$ are not delivered, $\forall\mathcal{S} \subset [K]$ such that $|\mathcal{S}| = s$. As a result, the required delivery load is

$$\begin{aligned} D(\mathbf{d}) &= \sum_{s=1}^{K} \left(\tfrac{M}{N}\right)^{s-1} \left(1-\tfrac{M}{N}\right)^{K-s+1} \left(\binom{K}{s} - \binom{K-N_{\mathbf{d}}}{s}\right) \\ &= \tfrac{N-M}{M}\left(1-(1-M/N)^{N_{\mathbf{d}}}\right). \end{aligned} \quad (3.22)$$

The worst-case demand combination happens when $N_{\mathbf{d}} = \min\{N, K\}$ leading to the following delivery load-cache size trade-off:

$$D(M) = \tfrac{N-M}{M}\left(1-(1-M/N)^{\min\{N,K\}}\right), \quad (3.23)$$

which is shown to be optimal under uncoded cache placement constraint [10].

3.2 Erasure BC

Having illustrated the fundamental caching and coded delivery techniques in the literature, we now consider a noisy BC during the delivery phase, and focus on the centralized caching scenario. In particular, a memoryless packet erasure BC is adopted, where each packet is separately channel coded, and it either arrives at a user correctly or is lost. In the previous section, our goal was to minimize the delivery load, which corresponds to the delay in satisfying all the user demands due to the error-free channel assumption. However, when there are errors over the channel, we have to introduce error correction codes to deal with these errors, and the communication rate becomes the appropriate measure of performance. Accordingly, each file in the library is assumed to be distributed uniformly over the set $\left[\lceil 2^{nR} \rceil\right]$, where R represents the rate of the file, and n denotes the channel coding blocklength.

3.2.1 Problem formulation

At each channel use, the server broadcasts an L-bit common message X from the alphabet $\mathcal{X} \triangleq \{0,1\}^L$, and the output alphabet is represented by $\mathcal{Y} \triangleq \mathcal{X} \cup \{\Delta\}$, where Δ denotes the erased packet. User k receives a packet correctly with probability $1-\delta_k$, and the symbol Δ with probability δ_k, $k \in [K]$; i.e., we have

$$P\left(Y_k = y_k \,\middle|\, X = x\right) = \begin{cases} 1-\delta_k, & \text{if } y_k = x, \\ \delta_k, & \text{if } y_k = \Delta. \end{cases} \quad (3.24)$$

Users are grouped into two disjoint sets of *weak* and *strong* users with respect to their erasure probabilities, where the users in the set of strong users have statistically better channel qualities than those in the set of weak users. Without loss of generality, the users are enumerated such that $\delta_1 \geq \delta_2 \geq \cdots \geq \delta_K$. The first K_w users are selected as weak users, denoted by $\mathcal{K}_w \triangleq [K_w]$, and the next $K_s = K - K_w$ users are selected as strong users, denoted by $\mathcal{K}_s \triangleq [K_w + 1 : K]$. To compensate for their worse channel quality, each weak user is equipped with a cache of size nM bits, as depicted in Figure 3.2.

Let $\boldsymbol{\delta} \triangleq \{\delta_1, \ldots, \delta_K\}$. Given a demand vector $\mathbf{d}$, a (R, M, n, F) caching and delivery code over the packet erasure BC consists of

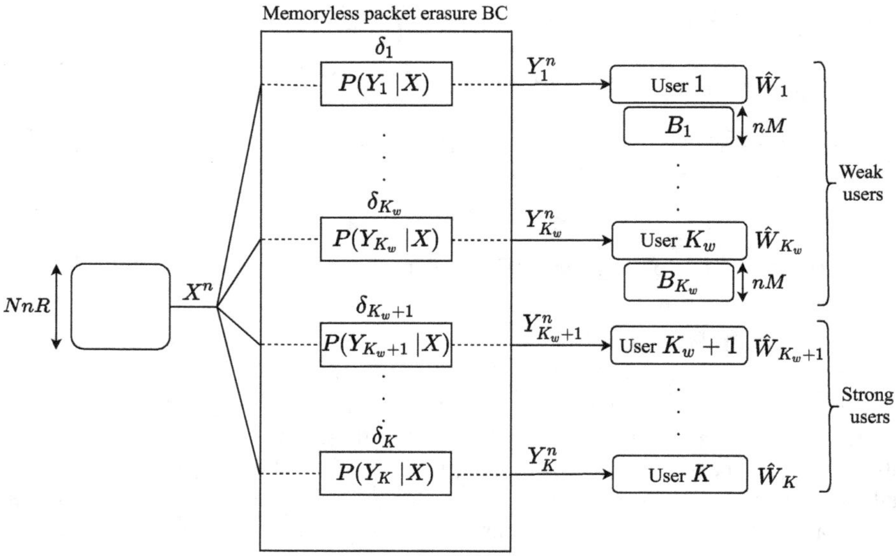

Figure 3.2 *Illustration of cache-aided memoryless packet erasure BC. Weak users have statistically worse channel qualities than the strong users, and each is equipped with a cache of size* nM *bits.*

i. K_w caching functions:

$$\phi_{\boldsymbol{\delta},k} : [\lceil 2^{nR} \rceil]^N \to [\lfloor 2^{nM} \rfloor], \quad k \in \mathcal{K}_w, \tag{3.25}$$

each of which maps the library to the cache content B_k at user k assuming the knowledge of $\boldsymbol{\delta}$, i.e., $B_k = \phi_{\boldsymbol{\delta},k}(\mathbf{W})$;

ii. Encoding function:

$$\psi_{\boldsymbol{\delta}} : [\lceil 2^{nR} \rceil]^N \times [N]^K \to \mathcal{X}^n, \tag{3.26}$$

which generates the length-n channel input X^n according to the library and demand vector $\mathbf{d}$ having the knowledge of $\boldsymbol{\delta}$, i.e., $X^n = \psi_{\boldsymbol{\delta}}(\mathbf{W}, \mathbf{d})$;

iii. K decoding functions:

$$\mu_k : \mathcal{Y}^n \times [\lfloor 2^{nM} \rfloor] \times [N]^K \to [\lceil 2^{nR} \rceil], \quad \forall k \in \mathcal{K}_w, \tag{3.27}$$

each of which recovers the file requested by weak user $k \in \mathcal{K}_w$ from its received signal Y_k^n, cache content B_k, and demand vector $\mathbf{d}$, i.e., $\hat{W}_k = \mu_k(Y_k^n, B_k, \mathbf{d})$, and

$$\mu_k : \mathcal{Y}^n \times [N]^K \to [\lceil 2^{nR} \rceil], \quad \forall k \in \mathcal{K}_s, \tag{3.28}$$

which recovers the file requested by strong user $k \in \mathcal{K}_s$ solely from its received signal Y_k^n and demand vector $\mathbf{d}$, i.e., $\hat{W}_k = \mu_k(Y_k^n, \mathbf{d})$.

The error probability P_e is defined as in (3.15). A rate-memory pair (R, M) is said to be *achievable* if there exists a caching and delivery code (R, M, n, L) such that, for large enough n, $P_e < \varepsilon$, for any $\varepsilon > 0$. Also, the capacity-memory trade-off for the caching and delivery code over the erasure BC is defined as

$$C(M) \triangleq \sup\{R : (R, M) \text{ is achievable}\}. \tag{3.29}$$

3.2.2 Preliminaries

Here we provide some background information on the memoryless packet erasure BCs, which will be used in presenting the achievable results later.

Memoryless packets erasure BC capacity. The capacity region of a memoryless packet erasure BC with K users and an independent message W_k with rate R_k for user k with erasure probability δ_k, for $k \in [K]$, is the closure of the set of non-negative rate tuples $(R_1, ..., R_K)$ satisfying [18]

$$\sum_{i=1}^{K} \frac{R_i}{(1-\delta_i)L} \leq 1, \tag{3.30}$$

where L is the length of the binary channel input per channel use.

Packet erasure BC with side information. Let W_1 and W_2 denote two messages distributed uniformly over $[2^{nR_1}]$ and $[2^{nR_2}]$, respectively, where we assume that 2^{nR_k} is an integer, for $k = 1, 2$. There are two users in the system with erasure probabilities $\delta_1 \geq \delta_2$. Messages W_1 and W_2 are targeted for users 1 and 2, respectively, and we assume that message W_2 is known by user 1 as side information. With the joint encoding technique introduced in [19], $2^{n(R_1+R_2)}$ codewords of length n are generated representing a pair of messages (w_1, w_2), for $w_1 \in [2^{nR_1}]$ and $w_2 \in [2^{nR_2}]$, where each entry of each codeword is generated independently according to a fixed distribution. To transmit particular messages $W_1 = w_1$ and $W_2 = w_2$ to user 1 and 2, respectively, the corresponding codeword is transmitted. From the joint typicality [20], for large enough n, user 2 can decode message $W_2 = w_2$ if

$$R_1 + R_2 \leq I(X; Y_2), \tag{3.31}$$

where $I(\cdot; \cdot)$ denotes the mutual information between its arguments, and user 1 can decode $W_1 = w_1$ using side information W_2, if

$$R_1 \leq I(X; Y_1). \tag{3.32}$$

For a memoryless packet erasure BC, the achievable rate region is given by

$$R_1 \leq (1 - \delta_1) L, \tag{3.33a}$$
$$R_1 + R_2 \leq (1 - \delta_2) L. \tag{3.33b}$$

This result can be generalized to multiple users. Consider a memoryless packet erasure BC with two disjoint sets of users $\mathcal{S}_1$ and $\mathcal{S}_2$, where all the users in set $\mathcal{S}_i$ have the same erasure probability δ_i, for $i = 1, 2$. A common message with rate W_i with rate R_i is to be transmitted to users in set $\mathcal{S}_i$, for $i = 1, 2$, where all the users in set $\mathcal{S}_1$

have access to message W_2 as side information. With the joint encoding technique, denoted by

$$\mathrm{JE}\left(\left(W_1\right)_{\mathcal{S}_1}, \left(W_2\right)_{\mathcal{S}_2}\right), \tag{3.34}$$

the rate pair (R_1, R_2) satisfying the following conditions is achievable

$$R_1 \quad \leq \left(1-\delta_1\right) L, \tag{3.35a}$$

$$R_1 + R_2 \leq \left(1-\delta_2\right) L, \tag{3.35b}$$

which is equivalent to

$$\max\left\{\frac{R_1}{(1-\delta_1)L}, \frac{R_1+R_2}{(1-\delta_2)L}\right\} \leq 1. \tag{3.36}$$

3.2.3 Achievable scheme

An achievable scheme providing a lower bound on the capacity-memory trade-off $C(M)$ is outlined in the following. The achievable scheme is characterized by a pair of parameters (p, q), where $p \in \left[0 : K_w\right]$ and $q \in \left[p : K_w\right]$.

Placement phase. File W_i, for $i \in [N]$, is divided into $(q-p+1)$ non-overlapping subfiles

$$W_i = \left(W_i^p, W_i^{p+1}, \ldots, W_i^q\right), \tag{3.37}$$

where subfile W_i^j, for $j \in \left[p : q\right]$, has rate

$$R^j \triangleq \frac{\gamma(p,\boldsymbol{\delta},j)}{\sum_{l=p}^{q}\gamma(p,\boldsymbol{\delta},l)} R, \tag{3.38}$$

with $\gamma(p, \boldsymbol{\delta}, j)$ defined as, for $j \in [p : q]$,

$$\gamma(p,\boldsymbol{\delta},j) \triangleq \frac{\binom{K_w}{j}}{\binom{K_w}{p}K_S^{j-p}} \prod_{i=p}^{j-1}\left(\frac{K_S}{\left(1-\delta_{K_w-i}\right)\sum_{l=K_w+1}^{K}\frac{1}{1-\delta_l}} - 1\right). \tag{3.39}$$

Note that $\sum_{j=p}^{q} R^j = R$. In the placement phase, subfiles $W_1^j, \ldots, W_N^j$ are cached by the K_w weak users using the MAN centralized cache placement described in Section 3.1.1 for a caching factor $t = j$, $\forall j \in \left[p : q\right]$; i.e., each subfile W_i^j is partitioned into $\binom{K_w}{j}$ equal-rate pieces,

$$W_i^j = \left(W^j_{i,\mathcal{S}^j_{K_w,1}}, W^j_{i,\mathcal{S}^j_{K_w,2}}, \ldots, W^j_{i,\mathcal{S}^j_{K_w,\binom{K_w}{j}}}\right), \ \forall i \in \left[N\right], \forall j \in \left[p : q\right]. \tag{3.40}$$

The piece $W^j_{i,\mathcal{S}^j_{K_w,l}}$, which has rate $R^j/\binom{K_w}{j}$, is cached by users in set $\mathcal{S}^j_{K_w,l}$, for $l \in \left[\binom{K_w}{j}\right]$. Therefore, the cache content at the weak user $k \in \left[K_w\right]$ is given by

$$B_k = \bigcup_{i\in[N]} \bigcup_{j\in[p:q]} \bigcup_{l\in\left[\binom{K_w}{j}\right]:k\in\mathcal{S}^j_{K_w,l}} W^j_{i,\mathcal{S}^j_{K_w,l}}. \tag{3.41}$$

Accordingly, $\binom{K_w-1}{j-1}$ pieces of subfile W^j_i, each of rate $R^j/\binom{K_w}{j}$, are cached by each weak user, for $j \in [p:q]$, $i \in [N]$. Thus, the normalized cache size at each weak user is given by

$$M = N\sum_{j=p}^{q}\binom{K_w-1}{j-1}\frac{R^j}{\binom{K_w}{j}} = \frac{N}{K_w}\sum_{j=p}^{q} jR^j = \frac{N\sum_{j=p}^{q} j\gamma(p,\delta,j)}{K_w\sum_{l=p}^{q}\gamma(p,\delta,l)}R. \tag{3.42}$$

Delivery phase. The delivery phase takes place over $q-p+2$ orthogonal time slots. During the i-th time slot a message is transmitted using $\beta_i n$ channel uses, such that $\sum_{i=1}^{q-p+2}\beta_i = 1$.

The first message of the delivery phase is targeted only for the weak users, and the goal is to deliver the missing bits of subfile $W^q_{d_k}$ to users in $\mathcal{K}_w$. Note that, for $q = K_w$, according to the cached content in (3.41), all the weak users have access to W^q_i, $\forall i \in [N]$; therefore, no message needs to be delivered. In the following, we consider $q < K_w$. The first message of the delivery phase is transmitted over $\binom{K_w}{q+1}$ orthogonal time slots, where in each slot, a common sub-message is delivered to a distinct set of $q+1$ weak users. Sub-message l is a codeword of length $\beta_{1,l}n$ channel uses, and is targeted to the users in set $\mathcal{S}^{q+1}_{K_w,l}$, for $l \in \left[\binom{K_w}{q+1}\right]$, such that $\sum_{l=1}^{\binom{K_w}{q+1}}\beta_{1,l} = \beta_1$. Following the coded delivery procedure of the MAN scheme described in Section 3.1.1, the coded packet delivered by the sub-message l is given by

$$W^q_{\mathcal{S}^{q+1}_{K_w,l}} \triangleq \bigoplus_{k\in\mathcal{S}^{q+1}_{K_w,l}} W^q_{d_k,\mathcal{S}^{q+1}_{K_w,l}\backslash\{k\}}, \quad \text{for } l \in \left[\binom{K_w}{q+1}\right]. \tag{3.43}$$

After receiving $W^q_{\mathcal{S}^{q+1}_{K_w,l}}$, user $k \in \mathcal{S}^{q+1}_{K_w,l}$ can recover $W^q_{d_k,\mathcal{S}^{q+1}_{K_w,l}\backslash\{k\}}$, for $l \in \left[\binom{K_w}{q+1}\right]$, having access to cache content B_k given in (3.41). Thus, together with its cache content, user $k \in [K_w]$ can recover subfile $W^q_{d_k}$.

The delivery technique transmitting messages $2, 3, \ldots, q-p+1$ follow the same procedure, described next. With the message $q-i+1$ of length $\beta_{q-j+1}n$ channel uses, the server delivers the missing bits of subfile $W^q_{d_{k_1}}$ to weak user k_1, $k_1 \in \mathcal{K}_w$, and $W^{j+1}_{d_{k_2}}$ to strong user k_2, $k_2 \in \mathcal{K}_s$, for $j = q-1, q-2, \ldots, p$.[a] Message $(q-j+1)$ is delivered through $\binom{K_w}{j+1}$ sub-messages, transmitted over orthogonal time slots, where sub-message l is of length $\beta_{q-j+1,l}n$ channel uses, such that $\sum_{l=1}^{\binom{K_w}{j+1}}\beta_{q-j+1,l} = \beta_{q-j+1}$. With the l-th sub-message, the coded packet

$$W^j_{\mathcal{S}^{j+1}_{K_w,l}} \triangleq \bigoplus_{k\in\mathcal{S}^{j+1}_{K_w,l}} W^j_{d_k,\mathcal{S}^{j+1}_{K_w,l}\backslash\{k\}}, \tag{3.44}$$

is delivered to the weak users in set $\mathcal{S}^{j+1}_{K_w,l}$, while packets

[a]For example, with the second message, subfile $W^q_{d_{k_2}}$ is delivered to strong user $k_2 \in \mathcal{K}_s$, and subfile $W^{q-1}_{d_{k_1}}$ to weak user $k \in \mathcal{K}_w$. With the third message, subfile $W^{q-2}_{d_{k_2}}$ is delivered to strong user $k_2 \in \mathcal{K}_s$, and subfile $W^{q-2}_{d_{k_1}}$ to weak user $k_1 \in \mathcal{K}_w$, and so on.

$$\left\{ W^{j+1}_{d_{K_w+1}, S^{j+1}_{K_w,l}}, \dots, W^{j+1}_{d_K, S^{j+1}_{K_w,l}} \right\}, \tag{3.45}$$

are delivered to the strong users in $\mathcal{K}_s$, for $j = q-1, \dots, p$, and $l \in \left[\binom{K_w}{j+1}\right]$. Observe that, after receiving sub-message $W^j_{S^{j+1}_{K_w,l}}$, weak user $k \in S^{j+1}_{K_w,l}$ can recover $W^j_{d_k, S^{j+1}_{K_w,l} \setminus \{k\}}$, for $l \in \left[\binom{K_w}{j+1}\right]$, for $j = q-1, \dots, p$. Note that, the packets in (3.45), which are targeted to the strong users, are all known by each weak user in set $S^{j+1}_{K_w,l}$ as side information in their caches. Accordingly, the l-th sub-message of message $q-j+1$, for $j = q-1, \dots, p$, and $l \in \left[\binom{K_w}{j+1}\right]$, is transmitted using the joint encoding technique:

$$\mathrm{JE}\left(\left(W^j_{S^{j+1}_{K_w,l}} \right)_{S^{j+1}_{K_w,l}}, \left(W^{j+1}_{d_{K_w+1}, S^{j+1}_{K_w,l}}, \dots, W^{j+1}_{d_K, S^{j+1}_{K_w,l}} \right)_{\mathcal{K}_s} \right). \tag{3.46}$$

To increase the efficiency of the delivery phase, the j-th sub-message is delivered via K_s orthogonal time periods, such that in the m-th period a codeword of length $\beta_{q-j+1,l,m} n$ channel uses is transmitted, where $\sum_{m=1}^{K_s} \beta_{q-j+1,l,m} = \beta_{q-j+1,l}$. To this end, coded packet $W^j_{S^{j+1}_{K_w,l}}$, targeted for users in $S^{j+1}_{K_w,l}$, is divided into K_s non-overlapping equal-rate pieces

$$W^j_{S^{j+1}_{K_w,l}} = \left(W^j_{S^{j+1}_{K_w,l},1}, \dots, W^j_{S^{j+1}_{K_w,l},K_s} \right), \text{ for } l \in \left[\binom{K_w}{j+1}\right], \tag{3.47}$$

and during the m-th time period the following packet is delivered through joint encoding:

$$\mathrm{JE}\left(\left(W^j_{S^{j+1}_{K_w,l},m} \right)_{S^{j+1}_{K_w,l}}, \left(W^{j+1}_{d_{K_w+m}, S^{j+1}_{K_w,l}} \right)_{\{K_w+m\}} \right), \text{ for } m \in [K_s]. \tag{3.48}$$

Note that, after receiving messages 2 to $q-p+1$, weak user $k_1 \in \mathcal{K}_w$ can decode subfiles $\left(W^{q-1}_{d_{k_1}}, W^{q-2}_{d_{k_1}}, \dots, W^p_{d_{k_1}} \right)$, while strong user $k_2 \in \mathcal{K}_s$ can decode subfiles $\left(W^q_{d_{k_2}}, W^{q-1}_{d_{k_2}}, \dots, W^{p+1}_{d_{k_2}} \right)$; therefore, together with message 1, weak users' demands are fully satisfied. However, strong user $k_2 \in \mathcal{K}_s$ only requires to receive subfile $W^p_{d_{k_2}}$.

During the last time slot, subfile $W^p_{d_k}$ is delivered to the strong user $k \in \mathcal{K}_s$ using the channel coding scheme for standard memoryless packet erasure BCs.

With the packet delivered in (3.48) through joint encoding, the strong users can also benefit from the caches available only at the weak users, without any additional cost.

Table 3.1 summarizes the packets delivered with each message in the delivery phase. Also, for clarification, Table 3.2 provides the packets delivered using the above achievable scheme for a simple example, where there are $K_w = 3$ weak and $K_s = 2$ strong users in the system. Please refer to [21] for a detailed presentation of the above example with performance analysis.

Table 3.1 Packets sent with messages 1 to q−p+2 in the delivery phase

Message index	Sub-message index	Delivered content
1	1	$W^q_{\mathcal{S}^{q+1}_{K_w,1}}$ to users in $\mathcal{S}^{q+1}_{K_w,1}$
	⋮	⋮
	$\binom{K_w}{q+1}$	$W^q_{\mathcal{S}^{q+1}_{K_w,\binom{K_w}{q+1}}}$ to users in $\mathcal{S}^{q+1}_{K_w,\binom{K_w}{q+1}}$
$t=2,\ldots,$ $q-p+1$	1	$\mathrm{JE}\left(\left(W^{q-t+1}_{\mathcal{S}^{q-t+2}_{K_w,1},m}\right)_{\mathcal{S}^{q-t+2}_{K_w,1}},\left(W^{q-t+2}_{d_{K_w+m},\mathcal{S}^{q-t+2}_{K_w,1}}\right)_{\{K_w+m\}}\right)$ in m-th time period, for $m\in[K_S]$
	⋮	⋮
	$\binom{K_w}{q-t+2}$	$\mathrm{JE}\left(\left(W^{q-t+1}_{\mathcal{S}^{q-t+2}_{K_w,\binom{K_w}{q-t+2}},m}\right)_{\mathcal{S}^{q-t+2}_{K_w,\binom{K_w}{q-t+2}}},\left(W^{q-t+2}_{d_{K_w+m},\mathcal{S}^{q-t+2}_{K_w,\binom{K_w}{q-t+2}}}\right)_{\{K_w+m\}}\right)$ in m-th time period $m\in[K_s]$
$q-p+2$		$W^p_{d_k}$ to user $k\in\mathcal{K}_s$

3.2.4 *Achievable rate-memory pair*

Here the above achievable scheme is analyzed, providing a lower bound on the capacity-memory trade-off $C(M)$. The rate of each coded packet targeted to a subset of weak users is chosen such that the weakest user among the intended subset of users can decode it.

With sub-message l of message 1 of blocklength $\beta_{1,l}n$, $W^q_{\mathcal{S}^{q+1}_{K_w,l}}$, given in (3.43), of rate $R^q/\binom{K_w}{q}$ is transmitted to the weak users in $\mathcal{S}^{q+1}_{K_w,l}$, for $l\in\left[\binom{K_w}{q+1}\right]$. The rate of $W^q_{\mathcal{S}^{q+1}_{K_w,l}}$ is set such that the weakest user in $\mathcal{S}^{q+1}_{K_w,l}$ can decode it, i.e.,

$$\frac{R^q}{\binom{K_w}{q}}\le\beta_{1,l}\left(1-\max_{r\in\mathcal{S}^{q+1}_{K_w,l}}\{\delta_r\}\right)L,\quad\text{for } l\in\left[\binom{K_w}{q+1}\right].\tag{3.49}$$

Summing over all possible subsets of users $\mathcal{S}^{(q+1)}_{K_w,l}$, $\forall l\in\left[\binom{K_w}{q+1}\right]$, it follows that

Table 3.2 Packets sent with messages 1 to 4 in the delivery phase when $K_w = 3$ and $K_s = 2$

Message 1	$W^2_{d_1,\{2,3\}} \oplus W^2_{d_2,\{1,3\}} \oplus W^2_{d_3,\{1,2\}}$ to receivers 1, 2, 3	
Message 2	Sub-message 1	$\mathrm{JE}\left(\left(W^1_{d_1,\{2\}} \oplus W^1_{d_2,\{1\}}\right)_{\{1,2\}}, \left(W^2_{d_4,\{1,2\}}, W^2_{d_5,\{1,2\}}\right)_{\{4,5\}}\right)$
	Sub-message 2	$\mathrm{JE}\left(\left(W^1_{d_1,\{3\}} \oplus W^1_{d_3,\{1\}}\right)_{\{1,3\}}, \left(W^2_{d_4,\{1,3\}}, W^2_{d_5,\{1,3\}}\right)_{\{4,5\}}\right)$
	Sub-message 3	$\mathrm{JE}\left(\left(W^1_{d_2,\{3\}} \oplus W^1_{d_3,\{2\}}\right)_{\{2,3\}}, \left(W^2_{d_4,\{2,3\}}, W^2_{d_5,\{2,3\}}\right)_{\{4,5\}}\right)$
Message 3	Sub-message 1	$\mathrm{JE}\left(\left(W^0_{d_1,\emptyset}\right)_{\{1\}}, \left(W^1_{d_4,\{1\}}, W^1_{d_5,\{1\}}\right)_{\{4,5\}}\right)$
	Sub-message 2	$\mathrm{JE}\left(\left(W^0_{d_2,\emptyset}\right)_{\{2\}}, \left(W^1_{d_4,\{2\}}, W^1_{d_5,\{2\}}\right)_{\{4,5\}}\right)$
	Sub-message 3	$\mathrm{JE}\left(\left(W^0_{d_3,\emptyset}\right)_{\{3\}}, \left(W^1_{d_4,\{3\}}, W^1_{d_5,\{3\}}\right)_{\{4,5\}}\right)$
Message 4	$W^0_{d_4,\emptyset}$ to user 4 and $W^0_{d_5,\emptyset}$ to user 5	

$$\frac{R^q}{\binom{K_w}{q}} \sum_{r=1}^{K_w-q} \frac{\binom{K_w-r}{q}}{1-\delta_r} \leq \sum_{l=1}^{\binom{K_w}{q+1}} \beta_{1,l} L = \beta_1 L. \tag{3.50}$$

The packet is given in (3,48), $W^j_{S^{j+1}_{K_w},l,m}$, targeted for the users in $\mathcal{S}^{j+1}_{K_w,l}$, is of rate $R^j/\left(K_s\binom{K_w}{j}\right)$, while $W^{j+1}_{d_{K_w+m},S^{j+1}_{K_w,l}}$, destined for user K_w+m, is of rate $R^{j+1}/\binom{K_w}{j+1}$, for $j = q-1, \ldots, p$, $l \in \binom{K_w}{j+1}$, and $m \in [k_s]$. These two packets are transmitted through joint encoding, where all the users in set $\mathcal{S}^{j+1}_{K_w,l}$ have access to $W^{j+1}_{d_{K_w+m},S^{j+1}_{K_w,l}}$ in their caches. Therefore, from (3.36), it follows that the codeword in (3.48) can be decoded correctly by the intended users if, for $m \in [k_s]$,

$$\max\left\{\frac{R^j/\left(K_s\binom{K_w}{j}\right)}{\left(1-\max_{r\in S^{j+1}_{K_w,l}}\{\delta_r\}\right)L}, \frac{R^j/\left(K_s\binom{K_w}{j}\right)+R^{j+1}/\binom{K_w}{j+1}}{(1-\delta_{K_w+m})L}\right\} \leq \beta_{q-j+1,l,m}, \tag{3.51}$$

where the rate of $W^j_{S^{j+1}_{K_w,l},m}$ is limited by the weakest receiver in $\mathcal{S}^{j+1}_{K_w,l}$, for $j = q-1, \ldots, p$, and $l \in \left[\binom{K_w}{j+1}\right]$. By summing up all the K_s inequalities in (3.51), we have

$$\max\left\{\frac{R^j/\binom{K_w}{j}}{\left(1-\max_{r\in S^{j+1}_{K_w,l}}\{\delta_r\}\right)L}, \left(\frac{R^j}{K_s\binom{K_w}{j}} + \frac{R^{j+1}}{\binom{K_w}{j+1}}\right)\sum_{m=1}^{K_s}\frac{1}{\left(1-\delta_{K_w+m}\right)L}\right\}$$
$$\leq \beta_{q-j+1,l}, \quad \text{for } j = q-1, \ldots, p, \text{ and } l \in \left[\binom{K_w}{j+1}\right]. \tag{3.52}$$

According to the choice of (3.38), and the fact that

$$\gamma\left(p,\boldsymbol{\delta},j+1\right)=\gamma\left(p,\boldsymbol{\delta},j\right)\frac{\binom{K_w}{j+1}}{\binom{K_w}{j}K_S}\left(\frac{K_S}{\left(1-\delta_{K_w-j}\right)\sum_{l=K_w+1}^{K}\frac{1}{1-\delta_l}}-1\right), \tag{3.53}$$

which follows from the definition in (3.39), the second term of the maximization in (3.52) is reduced to

$$\frac{R^j/\binom{K_w}{j}}{\left(1-\delta_{K_w-j}\right)L}. \tag{3.54}$$

Thus, (3.52) is simplified as follows:

$$\max\left\{\frac{R^j/\binom{K_w}{j}}{\left(1-\max_{r\in\mathcal{S}_{K_w,l}^{j+1}}\{\delta_r\}\right)L},\frac{R^j/\binom{K_w}{j}}{\left(1-\delta_{K_w-j}\right)L}\right\}\le\beta_{q-j+1,l},$$

$$\text{for } j=q-1,\dots,p,\text{ and } l\in\left[\binom{K_w}{j+1}\right]. \tag{3.55}$$

Note that $\left|\mathcal{S}_{K_w,l}^{j+1}\right|=j+1$; therefore, we have, for $j=q-1,\dots,p,$

$$\max_{r\in\mathcal{S}_{K_w,l}^{j+1}}\{\delta_r\}\ge\delta_{K_w-j},\quad\forall l\in\left[\binom{K_w}{j+1}\right]. \tag{3.56}$$

From (3.56), (3.55) is reduced to, for $j=q-1,\dots,p,$ and $l\in\left[\binom{K_w}{j+1}\right]$,

$$\frac{R^j/\binom{K_w}{j}}{\left(1-\max_{r\in\mathcal{S}_{K_w,l}^{j+1}}\{\delta_r\}\right)L}\le\beta_{q-j+1,l}, \tag{3.57}$$

After summing up the inequality in (3.57) over all values of l, each corresponds to a distinct $(j+1)$-element subset of weak users $\mathcal{K}_w$, it follows that, for $j=q-1,\dots,p,$

$$\frac{R^j}{\binom{K_w}{j}}\sum_{r=1}^{K_w-j}\frac{\binom{K_w-r}{j}}{1-\delta_r}\le\sum_{l=1}^{\binom{K_w}{j+1}}\beta_{q-j+1,l}L=\beta_{q-j+1}L. \tag{3.58}$$

During the last time slot, strong users are served, and subfile $W_{d_k}^p$ is delivered to user k, $\forall k\in\mathcal{K}_s$ following the channel coding technique for memoryless packet erasure BCs. According to (3.30), user $k\in\mathcal{K}_s$ can decode subfile $W_{d_k}^p$ of rate R^p correctly, if

$$R^p\sum_{k=K_w+1}^{K}\frac{1}{1-\delta_k}\le\beta_{q-p+2}L. \tag{3.59}$$

By combining inequalities (3.50), (3.58), and (3.59), we have

$$\sum_{j=p}^{q}\left(\frac{R^j}{\binom{K_w}{j}}\sum_{l=1}^{K_w-j}\frac{\binom{K_w-l}{j}}{1-\delta_l}\right)+R^p\sum_{j=K_w+1}^{K}\frac{1}{1-\delta_j}\le\sum_{j=p-1}^{q}\beta_{q-j+1}L=L. \tag{3.60}$$

By replacing R^j, for $j\in[p:q]$, with the expression in (3.38), it follows that

$$R \leq \frac{L\sum_{j=p}^{q}\gamma(p,\delta,i)}{\sum_{j=p}^{q}\left(\frac{\gamma(p,\delta,j)}{\binom{K_w}{j}}\sum_{l=1}^{K_w-j}\frac{\binom{K_w-l}{j}}{1-\delta_l}\right)+\sum_{j=K_w+1}^{K}\frac{1}{1-\delta_j}}, \tag{3.61}$$

which together with the normalized cache size M, given in (3.40), proves the achievability of a rate-memory trade-off pair presented in the following theorem.

Theorem 1.1: Consider caching and coded delivery of N files over a memoryless packet erasure BC with K_w weak and K_s strong users, where each weak user is equipped with a cache of normalized size M. Memory-rate pairs $\left(M_{(p,q)}, R_{(p,q)}\right)$, $\forall p \in [0 : K_w]$, $\forall q \in [p : K_w]$, are achievable, where

$$R_{(p,q)} \triangleq \frac{L\sum_{j=p}^{q}\gamma(p,\delta,j)}{\sum_{j=p}^{q}\left(\frac{\gamma(p,\delta,j)}{\binom{K_w}{j}}\sum_{l=1}^{K_w-j}\frac{\binom{K_w-l}{j}}{1-\delta_l}\right)+\sum_{j=K_w+1}^{K}\frac{1}{1-\delta_j}}, \tag{3.62a}$$

$$M_{(p,q)} \triangleq \frac{N\sum_{j=p}^{q}j\gamma(p,\delta,j)}{K_w\sum_{l=p}^{q}\gamma(p,\delta,l)}R_{(p,q)}. \tag{3.62b}$$

The upper convex hull of these $(K_w + 1)(K_w + 2)/2$ memory-rate pairs can also be achieved through memory-sharing.

3.2.5 Upper bound

Here we provide an upper bound on the capacity-memory trade-off $C(M)$. We first consider a general cache-aided degraded BC, and we then specify the results for the memoryless packet erasure BC model.

Consider a cache-aided BC where user k is equipped with a cache of size nM_k bits, $k \in [K]$, and a memoryless BC with an arbitrary transition probability

$$P\left(Y_K = y_K, \ldots, Y_1 = y_1 \,|X = x\right). \tag{3.63}$$

Assume that the BC is *physically degraded*; i.e., we have the following Markov chain:

$$X - Y_K - Y_{K-1} - \cdots - Y_1. \tag{3.64}$$

We modify the caching and coded delivery code defined in Section 3.2.1 by letting users have different cache sizes as described above and represent the new code with the tuple $(R, M_1, \ldots, M_K, n, L)$. A rate-memory tuple $(R, M_1, \ldots, M_K)$ is said to be *achievable*, if there exists a caching and delivery code $(R, M_1, \ldots, M_K, n, L)$ such that, for large enough n, $P_e < \varepsilon$, for any $\varepsilon > 0$. Also, the capacity-memory trade-off is defined as

$$C\left(M_1, \ldots, M_K\right) \triangleq \sup\left\{R : \left(R, M_1, \ldots, M_K\right) \text{ is achievable}\right\}. \tag{3.65}$$

Given a non-zero subset of users $\mathcal{S} = \left\{s_1, s_2, \ldots, s_{|\mathcal{S}|}\right\} \subset [K]$, such that $s_1 < s_2 < \cdots < s_{|\mathcal{S}|}$, we define

$$R_{\mathcal{S}} \triangleq \max\min\left\{I\left(U_1;Y_{s_1}\right),I\left(U_2;Y_{s_2}\left|U_1\right.\right),...,I\left(U_{|\mathcal{S}|-1};Y_{s_{|\mathcal{S}|-1}}\left|U_{|\mathcal{S}|-2}\right.\right),\right. \tag{3.66}$$

for auxiliary random variables $U_1, \ldots, U_{|\mathcal{S}|-1}$ forming the following Markov chain:

$$U_1 - U_2 - \cdots - U_{|\mathcal{S}|-1} - X - (Y_{s_1}, \ldots, Y_{s_{|\mathcal{S}|}}), \tag{3.67}$$

where the maximization is over the joint distribution $U_1, \ldots, U_{|\mathcal{S}|-1}, X$. We note that $R_{\mathcal{S}}$ corresponds to the highest achievable common rate for transmission to the receivers in $\mathcal{S}$ over a BC without receiver caching. The following theorem, whose proof is provided in [22, Appendix B], presents an upper bound on $C(M_1, ..., M_K)$.

Theorem 1.2: For the cache-aided degraded BC with a normalized cache size M_k at user k, $C(M_1, ..., M_K)$ is upper bounded as

$$C(M_1, ..., M_K) \leq \min_{\mathcal{S}\subset[K]} \left\{R_{\mathcal{S}} + \textstyle\sum_{k\in\mathcal{S}} M_k/N\right\}. \tag{3.68}$$

According to Theorem 1.2, the capacity-memory trade-off $C(M)$ for memoryless packet erasure BC with receiver caching only at the weak users is upper bounded as

$$C(M) \leq \min_{\mathcal{S}\subset[K]} \left\{L\left(\textstyle\sum_{k\in\mathcal{S}} \frac{1}{1-\delta_k}\right)^{-1} + \frac{M}{N}\left|\mathcal{S}\cap\mathcal{K}_w\right|\right\}. \tag{3.69}$$

3.3 Gaussian BC

We now consider a wireless scenario, in which the shared link from the server to the users during the delivery phase is modelled as a Gaussian BC, while the placement phase is performed in an error-free fashion without any knowledge of the CSI during the delivery phase. In a Gaussian channel the transmission rate depends on the transmit power. Accordingly, we will study the impact of receiver caches on the minimum required transmit power at the server. Assuming uniformly random demands across the files in the library, both the minimum average transmit power, averaged over all the demand combinations, and the minimum peak transmit power, required to satisfy all the demand combinations, are investigated for both centralized and decentralized cache placement.

3.3.1 *Problem formulation*

Similarly to the model described in Section 3.2, each file in the library is assumed to be uniformly distributed over the set $\left[\lceil 2^{nR}\rceil\right]$. We consider K users in the system, each equipped with a cache of size nMR bits. Cache placement takes place without any CSI knowledge, where the caching function for user k, $k \in [K]$, is given by

$$\phi_k : \left[\lceil 2^{nR}\rceil\right]^N \to \left[\lfloor 2^{nMR}\rfloor\right], \tag{3.70}$$

which maps the library to the cache content B_k.

Assume that the demands are independent and uniformly distributed over the files in the library; i.e., $\Pr\{d_k = i\} = 1/N$, for $i \in [N], \forall k \in [K]$. After revealing the demand vector $\mathbf{d}$, the delivery phase takes place over a Gaussian BC, characterized by

$$Y_{k,i}\left(\mathbf{W},\mathbf{d}\right) = h_k X_i\left(\mathbf{W},\mathbf{d}\right) + Z_{k,i}, \quad \text{for } i \in [n], k \in [K], \tag{3.71}$$

where $X_i\left(\mathbf{W},\mathbf{d}\right)$ is the channel input at time i, h_k is the real channel gain from the server to user k, $Z_{k,i}$ is zero-mean unit-variance noise at user k at time i, and $Y_{K,i}$ is the channel output at user k at time i. The additive noise is assumed to be independent and identically distributed (i.i.d.) across time and users. The users are enumerated such that $h_1^2 \leq h_2^2 \leq \cdots \leq h_K^2$, and let $\mathbf{h} \triangleq \{h_1, \ldots, h_K\}$. The length-$n$ channel input $X^n\left(\mathbf{W},\mathbf{d}\right)$ is generated by

$$\psi_{\mathbf{h}} : \left[\lceil 2^{nR} \rceil\right]^N \times [N]^K \to \mathbb{R}^n, \tag{3.72}$$

which maps the library and demand vector $\mathbf{d}$ to a real length-n vector, with an average transmit power $P\left(\mathbf{W},\mathbf{d}\right) \triangleq \frac{1}{n}\sum_{i=1}^{n} X_i^2\left(\mathbf{W},\mathbf{d}\right)$. The average power of the above encoding function for demand vector $\mathbf{d}$ is defined as

$$P\left(\mathbf{d}\right) \triangleq \max_{W_1,\ldots,W_N} P\left(\mathbf{W},\mathbf{d}\right), \tag{3.73}$$

where the maximization is over all possible realizations of the file library. Having known $\mathbf{h}$, user k recovers $\hat{W}_k$ using its channel output $Y_k^n\left(\mathbf{W},\mathbf{d}\right)$, cache content B_k, and demand vector $\mathbf{d}$ through the following decoding function:

$$\mu_{\mathbf{h},k} : \mathbb{R}^n \times \left[\lfloor 2^{nMR} \rfloor\right] \times [N]^K \to \left[\lceil 2^{nR} \rceil\right]. \tag{3.74}$$

The error probability is defined as in (3.15) considering all the demand combinations, where

$$P_e = \Pr\left\{ \bigcup_{\mathbf{d}\in[N]^K} \bigcup_{k=1}^{K} \left\{\hat{W}_k \neq W_{d_k}\right\}\right\}. \tag{3.75}$$

A (R, M, n) *caching and delivery code* consists of K caching functions $\phi_1, \ldots, \phi_K$, an encoding function $\psi_{\mathbf{h}}$, and K decoding functions $\mu_{\mathbf{h},1}, \ldots, \mu_{\mathbf{h},K}$. A $\left(R, M, \bar{P}, \hat{P}\right)$ tuple is *achievable* if, for every $\varepsilon > 0$, there exists a (n, R, M) code with sufficiently large n, which satisfies $P_e < \varepsilon$, $\mathrm{E}_{\mathbf{d}}\left[P(\mathbf{d})\right] \leq \bar{P}$, and $P(\mathbf{d}) \leq \hat{P}$, $\forall \mathbf{d}$. Given rate R and normalized cache size M, the average and peak power-memory trade-offs are defined, respectively, as

$$\bar{P}^*\left(R, M\right) \triangleq \inf\left\{\bar{P} : \left(R, M, \bar{P}, \infty\right) \text{ is achievable}\right\}, \tag{3.76a}$$

$$\hat{P}^*\left(R, M\right) \triangleq \inf\left\{\hat{P} : \left(R, M, \hat{P}, \hat{P}\right) \text{ is achievable}\right\}. \tag{3.76b}$$

3.3.2 Preliminaries

Gaussian BC without cache. Consider a K-user Gaussian BC without any caches, where a message of rate R_k is intended for user k, $k \in [K]$. The minimum transmit power P that is required to serve all the users reliably is achieved by the superposition coding with Gaussian codewords of power $\alpha_k P$ allocated to user k, $k \in [K]$, where [23]

$$\alpha_k P = \left(\frac{2^{2R_k}-1}{h_k^2}\right)\left(1 + h_k^2 \sum_{i=k+1}^{K}\left(\frac{2^{2R_i}-1}{h_i^2}\right)\prod_{j=k+1}^{i-1} 2^{2R_j}\right), \tag{3.77}$$

which results in a transmit power

$$P = \sum_{k=1}^{K}\left(\frac{2^{2R_k}-1}{h_k^2}\right)\prod_{i=1}^{k-1} 2^{2R_i}. \tag{3.78}$$

Set of leaders. Given a demand vector $\mathbf{d}$, the definition of set of leaders $\mathcal{U}_{\mathbf{d}}$, defined in Section 3.1.2 for the error-free BC, is updated such that it consists of the $N_{\mathbf{d}}$ users with distinct requests and the worst channel qualities, where we remind that $N_{\mathbf{d}}$ denotes the number of distinct requests of demand vector $\mathbf{d}$. Accordingly, a user is included in the set $\mathcal{U}_{\mathbf{d}}$ if it has the worst channel quality among all the users with the same demand. It is easy to verify that for any demand vector $\mathbf{d}$, $1 \in \mathcal{U}_{\mathbf{d}}$. Also, for given $\mathbf{d}$ and user k, $k \in [K]$, let $\mathcal{U}_{\mathbf{d},k}$ denote the set of users in $\mathcal{U}_{\mathbf{d}}$ with better channel qualities than user k; i.e.,

$$\mathcal{U}_{\mathbf{d},k} \triangleq \{i \in \mathcal{U}_{\mathbf{d}} : i > k\}, \quad k \in [K], \tag{3.79}$$

and let $N_{\mathbf{d},k} \triangleq |\mathcal{U}_{\mathbf{d},k}|$.

In the following, both upper and lower bounds on the minimum average and peak power-memory trade-offs for both centralized and decentralized caching are presented.

3.3.3 Centralized cache-aided broadcasting

Here the achievable scheme is described providing an upper bound on the minimum average and peak power-memory trade-offs for the centralized caching scenario. First, we focus on an integer caching factor $t = MK/N$, and the extension to any real t value will be outlined next.

Placement phase. Consider a caching factor $t \in [0 : K]$. The placement phase follows that of the MAN scheme. File W_i is split into $\binom{K}{t}$ equal-length non-overlapping subfiles, denoted by $W_{i,\mathcal{S}_{K,1}^t}, \ldots W_{i,\mathcal{S}_{K,\binom{K}{t}}^t}$, each of rate $R/\binom{K}{t}$. User k's cache content is given by

$$B_k = \bigcup_{i\in[N]} \bigcup_{l\in\left[\binom{K}{t}\right]: k\in\mathcal{S}_{K,l}^t} W_{i,\mathcal{S}_{K,l}^t}, \quad \text{for } k \in [K]. \tag{3.80}$$

Delivery phase. Given a demand vector $\mathbf{d}$, similarly to the delivery phase of the MAN scheme, the following coded packet is intended for users in set $\mathcal{S}_{K,l}^{t+1}$

$$W_{\mathcal{S}_{K,l}^{t+1}} = \bigoplus_{k\in\mathcal{S}_{K,l}^{t+1}} W_{d_k, \mathcal{S}_{K,l}^{t+1}\setminus\{k\}}, \quad l \in \left[\binom{K}{t+1}\right]. \tag{3.81}$$

However, as described in Section 3.1.2, demand vector $\mathbf{d}$ can be fully satisfied even without transmitting the coded packets targeted for users in set $\mathcal{S}_{K,l}^{t+1}$ if $\mathcal{S}_{K,l}^{t+1} \cap \mathcal{U}_{\mathbf{d}} = \emptyset$, $l \in \left[\binom{K}{t+1}\right]$. Accordingly, we utilize this technique for an efficient coded delivery designed for each specific demand combination, and note that for a demand vector $\mathbf{d}$ with $N_{\mathbf{d}}$ distinct demands, if $K - N_{\mathbf{d}} \geq t + 1$, not all the coded messages $W_{\mathcal{S}_{K,l}^{t+1}}$, $\forall l \in \left[\binom{K}{t}\right]$, need to be delivered. For demand vector $\mathbf{d}$, only the coded packets targeted for at least one user in $\mathcal{U}_{\mathbf{d}}$ are delivered, i.e., coded packets $W_{\mathcal{S}_{K,l}^{t+1}}$ such that $\mathcal{S}_{K,l}^{t+1} \cap \mathcal{U}_{\mathbf{d}} \neq \emptyset$, $\forall l \in \left[\binom{K}{t+1}\right]$. The transmit power of coded packet $W_{\mathcal{S}_{K,l}^{t+1}}$, targeted for users in $\mathcal{S}_{K,l}^{t+1}$, for transmission over the Gaussian BC is adjusted such that the worst among the users in $\mathcal{S}_{K,l}^{t+1}$ can decode it; and so can all the other users in $\mathcal{S}_{K,l}^{t+1}$ due to the degradedness of the Gaussian BC. As a result, the demand of every user in $\mathcal{U}_{\mathbf{d}}$ will be satisfied.

The goal is to design a coded delivery scheme by specifying the coded packets targeted for each user that will minimize the total transmit power, while guaranteeing the fulfilment of all the demands. The coded packets targeted for different users are superposed for transmission over the Gaussian BC. For this purpose, we start from the worst user, i.e., user 1, and deliver all the coded packets intended for user 1. We then target the second user and deliver the coded packets intended for it that have not already been delivered, keeping in mind that only the coded messages $W_{\mathcal{S}_{K,l}^{t+1}}$ for which $\mathcal{S}_{K,l}^{t+1} \cap \mathcal{U}_{\mathbf{d}} \neq \emptyset$ are delivered. We continue this procedure until the delivery of the coded packets targeted for the best user in $\mathcal{U}_{\mathbf{d}}$.

For demand vector $\mathbf{d}$, we denote the collection of coded packets designed for delivery to user k by $\tilde{W}_k$, and their sum rate by $R^{\mathrm{C}}_{\mathbf{d},t,k}$. All the coded packets targeted for user 1 are delivered, i.e.,

$$\tilde{W}_1 = \bigcup\nolimits_{l:1 \in \mathcal{S}_{K,l}^{t+1}} W_{\mathcal{S}_{K,l}^{t+1}}. \tag{3.82}$$

Note that there are $\binom{K-1}{t}$ coded packets targeted for user 1, and the total rate of the coded packets targeted for user 1 is

$$R^{\mathrm{C}}_{\mathbf{d},t,1} = \frac{\binom{K-1}{t}}{\binom{K}{t}} R = \frac{K-t}{K} R. \tag{3.83}$$

For users in $k \in [2:K]$, the coded packets targeted for user k that have not been sent through the transmissions to the previous $k-1$ users and are targeted for at least one user in $\mathcal{U}_{\mathbf{d}}$ are delivered. Accordingly, for $k \in [2:K]$, we deliver

$$\tilde{W}_k = \bigcup_{l:\mathcal{S}_{K,l}^{t+1} \cap \mathcal{U}_{\mathbf{d}} \neq \emptyset, \mathcal{S}_{K,l}^{t+1} \cap [k-1] = \emptyset, k \in \mathcal{S}_{K,l}^{t+1}} W_{\mathcal{S}_{K,l}^{t+1}}, \tag{3.84}$$

which is equivalent to

$$\begin{aligned} \tilde{W}_k &= \left(\bigcup_{l:\mathcal{S}_{K,l}^{t+1} \cap [k-1] = \emptyset, k \in \mathcal{S}_{K,l}^{t+1}} W_{\mathcal{S}_{K,l}^{t+1}} \right) \\ &\quad - \left(\bigcup_{l:\mathcal{S}_{K,l}^{t+1} \cap \mathcal{U}_{\mathbf{d}} = \emptyset, \mathcal{S}_{K,l}^{t+1} \cap [k-1] = \emptyset, k \in \mathcal{S}_{K,l}^{t+1}} W_{\mathcal{S}_{K,l}^{t+1}} \right). \end{aligned} \tag{3.85}$$

For each user $k \in [2:K]$, there are $\binom{K-k}{t}$ different $(t+1)$-element subsets $\mathcal{S}_{K,l}^{t+1}$, such that $\mathcal{S}_{K,l}^{t+1} \cap [k-1] = \emptyset$ and $k \in \mathcal{S}_{K,l}^{t+1}$, for $l \in \left[\binom{K}{t+1}\right]$. On the other hand, for each user $k \in [2:K] \backslash \mathcal{U}_{\mathbf{d}}$, there are $\binom{K-k-N_{\mathbf{d},k}}{t}$ different $(t+1)$-element subsets $\mathcal{S}_{K,l}^{t+1}$, such that $\mathcal{S}_{K,l}^{t+1} \cap (\mathcal{U}_{\mathbf{d}} \cup [k-1]) = \emptyset$ and $k \in \mathcal{S}_{K,l}^{t+1}$, for $l \in \left[\binom{K}{t+1}\right]$. Note that, if $k \in \mathcal{U}_{\mathbf{d}}$, the second term on the RHS of (3.3.3) includes no content. Thus, if $k \in [2:K] \cap \mathcal{U}_{\mathbf{d}}$, total rate targeted for user k is

$$R_{\mathbf{d},t,k}^{\mathrm{C}} = \frac{\binom{K-k}{t}}{\binom{K}{t}} R = \left(\prod_{i=0}^{k-1} \frac{K-t-i}{K-i} \right) R, \tag{3.86}$$

while the total rate targeted for user k, $k \in [2:K] \backslash \mathcal{U}_{\mathbf{d}}$, is

$$R_{\mathbf{d},t,k}^{\mathrm{C}} = \frac{\binom{K-k}{t} - \binom{K-k-N_{\mathbf{d},k}}{t+1}}{\binom{K}{t}} R. \tag{3.87}$$

In summary, the above achievable scheme intends to deliver contents of total rate $R_{\mathbf{d},t,k}^{\mathrm{C}}$ to user k, for $k \in [K]$, where

$$R_{\mathbf{d},t,k}^{\mathrm{C}} = \begin{cases} \frac{\binom{K-k}{t}}{\binom{K}{t}} R, & \text{if } k \in \mathcal{U}_{\mathbf{d}}, \\ \frac{\binom{K-k}{t} - \binom{K-k-N_{\mathbf{d},k}}{t}}{\binom{K}{t}} R, & \text{otherwise}. \end{cases} \tag{3.88}$$

The centralized caching and coded packet generation described above is summarized in Algorithm 1.

Algorithm 1 Centralized Caching and Coded Packet Generation

procedure PLACEMENT PHASE
 $W_i = \bigcup_{l=1}^{\binom{K}{t}} W_{i,\mathcal{S}_{K,l}^t}$, for $i \in [N]$
 $B_k = \bigcup_{i \in [N]} \bigcup_{l \in \left[\binom{K}{t}\right] : k \in \mathcal{S}_{K,l}^t} W_{i,\mathcal{S}_{K,l}^t}$, for $k \in [K]$
end procedure

procedure CODED PACKET GENERATION
 $W_{\mathcal{S}_{K,l}^{t+1}} = \bigoplus_{k \in \mathcal{S}_{K,l}^{t+1}} W_{d_k, \mathcal{S}_{K,l}^{t+1} \backslash \{k\}}$, for $l \in \left[\binom{K}{t+1}\right]$
 $\tilde{W}_k = \bigcup_{l : \mathcal{S}_{K,l}^{t+1} \cap \mathcal{U}_{\mathrm{d}} \neq \emptyset, \mathcal{S}_{K,l}^{t+1} \cap [k-1] = \emptyset, k \in \mathcal{S}_{K,l}^{t+1}} W_{\mathcal{S}_{K,l}^{t+1}}$, for $k \in [K]$
end procedure

Once the coded packets intended to each user are specified, the next step is to design the physical layer channel coding scheme to deliver these packets over the Gaussian BC. Given $\mathbf{d}$, K codebooks are generated, where the codebook $k \in [K]$ is designed to deliver the contents in $\tilde{W}_k$ to user k. The k-th codebook consists of $2^{nR_{\mathbf{d},t,k}^{\mathrm{C}}}$ i.i.d. Gaussian codewords $x_k^n(\mathbf{W}, \mathbf{d})$, generated according to the normal distribution $\mathcal{N}\left(0, \alpha_k P_{\mathrm{UB}}^{\mathrm{C}}(R, M, \mathbf{d})\right)$, where $\alpha_k \geq 0$ and $\sum_{i=1}^{K} \alpha_i = 1$. The transmission over the Guassian BC is performed through superposition coding; i.e., the channel input is given by $\sum_{k=1}^{K} x_k^n(\mathbf{W}, \mathbf{d}, \tilde{q}_k)$, where $\tilde{q}_k \in \left[2^{nR^{\mathrm{C}}_{\mathbf{d},t,k}}\right]$. User k decodes codewords $x_1^n(\mathbf{W}, \mathbf{d}), ..., x_k^n(\mathbf{W}, \mathbf{d})$ by successive decoding, considering all the codewords in higher levels as noise. User $k \in [K]$ can recover contents $\tilde{W}_1, ..., \tilde{W}_k$ if all the k codewords are decoded successfully. Accordingly, user $k \in [K]$ will receive all the coded contents targeted for it except those that are not intended for any of the users in $\mathcal{U}_{\mathbf{d}}$ (which have not been delivered); i.e., user k receives all the coded contents

$$\bigcup_{l:\mathcal{S}_{K,l}^{t+1} \cap \mathcal{U}_{\mathbf{d}} \neq \emptyset, k \in \mathcal{S}_{K,l}^{t+1}} W_{\mathcal{S}_{K,l}^{t+1}}. \tag{3.89}$$

Also, thanks to the degradation of the underlying Gaussian BC, it can decode the coded packets intended for user $[k-1] \cap \mathcal{U}_{\mathbf{d}}$, i.e., all the coded packets in

$$\bigcup_{l:\mathcal{S}_{K,l}^{t+1} \cap \mathcal{U}_{\mathbf{d}} \neq \emptyset, [k-1] \cap \mathcal{S}_{K,l}^{t+1} \neq \emptyset} W_{\mathcal{S}_{K,l}^{t+1}}. \tag{3.90}$$

Note that, if $k \in \mathcal{U}_{\mathbf{d}}$, (3.89) reduces to $\bigcup_{l:k \in \mathcal{S}_{K,l}^{t+1}} W_{\mathcal{S}_{K,l}^{t+1}}$, which shows that the demand of user $k \in \mathcal{U}_{\mathbf{d}}$ is fully satisfied. Next we show that the demand of users in $[K] \backslash \mathcal{U}_{\mathbf{d}}$ can also be satisfied without sending any extra messages. For any set of users $\mathcal{S}_{K,l}^{t+1}$ such that $\mathcal{S}_{K,l}^{t+1} \cap \mathcal{U}_{\mathbf{d}} = \emptyset$, we need to show that every user in $\mathcal{S}_{K,l}^{t+1}$ can decode all the coded packets $W_{\mathcal{C} \backslash \mathcal{G}}$, $\forall \mathcal{G} \in \mathcal{G}_{\mathcal{C}} \backslash \mathcal{U}_{\mathbf{d}}$, where $\mathcal{C} = \mathcal{S}_{K,l}^{t+1} \cup \mathcal{U}_{\mathbf{d}}$. As a result, they can also decode $W_{\mathcal{S}_{K,l}^{t+1}}$ through (3.18).

Assume that there exists a set of users $\mathcal{S}_{K,l}^{t+1}$ such that $\mathcal{S}_{K,l}^{t+1} \cap \mathcal{U}_{\mathbf{d}} = \emptyset$; set $\mathcal{C} = \mathcal{S}_{K,l}^{t+1} \cup \mathcal{U}_{\mathbf{d}}$. According to (3.19a), $(\mathcal{C} \backslash \mathcal{G}) \cap \mathcal{U}_{\mathbf{d}} \neq \emptyset$, $\forall \mathcal{G} \in \mathcal{G}_{\mathcal{C}} \backslash \mathcal{U}_{\mathbf{d}}$. Therefore, all the coded packets $W_{\mathcal{C} \backslash \mathcal{G}}$ have been delivered by the above-coded delivery scheme outlined, in Algorithm 1. We highlight that, for each user $k \in \mathcal{S}_{K,l}^{t+1}$ and $\forall \mathcal{G} \in \mathcal{G}_{\mathcal{C}} \backslash \mathcal{U}_{\mathbf{d}}$, either $k \in \mathcal{C} \backslash \mathcal{G}$ or $\exists k' \in \mathcal{C} \backslash \mathcal{G}$, where $d_{k'} = d_k$ and $k' \in \mathcal{U}_{\mathbf{d}}$, i.e., $h_{k'}^2 \leq h_k^2$. If $k \in \mathcal{C} \backslash \mathcal{G}$, since $\mathcal{C} \backslash \mathcal{G} \cap \mathcal{U}_{\mathbf{d}} \neq \emptyset$, according to (3.89), user k can obtain $W_{\mathcal{C} \backslash \mathcal{G}}$. On the other hand, if $k \notin \mathcal{C} \backslash \mathcal{G}$, then $\exists k' \in \mathcal{C} \backslash \mathcal{G}$ with $d_{k'} = d_k$ and $k' \in \mathcal{U}_{\mathbf{d}}$; so, user k' can decode $W_{\mathcal{C} \backslash \mathcal{G}}$, and since $h_{k'}^2 \leq h_k^2$, user k can also decode $W_{\mathcal{C} \backslash \mathcal{G}}$ due to the degradedness of the underlying Gaussian BC. Thus, all the users $k \in \mathcal{S}_{K,l}^{t+1}$ can decode $W_{\mathcal{S}_{K,l}^{t+1}}$ successfully, for any set of users $\mathcal{S}_{K,l}^{t+1}$ that satisfies $\mathcal{S}_{K,l}^{t+1} \cap \mathcal{U}_{\mathbf{d}} = \emptyset$. This confirms that, with the above coded delivery scheme, the demands of all the users in $[K] \backslash \mathcal{U}_{\mathbf{d}}$ are also satisfied.

A simple example elaborating the above-coded delivery scheme in detail is available in [24]. In the following, the above caching and coded delivery technique is extended to non-integer caching factor, t, values. First, the files in the library and the cache available at each user is divided into two disjoint parts, where the above caching and coded delivery scheme is performed for each part separately.

Placement phase. File W_i is divided into two non-overlapping subfiles, W_i^1 and W_i^2 with rates R_1 and R_2, respectively, for $i \in [N]$. We set

$$R_1 = \left(\lfloor t \rfloor + 1 - t\right) R, \tag{3.91a}$$

$$R_2 = \left(t - \lfloor t \rfloor\right) R, \tag{3.91b}$$

such that $R_1 + R_2 = R$. For subfiles $\{W_1^1, \ldots, W_N^1\}$, the placement phase presented in Algorithm 1 corresponding to a caching factor $\lfloor t \rfloor$ is performed, and it requires a cache size of $n\left(\lfloor t \rfloor N/K\right) R_1$ bits. On the other hand, for subfiles $\{W_1^2, \ldots, W_N^2\}$, the placement phase outlined in Algorithm 1 corresponding to a caching factor $\lfloor t \rfloor + 1$ is performed, and it requires a cache size of $n\left(\left(\lfloor t \rfloor + 1\right) N/K\right) R_2$ bits. Therefore, the total cache size is nMR, which satisfies the cache size constraint at each user.

Delivery phase. For a demand vector $\mathbf{d}$, the coded delivery presented in Algorithm 1 corresponding to caching factors $\lfloor t \rfloor$ and $\lfloor t \rfloor + 1$ is performed to deliver the missing bits of subfiles $\{W_1^1, \ldots, W_N^1\}$ and $\{W_1^2, \ldots, W_N^2\}$ to the intended users, respectively. Let define

$$W^i_{\mathcal{S}_{K,l}^{t_i+1}} \triangleq \bigoplus_{k \in \mathcal{S}_{K,l}^{t_i+1}} W^i_{d_k, \mathcal{S}_{K,l}^{t_i+1} \setminus \{k\}}, \qquad \text{for } i = 1, 2, \tag{3.92}$$

where $t_1 \triangleq \lfloor t \rfloor$ and $t_2 \triangleq \lfloor t \rfloor + 1$. According to the coded delivery scheme in Algorithm 1, for a caching factor t_i, to serve subfiles $\{W_1^i, \ldots, W_N^i\}$, each of rate R_i, user $k \in [K]$ should receive

$$\tilde{W}_k^i = \bigcup_{l: \mathcal{S}_{K,l}^{t_i+1} \cap \mathcal{U}_{\mathbf{d}} \neq \emptyset, \mathcal{S}_{K,l}^{t_i+1} \cap [k-1] = \emptyset, k \in \mathcal{S}_{K,l}^{t_i+1}} W^i_{\mathcal{S}_{K,l}^{t_i+1}}, \qquad \text{for } i = 1, 2, \tag{3.93}$$

of the following total rate obtained according to (3.88)

$$R^{\mathrm{C}}_{\mathbf{d},t_i,k} = \begin{cases} \dfrac{\binom{K-k}{t_i}}{\binom{K}{t_i}} R_i, & \text{if } k \in \mathcal{U}_{\mathbf{d}}, \\ \dfrac{\binom{K-k}{t_i} - \binom{K-k-N_{\mathbf{d},k}}{t_i}}{\binom{K}{t_i}} R_i, & \text{otherwise.} \end{cases} \tag{3.94}$$

The coded packets targeted for user k is $\tilde{W}_k = \left(\tilde{W}_k^1, \tilde{W}_k^2\right)$ with a total rate of $R^{\mathrm{C}}_{\mathbf{d},t,k} = R^{\mathrm{C}}_{\mathbf{d},t_1,k} + R^{\mathrm{C}}_{\mathbf{d},t_2,k}$, given by

$$R^{\mathrm{C}}_{\mathbf{d},t,k} = \begin{cases} \dfrac{\binom{K-k}{\lfloor t \rfloor}}{\binom{K}{\lfloor t \rfloor}} \left(\lfloor t \rfloor + 1 - t\right) R + \dfrac{\binom{K-k}{\lfloor t \rfloor + 1}}{\binom{K}{\lfloor t \rfloor + 1}} \left(t - \lfloor t \rfloor\right) R, & \text{if } k \in \mathcal{U}_{\mathbf{d}}, \\ \dfrac{\binom{K-k}{\lfloor t \rfloor} - \binom{K-k-N_{\mathbf{d},k}}{\lfloor t \rfloor}}{\binom{K}{\lfloor t \rfloor}} \left(\lfloor t \rfloor + 1 - t\right) R & \\ \quad + \dfrac{\binom{K-k}{\lfloor t \rfloor + 1} - \binom{K-k-N_{\mathbf{d},k}}{\lfloor t \rfloor + 1}}{\binom{K}{\lfloor t \rfloor + 1}} \left(t - \lfloor t \rfloor\right) R, & \text{otherwise.} \end{cases} \tag{3.95}$$

Transmit power analysis. After determining the packets targeted for all the users, $\tilde{W}_1, \ldots, \tilde{W}_K$, superposition coding is performed, and the channel input is given by $\sum_{k=1}^K x_k^n\left(\mathbf{W}, \mathbf{d}, \tilde{q}_k\right)$, where $\tilde{q}_k \in \left[2^{nR^{\mathrm{C}}_{\mathbf{d},t,k}}\right]$, and codeword $x_k^n\left(\mathbf{W}, \mathbf{d}\right)$ is generated according to the normal distribution $\mathcal{N}\left(0, \alpha_k P^{\mathrm{C}}_{\mathrm{UB}}\left(R, M, \mathbf{d}\right)\right)$, where $\alpha_k \geq 0$ and $\sum_{i=1}^K \alpha_i = 1$. User $k \in [K]$ decodes codewords $x_1^n\left(\mathbf{W}, \mathbf{d}\right), \ldots, x_k^n(\mathbf{W}, \mathbf{d})$ successively by considering all the other codewords in higher levels as noise. User k can decode its intended message successfully if, for $k \in [K]$,

$$R^{\mathrm{C}}_{\mathbf{d},t,k} \leq \frac{1}{2}\log_2\left(1 + \frac{\alpha_k h_k^2 P^{\mathrm{C}}_{\mathrm{UB}}(R, M, \mathbf{d})}{h_k^2 \sum_{i=k+1}^K \alpha_i P^{\mathrm{C}}_{\mathrm{UB}}(R, M, \mathbf{d}) + 1}\right). \tag{3.96}$$

From (3.78), the minimum transmit power of the above caching and coded delivery scheme for demand vector $\mathbf{d}$ is given by

$$P^{\mathrm{C}}_{\mathrm{UB}}\left(R,M,\mathbf{d}\right) \triangleq \sum_{i=1}^{K}\left(\frac{2^{2R^{\mathrm{C}}_{\mathbf{d},t,i}}-1}{h_i^2}\right)\prod_{j=1}^{i-1}2^{2R^{\mathrm{C}}_{\mathbf{d},j}}, \tag{3.97}$$

and the average power-memory trade-off is $\bar{P}^{\mathrm{C}}_{\mathrm{UB}}(R,M) \triangleq \mathrm{E}_{\mathbf{d}}\left[P^{\mathrm{C}}_{\mathrm{UB}}\left(R,M,\mathbf{d}\right)\right]$, which assuming uniformly distributed demands over the files, it follows that

$$\bar{P}^{\mathrm{C}}_{\mathrm{UB}}(R,M) = \frac{1}{N^K}\sum_{\mathbf{d}\in[N]^K}\left[\sum_{i=1}^{K}\left(\frac{2^{2R^{\mathrm{C}}_{\mathbf{d},t,i}}-1}{h_i^2}\right)\prod_{j=1}^{i-1}2^{2R^{\mathrm{C}}_{\mathbf{d},t,j}}\right], \tag{3.98}$$

where $R^{\mathrm{C}}_{\mathbf{d},t,i}$ is given in (3.95). To find the achievable peak power-memory trade-off, we need to find the worst-case demand for the caching and coded delivery scheme. Observe that, for the demand combinations with the same $\mathcal{U}_{\mathbf{d}}$, the required transmit power is the same. It is shown in [24, Appendix B] that the worst-case demand combination happens when $\mathcal{U}_{\mathbf{d}} = \left[\min\{N,K\}\right]$, i.e., when the first $\min\{N,K\}$ users have different demands. Accordingly, the peak power-memory trade-off is as follows:

$$\hat{P}^{\mathrm{C}}_{\mathrm{UB}}(R,M) \triangleq \sum_{i=1}^{K}\left(\frac{2^{2\hat{R}^{\mathrm{C}}_{\mathbf{d},t,i}}-1}{h_i^2}\right)\prod_{j=1}^{i-1}2^{2\hat{R}^{\mathrm{C}}_{\mathbf{d},t,j}}, \tag{3.99}$$

where

$$\hat{R}^{\mathrm{C}}_{\mathbf{d},t,k} \triangleq \begin{cases} \frac{\binom{K-k}{\lfloor t\rfloor}}{\binom{K}{\lfloor t\rfloor}}\left(\lfloor t\rfloor+1-t\right)R + \frac{\binom{K-k}{\lfloor t\rfloor+1}}{\binom{K}{\lfloor t\rfloor+1}}\left(t-\lfloor t\rfloor\right)R, & \text{if } k\in\left[\min\{N,K\}\right],\\ 0, & \text{otherwise.}\end{cases} \tag{3.100}$$

3.3.4 Decentralized cache-aided broadcasting

Placement phase. The decentralized placement phase follows the one introduced in [2], where each user caches nMR/N random bits of each file of length nR bits independently. The cache content at user k is the same as in (3.11), where the length of $W_{i,\mathcal{S}}$ is approximated by $\left(\frac{M}{N}\right)^{|\mathcal{S}|}\left(1-\frac{M}{N}\right)^{K-|\mathcal{S}|}nR$ bits.

Delivery phase. Similarly to the delivery phase for the centralized caching scenario, for demand vector $\mathbf{d}$, the coded packets

$$W_{\mathcal{S}} = \bigoplus_{k\in\mathcal{S}} W_{d_k,\mathcal{S}\backslash\{k\}}, \quad \forall\mathcal{S}\subset[K], \tag{3.101}$$

targeted for users in $\mathcal{S}$, are delivered if $\mathcal{S}\cap\mathcal{U}_{\mathbf{d}}\neq\emptyset$, and the remaining coded packets can be recovered through (3.18). For any such set $\mathcal{S}$ with $\mathcal{S}\cap\mathcal{U}_{\mathbf{d}}\neq\emptyset$, the transmit power is adjusted such that the worst user in $\mathcal{S}$ can decode the intended coded packet; and so all the other users in $\mathcal{S}$ can decode it, which is due to the degradedness of the Gaussian BC. It is worth noting that, for the centralized caching scheme presented in Algorithm 1, the coded packets are targeted for $t+1$ users, $t\in[0:K-1]$; however, in the decentralized caching scenario, the coded packets are targeted for all the non-empty subsets of users, i.e., for $(t+1)$-element subsets of users, $\forall t\in[0:K-1]$. Therefore, a similar delivery technique can be employed to guarantee that all the users' demands are fully satisfied.

For demand vector $\mathbf{d}$, the following contents are intended for delivery to user k, for $k \in [K]$:

$$\tilde{W}_k = \bigcup_{\mathcal{S} \subset [k:K]: \mathcal{S} \cap \mathcal{U}_{\mathbf{d}} \neq \emptyset, k \in \mathcal{S}} W_{\mathcal{S}}, \tag{3.102}$$

which are equivalent to

$$\tilde{W}_k = \bigcup_{\mathcal{S} \subset [k:K]: k \in \mathcal{S}} W_{\mathcal{S}} - \bigcup_{\mathcal{S} \subset [k:K] \backslash \mathcal{U}_{\mathbf{d},k}, k \in \mathcal{S}} W_{\mathcal{S}}. \tag{3.103}$$

For user k, $k \in [K]$, there are $\binom{K-k}{i}$ different $(i+1)$-element subsets of $[k : K]$ that include k, $i \in [0 : K-k]$. Thus, if $k \in \mathcal{U}_{\mathbf{d}}$, from (3.97), the total amount of rate targeted for user k, denoted by $R^{\mathrm{D}}_{\mathbf{d},k}$, is given by

$$R^{\mathrm{D}}_{\mathbf{d},k} = \sum_{i=0}^{K-k} \binom{K-k}{i} (M/N)^i (1 - M/N)^{K-i} R = (1 - M/N)^k R. \tag{3.104}$$

On the other hand, for user k, $k \in [K]$, there are $\binom{K-k-N_{\mathbf{d},k}}{i}$ different $(i+1)$-element subsets of $[k : K] \backslash \mathcal{U}_{\mathbf{d},k}$ that include k, $i \in [0 : K-k]$. Thus, if $k \notin \mathcal{U}_{\mathbf{d}}$, from (3.103), the total amount of rate targeted for user $k \in [K]$ is given by

$$\begin{aligned} R^{\mathrm{D}}_{\mathbf{d},k} = & \sum_{i=0}^{K-k} \binom{K-k}{i} (M/N)^i (1 - M/N)^{K-i} R \\ & - \sum_{i=0}^{K-k-N_{\mathbf{d},k}} \binom{K-k-N_{\mathbf{d},k}}{i} (M/N)^i (1 - M/N)^{K-i} R \\ = & (1 - M/N)^k \left(1 - (1 - M/N)^{N_{\mathbf{d},k}}\right) R. \end{aligned} \tag{3.105}$$

In summary, the rate of the contents targeted for user k, $k \in [K]$, is given by

$$R^{\mathrm{D}}_{\mathbf{d},k} = \begin{cases} \left(1 - \frac{M}{N}\right)^k R, & \text{if } k \in \mathcal{U}_{\mathbf{d}}, \\ \left(1 - \frac{M}{N}\right)^k \left(1 - \left(1 - \frac{M}{N}\right)^{N_{\mathbf{d},k}}\right) R, & \text{otherwise.} \end{cases} \tag{3.106}$$

For demand vector $\mathbf{d}$, the linear superposition of K codewords $x_1^n(\mathbf{W}, \mathbf{d}), \ldots, x_K^n(\mathbf{W}, \mathbf{d})$, each chosen from an independent codebook, is transmitted over the Gaussian BC. Codebook k consists of $2^{nR^{\mathrm{D}}_{\mathbf{d},k}}$ i.i.d. codewords $x_k^n(\mathbf{W}, \mathbf{d})$ generated according to $\mathcal{N}\left(0, \alpha_k P^{\mathrm{D}}_{\mathrm{UB}}(R, M, \mathbf{d})\right)$, where $\alpha_k \geq 0$ and $\sum_{i=1}^{K} \alpha_i = 1$, which satisfy the power constraint, for $k \in [K]$.

User k, $k \in [K]$, decodes codewords $x_1^n(\mathbf{W}, \mathbf{d}), \ldots, x_k^n(\mathbf{W}, \mathbf{d})$ through successive decoding, while considering all the codewords in the higher levels, $x_{k+1}^n(\mathbf{W}, \mathbf{d}), \ldots, x_K^n(\mathbf{W}, \mathbf{d})$, as noise. After decoding the first k codewords, user k can recover the contents $\tilde{W}_1, \ldots, \tilde{W}_k$, from which it can obtain all the coded contents $\bigcup_{\mathcal{S} \cap \mathcal{U}_{\mathbf{d}} \neq \emptyset, [k] \cap \mathcal{S} \neq \emptyset} W_{\mathcal{S}}$. Accordingly, user k can obtain all the coded contents targeted for it except those that are not intended for any of the users in $\mathcal{U}_{\mathbf{d}}$ (which have not been delivered). It can further obtain all the coded contents targeted for users $[k-1]$, which are also intended for at least one user in $\mathcal{U}_{\mathbf{d}}$, i.e., all the coded contents $\bigcup_{\mathcal{S} \cap \mathcal{U}_{\mathbf{d}} \neq \emptyset, [k-1] \cap \mathcal{S} \neq \emptyset} W_{\mathcal{S}}$. Similarly to the centralized caching scenario, it can be verified that this decentralized caching and coded delivery technique enables user

k to retrieve all the coded packets targeted for it, $\bigcup_{\mathcal{S}:k\in\mathcal{S}} W_{\mathcal{S}}$, $\forall k$. Therefore, all the users' demands are fulfilled.

Transmit power analysis. For demand vector $\mathbf{d}$, user k decodes the messages up to level k while considering messages in the higher levels as noise, for $k \in [K]$. For successful decoding at user k, it requires that

$$R^{\mathrm{D}}_{\mathbf{d},k} \leq \tfrac{1}{2}\log_2\left(1 + \frac{\alpha_k h_k^2 P^{\mathrm{D}}_{\mathrm{UB}}(R,M,\mathbf{d})}{h_k^2 \sum_{i=k+1}^{K} \alpha_i P^{\mathrm{D}}_{\mathrm{UB}}(R,M,\mathbf{d})+1}\right), \tag{3.107}$$

where, from (3.78), the total transmit power is given by

$$P^{\mathrm{D}}_{\mathrm{UB}}\left(R,M,\mathbf{d}\right) \triangleq \sum_{i=1}^{K}\left(\frac{2^{2R^{\mathrm{D}}_{\mathbf{d},i}}-1}{h_i^2}\right)\prod_{j=1}^{i-1} 2^{2R^{\mathrm{D}}_{\mathbf{d},j}}, \tag{3.108}$$

where $R^{\mathrm{D}}_{\mathbf{d},k}$, $\forall k$, is given in (3.106). Accordingly, the achievable average power-memory trade-off of the above decentralized caching and coded delivery scheme is $\bar{P}^{\mathrm{D}}_{\mathrm{UB}}(R,M) \triangleq \mathrm{E}_{\mathbf{d}}\left[P^{\mathrm{D}}_{\mathrm{UB}}\left(R,M,\mathbf{d}\right)\right]$, which for uniformly distributed demands over the library, it follows that

$$\bar{P}^{\mathrm{D}}_{\mathrm{UB}}(R,M) = \frac{1}{N^K}\sum_{\mathbf{d}\in[N]^K}\left[\sum_{i=1}^{K}\left(\frac{2^{2R^{\mathrm{D}}_{\mathbf{d},i}}-1}{h_i^2}\right)\prod_{j=1}^{i-1} 2^{2R^{\mathrm{D}}_{\mathbf{d},j}}\right]. \tag{3.109}$$

Also, it is shown in [24, Appendix C], that the worst-case demand for this scheme corresponds to the demands with $\mathcal{U}_{\mathbf{d}} = \left[\min\{N,K\}\right]$, in which case we achieve the following peak power-memory trade-off

$$\hat{P}^{\mathrm{D}}_{\mathrm{UB}}(R,M) \triangleq \sum_{i=1}^{\min\{N,K\}}\left(\frac{2^{2R\left(1-\frac{M}{N}\right)^i}-1}{h_i^2}\right)2^{2R\left(\frac{N}{M}-1\right)\left(1-\left(1-\frac{M}{N}\right)^{i-1}\right)}. \tag{3.110}$$

3.3.5 *Lower bound*

In this section, we present lower bounds on the average and peak power-memory trade-offs assuming an uncoded placement phase. The main challenge in deriving these bounds is that the cache-aided Gaussian BC is not degraded due to the presence of the caches. Degradedness is created by providing the requested files and cache contents of users in $[k-1]$ to users in $[k:K]$, $\forall k \in [2:K]$.

For any possible set of leaders $\mathcal{U}_{\mathbf{d}}$, let $\mathcal{D}_{\mathcal{U}_{\mathbf{d}}}$ denote the set of all demand vectors resulting the same $\mathcal{U}_{\mathbf{d}}$. Note that the union $\bigcup_{\mathcal{U}_{\mathbf{d}}} \mathcal{D}_{\mathcal{U}_{\mathbf{d}}}$ form the set of all possible demand vectors $[N]^K$. Therefore, the set of demand vectors can be broken into classes $\mathcal{D}_{\mathcal{U}_{\mathbf{d}}}$ corresponding to a specific $\mathcal{U}_{\mathbf{d}}$. We modify the error probability definition of the (R,M,n) caching and coded delivery code defined in Section 3.3.1, and customize it for any set of leaders $\mathcal{U}_{\mathbf{d}}$. We consider a particular $\mathcal{U}_{\mathbf{d}}$, and define the error probability as

$$P_e\left(\mathcal{U}_{\mathbf{d}}\right) \triangleq \Pr\left\{\bigcup_{\mathbf{d}\in\mathcal{D}_{\mathcal{U}_{\mathbf{d}}}}\bigcup_{k\in\mathcal{U}_{\mathbf{d}}}\left\{\hat{W}_k \neq W_{d_k}\right\}\right\}. \tag{3.111}$$

Let $P_{\mathcal{U}_{\mathbf{d}}}(\mathbf{d})$ denote the average transmit power at the server generated by this code. An $\left(R, M, \bar{P}, \hat{P}\right)$ tuple is said to be $\mathcal{U}_{\mathbf{d}}$-*achievable* if, for every $\varepsilon > 0$, there exists an (n, R, M) code with sufficiently large n, which satisfies $P_e(\mathcal{U}_{\mathbf{d}}) < \varepsilon$, $\mathrm{E}_{\mathbf{d}}\left[P_{\mathcal{U}_{\mathbf{d}}}(\mathbf{d})\right] \leq \bar{P}$, and $P_{\mathcal{U}_{\mathbf{d}}}(\mathbf{d}) \leq \hat{P}$, $\forall \mathbf{d} \in \mathcal{D}_{\mathcal{U}_{\mathbf{d}}}$. We can also define $\bar{P}^*_{\mathcal{U}_{\mathbf{d}}}(R, M)$ and $\hat{P}^*_{\mathcal{U}_{\mathbf{d}}}(R, M)$ in the same manner as (3.76a) and (3.76b), respectively, by considering $\mathcal{U}_{\mathbf{d}}$-achievable codes.

According to (3.111), we note that a $\mathcal{U}_{\mathbf{d}}$-achievable code satisfies only the demands of users in $\mathcal{U}_{\mathbf{d}}$. Comparing the error probability definitions in (3.75) and (3.111), it is easy to verify that an achievable $(R, M, \bar{P}, \hat{P})$ tuple is also $\mathcal{U}_{\mathbf{d}}$-achievable, since $P_e(\mathcal{U}_{\mathbf{d}}) \leq P_e$, for any $\mathcal{U}_{\mathbf{d}}$ set. Therefore, lower bounds on $\bar{P}^*_{\mathcal{U}_{\mathbf{d}}}(R, M)$ and $\hat{P}^*_{\mathcal{U}_{\mathbf{d}}}(R, M)$ also lower bound $\bar{P}^*(R, M)$ and $\hat{P}^*(R, M)$, respectively. In the following, lower bounds on $\bar{P}^*_{\mathcal{U}_{\mathbf{d}}}(R, M)$ and $\hat{P}^*_{\mathcal{U}_{\mathbf{d}}}(R, M)$ are provided.

In total, we have 2^{K-1} different $\mathcal{U}_{\mathbf{d}}$ sets considering all possible demand combinations. This follows since $N_{\mathbf{d}} \leq K$ and $1 \in \mathcal{U}_{\mathbf{d}}$, $\forall \mathbf{d}$. For any demand vector $\mathbf{d}$, let $\mathcal{U}_{\mathbf{d}} = \{u_1, u_2, \ldots, u_{N_{\mathbf{d}}}\}$, where $1 = u_1 \leq u_2 \leq \cdots \leq u_{N_{\mathbf{d}}}$. The number of demand vectors with the same $\mathcal{U}_{\mathbf{d}}$ is given by

$$N_{\mathcal{U}_{\mathbf{d}}} \triangleq \binom{N}{N_{\mathbf{d}}} N_{\mathbf{d}}! \left(\prod_{j=2}^{N_{\mathbf{d}}} j^{u_{j+1}-u_j-1}\right), \tag{3.112}$$

where we define $u_{N_{\mathbf{d}}+1} \triangleq K + 1$.

Let $(R, M, \bar{P}, \hat{P})$ be a $\mathcal{U}_{\mathbf{d}}$-achievable tuple. For uniformly distributed demands, we have

$$\bar{P} \geq \mathrm{E}_{\mathbf{d}}\left[P_{\mathcal{U}_{\mathbf{d}}}(\mathbf{d})\right] = \mathrm{E}_{\mathcal{U}_{\mathbf{d}}}\left[\frac{1}{N_{\mathcal{U}_{\mathbf{d}}}} \sum_{\mathbf{d} \in \mathcal{D}_{\mathcal{U}_{\mathbf{d}}}} P_{\mathcal{U}_{\mathbf{d}}}(\mathbf{d})\right], \tag{3.113}$$

which follows since the probability of each demand vector in $\mathcal{D}_{\mathcal{U}_{\mathbf{d}}}$ is equal. We divide the set of demand vectors $\mathcal{D}_{\mathcal{U}_{\mathbf{d}}}$ into different sub-classes based on the demands of leaders, $\mathcal{U}_{\mathbf{d}}$, where each sub-class consists of the demand vectors for which the demands of all the leaders are the same. It is easy to verify that there are $N_{\mathcal{U}_{\mathbf{d}},1} \triangleq \binom{N}{N_{\mathbf{d}}} N_{\mathbf{d}}!$ such sub-classes, denoted by $\mathcal{D}^l_{\mathcal{U}_{\mathbf{d}}}$, $\forall l \in [N_{\mathcal{U}_{\mathbf{d}},1}]$, i.e., $\mathcal{D}_{\mathcal{U}_{\mathbf{d}}} = \bigcup_{l=1}^{N_{\mathcal{U}_{\mathbf{d}},1}} \mathcal{D}^l_{\mathcal{U}_{\mathbf{d}}}$, and the number of demand vectors in each sub-class $\mathcal{D}^l_{\mathcal{U}_{\mathbf{d}}}$, denoted by $N_{\mathcal{U}_{\mathbf{d}},2}$, is the same, and is given by

$$N_{\mathcal{U}_{\mathbf{d}},2} = \prod_{j=2}^{N_{\mathbf{d}}} j^{u_{j+1}-u_j-1}. \tag{3.114}$$

Thus, $N_{\mathcal{U}_{\mathbf{d}}} = N_{\mathcal{U}_{\mathbf{d}},1} N_{\mathcal{U}_{\mathbf{d}},2}$, and (3.113) can be rewritten as follows:

$$\begin{aligned} \bar{P} &\geq \mathrm{E}_{\mathcal{U}_{\mathbf{d}}}\left[\frac{1}{N_{\mathcal{U}_{\mathbf{d}}}} \sum_{\mathbf{d} \in \mathcal{D}_{\mathcal{U}_{\mathbf{d}}}} P_{\mathcal{U}_{\mathbf{d}}}(\mathbf{d})\right] \\ &= \mathrm{E}_{\mathcal{U}_{\mathbf{d}}}\left[\frac{1}{N_{\mathcal{U}_{\mathbf{d}},1}} \sum_{l=1}^{N_{\mathcal{U}_{\mathbf{d}},1}} \left(\frac{1}{N_{\mathcal{U}_{\mathbf{d}},2}} \sum_{\mathbf{d} \in \mathcal{D}^l_{\mathcal{U}_{\mathbf{d}}}} P_{\mathcal{U}_{\mathbf{d}}}(\mathbf{d})\right)\right]. \end{aligned} \tag{3.115}$$

It is proved in [24] that, for a demand vector $\mathbf{d} \in \mathcal{D}^l_{\mathcal{U}_{\mathbf{d}}}$, $l \in [N_{\mathcal{U}_{\mathbf{d}},1}]$, by limiting to the uncoded cache placement, we have

$$P_{\mathcal{U}_\mathbf{d}}(\mathbf{d}) \geq f\left(c^l_{\pi_{\mathcal{U}_\mathbf{d}}(1)}, \ldots, c^l_{\pi_{\mathcal{U}_\mathbf{d}}(N_\mathbf{d})}\right) \triangleq \sum_{i=1}^{N_\mathbf{d}} \left(\frac{2^{2c^l_{\pi_{\mathcal{U}_\mathbf{d}}(i)}}-1}{h^2_{\pi_{\mathcal{U}_\mathbf{d}}(i)}}\right) \prod_{j=1}^{i-1} 2^{2c^l_{\pi_{\mathcal{U}_\mathbf{d}}(j)}}, \tag{3.116}$$

where, for $i \in [N_d]$,

$$c^l_{\pi_{\mathcal{U}_\mathbf{d}}(i)} \triangleq R - \frac{1}{n} I\left(W_{d^l_{\pi_{\mathcal{U}_\mathbf{d}}(i)}}; U_{\pi_{\mathcal{U}_\mathbf{d}}(1)}, \ldots, U_{\pi_{\mathcal{U}_\mathbf{d}}(i)}\right), \tag{3.117}$$

and $\pi_\mathcal{S}$ denotes the permutation over any subset of users $\mathcal{S} \subset [K]$ such that $h^2_{\pi_\mathcal{S}(1)} \leq h^2_{\pi_\mathcal{S}(2)} \leq \cdots \leq h^2_{\pi_\mathcal{S}(|\mathcal{S}|)}$. From the convexity of function f, as outlined in [24, Lemma 1], the following lower bound is obtained from averaging over all the demand combinations in $\mathcal{D}^l_{\mathcal{U}_\mathbf{d}}$, $l \in [N_{\mathcal{U}_\mathbf{d},1}]$:

$$\frac{1}{N_{\mathcal{U}_\mathbf{d},1}} \sum_{l=1}^{N_{\mathcal{U}_\mathbf{d},1}} f\left(c^l_{\pi_{\mathcal{U}_\mathbf{d}}(1)}, \ldots, c^l_{\pi_{\mathcal{U}_\mathbf{d}}(N_\mathbf{d})}\right) \geq \sum_{i=1}^{N_\mathbf{d}} \left(\frac{2^{2R(1-\min\{iM/N,1\})}-1}{h^2_{\pi_{\mathcal{U}_\mathbf{d}}(i)}}\right) \prod_{j=1}^{i-1} 2^{2R\left(1-\min\left\{jM/N,1\right\}\right)}. \tag{3.118}$$

By substituting (3.116) and (3.118) into (3.115), it follows that

$$\bar{P} \geq \bar{P}_{\text{LB}}(R,M) \triangleq \mathrm{E}_{\mathcal{U}_\mathbf{d}}\left[\sum_{i=1}^{N_\mathbf{d}} \left(\frac{2^{2R(1-\min\{iM/N,1\})}-1}{h^2_{\pi_{\mathcal{U}_\mathbf{d}}(i)}}\right) \prod_{j=1}^{i-1} 2^{2R\left(1-\min\left\{jM/N,1\right\}\right)}\right]. \tag{3.119}$$

As a result, $\bar{P}_{\text{LB}}(R,M)$ lower bounds $\bar{P}^*_{\mathcal{U}_\mathbf{d}}\left(R,M\right)$, and so

$$\bar{P}^*\left(R,M\right) \geq \bar{P}_{\text{LB}}(R,M). \tag{3.120}$$

It is further proved in [24] that

$$\hat{P}^*(R,M) \geq \hat{P}_{\text{LB}}(R,M) \triangleq \max_{\mathcal{S}\subset[\min\{N,K\}]} \left\{\sum_{i=1}^{|\mathcal{S}|} \left(\frac{2^{2R(1-\min\{iM/N,1\})}-1}{h^2_{\pi_\mathcal{S}(i)}}\right) \prod_{j=1}^{i-1} 2^{2R\left(1-\min\left\{jM/N,1\right\}\right)}\right\}. \tag{3.121}$$

3.4 Numerical experiments

In this section, we provide numerical results evaluating the upper and the lower bounds on peak and average power-memory trade-offs, and illustrating the gap between these bounds. We assume a fixed rate $R = 1$ for the files in the library, and channel gain $1/h_k^2 = 10 - \frac{9(k-1)}{K-1}$, $k \in [K]$.

The bounds on the average and peak power memory trade-offs are illustrated in Figures 3.3 and 3.4, respectively, for $K = 10$ users, $N = 15$ files in the library, and cache size $0 \leq M \leq 3$. As can be seen, even a small cache size can provide significant transmit power reduction. We observe that, in both scenarios, the gap between the centralized and decentralized caching schemes is relatively small, particularly

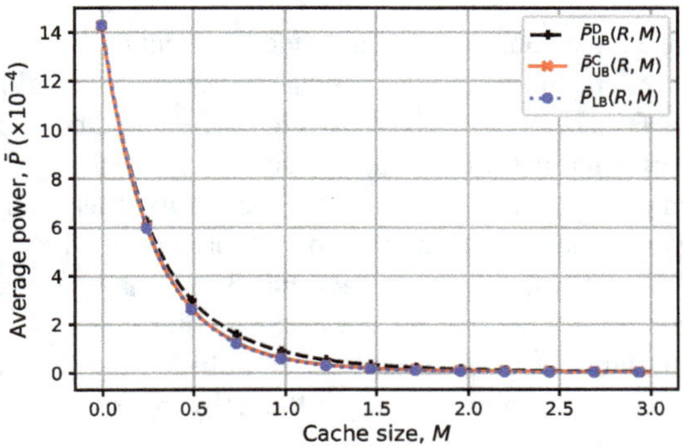

Figure 3.3 *Upper and lower bounds on the average power-memory trade-off for K=5 users, N=8 files with rate R=1, and channel gain $1/h_k^2 = 10 - \frac{9(k-1)}{K-1}$, $k \in [K]$*

for small and large cache sizes, which shows the robustness of the proposed coded caching scheme against the decentralization of the cache placement. It can be further seen that the gap between the centralized caching scheme and the lower bound is negligible, which illustrates the efficiency of the centralized caching scheme in energy saving and the tightness of the lower bound. The peak power-memory trade-off exhibits a similar behavior to the average case, but with significantly higher power values. Therefore, a considerable amount of energy can be saved by adapting the content delivery to the demand combination rather than designing it for only the worst-case demand.

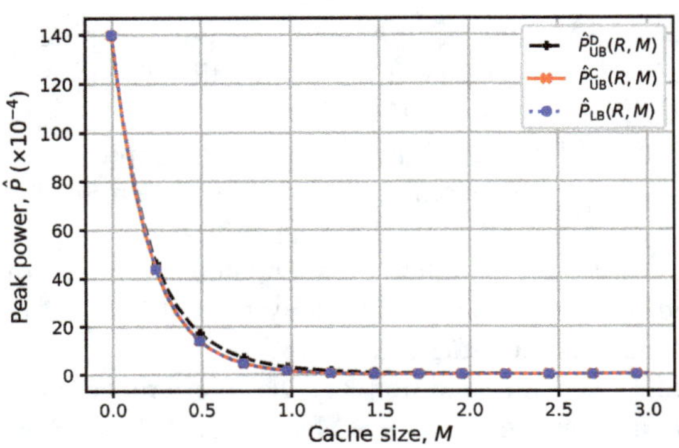

Figure 3.4 *Upper and lower bounds on the peak power-memory trade-off for K=5 users, N=8 files with rate R=1, and channel gain $1/h_k^2 = 10 - \frac{9(k-1)}{K-1}$, $k \in [K]$*

3.5 Summary

- In this chapter, we have considered cache-aided content delivery over noisy BCs, focusing on packet erasure and Gaussian channels. The physical layer characteristics of wireless BCs bring new challenges to cache-aided content delivery. The results in this chapter indicate that the coding schemes used for conventional broadcasting communication systems must be redesigned in order to exploit the availability of receiver caches across the network. More specifically, we have shown that, if properly exploited, even the users that do not have access to any cache memory can benefit from the presence of cache memories across the network.
- We have also shown that caching and coded delivery can improve the energy efficiency significantly, particularly when the delivery scheme is adapted to the demand combination.

References

[1] Maddah-Ali M.A., Niesen U. 'Fundamental limits of caching'. *IEEE Transactions on Information Theory*. 2014;**60**(5):2856–67.

[2] Maddah-Ali M.A., Niesen U. 'Decentralized coded caching attains order-optimal memory-rate tradeoff'. *IEEE/ACM Transactions on Networking*. 2014;**23**(4):1029–40.

[3] Chen Z., Fan P., Letaief K.B. 'Fundamental limits of caching: improved bounds for users with small buffers'. *IET Communications*. 2016;**10**(17):2315–18.

[4] Mohammadi Amiri M., Gündüz D. 'Improved delivery rate-cache capacity trade-off for centralized coded caching'. International Symposium on Information Theory and Its Applications (ISITA); Monterey, CA; Oct. 2016.

[5] Wan K., Tuninetti D., Piantanida P. 'On caching with more users than files'. Proceedings of IEEE International Symposium on Information Theory (ISIT); Barcelona, Spain; Jul. 2016. pp. 135–9.

[6] Mohammadi Amiri M., Yang Q., Gündüz D. 'Coded caching for a large number of users'. Proceedings of the IEEE Information Theory Workshop (ITW); Cambridge, UK; Sep. 2016. pp. 171–5.

[7] Sahraei S., Gastpar M. 'Users caching two files: an improved achievable rate'. Proceedings of the Annual Conference on Information Science and Systems (CISS); Princeton, NJ, USA; Mar. 2016. pp. 620–4.

[8] Mohammadi Amiri M., Gündüz D. 'Fundamental limits of coded caching: improved delivery rate-cache capacity tradeoff'. *IEEE Transactions on Communications*. Feb. 2017;**65**(2):806–15.

[9] Tian C., Chen J. 'Caching and delivery via interference elimination'. *IEEE Transactions on Information Theory*. Mar. 2018;**64**(3):1548–60.

[10] Yu Q., Maddah-Ali M.A., Avestimehr A.S. 'The exact rate-memory tradeoff for caching with uncoded prefetching'. *IEEE Transactions on Information Theory*. Feb. 2018;**64**(2):1281–96.

[11] Sengupta A., Tandon R. 'Improved approximation of storage-rate tradeoff for caching with multiple demands'. *IEEE Transactions on Communications*. May 2017;**65**(5):1940–55.

[12] Wang C.-Y., Saeedi Bidokhti S., Wigger M. 'Improved converses and gap results for coded caching'. *IEEE Transactions on Information Theory*. Nov. 2018;**64**(11):7051–62.

[13] Gregori M., Gomez-Vilardebo J., Matamoros J., Gündüz D. 'Wireless content caching for small cell and D2D networks'. *IEEE Journal on Selected Areas in Communications*. Mar. 2016;**34**(5):1222–34.

[14] Mohammadi Amiri M., Yang Q., Gündüz D. 'Decentralized caching and coded delivery with distinct cache capacities'. *IEEE Transactions on Communications*. 2017;**65**(11):4657–69.

[15] Niesen U., Maddah-Ali M.A. 'Coded caching with nonuniform demands'. *IEEE Transactions on Information Theory*. Feb. 2017;**63**(2):1146–58.

[16] Pedarsani R., Maddah-Ali M.A., Niesen U. 'Online coded caching'. *IEEE/ACM Transactions on Networking*. 2016;**24**(2):836–45.

[17] Yang Q., Hassanzadeh P., Gündüz D., Erkip E. 'Centralized caching and delivery of correlated contents over Gaussian broadcast channels'. *IEEE Transactions on Communications*. Jan. 2020;**68**(1):122–36.

[18] Urbanke R.L., Wyner A.D. 'Packetizing for the erasure broadcast channel with an internet application'. Proceeding of the International Conference on Combinatorics, Information Theory, and Statistics; Portland, ME; May 1997. p. 93.

[19] Tuncel E. 'Slepian-Wolf coding over broadcast channels'. *IEEE Transactions on Information Theory*. Apr. 2006;**52**(4):1469–82.

[20] El Gamal A., Kim Y.-H. *Network Information Theory*. Cambridge, UK: Cambridge University Press; 2012.

[21] Mohammadi Amiri M., Gündüz D. 'Cache-aided content delivery over erasure broadcast channels'. *IEEE Transactions on Communications*. Jan. 2018;**66**(1):370–81.

[22] Saeedi Bidokhti S., Wigger M., Timo R. 'Noisy broadcast networks with receiver caching'. *IEEE Transactions on Information Theory*. Nov. 2018;**64**(11):6996–7016.

[23] Bergmans P. 'Random coding theorem for broadcast channels with degraded components'. *IEEE Transactions on Information Theory*. 1973;**19**(2):197–207.

[24] Mohammadi Amiri M., Gunduz D. 'Caching and coded delivery over Gaussian broadcast channels for energy efficiency'. *IEEE Journal on Selected Areas in Communications*. Aug. 2018;**36**(8):1706–20.

Chapter 4
Multi-antenna cache-aided channels

Eleftherios Lampiris[1] *and Petros Elia*[2]

4.1 Abstract

This chapter discusses coded caching in multi-antenna systems and argues that when caching and multi-antenna delivery are treated jointly, multi-antenna systems can benefit from a powerful performance boost. Combining the two ingredients is only natural, given the promise of coded caching and the fact that multi-antenna technologies are currently the backbone of wireless systems.

While there has been an abundance of very interesting works on the combination of coded caching with multi-antenna systems, this particular chapter here places focus on how caching and multi-antenna systems – when fused in a joint, non-separable manner – have the potential to resolve each other's bottlenecks. While multi-antenna systems are limited by the well-known *feedback bottleneck* that requires channel-state information (CSI) costs that scale linearly with the degrees of freedom (DoF), single-antennacoded caching systems are also severely limited by the so-called *subpacketization bottleneck* that requires each file to be segmented into an astronomical number of packets that can scale exponentially with the number of users K. In the first case, finite coherence times can hardbound the feedback-constrained DoF, while in the second case, finite file sizes can hardbound the real *subpacketization-constrained DoF*. We review how a proper merging of the two ingredients can alleviate these constraints.

The chapter will first consider the L-antenna cache-aided Multiple-Input Single-Output (MISO) Broadcast Channel (BC) where a base station serves the demands of K cache-aided receivers. For this setting, we will discuss how switching from a single-antenna to a properly designed multi-antenna coded caching system can dramatically alleviate the subpacketization bottleneck of coded caching. This, in turn, will mean that going from 1 to L transmit antennas can boost the subpacketization-constrained (real) DoF (by boosting the real caching gain itself) by up to a multiplicative factor of L. This unprecedent improvement in the caching gain is due to

[1]Communications and Information Theory Group (CommIT), Technical University of Berlin, Berlin, Germany

[2]Communication Systems Department, EURECOM, Sophia Antipolis, France

a recent non-separable approach that will be reviewed here, where the designs of caching and multi-antenna delivery are treated jointly.

Then, we will see the other side of the story whose starting point is that in the cache-less MISO-BC, the real DoF is limited because the CSI cost generally matches the DoF gain. This unfortunate constraint remained in place – and was in fact aggravated – with the introduction of multi-antenna coded caching algorithms. We will review a recent non-separable multi-antenna coded caching approach that delivers the optimal DoF equal to the sum of the caching gain and the multiplexing gain, asking though only L of these users to feedback their CSI to the transmitter. This untangles the CSI costs from the coded caching gains of multi-antenna systems, in the sense that we pay for the multiplexing gain and we get the caching gain for free. Resolving the problems of simultaneously reducing subpacketization and CSI costs remains a big open problem.

4.2 The subpacketization bottleneck of coded caching

4.2.1 Coded caching with unbounded file sizes

We recall that the main idea of coded caching is to exploit receiver-side cached content in order to reduce multi-user interference. We also recall that in the standard single-stream coded caching scenario, a transmitter with access to a library of N files serves the demands of K users via a bottleneck link, where each user requests a single file and where the cache at each receiver can fit a fraction $\gamma = M/N \in [0, 1]$ of the library (Figure 4.1). For this setting, the work in [1] has shown that the theoretical worst-case delivery time takes the form

$$\mathscr{T}_{L=1}(K, \gamma) = \tfrac{K(1-\gamma)}{1+K\gamma} \overset{K\to\infty}{\longrightarrow} \tfrac{1-\gamma}{\gamma} \tag{4.1}$$

revealing that the overall delay remains fixed and bounded by the inverse of the individual cache size, irrespective of how many users are being served. This is achieved by transmitting multicast messages that serve $K\gamma + 1$ users at a time. This ability to deliver $K\gamma + 1$ DoF over a single-stream channel was a clear indication of the enormous potential of having modestly sized individual caches at the receivers.

4.2.2 Exponential subpacketization of coded caching

It is the case, though, that these gains are severely diminished by the fact that the algorithm may require files to have astronomical sizes. In particular, as we know from [2], the above optimal delay, and the corresponding DoF $K\gamma + 1$, requires – under the assumption of uncoded placement – that each file is split into

$$S = \tbinom{K}{K\gamma} \tag{4.2}$$

packets. This is the exact subpacketization of the algorithm in [1]. This means that if the file sizes are less than S, then naturally the coded caching algorithm cannot be applied in its full potential. The above subpacketization

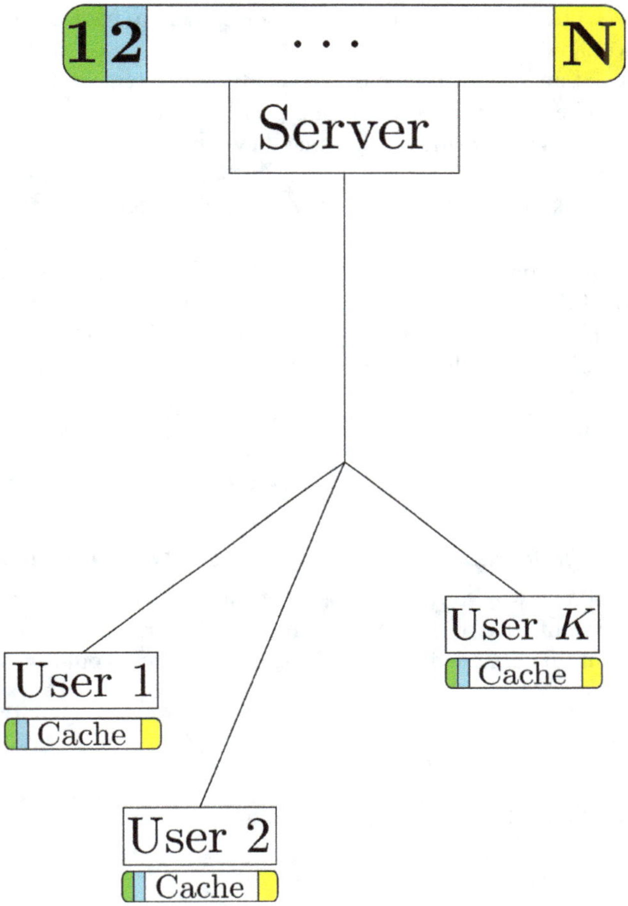

Figure 4.1 The single-stream, cache-aided bottleneck channel

$$S = \binom{K}{K\gamma} \geq \left(\frac{1}{\gamma}\right)^{K\gamma} \quad (4.3)$$

scales exponentially with K and can thus be prohibitive.

Observing the design in [1], we understand that this limitation is a direct outcome of the following observations.

- For each multicast message meant for $K\gamma + 1$ users, all unwanted subfiles must be cached by all unintended recipients, and this must be guaranteed for each possible collection of $K\gamma + 1$ users.
- In each multicast message meant for $K\gamma + 1$ users, each involved user must have cached the $K\gamma$ interfering subfiles of that message. This implies that any

set of $K\gamma + 1$ users that will be treated together need to have caches with some partial overlap.

- At the same time, this overlap has to be controlled. Users with identical caches cannot remove each other's interfering subfiles and thus cannot belong in the same multicast group, which means that they must be treated in different time-slots. In other words, full cache replication (rather than controlled overlap) would reduce subpacketization but would inevitably lead to reduced performance.

The above limitations hold for a general class of algorithms. As shown in [3] – in the context of decentralized coded caching (see [4]) and under the assumption of uncoded cache placement – any decentralized algorithm that seeks to achieve a coded caching gain of $\frac{4}{3}g$ will require a subpacketization level that scales at least as

$$S_{dec} = \mathcal{O}\left(\left(\tfrac{1}{\gamma}\right)^{g}\right) \tag{4.4}$$

which approximately matches the scaling in (4.3).

4.2.3 Algorithmic improvements for reducing subpacketization

As argued in [3], one way to reduce subpacketization would be to encode across a limited number Λ of users ($\Lambda < K$) at a time, thus generating a set of Λ different caches that would be assigned to the different users. This would yield a reduced subpacketization of

$$S = \binom{\Lambda}{\Lambda\gamma} \in \left[\left(\tfrac{1}{\gamma}\right)^{\Lambda\gamma}, \left(\tfrac{e}{\gamma}\right)^{\Lambda\gamma}\right] \tag{4.5}$$

at the expense of a much increased delay

$$\mathcal{T}_1(\Lambda, \gamma) = \tfrac{K(1-\gamma)}{1+\Lambda\gamma}. \tag{4.6}$$

This delay increase would be optimal [5], and it would be due to the fact that groups of users with identical caches must be treated in a Time Division Multiple Access (TDMA) manner.

Some progress came with the work in [6] and later in [7], which proposed new coded caching primitives that reduce subpacketization to $S = \left(\frac{1}{\gamma}\right)^{K\gamma-1}$ at the expense of achieving a slightly reduced DoF of $K\gamma$. Similarly, the work in [8] proposed new low-subpacketization algorithms based on hypergraph designs that achieve a coded caching gain that is approximately $\frac{K\gamma^2}{4}$. The same work also showed that no single-antenna algorithm can achieve a DoF that scales in any way with K while employing subpacketization that scales linearly with K. Other related approaches can be found in [9] with a design based on Rusza-Schemeredi graphs, as well as in [10–12].

While the above new algorithms provide exponential subpacketization reductions compared to the algorithm of [1] (MN algorithm), their improvement over the subpacketization-constrained DoF performance of MN can be modest when we focus on realistic values of γ and S_{max}. For example, when $\gamma \le 1/20$ and $S_{max} \le 10^5$,

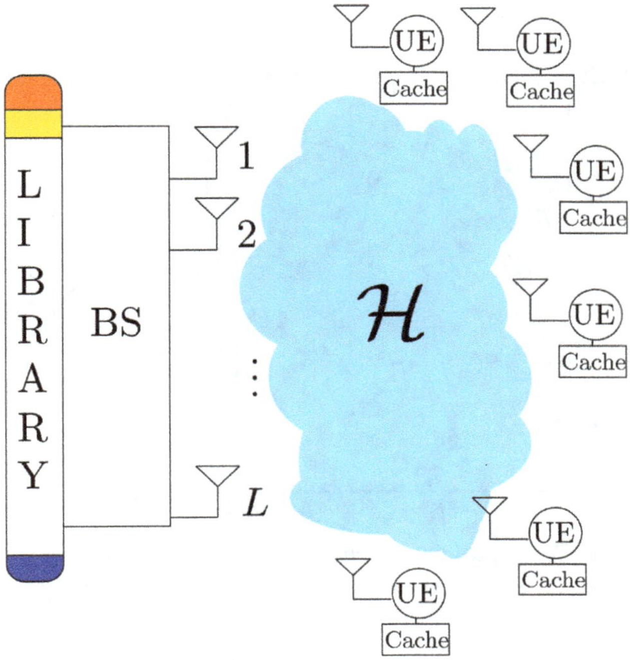

Figure 4.2 The cache-aided MISO BC

no known algorithm can improve over the MN algorithm's effective caching gain (and effective DoF) by more than two[a] additional users served at a time.

To the best of our knowledge, under the generous assumptions that $S_{max} \leq 10^5$, $\gamma \leq 1/50$ and $K \leq 10^5$, currently there exists no method that yields a DoF of more than 6 users served at a time.[b]

In the end, the subpacketization constraint remains the main factor that severely diminishes the real gains of single-stream coded caching.

4.2.4 Subpacketization bottleneck of separable multiple antenna coded caching

The notable works in [13, 14] have extended the ideas of [1] to the MISO setting with L transmit antennas[c] [13, 15] (see Figure 4.2) and showed that one can achieve a worst-case delivery time of

[a]This best-known improvement is due to Construction 6 in [8] ($a = b = 2, \lambda = 40$), which encodes over 3 160 users to give an effective (real) sum-DoF of 6, while the MN algorithm gives a DoF of 4.

[b]This corresponds to Construction 6 in [42] ($a = b = 2, \lambda = 100$), and it requires approximately 20 000 users.

[c]The same ideas apply to the setting where K_T cache-aided transmitters – each with access to some fraction γT of the library – are tasked with satisfying the requests of the users [14]. The exact association with the L-antenna MISO-BC setting can be made by setting $K_T \gamma_T = L$.

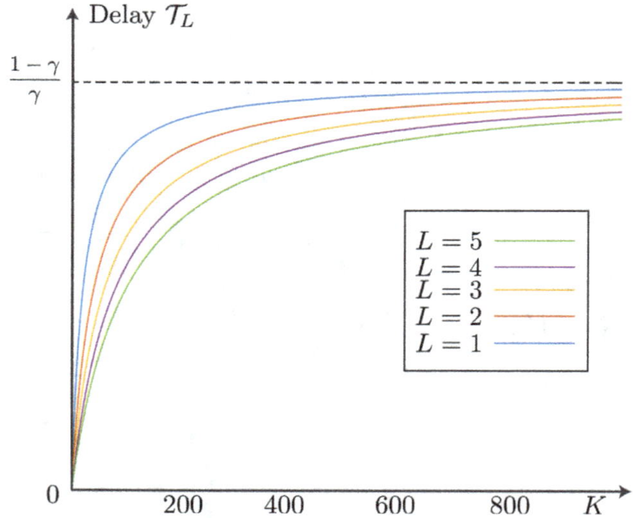

Figure 4.3 The effect of adding antennas in a cache-aided system. As the number of users increases, the delay asymptotically tends to $\frac{1-\gamma}{\gamma}$.

$$\mathcal{T}_L(K, \gamma) = \frac{K(1-\gamma)}{L+K\gamma} \tag{4.7}$$

thus also revealing that multiplexing and multicasting gains can combine in an additive manner to yield a sum-DoF of $L + K\gamma$. This performance – which is depicted in Figure 4.3 – could be achieved with a separable design that used the original placement and XOR generation algorithms of [1] and then employed multiple antennas to efficiently deliver these XORs to $L + K\gamma$ cache-aided receivers at a time.

It is the case though, that the subpacketization required for this performance further increased to

$$S_L = \binom{K}{K\gamma}\binom{K-K\gamma-1}{L-1} \tag{4.8}$$

thus making multi-antenna coded caching even more impractical under the assumption of realistic file and cache sizes.

4.3 Setting and notation

In reviewing different schemes, this chapter will focus on studying the DoF performance of the L-antenna MISO BC[d] with K cache-aided single-antenna users, where each user has the capacity to store a fraction γ of the library at its cache. Each user simultaneously asks for a single and different file from the library.

[d]We note that the K_T transmitter setting with caches at both transmitters and receivers exhibits the same DoF performance as the L-antenna MISO-BC when each transmitter caches fraction $\gamma_T = \frac{L}{K_T}$ of the library. Thus, the two settings could be used interchangeably.

We will use $\mathcal{T}_L(K,\gamma)$ to denote the normalized – by the file size and the link capacity – delivery time, and we will use $\mathcal{D}_L(K,\gamma)$ to describe the sum DoF of the same system. The relationship between the two quantities is also given by

$$\mathcal{D}_L(K,\gamma) = \frac{K(1-\gamma)}{\mathcal{T}_L(K,\gamma)}. \tag{4.9}$$

In order to satisfy the users' demands, the base station transmits the $L \times 1$ vector $\mathbf{x}$. The signal at each receiver $k \in [K]$ takes the form

$$y_k = \mathbf{h}_k^\dagger \mathbf{x} + w_k, \tag{4.10}$$

where $\mathbf{h}_k \in \mathbb{C}^{L\times 1}$ denotes the channel from the transmitter to receiver k, $\mathbf{x}$ is the $L \times 1$ vector message transmitted from the L antenna array, satisfying the power constraint $\mathbb{E}\{\|\mathbf{x}\|^2\} = P$, and $w_k \sim \mathcal{CN}(0,1)$ corresponds to the noise observed at user k. We assume that each node has all necessary CSI and that for a given signal-to-noise ratio (SNR), each link has the capacity of the form $\log(SNR) + o(log(SNR))$.

We will use $[K] \triangleq \{1,2,\ldots,K\}$ to denote the set of users, and for any subset $\lambda \subset [K]$ of size $|\lambda| = L$, we will denote with H_λ^{-1}, the $L \times L$ matrix formed as the normalized inverse of the channel between the L-antenna transmitter and the L users in set λ. In particular, this zero-forcing (ZF) matrix H_λ^{-1} can be written in the form

$$H_\lambda^{-1} \triangleq [\mathbf{h}_{\lambda\setminus\{\lambda(1)\}}^{\perp}, \ldots \mathbf{h}_{\lambda\setminus\{\lambda(L)\}}^{\perp}] \tag{4.11}$$

where

$$\mathbf{h}_k^\dagger \mathbf{h}_{\lambda\setminus\{\lambda(\ell)\}}^{\perp} = \begin{cases} 0, & k \in \lambda \setminus \{\lambda(\ell)\} \\ \neq 0, & k \in [K] \setminus \lambda \cup \{\lambda(\ell)\}. \end{cases} \tag{4.12}$$

4.4 Using multiple antennas to dramatically alleviate the subpacketization bottleneck of coded caching

We proceed to review the recent solution in [16] (see also [17]) that manages to employ multiple antennas to reduce the subpacketization constraint. We first summarize the main theorem of this work, noting that it assumes $\frac{K}{L} \in \mathbb{N}$ and $\frac{K\gamma}{L} \in \mathbb{N}$. The treatment of the other (non-integer) cases leads to a slightly reduced DoF and can be found in the appendices of [16].

Theorem 1: *In the cache-aided MISO-BC with L transmitting antennas and K receiving users, the delay of* $\mathcal{T}_L(K,\gamma) = \frac{K(1-\gamma)}{L+K\gamma}$ *and the corresponding DoF* $\mathcal{D}_L(K,\gamma) = L + K\gamma$ *can be achieved with subpacketization*

$$S_L = \binom{\frac{K}{L}}{\frac{K\gamma}{L}}. \tag{4.13}$$

4.4.1 Impact and insight

Before reviewing the schemes, we provide some insights on the above result.

Single-antenna system vs. multi-antenna system

Contrary to single-antenna algorithms that seek to reduce subpacketization, and which inevitably result in reduced DoF, what the above result shows is that – in addition of, course, to offering an increased multiplexing gain – increasing the number of transmitters also leads to an accelerated subpacketization reduction, without sacrificing the theoretical caching gains. This new subpacketization is approximately the L-th root of the equivalent single-antenna system with the same caching gain.

Multiplicative real-DoF gains

As we have seen, in the absence of file-size constraints, going from 1 to L antennas yields an additive DoF boost from $1 + K\gamma$ to $L + K\gamma$. What we note here, though, is that in the presence of such file-size constraints, this relationship can be of a much more powerful, multiplicative nature. To see this, let us assume that in the single-stream scenario, we use the MN algorithm in the presence of a subpacketization constraint $S_{max} = \binom{\Lambda}{\Lambda\gamma}$ for some integer Λ that divides K. We have seen that the real DoF performance of this system would be reduced to

$$\mathcal{D}_1(\Lambda, \gamma) = \Lambda\gamma + 1 \tag{4.14}$$

and that this could be achieved with cache replication (see Equation (4.6)). The above theorem directly says that – with the new algorithm that we will review below – the L-antenna equivalent system would exhibit a real DoF of up to

$$\mathcal{D}_L(K, \gamma) = L \cdot \Lambda\gamma + L = L(\Lambda\gamma + 1) \tag{4.15}$$

which is L times higher than that of the single-antenna system, as long as $L \leq K/\Lambda$.

4.4.2 Scheme description

We proceed to review the scheme's cache placement phase and then its delivery phase.

4.4.2.1 Placement phase

The placement phase is based on the algorithm of [1] and uses cache replication. Users are segmented into groups of L users each, where users belonging to the same group will cache *the exact same cache content*. This cache replication allows for a dramatic subpacketization reduction, but it conflicts with the aforementioned core idea of coded caching, where multicasting opportunities emerge only if the cache overlap is partial.

The first step involves organizing the users into $\frac{K}{L}$ groups, each of size L. We denote the $\Lambda = \frac{K}{L}$ groups by G_κ, $\kappa \in [\Lambda]$ while, for tractability, we assume the following placement of users $k \in [K]$ into groups

$$G_\kappa = \{\kappa, \kappa + \Lambda, \ldots, \kappa + (L-1) \cdot \Lambda\}, \quad \kappa \in [\Lambda]. \tag{4.16}$$

Group-based placement phase

The placement phase begins by segmenting each library file into $S_\Lambda = \binom{\Lambda}{\Lambda\gamma}$ disjoint parts, where, in particular, each file W^n, $n \in [N]$ is divided as

$$W^n \rightarrow \left\{W^n_\tau,\ \tau \subset [\Lambda],\ |\tau| = \Lambda\gamma\right\} \tag{4.17}$$

The cache $\mathcal{Z}_k$ of user $k \in G_\kappa$ is filled as follows

$$\mathcal{Z}_k = \{W^n_\tau\ : \kappa \in \tau, \forall n \in [N]\},\quad k \in G_\kappa. \tag{4.18}$$

Consequently here, index τ denotes the $\frac{K}{L}$ groups of users where this subfile is cached.

4.4.2.2 Group-based delivery of content

The delivery phase commences as soon as the users' requests are announced. Each user can ask for a different file. We will assume that feedback is perfect and instantaneous. For simplicity, the presented schemes will use ZF precoding, which is adequate for a DoF exposition of the problem. The extension to more efficient precoders is automatic. The steps are as follows.

1. Pick $\sigma \subseteq [\Lambda], |\sigma| = \Lambda\gamma + 1$ (select groups)
2. Let $G_m = \{g_{m(1)}, g_{m(2)}, \ldots, g_{m(L)}\}$
3. Transmit:

$$\mathbf{x}_\sigma = \sum\nolimits_{m\in\sigma} H^{-1}_{G_m} \cdot \begin{bmatrix} W^{d_{g_{m(1)}}}_{\sigma\setminus\{m\}} \\ \vdots \\ W^{d_{g_{m(L)}}}_{\sigma\setminus\{m\}} \end{bmatrix}. \tag{4.19}$$

The above algorithm selects in every transmission a new set σ of $\Lambda\gamma + 1$ groups that define the $L \cdot (\Lambda\gamma + 1)$ users to be served at that point. After creating the group-based precoders $H^{-1}_{G_m}$, $m \in \sigma$, the algorithm proceeds to form the message as a linear combination of $\Lambda\gamma + 1$ vectors, where each vector is intended for users of some group $\kappa \in \sigma$ and takes the form

$$H^{-1}_{G_\kappa} \cdot \begin{bmatrix} W^{d_{g_{\kappa(1)}}}_{\sigma\setminus\{\kappa\}} \\ \vdots \\ W^{d_{g_{\kappa(L)}}}_{\sigma\setminus\{\kappa\}} \end{bmatrix}. \tag{4.20}$$

We can see that every user of the same group will receive, each from their own different desired file, the part of that file that is indexed by $\sigma \setminus \{\kappa\}$.

4.4.2.3 Decodability

The received signal (after suppressing the noise for simplicity) of any user $k \in G_\kappa$, takes the form

$$y_k = \mathbf{h}_k^\dagger \cdot \sum_{m\in\sigma} H_{G_m}^{-1} \cdot \begin{bmatrix} W_{\sigma\backslash\{m\}}^{d_{g_m(1)}} \\ \vdots \\ W_{\sigma\backslash\{m\}}^{d_{g_m(L)}} \end{bmatrix}. \tag{4.21}$$

To decode, user k can first remove any subfile with any index $\tau \ni \kappa$ by using its cached content and the acquired CSI. Now, free of any intergroup interference, the remaining part of the received signal for that user takes the form

$$y_k^{\text{rem}} = \mathbf{h}_k^\dagger \cdot H_{G_\kappa}^{-1} \cdot \begin{bmatrix} W_{\sigma\backslash\{\kappa\}}^{d_{g_\kappa(1)}} \\ \vdots \\ W_{\sigma\backslash\{\kappa\}}^{d_{g_\kappa(L)}} \end{bmatrix} = W_{\sigma\backslash\{\kappa\}}^{d_k}. \tag{4.22}$$

This residual part in (4.21) constitutes a ZF-precoded signal which, by design, removes all intragroup interference such that finally each user receives their own subfile, interference free.

4.4.3 Intuition

As we have seen, adding antennas to a system allows more and more users to share the same cached content without affecting the caching gains. This transition is what allows the subpacketization reductions to emerge. In essence, a K-user L-antenna system can incur the same subpacketization as the single-stream equivalent but with $\frac{K}{L}$ users, without a sacrifice in the caching gain. This equivalency reveals how a multi-antenna setting can be *decomposed* into L parallel systems.

4.4.4 Extensions

The above result provides a way to dramatically reduce subpacketization in multi-antenna coded caching environments. Recent efforts sought to improve the above result in two ways. A first line of research aims to reduce the amount of CSI feedback that the algorithm requires (see Section 4.5), while a second line of research aims to design algorithms that achieve the maximal DoF of $K\gamma + L$ when $\frac{K}{L}$ and $\frac{K\gamma}{L}$ are non-integer. The second goal is addressed in [18] which, for the regime where $K\gamma > L$, designs reduced-subpacketization coded caching algorithms that retain the full coded caching and multiplexing gains.

4.5 The feedback bottleneck of multi-antenna systems

As noted before, in the MISO-BC, the real DoF is limited because the CSI cost generally matches the DoF gain. In such scenarios, feedback-related costs can severely reduce the fraction of the coherence time that can be used to send actual data (cf [19–21].), and thus these same costs are the reason why most multi-antenna solutions fail to scale.

4.5.1 Feedback bottleneck of multi-antenna coded caching

These high CSI costs remained in place with the introduction of multi-antenna coded caching algorithms, which would require all users that were treated in a single transmission, to deliver CSI to the transmitter. This is illustrated in the following example, which corresponds to the algorithm in [13].

Example 1: *Let us consider a system with* $K = 4$ *users, where each user is equipped with a cache of normalized size* $\gamma = \frac{1}{2}$. *In the single stream case, the DoF of* $\mathcal{D}_1(4, \frac{1}{2}) = 3$ *naturally requires no feedback costs. On the other hand, let us consider a characteristic separable, multi-antenna coded caching algorithm* [13] *for the case of* $L = 2$. *Such an algorithm can serve* $\mathcal{D}_2(4, \frac{1}{2}) = 4$ *users at a time but requires in each transmission that all four served users send their CSI. This can be deducted by observing the transmissions of the algorithm that take the form*

$$\begin{aligned} \mathbf{x} = \; & \mathbf{h}_4^{\perp}(A_{23} \oplus B_{13} \oplus C_{12}) + \mathbf{h}_3^{\perp}(A_{24} \oplus B_{14} \oplus D_{12}) \\ & + \mathbf{h}_2^{\perp}(A_{34} \oplus C_{14} \oplus D_{13}) + \mathbf{h}_1^{\perp}(B_{34} \oplus C_{24} \oplus D_{23}) \end{aligned} \tag{4.23}$$

From the above transmission, we can clearly see that the transmitter needs to know the channels of all four users in order to design the four precoder vectors ($\mathbf{h}_k^{\perp}$, $k \in [4]$). *Moreover, this feedback needs to be communicated to the users in order for them to decode their messages. Thus, the CSIT and the CSIR costs are both equal to* 4, *in the sense that* 4 *users need to communicate their CSI per transmission. Hence, here, adding a single DoF (by going from 1 to 2 antennas) increases the CSI cost from* 0 *to* 4. *This simple example reveals a rather peculiar disadvantage of existing separable multi-antenna coded caching designs that as soon as the number of antennas is increased by 1, the CSI cost goes from* 0 *to* 4, *which further means that adding a single DoF increases the CSI cost from* 0 *to* $2 + K\gamma$.

Multi-antenna coded caching with feedback considerations

Motivated by this feedback bottleneck, different works combined multiple antennas (or multiple transmitters) with coded caching and explored the possibility of reducing CSI costs. These CSI reductions, however, came at the expense of reduced performance. As an example, the works in [22, 23] consider reduced quality CSIT but yield a maximum DoF that is bounded close to $K\gamma + 1$, while the works in [24, 25] consider only statistical CSI but again achieve much lower DoF. Similarly, the interesting work in [26] uses ACK/NACK-type CSIT to ameliorate the negative effects of having users with unequal channel strengths, but the algorithm does not aim to achieve multiplexing gains.

4.6 Joint caching and multi-antenna delivery with low CSIT

In this section, we will review the recent multi-antenna algorithm in [27], which achieves the maximal known DoF performance, while requiring CSI cost equal to L. This means that the coded caching gain comes for free and the CSI cost is not dependent on the achieved DoF.

We will assume that feedback is perfect and instantaneous, while the feedback training process is divided into two phases: the uplink phase where the transmitter estimates the channels of some users and the downlink phase where receiver k estimates products $\mathbf{h}_k^{\dagger}\mathbf{h}_{\lambda}^{\perp}$, for some set $\lambda \subset [K], |\lambda| = L - 1$. The feedback training process, as described in [28], is such that feedback for C users will require C training slots in the uplink training phase (for CSIT) and C training slots in the downlink training phase (for local and global CSIR).

We proceed with the main result of the work in [27] (see also [28]).

Theorem 2: *In the MISO-BC setting with L transmit antennas and K single-antenna users, each equipped with cache of normalized size* γ, *the maximal DoF* $\mathcal{D}_L(K, \gamma) = L + K\gamma$ *can be achieved with CSI from only L users at a time, and thus with CSIT cost of L uplink training time-slots and CSIR cost of L downlink training time-slots.*

Intuition and small example

We begin by offering some intuition on this design. First, let us notice that the file segmentation and the cache placement draw directly from the placement algorithm of [1]. Contrarily, the design of the multicast messages will be fundamentally different. The steps are as follows.

1. Initially, we construct XORs composed of $\frac{K\gamma}{L} + 1$ subfiles, and then
2. a collection of L such XORs is communicated, thus the total number of users involved is $K\gamma + L$.
3. The set of "active users" (a set of $L + K\gamma$ served users) will be divided into two sets:
    - a set, λ, which consists of the L users that will be assisted by precoding and
    - a second set, π, consisting of $K\gamma$ users. These users will exclusively use their caches to remove interference since they will not be assisted by precoding.
4. Finally, the information vector comprised of the L XORs will be multiplied by the normalized inverse H_{λ}^{-1} of the channel matrix between the base station and the users in set λ.

This non-separable design, as we will see later, guarantees that all $K\gamma + L$ active users can decode their desired subfile. First, the users assisted by precoding will only "see" one of the XORs, due to the design of the precoder matrix. Since these XORs are smaller instances of the XORs designed in [1], it follows that these can be decoded by each user in set λ.

On the other hand, the remaining $K\gamma$ users, who belong to the set π, will receive a linear combination of the L XORs and thus need to be able to remove $K\gamma + L - 1$ subfiles in order to recover their respective subfiles. Thus, the main challenge of the algorithm is to design these XORs and then select subsets of L such XORs in such a manner that eventually allows everyone in set π to decode.

We proceed with illustrating a single transmission of the algorithm, using the same setting as Example 1. As discussed previously, we seek to achieve the same performance as before (all 4 users being able to decode their subfiles in each transmission), while using CSIT from only two users at a time ($L = 2$ training slots for the uplink and $L = 2$ training slots for the downlink).

Example 2: *In a MISO-BC setting with $L = 2$ transmit antennas, $K = 4$ users, and per-user cache of normalized size $\gamma = 1/2$ (see Example 1), one of the transmitted vectors from the proposed algorithm takes the form*[e]

$$\mathbf{x} = \mathbf{h}_2^{\perp}(A_{34} \oplus C_{14}) + \mathbf{h}_1^{\perp}(B_{34} \oplus D_{23}). \tag{4.24}$$

We remind that A, B, C, and D denote the files that are requested by users 1, 2, 3, and 4, respectively. Also, subfile W_{ij}, $W = \{A, B, C, D\}$ is stored at users i, j.

Then the received signals y_k, $k \in \{1, 2, 3, 4\}$ at the four users take the form

$$\begin{bmatrix} y_1 \\ y_2 \\ y_3 \\ y_4 \end{bmatrix} = \begin{bmatrix} \mathbf{h}_1^{\dagger}\left(\mathbf{h}_2^{\perp} \cdot (A_{34} \oplus C_{14}) + \mathbf{h}_1^{\perp} \cdot (B_{34} \oplus D_{23})\right) \\ \mathbf{h}_2^{\dagger}\left(\mathbf{h}_2^{\perp} \cdot (A_{34} \oplus C_{14}) + \mathbf{h}_1^{\perp} \cdot (B_{34} \oplus D_{23})\right) \\ \mathbf{h}_3^{\dagger}\left(\mathbf{h}_2^{\perp} \cdot (A_{34} \oplus C_{14}) + \mathbf{h}_1^{\perp} \cdot (B_{34} \oplus D_{23})\right) \\ \mathbf{h}_4^{\dagger}\left(\mathbf{h}_2^{\perp} \cdot (A_{34} \oplus C_{14}) + \mathbf{h}_1^{\perp} \cdot (B_{34} \oplus D_{23})\right) \end{bmatrix} \tag{4.25}$$

$$= \begin{bmatrix} A_{34} \oplus C_{14} \\ B_{34} \oplus D_{23} \\ \mathbf{h}_3^{\dagger}\left(\mathbf{h}_2^{\perp} \cdot (A_{34} \oplus C_{14}) + \mathbf{h}_1^{\perp} \cdot (B_{34} \oplus D_{23})\right) \\ \mathbf{h}_4^{\dagger}\left(\mathbf{h}_2^{\perp} \cdot (A_{34} \oplus C_{14}) + \mathbf{h}_1^{\perp} \cdot (B_{34} \oplus D_{23})\right) \end{bmatrix} \tag{4.26}$$

where we have suppressed noise for simplicity.

Hence, we see that user 1 *and user* 2 *only receive the first and second XOR, respectively (due to the design of the precoders), which means that each of these two users can decode their desired subfiles A_{34} and B_{34}, respectively, by "caching-out" the unwanted subfiles C_{14} and D_{23}, respectively.*

Examining the decoding process for the remaining users (users 3 *and* 4*), we can discern that user* 3 *must employ its cache to remove subfiles A_{34}, B_{34}, and D_{23} (which, by design, are cached at user* 3*) such as to decode the desired C_{14}. Further, user* 4 *needs to retrieve from its cache subfiles A_{34}, B_{34}, and C_{14} in order to decode the desired subfile D_{23}. Apart from employing their cached content, users* 3 *and* 4 *need some CSI knowledge; user* 3 *needs products $\mathbf{h}_3^{\dagger}\mathbf{h}_2^{\perp}$ and $\mathbf{h}_3^{\dagger}\mathbf{h}_1^{\perp}$, while user* 4 *needs $\mathbf{h}_4^{\dagger}\mathbf{h}_2^{\perp}$ and $\mathbf{h}_4^{\dagger}\mathbf{h}_1^{\perp}$.*

[e]We note that there is a notational discrepancy between the subfile indices of this example and the formal notation. In this example, we have kept the notation as simple as possible in order to more easily provide a basic intuition on the structure of the scheme.

4.6.1 Scheme description

4.6.1.1 Cache placement

The placement phase follows the scheme in [1], and as such it takes place without knowledge of the number of transmit antennas nor assumes any CSI knowledge. Each file $W^n, n \in [N]$ is initially split into $\binom{K}{K\gamma}$ subfiles as

$$W^n \rightarrow \{W^n_\tau,\ \tau \subset [K],\ |\tau| = K\gamma\} \tag{4.27}$$

and each cache $\mathcal{Z}_k$ is filled as

$$\mathcal{Z}_k = \{W^n_\tau : \forall \tau \ni k, |\tau| = K\gamma, \forall n \in [N]\} \tag{4.28}$$

Upon notification of the requests $\{W^{d_k},\ k \in [K]\}$, and after the number of antennas is revealed to be L, each requested subfile $W^{d_k}_\tau$ is further split twice as follows[f]

$$W^{d_k}_\tau \rightarrow \quad \{W^{d_k}_{\sigma,\tau},\ \sigma \subseteq [K] \setminus (\tau \cup \{k\}), |\sigma| = L-1\} \tag{4.29}$$

$$W^{d_k}_{\sigma,\tau} \rightarrow \{W^{\phi,d_k}_{\sigma,\tau},\ \phi \in [L+K\gamma]\} \tag{4.30}$$

4.6.1.2 CSI acquisition

The acquisition phase is split into the *uplink training phase* and the *downlink training phase*. The uplink phase is responsible for feeding the transmitter with the required CSIT from some set λ of L users. These users will transmit pilot signals, and the transmitter, in turn, can estimate their channel coefficients $\mathbf{h}_k$, $\forall k \in \lambda$, and thus construct precoders $\mathbf{h}^{\perp}_{\lambda\setminus\{k\}}$, $\forall k \in \lambda$. Further, the *downlink training phase* will inform the active users of the required CSIR knowledge. Hence, a set of signals (each comprised of Q vectors) for each of the L precoders will be broadcasted by the transmitter[g] as follows

$$\text{diag}\left(\mathbf{h}^{\perp}_{\lambda\setminus\{k\}}\right) \mathbf{S}_{L\times Q}, \quad \forall k \in \lambda \tag{4.31}$$

which, when multiplied by a user's individual channel, will allow for estimation at any user $p \in \pi \subset [K] \setminus \lambda$ (recall $|\pi| = K\gamma$) of the needed global-CSIR (channel-precoder) products $\mathbf{h}^{\dagger}_p \cdot \mathbf{h}^{\perp}_{\lambda\setminus\{k\}}$, $\forall k \in \lambda$. In the above, diag($\mathbf{h}$) refers to the square diagonal matrix with its elements being the entries of vector $\mathbf{h}$.

The above process consumes L training slots for CSIT acquisition because only the L users in set λ need to communicate their channels. Further, it occupies L slots for the CSIR phase because, for each fixed precoder, a single (training symbol) shot suffices to communicate the relevant composite channel-precoder product to any number of users.

[f]For clarity of exposition and to avoid many indices, the index φ of (4.29) will henceforth be suppressed; thus, any $W^{\phi,d_k}_{\sigma,\tau}$ will be denoted as $W^{d_k}_{\sigma,\tau}$ unless φ is explicitly needed.

[g]Q is the minimum number of vectors that are required for a perfect estimation of the intended CSI.

4.6.1.3 Delivery phase

We continue with the description of the algorithm, where in each different transmission, the transmitter creates a vector of L XORs and proceeds to precode these with the appropriate set of precoders.

Individual XOR design

For any two disjoint sets $\mu, \nu \subset [K]$ and $\mu \cap \nu = \emptyset$, where $|\mu| = \frac{K\gamma}{L} + 1, |\nu| = K\gamma\frac{L-1}{L}$, and for an arbitrary $\sigma \subseteq \left([K] \setminus (\mu \cup \nu)\right)$, $|\sigma| = L - 1$, we construct the XOR

$$X_{\mu}^{\nu,\sigma} = \bigoplus_{k\in\mu} W_{\sigma,(\nu\cup\mu)\setminus\{k\}}^{d_k} \tag{4.32}$$

which consists of $\frac{K\gamma}{L} + 1$ subfiles, where

- each subfile is requested by one user that belongs to set μ and
- all subfiles are known by all users in set ν.

The set $(\nu \cup \mu) \setminus \{k\}$, $k \in \mu$ acts as the set τ (see (4.33)), as it describes the set of users that have this subfile $W_{\tau}^{d_k}$ in their cache. Moreover, the set σ is a subset that reveals the $L - 1$ users (one from each of the $L - 1$ remaining XORs) which will be assisted by precoding.

Example 3: *Consider the K-user MISO-BC with $L = 2$ transmit antennas and a cumulative (normalized) cache-size of $K\gamma = 4$. Fix $\mu = \{1, 2, 3\}$ and $\nu = \{4, 5\}$ and pick any $\sigma \in [K] \setminus \{1, 2, 3, 4, 5\}$, $|\sigma| = 1$. Then, the XOR of Equation (4.28) becomes*

$$X_{\{123\}}^{45,\sigma} = W_{\{\sigma,\underbrace{2345}_{\tau}\}}^{d_1} \oplus W_{\{\sigma,1345\}}^{d_2} \oplus W_{\{\sigma,1245\}}^{d_3} \tag{4.33}$$

Communicating the above XOR would deliver subfiles requested by the users in set μ and at the same time, it can be completely cached-out by all users of set ν. Observing users 1, 2, and 3, we can see that they can cache out each others' subfiles (e.g., user 1 caches out $W_{\{\sigma,1345\}}^{d_2} \oplus W_{\{\sigma,1245\}}^{d_3}$ to get its own $W_{\{\sigma,2345\}}^{d_1}$). Users 4 and 5 can find in their cache all the subfiles of the above XOR and, thus, can completely remove it. We can easily verify that each index τ has size $|\tau| = K\gamma = 4$, which meets the available cache-size constraint.

4.6.1.4 Design of the transmitted vectors

As previously stated, the goal in each transmission is to serve different subfiles to some set of $K\gamma + L$ users, while requiring CSIT from only L users. The following scheme achieves this. The steps of this scheme are described below.

- In Step 1, a set λ of L users is chosen, which signifies the precoding-assisted users.

- In Step 2, a (ZF-type) precoder H_{λ}^{-1} is designed to spatially separate the L users who belong to set λ.
- In Step 3, the set $\pi \subseteq [K] \setminus \lambda$ of $K\gamma$ users is selected from the remaining users.

To construct the L XORs and to properly place them in the vector, the following steps take place.

- In Step 4, the set π of $K\gamma$ users is partitioned, arbitrarily, into L non-overlapping sets $\mathcal{F}_i$, $i \in [L]$, each having $\frac{K\gamma}{L}$ users.
- In Step 5, the L different XORs are created, each intended for $\frac{K\gamma}{L} + 1$ users.
  - Each user from set λ is associated to a distinct set $\mathcal{F}_i$, as a function of parameter s that takes values from $\{0, 1, ..., L-1\}$. For example, when $s = 0$, the first XOR of the vector will be intended for users in set $\lambda(1) \cup \mathcal{F}_1$ (set μ of (4.30)), while known by all users in $\pi \setminus \mathcal{F}_1$ (set ν of (4.30)). In the same manner, the second XOR will be intended for the users in the set $\mu = \lambda(2) \cup \mathcal{F}_2$, while known by all users in set $\nu = \pi \setminus \mathcal{F}_2$, and so on.
  - Further, when $s = 1$ the first XOR is intended for users in $\lambda(1) \cup \mathcal{F}_2$ (while known by all users in $\pi \setminus \mathcal{F}_2$), the second XOR is for users in $\lambda(2) \cup \mathcal{F}_3$ (while known by all users in $\pi \setminus \mathcal{F}_3$), and so on. In particular, Step 5 allows to iterate over all sets $\mathcal{F}_i$, associating every time a distinct set $\mathcal{F}_i$ to a distinct user from group λ, until all users from set λ have been associated with all sets $\mathcal{F}_i$.
- Then in the last step (Step 6), the vector of the L XORs is transmitted after being multiplied by precoder matrix H_{λ}^{-1}.

4.6.1.5 Decoding at the users

Due to the design of the XORs, which is described in (4.32), and the respective precoders, each vector guarantees that the users belonging in set λ can decode the XOR that they receive.

Moreover, the above XOR design also guarantees that each of the remaining users (users in set π) have cached all subfiles that are found in the entire vector, apart from their desired subfile. Furthermore, the training phase of Section 4.6.1.2 has provided the users of set π with all the necessary CSI estimates (specifically, it has provided the receivers with all the necessary precoder-channel composite scalars) to perform the decoding of the linear combination of the transmitted vector.

To see the above more clearly, let us look at the signal received and the decoding process at some of the users.

Let us first assume some user k belonging to set λ. The received message, due to the design of precoder matrix H_{λ}^{-1} (cf. (4.11)), takes the form

$$y_k = \mathbf{h}_k^\dagger \cdot H_\lambda^{-1} \begin{bmatrix} X_{\lambda(1)\cup\mathcal{F}_{r_1}}^{\pi\setminus\mathcal{F}_{r_1},\lambda\setminus\lambda(1)} \\ X_{\lambda(2)\cup\mathcal{F}_{r_2}}^{\pi\setminus\mathcal{F}_{r_2},\lambda\setminus\lambda(2)} \\ \vdots \\ X_{\lambda(L)\cup\mathcal{F}_{r_L}}^{\pi\setminus\mathcal{F}_{r_L},\lambda\setminus\lambda(L)} \end{bmatrix} = X_{\{k\}\cup\mathcal{F}_{r_k}}^{\pi\setminus\mathcal{F}_{r_k},\lambda\setminus\{k\}}. \tag{4.34}$$

The design of the remaining XOR (see (4.32)) means that all but one subfiles have been cached by user k; thus user k can remove the interfering subfiles in order to retrieve its desired subfile.

In contrast, decoding at some user $m \in \pi$ also requires access to feedback information. The received message at user $m \in \pi$ takes the form

$$y_m = \mathbf{h}_m^\dagger \cdot H_\lambda^{-1} \begin{bmatrix} X_{\lambda(1)\cup\mathcal{F}_{r_1}}^{\pi\setminus\mathcal{F}_{r_1},\lambda\setminus\lambda(1)} \\ X_{\lambda(2)\cup\mathcal{F}_{r_2}}^{\pi\setminus\mathcal{F}_{r_2},\lambda\setminus\lambda(2)} \\ \vdots \\ X_{\lambda(L)\cup\mathcal{F}_{r_L}}^{\pi\setminus\mathcal{F}_{r_L},\lambda\setminus\lambda(L)} \end{bmatrix} = \sum_{j=1}^{L} \mathbf{h}_m^\dagger \mathbf{h}_{\lambda\setminus\lambda(j)}^\perp X_{\lambda(j)\cup\mathcal{F}_{r_j}}^{\pi\setminus\mathcal{F}_{r_j},\lambda\setminus\lambda(j)}. \tag{4.35}$$

User m has estimated all products $\mathbf{h}_m^\dagger \mathbf{h}_{\lambda\setminus\lambda(j)}^\perp$, $\forall j \in [L]$ that appear in (4.35) due to the CSI acquisition phase of Section 4.6.1.2. Moreover, since sets $\mathcal{F}_{r_i}$ are non-overlapping, it follows that user m belongs in only one of those. Hence, user m has fully cached the content of all but one XORs (see (4.32)) and can completely remove them. The remaining XOR contains $\frac{K\gamma}{L}$ known subfiles and one unknown and desired subfile for user m, which can be decoded.

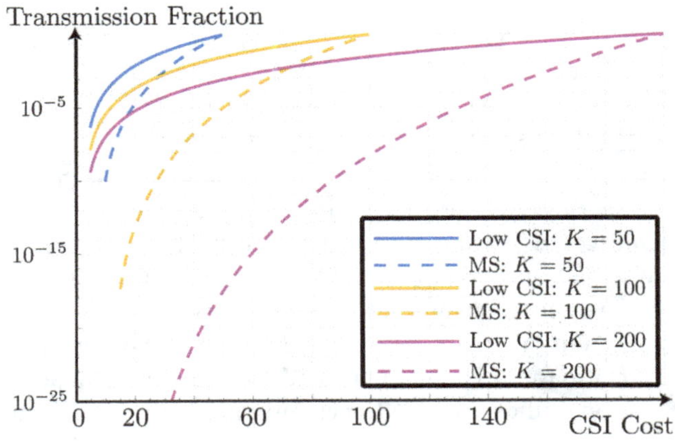

Figure 4.4 Accumulated feedback costs, in number of users needed to send CSI during a coherence period. Comparison between algorithms [13] and [27]. The coherence period is described as a function of the file fraction that can be transmitted.

Example 4: *Consider a transmitter with $L = 2$ antennas, serving $K = 4$ users whose caches allow for a cumulative caching redundancy $K\gamma = 2$. Each file is split into*

$$S_L = \overbrace{(K\gamma + L)}^{\phi} \overbrace{\binom{K - K\gamma - 1}{L-1}}^{\sigma} \overbrace{\binom{K}{K\gamma}}^{\tau} = 24 \tag{4.36}$$

sub-subfiles and the following are the $\binom{K}{L}\binom{K-L}{K\gamma}L = 12$ *transmissions that will satisfy all the users' requests.*

$$\begin{aligned}
\mathbf{x}_{12,1}^{34} &= H_{12}^{-1}\begin{bmatrix} A_{2,34}^{(1)} \oplus C_{2,14}^{(1)} \\ B_{1,34}^{(1)} \oplus D_{1,23}^{(1)} \end{bmatrix}, \mathbf{x}_{12,2}^{34} = H_{12}^{-1}\begin{bmatrix} A_{2,34}^{(2)} \oplus D_{2,13}^{(1)} \\ B_{1,34}^{(2)} \oplus C_{1,24}^{(1)} \end{bmatrix} \\
\mathbf{x}_{34,1}^{12} &= H_{34}^{-1}\begin{bmatrix} B_{4,13}^{(1)} \oplus C_{4,12}^{(1)} \\ A_{3,24}^{(1)} \oplus D_{3,12}^{(1)} \end{bmatrix}, \mathbf{x}_{34,2}^{12} = H_{34}^{-1}\begin{bmatrix} A_{4,23}^{(1)} \oplus C_{4,12}^{(2)} \\ B_{3,14}^{(1)} \oplus D_{3,12}^{(2)} \end{bmatrix} \\
\mathbf{x}_{24,1}^{13} &= H_{24}^{-1}\begin{bmatrix} A_{4,23}^{(2)} \oplus B_{4,13}^{(2)} \\ C_{2,14}^{(2)} \oplus D_{2,13}^{(2)} \end{bmatrix}, \mathbf{x}_{24,2}^{13} = H_{24}^{-1}\begin{bmatrix} B_{4,13}^{(3)} \oplus C_{4,12}^{(3)} \\ A_{2,34}^{(3)} \oplus D_{2,13}^{(3)} \end{bmatrix} \\
\mathbf{x}_{13,1}^{24} &= H_{13}^{-1}\begin{bmatrix} A_{3,24}^{(2)} \oplus B_{3,14}^{(2)} \\ C_{1,24}^{(2)} \oplus D_{1,23}^{(2)} \end{bmatrix}, \mathbf{x}_{13,2}^{24} = H_{13}^{-1}\begin{bmatrix} A_{3,24}^{(3)} \oplus D_{3,12}^{(2)} \\ B_{1,34}^{(3)} \oplus C_{1,24}^{(3)} \end{bmatrix} \\
\mathbf{x}_{14,1}^{23} &= H_{14}^{-1}\begin{bmatrix} A_{4,23}^{(3)} \oplus B_{4,13}^{(4)} \\ D_{1,23}^{(3)} \oplus C_{1,24}^{(4)} \end{bmatrix}, \mathbf{x}_{14,2}^{23} = H_{14}^{-1}\begin{bmatrix} A_{4,23}^{(4)} \oplus C_{4,12}^{(4)} \\ B_{1,34}^{(4)} \oplus D_{1,23}^{(4)} \end{bmatrix} \\
\mathbf{x}_{23,1}^{14} &= H_{23}^{-1}\begin{bmatrix} A_{3,24}^{(4)} \oplus B_{3,14}^{(3)} \\ C_{2,14}^{(3)} \oplus D_{2,13}^{(4)} \end{bmatrix}, \mathbf{x}_{23,2}^{14} = H_{23}^{-1}\begin{bmatrix} B_{3,14}^{(4)} \oplus D_{3,12}^{(4)} \\ C_{2,14}^{(4)} \oplus A_{2,34}^{(4)} \end{bmatrix}
\end{aligned} \tag{4.37}$$

As we see, the delay is $\mathscr{T}_2\left(4, \frac{1}{2}\right) = \frac{12}{24} = \frac{1}{2}$ and the DoF is

$$\mathscr{D}_2\left(4, \frac{1}{2}\right) = \frac{K(1-\gamma)}{T_2\left(4, \frac{1}{2}\right)} = 4 \tag{4.38}$$

4.6.2 Additional feedback-related open problems

The aforementioned CSI costs could further increase if the number of users involved during a single coherence period is larger than $K\gamma + L$. The above-reviewed scheme addresses this problem by utilizing the acquired CSI (from a fixed L-tuple λ) not only for one single transmission (to a fixed set $\lambda \cup \pi$ of $L + K\gamma$ users) but rather for many such transmissions. An early study on the corresponding gains from this reuse is presented in Figure 4.4, which describes the amount of information that is communicated for a given feedback cost. A fundamental exposition of the above CSI-accumulation issue remains an interesting open problem.

Another open problem would be to study the *joint* role (and limitations) of feedback and subpacketization in multi-antenna coded caching. It is natural to wonder if there exist a multi-antenna coded caching algorithm that minimizes the CSI costs

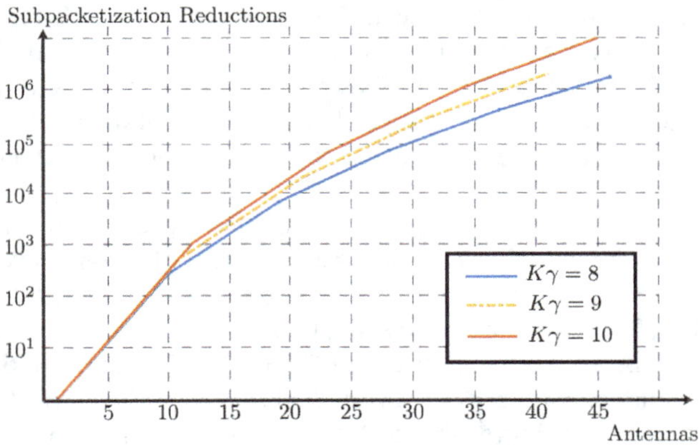

Figure 4.5 *Subpacketization reduction (multiplicative factor) achieved in [29], compared to the single-stream case, as a function of the number of antennas*

while also achieving the subpacketization savings that multiple transmit antennas are capable of. This remains a big open problem. Some recent progress can be found in [29, 30], which designed algorithms that achieve a near-optimal delay, with a CSI cost of $C = L$ and a subpacketization that is approximately e^{L-1} times smaller than that in [1], see (Figure 4.5).

4.7 Conclusions – further impact of combining multiple transmitters with coded caching

This chapter has highlighted the subpacketization and feedback-related benefits from combining coded caching with multiple antennas. Additional benefits can be found in abundance of other scenarios. One such scenario would involve hybrid systems where cache-aided and cache-less users coexist. For this setting, the recent work in [30, 31] has shown that – while in the single-antenna case, cache-less users cannot enjoy coded caching gains – introducing multiple antennas can allow both cache-aided and cache-less users to enjoy the full cache-aided DoF of $L + K\gamma$.

Further benefits from the synergy between coded caching and multiple antennas can be found when exploring coded caching in the low-to-moderate SNR regime. Little is known about this regime, but first indications suggest that the performance of coded caching can substantially deteriorate due to fading and path-loss, which introduce channel asymmetries and the "worst-user" phenomenon where the user with the weakest channel slows down the delivery of the entire multicast message.[h]

[h]The performance of coded caching under uneven channels has been studied under various scenarios such as the erasure broadcast channel [32–34], energy-aware transmissions [35], the benefits of cache-size asymmetry under uneven channels [36], lossy transmissions [37–39], etc. For the fundamental

Some very recent works that study such aspects in the finite SNR regime include [15, 26, 44–46]. As a testament to the powerful impact from joining multiple antennas and coded caching in the moderate SNR regime, the recent work in [44] displays how adding antennas can provide a super-linear increase in the rate, by exploiting the ability of multiple antennas to both alleviate the worst-user effect and multiplex to several users at a time.

It is also interesting to note that multi-transmitter coded caching environments without CSIT [24, 25], or with partial and/or delayed CSIT [47, 48], have also exhibited notable performance, despite this lack of good-quality feedback. For example, the works in [24, 25] showed how transmitter cooperation and channel unevenness can partially compensate for the absence of CSI knowledge in multi-transmitter coded caching. Further, works in [47, 48] studied channels with partial CSIT quality and/or delayed CSIT and showed that caching can provide significant improvements compared to the non-cache-aided channel.

4.7.1 From caching to distributed computing

Finally, similar benefits from fusing multiple antennas and coded caching can emerge in distributed computing systems that employ MapReduce-type computations. The equivalence between such systems and coded caching was revealed in [49, 50] with the introduction of the Coded MapReduce algorithm. The coded MapReduce framework forces computing nodes to perform more preprocessing computations, thus endowing them with redundant information that can be used to decode multicast messages in the communication phase of computational parallelization.

The benefits of implementing coded MapReduce in wireless systems have been demonstrated in various works [51, 52]. In this same context of the Coded MapReduce framework, having multiple transmitters can again be of high value. For example, the recent work in [53] shows that node cooperation can lead to reduced subpacketization, while [54] shows that exploiting the multi-transmitter nature of the device-to-device environment can mitigate the damage from having computing nodes with uneven computing capabilities.

References

[1] Amiri M.M., Gündüz D. 'Caching and coded delivery over Gaussian broadcast channels for energy efficiency'. *IEEE Journal on Selected Areas in Communications*. 2018;**36**(8):1706–20.

[2] Bayat M., Yapar C., Caire G. 'Spatially scalable lossy coded caching'. 2018 15th International Symposium on Wireless Communication Systems (ISWCS); 2018. pp. 1–6.

---

performance of single-antenna coded caching under uneven channel strengths and under the generalized DoF (GDoF) framework (cf. [40]) see works [41–43].

[3] Cacciapuoti A.S., Caleffi M., Ji M., Llorca J., Tulino A.M. 'Speeding up future video distribution via channel-aware caching-aided coded multicast'. *IEEE Journal on Selected Areas in Communications*. 2016;**34**(8):2207–18.

[4] Cheng M., Jiang J., Wang Q., Yao Y. 'A generalized grouping scheme in coded caching'. *IEEE Transactions on Communications*. 2019;**67**(5):3422–30.

[5] Cheng M., Yan Q., Tang X., Jiang J. 'Coded caching schemes with low rate and subpacketizations'. *arXiv preprint*. 2017:1703.01548.

[6] Chittoor H.H.S., Krishnan P. 'Low subpacketization coded caching via projective geometry for broadcast and D2D networks'. *arXiv preprint arXiv*. 2019:1902.08041.

[7] Destounis A., Kobayashi M., Paschos G., Ghorbel A. 'Alpha fair coded caching'. 15th International Symposium on Modeling and Optimization in Mobile, Ad Hoc, and Wireless Networks (WiOpt); 2017. pp. 1–8.

[8] Ghorbel A., Kobayashi M., Yang S. 'Content delivery in erasure broadcast channels with cache and feedback'. *IEEE Transactions on Information Theory*. 2016;**62**(11):6407–22.

[9] Ha S., Zhang J., Simeone O., Kang J. 'Wireless map-reduce distributed computing with full-duplex radios and imperfect CSI'. 2019 IEEE 20th International Workshop on Signal Processing Advances in Wireless Communications (SPAWC); 2019. pp. 1–5.

[10] Jafar S.A., Vishwanath S. 'Generalized degrees of freedom of the symmetric Gaussian user interference channel'. *IEEE Transactions on Information Theory*. 2010;**56**(7):3297–303.

[11] Joudeh H., Lampiris E., Elia P., Caire G. 'Fundamental limits of wireless caching under mixed cacheable and uncacheable traffic'. *IEEE Transactions on Information Theory*. 2020;**67**(7):4747–67.

[12] Kamel S., Sarkiss M., Wigger M. 'Decentralized joint cache-channel coding over erasure broadcast channels'. IEEE Middle East and North Africa Communications Conference (MENACOMM); 2018. pp. 1–6.

[13] Kobayashi M., Caire G. 'On the net DoF comparison between ZF and MAT over time-varying MISO broadcast channels'. 2012 IEEE International Symposium on Information Theory Proceedings; 2012. pp. 2286–90.

[14] Krishnan P. 'Coded caching via line graphs of bipartite graphs'. IEEE Information Theory Workshop (ITW); 2018. pp. 1–5.

[15] Lampiris E., Elia P., Caire G. 'Bridging the gap between multiplexing and diversity in finite snr multiple antenna coded caching'. 2019 53rd Asilomar Conference on Signals, Systems, and Computers; 2019. pp. 1272–7.

[16] Lampiris E., Elia P. 'Achieving full multiplexing and unbounded caching gains with bounded feedback resources'. IEEE International Symposium on Information Theory (ISIT); 2018. pp. 1440–4.

[17] Lampiris E., Elia P. 'Adding transmitters allows unbounded coded-caching gains with bounded file sizes'. IEEE International Symposium on Information Theory (ISIT); 2018. pp. 1435–9.

[18] Lampiris E., Elia P. 'Adding transmitters dramatically boosts coded-caching gains for finite file sizes'. *IEEE Journal on Selected Areas in Communications*. 2018;**36**(6):1176–88.

[19] Lampiris E., Elia P. 'Full coded caching gains for cache-less users'. IEEE Information Theory Workshop (ITW); 2018. pp. 1–5.

[20] Lampiris E., Elia P. 'Bridging two extremes: multi-antenna coded caching with reduced subpacketization and CSIT'. 2019 IEEE 20th International Workshop on Signal Processing Advances in Wireless Communications (SPAWC); 2019.

[21] Lampiris E., Jiménez Zorrilla D., Elia P. 'Mapping heterogeneity does not affect wireless coded MapReduce'. IEEE International Symposium on Information Theory (ISIT); 2019.

[22] Lampiris E., Zhang J., Elia P. 'Cache-aided cooperation with no CSIT'. IEEE International Symposium on Information Theory (ISIT); 2017. pp. 2960–4.

[23] Lampiris E., Elia P. 'Full coded caching gains for cache-less users'. *arXiv preprint arXiv*. 2018:1806.07800.

[24] Lampiris E., Elia P. 'Resolving a feedback bottleneck of multi-antenna coded caching'. *arXiv preprint arXiv*. 2018:1811.03935.

[25] Lampiris E., Zhang J., Simeone O., Elia P. 'Fundamental limits of wireless caching under uneven-capacity channels'. International Zurich Seminar on Information and Communication (IZS); 2020.

[26] Li F., Chen J., Wang Z. 'Wireless MapReduce distributed computing'. *IEEE Transactions on Information Theory*. 2019;**65**(10):6101–14.

[27] Li S., Maddah-Ali M.A., Avestimehr A.S. 'Coded MapReduce'. 53rd Annual Allerton Conference on Communication, Control, and Computing (Allerton); 2015. pp. 964–71.

[28] Li S., Maddah-Ali M.A., Yu Q., Avestimehr A.S. 'A fundamental tradeoff between computation and communication in distributed computing'. *IEEE Transactions on Information Theory*. 2018;**64**(1):109–28.

[29] Lozano A., Heath R.W Jr., Andrews J.G. 'Fundamental limits of cooperation'. *arXiv preprint arXiv*. 2012:1204.0011.

[30] Maddah-Ali M.A., Niesen U. 'Fundamental limits of caching'. *IEEE Transactions on Information Theory*. 2014;**60**(5):2856–67.

[31] Maddah-Ali M.A., Niesen U. 'Decentralized coded caching attains order-optimal memory-rate tradeoff'. *IEEE/ACM Transactions on Networking*. 2015;**23**(4):1029–40.

[32] Mohammadi Amiri M., Yang Q., Gündüz D. 'Decentralized caching and coded delivery with distinct cache capacities'. *IEEE Transactions on Communications*. 2017;**65**(11):4657–69.

[33] Naderializadeh N., Maddah-Ali M.A., Avestimehr A.S. 'Fundamental limits of cache-aided interference management'. *IEEE Transactions on Information Theory*. 2017;**63**(5):1–107.

[34] Nejad M.A.T., Shariatpanahi S.P., Khalaj B.H. 'On storage allocation in cache-enabled interference channels with mixed CSIT'. 2017 IEEE International

Conference on Communications Workshops (ICC Workshops); 2017. pp. 1177–82.

[35] Ngo K.-H., Yang S., Kobayashi M. 'Scalable content delivery with coded caching in multi-antenna fading channels'. *IEEE Transactions on Wireless Communications*. 2018;**17**(1):548–62.

[36] Parrinello E., Lampiris E., Elia P. 'Coded distributed computing with node cooperation substantially increases speedup factors'. IEEE International Symposium on Information Theory (ISIT); 2018. pp. 1291–5.

[37] Parrinello E., Ünsal A., Elia P. 'Fundamental limits of coded caching with multiple antennas, shared caches and uncoded prefetching'. *IEEE Transactions on Information Theory*. 2020;**66**(4):2252–68.

[38] Piovano E., Joudeh H., Clerckx B. 'On coded caching in the overloaded miso broadcast channel'. IEEE International Symposium on Information Theory(ISIT); 2017. pp. 2795–9.

[39] Piovano E., Joudeh H., Clerckx B. 'Generalized degrees of freedom of the symmetric cache-aided MISO broadcast channel with partial CSIT'. *IEEE Transactions on Information Theory*. 2019;**65**(9):5799–815.

[40] Salehi M., Tölli A., Shariatpanahi S.P. 'Coded caching with linear subpacketization is possible in multi-antenna communications'. *arXiv preprint arXiv*. 2019:1910.10384.

[41] Salehi M., Tölli A., Shariatpanahi S.P. 'Coded caching with uneven channels: a quality of experience approach'. *arXiv preprint arXiv*. 2020:2003.01928.

[42] Shangguan C., Zhang Y., Ge G. 'Centralized coded caching schemes: a hypergraph theoretical approach'. *IEEE Transactions on Information Theory*. 2018;**64**(8):5755–66.

[43] Shanmugam K., Ji M., Tulino A.M., Llorca J., Dimakis A.G. 'Finite-length analysis of caching-aided coded multicasting'. *IEEE Transactions on Information Theory*. 2016;**62**(10):5524–37.

[44] Shanmugam K., Tulino A.M., Dimakis A.G. 'Coded caching with linear subpacketization is possible using Ruzsa-Szeméredi graphs'. 2017 IEEE International Symposium on Information Theory (ISIT); 2017. pp. 1237–41.

[45] Shariatpanahi S.P., Caire G., Hossein Khalaj B. 'Physical-layer schemes for wireless coded caching'. *IEEE Transactions on Information Theory*. 2019;**65**(5):2792–807.

[46] Shariatpanahi S.P., Motahari S.A., Khalaj B.H. 'Multi-server coded caching'. *IEEE Transactions on Information Theory*. 2016;**62**(12):7253–71.

[47] Tang L., Ramamoorthy A. 'Coded caching schemes with reduced subpacketization from linear block codes'. *IEEE Transactions on Information Theory*. 2018;**64**(4):3099–120.

[48] Tolli A., Shariatpanahi S.P., Kaleva J., Khalaj B. 'Multicast beamformer design for coded caching'. IEEE International Symposium on Information Theory(ISIT); 2018. pp. 1914–18.

[49] Tölli A., Shariatpanahi S.P., Kaleva J., Babak K. 'Multi-antenna interference management for coded caching'. *arXiv preprint arXiv*. 2017:1711.03364.

[50] Yan Q., Cheng M., Tang X., Chen Q. 'On the placement delivery array design for centralized coded caching scheme'. *IEEE Transactions on Information Theory*. 2017;**63**(9):5821–33.

[51] Zhang J., Elia P. 'Fundamental limits of cache-aided wireless BC: interplay of coded-caching and csit feedback'. *IEEE Transactions on Information Theory*. 2017;**63**(5):1–60.

[52] Zhang J., Elia P. 'Wireless coded caching: a topological perspective'. IEEE International Symposium on Information Theory (ISIT); 2017. pp. 401–5.

[53] Zhang J., Engelmann F., Elia P. 'Coded caching for reducing csit-feedback in wireless communications'. 2015 53rd Annual Allerton Conference on Communication, Control, and Computing (Allerton); 2015. pp. 1099–105.

[54] Lizhong Zheng., Tse D.N.C. 'Communication on the Grassmann manifold: a geometric approach to the noncoherent multiple-antenna channel'. *IEEE Transactions on Information Theory*. 2002;**48**(2):359–83.

Chapter 5

Caching for interference networks: a separation architecture

Jad Hachem[1], *Urs Niesen, and Suhas Diggavi*[1]

5.1 Introduction

Content- or information-centric networking (CCN/ICN) is a proposed next-generation Internet architecture that has attracted significant interest from the wired Internet research communities [1, 2]. It makes use of the availability of caches in the network by routing requests and packets based on content addressing rather than traditional host addressing. By replicating content in several locations, CCN can alleviate communication bottlenecks in the network. The gains from CCN are largest when the last-hop communication link to the end user is not the communication bottleneck, and when local storage is large enough to cache most of the popular content [3].

The focus of this chapter is cache-enabled wireless networks, which are typically deployed for the last communication hop to the end user. Content distribution architectures designed for the wired Internet, including CCN/ICN, cannot be directly translated to this wireless setting. This is because the wireless last hop is a communication bottleneck link, and because the wireless cellular base station typically has limited storage. Therefore, last-hop content delivery schemes need to be designed cognizant of the wireless architecture being deployed.

Current wireless communication networks are heterogeneous: They consist of a dense deployment of wireless access points (AP), such as WiFi or small cells, with small coverage and high per-user data rates in combination with a sparse deployment of macro base stations (BS) with large coverage and lower per-user data rates. A benefit of this heterogeneous architecture is that an end user can connect to a BS as well as one or more APs. However, a corresponding downside is that the resulting wireless transmissions from those BS and AP interfere with one another. This situation is illustrated in Figure 5.1.

Coded caching [4] has emerged as a basis to create a content-centric wireless last-hop delivery system by harnessing the presence of cache memories at either end users or wireless AP. Work in the information-theory literature has demonstrated that

[1]University of California at Los Angeles, USA

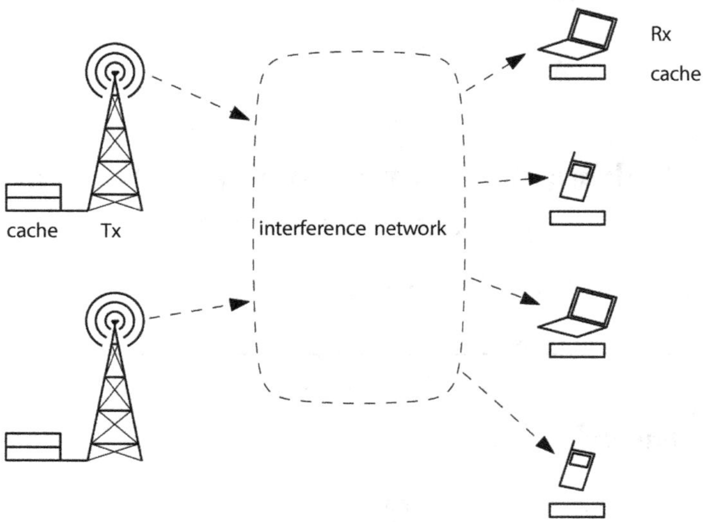

Figure 5.1 A wireless communication network with caches at the transmitters and receivers

this combination can yield substantial benefits (see [5–12] among several others). However, a significant amount of the work on this topic has been focused on systems with a single, noise-free, rate-limited broadcast link. In contrast, as mentioned in the previous paragraphs, a salient feature of wireless networks is interference from multiple concurrent transmitters.

How should we manage this interference within the coded caching framework? Further, the success of both the wired and wireless Internet architectures is arguably based on the separation approach (see [13, 14] and references therein). How can we integrate coded caching into such a layered network architecture?

Motivated by these questions, we review in this chapter our works published in [15, 16], where we propose and analyze a *separation-based* architecture for coded caching in wireless interference networks. The proposed architecture consists of a physical layer and a network layer. The physical layer creates appropriate multicast "bit pipes," which are then used in the network layer to implement a coded caching scheme. We demonstrate that this separation-based architecture is *approximately optimal*. Such a separation architecture could therefore be an attractive way to capture the advantages of coded caching in wireless interference networks.

This chapter is organized as follows. In Section 5.2, we set up the problem and define the notation used throughout the chapter. In Section 5.3, we state and explain the implications of the main results. Section 5.4 explains the subtleties in the design of the separation architecture and describes the physical and network (coded caching) layers. We conclude with a discussion of related work, putting our work in context, and an outlook to unresolved questions in Section 5.5.

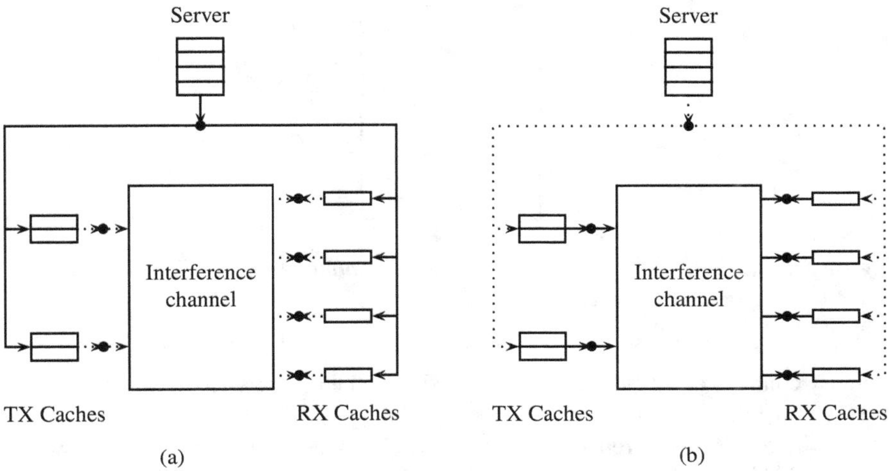

Figure 5.2 *Placement (a) and delivery phases (b) for a cache-aided interference network with* $K_t = 2$ *transmitters and* $K_r = 4$ *receivers. This models the wireless network scenario shown in Figure 5.1.*

5.2 Cache-aided interference networks

We start by describing a model for a cache-aided wireless interference network. Our model is sufficiently simple to allow formal analysis but at the same time sufficiently rich to capture the main system bottlenecks and trade-offs.

We have a content library consisting of N equal-sized files stored at a central server. Parts of this content library will be requested by K_r users (also referred to as receivers). These requests are to be fulfilled with the help of K_t transmitters, connected to the users via a wireless interference channel. Each of the K_t transmitters has a separate cache memory large enough to hold the equivalent of M_t files. Each of the K_r users has a separate cache memory large enough to hold the equivalent of M_r files. These cache memories can be used to aid the content delivery.

Content distribution proceeds in two phases: a content *placement phase* (taking place during a period of low network load) followed by a content *delivery phase* (taking place during a period of peak network load), see Figure 5.2.

During the placement phase, with network load low, the main constraints are the cache memory sizes. In this phase, the transmitter and receiver caches can be filled as an arbitrary function of the content library up to the memory constraint. In particular, we allow arbitrary intra-file and inter-file coding and impose no communication constraints. Crucially, during this phase the future user requests are unknown.

During the subsequent delivery phase, with network load high, the main constraint is the communication link. This phase starts with each of the users requesting one of the N files. The transmitters are informed of all these requests. Each transmitter responds by constructing a codeword as a function of only its local cache

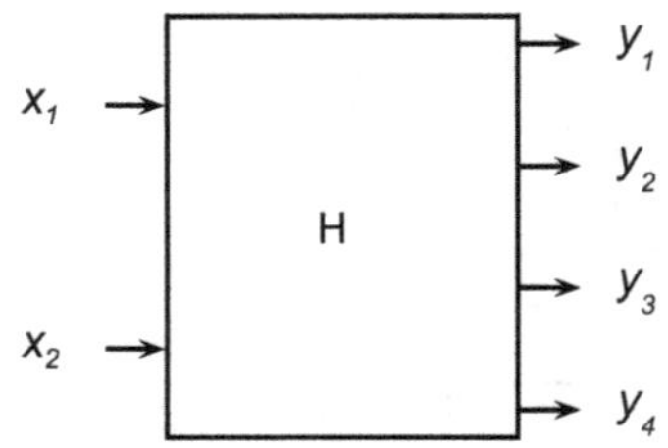

Figure 5.3 *Model for a wireless interference channel with* K_t = *2 transmitters and* K_r = *4 receivers*

memory contents and sending this codeword over the interference network connecting the transmitters with the receivers.

We model the interference network as a wireless interference channel as shown in Figure 5.3. The wireless medium is modeled as a (real) time-varying fading channel with additive Gaussian receiver noise. The received signal $y_i(\tau)$ at receiver i at time τ is given by a noisy linear superposition of the transmitted signals $x_1(\tau), \ldots, x_{K_t}(\tau)$. The linear superposition is dictated by the channel matrix $\boldsymbol{H}(\tau)$.

We assume full channel-state information $\boldsymbol{H}(\tau)$ is available at all transmitters and receivers. The transmitters are constrained by an average power constraint. A key system parameter is the ratio between the transmitter power constraint and the receiver noise variance denoted by SNR.

From the wireless channel output and its own cache memory contents, each user aims to recover its requested content file. The performance metric of interest is the per-user content delivery rate, which is the ratio of the file size divided by the number of uses of the wireless channel. We say that such a delivery rate is achievable if, for large enough file size, there exists a caching and transmission scheme that allows each user to recover its requested content file with vanishing probability of error.

The optimal transmission rate $R^{\star}(\mathrm{SNR})$ is the largest such achievable rate by the best caching and transmission schemes. As indicated by the notation, this optimal transmission rate depends on the signal-to-noise ratio (SNR).

Ideally, we would like to find an expression for the optimal transmission rate $R^{\star}(\mathrm{SNR})$ of the cache-aided interference network. Unfortunately, this turns out to be difficult. Indeed, even for the special case of an interference network without caches, the optimal transmission rate is unknown. Instead, it is common in wireless communication, to analyze the degrees of freedom (DoF) describing the behavior of the optimal transmission rate at high SNR through the relation $R^{\star}(\mathrm{SNR}) \approx \mathrm{DoF} \cdot \frac{1}{2}\log(\mathrm{SNR})$ for large SNR.

The remainder of this chapter studies this DoF quantity for the cache-aided interference network. Our results will take the form of a constant-factor approximation of DoF, i.e., upper and lower bounds on DoF as a function of all the system parameters that agree within a constant factor (in our case 13.5). As will be discussed in more detail in Section 5.5, finding tighter than constant-factor approximations of DoF appears difficult, since even for the well-studied special case of the cache-aided

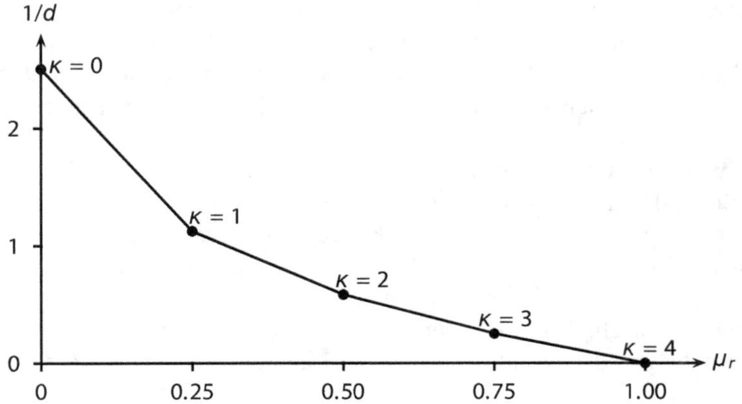

Figure 5.4 Approximate reciprocal degrees of freedom 1/d *as a function of normalized receiver cache size* $\mu_r \triangleq M_r/N$ *for the* 2 × 4 *cache-aided interference network with* N = 4 *files shown in Figure 5.2*

interference network with only a single transmitter, this type of approximation is the best known.

5.3 Performance limits

The main result of this chapter is a constant-factor approximation of the DoF of the cache-aided interference network. This approximation will be given in terms of the quantity $d = d(N, K_t, K_r, M_t, M_r)$ defined through

$$\frac{1}{d} \triangleq \frac{K_t - 1 + \min\{K_r/(\kappa+1), N\}}{K_t} \cdot \left(1 - \frac{\kappa}{K_r}\right), \tag{5.1}$$

with $M_r = \kappa N/K_r$. The definition (5.1) holds for M_r with integer $\kappa \in \{0, 1, \cdots, K_r\}$. The definition of d is extended to all $M_r \in [0, N]$ by constructing the lower convex envelope of $1/d$ from the integer points. An example of d as a function of M_r is shown in Figure 5.4.

We are now ready to state the main result of this chapter.

Theorem 1: The DoF of the cache-aided interference network is within a constant factor of d. More precisely,

$$d \leq \text{DoF} \leq 13.5 \cdot d. \tag{5.2}$$

Expressed in terms of the optimal rate $R^\star(\text{SNR})$, this result is equivalent to

$$d \cdot \tfrac{1}{2} \log \text{SNR} - o(\log \text{SNR}) \leq R^\star(\text{SNR}) \leq 13.5 \cdot d \cdot \tfrac{1}{2} \log \text{SNR} + o(\log \text{SNR}) \tag{5.3}$$

for large SNR.

Theorem 1 holds for all values of the number of transmitters K_t, number of receivers K_r, number of files N, transmitter cache sizes $M_t \in [N/K_t, N]$, and receiver cache sizes $M_r \in [0, N]$. The restriction on the transmitter cache size $M_t \geq N/K_t$ guarantees that the total transmitter memory K_tM_t is large enough to collectively hold the entire content library.

The upper bound on DoF is obtained through an information-theoretic cut-set argument. The lower bound is obtained by proposing a communication (i.e., physical layer) and caching (i.e., network layer) architecture and by analyzing its performance. The structure of this order-optimal architecture provides interesting engineering insights and is presented in more detail in Section 5.4.

Using this insight from the order-optimal communication architecture, we can decompose the sum degrees of freedom $K_r\text{DoF}$ into three factors or gains as

$$K_r\mathsf{DoF} \approx K_r d \approx \underbrace{\frac{K_tK_r}{K_t + K_r - 1}}_{\substack{\text{interference-alignment}\\ \text{gain}}} \cdot \underbrace{\frac{1}{1-\mu_r}}_{\substack{\text{local}\\ \text{caching gain}}} \cdot \underbrace{\frac{K_r\mu_r + 1}{\mu_r\left(\frac{1}{K_r} + \frac{1}{K_t - 1}\right)^{-1} + 1}}_{\substack{\text{global}\\ \text{caching gain}}}. \tag{5.4}$$

Here, we have defined the normalized cache size $\mu_r \triangleq M_r/N$. The approximation is valid assuming large content library so that the number of files N is larger than the number of receivers K_r. As indicated in (5.2), the three gains are the *interference-alignment gain*, the *local caching gain*, and the *global caching gain*.

The interference-alignment gain is a communication gain achieved by aligning the interference subspaces in the physical layer. The local caching gain arises at the network layer from the ability to store content locally at each receiver. Note that the local caching gain depends on the (normalized) local cache size μ_r. The global caching gain arises at the interface between the network and physical layers by creating coded multicasting opportunities that allow several receivers with different demands to be served by a single transmission. Note that the global caching gain depends on the (normalized) global cache size $K_r\mu_r$ that appears in the numerator.

We next explore how these three gains scale with various system parameters. By considering the logarithm of the sum DoF, i.e., $\log(K_r\text{DoF})$, the three gains become additive terms that can be easily visualized.

5.3.1 *Scaling with normalized transmitter cache size* $\mu_t \triangleq M_t/N$

Recall that the normalized transmitter cache size needs to satisfy $\mu_t \geq 1/K_t$ for Theorem 5.3 to hold. However, beyond this condition, Theorem 5.3 shows that the approximate DoF does not depend on the transmit memory size. Put differently, increasing transmitter memory can lead to at most a constant-factor gain in DoF. In particular, this implies that strategies such as transmit beamforming are not needed for order-optimal system operation.

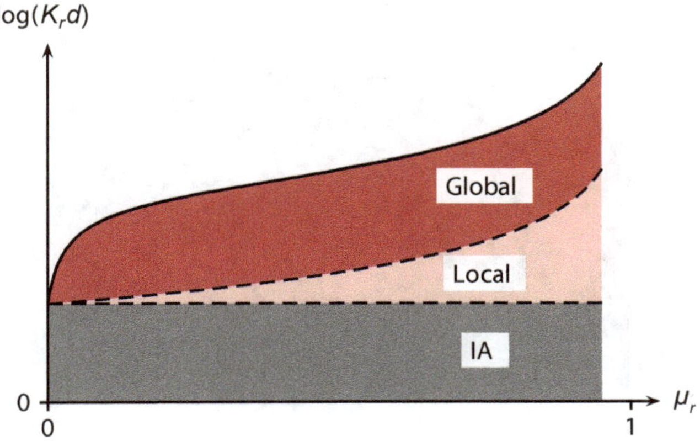

Figure 5.5 *Scaling of DoF gains with the normalized receiver cache size* μ_r

5.3.2 *Scaling with normalized receiver cache size* $\mu_r \triangleq M_r/N$

The scaling of the approximate sum DoF with the normalized receiver memory size is shown in Figure 5.5. Observe that the interference-alignment gain is independent of the receiver memory size. The local gain becomes significant whenever the normalized receiver cache size is large, say $\mu_r \geq 0.5$, and dominates as $\mu_r \to 1$. On the other hand, the global gain is significant already when $\mu_r \geq 1/K_r$, i.e., already for fairly small receiver cache sizes.

5.3.3 *Scaling with number of receivers* K_r

The scaling of the approximate sum DoF with the number of receivers K_r is shown in Figure 5.6. The local gain is independent of the number of receivers. The

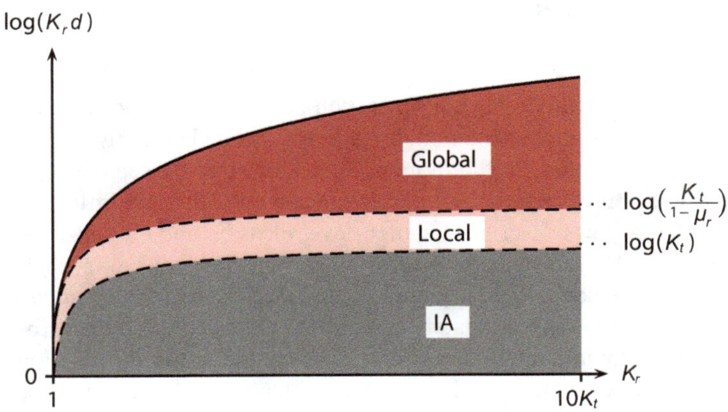

Figure 5.6 *Scaling of DoF gains with the number of receivers* K_r

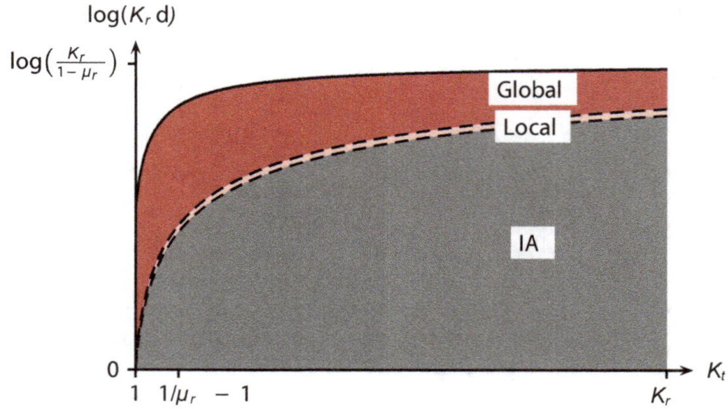

Figure 5.7 Scaling of DoF gains with the number of transmitters K_t

interference-alignment gain grows with K_r and approaches the limiting value of K_t. The global gain also grows with K_r, but (unlike the interference-alignment gain) it grows linearly without bounds as K_r increases. For large number of receivers, it is therefore the dominant gain.

5.3.4 Scaling with number of transmitter K_t

The scaling of the approximate sum DoF with the number of transmitters K_t is shown in Figure 5.7. As K_t increases, the sum DoF increases to a finite limiting value of $K_r/(1 - \mu_r)$. This limiting value is approximately achieved as soon as $K_t \approx 1/\mu_r - 1$, Therefore, increasing the number of transmitters beyond this value leads only to limited increase in system performance.

5.4 Separation architecture

As discussed in the introduction, we propose a separation architecture for coded caching in wireless networks. Such a separation achieves an attractive goal of any strategy: modularity of implementation, by layering a coded caching scheme on top of a channel coding scheme. Furthermore, we will show that the separation architecture we choose is approximately optimal; in other words, there is little cost to this modularity. What follows describes the high-level ideas behind the separation.

Consider first the special case with a single transmitter, i.e., a cache-aided Gaussian broadcast channel. In [4], the cache-aided *error-free* broadcast channel was studied, and an approximately optimal coded caching strategy for it was devised. This strategy uses the receiver caches to create shared information across receivers, providing opportunities for the transmitter to send coded messages that are useful for multiple receivers at once. A natural strategy for the Gaussian case is to leverage this coded caching strategy for the error-free case. In particular, by choosing the appropriate channel coding scheme to deliver broadcast messages from the transmitter to all the receivers, we would create an error-free broadcast link on top of which we can layer the coded caching strategy from [4]. Effectively, we are separating the problem into a *network layer* and a *physical layer*, and layering a *coded caching* scheme for the first on top of a *channel coding* scheme for the second.

The natural way to generalize this separation approach to the cache-aided interference network (with multiple transmitters) is to have the transmitters take turns broadcasting messages to all of the receivers. This is an especially attractive approach considering it reduces the problem to multiple instances of the broadcast case. However, a major drawback of this time-shared broadcast approach is that, at any given time, almost all transmitters are silent while only one broadcasts.

Alternatively, we can try to minimize this time-sharing by choosing a different separation architecture. Instead of each transmitter sending one message to all receivers, we let each transmitter send one separate message to every receiver. At the physical layer, we apply an interference alignment technique, first proposed in [17], to efficiently transmit these unicast messages. However, the restriction to unicast messages means that we are unable to take advantage of any coding opportunities created by the receiver caches.

The major feature these two approaches have in common is the separation of the network layer from the physical layer. These layers interface through a *message set*: the physical layer deals with transmitting these messages across the channel, while the network layer uses them as error-free *bit pipes*. In the first approach, the separation creates *broadcast* bit pipes, which allows coding opportunities at the expense of inefficient channel use. This is only useful when the receiver cache memory is large, enabling the transmitters to send messages that are useful to almost all users at once. In the second approach, the separation creates *unicast* bit pipes, which forgoes any coding opportunities in order to use the channel efficiently. This is only useful when the receiver cache memory is too small to enable any coding opportunities, meaning messages cannot be used by multiple users at once.

For intermediate receiver cache memory, it becomes important to find a compromise between these two approaches. The solution is to define the separation in terms of *multicast* messages: at the physical layer, each message is sent from one transmitter to be collectively decoded by multiple receivers; at the network layer, the messages are abstracted by a set of *non-overlapping multicast* bit pipes, each connecting one transmitter to a subset of receivers.[a] This multicast-based separation architecture turns out to be approximately optimal and exhibits the properties discussed in Section 5.3. In other words, we can achieve the goal of modularity with no cost to approximate optimality. Note that the multiple-unicast and multiple-broadcast approaches outlined above are special cases of this multicast approach; see Table 5.1. Figure 5.8 illustrates a simple example of a separation architecture with three multicast messages and the corresponding three multicast bit pipes.

[a]In some cases, a more sophisticated separation architecture that allows the bit pipes to overlap can achieve a better DoF. We expand on such an architecture in [16]. However, as follows from Theorem 1, this cannot yield more than a constant-factor benefit to the DoF.

Table 5.1 *Choice of the multicast size for the multicast messages used in the separation architecture*

Receiver cache size	Small	Medium	Large
Multicast size	1 (unicast)	2, 3, 4, . . .	All (broadcast)

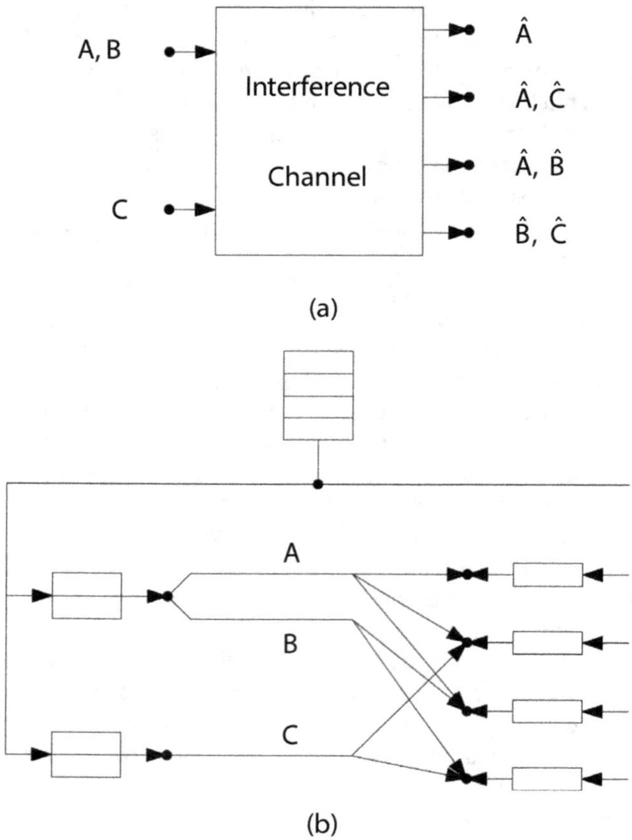

Figure 5.8 *A possible separation architecture for the interference network with* $K_t = 2$ *transmitters and* $K_r = 4$ *receivers shown in Figure 5.2. This specific separation uses a set of three multicast messages A, B, and C; each message is sent by one transmitter and intended for a subset of the receivers. The physical layer (a) ensures that such messages can be recovered at the receivers, while the network layer (b) can use them as multicast bit pipes linking the transmitters to the receivers.*

In this section, we will prove the approximate optimality of this separation architecture. We will first describe the separation architecture in detail, including the strategies for the two layers. We will then compute the achievable DoF, and finally derive information-theoretic outer bounds on the optimal DoF, proving the approximate optimality result.

5.4.1 Separating the physical and network layers

As discussed above, there are many ways to specify a separation architecture for this network, and we choose a class of architectures defined by a set of *multicast* messages. Let $V_{j,S}$ denote a multicast message intended from transmitter $j \in \{1, \ldots, K_t\}$ to all of the receivers in $S \subseteq \{1, \ldots, K_r\}$. For instance, in Figure 5.8, the multicast messages A, B, and C can be denoted, respectively as $V_{1,\{1,2,3\}}$, $V_{1,\{3,4\}}$, and $V_{2,\{2,4\}}$. Then, the message set that we choose is always of the form

$$\mathcal{V}_\sigma = \left\{ V_{j,S} : j \in \{1, \ldots, K_t\}, S \subseteq \{1, \ldots, K_r\}, |S| = \sigma \right\}, \tag{5.5}$$

for some $\sigma \in \{1, \ldots, K_r\}$, which we refer to as the *multicast size*. In other words, all the messages are intended for the same number of receivers σ, and conversely every set of σ receivers accept one message from every single transmitter.

As we will show, this choice of message set turns out to be without loss of approximate optimality. For now, we will give a high-level intuition behind it. First, there is a clear symmetry in the problem statement with respect to the transmitters and the receivers, and that symmetry will naturally translate to the choice of message set. As a result, receiver subsets of the same size should be represented equally. Second, the decision to choose messages of the same multicast size (as opposed to having different multicast sizes for different messages) follows from the properties of the caching problem itself: even in the simpler case of an error-free broadcast channel, an approximately optimal strategy is to send messages with a fixed multicast size that depends on the size of the receiver cache, because that maximizes the network-coding opportunities [4]. This property carries over to the more general case of interference networks.

With this message set in hand (parametrized by the multicast size σ), we can write the achievable rate of the full system as a function of the specifics of each of the network and physical layers. If the physical layer is able to transmit the message set with a per-user transmission rate of $\tilde{R}_\sigma(\text{SNR})$, and if the network layer can deliver all the requested files with messages of size at most $L_\sigma F$ bits in the worst case, then the end-to-end achievable rate is

$$R_\sigma(\text{SNR}) \geq \frac{\tilde{R}_\sigma(\text{SNR})}{L_\sigma}. \tag{5.6}$$

Since the multicast size σ directly influences the placement and delivery strategies, we will choose the value that makes the most sense for the network-layer strategy. This will be described further in Section 5.4.2. Note that both the network-layer scheme and the physical-layer scheme are described below for specific values of the receiver memory M_r, namely those such that $K_r M_r / N$ is an integer. All other values of M_r can be handled by time-and-memory-sharing between these values.

5.4.2 The network layer

At the network layer, we have a set of non-overlapping multicast bit pipes. Each pipe connects one transmitter to a set of σ receivers. The idea is to choose the multicast size σ that maximizes the number of coding opportunities. We motivate our choice by first describing the simplest cache-aided network: the error-free broadcast case [4].

In the error-free broadcast network, there is a single transmitter with access to the full content library, and a single multicast bit pipe connecting it to all the receivers. When the receiver cache size is such that $\kappa = K_r M_r / N$ is an integer, the cache placement is done so that every subset of κ users shares some piece of information. In the delivery phase, despite the fact that there is an error-free broadcast link to *all* users, the transmitter sends *multicast* messages to set of only $\kappa + 1$ users; nothing is gained by sending messages to all the users instead. In other words, there is a specific multicast size that maximizes the global caching gain, and this multicast size depends on the receiver cache memory. Using this strategy, the worst-case number of bits sent through the broadcast link is $F \cdot (K_r - \kappa)/(\kappa + 1)$.

This idea can be generalized to the cache-aided interference channel, using the network-layer scheme described below at a high level.

1. Each file is partitioned evenly across all the transmitter caches, effectively splitting the content library into K_t sub-libraries. Note that this takes up only $N/K_t \cdot F$ bits in each transmitter's cache, even if $M_t > N/K_t$, and that is sufficient for order-optimality.
2. Each receiver cache is also partitioned into K_t parts, each allocated exclusively to one sub-library. This effectively translates the interference problem at a high level into K_t separate broadcast problems.
3. Instead of creating a broadcast bit pipe from each transmitter to all receivers, we create multicast bit pipes of only the necessary multicast size $\sigma = \kappa + 1$, where $\kappa = K_r M_r / N$ is an integer.
4. The placement and delivery strategies then follow the broadcast case on each sub-library, where requesting a specific file translates to requesting all of its K_t parts from each of the transmitters.

Notice that the above strategy ends up sending the same number of bits through the network as the strategy in [4] does in the error-free broadcast cast. Consequently, we can achieve the sum network load in the following theorem.

Theorem 2: If $M_r = \kappa N/K_r$ for some $\kappa \in \{0, 1, \ldots, K_r\}$, a network-layer setup with a multicast size of $\kappa + 1$ has an achievable sum network load of

$$L = \frac{K_r - \kappa}{\kappa + 1} = \frac{K_r(1 - M_r/N)}{K_r M_r/N + 1}. \tag{5.7}$$

5.4.3 The physical layer

Recall that the physical layer deals with transmitting the messages in $\mathcal{V}_\sigma$ across the interference channel reliably. In particular, each transmitter j has a message $V_{j,S}$ for every subset S of receivers of size σ. The total number of messages is $K_t\binom{K_r}{\sigma}$, and hence a straightforward TDMA approach can achieve a per-message DoF of $1/K_t\binom{K_r}{\sigma}$ (for a sum DoF of one). This, however, is inefficient.

Consider the simpler case of a 2×2 interference channel, where each transmitter j has a unicast message V_{ji} for each receiver i (we abuse notation and write V_{ji} in place of $V_{j,\{i\}}$). Let us focus on receiver one. With a TDMA approach, receiver one is essentially capable of recovering not only its intended messages V_{11} and V_{21} but also all the other messages (V_{12} and V_{22}) which it does not need. Using the interference alignment technique introduced in [17], we can "align" the unwanted messages so that receiver one cannot recover them individually but only some combination of the two. Roughly speaking, it is as though the most that receiver one can decode, aside from its intended messages, is "$V_{12} + V_{22}$." Consequently, the transmission rate is increased, and in particular, an improved sum DoF of 4/3 can be achieved.

The above technique was established for the general *unicast* X-channel in [17] and was proved to achieve the optimal DoF in the unicast case. We here generalize this technique to a *multicast* X-channel of any multicast size σ. In this general setting, each receiver has $K_t\binom{K_r-1}{\sigma-1}$ messages intended for it and $K_t\binom{K_r-1}{\sigma}$ interfering messages. Since the interfering messages do not need to be decoded by the receiver, we can try to "align" them into a smaller subspace than is needed for their decoding, resulting in a more efficient transmission and a higher DoF. More precisely, the minimal subspace in which they can be aligned has a dimension of $\binom{K_r-1}{\sigma}$, i.e., its dimension is smaller by a factor of K_t, yielding the following result.

Theorem 3: In the $K_t \times K_r$ multiple-multicast X-channel with a multicast size of σ, the DoF is exactly characterized by

$$\tilde{d}_\sigma(K_t, K_r) = \frac{1}{K_t\binom{K_r-1}{\sigma-1}+\binom{K_r-1}{\sigma}}. \tag{5.8}$$

Notice that Theorem 3 provides an *exact* characterization of the DoF of this physical-layer problem. The rigorous analysis of this alignment is out of the scope of this chapter and can be found in [15], but we provide the high-level idea of the alignment strategy that achieves this DoF below.

Suppose we want to communicate the message set across the channel over T time slots. Every transmitter j will beamform each message $V_{j,S}$, for each receiver set S, along a particular T-dimensional vector. It will then send the sum of all of these vectors through the channel as its codeword. In effect, while all of the transmitter's messages belong to an overall T-dimensional vector space, each message occupies a specific 1-dimensional subspace. In order to maximize the transmission rate, we aim to align, at every receiver, the messages that are *not* intended for it in the smallest possible subspace.

To do so, we enforce the following three principles, for which we give intuitions below. At every receiver i:

1. Each message that is intended for receiver i (i.e., each $V_{j,S}$ such that $i \in S$) must be in a subspace of dimension 1 and not aligned with any other message;

2. Messages from the same transmitter must be in non-aligned subspaces;
3. For every set of receivers S with $i \notin S$, all messages intended for S must be aligned in the same subspace of dimension 1.

The intuition behind these principles is as follows. The first principle is meant to ensure that all messages intended for receiver i can in fact be decoded at receiver i. The second principle is in fact a necessary condition for the first principle to hold for all receivers and thus ensures decodability at *other* receivers. Indeed, if two messages originating from the same transmitter are aligned at any receiver, then they must have been aligned at the transmitter as well, because they are encoded into the same codeword and go through the same channels. As a result, they must necessarily be aligned at *all* receivers, including those that need to decode them. Finally, the third principle provides maximal alignment of the interfering messages without violating either of the other two principles.

Each receiver has $K_t\binom{K_r-1}{\sigma-1}$ intended messages and $K_t\binom{K_r-1}{\sigma}$ interfering messages. The first principle requires that each intended message must be in a subspace of dimension 1, while the third principle implies aligning each K_t interfering messages in a single subspace of dimension 1. This can be ensured if the dimension of the overall vector space is $T = K_t\binom{K_r-1}{\sigma-1} + \binom{K_r-1}{\sigma}$. Consequently, the interference alignment technique achieves a DoF of $1/T$, which gives the expression in the theorem.

5.4.4 Information-theoretic outer bounds

The approximate optimality of the separation architecture can be proved by deriving cut-set upper bounds on the DoF of the cache-aided interference channel. The cut-set bounds are stated in the following lemma, for which a proof sketch is given at the end of this section.

Lemma 1: The DoF of the cache-aided interference channel must satisfy

$$\frac{1}{\mathsf{DoF}} \geq \max_{s \in \{1,\ldots,K_r\}} \frac{s\left(1-\frac{M_r}{\lfloor N/s \rfloor}\right)}{\min\{K_t, s\}}. \tag{5.9}$$

With Lemma 1 providing the outer bounds on the optimal DoF, and Theorems 2 and 3 combined in (5.4) providing the achievable DoF, we can prove the approximate optimality of the separation architecture and the characterization of the optimal DoF given in Theorem 1.

Rigorously applying the lemma in order to prove approximate optimality requires considering four separate regimes of the problem parameters. This is done in [15], but we can here get an intuition of the techniques by focusing on a single regime. Consider the regime of most interest, in which the number of transmitters is small compared to the number of receivers and the number of files, and $N/K_r \leq M_r < 0.1N/K_t$, i.e., the receivers can collectively hold the entire content library while an individual receiver cannot hold more than a fraction of the contents of a single transmitter's cache. In this regime, the separation architecture achieves an inverse DoF that is approximately N/K_tM_r, and applying Lemma 5.4.4

with $s \approx N/2M_r$ yields $1/\text{DoF} \gtrapprox N/4K_tM_r$. This demonstrates that the gap between the achievable DoF and the cut-set bounds is no more than a constant factor in this regime. The remaining regimes are effectively edge cases, and applying the lemma in these regimes (with limiting values of s such as 1, K_r, or N) proves the constant-factor gap universally.

5.4.4.1 Proof sketch of lemma

Roughly speaking, the cut-set bounds focus on a subset of s users, and consider the $K_t \times s$ MIMO channel between all transmitters and these receivers. These s users are allowed to fully cooperate, while the remaining users are served by a genie (and henceforth ignored). From the problem requirements, these s users can request up to s files, and are able to decode them with a single set of their channel outputs along with their cache contents. Therefore, using their cache contents and approximately N/s sets of channel outputs, they should be able to decode all N files.

Consequently, the s receiver caches of size M_rRT bits each, along with the N/s sets of outputs of the $K_t \times s$ MIMO channel, must have at least enough information to decode all N files, whose total size is NRT bits,

$$NRT \lessapprox sM_rRT + (N/s) \cdot C(K_t \times s) \cdot T, \tag{5.10}$$

where $C(K_t \times s)$ is the capacity bound of the $K_t \times s$ MIMO channel. In the limit as the SNR grows, the DoF of the MIMO channel is $\min\{K_t, s\}$. Since the number of receivers s in the subset was arbitrary, we get

$$\text{DoF} \lessapprox \min_{s \in \{1,\ldots,K_r\}} \frac{\min\{K_t,s\}}{s(1-sM_r/N)}, \tag{5.11}$$

whose inverse yields the result.

5.4.5 *A note on the low-SNR regime*

While this chapter largely focuses on the high-SNR regime and the DoF characterization of the communication rate, it is worthwhile to briefly discuss separation architectures in the low-SNR regime. In particular, we propose in [18] a similar separation architecture for the low-SNR regime. Of note are the following results:

- Separation is order-optimal in at least two cases: the single-transmitter (broadcast) case and the single-receiver (multiple-access) case. We conjecture that order-optimality extends beyond these two cases.
- Unlike in the high-SNR regime, transmitter cooperation becomes necessary in the low-SNR regime. Indeed, the message set must include messages originating from multiple transmitters at once. These transmitters must hence have shared information in their caches, which they can use to beamform messages to the intended receivers.

We refer the reader to [18] for the complete results.

5.5 Related work and outlook

In this chapter, we reviewed our results for a separation-based architecture for cache-aided wireless interference networks. In this section, we briefly put this work in the context of related work and also give an outlook to several unresolved question in this topic.

5.5.1 Related work

An information-theoretic framework for coded caching was introduced in [4] in the context of the deterministic (noiseless) broadcast channel. This framework was also applied to hierarchical (tree) topologies in [8], and to non-uniform content popularities in [9, 19–23], among many other extensions (see [10–12] among many others). Coded caching for erasure (noisy) broadcast channels has also been studied [24–26].

As argued in this chapter, interference is central to wireless network. Motivated by this, coded caching has been extended to interference networks with caches only at the transmitters in [27–29], where there are caches only at the transmitters. When caches are used *both* at transmitters and at receivers works in [16, 30, 31], consider a setup similar to the one in this chapter. However, each of these works has some restrictions. The work in [30] focuses on one-shot linear schemes, while [31] prohibits inter-file coding during placement and limits the number of receivers to three. Our work in [16] studies the same setup but with only two transmitters and two receivers. The work reviewed in this chapter (summarizing [15]) differs from those above in that it considers an arbitrary number of transmitters and receivers and proves order-optimality using outer bounds that assume *no restrictions* on the scheme. The work in [32] discussed a similar setup as described in this chapter but with only two or three receivers, as well as an outer bound that prohibits inter-file coding during placement; this was similar to [31].[b] To the best of our knowledge, the low-SNR regime for cache-aided interference networks is studied only in [18], briefly described in Section 5.4.5.

(Approximate) information-theoretic optimality of separation architectures has a long history, starting with Shannon's optimality source-channel separation theorem for point-to-point communication [34, 35]. In extensions beyond point-to-point channels, forms of source-channel separation have been examined for specific classes of sources and channels (see, e.g., [36–39] and references therein). More recently [40] has shown approximate and exact optimality of separation in several more general network data compression scenarios.

[b]For context and the timeline of these works published in [33] and those published in [15], please see section I-A, in [15].

5.5.2 Outlook

In this chapter we advocated a separation-based architecture for wireless (interference) networks. We believe that such an architecture is appropriate for several wireless network configurations including noisy broadcast, potentially multi-hop wireless networks, etc., for which we have not developed precise results characterizing the gap between a separation architecture and those without such a restriction. An important open question is to develop a "meta-result" that enables us to answer when such a separation architecture still retains most of the benefits, and perhaps when it is significantly advantageous to break such a layering.

In terms of technical questions that are unresolved, exact characterization of the DoF is open; however, this might be challenging as the exact rate-memory trade-off is unknown even for the error-free broadcast case. Note also that for general interference channels, using interference alignment, still only DoF characterizations are known, even for just the communication problem.

There are some improvements that can certainly be made to the schemes described in this chapter (as also seen in Section VII of [15]), where at least a constant factor improvement can be obtained. Potential further improvements could be obtained by using transmit zero-forcing or by placing coded content in the caches. Another aspect which could result in better approximation or characterization could be through a derivation of tighter outer bounds. Another interesting direction is to characterize (exactly) the optimal strategy when the problem imposes a restriction of uncoded cache placement, in a similar manner done for the deterministic broadcast case [41, 42].

On the more applied side of the story line, combining these ideas with wireless networking protocols would be an interesting direction of research. These would require the design of not just the physical layer schemes but also scheduling techniques to gauge the benefits of coded caching in wireless interference networks particularly in the finite blocklength regime. Moreover, the combination of coded caching with the CCN/ICN architectures being deployed for the Internet would also be potentially fruitful directions for future work.

Acknowledgement

The work of Suhas Diggavi was supported in part by NSF grants # 2007714 and 1955632.

References

[1] Koponen T., Chawla M., Chun B.-G., *et al.* 'A data-oriented (and beyond) network architecture'. *ACM SIGCOMM Computer Communication Review*. 2007;**37**(4):181–92.

[2] Zhang L., Afanasyev A., Burke J., *et al.* 'Named data networking'. *ACM SIGCOMM Computer Communication Review*. 2014;**44**(3):66–73.

[3] Korupolu M.R., Plaxton C.G., Rajaraman R. Placement algorithms for hierarchical cooperative caching. In: Proceedings of the ACM-SIAM Symposium on Discrete Algorithms (SODA); 1999. pp. 586–95.

[4] Maddah-Ali M.A., Niesen U. 'Fundamental limits of caching'. *IEEE Transactions on Information Theory*. 2014;**60**(5):2856–67.

[5] Hachem J., Karamchandani N., Diggavi S.N. 'Content caching and delivery over heterogeneous wireless networks'. Proceedings of the IEEE INFOCOM. IEEE; 2015. pp. 756–64.

[6] Shanmugam K., Golrezaei N., Dimakis A.G., Molisch A.F., Caire G. 'FemtoCaching: wireless content delivery through distributed caching helpers'. *IEEE Transactions on Information Theory*. 2013;**59**(12):8402–13.

[7] Ji M., Caire G., Molisch A.F. 'Wireless device-to-device caching networks: basic principles and system performance'. *IEEE Journal on Selected Areas in Communications*. 2016;**34**(1):176–89.

[8] Karamchandani N., Niesen U., Maddah-Ali M.A., Diggavi S.N. 'Hierarchical coded caching'. *IEEE Transactions on Information Theory*. 2016;**62**(6):3212–29.

[9] Hachem J., Karamchandani N., Diggavi S. 'Coded caching for multi-level popularity and access'. *IEEE Transactions on Information Theory*. 2017;**63**(5):1–3141.

[10] Wang S., Li W., Tian X., Liu H. *Fundamental limits of heterogenous cache. CoRR. 2015;abs/1504.01123*. Available from http://arxiv.org/abs/1504.01123.

[11] Maddah-Ali M.A., Niesen U. 'Decentralized coded caching attains order-optimal memory-rate tradeoff'. *IEEE/ACM Transactions on Networking*. 2015;**23**(4):1029–40.

[12] Zhang J., Lin X., Wang C.C., Wang X. 'Coded caching for files with distinct file sizes'. Proceedings of the 2015 IEEE International Symposium on Information Theory (ISIT); 2015. pp. 1686–90.

[13] Clark D.D. 'The design philosophy of the DARPA internet protocols'. Proceedings of the ACM SIGCOMM; 1988. Reprinted in 18 ACM Computer Communication Review 106; 1988.

[14] Chiang M., Low S.H., Calderbank A.R., Doyle J.C. 'Layering as optimization decomposition: a mathematical theory of network architectures'. *Proceedings of the IEEE*. 2007;**95**(1):255–312.

[15] Hachem J., Niesen U., Diggavi S.N. 'Degrees of freedom of cache-aided wireless interference networks'. *IEEE Transactions on Information Theory*. 2018;**64**(7):5359–80.

[16] Hachem J., Niesen U., Diggavi S. 'A layered caching architecture for the interference channel'. Proceedings of the 2016 IEEE International Symposium on Information Theory (ISIT); 2016. pp. 415–19.

[17] Cadambe V.R., Jafar S.A. 'Interference alignment and the degrees of freedom of wireless X networks'. *IEEE Transactions on Information Theory*. 2009;**55**(9):3893–908.

[18] Hachem J., Niesen U., Diggavi S. 'Energy-efficiency gains of caching for interference channels'. *IEEE Communications Letters*. 2018;**22**(7):1434–7.

[19] Niesen U., Maddah-Ali M.A. 'Coded caching with nonuniform demands'. Computer Communications Workshops (INFOCOM WKSHPS) 2014 IEEEConference on; 2014. pp. 221–6.

[20] Niesen U., Maddah-Ali M.A. 'Coded caching with nonuniform demands'. *IEEE Transactions on Information Theory*. 2017;**63**(2):1146–58.

[21] Ji M., Tulino A.M., Llorca J., Caire G. 'Order-Optimal rate of caching and coded multicasting with random demands'. *IEEE Transactions on Information Theory*. 2017;**63**(6):3923–49.

[22] Zhang J., Lin X., Wang X. 'Coded caching under arbitrary popularity distributions'. *IEEE Transactions on Information Theory*. 2018;**64**(1):349–66.

[23] Hachem J., Karamchandani N., Diggavi S. 'Multi-level coded caching'. Proceedings of the 2014 IEEE International Symposium on Information Theory (ISIT); 2014. pp. 56–60.

[24] Timo R., Wigger M. 'Joint cache-channel coding over erasure broadcast channels'. Proceedings of the IEEE International Symposium on Wireless Communication Systems (ISWCS); 2015. pp. 201–5.

[25] Ghorbel A., Kobayashi M., Yang S. 'Content delivery in erasure broadcast channels with cache and feedback'. *IEEE Transactions on Information Theory*. 2016;**62**(11):6407–22.

[26] Mohammadi Amiri M., Gunduz D. 'Cache-aided content delivery over erasure broadcast channels'. *IEEE Transactions on Communications*. 2018;**66**(1):370–81.

[27] Maddah-Ali M.A., Niesen U. 'Cache-aided interference channels'. Proceedings of the 2015 IEEE International Symposium on Information Theory (ISIT); 2015. pp. 809–13.

[28] Maddah-Ali M.A., Niesen U. 'Cache-aided interference channels'. *IEEE Transactions on Information Theory*. 2019;**65**(3):1714–24.

[29] Sengupta A., Tandon R., Simeone O. 'Fog-aided wireless networks for content delivery: fundamental latency tradeoffs'. *IEEE Transactions on Information Theory*. 2017;**63**(10):6650–78.

[30] Naderializadeh N., Maddah-Ali M.A., Avestimehr A.S. 'Fundamental limits of cache-aided interference management'. *IEEE Transactions on Information Theory*. 2017;**63**(5):3092–107.

[31] Xu F., Liu K., Tao M. 'Cooperative Tx/Rx caching in interference channels: a storage-latency tradeoff study'. 2016 IEEE International Symposium on Information Theory (ISIT); 2016. pp. 2034–8.

[32] Xu F., Tao M., Liu K. 'Fundamental tradeoff between storage and latency in Cache-Aided wireless interference networks'. *arXiv:160500203v3 [csIT]*. 2017.

[33] Xu F., Tao M., Liu K. 'Fundamental tradeoff between storage and latency in cache-aided wireless interference networks'. *IEEE Transactions on Information Theory*. 2017;**63**(11):7464–91.

[34] Shannon C.E. 'A mathematical theory of communication'. *Bell System Technical Journal*. 1948;**27**(3):379–423.

[35] Shannon C.E. 'A mathematical theory of communication'. *Bell System Technical Journal*. 1948;**27**(4):623–56.

[36] Han T.S. 'Multicasting correlated multi-source to multi-sink over a network'. *IEEE Trans Information Theory*. 2011;**57**(1):4–13.

[37] Mohajer S., Tian C., Diggavi S.N. 'On source transmission over deterministic relay networks'. 2010 IEEE Information Theory Workshop on Information Theory (ITW 2010, Cairo); 2010. pp. 1–5.

[38] Xiao J.-J., Luo Z.-Q. 'Multiterminal source–channel communication over an orthogonal multiple-access channel'. *IEEE Transactions on Information Theory*. 2007;**53**(9):3255–64.

[39] Tuncel E. 'Slepian-Wolf coding over broadcast channels'. *IEEE Transactions on Information Theory*. 2006;**52**(4):1469–82.

[40] Tian C., Chen J., Diggavi S.N., Shamai S. 'Optimality and approximate optimality of source-channel separation in networks'. *IEEE Transactions on Information Theory*. 2014;**60**(2):904–18.

[41] Wan K., Tuninetti D., Piantanida P. 'An index coding approach to caching with uncoded cache placement'. *IEEE Transactions on Information Theory*. 2020;**66**(3):1318–32.

[42] Yu Q., Maddah-Ali M.A., Avestimehr A.S. 'The exact rate-memory tradeoff for caching with uncoded prefetching'. *IEEE Transactions on Information Theory*. 2018;**64**(2):1281–96.

Chapter 6

Scalable delivery of correlated video content over cache-aided broadcast networks

Parisa Hassanzadeh[1], Antonia M. Tulino[2], Jaime Llorca[3], and Elza Erkip[3]

Abstract

The increasingly dynamic, unpredictable, and personalized nature of the content that users consume challenges the efficiency of existing caching-based solutions in which only exact content reuse is explored. This calls for novel content delivery schemes that account for the additional gains that can be obtained from the joint compression of correlated content distributed throughout the network, such as updated versions of dynamic data. This chapter reviews state-of-the-art studies on the fundamental limits of cache-aided communication systems for the delivery of correlated content. Two scenarios are considered: (i) a *static* setting, in which the same correlated content library is used during both the caching and delivery phases and (ii) a *dynamic* setting, in which an updated version of the content library may become available during the delivery phase. For the static setting, a class of achievable schemes based on a two-step source coding approach is described. The content library is first jointly compressed, with resulting files cached and delivered using a combination of correlation-unaware cache-aided delivery schemes. For the dynamic setting, a second class of achievable schemes is described, in which, during the caching phase, content pieces are stored based on their correlation with the rest of the library, and during delivery, the requested content is jointly compressed while using cached information as references. With accompanying performance analyses and numerical results, this chapter illustrates that as we move towards real-time and/or personalized media dominated services, where exact cache hits are almost nonexistent (challenging the efficiency of traditional caching schemes), the high correlation present among increasingly dynamic content still enables exploiting network cached information as references for network compression.

[1]J. P. Morgan AI Research, USA
[2]DIETI, University of Naples Federico II, Italy
[3]ECE Department, New York University, USA

6.1 Introduction

6.1.1 Edge caching

Internet traffic has experienced dramatic growth in the past decade, reaching more than 45 000 GB/second in 2017, and is expected to nearly triple by 2022 [1]. With the advent of new applications such as augmented reality, social networking, and ultra-high-definition video streaming, it is projected that traffic from wireless and mobile devices will account for 71% of total traffic by 2022. Providing such services requires high data rates with gigabit speed, latency as low as 1 ms, and ultra-high reliability. In addition to making infrastructural improvements such as increasing base station density and offloading traffic to Wi-Fi, given the declining costs of storage capacity, wireless operators are considering *edge caching* as a technique to reduce traffic load by exploiting the high degree of asynchronous content reuse in wireless networks. Prefetching portions of popular content into cache memories that are distributed throughout the network is regarded as a highly effective technique for reducing wireless traffic load, since it not only provides local content delivery, but it also enables coded multicast transmissions that are useful for multiple receivers. During off-peak periods when network resources are abundant, popular content can be stored at the wireless edge (e.g., small-cell base stations, wireless access points, or end-user devices), so that peak hour demands can be met with reduced access latencies and bandwidth requirements.

The simplest form of caching is to prefetch and store the most popular content, such as the latest release of a Netflix series, at the local caches of all network elements. Requests for popular cached files can then be served locally, while cache misses need to be served by the base station, achieving what is referred to as a *local* caching gain. However, merely replicating content on many devices can result in an inefficient use of the aggregate cache capacity. Recently, it was shown that distributing different portions of the content among the caches across the network, and capitalizing on the spatial reuse of the cached information (with device-to-device communications or multicasting coded messages) can yield significant multiplicative gains, referred to as *global* caching gains [2–8]. These schemes are referred to in literature as *Cache-Aided Coded Multicast* (CACM) or more simply as coded caching. In all these works on cache-aided networks, it is assumed that the network operates in two phases: a caching (or placement) phase taking place at network setup followed by a delivery phase where the network is repeatedly used in order to satisfy receiver demands. The design of the caching and delivery phases forms what is referred to as a *caching scheme*. During the caching phase, caches are filled with content from the library according to a properly designed *caching distribution*. During the delivery phase, the sender compresses the set of requested files into a multicast codeword by computing an *index code* [9, 10].

6.1.2 Turning memory into bandwidth

The first information-theoretic formulation of content delivery over cache-aided wireless networks was given by Maddah-Ali and Niesen in [6], which studied the fundamental limits of a multi-user caching network where users are connected to a server through an error-free shared broadcast link. The authors proposed a coded caching scheme that, by jointly optimizing the placement and delivery phases, satisfies multiple distinct user demands via a single coded multicast transmission, achieving global caching gains in addition to the conventional local caching gains. They showed that such a scheme performs within a constant factor of the information-theoretic optimum in the worst-case scenario, i.e., when all users request different content. They also showed that the global caching gains are relevant in settings where the aggregate cache size, i.e., the cumulative cache available at all edge caches, is large enough compared to the total amount of popular content. The scheme proposed in [6] relied on a central server to carefully design and place the cached content. In order to make the content delivery robust to network changes, the authors proposed a random placement scheme in [5], still able to create coded multicast opportunities and achieve a delivery rate close to that of the optimal deterministic scheme. Several works including [11–22] have studied the fundamental limits in caching networks by improving the converse bounds and proposing more efficient caching schemes. Finally, in [8], the exact memory-rate trade-off for the worst-case and average-case demand was characterized for the class of caching schemes that fill caches with uncoded data in the placement phase, and in [23], the trade-off was characterized within a factor of two of an information-theoretic lower bound for uniformly popular files.

The idea of exploiting coded multicast opportunities to further reduce broadcast traffic was extended to various settings in [2, 4, 24–37]. The case of nonuniform demands according to an arbitrary popularity distribution was analyzed in [4, 33, 38], which proposed schemes that are able to achieve performance within a constant factor of the optimal. The work in [2] studied the multiple per-user demands scenario [35], studied online caching, the works in [27, 29] investigated hierarchical caching settings, and the works in [31, 32] analyzed cache-aided interference channels. Other settings include device-to-device caching [28], noisy broadcast channels [24, 26], C-RAN networks [25, 34], mixed time-scale problems coupling between the transmissions and the caching strategies [30], and caching for channels with delayed feedback with channel state information [36, 37]. A comprehensive review of existing work on caching networks can be found in [39].

6.1.3 Exploiting content correlations

While existing work on wireless caching is motivated by video applications, the majority do not exploit specific properties of video, such as correlated content reuse and scalability of video content, in the caching and delivery phases. Specifically, most cache-aided content delivery schemes available in literature:

i. treat network content as independent pieces of information, and exploit *exact* content reuse across the network without accounting for the additional gains that can be obtained from the joint compression of the correlated content distributed throughout the network.

ii. are designed based on fixed-to-variable source encoding, designed to minimize the aggregate rate on the shared link so that the requested files are recovered with zero distortion.

In this chapter, we will provide a comprehensive overview of the main results around the fundamental limits and the design of caching schemes that exploit the key correlation and scalability properties of video-based applications to push the limits of efficient and robust wireless content delivery. Motivated by the increasing dynamic and personalized nature of next-generation content services, we will illustrate some correlation-aware schemes mainly for two settings: (i) a *static* setting, in which we assume that the same (correlated) content library is used during both the caching and delivery phases and (ii) a *dynamic* setting, in which an updated version of the content library may become available during the delivery phase.

The chapter is organized as follows. Section 6.2 provides a high level overview of the latest studies around wireless delivery of correlated content, including lossless and lossy settings. The remaining sections then focus on providing a more detailed description of the lossless setting. Specifically, Section 6.3 presents an information-theoretic formulation for the problem of lossless delivery of static and dynamic content over a broadcast caching network. Sections 6.4 and 6.5 review state-of-the-art solutions for static content and for dynamic content, respectively. Finally, Section 6.6 provides illustrative numerical results and concluding remarks.

6.2 Delivering correlated content over cache-aided broadcast networks

Exploiting content correlation is especially relevant among content files of the same category, such as episodes of a TV show or same-sport recordings, which may share common backgrounds and scene objects, but becomes also particularly critical as we move from static content distribution toward real-time delivery of rapidly changing (and aging), but highly correlated, personalized data (as in news updates, social networks, remote sensing, augmented reality, etc.) where information rapidly loses relevance, requires timely updates and in which exact content reuse is almost nonexistent [40, 41]. Additionally, video can be coded and delivered in a scalable fashion rather than as a file with a predetermined size, and it can be compressed at different rates resulting in various playback qualities. In scalable video coding (SVC) [42], video files are encoded into layers such that the base layer contains the lowest quality level and additional enhancement layers allow successive improvement of the video streaming quality. SVC strategies are especially suitable for heterogeneous wired and wireless networks, since they encode video into a scalable bitstream such that videos of lower qualities, spatial resolutions, and temporal resolutions can be generated by simply truncating the scalable bitstream. This scalability accommodates network requirements such as bandwidth limitations, user device capability, and quality-of-service restrictions in video streaming applications [43].

This calls for new strategies that are able to efficiently exploit the correlation between multiple versions of dynamic data in order to enable both compressed delivery, as well as efficient updating of cached references. Existing studies in this regard have considered two main setups:

i. *lossless* setting, where requested content must be reconstructed with zero distortion at all receivers and
ii. *lossy* setting, where requested content can be reconstructed with nonzero distortion depending on each receiver's minimum rate requirement.

In both setups, fundamental information-theoretic limits have been investigated, and achievable schemes for correlation-aware content caching and delivery have been designed. In the first setting, existing works have focused on the design of what are known as *correlation-aware CACM* schemes, whose main goal is to exploit the additional gains that can be obtained when accounting for the correlations that may exist among different static files as well as among multiple versions of dynamic data. In the second setting, the main focus has been on using caching as a method to enhance file (e.g., video) reconstruction quality at user devices. The goal has been to characterize fundamental *distortion-memory trade-offs* by designing caching schemes that, for a given broadcast rate, minimize the average distortion experienced at user devices.

6.2.1 Correlation-aware cache-aided coded multicasting

Only a few recent works have studied the cache-aided broadcast network with a library composed of correlated files [44–49]. The works in [47] and [46] consider a single-receiver single-cache multiple-file network with lossy reconstructions, and characterize the trade-offs between rate, cache capacity, and reconstruction distortions. The analysis in [47] also considers two receivers and one cache, in which again only local caching gains can be explored. The authors determine the information common to all sources to be the most useful content to be placed in the cache. The work in [46] models the caching problem in a way that resembles the Gray-Wyner network [50]. The work in [45] studies a general network and proposes an achievable scheme that takes on a more practical approach to finding the information common to a set of files. To this end, the library files are grouped into sets of correlated files, and a subset of the files that are most representative of the library are selected as references, referred to as *I-files*, and the remaining files are inter-compressed with respect to the selected files and referred to as *P-files*, in resemblance to the use of *I-frames* and *P-frames* in local video compression algorithms. This results in a compressed library where each file is made up of an *I-file* and a *P-file* leading to a multiple-request caching problem. Differently from previous multiple-request schemes provided in [2, 51–54], the demand in [45] has a specific structure dictated by the configuration of the resulting compressed library in terms of I-files and P-files. In [48], the authors consider a library that has a special correlation structure such that each file is composed of multiple independent subfiles that are common among a fixed set of files in the library. This, as in [45], leads to a multiple-request caching problem where the demand has a particular configuration dictated by the specific library structure. They propose a caching scheme for an arbitrary number of files and receivers, and numerically compare the performance of their scheme with a lower bound on the worst-case rate-memory trade-off. They numerically compare the performance of their scheme with a lower bound on the worst-case rate-memory trade-off. The follow-up work in [49] considers a setting where a server holding a database of correlated content communicates with the users over a Gaussian broadcast channel, and evaluates the performance of this system in terms of the minimum transmission power required to satisfy all possible demands. The authors propose an achievable scheme based on superposition coding and power allocation. Specifically, for each demand realization, the server identifies the messages that need to be delivered to each user using the same scheme proposed in [49], and then employs superposition coding to reliably communicate coded messages over the Gaussian broadcast channel.

All previously cited works provide achievable caching schemes without analytically bounding the gap to optimality, except for the special case of two files and two receivers, and three files and two receivers presented in [55] and [56], respectively. In [57], by focusing on lossless reconstructions, the authors extend the information-theoretic analysis of the broadcast caching network done in [55] to arbitrary number of receivers, each equipped with its own cache. Differently from [44, 45, 48, 58], the authors in [57] characterize the peak and average rate-memory region for files generated by a discrete memoryless source with arbitrary joint distribution, and propose a class of optimal or near-optimal two-step schemes.[a] The main contributions derived in [57] are summarized as follows: The authors

- formulate the problem of efficient delivery of multiple correlated files over a broadcast caching network with arbitrary number of receivers via information-theoretic tools.
- propose a class of correlation-aware two-step schemes, in which the files are first encoded based on the Gray-Wyner network [50], and in the second step, they are cached and delivered through a combination of multiple-request caching schemes that do not exploit the correlations among the content, which is general enough to account for files of different length.
- discuss the optimality of the proposed two-step scheme in a two-file K-receiver network by characterizing an upper bound on the peak and average rate-memory trade-offs for this class of schemes, and by comparing it with a new lower bound on the optimal rate-memory trade-offs derived for correlated files. In [57], the authors identify a set of operating points in the achievable Gray-Wyner region [50] for which the proposed two-step scheme is optimal over a range of cache

[a]By near-optimality we refer to schemes with achievable rates that are within a constant additive factor from the optimal solution. The term *order-optimal*, on the other hand, is used when the achievable rate is within a constant multiplicative factor from the optimal solution.

capacities, and approximates the optimal rate in the two-file network within half of the conditional entropy for all cache sizes.

- extend the analysis to the three-file scenario which captures the essence of the multiple-file case without involving the exponential complexity of the multiple-file Gray-Wyner network. We show that for two receivers the proposed scheme is optimal for high memory sizes and its gap to optimality is less than half of the joint entropy of two of the sources conditioned on the third source for other memory.
- motivated by the second step of the proposed Gray-Wyner-based methodology for three files, they also present a novel near-optimal caching scheme for a network with two receivers and three independent files, where each receiver requests two of the files. The proposed scheme uses *coded* cache placement to achieve optimality for cache capacities up to half of the library size.

6.2.1.1 The case for dynamic correlated content

Even though jointly compressing the library prior to the caching phase, as in [45–48], is a natural solution to reducing the load on the shared link, this approach is mainly suitable for a library with static content, losing robustness and leading to significant performance degradation when content is highly dynamic. For example, if the library is updated and a new file is added, all files need to be jointly re-compressed and the content cached across the entire network needs to be updated based on the new compressed versions. On the other hand, schemes that are not based on a-priori library compression may only require updating a portion of the network cached content, which is especially critical for the delivery of dynamic data in next-generation services. In [44], the authors propose a correlation-aware scheme that is robust to changes in the library, where content is cached according to both the popularity of files and their correlation with the rest of the library. Cached information is then used as references for the compression of requested files during the delivery phase. In a subsequent work [58], the same authors design a practical scheme that is robust to the variations inherent to the delivery of real-time services over wireless networks. To this end, they propose an achievable scheme that stores individually (rather than jointly) compressed content pieces during the caching phase, which then serve as references for compression during the delivery phase, resulting in an efficient scheme that is robust to dynamic changes in the content library. Compared with previously proposed schemes [45, 47, 48, 55–57], where correlation-aware compression takes place before the caching phase, their scheme provides on-demand compression during the delivery phase and hence is able to adapt to system variations while continuing to exploit available content correlations. Specifically, in [58], the authors propose a correlation-aware scheme that consists of (i) storing individually compressed content pieces based on their popularity as well as on their correlation with the rest of the library in the caching phase and (ii) sending compressed versions of the requested files according to the information distributed throughout the network and their joint statistics during the delivery phase.

By using an individually compressed caching phase, and providing *on-demand* compression during the delivery phase, such a scheme is robust to changes in the content library, particularly relevant in next-generation dynamic content services. The authors analyze the performance of the proposed correlation-aware scheme for both static and dynamic settings. They characterize an upper bound on the expected rate of the proposed correlation-aware scheme in a network with arbitrary number of files and receivers, and numerically compare the achievable rate with that of existing correlation-unaware schemes, which confirm the additional available gains, especially for small memory sizes. Finally, for the special case of the two-file two-receiver network considered in [55, 57], the authors present a scheme based on a more structured cache placement that achieves a rate within half of the mutual information of the two files for all cache sizes.

6.2.2 Distortion-memory trade-offs

As discussed above, even though caching networks address the network congestion due to multimedia applications, most existing works do not take into account specific properties of such content such as their scalability. They assume that content are files with fixed sizes and their objective is to losslessly deliver the entire content files to users with the goal of minimizing the total rate transmitted over the shared link, in order to recover all requested fixed-size video files in whole [2, 4–6, 23, 52, 59]. In practice however, strategies such as scalable video encoding is used to allow for lossy content delivery and for adjusting to different network limitations such as quality-of-service and bandwidth requirements. There are only a few works that study the lossy cache-aided broadcast network [47, 60, 61]. In [47], the authors study the delivery rate, cache capacity and reconstruction distortion trade-offs in a network with arbitrarily correlated sources. They characterize the optimal trade-off for the single-user network and some special cases of a two-user problem. Similarly to this paper, [61] and [60] assume successively refinable sources in a setting where receivers have heterogeneous distortion requirements. In [61], the authors study the problem of minimizing the worst-case delivery rate for Gaussian sources and heterogeneous distortion requirement at the users. They characterize the optimal delivery rate for the two-file two-user case, and propose efficient centralized and decentralized caching schemes based on successive refinement coding for the general case. The work in [60] extends [61] to a setting where the server not only designs the users' cache contents but also optimizes their cache sizes subject to a total cache memory budget.

The work in [62] differs from [47, 60, 61] in a number of ways. Ref.[62] focuses on lossy delivery of multimedia content in a heterogeneous rate-limited caching network for a general setting with arbitrary number of users and files and formulates the problem using information-theoretic tools.

Compared to [47], which considers a single-cache network, [62] has a large network with arbitrary number of receivers, each equipped with a cache memory of different capacity. The works in [60, 61] minimize the worst-case rate transmitted over the broadcast link for a set of predetermined reconstruction distortion requirements at each user, while [62] minimizes the expected distortion experienced across all the users and demands subject to a given broadcast rate for a more general setting as elaborated below. Hassanzadeh *et al.* [62] propose two classes of achievable schemes based on scalable source coding where files are compressed into layers that are successively refinable: (i) a scheme that designs the placement phase only based on users' local information, and then uses unicast transmissions during the delivery phase and (ii) a scheme that jointly designs the content to be cached by all users and delivered using all network parameters, and uses coded mutlicasting in the delivery phase.

After formulating the problem of efficient lossy delivery of sources over a heterogeneous rate-limited broadcast caching network via information-theoretic tools, [62] studies the trade-off between user cache sizes, broadcast rate and the expected reconstruction distortion across users and demands. In [62], the authors allow for sources to have different distortion-rate functions and for users to have different cache sizes and different demand distributions. They propose a class of cache-aided delivery schemes referred as *Local Cache-aided Unicast (LC-U)* schemes, in which, to limit the computational complexity and reduce the communication overhead, the sender only takes into account users' local cached content during the delivery phase, and generates the transmit message independently for each receiver without exploring multicast coding opportunities. It is shown that the optimal caching policy in LC-U admits a reverse water-filling type solution, which can be implemented locally and independently across users, without the need of global coordination. Hassanzadeh *et al.* [62] also propose another class of schemes, referred to as the *Cooperative Cache-aided Coded Multicast* (*CC-CM*) scheme where the sender designs the caching and delivery phases jointly across all receivers based on global network knowledge (user cache contents and demand distributions, and file rate-distortion functions), and compresses the files accordingly. In this scheme, global network knowledge is used to fill user caches and to construct codes that fully exploit the multicast nature of a wireless system. The CC-CM proposed in [62] is based on two components, (i) a coded delivery scheme, adopted by CC-CM to implement the caching phase and to deliver a portion of the multicast message and referred to as *Random Fractional caching with Greedy Constrained Coloring (RF-GCC)* and (ii) unicast scheme. The RF-GCC is a generalization of the scheme proposed in [4] to a setting with heterogeneous cache sizes and demand distributions. Here the users are interested in receiving possibly degraded versions (different-length portions) of a given file in the library, as the *RF-GCC*. As mentioned above, the RF-GCC is used in CC-CM to deliver part of the requested file. The remaining part is delivered via unicast. Based on these two components, the rate-distortion-memory trade-off achieved with CC-CM is characterized. In particular the rate-distortion-memory trade-off characterization in [62] is obtained by evaluating upper bounds on the per-demand and average delivery rates achieved with RF-GCC. It is shown that RF-GCC allows the generalization of the problem studied in [61] where (i) files have different distortion-rate functions and (ii) users have different reconstruction distortion targets for each library file. When specialized to the setting in [61], RF-GCC achieves equal or better worst-case delivery rate compared to the decentralized scheme proposed in [61]. Combining the two components, [62] numerically shows that the CC-CM offers notable performance improvements (order improvements) over LC-U in terms of average file reconstruction distortion for the same available broadcast rate.

6.3 Caching networks with correlated sources

In this section, we present a general problem formulation for the lossless delivery of static and dynamic content. The formulation is general enough to specialize to a *static* setting, where the statistics of the correlated library content remain the same during the caching and delivery phases, studied in Section 6.4, as well as to a *dynamic* setting that accounts for the correlations between updated versions of dynamic data, considered in Section 6.5. We consider a network setup similar to [2–8], but we assume that the files in the library are generated according to a joint distribution and are therefore correlated. We also introduce two criteria, namely the peak and the average rate-memory trade-offs, that are used for studying the fundamental limits of broadcast caching networks in the sections that follow.

6.3.1 Problem formulation

We consider a broadcast caching network composed of one sender (e.g., base station) with access to a library made of N uniformly popular files, generated by an N-component discrete memoryless source (N-DMS). The N-DMS model consists of N finite alphabets $\mathcal{W}_1, \ldots, \mathcal{W}_N$ and a joint probability mass function (p.m.f.) $p(\mathbf{w}) = p(w_1, \ldots, w_N)$ over $\mathcal{W}_1 \times \cdots \times \mathcal{W}_N$. For a block length F, library file $n \in \{1, \ldots, N\}$ is represented by a sequence $W_n^F = (W_{n1}, \ldots, W_{nF})$, where $W_n^F \in \mathcal{W}_n^F$, and $(W_{1i}, \ldots, W_{Ni})$, $i \in \{1, \ldots, F\}$, is generated i.i.d. according to $p(\mathbf{w})$. The sender communicates with K receivers, $\{1, 2, \ldots, K\}$, over a shared error-free broadcast link. Each receiver is equipped with a cache of size MF bits, where $M \in \left[0, \sum_{n=1}^{N} H(W_n)\right]$ denotes the (normalized) cache capacity. The network operates in two phases:

i. A caching phase that takes place at the network setup and is assumed to happen during off-peak hours without consuming actual delivery rate. During this phase, the receivers fill their caches from $\{W_1^F, \ldots, W_N^F\}$, referred to as the *original library*.
ii. A delivery phase where the network is repeatedly used and library files can be dynamically updated. During this phase, when receiver demands are revealed and resources are limited, the sender broadcasts the shortest possible codeword that allows each receiver to losslessy recover its demand. The most recent versions of the files are denoted by $\{V_1^F, \ldots, V_N^F\}$. For each $i \in \{1, \ldots, F\}$, conditioned on $(W_{1i}, \ldots, W_{Ni})$, the vectors $(V_{1i}, \ldots, V_{Ni})$ are i.i.d. according to $(V_1, \ldots, V_N) \sim p(\mathbf{v}|\mathbf{w}) = p(v_1, \ldots, v_N|\mathbf{w})$. Without

loss of generality, we assume that the most recent versions have the same alphabets as the original files. Availability of the most recent content at the server is subject to network delays, and as a result, we assume that only with probability π_n the server observes the most recent version of file $n \in \{1, \ldots, N\}$. Hence the *updated library*, denoted by $\{U_1^F, \ldots, U_N^F\}$, is such that $U_n^F = V_n^F$ with probability π_n and $U_n^F = W_n^F$ with probability $1 - \pi_n$ for $n \in \{1, \ldots, N\}$, where availability of the most recent content at the server is modeled as an independent process across files. We also assume that in different uses of the network, the most recent versions are obtained conditionally i.i.d. with respect to the original library, and files are updated at the server (with respect to the cached versions) in an i.i.d fashion. The *library distribution* is then characterized by $p(\mathbf{w}, \mathbf{v})$ and $\boldsymbol{\pi} = (\pi_1, \ldots, \pi_N)$. Receivers request files in an i.i.d. manner according to a uniform demand distribution, and the sender satisfies the demands using the updated library. The demand realization is denoted by $\mathbf{d} = (d_1, \ldots, d_K)$, where $\mathbf{d} \in \mathcal{D} \equiv \{1, \ldots, N\}^K$.

A caching scheme for this network consists of:

- **Cache Encoder:** The cache encoder at the sender computes the content to be cached at receiver $k \in \{1, \ldots, K\}$, denoted by Z_k, using the set of functions $f_k^{\mathfrak{C}} : \mathcal{W}_1^F \times \cdots \times \mathcal{W}_N^F \to [1 : 2^{MF})$, as

$$Z_k = f_k^{\mathfrak{C}}\left(\{W_n^F\}_{n=1}^N\right). \tag{6.1}$$

  The encoder designs the cache configuration $\{Z_1, \ldots, Z_K\}$ jointly across receivers, taking into account available global system knowledge, such as the number of receivers and their cache sizes, the number of files and the p.m.f $p(\mathbf{w})$.
- **Multicast Encoder:** After the caches are populated, the network is repeatedly used. At each use, a demand $\mathbf{d}$ from the updated library, $\{U_1^F, \ldots, U_N^F\}$, is revealed to the sender. The multicast encoder uses a fixed-to-variable encoding function

$$f^{\mathfrak{M}} : \mathcal{D} \times \mathcal{W}_1^F \times \cdots \times \mathcal{W}_N^F \times [1 : 2^{MF})^K \to \mathcal{Y}^\star, \tag{6.2}$$

  to generate a multicast codeword $Y_{\mathbf{d}}$ according to the demand realization $\mathbf{d}$, the updated library $\{U_1^F, \ldots, U_N^F\}$, the cache configuration $\{Z_1, \ldots, Z_K\}$, and the library distribution $p(\mathbf{w}, \mathbf{v})$. Then,

$$Y_{\mathbf{d}} = f^{\mathfrak{M}}\left(\mathbf{d}, \{U_n^F\}_{n=1}^N, \{Z_k\}_{k=1}^K\right), \tag{6.3}$$

  where $\mathcal{Y}^\star$ denotes the set of finite length sequences.
- **Multicast Decoders:** Receivers recover their requested files using their cached content and the received multicast codeword. Receiver $k \in \{1, \ldots, K\}$ recovers $U_{d_k}^F$ using a decoding function $h_k^{\mathfrak{M}} : \mathcal{D} \times \mathcal{Y}^\star \times [1 : 2^{MF}) \to \mathcal{W}_{d_k}^F$, as

$$\widehat{U}^F_{d_k} = h^{\mathfrak{M}}_k(\mathbf{d}, Y_\mathbf{d}, Z_k). \tag{6.4}$$

We refer to the overall scheme as a *CACM* scheme.

6.3.2 Peak and the average rate-memory functions

The worst-case probability of error of the corresponding CACM scheme is defined as

$$P^{(F)}_e = \max_{\mathbf{d}\in\mathscr{D}} \mathbb{P}\left(\bigcup_{k\in\{1,\dots,K\}} \left\{\widehat{U}^F_{d_k} \neq U^F_{d_k}\right\} \right). \tag{6.5}$$

We consider two performance criteria:

i) The peak multicast rate, $R^{(F)}$, which corresponds to the worst-case demand,

$$R^{(F)} = \max_{\mathbf{d}\in\mathscr{D}} \frac{\mathbb{E}[L(Y_\mathbf{d})]}{F}, \tag{6.6}$$

where $L(Y)$ denotes the length (in bits) of the multicast codeword Y, and the expectation is over the library distribution $p(\mathbf{w}, \mathbf{v})$ used during caching and delivery phases.

ii) The average multicast rate, $\bar{R}^{(F)}$, over all possible demands

$$\bar{R}^{(F)} = \frac{\mathbb{E}[L(Y_\mathbf{d})]}{F}, \tag{6.7}$$

where the expectation is over the demands and the library distribution $p(\mathbf{w}, \mathbf{v})$.

Definition 1. A peak rate-memory pair (R, M) is *achievable* if there exists a sequence of CACM schemes for cache capacity M and increasing file size F, such that:

$$\lim_{F\to\infty} P^{(F)}_e = 0, \qquad \limsup_{F\to\infty} R^{(F)} \leq R. \tag{6.8}$$

Definition 2. The peak rate-memory region $\mathfrak{R}^*$ is the closure of the set of achievable peak rate-memory pairs (R, M) and the optimal peak rate-memory function, $R^*(M)$, is

$$R^*(M) = \inf\{R : (R, M) \in \mathfrak{R}^*\}. \tag{6.9}$$

Definition 3. An average rate-memory pair (R, M) is *achievable* if there exists a sequence of CACM schemes for cache capacity M and increasing file size F, such that $\lim_{F\to\infty} P^{(F)}_e = 0$, and $\limsup_{F\to\infty} \bar{R}^{(F)} \leq R$.

Definition 4. The average rate-memory region $\bar{\mathfrak{R}}^*$ is the closure of the set of achievable average rate-memory pairs (R, M) and the optimal average rate-memory function, $\bar{R}^*(M)$, is

$$\bar{R}^*(M) = \inf\{R : (R, M) \in \bar{\mathfrak{R}}^*\}. \tag{6.10}$$

The goal is to design CACM schemes that result in small achievable peak and average rates for a given memory size M. We consider the following two settings:

- **Static Setting**: In this setting, the library remains static throughout the network setup and its consecutive uses, such that content is requested and delivered from the same library that is used in the caching phase; i.e., $\pi_n = 0$, $n \in \{1, \dots, N\}$. Hence, in this scenario the library is composed of static correlated content, as in [45–49], and when particularized

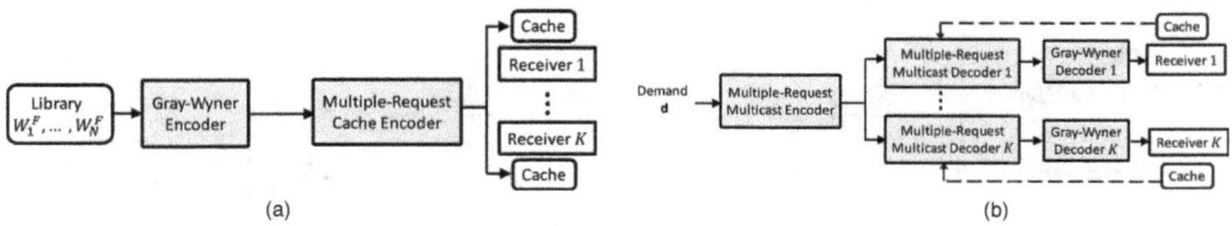

Figure 6.1 *The GW-MR scheme: a two-step correlation-aware scheme composed of a Gray-Wyner source coding step followed by a multiple-request CACM step [57]*

to independent files, the setting is equivalent to that of most prior works including [2, 4–7, 53].

- **Dynamic Setting:** In this case, for ease of exposition, we assume that the library used during the caching phase is originally composed of N independent files, i.e., $p(\mathbf{w}) = \prod_{n=1}^{N} p(w_n)$, and that the most recent versions of the files are generated such that $p(\mathbf{v}|\mathbf{w}) = \prod_{n=1}^{N} p(v_n|w_n)$, but otherwise the library distribution follows the general model described in this section.

In Sections 6.4 and 6.5, we study the static and dynamic settings, respectively. For each setting, we describe corresponding CACM schemes and analyze their performance by characterizing an upper bound on the achievable rate and comparing it with a lower bound on the optimal peak and average rate-memory trade-offs as well as with existing correlation-unaware schemes.

6.4 Caching networks with correlated sources – the static case

In this section, we particularize the system model introduced in Section 6.3 for a static library, where the same library is used during both caching and delivery phases. We describe a class of achievable correlation-aware schemes based on a two-step source coding approach introduced in [44] and depicted in Figure 6.1.

Library files are first jointly compressed, with resulting versions cached and delivered using a combination of correlation-unaware multiple-request CACM schemes. The first step uses Gray-Wyner source coding to represent the library via private descriptions and descriptions that are common to more than one file. The second step then becomes a multiple-request caching problem, where the demand structure is dictated by the configuration of the compressed library.

Such a two-step scheme exploits the correlation among the library content by first compressing the library using the Gray-Wyner encoder, described in detail in Section 6.4.1, and depicted for two and three files in Figure 6.2. In the first step, the Gray-Wyner encoder represents the library using N private descriptions and $\binom{N}{\ell}$ descriptions that are common to ℓ files for $\ell \in \{2, \dots, N\}$, thereby transforming the caching problem with correlated content and receivers requesting only one file, into a caching problem

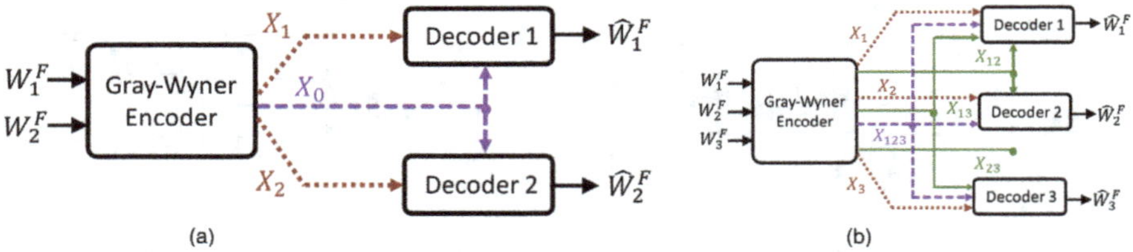

Figure 6.2 Gray-Wyner network for (a) two files and (b) three files [57]

with a larger number of files where receivers require multiple descriptions. Note that this multiple-request caching problem has a specific class of demands. In the second step, we assume that the multiple-request CACM scheme, described in detail in Section 6.4.3, is agnostic to the correlation among the content generated by the Gray-Wyner encoder, i.e., the second step is correlation-unaware. The two steps are jointly designed to optimize the performance of the overall scheme, referred to as the *Gray-Wyner Multiple-Request CACM* (GW-MR) scheme.

The rest of this section is organized as follows. Section 6.4.1 provides a brief review the Gray-Wyner network. Section 6.4.2 formally describes the two-step GW-MR scheme. The multiple-request caching problem arising after the Gray-Wyner step is discussed in Section 6.4.3 and an achievable MR scheme for the two-file network is described and analyzed in detail in Section 6.4.4.

6.4.1 Gray-Wyner network

The Gray-Wyner network was first introduced for two files in [15], in which a 2-DMS (W_1, W_2) is represented by three descriptions $\{X_0, X_1, X_2\}$, where $X_0 \in [1 : 2^{F\rho_0})$ is referred to as the common description, and $X_1 \in [1 : 2^{F\rho_1})$ and $X_2 \in [1 : 2^{F\rho_2})$ are the corresponding private descriptions as depicted in Figure 6.2(a). The descriptions are such that file W_1^F can be losslessly recovered from descriptions (X_0, X_1), and file W_2^F can be losslessly recovered from descriptions (X_0, X_2), both asymptotically, as block length $F \to \infty$. In [50], Gray and Wyner fully characterized the rate region for lossless reconstruction of both files, which is restated in the following theorem.

Theorem 1 [Gray-Wyner Rate Region]. The optimal rate region for the two-file Gray-Wyner network, $\mathfrak{S}_{GW}$, is

$$\mathfrak{S}_{GW} = cl\left\{\bigcup \left\{(\rho_0, \rho_1, \rho_2) : \rho_0 \geq I(W_1, W_2; \widetilde{\mathrm{U}}),\ \rho_1 \geq H(W_1|\widetilde{\mathrm{U}}),\ \rho_2 \geq H(W_2|\widetilde{\mathrm{U}})\right\}\right\} \tag{6.11}$$

where $cl\{S\}$ denotes the closure of set S, and the union is over all choices of $\widetilde{\mathrm{U}}$ for some $p(\tilde{u}|w_1, w_2)$ with $|\widetilde{\mathcal{U}}| \leq |\mathcal{W}_1||\mathcal{W}_2| + 2$.

The Gray-Wyner network can be extended to N files such that the Gray-Wyner encoder observes a N-DMS $(W_1, \ldots, W_N)$ and communicates W_i to decoder $i \in \{1, \ldots N\}$. The encoder is connected to the decoders through 2^{N-1} error-free links, such that there is a link connecting the encoder to any subset $s \subseteq \{1, \ldots, N\}$ of the decoders. In particular, there is one common link connecting the encoder to all N decoders, there are $\binom{N}{\ell}$ links common to any $\ell \in \{2, \ldots, N-1\}$ of the decoders, and finally, there are N private links connecting the encoder to each decoder. For any nonempty set $\mathscr{A} \subseteq \{1, \ldots, N\}$, description $W_{\mathscr{A}} \in [1 : 2^{F\rho_{\mathscr{A}}})$ is communicated to all decoders $i \in \mathscr{A}$. The Gray-Wyner rate region, $\mathfrak{S}_{GW}$, is represented by the set of all rate-tuples $\boldsymbol{\rho}$ for the $2^N - 1$ descriptions, for which file W_i^F, $i \in \{1, \ldots, N\}$ can be losslessly reconstructed from the descriptions $\left\{X_s : s \subseteq \{1, \ldots, N\},\ i \in s\right\}$, asymptotically, as $F \to \infty$. In general, N files are encoded into $2^N - 1$ descriptions, such that: (i) N of the descriptions contain information exclusive to only one file and (ii) the remaining descriptions comprise information common to more than one file. While the rate region of the N-file Gray-Wyner network has been studied in a number of papers [63–65], the optimal characterization for general sources is not known.

6.4.2 Gray-Wyner multiple-request CACM

The class of two-step schemes referred to as Gray-Wyner Multiple-Request CACM (GW-MR) consists of:

- **Gray-Wyner Encoder:** Given the library $\{W_1^F, \ldots, W_N^F\}$, the Gray-Wyner encoder at the sender computes descriptions $\{X_s : s \in \mathscr{S}\}$, where $\mathscr{S}$ is the set of all nonempty subsets of $\{1, \ldots, N\}$, using a mapping

$$f^{GW} : \mathscr{W}_1^F \times \cdots \times \mathscr{W}_N^F \to \prod_{s \in \mathscr{S}} \left[1 : 2^{F\rho_{\mathscr{S}}}\right), \tag{6.12}$$

  for $\boldsymbol{\varrho} \in \mathfrak{S}_{GW}$.
- **Multiple-Request Multicast Encoder:** For any demand realization $\mathbf{d} \in \mathscr{D}$ revealed to the sender, the correlation-unaware multicast encoder generates and transmits the multicast codeword $Y_{\mathbf{d}} = f^{\mathfrak{M}_{MR}}\left(\mathbf{d}, \{X_s : s \in \mathscr{S}\}, \{Z_k\}_{k=1}^K\right)$.
- **Multiple-Request Multicast Decoder:** Receiver $k \in \{1, \ldots, K\}$ decodes the descriptions corresponding to its requested file as

$$\left\{\hat{X}_s : s \in \mathscr{S}_{d_k}\right\} = g_k^{\mathfrak{m}_{MR}}\left(\mathrm{d}, \mathrm{Y}_{\mathrm{d}}, \mathrm{Z}_{\mathrm{k}}\right), \tag{6.13}$$

  where $\mathscr{S}_{d_k} \triangleq \left\{s \in \mathscr{S} : d_k \in s\right\}$
- **Gray-Wyner Decoder:** Receiver $k \in \{1, \ldots, K\}$ decodes its requested file using the descriptions recovered by the multicast decoder, as $\widehat{W}_{d_k}^F = g_k^{GW}\left(\left\{\widehat{X}_s : s \in \mathscr{S}_{d_k}\right\}\right)$, via the Gray-Wyner decoder

$$g_k^{GW} : \prod_{s \in \mathscr{S}_{d_k}} [1 : 2^{F\rho_s}) \to \mathscr{W}_{d_k}^F. \tag{6.14}$$

- **Multiple-Request Cache Encoder:** Given the compressed descriptions, the correlation-unaware cache encoder at the sender computes the cache content at receiver $k \in \{1, \ldots, K\}$, as $Z_k = f_k^{\mathfrak{C}_{MR}}\left(\{X_s : s \in \mathscr{S}\}\right)$.

The Gray-Wyner encoder and decoder correspond to the first step, namely the encoder and decoder of the Gray-Wyner network, and the multiple-request cache encoder, multiple-request multicast encoder, and multiple-request multicast decoder comprise the second multiple-request CACM, referred to as MR, step of the two-step scheme.

Note that the performance of the class of schemes described above depends on the operating point of the Gray-Wyner network $\boldsymbol{\rho} \in \mathfrak{S}_{GW}$. For a given $\boldsymbol{\rho}$, the performance of the overall two-step scheme is dictated by the peak and average multicast rates of the MR scheme, which is similar to (6.2) and (6.3), are defined as

$$R_{MR}^{(F)}(\boldsymbol{\rho}) = \max_{\mathbf{d}\in\mathscr{D}} \frac{\mathbb{E}[L(Y_{\mathbf{d}})]}{F}, \tag{6.15}$$

$$\bar{R}_{MR}^{(F)}(\boldsymbol{\rho}) = \frac{\mathbb{E}[L(Y_{\mathbf{d}})]}{F}, \tag{6.16}$$

respectively. Furthermore, the worst-case probability of error of the class of two-step schemes depends on the probability of error of the Gray-Wyner source coding step and the probability of error of the multiple-request CACM step. Since $\boldsymbol{\rho} \in \mathfrak{S}_{GW}$, and $\left\{X_s : s \in \mathscr{S}_{d_{r_k}}\right\}$ is a Gray-Wyner description of $W_{d_k}^F$ with $d_k \in \{1, \ldots, N\}$, it is guaranteed that Gray-Wyner decoding is asymptotically lossless with F. Hence, the probability of error of a two-step scheme is approximately upper bounded by the probability of error of the MR scheme, given by

$$P_{e,MR}^{(F)} = \max_{\mathbf{d}\in\mathscr{D}} \mathbb{P}\left(\bigcup_{k\in\{1,\ldots,K\}} \left\{\widehat{X}_s \neq X_s, \forall s \in \mathscr{S}_{d_k}\right\}\right). \tag{6.17}$$

6.4.2.1 Rate-memory functions for the GW-MR scheme

Definition 5. For a given rate-tuple $\boldsymbol{\rho} \in \mathfrak{S}_{GW}$, an MR peak rate-memory pair (R, M) is *achievable* if there exists a sequence of MR schemes with rate $R_{MR}^{(F)}(\varrho)$, for cache capacity M and increasing file size F such that $\lim_{F\to\infty} P_{e,MR}^{(F)} = 0$, and $\limsup_{F\to\infty} R_{MR}^{(F)}(\boldsymbol{\rho}) \leq R$.

Definition 6. For a given rate-tuple $\boldsymbol{\rho} \in \mathfrak{S}_{GW}$, the MR peak rate-memory region $\mathfrak{R}_{MR}^*(\boldsymbol{\rho})$ is the closure of the set of achievable MR peak rate-memory pairs (R, M), and the MR peak rate-memory function $R_{MR}^*(M, \boldsymbol{\rho})$ is defined as

$$R_{MR}^*(M, \boldsymbol{\rho}) = \inf\{R : (R, M) \in \mathfrak{R}_{MR}^*(\boldsymbol{\rho})\}. \tag{6.18}$$

In the class of two-step schemes, we refer to the scheme operating at the rate-tuple $\boldsymbol{\rho} \in \mathfrak{S}_{GW}$ that minimizes the MR peak rate-memory function as the *GW-MR scheme*. The peak rate-memory pair achieved by this scheme is the GW-MR peak rate-memory function defined below.

Definition 7. The GW-MR peak rate-memory function $R_{GW\text{-}MR}^*(M)$ is given by

$$\bar{R}_{GW\text{-}MR}^*(M) = \inf\{\bar{R}_{MR}^*(M, \boldsymbol{\rho}) : \boldsymbol{\rho} \in \mathfrak{S}_{GW}\}. \tag{6.19}$$

In line with Definitions 5-7, for the average rate criterion we have the following.

Definition 8. For a given rate-tuple $\boldsymbol{\rho} \in \mathfrak{S}_{GW}$, an MR average rate-memory pair (R, M) is *achievable* if there exists a sequence of MR schemes with rate $\bar{R}^{(F)}_{MR}(\boldsymbol{\rho})$, for cache capacity M and increasing file size F, such that $\lim_{F\to\infty} P^{(F)}_{e,MR} = 0$, and $\limsup_{F\to\infty} \bar{R}^{(F)}_{MR}(\boldsymbol{\rho}) \le R$.

Definition 9. For a given rate-tuple $\boldsymbol{\rho} \in \mathfrak{S}_{GW}$, the MR average rate-memory region $\bar{\mathfrak{R}}^*_{MR}(\boldsymbol{\rho})$ is the closure of the set of achievable MR average rate-memory pairs (R, M), and the MR average rate-memory function $\bar{R}^*_{MR}(M, \boldsymbol{\rho})$ is defined as

$$\bar{R}^*_{MR}(M, \boldsymbol{\rho}) = \inf\{R : (R, M) \in \bar{\mathfrak{R}}^*_{MR}(\boldsymbol{\rho})\}. \tag{6.20}$$

Definition 10. The GW-MR average rate-memory function $\bar{R}^*_{GW\text{-}MR}(M)$ is given by

$$\bar{R}^*_{GW\text{-}MR}(M) = \inf\{\bar{R}^*_{MR}(M, \boldsymbol{\rho}) : \boldsymbol{\rho} \in \mathfrak{S}_{GW}\}. \tag{6.21}$$

In the remainder of this chapter, we first present the MR scheme in the second step for a general setting in Section 6.4.3, and then describe and analyze the performance of the overall GW-MR scheme for the case of two files and K receivers in Section 6.4.4.

6.4.3 Multiple-request CACM

In this section, we focus on the second step of the GW-MR scheme depicted in Figure 6.1, namely the MR scheme when the Gray-Wyner network operates at $\boldsymbol{\rho} \in \mathfrak{S}_{GW}$. As mentioned previously, in the first step, the **Gray-Wyner Encoder** converts the N-file library into $2^N - 1$ descriptions, such that multiple descriptions are required to losslessly recover a file from the original library. In the second step, the **Multiple-Request CACM** scheme arranges these descriptions into N groups, $L_1, \ldots, L_N$, referred to as *sublibraries*, and delivers the subset of descriptions corresponding to the requested files. For a given choice of the MR scheme, the overall GW-MR scheme operates at the point $\boldsymbol{\rho} \in \mathfrak{S}_{GW}$ that minimizes the overall network rate.

Sublibrary $L_\ell = \left\{X_s : s \subseteq \{1, \ldots, N\}, |s| = \ell\right\}$ contains the descriptions that are common to exactly ℓ files. We refer to sublibrary $L_N = \{X_{12\ldots N}\}$ as the *common-to-all* (or simply common) sublibrary, and to $L_1 = \{X_1, \ldots, X_N\}$, which contains all the private descriptions, as the *private* sublibrary. The MR scheme accounts for populating the receiver caches with content from sublibraries $L_1, \ldots, L_N$ and serving the demand realizations placed in the original library, which translate into a new class of demands from the compressed sublibraries. More specifically, each receiver demand corresponds to a demand for a set of descriptions from each of the sublibraries (hence the term multiple-request), such that the original library file W_{d_k} requested by receiver k maps to descriptions $\left\{X_s \in L_\ell : d_k \in s\right\}$ from sublibrary L_ℓ, $\ell \in \{1, \ldots, N\}$. Even though receivers request single files from the original library independently and according to a uniform distribution, the Gray-Wyner encoding process leads to a nonuniform multiple-request demand for descriptions (files) from the compressed sublibraries. As described in Section 6.4.2, the MR scheme consists of three components:

- **MR Cache Encoder**, which populates the receiver caches following a specific and well-designed cache allocation policy.
- **MR Multicast Encoder**, which transmits a multicast codeword that is determined from the descriptions in the sublibraries according to a coded multicast scheme that accounts for the demand structure dictated by the configuration of the compressed library.
- **MR Multicast Decoder**, which is located at each receiver and uses the received multicast codeword and the receiver's cached content to recover the requested file.

One possible choice in designing an achievable MR scheme is to treat each sublibrary independently during both caching and delivery phases. Specifically, the descriptions from each sublibrary are cached and delivered as follows:

i. the description in the common-to-all sublibrary L_N, which is required for the reconstruction of all files, and hence requested by all receivers, is cached according to the least frequently used (LFU)[b] strategy and delivered through uncoded (naive) multicasting[c],
ii. sublibrary L_1 is cached and delivered according to any single-request correlation-unaware CACM scheme (such as [66–68] for unequal-length descriptions, and [4–6, 8, 13, 16, 20] for equal-length descriptions),
iii. sublibrary L_ℓ, $\ell \in \{N-1, \ldots, 2\}$ is cached and delivered according to any multiple-request correlation-unaware CACM scheme in which each receiver requests exactly $\binom{N-1}{\ell-1}$ descriptions.

The single-request and multiple-request correlation-unaware CACM schemes for each sublibrary are selected with respect to a given performance or complexity criteria. In the MR scheme, the fraction of each receiver's cache capacity that is allocated to prefetching content from each sublibrary is optimally determined using a cache allocation optimization problem described in the next section.

Multiple-request correlation-unaware CACM schemes where each receiver requests more than one file have been analyzed in previous works, e.g., [2, 51–54]. However, in addition to being limited to settings with equal-length files, they have been designed for arbitrary demand combinations and could

[b]The term LFU in the context of cache placement (as opposed to replacement) policies is used to describe a local policy that caches the most popular files.
[c]Uncoded multicasting refers to the transmission of a common not-network-coded message simultaneously received and decoded by multiple receivers.

therefore be suboptimal for the specific class of demands considered in our MR scheme.

An achievable multiple-request scheme, where the demand combinations reflect the specific class of demands of the MR scheme is proposed in [48]. This work considers a setting with *equal-length* files composed of independent subfiles that are common to a fixed set of the library files, which could model the descriptions generated by the Gray-Wyner network in the proposed two-step GW-MR scheme. One could adopt the scheme proposed in [48], which is designed for equal length files, for the MR scheme, by operating the Gray-Wyner network at a point $\boldsymbol{\rho} \in \mathfrak{S}_{GW}$ where all the $\binom{N-1}{\ell-1}$ descriptions in sublibrary $\ell \in \{1, \ldots N-1\}$ have equal length. Equivalently, the system operates at point $\boldsymbol{\rho} \in \mathfrak{S}_{GW}$, where for all $s_1, s_2 \subseteq \{1, \ldots, N\}$, $\rho_{s_1} = \rho_{s_2} = \rho_\ell$ if $|s_1| = |s_2| = \ell$[d]. However, as shown in [55–57], adopting this scheme for the MR CACM scheme can result in a suboptimal GW-MR scheme.

An approach for designing a MR scheme which is proved to lead to a near optimal GW-MR scheme, is proposed in [55–57], which also presents a new framework for deriving tight lower bounds on the optimal rate-memory trade-offs for the setting with correlated files. In [55–57], the authors resort to a MR scheme that properly combines near optimal correlation-unaware schemes available in the literature, generalized to unequal file lengths, as described in the next section to deliver the $\binom{N-1}{\ell-1}$ descriptions requested by each receiver from sublibrary L_ℓ, $\ell \in \{1, \ldots, N-1\}$. The resulting scheme is proved to be optimal for several regions of the cache capacity in the special case of a two-file multiple-receiver network, and in a three-file two-receiver network, and to be near-optimal in all the other regions of the memory, where by near-optimality we refer to schemes that achieve rates within a constant distance (or additive factor) from the optimal solution. The proposed scheme achieves optimality in some regions of the memory by using coded cache placement. In [55–57], the optimality of the proposed two-step GW-MR scheme was evaluated by comparing its achievable rate with a lower bound on the optimal peak and average rate-memory trade-offs.

6.4.3.1 Cache allocation policy

Let $\boldsymbol{\rho} = (\boldsymbol{\rho}_{L_1}, \ldots, \boldsymbol{\rho}_{L_N}) \in \mathfrak{S}_{GW}$ denote the point at which the Gray-Wyner network is operating, where $\boldsymbol{\rho}_{L_\ell} = (\rho_s, s \subseteq \{1, \ldots, N\}, |s| = \ell)$ denotes the vector of the description rates (lengths) in sublibrary ℓ, and ρ_s denotes the rate of description X_s. As described in Section 6.4.3, the MR scheme arranges the descriptions generated by the Gray-Wyner network into N sublibraries, such that each receiver demand from the original library corresponds to requesting 2^{N-1} descriptions: one from the common sublibrary $L_N = \{X_{12\ldots N}\}$, one from the private

[d]While the Gray-Wyner encoder can operate at any point in the region that satisfies the equal-length description constraint, without loss of generality, it can be assumed that the MR scheme operates at the boundary of this region.

sublibrary $L_1 = \{X_1, X_2, \ldots, X_N\}$, and $\binom{N-1}{\ell-1}$ descriptions from sublibrary L_ℓ, with $\ell = 2, \ldots, N-1$.

For a given operating point $\boldsymbol{\rho} \in \mathfrak{S}_{GW}$, the MR scheme described in Section 6.4.3 operates as follows: it (i) adopts a single-request correlation-unaware CACM scheme for sublibrary L_1, (ii) adopts a multiple-request correlation-unaware CACM scheme for each sublibrary L_ℓ, $\ell = \{2, \ldots, N-1\}$, (iii) divides the cache capacity M optimally among the N sublibraries, (iv) fills receiver caches independently from each sublibrary according to the adopted scheme, and (v) delivers the requested content from each sublibrary independently, i.e., with no further coding across them. For each sublibrary, the available memory is allocated to each description according to the adopted scheme, and the cache allocation policy of the MR scheme, i.e, the optimal memory allocation among the sublibraries, is determined from the following optimization problem.

Let $R_{ach}(M_l, \boldsymbol{\rho}_{Ll})$ denote the (peak or average) rate achieved by the correlation-unaware CACM scheme adopted for sublibrary when memory has been assigned to sublibrary L_l. The MR scheme determines the optimal cache allocation policy by minimizing the sum rate as

$$\begin{aligned}
&\min \sum_{l=1}^{N} R_{ach}(M_l, \boldsymbol{\rho}_{Ll}) \\
&S.t. \ \sum_{l=1}^{N} M_l = M \\
&0 \leq M_l \leq k_l
\end{aligned} \tag{6.22}$$

with $k_l = \sum s \in L_l \rho_s$, which leads to a memory allocation policy as $M_l^* = \min\left\{B_l^*, K_l\right\}$, where $\mathrm{B}_l^* = \min\left\{M : \left|\frac{\partial -}{\partial M} R_{ach}(M, \boldsymbol{\rho}_{Ll})\right| \leq \lambda\right\}$ for $\lambda > 0$ such that $\sum_{l=1}^{N} M_l^* = M$, where $\frac{\partial -}{\partial M}$ denotes the left partial derivative with respect to M.

We illustrate the optimization algorithm in the next section by focusing on the two-file-receiver setting, where the optimal solution admits a graphical "two leaky-buckets water-filling" interpretation. We refer the reader to [55–57] for more details.

6.4.4 *GW-MR for two files and* K *receivers*

This section describes more in detail the MR scheme introduced in Section 6.4.3 as well as the resulting GW-MR scheme for a network with two files and K receivers.

Assume that the Gray-Wyner encoder operates at a point $\boldsymbol{\rho} = (\rho_0, \boldsymbol{\rho}_{L_1}) \in \mathfrak{S}_{GW}$ with $\boldsymbol{\rho}_{L_1} = (\rho_1, \rho_2)$, where ρ_0 denotes the rate of the common description X_{12}, and ρ_1 and ρ_2 denote the rate of the private descriptions X_1 and X_2, respectively. As described in Section 6.4.3, the MR scheme arranges the three descriptions generated by the Gray-Wyner network into a common sublibrary $L_2 = \{X_{12}\}$, and a private sublibrary $L_1 = \{X_1, X_2\}$. Each receiver demand corresponds to requesting two descriptions: one from the common sublibrary L_2, and one from the private sublibrary L_1.

While the common description X_{12} in L_2 is cached according to the LFU strategy and delivered through uncoded multicasting, the private descriptions in L_1 can be cached and delivered according to any correlation-unaware CACM scheme available in literature (e.g., [4–6, 8, 13, 16, 20]), properly generalized to a setting with unequal-length files. In the following, we first describe how the MR scheme allocates the memory to each sublibrary according to the optimal cache allocation policy, we then characterize the peak and average rates achieved by the corresponding MR scheme, and finally, we present the GW-MR scheme and discuss its optimality.

6.4.4.1 MR Scheme for two files and K receivers

Let $R_{ach}(M, \boldsymbol{\rho}_{L_1})$ denote the lower convex envelope of the (peak or average) rate achieved by the scheme adopted for sublibrary L_1, where $\boldsymbol{\rho}_{L_1} = (\rho_1, \rho_2)$. Under the peak rate criterion, the cache allocation that minimizes the overall delivery rate (i.e., the sum delivery rate for each sublibrary) is as follows. Let

$$B^* \triangleq \min\left\{M : \left|\frac{\partial}{\partial M}R_{ach}(M, \boldsymbol{\rho}_{L_1})\right| < 1\right\} \tag{6.23}$$

denote the cache encoder threshold. For a given $\boldsymbol{\rho} \in \mathfrak{S}_{GW}$, the cache encoder allocates the memory to each sublibrary as follows:

- If $M \in \left[0, B^*\right)$, the common description X_{12} is not cached at either receiver, and descriptions $\{X_1, X_2\}$ from L_1 are cached according to the caching strategy of the adopted correlation-unaware CACM scheme.
- If $M \in \left[B^*, B^* + \rho_0\right)$, the first $F(M - B^*)$ bits of description X_{12} are cached at both receivers (as per LFU caching), and descriptions $\{X_1, X_2\}$ are cached according to the scheme adopted for sublibrary L_1 over the remaining memory B^*.
- If $M \in \left[B^* + \rho_0, \rho_0 + \rho_1 + \rho_2\right]$, the common description X_{12} is fully cached at both receivers, and descriptions $\{X_1, X_2\}$ are cached according to the adopted correlation-unaware CACM scheme over the remaining cache capacity $M - \rho_0$.

The cache allocation described above is proven to be optimal in [57], and its graphical representation is shown in Figure 6.3. This representation can be understood as "water-filling of two leaky buckets," which is related to the well-known water-filling optimization problem [69]. For small cache capacities up to B^*, all the memory is allocated to the private sublibrary. As the cache size increases, it is optimal to first store the common description until it is fully cached, and for cache capacities larger than B^*, the residual memory is allocated to storing the private descriptions.

Remark 1. Differently from the single-cache setting analyzed in [47], where it is always optimal to cache the common description, in this case, when the cache capacity is smaller than B^* it is optimal to first cache the private descriptions. This is due to the fact that for small cache sizes delivering the private descriptions with coded multicast transmissions is more effective in reducing the overall rate compared to mutlicasting the common description, and therefore it is preferable to prioritize caching from the private

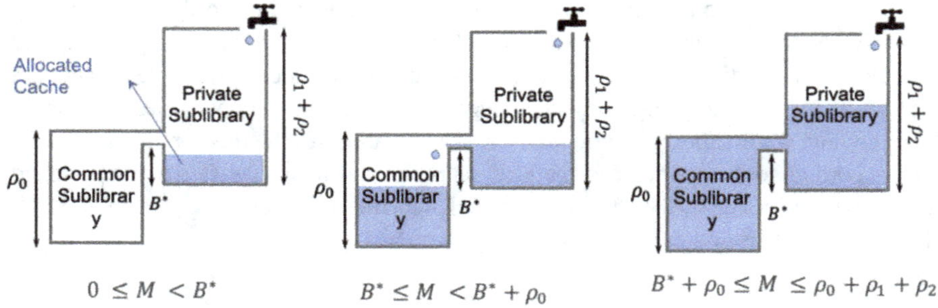

Figure 6.3 Optimal caching strategy in the MR scheme for two files with B^* defined in (6.6)

sublibrary. As the cache size increases, this difference in rate reduction diminishes to the extent that assigning memory to the common sublibrary is more effective in reducing the rate compared to the private sublibrary. Therefore, it is preferable to fully store the common description prior to allocating additional memory for storing the private descriptions.

The (peak and average) rate achieved by the MR scheme for the optimal cache allocation policy described above is given by

$$R_{MR}(M, \boldsymbol{\rho}) = \begin{cases} \rho_0 + R_{ach}(M, \boldsymbol{\rho}_{L_1}), & M \in \left[0, B^*\right) \\ \rho_0 + R_{ach}(B^*, \boldsymbol{\rho}_{L_1}) + B^* - M, & \\ & M \in \left[B^*, \rho_0 + B^*\right) \\ R_{ach}(M - \rho_0, \boldsymbol{\rho}_{L_1}), & \\ & M \in \left[\rho_0 + B^*, \rho_0 + \rho_1 + \rho_2\right] \end{cases} \tag{6.24}$$

where B^* is defined in (6.6) and $\boldsymbol{\rho}_{L_1} = (\rho_1, \rho_2)$.

6.4.4.2 Optimality of GW-MR

Equation (6.7) provides the (peak and average) rate achieved by the MR scheme for a given single-request correlation-unaware CACM scheme adopted for library L_1, which achieves a (peak and average) rate equal to $R_{ach}(M, \boldsymbol{\rho}_{L_1})$. The corresponding GW-MR scheme operates at the point $\boldsymbol{\rho}$ in the Gray-Wyner region that minimizes the rate achieved by the MR scheme and achieves the (peak and average) rate given by

$$R_{GW-MR}(M) = \inf\{R_{MR}(M, \boldsymbol{\rho}) : \boldsymbol{\rho} \in \mathfrak{S}_{GW}\}. \tag{6.25}$$

The performance of the two-step GW-MR depends on both the correlation-unaware CACM scheme adopted for the private sublibrary and the operating point $\boldsymbol{\rho} \in \mathfrak{S}_{GW}$, as seen from (6.7) to (6.8). In [55,57], the authors analyze both peak and average rates of the GW-MR scheme using the result in (6.7), and (i) adopt a scheme for the private sublibrary

L_1 that is based on memory sharing[e] among generalizations of the correlation-unaware CACM schemes proposed in [8] and [13] to files with unequal lengths and (ii) determine the operating point of the Gray-Wyner network to be $\boldsymbol{\rho}^* = \arg\max_{\boldsymbol{\rho}\in\mathfrak{S}_{GW}} \min\{\rho_1, \rho_2\}$, with the constraint that $\rho_0 + \rho_1 + \rho_2 = H(X_1, X_2)$. To satisfy this condition, it is sufficient to choose the rate-tuple $\boldsymbol{\rho} \in \mathfrak{S}_{GW}$ such that

$$\rho_0 = I(X_1, X_2; \widetilde{\mathsf{U}}), \quad \rho_1 = H(X_1|\widetilde{\mathsf{U}}) \quad \rho_2 = H(X_2|\widetilde{\mathsf{U}}) \tag{6.26}$$

as in (6.4.1), for the $\widetilde{\mathsf{U}}$ of the form $X_1 - \widetilde{\mathsf{U}} - X_2$ that achieves M_K given by

$$M_K \triangleq \max_{X_1 - \widetilde{\mathsf{U}} - X_2} \frac{1}{K} \min\left\{H(X_1|\widetilde{\mathsf{U}}), H(X_2|\widetilde{\mathsf{U}})\right\}. \tag{6.27}$$

Note that the operating point $\boldsymbol{\rho}^* \in \mathfrak{S}_{GW}$ is related to Wyner's common information. In fact, in Wyner's common information the goal is to minimize ρ_0 subject to $\rho_0 + \rho_1 + \rho_2 = H(X_1, X_2)$, while in our case the goal is to maximize $\min\{\rho_1, \rho_2\}$ subject to $\rho_0 + \rho_1 + \rho_2 = H(X_1, X_2)$. By comparing the achievable rate of the two-step GW-CACM scheme with a lower bound on the optimal peak and average rate-memory trade-offs [55, 57], show that in a network with two files and K receivers, the achievable rate matches the lower bound for a significant memory regime and it is within $\frac{K-2}{2K}$ of the conditional entropy of files for all other memory values. An interesting future direction would be to investigate *joint* compression and caching strategies or to design a different two-step approach for this region of the memory.

In [57], the authors show that for the special case of a 2-DMS with $X_1 = (X_1', \widetilde{\mathsf{V}})$ and $X_2 = (X_2', \widetilde{\mathsf{V}})$ such that X_1' and X_2' are conditionally independent given $\widetilde{\mathsf{V}}$, the optimal operating point is $\boldsymbol{\rho}^* = \left(I(X_1; X_2), H(X_1|X_2), H(X_2|X_1)\right)$. In this setting, the set of descriptions generated by the Gray-Wyner network operating at $\boldsymbol{\rho}^*$, is equivalent to the library considered in [48], where each file is composed of two independent subfiles, one of which is common among the two files. Hence, it follows that the GW-MR scheme provides an optimal solution to the two-file two-receiver setting considered in [48].

For the special case of three files and multiple receivers [56, 57], show that the two-step strategy achieves the lower bound for large cache capacities, and it is within half of the joint entropy of two of the sources conditioned on the third one for all other cache sizes.

6.5 Caching networks with correlated sources – the dynamic case

In this section, we consider the network setup presented in Section 6.3 for dynamic sources. Even though jointly compressing the library before the caching phase, as in [45–49, 55, 56], is a natural solution to reducing the load on the shared link, this approach is

[e] Specifically, for small cache sizes, the adopted scheme uses the scheme in [9] (first introduced in [57] for a specific cache capacity), which prefetches coded content during the caching phase, while for large cache sizes, it uses the scheme in [65] with uncoded cache placement.

mainly suitable for a library with static content, losing robustness and leading to significant performance degradation when content is highly dynamic. For example, if the library is updated and a new file is added, all files need to be jointly re-compressed and the content cached across the entire network needs to be updated based on the new compressed versions. On the other hand, schemes that are not based on a-priori library compression may only require updating a portion of the content cached across the network, which is especially critical for the delivery of dynamic data in next-generation services. Motivated by the increasing dynamic and personalized nature of next-generation content services, in this section, we consider the problem of efficient caching and delivery of correlated content in a general setting with multiple receivers and multiple files and for an arbitrary joint file distribution. The goal is to design a practical scheme that is robust to the variations inherent to the delivery of real-time services over wireless networks. To this end, Refs. [44, 58] propose a class of correlation-aware CACM schemes that stores individually (rather than jointly) compressed content pieces during the caching phase, which then serve as references for compression during the delivery phase, resulting in an efficient scheme that is robust to dynamic changes in the content library. Compared with previously proposed schemes [45–49, 55, 56], where correlation-aware compression takes place before the caching phase, as described in Section 6.3, the class of schemes proposed in [44, 58] provides on-demand compression during the delivery phase and hence is able to adapt to system variations while continuing to exploit available content correlations. This section describes this class of schemes and evaluates its achievable gains numerically. We adopt the network setting presented in Section 6.3.1, and without loss of generality, we assume that $H(W_n) = H(W)$ for $n \in \{1, \ldots, N\}$.

6.5.1 Dynamic correlation-aware scheme

The class of correlation-aware CACM schemes proposed in [44, 58] distributes the most relevant library content among the receiver caches such that during the delivery phase, the correlation among the aggregate cache and the demand can be exploited to send multicast codewords of compressed files that further improve the global caching gain. In the following, we refer to the overall scheme as DCA-CACM.

6.5.1.1 Dynamic correlation-aware CACM

A DCA-CACM scheme, as depicted in Figure 6.4, is composed of the following components, which we describe in more detail in the following subsections:

i. *Correlation-Aware Cache Encoder:* As in conventional CACM schemes, a fractional cache encoder is used to divide each file into packets and to determine the subset of each file's packets to be cached at each receiver. This can be done in a *deterministic* fashion, referred to as centralized caching in the literature, as in [2, 6, 8], or, in order to be more robust to system dynamics, in a *random* fashion, referred to as decentralized caching, as in [3–5, 7]. Differently from the random cache placement, where only the portion of packets to be cached from each file is stated, deterministic

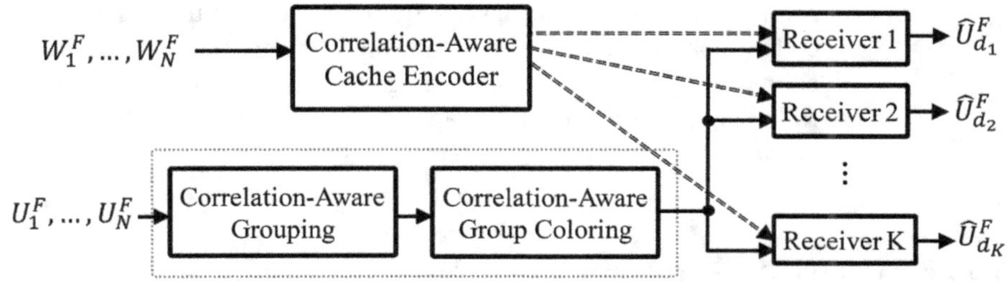

Figure 6.4 The correlation-aware DCA-CACM scheme

cache placement also specifies their identity and, therefore, results in better performance compared to random caching. For a detailed description deterministic caching, we refer the reader to [7, 12, 20, 31, 70]. In the next section, we briefly describe the random version of the cache encoder. In addition to providing increased robustness to system dynamics, this strategy also allows us to analytically derive an upper bound on the rate achieved by the DCA-CACM scheme, the details of which can be found in [58].

ii. *Correlation-Aware Multicast Encoder:* As in conventional CACM schemes [2–8], the multicast encoder generates a multicast codeword constructed as a linear combination of file packets. While in the conventional CACM schemes, the multicast codeword consists of the requested packets that are linearly combined according to the coloring of an index coding conflict graph (for more details see [58, 59]), the multicast codeword computed by the correlation-aware multicast encoder, is a linear combination of only a subset of the requested packets and includes other packets that are correlated with the requested ones as well as additional refinement packets. The packets included in the multicast codeword by the multicast encoder are determined following two steps:

- *Correlation-Aware Grouping*: Given a demand realization in the delivery phase, the multicast encoder identifies all the packets among the collective cache and demand that are correlated with each requested packet that is not locally available. Any of the identified correlated packets can be used in the multicast codeword in place of the requested packet, subject to transmitting additional refinement codewords that ensure lossless reconstruction of the requested packet.
- *Correlation-Aware Group Coloring*: The multicast encoder generates a multicast codeword constructed as a linear combination of a subset of the requested packets and a subset of their correlated packets. This is done by computing a *group*

coloring of a graph that results from augmenting the index coding conflict graph [9, 10] with the correlated ensembles computed by the correlation-aware grouping procedure. In the following, we refer to this graph as the *augmented conflict graph.*

6.5.1.2 Correlation-aware cache encoder

In the random version of the cache encoder, which we refer to as the random fractional caching strategy, the packets to be cached are selected according to a *caching distribution.* The caching distribution is optimized in order to minimize the achievable delivery rate with respect to the library distribution, as a function of not only file popularity (as in a correlation-unaware CACM) but also their cross-correlation.

Each file W_n^F is partitioned into B equal-size packets, where packet $b \in \{1, \dots, B\}$ of file $n \in \{1, \dots, N\}$ is denoted by $W_n^{(b)}$. The content to be cached at each receiver is determined according to a *caching distribution,* $\mathbf{p} = (p_1, \dots, p_N)$[f] with $0 \leq p_n \leq 1/M$, $\forall n \in \{1, \dots, N\}$ and $\sum_{n=1}^{N} p_n = 1$, which is optimized to minimize the rate of the corresponding index coding delivery scheme. For a given caching distribution $\mathbf{p}$, each receiver randomly selects and caches a subset of $p_n MB$ distinct packets from file $n \in \{1, \dots, N\}$. We denote by $\mathbf{C} = \{\mathbf{C}_1, \dots, \mathbf{C}_K\}$ the packet-level cache configuration, where $\mathbf{C}_k$ denotes the set of packets $W_n^{(b)}$, $n \in \{1, \dots, N\}$, $b \in \{1, \dots, B\}$, cached at receiver k. While the caching distribution of a correlation-unaware scheme prioritizes the caching of packets according to the aggregate popularity distribution (see [4]), the correlation-aware caching distribution accounts for both the aggregate popularity and the correlation among the library files when determining the number of packets to be cached from each file.

6.5.1.3 Correlation-aware multicast encoder

The correlation-aware coded multicast encoder capitalizes on the additional coded multicast opportunities that arise from incorporating cached packets that are not only equal to but also correlated with the requested packets into the multicast codeword. For a given demand realization $\mathbf{d}$, the packet-level demand realization is denoted by $\mathbf{Q} = \{\mathbf{Q}_1, \dots, \mathbf{Q}_K\}$, where $\mathbf{Q}_k$ denotes the packets of file $U_{d_k}^F$ requested, but not cached, by receiver k.

Correlation-Aware Grouping. For each requested packet $U_n^{(b)} \in \mathbf{Q}$, the correlation-aware grouping procedure computes a δ-ensemble $\Omega_{U_n^{(b)}}$, where $\Omega_{U_n^{(b)}}$ is the union of $U_n^{(b)}$ and the subset of all cached and requested packets in $\mathbf{C} \cup \mathbf{Q}$ that are δ-correlated with $U_n^{(b)}$, as per the following definition.

Definition 11. **(δ-Correlated Vectors)** Consider two random vectors $\mathbf{X} = [X_1, \dots, X_L]$ and $\mathbf{Y} = [Y_1, \dots, Y_L]$ with i.i.d. entries such that $(X_i, Y_i) \sim p_{X,Y}(x, y)$, and $H(X) = H(Y)$. For a given threshold $\delta \leq 1$, we say that $\mathbf{X}$ is δ-correlated with $\mathbf{Y}$ if $H(X, Y) \leq (1 + \delta)H(X)$.

[f]Not to be confused with the library distributions $p(w)$ and $p(w, v)$.

Note that δ is a system parameter that can be optimized as a function of the system parameters K, N, M, and the library distribution, in order to minimize the rate of the overall DCA-CACM scheme as quantified in (6.3).[g] In fact, based on Definition 11, higher δ, allows for more packets to be δ-correlated, but with higher refinements.

Correlation-Aware Group Coloring. After computing the δ-ensembles for each requested packet, the multicast encoder computes the multicast codeword by following a group coloring procedure on an *augmented* conflict graph. For completeness, we first describe the conventional index coding conflict graph. A conventional index coding conflict graph, which is the complement of the side information graph [9, 10], is constructed based on the packet-level cache configuration **C** and demand realization **Q** in such a way that there is one vertex for each requested packet in **Q**, and there is an edge between any two vertices if one of the packets is not cached by the receiver requesting the other packet. Such construction ensures decodability of an index code constructed by first computing a valid coloring of the conflict graph (see Definition 12), and then XORing the packets with the same color. The augmented conflict graph is constructed by augmenting the conventional index coding conflict graph with the δ-ensembles computed by the grouping procedure, such that for each requested packet in **Q**, there are a subset of vertices corresponding to its δ-ensemble. Therefore, any vertex in the augmented graph is either a requested packet or a packet δ-correlated with it. For any pair of vertices v_1 and v_2 in the augmented graph, there is an edge between v_1 and v_2 if both v_1 and v_2 are in the same group (δ-ensemble), or if the packet associated to one of the vertices is not cached by the receiver requesting the packet associated to the other vertex.

Definition 12. (Graph Coloring) A valid coloring of a graph is an assignment of colors to the vertices of the graph such that no two adjacent vertices are assigned the same color.

Given a valid coloring of the augmented graph, a valid *group coloring* of the augmented graph consists of assigning to each δ-ensemble one of the colors assigned to the vertices inside it. For each color in the group coloring, only the packets with the same color as the color assigned to their corresponding group are XORed together and multicasted to the receivers. Using its cached information and the received XORed packets, each receiver is able to reconstruct a (possibly) distorted version of its requested packets, due to the potential reception of a packet that is δ-correlated with its requested one. The encoder transmits refinement segments, when needed, to enable lossless reconstruction of the demand at each receiver. The coded multicast codeword results from concatenating: (1) for each color in the group coloring, the XOR of the packets with the same color and (2) for each receiver, if needed, the refinement segment. The multicast encoder selects the valid group coloring corresponding to the shortest coded multicast codeword.

For a more detailed description of the augmented graph the reader can refer to [58]. In the following we illustrate how correlation is exploited during the delivery phase through a simple example in a dynamic setting.

[g]In practice, the designer shall determine the level at which to compute and exploit correlations between files based on performance-complexity trade-offs.

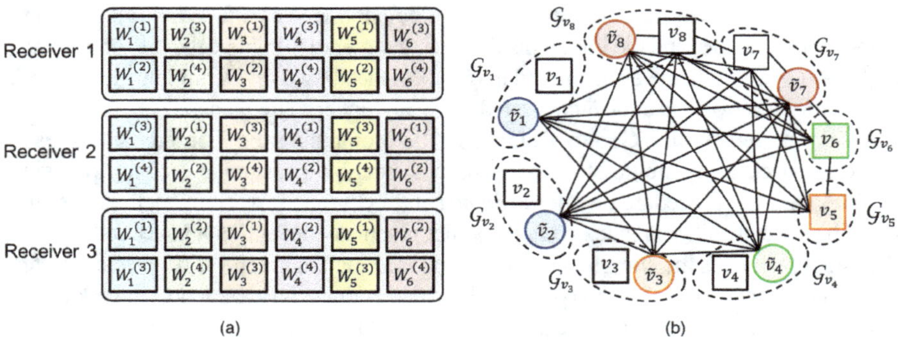

Figure 6.5 *Example 1 in Section 6.5.1.3: (a) Packet-level cache configuration. (b) Coloring of the augmented conflict graph*

Example 1. Consider a network with $K = 3$ receivers, cache capacity $M = 3$, and a dynamic library composed of $N = 6$ uniformly popular files $\{W_1^F, W_2^F, \ldots, W_6^F\}$, with $H(W) = 1$, and update probability vector $\pi = (1, 0, \ldots, 0)$. We assume that all files are independent except for the files within each pair $\{W_1^F, W_2^F\}$ and $\{W_5^F, W_6^F\}$ which are δ-correlated, such that packets $W_1^{(b)}$ and $W_2^{(b')}$ and $W_5^{(b)}$ and $W_6^{(b')}$ are δ-correlated for $b = b'$ and independent otherwise. The packet-level cache configuration is given in Figure 6.5 (a). During the delivery phase, an updated version of the first file becomes available which is correlated with W_1^F and independent of all other files.[h] For demand $\mathbf{d} = (1, 3, 5)$, the correlation-aware grouping component computes the following δ-ensembles:

$$\begin{aligned}&\Omega_{U_1^{(1)}} = \{U_1^{(1)}, W_1^{(1)}\},\ \Omega_{U_1^{(2)}} = \{U_1^{(2)}, W_1^{(2)}\},\ \Omega_{U_1^{(3)}} = \{U_1^{(3)}, W_1^{(3)}\},\\ &\Omega_{U_1^{(4)}} = \{U_1^{(4)}, W_1^{(4)}\}, \Omega_{U_5^{(2)}} = \{U_5^{(2)}, W_6^{(2)}\},\ \Omega_{U_5^{(4)}} = \{U_5^{(4)}, W_6^{(4)}\},\\ &\Omega_{U_3^{(1)}} = \{U_3^{(1)}\},\ \Omega_{U_3^{(2)}} = \{U_3^{(2)}\},\end{aligned} \tag{6.28}$$

Given the packet-level cache and demand configurations, the corresponding augmented conflict graph is shown in Figure 6.5 for which a valid group coloring based on a greedy coloring is also shown. δ-ensembles are delineated with dashed ovals, vertices corresponding to a requested packet are represented by squares, and circles have been used for those correlated with requested packets. Groups $\mathcal{G}_{v_1}, \ldots, \mathcal{G}_{v_4}$ correspond to receiver 1's demand, $\mathcal{G}_{v_5}$ and $\mathcal{G}_{v_4}$ correspond to receiver 2, and $\mathcal{G}_{v_7}$ and $\mathcal{G}_{v_8}$ correspond to receiver 3. Vertices v_1, v_2, v_3, and v_4 are connected to all other vertices, and all corresponding edges have been removed in the

[h]Hence, the updated library contains $\{U_1^F, U_2^F, \ldots, U_6^F\}$ with $U_i^F = W_i^F$ for all i = 2,...,6.

depicted graph. Each group is colored with the same color as the colored vertex shown in that group.

Based on the greedy group coloring, the groups are colored with 4 distinct colors, such that one color is assigned to nodes $\tilde{v}_1$ and $\tilde{v}_2$, one color is assigned to nodes $\tilde{v}_3$ and v_5, one color is assigned to nodes $\tilde{v}_4$ and node v_6, and finally one color is assigned to nodes $\tilde{v}_7$ and $\tilde{v}_8$. Since the packets corresponding to $\tilde{v}_1$ and $\tilde{v}_2$, and the packets of $\tilde{v}_7$ and $\tilde{v}_8$ are cached at receivers 1 and 3, respectively, they are not XORed together, and not included in the multicast codeword. Based on this correlation-aware group coloring, the sender transmits a coded segment $Y_{\mathbf{d}}^{CM} = \{W_1^{(3)} \oplus W_3^{(1)}, W_1^{(4)} \oplus W_3^{(2)}\}$ followed by a refinement segment $Y_{\mathbf{d}}^{UM}$, with rate $H(U_1|W_1) + \frac{1}{2}H(W_5|W_6) \leq \frac{3}{2}\delta$. Therefore, the overall load is at most $\frac{1}{2} + \frac{3}{2}\delta$. The correlation-unaware CACM scheme in [4] constructs the conventional index coding conflict graph, which only contains vertices $v_1, \ldots, v_8$, and multicasts $\{U_1^{(1)}, U_1^{(2)}, U_1^{(3)}, U_1^{(4)}\}$, with rate $\frac{7}{4}$.

6.5.2 Greedy group coloring

In the previous section, we described how to design a coded multicast codeword that is useful to a large number of receivers based on a group coloring of the augmented conflict graph. The multicast encoder determines the optimal multicast codeword across all valid graph colorings and their corresponding group colorings. Given that graph coloring, and by extension group coloring, is NP-Hard [58], proposes a *Greedy Group Coloring* (*GGC*), a polynomial-time approximation to the group coloring problem. GGC extends the existing GCC scheme [4] to account for file correlation in cache-aided networks. We refer the reader to [4, 58] for more details on the greedy coloring approach.

6.6 Illustration results and conclusions

In order to illustrate the significant gains that can be achieved by exploiting correlation, especially in the dynamic setting, we plot the expected achievable rate, in average number of transmissions normalized to the file size, versus memory size (cache capacity) M, normalized to the file size for a library and versus the number of users. We numerically compare the rates achieved by the state-of-the-art correlation-aware and correlation-unaware schemes. Specifically, we compare the DCA-CACM scheme of [58] described in Section 6.5.1.1 with the state-of-the-art correlation-unaware CACM scheme of [4]. In Figure 6.6, the schemes are compared for a library with $N = 1\,000$ files, while the number of receivers varies from $K = 10$ to 50 when each receiver's cache capacity is 10% of the entire library size. We remark that the results displayed here correspond to a given system parameter δ, and we do not optimize over the choice of δ.

Figure 6.7(a) and (b) displays the rate-memory trade-off in a static setting under the assumption that each file (and consequently each of its packets) is δ-correlated with G_δ other files. Figure 6.7(c) and (d) displays the rate-memory trade-off in a network with dynamically correlated content. We observe that the correlation-aware

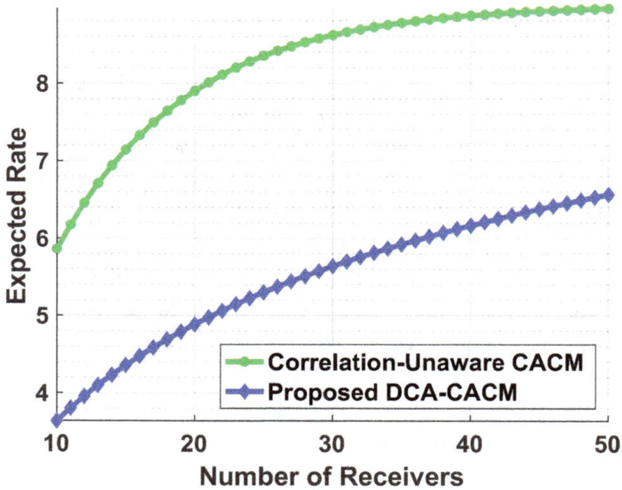

Figure 6.6 *Rate-memory trade-off in a network with* $N = 1\,000$, M, $\delta = 0.1$, $G_\delta = 5$ *for number of receivers from* 10 *to* 50

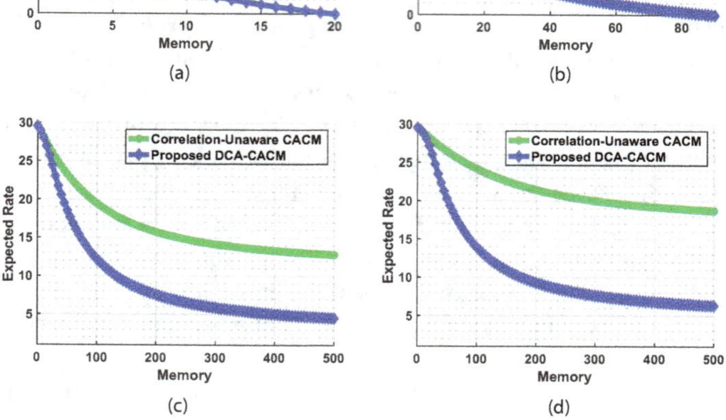

Figure 6.7 *Rate-memory trade-off for a library with: (a), (b) static correlated content, and (c), (d) dynamic correlated content*

scheme is able to achieve rate reductions that go well beyond the state-of-the art correlation-unaware counterpart with a 1.6× reduction in the expected rate compared to the correlation-unaware scheme for a cache capacity equal to 10% of the library size when $N = 20$.

When new versions of files become available at the sender with a probability of $\pi = 0.4$, the DCA-CACM scheme performs 2.8× better compared to the correlation-unaware scheme for a cache size equal to half the library, and its performance gap becomes 3× when updated versions become available with $\pi = 0.6$. The DCA-CACM scheme is steadily effective in exploring global caching gains, across all cache capacities, among the files that have not been updated as well as the files for which a new version is available. However, since the new content has not been prefetched in the receiver caches, the correlation-unaware scheme falls short from constructing coded messages for delivering the requested content for which a new version is available, and can only use coding for the files that have not changed. The numerical results confirm that the network cached information can still be used to create coding opportunities in settings with dynamic and personalized data, where contents are rapidly evolving but have high-level correlations.

As a promising research direction, one could investigate how the inherent correlations in dynamic settings can be exploited for efficiently updating the caches in settings where information requires timely updates. That is, to understand the trade-off between allocating the available network resources to satisfy current user demands and allocating the resources for edge-cache updates, which, while not relevant for the current demand, can be advantageous for future network uses. Additionally, lossy cache-aided content delivery can be extended to settings with nonideal channel network conditions, e.g., when the wireless link is not error-free and deterministic, in which strategies based on joint source-channel coding for caching and delivery can potentially improve the expected distortion experienced at user devices.

References

[1] Ji M., Tulino A., Llorca J., Caire G. Cisco visual networking index: Forecast and trends, 2017-2022. Available from https://www.cisco.com/c/en/us/solutions/collateral/service-provider/visual-networking-index-vni/white-paper-c11-741490.html [Accessed 2018].

[2] Ji M., Tulino A.M., Llorca J., Caire G. 'Caching and coded multicasting: multiple groupcast index coding'. Proceedings of IEEE Global Conference on Signal and Information Processing (GlobalSIP); 2014. pp. 881–5.

[3] Ji M., Tulino A., Llorca J., Caire G. 'Caching-aided coded multicasting with multiple random requests'. Proceedings of IEEE Information Theory Workshop (ITW); 2015. pp. 1–5.

[4] Ji M., Tulino A.M., Llorca J., Caire G. 'Order-optimal rate of caching and coded multicasting with random demands'. *IEEE Transactions on Information Theory*. 2017;**63**(6):3923–49.

[5] Maddah-Ali M.A., Niesen U. 'Decentralized coded caching attains order-optimal memory-rate tradeoff'. *IEEE/ACM Transactions on Networking*. 2015;**23**(4):1029–40.
[6] Maddah-Ali M.A., Niesen U. 'Fundamental limits of caching'. *IEEE Transactions on Information Theory*. 2014;**60**(5):2856–67.
[7] Shanmugam K., Ji M., Tulino A.M., Llorca J., Dimakis. A.G. 'Finite-length analysis of caching-aided coded multicasting'. *IEEE Transactions on Information Theory*. 2016;**62**(10):5524–37.
[8] Yu Q., Maddah-Ali M.A., Avestimehr A.S. 'The exact rate-memory tradeoff for caching with uncoded prefetching'. *IEEE Transactions on Information Theory*. 2018;**64**(2):1281–96.
[9] Bar-Yossef Z., Birk Y., Jayram T.S., Kol T. 'Index coding with side information'. *IEEE Transactions on Information Theory*. 2011;**57**(3):1479–94.
[10] Birk Y., Kol T. 'Informed-source coding-on-demand (ISCOD) over broadcast channels'. Proceedings of IEEE INFOCOM'98. Seventeenth Annual Joint Conference of the IEEE Computer and Communications Societies; 1998. pp. 1257–64.
[11] Mohammadi Amiri M., Gunduz D. 'Fundamental limits of coded caching: improved delivery rate-cache capacity tradeoff'. *IEEE Transactions on Communications*. 2017;**65**(2):806–15.
[12] Amiri M.M., Yang Q., Gündüz D. 'Coded caching for a large number of users'. Proceedings of IEEE Information Theory Workshop (ITW). IEEE; 2016. pp. 171–5.
[13] Chen Z., Fan P., Letaief K.B. 'Fundamental limits of caching: improved bounds for small buffer users'. *arXiv preprint arXiv:1407.1935*. 2014.
[14] Ghasemi H., Ramamoorthy A. 'Improved lower bounds for coded caching'. *IEEE Transactions on Information Theory*. 2017;**63**(7):4388–413.
[15] Gómez-Vilardebó J. 'Fundamental limits of caching: improved bounds with coded prefetching'. *arXiv preprint arXiv:1612.09071*. 2016.
[16] Lim S.H., Wang C.-Y., Gastpar M. 'Information theoretic caching: the multi-user case'. Proceedings of IEEE International Symposium on Information Theory (ISIT); 2016. pp. 525–9.
[17] Sahraei S., Gastpar M. 'K users caching two files: an improved achievable rate'. Proceedings of of Annual Conference on Information Science and Systems (CISS). IEEE; 2016. pp. 620–4.
[18] Sengupta A., Tandon R., Clancy T.C. 'Improved approximation of storage-rate tradeoff for caching via new outer bounds'. Proceedings of IEEE International Symposium on Information Theory (ISIT). IEEE; 2015. pp. 1691–5.
[19] Tian C. 'Symmetry, outer bounds, and code constructions: a computer-aided investigation on the fundamental limits of caching'. *Entropy*. 2018;**20**(8):603.
[20] Tian C., Chen J. 'Caching and delivery via interference elimination'. *IEEE Transactions on Information Theory*. 2018;**64**(3):1548–60.
[21] Wan K., Tuninetti D., Piantanida P. 'On caching with more users than files'. Proceedings of IEEE International Symposium on Information Theory (ISIT). IEEE; 2016. pp. 135–9.

[22] Wang C.-Y., Lim S.H., Gastpar M. 'A new converse bound for coded caching'. Proceedings of IEEE Information Theory and Applications Workshop (ITA). IEEE; 2016. pp. 1–6.
[23] Yu Q., Maddah-Ali M.A., Avestimehr A.S. 'Characterizing the rate-memory tradeoff in cache networks within a factor of 2'. *IEEE Transactions on Information Theory*. 2019;**65**(1):647–63.
[24] Amiri M.M., Gündüz D. 'Caching and coded delivery over Gaussian broadcast channels for energy efficiency'. *IEEE Journal on Selected Areas in Communications*. 2018;**36**(8):1706–20.
[25] Azimi S.M., Simeone O., Sengupta A., Tandon R. 'Online edge caching and wireless delivery in fog-aided networks with dynamic content popularity'. *IEEE Journal on Selected Areas in Communications*. 2018;**36**(6):1189–202.
[26] Saeedi Bidokhti S., Wigger M., Timo R. 'Noisy broadcast networks with receiver caching'. *IEEE Transactions on Information Theory*. 2018;**64**(11):6996–7016.
[27] Hachem J., Karamchandani N., Diggavi S. 'Multi-level coded caching'. Proceedings of IEEE International Symposium on Information Theory (ISIT). IEEE; 2014. pp. 56–60.
[28] Ji M., Caire G., Molisch A.F. 'Fundamental limits of caching in wireless D2D networks'. *IEEE Transactions on Information Theory*. 2016;**62**(2):849–69.
[29] Karamchandani N., Niesen U., Maddah-Ali M.A., Diggavi S.N. 'Hierarchical coded caching'. *IEEE Transactions on Information Theory*. 2016;**62**(6):3212–29.
[30] Liu A., Lau V.K.N. 'Mixed-timescale precoding and cache control in cached mimo interference network'. *IEEE Transactions on Signal Processing*. 2013;**61**(24):6320–32.
[31] Maddah-Ali M.A., Niesen U. 'Cache-aided interference channels'. *IEEE Transactions on Information Theory*. 2019;**65**(3):1714–24.
[32] Naderializadeh N., Maddah-Ali M.A., Avestimehr A.S. 'Fundamental limits of cache-aided interference management'. *IEEE Transactions on Information Theory*. 2017;**63**(5):1–107.
[33] Niesen U., Maddah-Ali M.A. 'Coded caching with nonuniform demands'. *IEEE Transactions on Information Theory*. 2017;**63**(2):1146–58.
[34] Park S.-H., Simeone O., Shamai Shitz S., Shitz S.S. 'Joint optimization of cloud and edge processing for FOG radio access networks'. *IEEE Transactions on Wireless Communications*. 2016;**15**(11):7621–32.
[35] Pedarsani R., Maddah-Ali M.A., Niesen U. 'Online coded caching'. *IEEE/ACM Transactions on Networking*. 2016;**24**(2):836–45.
[36] Zhang J., Elia P. 'Fundamental limits of cache-aided wireless BC: interplay of coded-caching and CSIT feedback'. *IEEE Transactions on Information Theory*. 2017;**63**(5):1–60.
[37] Zhang J., Engelmann F., Elia P. 'Coded caching for reducing CSIT-feedback in wireless communications'. Proceedings of 53rd Annual Allerton Conference on Communication, Control, and Computing (Allerton). IEEE; 2015. pp. 1099–105.

[38] Zhang J., Lin X., Wang X. 'Coded caching under arbitrary popularity distributions'. *IEEE Transactions on Information Theory*. 2018;**64**(1):349–66.

[39] Paschos G.S., Iosifidis G., Tao M., Towsley D., Caire G. 'The role of caching in future communication systems and networks'. *IEEE Journal on Selected Areas in Communications*. 2018;**36**(6):1111–25.

[40] Kam C., Kompella S., Nguyen G.D., Wieselthier J.E., Ephremides A. 'Information freshness and popularity in mobile caching'. Proceedings of IEEE International Symposium on Information Theory (ISIT); 2017. pp. 136–40.

[41] Miorandi D., Sicari S., De Pellegrini F., Chlamtac I. 'Internet of things: vision, applications and research challenges'. *Ad Hoc Networks*. 2012;**10**(7):1497–516.

[42] Schwarz H., Marpe D., Wiegand T. 'Overview of the scalable video coding extension of the H.264/AVC standard'. *IEEE Transactions on Circuits and Systems for Video Technology*. 2007;**17**(9):1103–20.

[43] Sun H., Vetro A., Xin J. 'An overview of scalable video streaming'. *Wireless Communications and Mobile Computing*. 2007;**7**(2):159–72.

[44] Hassanzadeh P., Tulino A., Llorca J., Erkip E. 'Cache-aided coded multicast for correlated sources'. Proceedings of IEEE International Symposium on Turbo Codes and Iterative Information Processing (ISTC); 2016. pp. 360–4.

[45] Hassanzadeh P., Tulino A., Llorca J., Erkip E. 'Correlation-aware distributed caching and coded delivery'. Proceedings of IEEE Information Theory Workshop; 2016. pp. 166–70.

[46] Op't Veld G., Gastpar M.C. 'Caching of bivariate Gaussians with non-uniform preference probabilities'. Proceedings of Symposium on Information Theory and Signal Processing in the Benelux; 2017. pp. 176–83.

[47] Timo R., Saeedi Bidokhti S., Wigger M., Geiger B.C. 'A rate-distortion approach to caching'. *IEEE Transactions on Information Theory*. 2018;**64**(3):1957–76.

[48] Yang Q., Gündüz D. 'Centralized coded caching of correlated contents'. Proceedings of IEEE International Conferece on Communications; 2018. pp. 1–6.

[49] Yang Q., Hassanzadeh P., Gündüz D., Erkip E. 'Centralized caching and delivery of correlated contents over Gaussian broadcast channels'. *IEEE Transactions on Communications*. 2020;**68**(1):122–36.

[50] Gray R.M., Wyner A.D. 'Source coding for a simple network'. *Bell System Technical Journal*. 1974;**53**(9):1681–721.

[51] Daniel A.M., Yu W. 'Optimization of heterogeneous coded caching'. *arXiv preprint arXiv:1708.04322*. 2017.

[52] Ji M., Shanmugam K., Vettigli G., Llorca J., Tulino A., Caire G. 'An efficient multiple-groupcast coded multicasting scheme for finite fractional caching'. Proceedings of IEEE International Conference on Communications (ICC); 2015. pp. 3801–6.

[53] Ji M., Tulino A., Llorca J., Caire G. 'Caching-aided coded multicasting with multiple random requests'. Proceedings of IEEE Information Theory Workshop; 2015. pp. 1–5.

[54] Sengupta A., Tandon R. 'Improved approximation of storage-rate tradeoff for caching with multiple demands'. *IEEE Transactions on Communications*. 2017;**65**(5):1940–55.

[55] Hassanzadeh P., Tulino A.M., Llorca J., Erkip E. 'Rate-memory trade-off for the two-user broadcast caching network with correlated sources'. Proceedings of IEEE International Symposium on Information Theory (ISIT); 2017. pp. 1227–31.

[56] Hassanzadeh P., Tulino A.M., Llorca J., Erkip E. 'Broadcast caching networks with two receivers and multiple correlated sources'. Proceedings of IEEE Asilomar conference on Signals, Systems and Computers; 2017. pp. 1449–54.

[57] Hassanzadeh P., Tulino A.M., Llorca J., Erkip E. 'Rate-memory tradeoff for caching and delivery of correlated sources'. *IEEE Transactions on Information Theory*. 2020;**66**(4):2219–51.

[58] Hassanzadeh P., Tulino A.M., Llorca J., Erkip E. 'On coding for cache-aided delivery of dynamic correlated content'. *IEEE Journal on Selected Areas in Communications*. 2018;**36**(8):1666–81.

[59] Ji M., Tulino A.M., Llorca J., Caire G. 'On the average performance of caching and coded multicasting with random demands'. Proceedings of IEEE International Symposium on Wireless Communications Systems (ISWCS); 2014. pp. 922–6.

[60] Ibrahim A.M., Zewail A.A., Yener A. 'On coded caching with heterogeneous distortion requirements'. Proceedings of Information Theory and Applications Workshop (ITA). IEEE; 2018. pp. 1–9.

[61] Yang Q., Gündüz D. 'Coded caching and content delivery with heterogeneous distortion requirements'. *IEEE Transactions on Information Theory*. 2018;**64**(6):4347–64.

[62] Hassanzadeh P., Tulino A.M., Llorca J., Erkip E. 'Rate-distortion-memory trade-offs in heterogeneous caching networks'. *IEEE Transactions on Wireless Communications*. 2020;**19**(5):3019–33.

[63] Liu W., Xu G., Chen B. 'The common information of N dependent random variables. Proceedings of 48th Annual Allerton Conference on Communication, Control, and Computing (Allerton). IEEE; 2010. pp. 836–43.

[64] Viswanatha K., Akyol E., Rose K. 'Subset typicality lemmas and improved achievable regions in multiterminal source coding'. *arXiv preprint arXiv:1205.1173*. 2012.

[65] Tandon R., Sankar L., Poor H.V. 'Multi-user privacy: the Gray-Wyner system and generalized common information'. Proceedings of IEEE International Symposium on Information Theory (ISIT); 2011. pp. 563–7.

[66] Cheng H., Li C., Xiong H., Frossard P. 'Optimal decentralized coded caching for heterogeneous files'. Proceedings of IEEE European Signal Processing Conference; 2017. pp. 2531–5.

[67] Li C. 'On rate region of caching problems with non-uniform file and cache sizes'. *IEEE Communications Letters*. 2017;**21**(2):238–41.

[68] Zhang J., Lin X., Wang C.-C., Wang X. 'Coded caching for files with distinct file sizes'. Proceedings of IEEE International Symposium on Information Theory (ISIT); 2015. pp. 1686–90.

[69] Cover T.M., Thomas J.A. *Elements of Information Theory*. John Wiley & Sons; 2012.

[70] Shanmugam K., Tulino A.M., Dimakis A.G. 'Coded caching with linear subpacketization is possible using Ruzsa-Szemèredi graphs'. 2017 IEEE International Symposium on Information Theory (ISIT); 2017. pp. 1237–41.

Chapter 7
Rate-distortion theory for caching

Shirin Saeedi Bidokhti[1], Roy Timo[2], and Michèle Wigger[3]

7.1 Introduction

Caching is an appealing solution to reduce network congestion, which is often due to media traffic. Indeed, media streaming is an important application of caching, because video and music files are resource-demanding, have stable contents, and are usually requested by several users simultaneously. Video and music files often encode details that human auditory or visual system cannot distinguish, and, therefore, most users require only a lossy version of them. In fact, the required level of detail often depends on the user's device. For example, if a song is to be played on a simple mobile phone, the distortions created by the phone's audio equipment (e.g., loudspeakers) render a high-resolution music file useless as opposed to the case of a high-end surround system. Similarly, different video qualities are desired depending on the replay type (HD or SD) and the screen's pixel size. In a setup with multiple users, the desired quality of video or music files thus often vary across different users and files.

In an information-theoretic framework, requirements on quality and resolution of files are typically phrased in terms of distortion measures and maximum allowed distortions. An example of a distortion measure for digital sources often considered in the information theory literature is the *Hamming distortion*, which counts the number of pixels or bits where the reconstruction differs from the original source. For continuous sources, the *squared-error distortion* measure is frequently used, which sums the squares of the differences between any sample and its reconstruction. We are interested in understanding the tradeoff between the rate at which a file is transmitted and the minimum distortion with which it can be reconstructed. A *rate-distortion (RD)* function characterizes the smallest compression rate that is needed in order to reproduce a file satisfying the maximum allowed distortion criteria (under a given distortion measure).

[1]University of Pennsylvania, USA
[2]Ericsson Research, Sweden
[3]Telecom Paris, IP Paris, France

In this chapter, we consider a single-server K-user centralized caching setup with a library of correlated files. In the first communication phase, *the placement phase*, the library prefetches and stores content in the users' cache memories. Later, each user demands one specific file from the library and the demands are revealed to the server and all the users in the system. During the second communication phase, *the delivery phase*, the server sends a common delivery message that is observed by all the users in the system and allows the users to reconstruct, using the contents in their local cache memory, their requested file with the desired distortion. The main focus of this chapter is on the minimum rate of this delivery message given a set of distortion criteria and cache sizes at the users. We call this quantity the *rate-distortion-memory (RDM) function* of the system.

The RDM problem for the single-server $K \geq 1$-user setup was introduced by Timo *et al.* in [1, 2]. In their setup (reproduced here in Section 7.3) the distortion measure can differ over files and users. This assumption is motivated by the different types of the source files (audio or video files) and by the different kinds of user terminals (mobile phones or high-end surround systems). The RDM function for this model is still open, so in this chapter we focus on interesting special cases.

The first special case is the single-user model (Section 7.4), for which Timo *et al.* [1, 2] determined the optimal RD function and the optimal placement and delivery strategies for arbitrary sources and distortion measures (under some technical conditions that we will explain). The single-user setup with bivariate Gaussian sources and squared-error distortion measures was also treated by Op't Veld and Gastpar in [3, 4]. Both lines of work exploit the connection of the single-user RD caching problem to the Gray-Wyner source coding network and draw connections to known common information (CI) measures such as lossy Wyner [5–7] and Gács-Körner [7–10] CIs.

More involved is the two-user case, which we treat in Section 7.5. Here, a complete solution seems out of reach and the literature has focused on various special cases. The work in [1] finds lower and upper bounds for the case where only one of the two users has a cache memory, and [11, 12] focus on lossless caching and propose two-step strategies in which source coding and coded caching are separated.

Finally, due to the complexity of the problem, for the general K-user scenario the literature has focused on *uncorrelated* Gaussian sources and mean squared-error distortion. As we explain in Section (7.6) of this chapter, under this assumption, it is natural to compress each file individually into different layers, and upon request deliver each user the number of layers allowing it to attain the desired distortion. Scalable source coding [13–15] is typically applied as a framework for this approach. Yang and Gündüz [16, 17], Hassanzadeh *et al.* [18], and Zewaili and Yener [19] considered such a scalable source-coding approach for uncorrelated Gaussian sources (files) under mean-squared error distortion measures. More specifically, Yang and Gündüz [16, 17] proposed practical cache allocation algorithms to identify what percentage of the cache memory each user should allocate to each layer. In their approach, placement and delivery are performed separately for each layer. This per-layer approach was proved by Zewaili and Yener [19] to be generally suboptimal under uncoded placement and linear delivery, but optimal under these assumptions when one can optimize the assignment of the cache sizes to the different users subject to a total cache memory budget. (The problem of cache assignment was first studied in [20, 21] for the standard coded caching setup over a noisy broadcast channel (BC).) Hassanzadeh *et al.* [22] considered a different global optimization problem across users, where the maximum permissible distortion levels are not fixed for each user and file. Instead, in their approach the goal is to minimize the *average* mean squared-error distortion across all the K users and all files (which are requested according to a given file popularity distribution).

Related, but slightly beyond the scope of this chapter are also the "caching for computing" framework of [23, 24] and the RD formulation of the caching problem over noisy BCs. Wang *et al.* [23, 24] formulated the "caching for computing" problem that is more resemblant of traditional information-theoretic source coding with side-information. In their scenario a user does not emit a single request to indicate the file that it wishes to download, but it makes a new request for each symbol. The request sequence is then modelled as the side-information sequence of a standard source-coding problem. Since the focus of [23] is on *lossless compression*, we will not treat it in detail in this chapter. It is however obvious that similar results can also be obtained for a RD formulation of their problem.

So far, delivery communication was assumed to take place over a noise-less rate-limited link. Timo *et al.* introduced a variation of the coded caching problem where delivery communication is over a noisy BC [25, 26]. They showed at hand of erasure BCs that joint source-channel coding (also termed *joint cache-channel coding* in this context) can greatly improve the performance of lossless caching systems when delivery is over a noisy BC. Similar results were also obtained for other channels such as the K-user Gaussian BC [20, 21]. The RD caching problem with correlated files and delivery over noisy BCs still remains an open research problem. Simple results can be obtained by appropriately combining previous results on the lossless caching over noisy BCs with the RD results presented in this chapter. The lossless counterpart of the RD caching problem over a Gaussian BC has been studied by Yang *et al.* [27].

7.2 Rate-distortion functions

We provide a quick overview of the basic quantities in RD theory used in this chapter: the single-source RD function, the joint RD function and the conditional RD function. More details can be found in standard textbooks, see [28, 29].

Given a source V following some distribution P_V over an alphabet $\mathcal{V}$, a reconstruction alphabet $\hat{\mathcal{V}}$, a nonnegative and bounded per-letter distortion measure $\delta\colon \hat{\mathcal{V}} \times \mathcal{V} \to \mathbb{R}^+$ (where the latter set denotes the nonnegative reals), and a nonnegative distortion $D \geq 0$, the *single-source RD function* is defined as

$$R_V(D) := \min I(V; \hat{V}), \tag{7.1}$$

where the minimum is over all auxiliary random variables $\hat{V}$ over the set $\hat{\mathcal{V}}$ jointly distributed with V and satisfying:

$$\mathbb{E}[\delta(\hat{V}, V)] \le D. \tag{7.2}$$

The single-source RD function determines the minimum number of bits *per source symbol* required to describe an *independent and identically distributed (i.i.d.)* source sequence $V^n := (V_1, \ldots, V_n)$ distributed according to P_V such that a decoder that has access to this description can produce a reconstruction $\hat{V}^n \in \hat{\mathcal{V}}^n$ that matches the original source up to average per-letter distortion D in the asymptotic limit as the blocklength n tends to infinity, i.e., so that

$$\overline{\lim_{n\to\infty}} \frac{1}{n}\sum_{i=1}^{n} \delta(\hat{V}_i, V_i) \le D. \tag{7.3}$$

The joint RD function has an analogous meaning when the single source V is replaced by a tuple $(V_1, \ldots, V_t) \sim P_{V_1 \cdots V_t}$. In particular, it determines the minimum number of bits *per source tuple* required to *jointly describe* the i.i.d. source sequences $V_1^n, \ldots, V_t^n$, for $V_j^n := (V_{j,1}, \ldots V_{j,n})$ and $(V_{1,i}, \ldots, V_{t,i}) \sim P_{V_1 \cdots V_t}$, with a common message index J, such that a decoder that has access to J can produce reconstructions $\hat{V}_1^n, \ldots, \hat{V}_t^n$ that individually match the original sources up to average per-letter distortion $D_1, \ldots, D_t$ in the asymptotic limit as the blocklength n tends to infinity. Formally, the joint RD function is defined as

$$R_{V_1 \cdots V_L}(D) := \min I(V_1, \ldots, V_t; \hat{V}_1, \ldots, \hat{V}_t), \tag{7.4}$$

where the minimum is over all auxiliary random variables $\hat{V}_1, \ldots, \hat{V}_t$ over the given reconstruction sets $\hat{\mathcal{V}}_1, \ldots, \hat{\mathcal{V}}_t$ jointly distributed with $V_1, \ldots, V_t$ and satisfying:

$$\mathbb{E}[\delta_i(\hat{V}_i, V_i)] \le D_i, \qquad i \in \{1, \ldots, t\}. \tag{7.5}$$

Finally, the *conditional RD function* has a similar operational meaning as the single-file RD function, but when the encoder and the decoder (who produces the reconstruction) have access to a random sequence $S^n := (S_1, \ldots, S_n)$ where for each $i \in \{1, \ldots, n\}$ the pair (S_i, V_i) follows a given joint distribution P_{SV}. Formally, the conditional RD function is defined as

$$R_{V|S}(D) := \min I(V; \hat{V}|S), \tag{7.6}$$

where the minimum is over all auxiliary random variables $\hat{V}$ over the given reconstruction set $\hat{\mathcal{V}}$ jointly distributed with the pair (S, V) and satisfying:

$$\mathbb{E}[\delta(\hat{V}, V)] \le D. \tag{7.7}$$

7.3 Description of the system model

We assume throughout this chapter that the library consists of $(L \ge 1)$ different source files, denoted by

$$W_1^n, \ldots, W_L^n. \tag{7.8}$$

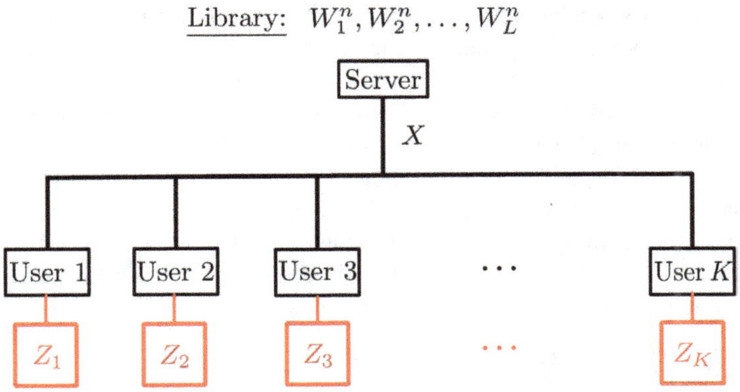

Figure 7.1 *Illustration of the general model with a single server and* K *cache-aided users*

We further assume that the ℓ-th source file, W_ℓ^n, where $\ell \in \mathcal{L} := \{1, 2, \ldots, L\}$, consists of n symbols chosen from a discrete and finite alphabet[a] $\mathcal{W}_\ell$. That is, we can write the ℓ-th source file as

$$W_\ell^n := (W_{\ell,1} \ldots, W_{\ell,n}), \tag{7.9}$$

for $W_{\ell,1}, \ldots, W_{\ell,n} \in \mathcal{W}_\ell$. The number of source symbols n (also called the blocklength) will be allowed to grow without bound so as to enable an information-theoretic analysis. Thus, we are interested in libraries consisting of a fixed number of large source files.

We further assume that the library is randomly generated by a discrete memoryless source (DMS), i.e., $W_1^n, \ldots, W_L^n$ is the sequence of L-tuples $\{(W_{1,i}, \ldots, W_{L,i})\}_{i=1}^n$ and is i.i.d. according to a given, but arbitrary, joint distribution $P_{W_1 \cdots W_L}$ defined on $\mathcal{W}_1 \times \cdots \times \mathcal{W}_L$:

$$(W_1, \ldots, W_L) \sim P_{W_1 \cdots W_L}, \tag{7.10}$$

This assumption is quite common in the multi-terminal RD theory literature, as it admits rigorous proofs and gives some insight to more complicated models. Although it is somewhat restrictive, some important transformations (e.g. Burrows-Wheeler) are known to emit almost memoryless processes [30, 31].

We consider a communication system that consists of a single server and K users, which we enumerate from 1 to K (see Figure 7.1). The server has access to the entire library $W_1^n, \ldots, W_L^n$ and each user k is interested in learning one of these files. It is however à priori unknown which files the users are interested in. User k

[a] We will drop the finite alphabet assumption for some Gaussian source examples.

has access to a local cache memory whose content is denoted by Z_k and that can fit up to $n \cdot M_k$ bits. We thus say that the cache memory has a capacity of M_k bits per source symbol.

As in the previous chapters of this book, we assume that the server can prefetch arbitrary contents in all cache memories during an idle-network period when the users have not yet decided (or communicated) which files they wish to download. In this chapter we mostly concentrate on the so-called *centralized caching scenario* where the server centrally decides on all cache contents, as opposed to a *decentralized caching scenario* where each user locally decides on how to fill its own cache memory. The cache memories $Z_1, \ldots, Z_K$ are thus filled according to

$$Z_k = \phi_k(W_1^n, \ldots, W_L^n), \tag{7.11}$$

for arbitrary placement functions $\phi_1, \ldots, \phi_K$ on appropriate domains.

Following the placement phase where the users' cache memories are filled with contents, the users communicate to the server and the other users in the system which files they wish to learn.[b] It will be interesting to allow for restricted demand sets at the various users. That means, assume given subsets

$$\mathcal{L}_1, \ldots, \mathcal{L}_K \subseteq \mathcal{L}, \tag{7.12}$$

where for each k the set $\mathcal{L}_k$ contains the indices of all files that can be requested by user k. In most of the caching literature, all users can potentially demand all files, i.e., $\mathcal{L}_1 = \cdots = \mathcal{L}_K = \mathcal{L}$. Certain situations, e.g., systems with different kinds of subscriptions, are however more adequately modeled with unequal demand sets $\mathcal{L}_1, \ldots, \mathcal{L}_K$. In any case, at the end of the placement phase, each user k produces a demand

$$\ell_k \in \mathcal{L}_k, \tag{7.13}$$

and in the subsequent *delivery* phase the server conveys a message X taking value in a finite index set $\mathcal{X}$ over a noise- and error-free link to all the users. Mathematically speaking, the message X is of the form

$$X = f(W_1^n, \ldots, W_L^n, \ell_1, \ldots, \ell_K, Z_1, \ldots, Z_K), \tag{7.14}$$

for some delivery encoding function f on appropriate domains.

The goal is to design the message X in a way that each user can reconstruct its desired file up to a maximum admissible distortion. In practice, a given user might decide on this maximum admissible distortion and the specific distortion measure at the moment where it requests the file, e.g., in function of the current channel conditions or on the specific application running on the device. This, however, makes the system more complicated. For simplicity and to better illustrate the impact of lossy

[b]It suffices that the server learns the requests, because it can communicate it using KlogL bits to all users, which requires 0 delivery rate.

reconstructions on a caching system, we assume throughout this chapter that for each file W_ℓ^n each user k has a fixed maximum admissible distortion level $D_{k,\ell}$ under a given distortion measure $\delta_{k,\ell}^{(n)}$. Moreover, both sets $\{D_{k,\ell}\}$ and $\{\delta_{k,\ell}^{(n)}\}$ are known to the server and all the users in advance, before the cache memories are filled with contents.

Each user k is thus supposed to apply a reconstruction function g_k on appropriate domains to the received message X and its local cache content Z_k so that in the limit of large blocklengths n, the reconstruction

$$\hat{W}_k^n = g_k(X, Z_k, \ell_1, \ldots, \ell_K) \tag{7.15}$$

satisfies the distortion constraint

$$\overline{\lim_{n\to\infty}}\, \mathbb{E}\left[\delta_{k,\ell_k}^{(n)}(\hat{W}_k^n, \hat{W}_{\ell_k}^n)\right] \leq D_{k,\ell_k}, \quad \forall k \in \{1,\ldots,K\}. \tag{7.16}$$

In this chapter, we consider *separable distortion functions* that take on the form

$$\delta_{k,\ell}^{(n)}(\hat{W}_\ell^n, W_\ell^n) = \frac{1}{n}\sum_{i=1}^{n} \delta_{k,\ell}(\hat{W}_{\ell,i}, W_{\ell,i}) \tag{7.17}$$

for some per-letter distortion function $\delta_{k,\ell}\colon \hat{\mathcal{W}}_\ell \times \mathcal{W}_\ell \to [0,\infty)$. We allow for arbitrary (not only squared-error) per-letter distortion measures $\{\delta_{k,\ell}\}$ subject to the following two conditions:[c]

1. For each source symbol $w_\ell \in \mathcal{W}_\ell$ there exists a reconstruction symbol $\hat{w}_\ell \in \hat{\mathcal{W}}_\ell$ such that $\delta_{k,\ell}(\hat{w}_\ell, w_\ell) = 0$; and
2. there exists a finite $D_{\max} > 0$ such that $\delta_{k,\ell}(\hat{w}_\ell, w_\ell) \leq D_{\max}$ for all $w_\ell \in \mathcal{W}_\ell$ and $\hat{w}_\ell \in \hat{\mathcal{W}}_\ell$, and all $k \in \{1,\ldots,K\}$ and $\ell \in \mathcal{L}$.

For example, $\delta_{k,\ell}$ can be the *Hamming distortion function* where $\hat{\mathcal{W}}_\ell = \mathcal{W}_\ell$ and for all $\hat{w}_\ell, w_\ell \in \mathcal{W}_\ell$:

$$\delta_{k,\ell}(\hat{w}_\ell, w_\ell) = \begin{cases} 1 & \text{if } \hat{w}_\ell \neq w_\ell \\ 0 & \text{if } \hat{w}_\ell = w_\ell. \end{cases} \tag{7.18}$$

We will further restrict to a maximum *expected* distortion criterion as is more common both in the general information-theoretic literature and in the caching literature.

Alternatively, one could also consider a maximum *excess* distortion criterion instead of the abovementioned maximum expected distortion criterion, or allow for the more general f-*separable distortion measures*. Under an f-separable distortion measure the n-symbol distortion is calculated as [32]

[c]An exception is made for Gaussian sources with squared-error distortion measures where assumption 2. is not valid.

$$\delta^{(n)}(\hat{w}^n, w^n) := \mathsf{f}_\ell^{-1}\left(\frac{1}{n}\sum_{i=1}^{n}\mathsf{f}_\ell\big(\mathsf{d}_\ell(\hat{w}_i, w_i)\big)\right), \tag{7.19}$$

for some given continuous and strictly increasing function $\mathsf{f}_\ell\colon [0,\infty) \to [0,\infty)$. For example, f_ℓ can be chosen to assign appropriate (possibly nonlinear) frequency costs to different quantization error events. Results for f-separable distortion measures and maximum excess distortion constraints were presented in [2].

As mentioned previously, in this chapter, we focus on separable distortion measures and the maximum expected distortion criterion.

Let $\mathbf{D} := (D_{1,1}, \ldots, D_{1,L}, D_{2,1}, \ldots, D_{2,L}, \ldots, D_{K,1}, \ldots, D_{K,L})$ and $\mathbf{M} := (M_1, \ldots, M_K)$ be arbitrary nonnegative allowed distortions and cache sizes at the users, and $\{\delta_{k,\ell}\}$ arbitrary per-letter distortion measures. We say that a delivery rate $R \geq 0$ is $(\mathbf{D}, \mathbf{M})$*-admissible* if for every $\varepsilon > 0$ there exists a sufficiently large blocklength n and placement, delivery encoding, and reconstruction functions $\phi_1, \ldots, \phi_K, f, g_1, \ldots, g_K$ such that for all K-tuples $(\ell_1, \ldots, \ell_K) \in \mathcal{L}_1 \times \cdots \times \mathcal{L}_K$:

$$\mathbb{E}\left[\frac{1}{n}\sum_{i=1}^{n}\delta_{k,\ell_k}(\hat{W}_{k,i}, W_{\ell_k,i})\right] \leq D_{k,\ell_k} + \varepsilon, \qquad \forall k \in \{1, \ldots, K\}, \tag{7.20}$$

and the cardinality of the delivery message is bounded as

$$\frac{1}{n}\log_2|\mathscr{X}| \leq R + \varepsilon, \quad \forall(\ell_1, \ldots, \ell_K) \in \mathscr{L}^k. \tag{7.21}$$

We thus employ a *worst-case* perspective where the delivery rate needs to suffice to deliver the required information irrespective of the actual demands. Alternatively, one can also focus on the *average delivery rate* in which case the alphabet of the delivery message X can depend on the demand vector

$$\boldsymbol{\ell} := (\ell_1, \ldots, \ell_K). \tag{7.22}$$

Under the *average-case* perspective, we write $\mathscr{X}_{\boldsymbol{\ell}}$ for the alphabet of X under the demand vector $\boldsymbol{\ell} = (\ell_1, \ldots, \ell_K)$, and replace (7.21) by

$$\frac{1}{|\mathscr{L}|^k}\sum_{\boldsymbol{\ell}=(\ell_1,\ldots,\ell_K)\in\mathscr{L}^k}\frac{1}{n}\log_2|\mathscr{X}_{\boldsymbol{\ell}}| \leq R + \varepsilon, \tag{7.23}$$

Through a slight change, it is also possible to account for file popularities and to focus on the *expected* delivery rate. Under this perspective, we let $p_{k,\ell}$ denote the probability that user k demands file W_ℓ, and replace constraint (7.23) by

$$\sum_{\boldsymbol{\ell}=(\ell_1,\ldots,\ell_K)\in\mathscr{L}_k}\left(\prod_{k=1}^{K} p_{k,\ell_k}\right)\cdot\frac{1}{n}\log_2|\mathscr{X}_{\boldsymbol{\ell}}| \leq R + \varepsilon. \tag{7.24}$$

Under any of the three perspectives, we call $\mathbf{M}$ the *cache memory vector* and $\mathbf{D}$ the *distortion constraint vector*. The optimal RDM tradeoff in the asymptotic regime of infinite blocklengths $n \to \infty$ is characterized by the following function.

The *worst-case RDM function* is

$$\mathsf{R}_{\text{worst}}(\mathbf{D}, \mathbf{M}) := \inf\{R \geq 0 : R \text{ is } (\mathbf{D}, \mathbf{M})\text{-admissible}\}. \tag{7.25}$$

The *average RDM function* $\mathsf{R}_{\text{avg}}(\mathbf{D},\mathbf{M})$ and the *expected RDM function* $\mathsf{R}_{\text{exp}}(\mathbf{D},\mathbf{M})$ are defined analogously.

Our main focus will be on the worst-case RDM function, but we will also discuss results on the average and expected RDM functions. Some simple properties of the worst-case RDM function $\mathsf{R}_{\text{worst}}(\mathbf{D},\mathbf{M})$ have been proved in [2] for $K = 1$ or $K = 2$. We first note that the worst-case RDM function $\mathsf{R}_{\text{worst}}(\mathbf{D},\mathbf{M})$ is jointly convex, continuous and non-increasing in each component of the vectors $\mathbf{D}$ and $\mathbf{M}$. Notice further that without cache memories, $\mathbf{M} = (0, 0, \dots, 0)$, all information has to be sent in the delivery phase. By the broadcast nature of the delivery channel (every user observes the entire transmission), the worst-case RDM function for the *all-zero* cache memory vector $\mathbf{M} = \mathbf{0} := (0, 0, \dots, 0)$ is:

$$\mathsf{R}_{\text{worst}}(\mathbf{D},\mathbf{0}) = \max_{(\ell'_1,\dots,\ell'_K)\in\mathcal{L}_k} \mathsf{R}_{W_{\ell'_1}\cdots W_{\ell'_K}}(D_{1,\ell'_1},\dots,D_{K,\ell'_K}), \tag{7.26}$$

where recall that (see Section 7.2) $\mathsf{R}_{W_{\ell'_1}\cdots W_{\ell'_K}}(D_{1,\ell'_1},\dots,D_{K,\ell'_K})$ denotes the standard *joint RD function* for the file tuple $W_{\ell'_1},\dots,W_{\ell'_K} \sim P_{W_{\ell'_1}\cdots W_{\ell'_K}}$.

In the other extreme, when the cache memories are sufficiently large so that all relevant information can be stored at every single user, then no delivery message is required. That means, for any cache memory vector $\mathbf{M} = (M_1, \dots, M_K)$ and distortion vector $\mathbf{D}$ satisfying:

$$M_k \geq \mathsf{R}_{W_1\cdots W_L}(D_{k,1},\dots,D_{k,L}), \quad \forall k \in \{1,\dots,K\}, \tag{7.27}$$

the RDM function vanishes:

$$\mathsf{R}_{\text{worst}}(\mathbf{D},\mathbf{M}) = 0. \tag{7.28}$$

Determining the worst-case RDM function in general is challenging. In fact, even for lossless reconstructions the problem remains unsolved. We will therefore focus on interesting special cases to get some insight on the general problem.

7.4 Single-user systems

We start by considering a single user ($K = 1$), see Figure 7.2, and without ambiguity drop the user index 1 to simplify notation. That means, we denote the single user's cache content by Z, its size by M, the single-user's request by ℓ, the demand set by $\mathscr{L}$, the per-letter distortion measure for a given file W_ℓ^n by δ_ℓ, and its maximum permissible distortion by D_ℓ.

The results in this section on the worst-case RDM function are based on [1, 2], and the results on the average-case RDM function in the Gaussian case are based on [3, 4].

7.4.1 The RDM functions of the single-user system

We start by deriving the RDM functions of the single-user caching system, by making use of an operational equivalence between the caching system and the Gray-Wyner

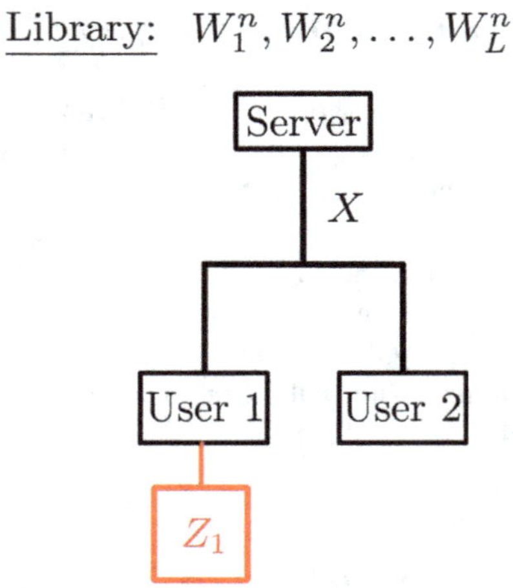

Figure 7.2 Illustration of single-user caching model

source coding network [33], as well as the tight characterization of the RD region for this problem found by Gray and Wyner in their original manuscript.

7.4.1.1 The Gray-Wyner source coding system

The single-user caching system is closely related to the Gray-Wyner source coding network [33] in Figure 7.3. The Gray-Wyner network consists of a single encoder observing the correlated sources $W_1^n, \ldots, W_L^n$ and L decoders $\ell = 1, \ldots, L$, each wishing to reconstruct their corresponding source W_ℓ with a prescribed maximum distortion level $\tilde{D}_\ell$ under a given per-letter distortion measure δ_ℓ. The Gray-Wyner network is comprised of a common rate-R_c communication link from the encoder

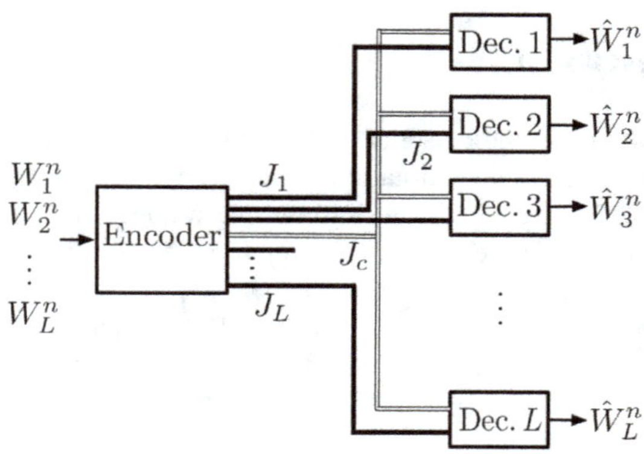

Figure 7.3 Illustration of the Gray-Wyner source coding network with a common message J_c of rate R_c and individual messages $J_1, \ldots, J_L$ of rates $R_1, \ldots, R_L$

to *all decoders* and L individual communication links of rates $R_1, \ldots, R_L$ from the encoder to each of the decoders. Let $\mathbf{R}_{\text{ind}} := (R_1, \ldots, R_L)$ denote the individual-rates vector and $\tilde{\mathbf{D}} := (\tilde{D}_1, \ldots, \tilde{D}_L)$ the distortions vector. By standard convention, a RD tuple $(R_c, \mathbf{R}_{\text{ind}}, \tilde{\mathbf{D}})$ is called achievable in the Gray-Wyner network if it is possible to find a sequence of encoders and decoders satisfying the rate constraints and the maximum allowed distortions [6, 7].

The set of all achievable RD tuples $(R_c, \mathbf{R}_{\text{ind}}, \tilde{\mathbf{D}})$ for the L-decoder Gray-Wyner network is given by [6, 7]

$$\mathscr{R}_{\mathrm{GW}}(\tilde{\boldsymbol{D}}) := \bigcup_{P_{U|W_1\cdots W_L}} \left\{ (R_c, \mathbf{R}_{ind}) : \begin{array}{rcl} R_c & \geq & I(W_1 \cdots W_L; U) \\ R_{l'} & \geq & \mathsf{R}_{W_{l'}|U}(\tilde{D}_{l'}) \quad \forall l' \in \mathscr{L} \end{array} \right\}, \tag{7.29}$$

where recall that (see Section 7.2) $\mathsf{R}_{W_{\ell'}|U}(\tilde{D}_{\ell'})$ is the conditional RD function of file $W^n_{\ell'}$ given a sequence U^n that is jointly i.i.d. with $W_{\ell'}$ according to $P_{UW_{\ell'}}$ and known both to the encoder and the decoder.

The following scheme achieves the RD tuples in $\mathscr{R}_{\mathrm{GW}}(\tilde{\mathbf{D}})$. The encoder quantizes the n-tuple source $W^n_1, \ldots, W^n_L$ into a codeword U^n and sends the description using the common index J_c to all the decoders. Subsequently, the encoder applies a separate conditional RD code to each file W^n_ℓ, for $\ell \in \mathscr{L}$, assuming side-information U^n at both the encoder and decoder and for the target distortion $\tilde{D}_\ell$. It finally sends the resulting description $\hat{W}_\ell$ using the private index J_ℓ to Decoder ℓ, which produces $\hat{W}_\ell$ as its reconstruction.

7.4.1.2 An operational equivalence and the RDM function

We now present an observation that is key to solving the single-user RD caching problem of this section. The Gray-Wyner RD problem is equivalent to the single-user caching system, as seen through the following associations.

- The cache memory Z in the caching setup corresponds to the index J_c sent over the common link in the Gray-Wyner source coding setup, and the cache size M to the common rate R_c in the Gray-Wyner setup.
- If the single user in the caching system demands file W_ℓ, then it can be associated with User ℓ in the Gray-Wyner network. Under demand ℓ, the delivery message X in the caching setup then corresponds to the private index J_ℓ in the Gray-Wyner network and the delivery rate $\frac{1}{n}\log|\mathscr{X}|$ corresponds to Gray-Wyner's individual rate R_ℓ.
- Consequently, the worst-case, average, and expected RDM functions of the caching setup are given by the maximum, average, and expected Gray-Wyner individual rates. Applying the expression for the set of achievable RD tuples in (7.29) then yields the following worst-case RDM function:

$$\mathsf{R}_{(\text{worst})}(\mathbf{D}, M) = \min_{\mathbf{R}:(M,\mathbf{R})\in\mathscr{R}_{\mathrm{GW}}(\mathbf{D})} \max_{\ell\in\mathscr{L}} R_\ell \tag{7.30}$$

$$= \min_{U:\, I(W_1,\ldots,W_L;U)\leq M} \max_{\ell\in\mathscr{L}} \mathsf{R}_{W_\ell|U}(D_\ell), \tag{7.31}$$

and for the average RDM function:

$$\mathsf{R}_{\text{avg}}(\mathbf{D}, M) = \min_{\mathbf{R}:\,(M,\mathbf{R})\in\mathscr{R}_{\text{GW}}(\mathbf{D})} \frac{1}{L}\sum_{\ell\in\mathscr{L}} R_\ell \tag{7.32}$$

$$= \min_{U:\, I(W_1,\ldots,W_L;U)\le M} \frac{1}{L}\sum_{\ell\in\mathscr{L}} \mathsf{R}_{W_\ell|U}(D_\ell). \tag{7.33}$$

The expression for the *expected* RDM function $\mathsf{R}_{\text{avg}}(\mathbf{D}, M)$ is analogous and omitted.

7.4.1.3 Two examples

Even though (7.31) and (7.33) characterize the worst-case and average RDM functions exactly, an analytical evaluation of these formulas for specific examples can be tricky because it is unclear how to choose the auxiliary random variable U. We present the bounds on the worst-case RDM function $\mathsf{R}_{\text{worst}}(\mathbf{D}, M)$ found in [2] for two canonical examples: the *doubly symmetric binary source* (DSBS) under Hamming distortions with $D_1 = D_2 = 0$ and a centered bivariate Gaussian source under squared-error distortion. For the Gaussian example, we also provide the bound on the average-case RDM function $\mathsf{R}_{\text{avg}}(\mathbf{D}, M)$ found in [4].

Example 1 (Doubly-Symmetric Binary Source with Lossless Reconstructions) *Consider a source with only two files* (W_1^n, W_2^n) *whose symbols are binary and follow the distribution*

$$P_{W_1W_2}(w_1, w_2) = \tfrac{1}{2}(1-\rho)\cdot 1\{w_1 = w_2\} + \tfrac{1}{2}\rho\cdot 1\{w_1 \neq w_2\}, \quad w_1, w_2 \in \{0, 1\},$$

for a given parameter $\rho \in [0, 1/2]$. *Consider the same Hamming distortion measure for all files:* $\delta_\ell(\hat{w}, w) = 1\{w \neq w\}$, *for all* $w, \hat{w} \in \{0, 1\}$ *and* $\ell \in \mathcal{L}$. *Let* h_b *denote the binary entropy function:*

$$h_b(t) := -t\log t - (1-t)\log(1-t), \quad t \in [0, 1], \tag{7.34}$$

and define $\rho^* := \frac{1}{2} - \frac{1}{2}\sqrt{1-2\rho}$.

Under zero distortions $D_1 = D_2 = 0$, depending on the available cache size M, the worst-case RDM function is bounded as follows:

- If $1 + h_b(\rho) - 2h_b(\rho^*) \le M \le 1 + h_b(\rho)$, then

$$\mathsf{R}_{\text{worst}}(\mathbf{0}, M) = \frac{1+h_b(\rho)-M}{2}. \tag{7.35}$$

- If $0 \le M < 1 - h_b(\rho)$, then

$$\mathsf{R}_{\text{worst}}(\mathbf{0}, M) \ge [1 - M]^+. \tag{7.36}$$

- If $0 \le M \le 1 + h_b(\rho) - 2h_b(\rho^*)$, then

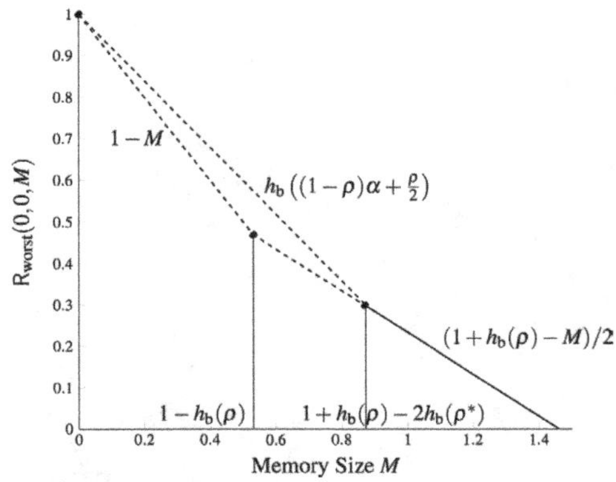

Figure 7.4 Illustration of the upper (achievable) and lower (converse) bounds on the RDM function of DSBS sources with correlation coefficient $\rho = 0.1$

$$\frac{1+h_b(\rho)-M}{2} \leq \mathsf{R}_{\text{worst}}(\mathbf{0}, M) \leq h_b\left((1-\rho)\alpha + \tfrac{\rho}{2}\right), \tag{7.37}$$

where $\alpha := h_b^{-1}\left(\frac{1-\rho-M}{1-\rho}\right)$ and h_b^{-1} denotes the inverse binary entropy function that takes values in the interval [0, 1/2].

Above bounds can be proved following similar steps as in the DSBS examples in ([33] Sec. 1.5), ([34] Sec. III.C), and ([23] Ex. 1). The same steps can also be used to show that the bounds remain valid for the average-case RDM function $\mathsf{R}_{\text{avg}}(\mathbf{D}, M)$ (which by definition is always smaller than or equal to the worst-case RDM function).

Bounds (7.35)–(7.37) are illustrated in Figure 7.4. We use a solid line to indicate the exact worst-case RDM function in the regime where it is known, and dashed lines in the regime where only upper and lower bounds are known.

In the second example, we consider a bivariate Gaussian source and squared-error distortion measures. For more than two sources, a lower bound on the worst-case RDM-function is presented in [2].

Example 2. Consider a library with two files $\{W_1^n, W_2^n\}$ generated by a centered bivariate Gaussian distribution with covariance matrix

$$\mathsf{K}\_{W_1 W_2} = \begin{bmatrix} 1 & \rho \\ \rho & 1 \end{bmatrix}, \tag{7.38}$$

for a given correlation coefficient $\rho \in (0, 1)$. Further, assume squared-error distortion functions for both files, i.e., $\delta_1(\hat{w}, w) = \delta_2(\hat{w}, w) = (\hat{w} - w)^2$, for all $w, \hat{w} \in \mathbb{R}$, and symmetric maximum distortions $D_1 = D_2 = D \geq 0$. The joint RD function $\mathsf{R}_{G,X_1X_2}(D)$ for this symmetric bivariate Gaussian source is ([35, 36], Thm. III.1):

i. If $0 < D \leq 1 - \rho$, then

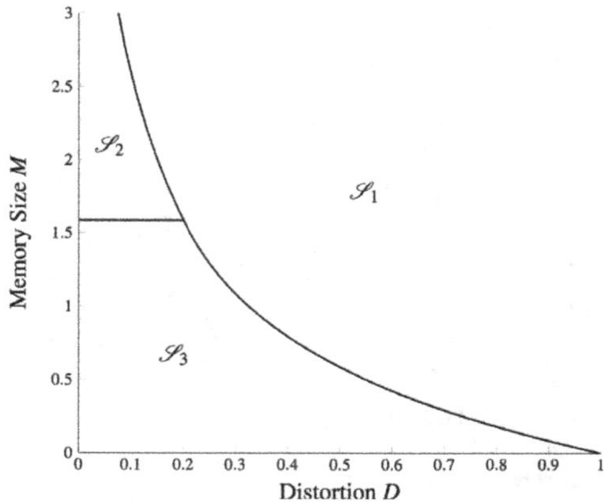

Figure 7.5 Illustration of the distortion-memory regions $\mathcal{S}_1$, $\mathcal{S}_2$, and $\mathcal{S}_3$ when$\rho = 0.8$

$$\mathsf{R}_{G,X_1X_2}(D) = \frac{1}{2}\log\frac{1-\rho^2}{D^2}. \tag{7.37}$$

ii. If $1-\rho \leq D \leq 1$, then

$$\mathsf{R}_{G,X_1X_2}(D) = \frac{1}{2}\log\frac{1+\rho}{2D-(1-\rho)}. \tag{7.40}$$

iii. If $D > 1$, then

$$\mathsf{R}_{G,X_1X_2}(D) = 0. \tag{7.41}$$

The worst-case RDM function shows a different behavior on the following three regions $\mathscr{S}_1, \mathscr{S}_2, \mathscr{S}_3$, illustrated in Figure 7.5:

$$\mathscr{S}_1 := \left\{(D, M) : M \geq R_{G,X_1X_2}(D)\right\}, \tag{7.42}$$

$$\mathscr{S}_2 := \left\{(D, M) : \frac{1}{2}\log\frac{1+\rho}{1-\rho} \leq M \leq \mathsf{R}_{G,X_1X_2}(D)\right\}, \tag{7.43}$$

and

$$\mathscr{S}_3 := \left\{(D, M) : D \leq 1-\rho,\ \ M \leq \frac{1}{2}\log\frac{1+\rho}{1-\rho}\right\} \\ \bigcup \left\{(D, M) : 1-\rho \leq D \leq 1,\ \ M \leq \mathsf{R}_{G,X_1X_2}(D)\right\}. \tag{7.44}$$

In fact, the worst-case RDM function satisfies [2]:

$$\mathsf{R}_{\text{worst}}(D, D, M) = \begin{cases} 0, & (D, M) \in \mathscr{S}_1, \\ \dfrac{1}{4}\log\dfrac{1-\rho^2}{D^2} - \dfrac{M}{2}, & (D, M) \in \mathscr{S}_2, \end{cases} \tag{7.45}$$

and if $(D, M) \in \mathscr{S}_3$:

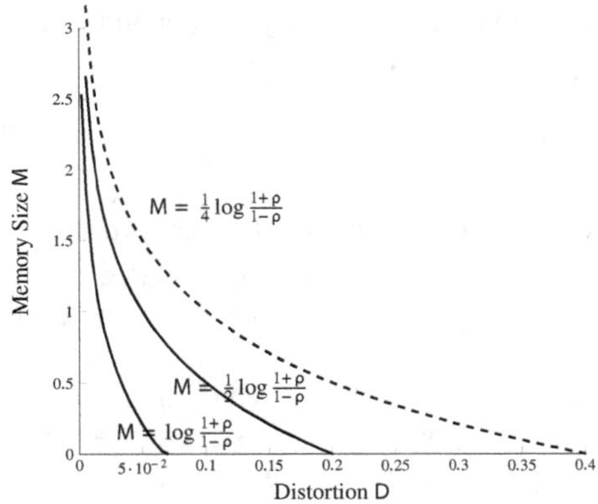

Figure 7.6 *Bounds on the worst-case RDM functions for a centered bivariate Gaussian source with correlation coefficient* $\rho = 0.8$ *and unit variances, and for different cache sizes*

$$\mathsf{R}_{\text{worst}}(D, D, M) \leq \frac{1}{2} \log \frac{1 - \frac{1}{2}(1+\rho)(1 - 2^{-2M})}{D}. \tag{7.46}$$

Given the symmetry of the problem, the same bounds apply also to the average-case RDM function $\mathsf{R}_{\text{avg}}(D, D, M)$, see [4].

Figure 7.6 illustrates the above bounds when $\rho = 0.8$. For cache sizes $M = \frac{1}{2} \log \frac{1+\rho}{1-\rho}$ and $M = \log \frac{1+\rho}{1-\rho}$ all relevant (D, M) pairs are in $\mathscr{S}_2$ (for small cache memories M) or in $\mathscr{S}_1$ (for large cache memories M), and thus the exact RDM function is known. For cache size $M = \frac{1}{4} \log \frac{1+\rho}{1-\rho}$, the considered (M, D) pairs are in $\mathscr{S}_3$, and only an explicit lower bound on the worst-case RDM function is known. This lower bound is dashed in Figure 7.6 to emphasize that it does not necessarily indicate the exact RDM function.

7.4.2 *Lower bounds, critical cache sizes, and common informations*

As we elaborate in this subsection, the RDM function for the single-user caching setup with correlated files is also related to (lossy versions of) *Wyner's common information (CI)* and *Gác's-Körner's CI* [5–8]. This does not come as a surprise because both CIs are known for their operational meanings on the Gray-Wyner source coding network. In the Gray-Wyner network, Wyner's CI characterizes the smallest common rate R_c for which there exist achievable individual rates so that the sum of all rates $R_c + R_1 + \cdots + R_L$ coincides with the joint RD-function $\mathsf{R}_{W_1 \cdots W_L}(\tilde{\mathbf{D}})$ of a super-receiver that observes all the indices $J_c, J_1, \ldots, J_L$. In the same network, the Gács-Körner CI designates the largest common rate R_c for which there exist achievable individual rates so that the sum of the rates $R_c + R'_\ell$ to any given user ℓ' coincides with the individual RD-function $\mathsf{R}_{W_{\ell'}}(\tilde{D}_{\ell'})$ to this user.

In our single-user caching setup the lossy Wyner and Gács-Körner CIs show up when analyzing the tightness of two simple lower bounds on the worst-case RDM function.

7.4.2.1 Genie-aided lower bound and Gács-Körner CI

The first lower bound is obtained by considering a related system where a genie tells the server in advance which file will be requested by the user during the delivery phase. The optimal caching strategy for this hypothetical *genie-aided* system is to compress the requested ℓ-th source W_ℓ^n using an optimal RD code for that source, to place any nM bits of the code's output in the user's cache memory, and to then send the remaining bits during the delivery phase. The RDM function of this system is easily found to be

$$\mathsf{g}(\mathbf{D}, M) = \left[\max_{\ell \in \mathcal{L}} \mathsf{R}_{W_\ell}(D_\ell) - M\right]^+, \tag{7.47}$$

where $[a]^+$ stands for $\max\{0, a\}$. In general, the genie-aided RDM function $\mathsf{g}(\mathbf{D}, M)$ provides a lower bound on the worst-case RDM function $\mathsf{R}_{\text{worst}}(\mathbf{D}, M)$ of the caching system under consideration:

$$\mathsf{R}_{\text{worst}}(\mathbf{D}, M) \geq \mathsf{g}(\mathbf{D}, M), \qquad \forall M \geq 0. \tag{7.48}$$

Moreover, in the absence of any cache memory, i.e., for $M = 0$, the lower bound is tight, $\mathsf{g}(\mathbf{D}, 0) = \mathsf{R}_{\text{worst}}(\mathbf{D}, 0)$. It can be shown that in some setups the lower bound is also tight for positive cache memory sizes $M > 0$. We define the *critical cache size* M_c^* as the largest $M \geq 0$ such that $\mathsf{g}(\mathbf{D}, M) = \mathsf{R}_{\text{worst}}(\mathbf{D}, M)$. Since $\mathsf{g}(\mathbf{D}, M)$ decreases linearly in M and $\mathsf{R}_{\text{worst}}(\mathbf{D}, M)$ is convex, this implies that the genie-aided lower bound is tight for all cache sizes below M_c^*. A single-letter expression for the critical cache size M_c^* was provided in [2], where it was also shown that it is lower bounded by the L-variable lossy Gács-Körner CI:

$$\mathsf{M}_\mathsf{c}^* \geq \mathsf{K}_{\text{GK}}(W_1, \ldots, W_L; D_1, \ldots, D_L), \tag{7.49}$$

where equality holds whenever the single-source RD functions are all the same: $\mathsf{R}_{W_1}(D_1) = \cdots = \mathsf{R}_{W_L}(D_L)$. The lossy Gács-Körner CI $\mathsf{K}_{\text{GK}}(W_1, \ldots, W_L; D_1, \ldots, D_L)$ for L variables $W_1, \ldots, W_L \sim P_{W_1 \cdots W_L}$ is a straightforward extension of the two-variable lossy Gács-Körner CI in [7] and was formally defined in [2]:

$$\mathsf{K}_{\text{GK}}(W_1, \ldots, W_L; D_1, \ldots, D_L) := \max_{(U, \hat{W}_1, \ldots, \hat{W}_L)} I(W_1, \ldots, W_L; U), \tag{7.50}$$

where the maximum is taken over all tuples $(U, \hat{W}_1, \ldots, \hat{W}_L)$ on $\mathcal{U} \times \hat{\mathcal{W}}_1 \times \cdots \hat{\mathcal{W}}_L$ jointly distributed with $W_1, \ldots, W_L$ and satisfying

i. $\forall\, \ell' \in \mathcal{L} : U \leftrightarrow W_{\ell'} \leftrightarrow \{W_{\ell^c}\}_{\ell^c \in \mathcal{L} \setminus \{\ell'\}}$
ii. $\forall\, \ell' \in \mathcal{L} : U \leftrightarrow \hat{W}_{\ell'} \leftrightarrow W_{\ell'}$
iii. $\forall\, \ell' \in \mathcal{L} : \mathbb{E}[\delta_{\ell'}(\hat{W}_{\ell'}, W_{\ell'})] \leq D_{\ell'}$

iv. $\forall\, \ell' \in \mathcal{L} : I(W_{\ell'}; \hat{W}_{\ell'}) = \mathsf{R}_{W_{\ell'}}(D_{\ell'})$.

When specialized to Hamming distortions and zero maximum distortions, the lossy Gács-Körner CI specializes to the original Gács-Körner CI [8] which we denote by $\mathsf{K}_{\mathrm{GK}}(W_1, \ldots, W_L)$. This latter CI can also be expressed as the largest entropy of a common random variable that can be extracted individually from each of the random variables $W_1, \ldots, W_L$. That means, the largest entropy of a random variable T for which there exist functions $f_1, \ldots, f_L$ so that with probability 1:

$$T = f_1(W_1) = \cdots = f_L(W_L). \tag{7.51}$$

7.4.2.2 Super-user lower bound and Wyner CI

The second lower bound is obtained by the RDM function of a setup with a super-user that observes delivery messages $X_1, \ldots, X_L$ corresponding to all possible L demands and, with these messages and the cache content, reconstructs the entire library $W_1^n, \ldots, W_L^n$ up to the desired distortions $D_1, \ldots, D_L$. The optimal caching strategy for this super-user setup is to take an optimal joint RD code that jointly describes all the files $W_1^n, \ldots, W_L^n$, to then place nM bits of the code's output in the user's cache memory, and to finally distribute the remaining bits equally over the L links in the delivery phase. The RDM function $\mathsf{s}(\mathbf{D}, M)$ of this super-user problem is

$$\mathsf{s}(\mathbf{D}, M) = \left[\frac{\mathsf{R}_{W_1 \cdots W_L}(\mathbf{D}) - M}{L}\right]^+, \tag{7.52}$$

and it lower bounds the RDM function of our single-user caching problem:

$$\mathsf{R}_{\mathrm{worst}}(\mathbf{D}, M) \geq \mathsf{s}(\mathbf{D}, M), \quad \forall M \geq 0. \tag{7.53}$$

As already mentioned in the previous section, the RDM function of our single-user caching setup is trivially 0 when the cache memory is sufficiently large to store all the bits of the joint RD code:

$$\mathsf{R}_{\mathrm{worst}}(\mathbf{D}, M) = 0, \qquad M \geq \mathsf{R}_{W_1 \cdots W_L}(\mathbf{D}). \tag{7.54}$$

Since $\mathsf{s}(\mathbf{D}, M)$ is nonnegative, the worst-case RDM trivially coincides with the super-user lower bound $\mathsf{s}(\mathbf{D}, M)$ for all cache sizes above $\mathsf{R}_{W_1 \cdots W_L}(\mathbf{D})$. It can be shown that in some problems even for smaller cache sizes the RDM functions $\mathsf{s}(\mathbf{D}, M)$ and $\mathsf{R}_{\mathrm{worst}}(\mathbf{D}, M)$ coincide. One can thus define the *critical cache size* $\mathsf{M}_{\mathrm{s}}^*(\mathbf{D})$ as the smallest cache size where the two bounds coincide. Since the RDM function $\mathsf{s}(\mathbf{D}, M)$ is linear and the RDM function $\mathsf{R}_{\mathrm{worst}}(\mathbf{D}, M)$ is convex, the bounds indeed coincide for all cache memory sizes above $\mathsf{M}_{\mathrm{s}}^*(\mathbf{D})$.

A single-letter expression for the critical cache size $\mathsf{M}_{\mathrm{s}}^*(\mathbf{D})$ was presented in [2], where it was also shown that it is always lower bounded by the L-variable lossy Wyner CI $\mathsf{K}_{\mathrm{W}}(W_1, \ldots, W_L; D_1, \ldots, D_L)$:

$$\mathsf{M}_{\mathrm{s}}^*(\mathbf{D}) \geq \mathsf{K}_{\mathrm{W}}(W_1, \ldots, W_L; D_1, \ldots, D_L). \tag{7.55}$$

The L-variable lossy Wyner CI of $W_1, \ldots, W_L$ is defined as [7]:

$$\mathsf{K_W}(W_1,\ldots,W_L;D_1,\ldots,D_L) := \min_{(U,\hat{W}_1,\ldots,\hat{W}_L)} I(W_1,\ldots,W_L;U), \tag{7.56}$$

where the minimum is taken over all tuples $(U,\hat{W}_1,\ldots,\hat{W}_L)$ on $\mathcal{U}\times\hat{\mathcal{W}}_1\times\cdots\times\hat{\mathcal{W}}_L$ such that the following four properties hold

i. $(W_1,\ldots,W_L) \leftrightarrow (\hat{W}_1,\ldots,\hat{W}_L) \leftrightarrow U$
ii. $\forall\, \ell' \in \mathcal{L}{:}\hat{W}_{\ell'} \leftrightarrow U \leftrightarrow \{\hat{W}_{\ell^c}\}_{\ell^c\in\mathscr{L}\setminus\{\ell'\}}$
iii. $\forall\, \ell' \in \mathcal{L}: \mathbb{E}[\delta_{\ell'}(\hat{W}_{\ell'},W_{\ell'})] \leq D_{\ell'}$
iv. $I(W_1,\ldots,W_L;\hat{W}_1,\ldots,\hat{W}_L) = \mathsf{R}_{W_1\cdots W_L}(\mathbf{D})$.

In the lossless case, i.e., under Hamming distortion measures and $\mathbf{D} = (0,\ldots,0)$, lossy Wyner CI coincides with Wyner's CI [5] that we denote $\mathsf{K_W}(W_1,\ldots,W_L)$. Moreover, in this case inequality (7.40) holds with equality whenever the caching problem is symmetric in the sense that $\mathsf{K_W}(W_1,\ldots,W_L) = I(\boldsymbol{W};U^*)$ for some U^* satisfying $H(W_1|U^*) = \cdots = H(W_L|U^*)$ and the Markov chain $W_{\ell'} \leftrightarrow U^* \leftrightarrow \{U^*_{\ell^c}\}_{\ell^c\in\mathscr{L}\setminus\{\ell'\}}$ for all $\ell' \in \mathscr{L}$.

Alternatively, the critical cache capacity $\mathsf{M}^*_\mathsf{s}(\mathbf{D})$ can also be defined in terms of the *average* RDM function instead of the *worst-case* RDM function. Under this average-case framework the critical cache capacity $\mathsf{M}^*_\mathsf{s}(\mathbf{D})$ always coincides with the lossy Wyner CI, see [3, 4].

7.4.2.3 Typical properties of the single-user RDM Function

Figure 7.7 provides an overview of the discussion in this subsection. It depicts the worst-case RDM function $\mathsf{R}_{\text{worst}}(\mathbf{D},M)$ of our single-user caching problem and the genie-aided and super-user lower bounds $\mathsf{g}(\mathbf{D},M)$ and $\mathsf{s}(\mathbf{D},M)$. These bounds are tight for cache sizes smaller than $\mathsf{M}^*_\mathsf{c}(\mathbf{D})$ or larger than $\mathsf{M}^*_\mathsf{s}(\mathbf{D})$, which in their turn are lower bounded by the L-variable lossy Wyner and Gács-Körner CI, and coincide with them in some special cases. Notice in particular that the worst-case RDM

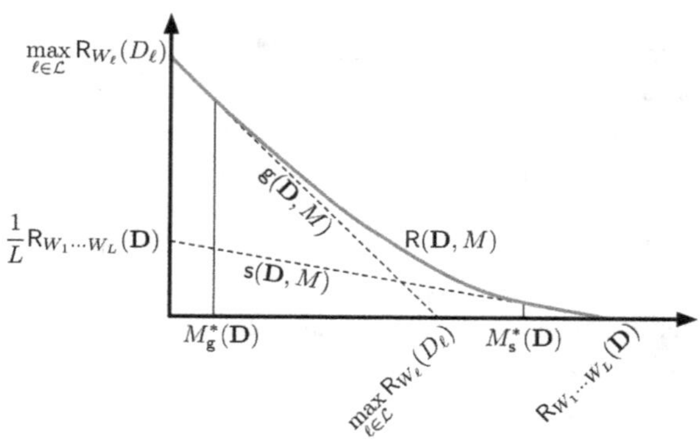

Figure 7.7 An illustration of some typical characteristics of the RDM function $\mathsf{R}_{\text{worst}}(\mathbf{D},M)$ *for a fixed distortion tuple* $\mathbf{D} = (D_1,\ldots,D_L)$

function $\mathsf{R}_{\text{worst}}(\mathbf{D}, M)$ decays linearly for cache sizes $M \in [0, \mathsf{M}_c^*(\mathbf{D})] \cup [\mathsf{M}_s^*(\mathbf{D}), \infty)$ because both lower bounds $\mathsf{g}(\mathbf{D}, M)$ and $\mathsf{s}(\mathbf{D}, M)$ decay linearly in the cache size M.

We can also interpret our findings for the DSBS source with zero Hamming distortions in Example 7.4.1.3 in the sense discussed above. For this source, the Gács-Körner CI is 0 because it is not possible to extract a common random variable individually from the pair $(W_1, W_2) \sim P_{W_1 W_2}$. The standard Wyner CI for the DSBS source is well known and equals

$$\mathsf{K}_\mathsf{W}(W_1, W_2) = 1 + h_b(\rho) - 2h_b(\rho^*). \tag{7.57}$$

Since the problem is completely symmetric, the critical rate M_c^* coincides with Wyner's CI $\mathsf{K}_\mathsf{W}(W_1, W_2)$ on this problem. In fact, as can be seen by inspecting Figure 7.4, the worst-case RDM function decays linearly for all cache sizes M above $\mathsf{K}_\mathsf{W}(W_1, W_2)$ where it coincides with the RDM function $\mathsf{s}(\mathbf{D})$ of the super-user caching problem.

7.5 Two-user systems

The mulit-user caching problem is more challenging than the single-user problem. Before jumping into the K–user problem in the next section, we illustrate some of these challenges for the two-user problem in this section. The full solution for the two-user case is still unknown; nevertheless, lower and upper bounds and special cases are studied in [1, 11, 12].

We start this section with a simple genie-aided lower bound for the general two-user case, similar to the lower bound presented in the previous section. We then present two core caching ideas for two specific two-user systems: In Section 7.5.2, which is based on [1], we consider the scenario in Figure 7.8 (left picture) where only one user has a local cache memory, and in Section 7.5.3, which is based on [11], we consider the scenario in Figure 7.8 (right picture) with two cache memories, but only two files in the library. (A generalization to multiple files and arbitrary number of users can be found in [12].) For the two-cache scenario, the authors in

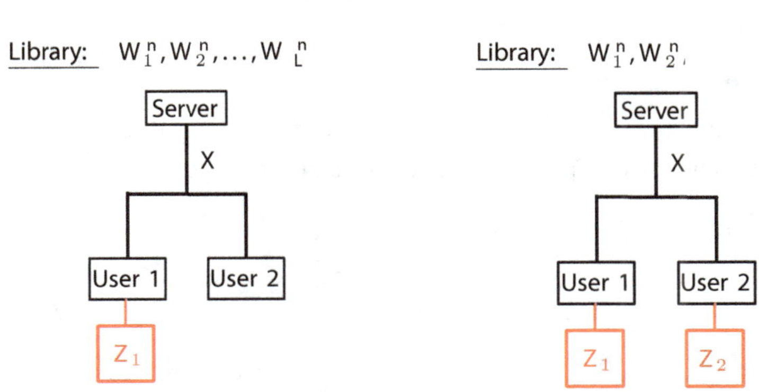

Figure 7.8 Two-user RD cache problem

[11] have focused on (almost) lossless compression. We follow their approach here; an extension to lossy reconstructions can be obtained using standard tools from RD theory.

In this section we follow the general notation introduced in Section 7.3. This means, in particular, that at user k file W_ℓ^n is measured with the per-letter distortion function $\delta_{k,\ell}$ and is subject to a maximum distortion $D_{k,\ell}$. The demand sets of the two users are denoted by $\mathscr{L}_1$ and $\mathscr{L}_2$, and their sizes by L_1 and L_2. We further use the abbreviation $\mathbf{W} := (W_1, \ldots, W_L)$ and write $\mathbf{W} \sim P_{W_1 \cdots W_L}$ to describe the underlying joint description of $\mathbf{W}$.

7.5.1 A genie-aided lower bound

Let $\mathsf{g}(\mathbf{D}, M_1, M_2)$ denote the RDM function of a two-user system with a genie that reveals the demands ℓ_1 and ℓ_2 to the server ahead of time (prior to the placement phase). For each realization of the demand pair ℓ_1, ℓ_2, where $\ell_1 \neq \ell_2$, this genie-information transforms the problem into a two-user two-file ($W_{\ell_1}^n$ and $W_{\ell_2}^n$) Gray-Wyner source coding problem where the common message J_c describes the delivery message X and the two private messages J_1 and J_2 describe the cache contents Z_1 and Z_2. The worst-case genie-aided RDM function $\mathsf{g}(\mathbf{D}, M_1, M_2)$ is then obtained from the characterization of the Gray-Wyner RD region $\mathscr{R}_{\mathrm{GW}}(\tilde{\mathbf{D}})$ in (7.27), and by taking the maximum over all demand pairs (ℓ_1, ℓ_2). In particular, we have

$$\mathsf{g}(\mathbf{D}, M_1, M_2) = \max_{(\ell_1,\ell_2)\in\mathscr{L}_1\times\mathscr{L}_2} \min I(U; \mathbf{W}), \tag{7.58}$$

where the minimum is over all auxiliary random variables U satisfying

$$\mathsf{R}_{W_{\ell_k}|U}(D_{k,\ell_k}) \leq M_k, \quad k \in \{1, 2\}. \tag{7.59}$$

In the special case where only user 1 has a cache memory, i.e., $M_2 = 0$, the above Gray-Wyner source-coding problem reduces to the successive-refinement problem introduced and studied by Equitz and Cover [13] and by Rimold [14], and the worst-case genie-aided RDM function is directly obtained from the RD region of the successive refinement problem.[d]

$$\begin{aligned} &\mathsf{g}(\mathbf{D}, M_1, M_2 = 0) \\ &= \max_{(\ell_1,\ell_2)\in\mathscr{L}_1\times\mathscr{L}_2} \min \max \left\{ I(\mathbf{W}; \hat{W}_{2,\ell_2}),\ I(\mathbf{W}; \hat{W}_{1,\ell_1}, \hat{W}_{2,\ell_2}) - M_1 \right\}, \end{aligned} \tag{7.60}$$

where the minimum is taken over all tuples $\{\hat{W}_{1,\ell_1}\}_{\ell_1\in\mathcal{L}_1}$ and $\{\hat{W}_{2,\ell_2}\}_{\ell_2\in\mathcal{L}_2}$ that are jointly distributed with $\mathbf{W}$ and satisfy for all $(\ell_1, \ell_2) \in \mathcal{L}_1 \times \mathcal{L}_2$:

$$\mathbb{E}\left[\delta_{k,\ell_k}(\hat{W}_{k,\ell_k}, W_{\ell_k})\right] \leq D_{k,\ell_k}, \quad k \in \{1, 2\}. \tag{7.61}$$

[d] The bound in (7.44) can also be obtained directly from (7.42) by rate-transfer arguments and exploiting properties of the optimal choice of U in (7.43) when $M_2 = 0$.

Notice that the outer maximization and the minimization can be exchanged because the objective function only depends on the joint distributions of each pair $(\hat{W}_{1,\ell_1}, \hat{W}_{2,\ell_2})$ with $\mathbf{W}$.

The genie-aided RDM function $g(\mathbf{D}, M_1, M_2)$ lower bounds the worst-case RDM function of our general two-user caching problem:

$$\mathbf{R}_{\text{worst}}(\mathbf{D}, M_1, M_2) \geq g(\mathbf{D}, M_1, M_2), \tag{7.62}$$

because genie-information can only reduce the (worst-case) RDM function. In the following two subsections, we present upper bounds on $\mathbf{R}_{\text{worst}}(\mathbf{D}, M_1, M_2)$ when $M_2 = 0$ and show special cases where they coincide with the genie-aided lower bound.

7.5.2 Two users and one cache

Consider the setup in Figure 7.8 (left picture). User 1 has a cache memory Z_1 with capacity M_1 bits per source symbol and user 2 does not have a cache ($M_2 = 0$). We first consider two lossy setups and then an (almost) lossless setup.

7.5.2.1 Lossy reconstructions with $\mathcal{L}_2$ a singleton

Consider first the special case where $\mathcal{L}_2$ is a singleton, i.e., $\mathcal{L}_2 = \{\ell_2^\star\}$ for some $\ell_2^\star \in \mathcal{L}$. In this case, the following placement and delivery strategies are optimal:

- The server quantizes $W^n_{\ell_2^\star}$ into user 2's reconstruction $\hat{W}^n_2$ such that it satisfies the desired distortion constraint $D_{2,\ell_2^\star}$. The server will send the description of $\hat{W}^n_2$ in the delivery phase as part of the message X, letting user 2 reconstruct $\hat{W}^n_2$ directly from the delivery message.
- Since user 2 can directly reproduce $\hat{W}^n_2$ from the delivery message (without help from the local cache memory), so can user 1. Given this side information (SI), the server applies an optimal placement and delivery strategy for a single-user RD caching problem where both the encoder (the server) and the decoder (user 1) have access to the SI sequence $\hat{W}^n_2$. That is, the server applies a modified Gray-Wyner source code with SI where it first quantizes the entire library $W^n_1, \ldots, W^n_L$ into an auxiliary codeword U^n, assuming SI $\hat{W}^n_2$ at the encoder and decoder, and places the description of the quantized codeword into user 1's cache memory. After receiving user 1's demand ℓ_1[e], the server applies a conditional RD-code (with the target distortion $D_{1,\ell}$) to the file $W^n_{\ell_1}$ requested by user 1, assuming that SI $(U^n, \hat{W}^n_2)$ is known to the encoder and the decoder. The description of $\hat{W}^n_1$ is sent along with the description of $\hat{W}^n_2$ in the delivery phase.
- User 1 first recovers $\hat{W}^n_2$ from the delivery message X. Using $\hat{W}^n_2$ as SI, it then recovers the auxiliary codeword U^n from the cache memory. Finally, given

[e]User 2's demand has been known because L_2 is a singleton.

$(U^n, \hat{W}_2^n)$ as SI and using the partial description of $\hat{W}_1^n$ via the delivery message, user 1 reconstructs $\hat{W}_1^n$.

In summary, during the placement phase, the server puts a description of codeword U^n into user 1's cache memory using SI $\hat{W}_2^n$ (which is user 2's reconstruction) and assuming that this SI is also available at the decoder. During the subsequent delivery phase, the server sends a description of $\hat{W}_2^n$ as well as a description of $\hat{W}_1^n$ using the SI pair $(U^n, \hat{W}_2^n)$. Notice that since user 1 only learns the SI $\hat{W}_2^n$ with the delivery message X, it can only make use of its cache content at the end of the delivery phase.

For any demand $\ell_1 \in \mathcal{L}_1$, the described strategy achieves the desired distortion constraints $(D_{1,\ell_1}, D_{2,\ell_2^\star})$ if the cache size M_1 and the delivery rate R satisfy

$$M_1 > I(\mathbf{W}; U|\hat{W}_2) \tag{7.63}$$

$$R > I(\mathbf{W}; \hat{W}_2) + I(\mathbf{W}; \hat{W}_1|U, \hat{W}_2), \tag{7.64}$$

for some triple $(\hat{W}_1, \hat{W}_2, U)$ jointly distributed with $\mathbf{W}$ satisfying

$$\mathbb{E}\left[\delta_{1,\ell_1}(\hat{W}_1, W_{\ell_1})\right] \leq D_{1,\ell_1} \quad \textit{and} \quad \mathbb{E}\left[\delta_{2,\ell_2^*}(\hat{W}_2, W_{\ell_2^*})\right] \leq D_{2,\ell_2^*}. \tag{7.65}$$

The achieved distortions remain unchanged if instead of storing some of the bits in user 1's cache memory, the server sends them as additional bits in the delivery message to both users. This classical rate-transfer argument and the above two conditions (7.63) and (7.64) prove that the worst-case RDM function of the described setup is upper-bounded by

$$\begin{aligned}&\mathsf{R}_{\text{worst}}(\mathbf{D}, M_1, M_2 = 0)\\ &\leq \min \max_{\ell_1 \in \mathcal{L}_1} \max \Big\{ I(\mathbf{W}; \hat{W}_{2,\ell_2^*}) + I(\mathbf{W}; \hat{W}_{1,\ell_1}|U, \hat{W}_{2,\ell_2^*}),\ I(\mathbf{W}; U, \hat{W}_{1,\ell_1} \hat{W}_{2,\ell_2^*}) -\end{aligned} \tag{7.66}$$

where the minimum is taken over all tuples $(\{\hat{W}_{1,\ell_1}\}_{\ell_1 \in \mathcal{L}_1}, \hat{W}_{2,\ell_2^\star})$ jointly distributed with $\mathbf{W}$ and satisfying (7.65) for all $\ell_1 \in \mathcal{L}_1$.

Using standard converse steps, it can be shown that the inequality in (7.66) is actually an equality, and the above described scheme achieves the optimal RDM function when $\mathcal{L}_2^\star$ is a singleton.

7.5.2.2 Lossy reconstructions and general demand sets

We present a general upper bound on the RDM function $\mathsf{R}_{\text{worst}}(\mathbf{D}, M_1, M_2 = 0)$ by extending the placement and delivery strategies described in the previous subsection to demand sets $\mathcal{L}_2$ that are not singletons. First recall that when $\mathcal{L}_2$ is a singleton, the server naturally knows user 2's demand *a priori* in the placement phase. It can, therefore, treat user 2's reconstruction $\hat{W}_2^n$ as SI known to both the encoder and the decoder and exploit it in designing user 1's cache content (described by the auxiliary codeword U^n).

For general demand sets $\mathcal{L}_2$, SI $\hat{W}_2^n$ is still available at user 1 to recover the codeword U^n from its cache memory. (Recall that this holds because user 2 does not have a cache memory and $\hat{W}_2^n$ must thus be entirely described by the delivery message X.) The server however does not know the SI $\hat{W}_2^n$ when placing its description of U^n in the cache memory, because it has not yet learned the demand ℓ_2. It can thus not apply a standard conditional RD code to generate U^n.

Instead [1] proposed that the server generates the auxiliary codeword U^n so that under any possible demand $\ell_2 \in \mathscr{L}_2$ the corresponding reconstruction sequence $\hat{W}_{2,\ell_2}^n$ can serve as decoder SI. The server then places a coarse bin index for the auxiliary codeword U^n in user 1's cache memory, and refines the bin *a posteriori* after learning the demand ℓ_2. The refined bin index is sent as part of the delivery message X and is chosen to be of exactly the right size (depending on user 2's demand ℓ_2) so as to allow user 1 to recover U^n using SI $\hat{W}_{\ell,2}^n$. This technique leads to the following upper bound on the worst-case RDM function:

$$\mathsf{R}_{\text{worst}}(\mathbf{D}, M_1) \leq \min \max_{(\ell_1,\ell_2)\in\mathcal{L}_1\times\mathcal{L}_2} \max \left\{ \begin{array}{l} I(\mathbf{W}; \hat{W}_{2,\ell_2}) + I(\mathbf{W}; \hat{W}_{1,\ell_1}|U, \hat{W}_{2,\ell_2}), \\ I(\mathbf{W}; U, \hat{W}_{1,\ell_1}, \hat{W}_{2,\ell_2}) - M_1 \end{array} \right\} \tag{7.67}$$

where the minimum is taken over all tuples $\left(U, \{\hat{W}_{1,\ell_1}\}_{\ell_1\in\mathscr{L}_1}, \{\hat{W}_{2,\ell_2}\}_{\ell_2\in\mathscr{L}_2}\right)$ that are jointly distributed with $\mathbf{W}$ and satisfy for all demand pairs $(\ell_1, \ell_2) \in \mathscr{L}_1 \times \mathscr{L}_2$:

$$\mathbb{E}\left[\delta_{k,\ell_k}(\hat{W}_{k,\ell_k}, W_{\ell_k})\right] \quad \leq \quad D_{k,\ell_k}, \quad k \in \{1,2\}. \tag{7.68}$$

This bound matches the genie-aided lower bound in (7.60) in some special cases. For example, when $\mathscr{L}_1 = \{\ell_1^\star\}$ is a singleton, in which case it is optimal to set $U = \hat{W}_{1,\ell_1^\star}$.

7.5.2.3 (Almost) lossless reconstructions

Consider the case where both users wish to reconstruct their demanded sources $W_{\ell_1}^n$ and $W_{\ell_2}^n$ (almost) losslessy. That is, we have

$$\delta_{k,\ell_k}(w_{\ell_k}, \hat{w}_{k,\ell_k}) = \mathbb{1}\left\{\hat{w}_{k,\ell_k} \neq w_{\ell_k}\right\}, \qquad \forall k \in \{1,2\}, \quad \forall(\ell_1, \ell_2) \in \mathscr{L}_1 \times \mathscr{L}_2$$

and $\mathbf{D} = \mathbf{0}$. The bounds in (7.60) and (7.67) specialize to:

$$\mathrm{R}\_{\text{worst}(\mathbf{0},M_1)\geq} \max_{(\ell_1,\ell_2)\in\mathcal{L}_1\times\mathcal{L}_2} \max \left\{ \begin{array}{l} H(W_{\ell_2}), \\ H(W_{\ell_1}, W_{\ell_2}) - M_1 \end{array} \right\}, \tag{7.69}$$

$$\mathrm{R}\_{\text{worst}(\mathbf{0},M_1)\leq\min} \max_{(\ell_1,\ell_2)\in\mathcal{L}_1\times\mathcal{L}_2} \max \left\{ \begin{array}{l} H(W_{\ell_2}) + H(W_{\ell_1}|U, W_{\ell_2}), \\ H(U, W_{\ell_1}, W_{\ell_2}) - M_1 \end{array} \right\} \tag{7.70}$$

where the minimisation in (7.69) is over all auxiliaries U. The two bounds coincide for:

1. $\mathcal{L}_1 = \{\ell\}$for some $\ell \in \mathcal{L}$ (set $U = W_\ell$); and
2. $\mathcal{L}_1 = \mathcal{L}_2 = \{\ell, \ell'\}$for $\ell, \ell' \in \mathcal{L}$ (set $U = (W_\ell, W_{\ell'})$).

The lower bound above (7.69) is derived from the genie-aided bound in (7.62) and its tightness implies that there is no penalty due to the lack of information about ℓ_1, ℓ_2 at the server in the placement phase.

7.5.3 Two users and two files

To develop further understanding on the caching of correlated sources, we examine the lossless caching scenario illustrated in Figure 7.8 (right picture) where both users have the same demand sets:

$$\mathcal{L}_1 = \mathcal{L}_2 = \mathcal{L} = \{1, 2\}, \tag{7.71}$$

and both users have identical cache sizes:

$$M_1 = M_2 = M. \tag{7.72}$$

An approximate solution to this problem can be found by a scheme that separates source coding (which is correlation aware) from coded caching (which is agnostic to correlation).

In particular, set $L = 2$ and $\mathbf{D} = 0$ in the Gray-Wyner network of Figure 7.3 and its rate-region in (7.29). The descriptions J_0, J_1, and J_2 (of rates R_c, R_1, R_2, respectively) are thus common and individual encoded messages and are designed such that W_ℓ, $\ell \in \{1, 2\}$, can be recovered from (J_0, J_ℓ) almost losslessly. Once the library is encoded, the cache contents and the delivery message are designed solely based on J_0, J_1, J_2. Let $R_1 = R_2 = \tilde{R}$ such that the rate tuple $(R_c, \tilde{R}, \tilde{R})$ satisfies the inequalities in (7.29) for some choice of U with $P_{W_1|U} = P_{W_2|U}$. We will optimize over the rate tuple and the corresponding random variable U later.

The cache encoder operates as follows:

- If $M < \tilde{R}$, only the individual messages are cached (e.g., according to the scheme of [37] described in the earlier chapters).
- If $\tilde{R} \leq M < R_c + \tilde{R}$, the first $n(M - \tilde{R})$ bits of J_0 are cached at both users and the remaining cache memory is used for the caching of J_1, J_2 (e.g., according to [37]).
- If $R_c + \tilde{R} \leq M < R_c + 2\tilde{R}$, the common description J_0 is cached at both users and the remaining cache memories are used for the caching of J_1, J_2 (e.g., according to [37]).
- If $M \geq R + 2\tilde{R}$, all three descriptions can be fully stored at both users and no delivery message will be needed.

The delivery message then consists of the portion of J_0 that was not cached and a coded multicast message encoding the pair (J_1, J_2) (e.g. following [37]). Such a two-step caching scheme achieves the following upper bound on R_{worst}:

$$R_{\text{worst}}(\mathbf{0},\mathbf{0},M,M) \le \begin{cases} R_0 + 2\tilde{R} - 2M \\ R_0 + \frac{3}{2}\tilde{R} - M \\ \frac{1}{2}R_0 + \tilde{R} - \frac{1}{2}M. \end{cases} \tag{7.73}$$

The above upper bound can be optimized by minimizing over U and $(R_0, \tilde{R})$. Recall that U is characterized by $P_{U|W_1W_2}$, $P_{W_1|U} = P_{W_2|U}$, and the rate tuple $(R_0, \tilde{R}, \tilde{R})$ satisfies the constraints in (7.29) for the given choice of U.

Lower bounds on the RDM function can also be found using the genie lower bound in (7.58)–(7.59):

$$R_{\text{worst}}(\mathbf{0},\mathbf{0},M,M) \ge \min \left\{ \begin{array}{l} H(W_1, W_2) - 2M, \\ \frac{1}{2}\left(H(W_1, W_2) - M\right), \\ \frac{1}{2}\left(H(W_1, W_2) + \max(H(W_1), H(W_2)) - M\right) \end{array} \right\}. \tag{7.74}$$

The RDM function can now be approximately found. In particular, the bounds in (7.74) and (7.75) coincide for the following two ranges of the cache memory size M (i) $M \le \tilde{M}$ and (ii) $H(W_1, W_2) - 2\tilde{M} \le M \le H(W_1, W_2)$, where

$$\tilde{M} = \max\ \tfrac{1}{2}\min\{H(X_1|U), H(X_2|U)\}, \tag{7.75}$$

and the max is over all choices of U such that $P_{W_1|U} = P_{W_2|U}$ and $W_1 \to U \to W_2$ forms a Markov chain. In other regimes of M, the gap between the upper bound in (7.74) and the lower bound in (7.75) is at most $\frac{1}{2}\min\{H(W_1|W_2), H(W_2|W_1)\} - \tilde{M}$.

7.6 *K*-user systems with uncorrelated sources

The K-user problem is particularly challenging and we restrict to the special case where the files $W_1, \ldots, W_L$ are uncorrelated, i.e., the joint probability distribution $P_{W_1 \cdots W_L}$ factorizes as:

$$P_{W_1 \cdots W_L} = P_{W_1} \cdot P_{W_2} \cdots P_{W_L}. \tag{7.76}$$

We further assume full demand sets $\mathcal{L}_1 = \cdots = \mathcal{L}_K = \mathcal{L}$ where all users potentially demand all files in the library. This setup was studied in [17–19] for Gaussian sources and squared-error distortion measures, so in this section we assume that each source distribution is

$$W_\ell \sim \mathcal{N}(0, \sigma_\ell^2), \tag{7.77}$$

for some positive variance σ_ℓ^2, and

$$\delta_{k,\ell}(\hat{w}, w) = (\hat{w} - w)^2, \quad \forall k, \ell. \tag{7.78}$$

The works in [17–19] are based on the *scalable source-coding* idea of [13–15], which we review next.

7.6.1 Scalable source coding

Scalable source coding, also called *successive refinement*, was introduced and studied in [13–15]. A source , is called *successively refinable* with respect to a distortion measure δ_ℓ, if for any $\epsilon > 0$ and sufficiently large blocklength n, and any $\ell \in \mathcal{L}$, positive integer ν_ℓ and decreasing sequence of distortion levels

$$\tilde{D}_{\ell,1} > \tilde{D}_{\ell,2} > \cdots > \tilde{D}_{\ell,\nu_\ell} \geq 0, \tag{7.79}$$

it is possible to encode the i.i.d. P_{W_ℓ}-source W_ℓ^n into a sequence of refinement layers

$$\bar{\Omega}_\ell := (\Omega_{\ell,1}, \ldots, \Omega_{\ell,\nu_\ell}) \tag{7.80}$$

of rates $r_{\ell,1}, \ldots, r_{\ell,\nu_\ell}$ by means of a scalable encoder

$$\bar{\Omega}_\ell = \text{sc\_enc}_\ell(W_\ell) \tag{7.81}$$

so that a user that observes the first $\lambda \leq \nu_\ell$ layers $\Omega_{\ell,1}, \ldots, \Omega_{\ell,\lambda}$ can produce a reconstruction

$$\hat{W}_\ell^n = \text{sc\_dec}(\Omega_{\ell,1}, \ldots, \Omega_{\ell,\lambda}) \tag{7.82}$$

of file W_ℓ^n that satisfies the predetermined distortion level $\tilde{D}_{\ell,\lambda}$,

$$\mathbb{E}\left[\sum_{i=1}^{n} \delta_\ell(\hat{W}_{\ell,i}, W_{\ell,i})\right] \leq \tilde{D}_{\ell,\lambda} + \epsilon, \tag{7.83}$$

and so that the cumulated rate of layers $\Omega_{\ell,1}, \ldots, \Omega_{\ell,\lambda}$, equals the minimum rate required to attain this specific distortion level $\tilde{D}_{\ell,\lambda}$:

$$\sum_{j=1}^{\lambda} r_{\ell,j} = R_{W_\ell}(\tilde{D}_{\ell,\lambda}). \tag{7.84}$$

It is known that Gaussian sources are successively refinable under squared-error distortion measures.

7.6.2 The layered approach in caching

The works in [17–19] propose to first compress each file W_ℓ^n by means of a scalable source encoder tailored to the (reordered) permissible distortion levels $D_{1,\ell}, \ldots, D_{K,\ell}$, and to then design the placement and delivery schemes in a way that each user k learns sufficiently many layers of its desired file W_{ℓ_k} to meet the required distortion level D_{k,ℓ_k}. Once the set of all possibly demanded layers is identified, the problem resembles a standard coded caching problem with a library that consists of the set of all layers and where each user wishes to download multiple files (corresponding here to multiple layers).

The described setup is rather involved in general, but simplifies significantly when the maximum permissible distortion levels only depend on the user or the file but not on both. When the maximum distortions only depends on the file, i.e., $D_{k,\ell} = D_\ell$, $\forall k \in \{1, \ldots, K\}$, then the setup even specializes to the standard coded caching setup because each file is encoded into a single layer. Technically more involved is the case where the maximum distortions only depend on the users but not on the files, i.e.,

$$D_{k,\ell} = D_k, \quad \forall \ell \in \{1, \dots, L\}. \tag{7.85}$$

This assumption is realistic in video streaming because video quality often depends only on the streaming terminal: a laptop requires a higher-quality video than a smaller mobile phone. In this case, (without loss of generality) we assume that the distortions are ordered:

$$D_1 \geq D_2 \geq \cdots \geq D_K. \tag{7.86}$$

Under above assumptions (7.46)–(7.83), the problem simplifies, because now the j-th layers of all files,

$$\boldsymbol{\Omega}^{(j)} := \{\Omega_{1,j}, \dots, \Omega_{L,j}\} \tag{7.87}$$

are all of the same rate $r_{1,j} = \cdots r_{L,j}$. We can thus define the layer-j rate

$$r_j := r_{\ell',j}, \qquad \ell' \in \mathcal{L}, \tag{7.88}$$

because the j-th layers of all files are intended (if the corresponding files are requested) to the same subset of users

$$\mathcal{U}^{(j)} := \{j, \dots, K\}. \tag{7.89}$$

A natural solution is then to time- and memory-share the placement and delivery schemes of K coded caching schemes, where the j-th scheme, for $j = 1, \dots, K$, considers only the users in $\mathcal{U}^{(j)}$ and the sublibrary $\Omega^{(j)}$. In fact, it can be shown that this approach achieves the optimal RDM-function for two users [38]. As we will see shortly, it is however suboptimal for $K \geq 3$ users. Nevertheless, the approach leads to interesting conclusions, and we will discuss the related literature in the next two subsections.

7.6.2.1 Cache allocation policies and coding within layers

Once the layers have been identified, one has to design a *cache-allocation policy* that at any given user assigns a given cache memory size to each layer. Two such algorithms were proposed in [17]. In the *Ordered Cache Allocation (OCA)* algorithm [17], each user k entirely stores the lowest $j_k^\star$ layers of all files in its cache memory, where the value of the index $j_k^\star$ depends on the user's cache size, it stores a part of the file-layers of layer $j_k^\star + 1$, and it stores no information about any of the higher layers. In the *Proportional Cache Allocation (PCA)* algorithm [17], any given user k allocates to each layer j a cache memory size that is proportional to the rate of this layer r_j. Thus, denoting by $M_{k,j}$ the cache-memory rate allocated to a given layer j at a user k,

$$M_{k,j} = \frac{r_j}{\sum\limits_{j' : k \in \mathcal{U}^{(j')}} r_{j'}} \cdot \frac{M}{L} \cdot \mathbb{1}\{k \in \mathcal{U}^{(j)}\}, \tag{7.90}$$

where $\mathbb{1}\{\cdot\}$ denotes the indicator function that equals 1 if the statement is correct and 0 otherwise. Other cache allocation strategies are of course possible. Finding the optimal allocation policy, however, seems difficult in general.

Once the cache allocation strategy is fixed, one has to design the coded caching algorithm for each layer $\Omega^{(j)}$ and the corresponding set of users $\mathscr{U}^{(j)}$. This task is more involved than in the standard coded caching setup of Chapter 2, because typically (as the case for OCA and PCA) the cache memory allocated for a given layer j differs across users, and thus standard coded caching algorithms designed under the assumption of equal cache sizes cannot be applied directly. This problem can be circumvented by splitting each layer into sublayers, and each sublayer l is then associated the same cache memory size at a subset of users in $\mathscr{U}_l^{(j)} \subseteq \mathscr{U}^{(j)}$ and no cache memory at the remaining users of $\mathscr{U}^{(j)} \backslash \mathscr{U}_l^{(j)}$ (and thus no cache memory at all users in $\{1, \ldots, K\} \backslash \mathscr{U}_l^{(j)}$). Standard coded caching schemes as presented in Chapter 2 can then be applied to each sublayer l of layer j and to the corresponding set of users $\mathscr{U}_l^{(j)}$. For the other users in $\mathscr{U}^{(j)}$, each requested sublayer l of layer j is transmitted by simple uncoded transmission as in the absence of cache memories.

For two users and assuming Gaussian sources and squared-error distortion measures, the described approach with scalable source coding and separately coding over the layers is optimal as shown in [38]. This optimal RDM function is achieved by employing a coded caching scheme for the lower layer (intended to both users) and a simple uncoded placement/uncoded delivery scheme for the higher layer. Depending on the cache sizes uncoded or coded placement have to be combined with coded delivery for the lower layer to attain the optimal RDM function.

7.6.2.2 Coding over layers

While appealingly simple, the proposed approach with scalable coding and separate coding over layers can be suboptimal for $K \geq 3$ users. In [19], it is proved that when one restricts to uncoded placements and linear delivery, then the delivery rate can be reduced by coding over multiple layers. The following example is taken from [19] and shows how coding across layers can be employed.

(Example 1 in [19]) Consider $K = 3$ users 1, 2, and 3 with cache sizes $M_1 = 0.1 \cdot L$, $M_2 = 0.2 \cdot L$, and $M_3 = 0.6 \cdot L$, where $L := |\mathcal{L}||$ denotes the number of files. As described above, a scalable encoder is applied to each file individually for the given maximum distortion levels $D_1 \geq D_2 \geq D_3$, and the j-th layer of file W_ℓ^n is denoted $\Omega_{\ell,j}$, and its rate by r_j. In this example, we assume that the maximum distortion levels $D_1 \geq D_2 \geq D_3$ are so that the layer-1, 2, 3 rates satisfy

$$r_1 = 0.2, \quad r_2 = 0.1, \quad \text{and} \quad r_3 = 0.5. \tag{7.91}$$

To describe the placement phase, the lowest layer of each file is split into two sublayers of equal rates: $\Omega_{\ell',1} = (\Omega_{\ell',1}^{(1)}, \Omega_{\ell',1}^{(2)})$. The first sublayers $\{\Omega_{\ell',1}^{(1)}\}_{\ell' \in \mathcal{L}}$ are all placed at users 1 and 2 and the second sublayers $\{\Omega_{\ell',1}^{(2)}\}_{\ell' \in \mathcal{L}}$ are placed at user 3. All second layers $\{\Omega_{\ell',2}\}_{\ell' \in \mathcal{L}}$ are placed at user 2 and all third layers $\{\Omega_{\ell',3}\}_{\ell' \in \mathcal{L}}$ at user 3. It can be verified that the described placement satisfies the cache memory constraints.

During the delivery, the server sends the two signals

$$X = (\Omega^{(1)}_{\ell_{3,1}} \oplus \Omega^{(2)}_{\ell_{1,1}}, \ \Omega_{\ell_{3,2}} \oplus \Omega^{(2)}_{\ell_{2,1}}), \tag{7.92}$$

where $\oplus$ denotes the component-wise XOR operation between the bits of the corresponding (sub)layers. With this delivery signal and its own cache content, user 1 can recover the missing sublayer $\Omega^{(2)}_{\ell_{1,1}}$, and combined with sublayer $\Omega^{(1)}_{\ell_{1,1}}$ in its cache memory, it can reconstruct $W^n_{\ell_1}$ with the desired distortion D_1. User 2 recovers sublayer $\Omega^{(2)}_{\ell_{2,1}}$ and with $\Omega^{(1)}_{\ell_{2,1}}$ and $\Omega_{\ell_{2,2}}$ from its cache memory it can reconstruct $W^n_{\ell_2}$ with distortion D_2. Finally, user 3 recovers both $\Omega^{(1)}_{\ell_{3,1}}$ and $\Omega_{\ell_{3,2}}$ from the delivery signal, and with $\Omega^{(2)}_{\ell_{3,1}}$ and $\Omega_{\ell_{3,3}}$ in its cache memory it can reconstruct its desired file $W^n_{\ell_3}$ up to distortion D_3.

The delivery-rate in the proposed scheme is 0.2. If one insists on coding over each layer separately, then under any uncoded placement and linear delivery strategy, the delivery rate cannot be smaller than 0.2167 [19].

7.6.3 Optimal cache allocation and uncoded placement – linear delivery schemes

As previously mentioned, it seems challenging to find the optimal cache-allocation to the different layers in general. The problem becomes tractable when restricting to uncoded placement and linear delivery schemes, where delivery can code across layers. In fact [19], showed that under these assumptions, even when one allows for coding over the layers (during the delivery phase), the cache allocation problem turns into a linear programming problem (LP).

Currently there are thus two approaches for the K-user Gaussian RD-problem with uncorrelated sources, both based on scalable coding with layers. The approach in [19] restricts to uncoded placement and linear delivery, but optimizes over the cache allocation to the various layers. The approach in [17] fixes the cache allocation algorithms (PCA or OCA) and depending on the cache size employs uncoded or coded placement. Which of these two approaches performs better depends on the cache sizes at the users.

7.6.4 Cache allocation across users

Reference [17] also considered the cache allocation problem across users, where the total cache memory budget M_{tot} is fixed and one optimizes the cache allocation to the different layers *and users*.[f] The optimization is again performed under the assumption of uncoded placement and coded delivery, possibly across layers. More specifically, an LP optimization problem is phrased to identify the optimal cache allocation to users and layers as well as the optimal linear delivery scheme. It

[f]The cache-allocation problem was also considered for the standard lossless coded caching setup with uncorrelated files when delivery is over noisy broadcast channels [20, 21].

is proved that without loss in optimality[g] delivery can be performed separately for each layer and that if the total cache budget M_{tot} permits, the optimal cache allocation assigns cache size

$$M_{k,j}^{(\text{LP})} = \frac{t_j r_j}{K-j+1} \cdot \mathbb{1}\{k \in \mathcal{U}^{(j)}\}, \qquad k \in \mathcal{U}^{(j)}, \tag{7.93}$$

to layer j and user k, for some decreasing sequence of positive integers $t_1, \ldots, t_K$ for which the cache allocation in (7.92) respects the total cache memory budget:

$$\sum_{k=1}^{K}\sum_{j=1}^{K} M_{k,j}^{(\text{LP})} = M_{\text{tot}}. \tag{7.94}$$

Notice that the cache allocation strategy in (7.93) allocates the same cache size for a given layer j to all users $k \in \mathcal{U}^{(j)}$, and therefore the standard coded caching schemes described in Chapter 2 are optimal on each layer. For cache memory budgets M_{tot} where no appropriate sequence of positive integers $t_1, \ldots, t_K$ exists, the optimal cache allocation is obtained by time- and memory-sharing two appropriate cache allocations as in (7.93).

7.6.5 Minimizing the average distortions

A similar setup to [17] and [19] was considered in [18], but with the goal to minimize the average *expected* distortion over all sources, all demands (which follow a given popularity distribution), and all users under a given constraint on the *expected* total delivery rate. The previously explained scalable coding approach is used, and uncoded placement and uncoded or coded delivery schemes are designed. Coding is even allowed over layers, as in [19].

For purely uncoded delivery, [18] proposes a water-filling cache allocation algorithm where user k allocates cache size

$$M_{k,j} = \max\left\{0,\ \frac{1}{2}\log_2\left(\sqrt{\frac{2\ln(2p_{k,j}\sigma_j^2)}{\lambda_k^*}}\right)\right\}, \tag{7.95}$$

to layer j, where recall that $p_{k,j}$ denotes the probability that user k wishes to download file W_j, and λ_k^* is a positive normalization parameter that ensures the constraint on the total cache size:

$$\sum_{j=1}^{L} M_{k,j} = M_k, \quad j \in \{1, \ldots, K\}. \tag{7.96}$$

A smaller expected distortion can be achieved by a hybrid uncoded/coded delivery scheme, where files are split into two parts and uncoded delivery is applied to the

[g]This holds under the total cache memory budget constraint, but not under individual cache size constraints.

first part and coded delivery to the second part. More specifically, for the coded delivery scheme [18] proposes to extend *Random Aggregate Popularity caching with Greedy Constrained Coloring (RAP-GCC)*, a decentralized caching algorithm proposed in [39], to account for the variable cache sizes across users and for the fact that different users request a different number of layers. The main novelty in the new algorithm, termed *Greedy Constrained Coloring (RF-GCC)* algorithm, is that users randomly place bits from the different layers that they are interested in, and the random conflict graph describing the delivery is designed across all layers.

7.7 Summary and conclusions

We introduced an RD framework for lossy caching to quantify tradeoffs between delivery rate, distortion, and cache memory sizes. In the special case where the various files are independent of each other, a layered approach based on scalable source coding is proposed. In this approach, each file W^n_ℓ is represented using a set of layers in a way that a user k can reconstruct this ℓ-th file with the desired distortion $D_{k,\ell}$ when learning the first $j_{k,\ell}$ layers. Different strategies have been proposed to place and deliver the various layers to the users, including specific cache allocation algorithms and delivery schemes that code within layers or across layers. Finding the optimal strategy remains an open problem.

In the case where the files in the library are correlated, "common parts" of the various files should be placed in the cache memories of the users, because they will be useful under various demands. This can be made more precise for the single-user caching system. Timo *et al.* showed that under certain symmetry assumptions and for relatively small cache sizes, it is optimal to place the lossy Gács-Körner common part in the cache memory. For sufficiently large cache sizes, they showed that it is optimal to place the lossy Wyner common part. These observations can also be rephrased in terms of CI. The lossy Gács-Körner CI characterizes the largest cache size for which there is no penalty incurred because the server learns the demands only *after* the placement phase. Similarly, the lossy Wyner CI characterizes the smallest cache size where the cache content is *jointly* maximally useful for all possible demands.

The general RD caching problem seems hard. For single-user systems, exploiting connections to the lossy Gray-Wyner source coding network, it is possible to characterize the RDM function. In the context of two users, two specific examples have been considered in the literature: one were only the first user has a cache memory and one where the library contains only two files. For the latter problem, Hassanzadeh *et al.* proposed to perform Gray-Wyner source coding and coded caching separately, and they showed that this approach attains near-optimal RDM tradeoffs. For the setup with a single cache memory, Timo *et al.* proposed a joint architecture, where in particular the delivery information addressed to the cache-free user can be used by the other user to decode its cache content. The proposed scheme achieves the optimal RDM function in some special cases, and it applies adaptive binning and rate-transfer from the placement to the delivery phase.

To summarize, the RD caching problem embraces both correlated files and lossy reconstructions, and remains widely open, especially for larger number of users. Solutions were found in interesting special cases. The main technical ingredients used in RD caching are lossy Gray-Wyner source coding, lossy Wyner CI and lossy Gács-Körner CI, joint source coding and caching, adaptive binning and rate-transfer, scalable coding, and coding within and across layers.

References

[1] Timo R., Saeedi Bidokhti S., Wigger M., Geiger B.C. 'A Rate-Distortion approach to Caching'. *IEEE Transactions on Information Theory*. 2016;**64**(3):1957–76.

[2] Timo R., Saeedi Bidokhti S., Wigger M., Geiger B.C. A Rate-Distortion approach to Caching. *International Zurich Seminar on Communications*; Zurich, March 2 - 4; 2016.

[3] Gastpar M., Op 't Veld G. 'Caching Gaussians: minimizing total correlation on the Gray-Wyner network'. *Annual Conference on Information Science and Systems (CISS)*; 2016. pp. 478–83.

[4] Gastpar M., Op 't Veld G. Caching two Gaussians. *WIC Symposium on Information Theory*; *Benelux*; 2015.

[5] Wyner A. 'The common information of two dependent random variables'. *IEEE Transactions on Information Theory*. 1975;**21**(2):163–79.

[6] Xu G., Liu W., Chen B. 'A Lossy Source Coding Interpretation of Wyner's Common Information'. *IEEE Transactions on Information Theory*. 2016;**62**(2):754–68.

[7] Viswanatha K.B., Akyol E., Rose K. 'The lossy common information of correlated sources'. *IEEE Transactions on Information Theory*. 2014;**60**(6):3238–53.

[8] Gács P., Körner J. 'Common information is far less than mutual information'. *Problems of Control and Information Theory*. 1973;**2**(2):149–62.

[9] Kamath S., Anantharam V. 'A new dual to the Gács-Körner common information defined via the Gray-Wyner system'. *In: 2010 48th annual Allerton conference on communication, control, and computing (Allerton)*; 2010. pp. 1340–6.

[10] Ahlswede R., Körner J. On common information and related characteristics of correlated information sources. *The 7th Prague conference on Information Theory*; *Prague*; 1974.

[11] Hassanzadeh P., Tulino A.M., Llorca J., Erkip E. 'Rate-memory trade-off for the two-user broadcast caching network with correlated sources'. *In: IEEE*

International Symposium on Information Theory (ISIT) Aachen, Germany; 2017. pp. 1227–31.

[12] Hassanzadeh P., Tulino A.M., Llorca J., Erkip E. 'Rate-Memory trade-off for Caching and delivery of correlated sources'. *IEEE Transactions on Information Theory*. 2020;**66**(4):2219–51.

[13] Equitz W.H.R., Cover T.M. 'Successive refinement of information'. *IEEE Transactions on Information Theory*. 1991;**37**(2):269–75.

[14] Rimoldi B. 'Successive refinement of information: characterization of the achievable rates'. *IEEE Transactions on Information Theory*. 1994;**40**(1):253–9.

[15] Diggavi S.N., Chao Tian. 'On multistage successive refinement for Wyner–Ziv source coding with degraded side Informations'. *IEEE Transactions on Information Theory*. 2007;**53**(8):2946–60.

[16] Yang Q., Gündüz D., ISIT. 'Centralized coded caching for heterogeneous lossy requests'. *IEEE International Symposium on Information Theory (ISIT). Barcelona, Spain*; 2016. pp. 405–9.

[17] Yang Q., Gunduz D. 'Coded Caching and content delivery with heterogeneous distortion requirements'. *IEEE Transactions on Information Theory*. 2018;**64**(6):4347–64.

[18] Hassanzadeh P., Tulino A.M., Llorca J., Erkip E. 'Rate-Distortion-Memory trade-offs in heterogeneous Caching networks'. *IEEE Transactions on Wireless Communications*. 2020;**19**(5):3019–33.

[19] Ibrahim A.M., Zewail A.A., Yener A., ITA. 'On Coded Caching with Heterogeneous Distortion Requirements'. *Information Theory and Applications Workshop*; 2018. pp. 1–9.

[20] Saeedi Bidokhti S., Wigger M., Yener A., ICC. 'Gaussian broadcast channels with receiver cache assignment'. *IEEE International Conference on Communications*; 2017. pp. 1–6.

[21] Saeedi Bidokhti S., Wigger M., Yener A. 'Benefits of Cache assignment on degraded broadcast channels'. *IEEE Transactions on Information Theory*. 2019;**65**(11):6999–7019.

[22] Hassanzadeh P., Erkip E., Llorca J. 'Distortion-memory tradeoffs in cache-aided wireless video delivery'. *In: 53rd Annual Allerton Conference on Communication, Control, and Computing. Monticello, IL*; 2015. pp. 1150–7.

[23] Wang C.-Y., Lim S.H., Gastpar M. 'Information-Theoretic Caching: sequential coding for computing'. *IEEE Transactions on Information Theory*. 2016;**62**(11):6393–406.

[24] Lim S.H., Wang C.-Y., Gastpar M. 'Information-Theoretic Caching: the Multi-User case'. *IEEE Transactions on Information Theory*. 2017;**63**(11):7018–37.

[25] Timo R., Wigger M. 'Joint cache-channel coding over erasure broadcast channels'. *In: International Symposium on Wireless Communication Systems*; 2015. pp. 201–5.

[26] Saeedi Bidokhti S., Wigger M., Timo R. 'Noisy broadcast networks with receiver Caching'. *IEEE Transactions on Information Theory*. 2018;**64**(11):6996–7016.

[27] Yang Q., Hassanzadeh P., Gündüz D., Erkip E. 'Centralized Caching and delivery of correlated contents over Gaussian broadcast channels'. *IEEE Transactions on Communications*. 2020;**68**(1):122–36.
[28] Cover T.M., Thomas J.A. *Elements of information theory (Wiley series in telecommunications and signal processing)*. USA: Wiley-Interscience; 2006.
[29] El Gamal A., Kim Y.H. *Network information theory*. Cambridge: Cambridge University Press; 2011.
[30] Visweswariah K., Kulkarni S., Verdú S. 'Output distribution of the Burrows-Wheeler transform'. *In: IEEE International Symposium on information theory (ISIT). Sorrento, Italy*; 2000. p. 53.
[31] Effros M., Visweswariah K., Kulkarni S.R., Verdu S. 'Universal lossless source coding with the Burrows Wheeler transform'. *IEEE Transactions on Information Theory*. 2002;**48**(5):1061–81.
[32] Shkel Y., Verdú S. 'A coding theorem for f-separable distortion measures'. *In: Information Theory and Applications Workshop (ITA). San Diego, CA*; 2016. pp. 1–7.
[33] Gray R.M., Wyner A.D. 'Source coding for a simple network'. *Bell System Technical Journal*. 1974;**53**(9):1681–721.
[34] Timo R., Grant A., Kramer G. 'Lossy Broadcasting with complementary side information'. *IEEE Transactions on Information Theory*. 2013;**59**(1):104–31.
[35] Lapidoth A., Tinguely S. 'Sending a bivariate Gaussian over a Gaussian MAC'. *IEEE Transactions on Information Theory*. 2010;**56**(6):2714–52.
[36] Xiao J.J., Luo Z.Q. 'Compression of correlated Gaussian sources under individual distortion criteria'. *In: 43rd Annual Allerton Conference on Communication, Control, and Computing. Monticello, IL*; 2005. pp. 438–47.
[37] Tian C., Chen J. 'Caching and delivery via interference elimination'. *IEEE Transactions on Information Theory*. 2018;**64**(3):1548–60.
[38] Cao D., Zhang D., Chen P., Liu N., Kang W., Gunduz D. 'Coded Caching with asymmetric Cache sizes and link qualities: the Two-User case'. *IEEE Transactions on Communications*. 2019;**67**(9):6112–26.
[39] Ji M., Tulino A.M., Llorca J., Caire G. 'Order-Optimal rate of Caching and coded Multicasting with random demands'. *IEEE Transactions on Information Theory*. 2017;**63**(6):3923–49.

Chapter 8
Secrecy of edge caching

Mohamed Nafea[1] *and Aylin Yener*[2]

Wireless networks are naturally vulnerable to a variety of security threats. The common thread is to complement developments in any aspect of wireless technologies with security-preserving mechanisms. To this end, we study in this chapter various types of confidentiality concerns and adversarial attacks in caching networks. The underlying model assumptions and network topology are decidedly chosen to abstract the security concern in interest in its most basic formulation. The defining problem, in most of the considered settings, is to study the fundamental trade-off between the cache-memory sizes in the network and the delivery load such that the user requests and the desired security requirements are satisfied. Characterizing such trade-offs is approached by constructing achievability algorithms and deriving impossibility bounds. A particular emphasis is given to a setting in which a powerful strategic adversary optimizes its tapping capability over the two phases of communication in a cache-aided system, i.e., cache placement and delivery.

8.1 Introduction

Coded caching is a promising paradigm for content-centric wireless networks. The challenge in caching systems stems from the fact that the server, that has access to a database of files, has no knowledge about future demands of end users when designing their cache contents. In coded caching, the server jointly optimizes cache placement and content delivery. In particular, in an error-free broadcast system, the users' cache memories not only provide local access to contents but also enable multicasting opportunities during peak traffic times [1]. This leads to a *global caching gain,* achieved by serving multiple requests via multicast signals. As the storage capabilities increase, the required delivery transmission rates during high traffic can decrease, inducing a trade-off between the cache memory sizes and the transmission rate. The proposed scheme in [1] is shown to be order-optimal, i.e., the multiplicative gap between the achievable rate and the lower bound is bounded by a constant.

[1]Electrical and Computer Engineering Department, University of Detroit Mercy, USA
[2]Electrical and Computer Engineering Department, Ohio State University, USA

Caching coded content brings legitimate concerns about confidentiality and privacy of information. External wiretapping adversaries may have access to the delivery transmission and/or cached contents. Further, one may need to ensure that the end user, and only this end user, is able to access the file it requests. This presents a challenge since users who are not requesters of the file store "pieces" of the file. In a different network topology, the server maybe concerned about placing content in intermediate nodes, i.e., relays, that may not be authorized to access content. In all these scenarios, cache placement and delivery must also accompany the security guarantee that no information leakage to an unauthorized party occurs.

In this chapter, we study coded caching under various confidentiality and privacy requirements. First, we consider the presence of a wiretapping adversary which has full access to the delivery transmission [2]. We study the memory-rate trade-off such that the external wiretapper is ignorant of the database files. This requirement is known as *secure delivery* [2, 3]. Next, we consider a privacy setting in which each user is only allowed to recover its requested files, yet gains no information about the remaining files. This requirement is referred to as *private coded caching* (or alternatively, secure caching) [4]. In particular, since user requests are unknown during cache placement, secure caching requirement implies that no user is able to utilize its cache memory to recover any file. Thus, an external adversary with full access to one user's cache contents gains no information about any of the files.

Coded caching has been studied under various modeling assumptions and network topologies, see, for example [5–14]. It is of interest to investigate security concerns in more complex network topologies empowered with coded caching paradigm. To this end, we consider coded caching in combination networks introduced in [10, 11, 15] under confidentiality requirements. In particular, a combination network is a symmetric layered network where the server is connected to a set of h relay nodes communicating with $\binom{h}{r}$ users, and each user is connected to a distinct set of r relay nodes. We study a general combination network equipped with cache memories both at the relays and the end users under the aforementioned secure delivery and private caching requirements [11]. This model enables cooperation between caches from different layers to aid the server.

A common theme among the aforementioned works is the assumption of secure cache placement, i.e., the adversary cannot tap into the cache contents (of more than one user) or into the communication that performs cache placement. At the other extreme, if cache placement were to be public, i.e., the adversary were to have perfect access to all the cache contents in the network, it follows in a straightforward fashion from [16, 17] that the presence of cache memories cannot increase the secrecy capacity. Given the results of these two extreme settings, we consider an intermediate scenario in which the adversary has *partial access* to cache placement [18, 19].

The wiretap channel (WTC) II [20] provides a model for an adversary with partial access to the legitimate communication, in the form of a threshold on the time fraction during which the adversary is capable of overhearing the communication. In particular, the model considers a noiseless legitimate channel and an adversary who *chooses* a *fixed-size* subset of the transmitted symbols to noiselessly observe. Ozarow and Wyner [20] have shown that, despite this ability to choose which symbols to tap into, with proper coding, the adversary can be made no more powerful than nature, i.e., the secrecy capacity of the WTC II is identical to that of a binary erasure WTC with the same fraction of erasures. Nafea and Yener [21] have generalized the model to one with a discrete memoryless, i.e., noisy, legitimate channel. Nafea and Yener [22] have introduced a generalized wiretap model that subsumes both the classical wiretap [23] and wiretap II [21] models as its extreme special cases. This generalized model has been extended to multi-transmitter and multi-receiver networks in [24–26]. In all these models, the common theme is the robustness of stochastic wiretap coding [23] against type II adversaries.

We consider an adversary model of type II as in [20–22, 24–27], but in a cache-aided communication setting [18, 19]. The adversary noiselessly observes a partial subset of its choosing of the transmitted symbols over the cache placement and/or delivery phases. Thus we term this model the caching broadcast channel with a *wire and cache tapping* adversary of type II (CBC-WCT II). The legitimate parties do not know whether cache placement, delivery, or both transmissions are tapped, the relative fractions of tapped symbols in each, nor their positions. Only the knowledge of the overall size of the tapped set is available to the legitimate terminals. We show that even under these unfavorable conditions, strong secrecy guarantees can be provided that are invariant to the positions of the tapped symbols varying between cache placement, delivery, or both.

In coded caching literature up to date, the physical communication that populates the cache memories at end users is not considered in the problem formulation, due to the assumption of secure cache placement. By contrast, in order to model cache placement that is tapped by an adversary, we consider a length-n communication block over a two-user broadcast WTC II [26]. The sizes of cache memories at the users are fixed in this model. We note that introducing variable memory sizes for which a rate-memory trade-off can be characterized, as in the usual setup for caching, requires considering additional communication blocks for cache placement.

Notation: We remark the notation we utilize throughout the chapter. $\mathbb{N}$, $\mathbb{Z}$, and $\mathbb{R}$ denote the sets of natural, integer, and real numbers, respectively. For $a, b \in \mathbb{R}$, $[a : b]$ is the set of integers $\{i \in \mathbb{N} : a \leq i \leq b\}$. For two sets $\mathcal{A}_1$, $\mathcal{A}_2$, $\mathcal{A}_1 \times \mathcal{A}_2$ is their Cartesian product. $\mathcal{A}^T$ denotes the T-fold Cartesian product of $\mathcal{A}$. For $W_1, W_2 \in [1 : M]$, $W_1 \oplus W_2$ denotes the bit-wise XOR on the binary bit strings that correspond to W_1, W_2. $\mathbb{1}_{\mathcal{A}}$ denotes the indicator function for the event $\mathcal{A}$. $\mathbb{D}(p_x \| q_x)$ denotes the Kullback-Leibler divergence between the probability distributions p_x, q_x, defined on the same probability space. $\{\epsilon_n\}_{n \geq 1}$ denotes a sequence of positive real numbers such that $\epsilon_n \to 0$ as $n \to \infty$.

8.2 Coded caching with secure delivery

We study the fundamental security aspects of the caching problem in the presence of an external wiretapping adversary that noiselessly observes the delivery

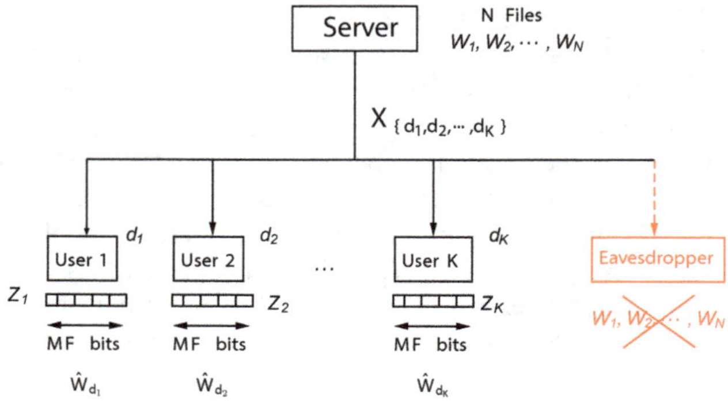

Figure 8.1 System model for caching with secure delivery

transmission. Specifically, the multicast communication between the server and the users occurs over a public insecure error-free channel. By contrast, cache placement occurs over an error-free secure channel, i.e., the adversary is oblivious about the placement phase. This assumption may be justified as cache placement, traditionally, is carried out at off-peak times. This, for instance, allows for secure unicasting at lower transmission rates. This security requirement is termed as secure delivery [2].

Consider the caching model in Figure 8.1. The server has a database with N independent files, $W_{[1:N]} = \{W_1, \cdots, W_N\}$, each is uniform over $[1 : 2^F]$, i.e., each file is of size F bits, $F \in \mathbb{N}$. The server wishes to send K files to the K users according to their demands. Each user has a cache memory Z_k of size MF bits; $1 \leq M \leq N$ is the normalized memory size per user. The communication occurs over two phases:

Cache Placement Phase: The server populates the cache memory of user k, Z_k, with a (possibly randomized) function of its N files, through a unicast transmission. During this phase, the server is unaware of the future demands by users. Thus

$$Z_k \triangleq \mu_k(W_{[1:N]}), \quad k \in [1 : K], \tag{8.1}$$

where $\mu_k(W_{[1:N]}) : [1 : 2^F]^N \mapsto [1 : 2^{MF}]$.

Delivery Phase: At the beginning of the delivery phase, the K users announce their demands $\mathbf{d} = \{d_1, d_2, \cdots, d_K\}$ to the server. In order to satisfy these demands, the server sends a multicast transmission over a noiseless link, which depends on its N files and the users requests. Thus, the multicast transmission can be written as

$$X_{\mathbf{d}} \triangleq \phi_{\mathbf{d}}(W_{[1:N]}), \tag{8.2}$$

where $X_{\mathbf{d}}$ is of size R_sF bits; R_s is the normalized delivery rate. That is, for $\phi_{\mathbf{d}}(W_{[1:N]}) : [1 : 2^F]^N \times [1 : N]^K \mapsto [1 : 2^{R_sF}]$. Using its cached contents, Z_k, and its received signal, $X_{\mathbf{d}}$, user k outputs the estimate $\hat{W}_{d_k}$ of its requested file:

$$\hat{W}_{d_k} \triangleq \psi_{\mathbf{d},k}(X_{\mathbf{d}}, Z_k). \tag{8.3}$$

An external eavesdropper observes the delivery transmission, $X_{\mathbf{d}}$, over the noiseless insecure broadcast link, and is to be kept ignorant of the files $W_{[1:N]}$. A caching scheme thus consists of: (i) K random caching functions $\{\mu_k : k \in [1:K]\}$; (ii) N^K encoding functions $\{\phi_{\mathbf{d}} : \mathbf{d} \in [1:N]^K\}$; and (iii) KN^K decoding functions $\{\psi_{\mathbf{d},k} : k \in [1:K], \mathbf{d} \in [1:N]^K\}$. The memory-rate pair (M, R_s) is said to be securely achievable if there exists a caching scheme such that (s.t.), for any $\epsilon > 0$,

$$P_e = \max_{\mathbf{d}\in[1:N]^K} \max_{k\in[1:K]} \mathbb{P}(\hat{W}_{d_k} \neq W_{d_k}) \leq \epsilon, \tag{8.4}$$

$$I(X_{\mathbf{d}}; W_{[1:N]}) \leq \epsilon. \tag{8.5}$$

The error probability P_e is calculated for the worst-case demands. The security condition in (8.5) implies that the delivery transmission is independent from the database files. For a memory size M, the minimum securely achievable delivery rate is

$$R_s^*(M) = \inf\{R_s : (M, R_s) \text{ is securely achievable}\}. \tag{8.6}$$

The following theorem provides an upper bound on $R_s^*(M)$:
Theorem 1: [2, Theorem 1] For a database with N files, and K users, each having a cache memory of size $M \in \left\{\frac{N-1}{K}t + 1 : t \in [0:K]\right\}$, we have

$$R_s^*(M) \leq R_s^C(M) = K\Big(1 - \frac{M-1}{N-1}\Big)\Big(\frac{1}{1 + K(\frac{M-1}{N-1})}\Big). \tag{8.7}$$

For M s.t. $1 \leq M \leq N$, the lower convex envelop of the points in (8.7) is achievable.

We state the following remarks about Theorem 1.

Remark 1: The upper bound on the achievable delivery rate without the security constraint is $R_s^*(M) \leq K\left(1 - \frac{M}{N}\right)\left(\frac{1}{1+K(\frac{M}{N})}\right)$[1]. Comparing this bound to (8.7), the terms $\frac{M-1}{N-1}$ are replaced with $\frac{M}{N}$, which represent the increase in the minimum delivery rate due to the security constraint, i.e., the security cost.

Remark 2: Unlike the scheme for the case without security in [1], in order to secure the delivery transmission against the eavesdropper, randomness must be introduced into the delivery phase. To do so, the server populates a portion of the cache memory of each user with key bits. During delivery, the server uses these keys to encode the delivery transmission. For instance, when $M = 1$, the server stores a unique key in the cache memory of each user (of the same size as a single file). The server then sends the requested file by each user XOR-ed with its stored key. Thus, the extreme point $(0, N)$ is not feasible under the security constraint. The worst-case delivery rate, N, is achieved at $M = 1$, i.e., the pair $(1, N)$ is an extreme point. On the other extreme, the pair $(N, 0)$ is achievable by storing the whole database at each user.

The proof for Theorem 1 is given in [2, Appendix A]. We present the following example to explain the intuition behind the achievability of the rate in (8.7).

Example 1: Consider $N = K = 2$. Theorem 1 states that, for $M = 1, \frac{3}{2}, 2$, the rates $2, \frac{1}{2}, 0$ are achievable, respectively. The pairs $(1, 2)$ and $(2, 0)$ represent the extreme points explained in Remark 2. It remains to show the pair $(3/2, 1/2)$ is achievable. The server divides the file W_k into the subfiles $\{W_k^{(1)}, W_k^{(2)}\}$, $k = 1, 2$, which are independent and uniform over $[1:2^{\frac{F}{2}}]$. The server generates the key K, independently from W_1, W_2, and uniformly over $[1:2^{\frac{F}{2}}]$. The server populates the cache memory of user k with $Z_k = \{W_1^{(k)}, W_2^{(k)}, K\}$, $k = 1, 2$. During delivery, for the demand $\{d_1, d_2\} \in [1:2]^2$, the server sends $X_{\{d_1,d_2\}} = W_{d_1}^{(1)} \oplus W_{d_2}^{(2)} \oplus K$. Using Z_k and $X_{\{d_1,d_2\}}$, user k successfully recovers its desired file W_{d_k}.

The next theorem provides a lower bound on the minimum delivery rate:

Theorem 2: [2, Theorem 2] For $1 \leq M \leq N$, we have

$$R_s^*(M) \geq \max_{s\in[1:\min\{N,K\}]} \left(s - \frac{s(M-1)}{\lfloor \frac{N}{s} \rfloor - 1}\right). \tag{8.8}$$

The proof for Theorem 2 extends the cut-set bound in [1] to incorporate the security constraint and is given in [2, Appendix B]. The upper and lower bounds in Theorems 1, 2 are within a constant gap, for a certain range of the model parameters.

Theorem 3: [2, Theorem 3] For $\max\{\frac{(K-N)(N-1)}{KN} + 1, 1\} \leq M \leq N$, we have

$$1 \leq \frac{R_s^C(M)}{R_s^*(M)} \leq 17. \tag{8.9}$$

8.3 Private coded caching: secure caching

In the previous section, we have considered an external eavesdropper with full access to the delivery transmission and no access to the cached contents. Here, we consider a confidentiality condition imposed at each user. In particular, each user is only allowed to decode its requested file and should not gain any information about the remaining files, a requirement termed as *private (or secure) caching*. To do so, the cache contents at user k, as well as the delivery transmission, should be independent from all files except the requested file by user k, i.e., W_{d_k} [4]. In addition, user k's cache contents and the delivery transmission should be sufficient for successful recovery of W_{d_k}. This formulation naturally suggests utilizing the idea of secret sharing [28]. In secret sharing, each file is mapped into a set of shares such that the file can be recovered using all the corresponding shares, however, any subset of the shares, of size less than a certain threshold, reveals no information about the file.

We consider the caching system in the previous section without the external eavesdropper and under the private caching requirement, see Figure 8.2. The memory-rate pair (M, R_p), where M is the normalized memory size per user and R_p is the normalized delivery rate, is said to be *privately* achievable if there exists a caching scheme such that the reliability condition in (8.4) is satisfied and, for all $\epsilon > 0$,

$$\max_{\mathbf{d}\in[1:N]^K} \max_{k\in[1:K]} I(X_{\mathbf{d}}, \mathbf{W}_{-k}; W_{d_k}) \leq \varepsilon, \tag{8.10}$$

$\mathbf{W}_{-k} = \{W_{[1:N]}\} \setminus W_{d_k}$. Equation (8.10) represents the private caching requirement. As in (8.6), the minimum privately achievable delivery rate, for a memory size M, is

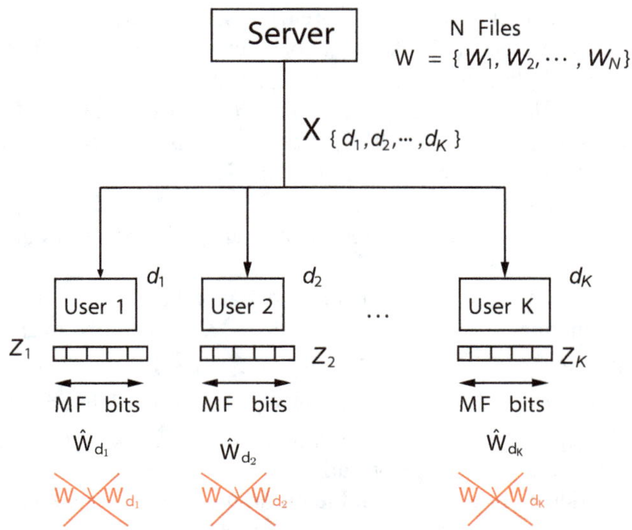

Figure 8.2 *System model for private caching*

$$R_p^*(M) = \inf\{R_p : (M, R_p) \text{ is privately achievable}\}. \tag{8.11}$$

The theorem below provides an upper bound on $R_p^*(M)$:

Theorem 4: [4, Theorem 1] For $M \in \{\frac{Nt}{K-t} + 1 : t \in [0 : K-2]\}$, the rate

$$R_p^*(M) \leq R_p^C(M) = \frac{K(N+M-1)}{N+(K+1)(M-1)}, \tag{8.12}$$

is privately achievable. For $M = N(K-1)$, $R_p^C(M) = 1$ is achievable. For any M s.t. $1 \leq M \leq N(K-1)$, the lower convex envelop of the points in (8.12) is achievable.

Remark 3: Not surprisingly, the minimum privately achievable delivery rate for any private caching scheme with $M \geq 1$ is $R_p^C(M) = 1$, which is achieved at $M = N(K-1)$, i.e., $R_p^*(M) \geq 1$ for any $M \geq 1$. The private caching requirement in (8.10) constraints the cache contents at each user not to reveal any information about the files not requested by this user. Since the user demands are unknown during placement, a delivery transmission of positive rate is necessary for each user to recover its file. Further, $M \geq 1$ is a necessary condition for the existence of a private caching scheme.

The achievability proof for Theorem 4 is given in [4, Section V]. As we mentioned earlier, the achievability under private caching requirement necessitates utilizing a secret sharing scheme [28]. We present the following example to provide the idea behind the achievability of Theorem 4. Let us first state the following definition.

Definition 1: A (m, n) secret sharing scheme, where $m < n$, is a randomized mapping from a uniformly distributed secret W to n equal-sized shares $S_1, S_2, \cdots, S_n$ such that any m shares reveal no information about W and access to all the n shares completely reveals W. That is, $I(W; S_{\mathcal{A}}) = 0$, $\forall \mathcal{A} \subseteq [1 : n]$ s.t. $|\mathcal{A}| \leq m$ and $H(W|S_1, \cdots, S_n) = 0$.

Example 2: Consider $N = K = 2$. Theorem 4 states that, for $M = 1$, the rate $R_p^C(1) = 2$ is achievable. Let K_1, K_2 be two independent keys, each is independent from the files W_1, W_2, and uniform over $[1 : 2^F]$. A simple $(1, 2)$ secret sharing scheme for the file $W_k, k = 1, 2$, is mapping W_k into the shares $\{W_k \oplus K_k,\ K_k\}$. Using both shares, one can recover W_k, yet each share is independent from W_k. The server (i) populates the cache memory of user k with $Z_k = K_k$, $k = 1, 2$, and (ii) during delivery, for $\{d_1, d_2\} \in [1 : 2]^2$, sends $X_{\{d_1,d_2\}} = \{W_{d_1} \oplus K_1, W_{d_2} \oplus K_2\}$. Using Z_k and $X_{\{d_1,d_2\}}$, user k recovers its desired file W_{d_k}, and gains no information about W_j, $j \neq d_k$.

For $M = N(K - 1)$, i.e., $M = 2$, Theorem 4 states that $R_p^C(2) = 1$ is achievable. The server (i) generates the independent keys K_1, K_2 uniformly over $[1 : 2^F]$; (ii) maps W_k into the shares $\{S_k^{(1)} = K_k,\ S_k^{(2)} = W_k \oplus K_k\}$, $k = 1, 2$, and (iii) populates the cache memory of user k with $Z_k = \{S_1^{(j)}, S_2^{(j)}\}, j \neq k$. During delivery, for d_1, d_2, the server sends $X_{\{d_1,d_2\}} = S_{d_1}^{(1)} \oplus S_{d_2}^{(2)}$. Using Z_k and $X_{\{d_1,d_2\}}$, user k successfully recovers its desired file W_{d_k}, and gains no information about W_j, $j \neq d_k$.

We next provide a lower bound on the minimum privately achievable rate:

Theorem 5: [4, Theorem 2] For $1 \leq M \leq N(K - 1)$, we have

$$R_p^*(M) \geq \max_{s \in [1:\min\{\frac{N}{2},K\}]} \left(\frac{s\lfloor \frac{N}{s} \rfloor - 1 - (s-1)M}{\lfloor \frac{N}{s} \rfloor - 1} \right). \tag{8.13}$$

The proof for Theorem 5 utilizes cut-set bound arguments [4, Section V].

The lower bound in (8.13) suggests that for, $N = K = 2$ and $M = 1$, $R_p^*(1) \geq 1$. In Example 2, we have seen an achievability scheme for the upper bound guaranteed by Theorem 4, $R_p^C(1) = 2$. However, the lower bound in (8.13) is in fact achievable, i.e., tight. The optimal scheme is explained as follows:

The server (i) divides $W_k \in [1 : 2^F]$ into the subfiles $\{W_k^{(1)}, W_k^{(2)}\}$, $k = 1, 2$, where $\{W_1^{(1)}, W_1^{(2)}, W_2^{(1)}, W_2^{(2)}\}$ are independent and uniform over $[1 : 2^{\frac{F}{2}}]$; (ii) generates the keys K_1, K_2, independent from one another and from W_1, W_2 and uniform over $[1 : 2^{\frac{F}{2}}]$; (iii) populates the cache memory of user k with $Z_k = \{\{W_1^{(k)} \oplus W_2^{(k)} \oplus K_j\}, K_k\}$, $k, j = 1, 2$ and $j \neq k$. During delivery, for $\{d_1, d_2\}$ where $d_1 \neq d_2$, the server sends $X_{\{d_1,d_2\}} = \{W_{d_1}^{(2)} \oplus K_1, W_{d_2}^{(1)} \oplus K_2\}$. Using Z_k and $X_{\{d_1,d_2\}}$, user k recovers W_{d_k} and gains no information about W_j, $j \neq d_k$ and $j, d_k \in [1, 2]$.

For a certain range of M, the bounds in (8.12) and (8.13) are within a constant gap.

Theorem 6: [4, Theorem 3] For $1 + \max\{0, \frac{N(K-N)}{(N-1)K+N}\} \leq M \leq N(K - 1)$, we have

$$1 \leq \frac{R_p^C(M)}{R_p^*(M)} \leq 16. \tag{8.14}$$

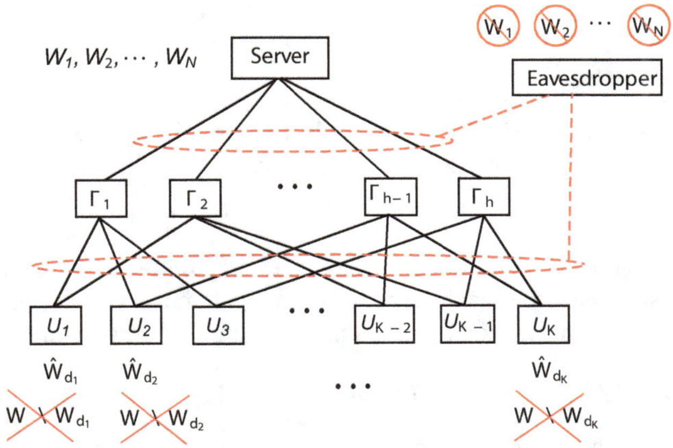

Figure 8.3 *A combination network with secrecy requirements*

8.4 Combination network with security requirements

We consider a cache-aided symmetric two-hop network, known as the combination network [15], under the security requirements presented in Sections 8.2, 8.3 [11, 29]. In such a network, the server is connected to a set of h relay nodes that communicate with $\binom{h}{r}$ users, while each user is connected to a distinct set of r relays. Both the relays and the end users are equipped with cache memories. For the same model without security guarantees, maximum distance separable (MDS) codes are utilized in order to decompose the combination network into h virtual sub-networks, along with jointly optimizing the cache placement and delivery policies [30]. When secure delivery and/or secure caching are considered, the use of MDS codes should be integrated with storing key bits in the user caches and/or secret sharing schemes.

Consider the combination network in Figure 8.3. A server is connected to K users through a layer of h relays. Each user is connected to a distinct set of r relays, where $r < h$. The number of distinct sets of r relays is $\binom{h}{r}$, and thus we consider $K = \binom{h}{r}$. Each relay is connected to $\hat{K} = \binom{h-1}{r-1} = \frac{rK}{h}$ users. All network links are assumed to be noiseless. Let $\mathscr{R} = \{\Gamma_1, \cdots, \Gamma_h\}$, $\mathscr{U} = \{U_1, \cdots, U_K\}$ be the sets of all relays and all users, respectively. $\mathcal{N}(\Gamma_i)$ and $\mathcal{N}(U_k)$ denote the set of end users connected to relay Γ_i and the set of relays connected to user U_k, $i \in [1 : h]$, $k \in [1 : K]$.

A caching scheme is described as follows. The server has a database with N independent files $W_{[1:N]}$, each is uniform over $[1 : 2^F]$. Each relay has a cache memory of size M_rF bits, and each user has a cache memory of size M_uF bits. That is, M_r and M_u are the normalized memory sizes per relay and per user, respectively. For this model, cache placement and delivery are described as follows:

Cache placement: The server populates the caches of relay i and user k with

$$V_i \triangleq \nu_i(W_{[1:N]}), \quad Z_k \triangleq \mu_k(W_{[1:N]}), \quad i \in [1 : h], \quad k \in [1 : K]. \tag{8.15}$$

Delivery: The users announce their demands $\mathbf{d} = \{d_1, \cdots, d_K\}$. The server sends

$$X_{i,\mathbf{d}} \triangleq \phi_{\mathbf{d},i}(W_{[1:N]}), \qquad i \in [1:h] \tag{8.16}$$

to relay i, where $X_{i,\mathbf{d}}$ is of size R_1F bits; R_1 is the normalized rate of the transmitted signal from the server to each relay. Subsequently, relay Γ_i multicasts:

$$Y_{i,\mathbf{d},k} \triangleq \varphi_{\mathbf{d},k}(X_{i,\mathbf{d}}, V_i), \qquad k \in \mathcal{N}(\Gamma_i), \tag{8.17}$$

to $\mathcal{N}(\Gamma_i)$, i.e., its connected users. $Y_{i,\mathbf{d},k}$ is of size R_2F bits; R_2 is the normalized rate of the multicast signal from each relay to its connected users. Using its cache contents and received signals, user k outputs the estimate $\hat{W}_{d_k}$ of its requested file:

$$\hat{W}_{d_k} \triangleq \psi_{\mathbf{d},k}\left(Z_k, \{Y_{i,\mathbf{d},k} : i \in \mathcal{N}(U_k)\}\right), \qquad k \in [1:K]. \tag{8.18}$$

Let $\mathscr{X}, \mathscr{Y}$ denote the sets of all transmitted signals by the server and relays, respectively. The secure delivery and private caching requirements are: For any $\epsilon > 0$,

$$I(\mathscr{X}, \mathscr{Y}; W_{[1:N]}) \leq \epsilon \qquad \text{(Secure Delivery)} \tag{8.19}$$

$$\max_{\mathbf{d}} \max_{k\in[1:K]} I\left(\mathbf{W}_{-k}; Z_k, \{Y_{i,\mathbf{d},k} : i \in \mathcal{N}(U_k)\}\right) \leq \epsilon \quad \text{(Private Caching)} \tag{8.20}$$

The tuple $(M_r, M_u, R_{s,1}, R_{s,2})$ is achievable under secure delivery (against an external eavesdropper) if there is a caching scheme, with relay and end user normalized memory sizes M_r, M_u, and first and second hop normalized rates $R_1 = R_{s,1}$, $R_2 = R_{s,2}$, such that the reliability and security conditions in (8.4) and (8.19) are satisfied. Similarly, a tuple $(M_r, M_u, R_{p,1}, R_{p,2})$ is achievable under private caching if there is a caching scheme, with $R_1 = R_{p,1}$, $R_2 = R_{p,2}$, such that the conditions in (8.4) and (8.20) are satisfied. The following theorems provide upper bounds on the securely and privately achievable rates, for given relay and user memory sizes.

Theorem 7: [11, Theorem 3] For $M_r \geq 0$ and $M_u = 1 + \frac{t_2(N-1)}{\hat{K}} + \frac{(t_1-t_2)r(N-1)M_r}{\hat{K}(N+\hat{K}-t_1)}$, where $t_1, t_2 \in [0:\hat{K}]$ and $\frac{t_1}{\hat{K}} \leq \frac{M_r}{N+\hat{K}-t_1}$, the transmission rates under secure delivery are upper bounded as

$$R_{s,1} \leq \frac{\hat{K}-t_2}{r(t_2+1)}\left(1 - \frac{rM_r}{N+\hat{K}-t_1}\right), \quad R_{s,2} \leq \frac{1}{r}\left(1 - \frac{M_u-1}{N-1}\right). \tag{8.21}$$

The convex envelop of the points in (8.21) is achievable.

Theorem 8: [11, Theorem 4] For $0 \leq M_r \leq \frac{N}{r}$ and $M_u = \frac{tN}{\hat{K}-t}\left(1 - \frac{rM_r}{N}\right)$, where $t \in [0:\hat{K}-1]$, the transmission rates under private caching are upper bounded as

$$R_{p,1} \leq \frac{\hat{K}(N+M_u-rM_r)}{r\left((\hat{K}+1)M_u+N-rM_r\right)}\left(1 - \frac{rM_r}{N}\right), \quad R_{p,2} \leq \frac{1}{r}. \tag{8.22}$$

The convex envelop of the points in (8.22) is achievable.

The proofs for Theorems 7 and 8 are given in [11, Sections IV,V]. Without the security requirements, the main idea behind achievability is to encode each file using an (h, r) MDS code [31]. In particular, using (h, r) MDS codes, each file is mapped into h symbols such that the file can be recovered from any r symbols. Then, each relay acts as a server for one of the resulting encoded symbols. Since each user is connected to r different relay nodes, by the end of the delivery phase, it will be able to obtain r different encoded symbols that can be used to recover its requested file. Under the secure delivery requirement, along with using MDS codes, the server store key bits in the end users caches, and use these keys to encrypt the transmitted signals. For secure caching, the use of MDS codes is complemented with a secret sharing scheme.

8.5 The caching broadcast channel with a wire and cache tapping adversary

So far, we have considered the settings in which cache placement occurs over a secure channel. Here, we introduce the notion of *cache-tapping* into the information theoretic models of coded caching. In particular, an adversary of type II is able to tap into a fixed-size subset of symbols of its choosing either from cache placement, delivery, or both transmissions. We characterize the strong secrecy capacity of the model, i.e., the maximum achievable file rate that keeps the overall database strongly secure, for the instance of a server's database with two files. In particular, we devise an achievability scheme which integrates wiretap coding [20], security embedding codes [32, 33], one-time pad keys [16], *coded* cache placement, and *uncoded* delivery [1]. In addition, we propose a genie-aided upper bound, in which a genie provides the server with user demands before cache placement, rendering the model to a two-user broadcast wiretap channel (WTC) II [26], in order to establish the converse.

For a server's database with more than two files, we derive lower and upper bounds on the strong secrecy file rate. We utilize a coding scheme similar to the scheme we used for two files, however, the cache placement and delivery schemes are different. In particular, we utilize here *uncoded* cache placement and a *partially coded* delivery. We derive the upper bound in three steps. First, we consider a transformed channel with an adversary who can tap an equal fraction of symbols to our model, but is only allowed to tap into the delivery phase. Since this adversary has a more restricted strategy space than the original one, the corresponding secrecy capacity is at least as large as our original model. Next, we further upper bound the secrecy capacity for the restricted adversary model by the secrecy capacity when the adversary encounters a discrete memoryless binary erasure channel. Finally, the secrecy capacity of the discrete memoryless model is upper bounded by the secrecy capacity of a single user setting in which the user requests two files from the database.

8.5.1 System model

Consider the cache-aided communication system depicted in Figure 8.4. The adversary taps into both cache placement and delivery transmissions. The server has a database with $N \geq 2$ independent files, $W_{[1:N]} = W_1, \cdots, W_N$, each is uniform over $[1 : 2^{nR_s}]$. Each user has a cache memory of size $M = \frac{n}{2}$ bits. As in Section 8.2,

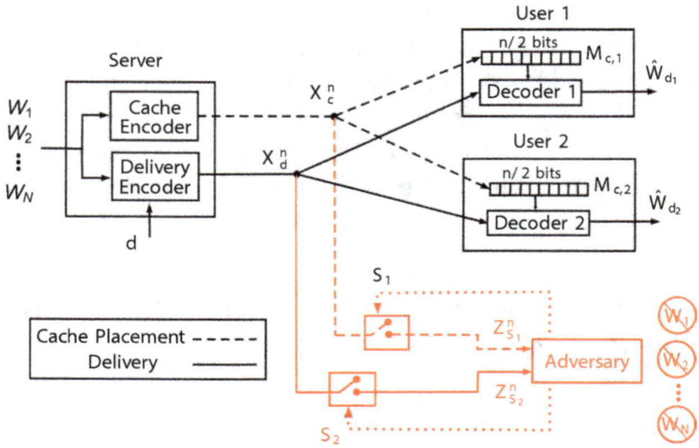

Figure 8.4 *The caching broadcast channel with a wire and cache tapping adversary of type II (CBC-WCT II). The adversary chooses tapping sets* S$_1$ *and* S$_2$ *in cache placement and delivery*

communication occurs over two phases: placement and delivery, and the broadcast channel is noiseless during both phases. The communication model is described as:

Cache placement phase: The server broadcasts a length-n binary signal, $\mathbf{X}_c^n \in \{0,1\}^n$, to both users. The codeword $\mathbf{X}_c^n$ is a function of the database files, i.e., $\mathbf{X}_c^n \triangleq f_c(W_{[1:N]})$. The server does not know the users' demands during cache placement [1]. Each user has a cache memory of size $\frac{n}{2}$ bits in which they store a function of $\mathbf{X}_c^n$, $M_{c,j} \triangleq f_{c,j}(\mathbf{X}_c^n)$, where $f_{c,j} : \{0,1\}^n \mapsto [1:2^{\frac{n}{2}}]$ and $j = 1, 2$.

Delivery phase: At the beginning of the delivery phase, the two users announce their demands $\mathbf{d} \triangleq (d_1, d_2) \in [1:N]^2$. The server, in order to satisfy these demands, encodes $W_{[1:N]}$ and $\mathbf{d}$ into the binary codeword $\mathbf{X}_\mathbf{d}^n \in \{0,1\}^n$. In particular, for each $\mathbf{d}$, the server uses the encoder $f_\mathbf{d} : [1:2^{nR_s}]^N \mapsto \{0,1\}^n$ and sends $\mathbf{X}_\mathbf{d}^n \triangleq f_\mathbf{d}(W_{[1:N]})$.

Decoding: User j uses the decoder $g_{\mathbf{d},j} : [1:2^{\frac{n}{2}}] \times \{0,1\}^n \mapsto [1:2^{nR_s}]$, to output the estimate $\hat{W}_{d_j} \triangleq g_{\mathbf{d},j}(f_{c,j}(\mathbf{X}_c^n), \mathbf{X}_\mathbf{d}^n)$ of its requested file W_{d_j}, $j = 1, 2$.

Adversary model: The adversary chooses two subsets $S_1, S_2 \subseteq [1:n]$. The size of the sum of cardinalities of S_1 and S_2 is fixed, i.e., for $|S_1| = \mu_1$, $|S_2| = \mu_2$, $\mu_1, \mu_2 \leq n$, we have $\mu_1 + \mu_2 = \mu$. The subsets S_1, S_2 indicate the positions tapped by the adversary during cache placement and delivery, respectively. Over the two phases, the adversary observes the length-$2n$ sequence $\mathbf{Z}_S^{2n} = [\mathbf{Z}_{S_1}^n, \mathbf{Z}_{S_2}^n] \in \mathcal{Z}^{2n}$, $\mathbf{Z}_{S_j}^n \triangleq [Z_{S_j,1}, Z_{S_j,2}, \ldots, Z_{S_j,n}] \in \mathcal{Z}^n$, $j = 1, 2$,

$$Z_{S_1,i} = \begin{cases} X_{c,i}, & i \in S_1 \\ ?, & i \notin S_1 \end{cases} \quad \text{and} \quad Z_{S_2,i} = \begin{cases} X_{\mathbf{d},i}, & i \in S_2 \\ ?, & i \notin S_2. \end{cases} \tag{8.23}$$

The alphabet is $\mathcal{Z} = \{0, 1, ?\}$, where "?" denotes an erasure.

The legitimate terminals know neither the realizations of S_1, S_2 nor the values of μ_1, μ_2. Only μ is known. Let $\alpha_1 = \frac{\mu_1}{n}$, $\alpha_2 = \frac{\mu_2}{n}$ denote the fractions of the tapped symbols in cache placement and delivery, and let $\alpha = \alpha_1 + \alpha_2$ be the overall tapped ratio: $\alpha_1, \alpha_2 \in [0,1]$ and $\alpha \in [0,2]$. Notice that α is strictly greater than zero, i.e., the adversary is present. For $\alpha = 0$, i.e., no adversary, the problem reduces to [1].

A channel code $\mathscr{C}_{2n}$ for this model consists of: (i) N message sets, $\mathscr{W}_l \triangleq [1:2^{nR_s}]$, $l \in [1:N]$; (ii) Cache encoder, $f_c : [1:2^{nR_s}]^N \mapsto \{0,1\}^n$; (iii) Cache decoders, $f_{c,j} : \{0,1\}^n \mapsto [1:2^{\frac{n}{2}}]$, $j = 1,2$; (iv) Delivery encoders, $\{f_{\mathbf{d}} : \mathbf{d} \in [1:N]^2\}$, $f_{\mathbf{d}} : [1:2^{nR_s}]^N \mapsto \{0,1\}^n$; and (v) Decoders, $\{g_{\mathbf{d},j} : j = 1,2,\ \mathbf{d} \in [1:N]^2\}$, $g_{\mathbf{d},j} : [1:2^{\frac{n}{2}}] \times \{0,1\}^n \mapsto [1:2^{nR_s}]$. The file rate R_s is said to be achievable with strong secrecy if there exists a sequence $\{\mathscr{C}_{2n}\}_{n \geq 1}$ satisfying

$$\lim_{n\to\infty} \max_{\mathbf{d}\in[1:N]^2} \mathbf{P}\Big(\bigcup_{j=1,2}(\hat{W}_{d_j} \neq W_{d_j})\Big) = 0, \tag{8.24}$$

and

$$\lim_{n\to\infty} \max_{\substack{S_1,S_2\subseteq[1:n]:\\ |S_1|+|S_2|\leq\mu}} I(W_{[1:N]}; \mathbf{Z}^n_{S_1}, \mathbf{Z}^n_{S_2}) = 0. \tag{8.25}$$

That is, R_s is the symmetric secure file rate under any demand vector and adversarial strategy. The strong secrecy capacity C_s is the supremum of all achievable R_s. The condition in (8.25) guarantees strong secrecy against all possible strategies for the adversary, i.e., choices of the subsets S_1, S_2 such that $|S_1| + |S_2| \leq \mu$.

8.5.2 *Main results*

We first study the model in Section 8.5.1 when the server has two files; $N = 2$. We then extend the ideas and the analysis to $N > 2$. For $N > 2$, we utilize a channel coding scheme similar to the scheme we construct for $N = 2$, but the cache placement and delivery schemes which achieve the best rates are different from those for $N = 2$.

The following theorem presents the strong secrecy capacity for $N = 2$.

Theorem 9: For $0 < \alpha \leq 2$, $N = 2$, the strong secrecy capacity for the caching broadcast channel with a wire and cache tapping type II adversary (CBC-WCT II) is

$$C_s(\alpha) = 1 - \frac{\alpha}{2}. \tag{8.26}$$

Proof. The proof is provided in Section 8.5.3.

Theorem 10 presents an achievable strong secrecy file rate for $N > 2$.

Theorem 10: For $0 < \alpha \leq 2$ and $N > 2$, the achievable strong secrecy file rate for the CBC-WCT II is

$$R_s(\alpha) \geq \begin{cases} \frac{1}{2} + \frac{3(1-\alpha)}{4N}, & 0 < \alpha < 1 \\ 1 - \frac{\alpha}{2}, & 1 \leq \alpha \leq 2. \end{cases} \tag{8.27}$$

Proof. The proof is provided in Section 8.5.4.

The following theorem upper bounds the secure file rate when $N > 2$.

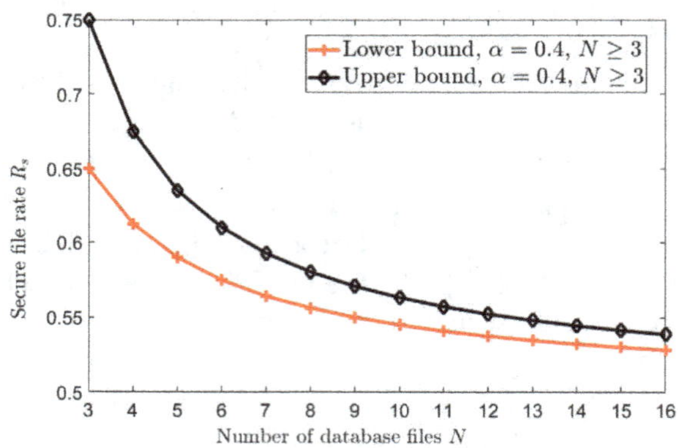

Figure 8.5 *Bounds on the achievable strong secrecy file rate* R_s, $\alpha = 0.4$, $N \geq 3$

Theorem 11: For $0 < \alpha \leq 2$ and $N > 2$, the achievable strong secrecy file rate for the CBC-WCT II is upper bounded as

$$R_s(\alpha) \leq \begin{cases} \frac{1}{2} + \frac{2N-1}{2N(N-1)}(1-\alpha), & 0 < \alpha < 1 \\ 1 - \frac{\alpha}{2}, & 1 \leq \alpha \leq 2. \end{cases} \tag{8.28}$$

Proof. The proof is provided in Section 8.5.5.

The following corollary is immediate from Theorems 9, 10, and 11.

Corollary 1: For $1 \leq \alpha \leq 2$, that is when the adversary can tap longer than one phase of communication, the strong secrecy capacity for the CBC-WCT II is

$$C_s(\alpha) = 1 - \frac{\alpha}{2}. \tag{8.29}$$

Remark 4: When $\alpha \in [1,2]$, i.e., $n \leq \mu \leq 2n$, two possible strategies for the adversary are $\{S_1 = [1:n], S_2 \subset [1:n]\}$ or $\{S_1 \subset [1:n], S_2 = [1:n]\}$: The adversary can tap into all symbols in one phase and a subset of symbols in the other phase. Interestingly, a positive strong secrecy capacity is achievable against such an adversary.

Unlike for $1 \leq \alpha \leq 2$, for $0 < \alpha < 1$, the lower and upper bounds in (8.27), (8.28), have a gap. These bounds are plotted for $\alpha = 0.4$ in Figure 8.5.

Remark 5: When $\alpha = 0$, i.e., no adversary, our scheme for $N > 2$ in Section 8.5.4 reduces to the scheme in [1], which achieves the optimal rate-memory trade-off for two users and a database size of three or larger [34, 35]. The upper bound for $N > 2$ derived in this work is to address the intricacies of the adversarial model and is useful only when the adversary is present ($\alpha > 0$), i.e., (8.28) is loose when $\alpha = 0$.

8.5.3 Proof of Theorem 9

8.5.3.1 Converse

When the demand vector $\mathbf{d} = (d_1, d_2)$ is known to the server during cache placement, the model in Theorem 9 reduces to a broadcast WTC II over a length-$2n$ communication block, whose strong sum secrecy rate is upper bounded as [26, Theorem 1]

$$2R_s \leq 2 - \alpha. \tag{8.30}$$

Equation (8.30) holds for any $\mathbf{d}$ s.t. $d_1 \neq d_2$, i.e., the worst-case demands. Since $\mathbf{d}$ is unknown for the original model, $1 - \frac{\alpha}{2}$ is an upper bound for its strong secrecy capacity.

8.5.3.2 Restricted adversary models as building blocks

Before proceeding with the achievability proof, it is relevant to investigate the secrecy capacity when a known fraction of cache placement, a known fraction of delivery, or both, is tapped. In particular, we consider that the adversary taps into (i) cache placement only, (ii) delivery only, or (iii) both and the relative fractions of tapped symbols in each are known. For these models, we show that the strong secrecy file rate in (8.26), i.e., $1 - \frac{\alpha}{2}$, is achievable, and hence determines their strong secrecy capacities. We then use these models as building blocks for when the relative fractions are *unknown* and provide the proof in Sections 8.5.3.3 and 8.5.3.4.

Setting 1: The adversary taps into cache placement only: This setting corresponds to $\alpha_1 = \alpha$ ($\alpha_2 = 0$) and $|S_1| = \mu$ ($S_2 = \varnothing$). The server knows that $\alpha_1 = \alpha$. The server divides the file W_l, $l = 1, 2$, into three independent subfiles, $W_l^{(1)}$, $W_l^{(2)}$, $W_{l,s}$: $W_l^{(1)}$, $W_l^{(2)}$ are uniform over $[1 : 2^{n\frac{1-\alpha-\epsilon_n}{2}}]$; $W_{l,s}$ is uniform over $[1 : 2^{n\frac{\alpha+\epsilon_n}{2}}]$. Let

$$M_c = \{M_{c,1}, M_{c,2}\}; \quad M_{c,1} = W_1^{(1)} \oplus W_2^{(1)}, \quad M_{c,2} = W_1^{(2)} \oplus W_2^{(2)}, \tag{8.31}$$

$$M_{\mathbf{d}} = \{W_{d_1}^{(2)}, W_{d_2}^{(1)}, W_{d_1,s}, W_{d_2,s}\}. \tag{8.32}$$

During placement, the server maps M_c into $\mathbf{X}_c^n$ using wiretap coding [23]. Since the rate of M_c is less than $1 - \alpha$, M_c is strongly secure from the adversary who observes $n\alpha$ symbols of $\mathbf{X}_c^n$ [22, 27]. During delivery, the server sends $\mathbf{X}_{\mathbf{d}}^n$ as the binary representation of $M_{\mathbf{d}}$ which is of length n bits, since the delivery phase is noiseless and secure. Using $\mathbf{X}_c^n$, user j, $j = 1, 2$, recovers $M_{c,j}$ and stores it in its cache. The size of $M_{c,j}$, for $j = 1, 2$, is smaller than $\frac{n}{2}$ bits (the cache size at each user). Using $\mathbf{X}_{\mathbf{d}}^n$, both users recover $M_{\mathbf{d}}$. Using $M_{\mathbf{d}}$ and $M_{c,j}$, and for n sufficiently large,[a] user j recovers W_{d_j}. The secrecy condition in (8.25) is satisfied [36, Appendix A]. Since $\epsilon_n \to 0$ as $n \to \infty$, the achievable strong secrecy file rate is

$$R_s(\alpha) = 2 \times \frac{1-\alpha}{2} + \frac{\alpha}{2} = 1 - \frac{\alpha}{2}. \tag{8.33}$$

[a]Large block-length n is needed in order to ensure a valid subpacketization of the file into the subfiles $\{W_1^{(1)}, W_1^{(2)}, W_{1,s}\}$, for l=1,2. That is, a bijective map between the file and its subfiles is preserved.

Setting 2: The adversary taps into the delivery only: This setting corresponds to $\alpha_1 = 0$. The server performs the same division of W_1, W_2 as in Setting 1 and generates the keys K_1, K_2, each is uniform over $[1 : 2^{n\frac{\alpha+\epsilon_n}{2}}]$, independent from one another and from W_1, W_2. Define $M_c = \{M_{c,1}, M_{c,2}\}$, $M_{\mathbf{d}}$, and $\tilde{M}_{\mathbf{d}}$ as:

$$M_{c,1} = \{W_1^{(1)} \oplus W_2^{(1)}, K_1\},\ M_{c,2} = \{W_1^{(2)} \oplus W_2^{(2)}, K_2\}, \tag{8.34}$$

$$M_{\mathbf{d}} = \{W_{d_1}^{(2)}, W_{d_2}^{(1)}\},\ \ \tilde{M}_{\mathbf{d}} = \{W_{d_1,s} \oplus K_1, W_{d_2,s} \oplus K_2\}. \tag{8.35}$$

During placement, the server sends $\mathbf{X}_c^n$ as the binary representation of M_c, and user j stores $M_{c,j}$ in its cache. During delivery, the server encodes $M_{\mathbf{d}}$ into $\mathbf{X}_{\mathbf{d}}^n$ using wiretap coding, while using $\tilde{M}_{\mathbf{d}}$ as the randomization message. User j recovers $M_{\mathbf{d}}$ and $\tilde{M}_{\mathbf{d}}$, using which, along with $M_{c,j}$, it decodes W_{d_j}, for sufficiently large n. By contrast, the adversary can only retrieve $\tilde{M}_{\mathbf{d}}$ using which it gains no information about W_1 and W_2. The secrecy condition in (8.25) is satisfied [36, Appendix B]. The achievable strong secrecy file rate is again $1 - \frac{\alpha}{2}$.

Setting 3: The legitimate terminals know the values of α_1 and α_2: For this setting, neither $\alpha_1 = 0$ nor $\alpha_2 = 0$. However, the server and the users know the values of α_1 and α_2. Under these assumptions, the scheme which achieves the strong secrecy rate of $1 - \frac{\alpha}{2}$ depends on whether $\alpha_1 \geq \alpha_2$. In particular, for $\alpha_1 \geq \alpha_2$, we utilize an achievability scheme similar to Setting 1, and for $\alpha_1 < \alpha_2$, we utilize an achievability scheme similar to Setting 2, along with wiretap coding in both phases.

With the aforementioned settings, we showed that the same secrecy file rate, $1 - \frac{\alpha}{2}$, is achievable irrespective of where the adversary taps as long as α_1, α_2 are known. The question then arises whether the lack of knowledge about α_1, α_2 would decrease the secrecy capacity. The following setting provides a hint on the answer.

Setting 4: Either $\alpha_1 = 0$ or $\alpha_2 = 0$, and the legitimate terminals do not know which is zero. That is, the adversary taps into either cache placement or delivery, but not both, and the legitimate terminals *do not* know which phase is tapped. We show that the strong secrecy rate $1 - \frac{\alpha}{2}$ is again achievable.

The server performs the same division of W_1, W_2 as in Settings 1, 2, and generates independent keys K_1, K_2 as in Setting 2. Let $M_c = \{M_{c,1}, M_{c,2}\}$, where

$$M_c = \{M_{c,1}, M_{c,2}\}, \qquad M_{c,1} = W_1^{(1)} \oplus W_2^{(1)}, \quad M_{c,2} = W_1^{(2)} \oplus W_2^{(2)} \tag{8.36}$$

$$\tilde{M}_c = \{\tilde{M}_{c,1}, \tilde{M}_{c,2}\}; \qquad \tilde{M}_{c,1} = K_1, \qquad \tilde{M}_{c,2} = K_2 \tag{8.37}$$

$$M_{\mathbf{d}} = \{W_{d_1}^{(2)}, W_{d_2}^{(1)}\} \tag{8.38}$$

$$\tilde{M}_{\mathbf{d}} = \{\tilde{M}_{\mathbf{d},1}, \tilde{M}_{\mathbf{d},2}\}, \qquad \tilde{M}_{\mathbf{d},1} = W_{d_1,s} \oplus K_1, \quad \tilde{M}_{\mathbf{d},2} = W_{d_2,s} \oplus K_2. \tag{8.39}$$

During placement, the server encodes M_c into $\mathbf{X}_c^n$ using wiretap coding, while using $\tilde{M}_c$ as the randomization message. User $j = 1, 2$, stores $M_{c,j}$, $\tilde{M}_{c,j}$, in its cache. During delivery, the server encodes $M_{\mathbf{d}}$ into $\mathbf{X}_{\mathbf{d}}^n$ using wiretap coding, while using $\tilde{M}_{\mathbf{d}}$ as the randomization message. Using $M_{c,j}$, $M_{\mathbf{d}}$, $\tilde{M}_{\mathbf{d}}$, user j recovers W_{d_j}. By contrast, the adversary can only retrieve either $\{K_1, K_2\}$ or $\{W_{d_1,s} \oplus K_1, W_{d_2,s} \oplus K_2\}$, but not both, using which it gains no information about W_1, W_2. The secrecy condition (8.25) is satisfied [36, Appendix E]. The achievable strong secrecy file rate is $1 - \frac{\alpha}{2}$.

8.5.3.3 Achievability for $\alpha \in (0,1)$

In Setting 4, the lack of knowledge about which phase is tapped is countered by encrypting pieces of information, $\{W_{d_1,s}, W_{d_2,s}\}$, with one-time pad keys, while ensuring that the adversary only retrieves either the keys or the encrypted bits but not both, using which it gains no information about the files. We generalize this idea to tackle the case when the adversary taps into both phases with no knowledge about the relative fractions of tapped symbols in each, i.e., the model in Figure 8.4. In particular, similar to [33], in which the uncertainty about the wiretapper's channel is treated by using a security embedding code [32], here, in each phase, we construct an embedding code in which $n\alpha$ single-bit layers are embedded into one another. Doing so, we ensure that, no matter what the values for α_1, α_2 are, the adversary can retrieve no more than $n\alpha_1$ bits from cache placement and $n\alpha_2$ bits from delivery. By designing what the adversary retrieves to be either a set of key bits and/or information bits encrypted with a distinct set of key bits, we guarantee no information on the files is asymptotically leaked to the adversary. We thus prove the lack of knowledge about relative fractions of tapped symbols *does not decrease* the secrecy capacity.

We are now ready to present the achievability proof for $N = 2$. Consider first $\alpha \in (0,1)$. For simplicity, assume that $n\frac{\alpha_1}{2} = \frac{\mu_1}{2}$ and $n\frac{\alpha_2}{2} = \frac{\mu_2}{2}$ are integers. A minor modification to the analysis can be adopted otherwise. The server (i) divides $W_l, l = 1,2$, into the independent subfiles $W_l^{(1)}, W_l^{(2)}, W_{l,s} : W_l^{(1)}, W_l^{(2)}$ are uniform over $[1 : 2^{n\frac{1-\alpha}{2}}]$, $W_{l,s}$ is uniform over $[1 : 2^{n\frac{\alpha}{2}}]$, and (ii) generates the independent keys K_1, K_2, uniform over $[1 : 2^{n\frac{\alpha}{2}}]$, and independent from W_1, W_2. For simplicity, we have ignored the small rate reduction ϵ_n at this stage, and will introduce this later into the security analysis. The main ideas of the achievability proof are:

1. The server uses wiretap coding with a randomization message of size $n(\alpha_1 + \alpha_2) = n\alpha$ bits in *both cache placement and delivery*. As the adversary does not tap into more than $n\alpha$ bits in each phase, a secure transmission rate of $1 - \alpha$ is achievable in each phase, as long as the randomization messages in the two phases are independent. Using coded placement for $W_1^{(1)}, W_1^{(2)}, W_2^{(1)}, W_2^{(2)}$, a secure file rate of $1 - \alpha$ can be achieved.
2. The randomization messages over the two phases can deliver additional secure information, of rate $\frac{\alpha}{2}$ per file, via encryption. The overall achievable file rate is thus $R_s = 1 - \frac{\alpha}{2}$. In particular, we utilize the keys K_1, K_2, as the randomization message for cache placement. Along with wiretap coding, we employ a security embedding code by using bits of K_1, K_2, in a manner that allows the adversary to be able to retrieve only the last $n\frac{\alpha_1}{2}$ bits from each. In the delivery phase, we encrypt additional pieces of information, $W_{d_1,s}$ and $W_{d_2,s}$, with the keys K_1, K_2, and utilize this encrypted information as the randomization message. We employ again a security embedding code, in the *reverse order*, such that the adversary can only retrieve the first $n\frac{\alpha_2}{2}$ bits from each of $W_{d_1,s} \oplus K_1, W_{d_2,s} \oplus K_2$.
3. With the aforementioned construction, the adversary, for any values of α_1, α_2 it chooses, can only retrieve a set of key bits and/or information bits

encrypted with other key bits. Due to the *reversed embedding order*, the adversary does not obtain, in the delivery phase, any file bits encrypted with key bits it has seen during cache placement. Since $\{K_1, K_2\}$ is independent from $\{W_{d_1,s} \oplus K_1, W_{d_2,s} \oplus K_2\}$, and is an independent sequence, the adversary cannot use the revealed key bits in cache placement to obtain any information about the bits of $W_{d_1,s} \oplus K_1, W_{d_2,s} \oplus K_2$ that need to be securely transmitted in the delivery.

Let M_c, $\tilde{M}_c$ be as in (8.36) and (8.37). M_c is the message to be securely sent during placement regardless of the adversary's choice of α_1. $\tilde{M}_c$ is the randomization message used for wiretap coding. The server further divides $\tilde{M}_{c,1}$, $\tilde{M}_{c,2}$ into sequences of independent binary bits $\{\tilde{M}_{c,1}^{(1)}, \cdots, \tilde{M}_{c,1}^{(n\frac{\alpha}{2})}\}$, $\{\tilde{M}_{c,2}^{(1)}, \cdots, \tilde{M}_{c,2}^{(n\frac{\alpha}{2})}\}$, and generates $\mathbf{X}_c^n$:

Cache Placement Codebook Generation: Let m_c, $\tilde{m}_{c,1} = \{\tilde{m}_{c,1}^{(1)}, \cdots, \tilde{m}_{c,1}^{(n\frac{\alpha}{2})}\}$, and $\tilde{m}_{c,2} = \{\tilde{m}_{c,2}^{(1)}, \cdots, \tilde{m}_{c,2}^{(n\frac{\alpha}{2})}\}$ be the realizations of M_c, $\tilde{M}_{c,1}$ and $\tilde{M}_{c,2}$ in (8.36) and (8.37). We construct the cache placement codebook $\mathcal{C}_{c,n}$, from which $\mathbf{X}_c^n$ is drawn, as follows. We randomly and independently distribute all the possible 2^n length-n binary sequences into $2^{n(1-\alpha)}$ bins, indexed by $m_c \in [1 : 2^{n\frac{1-\alpha}{2}}]^2$. Each bin m_c contains $2^{n\alpha}$ binary sequences (codewords). Further, we randomly and independently divide each bin m_c into two sub-bins, indexed by $\tilde{m}_{c,1}^{(1)}$, and each contains $2^{n\alpha-1}$ codewords. The two sub-bins $\tilde{m}_{c,1}^{(1)}$ are further divided into smaller bins, indexed by $\tilde{m}_{c,2}^{(1)}$, and each contains $2^{n\alpha-2}$ codewords. The process continues, going over $\tilde{m}_{c,1}^{(2)}$, $\tilde{m}_{c,2}^{(2)}, \cdots,$ $\tilde{m}_{c,1}^{(n\frac{\alpha}{2}-1)}$, $\tilde{m}_{c,2}^{(n\frac{\alpha}{2}-1)}$, $\tilde{m}_{c,1}^{(n\frac{\alpha}{2})}$, until the remaining two codewords, after each sequence of divisions, are indexed by $\tilde{m}_{c,2}^{(n\frac{\alpha}{2})}$. The codebook $\mathscr{C}_{c,n}$ is described in Figure 8.6.

Cache Encoder: Given w_1, w_2, i.e., $\{w_1^{(1)}, w_1^{(2)}, w_{1,s}\}$, $\{w_2^{(1)}, w_2^{(2)}, w_{2,s}\}$, the server generates m_c, $\tilde{m}_c = \{\tilde{m}_{c,1}, \tilde{m}_{c,2}\}$ as in (8.36), (8.37). Using $\mathcal{C}_{c,n}$, the server sends $\mathbf{x}_c^n$ which corresponds to m_c, $\tilde{m}_{c,1}$, $\tilde{m}_{c,2}$, i.e., $\mathbf{x}_c^n\left(m_c, \tilde{m}_{c,1}^{(1)}, \tilde{m}_{c,2}^{(1)}, \cdots, \tilde{m}_{c,1}^{(n\frac{\alpha}{2})}, \tilde{m}_{c,2}^{(n\frac{\alpha}{2})}\right)$.

For the delivery phase, let $M_\mathbf{d}$ and $\tilde{M}_\mathbf{d}$ be defined as in (8.38) and (8.39). $M_\mathbf{d}$ is the message to be securely sent during delivery regardless the adversary's choice of α_2. $\tilde{M}_\mathbf{d}$ is the randomization message used for the wiretap coding. Similar to placement, the server further divides $\tilde{M}_{\mathbf{d},1}$, $\tilde{M}_{\mathbf{d},2}$ into sequences of independent binary bits, $\{\tilde{M}_{\mathbf{d},1}^{(1)}, \cdots, \tilde{M}_{\mathbf{d},1}^{(n\frac{\alpha}{2})}\}$, $\{\tilde{M}_{\mathbf{d},2}^{(1)}, \cdots, \tilde{M}_{\mathbf{d},2}^{(n\frac{\alpha}{2})}\}$, and generates $\mathbf{X}_\mathbf{d}^n$ as follows.

Delivery Codebook Generation: Let $m_\mathbf{d}$, $\tilde{m}_{\mathbf{d},1} = \{\tilde{m}_{\mathbf{d},1}^{(1)}, \cdots, \tilde{m}_{\mathbf{d},1}^{(n\frac{\alpha}{2})}\}$, and $\tilde{m}_{\mathbf{d},2} = \{\tilde{m}_{\mathbf{d},2}^{(1)}, \cdots, \tilde{m}_{\mathbf{d},2}^{(n\frac{\alpha}{2})}\}$ be the realizations of $M_\mathbf{d}$, $\tilde{M}_{\mathbf{d},1}$, $\tilde{M}_{\mathbf{d},2}$ in (8.38) and (8.39). We construct the delivery codebook $\mathscr{C}_{\mathbf{d},n}$, from which $\mathbf{X}_\mathbf{d}^n$ is drawn, in a similar fashion as $\mathscr{C}_{c,n}$, but with a reversed indexing of the sub-bins. In particular, we randomly and independently divide all the 2^n binary sequences into $2^{n(1-\alpha)}$ bins, indexed by $m_\mathbf{d} \in [1 : 2^{n\frac{1-\alpha}{2}}]^2$, and each contains $2^{n\alpha}$ codewords. We further randomly and independently divide each bin $m_\mathbf{d}$ into two sub-bins, indexed by $\tilde{m}_{\mathbf{d},1}^{(n\frac{\alpha}{2})}$, and each contains $2^{n\alpha-1}$ codewords. The process continues, going in reverse order

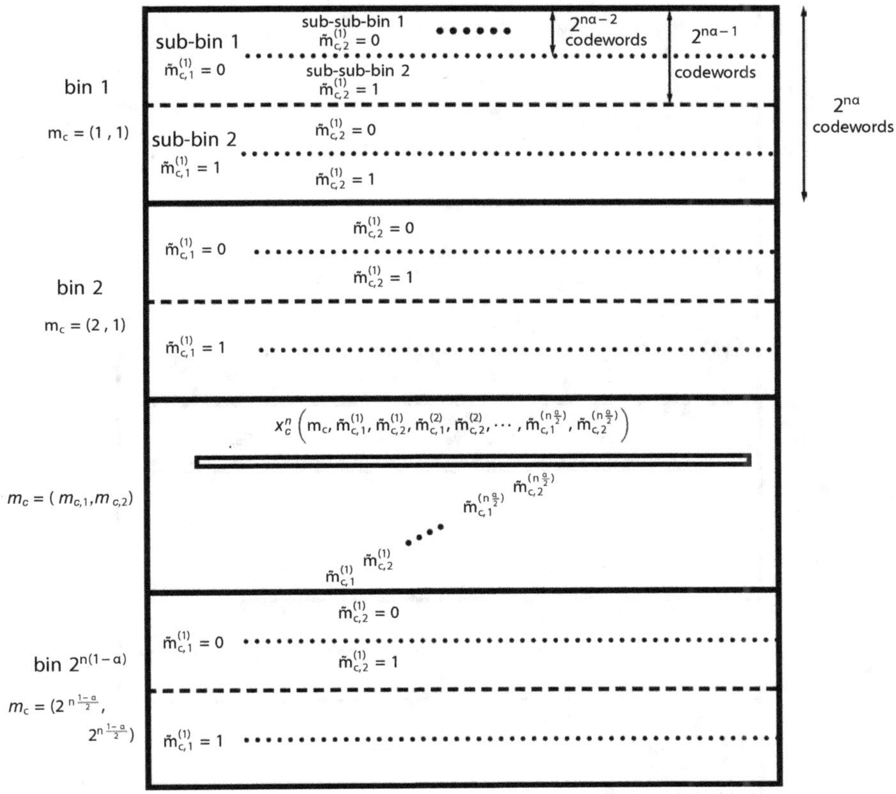

Figure 8.6 Codebook construction for the cache placement phase, $\mathscr{C}_{c,n}$

over $\tilde{m}_{\mathbf{d},2}^{(n\frac{\alpha}{2})}, \tilde{m}_{\mathbf{d},1}^{(n\frac{\alpha}{2}-1)}, \tilde{m}_{\mathbf{d},2}^{(n\frac{\alpha}{2}-1)}, \ldots, \tilde{m}_{\mathbf{d},1}^{(1)}$, until the remaining two codewords, after each sequence of divisions, are indexed by $\tilde{m}_{\mathbf{d},2}^{(1)}$. The codebook $\mathscr{C}_{\mathbf{d},n}$ is described in Figure 8.7.

Delivery Encoder: Given w_1, w_2, and $\mathbf{d} = (d_1, d_2)$, the server generates $m_{\mathbf{d}}$, $\tilde{m}_{\mathbf{d}} = \{\tilde{m}_{\mathbf{d},1}, \tilde{m}_{\mathbf{d},2}\}$ as in (8.38) and (8.39). The server sends $\mathbf{x}_{\mathbf{d}}^n$, from $\mathscr{C}_{\mathbf{d},n}$, which corresponds to $m_{\mathbf{d}}$, $\tilde{m}_{\mathbf{d},1}$, and $\tilde{m}_{\mathbf{d},2}$, i.e., $\mathbf{x}_{\mathbf{d}}^n\left(m_{\mathbf{d}}, \tilde{m}_{\mathbf{d},1}^{(n\frac{\alpha}{2})}, \tilde{m}_{\mathbf{d},2}^{(n\frac{\alpha}{2})}, \cdots, \tilde{m}_{\mathbf{d},1}^{(1)}, \tilde{m}_{\mathbf{d},2}^{(1)}\right)$.

Decoding: Using $\mathbf{X}_c^n$, user j, $j = 1, 2$, recovers $M_{c,j}$, $\tilde{M}_{c,j}$, and stores them in its cache memory. For $j = 1, 2$, the combined size of $M_{c,j}$ and $\tilde{M}_{c,j}$ does not exceed $\frac{n}{2}$ bits. Using $\mathbf{X}_{\mathbf{d}}^n$, both users recover $M_{\mathbf{d}}$, $\tilde{M}_{\mathbf{d}}$. Using $M_{\mathbf{d}}$, $\tilde{M}_{\mathbf{d}}$, $M_{c,j}$, $\tilde{M}_{c,j}$, and for n sufficiently large, user j correctly decodes W_{d_j}.

Security Analysis: Let us slightly modify the construction above as follows. $\{\epsilon_n\}_{n\geq 1}$ is a sequence of positive real numbers such that $\epsilon_n \to 0$ as $n \to \infty$. Define

$$\alpha_\epsilon = \alpha + 2\epsilon_n, \qquad \alpha_{1,\epsilon} = \alpha_1 + \epsilon_n, \qquad \alpha_{2,\epsilon} = \alpha_2 + \epsilon_n. \tag{8.40}$$

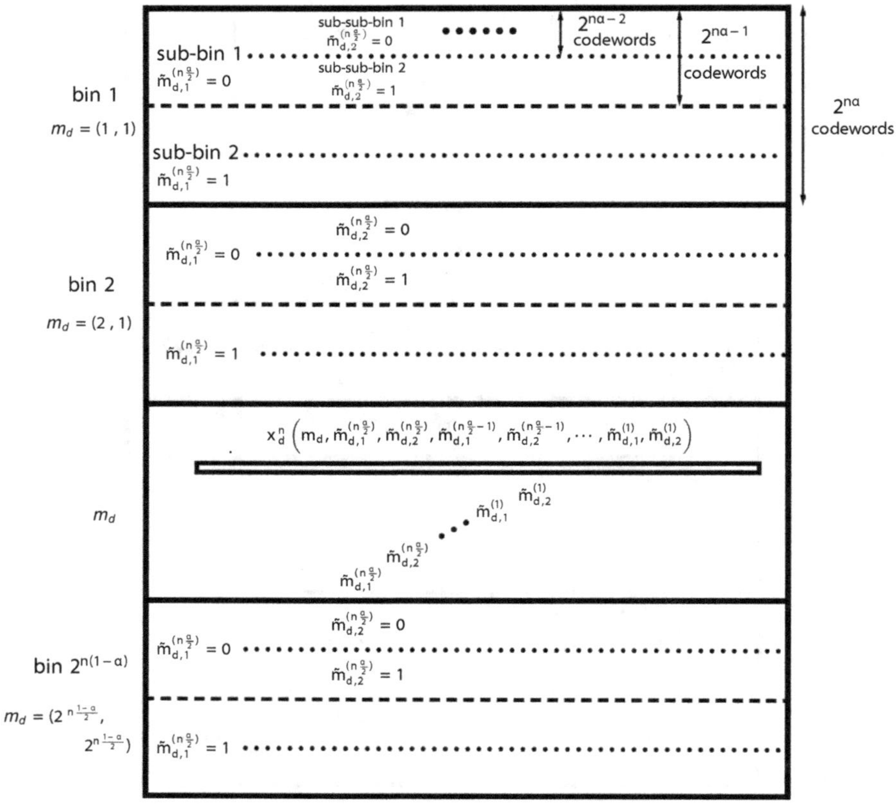

Figure 8.7 Codebook construction for the delivery phase, $\mathscr{C}_{\mathbf{d},n}$

That is, $\alpha_{1,\epsilon} + \alpha_{2,\epsilon} = \alpha_\epsilon$. We increase the sizes of K_1, K_2 into $\frac{n\alpha_\epsilon}{2}$ bits, from $n\frac{\alpha}{2}$, and zero-pad the bit strings of $W_{d_1,s}, W_{d_2,s}$ accordingly. We also decrease the sizes of $W_l^{(1)}, W_l^{(2)}$, $l = 1, 2$, to $n\frac{1-\alpha_\epsilon}{2}$ bits, instead of $n\frac{1-\alpha}{2}$. Once again, we assume that $\frac{n\alpha_\epsilon}{2}$ and $\frac{n\alpha_{1,\epsilon}}{2}$ are integers; as minor modifications can be adopted otherwise.

Fix $S_1, S_2 \subseteq [1:n]$. For the corresponding values of α_1, α_2, the placement codebook $\mathscr{C}_{c,n}$ can be viewed as a wiretap code with $2^{n(1-\alpha_{1,\epsilon})}$ bins, each is indexed by

$$w_c = (m_c, \tilde{m}_{c,1}^{(1)}, \tilde{m}_{c,2}^{(1)}, \tilde{m}_{c,1}^{(2)}, \tilde{m}_{c,2}^{(2)}, \cdots, \tilde{m}_{c,1}^{(n\frac{\alpha_{2,\epsilon}}{2})}, \tilde{m}_{c,2}^{(n\frac{\alpha_{2,\epsilon}}{2})}). \tag{8.41}$$

Each bin w_c contains $2^{n\alpha_{1,\epsilon}}$ codewords, indexed by the randomization message

$$\tilde{w}_c = (\tilde{m}_{c,1}^{(n\frac{\alpha_{2,\epsilon}}{2}+1)}, \tilde{m}_{c,2}^{(n\frac{\alpha_{2,\epsilon}}{2}+1)}, \tilde{m}_{c,1}^{(n\frac{\alpha_{2,\epsilon}}{2}+2)}, \tilde{m}_{c,2}^{(n\frac{\alpha_{2,\epsilon}}{2}+2)}, \cdots, \tilde{m}_{c,1}^{(n\frac{\alpha_\epsilon}{2})}, \tilde{m}_{c,2}^{(n\frac{\alpha_\epsilon}{2})}). \tag{8.42}$$

Similarly, the delivery codebook $\mathcal{C}_{\mathbf{d},n}$ can be seen as a wiretap code with $2^{n(1-\alpha_{2,\epsilon})}$ bins, each of which is indexed by the message

$$w_{\mathbf{d}} = (m_{\mathbf{d}}, \tilde{m}_{\mathbf{d},1}^{(n\frac{\alpha_\epsilon}{2})}, \tilde{m}_{\mathbf{d},2}^{(n\frac{\alpha_\epsilon}{2})}, \tilde{m}_{\mathbf{d},1}^{(n\frac{\alpha_\epsilon}{2}-1)}, \tilde{m}_{\mathbf{d},2}^{(n\frac{\alpha_\epsilon}{2}-1)}, \cdots, \tilde{m}_{\mathbf{d},1}^{(n\frac{\alpha_{2,\epsilon}}{2}+1)}, \tilde{m}_{\mathbf{d},2}^{(n\frac{\alpha_{2,\epsilon}}{2}+1)}). \tag{8.43}$$

Each bin $w_{\mathbf{d}}$ contains $2^{n\alpha_{2,\epsilon}}$ codewords, indexed by the randomization message

$$\tilde{w}_{\mathbf{d}} = (\tilde{m}_{\mathbf{d},1}^{(n\frac{\alpha_{2,\epsilon}}{2})}, \tilde{m}_{\mathbf{d},2}^{(n\frac{\alpha_{2,\epsilon}}{2})}, \tilde{m}_{\mathbf{d},1}^{(n\frac{\alpha_{2,\epsilon}}{2}-1)}, \tilde{m}_{\mathbf{d},2}^{(n\frac{\alpha_{2,\epsilon}}{2}-1)}, \cdots, \tilde{m}_{\mathbf{d},1}^{(1)}, \tilde{m}_{\mathbf{d},2}^{(1)}). \tag{8.44}$$

Let $\{\mathscr{B}_{w_c} : w_c \in [1 : 2^{n(1-\alpha_{1,\epsilon})}]\}$, $\{\mathscr{B}_{w_{\mathbf{d}}} : w_{\mathbf{d}} \in [1 : 2^{n(1-\alpha_{2,\epsilon})}]\}$ denote the partition, i.e., bins, of $\mathscr{C}_{c,n}$, $\mathscr{C}_{\mathbf{d},n}$, which correspond to the messages w_c, $w_{\mathbf{d}}$ in (8.41) and (8.43), respectively. Let $\mathbf{x}^{2n} \triangleq (\mathbf{x}_c^n, \mathbf{x}_{\mathbf{d}}^n)$ denote the concatenation of the two length-n binary codewords $\mathbf{x}_c^n$, $\mathbf{x}_{\mathbf{d}}^n$. Define the Cartesian product of the bins $\mathscr{B}_{w_c}$ and $\mathscr{B}_{w_{\mathbf{d}}}$, as

$$\mathscr{B}_{w_c,w_{\mathbf{d}}} \triangleq \{\mathbf{x}^{2n} = (\mathbf{x}_c^n, \mathbf{x}_{\mathbf{d}}^n) : \mathbf{x}_c^n \in \mathcal{B}_{w_c}, \mathbf{x}_{\mathbf{d}}^n \in \mathscr{B}_{w_{\mathbf{d}}}\}. \tag{8.45}$$

Since the partitioning of $\mathscr{C}_{c,n}, \mathscr{C}_{\mathbf{d},n}$ is random, for every $w_c, w_{\mathbf{d}}$, $\mathscr{B}_{w_c,w_{\mathbf{d}}}$ is a random codebook, which results from the Cartesian product of the random bins $\mathscr{B}_{w_c}$, $\mathscr{B}_{w_{\mathbf{d}}}$. Recall that $\mathscr{B}_{w_c}$ contains $2^{n\alpha_{1,\epsilon}}$ and $\mathscr{B}_{w_{\mathbf{d}}}$ contains $2^{n\alpha_{2,\epsilon}}$ length-n binary codewords. Thus, the product $\mathscr{B}_{w_c,w_{\mathbf{d}}}$ contains $2^{n\alpha_\epsilon}$ length-$2n$ binary codewords.

Let $\{W_{d_l,s}^{(1)}, W_{d_l,s}^{(2)}, \cdots, W_{d_l,s}^{(n\frac{\alpha_\epsilon}{2})}\}$ and $\{K_l^{(1)}, K_l^{(2)}, \cdots, K_l^{(n\frac{\alpha_\epsilon}{2})}\}$ denote the binary bit strings of $W_{d_l,s}$ and K_l, $l = 1, 2$. In addition, for notational simplicity, define

$$\mathbf{W}_s^{(1)} = \{W_{d_1,s}^{(1)}, W_{d_2,s}^{(1)}, \cdots, W_{d_1,s}^{(n\frac{\alpha_{2,\epsilon}}{2})}, W_{d_2,s}^{(n\frac{\alpha_{2,\epsilon}}{2})}\}, \tag{8.46}$$

$$\mathbf{W}_s^{(2)} = \{W_{d_1,s}^{(n\frac{\alpha_{2,\epsilon}}{2}+1)}, W_{d_2,s}^{(n\frac{\alpha_{2,\epsilon}}{2}+1)}, \cdots, W_{d_1,s}^{(n\frac{\alpha_\epsilon}{2})}, W_{d_2,s}^{(n\frac{\alpha_\epsilon}{2})}\}, \tag{8.47}$$

$$\mathbf{K}^{(1)} = \{K_1^{(1)}, K_2^{(1)}, \cdots, K_1^{(n\frac{\alpha_{2,\epsilon}}{2})}, K_2^{(n\frac{\alpha_{2,\epsilon}}{2})}\}, \tag{8.48}$$

$$\mathbf{K}^{(2)} = \{K_1^{(n\frac{\alpha_{2,\epsilon}}{2}+1)}, K_2^{(n\frac{\alpha_{2,\epsilon}}{2}+1)}, \cdots, K_1^{(n\frac{\alpha_\epsilon}{2})}, K_2^{(n\frac{\alpha_\epsilon}{2})}\}, \tag{8.49}$$

$$\mathbf{W}_{\oplus\mathbf{K}}^{(1)} = \{W_{d_1,s}^{(i)} \oplus K_1^{(i)}, W_{d_2,s}^{(i)} \oplus K_2^{(i)} : \ i = 1, 2, \cdots, n\frac{\alpha_{2,\epsilon}}{2}\}, \tag{8.50}$$

$$\mathbf{W}_{\oplus\mathbf{K}}^{(2)} = \{W_{d_1,s}^{(i)} \oplus K_1^{(i)}, W_{d_2,s}^{(i)} \oplus K_2^{(i)} : \ i = n\frac{\alpha_{2,\epsilon}}{2} + 1, \ n\frac{\alpha_{2,\epsilon}}{2} + 2, \cdots, n\frac{\alpha_\epsilon}{2}\}. \tag{8.51}$$

Let W_c, $\tilde{W}_c$, $W_{\mathbf{d}}$, $\tilde{W}_{\mathbf{d}}$ denote the random variables that correspond to the realizations defined in (8.41)–(8.44). Using (8.36)–(8.39), (8.41)–(8.44), and (8.48)–(8.51),

$$W_c = \{M_c, \mathbf{K}^{(1)}\} = \{W_1^{(1)} \oplus W_2^{(1)}, W_1^{(2)} \oplus W_2^{(2)}, \mathbf{K}^{(1)}\}, \qquad \tilde{W}_c = \mathbf{K}^{(2)} \tag{8.52}$$

$$W_{\mathbf{d}} = \{M_{\mathbf{d}}, \mathbf{W}_{\oplus\mathbf{K}}^{(2)}\} = \{W_{d_1}^{(2)}, W_{d_2}^{(1)}, \mathbf{W}_{\oplus\mathbf{K}}^{(2)}\}, \qquad \tilde{W}_{\mathbf{d}} = \mathbf{W}_{\oplus\mathbf{K}}^{(1)}. \tag{8.53}$$

$\tilde{W}_c$, $\tilde{W}_{\mathbf{d}}$ are independent, and each is uniform. Thus, $\{\tilde{W}_c, \tilde{W}_{\mathbf{d}}\}$ is jointly uniform. $\{\tilde{W}_c, \tilde{W}_{\mathbf{d}}\}$ is independent from $\{W_c, W_{\mathbf{d}}\}$. Thus, we can apply [26, 94–103] to show that, for every S_1, S_2, w_c, and $w_{\mathbf{d}}$, and every $\epsilon > 0$, there exists $\gamma(\epsilon) > 0$ s.t.

$$\mathbb{P}_{\mathscr{B}_{w_c,w_{\mathbf{d}}}}(\mathbb{D}(P_{\mathbf{Z}^n_{S_1}\mathbf{Z}^n_{S_2}|W_c=w_c,W_{\mathbf{d}}=w_{\mathbf{d}}}||P_{\mathbf{Z}^n_{S_1}\mathbf{Z}^n_{S_2}})>\epsilon)\leq\exp(-e^{n\gamma(\epsilon)}). \tag{8.54}$$

$P_{\mathbf{Z}^n_{S_1}\mathbf{Z}^n_{S_2}|W_c=w_c,W_{\mathbf{d}}=w_{\mathbf{d}}}$ is the induced distribution at the adversary when $\mathbf{x}^n_c(w_c,\tilde{w}_c)$ and $\mathbf{x}^n_{\mathbf{d}}(w_{\mathbf{d}},\tilde{w}_{\mathbf{d}})$ are the transmitted codewords over cache placement and delivery phases. $P_{\mathbf{Z}^n_{S_1}\mathbf{Z}^n_{S_2}}$ is the output distribution at the adversary.

The number of the messages $\{w_c,w_{\mathbf{d}}\}$ is $2^{n(2-\alpha_\epsilon)}$, and the number of the subsets S_1,S_2 is $\binom{2n}{\alpha n}<2^{2n}$. The combined number of the messages and the subsets is at most exponential in n. Using (8.54) and the union bound, as in [22, 27], we have

$$\lim_{n\to\infty}\max_{S_1,S_2} I(W_c,W_{\mathbf{d}};\mathbf{Z}^n_{S_1},\mathbf{Z}^n_{S_2})=0. \tag{8.55}$$

We also have, for any $\mathbf{d}=(d_1,d_2)$, $d_1,d_2\in\{1,2\}$,

$$I(W_1,W_2;\mathbf{Z}^n_{S_1},\mathbf{Z}^n_{S_2})=I(W_1^{(1)},W_1^{(2)},W_2^{(1)},W_2^{(2)},W_{1,s},W_{2,s};\mathbf{Z}^n_{S_1},\mathbf{Z}^n_{S_2}) \tag{8.56}$$

$$=I(W_1^{(1)},W_1^{(2)},W_2^{(1)},W_2^{(2)},\mathbf{W}_s^{(1)},\mathbf{W}_s^{(2)};\mathbf{Z}^n_{S_1},\mathbf{Z}^n_{S_2}) \tag{8.57}$$

$$=I(W_1^{(1)}\oplus W_2^{(1)},W_1^{(2)}\oplus W_2^{(2)},W_{d_1}^{(2)},W_{d_2}^{(1)},\mathbf{W}_s^{(1)},\mathbf{W}_s^{(2)};\mathbf{Z}^n_{S_1},\mathbf{Z}^n_{S_2}) \tag{8.58}$$

$$=I(M_c,M_{\mathbf{d}},\mathbf{W}_s^{(1)},\mathbf{W}_s^{(2)};\mathbf{Z}^n_{S_1},\mathbf{Z}^n_{S_2}) \tag{8.59}$$

$$\leq I(M_c,M_{\mathbf{d}},\mathbf{W}_s^{(1)},\mathbf{W}_{\oplus\mathbf{K}}^{(2)};\mathbf{Z}^n_{S_1},\mathbf{Z}^n_{S_2}) \tag{8.60}$$

$$=I(M_c,\mathbf{W}_s^{(1)},W_{\mathbf{d}};\mathbf{Z}^n_{S_1},\mathbf{Z}^n_{S_2}) \tag{8.61}$$

$$=H(\mathbf{Z}^n_{S_1},\mathbf{Z}^n_{S_2})-H(\mathbf{Z}^n_{S_1},\mathbf{Z}^n_{S_2}|M_c,\mathbf{W}_s^{(1)},W_{\mathbf{d}}). \tag{8.62}$$

Equation (8.57) follows since $\mathbf{Z}^n_{S_1},\mathbf{Z}^n_{S_2}$ depend only on $W_1^{(1)},W_1^{(2)},W_2^{(1)},W_2^{(2)},W_{d_1,s},W_{d_2,s}$, and by using (8.46) and (8.47). Equation (8.58) follows because there is a bijection between $\{W_1^{(1)},W_1^{(2)},W_2^{(1)},W_2^{(2)}\}$ and $\{W_1^{(1)}\oplus W_2^{(1)},W_1^{(2)}\oplus W_2^{(2)},W_{d_1}^{(2)},W_{d_2}^{(1)}\}$. The inequality in (8.60) follows due to the Markov chain $\mathbf{W}_s^{(2)}-\{M_c,M_{\mathbf{d}},\mathbf{W}_s^{(1)},\mathbf{W}_{\oplus\mathbf{K}}^{(2)}\}-\{\mathbf{Z}^n_{S_1},\mathbf{Z}^n_{S_2}\}$, and the data processing inequality. This Markov chain holds because $\{M_c,M_{\mathbf{d}},\mathbf{W}_s^{(1)}\}$ are independent from $\{\mathbf{W}_s^{(2)},\mathbf{K}^{(2)}\}$, and only the encrypted information $\mathbf{W}_{\oplus\mathbf{K}}^{(2)}$ is transmitted. Equation (8.61) follows from (8.53).

The second term on the right hand side of (8.62) can be lower bounded as

$$H(\mathbf{Z}^n_{S_1},\mathbf{Z}^n_{S_2}|M_c,\mathbf{W}_s^{(1)},W_{\mathbf{d}})=H(\mathbf{Z}^n_{S_1},\mathbf{Z}^n_{S_2},\mathbf{W}_s^{(1)}|M_c,W_{\mathbf{d}})-H(\mathbf{W}_s^{(1)}|M_c,W_{\mathbf{d}}) \tag{8.63}$$

$$=H(\mathbf{Z}^n_{S_1},\mathbf{Z}^n_{S_2},\mathbf{W}_s^{(1)},\mathbf{W}_{\oplus K}^{(1)}|M_c,W_{\mathbf{d}})-H(\mathbf{W}_{\oplus K}^{(1)}|M_c,W_{\mathbf{d}},\mathbf{W}_s^{(1)},\mathbf{Z}^n_{S_1},\mathbf{Z}^n_{S_2})-H(\mathbf{W}_s^{(1)}) \tag{8.64}$$

$$=H(\mathbf{Z}^n_{S_1},\mathbf{Z}^n_{S_2},\mathbf{K}^{(1)},\mathbf{W}_{\oplus K}^{(1)}|M_c,W_{\mathbf{d}})-H(\mathbf{K}^{(1)}|M_c,W_{\mathbf{d}},\mathbf{W}_s^{(1)},\mathbf{Z}^n_{S_1},\mathbf{Z}^n_{S_2})-H(\mathbf{W}_s^{(1)}) \tag{8.65}$$

$$\geq H(\mathbf{Z}^n_{S_1},\mathbf{Z}^n_{S_2},\mathbf{K}^{(1)},\mathbf{W}_{\oplus K}^{(1)}|M_c,W_{\mathbf{d}})-H(\mathbf{W}_s^{(1)})-\epsilon_n' \tag{8.66}$$

$$\geq H(\mathbf{K}^{(1)}|M_c, W_{\mathbf{d}}) + H(\mathbf{Z}^n_{S_1}, \mathbf{Z}^n_{S_2}|M_c, \mathbf{K}^{(1)}, W_{\mathbf{d}}) - H(\mathbf{W}^{(1)}_s) - \epsilon'_n \tag{8.67}$$

$$= H(\mathbf{Z}^n_{S_1}, \mathbf{Z}^n_{S_2}|W_c, W_{\mathbf{d}}) + H(\mathbf{K}^{(1)}) - H(\mathbf{W}^{(1)}_s) - \epsilon'_n \tag{8.68}$$

$$= H(\mathbf{Z}^n_{S_1}, \mathbf{Z}^n_{S_2}|W_c, W_{\mathbf{d}}) - \epsilon'_n, \tag{8.69}$$

where $\epsilon'_n \to 0$ as $n \to \infty$. Equation (8.64) follows since $\mathbf{W}^{(1)}_s$ is independent from $\{M_c, W_{\mathbf{d}}\}$. Equation (8.65) follows because there is a bijection between $\{\mathbf{W}^{(1)}_s, \mathbf{W}^{(1)}_{\oplus K}\}$ and $\{\mathbf{K}^{(1)}, \mathbf{W}^{(1)}_{\oplus K}\}$. Equation (8.68) follows from (8.52), and since $\mathbf{K}^{(1)}$ is independent from $\{M_c, W_{\mathbf{d}}\}$. Equation (8.69) follows since $\mathbf{K}^{(1)}$ and $\mathbf{W}^{(1)}_s$ are independent and identically distributed. The inequality in (8.66) follows because, given $\{M_c, \mathbf{W}^{(1)}_s, W_{\mathbf{d}}\}$, and for n sufficiently large, the adversary can decode $\mathbf{K}^{(1)}$ using its observations $\mathbf{Z}^n_{S_1}, \mathbf{Z}^n_{S_2}$. In particular, $\{M_c, \mathbf{W}^{(1)}_s, W_{\mathbf{d}}\}$ determines a partition of the codebook into bins, each of which contains $2^{n\alpha_\epsilon}$ binary codewords. For n sufficiently large, and given the values of M_c, $\mathbf{W}^{(1)}_s$, $W_{\mathbf{d}}$, i.e., the bin index, the adversary can determine the codeword index inside the bin, and hence decode $\mathbf{K}^{(1)}$. We conclude that, $H(\mathbf{K}^{(1)}\big|M_c, W_{\mathbf{d}}, \mathbf{W}^{(1)}_s, \mathbf{Z}^n_{S_1}, \mathbf{Z}^n_{S_2}) \leq \epsilon'_n$, where $\epsilon'_n \to 0$ as $n \to \infty$.

Substituting (8.69) in (8.62) gives

$$I(W_1, W_2; \mathbf{Z}^n_{S_1}, \mathbf{Z}^n_{S_2}) \leq I(W_c, W_{\mathbf{d}}; \mathbf{Z}^n_{S_1}, \mathbf{Z}^n_{S_2}) + \epsilon'_n. \tag{8.70}$$

Using (8.55) and (8.70), the secrecy constraint in (8.25) is satisfied. Since $\epsilon_n \to 0$ as $n \to \infty$, we conclude that, the achievable strong secrecy file rate is

$$R_s(\alpha) = 2 \times \frac{1-\alpha}{2} + \frac{\alpha}{2} = 1 - \frac{\alpha}{2}. \tag{8.71}$$

Remark 6: Although the cache placement and delivery codebooks, $\mathcal{C}_{c,n}$, $\mathcal{C}_{\mathbf{d},n}$, are designed and generated disjointly, in the security analysis, we consider the Cartesian products of the individual bins of the two codebooks. We are able to do so since the input distributions for generating the two codebooks are identical.

8.5.3.4 Achievability for $\alpha \in [1,2]$

For $\alpha \in [1, 2]$, we adapt the scheme in Section 8.5.3.3 as follows. The files W_1, W_2 are uniform over $[1 : 2^{n\frac{2-\alpha_\epsilon}{2}}]$; α_ϵ is defined in (8.40). The server generates the independent keys K_1, K_2, uniform over $[1 : 2^{n\frac{2-\alpha_\epsilon}{2}}]$, and independent from W_1, W_2. In addition, the server, independently from W_1, W_2, K_1, K_2, generates the independent randomization messages $\tilde{W}$ and $\tilde{W}_K$, uniformly over $[1 : 2^{n(\alpha_\epsilon - 1)}]$.

The messages for cache placement at users 1 and 2 are

$$M_{c,1} = K_1, \qquad M_{c,2} = K_2. \tag{8.72}$$

That is, user j, $j = 1, 2$, stores the key K_j in its cache memory. The message to be securely transmitted during delivery is

$$M_{\mathbf{d}} = \{M_{\mathbf{d},1}, M_{\mathbf{d},2}\}; \;\; M_{\mathbf{d},1} = W_{d_1} \oplus K_1, \;\; M_{\mathbf{d},2} = W_{d_2} \oplus K_2. \tag{8.73}$$

Let $\{W_{d_l}^{(1)}, \cdots, W_{d_l}^{(n\frac{2-\alpha_\epsilon}{2})}\}$, $\{K_l^{(1)}, \cdots, K_l^{(n\frac{2-\alpha_\epsilon}{2})}\}$, and $\{M_{\mathbf{d},l}^{(1)}, \cdots, M_{\mathbf{d},l}^{(n\frac{2-\alpha_\epsilon}{2})}\}$ denote the bit strings of W_{d_l}, K_l, and $M_{\mathbf{d},l}$, $l = 1, 2$.

For $\alpha \in [1, 2]$, the adversary can see all symbols in at least one of the phases. Thus, unlike Section 8.5.3.3, we cannot utilize randomization messages, $\tilde{W}$, $\tilde{W}_K$, to carry any information; only keys are stored in the cache memories of the users. Additionally, the cache placement and delivery codebooks for this case have a different embedding structure than for $\alpha \in (0, 1)$ in Section 8.5.3.3. In particular, the embedding here is performed on the bits of the messages M_c, $M_{\mathbf{d}}$, while the embedding in Section 8.5.3.3 is performed on the bits of the randomization messages $\tilde{M}_c$, $\tilde{M}_{\mathbf{d}}$.

Cache Placement Codebook: During cache placement, the server generates $\mathscr{C}_{c,n}$ as follows. The server randomly and independently divides all the 2^n length-n binary sequences into 2 bins, indexed by $K_1^{(1)}$, and each contains 2^{n-1} codewords. These two bins are further randomly and independently divided into two sub-bins, indexed by $K_2^{(1)}$, and each contains 2^{n-2} codewords. The process continues, going over $K_1^{(2)}$, $K_2^{(2)}, \ldots, K_1^{(n\frac{2-\alpha_\epsilon}{2})}$, $K_2^{(n\frac{2-\alpha_\epsilon}{2})}$, until the remaining $2^{n(\alpha_\epsilon - 1)}$ codewords, after each sequence of divisions, are indexed by $\tilde{W}_K$.

Cache Encoder: The server sends $\mathbf{X}_c^n$ which corresponds to the keys K_1, K_2, and the randomization message $\tilde{W}_K$, i.e., $\mathbf{X}_c^n(K_1^{(1)}, K_2^{(1)}, \cdots, K_1^{(n\frac{2-\alpha_\epsilon}{2})}, K_2^{(n\frac{2-\alpha_\epsilon}{2})}, \tilde{W}_K)$.

Delivery Codebook: During delivery, the server generates $\mathscr{C}_{\mathbf{d},n}$ as follows. The server randomly and independently divides all the 2^n length-n binary sequences into two bins, indexed by $M_{\mathbf{d},1}^{(n\frac{2-\alpha_\epsilon}{2})}$, and each contains 2^{n-1} codewords. These two bins are further randomly and independently divided into two sub-bins, indexed by $M_{\mathbf{d},2}^{(n\frac{2-\alpha_\epsilon}{2})}$, and each contains 2^{n-2} codewords. The process continues, going in reverse order over $M_{\mathbf{d},1}^{(n\frac{2-\alpha_\epsilon}{2}-1)}, M_{\mathbf{d},2}^{(n\frac{2-\alpha_\epsilon}{2}-1)}, \cdots, M_{\mathbf{d},1}^{(1)}, M_{\mathbf{d},2}^{(1)}$, until the remaining $2^{n(\alpha_\epsilon - 1)}$ codewords, after each sequence of divisions, are indexed by $\tilde{W}$.

Delivery Encoder: Given W_1, W_2, K_1, K_2, $\tilde{W}$, and $\mathbf{d} = (d_1, d_2)$, the server forms $M_{\mathbf{d},1}$ and $M_{\mathbf{d},2}$ as in (8.73) and sends $\mathbf{X}_{\mathbf{d}}^n$ which corresponds to $M_{\mathbf{d},1}$, $M_{\mathbf{d},2}$, and $\tilde{W}$, i.e., $\mathbf{X}_{\mathbf{d}}^n\left(M_{\mathbf{d},1}^{(n\frac{2-\alpha_\epsilon}{2})}, M_{\mathbf{d},2}^{(n\frac{2-\alpha_\epsilon}{2})}, \cdots, M_{\mathbf{d},1}^{(1)}, M_{\mathbf{d},2}^{(1)}, \tilde{W}\right)$.

The remainder of the analysis follows similar steps as when $\alpha \in [0, 1)$ and can be found in [36, Section IV-D]. The achievable secrecy file rate is $R_s = 1 - \frac{\alpha}{2}$. This completes the proof for Theorem 9.

8.5.4 Proof of Theorem 10

We extend the scheme in Section 8.5.3 to derive a lower bound on the achievable strong secrecy file rate when $N > 2$. As in Section 8.5.3, we divide the proof into two cases for the ranges $\alpha \in (0, 1)$ and $\alpha \in [1, 2]$.

8.5.4.1 $\alpha \in [1, 2]$

We use the same exact scheme in Section 8.5.3.4. The reason behind this is, for this range of α, only the keys K_1, K_2 are transmitted in the cache placement and stored in the users cache memories. That is, no files are stored in the caches, and the user demands are known during delivery. The achievable strong secrecy file rate is $1 - \frac{\alpha}{2}$.

8.5.4.2 $\alpha \in (0, 1)$

The achievability scheme for this case has the same channel coding structure as in Section 8.5.3.3. The difference, however lies in generating the messages to be securely communicated over cache placement and delivery, i.e., M_c, $M_{\mathbf{d}}$. In particular, we utilize here uncoded placement for designing the cache contents, and a partially coded delivery transmission that is simultaneously useful for both users.

The server divides W_l, $l \in [1:N]$, into the independent subfiles $\{W_l^{(1)}, W_l^{(2)}, W_{l,t}, W_{l,s}\}$. $W_l^{(1)}, W_l^{(2)}$ are uniform over $[1 : 2^{n\frac{1-\alpha_\epsilon}{2N}}]$; α_ϵ is defined in (8.40). $W_{l,t}$ is uniform over $[1 : 2^{n\frac{(2N-1)(1-\alpha_\epsilon)}{4N}}]$, and $W_{l,s}$ is uniform over $[1 : 2^{n\frac{\alpha_\epsilon}{2}}]$. The server, independently from $W_{[1:N]}$, generates the independent keys K_1, K_2, uniform over $[1 : 2^{n\frac{\alpha_\epsilon}{2}}]$.

Let $M_c = \{M_{c,1}, M_{c,2}\}$. Unlike (8.36), we use here *uncoded placement*:

$$M_{c,1} = \{W_1^{(1)}, W_2^{(1)}, \cdots, W_N^{(1)}\}, \tag{8.74}$$

$$M_{c,2} = \{W_1^{(2)}, W_2^{(2)}, \cdots, W_N^{(2)}\}. \tag{8.75}$$

The randomization message for cache placement, $\tilde{M}_c = \{\tilde{M}_{c,1}, \tilde{M}_{c,2}\}$, is identical to (8.37). User j stores $M_{c,j}$ and $\tilde{M}_{c,j}$ in its cache memory.

Unlike (8.36), we use here *partially coded* delivery:

$$M_{\mathbf{d}} = \{W_{d_2}^{(1)} \oplus W_{d_1}^{(2)}, W_{d_1,t}, W_{d_2,t}\}. \tag{8.76}$$

The randomization message for delivery, $\tilde{M}_{\mathbf{d}}$, is identical to (8.39).

Remark 7: Notice that the sizes of M_c, $M_{\mathbf{d}}$, $\tilde{M}_c$, $\tilde{M}_{\mathbf{d}}$, are the same as in Section 8.5.3.3. In particular, the sizes of $\tilde{M}_c$, $\tilde{M}_{\mathbf{d}}$ are both $n\alpha_\epsilon$ bits. The size of M_c is

$$2 \times N \times n\frac{1-\alpha_\epsilon}{2N} = n(1-\alpha_\epsilon) \quad \text{bits}, \tag{8.77}$$

and the size of $M_{\mathbf{d}}$ is given by

$$n\frac{1-\alpha_\epsilon}{2N} + 2 \times n\frac{(2N-1)(1-\alpha_\epsilon)}{4N} = n(1-\alpha_\epsilon) \quad \text{bits}. \tag{8.78}$$

Codebooks Generation and Encoders: For M_c, $M_{\mathbf{d}}$, $\tilde{M}_c$, $\tilde{M}_{\mathbf{d}}$ defined above, the cache placement and delivery codebooks, $\mathscr{C}_{c,n}$, $\mathscr{C}_{\mathbf{d},n}$, and the cache and delivery encoders, are designed in the same exact manner as in Section 8.5.3.3.

Decoding: As in Section 8.5.3.3, using $M_{\mathbf{d}}$, $\tilde{M}_{\mathbf{d}}$, $M_{c,j}$, $\tilde{M}_{c,j}$, and for n sufficiently large, user j correctly decodes W_{d_j}, $j = 1, 2$.

Security analysis: Let W_c, $\tilde{W}_c$, $W_{\mathbf{d}}$, and $\tilde{W}_{\mathbf{d}}$, be defined as in (8.41)–(8.44), (8.52) and (8.53). $\tilde{W}_c$ and $\tilde{W}_{\mathbf{d}}$ are independent and uniform, and hence $\{\tilde{W}_c, \tilde{W}_{\mathbf{d}}\}$ is jointly uniform. In addition, $\{W_c, W_{\mathbf{d}}\}$ are independent from $\{\tilde{W}_c, \tilde{W}_{\mathbf{d}}\}$. Thus, (8.55) holds for this case. For any $\mathbf{d} = (d_1, d_2)$,

$$I(W_{[1:N]}; \mathbf{Z}^n_{S_1}, \mathbf{Z}^n_{S_2}) = I(\{W_l^{(1)}, W_l^{(2)}, W_{l,t}, W_{l,s}\}_{l=1}^N; \mathbf{Z}^n_{S_1}, \mathbf{Z}^n_{S_2}) \tag{8.79}$$

$$\leq I(M_c, \{W_l^{(1)}, W_l^{(2)}, W_{l,t}, W_{l,s}\}_{l=1}^N; \mathbf{Z}^n_{S_1}, \mathbf{Z}^n_{S_2}) \tag{8.80}$$

$$\leq I(M_c, W_{d_2}^{(1)} \oplus W_{d_1}^{(2)}, W_{d_1,t}, W_{d_2,t}, W_{d_1,s}, W_{d_2,s}; \mathbf{Z}^n_{S_1}, \mathbf{Z}^n_{S_2}) \tag{8.81}$$

$$= I(M_c, M_{\mathbf{d}}, W_{d_1,s}, W_{d_2,s}; \mathbf{Z}^n_{S_1}, \mathbf{Z}^n_{S_2}) \tag{8.82}$$

$$\leq I(W_c, W_{\mathbf{d}}; \mathbf{Z}^n_{S_1}, \mathbf{Z}^n_{S_2}) + \epsilon'_n, \tag{8.83}$$

Equation (8.81) follows since $W_{[1:N]} - \{M_c, W_{d_2}^{(1)} \oplus W_{d_1}^{(2)}, W_{d_1,t}, W_{d_2,t}, W_{d_1,s}, W_{d_2,s}\}$ $-\{\mathbf{Z}^n_{S_1}, \mathbf{Z}^n_{S_2}\}$ is a Markov chain; (8.82) follows from (8.76), and (8.83) follows using similar steps as in (8.58)–(8.69). Using (8.55) and (8.83), the secrecy constraint in (8.25) is satisfied.

With $\varepsilon_n \to 0$ as $n \to \infty$, the achievable strong secrecy file rate is

$$R_s(\alpha) = 2 \times \frac{1-\alpha}{2N} + \frac{(2N-1)(1-\alpha)}{4N} + \frac{\alpha}{2} \tag{8.84}$$

$$= \frac{1}{2} + \frac{3(1-\alpha)}{4N}. \tag{8.85}$$

This completes the proof for Theorem 10.

Remark 8: For $N = 2$, the achievable secrecy rate in (8.85) is strictly smaller than the secrecy rate obtained by *coded* placement and *uncoded* delivery in Section 8.5.3.3, i.e., $1 - \frac{\alpha}{2}$.

Remark 9: An achievability scheme that utilizes coded placement and uncoded delivery, as in Section 8.5.3.3, achieves the same secure file rate as (8.85) for $N = 3$. However, this scheme achieves a strictly smaller secure file rate for $N \geq 4$. In this scheme, $W_l^{(1)}$, $W_l^{(2)}$ are uniform over $[1 : 2^{n\frac{1-\alpha_\epsilon}{2(N-1)}}]$. $W_{l,t}$ is uniform over $[1 : 2^{n\frac{(N-2)(1-\alpha_\epsilon)}{2(N-1)}}]$. $W_{l,s}, K_1, K_2$ are uniform over $[1 : 2^{n\frac{\alpha_\epsilon}{2}}]$. $M_c = \{M_{c,1}, M_{c,2}\}$, $M_{\mathbf{d}}$ are given by

$$M_{c,1} = \{W_1^{(1)} \oplus W_2^{(1)}, W_2^{(1)} \oplus W_3^{(1)}, \cdots, W_{N-1}^{(1)} \oplus W_N^{(1)}\}, \tag{8.86}$$

$$M_{c,2} = \{W_1^{(2)} \oplus W_2^{(2)}, W_2^{(2)} \oplus W_3^{(2)}, \cdots, W_{N-1}^{(2)} \oplus W_N^{(2)}\}, \tag{8.87}$$

$$M_{\mathbf{d}} = \{W_{d_2}^{(1)}, W_{d_1}^{(2)}, W_{d_1,t}, W_{d_2,t}\}. \tag{8.88}$$

Without loss of generality, let $d_1 < d_2$. For $\mathbf{d} = (d_1, d_2)$, using $M_{c,j}$ in (8.86) and (8.87), user j, restores $W_{d_1}^{(j)} \oplus W_{d_2}^{(j)}$ by XOR-ing $\{W_{d_1}^{(j)} \oplus W_{d_1+1}^{(j)}\}, \{W_{d_1+1}^{(j)} \oplus W_{d_1+2}^{(j)}\}, \quad , \{W_{d_2-1}^{(j)} \oplus W_{d_2}^{(j)}\}$. The achievable secrecy file rate using this scheme is $R_s(\alpha) = \frac{1}{2} + \frac{1-\alpha}{2(N-1)}$.

8.5.5 *Proof of Theorem 11*

When $\alpha \in [1, 2]$, the upper bound on R_s follows as in Section 8.5.3.1. It remains to prove the upper bound for $\alpha \in (0, 1)$. The proof is divided into three steps.

Step 1: We first upper bound R_s by the secrecy capacity when the adversary is restricted to tap into the delivery transmission only, denoted as C_s^{Res} : C_s^{Res} is the maximum achievable file rate when $\alpha_1 = 0, \alpha_2 = \alpha$. Restricting the adversary to only tap into the delivery phase cannot decrease the secrecy capacity, i.e., $R_s \leq C_s^{\text{Res}}$, since this setting is included in the feasible strategy space for the adversary. Cache placement is secure, and each user has a secure cache memory of size $\frac{n}{2}$ bits.

Step 2: We upper bound C_s^{Res} by the secrecy capacity, i.e., the maximum achievable file rate, when the delivery channel to the adversary is replaced by a discrete memoryless binary erasure channel, with erasure probability $1 - \alpha$, denoted as C_s^{DM}. The proof for this step follows the same lines as in [22, Section V].

Step 3: From Step 1, each user has a secure cache of size $\frac{n}{2}$ bits. Since increasing the cache sizes cannot decrease the achievable file rate, we further upper bound C_s^{DM} by the maximum achievable file rate when each user has a cache memory of size n bits, in which it stores $\mathbf{X}_c^n$. User j, $j = 1, 2$, utilizes both $\mathbf{X}_c^n$ and $\mathbf{X}_{\mathbf{d}}^n$ to decode its desired file W_{d_j}, i.e., $\hat{W}_{d_j} = g_{\mathbf{d},j}(\mathbf{X}_c^n, \mathbf{X}_{\mathbf{d}}^n)$, $\mathbf{d} = (d_1, d_2)$. This setup is equivalent to a single user, with a cache of size n bits, who demands two files W_{d_1}, W_{d_2}, and utilizes the decoder $g_{\mathbf{d}} \triangleq \{g_{\mathbf{d},1}, g_{\mathbf{d},2}\}$. Denoting the maximum achievable file rate for this single user model as C_s^{SR}, $C_s^{\text{DM}} \leq C_s^{\text{SR}}$. In the following, we upper bound C_s^{SR}.

Let M_{D} denote the fraction of the size-n bits cache memory dedicated to store (coded or uncoded) information bits, and let M_{K} denote the fraction dedicated to store key bits. That is, $M_{\text{D}} + M_{\text{K}} = 1$. Let S_{D} denote the information bits stored in this memory, i.e., $S_{\text{D}} = f_{\text{D}}(W_{[1:N]})$ and $H(S_{\text{D}}) = nM_{\text{D}}$. We utilize the following lemma in order to upper bound C_s^{SR}.

Lemma 1: [37, Lemma 1] For a fixed allocation of M_{D}, M_{K}, and a user who demands W_{d_1}, W_{d_2}, the secrecy rate for the single user model is upper bounded as

$$2R_s^{\text{SR}} \leq \min\{1, 1 - \alpha + M_{\text{K}}\} + \frac{1}{n} I\left(W_{d_1}, W_{d_2}; S_{\text{D}}\right). \tag{8.89}$$

Notice that (8.89) holds for any demand pair $\mathbf{d} = (d_1, d_2)$ such that $d_1 \neq d_2$, i.e., the worst-case demands. Summing over all such demands, we have

$$2R_s^{\text{SR}} \leq \min\{1, 1 - \alpha + M_{\text{K}}\} + \frac{1}{nN(N-1)} \sum_{d_1, d_2 \in [1:N],\, d_1 \neq d_2} I\left(W_{d_1}, W_{d_2}; S_{\text{D}}\right). \tag{8.90}$$

The second term on the right-hand side of (8.90) can be written as

$$\frac{1}{nN(N-1)}\sum_{d_1,d_2\in[1:N],\,d_1\neq d_2} I\left(W_{d_1},W_{d_2};S_{\mathrm{D}}\right)$$
$$=\frac{1}{nD}\sum_{d_1\in[1:N]} I\left(W_{d_1};S_{\mathrm{D}}\right) \tag{8.91}$$
$$+\frac{1}{nN(N-1)}\sum_{d_1,d_2\in[1:N],\,d_1\neq d_2} I\left(W_{d_2};S_{\mathrm{D}}|W_{d_1}\right)$$
$$\leq\frac{1}{nD}\sum_{d_1\in[1:N]} I\left(W_{d_1};S_{\mathrm{D}}\right)+\frac{1}{nN(N-1)}\sum_{d_1\in[1:N]}\left(\sum_{d_2\in[1:N]} I\left(W_{d_2};S_{\mathrm{D}}|W_{d_1}\right)\right). \tag{8.92}$$

For any $d_1\in[1:N]$, we have

$$\sum_{d_2\in[1:N]} I\left(W_{d_2};S_{\mathrm{D}}|W_{d_1}\right)=\sum_{d_2=1}^{N}\left[H\left(W_{d_2}|W_{d_1}\right)-H\left(W_{d_2}|W_{d_1},S_{\mathrm{D}}\right)\right] \tag{8.93}$$

$$\leq\sum_{d_2=1}^{N}\left[H\left(W_{d_2}|W_1,\cdots,W_{d_2-1},W_{d_1}\right)-H\left(W_{d_2}|W_1,\cdots,W_{d_2-1},W_{d_1},S_{\mathrm{D}}\right)\right] \tag{8.94}$$

$$=I\left(W_1,\cdots,W_N;S_{\mathrm{D}}|W_{d_1}\right) \tag{8.95}$$
$$\leq H(S_{\mathrm{D}})=nM_{\mathrm{D}}, \tag{8.96}$$

Equation (8.94) follows because when $d_2=d_1$, $H\left(W_{d_2}|W_{d_1}\right)=H\left(W_{d_2}|W_1,\cdots,W_{d_2-1},W_{d_1}\right)=0$, and when $d_2\neq d_1$, $H\left(W_{d_2}|W_{d_1}\right)=H\left(W_{d_2}|W_1,W_2,\cdots,W_{d_2-1},W_{d_1}\right)=H(W_{d_2})$.

Similarly, we have

$$\sum_{d_1\in[1:N]} I\left(W_{d_1};S_{\mathrm{D}}\right)\leq H(S_{\mathrm{D}})=nM_{\mathrm{D}}. \tag{8.97}$$

Substituting (8.96) and (8.97) in (8.92) gives

$$\frac{1}{nN(N-1)}\sum_{d_1,d_2\in[1:N],\,d_1\neq d_2} I\left(W_{d_1},W_{d_2};S_{\mathrm{D}}\right)\leq\frac{2N-1}{N(N-1)}M_{\mathrm{D}}. \tag{8.98}$$

Thus, using (8.90) and (8.98), R_s^{SR} is further upper bounded as

$$R_s^{\mathrm{SR}}\leq\frac{1}{2}\left[\min\{1,1-\alpha+M_{\mathrm{K}}\}+\frac{2N-1}{N(N-1)}M_{\mathrm{D}}\right]. \tag{8.99}$$

Finally, by maximizing over all possible allocations for M_{D} and M_{K} such that $M_{\mathrm{D}}+M_{\mathrm{K}}=1$, we obtain

$$C_s^{\mathrm{SR}}\leq\frac{1}{2}\max_{M_{\mathrm{D}},M_{\mathrm{K}}:\,M_{\mathrm{D}}+M_{\mathrm{K}}=1}\left\{\min\{1,1-\alpha+M_{\mathrm{K}}\}+\frac{2N-1}{N(N-1)}M_{\mathrm{D}}\right\} \tag{8.100}$$
$$=\frac{1}{2}\left[1+\frac{2N-1}{N(N-1)}(1-\alpha)\right]. \tag{8.101}$$

Equation (8.101) follows because, for $N \geq 3$, the maximum occurs at $M_K = \alpha$ and $M_D = 1 - \alpha$. This completes the proof for Theorem 11.

Remark 10: An upper bound considering uncoded placement only can be derived as follows. The same analysis as in (8.89)–(8.101) carries through with $I(W_{d_2}; S_D | W_{d_1})$ in (8.92) is equal to $I(W_{d_2}; S_D)$. Hence the right-hand side of (8.98) is replaced by $\frac{2M_D}{N}$. The resulting bound $R_s \leq \frac{1}{2} + \frac{(1-\alpha)}{N}$ is tighter than (8.101).

8.6 Conclusion

In this chapter, we discussed security and privacy aspects in coded caching networks. We presented the fundamental rate-memory trade-off when security against an external wiretapper with full access to the delivery phase is required. We next studied the rate-memory trade-off when the files requested by the end users are confidential, i.e., each user is allowed to recover only its desired file. We then considered the aforementioned security criteria in a layered network topology termed as the combination network. Finally, we studied a setting in which both cache placement and delivery phases are compromised by an adversary who optimizes its tapping capabilities.

References

[1] Maddah-Ali M.A., Niesen U. 'Fundamental limits of caching'. *IEEE Transactions on Information Theory*. 2014;**60**(5):2856–67.

[2] Sengupta A., Tandon R., Clancy T.C. 'Fundamental limits of caching with secure delivery'. *IEEE Transactions on Information Forensics and Security*. 2015;**10**(2):355–70.

[3] Awan Z.H., Sezgin A. 'Fundamental limits of caching in D2D networks with secure delivery'. IEEE International Conference on Communication Workshop; 2015. pp. 464–9.

[4] Ravindrakumar V., Panda P., Karamchandani N., Prabhakaran V.M. 'Private coded caching'. *IEEE Transactions on Information Forensics and Security*. 2017;**13**(3):685–94.

[5] Pedarsani R., Maddah-Ali M.A., Niesen U. 'Online coded caching'. *IEEE/ACM Transactions on Networking*. 2016;**24**(2):836–45.

[6] Karamchandani N., Niesen U., Maddah-Ali M.A., Diggavi S.N. 'Hierarchical coded caching'. *IEEE Transactions on Information Theory*. 2016;**62**(6):3212–29.

[7] Niesen U., Maddah-Ali M.A. 'Coded caching with nonuniform demands'. *IEEE Transactions on Information Theory*. 2017;**63**(2):1146–58.

[8] Ibrahim A.M., Zewail A.A., Yener A. 'Centralized coded caching with heterogeneous cache sizes'. IEEE Wireless Communications and Networking Conference. IEEE; 2017. pp. 1–6.

[9] Maddah-Ali M.A., Niesen U. 'Cache-aided interference channels'. IEEE International Symposium on Information Theory; 2015. pp. 809–13.

[10] Ji M., Wong M.F., Tulino A.M. 'On the fundamental limits of caching in combination networks'. IEEE International Workshop on Signal Processing Advances in Wireless Communications; 2015. pp. 695–9.
[11] Zewail A.A., Yener A. 'Combination networks with or without secrecy constraints: the impact of caching relays'. *IEEE Journal on Selected Areas in Communications*. 2018;**36**(6):1140–52.
[12] Ji M., Caire G., Molisch A.F. 'Fundamental limits of Caching in wireless D2D networks'. *IEEE Transactions on Information Theory*. 2016;**62**(2):849–69.
[13] Ibrahim A., Zewail A.A., Yener A. 'Device-to-device coded caching with heterogeneous cache sizes'. IEEE International Conference on Communications; 2018. pp. 1–6.
[14] Ghorbel A., Kobayashi M., Yang S. 'Content delivery in erasure broadcast channels with cache and feedback'. *IEEE Transactions on Information Theory*. 2016;**62**(11):6407–22.
[15] Ngai C.K., Yeung R.W. 'Network coding gain of combination networks'. *in Information Theory Workshop. IEEE*. 2004:283–7.
[16] Shannon C.E. 'Communication theory of secrecy systems'. *Bell System Technical Journal*. 1949;**28**(4):656–715.
[17] Ahlswede R., Csiszár I. 'Common randomness in information theory and cryptography. I. Secret sharing'. *IEEE Transactions on Information Theory*. 1993;**39**(4):1121–32.
[18] Nafea M., Yener A. 'The caching broadcast channel with a wire and cache tapping adversary of type II'. 2018 IEEE Information Theory Workshop (ITW). IEEE; 2018. pp. 1–5.
[19] Nafea M., Yener A. 'The caching broadcast channel with a wire and cache tapping adversary of type II: multiple library files'. 56th Annual Allerton Conference on Communication, Control, and Computing (Allerton). IEEE; 2018. pp. 989–96.
[20] Ozarow L.H., Wyner A.D. 'Wire-tap channel II'. *AT&T Bell Laboratories Technical Journal*. 1984;**63**(10):2135–57.
[21] Nafea M., Yener A. 'Wiretap channel II with a noisy main channel'. IEEE International Symposium on Information Theory; 2015. pp. 1159–63.
[22] Nafea M., Yener A. 'A new wiretap channel model and its strong secrecy capacity'. *IEEE Transactions on Information Theory*. 2018;**64**(3):2077–92.
[23] Wyner A.D. 'The wire-tap channel'. *Bell System Technical Journal*. 1975;**54**(8):1355–87.
[24] Nafea M., Yener A. 'Generalizing multiple access wiretap and wiretap II channel models: achievable rates and cost of strong secrecy'. *IEEE Transactions on Information Theory*. 2019;**65**(8):5125–43.
[25] Nafea M., Yener A. 'New models for the interference and broadcast channels with confidential messages'. IEEE International Symposium on Information Theory; 2017. pp. 1808–12.
[26] Nafea M., Yener A. 'A new broadcast wiretap channel model'. IEEE International Symposium on Information Theory; Aachen, Germany, June; 2017. pp. 2128–32.

[27] Goldfeld Z., Cuff P., Permuter H.H. 'Semantic-security capacity for wiretap channels of type II'. *IEEE Transactions on Information Theory*. 2016;**62**(7):3863–79.

[28] Shamir A. 'How to share a secret'. *Communications of the ACM*. 1979;**22**(11):612–13.

[29] Zewail A.A., Yener A. 'Cache-aided combination networks with secrecy guarantees'. 2018 52nd Annual Conference on Information Sciences and Systems (CISS). IEEE; 2018. pp. 1–6.

[30] Zewail A.A., Yener A. 'Coded caching for combination networks with cache-aided relays'. 2017 IEEE International Symposium on Information Theory (ISIT). IEEE; 2017. pp. 2433–7.

[31] Lin S., Costello D.J. *Error Control Coding*. **2**. Hoboken: Prentice Hall; 2001.

[32] Ly H.D., Liu T., Blankenship Y. 'Security embedding codes'. *IEEE Transactions on Information Forensics and Security*. 2012;**7**(1):148–59.

[33] Liang Y., Lai L., Poor H.V., Shamai S. 'A broadcast approach for fading wiretap channels'. *IEEE Transactions on Information Theory*. 2014;**60**(2):842–58.

[34] Tian C. 'Symmetry, demand types and outer bounds in caching systems'. IEEE International Symposium on Information Theory; 2016. pp. 825–9.

[35] Cao D., Zhang D., Chen P., Liu N., Kang W., Gündüz D. 'Coded caching with heterogeneous cache sizes and link qualities: the two-user case'. IEEE International Symposium on Information Theory; 2018. pp. 1545–9.

[36] Nafea M., Yener A. 'Coded caching in the presence of a wire and a cache tapping adversary of type II'. *IEEE Journal on Selected Areas in Information Theory*. 2018;**2**(1):65–81.

[37] Zewail A.A., Yener A. 'The wiretap channel with a cache'. 2018 IEEE International Symposium on Information Theory (ISIT). IEEE; 2018. pp. 1720–4.

Chapter 9
Scaling laws for cache-aided device-to-device networks for wireless video

Ming-Chun Lee[1], Mingyue Ji[2], and Andreas F Molisch[3]

9.1 Introduction

9.1.1 Motivation for D2D-based video distribution

Wireless data traffic has been increasing by 50%–100% per year over the past decade and is expected to continue this spectacular growth. The majority of this video traffic (currently > 65% and growing) stems from wireless video, in particular video streaming, i.e., services like Netflix, Disney+, and YouTube. Essentially, video streaming has emerged as the "killer application" of 4G and early 5G wireless systems.

At the same time, the large and increasing data traffic strains the resources of the networks. Traditional methods for increasing network capacity include addition of new spectrum, network densification, heterogeneous networks, and use of massive antenna arrays. However, all of these solutions are expensive as they require either license fees or new infrastructure. Furthermore, they are *not scalable* – both spectrum, and locations at which base stations (BSs) (and their associated backhaul) can be placed are limited.

Yet, wireless video has a special property that allows a much more efficient distribution than most other types of data files: it has a very concentrated popularity distribution, i.e., a relatively small number of popular movies/clips accounts for the majority of the data traffic. In traditional TV, this fact was exploited by the broadcast effect, such that the system throughput per user was independent of the number of users. Of course, this was achieved by forcing users to watch a pre-determined video at a pre-determined time. *Asynchronous content reuse*, which normally requires transmission at unpredictable times, to each user separately, can be made more efficient by caching of content at the devices; the downloading to the devices can then

[1]Institute of Communications Engineering, National Yang Ming Chiao Tung University, Taiwan
[2]Department of Electrical and Computer Engineering, University of Utah, USA
[3]Ming-Hsieh Department of Electrical and Computer Engineering, University of Southern California, USA

occur at times when it does not affect overall network performance, e.g., during the night. In its simplest form, called self-caching, a device would cache the videos that it anticipates its owner to watch (this is implemented, e.g., in Netflix's "smart download"). However, due to the limited storage space on cellphones and similar devices, the percentage of files that can be downloaded is small and does not lead to a significant relief of the network.

A solution for this problem was first suggested in [1–3]: different devices, which are in close proximity, and which have different videos cached, exchange (upon request) files through a spectrally efficient device-to-device communications process. If at the time of caching it is known which devices will be close together during the file exchange, a deterministic choice of which device caches which video can be made; otherwise, each device can determine, independently and randomly (but according to a given probability density function), which videos it will cache. A group of close-together devices can then be seen as "pooling" their caching resources, while efficiently communicating from the device that has the video stored, to the device that requests it. The more caching space is available on each device, the less bandwidth is required for communication (desired files are closer to the requesting user, so that area spectral efficiency (ASE) of the communication is increased). This thus allows to "turn memory into bandwidth." Considering that memory is a relatively cheap resource, and bandwidth is the most expensive one in wireless systems, this is a very attractive tradeoff. Simulations under realistic settings indicated that network throughput can be increased by one to two orders of magnitude. Consequently, this topic has been investigated extensively over the past 10 years, and some 1 000 papers have been published – various aspects are discussed in other parts of this book and the reviews [4–6].

A fundamental question is whether the approach is scalable: in particular, as the density of users increases, are the benefits of caching retained, or do they peter out? From intuition, one might hope that caching benefits are retained: the number of file requesters, and the number of available caches, both increase proportionately as the user density increases. Yet, other effects, such as the increase of the file library (i.e., the collection of videos that the users might want to see) might create problems. Scaling laws aim to obtain the fundamental behavior of throughput and other relevant quantities as user density becomes very large. By providing lower and upper bounds (achievable bound and converse), we can judge not only the potential of a scheme, but also whether an implementable scheme might be close to the theoretical optimum.

In this chapter, we will first review scaling laws for "standard" (non-caching) D2D networks, as well as other caching approaches that do not make use of D2D communications (see Section 9.1.2). Section 9.2 summarizes the system model that underlies most scaling law investigations. This is followed by a review of scaling laws for single-hop and multihop networks in Sections 9.3 and 9.4, respectively. Next, we discuss the scaling laws when coded caching is combined with D2D communications in Section 9.5. A discussion of how the scaling laws relate to practical implementation rounds off the chapter.

9.1.2 D2D scaling laws without caching

Cache-aided D2D networks show considerable similarity to ad hoc networks. In both cases, communications occur not through a central infrastructure node but directly between devices. The main difference is that in ad hoc networks a device might want to talk to a specific other device, while in cache-aided D2D networks, a device wants to receive a *content*, and does not care which device sends it. Still, scaling laws for ad hoc networks are an important reference point and many of the techniques developed for them can be adapted for cache-aided D2D networks.

A milestone in scaling laws for ad hoc networks was the work by Gupta and Kumar [7], which investigated the transport capacity when multihop relaying was allowed. Placing N users that are either placed arbitrarily or randomly, both a lower (achievable) bound on the throughput per user, of the order $\Theta\left(\frac{1}{\sqrt{N \log N}}\right)$ and an upper bound (under some conditions) of $\Theta\left(\frac{1}{\sqrt{N}}\right)$ were derived. In [8], a similar analysis was conducted with a more general physical model and the upper bound $\Theta\left(\frac{1}{\sqrt{N}}\right)$ was validated under general conditions. The $\Theta(\sqrt{\log N})$ gap between the achievable throughput and the upper bound was closed in [9], however, with a slightly different model where the user distribution is described by a Poisson Point Process (PPP). A number of other schemes and channel models were investigated in other papers as well.

While all the above techniques were based on single-hop or multihop relaying, a more complicated transmission scheme can actually increase the throughput in a scaling law sense. Reference [10] introduced a hierarchical cooperation scheme, where the cooperation between users is employed to form a distributed multiple-input multiple-output (MIMO) system among the users. The resulting scaling of the throughput per user is almost $\Theta(1)$, at the price of very complicated cooperation among user nodes. Finally, besides the scaling laws for throughput also those for the throughput-delay tradeoff have been investigated [11–13].

Other scaling laws were developed in the computer science literature, with content delivery networks in mind. Assuming *wired* connections, [14] proposed a caching policy (square-root replication policy) that provides the optimum design in terms of the *expected number of nodes* to visit until finding the desired content.

An important alternative to D2D-based caching networks is the so-called *coded caching*, in which devices locally cache part of the files. Then, the BS broadcasts a network-coded transmission such that each device can recover the desired files from the stored parts of the files and the broadcast transmission; thus all devices can recover the desired files simultaneously. The seminal paper of [15] provided the scaling laws which are – as will be discussed more in Section 9.5 – comparable to those achievable with cache-aided D2D networks, though the latter perform better in realistic circumstances.

Of course, the approach that is used in currently implemented networks, namely treating each file transmission as a separate unicast transmission, has a scaling law for the throughput as well – the throughput per user decreases as $\Theta(1/N)$.

9.2 System model

9.2.1 Popularity model

The scaling laws discussed in this chapter assume a certain popularity distribution of the files. More precisely, it is assumed that user $u \in \mathcal{U}$ requests a file $\mathsf{f}_u \in \mathcal{F} = \{1, \ldots, M\}$ in an independent and identically distributed (i.i.d.) manner, according to a given request probability mass function (PMF) $\{P_r(f) : f \in \mathcal{F}\}$. Here $\mathcal{F} = \{1, \ldots, M\}$ denotes the "file library," from which the requests are made. Note that this library contains only the files the users might be interested in, and thus M may change as the number of users changes (see below). We note that the assumption of i.i.d. requests might be violated in practice, either because different users have different tastes and thus different request probability, and/or because different users influence each other, e.g., through social networks. Yet, the i.i.d. model is used in essentially all scaling law investigations, because it enables much better mathematical tractability. Moreover, it is assumed that different users making the requests on the same file would request different segments of the file (e.g., due to different start times for streaming a video), which means that naive multicasting does not provide any gain. It is furthermore commonly assumed that all files have the same size; if large files exist, they can be broken into smaller chunks.

The most common assumption for the PMF of the file popularity is that P_r is a Zipf distribution with parameter γ [16], i.e.,

$$P_r(f) = \frac{f^{-\gamma}}{\sum_{j=1}^{M} j^{-\gamma}} \tag{9.1}$$

for $f = 1, \ldots, M$. This distribution is not only well-suited for closed-form analysis but also was shown to be in agreement with early investigations into the popularity of YouTube videos [16]. The Zipf parameter γ can be interpreted as a measure for the popularity concentration – the larger γ, the more a few popular videos determine traffic. It also has important mathematical consequences: for $\gamma < 1$, the distribution is heavy-tailed, and for $M \to \infty$, $P_r(f) \to 0$ for all f. For $\gamma > 1$, the sum in the denominator of (9.1) sums to a finite value for $M \to \infty$, i.e., the distribution has a fast-decaying tail.

A more general distribution is the MZipf distribution [17] :

$$P_r(f) = \frac{(f+q)^{-\gamma}}{\sum_{j=1}^{M} (j+q)^{-\gamma}}, \quad f = 1, 2, \ldots, M. \tag{9.2}$$

The parameter q denotes the "plateau" factor, such that the q most popular files have almost the same popularity, while beyond that the PMF decays, essentially, like a Zipf distribution. The authors of [18] showed that for an extensive real-world dataset of long-form videos (TV shows and movies), the MZipf distribution better fits the requesting behaviors of mobile users. In practice, the dataset in [18] showed that $q \ll M$.

9.2.2 Network model

The scaling analysis generally considers a network deployed over a unit-area square region and formed by N nodes $\mathcal{U} = \{1, \ldots, N\}$. Some papers place those nodes on a regular grid with minimum node distance $1/\sqrt{N}$, while others assume that the nodes are distributed according to either a binomial point process (BPP) or a homogeneous PPP; Sections 9.3–9.6 will specify which results are derived under which of those assumptions. Each node has a cache of size S (in units of files).

A scaling analysis can generally consider two approaches for the caching of files on the nodes: deterministic and stochastic. In the deterministic approach, it is assumed that during the distribution of the files to the caches, the location of all devices and the file transmission scheme during the demand phase will be known; the caching content for each node can then be deterministically optimized. In a "clustering scheme" (see below for details), this can be done, e.g., such that each file is cached exactly once in each cluster. In the most common random approach, namely a simple "decentralized" random caching strategy, each user caches S files selected independently from the library $\mathcal{F}$ with probability $P_c(f)$, where $0 \leq P_c(f) \leq 1$ and $\sum_{f=1}^{M} P_c(f) = 1$. The caching probability distribution, $P_c(f)$, which is by definition common to all users, can be optimized by the network designer. Note that when using this policy, a user might cache the same file multiple times, and this policy is used for the sake of analysis. To avoid caching the same file multiple times, an improved random caching strategy that considers $\sum_{f=1}^{M} P_c(f) = S$ and implemented via the approach proposed in [19] can be adopted.

For the delivery of the files to the requesting user, the cluster model is the most common model for the achievability bounds. The network is divided into clusters of equal size, such that the number of nodes in each cluster is $g_c(M)$, which are independent of the users' requests and cache placement realization. A user can look for the requested file only inside its own cluster. If a user can find the requested file inside the cluster, we say there is one *potential link* in this cluster. An *interference avoidance* scheme ensures that at most one transmission is allowed in each cluster on any time-frequency slot (transmission resource), while inter-cluster interference is avoided by a time-frequency reuse scheme ([20], Ch. 17) with parameter K, as shown in Figure 9.1. In particular, we can pick $K = \left(\left\lceil \sqrt{2}(1 + \Delta)\right\rceil + 1\right)^2$, where Δ is the interference parameter in the protocol model.

A system admission control scheme decides whether to serve potential links or ignore them. The served potential links in the same cluster are scheduled with equal probability (or, equivalently, in round robin), such that all admitted user requests have the same average throughput $\mathbb{E}[T_u] = \overline{T}_{\min}$ (see Section 9.2.3), for all users u, where expectation is with respect to the random user requests, random caching, and the link scheduling policy (which may be randomized or deterministic, as a special case).

9.2.3 Target functions for scaling

Ultimately, the goal of the scaling analysis is to provide the asymptotic behavior of the target function when $N \to \infty$, $M \to \infty$. The simplest target function is the

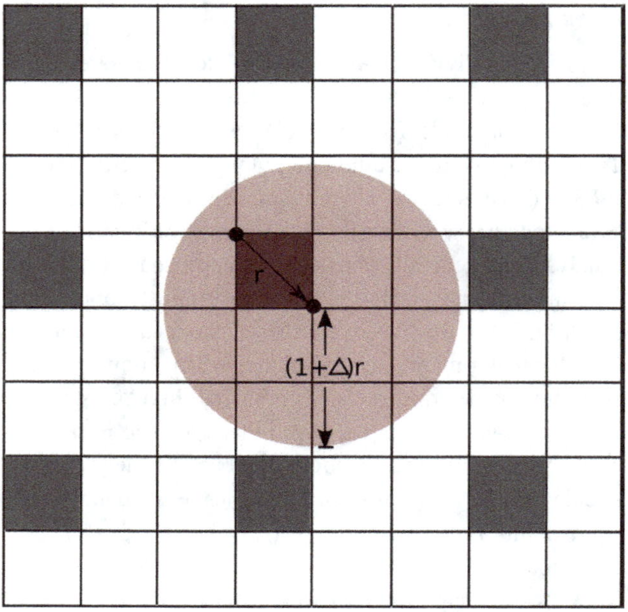

Figure 9.1 Single-cell layout and the interference avoidance TDMA scheme. Each square represents a cluster with gray squares representing concurrently transmitting clusters. The red area is the disk where the protocol model allows no other concurrent transmission.

sum throughput (or equivalently, the average per-user throughput), which was used in some of the earlier scaling analyses. However, this quantity has the drawback that it can take on very large values while leading to fundamental unfairness in the system – essentially, it might induce the caching distribution to concentrate on the most popular files and thus boost throughput but neglecting to serve users with other requests.

For this reason, later papers concentrated on the throughput-outage tradeoff. Qualitatively (for formal definition see [21]), we say that a user is in outage if the user cannot be served by the D2D network. This can be caused by: (i) the file requested by the user is not in the user's own cluster, (ii) the system admission control decides to ignore the request. We define the outage probability p_o as the average fraction of users in outage.

More precisely, it is common to define for a given network and request PMF, an outage-throughput pair (p, t) as *achievable* if there exists a cache placement scheme and an admission control and transmission scheduling policy with outage probability $p_o \leq p$ and minimum per-user average throughput $\overline{T}_{\min} \geq t$. The outage-throughput achievable region $\mathcal{T}(P_r, n, m)$ is the closure of all achievable outage-throughput pairs (p, t). In particular, we let $T^*(p) = \sup\{t : (p, t) \in \mathcal{T}(P_r, n, m)\}$.

Notice that $T^*(p)$ is the result of the optimization problem:

Table 9.1 *Scaling law order notations*

Scaling law notations	**Mathematical definitions**	**Asymptotic interpretations**
$f(n) = \mathcal{O}(g(n))$	$\exists$ a constant c and integer N such that $f(n) \leq cg(n)$ for $n > N$	$f(n) \leq g(n)$
$f(n) = o(n)$	$\lim_{n\to\infty} \frac{f(n)}{g(n)} = 0$	$f(n) \ll g(n)$
$f(n) = \Omega(g(n))$	$g(n) = \mathcal{O}(f(n))$	$f(n) \geq g(n)$
$f(n) = \omega(g(n))$	$g(n) = o(f(n))$	$f(n) \gg g(n)$
$f(n) = \Theta(g(n))$	$f(n) = \mathcal{O}(g(n))$ *and* $g(n) = \mathcal{O}(f(n))$	$f(n) = g(n)$

$$\begin{aligned} &\text{maximize} \quad \bar{T}_{\min} \\ &\text{subject to} \quad p_o \leq p, \end{aligned} \tag{9.3}$$

where the maximization is with respect to the cache placement and transmission policies. Since the scaling law analysis is essentially to characterize the asymptotic behavior of the networks, the scaling law order notations defined in Table 9.1 will be intensively used below.

9.3 Single-hop networks

In this section, we discuss scaling law results of cache-aided D2D networks considering uncoded single-hop communications.[a] The main benefit of considering such networks is that the implementation can remain simple when files are delivered. Additionally, single-hop communications are more plausible in current cellular systems [22, 23]. We will, in the following, first discuss the theoretical studies, and then provide numerical results under realistic setups to support the theories.

[a]"Uncoded" here means that no inter-file coded schemes will be used.

9.3.1 Scaling law results

The first work investigating the scaling law of cache-aided single-hop uncoded D2D networks was [24]. This paper investigated the scaling law of the maximum expected throughput, under the assumptions that the popularity model is characterized by the Zipf distribution, the protocol model is adopted for communications, and N users are distributed within the network uniformly at random. The results indicate that the throughput per user can scale with $\Theta(1)$ when $\gamma > 1$ and scale with $\Theta\left(\frac{1}{M^\eta}\right)$, where $\eta = \frac{1-\gamma}{2-\gamma}$, when $\gamma < 1$.[b] These scaling laws are achieved via adopting a caching policy that caches files according to a Zipf distribution and then optimizes the collaboration distance in order to pursue the maximum expected number of links in the network. These scaling law results indicate that the expected number of links, and thus the throughput, that can be offloaded from the BSs to the D2D network is significant as the expected throughput scales linearly with N when $\gamma > 1$ and scales with $\Theta\left(\frac{N}{M^\eta}\right)$ when $\gamma < 1$.

Although results in [24] showed promising offloading performance when using cache-aided D2D networks, the outage probability was ignored. In fact, it was found later that when $\gamma \in (0, 1)$ (heavy-tail popularity distribution), this maximum expected throughput can only be achieved when the outage probability goes to 1 as $N \to \infty$. As a consequence, only a small portion of the users are served, while the majority of the users are left without having services (i.e., in outage). Reference [21] discovered this problem and investigated the scaling behavior in terms of throughput-outage tradeoff. It provides a tight characterization for this tradeoff in the order sense.

Reference [21] assumes a grid network for the user distribution and a protocol model for the communication links. The discussions in [21] focused on results under the assumption that the popularity distribution is Zipf distribution. The model was later generalized in [18] to the MZipf distribution, as it is of practical interest how the plateau factor q of a MZipf distribution influences the performance. For q is a constant while $M \to \infty$, the MZipf distribution would behave identical to a Zipf distribution, while consideration of $q = \mathcal{O}(M) \to \infty$ reveals the influence of the plateau factor q. In addition, the MZipf distribution can converge to simple uniform distribution when $q = \omega(M)$.

To obtain the lower performance bound, Ji *et al.* [21] and Lee *et al.* [18] consider the following achievable scheme. The transmission policy is based on *clustering* described in Section 9.2, and the caching policy on *independent random caching*. Only a single link can be activated at a time in the cluster because single-hop D2D is considered. The independent random caching policy that minimizes the outage probability in a cluster is as follows (Theorem 1 in [18]):

[b]To be more specific, the $\Theta\left(\frac{1}{M^\eta}\right)$ scaling law when $\gamma < 1$ is a performance outer bound, and there is a small gap between the achievable performance and this value.

Theorem: Define $c_2 = qd'$, where $d' = \frac{\gamma}{S(g_c(M)-1)-1}$. Let $c_1 \geq 1$ be the solution of the equality $c_1 = 1 + c_2 \log\left(1 + \frac{c_1}{c_2}\right)$. Let $M \to \infty$ and $N \to \infty$. Suppose $g_c(M) \to \infty$ as $M \to \infty$, and denote m^* as the smallest index such that $P_c^*(m^* + 1) = 0$. The caching distribution $P_c^*(\cdot)$ that maximizes the probability that any user u finds its requested file inside the corresponding cluster is:

$$P_c^*(f) = \left[1 - \frac{\nu}{z_f}\right]^+, f = 1, \dots, M, \tag{9.4}$$

where $m^* = \Theta\left(\min\left(\frac{c_1 S g_c(M)}{\gamma}, M\right)\right)$,,$z_f = (P_r(f))^{\frac{1}{S(g_c(M)-1)-1}}$, $[x]^+ = \max(x, 0)$, and

$$m^* = \Theta\left(\min\left(\frac{c_1 S g_c(M)}{\gamma}, M\right)\right). \tag{9.5}$$

From the theorem, we observe that $P_c^*(f)$ is a monotonically decreasing function, indicating that the optimal policy generally is to cache the file requested more times with high probability. Besides, when $q = \Theta(1)$ is a constant, $c_1 \to 1$ can be observed, and the theorem applies to networks that consider Zipf distributions (Theorem 4 in [21]). The influence of q on the optimal caching policy is realized through c_1 and c_2. We observe that $c_1 = \mathcal{O}(c_2)$, when $c_2 = \Omega(1)$. Thus, when considering $q = \Omega\left(\frac{S g_c(M)}{\gamma}\right)$ and $\frac{c_1 S g_c(M)}{\gamma} < M$, we obtain $m^* = \Theta(\frac{c_1 S g_c(M)}{\gamma}) = \Theta\left(\sqrt{\frac{S q g_c(M)}{\gamma}}\right)$. Combining the above results, this theorem indicates that the number of files that the caching policy should cover not only is relevant to the number of nodes in a cluster but also is relevant to the rank q (orderwise) in the library. This is intuitive because the MZipf distribution has a relatively flat plateau regime and q characterizes the break point between the plateau and rolloff regimes. The following expositions assume that $P_c(f) = P_c^*(f)$.

The practically most interesting cases occur for the regime with small outage probability. Considering first the situation with $\gamma < 1$, it is found

Theorem (Theorem 2 in [18]): Let $M \to \infty$ and $N \to \infty$. Suppose $g_c(M) \to \infty$ as $M \to \infty$. Consider $M = \mathcal{O}(N)$, $SN \gg M$, $q = \mathcal{O}\left(\frac{S g_c(M)}{\gamma}\right)$, and $\gamma < 1$. Define $D = \frac{q}{M}$. When $g_c(M) = \frac{\rho M}{c_1 S}$, where $\rho \geq \gamma$, the achievable throughput-outage tradeoff is:

$$\begin{aligned} T(P) &= \frac{C}{K}\frac{Sc_1}{\rho M}, \\ p &= \frac{(1-\gamma)e^{-(\rho/c_1-\gamma)}}{(1+D)^{1-\gamma} - (D)^{1-\gamma}} \\ &\cdot \left[(1+D)^{\frac{\gamma}{S(g_c(M)-1)-1}+1} - (D)^{\frac{\gamma}{S(g_c(M)-1)-1}+1}\right]^{-(S(g_c(M)-1)-1)}. \end{aligned} \tag{9.6}$$

By choosing $g_c(M) = \beta M$ for some $\beta > 0$, it is apparent from the theorem that p strictly bounded away from 1. By fixing a small but positive target outage probability, the per-user average throughput of the D2D one-hop caching network with random (decentralized) caching scales as $T(p) = \Theta\left(\max\left\{\frac{1}{N}, \frac{S}{M}\right\}\right)$, where the scaling $\Theta\left(\frac{1}{N}\right)$ can be trivially achieved by letting the whole network to be a single cluster and serving one demand per unit time. This scaling is equivalent to conventional unicasting from a single omniscient node, which can be regarded as the state of the art of today's (single cell) systems, with a BS or access point serving individual requests without exploiting the asynchronous content reuse. We notice that when $SN \gg M$, the throughput of the D2D caching network achieves per-user throughput that increases linearly with S. This indicates that caching in the user nodes and exploiting the spatial reuse of the D2D network is a very attractive approach in dense networks, since storage space is much "cheaper" than scarce resources such as bandwidth or dense BS deployment. Since the benefits of uncoded caching is significant under the assumption that $SN \gg M$, the reminder of Sections 3 and 4 considers such an assumption unless otherwise stated. Finally, we note that when considering a Zipf (i.e., $q = \Theta(1)$), we have $p = (1 - \gamma)e^{\gamma - \rho}$, corresponding to the results in Theorem 5 of [21].

Apart from the achievable scheme, Ji *et al.* [21] also showed that the scaling law outer bound of the considered network is identical to the achievable performance discussed above, i.e., $T(p) = \Theta\left(\frac{S}{M}\right)$ when the outage probability is negligible. This validates the optimality of the scaling law performance here. Note that although the outer bound in [21] was derived considering of Zipf distribution, it is also valid for the MZipf distribution, as increasing q can only degrade the performance for a given random caching policy (see Appendix D in [18]).

To understand the potential of cache-aided D2D networks, it is instructive to compare the scaling laws achieved by the D2D caching network with those achievable by other possible approaches. We have already discussed conventional unicast, achieving $\Theta(\frac{1}{N})$ throughput. When the number of files M is less than the number of users N, an alternative and well-known approach is Harmonic Broadcasting [25]. In brief, Harmonic Broadcasting works as follows: fix the maximum waiting delay of τ "chunks" (from the time a streaming session is initiated to the time playback is started), and let L denote the total length of the video file, expressed in chunks. In Harmonic Broadcasting, the video file is split into successive blocks such that for $i = 1, \ldots, \lceil L/\tau \rceil$, there are i blocks of length τ/i. Then, each ith set of blocks of length τ/i is repeated periodically on a (logical) subchannel of throughput R/i, where R is the transmission rate (in bit/s) of the video playback. Users receive these channels in parallel. In this way, each file requires a downlink rate of $R\log(L/\tau)$. Hence, the total number of files that can be sent in the common downlink stream is $m' = \min\left\{\frac{C_{r0}}{R\log(L/\tau)}, M\right\}$, yielding an average throughput per user of $R\left(1 - p_o\right)$ with outage probability $p_o = \sum_{f=m'+1}^{M} P_r(f)$, since all requests to files not included in the common downlink stream are necessarily in outage. Accordingly, throughput per user of Harmonic Broadcasting scales as $\Theta\left(\frac{1}{m' \log\frac{L}{\tau}}\left(1 - \sum_{f=m'+1}^{M} P_r(f)\right)\right)$, where $m' \leq M$. It follows that when $\gamma < 1$, the throughput of Harmonic Broadcasting scales as $\Theta(\frac{1}{M\log L})$ for an outage probability that is bounded away from 1. From a strictly technical viewpoint, since in the system assumptions we study the system performance by first letting $L \to \infty$, and then considering N, M that simultaneously grow in some relation, the throughput of Harmonic Broadcasting under our assumption is identically zero. In practice, for large but finite L, the gain of D2D caching over Harmonic Broadcasting can be appreciated by comparing the term $\frac{S}{M}$ with the term $\frac{1}{M \log L}$ in the per-user throughput. It is clear that Harmonic Broadcasting does not take advantage of the user nodes storage memory and in addition suffers an arbitrarily large multiplicative penalty as the length of the files L increases.

It is interesting to compare cache-aided D2D with the coded multicasting (coded caching) scheme in [15], which represents another example of one-hop network with caching in the user nodes, able to make efficient (and in fact, near-optimal in an information theoretic sense) use of caching. The rate analysis provided in [15] shows that for large M, N and finite S, the scaling of the per-user throughput of the coded multicasting is $\Theta\left(\max\left\{\frac{1}{N}, \frac{S}{M}\right\}\right)$, where the two terms inside the max are realized depending on whether $NS \gg M$ or not. Interestingly and somehow surprisingly, this is the same scaling behavior of the D2D caching network derived above. However, coded multicasting requires higher implementation complexity due to the need for network coding and is vulnerable to fading effects, as shown in the simulations below.

The above results are based on the assumption that $\gamma < 1$. The following theorem characterizes the scaling law under the assumption that $\gamma > 1$.

Theorem (Theorem 4 in [18]): Let $M \to \infty$, $N \to \infty$, and $q \to \infty$. Suppose $g_c(M) \to \infty$ as $M \to \infty$. Consider $\gamma > 1$, $g_c(M) = o(M) \leq N$, and $q = \mathcal{O}\left(\frac{Sg_c(M)}{\gamma}\right)$. Define $c_6 = \frac{q}{g_c(M)}$. The achievable throughput-outage tradeoff is

$$T(p) = \frac{C}{K}\frac{1}{g_c(M)} + o\left(\frac{1}{g_c(M)}\right), \quad p = (c_6)^{\gamma-1}\frac{Sc_1+c_6}{\left(\frac{Sc_1}{\gamma}+c_6\right)^{\gamma}}. \tag{9.7}$$

It should be noted that this theorem considers $g_c(M) = o(M)$ and $q = \mathcal{O}(g_c(M))$. This is because such a consideration leads to a more interesting scaling law, as it breaks the $T(p) = \Theta\left(\frac{S}{M}\right)$ limit as compared to the case that $\gamma < 1$. As a matter of practice, we observe that $q \ll M$ from the dataset in [18]. From the theorem, we observe that if we let $g_c(M) = \Theta(q)$, the achievable scaling law is $\Theta(\frac{S}{q})$ when $NS \gg q$. This implies that when $\gamma > 1$ and the aggregate memory is larger than the order of q, the improvement of the cache-aided D2D is significant even if we have a large M. This relaxes the condition that applies for $\gamma < 1$, namely that a small library (i.e., $NS \gg M$) is required to have significant benefits. Furthermore, the fact that the real dataset in [18] has $q \ll M$ and $\gamma > 1$ for mobile users strongly supports the practical significance of such a result. Finally, it should be noted that the case considering a Zipf distribution with $\gamma > 1$ can be viewed as either a special case of the MZipf distribution with $q = \mathcal{O}(1)$ and $\gamma > 1$ or a special case of the Zipf distribution with $M = \mathcal{O}(1)$ and $\gamma < 1$. The latter one holds because when considering a Zipf distribution with $\gamma > 1$, a finite number of files collects essentially all the request probability mass. Accordingly, we can conclude that the scaling law with $\Theta\left(\frac{S}{M^{\epsilon}}\right)$, where ϵ is arbitrarily small, is achievable with negligibly small outage probability when a Zipf distribution with $\gamma > 1$ is considered.

9.3.2 Numerical studies under realistic setups

Ji *et al.* [26] and Lee *et al.* [18] provided abundant simulations under realistic setups to show the advantages of cache-aided D2D networks. As an example, we discuss here the "mixed-deployment" scenario described in [26]. A 0.36 km^2 (600 m × 600 m) area that contains buildings as well as streets/outdoor environments is considered. $N = 10\,000$ users are uniformly and independently distributed in the cell, i.e., on average, there are $2 \sim 3$ nodes in each square of 10×10 m. The cell contains a Manhattan grid of square buildings with side length of 50 m, separated by the streets of width 10 m. Each building is made up of offices of size 6.2 m × 6.2 m. Within the cell, users (devices) are distributed at random according to a uniform distribution. Each node is assigned to be outdoors or indoors, and in the latter case in a particular office, according to where its location falls on the map of the environment. D2D communications in the 2.4 GHz band is considered. Depending on the locations of the two users in the D2D communication pair, the employed channel model is for indoor communication (Winner model A1), outdoor-to-indoor communication (B4), indoor-to-outdoor communication (A2), and outdoor communication (B1).

In particular, the simulations directly use the appropriate Winner II channel models with antenna heights of 1.5 m, and the corresponding probabilistic line of sight (LOS) and non line of sight (NLOS) models. A probabilistic body shadowing loss (σ_{L_b}) with a log normal distribution accounts for the blockage of radiation by the person holding the device, where for LOS, $\sigma_{L_b} = 4.2$ and for NLOS, $\sigma_{L_b} = 3.6$ [27].

Figure 9.2 compares throughput/outage for cache-aided D2D, conventional unicasting, harmonic broadcasting, and coded multicasting. The number of files in the library is $M = 300$. The user cache size is $S = 20$ files, which even with high definition (HD) quality requires less than the (nowadays) ubiquitous 64 GByte of storage space. Each user independently makes a request by sampling from a Zipf distribution with $\gamma = 0.4$; this value is at the lower edge of the range of values that have been measured in practice [16]; note that the advantages of caching would be *more* pronounced for larger γ. The interference between concurrent D2D links sharing the same frequency band is treated as noise. The harmonic broadcasting uses a video file size of $L = 5,400$ chunks and $\tau = 10$ chunks, so that the number of blocks is $L_N = \lceil \frac{L}{\tau} \rceil = 540$ [28].

We can see that the throughput of the D2D scheme is markedly (orders of magnitude at low outages) higher than the conventional unicasting, harmonic broadcasting, and even coded multicast. This shows that in practical situations, the "scaling law" is not the only aspect of importance. Rather, the higher capacity of the short-distance links plays a significant role, and a good throughput-outage tradeoff can be achieved even without the use of a BS connection as "backstop." The main reason

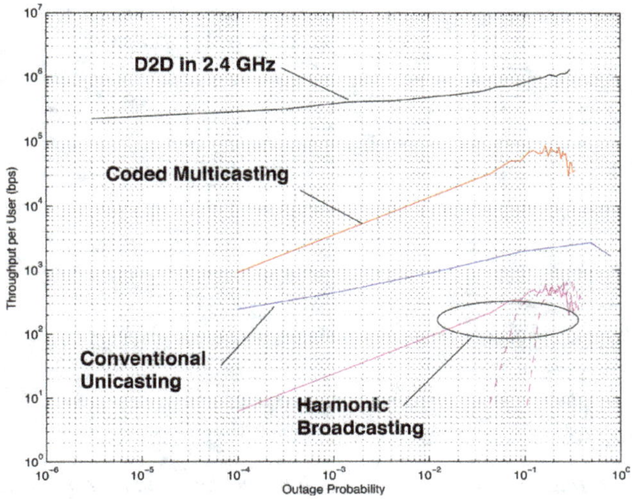

Figure 9.2 *Simulation results for the throughput-outage tradeoff for conventional unicasting, coded multicasting, harmonic broadcasting, and the* 2.4 *GHz D2D communication scheme under indoor office channel models. For harmonic broadcasting with only the m' most popular files, solid line: m'* = 300*; dash-dot line: m'* = 280*; dash line: m'* = 250*. We have N* = 10 000*, M* = 300*, S* = 20 *and γ* = 0.4.

lies in the fact that for the coded multicasting or harmonic broadcasting scheme, outage is determined by bad channel conditions, and no diversity is built into the system. For D2D the channel diversity plays a more importance role.

To understand the impact of a different popularity distribution, Figure 9.3 shows results in similar settings, but for an MZipf distribution. Specifically, the MZipf distribution with $\gamma = 1.16$, $q = 22$, and $M = 7\,345$ is considered, where these parameters are extracted from the July Metro region 1 dataset defined in [18]. Figure 9.3 shows the throughput-outage tradeoff for different cache sizes on each device. We observe that the throughput of 10^5 bps can be achieved if the cache size of each user is up to 1/10 of the library size, showing significant improvement as compared to the conventional unicasting. Even for $S = M/50$, i.e., approximately 100 files, the advantage compared to the conventional unicasting is two orders of magnitude. Even just caching of 30 files ($M/200$) also provides significant throughput gains, though only for outage probabilities >0.01.

Combining results in Figures 9.2 and 9.3, we can conclude that cache-aided D2D networks show significant improvement, as compared to other video distribution schemes. Besides, by comparing Figures 9.2 and 9.3, we observe that even though the scaling law changes when changing the popularity distribution from one to another, the performance of the cache-aided D2D network is not very sensitive to such a change, as long as the total cache space in the network is sufficient. This

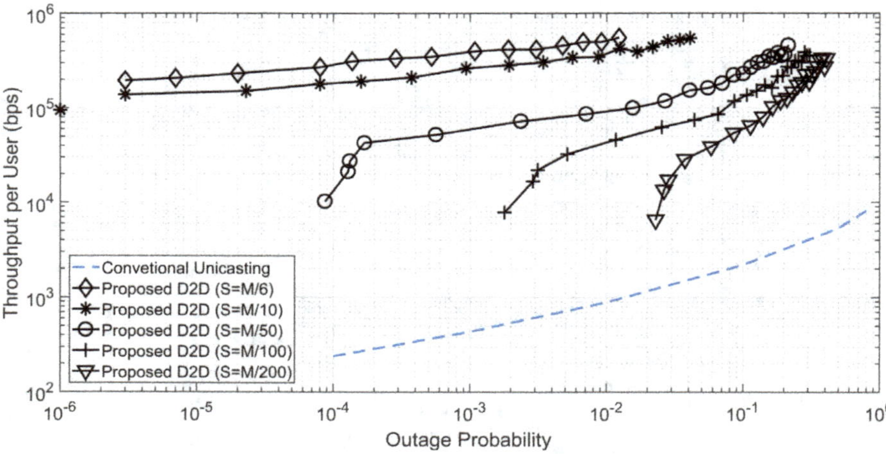

Figure 9.3 Throughput-outage tradeoff in networks assuming office scenario for propagation channel; varying local storage size; and $\gamma = 1.16$, $q = 22$, *and* $M = 7\,345$

again indicates that the "scaling law" is not the only aspect of importance in practical systems.

9.4 Multihop networks

Although the implementation is more difficult than single-hop D2D, adopting multihop delivery in cache-aided D2D networks can significantly improve scaling laws. We thus in this section discuss scaling laws of cache-aided multihop D2D networks.

Two multihop file delivery schemes have been used in the literature for scaling law analysis. The first scheme, proposed in [29], uses local multihop delivery with clustering. Specifically, the network is first split into clusters where interference between clusters is avoided by frequency reuse, just as in the schemes described in Section 9.3. Then, within each cluster, users obtain (if possible) files from other users in the same cluster via a multihop delivery scheme; a user in a cluster can only access files cached by users in the same cluster. The local multihop delivery scheme within each cluster operates as follows. First, the cluster is split into small equally sized square hopping cells. Then, users requesting files are paired with other users who cache their desired files, constructing source-destination pairs. Each source-destination pair defines their vertical and horizontal hopping path that connects the source and destination. To deliver the file, the source user first transmits the desired file through the horizontal path hopping along adjacent hopping cells, and then hopping through the vertical path, as illustrated in Figure 9.4. When several source-destination pairs in the cluster transmit files, a hopping cell might need to relay multiple files to its adjacent cells. In this case, a round-robin approach is used for all files in the same hopping cell. Note that to avoid interference between

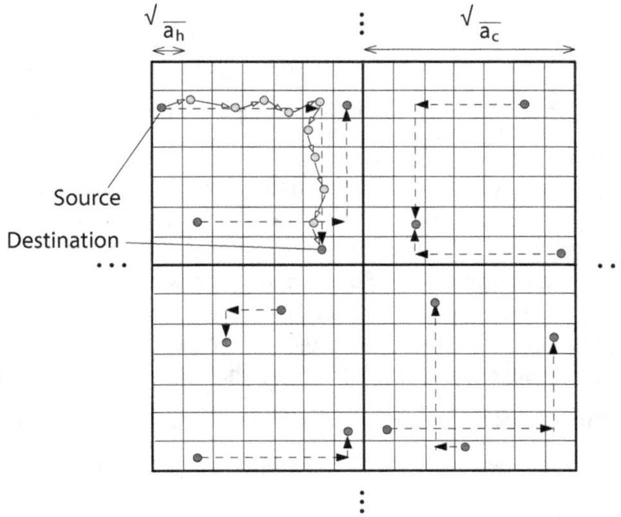

Figure 9.4 *The multihop delivery after the source node selection*

different hopping cells, a TDMA scheme with reuse factor J is used as the interference avoidance scheme for which a hopping cell is activated every J time-slots. The throughput-outage tradeoff is then obtained by adjusting the sizes of clusters and hopping cells, leading to the desired scaling law.

The second scheme, which was proposed in [30], also adopts the hybrid clustering and multihop delivery structure, while a different multihop delivery approach is used within each cluster. Specifically, the multihop delivery scheme proposed in [9] is exploited within each cluster for file delivery in this case. Denote $\mathcal{V}_f$ as the set of users in a cluster that cache file f. Then for each user u in the cluster, if the requested file f can be found in the caches of users in $\mathcal{V}_f$, then a user $v_{f'}$ randomly selected from $\mathcal{V}_f$, is chosen as the source to deliver the requested (real) file f to user u. If the requested file cannot be found from the caches of any users in the cluster, user u is matched with a randomly selected user v from all users in the cluster, and then user v is set as the source for delivering a *virtual* file to user u. Note that it does not matter what file is delivered in this case, as the user is actually in outage. After the establishment of the matching of the sources and destinations, to deliver (both real and virtual) files, the multihop approach proposed in [9] directly applies. Such scheme is suboptimal as the delivery of virtual files cannot generate any effective throughput. Nevertheless, when the outage probability is either negligibly small or converging to zero, this scheme will be orderwise optimal because the performance degradation caused by delivering virtual files is negligible.

Besides assuming the first of the two multihopping schemes, Jeon *et al.* [29] assume (i) the protocol model (ii) N users distributed according to a BPP (i.e., N users uniformly and randomly distributed within the network), (iii) a heavy-tailed popularity distribution (i.e., Zipf distribution with $\gamma < 1$), (iv) $SN \gg M$, (v) a uniformly random caching policy, in which each user caches S distinct files that are selected uniformly at random from the library $\mathcal{F}$. Then the throughput scaling law $\Theta\left(\sqrt{\frac{S}{M}}N^{-\varepsilon}\right)$ is achievable with negligible outage probability, where ε is an arbitrarily small number.

Jeon *et al.* [29] also characterized the outer bound of the scaling law when the outage probability converges to zero. The key of the outer bound characterization is to obtain the minimum distances the source-destination pairs need to traverse when delivering files. Such a distance characterization is relevant to the interference regime created by a successful file delivery of a source-destination pair. Given the interference regimes of source-destination pairs, the maximum number of concurrent transmissions can be upper bounded, leading to the outer bound of the throughput. The resulting throughput scaling law outer bound is $\mathcal{O}\left(\sqrt{\frac{S}{M}}N^{\varepsilon}\right)$. By comparing between the achievable performance and the outer bound, we can see that there is a $\Theta\left(N^{2\varepsilon}\right) = \Theta\left(\log(N)\right)$ gap between them.

Lee *et al.* [30] adopted a somewhat more general model, in particular assuming an MZipf distribution for the file popularity, PPP for user distribution, and a physical model (interference) model for communications. It obtained achievable scaling law and outer bound for the regimes that the outage probability is either negligibly small or converging to zero and showed that they are tight. Since the Zipf distribution model is only a special case for the MZipf model, such an optimality can also be applied to network considering Zipf distributions. This thus closes the gap mentioned above.

Specifically, when the average number of users in the network is $N \to \infty$, the scaling law analysis is conducted under the conditions that $M = o(N)$ and $q = \mathcal{O}(M)$ when $\gamma < 1$ and $q = o(M)$ when $\gamma > 1$. The plateau factor q can either go to infinity or remain as a constant, regarded as MZipf and Zipf distributions, respectively, from the asymptotic analysis perspective. To analyze the scaling law under the above assumptions, the following caching policy minimizes the outage probability of users in a cluster:

Theorem (Theorem 1 in [30]): Let $N \to \infty$, $M \to \infty$, $q \to \infty$, and $g_c(M) \to \infty$. Denote m^* as the smallest index such that $P_c^*(m^*+1) = 0$. Let $C_2 = \frac{q\gamma}{Sg_c(M)}$; C_1 is the solution of the equation: $C_1 = 1 + C_2 \log\left(1 + \frac{C_1}{C_2}\right)$. The caching distribution $P_c^*(\cdot)$ that minimizes the outage probability p_o is as follows:

$$P_c^*(f) = \left[\log\left(\tfrac{z_f}{\nu}\right)\right]^+, \; f = 1, \ldots, M, \tag{9.8}$$

where$m^* = \Theta\left(\min\left(\frac{C_1 Sg_c(M)}{\gamma}, M\right)\right)$.,$z_f = (P_r(f))^{\frac{1}{g_c(M)}}$, $[x]^+ = \max(x, 0)$, and

$$m^* = \Theta\left(\min\left(\frac{c_1 Sg_c(M)}{\gamma}, M\right)\right). \tag{9.9}$$

This caching policy, implemented via the approach proposed in [19], has a similar structure to the one introduced in [18], while the mathematical details are different because different user distribution and communication models are used. However, the interpretation is the same – the most popular q files (orderwise) have similar request probabilities, and we need to cache them to minimize the outage probability.

When the second multihop file delivery scheme is considered to deliver files and the aforementioned caching policy is adopted, the network has the following achievable throughput-outage tradeoffs:

Tradeoff for $\gamma < 1$ (Corollary 1 in [30]): Let $M \to \infty$, $N \to \infty$, and $q \to \infty$. Suppose $\gamma < 1$ and $g_c(M) = \frac{\rho M}{C_1 S} = o(N)$. When considering $\rho = \Theta(1) \geq \gamma$, the following throughput-outage performance is achievable:

$$T(p) = \Omega\left(\sqrt{\frac{S}{\rho M}}\right), \quad p = \varepsilon_1(\rho), \tag{9.10}$$

where $\varepsilon_1(\rho) > 0$ can be arbitrarily small. Furthermore, when considering $\rho \to \infty$, i.e., $\rho = \omega(1)$, we obtain the following achievable throughput-outage performance:

$$T(p) = \Omega\left(\sqrt{\frac{S}{\rho M}}\right), \quad p = \Theta\left(e^{-\rho}\right) = o(1). \tag{9.11}$$

Tradeoff for $\gamma > 1$ (Corollary 4 in [30]): Let $M \to \infty$, $N \to \infty$, and $q \to \infty$. Suppose $\gamma > 1$ and $g_c(M) \to \infty$. Consider $g_c(M) = o(M)$, $q = o(M)$, and $g_c(M) = \frac{\alpha_1 q}{S}$. When considering $\alpha_1 = \Theta(1)$ to be large enough, the following throughput-outage performance is achievable:

$$T(p) = \Omega\left(\sqrt{\frac{S}{\alpha_1 q}}\right), \quad p = \varepsilon_2(\alpha_1), \tag{9.12}$$

where $\varepsilon_2(\alpha_1) > 0$ can be arbitrarily small. Furthermore, when considering $\alpha_1 = \omega(1) \to \infty$, we obtain the following throughput-outage performance:

$$T(p) = \Omega\left(\sqrt{\frac{S}{\alpha_1 q}}\right), \quad p = \Theta\left(\frac{1}{(\alpha_1)^{\gamma-1}}\right) = o(1). \tag{9.13}$$

From these results, we understand that when $\gamma < 1$ and the outage probability is negligibly small, the achievable throughput per user is $\Omega\left(\sqrt{\frac{S}{M}}\right)$. Additionally, the tradeoff between the throughput and outage probability is controlled by the parameter ρ. When considering $\gamma > 1$ and the outage probability is negligibly small, we see that the throughput per user is $\Omega\left(\sqrt{\frac{S}{q}}\right)$, and the tradeoff between throughput and outage probability is controlled by the parameter α_1. By comparing between the results of $\gamma < 1$ and $\gamma > 1$, we understand that when the popularity distribution has a light tail ($\gamma > 1$), we can improve the scaling law as $q \ll M$ holds in practice. Finally, we note that the outer bound analysis in Theorems 4 and 5 in [30] indicates that above achievable throughput-outage performance is (orderwise) optimum as the multiplicative gap between the achievable performance and corresponding outer bound can be upper bounded by a constant. The outer bound is obtained by first characterizing the upper bound of the source-destination distances in the network when given an outage probability lower bound, and then obtain the upper bound of the throughput per user in this context. As a final remark, we note that although the achievable scheme has the assumption that a user may get a desired file from only its own cluster, the fact that this scheme can achieve (in the order sense) the outer bound shows that inclusion of inter-cluster communication does not change the scaling law.

The above achievable scheme and outer bound analysis can be conducted following the similar procedure for the case that $q = \mathcal{O}(1)$ is a constant. Such an analysis leads to scaling laws of networks considering the Zipf distribution. The results lead to the following optimal scaling laws: when $\gamma < 1$, the throughput-outage tradeoff is

$$T(p) = \Theta\left(\sqrt{\frac{S}{\rho_{\text{zip}}M}}\right), \quad p = \Theta\left(e^{-\rho_{\text{zip}}}\right), \tag{9.14}$$

where $\rho_{\text{zip}} = \Omega(1)$; when $\gamma > 1$, the throughput-outage tradeoff is

$$T(p) = \Theta\left(\sqrt{\frac{S}{\alpha'_{1,\text{zip}}}}\right), \quad p = \Theta\left(\frac{1}{(\alpha'_{1,\text{zip}})^{\gamma-1}}\right), \tag{9.15}$$

where $\alpha'_{1,\text{zip}} = o(M)$ is any function that goes to infinity as $M \to \infty$.

Discussions in Section 9.4 to this point characterize the throughput-outage scaling laws considering decentralized caching policies. Yet, there exist some papers investigating cache-aided multihop D2D networks adopting centralized caching policies. In [31], the scaling law of the average throughput per node for the cache-aided multihop D2D network was characterized with the assumption of user locations on a grid and with a centralized caching policy. It shows that the scaling laws of $\Theta\left(\sqrt{\frac{S}{M}}\right)$ for $\gamma < 1$, of $\Theta\left(\frac{\sqrt{S}}{M^{\frac{3}{2}-\gamma}}\right)$ for $1 < \gamma < \frac{3}{2}$, and of $\Theta\left(\sqrt{S}\right)$ for $\gamma > \frac{3}{2}$ can be achieved, respectively, when a Zipf distribution is used as popularity model. Qiu and Cao [32] proposed outer bounds considering also the Zipf distribution. It showed that the outer bounds are tight to the achievable performance in [31], indicating the potential optimum scaling laws for networks adopting centralized caching policy.[c] It should be noted that when a centralized caching policy is adopted and the total cache space in the network is large enough, since the cache space can be fully controlled, the outage probability can be exactly zero, i.e., completely no outage occurs. This is different from the decentralized random caching policy that the outage probability cannot be completely eliminated. Accordingly, the comparisons between the scaling law results of the centralized caching policy and the randomized caching policy might not be completely fair. This might also explain the additional regime that

[c]Note that [31] and [32] actually considered slightly different network models, while their insights are essentially the same.

$1 < \gamma < \frac{3}{2}$ in the scaling law of the centralized caching policy, as compared to the scaling law of the randomized caching policy – the allowance of a tiny outage (either negligibly small or converging to zero) in the scaling law of the randomized caching policy eliminates such regime.

By comparing the scaling laws of networks adopting centralized policy with those adopting decentralized caching policy, we observe that they have the same scaling law when a negligibly small outage probability is allowed for the network adopting a decentralized policy. Therefore, as both centralized and decentralized caching policies have almost identical scaling laws, the use of a randomized policy is more attractive, as it is easier to implement and more robust to user mobility [33].

Finally, we would like to point out that the multihop schemes discussed above in this section are based on simple multihop transmissions without complicated physical layer processing. On the other hand, there exist papers combining caching with hierarchical cooperation [10] which enables the distributed MIMO gain during multihop transmissions. These papers [34, 35] can lead to better scaling laws than those introduced above at the price of very complicated cooperation among user nodes.

9.5 D2D-coded caching

In this section, we will discuss the application of network coding in D2D caching networks. In particular, we will first discuss the original single-hop D2D-coded caching scheme introduced in [36] and then introduce an improved scheme proposed in [37]. In the end, we will discuss the extension of D2D-coded caching to multihop networks.

9.5.1 Single-hop-coded D2D caching

One important constraint of the schemes described in previous sections is that both the cache placement and the delivery schemes exploit an "uncoded" approach in the delivery phase. The throughput gain is mainly obtained by effective cache placement and spatial reuse (TDMA/FDMA) if applicable. A natural question to ask is whether coded delivery (or coded multicasting) for D2D transmissions can provide an additional gain, or whether the coding gain and the spatial reuse gain can be combined. Ji *et al.* [36] (see also [38]) design a deterministic subpacketized caching and a network-coded delivery scheme for the single-hop D2D caching networks. The scheme is explained by the example shown in Figure 9.5, where we assume no spatial reuse can be used, or only one transmission per time-frequency slot is allowed but the transmission range can cover the entire network. This scheme can be generalized to any N, M, S. Without using spatial reuse, for zero outage, the achievable normalized communication load (by the file size) such that every user

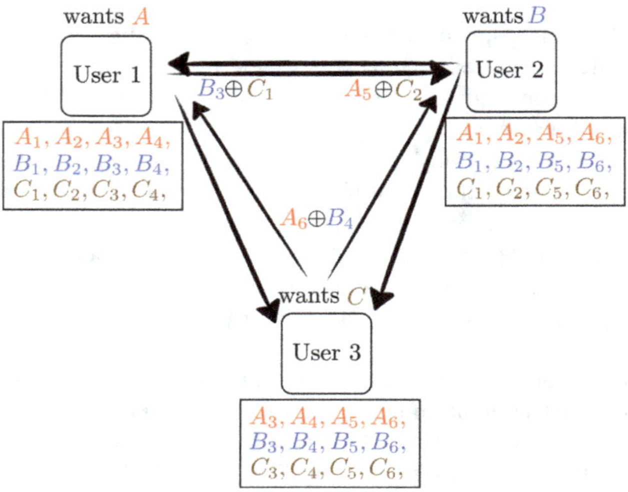

Figure 9.5 *Illustration of the example of* 3 *users,* 3 *files, and* $S = 2$*, achieving* 1/2 *transmission in terms of file. We divide each file into* 6 *packets (e.g., A is divided into* $A_1, \cdots, A_6$*). User 1 requests A; user 2 requests B; and user 3 requests C. The cached packets are shown in the rectangles under each user. For the delivery phase, user* 1 *transmits* $B_3 \oplus C_1$*; user 2 transmits* $A_5 \oplus C_2$*, and user 3 transmits* $A_6 \oplus B_4$*. The normalized number of transmissions is* $3 \cdot \frac{1}{6} = \frac{1}{2}$*, which is also information theoretically optimal for this network.*

can successfully decode is $R = \frac{M}{S}\left(1 - \frac{S}{M}\right) = \frac{M}{S} - 1$,[d] which is surprisingly almost the same as the result shown by [15], where instead of D2D communications, one central server (BS) which has access to all the video files, multicasts coded messages.

Suppose that (M, R) is achievable and that under the worst-case scenario, there exists a transmission policy that can deliver all coded messages necessary to decode the requested file of each user in no more than D channel uses. Then, the per-user throughput, measured in useful information bits per channel use, is given by $T = \frac{F}{D}$, where F denotes the file size in bits. Hence, we can see that the per-user throughput $\Theta\left(\frac{S}{M}\right)$ has the same scaling law as the throughput by using the decentralized random caching and uncoded delivery scheme described in Section 9.3. In addition, it can be shown that there is no further gain when spatial reuse is also exploited. In other words, *the gains of spatial reuse and coded delivery cannot accumulate.* Intuitively, if spatial reuse is not allowed, a complicated caching scheme can be designed such that one transmission can be useful for as many users as possible. If

[d]It can be seen that the per-user throughput T is inversely proportional to R under protocol model.

the transmission range is reduced and coding is done in one cluster, then the number of users benefited by one transmission is reduced but the D2D transmissions can operate simultaneously at a higher rate. Further, the complexity of caching subpacketization and coding can also be reduced. Therefore, the benefit of coding depends on the actual physical layer throughput (bits/s/Hz) and the caching/coding complexity rather than throughput scaling laws.[e]

The main drawback of the deterministic caching placement in the above proposed scheme is that, in practice, a tight control on the users caches must be implemented in order to make sure that, at any point in time, the files subpackets are stored into the caches in the required way. While this is conceptually possible, such approach is not robust to events such as user mobility and nodes turning on and off, as it may happen in a D2D wireless network with caching in the user devices. In [36], the authors proposed a decentralized random caching and coded delivery scheme that allows for more system robustness. In this scheme, each file of F bits is divided into L blocks or packets of F/L bits each. These packets are interpreted as the elements of the binary extension field $\mathbb{F}_{2^{F/L}}$, and are encoded using a $(L, L/\rho)$-MDS code, for some $\rho < 1$ the choice of which is essential to guarantee enough coded symbols per file are cached in the network such that the entire library is cached with high probability as $L \to \infty$. Notice that this expands the size of each packet from F to F/ρ. The resulting L/ρ encoded blocks of F/L bits each will be referred to as "MDS-coded symbols." Each user independently of the other users, uniformly chooses $\frac{ML}{m}$ elements from $\{1, 2, \cdots, L/\rho\}$ at random. In the delivery, similar to [40], each user seeks for possible coded multicasting opportunities with different group size. It can be shown that the achievable per user throughput of the proposed decentralized D2D-coded caching scheme is the same as the deterministic case.

9.5.2 An improved coded D2D caching schemes

The scheme in [36] can be understood as dividing the D2D caching network into N independent shared-link caching networks, in each of which one device serves as the BS and the rest $N-1$ devices are users. In each of the independent shared-link caching network, the coded multicasting scheme proposed in [15] is applied. In [37], an improved coded D2D caching scheme is proposed. In this scheme, under the same cache placement as in [15, 36], the delivery phase also divides the D2D network into N shared link caching networks. However, instead of applying the coded multicasting in each shared-link caching network in [15], it applies the *leader-based* coded multicasting scheme proposed in [41]. Consider a shared-link caching network with user demand vector $\mathbf{d}$ and denote $\mathcal{N}_e(\mathbf{d})$ as the number of distinct files in a request vector $\mathbf{d}$, in the *leader-based* coded multicasting scheme, any subset of users that request $\mathcal{N}_e(\mathbf{d})$ distinct files is referred to as leaders and the BS only

[e]An extensive analysis of the performance for D2D multicast is given by [39].

needs to multicast the codewords involving the requests of the leaders. Based on this scheme, the improved worst-case communication load for the coded D2D caching scheme is given by

$$R_{\text{Improved}} = \max_{\mathbf{d}} \frac{\binom{N-1}{t} - \frac{1}{N}\sum_{i=1}^{N}\binom{N-1-\mathcal{N}_e(\mathbf{d}\setminus\{i\})}{t}}{\binom{N-1}{t-1}}, \tag{9.16}$$

where $t = NS/M$ and $\mathbf{d} \setminus i$ denote all the elements in $\mathbf{d}$ except of i. It can be shown that when $N > M$, R_{Improved} strictly reduces the communication load in [36], which is $R = \frac{M}{S}\left(1 - \frac{S}{M}\right) = \frac{\binom{N-1}{t}}{\binom{N-1}{t-1}}$. Moreover, it can also be shown that R_{Improved} is optimal under the constraint of uncoded cache placement and one-shot delivery.[f]

9.5.3 Extension to multihop coded D2D caching networks

In Section 9.4, under uncoded cache placement and delivery constraint, we extended the uncoded D2D single-hop transmission scheme to the multihop case. For small outage probability, the per-user throughput of multihop delivery can be improved to $\Theta\left(\sqrt{\frac{S}{M}}\right)$, which shows an order gain compared to that for single-hop D2D communications. In Section 9.5.1, we allow coding for both cache placement phase, delivery (transmission) phase but focus on only single hop transmissions. Therefore, it is natural to consider the extension that relaxes all the constraints on cache placement, delivery and transmission phases under protocol channel model. It means that we can allow coding on both cache placement and delivery phases, and also allow multihop communications. Ji *et al.* [42] consider this scenario and propose an intra-file-coded cache placement based on deterministic assignment of minimum distance separable (MDS)-coded packets of the files, a coded multicast delivery scheme where the users send linearly coded messages (similar to the example in Figure 9.5) to each other in order to collectively satisfy their demands, and a randomized Euclidean Minimum Spanning Tree (EMST)-based routing strategy for transmission, where each hop in the EMST-based routing is implemented via the Manhattan Routing approach. These are illustrated in Figure 9.6. Under the worst-case demands, with some constant C, it shows that this approach actually achieves the throughput scaling law of $T = C\sqrt{\frac{\frac{S}{M}}{1-\frac{S}{M}}} = \Theta\left(\sqrt{\frac{S}{M}}\right)$, which has the same scaling law as the per-user throughput achieved by the multihop D2D-uncoded scheme under the PPP model. This throughput scaling law in [42] also achieves information theoretic outer bound within a multiplicative constant factor in practical parameter regimes.

[f] One-shot delivery means that each user can recover each of its needed bits from its own cache and the transmission of a single other device.

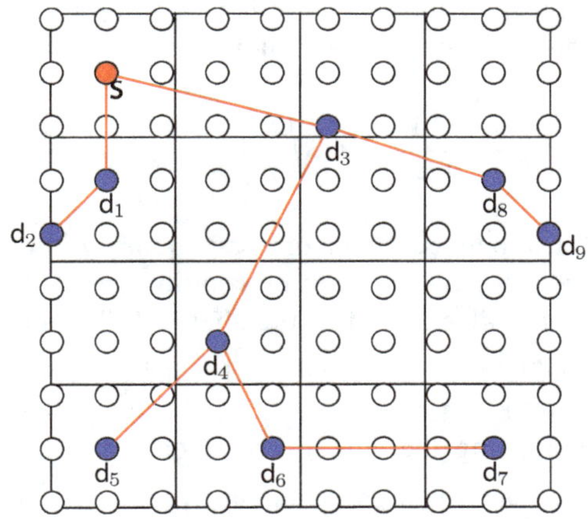

(a)

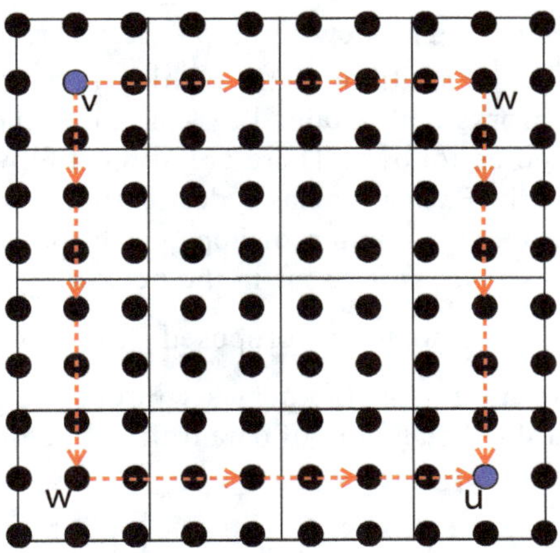

(b)

Figure 9.6 *(a) The D2D caching network is partitioned into square clusters with side length of* $\frac{r}{\sqrt{5}}$*. The red nodes represent the source node and all the blue nodes represent its* $t = 9$ *multicast destinations* $d_1, \cdots, d_9$*. The constructed EMST for this multicast session is shown by the red, blue nodes, and the red solid lines. (b) A illustration of the Manhattan routing approach from node* v *(transmitter) to node* u *(receiver). The red dashed line represents two Manhattan routing paths. In this example, the transmission range is assumed to be* $\frac{3}{\sqrt{N}}$*.*

Moreover, the achievable throughput in the multihop D2D-coded caching can be written as the product of four terms as follows.

$$T = C\frac{\sqrt{\frac{S}{M}}}{1-\frac{S}{M}} = \Theta\left(\frac{1}{N}\cdot\frac{1}{1-\frac{S}{M}}\cdot t\cdot\sqrt{\frac{N}{t}}\right), \tag{9.17}$$

where $t = \frac{NS}{M}$ is the ratio between the aggregate memory in the network and the library size. We have the following interpretation for (9.17). The first three terms $\frac{1}{N}$, $\frac{1}{1-\frac{S}{M}}$, and t are also found in the shared link caching networks [15] and the single-hop D2D caching networks [36]. In particular, the term $\frac{1}{N}$ is the per-user throughput by using a conventional scheme that serves individual demands without exploiting the demand redundancy; $\frac{1}{1-\frac{S}{M}}$ can be viewed as the *local caching gain*, any user can cache a fraction S/M of each file, hence it needs to receive only the remaining part; t is referred to as *global caching gain*, which is the gain due to the aggregate memory in the network rather than individual user's memory.[g] The new term $\sqrt{\frac{N}{t}}$ is the *multihop transmission gain*, which is the gain of multihop D2D caching networks over the single-hop D2D caching networks and obtained by using multihop. Intuitively, this term can be interpreted as follows. There are at most $\Theta(N)$ concurrent transmissions in the network. Each user has t destinations uniformly selected at random such that the average number of hops per bit is $\Theta(\sqrt{Nt})$.[h] Hence, the average number of bits that can be sent in the network per time slot is given by $\Theta\left(\frac{N}{\sqrt{Nt}}\right) = \Theta\left(\sqrt{\frac{N}{t}}\right)$. Although the proposed D2D-coded multihop caching scheme does not gain in terms of the per-user throughput, it may have some other gains. For example, the amount of data traffic N_S generated at each node from its memory is given by

$$N_S = \frac{F}{(1-\varepsilon)t\binom{N}{t}}\cdot\binom{N-1}{t} = \frac{F}{1-\varepsilon}\left(1-\frac{S}{M}\right)\frac{1}{t} = \Theta\left(\left(1-\frac{S}{M}\right)\frac{F}{t}\right), \tag{9.18}$$

where ε is an arbitrarily small constant. In contrast, due to the constraint of caching the entire files, the data traffic generated at each user from its own memory for the scheme discussed in Section 9.4 is given by $\Theta(F)$ bits for fixed S/M. Hence, in terms of the data generated at each node, the proposed scheme in this paper has a gain of t, which is the global caching gain.

[g]Note that the global caching gain for the shared link caching network is $t+1$ [15]. For $NS \gg M$ $(t \geq 1)$, these factors are almost identical.

[h]Note that for unicast network, where $t = 1$, the number of hops per bit reduces to $\Theta(\sqrt{n})$ as shown in the seminal paper [7].

9.6 Practical implementation considerations for D2D caching networks

In this section, we will discuss some practical consideration regarding the implementations of D2D caching networks. In particular, we will first discuss D2D caching under realistic channel conditions. Second, we will focus on the mobile D2D caching networks. Finally, we will discuss the complexity issues regarding D2D-coded caching.

9.6.1 D2D caching under realistic channel model

The results based on protocol channel model developed in previous sections reveal the potential of D2D caching networks in a fundamental manner without much practical considerations. In the literature, there are extensive studies on relaxing the constraints of the simplified protocol channel model via the consideration of much more realistic physical layer channel models. It turns out that coupling the realistic physical layer channel models significantly increase the difficulty in terms of theoretical analysis. Hence, one needs to find good models that can capture the main features of wireless D2D caching networks and are analytically tractable. A significant amount of work in this domain has been developed in the past few years, see [43–52] for a few examples.

When interferences are treated as noise, in general, mainly three representative models are used. The first two models consider the sub-6G Hz channel model and the location of the devices are based on various PPPs and Poisson Cluster Processes (PCPs), where the former does not consider the correlation of the user devices while the latter considers this phenomenon. The third model focuses on the mm-wave channel model, which is very attractive for D2D transmissions due to the high available bandwidth. The location distributions of the devices also follow PPPs. For each model, due to the difficulty of analysis, rather than the scaling laws of throughput-outage tradeoff, different performance metrics including density of successful receptions (DSR) [48], coverage probability, ASE [53] and offloading factor [46] are defined and analyzed. Although the caching polices can be different among distinct physical layer models, the analysis and simulation results using these models confirm the significant advantages of D2D caching networks in terms of different throughput related metrics.

When interferences are not treated as noise, or cooperative D2D communication is used, it turns out that cached information can be used to mitigate interference. In [34], it assumes a random phase fading proposed a delivery scheme based on the Hierarchical MIMO Cooperation scheme originally proposed in [10]. Let all nodes be uniformly and independently distributed in a unit square, let $SN > M$, the network is divided into square clusters of area $n^{-\nu}$ for some $\nu > 0$. The user request follows a uniform distribution (Zipf distribution with $\gamma = 0$). The cache placement is similar to the random decentralized cache placement that discussed in Section 9.5.1, where no MDS coding is used ($\rho = 1$). In the delivery phase, each user will first identify L users holding distinct packets requested by this user as the source nodes. Then, to serve the targeted user, the hierarchical MIMO cooperation scheme in [10] can be applied. In particular, a virtual MIMO transmission can be formed from these L source nodes to the nodes in the corresponding cluster of the targeted user. Then each user in the corresponding cluster quantizes the received signal and sends it to the targeted user which will jointly process the all copies of superimposed signals received from previous virtual MIMO transmissions. For serving all users in the network, the same transmission policy is applied in a TDMA manner. The per-user throughput of this scheme is $\Theta\left(n^{-\frac{1}{\tau+1}}\right)$ with a vanishing outage probability, where τ is an integer depending on the number of hierarchies but independent of N. It can be seen that in this case, the per-user throughput improves significantly compared to that for both single-hop and multihop D2D communications under protocol model (interference avoidance). As an independent work, Liu *et al.* [35] consider the extended network model with the similar physical layer assumptions.[i] In particular, a hierarchical cache content placement based on the content popularity and a tree-graph-based content delivery based on both multihop and hierarchical MIMO cooperation are proposed. Under the same extended network condition and a Zipf popularity distribution, it can be shown that this proposed scheme achieves zero outage probability and has an order gain in terms of per-user throughput when the path loss parameter is less than 3 and the Zipf parameter $\gamma < 3/2$.

From the previous discussions, it can be seen that there is indeed no order gain in terms of throughput scaling laws of D2D-coded caching in both single-hop and multihop compared to the counterpart of D2D uncoded caching schemes. The achievable throughput of D2D-caching in practice will depend on the realistic physical layer channel conditions.

9.6.2 D2D caching in mobile networks

The previous sections assumed all D2D users are static or do not assume explicit models for their mobility. In this section, we discuss how mobility can affect the behavior and the performance of D2D caching networks.

The effect of user mobility on D2D-uncoded caching is investigated via simulations in [3], which shows that user mobility does not have a significant impact on a random caching scheme. In addition, as discussed in Section 9.5.1, in [36], a decentralized coded caching scheme based on random cache placement is proposed and this scheme can also be robust for user mobility since the distribution of cache placement may not change when user moves. However, such caching schemes may not take full advantages of the specific user mobility pattern. The authors in [54] show that user mobility has positive effect on D2D caching and the authors in [55] consider the case where mobile users can update cache placement based on the demand and user mobility. However, it is assumed that one complete file can be transmitted via any D2D link when two users contact, which may not be not practical. A random

[i]All model considered in this chapter is a dense network model, whose area is a constant while the area of extended networks can grow linearly with the number of users and the nearest distance between two users are above a constant.

work-based user mobility model is introduced in [56], where user moves follow a discrete Markov process. Furthermore, another interesting user mobility models based on the distribution of contact time/inter-contact time is introduced in [56, 57], which will be discussed in the following.

In this model, mobile users may have contact with each other when they are within the transmission range. Correspondingly, the *contact time* for two mobile users is defined as the time slot that they are able to serve each other. The duration of each contact of user i and j is $t^c_{i,j}$ seconds. The *inter-contact time* for two mobile users is defined as the time duration between two consecutive contact times. In particular, the locations of contact times in the timeline for any two users i and j are modeled as a Poisson process with intensity $\lambda_{i,j}$, which represent the average number of contacts per unit time slot. For simplicity, the timelines for different user pairs are independent. Similar to the decentralized D2D-coded caching discussed in Section 9.5.1, an MDS type coded file cache placement is considered. A stochastic mixed integer nonlinear programming with the consideration of the contact-time and inter-contact time of mobile users is formulated with average data offloading ratio as the objective function and the cache size as the constraint. Sub-optimal solutions based on dynamic programming and sub-modular optimization are evaluated. Simulation results show the significant advantage of taking the consideration of the mobility pattern compared to the commonly used caching strategies including popular caching and random caching placement. Interestingly, it can be observed that users with very low velocity tend to rarely be in contact with others, and thus they need to cache the most popular files to meet their own requests. On the other hand, the users with high velocity contact others frequently, and they need to cache the most popular files as well, in order to help others to download the requested files. This work is generalized by the same authors in [58], which relaxes the assumption of constant contact durations and provides analytical performance evaluations.

9.6.3 Complexity of D2D-coded caching

When D2D-coded caching is considered, one of the most significant practical constraints is the requirement of extremely large subpacketization levels. This will lead to high complexity of D2D-coded caching schemes. For example, in [36], each file needs to be partitioned into $\binom{N}{t}t$ subfiles, where $t = \frac{NS}{M}$ represents the aggregate storage capacity in the network. When $\mu = \frac{S}{M}$ is fixed, $\binom{N}{t}t = t2^{NH_b(\mu)}$ grows exponentially with n. To address this concern, Woolsey *et al.* [59–61] propose several approaches to design coded caching networks with reduced subpacketization levels. In particular, two combinatorial design approaches were proposed for centralized D2D caching networks that have reduced subpacketization compared to [36]. The first approach is based on a "hypercube" design to define the cache placement. It demonstrates how the geometry of this hypercube can help to exploit coded multicasting opportunities for the delivery. One example of the 3-dimension hypercube (or cube) with 9 users and $t = 3$ is shown in Figure 9.7. It can be observed that the hypercube approach is designed and optimized specifically for D2D caching networks as opposed to adapting an existed caching scheme designed for shared-link

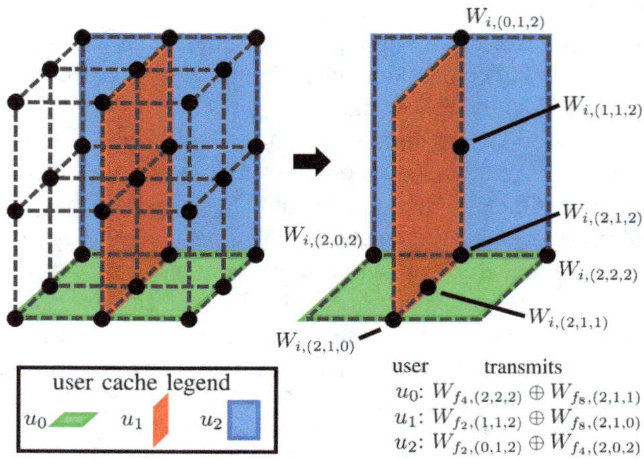

Figure 9.7 Each point on the 3D-lattice represents a set of packets. Each user's cache is represented by a plane of lattice points. The right figure represents a multicasting group, or a collection of 3 planes, where any two are orthogonal. The points on the intersection of two or more of these planes are highlighted. Every two users in this group share two packets that are requested by the third user. $W_{i,(j,k,\ell)}, i \in \{1, \ldots, M\}, j, k, \ell \in \{0, 1, 2\}$ *represent the packets of file i. In this example,* $N = 9$ *and* $t = 3$*. The required number of packets per file is 27. The required number of packets using the scheme in [36] is 252.*

caching network as the approach proposed in [42]. The number of required subfiles per file is reduced to $\left(\frac{M}{S}\right)^t$ from $\binom{N}{t}t$ while the communication load is $\frac{M}{S}$ which is almost the same as $\frac{N}{S} - 1$ as in [36]. In addition, by adopting the idea recently proposed in [62], this scheme can also be extended to a decentralized coded D2D caching scheme, which leads to a much more flexible design for given network parameters. Meanwhile, the advantage of the reduced subpacketizations of the hypercube approach still remains in the decentralized designs of the D2D caching networks. The second approach is based on an application of the Ruzsa-Szemérédi graph proposed in [63, 64], which is firstly used for the design in shared-link caching networks in [65]. Woolsey *et al.* [59] apply the Ruzsa-Szemérédi graph-based design in coded D2D caching networks and show that the requirement of file subpacketization is at most sub-quadratic in terms of the number of users if no spatial reuse is allowed while the per-user throughput scales as $\Theta(N^{-\delta})$ for some arbitrarily small δ under some parameter regimes. Both D2D combinatorial designs sustain the significant throughput gain compared to conventional uncoded unicast [26] and the required packetizations are reduced exponentially compared to [36] with respect to the number of users N while keeping the library size M and memory size S fixed. Finally, the impact of enabling spatial reuse in these caching network designs is also studied and shows this can further reduce the packetizations levels, while also improving the per-user throughput significantly for some parameter regimes, in contrast to the case in [36]. All these work show gives a hint that although the D2D caching schemes can be transferred from the shared-link caching schemes without much loss in terms of throughput as shown in [36], when subpacketizations are concerned, the D2D caching networks need a specific design approach that is different from the shared-link caching network. This idea is formalized in [66], in which the authors show that under some parameter regimes for D2D caching networks, there indeed exists a different design that achieves the exact optimal communication load as in [36] while requiring an orderwise smaller number of packets per file.

9.7 Conclusions and further readings

The area of D2D caching networks has attracted significant attentions in recent years. In this chapter, we discussed a few fundamental approaches that are typical for analyzing the scaling laws of D2D-caching networks including single-hop D2D uncoded/coded caching and multihop D2D-unocded/coded caching approaches. Tight scaling laws are characterized and show great potential of D2D coded caching network. In particular, the per-user throughput gain of caching is multiplicative under different network models.

There are a number of other important and interesting directions for this topic. We will illustrate a few of them as follows. First, under the condition of treating interference as noise, unlike the work discussed in Section 9.6.1, where only simple scheduling schemes are considered, reference [49] considers the ITLinQ scheduling approach [67], which introduced the optimal condition of treating interference as noise. Second, it is also possible to align or cancel interference, or use cooperation approaches rather than treating interference as noise. For example, Hachem *et al.* [68] and Naderializadeh *et al.* [69] focus on the interference channel[j] and used zero forcing and cache cancellation or interference alignment to remove the effect of interference. Finally, in the area of coded D2D caching, besides the paper discussed in Section 9.5.1, there are other coded D2D caching approaches such as [70] for distinct cache sizes and [71] for secure D2D caching.

References

[1] Golrezaei N., Dimakis A.G., Molisch A.F., Caire G. 'Wireless video content delivery through distributed caching and peer-to-peer gossiping'. 2011 Conference Record of the Forty Fifth Asilomar Conference on Signals, Systems and Computers (ASILOMAR), IEEE; 2011. pp. 1177–80.

[j] Note that the interference channel can be understood as D2D communications when the transmitters and receivers are fixed.

[2] Golrezaei N., Molisch A.F., Dimakis A.G., Caire G. 'Femtocaching and device-to-device collaboration: a new architecture for wireless video distribution'. *IEEE Communications Magazine*. 2013;**51**(4):142–9.

[3] Golrezaei N., Mansourifard P., Molisch A.F., Dimakis A.G. 'Base-station assisted device-to-device communications for high-throughput wireless video networks'. *IEEE Transactions on Wireless Communications*. 2014;**13**(7):3665–76.

[4] Ahmed I., Ismail M.H., Hassan M.S. 'Video transmission using device-to-device communications: a survey'. *IEEE Access*. 2019;**7**:131019–38.

[5] Li L., Zhao G., Blum R.S. 'A survey of caching techniques in cellular networks: research issues and challenges in content placement and delivery strategies'. *IEEE Communications Surveys & Tutorials*. 2018;**20**(3):1710–32.

[6] Mehrabi M., You D., Latzko V., Salah H., Reisslein M., Fitzek F.H.P. 'Device-enhanced MEC: multi-access edge computing (MEC) aided by end device computation and caching: a survey'. *IEEE Access*. 2019;**7**:166079–108.

[7] Gupta P., Kumar P.R. 'The capacity of wireless networks'. *IEEE Transactions on Information Theory*. 2000;**46**(2):388–404.

[8] Agarwal A., Kumar P.R. 'Capacity bounds for ad hoc and hybrid wireless networks'. *ACM SIGCOMM Computer Communication Review*. 2004;**34**(3):71–81.

[9] Franceschetti M., Dousse O., Tse D.N.C., Thiran P. 'Closing the gap in the capacity of wireless networks via percolation theory'. *IEEE Transactions on Information Theory*. 2007;**53**(3):1009–18.

[10] Ozgur A., Leveque O., Tse D.N.C. 'Hierarchical cooperation achieves optimal capacity scaling in ad hoc networks'. *IEEE Transactions on Information Theory*. 2007;**53**(10):3549–72.

[11] El Gamal A., Mammen J., Prabhakar B., Shah D. 'Optimal throughput–delay scaling in wireless networks—Part II: Constant-size packets'. *IEEE Transactions on Information Theory*. 2006;**52**(11):5111–16.

[12] El Gamal A., Mammen J., Prabhakar B., Shah D. 'Optimal throughput-delay scaling in wireless networks – Part I: the fluid model'. *IEEE Transactions on Information Theory*. 2006;**52**(6):2568–92.

[13] El Gamal A., Mammen J., Prabhakar B., Shah D. 'Throughput-delay trade-off in wireless networks'. IEEE INFOCOM 2004. IEEE; 2004. p. 1.

[14] Cohen E., Shenker S. 'Replication strategies in unstructured peer-to-peer networks'. *ACM SIGCOMM Computer Communication Review*. 2002;**32**(4):177–90.

[15] Maddah-Ali M.A., Niesen U. 'Fundamental limits of caching'. *IEEE Transactions on Information Theory*. 2014;**60**(5):2856–67.

[16] Breslau L., Cao P., Fan L., Phillips G., Shenker S. 'Web caching and zipf-like distributions: evidence and implications'. INFOCOM'99. Eighteenth Annual Joint Conference of the IEEE Computer and Communications Societies. Proceedings. IEEE; 1999. pp. 126–34.

[17] Hefeeda M., Saleh O. 'Traffic modeling and proportional partial caching for peer-to-peer systems'. *IEEE/ACM Transactions on Networking.* 2008;**16**(6):1447–60.

[18] Lee M.-C., Ji M., Molisch A.F., Sastry N. 'Throughput–outage analysis and evaluation of cache-aided D2D networks with measured popularity distributions'. *IEEE Transactions on Wireless Communications.* 2019;**18**(11):5316–32.

[19] Blaszczyszyn B., Giovanidis A. 'Optimal geographic caching in cellular networks'. 2015 IEEE International Conference on Communications (ICC); 2015. pp. 3358–63.

[20] Molisch A.F. *Wireless Communications*. 2nd ed. Wiley-IEEE Press; 2011.

[21] Ji M., Caire G., Molisch A.F. 'The throughput-outage tradeoff of wireless one-hop caching networks'. *IEEE Transactions on Information Theory.* 2015;**61**(12):6833–59.

[22] Lin X., Andrews J., Ghosh A., Ratasuk R. 'An overview of 3GPP device-to-device proximity services'. *IEEE Communications Magazine.* 2014;**52**(4):40–8.

[23] Nishiyama H., Ito M., Kato N. 'Relay-by-smartphone: realizing multihop device-to-device communications'. *IEEE Communications Magazine.* 2014;**52**(4):56–65.

[24] Golrezaei N., Dimakis A.G., Molisch A.F. 'Scaling behavior for device-to-device communications with distributed caching'. *IEEE Transactions on Information Theory*. 2014;**60**(7):4286–98.

[25] Juhn L.-S., Tseng L.-M. 'Harmonic broadcasting for video-on-demand service'. *IEEE transactions on broadcasting*. 1997;**43**(3):268–71.

[26] Ji M., Caire G., Molisch A.F. 'Wireless device-to-device caching networks: basic principles and system performance'. *IEEE Journal on Selected Areas in Communications*. 2016;**34**(1):176–89.

[27] Karedal J., Johansson A., Tufvesson F., Molisch A. 'A measurement-based fading model for wireless personal area networks'. *IEEE Transactions on Wireless Communications*. 2008;**7**(11):4575–85.

[28] Sánchez de la Fuente Y., Schierl T., Hellge C. 'iDASH: improved dynamic adaptive streaming over HTTP using scalable video coding'. proceedings of the Second Annual ACM Conference on Multimedia Systems, ACM; 2011. pp. 257–64.

[29] Jeon S.-W., Hong S.-N., Ji M., Caire G., Molisch A.F. 'Wireless multihop device-to-device caching networks'. *IEEE Transactions on Information Theory*. 2017;**63**(3):1662–76.

[30] Lee M.-C., Ji M., Molisch A.F. 'Optimal throughput–outage analysis of cache-aided wireless multi-hop D2D networks – derivations of scaling laws'. *arXiv preprint arXiv:2005.05149*. 2020.

[31] Gitzenis S., Paschos G.S., Tassiulas L. 'Asymptotic laws for joint content replication and delivery in wireless networks'. *Arxiv preprint arXiv:1201.3095*. 2012.

[32] Qiu L., Cao G. 'Popularity-aware caching increases the capacity of wireless networks'. *IEEE Transactions on Mobile Computing*. 2019;**19**(1):173–87.

[33] Molisch A.F., Caire G., Ott D., Foerster J.R., Bethanabhotla D., Ji M. 'Caching eliminates the wireless bottleneck in video aware wireless networks'. *Advances in Electrical Engineering*. 2014;**2014**(261390):1–13.

[34] Guo J., Yuan J., Zhang J. 'An achievable throughput scaling law of wireless device-to-device caching networks with distributed MIMO and hierarchical cooperations'. *IEEE Transactions on Wireless Communications*. 2018;**17**(1):492–505.

[35] Liu A., Lau V.K.N., Caire G. 'Cache-induced hierarchical cooperation in wireless device-to-device caching networks'. *IEEE Transactions on Information Theory*. 2018;**64**(6):4629–52.

[36] Ji M., Caire G., Molisch A.F. 'Fundamental limits of caching in wireless D2D networks'. *IEEE Transactions on Information Theory*. 2016;**62**(2):849–69.

[37] Yapar Ç., Wan K., Schaefer R.F., Caire G. 'On the optimality of D2D coded caching with uncoded cache placement and one-shot delivery'. *IEEE Transactions on Communications*. 2019;**67**(12):8179–92.

[38] Ji M., Caire G., Molisch A.F. 'Fundamental limits of distributed caching in D2D wireless networks'. Information Theory Workshop (ITW), 2013 IEEE, IEEE; 2013. pp. 1–5.

[39] Lin X., Ratasuk R., Ghosh A., Andrews J.G. 'Modeling, analysis, and optimization of multicast device-to-device transmissions'. *IEEE Transactions on Wireless Communications*. 2014;**13**(8):4346–59.

[40] Maddah-Ali M.A., Niesen U. 'Decentralized caching attains order-optimal memory-rate tradeoff'. *arXiv preprint arXiv:1301.5848*. 2013.

[41] Yu Q., Maddah-Ali M.A., Avestimehr A.S. 'The exact rate-memory tradeoff for caching with uncoded prefetching'. *IEEE Transactions on Information Theory*. 2018;**64**(2):1281–96.

[42] Ji M., Chen R., Caire G., Molisch A.F. 'Fundamental limits of distributed caching in multihop D2D wireless networks'. 2017 IEEE International Symposiumon Information Theory (ISIT); 2017. pp. 2950–4.

[43] Afshang M., Dhillon H.S. 'Optimal geographic caching in finite wireless networks'. *arXiv preprint arXiv:1603.01921*. 2016.

[44] Afshang M., Dhillon H., Chong P., Joo Chong P.H. 'Modeling and performance analysis of clustered device-to-device networks'. *IEEE Transactions on Wireless Communications*. 2016;**15**(7):1–72.

[45] Amer R., Elsawy H., Butt M.M., Jorswieck E.A., Bennis M., Marchetti N. 'Optimizing joint probabilistic caching and communication for clustered D2D networks'. *arXiv preprint arXiv:1810.05510*. 2018.

[46] Giatsoglou N., Ntontin K., Kartsakli E., Antonopoulos A., Verikoukis C. 'D2D-aware device caching in mmwave-cellular networks'. *IEEE Journal on Selected Areas in Communications*. 2017;**35**(9):2025–37.

[47] Karasik R., Simeone O., Shamai S. 'Information-theoretic analysis of D2D-aided pipelined content delivery in fog-ran'. 2018 15th International Symposium on Wireless Communication Systems (ISWCS), IEEE; 2018. pp. 1–5.

[48] Malak D., Shalash M., Andrews J.G. 'Optimizing content caching to maximize the density of successful receptions in device-to-device networking'. *IEEE Transactions on Communications*. 2016;**64**(10):1–80.

[49] Naderializadeh N., Kao D.T.H., Avestimehr A.S. 'How to utilize caching to improve spectral efficiency in device-to-device wireless networks'. 2014 52nd Annual Allerton Conference on Communication, Control, and Computing (Allerton); 2014. pp. 415–22.

[50] Nuradha T., Samarasinghe T., Hemachandra K.T. 'A novel content caching and delivery scheme for millimeter wave device-to-device communications'. *arXiv preprint arXiv:1911.06517*. 2019.

[51] Vuppala S., Vu T.X., Gautam S., Chatzinotas S., Ottersten B. 'Cache-aided millimeter wave ad hoc networks with contention-based content delivery'. *IEEE Transactions on Communications*. 2018;**66**(8):3540–54.

[52] Wu W., Zhang N., Cheng N., Tang Y., Aldubaikhy K., Shen X. 'Beef up mmwave dense cellular networks with D2D-assisted cooperative edge caching'. *IEEE Transactions on Vehicular Technology*. 2019;**68**(4):3890–904.

[53] Afshang M., Dhillon H.S., Chong P.H.J. 'Fundamentals of cluster-centric content placement in cache-enabled device-to-device networks'. *IEEE Transactions on Communications*. 2016;**64**(6):2511–26.

[54] Krishnan S., Dhillon H.S. 'Effect of user mobility on the performance of device-to-device networks with distributed caching'. *IEEE Wireless Communications Letters*. 2017;**6**(2):194–7.

[55] Lan R., Wang W., Huang A., Shan H. 'Device-to-device offloading with proactive caching in mobile cellular networks'. 2015 IEEE Global Communications Conference (GLOBECOM); 2015. pp. 1–6.

[56] Hosny S., Eryilmaz A., Abouzeid A.A., El Gamal H. 'Mobility-aware centralized D2D caching networks'. 2016 54th Annual Allerton Conference on Communication, Control, and Computing (Allerton); 2016. pp. 725–32.

[57] Wang R., Zhang J., Song S.H., Letaief K.B. 'Mobility-aware caching in D2D networks'. *IEEE Transactions on Wireless Communications*. 2017;**16**(8):5001–15.

[58] Wang R., Zhang J., Song S.H., Letaief K.B. 'Exploiting mobility in cache-assisted D2D networks: performance analysis and optimization'. *IEEE Transactions on Wireless Communications*. 2018;**17**(8):5592–605.

[59] Woolsey N., Chen R., Ji M. 'Device-to-device caching networks with sub-quadratic subpacketizations'. GLOBECOM 2017–2017 IEEE Global Communications Conference; 2017. pp. 1–6.

[60] Woolsey N., Chen R., Ji M. 'Coded caching in wireless device-to-device networks using a hypercube approach'. 2018 IEEE International Conference on Communications Workshops (ICC Workshops); 2018. pp. 1–6.

[61] Woolsey N., Chen R.-R., Ji M. 'Towards practical file packetizations in wireless device-to-device caching networks'. *arXiv preprint arXiv:1712.07221*. 2017.

[62] Jin S., Cui Y., Liu H., Caire G. 'New order-optimal decentralized coded caching schemes with good performance in the finite file size regime'. *arXiv preprint arXiv:1604.07648*. 2016.

[63] Alon N., Moitra A., Sudakov B. 'Nearly complete graphs decomposable into large induced matchings and their applications'. proceedings of the forty-fourth annual ACM Symposium on theory of computing, ACM; 2012. pp. 1079–90.

[64] Ruzsa I., Szemerédi E. 'Triple systems with no six points carrying three triangles combinatorics (Keszthely, 1976)'. *Colloquia Mathematica Societatis Janos Bolyai*. 1978;**18**:939–45.

[65] Shanmugam K., Tulino A.M., Dimakis A.G. 'Coded caching with linear subpacketization is possible using Ruzsa-Szeméredi graphs'. *arXiv preprint arXiv:1701.07115*. 2017.

[66] Zhang X., Yang X.F., Ji M. 'A new design framework on device-to-device coded caching with optimal rate and significantly less subpacketizations'. *2020 IEEE International Symposium on Information Theory (ISIT)*; 2020. pp. 1699–1704.

[67] Naderializadeh N., Avestimehr A.S. 'ITLinQ: a new approach for spectrum sharing in device-to-device communication systems'. *IEEE Journal on Selected Areas in Communications*. 2014;**32**(6):1139–51.

[68] Hachem J., Niesen U., Diggavi S.N. 'Degrees of freedom of cache-aided wireless interference networks'. *IEEE Transactions on Information Theory*. 2018;**64**(7):5359–80.

[69] Naderializadeh N., Maddah-Ali M.A., Avestimehr A.S. 'Fundamental limits of cache-aided interference management'. *IEEE Transactions on Information Theory*. 2017;**63**(5):1–107.

[70] Ibrahim A.M., Zewail A.A., Yener A. 'Device-to-device coded caching with distinct cache sizes'. *arXiv preprint arXiv:1903.08142*. 2019.

[71] Zewail A.A., Yener A. 'Device-to-device secure coded caching'. *IEEE Transactions on Information Forensics and Security*. 2020;**15**:1513–24.

Chapter 10

Caching increases the capacity of wireless networks

Li Qiu[1] *and Guohong Cao*[1]

Abstract Content Caching at the network edge, e.g., base stations and mobile devices, is a promising solution to meet the demand of massive mobile traffic growth. In this chapter, we focus on edge caching in a wireless ad hoc network without any infrastructure support, and study how content caching at mobile devices can help improve the network capacity. In wireless ad hoc networks, due to the interference between concurrent transmissions, the per-node capacity generally decreases with the increasing number of nodes in the network. Caching can help improve the network capacity, as it shortens the content transmission distance and reduces the communication interference. However, the fundamental performance limits of caching in wireless ad hoc networks have rarely been studied in an analytical manner. In this chapter, we first illustrate the benefit of caching under uniform content popularity, i.e., the capacity of wireless ad hoc networks with caching can remain constant even as the number of nodes in the network increases. Then, we generalize the results, evaluate how the distribution of the content popularity affects the per-node capacity, and derive different capacity scaling laws based on the skewness of the content popularity. Our results suggest that for wireless networks with caching, when contents have skewed popularity, increasing the number of nodes monotonically increases the per-node capacity.

10.1 Introduction

Due to the popularity of mobile devices and content-centric nature of many Web applications [1], user requests for content files are explosively growing. Caching at the network edge (e.g., base stations and mobile devices) has been proposed to accommodate these increasing user requests, as such technique brings the content closer to the end users and effectively improves network performance. Although there is a large amount of existing work on edge caching, most of them assume

[1]The Pennsylvania State University, United States

that the mobile devices [2–4] have well-connected communication links to the edge server that holds the content. However, this is not necessarily true for many mobile edge networks in battlefield, disaster recovery, or mobile environments without infrastructure support. In these mobile edge networks or wireless ad hoc networks, contents are generally cached on the mobile nodes, and the effects of caching on network capacity in such networks should also be studied.

Wireless ad hoc networks enable communications among mobile nodes without any infrastructure support, as nodes themselves relay and forward packets for each other. Due to the interference between concurrent transmissions, the per-node capacity generally decreases with the increasing number of nodes in the network. This was first found by Gupta and Kumar [5]. They showed that for a wireless ad hoc network with n nodes, each node can at most transmit at a rate of $\Theta(\frac{1}{\sqrt{n}})$ to its destination, even with optimal scheduling of transmissions from various nodes.

Later, Grossglauser and Tse [6] examined the capacity of a mobile ad hoc network. They proved that with node mobility, the per-node capacity can be kept constant even when the number of nodes in the network grows. In their proposed scheme, two nodes wait until they move close enough to transmit data. In this way, the per-node capacity can be significantly improved, since the transmissions are limited to nearby neighbors and the consequent interference will affect fewer number of nodes. However, the delay of their proposed scheme is extremely long, because nodes need to wait until they are close enough to transmit.

Caching can be used to improve the network capacity. By caching contents at multiple nodes, network traffic and content access delay can be reduced as users can access contents from near neighbors instead of the faraway server. Although various caching techniques [7–10] have been proposed to improve the performance of wireless ad hoc networks, these studies are limited to protocol design and the associated performance evaluations are based on simulations. The fundamental performance limits of caching in wireless ad hoc networks have rarely been studied in an analytical manner. Although there are some analytic performance studies on caching in Internet such as modeling the cache hit ratio [11–13], modeling the steady-states of cache networks [14], performance analysis of optimal routing and content caching [15], etc., these results cannot be directly applied to wireless ad hoc networks. Although the network capacity of wireless ad hoc network has been analyzed in [5, 6], caching is not considered in their studies and it is not clear how caching affects the capacity of wireless ad hoc networks.

In this chapter, we study the effect of caching on the capacity of wireless ad hoc networks. We first illustrate the benefit of caching under uniform content popularity, i.e., the capacity of wireless ad hoc networks with caching can remain constant even as the number of nodes in the network increases. Then, we generalize the results to popularity-aware caching. To maximize the per-node capacity given the content popularity, we first study the optimal caching strategy; i.e., how frequently contents with various popularity should be cached so that the per-node capacity is maximized. Based on the optimal caching strategy, we then evaluate the effect of content popularity on per-node capacity, and derive different capacity scaling laws for networks with different content popularity skewness. For all the different capacity scaling laws, we analytically investigate how the per-node capacity is affected by various parameters, including the number of nodes (n), the cache size (s), and the number of unique contents (m). The main results of the chapter are summarized as follows:

- For contents with uniform popularity, the per-node capacity is upper bounded by $O\left(\sqrt{\frac{s}{m}}\right)$, and lower bounded by $\Theta\left(\frac{s}{m}\right)$, i.e., the capacity of wireless ad hoc networks with caching can remain constant even as the number of nodes in the network increases.
- We derive different capacity scaling laws based on the skewness of the content popularity. Basically, as the distribution of the content popularity changes from uniform distribution to more skewed distributions, the per-node capacity increases from $\Theta\left(\sqrt{\frac{s}{m}}\right)$ to roughly $\Theta\left(\sqrt{s}\right)$.
- We find that for wireless ad hoc networks with caching, when contents have skewed popularity, increasing the number of nodes monotonically increases the per-node capacity.
- We propose a distributed caching algorithm which enables nodes to optimally cache contents and maximize the per-node capacity only based on local knowledge.

The rest of the chapter is organized as follows. Section 10.2 reviews existing work. Section 10.3 introduces the network model, and Section 10.4 presents the capacity results when content popularity follows a uniform distribution. We derive the capacity scaling laws and analyze the effects of various parameters for the more general Zipf distribution in Section 10.5. Section 10.6 presents the distributed caching algorithm, and Section 10.7 concludes the chapter.

10.2 Related work

Caching in wireless networks is a traditional topic that has been studied from various aspects. In [7], Yin and Cao have designed and evaluated cooperative caching schemes to support data access in wireless ad hoc networks. In [8], the authors have implemented and examined a caching scheme for wireless P2P networks. Fiore *et al.* [9] have proposed an algorithm to help users decide whether a content should be cached, so as to reduce the data redundancy among neighbors. Recently, various content caching and cache replacement algorithms have also been proposed for information-centric networks [16, 17]. However, these algorithms are only based on heuristic and the associated performance evaluations are based on simulations. None of them has analytically investigated the fundamental performance limits of caching in wireless networks.

The problem of optimal caching has only been investigated in a few existing works. When the content popularity is known, Cohen and Shenker [18] have studied the problem of how the contents with various popularity should be cached in P2P networks, so that the number of searches to retrieve the content is minimized. Besides, Jin and Wang [19] have proposed techniques to determine how contents should be cached at various nodes in wireless networks. When the content popularity is unknown, content caching at the base stations has been examined in [20, 21]. The authors have proposed learning-based approaches, where the base station learns the content popularity based on user requests or context information, and then caches the contents according to the observed popularity distribution. However, since these algorithms require the exact content popularity, they may not work well in a mobile ad hoc network, which does not have a base station to observe and process all content requests. Furthermore, these works [18–21] only proposed caching schemes, and the effect of caching on network capacity was not investigated.

Recently, researchers have become more interested in studying the fundamental property of caching in wireless networks, based on different models and assumptions. In [22, 23], Ji *et al.* presented theoretical bounds as well as simulations results on per-node capacity, but their results were limited to single-hop data transmission. In [24], the authors presented theoretical results on network throughput and content access delay for a heterogenous network, which consists of base stations, relays, and device-to-device pairs. Niesen *et al.* [25] focused on deriving the feasible request serving rate by knowing where and what data has been cached. In [26], Gitzenis *et al.* examined the asymptotic laws for joint replication and delivery in wireless networks. They derived the minimum throughput on each link so that every node can satisfy one request per second. However, these works have assumed different communication or content access models, thus leading to results different from ours.

10.3 Preliminaries

10.3.1 Network model

We consider a wireless network consisting of n nodes that are independently and uniformly distributed on the surface of a unit sphere. As in [5], we analyze the per-node capacity when nodes are located on the surface of sphere S^2 rather than on a disk so as to eliminate the edge effects; i.e., nodes near the edge have much fewer neighbors than nodes near the center.

Assume all nodes employ the same amount of power for transmission, and they transmit over a common wireless channel which can support W bits per second. According to the physical model in [5], when a node i sends data to node j, the transmission rate can reach W bits per second, if the signal-to-interference-plus-noise ratio (SINR) at node j is greater than or equal to β, where β is the minimum SINR for successful reception.

10.3.2 Content access and Zipf distribution

Let m denote the number of unique contents throughout the network. To simplify the analysis, we assume each content has one bit. These m contents are cached by various nodes. Suppose each node has a cache size of s. For any node i, it caches s out of m contents locally. In case each content has b bits rather than one bit, the capacity results can be obtained by replacing s with $\frac{s}{b}$. The per-node capacity decreases as the content size b increases, since a larger content size results in a longer average distance for the nodes to retrieve the content.

We assume the three parameters n, s and m are independent. In wireless networks with caching, generally each node can only cache a small portion of the m unique contents, in this way, $m \gg s$. On the other hand, to guarantee that all content requests can be served by the nodes in the network, the total cache size of all nodes should be greater than or equal to m, i.e., $n * s \geq m$. Thus, we assume $\frac{m}{n} \leq s \ll m$.

At each node, there is always a content request, and a new request will arrive after the previous request has been served. If the requested content has been cached locally, the request can be directly served. Otherwise, the node will retrieve the requested content from other nodes through the shortest path via multi-hop transmission. We assume the content requests are independent. The popularity of the contents follows a Zipf distribution, i.e., the probability that the i-th most popular content being requested is proportional to $\frac{1}{i^\gamma}$. This content access pattern has been used by previous studies, and existing works [27, 28] have shown that the content access pattern in the Internet follows a Zipf distribution. We focus on the scenario when the content popularity changes slowly or even remains unchanged. Typical examples include music files, and uploaded videos, which may remain popular for a long period of time. There are m unique contents, and the probability that the i-th most popular content will be requested, denoted as ρ_i, will be $\rho_i = \frac{1}{i^\gamma H_{m,\gamma}}$, where $H_{m,\gamma}$ is the generalized harmonic number, given by

$$H_{m,\gamma} = \sum_{j=1}^{m} \frac{1}{j^\gamma}. \tag{10.1}$$

For $H_{m,\gamma}$, it must satisfy

$$\int_1^{m+1} \frac{1}{x^\gamma} \mathrm{d}x \leq H_{m,\gamma} = \sum_{j=1}^{m} \frac{1}{j^\gamma} \leq 1 + \int_1^{m} \frac{1}{x^\gamma} \mathrm{d}x. \tag{10.2}$$

Accordingly, we have

$$\begin{cases} \log(m+1) \leq H_{m,\gamma} \leq \log m + 1, & \text{if } \gamma = 1 \\ \dfrac{(m+1)^{1-\gamma} - 1}{1-\gamma} \leq H_{m,\gamma} \leq \dfrac{m^{1-\gamma} - 1}{1-\gamma} + 1, & \text{otherwise} \end{cases} \tag{10.3}$$

As $m \to \infty$, $H_{m,\gamma}$ converges if and only if $\gamma > 1$.

10.3.3 Capacity

As in [5, 6], the per-node capacity describes each node's capability to transmit or retrieve contents. In a wireless network with caching, we define the *per-node capacity* as the number of bits each node utilizes per second to satisfy content requests, which includes the amount of contents it receives from others, and the contents in the local cache that have been used to serve requests.

We study the scaling behavior of the per-node capacity C based on four parameters: the number of nodes n, the number of unique contents m, the cache size s, and the Zipf parameter γ. The per-node capacity is studied under the assumption, $n \to \infty$, $m \to \infty$ and $\frac{m}{n} \leq s \ll m$.

10.4 Capacity under uniform content popularity

Before presenting the analytical results of caching under the more general Zipf content access, we first analyze a specific scenario where the content popularity follows a uniform distribution ($\gamma = 0$), to help understand the benefit of caching. Under uniform content popularity, we study the fundamental property of wireless networks with caching, i.e., the scaling laws of the network capacity based on cache size of individual node, the total size of unique content and the number of nodes in the network. We present an upper bound on network capacity, and present an achievable capacity lower bound, where we propose a caching scheme to show what capacity can actually be achievable.

10.4.1 Capacity upper bound

For a node i, after receiving requests for content ρ_j not cached locally, it has to retrieve ρ_j from the node that caches the content. Under different caching schemes, i may contact nearby nodes or faraway nodes to retrieve ρ_j, and then lead to different performance. To derive a capacity upper bound for all possible schemes, we first discuss what may affect the capacity.

In a wireless network with n nodes, where each node transmits to an arbitrarily chosen destination, Gupta and Kumar [5] prove that the per-node capacity is upper bounded by $\frac{W}{L\sqrt{n}}$, where L is the average distance between the source and destination pairs. This result indicates that as nodes communicate with more distant nodes, the capacity decreases. A similar result has also been found in [6, 29], where the authors suggest that the capacity can be improved by shortening the transmission distance between the source node and the destination node. This is because as the transmission distance increases, more intermediate nodes will be affected and then less number of simultaneous transmissions are allowed.

Based on the above results, each node prefers the other $m - s$ contents (s contents can be cached locally) to be cached at nodes as close as possible to achieve higher capacity. Thus, to find an upper bound, we assume each node i caches s contents locally, and the closest $\lceil \frac{m-s}{s} \rceil$ nodes cache the remaining $m - s$ contents. Under such a scenario, the average distance to get a content is:

$$\mathrm{E}(L) \geq \frac{2}{3} \arccos\left(1 - \frac{2}{n}\left(\left\lceil \frac{m}{s} \right\rceil - 1\right)\right). \tag{10.4}$$

Combining the above equation with the fact that per-node capacity is upper-bounded by $\frac{W}{L\sqrt{n}}$, we have the following theorem.

Theorem 1: Under our network and content access model, the per-node capacity is upper bounded by

$$C = O\left(W\sqrt{\tfrac{s}{m}}\right). \tag{10.5}$$

Proof. For nodes uniformly distributed on the surface of the sphere, the per-node capacity can not be larger than $\frac{\sqrt{8}}{\sqrt{\pi}}\frac{W}{\beta^{\frac{1}{\alpha}}-1}\frac{1}{\mathrm{E}(L)\sqrt{n}}$ (α is the signal scaling factor) [5], hence we have:

$$\begin{aligned} C &\leq \frac{\sqrt{8}}{\sqrt{\pi}}\frac{W}{\beta^{\frac{1}{\alpha}}-1}\frac{1}{\mathrm{E}(L)\sqrt{n}} \\ &\leq \frac{\sqrt{8}}{\sqrt{\pi}}\frac{W}{\beta^{\frac{1}{\alpha}}-1}\frac{1}{\sqrt{n}}\frac{3}{2\arccos(1-\frac{2}{n}(\lceil\frac{m}{s}\rceil-1))} \\ &= O\left(W\sqrt{\tfrac{s}{m}}\right) \end{aligned} \tag{10.6}$$

Note the last inequality is due to $m \gg s$ and $\arccos(1-2x) \approx 2\sqrt{x}$.

10.4.2 Capacity lower bound

In this subsection, we construct a caching scheme to obtain an achievable capacity lower bound.

10.4.2.1 Cell construction

We partition the surface of the sphere S^2 into cells with similar sizes. More specifically, each cell contains a disk of radius ρ, and is contained in a disk of radius 2ρ, where ρ is the radius of the disk with area $\frac{8\pi}{n}\lceil\frac{m}{s}\rceil$ on S^2. In this way, the area of each cell is at least $\frac{8\pi}{n}\lceil\frac{m}{s}\rceil$, and at most $c_1\frac{32\pi}{n}\lceil\frac{m}{s}\rceil$, where c_1 is a constant close to 1 (note the area of a disk with radius 2ρ on the surface of the sphere is different from $4\pi\rho^2$).

Let v_i denote the number of nodes in cell V_i, and $\eta = \lceil\frac{m}{s}\rceil$. The probability for any node j falls in cell V_i is at least $\frac{2}{n}\lceil\frac{m}{s}\rceil$. Then, based on Chernoff bound [30], the probability for any cell V_i to have at least η nodes is:

$$\Pr(v_i \geq \eta) \geq 1 - \Pr(v_i \leq \eta) \geq 1 - e^{-\frac{1}{4}}. \tag{10.7}$$

Hence, there is a high probability (at least $1 - e^{-\frac{1}{4}}$) that any cell can cache all m contents.

10.4.2.2 Transmission strategy

Within each cell, we only allow one pair of nodes to transmit simultaneously. As a result, the transmission in each cell will only affect a limited number of neighboring cells. Formally, for any cell V_i, V_j, we say they are Interfering Neighbors if the transmission in one cell will interfere with the transmission in the other.

Consider coloring all the cells such that no two interfering neighbors have the same color. Then, for any period T, the cells of each color can have a period of $\frac{T}{c_2}$ for transmission, where c_2 is the total number of colors (see [31] for details).

10.4.2.3 Achievable lower bound

Based on the above results, we now present achievable lower bounds on network capacity.

Theorem 2: Under our network and content access model, the per-node capacity is lower bounded by

$$C = \Theta\left(W\frac{s}{m}\right). \tag{10.8}$$

Proof. For a time period T, each cell can have a period of $\frac{T}{c_2}$ during which the transmission in the cell is always successfully. Then for cells that can cache all contents, during the period of $\frac{T}{c_2}$, nodes within the cell can receive $W\frac{T}{c_2}$ bits of contents regardless of which contents have been requested. Furthermore, at least $c_3\frac{n}{\eta}$ cells can cache all contents. Summing over all the cells that can cache all the contents, the total number of bits received is

$$c_3\frac{n}{\eta} \times W\frac{T}{c_2} = c_4\frac{WnT}{\eta}. \tag{10.9}$$

Then on average each node can achieve a capacity of

$$C = c_4\frac{WnT}{\eta}/nT = c_4\frac{W}{\eta} = \Theta\left(W\frac{s}{m}\right), \tag{10.10}$$

as $\eta = \lceil \frac{m}{s} \rceil$.

10.4.3 Numerical results

In this subsection, we compare our results with the analytical results presented in [5, 6], and show that our results are consistent with theirs when our caching scenario matches theirs.

(i) In case of $m = s * n$, the total cache size is just enough to store all contents. To find the capacity upper bound, each content is only cached at one node. For any node i receiving a request, the requested content may be cached at any node with equal probability. Thus, if the requested content is not cached locally, it is equally likely for i to retrieve the requested content from any other node. This scenario is similar to the scenario of random network in [5], where all nodes randomly choose a destination for transmission. Our derived upper bound does conform to their results, as they present an upper bound of $O\left(\frac{W}{\sqrt{n}}\right)$ while our capacity is also upper bounded by $W\sqrt{\frac{s}{m}} = \frac{W}{\sqrt{n}}$ when $m = s * n$.

(ii) In case of $\frac{m}{2} \leq s < m$, two nodes will be enough to cache all the contents. If contents are carefully cached, each node only need to retrieve contents from its closest neighbor. This scenario is similar to the scenario constructed in [6], as communications are restricted to closest neighbors. Under this special case, we are also able to obtain an achievable capacity of $\Theta(W)$, which conforms to their results.

(iii) The most important implication of our results is that the network capacity and the request satisfaction rate (averaged over all nodes) can remain constant even if the number of nodes increase. This is because although more nodes bring in more interferences, high node density also helps each node retrieve contents from closer neighbors.

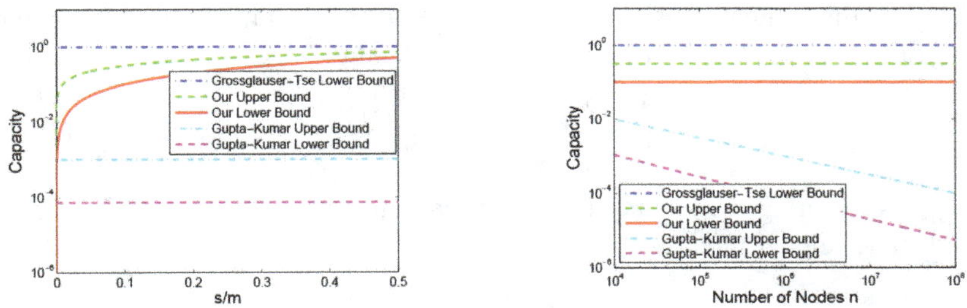

(a) Capacity vs. cache size ($n = 10^6$)

(b) Capacity vs. the number of nodes ($\frac{s}{m} = \frac{1}{10}$)

Figure 10.1 Comparisons to existing work

Figure 10.1 compares our results with [5, 6] based on numerical results. Figure 10.1a shows how the capacity changes with $\frac{s}{m}$ when $n = 10^6$. Apparently, the capacity results of Gupta-Kumar [5] and Grossglauser-Tse [6] will not change with $\frac{s}{m}$. When $\frac{s}{m}$ is very small, our result is comparable to Gupta-Kumar. However, as $\frac{s}{m}$ increases, our capacity result quickly approaches to Grossglauser-Tse's result. Figure 10.1b shows the capacity as a function of the number of nodes when $\frac{s}{m} = \frac{1}{10}$. Grossglauser-Tse shows that the network capacity can remain constant through node mobility. Similarly, caching also helps achieve constant network capacity. For wireless networks without considering mobility and caching, as shown in Gupta-Kumar, the capacity decreases quickly as the number of nodes increases. Note that although Grossglauser-Tse can achieve higher network capacity than ours, they are at the cost of long delay. On the other hand, caching can increase the network capacity, and reduce the content access delay.

10.5 Capacity under Zipf content popularity

In this section, we quantify the effect of popularity-aware caching on the capacity of wireless networks. To maximize the per-node capacity given the content popularity, we first study the optimal caching strategy; i.e., how frequently contents with various popularity should be cached so that the per-node capacity is maximized. Based on the optimal caching strategy, we then evaluate the effect of content popularity on per-node capacity, and derive different capacity scaling laws for networks with different content popularity skewness.

10.5.1 Problem formulation

When content popularity is known, the per-node capacity depends on the caching strategy, i.e., for content i with popularity ρ_i, how many nodes in the network should cache that content? Let $P = [p_1, p_2, \ldots, p_m]$ denote the densities of the contents, where p_i is the density of content i (i.e., the fraction of content i among all ns contents). We aim to find the optimal content densities, such that when the contents are cached according to the optimal densities, the per-node capacity is maximized.

Note that the content density P has some practical constraints. First, the sum of the densities of all contents must be less than or equal to 1:

$$\sum_{i=1}^{m} p_i \leq 1. \tag{10.11}$$

Second, it is meaningless for any node to cache more than one copy of the same content locally, thus p_i must satisfy

$$p_i \leq \frac{1}{s}, \text{ for } i = 1, \ldots, m. \tag{10.12}$$

Third, to guarantee that all content requests can be served, each content must have at least one copy in the network, that is:

$$p_i \geq \frac{1}{ns}, \text{ for } i = 1, \ldots, m. \tag{10.13}$$

To compute the per-node capacity based on the densities P, we review one important result from [5]. In that paper, the authors proved that for n nodes uniformly distributed on the surface of a sphere, the per-node capacity is upper bounded by $\frac{W}{L\sqrt{n}}$, where L is the average transmission distance between the source nodes and destination nodes. This result implies that the per-node capacity will increase as nodes retrieve contents from closer neighbors. Hence, to obtain the highest capacity, we should find the densities P that can minimize the average transmission distance L, where P is subject to constraints (10.11), (10.12), and (10.13).

Consider a content i with popularity ρ_i and density p_i. Since n nodes with cache size s can in total cache ns contents, i is cached at nsp_i nodes in the network (note that each node caches at most one copy of i locally). For the remaining $n - nsp_i$ nodes that have not cached i, they have to retrieve i from their neighbors. Thus, each node that caches i will on average be responsible for requests of i from $\frac{1}{p_i s} - 1$ other nodes. As all n nodes are uniformly and independently distributed, for any node j that caches i, based on [32], its average distance to itself and its closest $\frac{1}{p_i s} - 1$ neighbors is approximately:

$$L_i \approx \frac{2}{3}\arccos\left(1 - \frac{2}{n}\left(\frac{1}{p_i s} - 1\right)\right) \geq \sqrt{\frac{1}{p_i sn} - \frac{1}{n}} \geq \sqrt{\frac{1}{p_i sn}} - \sqrt{\frac{1}{n}}. \tag{10.14}$$

Therefore, the average transmission distance of content i is at least $\sqrt{\frac{1}{p_i sn}} - \sqrt{\frac{1}{n}}$. Applying the inequality to all m contents, the transmission distance L averaged over all contents will be

$$L = \sum_{i=1}^{m} L_i \cdot \rho_i \geq \sum_{i=1}^{m} \left(\sqrt{\frac{1}{np_i s}} - \sqrt{\frac{1}{n}}\right) \rho_i. \tag{10.15}$$

Based on the above inequality, the optimal value of the following nonlinear program L^*, will be a lower bound on the average transmission distance L.

$$\min_{\mathrm{P}\in\Omega} L = f(\mathrm{P}) = \sum_{i=1}^{m}\left(\frac{1}{\sqrt{p_i sn}} - \frac{1}{\sqrt{n}}\right)\rho_i$$
$$s.t.\ g(\mathrm{P}) = \sum_{i=1}^{m} p_i - 1 \le 0 \tag{10.16}$$

where $\Omega = \left[\frac{1}{ns}, \frac{1}{s}\right]^m$. The above convex nonlinear program is referred to as the primal problem, and it has the same constraints (10.11), (10.12) and (10.13).

To obtain L^*, let us consider the dual of the above nonlinear program:

$$\max\ d = D(u)$$
$$s.t.\ u \ge 0, \tag{10.17}$$

where the dual objective function $D(u)$ is given by

$$D(u) = \min_{\mathrm{P}\in\Omega}\left(f(\mathrm{P}) + u\cdot g(\mathrm{P})\right) = \min_{\mathrm{P}\in\Omega}\left(\sum_{i=1}^{m}\left(\frac{1}{\sqrt{p_i sn}} - \frac{1}{\sqrt{n}}\right)\rho_i + u\left(\sum_{i=1}^{m} p_i - 1\right)\right). \tag{10.18}$$

The above nonlinear program is referred to as the dual problem.

Definition 1: A pair (P^*, u^*) with $\mathrm{P}^* \in \Omega$ and $u^* \ge 0$ satisfies the global optimality conditions for the primal problem, if

$$(i)\ f(\mathrm{P}^*) + u^* g(\mathrm{P}^*) = \min_{\mathrm{P}\in\Omega}\left(f(\mathrm{P}) + u^* g(\mathrm{P})\right) \tag{10.19}$$
$$(ii)\ u^* g(\mathrm{P}^*) = 0 \tag{10.20}$$
$$(iii)\ g(\mathrm{P}^*) \le 0 \tag{10.21}$$

Lemma 1: If a pair (P^*, u^*) satisfies the global optimality condition given in Definition 1, P^* is optimal in the primal problem.

Proof. The detailed proof can be found in [33]. Basically, given Eq. (10.20) , $f(\mathrm{P}^*) = D(u^*)$. Weak duality theorem [33] states that $f(\mathrm{P}) \ge D(u)$ for any feasible P and u, therefore P^* optimizes the primal problem.

In the above lemma, P^* optimizes the primal problem, i.e., P^* represents the optimal content density that maximizes the per-node capacity. The above lemma states that the optimal content density P^* can be obtained by solving (10.19), (10.20), and (10.21). Then, the minimum transmission distance is $L^* = f(\mathrm{P}^*)$, and accordingly the per-node capacity is upper bounded by $\Theta\left(\frac{W}{L^*\sqrt{n}}\right)$.

10.5.2 Capacity upper bound

10.5.2.1 Optimal content densities

We discuss the optimal content densities. Based on Eq. (10.19), the optimal P^* minimizes $f(\mathrm{P}) + u^* g(\mathrm{P})$. We compute the partial derivative of $f(\mathrm{P}) + u^* g(\mathrm{P})$ over any p_i, which is

$$\frac{\partial\left(f(\mathrm{P}) + u^* g(\mathrm{P})\right)}{\partial p_i} = u^* - \frac{1}{2}\frac{\rho_i p_i^{-\frac{3}{2}}}{\sqrt{sn}}. \tag{10.22}$$

The above derivative monotonically increases with p_i, and as $u^* \geq 0$ (the constraint of the dual), the function $f(\mathrm{P}) + u^* g(\mathrm{P})$ will first decrease with p_i, and then increase with p_i when p_i is larger than $\left(\frac{\rho_i}{2u^*}\right)^{\frac{2}{3}}$. Thus, the function $f(\mathrm{P}) + u^* g(\mathrm{P})$ is minimized, when $p_i = \left(\frac{\rho_i}{2u^*}\right)^{\frac{2}{3}}$. Recall that p_i is restricted by $\frac{1}{sn} \leq p_i \leq \frac{1}{s}$, hence the optimal content densities are given as follows:

$$\mathrm{pi}^* = \begin{cases} \frac{1}{s}, & \text{if } \left(\frac{\rho_i}{2u^*\sqrt{sn}}\right)^{\frac{2}{3}} \geq \frac{1}{s} \\ \left(\frac{\rho_i}{2u^*\sqrt{sn}}\right)^{\frac{2}{3}}, & \text{if } \frac{1}{sn} < \left(\frac{\rho_i}{2u^*\sqrt{sn}}\right)^{\frac{2}{3}} < \frac{1}{s} \\ \frac{1}{sn}, & \text{if } \left(\frac{\rho_i}{2u^*\sqrt{sn}}\right)^{\frac{2}{3}} \leq \frac{1}{sn} \end{cases} \tag{10.23}$$

Basically, the above equation indicates that: popular contents (with densities equal to $\frac{1}{s}$) are cached by all nodes in the network, and unpopular contents (with densities equal to $\frac{1}{sn}$) are only cached at one node. For the remaining contents, their densities are proportional to $\rho_i^{\frac{2}{3}}$.

Due to the convexity of the primal problem and the condition $\frac{m}{n} \leq s \ll m$, P* exists and is unique. To simplify the notations, let μ denote the number of contents with density $\frac{1}{s}$, and let λ denote the number of contents with density $\frac{1}{sn}$. Then, among all m contents, the most popular μ contents each has a density of $\frac{1}{s}$, the most unpopular λ contents each has a density of $\frac{1}{sn}$, and the density for each of the remaining contents is $\left(\frac{\rho_i}{2u^*\sqrt{sn}}\right)^{\frac{2}{3}}$. The constraint of the primal problem must be tight, thus the densities of all contents add up to 1, which leads to:

$$\frac{\mu}{s} + \left(\frac{1}{2u^*\sqrt{sn}}\right)^{\frac{2}{3}} \sum_{i=\mu+1}^{m-\lambda} \left(\frac{\frac{1}{i^\gamma}}{H_{m,\gamma}}\right)^{\frac{2}{3}} + \frac{\lambda}{sn} = 1. \tag{10.24}$$

Specifically, for the value of μ, when $\left(\frac{\rho_1}{2u^*\sqrt{sn}}\right)^{\frac{2}{3}} \geq \frac{1}{s}$, μ is the non-negative integer satisfying

$$\left(\frac{\rho_{\mu+1}}{2u^*\sqrt{sn}}\right)^{\frac{2}{3}} < \frac{1}{s} \leq \left(\frac{\rho_\mu}{2u^*\sqrt{sn}}\right)^{\frac{2}{3}}. \tag{10.25}$$

When μ is reasonably large, $\left(\frac{\rho_\mu}{2u^*\sqrt{sn}}\right)^{\frac{2}{3}} \approx \frac{1}{s}$, and

$$\mu \approx \left(\frac{1}{\sqrt{n}} \frac{s}{2u^* H_{m,\gamma}}\right)^{\frac{1}{\gamma}}. \tag{10.26}$$

While for the value of λ, when $\left(\frac{\rho_m}{2u^*\sqrt{sn}}\right)^{\frac{2}{3}} \leq \frac{1}{sn}$, λ is the non-negative integer that satisfies

$$\left(\frac{\rho_{m-\lambda+1}}{2u^*\sqrt{sn}}\right)^{\frac{2}{3}} \leq \frac{1}{sn} < \left(\frac{\rho_{m-\lambda}}{2u^*\sqrt{sn}}\right)^{\frac{2}{3}}. \tag{10.27}$$

If λ is not too close to m, $\left(\frac{\rho_{m-\lambda}}{2u^* \sqrt{sn}}\right)^{\frac{2}{3}} \approx \frac{1}{sn}$, and

$$m - \lambda \approx \left(\frac{sn}{2u^* H_{m,\gamma}}\right)^{\frac{1}{\gamma}}. \tag{10.28}$$

10.5.2.2 Minimum transmission distance

We solve the primal problem based on various values of Zipf parameter γ : $\gamma < \frac{3}{2}$, $\gamma = \frac{3}{2}$ and $\gamma > \frac{3}{2}$. Specifically, we have the following three propositions:

Proposition 1: When $\gamma < \frac{3}{2}$, a lower bound on the average transmission distance is

$$L^* = \begin{cases} \dfrac{1}{\sqrt{sn}} \dfrac{\left(H_{m,\frac{2\gamma}{3}}\right)^{\frac{3}{2}}}{H_{m,\gamma}}, & \text{if } m < \left(1 - \dfrac{2}{3}\gamma\right) ns \\ \dfrac{(m-\lambda)^{1-\gamma}}{(1-\frac{2}{3}\gamma)H_{m,\gamma}} + 1 - \dfrac{H_{m-\lambda,\gamma}}{H_{m,\gamma}}, & \text{otherwise} \end{cases} \tag{10.29}$$

where $m - \lambda = (sn - m)(\dfrac{3}{2\gamma} - 1)$.

Proposition 2: When $\gamma = \frac{3}{2}$, a lower bound on the average transmission distance is

$$L^* = \begin{cases} \dfrac{1}{\sqrt{sn}} \dfrac{(H_{m,1})^{\frac{3}{2}}}{H_{m,\frac{3}{2}}}, & \text{if } m \ln m < ns \\ \dfrac{2(\kappa^{-\frac{1}{2}} - m^{-\frac{1}{2}}) + \kappa^{-\frac{1}{2}} \ln \kappa}{H_{m,\frac{3}{2}}}, & \text{otherwise} \end{cases} \tag{10.30}$$

where κ satisfies $\kappa(\ln \kappa - 1) = sn - m$.

Proposition 3:When $\gamma > \frac{3}{2}$, a lower bound on the average transmission distance is

$$L^* = \begin{cases} \dfrac{1}{\sqrt{sn}} \dfrac{\left(H_{m,\frac{2\gamma}{3}}\right)^{\frac{3}{2}}}{H_{m,\gamma}}, & \text{if } s < \chi_1 \text{ and } sn > \chi_2 \\ \dfrac{1}{\sqrt{sn-m}} \dfrac{\left(H_{m,\frac{2\gamma}{3}}\right)^{\frac{3}{2}}}{H_{m,\gamma}}, & \text{if } s < \chi_3 \text{ and } sn \le \chi_2 \\ \dfrac{\left(\frac{2\gamma-3}{2\gamma}\left(s - \frac{m}{n}\right)\right)^{1-\gamma}}{\sqrt{n}H_{m,\gamma}(\frac{2}{3}\gamma - 1)}, & \text{if } s \ge \chi_3 \text{ and } sn \le \chi_4 \\ \dfrac{\left(\frac{2\gamma-3}{2\gamma}s\right)^{1-\gamma}}{\sqrt{n}H_{m,\gamma}(\frac{2}{3}\gamma - 1)}, & \text{if } s \ge \chi_1 \text{ and } sn > \chi_4 \end{cases} \tag{10.31}$$

where $\chi_1 = H_{m,\frac{2\gamma}{3}}$, $\chi_2 = m^{\frac{2\gamma}{3}} H_{m,\frac{2\gamma}{3}}$, $\chi_3 = \frac{m}{n} + \frac{2\gamma}{2\gamma-3}$ and $\chi_4 = \frac{2\gamma}{2\gamma-3} \frac{m}{n^{\frac{3}{2\gamma}}} + m$.

The proof of the propositions are emitted due to space limitations.

10.5.2.3 Per-node capacity

Based on previous results on the average transmission distance, we now present an upper bound on the per-node capacity. Recall that given the average transmission distance L, the per-node capacity is upper bounded by $\frac{\sqrt{8}}{\sqrt{\pi}}\frac{W}{\beta^{\frac{1}{\alpha}}-1}\frac{1}{L\sqrt{n}}$ [5]. Combining this result with the lower bounds on average transmission distance given in propositions 1, 2 and 3, we have the following theorem.

Theorem 3: Under our network and content access model, an upper bound C on the per-node capacity (with constants $\frac{\sqrt{8}}{\sqrt{\pi}}\frac{W}{\beta^{\frac{1}{\alpha}}-1}$ omitted) is given as follows.

(i) $\gamma < \frac{3}{2}$

If $m < (1-\frac{2}{3}\gamma)sn$:

$$C=\begin{cases}\sqrt{\frac{s}{m}}\frac{\left(1-\frac{2}{3}\gamma\right)^{\frac{3}{2}}}{1-\gamma}, & \text{if } \gamma<1\\ 3^{-\frac{3}{2}}\sqrt{\frac{s}{m}}\ln m, & \text{if } \gamma=1\\ H_{m,\gamma}\frac{\sqrt{s}}{m^{\frac{3}{2}-\gamma}}\left(1-\frac{2}{3}\gamma\right)^{\frac{3}{2}}, & \text{if } 1<\gamma<\frac{3}{2}\end{cases} \tag{10.32}$$

If $m \ge (1-\frac{2}{3}\gamma)sn$:

$$C=\begin{cases}\frac{1}{\sqrt{n}}\frac{1}{1-\frac{1}{2}(\frac{3}{2\gamma}-1)^{-\gamma}\left(\frac{sn}{m}-1\right)^{1-\gamma}}, & \text{if } \gamma<1\\ \frac{1}{\sqrt{n}}\frac{\ln m}{\ln\frac{2m}{sn-m}+3}, & \text{if } \gamma=1\\ \frac{1}{\sqrt{n}}\frac{H_{m,\gamma}\frac{\gamma-1}{m^{1-\gamma}}}{\frac{1}{2}(\frac{3}{2\gamma}-1)^{-\gamma}(\frac{sn}{m}-1)^{1-\gamma}-1}, & \text{if } 1<\gamma<\frac{3}{2}\end{cases} \tag{10.33}$$

(ii) $\gamma = \frac{3}{2}$

$$C=\begin{cases}\frac{\sqrt{s}}{(\ln m)^{\frac{3}{2}}}H_{m,\frac{3}{2}}, & \text{if } m\ln m<sn\\ \frac{1}{\sqrt{n}}\frac{H_{m,\frac{3}{2}}}{2(\kappa^{-\frac{1}{2}}-m^{-\frac{1}{2}})+\kappa^{-\frac{1}{2}}\ln\kappa}, & \text{if } m\ln m\ge sn\end{cases} \tag{10.34}$$

where κ satisfies $\kappa(\ln\kappa-1)=sn-m$.

(iii) $\gamma > \frac{3}{2}$

$$C=\begin{cases}\sqrt{s}\dfrac{H_{m,\gamma}}{\left(H_{m,\frac{2\gamma}{3}}\right)^{\frac{3}{2}}}, & \text{if } s<\chi_1 \text{ and } sn>\chi_2\\ \sqrt{s-\dfrac{m}{n}}\dfrac{H_{m,\gamma}}{\left(H_{m,\frac{2\gamma}{3}}\right)^{\frac{3}{2}}}, & \text{if } s<\chi_3 \text{ and } sn\le\chi_2\\ \dfrac{2}{3}\left(s-\dfrac{m}{n}\right)^{\gamma-1}\dfrac{\gamma H_{m,\gamma}}{\left(\frac{2\gamma-3}{2\gamma}\right)^{-\gamma}}, & \text{if } s\ge\chi_3 \text{ and } sn\le\chi_4\\ \dfrac{2}{3}s^{\gamma-1}\dfrac{\gamma H_{m,\gamma}}{\left(\frac{2\gamma-3}{2\gamma}\right)^{-\gamma}}, & \text{if } s\ge\chi_1 \text{ and } sn>\chi_4\end{cases} \tag{10.35}$$

where the values of χ_1, χ_2, χ_3 and χ_4 are given in Proposition 3.

- *Proof.* Recall p_i is the density of content i, then $1 - sp_i$ is the fraction of nodes that have not cached content i locally. Let $\mathscr{L}_i$ represent the average distance for those nodes to retrieve i from others. Then, the average distance for nodes to retrieve contents from others (averaged over all contents), denoted by $\mathscr{L}$, is

$$\mathscr{L} = \frac{\sum_{i=1}^{m} \rho_i(1 - sp_i)\mathscr{L}_i}{\sum_{i=1}^{m} \rho_i(1 - sp_i)}. \tag{10.36}$$

According to [5], the number of bits each node can receive from others per second is upper bounded by $\frac{1}{\mathscr{L}\sqrt{n}}$. The cache hit ratio (i.e., the probability that a request is served by local cache) is $\mathscr{P} = \sum_{i=1}^{m} \rho_i sp_i$. Then, besides the bits received from others, $\frac{\mathscr{P}}{1-\mathscr{P}}\frac{1}{\mathscr{L}\sqrt{n}}$ bits from local cache are also used to serve requests. Combining the bits from others and the bits from local cache, we can get an upper bound on per-node capacity as follows:

$$C = \frac{1}{\mathscr{L}\sqrt{n}} \cdot \left(1 + \frac{\mathscr{P}}{1 - \mathscr{P}}\right) = \frac{1}{\sqrt{n}\sum_{i=1}^{m} \rho_i(1 - sp_i)\mathscr{L}_i}. \tag{10.37}$$

Recall that L_i is the transmission distance of content i averaged over all nodes (including the nodes that have cached i locally), then $L_i = \mathscr{L}_i \cdot (1 - sp_i) + 0 \cdot sp_i$, for $i = 1, \ldots, m$. The transmission distance averaged over all contents is $L = \sum_{i=1}^{m} L_i\rho_i = \sum_{i=1}^{m} \mathscr{L}_i(1 - sp_i)\rho_i$. Then, Eq. (10.37) can be rewritten as

$$C = \frac{1}{L\sqrt{n}}. \tag{10.38}$$

Combining the above equation with propositions 10.5.2.2, 2 and 3 yields the results in the theorem.

10.5.3 Analytic results

10.5.3.1 Influence of different parameters

If we ignore insignificant terms in above capacity results, simplified capacity scaling laws can be obtained as follows:

Table 10.1 *Per-node capacity under different* γ

γ	C
$\gamma < 1$	$O\left(\sqrt{\frac{s}{m}}\right)$
$\gamma = 1$	$O\left(\sqrt{\frac{s}{m}} \ln m\right)$
$1 < \gamma < \frac{3}{2}$	$O\left(\frac{\sqrt{s}}{m^{3/2-\gamma}}\right)$
$\gamma = \frac{3}{2}$	$O\left(\frac{\sqrt{s}}{(\ln m)^{3/2}}\right)$
$\gamma > \frac{3}{2}$	$O\left(\sqrt{s - \frac{m}{n}}\right)$, *if* $s < H_{m,2\gamma/3}$
	$O\left(\left(s - \frac{m}{n}\right)^{\gamma-1}\right)$, *if* $s \geq H_{m,2\gamma/3}$

We have the following observations. First, the per-node capacity will not decrease when the number of nodes in the network increases. Specifically, for small γ (i.e., $\gamma \leq \frac{3}{2}$), per-node capacity remains constant even when n increases; while for large γ (i.e., $\gamma > \frac{3}{2}$), it is possible for the per-node capacity to increase as n grows. This is in sharp contrast to previous Guptar-Kumar's result, where the per-node capacity without caching scales like $\frac{1}{\sqrt{n}}$, which decreases quickly as the number of nodes in the network grows.

Second, as cache size s grows, the per-node capacity increases. This is because when the cache size increases, more requests can be served by local cache or nearby nodes. Our results also show that the capacity scales like $s^{\gamma-1}$ with cache size s when $\gamma > \frac{3}{2}$, and scales like $\sqrt{s}$ when $\gamma <= \frac{3}{2}$. Thus, when γ is large, increasing the cache size can significantly improve the per-node capacity.

Third, the per-node capacity decreases when the number of unique contents m increases, since a larger m reduces the probability for the requests to be served locally or by close neighbors. However, as γ increases from 0 to more than $\frac{3}{2}$, m has a smaller effect on per-node capacity. When $\gamma < 1$, m is somewhat equally important as s, as s needs to increase at the same speed with m to keep the capacity constant. On the other hand, when $\gamma > \frac{3}{2}$, for some cases the capacity is almost irrelevant to m.

Forth, γ also plays an important role in per-node capacity. When $\gamma = 0$, contents follow uniform distribution. As γ grows from 0 to more than $\frac{3}{2}$, the per-node capacity is increased drastically as the exponent of m grows from $-\frac{1}{2}$ to 0. This result suggests that caching is more effective when contents have skewed popularity.

10.5.3.2 Effects of local cache

We analyze the cache hit ratio $\mathscr{P}$, i.e., the probability that a request is served locally, when contents have been optimally cached.

Lemma 2: When contents have been optimally cached to maximize the per-node capacity, for any $\gamma \leq \frac{3}{2}$, $\mathscr{P} \rightarrow 0$.

Proof. Given P*, the probability that requests being served by local cache is $\mathscr{P} = \sum_{i=1}^{m} \rho_i p_i^* s$. Note that when $\gamma \leq \frac{3}{2}$, $\mu = 0$, and accordingly Eq. (10.24) is

$$1 = \sum_{i=1}^{m-\lambda} \left(\frac{\rho_i}{2u^* \sqrt{sn}} \right)^{\frac{2}{3}} + \frac{\lambda}{sn} \geq \sum_{i=1}^{m} \left(\frac{\rho_i}{2u^* \sqrt{sn}} \right)^{\frac{2}{3}}. \tag{10.39}$$

Since $\sum_{i=1}^{m} \rho_i^{\frac{2}{3}} = \frac{H_{m,\frac{2\gamma}{3}}}{(H_{m,\gamma})^{\frac{2}{3}}}$, the above inequality (10.39) is equivalent to $\frac{(H_{m,\gamma})^{\frac{2}{3}}}{H_{m,\frac{2\gamma}{3}}} \geq (\frac{1}{2u^* \sqrt{sn}})^{\frac{2}{3}}$. Combining the previous inequality and the value of p_i^* given in Eq. (10.23), and let $\tilde{p_i^*} = \frac{(\rho_i H_{m,\gamma})^{\frac{2}{3}}}{H_{m,\frac{2\gamma}{3}}}$, we can get

$$\tilde{p_1^*} - p_1^* \geq \tilde{p_2^*} - p_2^* \geq \ldots \geq \tilde{p_m^*} - p_m^*. \tag{10.40}$$

Based on the definition of Zipf distribution, the popularity of contents satisfies $\rho_1 \geq \rho_2 \geq \ldots \geq \rho_m > 0$. Combining this content popularity inequality with inequality (10.40), and as $\sum_{i=1}^{m} \tilde{p_i^*} - p_i^* = 0$, we have $\sum_{i=1}^{m} \rho_i(\tilde{p_i^*} - p_i^*)s \geq 0$, which leads to:

$$\sum_{i=1}^{m} \rho_i \tilde{p_i^*} s \geq \sum_{i=1}^{m} \rho_i p_i^* s. \tag{10.41}$$

The left-hand side of the above inequality is

$$\sum_{i=1}^{m} \rho_i \tilde{p_1^*} s = \sum_{i=1}^{m} \frac{s}{i^\gamma H_{m,\gamma}} \frac{1}{i^{\frac{2}{3}\gamma} H_{m,\frac{2\gamma}{3}}} = \frac{sH_{m,\frac{5\gamma}{3}}}{H_{m,\gamma} H_{m,\frac{2\gamma}{3}}}. \tag{10.42}$$

Since $\frac{sH_{m,\frac{5\gamma}{3}}}{H_{m,\gamma} H_{m,\frac{2\gamma}{3}}} \rightarrow 0$ for any $\gamma \leq \frac{3}{2}$, $\sum_{i=1}^{m} \rho_i p_i^* s = \mathscr{P} \rightarrow 0$.

The above lemma states that when $\gamma \leq \frac{3}{2}$, only a small portion of requests is served locally, and the effect of the local cache is negligible. On the other hand, when $\gamma > \frac{3}{2}$, the content access is mainly focused on a few popular contents that may have been cached locally, and the local cache will have a more significant effect on per-node capacity. In the following lemma, we give two sufficient conditions for a constant portion of requests being served by local cache.

Lemma 3: When contents have been optimally cached to maximize the per-node capacity, for any $\gamma > \frac{3}{2}$, if $\mu \neq 0$ or $\mu = \lambda = 0$, there exists a constant c independent of m, s and n, such that $\mathscr{P} \geq c$.

Proof. (i)$\mu \neq 0$ In case of $\mu \geq 1$, $p_1 = \frac{1}{s}$ and every node caches the most popular content locally. Then, requests for that content is always served by the local cache, and $\mathscr{P}$ must satisfy

$$\mathscr{P} > \rho_1 = \frac{1}{H_{m,\gamma}} > \frac{1}{1+\frac{1}{\gamma-1}} = \frac{\gamma-1}{\gamma}. \tag{10.43}$$

(ii)$\boldsymbol{\mu} = \boldsymbol{\lambda} = \mathbf{0}$

When both λ and μ are 0, we have

$$\mathscr{P} = \sum_{i=1}^{m} \rho_i p_i^* s = \frac{sH_{m,\frac{5\gamma}{3}}}{H_{m,\gamma}H_{m,\frac{2\gamma}{3}}} \geq \frac{H_{m,\frac{5\gamma}{3}}}{H_{m,\gamma}H_{m,\frac{2\gamma}{3}}}. \tag{10.44}$$

As $m \to \infty$, $\frac{H_{m,\frac{5\gamma}{3}}}{H_{m,\gamma}H_{m,\frac{2\gamma}{3}}}$ converges to a constant which only depends on γ.

Therefore, $\mathscr{P}$ is larger than a constant when $\lambda = \mu = 0$.

10.5.4 Numerical results

10.5.4.1 Effects of various parameters

Based on above theoretical analysis, we now present some numerical results to illustrate how various parameters affect the capacity. Figure 10.2a shows how the per-node capacity varies with the number of nodes (n) given different values of γ, where $s = 10$ and $m = 10^4$. As shown in the figure, the per-node capacity increases as γ increases. When γ is small, the content access is more like a uniform distribution (note that $\gamma = 0$ corresponds to the uniform distribution). When γ is large, the content access is focused on some hot (frequently accessed) content that may have been cached, and then improving the per-node capacity. Under all five values of γ, the per-node capacity first increases with n, and then remains constant. This is different from Gupta-Kumar's result where cache is not considered and the per-node capacity decreases with increasing n. With caching, the per-node capacity will not decrease as n increases. Moreover, it shows that when n is relatively small, it is possible that increasing n will improve the per-node capacity. This is because when n is relatively small, most contents only have one replica, and nodes need to traverse the whole network to obtain the contents. On the other hand, when n is relatively large, some of the contents could be cached by nearby nodes, which makes content retrieval easier and then improves the per-node capacity.

Figure 10.2b illustrates how the number of unique contents m affects the per-node capacity. Generally, increasing m results in a reduction in per-node capacity. This is because with more unique contents, nodes will be more likely to retrieve contents from further away nodes, which leads to more interference and less capacity. More interestingly, when γ increases, m has a diminishing effect on capacity. For example, when $\gamma = 0$, the capacity decreases quickly with m; while for $\gamma = 2$, the capacity almost remains constant for small m. This is because the unpopular contents will be less frequently requested when γ increases. When γ is relatively large, increasing m will only add a few extremely unpopular contents, and will hardly affect what contents will be cached and requested.

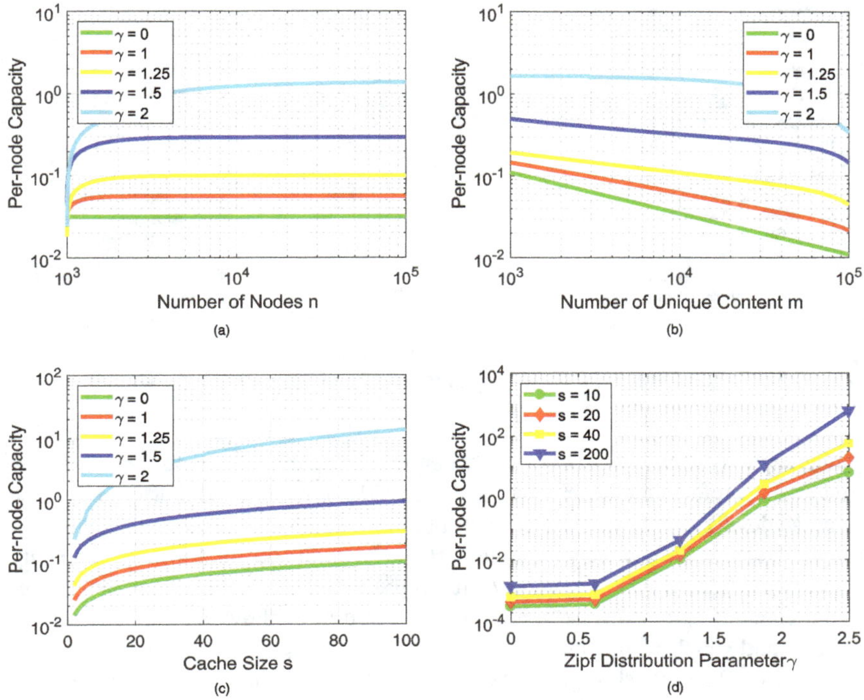

Figure 10.2 Effects of various parameters on capacity

Figure 10.2c shows how the cache size s affects per-node capacity. As can be seen, as the cache size increases, the per-node capacity increases. When γ is relatively small, the contents have similar popularity, and caching a few more contents will not result in a large increase in the network capacity. When γ is relatively large, the popular contents are requested more frequently, and only caching a few more popular contents can significantly improve the per-node capacity.

Figure 10.2d illustrates how the Zipf parameter γ affects the per-node capacity. Compared to the three parameters (s, m, n) discussed above, γ has the largest impact on per-node capacity. As γ grows from 0 to 2.5, the capacity dramatically increases from 10^{-3} to roughly 10^2. Increasing γ can significantly improve the capacity, since a larger γ results in more skewed content popularity where a few popular contents are more frequently requested, which makes caching more effective.

10.5.4.2 Comparisons to existing work

In Figure 10.3, we compare our capacity results with previous work [5, 6, 31] based on numerical results. From [5], for wireless networks without caching, the per-node capacity scales like $\frac{1}{\sqrt{n}}$, which is shown in the figure by the green dashed line. Grossglauser and Tse [6] have proved that when nodes are mobile, the

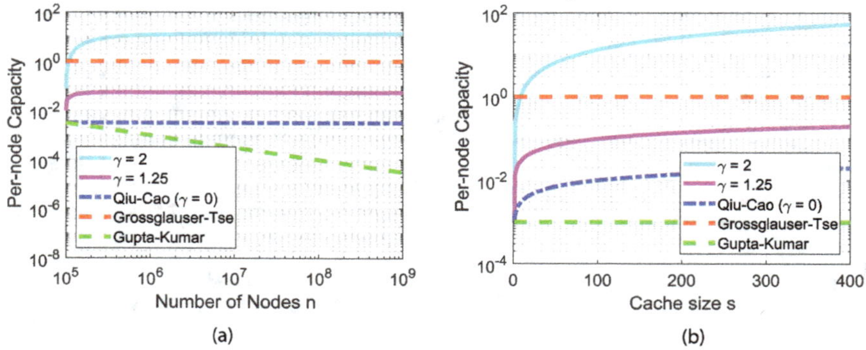

Figure 10.3 Comparisons to existing work

per-node capacity can remain constant when n grows, and their result is shown by the red dashed line in the figure. Qiu and Cao [31] have derived that the per-node capacity scales like $\sqrt{\frac{s}{m}}$ when the content popularity follows a uniform distribution (i.e., $\gamma = 0$), which is shown by the blue dashed line in the figure. Since our capacity result at $\gamma = 0$ (i.e., $C = \Theta(\sqrt{\frac{s}{m}})$) conforms to their result, this blue line also represents our capacity at $\gamma = 0$. The remaining two solid lines show our capacity results at $\gamma = 1.25$ and $\gamma = 2$, respectively.

In Figure 10.3a, we show how the per-node capacity changes with the number of nodes, where $s = 100$ and $m = 10^7$. As shown in the figure, when the number of nodes n increases, the capacity result of Gupta-Kumar drops quickly, and the capacity of Grossglauser-Tse and Qiu-Cao remain unchanged. In our approaches, the per-node capacity will not decrease when n increases. This is a significant improvement compared to the Gupta-Kumar's result; i.e., with caching, increasing the number of nodes will not reduce the per-node capacity.

Figure 10.3b illustrates the per-node capacity as a function of the cache size, where $n = m = 10^6$. Since Gupta-Kumar and Grossglauser-Tse do not consider caching, their results do not change with the cache size. When the cache size is extremely small ($s = \frac{m}{n}$), under all three values of γ, our per-node capacity is comparable with Gupta-Kumar results. When s approaches $\frac{m}{n}$, each content only has one replica in the network. Then, retrieving a content is like to communicate with a random node, which is identical to the communication scenario in [5].

As the cache size increases, the per-node capacity of our approach with various γ increases quickly, and significantly higher than Gupta-Kumar. When compared to Grossglauser-Tse, the result depends on γ. When γ is large (i.e., $\gamma = 2$), the content access is focused on some hot (frequently accessed) content that may have been cached, and then improving the per-node capacity. As a result, the per-node capacity of our approach with $\gamma = 2$ significantly outperforms that of Grossglauser-Tse. When γ is relatively small (i.e., $\gamma = 0$), the content access is more like a uniform distribution and the caching advantage is not very high, and our per-node capacity is lower. However, Grossglauser-Tse has much longer delay since nodes have to wait until they move close to the destination. While compared to Qiu-Cao, our capacity conforms to theirs at $\gamma = 0$. As γ increases (e.g., $\gamma = 1.25$), caching becomes more effective due to more skewed content popularity, and our capacity grows much higher than Qiu-Cao.

10.6 Caching algorithms

In the previous sections, we have presented the optimal content densities, and derived an upper bound on the per-node capacity when contents have been optimally cached. In practice, the exact content popularity is usually unknown a priori. To show that the optimal content densities are achievable in real scenarios, we present a distributed algorithm that achieves the optimal content densities only based on each node's local information.

10.6.1 Algorithm description

According to Eq. (10.23), each of the most unpopular λ contents only has one copy in the network. On the other hand, each of the most popular μ contents has n copies and is cached at all the nodes in the network. For any remaining content i, its popularity is proportional to $\rho_i^{\frac{2}{3}}$, where ρ_i is the popularity of content i.

To guarantee that all content requests can be served, a permanent copy for each content is randomly cached at some node in the network, before each node runs the distributed caching algorithm. To achieve the $p_i \propto \rho_i^{\frac{2}{3}}$ law for contents with moderate popularity, we use the GreedyDual-Size algorithm [34, 35]. The algorithm was originally proposed to reduce the access cost in Web caching, and it had been proved to be k-competitive for Web caching (where k is defined to be ratio of the cache size to the content size). However, we found that for caching in wireless networks, such an algorithm does not perform well especially when contents have very skewed popularity (i.e., γ is large). Thus, we propose a modified version of the GreedyDual-Size algorithm, so as to achieve the desired densities.

In our algorithm, each node maintains a value $T(i)$ for any content i cached locally. Each node also maintains an "inflation" value V, which is initially set to 0. For any content request, if the requested content i is cached locally, then $T(i)$ is increased by $l(i)$, where $l(i)$ is the latest physical distance to retrieve content from others. On the other hand, if content i is not in the local cache, i will be cached locally, and $T(i)$ is set to $V + l(i)$ ($l(i)$ is the physical distance to retrieve). For the value of V, if the memory is full, V equals to the smallest $T(j)$; otherwise, V equals to 0. Note that when the memory is full, to cache locally, the content with the smallest $T(j)$ will be evicted. Let S denote the set of contents cached locally, the *Distributed Caching Algorithm* running at each node is given as follows.

Algorithm 1 Distributed Caching Algorithm

1: $V \leftarrow 0$
2: for content request for any content i do
3:   if i is in local cache then
4:     $T(i) \leftarrow T(i) + l(i)$ $\{l(i)$ is the latest physical distance to retrieve content $i\}$
5:   else
6:     if the memory is full then
7:       $V \leftarrow \min\{T(j) | j \in S\}$
8:       evict content j such that $j = \operatorname{argmin}_{k \in S} T(k)$
9:     end if
10:     cache i locally, $T(i) \leftarrow V + l(i)$
11:   end if
12: end for

In practice, the value of $l(i)$, i.e., the physical distance to retrieve a content, can be obtained based on the geographic coordinates of the source node and the destination node. Since each node only maintains $T(i)$ and $l(i)$ for contents cached locally, the space complexity of our algorithm is $O(s)$. We also have the following theorem on the time complexity of the proposed algorithm.

Theorem 4: For the Distributed Caching Algorithm, the time complexity for processing each content request is $O(\log s)$.

Proof. In Line 3, it takes $O(1)$ time to determine whether a content has been cached locally, if a hash table is used. In Line 7, it takes $O(\log s)$ time to find the minimum $T(i)$, if the value $T(i)$ are stored in a min-heap. All the other operations inside the for loop take constant time. Hence, it takes $O(\log s)$ time to perform the operations inside the for loop (i.e., process each content request).

The proposed distributed caching algorithm converges to and achieves the optimal content density P*. A rigorous proof can be found in [36].

10.6.2 Evaluation results

In this subsection, we evaluate the performance of our *Distributed Caching Algorithm*. In the simulations, n nodes are uniformly and independently placed on the surface of a unit sphere, where each node can cache s contents locally. There are m unique contents in the network, where the content popularity follows a Zipf distribution with parameter γ. Suppose there is always a content request at each node. Nodes send requests sequentially to the nearest node that caches the requested content. After the content has been retrieved, it will be cached locally. If the memory is full, one content will be evicted based on the Distributed Caching Algorithm.

Figure 10.4compares the average transmission distance in the simulations with the theoretical minimum transmission distance. The theoretical distance is shown by the solid line, while the distance based on simulations is shown by the dashed line.

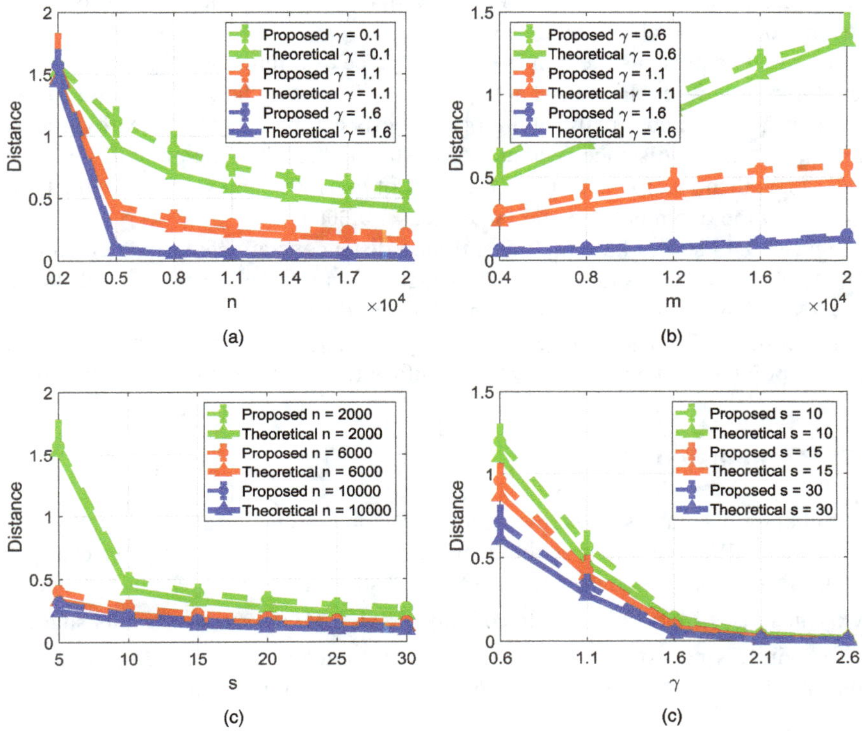

Figure 10.4 Results of the Distributed Caching Algorithm vs. theoretical value

Recall that the per-node capacity scales like $O(\frac{1}{\sqrt{L}})$ with the transmission distance L, hence, a smaller distance implies higher capacity. Each experiment is repeated for 50 times, and the 10% and 90% confidence intervals are also reported in the figure. Figure 10.4a shows the distance as a function of the number of nodes, where γ ranges from 0.1 to 1.6, $m = 10000$, and $s = 5$. When $ns = m$ (i.e., $n = 2000$), both the theoretical approach and the proposed algorithm require that each content has exactly one copy in the network. Therefore, the simulation results conform quite well to the theoretical results. As n increases, the simulation results slightly deviate from the theoretical optimal value. One possible explanation is that the content density p_i in the theoretical analysis can take any value between $\frac{1}{sn}$ and $\frac{1}{s}$, while in simulations sp_i has to be an integer (since the number of copies has to be an integer). Thus, the proposed algorithm performs a bit worse than the theoretical value. Besides, as γ grows, the transmission distance decreases.

Figure 10.4b shows the effect of m on the transmission distance, given different γ, when $n = 2500$ and $s = 10$. As the number of unique contents grows, the transmission distance increases, and then the per-node capacity decreases. This is because with more unique contents, it becomes less likely for the content requests to be served by local cache or nearby nodes.

Figure 10.4c presents the transmission distance as a function of the cache size, given different n, when $m = 10000$ and $\gamma = 1.1$. When the cache size increases, more contents can be cached locally, and more requests can be served by nearby nodes, and the transmission distance decreases. Note that the result at the point of $n = 2000$ and $s = 25$, is identical to the result at the point of $n = 10000$ and $s = 5$, since the total cache size at the two points are the same (i.e., $ns = 50000$).

Figure 10.4d shows how the transmission distance is affected by γ, given different s, when $n = 2000$ and $m = 10000$. Not surprisingly, the transmission distance decreases with γ and s. When $\gamma = 2.6$, all three cases (i.e., $s = 10$, $s = 15$ and $s = 30$) perform similarly. This is because for large γ, increasing the cache size will only add a few less popular contents, and may not significantly affect the system performance.

10.7 Conclusions

In this chapter, we have studied the effect of caching on the capacity of wireless ad hoc networks. We found that as the distribution of the content popularity changes from uniform distribution to more skewed distributions, the per-node capacity quickly increases from $\Theta\left(\sqrt{\frac{s}{m}}\right)$ to roughly $\Theta\left(\sqrt{s}\right)$. Moreover, our results suggest that for wireless networks with caching, when contents have skewed popularity, increasing the number of nodes monotonically increases the per-node capacity.

References

[1] Index C.V.N. 'Global mobile data traffic forecast update'. *2016–2021. white paper*. 2017.

[2] Bastug E., Bennis M., Debbah M. 'Living on the edge: the role of proactive caching in 5G wireless networks'. *IEEE Communications Magazine*. 2014;**52**(8):82–9.

[3] Jiang W., Feng G., Qin S. 'Optimal cooperative content caching and delivery policy for heterogeneous cellular networks'. *IEEE Transactions on Mobile Computing*. 2016;**16**(5):1382–93.

[4] Li X., Wang X., Wan P.-J., Han Z., Leung V.C.M. 'Hierarchical edge caching in device-to-device aided mobile networks: modeling, optimization, and design'. *IEEE Journal on Selected Areas in Communications*. **36**; 2018. pp. 1768–85.

[5] Gupta P., Kumar P.R. 'The capacity of wireless networks'. *IEEE Transactions on Information Theory*. 2000;**46**(2):388–404.

[6] Grossglauser M., Tse D. 'Mobility increases the capacity of ad-hoc wireless networks'. In: IEEE INFOCOM; 2001.

[7] Yin L., Cao G. 'Supporting cooperative caching in AD hoc net works'. In: IEEE INFOCOM; 2004.

[8] Zhao J., Zhang P., Cao G., Das C.R. 'Cooperative caching in wireless p2p networks: design, implementation, and evaluation'. *IEEE Transactions on Parallel and Distributed Systems*. 2010;**21**(2):229–41.

[9] Fiore M., Mininni F., Casetti C., Chiasserini C.-F. 'To cache or not to cache?'. In: IEEE INFOCOM; 2009.

[10] Tang B., Gupta H., Das S.R. 'Benefit-based data caching in AD hoc networks'. *IEEE transactions on mobile computing*. 2008;**7**(3):289–304.

[11] Leonardi E., Torrisi G.L. 'Least recently used caches under the shot noise mode'. In:IEEE INFOCOM; 2015.

[12] Fricker C., Robert P., Roberts J. 'A versatile and accurate approximation for LRU cache performance'. *In: International Teletraffic Congress (ITC)*. 2012.

[13] Choungmo Fofack N., Nain P., Neglia G., Towsley D. 'Performance evaluation of hierarchical TTL-based Cache networks'. *Computer Networks*. 2014;**65**(7):212–31.

[14] Rosensweig E.J., Menasche D.S., Kurose J. 'On the steady-state of cache networks'. In: IEEE INFOCOM; 2013.

[15] Dehghan M., Seetharam A., Jiang B. 'On the complexity of optimalrouting and content caching in heterogeneous networks'. *In: IEEE INFOCOM*; 2015.

[16] Dabirmoghaddam A., Barijough M.M., Garcia-Luna-Aceves J. 'Understanding optimal caching and opportunistic caching at the edge of information-centric networks'. In: Proceedings of the 1st ACM conference on information-centric networking; 2014. pp. 47–56.

[17] Zhang G., Li Y., Lin T. 'Caching in information centric networking: a survey'. *Computer Networks*. 2013;**57**(16):3128–41.

[18] Cohen E., Shenker S. 'Replication strategies in unstructured peer-to-peer networks'. *In: ACM SIGCOMM Computer Communication Review*. **32**; 2002. pp. 177–90.

[19] Jin S., Wang L. 'Content and service replication strategies in multi-hop wireless mesh networks'. In: ACM MSWiM; 2005.

[20] Song J., Sheng M., Quek T.Q.S., Xu C., Wang X. 'Learning based content Caching and sharing for wireless networks'. *IEEE Transactions on Communications*. 2017;**65**(10):1–24.

[21] Müller S., Atan O., van der Schaar M., Klein A. 'Context-aware proactive content caching with service differentiation in wireless networks'. *IEEE Transactions on Wireless Communications*. 2017;**16**(2):1024–36.

[22] Ji M., Caire G., Molisch A.F. 'Fundamental limits of caching in wireless D2d networks'. *IEEE Transactions on Information Theory*. 2016;**62**(2):849–69.

[23] Ji M., Caire G., Molisch A.F. 'Wireless device-to-device caching networks: basic principles and system performance'. *IEEE Journal on Selected Areas in Communications*. 2016;**34**(1):176–89.

[24] Yang C., Yao Y., Chen Z., Xia B. 'Analysis on cache-enabled wireless heterogeneous networks'. *IEEE Transactions on Wireless Communications*. 2016;**15**(1):131–45.

[25] Niesen U., Shah D., Wornell G.W. 'Caching in wireless networks'. *IEEE Transactions on Information Theory*. 2012;**58**(10):6524–40.

[26] Gitzenis S., Paschos G.S., Tassiulas L. 'Asymptotic laws for joint content replication and delivery in wireless networks'. *IEEE Transactions on Information Theory*. 2013;**59**(5):2760–76.

[27] Breslau L., Cao P., Fan L., Phillips G., Shenker S. 'Web caching and Zipf-like distributions: Evidence and implications'. In: IEEE INFOCOM; 1999.

[28] Cha M., Kwak H., Rodriguez P., Ahn Y.Y., Moon S. 'I tube, you tube, everybody tubes: analyzing the world's largest user generated content video system'. In: ACM SIGCOMM conference on Internet measurement; 2007.

[29] Li J., Blake C., De Couto D.S., Lee H.I., Morris R. 'Capacity of ad hoc wireless networks'. In: ACM MobiCom; 2001.

[30] Chernoff H. 'A measure of asymptotic efficiency for tests of a hypothesis based on the sum of observations'. *The Annals of Mathematical Statistics*. 1952;**23**(4):493–507.

[31] Qiu L., Cao G. 'Cache increases the capacity of wireless netw orks'. In: IEEE INFOCOM; 2016.

[32] David H.A., Nagaraja H.N. 'Order statistics'. *Wiley Online Library*. 1970.

[33] Bertsekas D.P. 'Nonlinear programming'. *Athena scientific Belmont*. 1999.

[34] Cao P., Irani S. 'Cost-aware www proxy caching algorithms'. *In: Usenix symposium on internet technologies and systems*. **12**; 1997. pp. 193–206.

[35] Zhang Y., Zhao J., Cao G. 'Roadcast: A popularity aware content sharing scheme in VANETs'. In: IEEE ICDCS; 2009.

[36] Qiu L., Cao G. 'Popularity-aware caching increases the capacity of wireless networks'. *IEEE Transactions on Mobile Computing*. 2019;**19**(1):173–87.

Chapter 11
Cache-induced hierarchical cooperation in wireless D2D networks

An Liu[1], Vincent Lau[2], and Giuseppe Caire[3]

11.1 Introduction

An increase of 1 000× in wireless data traffic is expected in the near future. More than 50% of this will be generated by high-definition video and content delivery applications. Many recent works have shown that wireless caching is one of the most promising solutions to handle the high traffic load caused by content delivery applications [1–3]. By exploiting the fact that content is "cachable," wireless nodes can cache some popular content during off-peak hours (*cache initialization phase*), in order to reduce the traffic rate at peak hours (*content delivery phase*). Early works focused on caching at the network side, such as at base stations (BSs). Recently, however, caching at the user/device side has also gained increasing interest, due to the number of wireless devices increasing faster than the number of BSs, and wireless device storage being arguably the cheapest and most rapidly growing network resource. It has been shown in [4–7] that combining wireless device caching with short-range device-to-device (D2D) communications can significantly improve the throughput of wireless networks. After that, many efficient wireless caching schemes have been proposed for *wireless D2D caching networks*.

One of the most important lesson that Shannon, the inventor of information theory, has taught the communication engineers is that the answers to the following two questions play a fundamental role in designing a practical communication system: what is the fundamental limit of a given communication network and what are the techniques to achieve such fundamental limit? For a point-to-point communication system, the fundamental limit is characterized by the channel capacity, which is the maximum transmission rate (in bits/second) that can be achieved by any communication scheme with a vanishing error probability. For a network with multiple transmitter–receiver pairs, the fundamental limit is usually characterized by the per-node capacity, which is the maximum rate that is achievable for all transmitter-receiver

[1]College of Information Science and Electronic Engineering, Zhejiang University, Hangzhou, China
[2]Department of ECE, The Hong Kong University of Science and Technology, Hong Kong, China
[3]Department of Telecommunication Systems, Technical University of Berlin, Berlin, Germany

pairs. Unfortunately, the fundamental limit of *wireless D2D caching networks* and the associated optimal caching scheme remains an open problem. In this chapter, we will provide a partial solution to this open problem.

11.1.1 Related work

11.1.1.1 Capacity scaling law in wireless ad hoc networks

It is extremely hard to characterize the exact per-node capacity of general wireless networks. For large wireless networks, scaling laws* provide a useful way to characterize the behavior of the capacity order. The capacity scaling law of wireless ad hoc networks was first studied by Gupta and Kumar in the seminal paper [8], where they showed that in a large wireless ad hoc network with n randomly located nodes, the per-node throughput of the classical multihop scheme scales at most as $\Theta(n^{-\frac{1}{2}})$ under a *protocol model*. In the protocol model, the transmission between a transmitter and receiver pair is successful if and only if the distance from any other interfering transmitter is $(1+\Delta)$ times larger than the distance between this transmitter and receiver pair, where $\Delta > 0$ models situations where a guard zone is specified by the protocol to prevent a neighboring node from transmitting at the same time. Since then, a number of works [9–11] have studied the information theoretic capacity scaling law under a more realistic *physical model* that includes distance-dependent propagation path-loss, fading, Gaussian noise, and signal interference. In this case, the capacity scaling law depends on whether the network is "extended" (constant node density, with the network area growing as $\Theta(n)$), or "dense" (constant network area, with the node density growing as $\Theta(n)$). Specifically, it was shown in [12] that under a physical model with path loss exponent $\alpha \geq 2$, the per-node capacity of a *dense network* scales as $\Theta(1)$ and that of an *extended network* scales as $\Theta\left(n^{1-\frac{\min(3,\alpha)}{2}}\right)$, both of which are orders better than the $\Theta(n^{-\frac{1}{2}})$ scaling law achieved by the classical multihop scheme. Moreover, this capacity scaling is achieved by *hierarchical cooperation*, with the number of hierarchical stages going to infinity. In [13], the authors studied the capacity scaling in ad hoc networks with arbitrary node placement, and the capacity regions of ad hoc networks with the more complicated unicast or multicast traffic model were studied in [14].

Note that the results in [12–14] depended heavily on the physical channel model, which assumes independent fading coefficients between different nodes. In contrast, the authors in [15] showed that the total network capacity of a wireless network with area $\mathscr{A}$ is fundamentally limited by $\Theta\left(\frac{\sqrt{\mathscr{A}}}{\lambda}\right)$ using Maxwell's equations, where λ is the wavelength of the operating carrier frequency. The results in [15] imply that for practical dense networks, the assumption of independent fading coefficients may only be valid when $n \leq \frac{\sqrt{\mathscr{A}}}{\lambda} = \Theta\left(\frac{1}{\lambda}\right)$. Since $\Theta\left(\frac{1}{\lambda}\right)$ is usually

*Scaling law order notation: given two functions f and g, we say that: (1) $f(n) = O(g(n))$ if there exists a constant C and integer N such that $f(n) \leq cg(n)$ for $n > N$. (2) $f(n) = \Omega(g(n))$ if $g(n) = O(f(n))$. (3) $f(n) = \Theta(g(n))$ if $f(n) = O(g(n))$ and $g(n) = O(f(n))$.

not large enough to be considered as an asymptotic regime, the scaling law for dense networks is less interesting in practice, as pointed out in [16]. For extended networks, $\mathscr{A}$ scales linearly with n, and thus the scaling law analysis is more relevant in practice. Therefore, in this chapter, we will only study scaling laws for extended networks. For clarity, we will assume rich scattering and focus on the case with independent fading coefficients. However, we will also discuss the extension of the scaling law results to the case when the assumption of independent fading coefficients is invalid and $\frac{\sqrt{\mathscr{A}}}{\lambda}$ becomes the limiting factor of the capacity.

11.1.1.2 Capacity scaling law in D2D caching networks

D2D caching combines caching of files on the user devices with short-range D2D communications [4–7]. In this way, the caches of multiple devices form a common virtual cache that can store a large number of files, even if the cache on each separate device is not necessarily very large. In particular, as the number of users in the network grows, also their aggregate cache capacity grows accordingly. This is the key property that allows for significant gains of D2D caching with respect to the other caching methods where the content is only stored in the network infrastructure (e.g., in the serving cellular base station [17, 18] or in the helper nodes in Femtocaching [1]).

A number of works have studied the fundamental limit of D2D caching networks under different network configurations. Gitzenis *et al.* [19] studied the joint optimization of cache content replication and routing in a *regular network* and identified the throughput scaling laws for various regimes. Single-hop D2D caching networks, where the content delivery scheme is restricted to single-hop transmission, were considered in [4, 5]. Under a Zipf popularity distribution [20] with skewness parameter less than one, and the protocol model, it was shown in [4, 5] that the per-node capacity scales as $\Theta\left(L_C/L\right)$, where L_C is the number of files that each node can cache (cache capacity in the unit of file size) and L is the total number of files in the content library. Multihop D2D caching networks with the protocol model were considered in [7]. By allowing multihop transmission, the per-node capacity scales as $\Theta\left(\sqrt{L_C/L}\right)$ when the popularity distribution has the "heavy tail" property, which is much better than the single-hop case.

In [2], the authors studied a different caching network topology, where a single transmitter serves n user nodes through a common noiseless link of fixed capacity (bottleneck link). Coded caching schemes were proposed for this scenario to create coded multicast gain. Specifically, in the cache initialization phase, each file is partitioned into packets and each node stores subsets of packets from each file. In the content delivery phase, the BS can compute a multicast network-coded message (transmitted via the common link) such that each node can decode its own requested file from the multicast message and its cached file packets (side information). Under the worst-case arbitrary demands model, the per-node throughput scaling is again given by $\Theta\left(L_C/L\right)$, which is the same scaling law as achieved by single-hop D2D caching networks. A number of extensions under different user demands and network structures can be found in [21–23].

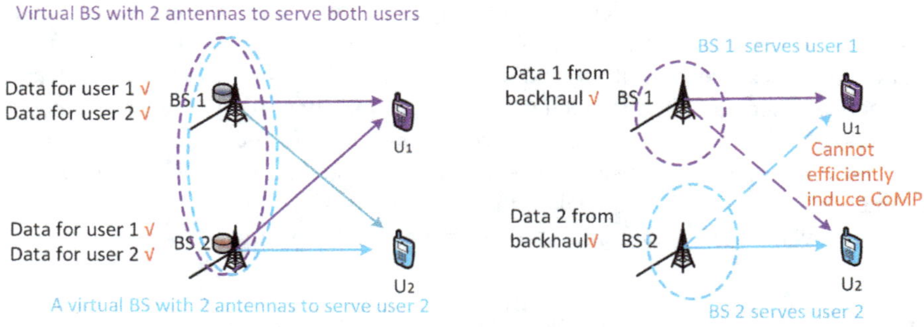

(a) When the data requested by the 2 users lies in the caches of both BSs, the RAN is a MIMO BC and CoMP is used to achieve a larger rate region.

(b) Without cache, the two BSs cannot induce CoMP if they do not fetch additional data from the backhaul.

Figure 11.1 *An illustration of PHY caching for a radio access network with two BS-user pairs. Each BS/user is equipped with a single antenna.*

11.1.1.3 Physical layer caching

A key feature of wireless networks is that interference can be handled at the physical layer (PHY) beyond the simple exclusion principle built into the previously mentioned protocol model. In particular, caching can also be exploited to mitigate interference and enable cooperative transmission at the PHY. For example, in cellular networks, when the user requested data exist in the BS cache (cache hits), they induce *dynamic side information* to the BSs, which can be further exploited to enhance the capacity of the radio interface. The concept of *cache-induced opportunistic MIMO cooperation*, or *PHY caching*, was first introduced in [17, 18] to achieve significant spectral efficiency gain without consuming BS backhaul. Figure 11.1 illustrates the key idea of PHY caching for a radio access network (RAN) with two BS-user pairs. When the data requested by the two users lie in the caches of both BSs, the RAN is a MIMO broadcast channel (BC) with the two BSs forming a virtual two-antenna transmitter, and Coordinated Multi-Point (CoMP) is employed to achieve a larger rate region, without consuming any backhaul resource. On the other hand, if there is no cache at the BSs and each BS only fetches the data for its serving user, the two BSs cannot induce CoMP and a smaller rate region is achieved. Since then, there have been many works on PHY caching, and they can be classified into two major classes, as discussed below.

High-SNR and fixed-size network regimes: These works focus on the degrees of freedom (DoF), i.e., the coefficient of the $O(\log \mathrm{SNR})$ leading term of the network sum capacity as SNR grows, but the network has a fixed number of nodes. For example, [24, 25] studied the average sum DoF (averaged over the user demands) for relay and interference channels with BS caching, respectively, under some achievable scheme. On the other hand, [26] studied the max-min sum DoF (i.e., maximizing the worst-case sum DoF over the user demands) of one-hop interference networks with caching at both the transmitters and receivers, and the impact of caching on the DoF of a Gaussian vector BC with delayed channel state information at the transmitter (CSIT) was also investigated in [27].

Large network and fixed SNR regimes: These works focus on studying the capacity/throughput scaling laws as the number of nodes grows, but with fixed SNR. In [28], PHY caching was used to exploit both the *cache-induced MIMO cooperation gain* and the *cache-assisted multihopping gain* (i.e., reducing the number of hops from the source to the destination) in backhaul-limited multihop wireless networks, and the throughput scaling laws achievable by PHY caching were identified for extended networks under Zipf popularity distributions (see also [3]). It was shown in [3, 28] that by exploiting the cache-induced MIMO cooperation, PHY caching can achieve significant throughput gain over conventional caching, which purely exploits cache-assisted multihopping gain. However, exploiting the cache-induced MIMO cooperation does not provide order gain in terms of throughput scaling laws.

11.1.2 Summary of the main results

As discussed above, capacity scaling laws have been obtained under the protocol model for one-hop and multihop wireless caching networks. For multihop wireless caching networks under the physical model, *achievable* scaling laws have been obtained and cache-induced MIMO cooperation has been shown to provide gains in terms of throughput (but not in terms of scaling laws). However, the question of the capacity scaling laws of wireless caching networks under the physical model has been unanswered so far. In this chapter, we provide an answer to this open question under the assumption of Zipf popularity distribution. Table 11.1 summarizes the existing capacity scaling law results for wireless D2D networks with and without

Table 11.1 Summary of the per-node capacity scaling laws *for wireless D2D networks with and without caching, where "H-Coop." stands for hierarchical cooperation. The new result of this chapter is highlighted with a star symbol.*

		Protocol model	**Physical model extended network**
Without cache	Capacity scaling	$\Theta\left(n^{-\frac{1}{2}}\right)$ [8]	$\Theta\left(n^{1-\frac{\min(3,\alpha)}{2}}\right)$ [12]
	Achievable scheme	Multihop	H-Coop. [12]
With cache (heavy tail popularity)	Capacity scaling	$\Theta\left(\sqrt{L_C/L}\right)$ [7]	$*\Theta\left((L_C/L)^{\frac{\min(3,\alpha)}{2}-1}\right)$
	Achievable scheme	Random caching + Multihop [7]	*Cache-induced H-Coop.
With cache (general popularity)	Capacity scaling	Unknown	*Known for Zipf
	Achievable scheme	Unknown	*Known for Zipf

caching, as well as the scaling law results from our work (highlighted with a star symbol).

In this chapter, we address the fundamental capacity scaling in extended wireless D2D caching networks under the physical model, and propose an associated order-optimal caching and content delivery scheme. With respect to the previous work on cache-induced MIMO cooperation, we shall design a more advanced cooperation scheme that can achieve an order gain in the throughput scaling law. As explained in Section 11.1.1.1, the capacity scaling law is less interesting for dense networks, and thus we will only focus on extended networks for the scaling law analysis. While scaling law analysis yields clean results, it cannot accurately reflect how a large network with a finite number of nodes really performs in terms of throughput. For example, as shown in the simulations in [16], the original hierarchical cooperation scheme in [12] performs even worse than the multihop scheme for networks of practical size. Therefore, in this chapter, we will also analyze the throughput of the proposed caching and content delivery scheme to verify its performance gain for such networks. The main new contributions of the chapter are summarized below.

- **Cache-induced hierarchical cooperation:** In this chapter, we combine the ideas of PHY caching (cache-induced MIMO cooperation) and hierarchical cooperation, and propose a novel caching and content delivery scheme called cache-induced hierarchical cooperation, which can achieve both a higher scaling law in extended networks and huge throughput gain in networks of practical size.
- **Cache content placement optimization:** We propose a low complexity cache content placement algorithm to optimize the parameters of the cache-induced hierarchical cooperation scheme and establish the order optimality of the proposed algorithm.
- **Throughput analysis:** We analyze the throughput performance of the proposed cache-induced hierarchical cooperation and show that it can achieve significant throughput gain over conventional caching, which purely exploits cache-assisted multihopping gain.
- **Capacity scaling laws in extended wireless D2D caching networks:** For the extended network model under a Zipf popularity distribution, we derive both the achievable throughput scaling laws of the proposed cache-induced hierarchical cooperation and an information-theoretic upper bound of the throughput scaling law. The scaling laws of the achievability and converse coincide, so that we can establish the capacity scaling law for the Zipf popularity distribution with the general skewness parameter τ. For the case of a "heavy tail" Zipf popularity distribution (i.e., the skewness parameter $\tau \leq 1$), the per-node capacity scales as $\Theta\left(\left(L_C/L\right)^{\frac{\min(3,\alpha)}{2}-1}\right)$. When $\alpha < 3$ and $L_C/L \ll 1$, this per-node capacity scaling law is much better than the $\Theta\left(\left(L_C/L\right)^{1/2}\right)$ per-node capacity scaling law of the cache-assisted multihop scheme under both the protocol model [7] and physical model [3].

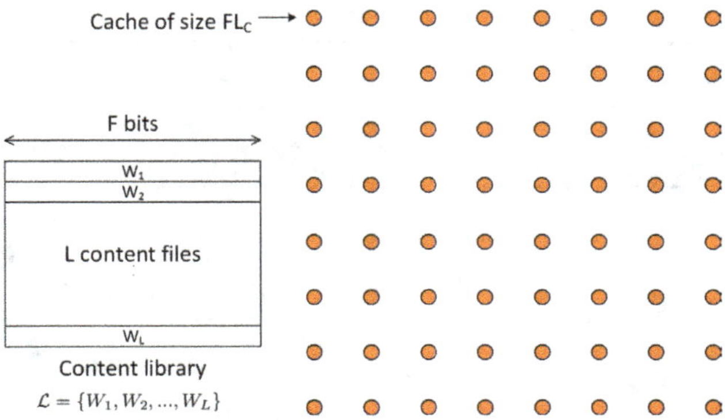

Figure 11.2 An illustration of a wireless D2D caching network

11.1.3 Organization

In Section 11.2, we introduce the architecture of wireless D2D caching networks and the channel model. In Section 11.3, we discuss some preliminary results on the improved hierarchical cooperation scheme in [16] and the classical multihop scheme, which are designed for wireless ad hoc/D2D networks without caching. In Sections 11.4 and 11.5, we describe the proposed cache-induced hierarchical cooperation scheme and the associated cache content placement optimization algorithm, respectively. The throughput performance of the proposed scheme is analyzed and compared in Section 11.6. The achievable scaling law of the cache-induced hierarchical cooperation and the converse proof are given in Section 11.7 for extended networks. The limitations and possible extensions of the scaling laws in this chapter are discussed in Section 11.8. Finally, the conclusion is given in Section 11.9.

11.2 System model

11.2.1 Wireless D2D caching networks

Consider a wireless D2D caching network with n nodes placed on a regular grid of area $A(n)$, as illustrated in Figure 11.2. For clarity, we focus on networks with $n = 4^M$ nodes, where M is some positive integer, and let $V(n)$ denote the set of all nodes in the network. The scaling law results can be easily generalized to the case when $n \neq 4^M$ as explained in Section 11.8.1.

In a wireless D2D caching network, the nodes request data (e.g., music or video) from a content library $\mathscr{L} = \{W_1, W_2, ..., W_L\}$ of $L = |\mathscr{L}|$ files (information messages), where W_l are drawn at random and independently with a uniform distribution over a message set $\mathbb{F}_2^F$ (binary strings of length F). Each node has a cache of size FL_C bits, which can be used to store a portion of the content files to serve the requests generated by the nodes in the network. We assume that L_C to avoid the trivial case when every node has enough cache capacity to store the whole content library $\mathscr{L}$. Furthermore, we assume $nL_C \geq L$, which is necessary and sufficient for the existence

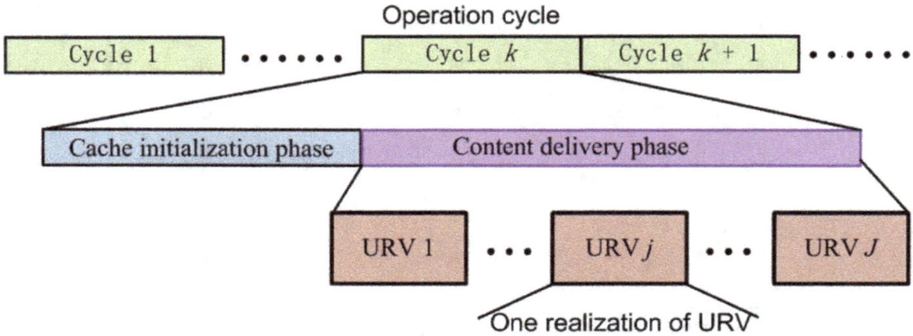

Figure 11.3 Time line of the wireless D2D caching network

of a caching policy such that there is at least one complete copy of each content file in the caches of the entire network.†

In this chapter, time is divided into operation cycles, where each operation cycle consists of one *cache initialization phase* followed by one *content delivery phase*, and each content delivery phase consists of many realizations of user requests, as illustrated in Figure 11.3. Without loss of generality, we focus on a single operation cycle.

In the cache initialization phase, each node caches a portion of the (possibly encoded) content files. In general, the *caching scheme* is defined as a collection of n mappings $\mathscr{B}_i : \mathbb{F}_2^{FL} \rightarrow \mathbb{F}_2^{FL_C}, i = 1, ..., n$ from the content library $\mathscr{L}$ to the content $B_i = \mathscr{B}_i(\mathscr{L})$ cached at node i. Since the popularity of content files changes very slowly (e.g., new movies are usually posted on a weekly or monthly timescale), the cache update overhead in the cache initialization phase is usually small. This is a reasonable assumption widely used in the literature [2–5, 19, 28].

The above general caching scheme includes many caching schemes as special cases. Some examples of caching schemes are given below.

- **Binary caching:** In the binary caching scheme, each node either caches a file entirely or does not cache it at all. In this case, each node will simply choose $\lceil L_C \rceil$ files from the content library to cache. Let $\mathscr{I}_i$ denote the index set of the files cached at node i. Then the mapping $\mathscr{B}_i$ is given by $B_i = \cup_{l \in \mathscr{I}_i} W_l$, and $\mathscr{I}_i$'s must satisfy the cache size constraint: $|\mathscr{I}_i| \leq L_C$.
- **Fractional caching:** In the fractional caching scheme, each node can cache a portion of each file. Let $B_{l,i}$ denote the portion of file l cached at node i. In this case, the mapping $\mathscr{B}_i$ is given by $B_i = \cup_{l \in \{1,...,L\}} B_{l,i}$, and $B_{l,i}$'s must satisfy the cache size constraint: $\sum_{l=1}^{L} |B_{l,i}| \leq FL_C$.

†The necessity is obvious: if the condition $nL_c \geq L$ is not satisfied, the library cannot be cached entirely and therefore some segment of some file is missing, since files are noncompressible i.i.d. bits. The sufficiency follows from the explicit construction of the proposed hierarchical cache content placement scheme in Section 11.4.

- **MDS-coded caching:** In MDS-coded caching scheme, each file is encoded using an ideal MDS rateless code. An MDS rateless code generates an arbitrarily long sequence of parity bits from a content file of F bits, such that if the decoder obtains any F parity bits, it can recover the original F information bits [29]. With MDS coding, each node only needs to determine how many parity bits should be cached for each file without having to specify which parity bits are cached. Such a good property provides more flexibility for the design of content delivery scheme.

In the content delivery phase, time is divided into time slots and each node independently requests the l-th content file with probability p_l, where the probability mass function $\mathbf{p} = [p_1, ..., p_L]$ represents the popularity of the content files. Without loss of generality, we assume $p_1 \geq p_2 \cdots \geq p_L$. Each node requests files one after another. When a requested file is delivered to node i, node i will request the next file immediately according to the popularity distribution $\mathbf{p}$.

Let $l_i(t)$ denote the content file requested by node i at time slot t, and let $\boldsymbol{l}(t) = [l_1(t), ..., l_n(t)]^T$ denote the user request vector (URV). Let t_i^j denote the time slot when $l_i(t)$ changes for the j-th time. In other words, node i starts to request file $l_i(t_i^j)$ at time slot t_i^j, and the delivery of file $l_i(t_i^j)$ to node i is finished at time slot $t_i^{j+1} - 1$. If the content $\mathscr{B}_i(\mathscr{L})$ cached at node i is sufficient to decode the requested content file $l_i(t_i^j)$, node i can obtain the requested file $l_i(t_i^j)$ immediately. Otherwise, node i has to obtain more information about the content file $l_i(t_i^j)$ from the other nodes in the network. Specifically, at time slot $t \in [t_i^j, ..., t_i^{j+1} - 1]$, each node $i' \neq i$ generates an *information message* $U_{i,i'}(t) = \mathscr{U}_{i,i'}(B_{i'}, t)$ for node i using a content delivery encoder $\mathscr{U}_{i,i'}(\cdot, t) : \mathbb{F}_2^{FLC} \rightarrow \mathbb{F}_2^{|U_{i,i'}(t)|}$.[‡] Let $U_i^j = \cup_{i' \neq i} \cup_{t \in [t_i^j, ..., t_i^{j+1} - 1]} U_{i,i'}(t)$ denote the aggregate information message for the j-th request of node i. The content delivery scheme treats the aggregate information messages of different users as independent messages and delivers each aggregate information message to the desired node. To be more specific, the content delivery scheme ensures that node i can successfully receive the aggregate information message U_i^j within the time window $[t_i^j, ..., t_i^{j+1} - 1]$ for any i and j. Note that t_i^j is a random variable depending on the specific content delivery scheme, the random URV process $\boldsymbol{l}(t)$, and other underlying random processes in the network, such as the fading channel and noise. When node i obtains U_i^j at time $t_i^{j+1} - 1$,

[‡]Note that $U_{i,i'}(t)$ can be empty, i.e., node i' does not generate any information message for node i at time slot t.

node i will apply a decoding function $\hat{W}_{l_i} = \phi_i^j\left(U_i^j, B_i\right)$ to obtain the estimated file $\hat{W}_{l_i}$, where $\phi_i^j : \mathbb{F}_2^{\left|U_i^j\right|} \times \mathbb{F}_2^{FL_C} \to \mathbb{F}_2^F$. A content delivery scheme is *feasible* if

$$\lim_{F\to\infty} \Pr\left[\phi_i^j\left(U_i^j, B_i\right)_{l_i}\right] = 0, \forall i, j. \tag{11.1}$$

Fano's inequality implies that a necessary condition for a content delivery scheme to be feasible is

$$H\left(W_{l_i}|U_i^j, B_i\right) \leq \varepsilon_F F, \tag{11.2}$$

where ε_F is a vanishing quantity as $F \to \infty$.

Similar to [3, 19], we assume a symmetric traffic model where all users have the same *average throughput requirement* R (averaged over all possible realizations of user requests). To be more specific, the average data rate of node i is defined as $R_i = \frac{F}{\overline{T}_i}$, where

$$\overline{T}_i = \lim_{J\to\infty} \frac{1}{J} \sum_{j=1}^{J} \mathbb{E}\left[t_i^{j+1} - t_i^j\right] = \lim_{J\to\infty} \frac{1}{J} \mathbb{E}\left[t_i^{J+1} - t_i^1\right] \tag{11.3}$$

is the average delivery time of one file to node i. A *symmetric per-node throughput* R is achievable if there exists a feasible caching and content delivery scheme with $F \to \infty$, such that $R_i \geq R, \forall i$.

11.2.2 Wireless channel model

We use a similar channel model to that in [12, 14]. The channel coefficient between a transmitter node j and a receiver node i is

$$h_{i,j} = \left(r_{i,j}\right)^{-\alpha/2} \exp\left(\sqrt{-1}\theta_{i,j}\right), \tag{11.4}$$

where $r_{i,j}$ is the distance between node j and i, $\theta_{i,j}$ is the random phase with uniform distribution on $(0, 2\pi]$, and $\alpha > 2$ is the path loss exponent. At each node, the received signal is also corrupted by a circularly symmetric Gaussian noise with zero mean and unit variance.

11.3 Preliminaries on hierarchical cooperation and classical multihop schemes

In the proposed cache-induced hierarchical cooperation scheme in Section 11.4, there are two PHY transmission modes, namely, the *hierarchical cooperation* mode and *multihop* mode. These two PHY transmission modes are based on the hierarchical cooperation scheme in [16] and the classical multihop scheme, respectively. The original hierarchical cooperation and multihop schemes are designed for wireless ad hoc/D2D networks without caching. In this case, the per-node throughput R depends on the *traffic pattern*, i.e., the number of source-destination pairs and the locations of each source-destination pair. In [16], the throughput performance of the hierarchical cooperation and multihop schemes are analyzed and compared for a wireless D2D network with n nodes under *uniform permutation traffic*, where the network consists of n source-destination pairs with the same throughput requirement R, such that each node is both a source and a destination, and pairs are selected at random over the set of n-permutations π that do not fix any elements (i.e., for which $\pi(i) \neq i$ for all $i = 1, ..., n$). In this section, we review some preliminary results from [16], which will be useful in the later sections.

11.3.1 Throughput performance of hierarchical cooperation under uniform permutation traffic

We first describe the hierarchical cooperation scheme for wireless D2D networks with area $A(n) = 1$. The hierarchical cooperation is based on a three-phase cooperative transmission scheme. The basic hierarchical cooperation scheme was first proposed in [12]. In such a scheme, the network is first divided into n/N clusters of N nodes each and then the following three phases are used to achieve cooperation gain.

- **Phase 1 (Information Dissemination):** Each source distributes N distinct subpackets of its message to the N neighboring nodes in the same cluster. One transmission is active per each cluster, in a round robin fashion, and clusters are active simultaneously in order to achieve some spatial spectrum reuse. The inter-cluster interference is controlled by a TDMA scheme with the reuse factor T_r, i.e., each cluster has one transmission opportunity every T_r^2 time slots (see Figure 11.4).
- **Phase 2 (Long-Range MIMO Transmission):** One cluster at a time is active, and when a cluster is active it operates as a single N-antenna MIMO transmitter, sending N independently encoded data streams to a destination cluster. Each node in the cooperative receiving cluster stores its own received signal.
- **Phase 3 (Cooperative Reception):** All receivers in each cluster share their own received and quantized signals so that each destination in the cluster decodes its intended message on the basis of the (quantized) N-dimensional observation. Each destination performs joint typical decoding to obtain its own desired message based on the quantized signals.

The basic hierarchical cooperation scheme employs the above three-phase cooperative transmission scheme as a recursive building block applied for local communication of a higher stage, i.e., at a larger space scale in the network, as illustrated in Figure 11.5 for the case of two stages. This scheme was improved in [16, 30]. Specifically [30] proposed an improvement where the local communication phase is formulated as a network multiple access, problem instead of being decomposed into a number of unicast network problems. Hong and Caire [16] further improved the throughput performance by using more efficient TDMA scheduling. The differences between the TDMA schemes in the conventional scheme in [12] and the improved scheme in [16] are illustrated in Figures 11.5 and 11.6. In this chapter, we will use

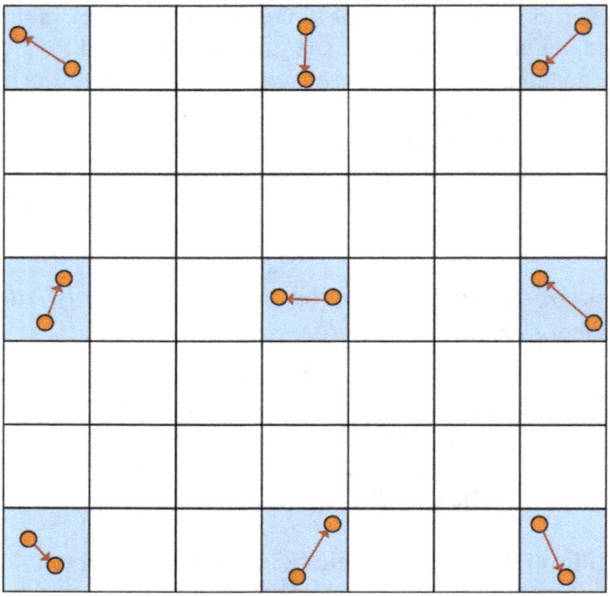

Figure 11.4 Grouping of interfering clusters in the TDMA scheme with reuse factor L = 3. Each square represents a cluster and the gray squares represent the concurrent transmitting clusters.

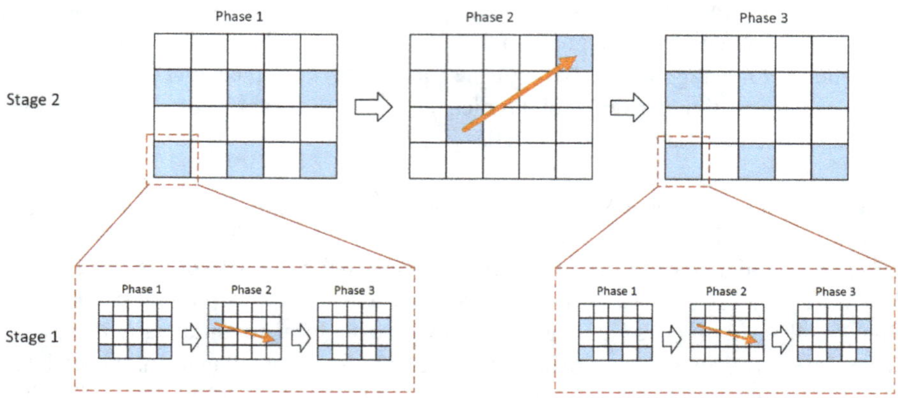

Figure 11.5 The silent features of the conventional scheme in [12] when 2-stage hierarchical cooperation architecture is applied, where the gray squares represent the concurrent transmitting clusters

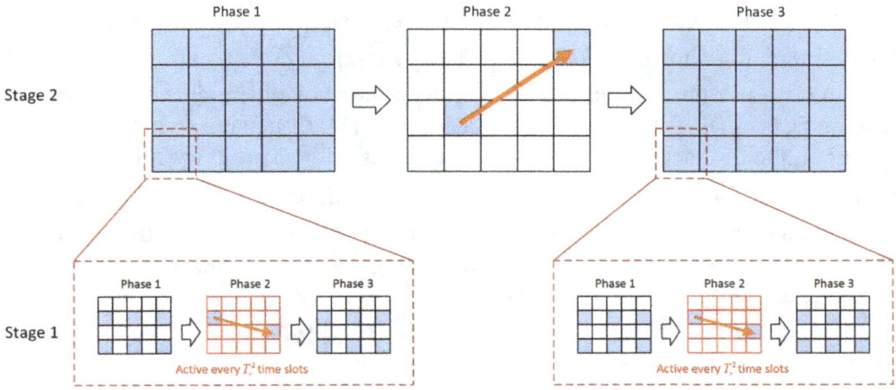

Figure 11.6 *The silent features of the improved TDMA schemes when 2-stage hierarchical cooperation architecture is applied, where the gray squares represent the concurrent transmitting clusters*

the hierarchical cooperation "method 4" from [16], with both improvements, as a building block for the proposed cache-induced hierarchical cooperation. For convenience, we will call "method 4" from [16] the *improved hierarchical cooperation* scheme, and its throughput performance is summarized in the following theorem.

Theorem 1 (per-node throughput of hierarchical cooperation): Consider a wireless D2D network of size n and area $A(n) = 1$ under uniform permutation traffic. Suppose each node has a sufficiently large transmit power with a uniform bound P_{max} that does not scale with n. The improved hierarchical cooperation scheme with s stages achieves a per-node throughput of

$$\mathrm{R}_H^{(s)}(n, P_I) = \begin{cases} \log\left(1 + \frac{\mathrm{SNR}}{1+P_I}\right) \frac{n^{\frac{-1}{2}}}{2\sqrt{2Tr}} & s = 1 \\ R_c(\alpha, P_I) \frac{n^{\frac{-1}{s+1}}}{(1+s)T_r^{\frac{2s}{s+1}}(3\cdot 2^s - 1)^{\frac{s}{2(s+1)}}}, & s \geq 2, \end{cases} \tag{11.5}$$

where

$$\begin{aligned} \mathrm{SNR} &= 2^{2(3+\alpha/2)} \\ T_r &= \left\lceil \sqrt{\mathrm{SNR}^{1/\alpha}} + 1 \right\rceil \\ P_I &= \sum_{i=1}^{\sqrt{n}} 8iSNR(T_r i - 1)^{-\alpha}. \end{aligned} \tag{11.6}$$

and $R_c(\alpha, P_I)$ is called the common message coding rate which is given in Appendix A.

Note that $R_c(\alpha, P_I) \leq \log\left(1 + \frac{\mathrm{SNR}}{1+P_I}\right)$, and refer to Appendix A for more details about the improved hierarchical cooperation scheme in [16] and how to determine the exact value of the common message coding rate $R_c(\alpha, P_I)$. Let $s_n^\star = \mathrm{argmax}_s R_H^{(s)}(n, P_I)$ denote the optimal number of stages. There is no closed form for $s_n^\star$. However, $s_n^\star$ can be easily found by a simple one-dimensional search. Furthermore, it is shown in [31] that $s_n^\star = \Theta\left(\sqrt{\ln n}\right)$.

Now we use the method in [12] to extend the above hierarchical cooperation scheme from $A(n) = 1$ to an arbitrary $A(n) \geq \Theta(1)$. Compared to networks with $A(n) = 1$, the distance between nodes in networks with an arbitrary $A(n)$ is scaled by a factor of $\sqrt{A(n)}$, and hence for the same transmit powers, the received powers are all scaled by a factor of $A(n)^{-\alpha/2}$. The hierarchical scheme for fixed peak power per node ($O(1)$ power per node) yields an *average* power per node of $O(1/n)$ since nodes are active only a fraction of $O(1/n)$ the time [12]. For a network with arbitrary $A(n)$, we need to scale the peak power up by a factor $A(n)^{\alpha/2}$ in order to compensate for the path loss. Imposing an average power per node $O(1)$, this yields that we can operate the network under the hierarchical cooperation scheme for a fraction of time $\min\left(nA(n)^{-\alpha/2}, 1\right)$. In this way, the hierarchical cooperation scheme for arbitrary $A(n)$ can achieve a per-node throughput of

$$\widetilde{R}_H^{(s)}(n, P_I) = R_H^{(s)}(n, P_I)\min\left(nA(n)^{-\alpha/2}, 1\right). \tag{11.7}$$

11.3.2 *Throughput performance of multihop scheme under uniform permutation traffic*

The multihop scheme is a classical communication architecture that has been widely used in practice. In this scheme, for a given source-destination pair, a routing path is first formed from the source to the destination. Then, on each routing path, packets are relayed from node to node. On each link of the routing path, each packet is fully decoded using conventional single-user decoding with all interference treated as noise.

In [16], the performance of the multihop scheme is compared with that of the hierarchical cooperation scheme for dense wireless D2D networks, under the following assumptions. The routing between each source-destination pair is to first proceed horizontally and then vertically in the network grid, as illustrated in Figure 11.7. Distance-dependent power control is applied and the interference is controlled by the reuse factor T_r, chosen to enforce the optimality condition of TIN as $T_r = \left\lceil \sqrt{\text{SNR}}^{1/\alpha} + 1 \right\rceil$. Under these assumptions, the per-node throughput for uniform permutation traffic is given by [16]

$$R_M(n, P_1) = \log\left(1 + \frac{\text{SNR}}{1+P_1}\right)\frac{n-\frac{1}{2}}{\left\lceil \sqrt{\text{SNR}}^{1/\alpha} + 1 \right\rceil^2}, \tag{11.8}$$

$$\text{SNR} = 2^{2(3+\alpha/\ln 4)}. \tag{11.9}$$

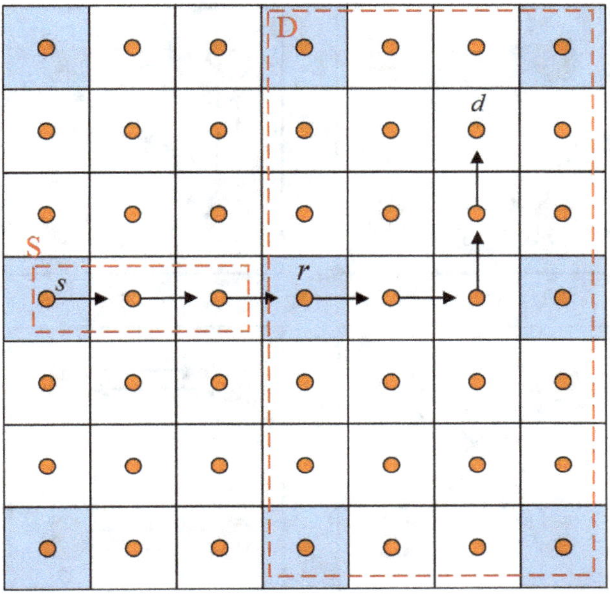

Figure 11.7 *Multihop routing over the grid network. Red arrow-lines represent a routing path from the source* s *to the destination* d.

11.3.3 Extension to per cluster uniform permutation traffic

In the proposed cache-induced hierarchical cooperation scheme, the traffic induced by the requests of all nodes will be grouped into *per cluster uniform permutation traffic* over clusters (subnetworks) of different sizes. The above hierarchical cooperation scheme or multihop scheme can be used to handle per cluster uniform permutation traffic over the n/N non-overlapping clusters with the same cluster size N, where there is uniform permutation traffic within each cluster but there is no traffic among clusters, as illustrated in Figure 11.8. Specifically, each cluster of size N simultaneously employs the hierarchical cooperation scheme or multihop scheme to serve the uniform permutation traffic within the cluster. Note that there is no need to apply TDMA among clusters to control the inter-cluster interference because the TDMA scheme within each cluster with reuse factor T_r already guarantees that the received power of the interference is upper bounded by $P_I = \sum_{i=1}^{\sqrt{n}} 8i\mathrm{SNR}\left(T_r i - 1\right)^{-\alpha}$. A similar idea is also used in [16] to improve the TDMA scheduling for hierarchical cooperation. The choice of hierarchical cooperation or multihop scheme depends on which will achieve higher throughput. If $\widetilde{R}_H^{(s_N^\star)}\left(N, P_I\right)_M\left(N, P_I\right)$, we will use the hierarchical cooperation scheme; otherwise, we will use the multihop scheme. In this case, the achievable per-node throughput is given by

$$R_u\left(N\right) = \max\left(\widetilde{R}_H^{(s_N^\star)}\left(N, P_I\right), R_M\left(N, P_I\right)\right). \tag{11.10}$$

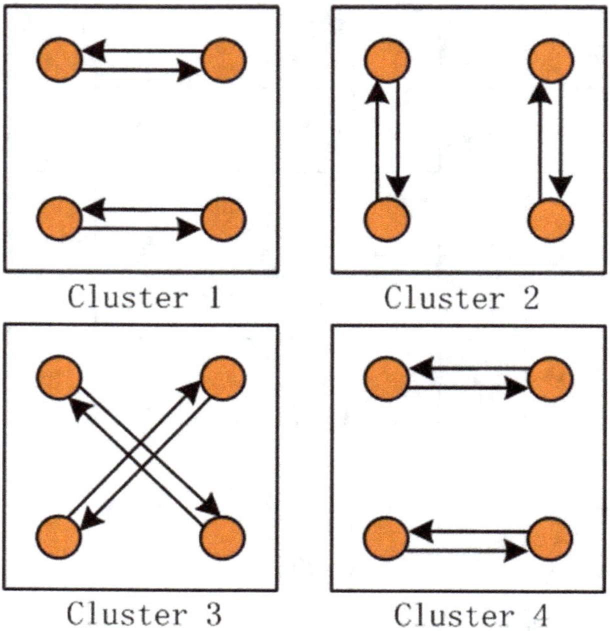

Figure 11.8 *An illustration of per cluster uniform permutation traffic over clusters of size* N = *4 in a network of size* n = *16*

11.4 Cache-induced hierarchical cooperation

In this section, we elaborate the proposed achievable scheme, called cache-induced hierarchical cooperation, which works for both dense and extended wireless D2D caching networks.

11.4.1 Key components of the cache-induced hierarchical cooperation scheme

The components of the proposed cache-induced hierarchical cooperation scheme and their inter-relationship are illustrated in Figure 11.9. There are two major components: the *hierarchical cache content placement*, working in the cache initiation phase, and the *tree-graph-based content delivery*, working in the content delivery phase. The hierarchical cache content placement decides how to distribute the content files into caches of different nodes (or mathematically decides the n mappings $\mathcal{B}_i, \forall i$). The tree-graph-based content delivery exploits the cached content at each node to serve the user requests, and it consists of four layers: the source determination layer, routing layer, cooperation layer, and PHY. In this content delivery scheme, the original network is abstracted as a tree graph and the source determination and routing are based on this graph. Specifically, each node is a leaf node in the tree graph and the set of source nodes for a leaf node is an internal node in the tree.

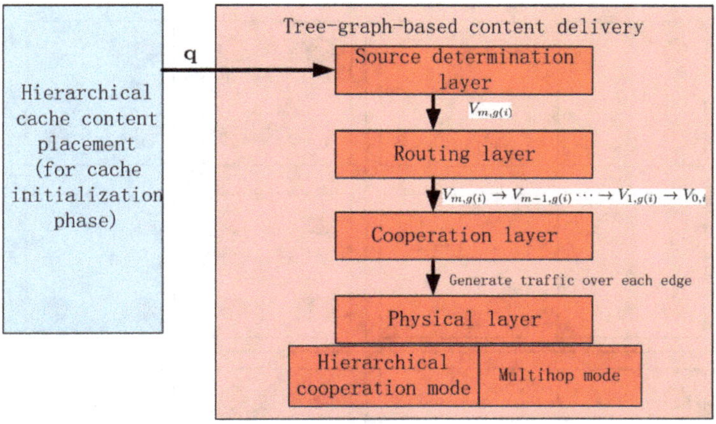

Figure 11.9 Components of the cache-induced hierarchical cooperation and their inter-relationship

The routing layer routes messages between the source nodes and destination node. The cooperation layer provides this tree abstraction to the routing layer by appropriately concentrating traffic over the network. Finally, the PHY implements this concentration of messages in the wireless network based on two PHY transmission modes, namely, *hierarchical cooperation* mode and multihop mode. The details of the components are elaborated in the following subsections.

11.4.2 Hierarchical cache content placement

In the proposed hierarchical cache content placement, nodes are partitioned into clusters of different levels. In the m-th level, $A(n)$ is partitioned into 4^{M-m} squares of equal size, as illustrated in Figure 11.10. Then the 4^m nodes in the same square form a cluster in the m-th level. Let $V_{m,i} \subseteq V(n)$ be the i-th cluster in the m-th level for $i \in \{1, ..., 4^{M-m}\}$. In the hierarchical cache content placement, all nodes cache the same number of $q_l F$ bits for the l-th file. Moreover, q_l can only take values from the discrete set $\left\{0, \frac{1}{4^M}, \frac{1}{4^{M-1}}, ..., \frac{1}{4}, 1\right\}$. If $q_l = \frac{1}{4^m}$, during the cache initiation phase, the l-th file will be equally distributed over the nodes in $V_{m,i}$ for any $i \in \{1, ..., 4^{M-m}\}$. In other words, each node in $V_{m,i}$ caches a portion of the $4^{-m}F$ bits for file l such that the l-th file can be reconstructed by collecting all portions from the caches of nodes in $V_{m,i}$. For convenience, we say that file l is cached at the m-th level if $q_l = \frac{1}{4^m}$. Such hierarchical cache content placement with parameter $q_l \in \left\{0, \frac{1}{4^M}, \frac{1}{4^{M-1}}, ..., \frac{1}{4}, 1\right\}$ can also be specified by another set of parameters $\mathbf{x} = \left[x_0, x_1, ..., x_M\right] \in \mathbb{Z}_+^{M+1}$, where

$$x_m = \sum_{l=1}^{L} 1\left(q_l = 4^{-m}\right) \tag{11.11}$$

is the number of files cached at the m-th level, and $1(\cdot)$ is the indication function. Note that $\mathbf{x}$ must satisfy the constraint $\sum_{m=0}^{M} x_m = L$ so that there is at least one

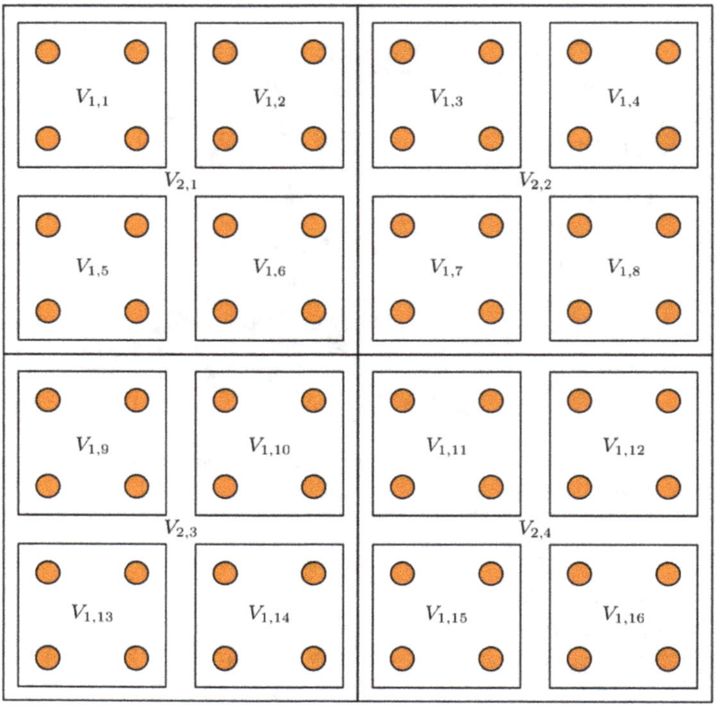

Figure 11.10 Illustration of clusters at different levels for a network with n = *64 nodes*

complete copy of each content file in the caches of the entire network. Moreover, $\mathbf{x}$ must also satisfy the cache size constraint $\sum_{m=0}^{M} x_m 4^{-m} \leq L_C$. Clearly, if file l is more popular than file l' (i.e., $p_l \geq p_{l'}$), file l should be replicated more frequently than file l'. Therefore, without loss of optimality, we let $q_1 \geq q_2 \cdots \geq q_L$. In other words, the more popular files are cached at lower levels and the less popular files are cached at higher levels.

11.4.3 Capacitated graph for content delivery

For a given hierarchical cache content placement with parameter $\mathbf{x}$, the content delivery scheme is based on a capacitated graph $\mathcal{G}$, which is similar to the communication schemes considered in [14, 32]. Specifically, the D2D networks are represented by a tree graph $\mathcal{G}$ whose leaf nodes are the nodes in $V(n)$ and whose internal nodes are node clusters. There are $M+1$ levels in the tree graph $\mathcal{G}$, where the lowest level is called the 0-th level, the next lowest level is called the first level, and the highest level is called the M-th level. With a slight abuse of notation, let $V_{0,i} \subseteq V(n)$ also denote the i-th leaf node at the 0-th level of $\mathcal{G}$, which represents the i-th node in the network. For $m \in \{1, ..., M\}$, let $V_{m,i}$ also denote the i-th internal node at the m-th level of $\mathcal{G}$. There are only edges between the nodes in adjacent levels. For

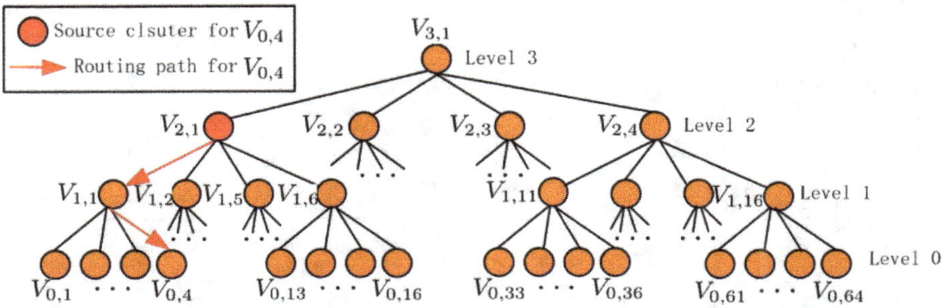

Figure 11.11 Illustration of the capacitated graph $\mathcal{G}$ for the network in Figure 11.10 with n = 64 nodes. Suppose node $V_{0,4}$ requests a file cached at the second level. Then the set of source nodes is $V_{2,1}$ (red node) and the routing path from the source cluster $V_{2,1}$ to the destination $V_{0,4}$ is illustrated with red arrows.

$m \in \{2, ..., M\}$, there is an edge between an internal node $V_{m-1,j}$ and an internal node $V_{m,i}$ if $V_{m-1,j} \subseteq V_{m,i}$. Similarly, there is an edge between a leaf node $V_{0,j}$ and an internal node $V_{1,i}$ at the first level if $V_{0,j} \subseteq V_{1,i}$. An example of the capacitated graph $\mathcal{G}$ is given in Figure 11.11.

11.4.3.1 Source determination layer

Let $V_{m,g(i)}$ denote the internal node at the m-th level in $\mathcal{G}$ that contains the leaf node $V_{0,i}$ (i.e., $V_{0,i} \subseteq V_{m,g(i)}$). Then for a leaf node $V_{0,i}$ requesting the l-th content cached at the m-th level (i.e., $q_l = 4^{-m}$), the set of source nodes is given by $V_{m,g(i)}$, which is the cluster at the m-th level that contains $V_{0,i}$, as illustrated in Figure 11.11 for $V_{0,4}$. Since the cache of each node in $V_{m,g(i)}$ stores a different portion of the $4^{-m}F$ bits of the l-th file, the leaf node $V_{0,i}$ can reconstruct a complete copy of the l-th file by collecting all the portions (*subfiles*) from the nodes in$V_{m,g(i)}$.

11.4.3.2 Routing layer

When a leaf node $V_{0,i}$ requests the l-th content cached at the m-th level, the requested content is sent from the source set $V_{m,g(i)}$ to the destination $V_{0,i}$ via the path $V_{m,g(i)} \rightarrow V_{m-1,g(i)} \cdots \rightarrow V_{1,g(i)} \rightarrow V_{0,i}$ in the capacitated graph $\mathcal{G}$, as illustrated in Figure 11.11 for $V_{0,4}$. This corresponds to the concentration of the content to smaller and smaller clusters until it finally concentrates to the single leaf node $V_{0,i}$ that requests the content.

11.4.3.3 Cooperation layer

To send information along an edge from a parent node to a child node, the routing layer calls upon the cooperation layer. Specifically, suppose the routing layer calls the cooperation layer to send a message from a parent node $V_{m,i}$ to a child node $V_{m-1,j}$. Assume each node in $V_{m,i}$ has access to a distinct 4^{-m} fraction of the message to be

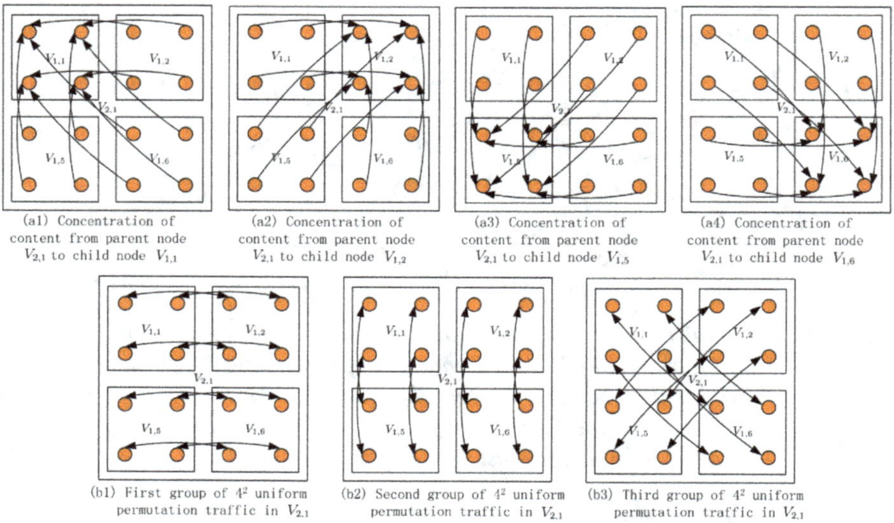

(a1) Concentration of content from parent node $V_{2,1}$ to child node $V_{1,1}$
(a2) Concentration of content from parent node $V_{2,1}$ to child node $V_{1,2}$
(a3) Concentration of content from parent node $V_{2,1}$ to child node $V_{1,5}$
(a4) Concentration of content from parent node $V_{2,1}$ to child node $V_{1,6}$
(b1) First group of 4^2 uniform permutation traffic in $V_{2,1}$
(b2) Second group of 4^2 uniform permutation traffic in $V_{2,1}$
(b3) Third group of 4^2 uniform permutation traffic in $V_{2,1}$

Figure 11.12 *Subfigures (a1)–(a4) illustrate the concentration of content from parent node $V_{2,1}$ to child nodes $V_{1,1}$, $V_{1,2}$, $V_{1,5}$, and $V_{1,6}$, respectively, for the network in Figure 11.10. The PHY partitions all the $3 \times 4^2 = 48$ subfile transmissions induced by the concentration of content from parent node $V_{2,1}$ to its child nodes into three groups of uniform permutation traffic. At each time, the PHY schedules one group for transmission using either the hierarchical cooperation or multihop scheme, as illustrated in subfigures (b1)–(b3).*

sent. Then each node in $V_{m,i} \backslash V_{m-1,j}$ sends its part of the message to a node in $V_{m-1,j}$ such that after the transmission, each node in $V_{m-1,j}$ will have access to a distinct $4^{-(m-1)}$ fraction of the message, as illustrated in Figure 11.12 for $m = 2$.

11.4.3.4 Physical layer

The PHY groups the traffic induced by the cooperation layer into per cluster uniform permutation traffic within different clusters at different levels so that we can use the existing hierarchical cooperation scheme or multihop scheme described in Section 11.3 as building blocks to handle the traffic induced by the cooperation layer. Specifically, the choice of PHY transmission mode for level m with cluster size 4^m depends on which PHY mode achieves a higher throughput, as described in Section 11.3.3.

To achieve this, the PHY needs to properly partition the available resources between different levels and different clusters, and schedule the transmissions. Specifically, the resource partitioning and scheduling at different levels/clusters are elaborated below.

The PHY time shares between the transmissions of M_b active levels, where $M_b = \max m$, s.t. $x_m > 0$. (Note that all levels higher than M_b are inactive since no content files are cached at these levels.) Note that $M_b \geq 1$ since L_C. For simplicity, in our achievability strategy, we choose to serve the levels in a round robin manner with the same fraction of time per level. This turns out to be sufficient in terms of scaling laws.

Within the m-th level for $m > 0$, there is no communication between clusters at the m-th level and the communications within each cluster of the m-th level occur simultaneously to achieve spatial reuse gain. This is exactly the per cluster uniform permutation traffic with cluster size $N = 4^m$ described in Section 11.3.3. Therefore, we can use the hierarchical cooperation or multihop scheme described in Section 11.3.3 to handle the traffic at the m-th level.

Within the i-th cluster at the m-th level $V_{m,i}$ for $m > 0$, there are a total number of 3×4^m subfile transmissions that need to be scheduled since each node in $V_{m,i}$ needs to collect a portion of the message from the other three nodes in $V_{m,i}$, as illustrated in Figure 11.12. At each time, the PHY can schedule 4^m subfile transmissions into a group of uniform permutation traffic for the nodes in $V_{m,i}$. Therefore, the PHY needs to further time share between the transmissions of the three groups of uniform permutation traffic, as illustrated in Figure 11.12.

As a result, the per traffic rate at the m-th level for $m > 0$ can be calculated as in the following lemma.

Lemma 1 (per traffic rate at different levels): For any cluster $V_{m,i}, i \in \{1, ..., 4^{M-m}\}$ at the m-th level, a per-node rate of R_m is achievable from one node in $V_{m,i}$ to another node in $V_{m,i}$, where

$$R_m = \frac{R_u(4^m)}{3M_b}, m = 1, ..., M_b. \tag{11.12}$$

$R_u(N)$ is the per-node rate for a regular grid network with n nodes under per cluster uniform permutation traffic with cluster size N, as given in (11.9).

Remark 1: Since there are a total number of 4^m transmissions from a parent node $V_{m,i}$ to a child node $V_{m-1,j}$, the edge between $V_{m,i}$ and $V_{m-1,j}$ can provide an achievable throughput of $C_m = 4^m R_m$.

Note that in the achievable scheme, we have implicitly assumed that each node knows which node stores which subfile of each file in advance before the content delivery phase. This is possible for a *semi-static wireless D2D caching network* where the node locations are fixed within one operation cycle (but can change over different operation cycles). When the nodes are allowed to move within one operation cycle, it is difficult to maintain the information about which node stores which subfile at each node. One potential method to address this problem is to use the MDS-coded caching [1, 3, 28]. An MDS rateless code generates an arbitrarily long sequence of parity bits from a content file of F bits, such that if the decoder obtains any F parity bits, it can recover the original F information bits [29]. Therefore, if we use the MDS-coded caching scheme where each content file is encoded using an MDS rateless code, we can modify the proposed hierarchical cache content placement scheme as follows to address the above problem. When $q_l = \frac{1}{4^m}$, each node in the network caches $4^{-m}F$ parity bits for file l during the cache initiation phase. Under such MDS-coded caching scheme, a node can always decode the requested content file l by collecting F parity bits from the caches of any nearby 4^m nodes without having to know which node stores which parity bits of each file in advance. Note that MDS coding is a well-known general technique widely used in distributed caching [1, 5, 33]. It turns the delivery of a file into a "fluid" problem. In this case, only the amount of data matters for retrieving a file and the identity of the data is no longer required.

11.5 Cache content placement optimization

In this section, we aim at finding the optimal cache content placement parameter $\mathbf{x}$ to maximize the per-node throughput R. We first derive the per-node throughput R for given cache content placement parameter $\mathbf{x}$ and formulate the cache content placement optimization problem. Then we propose a low-complexity cache content placement algorithm.

11.5.1 Problem formulation

We first analyze the total average traffic rate over an edge $e_{m,i,j}$ between a parent node $V_{m,i}$ to a child node $V_{m-1,j}$ at the m-th level, when the per-node throughput requirement is R. Whenever a user in $V_{m-1,i}$ requests a file that is cached at the m'-th level with $m' - 1$, it will induce a traffic rate of R on the edge $e_{m,i,j}$. Under the hierarchical cache content placement scheme, files with indices $\left\{\sum_{i=0}^{m-1} x_i + 1, \sum_{i=0}^{m-1} x_i + 2, ..., L\right\}$ are cached at levels higher than the $(m-1)$-th level. Therefore, for given cache content placement parameter $\mathbf{x}$ and per-node rate requirement R, the total average traffic rate over the edge $e_{m,i,j}$ is $4^{m-1}\sum_{l=\sum_{i=0}^{m-1} x_i+1}^{L} p_l R$. Clearly, a per-node throughput R is achievable if and only if the induced total average traffic rate does not exceed the capacity of the edges at all levels. Hence, for given cache content placement parameter $\mathbf{x}$, the maximum achievable per-node throughput is $\max R, \text{s.t.} \sum_{l=\sum_{i=0}^{m-1} x_i+1}^{L} p_l R \le C_m 4^{-(m-1)}$. Consequently, the cache content placement optimization problem to maximize the per-node throughput can be formulated as

$$\max_{\mathbf{x}\in\mathbb{Z}_+^m, R} \; R \tag{11.13}$$

$$\text{s.t.} \; \sum_{l=\sum_{i=0}^{m-1} x_i+1}^{L} p_l R \le C_m 4^{-(m-1)}, \forall m \tag{11.14}$$

$$\sum_{m=0}^{M} x_m 4^{-m} \le L_C, \tag{11.15}$$

$$\sum_{m=0}^{M} x_m = L, \tag{11.16}$$

where Constraint (11.14) is the cache size constraint. Note that for convenience, we have extended the definition of C_m from $m \in \{1, ..., M_b\}$ to $m \in \{1, ..., M\}$, where $C_m = \frac{4^m R_u(4^m)}{3M_b}, \forall m \in \{M_b + 1, ..., M\}$. Since $\sum_{l=\sum_{i=0}^{m-1} x_i+1}^{L} p_l R = 0$ for $m > M_b$, Constraint

(11.13) is equivalent to $\sum_{l=\sum_{i=0}^{m-1} x_i+1}^{L} p_l R \leq C_m 4^{-(m-1)}, m = 1, ..., M_b$, which is the link capacity constraint for the M_b active levels.

When $nL_C < L$, the condition $\sum_{m=0}^{M} x_m = L$ can never be satisfied. In this case, Problem (11.12) is infeasible. In fact, when $nL_C < L$, no cache content placement scheme can guarantee that all user requests are satisfied with probability 1 because in this case, the content library $\mathscr{L}$ cannot be stored in a lossy way in the network. Since we have assumed that $nL_C \geq L$ to avoid such a degenerate case, Problem (11.12) is always feasible.

The cache content placement optimization problem in (11.12) is an integer optimization problem, and an explicit (closed-form) solution amenable to the order-optimality analysis of the resulting throughput scaling law seems difficult to obtain. In the next section, we propose a low-complexity algorithm that can find an order-optimal solution for (11.12).

11.5.2 Low-complexity cache content placement algorithm

The capacity C_m of the edge $e_{m,i,j}$ depends on the number of active levels M_b, which is a complicated function of x_m. Clearly, C_m can be bounded as $\frac{1}{M}\overline{C}_m 4^{(m-1)} \leq C_m \leq \overline{C}_m 4^{(m-1)}$, where

$$\overline{C}_m = \frac{4R_u(4^m)}{3}, m = 1, ..., M. \tag{11.17}$$

Define $f(x) = (\lceil x \rceil - x) p_{\lfloor x \rfloor} + \sum_{l=\lceil x \rceil}^{L} p_l, x \in [1, L+1]$. Then we have $f\left(\sum_{i=0}^{m-1} x_i + 1\right) = \sum_{l=\sum_{i=0}^{m-1} x_i+1}^{L} p_l$ for $\sum_{i=0}^{m-1} x_i + 1 \in \mathbb{Z}_{++}$. Consider the following simplified cache content placement optimization problem:

$$\begin{aligned} &\max_{\mathbf{x} \in \mathbb{R}_+^m, R} \; R \\ &\text{s.t.} f\left(\sum_{i=0}^{m-1} x_i + 1\right) R \leq \overline{C}_m, \; m = 1, ..., M \\ &\sum_{m=0}^{M} x_m 4^{-m} \leq L_C, \; \sum_{m=0}^{M} x_m = L, \end{aligned} \tag{11.18}$$

where we have replaced C_m with its upper bound $\overline{C}_m 4^{(m-1)}$ and relaxed the integer optimization variables $\mathbf{x}$ to real variables. Note that the optimal solution of (11.17) would be the same if we were to replace C_m with its lower bound $\overline{C}_m 4^{(m-1)}/M$, although the optimal objective value would be scaled by $1/M$.

Based on the above analysis, the low-complexity cache content placement algorithm first solves the optimal solution of the simplified problem in (11.18) and then (approximately) projects the solution to the feasible set of the original problem in (11.13). In general, the function $f(x)$ depends on the content popularity distributions $p_1, ..., p_L$ and may not be convex. Therefore, Problem (11.18) may not be convex. In the following theorem, we prove that the optimal solution of Problem (11.18) must satisfy certain sufficient and necessary optimality conditions, from which a low complexity algorithm can be derived.

Theorem 2 (optimality condition of (11.17)): $(\mathbf{x}^*, R^*)$ is the optimal solution of Problem (11.18) if and only if

$$f\left(\sum_{i=m^*}^{m-1} x_i^* + 1\right) R^* = \overline{C}_m, m = m^* + 1, \ldots, M, \tag{11.19}$$

$$R^* \leq \overline{C}_{m^*},$$

$$\sum_{m=0}^{M} x_m^* = L, \tag{11.20}$$

$$\sum_{m=0}^{M} x_m^* 4^{-m} = L_C, \tag{11.21}$$

where $m^* = \min m$, s.t. $x_m^* > 0$, and $\overline{C}_0 = +\infty$ when $m^* = 0$.

Please refer to Appendix C for the proof.

From the optimality condition (11.18) in Theorem 2, the optimal cache content placement is to balance the traffic loading of the active levels. In Theorem 2, m^* is the lowest level at which a file can be cached and it depends on the cache size. The larger the cache size, the smaller m^* is. For example, when $L_C = L/n$, which is the minimum cache size to make the problem feasible, we have $m^* = M$, i.e., we can only cache all files at the highest level M. On the other hand, when $L_C = L$, we have $m^* = 0$; i.e., the cache size is enough to cache all files at the lowest level. As L_C increases from L/n to L, m^* decreases from M to 0. Motivated by this observation, we propose a bisection algorithm to find m^* and the optimal solution $\mathbf{x}^*$.

Specifically, for a given m^*, the solution of (11.18) and (11.19) is

$$\begin{aligned} x_{m^*}^* &= f^{-1}\left(\frac{\overline{C}_{m^*+1}}{R}\right) - 1, \\ x_m^* &= f^{-1}\left(\frac{\overline{C}_{m+1}}{R}\right) - f^{-1}\left(\frac{\overline{C}_m}{R}\right), \\ m &= m^* + 1, \ldots, M-1 \\ x_M^* &= L - f^{-1}\left(\frac{\overline{C}_M}{R}\right) + 1. \end{aligned} \tag{11.22}$$

Substituting (11.21) into (11.20), we have

$$L_{m^*}(R) \triangleq \sum_{m=m^*+1}^{M} \frac{3f^{-1}\left(\frac{\overline{C}_m}{R}\right)}{4^m} + \frac{L+1}{4^M} - \frac{1}{4^{m^*}} = L_C. \tag{11.23}$$

If we can find a solution R^* of $L_{m^*}(R) = L_C$ for $R \in \left(\overline{C}_{m^*+1}, \overline{C}_{m^*}\right]$, then m^* and R^* satisfy all the optimality conditions in Theorem 2.

For a given m^*, if $L_{m^*}\left(\overline{C}_{m^*+1}\right) \geq L_C$, it implies the cache size is insufficient to cache files at level m^* and thus m^* should be increased. If $L_{m^*}\left(\overline{C}_{m^*}\right)_C$, it implies the cache size is still sufficient to cache files at the lower level and thus m^* should be decreased. If $L_{m^*}\left(\overline{C}_{m^*+1}\right)_C$ and $L_{m^*}\left(\overline{C}_{m^*}\right) \geq L_C$, (11.23) must have a unique solution R^* for $R \in \left(\overline{C}_{m^*+1}, \overline{C}_{m^*}\right]$ because $L_{m^*}(R)$ is a strictly increasing function of R. In this case, $(\mathbf{x}^*, R^*)$ is the optimal solution of Problem (11.18) according to Theorem 2.

Based on the above analysis, the overall cache content placement algorithm is summarized in Algorithm 11.1. In Algorithm 11.1, Step 1 is the bisection algorithm to find m^* and the optimal solution $\mathbf{x}^*$ of Problem (11.18). After Step 1, we can determine the cache allocated to each level, e.g., the optimal cache allocated to level m (i.e., the amount of the cache used to store the files cached at the m-th level) is $x_m^* 4^{-m} F$. However, such a cache allocation scheme may not be feasible because x_m^* may not be integer. Therefore, Step 2 is to find a feasible solution $\mathbf{x}^o$ of (11.13) that is close to $\mathbf{x}^*$ (or equivalently, find a cache allocation scheme such that the cache allocated to the m-th level is close to $x_m^* 4^{-m} F$ and can be divided by $4^{-m}F$). Specifically, when $m = 0$, the available cache size is $x_0^* 4^{-0} F$ and thus the cache allocated to the 0-th level is $\lfloor x_0^* \rfloor 4^{-0} F$. Correspondingly, the number of files stored at the 0-th level is $x_0^o = \lfloor x_0^* \rfloor$. The released cache size from the 0-th level is $b_0 F = \left(x_0^* 4^{-0} - x_0^o 4^{-0}\right) F$. When $m = 1$, the available cache size (including the cache size released from the 0-th level) is $\left(x_1^* 4^{-1} F + b_0 F\right)$, and thus the cache allocated to the 1-th level is $\left\lfloor \left(x_1^* 4^{-1} F + b_0 F\right) / \left(4^{-1} F\right) \right\rfloor 4^{-1} F = \left\lfloor x_1^* + b_0 4^1 \right\rfloor 4^{-1} F$. Correspondingly, the number of files stored at the 1-th level is $x_1^o = \left\lfloor x_1^* + b_0 4^1 \right\rfloor$. The released cache size from the first level is $b_1 = \left(x_1^* 4^{-1} + b_0 - x_1^o 4^{-1}\right) F$. Similarly, when $m > 1$, the available cache size (including the cache size released from the $(m-1)$ -th level) is $\left(x_m^* 4^{-m} F + b_{m-1} F\right)$ and thus the cache allocated to the m-th level is $\left\lfloor \left(x_m^* 4^{-m} F + b_{m-1} F\right) / \left(4^{-m} F\right) \right\rfloor 4^{-1} F = \left\lfloor x_m^* + b_{m-1} 4^m \right\rfloor 4^{-m} F$. Correspondingly, the number of files stored at the m-th level is $x_m^o = \left\lfloor x_m^* + b_{m-1} 4^m \right\rfloor$. The released cache size from the m-th level is $b_m = \left(x_m^* 4^{-m} + b_{m-1} - x_m^o 4^{-m}\right) F$. The above cache allocation process continues until all L files have been cached. Finally, Steps 2c–2g are to balance the traffic loading of different levels.

Despite various relaxations and approximations, we will show that the proposed low-complexity cache content placement algorithm is order optimal in Section 11.6; i.e., it achieves the same order of throughput as the optimal solution of the cache content placement optimization problem in (11.13).

Remark 2: In Appendix B, we reformulate Problem (11.13) with fixed R as a zero-one linear programming (ZOLP) feasibility problem. Based on the ZOLP reformulation in (B.1), it is possible to find the optimal solution of Problem (11.1) by a bisection search over R, where for each fixed R, the ZOLP feasibility problem (B.1) is solved using standard ZOLP solvers. Although it is difficult to analyze the performance of the optimal solution, the ZOLP reformulation in (B.1) is elegant and may potentially achieve a better throughput performance. Readers interested in algorithm design may refer to Appendix B for the details.

11.6 Throughput performance of cache-induced hierarchical cooperation

For general content popularity distributions, it is very difficult to analyze the performance of the proposed cache-induced hierarchical cooperation scheme with the cache content placement parameter **x** determined by Algorithm 11.1. In this section, we assume the content popularity follows Zipf distribution [20] and analyze the throughput performance of the proposed scheme. Under the Zipf popularity distribution, the probability of requesting the l-th file is given by

$$p_l = \frac{1}{Z_\tau(L)} l^{-\tau}, l = 1, ..., L, \tag{11.24}$$

where τ is the *popularity skewness parameter* and $Z_\tau(L) = \sum_{l=1}^{L} l^{-\tau}$ is a normalization factor.

Algorithm 9.1 Cache content placement Algorithm

Step 1 (Bisection for solving (9.8)): Let $m_L = 0$. If $L_0(\overline{C}_1) < L_C$, let $m^* = m_L$ and goto Step 1c.
Let $m_H = M$. If $L4^{-M} \geq L_C$, let $m^* = m_H, x_m^* = 0, \forall m < m^*, x_M^* = L$ and goto Step 2; otherwise let $m^* = \lfloor \frac{m_L + m_H}{2} \rfloor$.
1a: If $L_{m^*}(\overline{C}_{m^*+1}) \geq L_C$, let $m_L = m^*$; elseif $L_{m^*}(\overline{C}_{m^*}) < L_C$, let $m_H = m^*$; else goto Step 1c.
1b: If $m_H - m_L = 1$, let $m^* = m_H$ and goto Step 1c; otherwise let $m^* = \lfloor \frac{m_L + m_H}{2} \rfloor$ and goto Step 1a.
1c: Let $x_m^* = 0, \forall m < m^*$ and $\{x_m^*, m = m^*, ..., M\}$ be the solution of (9.9) and (9.10) as given in (9.12).
Step 2 (Find a feasible solution close to x*): Let $b_{m^*-1} = 0$, $x_m^o = 0, m = 0, ..., m^* - 1$ and $m = m^*$.
2a: Let $x_m^o = \lfloor x_m^* + b_{m-1} 4^m \rfloor$. $b_m = x_m^* 4^{-m} + b_{m-1} - x_m^o 4^{-m}$.
2b: If $\sum_{i=0}^{m} x_i^o \geq L$, let $x_m^o = L - \sum_{i=0}^{m-1} x_i^o$ and goto Step 2c; otherwise let $m = m + 1$ and goto Step 2a.
2c: Let $M^\circ = \text{argmax}_m x_m^\circ$, s.t. $x_m^\circ > 0$ and $m^\circ = \text{argmin}_m x_m^\circ$, s.t. $x_m^\circ > 1$. If $M^\circ - m^\circ \leq 2$, goto Step 3.
2d: Let $x'_{m^\circ} = x^\circ_{m^\circ} - 1, x'_{m^\circ+1} = x^\circ_{m^\circ+1} + 1$ and $x'_m = x^\circ_m, \forall m \notin \{m^\circ, m^\circ + 1\}$.
2e: Let $M' = \text{argmax}_m x'_m$, s.t. $x'_m > 0$.
2f: While $M' > m^\circ + 1$ and $x'_{M'} > 0$ and $\sum_{m=0}^{M} x'_m 4^{-m} - 4^{-M'} + 4^{-(m^\circ+1)} \leq L_C$
    Let $x'_{M'} = x'_{M'} - 1, x'_{m^\circ+1} = x'_{m^\circ+1} + 1$,
    Let $M' = \text{argmax}_m x'_m$, s.t. $x'_m > 0$.
2g: Let $R' = \min_{m \in \{m^\circ, ..., M'\}} \overline{C}_m / f\left(\sum_{i=m^\circ}^{m-1} x'_i + 1\right)$ and $R^\circ = \min_{m \in \{m^\circ, ..., M^\circ\}} \overline{C}_m / f\left(\sum_{i=m^\circ}^{m-1} x_i^\circ + 1\right)$. If $R' > R^\circ$, let $x_m^\circ = x'_m, \forall m$, goto Step 2c.
Step 3 (Termination): Output $\mathbf{x}^o$.

11.6.1 Throughput bounds under Zipf popularity distribution

In this subsection, we derive the upper and lower bounds for the throughput of the proposed scheme under the Zipf popularity distribution. To achieve this, we first give upper and lower bounds for the $\overline{C}_m$'s in (11.16).

Lemma 2: The upper and lower bounds of $\overline{C}_m$ can be expressed in a unified form as $c_n 4^{-m\gamma_n}$ for some coefficient c_n and γ_n that depends on $A(n)$. Specifically, for a wireless D2D network with area $A(n) = n^{\kappa}$, with $\kappa \geq 0$, $\overline{C}_m$ can be bounded as $c_n^L(\kappa)\, 4^{-m\gamma_n^L(\kappa)} \leq \overline{C}_m \leq c_n^U(\kappa)\, 4^{-m\gamma_n^U(\kappa)}$, where

$$\begin{aligned}
\gamma_n^U(\kappa) &= \min\left(\frac{1}{2s_M+1} + \left(\frac{\alpha\kappa}{2} - 1\right)^+, \frac{1}{2}\right),\\
c_n^U(\kappa) &= \frac{2}{T_r 3^{\frac{5}{4}}} \log\left(1 + \frac{\mathrm{SNR}}{1+P_I}\right), s_M = \sqrt{M \ln 4},\\
\gamma_n^L(\kappa) &= \min\left(\frac{1}{s_M+1} + \left(\frac{\alpha\kappa}{2} - 1\right)^+, \frac{1}{2}\right),\\
c_n^L(\kappa) &= \frac{4R_c(\alpha, P_I)}{3\left(1+s_M\right) T_r^2 \left(3 \cdot 2^{s_M - 1}\right)^{\frac{s_M}{2(s_M+1)}}}.
\end{aligned} \tag{11.25}$$

The proof follows straightforwardly from the definition of $\overline{C}_m$ and the results in Section 11.3. The detailed derivations are omitted for conciseness.

One key challenge to derive the throughput lower bound is to quantify the throughput loss due to various relaxations and approximations in Algorithm 11.1. This challenge is addressed in the following lemma. Let $(\mathbf{x}^*, R^*)$ denote the optimal solution of the relaxed cache content placement optimization problem in (11.18) obtained in Step 1 of Algorithm 11.1, and $\mathbf{x}^o$ denote the feasible cache content placement parameter found by Step 2 of Algorithm 11.1. Then $\left(\mathbf{x}^o, \frac{1}{M(1+2^{\tau})} R^*\right)$ must be a feasible solution of the original cache content placement optimization problem in (11.12).

Please refer to Appendix C.2 for the proof.

Clearly, the optimal objective of Problem (11.18) provides an upper bound for the throughput achievable with the proposed cache-induced hierarchical cooperation. However, it is highly nontrivial to obtain the closed-form expression for the optimal objective of Problem (11.18) since there is no closed-form solution for Problem (11.18). To overcome this challenge, we first derive closed-form upper and lower bounds R_U and R_L for the optimal objective of Problem (11.18). Then R_U and $\frac{1}{M(1+2^{\tau})} R_L$ provide an upper bound and a lower bound for the achievable throughput, respectively. The detailed analysis is given in Appendix C.3, and the final results are summarized in the following theorem.

Theorem 3 (throughput bounds): Consider a wireless D2D network with area $A(n) = n^{\kappa}$, $\kappa \geq 0$. Let $R^{\star}$ and R° denote the per-node throughput achieved by the cache-induced hierarchical cooperation scheme with the optimal cache content placement parameter $\mathbf{x}^{\star}$ (i.e., the optimal solution of (11.13)) and with the low-complexity cache content placement solution $\mathbf{x}^{\circ}$ in Algorithm 11.1, respectively. Then both $R^{\star}$ and R° are lower bounded as $R^{\star} \geq R^{\circ} \geq \frac{1}{M(1+2^{\tau})} R_L$, where

$$R_L = \begin{cases} \overline{R}_{L|\tau<1} & \tau \in [0,1) \\ \left(\frac{\left(e^2 L - L - 1\right) 4^{-M + L_C}}{4\left(e^2 L - 1\right)} \right)^{\gamma_n} c_n & \tau = 1 \\ \left(\frac{4^{\frac{1+\gamma_n-\tau}{\tau-1}} - 1}{4^{\frac{\gamma_n+\tau-1}{\tau-1}} - 4} \right)^{\gamma_n} \frac{c_n L_C^{\gamma_n} L^{\tau-1-\gamma_n}}{\tau} & \tau \in (1, \gamma_n + 1) \\ \left(3 \log_4 L + 4\right)^{-\gamma_n} \frac{c_n}{\tau} L_C^{\tau-1} & \tau = \gamma_n + 1 \\ \frac{c_n L_C^{\tau-1}}{\left(\frac{3\tau^{\frac{1}{\tau-1}} 4^{\frac{\gamma_n+1-\tau}{\tau-1}}}{1-4^{\frac{\gamma_n+1-\tau}{\tau-1}}} + 4\tau^{\frac{1}{\gamma_n}} \right)^{\tau-1}} & \tau > \gamma_n + 1 \end{cases} \tag{11.26}$$

$$R_{L|\tau<1} = c_n \min \left(\left(\frac{4^{\gamma_n+1}-1}{4^{\gamma_n+2}-16} \frac{L_C}{L} \right)^{\gamma_n}, \frac{3\left(1 - \frac{L_C}{L}\right)^{-1}}{4^{\gamma_n+1}-1} \right), \tag{11.27}$$

$c_n = c_n^L(\kappa)$ and $\gamma_n = \gamma_n^L(\kappa)$. Moreover, both $R^{\star}$ and R° are upper bounded as $R_U \geq R^{\star} \geq R^{\circ}$, where

$$R_U = \begin{cases} \frac{c_n}{1-\tau} \left(\frac{4^{\gamma_n+1}-1}{4^{\gamma_n}-1} \right)^{\gamma_n} \left(\frac{L_C}{L} \right)^{\gamma_n} & \tau \in [0,1) \\ c_n \left(\frac{4e\left(4^M L_C + L + 1\right)}{3 \cdot 4^M L^{1-\frac{1}{\ln L}}} \right)^{\gamma_n} \ln L & \tau = 1 \\ \frac{L_C^{\gamma_n}}{\left(\frac{L^{\frac{1+\gamma_n-\tau}{\gamma_n}} \left(\frac{\tau}{c_n}\right)^{\frac{1}{\gamma_n}}}{2^{\tau-1}} - 4 c_n^{-\frac{1}{\gamma_n}} \right)^{\gamma_n}} & \tau \in (1, \gamma_n + 1) \\ \frac{\left(\tau - L^{1-\tau}\right) c_n 4^{-\gamma_n}}{\left(L_C+1\right)^{1-\tau} - (L+1)^{1-\tau}} & \tau \geq \gamma_n + 1 \end{cases} \tag{11.28}$$

$$c_n = c_n^U(\kappa) \text{ and } \gamma_n = \gamma_n^U(\kappa).$$

Note that in Theorem 3, R° is not the optimal objective of Problem (11.18). It is the objective value of the original problem in (11.13) achieved by the low-complexity cache content placement solution $\mathbf{x}^{\circ}$. Therefore, we must have $R^{\star} \geq R^{\circ}$.

11.6.2 Comparison with cache-assisted multihop scheme

In this subsection, we compare the per-node throughput of the cache-induced hierarchical cooperation scheme with that of a cache-assisted multihop scheme, which only has multihop PHY mode. Following a similar analysis, it can be shown that the lower and upper bounds of the per-node throughput of the cache-assisted multihop scheme is given in the same form as R_L and R_U in Theorem 3, but with different coefficients $c_n = c_n^M$ and $\gamma_n = \gamma_n^M$, where

$$\gamma_n^M = \frac{1}{2}, c_n^M = \frac{4}{3 T_r^2} \log \left(1 + \frac{\text{SNR}}{1 + P_I} \right). \tag{11.29}$$

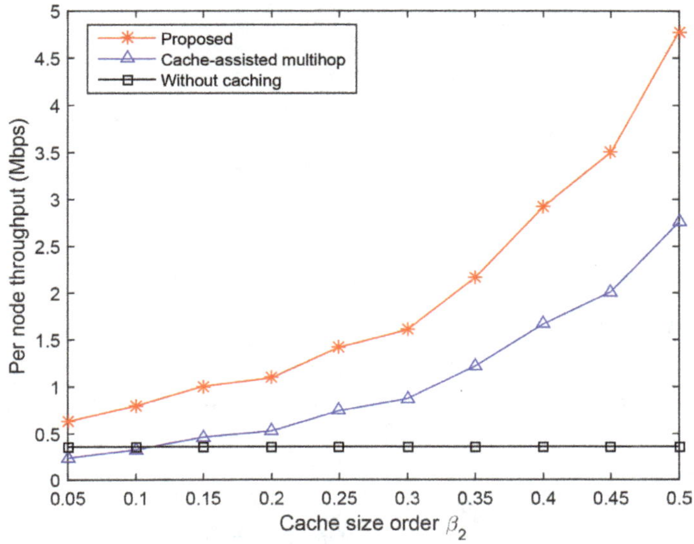

Figure 11.13 *Per-node throughput versus the cache size order β_2 for a dense network with* n = 4^9 *nodes. The content popularity skewness is* $\tau = 1$ *and the path loss exponent is* $\alpha = 4$.

Note that both R_L and R_U increase with c_n and decrease with γ_n, where γ_n determines the scaling of the throughput bounds w.r.t. n, as will be shown later in Theorem 11.7.2, and c_n determines the constant coefficients in the scaling law. Both $\gamma_n^U(\kappa)$ and $\gamma_n^L(\kappa)$ of the proposed scheme are smaller than the $\gamma_n^M = 1/2$ of the cache-assisted multihop scheme. As a result, the proposed scheme has huge throughput gain over the cache-assisted multihop scheme, especially when κ is smaller (i.e., denser networks), as will be shown in the simulations.

In Figures 11.13–11.15, we illustrate the throughput gain of the proposed cache-induced hierarchical cooperation for a dense wireless D2D network with $n = 4^9$ nodes, area $A(n) = 1$, and 200 MHz system bandwidth. There are $L = \lfloor n^{\beta_1} \rfloor$ content files on the content server, and the cache capacity at each node is $L_C = n^{\beta_2}$, where $\beta_1 = 0.9$ and $\beta_2 \in [0, \beta_1]$. The throughput of the *network without caching* is also given for comparison. The network without caching refers to arbitrary (random uniform permutation) source-destination traffic, as in [12], using the improved hierarchical cooperation in [16]. This comparison is just to give an idea of the advantage of caching when the demands are restricted to being in a given library of messages, rather than random source-destination traffic.

In Figure 11.13, we plot the per-node throughput versus the cache size order β_2. The total number of content files is $L = n^{0.9}$, and the content popularity skewness τ is fixed as 1. It can be seen that the throughput of both the proposed scheme and cache-assisted multihop scheme increases with the cache size order β_2. Moreover, the proposed scheme achieves large throughput gain over the two baseline schemes.

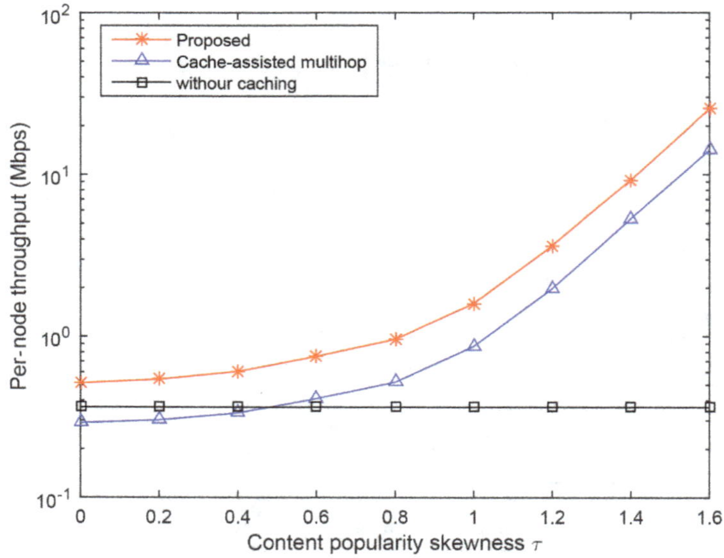

Figure 11.14 *Per node throughput versus the content popularity skewness τ for a dense network with* n = 4^9 *nodes. The cache size order is* $\beta_2 = 0.3$, *and the path loss exponent is* $\alpha = 4$.

We then simulate the case when the BS cache size is much smaller than the total content size. In Figure 11.14, we plot the per-node throughput versus the content popularity skewness τ. The total number of content files is $L = n^{0.9}$, and the cache size at each node is fixed as $L_C = n^{0.3}$. The results in Figure 11.14 show that the throughput of both the proposed scheme and cache-assisted multihop scheme increases with the content popularity skewness τ. Again, the proposed hierarchical cooperation achieves a large throughput gain over the two baseline schemes.

In Figure 11.15, we plot the per-node throughput versus the path loss exponent α. It can be seen that the throughput gain of the proposed scheme increases with the path loss exponent α for dense networks.

11.7 Scaling laws in extended networks

11.7.1 System scaling regime

In order to study the throughput scaling of extended wireless D2D caching networks (i.e., $A(n) = n$) for asymptotically large n, we consider that L and L_C scale with n according to the following functions:

$$L = a_1 n^{\beta_1} \text{ and } L_C = a_2 n^{\beta_2}, \tag{11.30}$$

where $\beta_1, a_1, a_2 > 0$ and $\beta_2 \in [0, \beta_1]$. When $\beta_1 = \beta_2$, we assume $a_1 > a_2$ to avoid the trivial case when each node has enough cache capacity to store the entire library $\mathscr{L}$.

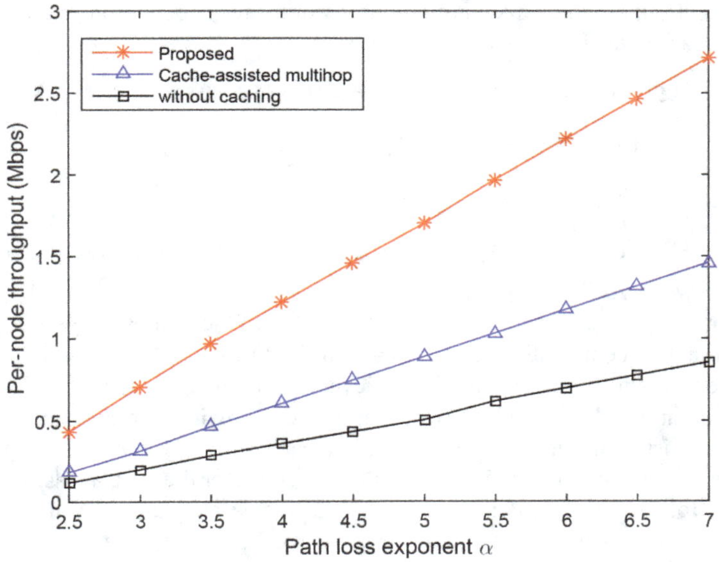

Figure 11.15 *Per-node throughput versus the path loss exponent* α *for a dense network with* n = 4^9 *nodes. The cache size order is* $\beta_2 = 0.3$*, and the content popularity skewness* $\tau = 1$.

Moreover, since nL_C, we have $\beta_1 - \beta_2 \le 1$ and when $\beta_1 - \beta_2 = 1$, we have $a_1 \le a_2$. Note that a similar scaling regime was also considered in [7].

Depending on the relative cache capacity $\frac{L_C}{L}$ at each node, the entire parameter space can be partitioned into two regimes as follows:

- Regime I: $\beta_1 - \beta_2 = 0, a_1 > a_2$
- Regime II: $\beta_1 - \beta_2 \in (0, 1)$, or $\beta_1 - \beta_2 = 1, a_1 \le a_2$.

11.7.2 Throughput scaling laws of cache-induced hierarchical cooperation

In this subsection, we obtain the throughput scaling laws of the proposed scheme. From the throughput bounds in Theorem 3, we can obtain the following achievable throughput scaling law.

Theorem 4 (Achievable Scaling Law in Extended Networks): For extended networks with $A(n) = n$, the achievable throughput $R^\star$ of the cache-induced hierarchical cooperation satisfies the following scaling law. In Regime I, we have

$$R^\star = \begin{cases} \Omega(1), & \tau \in [0, 1] \\ \Omega\left(n^{\beta_2(\tau-1)}\right), & \tau > 1 \end{cases}. \tag{11.31}$$

In Regime II, the achievable throughput scaling law depends on the popularity skewness parameter τ, summarized as follows:

$$R^\star = \begin{cases} \Omega\left(n^{(\beta_2-\beta_1)(\tau_\alpha-1)-\epsilon_\alpha}\right) & \tau \in [0,1] \\ \Omega\left(n^{\beta_1(\tau-\tau_\alpha)+\beta_2(\tau_\alpha-1)-\epsilon_\alpha}\right) & \tau \in (1,\tau_\alpha] \\ \Omega\left(n^{\beta_2(\tau-1)-\epsilon_\alpha}\right) & \tau > \tau_\alpha \end{cases}, \tag{11.32}$$

where $\tau_\alpha = \frac{\min(3,\alpha)}{2}, \epsilon_\alpha = \Theta\left(\frac{1}{\sqrt{\log n}}\right) \rightarrow 0$ as $n \rightarrow \infty$ for $\alpha \in (2,3)$, and $\epsilon_\alpha = 0$ for $\alpha \geq 3$. Moreover, we have $R^\circ = \Theta\left(R^\star\right)$.

Theorem 3 also establishes the order optimality of the proposed low-complexity cache content placement algorithm (Algorithm 11.1).

In the following, we compare the achievable scaling law of the proposed cache-induced hierarchical cooperation with that of the following two baseline schemes: the cache-assisted multihop scheme and PHY caching in [3]. Liu and Lau [3] only studied the achievable scaling law of the PHY caching for the special case of $\beta_2 = 0$. However, following a similar analysis to that in this chapter, we can extend the achievable scaling law in [3] to the more general case considered in this chapter, as summarized in the following theorem.

Theorem 5 (Achievable Scaling Law of Baseline Schemes): For extended networks, the achievable throughput R_{PHY} of the PHY caching scheme in [3] satisfies the following scaling law. In Regime I, we have

$$R_{PHY} = \begin{cases} \Omega\left(1\right) & \tau \in [0,1] \\ \Omega\left(n^{\beta_2(\tau-1)}\right) & \tau > 1 \end{cases}. \tag{11.33}$$

In Regime II, the achievable throughput scaling law depends on the popularity skewness parameter τ, summarized as follows:

$$R_{PHY} = \begin{cases} \Omega\left(n^{\frac{\beta_2-\beta_1}{2}-\epsilon}\right) & \tau \in [0,1] \\ \Omega\left(n^{\beta_1\left(\tau-\frac{3}{2}\right)+\frac{\beta_2}{2}-\epsilon}\right) & \tau \in \left(1,\frac{3}{2}\right] \\ \Omega\left(n^{\beta_2(\tau-1)-\epsilon}\right) & \tau > \frac{3}{2} \end{cases}, \tag{11.34}$$

where $\epsilon > 0$ is arbitrarily small. Moreover, the achievable throughput of the cache-assisted multihop scheme R_M satisfies the same scaling law as that of R_{PHY}, i.e., $R_M = \Theta\left(R_{PHY}\right)$.

For the special case of Regime I or Regime II with $\alpha \geq 3$, Theorem 4 reduces to the achievable scaling law of the two baseline schemes in Theorem 5. In Regime II with $\alpha < 3$, the scaling law achieved by the cache-induced hierarchical cooperation in Theorem 4 is better than that achieved by the two baseline schemes.

As shown in Figure 11.16, the achievable scaling laws of all schemes exhibit some phase transition phenomena as the content popularity skewness τ increases. Specifically, in Regime II, there are two *critical popularity skewness* points: $\tau = \tau_a$

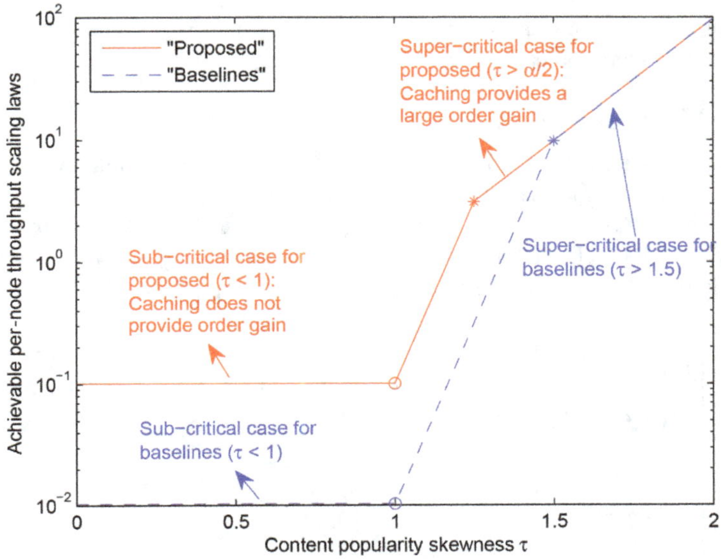

Figure 11.16 *Illustration of achievable throughput scaling laws in Theorem 4 for cache-induced hierarchical cooperation (solid curve) and Theorem 5 for baseline schemes (dashed curve). The number of nodes in the extended network is $n = 4^{11}$. The total content size order $\beta_1 = 0.9$ and the cache size order is $\beta_2 = 0.3$ (i.e., Regime II). The path loss exponent is $\alpha = 2.5$. In the figure, the circle and star symbols indicate the first and second critical popularity skewness points τ_a and τ_b, respectively.*

and $\tau = \tau_b$, where $\tau_a = 1$, $\tau_b = 1.5$ for the baseline schemes and $\tau_a = 1$, $\tau_b = \frac{\min(3,\alpha)}{2}$ for the cache-induced hierarchical cooperation scheme. For the *subcritical case* when $\tau < \tau_a$, the per-node throughput scales with n as $\Omega\left(n^{\eta_a}\right)$ with a smaller order η_a. For example, when $\beta_1 - \beta_2 = 1$ (i.e., $L_C \ll L$), $\eta_a = 1 - \frac{\min(3,\alpha)}{2}$ for the cache-induced hierarchical cooperation, which is the same as the scaling law of the hierarchical cooperation without caching; and $\eta_a = -0.5$ for the baseline schemes, which is the same as the Gupta–Kumar law [8]. Therefore, when $\tau < \tau_a$ and $\beta_1 - \beta_2 = 1$, caching does not provide order gain. For the *super-critical case* when $\tau > \tau_b$, on the other hand, the per-node throughput scales with n as $\Omega\left(n^{\eta_b}\right)$ with a much larger order η_b. For example, when $\beta_1 - \beta_2 = 1$ (i.e., $L_C \ll L$), we still have $\eta_b = \beta_2\left(\tau - 1\right) > 0$ for all schemes. In this case, caching provides a large order gain even when $L_C \ll L$.

From Figure 11.16, there are two advantages of the proposed cache-induced hierarchical cooperation over the PHY caching. First, when $\alpha < 3$, the second critical popularity skewness point τ_b of the cache-induced hierarchical cooperation is smaller than that of the baseline schemes. This implies that the cache-induced hierarchical cooperation can enjoy the large order gain $\eta_b = \beta_2\left(\tau - 1\right)$ under weaker conditions on the popularity distribution. Second, when $\alpha < 3$, the cache-induced hierarchical cooperation can achieve a better scaling law for $\tau < 1.5$.

11.7.3 Main converse results

In this section, we establish upper bounds on the throughput scaling laws. The main converse results are summarized in the following theorem.

Theorem 6 (Upper Bound of Scaling Laws in Extended Networks): In an extended wireless D2D caching network, the per-node throughput R of any feasible content delivery scheme must satisfy the following scaling laws. In Regime I, we have

$$R = \begin{cases} O\left(n^{\epsilon}\right), & \tau \in [0,1] \\ O\left(n^{\beta_2(\tau-1)+\epsilon}\right), & \tau > 1 \end{cases}. \tag{11.35}$$

In Regime II, we have

$$R = \begin{cases} O\left(n^{(\beta_2-\beta_1)(\tau_\alpha-1)+\epsilon}\right), & \tau \in [0,1] \\ O\left(n^{\beta_1(\tau-\tau_\alpha)+\beta_2(\tau_\alpha-1)+\epsilon}\right), & \tau \in (1,\tau_\alpha], \\ O\left(n^{\beta_2(\tau-1)+\epsilon}\right), & \tau > \tau_\alpha \end{cases} \tag{11.36}$$

where $\epsilon > 0$ is arbitrarily small.

Please refer to Section 11.7.4 for the proof.

In both Regimes I and II, the multiplicative gap between the achievable per-node throughput in Theorem 4 and its upper bound in Theorem 6 is within n^{ϵ} for $\epsilon > 0$ that can be arbitrarily small as $n \to \infty$. Therefore, the throughput scaling law depicted in Theorem 4 is order-optimal in the information theoretic sense for the Zipf popularity distribution.

11.7.4 Converse proof

11.7.4.1 Regime II with $\tau > \frac{\min(3,\alpha)}{2}$

The converse result for this case can be proved by considering the cut set bound between a reference node i and the rest of the network as follows. Let $q_l = I\left(B_i; W_l\right)/F$. Since B_i is a function of $W_1, \ldots, W_L$, we have

$$\begin{aligned} H\left(B_i\right) &= H\left(B_i\right) - H\left(B_i | W_1, \ldots, W_L\right) \\ &= I\left(B_i; W_1, \ldots, W_L\right) \\ &\overset{a}{=} \textstyle\sum_{l=1}^{L} I\left(B_i; W_l | W_1, \ldots, W_{l-1}\right) \\ &\overset{b}{=} \textstyle\sum_{l=1}^{L} I\left(B_i, W_1, \ldots, W_{l-1}; W_l\right) \\ &\geq \textstyle\sum_{l=1}^{L} I\left(B_i; W_l\right) = \sum_{l=1}^{L} q_l F, \end{aligned} \tag{11.37}$$

where (11.36-a) follows from the chain rule and (11.36-b) follows from the fact that the messages $W_1, ..., W_L$ are mutually independent. Hence, the q_l's must satisfy the cache capacity constraint $\sum_{l=1}^{L} q_l F \leq H\left(B_i\right) \leq L_C F$.

Under any feasible content delivery scheme, the amount of information u_i^j transmitted from the rest of the network to node i during the time window $\left[t_i^j, ..., t_i^{j+1} - 1\right]$ must satisfy

$$u_i^j \geq H\left(U_i^j|B_i\right) \tag{11.38}$$

$$\begin{aligned} &= H\left(W_{l_i}, U_i^j|B_i\right) - H\left(W_{l_i}|U_i^j, B_i\right) \\ &\geq H\left(W_{l_i}|B_i\right) - \varepsilon_F F \\ &= H\left(W_{l_i}\right) - I\left(W_{l_i}; B_i\right) - \varepsilon_F F \\ &= F\left(1 - q_{l_i} - \varepsilon_F\right), \end{aligned} \tag{11.39}$$

where $\varepsilon_F \to 0$ as $F \to \infty$, (11.38) is to ensure that node i can successfully receive the aggregate information message U_i^j, and (11.38) follows from the necessary condition in (11.1). As a result, for given q_l's and per-node rate requirement R, the total average traffic rate over the cut from the rest of the network to node i is $\left(\sum_{l=1}^{L} p_l \left(1 - q_l\right) - \varepsilon_F\right) R$. Clearly, a per-node throughput R is achievable only if the induced total average traffic rate does not exceed the sum capacity Γ_i of the MISO channel between the rest of the network and node i, which is upper bounded by $K \log n$ for some constant K [12]. Hence, the achievable per-node throughput is upper bounded by

$$R \leq \max_{\{q_l\}} \frac{\Gamma_i}{\sum_{l=1}^{L} p_l \left(1 - q_l\right) - \varepsilon_F}, \quad \text{s.t.} \sum_{l=1}^{L} q_l \leq \lceil L_C \rceil. \tag{11.40}$$

It is easy to see that the optimal q_l's to maximize the achievable per-node throughput upper bound is to cache the most popular $\lceil L_C \rceil$ files, i.e., $q_l^\star = 1, l = 1, ..., \lceil L_C \rceil$ and $q_l^\star = 0, l > \lceil L_C \rceil$. As a result, we have

$$R \leq \frac{\Gamma_i}{\sum_{l=1}^{L} p_l \left(1 - q_l^\star\right) - \varepsilon_F} = \frac{\Gamma_i}{\sum_{l=\lceil L_C \rceil + 1}^{L} p_l^\star - \varepsilon_F}. \tag{11.41}$$

In Regime II with $\tau > \frac{\min(3,\alpha)}{2}$, we have

$$\sum_{l=\lceil L_C \rceil + 1}^{L} p_l^\star = \Omega\left(n^{\beta_2(1-\tau)}\right). \tag{11.42}$$

Finally, it follows from $\Gamma_i \leq K \log n$, (11.41), and (11.42) that $R = O\left(n^{\beta_2(\tau-1)+\epsilon}\right)$ as $F \to \infty$.

11.7.4.2 Regime II with $\tau \in \left[0, \frac{\min(3,\alpha)}{2}\right]$ or Regime I

Draw a reference square at the center of the network with side length $\sqrt{\frac{1}{2}} n^{\frac{\beta_1 - \beta_2}{2}}$, as illustrated in Figure 11.17. Let $\mathcal{S}_c$ denote the set of nodes outside the reference

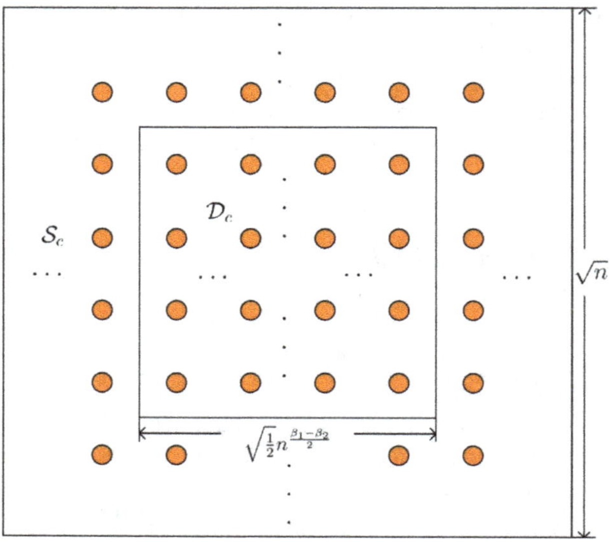

Figure 11.17 Illustration of reference square to construct the cut set bound in Lemma 3

square and $\mathcal{D}_c$ denote the set of nodes inside the reference square. The converse result for this case can be proved by considering the cut set bound between $\mathcal{S}_c$ and $\mathcal{D}_c$ as follows. Let $q_l = I\left(\cup_{i\in\mathcal{D}_c}B_i; W_l\right)/F$. Following similar analysis to that in Section 11.7.4.1, the q_l's must satisfy the total cache capacity constraint $\sum_{l=1}^{L} q_l \leq |\mathcal{D}_c| L_C$, where $|\mathcal{D}_c| = \frac{1}{2}n^{\beta_1-\beta_2}$. Moreover, under any feasible content delivery scheme, the amount of information u_i^j transmitted from the nodes in $\mathcal{S}_c$ to node i during the time window $\left[t_i^j, ..., t_i^{j+1} - 1\right]$ must satisfy $u_i^j \geq \left(1 - q_{l_i} - \varepsilon_F\right)F$. As a result, for given q_l's and per-node rate requirement R, the total average traffic rate over the cut from $\mathcal{S}_c$ to $\mathcal{D}_c$ is $|\mathcal{D}_c|\left(\sum_{l=1}^{L} p_l\left(1 - q_l\right) - \varepsilon_F\right)R$. Clearly, a per-node throughput R is achievable only if the induced total average traffic rate does not exceed the sum capacity Γ_c of the MIMO channel between the $\mathcal{S}_c$ and $\mathcal{D}_c$. Hence, the achievable per-node throughput is upper bounded by

$$R \leq \max_{\{q_l\}} \frac{\Gamma_c}{|\mathcal{D}_c|\left(\sum_{l=1}^{L} p_l\left(1 - q_l\right) - \varepsilon_F\right)},$$
$$\text{s.t.} \sum_{l=1}^{L} q_l \leq \left\lceil |\mathcal{D}_c| L_C \right\rceil. \tag{11.43}$$

It is easy to see that the optimal q_l's to maximize the achievable per-node throughput upper bound is to cache the most popular $\left\lceil |\mathcal{D}_c| L_C \right\rceil$ files, i.e., $q_l^\star = 1, l = 1, ..., \left\lceil |\mathcal{D}_c| L_C \right\rceil$ and $q_l^\star = 0, l > \left\lceil |\mathcal{D}_c| L_C \right\rceil$. As a result, we have

$$R \leq \frac{\Gamma_c}{p_c |\mathscr{D}_c|}, \tag{11.44}$$

where $p_c \triangleq \sum_{l=1}^{L} p_l \left(1 - q_l^{\star}\right) = \sum_{l=\lceil |\mathscr{D}_c| L_C \rceil + 1}^{L} p_l$ satisfies

$$p_c = \begin{cases} \Omega\left(1\right), & \tau \in \left[0, 1\right) \\ \Omega\left(\frac{1}{\log n}\right) & \tau = 1 \\ \Omega\left(n^{\beta_1(1-\tau)}\right), & \tau > 1 \end{cases} . \tag{11.45}$$

The following lemma bounds the sum capacity Γ_c between the $\mathscr{S}_c$ and $\mathscr{D}_c$.

Lemma 3: The sum capacity Γ_c of the MIMO channel between the $\mathscr{S}_c$ and $\mathscr{D}_c$ is bounded as

$$\Gamma_c \leq \Theta\left(n^{(\beta_1 - \beta_2)\left(2 - \frac{\min(3,\alpha)}{2}\right) + \epsilon}\right), \tag{11.46}$$

where $\epsilon < 0$ is arbitrarily small.

Please refer to Appendix C.4 for the proof.

Substituting the upper bound of Γ_c in (11.46) into (11.44) and letting $F \to \infty$, we obtain the desired results in Theorem 6 for Regime II with $\tau \in \left[0, \frac{\min(3,\alpha)}{2}\right]$ as well as Regime I.

11.8 Discussions on the scaling laws

In this section, we point out the limitations and possible extensions of the scaling laws in this chapter.

11.8.1 Extension for n $\neq$ 4^{M}

We have focused on the case when there are $n = 4^M$ nodes in the network for some positive integer M to simplify the description of the achievable scheme. In the following, we show that the proposed achievable scheme can be easily extended to the case when $n \neq 4^M$, and the scaling laws in this chapter remain valid.

11.8.1.1 Hierarchical cache content placement for $n \neq 4^M$

When $n \neq 4^M$, we can still define $M \triangleq \lfloor \log_4 n \rfloor$. In this case, n may not be divisible by 4^{M-m}, and the number of nodes in different clusters in the m-th level may be different. However, we can slightly change the hierarchical cache content placement scheme as follows to handle this case. In the m-th level for $m = 0$ to M, partition the nodes into 4^{M-m} clusters such that the number of nodes in the i-th cluster $\left|V_{m,i}\right|$ satisfies $\lfloor n4^{m-M} \rfloor \leq \left|V_{m,i}\right| \leq \lfloor n4^{m-M} \rfloor + 1$. In this case, the cache content placement parameter q_l only takes values from the discrete set $\left\{0, \frac{1}{n}, \frac{1}{\lfloor n4^{-1} \rfloor}, \ldots, \frac{1}{\lfloor n4^{-M} \rfloor}, 1\right\}$. When $q_l = \frac{1}{\lfloor n4^{m-M} \rfloor}$, each of the first $\lfloor n4^{m-M} \rfloor$ nodes in $V_{m,i}$ caches a portion of the

$\frac{F}{\lfloor n4^{m-M} \rfloor}$ bits for file l such that the l-th file can be reconstructed by collecting all portions from the caches of the first $\lfloor n4^{m-M} \rfloor$ nodes in $V_{m,i}$. If $V_{m,i}$ has $\lfloor n4^{m-M} \rfloor + 1$ nodes, then the $(\lfloor n4^{m-M} \rfloor + 1)$-th node in $V_{m,i}$ simply caches the same content as that in the first node in $V_{m,i}$.

11.8.1.2 Tree-graph-based content delivery for $n \neq 4^M$

Similar to the communication schemes considered in [14, 32], the proposed tree-graph-based content delivery scheme still works even when $n \neq 4^M$. In fact, the tree-graph-based content delivery scheme only requires that every cluster $V_{m,i}$ in the m-th level has the same order of $O\left(n4^{m-M}\right)$ nodes, which is still satisfied when $n \neq 4^M$.

11.8.1.3 Scaling laws for $n \neq 4^M$

As shown above, compared to the case with $n = 4^M$, the main difference when $n \neq 4^M$ is that the number of nodes in different clusters in the same level is no longer identical. However, since different clusters in the same level still have the same order of nodes even when $n \neq 4^M$, the achievable throughput scaling laws are not affected. On the other hand, the converse proof does not rely on the assumption of $n = 4^M$. Therefore, the scaling laws in this chapter are still valid for the case when $n \neq 4^M$.

11.8.2 Extension to coded-caching-enabled multicasting

In this chapter, we have considered an *independent delivery mechanism* in which messages of different users are treated as independent messages. Such independent delivery mechanism may not be throughput-optimal in some scenarios. For example, when multicasting opportunities are available, it is possible to further improve the throughput performance [2, 34].

There are two types of multicasting opportunities: *"naive" multicasting* due to repeated requests of the same segment of the same file at the same time from different users [34], and *coded-caching-enabled multicasting* as those considered in [2]. In practice, each file consists of a large number of segments and the probability of two users requesting the same segment at the same time is small. As pointed out in [5], even though users keep requesting the same few popular files, the asynchronism of their requests is usually large with respect to the duration of the file (e.g., video) itself, such that the probability that a single transmission from the source nodes is useful for more than one user is essentially zero. This phenomenon is called *"asynchronous content reuse"* in [5]. As a result, the "naive" multicasting due to repeated requests is unlikely to occur in practice.

On the other hand, it is possible to exploit the coded-caching-enabled multicasting to further improve the throughput. However, the coded-caching-enabled multi casting may not improve the scaling laws for wireless D2D caching networks.

Indeed, at least for networks with the protocol model, it has been shown in [33, 35] that the optimal scaling laws can be achieved by independent delivery mechanism without coded-caching-enabled multicasting. We conjecture that this is also true for networks with the physical model considered in this chapter. The reason is as follows. Note that the converse proof for Regime II with $\tau > \frac{\min(3,\alpha)}{2}$ in Section 11.7.4.1 does not rely on the assumption of independent delivery, and thus the scaling laws for Regime II with $\tau > \frac{\min(3,\alpha)}{2}$ in this chapter still hold even with the consideration of multicasting. In the converse proof for the other cases in Section 11.7.4.2, there is only one argument that uses the assumption of independent delivery, that is the total average traffic rate induced by all users in $\mathcal{D}_c$ over the cut from $\mathcal{S}_c$ to $\mathcal{D}_c$ is equal to the sum of the average traffic rate induced by each user in $\mathcal{D}_c$. This argument may no longer hold if multicasting is allowed. One potential technique to avoid using this argument in the converse proof is to treat all users in $\mathcal{D}_c$ as a virtual user with a cache of size $|\mathcal{D}_c| L_C F$ bits requesting data object $\cup_{i\in\mathcal{D}_c} W_{l_i}$. If "naive" multicasting is forbidden (but coded-caching-enabled multicasting is allowed), the data objects $W_{l_i}, i \in \mathcal{D}_c$ requested by different users in $\mathcal{D}_c$ are independent. In this case, we can use similar argument as in Section 11.7.4.1 together with the cut set bound in Lemma 11.7.4.2 to prove the converse and establish the same scaling laws as those in this chapter. Therefore, we conjecture that, as far as scaling laws are concerned, simple independent delivery mechanism is good enough.

However, a rigorous study of the impact of coded-caching-enabled multicasting on scaling laws requires a completely different user request model (such as the user request model considered in [5]) that can model the asynchronous content reuse and forbid any form of "naive" multicasting in a mathematically rigorous way, which is beyond the scope of this chapter and is left as future work.

11.8.3 Extension to more general PHY model

For clarity, we have assumed an independent phase fading channel model. However, the achievable throughput analysis in Theorem 3, and the capacity scaling law for extended networks in Theorems 4 and 6 can be readily extended to the more general PHY model. For an arbitrary PHY model, Theorems 3 and 6 still hold if we replace the coefficients c_n and γ_n in Theorem 3 (achievable throughput bounds) and the cut set bounds of the sum capacities Γ_i and Γ_c in the converse proof in Section 11.7.4 with proper expressions under the specific PHY model. For example, for an extended network under the free propagation model with $\alpha = 2$, the results in [36] show that $\gamma_n = 1 - \log\left(\frac{\sqrt{n}}{\lambda}\right)/\log n$, $\Gamma_i = O(\log n)$, and $\Gamma_c = \Theta\left(\left(\frac{\sqrt{n}}{\lambda}\right)^{\beta_1-\beta_2}\right)$ as $n \to \infty$, and thus the capacity scaling law in Regime II is given by

$$R=\begin{cases}\Theta\left(\lambda^{\beta_2-\beta_1}n^{\frac{\beta_2-\beta_1}{2}}\right) & \tau\in[0,1]\\ \Theta\left(\lambda^{\beta_2-\beta_1}n^{\beta_1\left(\tau-\frac{3}{2}\right)+\frac{\beta_2}{2}}\right) & \tau\in\left(1,\frac{\min(3,\alpha)}{2}\right],\\ \Theta\left(n^{\beta_2(\tau-1)}\right), & \tau>\frac{\min(3,\alpha)}{2}\end{cases} \tag{11.47}$$

where λ is the wavelength of the carrier frequency. In Regime I, the capacity scaling law is still given by Theorems 4 and 6.

11.8.4 Limitation of the user request model

In this chapter, we assume that when the requested file is delivered to a node, this node will immediately request another file with the same popularity distribution $\mathbf{p}$. Similar user request models have also been considered in many existing works on wireless caching to make the analysis tractable, see e.g., [3, 5, 19, 28]. In practice, a node usually will not request the same file immediately after receiving this file successfully. Therefore, a more practical user request model is that a node only requests one out of the remaining files that have not been delivered before (with an updated popularity distribution $\mathbf{p}$). The user request model in this chapter is a good approximation of this more practical user request model when there are a large number of content files L with less concentrated popularity distribution (i.e., smaller popularity skewness parameter τ). An interesting future work is to study the impact of more practical user request models on the scaling laws.

11.9 Conclusion

In this chapter, we combine wireless device caching and hierarchical cooperation to significantly improve the capacity of wireless D2D networks. Specifically, we propose a cache-induced hierarchical cooperation scheme where the network is abstracted as a tree graph with each virtual node in the graph representing a cluster of nodes in the network, and the content files are cached at different levels of the tree graph according to their popularities. The PHY has two possible modes: hierarchical cooperation mode or multihop mode, depending on which mode yields a better throughput. The corresponding optimal cache content placement is formulated as an integer programming problem. We propose a low-complexity cache content placement algorithm to solve the integer programming problem and bound the gap w.r.t. the optimal solution. Then we analyze the throughput performance of the cache-induced hierarchical cooperation scheme and show that the proposed scheme achieves significant throughput gain over the cache-assisted multihop scheme, which only supports multihop mode in the PHY. Finally, for extended networks under Zipf popularity distribution, we establish the per-node capacity scaling law by showing that the multiplicative gap between the achievable per-node throughput of the cache-induced hierarchical cooperation and an upper bound of the per-node throughput is within n^{ϵ} for $\epsilon > 0$ that can be arbitrarily small. When the path loss exponent $\alpha < 3$, the optimal per-node capacity scaling law in this chapter can be significantly better than that achieved by the existing state-of-the-art schemes. To the best of our knowledge, this is the first work that completely characterizes the per-node capacity scaling law for wireless caching networks under the physical model and Zipf distribution.

Appendix A

Overview of the improved hierarchical cooperation

In this section, we give an overview of the improved hierarchical cooperation scheme in [16]. We first provide more details on the basic three-phase cooperative transmission scheme outlined in Section 11.3.1. Then we review the improved hierarchical cooperation scheme and explain how to calculate the corresponding common message coding rate $R_c(\alpha, P_I)$.

Before getting into the details, we first explain how the per-node throughput is calculated for a (hierarchical) cooperation scheme. Let $R_c(\alpha, P_I)$ denote the common message coding rate for all users, expressed in bits per codeword symbol. The scheme delivers n messages using a certain number of time slots, each of which corresponds to the duration of a codeword. Hence, the network sum throughput is given by

$$R_{\text{sum}}(n,\alpha) = R_c(\alpha, P_I)\,\mathrm{T}(n,\alpha),$$

where

$$\mathrm{T}(n,\alpha) = n/(\text{required number of time slots}),$$

is the number of delivered source-destination messages per time slot ratio (referred to as packet throughput in the following). Then the per-node throughput is $R_{\text{sum}}(n,\alpha)/n$.

A.1 Basic three-phase cooperative transmission scheme

Phases 1 and 3 of the scheme employ intra-cluster TDMA and spatial reuse across the clusters. Since only one source-destination pair is active per cluster and there are n/N cluster, Phases 1 and 3 can be modeled as a n/N-user interference channel. A simple "Gaussian" single-user capacity achieving code is used in Phases 1 and 3 by treating interference as noise (TIN). The transmit power and reuse factor are chosen as $P = \mathrm{SNR}A(n)^{\alpha/2}$ and $T_r = \left\lceil \sqrt{\mathrm{SNR}}^{1/\alpha} + 1 \right\rceil$, respectively. Then, the local communication rate of $R^{(1)}(\mathrm{SNR}) = \log\left(1 + \frac{\mathrm{SNR}}{1+P_I}\right)$ is achievable by TIN, where the inter-cluster total interference is upper bounded by P_I. Reliable local communication is ensured by letting:

$$R_c(\alpha, P_I) \le R^{(1)}(\mathrm{SNR}).$$

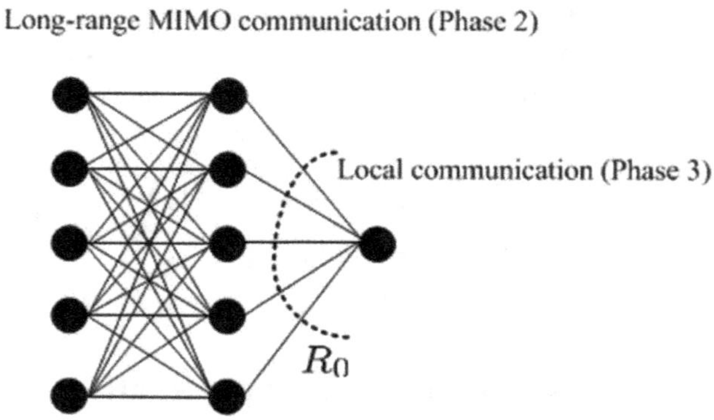

Figure A.1 *Distributed MIMO channel with finite backhaul capacity of rate* R_0, *where* R_0 *is determined by local communication rate*

By concatenating Phases 2 and 3 of the cooperative scheme, we obtain an equivalent distributed MIMO channel with finite backhaul capacity of rate R_0 as shown in Figure A.1, where the N transmit (resp., N receiver) antennas correspond to the N nodes in the source cluster (resp., destination cluster). The quantize-remap-and-forward (QMF) scheme (also referred to as noisy network coding) proposed in [37] is used for this equivalent distributed MIMO channel, where each destination performs joint typical decoding to obtain its own desired message based on the quantized signals. For the MIMO transmission in Phase 2, the transmit power is chosen to be

$$P_{\text{MIMO}} = \frac{\text{SNR}'}{N} A\left(n\right)^{\alpha/2},$$

where SNR' can be arbitrary chosen with a uniform bound SNR_{max} that does not scale with n. The local communication of Phase 3 can be expanded over Q time slots for some integer Q, in order to obtain more flexibility in the quantization rate of the underlying QMF scheme. This yields the backhaul capacity of the "equivalent" model as $R_0 = QR^{(1)}\left(\text{SNR}\right)$.

Let N_0 denote the variance of the additive noise plus inter-cluster interference. It is shown in [16] that the QMF achieves the symmetric rate of

$$\begin{aligned} &R_{\text{QMF}}\left(R_0, N_0, \text{SNR}\right) \\ &= \min\left\{R_0 - \log\left(1 + \frac{N_0}{\sigma_q^2}\right), \mathscr{C}\left(\frac{\text{SNR}}{N_0 + \sigma_q^2}\right)\right\}, \end{aligned}$$

for some quantization level $\sigma_q^2 \geq 0$, where

$$\mathscr{C}\left(x\right) \triangleq 2\log\left(\frac{1+\sqrt{1+4x}}{2}\right) - \frac{\log e}{4x}\left(\sqrt{1+4x} - 1\right)^2.$$

The optimal value of σ_q^2 can be easily attained by solving

$$R_0 - \log\left(1 + \frac{N_0}{\sigma_q^2}\right) = \mathscr{C}\left(\frac{\text{SNR}}{N_0+\sigma_q^2}\right)$$

using bisection method. In the rest of this section, we use $R_{\text{QMF}}\left(R_0, N_0, \text{SNR}\right)$ to denote the achievable rate of QMF with the optimal quantization level. Since the common message coding rate is limited by the achievable rates of both Phase 1 and QMF, we have

$$\begin{aligned} & R_c\left(\alpha, P_I\right) \\ & = \min\left\{R^{(1)}\left(\text{SNR}\right), R_{\text{QMF}}\left(QR^{(1)}\left(\text{SNR}\right), 1, \text{SNR}'\right)\right\} \end{aligned}$$

for the basic three-phase cooperative transmission scheme.

Finally, by counting the total number of time slots consumed by the three phases and optimizing the cluster size N, we have

$$\text{T}\left(n, \alpha\right) = \frac{\sqrt{n}}{2T_r\sqrt{1+Q}}$$

and the per-node throughput can be calculated accordingly.

A.2 Improved hierarchical cooperation scheme and the corresponding $R_c\left(\alpha, P_I\right)$

In the improved hierarchical cooperation scheme, we use the symmetric coding rate $R_c\left(\alpha, P_I\right)$ regardless of the number of hierarchical stages s. Choose the parameters as $P_i = \text{SNR}A_i\left(n\right)^{\alpha/2}$, $T_r = \left[\sqrt{SNR}^{1/\alpha} + 1\right]$ and $\text{SNR} = 2^{2(3+\alpha/\ln 2)}$ for stages $i = 1, \ldots, s$, where P_i is the transmit power for Phases 1 and 3 in stage i, and $A_i\left(n\right)$ denotes the cluster area of stage i. Such choice guarantees that, regardless of the hierarchical stage i, the received power of inter-cluster interference is upper bounded by P_I. For the MIMO transmission phase in stage i, the transmit power is chosen to be

$$P_{\text{MIMO},i} = \frac{\text{SNR}}{M}A_i\left(n\right)^{\alpha/2},$$

which also makes the interference power to be not larger than P_I.

From Section A, we have the rate-constraint of

$$R_c\left(\alpha, P_I\right) \le R^{(1)} \triangleq \log\left(1 + \frac{\text{SNR}}{1+P_I}\right),$$

for reliable local communication at the bottom stage (i.e., stage 1). Concatenating Phases 2 and 3 of stage 1, we can produce an equivalent distributed MIMO channel with backhaul capacity of $QR^{(1)}$. Then, the coding rate should satisfy the constraint

$$R_c\left(\alpha, P_I\right) \le R_{\text{QMF}}\left(QR^{(1)}, P_I + 1, \text{SNR}\right) \triangleq R^{(2)}.$$

It is shown in [16] that $R^{(2)} \le R^{(1)}$. Since $R^{(2)}$ is the local communication rate of stage 2, we can produce a degraded distributed MIMO channel with backhaul capacity $QR^{(2)} \le QR^{(1)}$, resulting in the rate-constraint

$$R_c\left(\alpha, P_I\right) \le R_{\text{QMF}}\left(QR^{(2)}, P_I + 1, \text{SNR}\right) \triangleq R^{(3)}.$$

Due to the smaller backhaul capacity, we have that $R^{(3)} \leq R^{(2)}$. Repeating the above procedure, we obtain that

$$R^{(s+1)} \triangleq R_{\text{QMF}}\left(QR^{(s)}, P_I + 1, \text{SNR}\right) \leq R^{(s)},$$

implying that $\{R^{(s)}\}$ is monotonically non-increasing. Hence, there exists a limit

$$\lim_{s\to\infty} R^{(s)} = R^{\star}\left(\alpha, P_I, Q\right).$$

All rate-constraints are satisfied by choosing $R_c\left(\alpha, P_I\right) = R^{\star}\left(\alpha, P_I, Q\right)$. It is observed in [16] that $Q = 2$ is the best choice and thus we have $R_c\left(\alpha, P_I\right) = R^{\star}\left(\alpha, P_I, 2\right)$, which can be easily calculated numerically.

To obtain the per-node throughput, we still need to calculate the packet throughput. By counting the total number of time slots and optimizing the cluster size N_i for stage $i = 1, ..., s$, we have

$$\mathrm{T}^{(s)}\left(n, \alpha\right) = \frac{n^{\frac{s}{s+1}}}{(1+s)\left(T_r\sqrt{2}\right)^{s}}, \tag{11.A.1}$$

for the basic hierarchical cooperation scheme in [12]. By using the network multiple access approach in [30] and the more efficient TDMA scheduling in [16], the packet throughput can be improved from (A.1) to

$$\mathrm{T}^{(s)}\left(n, \alpha\right) = \frac{n^{\frac{s}{s+1}}}{(1+s)T_r^{\frac{2s}{s+1}}\left(3\cdot 2^{s}-1\right)^{\frac{s}{2(s+1)}}},$$

for $s \geq 2$. As a result, the per-node throughput of the improved hierarchical cooperation scheme is given by $R_H^{(s)}\left(n, P_I\right)$ in Theorem 1.

Appendix B

Reformulation of the Problem

We reformulate Problem (11.12) as a ZOLP for fixed R as follows. Define the binary variables $\delta_{m,l} \in \{0, 1\}$ and the $(M+2) \times L$ matrix $\boldsymbol{\Delta} = [\delta_{m,l}]$ formed as follows: the first row is the all-one vector, i.e., $\delta_{m,l} = 1$ for all $l = 1, \ldots, L$. The last row is the all-zero vector, i.e., $\delta_{M+1,l} = 0$ for all $l = 1, \ldots, L$. The remaining rows between $m = 1$ and $m = M$ satisfy the following monotonicity conditions on the rows and on the columns:

$$\delta_{m,l} \leq \delta_{m,l+1}, = 1, ..., M, = 1, ..., L,$$
$$\delta_{m,l} \geq \delta_{m+1,l}, = 1, ..., M, = 1, ..., L.$$

In words, the matrix $\boldsymbol{\Delta}$ has monotonically non-decreasing rows, and monotonically non-increasing columns. Since the δ-variables are binary, this means that each row of $\boldsymbol{\Delta}$ is formed by a leading block of zeros followed by a block of ones, and each column of $\boldsymbol{\Delta}$ is formed by a leading block of ones followed by a block of zeros.

Observation 1: For $m = 0, \ldots, M$, we let

$$x_m = \sum_{l=1}^{L} \delta_{m,l} - \sum_{l=1}^{L} \delta_{m+1,l}.$$

Hence, the link capacity constraint (11.13) can be written as a linear constraint with respect to the variables $\delta_{m,l}$ for fixed R as follows

$$\mathbf{\Delta p}R \leq \mathbf{c},$$

where $\mathbf{p}$ is the $L \times 1$ vectors containing the file request distribution $\{p_l\}$, and $\mathbf{c}$ is a $(M+2) \times 1$ vector with the first element $c_0 = R$, the last element $c_{M+1} = 0$, and elements $m = 1, \ldots, M$ equal to $c_m = C_m 4^{-(m-1)}$.

Observation 2: We define row-differential matrix of dimensions $(M+1) \times (M+2)$ given by

$$\mathrm{D} = \begin{bmatrix} 1 & -1 & 0 & \cdots & & 0 \\ 0 & 1 & -1 & 0 & \cdots & 0 \\ \vdots & & \ddots & \ddots & & \vdots \\ 0 & \cdots & 0 & 1 & -1 & 0 \\ 0 & & \cdots & 0 & 1 & -1 \end{bmatrix}.$$

Then, it is not difficult to see that the cache size constraint (11.15) can be rewritten as

$$\mathbf{w}^T \mathbf{D\Delta 1} \leq L_C,$$

where $\mathbf{w}$ is a $(M+1) \times 1$ weight vector with elements $w_m = 4^{-m}$ and $\mathbf{1}$ is the all-one column vector of dimension $L \times 1$.

Observation 3: Constraint (11.15) with this new parameterization of the problem becomes irrelevant since it is automatically imposed by the side of the matrix $\mathbf{\Delta}$.

It follows that the problem re-parameterized in the binary variables $\delta_{m,l}$ can be written as

$$\begin{aligned} &\max_{R,\mathbf{\Delta}} R \\ &\text{s.t.} \mathbf{\Delta p}R \leq \mathbf{c}, \\ &\mathbf{w}^T \mathbf{D\Delta 1} \leq L_C, \\ &\delta_{m,l} \leq \delta_{m,l+1}, = 1, ..., M, = 1, ..., L, \\ &\delta_{m,l} \geq \delta_{m+1,l}, = 1, ..., M, = 1, ..., L, \\ &\delta_{0,l} = 1, = 1, ..., L, \\ &\delta_{m+1,l} = 0, = 1, ..., L. \end{aligned} \tag{11.B.1}$$

This is a ZOLP feasibility problem for any fixed value of R. Therefore, it is possible to use standard ZOLP solvers for fixed R, and perform a bisection search over $R \in [0, R_{\max}]$, where $R_{\max}$ is some upper bound on the per-node throughput that can be found by, e.g., relaxing the binary constraint on $\delta_{m,l}$ to $\delta_{m,l} \in \left[0, 1\right]$.

Appendix C

Proofs

C.1 Proof of Theorem 2

First, we show that if $(\mathbf{x}^*, R^*)$ is the optimal solution of Problem (11.18), it must satisfy the conditions in Theorem 2. The conditions $R \le \overline{C}_{m^*}$ and (11.20–11.21) are the constraints in Problem (11.18). Therefore, we only need to prove that $(\mathbf{x}^*, R^*)$ satisfy the first condition in (11.19). Suppose there exist $m' \in \{m^*+1, ..., M\}$ such that $f\left(\sum_{i=m^*}^{m'-1} x_i^* + 1\right) R^* < \overline{C}_{m'}$. Then we can find another feasible solution $\mathbf{x}$ to strictly improve the objective function R as follows. Let $x_{m'} = x^*_{m'} + \varepsilon$, where $\varepsilon > 0$ is a sufficiently small number. Let $x_{m'-1} = x^*_{m'-1} - \varepsilon$, $x_i = x_i^* - \varepsilon', \forall i \notin \{0, 1, ..., m^*-1\} \cup \{m'-1, m'\}$, and $x_0 = x_0 + \sum_{i \notin \{0,1,...,m^*-1\} \cup \{m'-1,m'\}} \varepsilon'$, where $\varepsilon' > 0$ is a sufficiently small number compared to ε. It can be verified that $\sum_{i=m^*}^{m-1} x_i > \sum_{i=m^*}^{m-1} x_i^*, \forall m \in \{m^*, ..., M\} \setminus \{m'\}$ and $\sum_{i=m^*}^{m'-1} x_i > \sum_{i=m^*}^{m'-1} x_i^* - \varepsilon$. Since $f(x)$ is a strictly decreasing function of x with a bounded derivative, for sufficiently small ε, we have

$$f\left(\sum_{i=m^*}^{m-1} x_i + 1\right) R^* < f\left(\sum_{i=m^*}^{m-1} x_i^* + 1\right) R^* \le \overline{C}_m, \forall m \in \{m^*, ..., M\} \backslash m',$$

and $f\left(\sum_{i=m^*}^{m'-1} x_i + 1\right) R^* \left(\sum_{i=m^*}^{m'-1} x_i + 1 - \varepsilon\right) R^* < \overline{C}_{m'}$. Therefore, we can strictly increase R without violating any constraints. Hence, at the optimal solution $(\mathbf{x}^*, R^*)$, the condition in (11.19) must be satisfied.

In the following, we show that the solution to the conditions in Theorem 2 is unique, which implies that these conditions are also sufficient for $(\mathbf{x}^*, R^*)$ to be the optimal solution of Problem (11.18). Specifically, it can be shown that

$$\begin{aligned} L_m\left(\overline{C}_m\right) &> L_m\left(\overline{C}_{m+1}\right) = L_{m+1}\left(\overline{C}_{m+1}\right) \\ L_m\left(\overline{C}_m\right) &= L_{m-1}\left(\overline{C}_m\right) < L_{m-2}\left(\overline{C}_{m-1}\right). \end{aligned} \tag{11.C.1}$$

If m^* is the solution to the optimality conditions, we must have $L_{m^*}\left(\overline{C}_{m^*+1}\right)_C$ and $L_{m^*}\left(\overline{C}_{m^*}\right) \ge L_C$. Then it follows from (14.24) that $L_m\left(\overline{C}_m\right)_C, \forall m^*$ and $L_m\left(\overline{C}_{m+1}\right) \ge L_C, \forall m^*$, which implies any $m \ne m^*$ cannot satisfy the two conditions $L_m\left(\overline{C}_{m+1}\right)_C$ and $L_m\left(\overline{C}_m\right) \ge L_C$ simultaneously. Therefore, the solution to the conditions in Theorem 2 is unique.

C.2 Proof of Lemma 11.6.1

After Steps 2a and 2b of Algorithm 11.1, at the m^* level, we have $x^o_{m^*} = \lfloor x^*_{m^*} \rfloor^*_{m^*}$, and thus

$$\sum_{i=m^*}^{M} x_i^\circ = \sum_{i=m^*}^{M} x_i^* = L,$$

$$\sum_{i=m^*+1}^{M} x_i^\circ = L - \lfloor x^*_{m^*} \rfloor < \sum_{i=m^*+1}^{M} x_i^* + 1.$$

At the m^*+1 level, we have $x^o_{m^*+1} = \left\lfloor x^*_{m*+1} + b_{m*}4^{m^*+1} \right\rfloor > \left\lfloor x^*_{m*+1} + 4\left(x^*_{m*} - x^o_{m*}\right) \right\rfloor$, and thus

$$\begin{aligned}\sum_{i=m^*+2}^{M} x_i^\circ &= L - \left\lfloor x^*_{m*+1} + 4\left(x^*_{m*} - x^o_{m*}\right) \right\rfloor - \left\lfloor x^*_{m*} \right\rfloor \\ &< \sum_{i=m^*+2}^{M} x_i^* + 1.\end{aligned}$$

Similarly, it can be shown that

$$\sum_{i=m}^{M} x_i^\circ < \sum_{i=m}^{M} x_i^* + 1, \forall m = m^*, ..., M. \tag{11.C.2}$$

Let $M^\circ = \mathrm{argmax}_m x_m^\circ$, s.t. $x_m^\circ > 0$ and $m^\circ = \mathrm{argmin}_m x_m^\circ$, s.t. $x_m^\circ \geq 1$, where x_m° is the cache content placement parameter after Steps 2a and 2b. It follows from (14.25) that $\sum_{i=m^\circ}^{M} x_i^* > \sum_{i=m^\circ}^{M} x_i^\circ - 1 = L - 1$, and thus

$$R^* = \overline{C}_{m^\circ}/f\left(\sum_{i=m^*}^{m^\circ-1} x_i^* + 1\right) \leq \overline{C}_{m^\circ}/f(2). \tag{11.C.3}$$

There are three cases as follows:

Case 1: $M^\circ - m^\circ = 0$

In this case, the achievable throughput after Steps 2a and 2b is

$$R^\circ = \overline{C}_{M^\circ}/f(1).$$

It follows from (14.25) that $\sum_{i=M^\circ}^{M} x_i^* > \sum_{i=M^\circ}^{M} x_i^\circ - 1 = L - 1$, and thus

$$R^* = \overline{C}_{M^\circ}/f\left(\sum_{i=m^*}^{M^\circ-1} x_i^* + 1\right) \leq \overline{C}_{M^\circ}/f(2).$$

Therefore,

$$R^\circ/R^* = f(2)/f(1) \geq 1/\left(1+2^\tau\right). \tag{11.C.4}$$

Case 2: $M^\circ - m^\circ = 1$

In this case, the achievable throughput after Steps 2a and 2b is

$$\begin{aligned}R^\circ &= \min\left(\frac{\overline{C}_{m^\circ}}{f(1)}, \frac{\overline{C}_{m^\circ+1}}{f\left(\sum_{i=m^*}^{m^\circ} x_i^\circ+1\right)}\right) \\ &\geq \frac{\overline{C}_{m^\circ+1}}{f(1)}.\end{aligned} \tag{11.C.5}$$

From (C.3), we have

$$\frac{\overline{C}_{m^\circ}}{f(1)\,R^*} \geq \frac{f(2)}{f(1)} \geq 1/\left(1+2^\tau\right).$$

If $x^{\circ}_{m^{\circ}} = L - 1$, we have $\sum_{i=m^*}^{m^{\circ}} x_i^{\circ} + 2 \le L$, and thus

$$\frac{\overline{C}_{m^{\circ}+1}}{f\left(\sum_{i=m^*}^{m^{\circ}} x_i^{\circ} + 1\right) R^*} = \frac{f\left(\sum_{i=m^*}^{m^{\circ}} x_i^* + 1\right)}{f\left(\sum_{i=m^*}^{m^{\circ}} x_i^{\circ} + 1\right)} \ge \frac{f\left(\sum_{i=m^*}^{m^{\circ}} x_i^{\circ} + 2\right)}{f\left(\sum_{i=m^*}^{m^{\circ}} x_i^{\circ} + 1\right)} \ge \frac{1}{1+2^{\tau}}.$$

If
for $L \ge 2$. From the above analysis, we have

$$R^{\circ}/R^* = \tfrac{1}{4} f(2)/f(1) \ge 1/\left(1 + 2^{\tau}\right). \tag{11.C.6}$$

Case 3: $M^{\circ} - m^{\circ} > 1$

In this case, after Step 2f is performed for the first time, the achievable throughput under the cache content placement parameter $\mathbf{x}'$ is given by

$$R' = \min_{m^{\circ} \le m \le M^{\circ}} \left(\frac{\overline{C}_m}{f\left(\sum_{i=m^*}^{m-1} x_i' + 1\right)} \right).$$

If $x^{\circ}_{M^{\circ}} < 3$, we have $x^{\circ}_{M^{\circ}} = 0$, and thus $\frac{\overline{C}_{M^{\circ}}}{f\left(\sum_{i=m^*}^{M^{\circ}-1} x_i' + 1\right)} = \frac{\overline{C}_{M^{\circ}}}{f(L+1)} = +\infty$. Otherwise, $x^{\circ}_{M^{\circ}} \ge 3$ and

$$\frac{\overline{C}_{M^{\circ}}}{f\left(\sum_{i=m^*}^{M^{\circ}-1} x_i' + 1\right) R^*} \ge \frac{\overline{C}_{M^{\circ}}}{f\left(\sum_{i=m^*}^{M^{\circ}-1} x_i^{\circ} + 1\right) R^*} = \frac{f\left(\sum_{i=m^*}^{M^{\circ}-1} x_i^* + 1\right)}{f\left(\sum_{i=m^*}^{M^{\circ}-1} x_i^{\circ} + 1\right)} \ge \frac{f\left(L - x^{\circ}_{M^{\circ}} + 2\right)}{f\left(L - x^{\circ}_{M^{\circ}} + 1\right)} \ge \frac{1}{1+2^{\tau}}. \tag{11.C.7}$$

For $m = m^{\circ} + 1$, we have

$$\frac{\overline{C}_{m^{\circ}+1}}{f\left(\sum_{i=m^*}^{m^{\circ}} x_i' + 1\right) R^*} = \frac{\overline{C}_{m^{\circ}+1}}{f\left(\sum_{i=m^*}^{m^{\circ}} x_i^{\circ}\right) R^*} = \frac{f\left(\sum_{i=m^*}^{m^{\circ}} x_i^* + 1\right)}{f\left(\sum_{i=m^*}^{m^{\circ}} x_i^{\circ}\right)} \ge \frac{f\left(\sum_{i=m^*}^{m^{\circ}} x_i^{\circ} + 2\right)}{f\left(\sum_{i=m^*}^{m^{\circ}} x_i^{\circ}\right)} \ge \frac{1}{1+2^{\tau}}. \tag{11.C.8}$$

Similarly, it can be shown that

$$\frac{\overline{C}_m}{f\left(\sum_{i=m^*}^{m-1} x_i' + 1\right)} > \frac{1}{1+2^{\tau}}, m = m^{\circ}, \ldots, M^{\circ},$$

from which it follows that $R'/R^* \geq \frac{1}{1+2^{\tau}}$.

Since Steps 2c to 2g do not decrease the achievable throughput, it follows from the above analysis that $R^{\circ}/R^* \geq 1/\left(1+2^{\tau}\right)$ also holds after the termination of the algorithm. Finally, the additional factor of $\frac{1}{M}$ is because $\frac{1}{M}\overline{C}_m 4^{(m-1)} \leq C_m \leq \overline{C}_m 4^{(m-1)}$ and we have used the upper bound $\overline{C}_m 4^{(m-1)}$ in the relaxed cache content placement optimization problem in (11.18).

C.3 Proof of Theorem 3

We first give some useful lemmas. The following lemma follows immediately from the optimality condition in Theorem 2. Let $0 \leq \bar{f}_L\left(R, m^*\right) \leq L_{m^*}\left(R\right), R \in \left(\overline{C}_{m^*+1}, \overline{C}_{m^*}\right]$ and $\bar{f}_U\left(R, m^*\right) \geq L_{m^*}\left(R\right), R \in \left(\overline{C}_{m^*+1}, \overline{C}_{m^*}\right]$ be some lower bound and upper bound of $L_{m^*}\left(R\right)$, respectively. For any R_U that satisfies $\bar{f}_L\left(R_U, m^*\right) \geq L_C, R_U \in \left(\overline{C}_{m^*+1}, \overline{C}_{m^*}\right]$ for some $m^* \in \mathbb{Z}_+$, we have $R_U \geq R^*$, where R^* is the optimal solution of the relaxed cache content placement optimization problem in (11.18). And for any R_L that satisfies $\bar{f}_U\left(R_L, m^*\right) \leq L_C, R_L \in \left(\overline{C}_{m^*+1}, \overline{C}_{m^*}\right]$ for some $m^* \in \mathbb{Z}_+$, we have $R_L \leq R^*$.

The following lemma gives closed-form bounds for $f^{-1}\left(\frac{\overline{C}m}{R}\right)$. For different regions of τ, $f^{-1}\left(\frac{\overline{C}m}{R}\right)$ can be lower bounded as

$$
\begin{aligned}
f^{-1}\left(\tfrac{\overline{C}m}{R}\right) &\geq L\left(1-\tfrac{\overline{C}m}{(1-\tau)R}\right)^{+}, \tau \in [0,1) \\
f^{-1}\left(\tfrac{\overline{C}m}{R}\right) &\geq e^{-1}L^{1-\frac{\overline{C}m}{R}}, \tau = 1 \\
f^{-1}\left(\tfrac{\overline{C}m}{R}\right) &\geq 2^{1-\tau}\min\left(\left(\tfrac{\overline{C}m\tau}{R}\right)^{\frac{1}{1-\tau}}, L+1\right), \tau > 1.
\end{aligned}
$$

and upper bounded as

$$
\begin{aligned}
f^{-1}\left(\tfrac{\overline{C}m}{R}\right) &\geq (L+1)\left(1-\tfrac{\overline{C}m}{R}\right), \tau \in [0,1) \\
f^{-1}\left(\tfrac{\overline{C}m}{R}\right) &\geq e^{2}L, \tau = 1 \\
f^{-1}\left(\tfrac{\overline{C}m}{R}\right) &\leq \min\left(\left(\tfrac{\overline{C}m\tau}{R}\right)^{\frac{1}{1-\tau}}, L+1\right), \tau > 1.
\end{aligned}
$$

Proof: The upper bound follows from the fact that $f(x) \geq \frac{\int_x^{L+1} z^{-\tau}dz}{\int_1^{L} z^{-\tau}dz+1}$ and the lower bound follows from the fact that $f(x) \leq \frac{\int_{\lfloor x \rfloor}^{L+1} z^{-\tau}dz+\lfloor x \rfloor^{-1}}{\int_1^{L+1} z^{-\tau}dz}$. The detailed calculations are omitted for conciseness.

With the above lemmas, we are ready to prove Theorem 3. The proof contains five cases depending on the value of τ.

Case 1: $\tau \in [0,1)$

We first prove the lower bound. Replace $f^{-1}\left(\frac{\overline{C}m}{R}\right)$ in $L_{m^*}\left(R\right)$ with the upper bound of $f^{-1}\left(\frac{\overline{C}m}{R}\right)$ for $t \in [0, 1)$ in Lemma C.3, and we obtain an upper bound of $L_{m^*}\left(R\right)$ as

$$\bar{f}_U(R, m^*) = L\left(4 - \frac{3c_n 4^{\gamma_n+1} 4^{-\gamma_n(m^*+1)}}{R(4^{\gamma_n+1}-1)}\right) 4^{-(m^*+1)}.$$

If $\frac{L_C}{L} > 1 - \frac{3\cdot 4^{\gamma_n}}{4^{\gamma_n+1}-1}$, it can be verified that $R_L^a = \frac{3c_n}{(4^{\gamma_n+1}-1)\left(1-\frac{L_C}{L}\right)}$ and $m^* = 0$ satisfies $t \in [0, 1)$. On the other hand, if $\frac{L_C}{L} \leq 1 - \frac{3\cdot 4^{\gamma_n}}{4^{\gamma_n+1}-1}$, it can be verified that $R_L^b = c_n\left(4 - \frac{12}{4^{\gamma_n+1}-1}\right)^{-\gamma_n}\left(\frac{L_C}{L}\right)^{\gamma_n}$ and $m^* = \left\lfloor \frac{1}{\gamma_n}\log_4 \frac{c_n}{R_L}\right\rfloor$ satisfies $\bar{f}_U(R_L^b, m^*) \leq L_C, R_L^b \in \left(\overline{C}_{m^*+1}, \overline{C}_{m^*}\right]$. Then from Lemma C.3, the lower bound given in Theorem 3 is valid for $\tau \in [0, 1)$.

Then we prove the upper bound. Replace $f^{-1}\left(\frac{\overline{C}_m}{R}\right)$ in $L_{m^*}(R)$ with the lower bound of $f^{-1}\left(\frac{\overline{C}_m}{R}\right)$ for $\tau \in [0, 1)$ in Lemma C.3, and we obtain a lower bound of $L_{m^*}(R)$ as

$$\bar{f}_L(R, m^*) \quad \geq L\left(1 - \frac{3\cdot 4^{\gamma_n}}{(4^{\gamma_n+1}-1)}\right)(1-\tau)^{\frac{1}{\gamma_n}}\left(\frac{R}{c_n}\right)^{\frac{1}{\gamma_n}}.$$

Let $L\left(1 - \frac{3\cdot 4^{\gamma_n}}{(4^{\gamma_n+1}-1)}\right)(1-\tau)^{\frac{1}{\gamma_n}}\left(\frac{R_U}{c_n}\right)^{\frac{1}{\gamma_n}} = L_C$, and we have

$$R_U = \frac{c_n}{1-\tau}\left(1 - \frac{3\cdot 4^{\gamma_n}}{(4^{\gamma_n+1}-1)}\right)^{-\gamma_n}\left(\frac{L_C}{L}\right)^{\gamma_n}.$$

Clearly, the above R_U and $m^* = \left\lfloor \frac{1}{\gamma_n}\log_4 \frac{c_n}{R_U}\right\rfloor^+$ satisfy

$$\bar{f}_L(R_U, m^*) > L_C, R_U \in \left(\overline{C}_{m^*+1}, \overline{C}_{m^*}\right].$$

Then from Lemma C.3, the upper bound R_U given in Theorem 3 is valid for $\tau \in [0, 1)$.

Case 2: $\tau = 1$

We first prove the lower bound. Replace $f^{-1}\left(\frac{\overline{C}_m}{R}\right)$ in $L_{m^*}(R)$ with the upper bound of $f^{-1}\left(\frac{\overline{C}_m}{R}\right)$ $\tau \in [0, 1)$ for $\tau = 1$ in Lemma C.3, and we obtain an upper bound of $L_{m^*}(R)$ as

$$\begin{aligned}\bar{f}_U(R, m^*) \quad &= (e^2L - 1)\,4^{-m^*} - (e^2L - L - 1)\,4^{-M} \\ &\leq (e^2L - 1)\,4\left(\frac{R}{c_n}\right)^{\frac{1}{\gamma_n}} - (e^2L - L - 1)\,4^{-M},\end{aligned}$$

where the last inequality follows from $4^{-m^*} < 4\left(\frac{R}{c_n}\right)^{\frac{1}{\gamma_n}}$. Let $(e^2L - 1)\,4\left(\frac{R_L}{c_n}\right)^{\frac{1}{\gamma_n}} - (e^2L - L - 1)\,4^{-M} = L_C$, and we have

$$R_L = c_n\left(\frac{(e^2L-L-1)4^{-M}+L_C}{4(e^2L-1)}\right)^{\gamma_n}.$$

Clearly, the above R_L and $m^* = \left\lfloor \frac{1}{\gamma_n}\log_4 \frac{c_n}{R_L}\right\rfloor^+$ satisfy

$$\bar{f}_U(R_L, m^*) \le L_C, R_L \in \left(\overline{C}_{m^*+1}, \overline{C}_{m^*}\right].$$

Then from Lemma C.3, the lower bound $\frac{1}{M(1+2^{\tau})}R_L$ given in Theorem 3 is valid for $\tau = 1$.

Then we prove the upper bound. Replace $f^{-1}\left(\frac{\overline{C}_m}{R}\right)$ in $L_{m^*}(R)$ with the lower bound of $f^{-1}\left(\frac{\overline{C}_m}{R}\right)$ for $\tau = 1$ in Lemma C.3, and we obtain a lower bound of $L_{m^*}(R)$ as

$$\begin{aligned} \bar{f}_L(R, m^*) &= \frac{3L}{e}\sum_{m=m^*+1}^{M} \frac{L^{-\frac{c_n}{R}4^{-m\gamma_n}}}{4^m} - \frac{(L+1)}{4^M} \\ &\ge \frac{3L}{e}\left(\frac{R}{c_n}\right)^{\frac{1}{\gamma_n}} \sum_{m=1}^{M-m} \frac{L^{-4^{-\gamma_n(m-1)}}}{4^m} - \frac{(L+1)}{4^M}, \end{aligned} \tag{11.C.9}$$

where the last inequality follows from $\frac{c_n}{R} < 4^{(m^*+1)\gamma_n}$ and $4^{-m^*} \ge \left(\frac{R}{c_n}\right)^{\frac{1}{\gamma_n}}$ since $R \in \left(\overline{C}_{m^*+1}, \overline{C}_{m^*}\right]$. It can be shown that $\max_m 4^{-m}L^{-4^{-\gamma_n(m-1)}} \ge \frac{(\ln L)^{-\frac{1}{\gamma_n}} L^{-\frac{1}{\ln L}}}{4}$, from which it follows that

$$\bar{f}_L(R, m^*) \ge \frac{3L}{e}\left(\frac{R}{c_n}\right)^{\frac{1}{\gamma_n}} \frac{(\ln L)^{-\frac{1}{\gamma_n}} L^{-\frac{1}{\ln L}}}{4} - \frac{(L+1)}{4^M}. \tag{11.C.10}$$

Let $\frac{3L}{e}\left(\frac{R_U}{c_n}\right)^{\frac{1}{\gamma_n}} \frac{(\ln L)^{-\frac{1}{\gamma_n}} L^{-\frac{1}{\ln L}}}{4} - (L+1)\,4^{-M} = L_C$, and we have

$$R_U = c_n\left(\frac{4e}{3}\right)^{\gamma_n} L^{\frac{\gamma_n}{\ln L}} \ln L\left(\frac{L_C}{L} + \frac{1}{4^M}\left(1 + \frac{1}{L}\right)\right)^{\gamma_n}.$$

Clearly, the above R_U and $m^* = \left\lfloor \frac{1}{\gamma_n}\log_4 \frac{c_n}{R_U}\right\rfloor^+$ satisfy

$$\bar{f}_L(R_U, m^*) > L_C, R_U \in \left(\overline{C}_{m^*+1}, \overline{C}_{m^*}\right].$$

Then from Lemma C.3, the upper bound R_U given in Theorem 3 is valid for $\tau = 1$.

Case 3: $\tau \in (1, \gamma_n + 1)$

We first prove the lower bound. Replace $f^{-1}\left(\frac{\overline{C}_m}{R}\right)$ in $L_{m^*}(R)$ with the lower bound of $f^{-1}\left(\frac{\overline{C}_m}{R}\right)$ for $t > 1$ in Lemma C.3, and we obtain a lower bound of $L_{m^*}(R)$ as

$$\bar{f}_U(R, m^*) = \frac{\left(4^{\frac{\gamma_n+\tau-1}{\tau-1}} - 4\right) L^{\frac{1+\gamma_n-\tau}{\gamma_n}} \left(\frac{\tau R}{c_n}\right)^{\frac{1}{\gamma_n}}}{\left(4^{\frac{1+\gamma_n-\tau}{\tau-1}} - 1\right)}.$$

Let $\bar{f}_U(R_L, m^*) = L_C$, and we have

$$R_L = \frac{c_n}{\tau}\left(\frac{4^{\frac{1+\gamma_n-\tau}{\tau-1}} - 1}{4^{\frac{\gamma_n+\tau-1}{\tau-1}} - 4}\right)^{\gamma_n} L_C^{\gamma_n} L^{\tau-1-\gamma_n}.$$

Clearly, the above R_L and $m^* = \left\lfloor \frac{1}{\gamma_n}\log_4 \frac{c_n}{R_L}\right\rfloor^+$ satisfy

$$\bar{f}_U(R_L, m^*) \le L_C, R_L \in \left(\overline{C}_{m^*+1}, \overline{C}_{m^*}\right].$$

Then from Lemma C.3, the lower bound $\frac{1}{M(1+2^{\tau})}R_L$ given in Theorem 3 is valid for $\tau \in (1, \gamma_n + 1)$.

Then we prove the upper bound. Replace $f^{-1}\left(\frac{\overline{C}_m}{R}\right)$ in $L_{m^*}(R)$ with the lower bound of $f^{-1}\left(\frac{\overline{C}_m}{R}\right)$ for $\tau > 1$ in Lemma C.3, and we obtain a lower bound of $L_{m^*}(R)$ as

$$\begin{aligned}\bar{f}_L(R, m^*) &= \frac{L^{\frac{1+\gamma_n-\tau}{\gamma_n}}\left(\frac{\tau R}{c_n}\right)^{\frac{1}{\gamma_n}}}{2^{\tau}-1} - 4^{-m^*} \\ &\geq \frac{L^{\frac{1+\gamma_n-\tau}{\gamma_n}}\left(\frac{\tau R}{c_n}\right)^{\frac{1}{\gamma_n}}}{2^{\tau}-1} - 4\left(\frac{R}{c_n}\right)^{\frac{1}{\gamma_n}},\end{aligned}$$

where the last inequality follows from $4^{-m^*} < 4\left(\frac{R}{c_n}\right)^{\frac{1}{\gamma_n}}$ since $R > \overline{C}_{m^*+1}$. Let $\frac{L^{\frac{1+\gamma_n-\tau}{\gamma_n}}\left(\frac{\tau R_U}{c_n}\right)^{\frac{1}{\gamma_n}}}{2^{\tau}-1} - 4\left(\frac{R_U}{c_n}\right)^{\frac{1}{\gamma_n}} = L_C$, and we have

$$R_U = L_C^{\gamma_n}\left(\frac{L^{\frac{1+\gamma_n-\tau}{\gamma_n}}\left(\frac{\tau}{c_n}\right)^{\frac{1}{\gamma_n}}}{2^{\tau}-1} - 4c_n^{-\frac{1}{\gamma_n}}\right)^{-\gamma_n}.$$

Clearly, the above R_U and $m^* = \left\lfloor \frac{1}{\gamma_n}\log_4\frac{c_n}{R_U}\right\rfloor^+$ satisfy

$$\bar{f}_L(R_U, m^*) > L_C, R_U \in \left(\overline{C}_{m^*+1}, \overline{C}_{m^*}\right].$$

Then from Lemma C.3, the upper bound R_U given in Theorem 3 is valid for $\tau \in (1, \gamma_n + 1)$.

Case 4: $\tau = \gamma_n + 1$

Replace $f^{-1}\left(\frac{\overline{C}_m}{R}\right)$ in $L_{m^*}(R)$ with the upper bound of $f^{-1}\left(\frac{\overline{C}_m}{R}\right)$ for $\tau > 1$ in Lemma E, and we obtain an upper bound of $L_{m^*}(R)$ as

$$\bar{f}_U(R, m^*) = (3\log_4 L + 4)\left(\frac{\tau R}{c_n}\right)^{\frac{1}{\gamma_n}}.$$

Let $\bar{f}_U(R_L, m^*) = L_C$, and we have

$$R_L = (3\log_4 L + 4)^{-\gamma_n}\frac{c_n}{\tau}L_C^{\gamma_n}.$$

Clearly, the above R_L and $m^* = \left\lfloor \frac{1}{\gamma_n}\log_4\frac{c_n}{R_L}\right\rfloor^+$ satisfy

$$\bar{f}_U(R_L, m^*) \leq L_C, R_L \in \left(\overline{C}_{m^*+1}, \overline{C}_{m^*}\right].$$

Then from Lemma C.3, the lower bound $\frac{1}{M(1+2^{\tau})}R_L$ given in theorem 3 is valid for $\tau = \gamma_n + 1$.

For the throughput upper bound, consider a scheme which caches the most popular files at the 0-th level and the remaining files at the 1-th level. Clearly, the throughput achieved by such a scheme must be larger than and is given by

$$\frac{c_n 4^{-\gamma_n}}{f(L-L_C+1)} \leq \frac{\left(\tau - L^{1-\tau}\right)c_n 4^{-\gamma_n}}{(L_C+1)^{1-\tau} - (L+1)^{1-\tau}}.$$

Case 5: $\tau > \gamma_n + 1$

Replace $f^{-1}\left(\frac{\overline{C}_m}{R}\right)$ in $L_{m*}(R)$ with the upper bound of $f^{-1}\left(\frac{\overline{C}_m}{R}\right)$ for $\tau > 1$ in Lemma E, and we obtain an upper bound of $L_{m*}(R)$ for $R \le c_n L^{\tau-1}$ as

$$\bar{f}_U(R, m^*) = \left(\frac{3\tau^{\frac{1}{\tau-1}} 4^{\frac{\gamma_n+1-\tau}{\tau-1}}}{1-4^{\frac{\gamma_n+1-\tau}{\tau-1}}} + 4\tau^{\frac{1}{\gamma_n}}\right)\left(\frac{R}{c_n}\right)^{\frac{1}{\tau-1}}.$$

Let $\bar{f}_U(R_L, m^*) = L_C$, and we have

$$R_L = c_n\left(\frac{3\tau^{\frac{1}{\tau-1}} 4^{\frac{\gamma_n+1-\tau}{\tau-1}}}{1-4^{\frac{\gamma_n+1-\tau}{\tau-1}}} + 4\tau^{\frac{1}{\gamma_n}}\right)^{1-\tau} L_C^{\tau-1}.$$

Clearly, the above R_L and $m^* = \left\lfloor \frac{1}{\gamma_n}\log_4 \frac{c_n}{R_L}\right\rfloor^+$ satisfy

$$\bar{f}_U(R_L, m^*) \le L_C, R_L \in \left(\overline{C}_{m^*+1}, \overline{C}_{m^*}\right].$$

Since $R_L \le c_n L^{\tau-1}$, from Lemma C.3, the lower bound $\frac{1}{M(1+2^{\tau})}R_L$ given in Theorem 3 is valid for $\tau > \gamma_n + 1$.

The throughput upper bound is the same as in case 4. This completes the proof.

C.4 Proof of Lemma 3

Lemma 3 can be proved using similar a technique to that in the proof of theorem 5.2 in [12]. With some bounded per-node power constraint P, the sum capacity of the MIMO channel between the $\mathcal{S}_c$ and $\mathcal{D}_c$ is

$$\Gamma_c = \max_{\substack{Q(H)\succeq 0 \\ \mathbb{E}\left(Q_{j,j}(H)\right)\le P, \forall j\in\mathcal{S}_c}} \mathbb{E}\left(\log\left|I + HQ(H)H^H\right|\right), \tag{11.C.11}$$

where $H = [h_{i,j}]_{i\in\mathcal{D}_c, j\in\mathcal{S}_c}$. Let $\overline{V}_c$ denote the set of nodes inside the square at the center of the network with area $(\sqrt{n_c}-2)^2$, where $n_c = n^{\beta_1-\beta_2}$, and let $V_c = \mathcal{D}_c\backslash\overline{V}_c$. By the generalized Hadamard's inequality, we have

$$\begin{aligned}\log\left|I + HQ(H)H^H\right| &\le \log\left|I + H^aQ(H)H^{aH}\right| \\ &+ \log\left|I + H^bQ(H)H^{bH}\right|,\end{aligned}$$

where $H^a = [h_{i,j}]_{i\in V_c, j\in\mathcal{S}_c}$ is the channel between the $\mathcal{S}_c$ and V_c, and

$$H^a = [h_{i,j}]_{i\in V_c, j\in\mathcal{S}_c}$$

is the channel between the $\mathcal{S}_c$ and $\overline{V}_c$, and thus (11.C.11) is bounded above by

$$\begin{aligned}\Gamma_c \le &\max_{\substack{Q(H^{(a)})\succeq 0 \\ \mathbb{E}\left(Q_{j,j}(H^a)\right)\le P, \forall j\in\mathcal{S}_c}} \mathbb{E}\left(\log\left|I + H^aQ(H^a)H^{aH}\right|\right) \\ &+\max_{\substack{Q(H^b)\succeq 0 \\ \mathbb{E}\left(Q_{j,j}(H^b)\right)\le P, \forall j\in\mathcal{S}_c}} \mathbb{E}\left(\log\left|I + H^bQ(H^b)H^{bH}\right|\right).\end{aligned} \tag{11.C.12}$$

Applying Hadamard's inequality once more, the first term in (11.C.12) can be upper-bounded by the sum of the capacities of the individual MISO channels between nodes in $\mathscr{S}_c$ and each node in V_c. Following a similar analysis to that in the proof of theorem 5.2 in [12], the first term in (11.C.12) is upper bounded by $K' \sqrt{n_c}(\log n)^2$, where K' is a constant independent of n.

To bound the second term in (11.C.12), we introduce the concept of the total power received by all the nodes in $\overline{V}_c$, when the nodes in $\mathcal{S}_c$ are transmitting independent signals with power P. Specifically, let P_j denote the total received power in $\overline{V}_c$ of the signal sent by $j \in \mathcal{S}_c$. Let $P_{tot}(n_c) = \sum_{j \in \mathscr{S}_c} P_j$ and define $\tilde{\boldsymbol{H}} = [h_{i,j}/\sqrt{d_j}]_{i \in \overline{V}_c, j \in \mathscr{S}_c}$, where $d_j = \sum_{i \in \overline{V}_c} r_{i,j}^{-\alpha}$. Then the second term is equal to

$$
\begin{aligned}
&\max_{\substack{Q(\tilde{H}) \succeq 0 \\ \mathbb{E}\left(\tilde{Q}_{j,j}(\tilde{H})\right) \leq P_j, \forall j \in \mathscr{S}_c}} \mathbb{E}\left(\log\left|\boldsymbol{I} + \tilde{\boldsymbol{H}}\tilde{\boldsymbol{Q}}(\tilde{\boldsymbol{H}})\tilde{\boldsymbol{H}}^H\right|\right) \\
&\leq \max_{\substack{Q(\tilde{H}) \succeq 0 \\ \mathbb{E}(\mathrm{Tr}(\tilde{Q}(\tilde{H}))) \leq P_{tot}(n_c)}} \mathbb{E}\left(\log\left|\boldsymbol{I} + \tilde{\boldsymbol{H}}\tilde{\boldsymbol{Q}}(\tilde{\boldsymbol{H}})\tilde{\boldsymbol{H}}^H\right|\right) \\
&\leq \max_{\substack{Q(\tilde{H}) \succeq 0 \\ \mathbb{E}(\mathrm{Tr}(\tilde{Q}(\tilde{H}))) \leq P_{tot}(n_c)}} \mathbb{E}\left(\log\left|\boldsymbol{I} + \tilde{\boldsymbol{H}}\tilde{\boldsymbol{Q}}(\tilde{\boldsymbol{H}})\tilde{\boldsymbol{H}}^H\right| 1_{B_{n,\epsilon}}\right) \\
&+ \max_{\substack{Q(\tilde{H}) \succeq 0 \\ \mathbb{E}(\mathrm{Tr}(\tilde{Q}(\tilde{H}))) \leq P_{tot}(n_c)}} \mathbb{E}\left(\mathrm{Tr}\left(\tilde{\boldsymbol{H}}\tilde{\boldsymbol{Q}}(\tilde{\boldsymbol{H}})\tilde{\boldsymbol{H}}^H\right) 1_{B_{n,\epsilon}^c}\right),
\end{aligned}
\tag{11.C.13}
$$

where the set $B_{n,\epsilon} = \left\{\|\tilde{\boldsymbol{H}}\|^2 \; \epsilon\right\}$.

For the first term in (11.C.13), denoted as $C_{B_{n,\epsilon}}$, using a similar analysis to equation (11) in [12], it can be shown that

$$
C_{B_{n,\epsilon}} \leq \tfrac{1}{2} n^{\beta_1 - \beta_2} \log\left(1 + \tfrac{nP_{tot}(n_c)}{\Pr(B_{n,\epsilon})}\right) \Pr\left(B_{n,\epsilon}\right). \tag{11.C.14}
$$

Furthermore, following a similar analysis to Lemma 5.3 in [12], it can be shown that for any $\epsilon > 0$ and $p \geq 1$, there exists $K_1' > 0$ such that for all n,

$$
\Pr\left(B_{n,\epsilon}\right) \leq \tfrac{K_1'}{n^p}. \tag{11.C.15}
$$

It follows from (11.C.14), (11.C.15), and $P_{tot}(n_c) \leq Pn^2$ that

$$
C_{B_{n,\epsilon}} \leq K_1' n^{\beta_1 - \beta_2 - p} \log\left(1 + \tfrac{n^{3+p}}{K_1'}\right), \tag{11.C.16}
$$

which decays to zero with an arbitrary exponent as n tends to infinity.

For the second term in (11.C.13), denoted as $C_{B_{n,\epsilon}^c}$, we have

$$
C_{B_{n,\epsilon}^c} \leq \max_{\substack{Q(\tilde{H}) \succeq 0 \\ \mathbb{E}(\mathrm{Tr}(\tilde{Q}(\tilde{H}))) \\ \leq P_{tot}(n_c)}} \mathbb{E}\left(\|\tilde{\boldsymbol{H}}\|^2 \mathrm{Tr}\left(\tilde{\boldsymbol{Q}}(\tilde{\boldsymbol{H}})\right) 1_{B_{n,\epsilon}^c}\right) \leq n^{\epsilon} P_{tot}(n_c). \tag{11.C.17}
$$

Moreover, it can be verified that

$$P_{tot}(n_c) = \begin{cases} K' n_c^{2-\alpha/2}, \alpha \in (2,3) \\ K' \sqrt{n_c}, \alpha \geq 3 \end{cases} \tag{11.C.18}$$

Finally, (11.39–11.41) complete the proof of Lemma 3.

References

[1] Golrezaei N., Shanmugam K., Dimakis A.G., *et al.*. 'FemtoCaching: Wireless video content delivery through distributed caching helpers'. Proceedings of IEEE INFOCOM; 2012. pp. 1107–15.

[2] Maddah-Ali M.A., Niesen U. 'Fundamental limits of caching'. *IEEE Transactions on Information Theory*. 2014;**60**(5):2856–67.

[3] Liu A., Lau V.K.N. 'Asymptotic scaling laws of wireless ad hoc network with physical layer caching'. *IEEE Transactions on Wireless Communications*. 2016;**15**(3):1657–64.

[4] Ji M., Caire G., Molisch A.F. 'Fundamental limits of distributed caching in D2D wireless networks'. 2013 IEEE Information Theory Workshop (ITW); 2013. pp. 1–5.

[5] Ji M., Caire G., Molisch A.F. 'The throughput-outage tradeoff of wireless one-hop caching networks'. *IEEE Transactions on Information Theory*. 2015;**61**(12):6833–59.

[6] Altieri A., Piantanida P., Vega L.R., Galarza C.G. 'On fundamental trade-offs of device-to-device communications in large wireless networks'. *IEEE Transactions on Wireless Communications*. 2015;**14**(9):4958–71.

[7] Jeon S.W., Hong S.N., Ji M., *et al.*. 'Caching in wireless multihop device-to-device networks'. Proceedings of IEEE International Conference on Communications 2015; 2015. pp. 6732–7.

[8] Gupta P., Kumar P.R. 'The capacity of wireless networks'. *IEEE Transactions on Information Theory*. 2000;**46**(2):388–404.

[9] Liang-Liang X., Kumar P.R. 'A network information theory for wireless communication: scaling laws and optimal operation'. *IEEE Transactions on Information Theory*. 2004;**50**(5):748–67.

[10] Jovicic A., Viswanath P., Kulkarni S.R. 'Upper bounds to transport capacity of wireless networks'. *IEEE Transactions on Information Theory*. 2004;**50**(11):2555–65.

[11] Xie L.L., Kumar P.R. 'On the path-loss attenuation regime for positive cost and linear scaling of transport capacity in wireless networks'. *IEEE Transactions on Information Theory*. 2006;**52**(6):2313–28.

[12] Ozgur A., Leveque O., Tse D.N.C. 'Hierarchical cooperation achieves optimal capacity scaling in ad hoc networks'. *IEEE Transactions on Information Theory*. 2007;**53**(10):3549–72.

[13] Niesen U., Gupta P., Shah D. 'On capacity scaling in arbitrary wireless networks'. *IEEE Transactions on Information Theory*. 2009;**55**(9):3959–82.

[14] Niesen U., Gupta P., Shah D. 'The balanced unicast and multicast capacity regions of large wireless networks'. *IEEE Transactions on Information Theory*. 2010;**56**(5):2249–71.

[15] Franceschetti M., Migliore M.D., Minero P. 'The capacity of wireless networks: information-theoretic and physical limits'. *IEEE Transactions on Information Theory*. 2009;**55**(8):3413–24.

[16] Hong S.-N., Caire G. 'Beyond scaling laws: on the rate performance of dense device-to-device wireless networks'. *IEEE Transactions on Information Theory*. 2015;**61**(9):4735–50.

[17] Liu A., Lau V.K.N. 'Cache-enabled opportunistic cooperative MIMO for video streaming in wireless systems'. *IEEE Transactions on Signal Processing*. 2014;**62**(2):390–402.

[18] Liu A., Lau V.K.N. 'Mixed-timescale precoding and cache control in cached MIMO interference network'. *IEEE Transactions on Signal Processing*. 2013;**61**(24):6320–32.

[19] Gitzenis S., Paschos G.S., Tassiulas L. 'Asymptotic laws for joint content replication and delivery in wireless networks'. *IEEE Transactions on Information Theory*. 2013;**59**(5):2760–76.

[20] Yamakami T. 'A Zipf-Like distribution of popularity and hits in the mobile web Pages with short life time'. Proceedings of IEEE International Conference on Parallel & Distributed Computing; 2006. pp. 240–3.

[21] Ji M., Tulino A.M., Llorca J., *et al.*. 'Caching and coded multicasting: multiple groupcast index coding'. Proceedings of 2014 IEEE Global Conference on Signal and Information Processing; 2014. pp. 881–5.

[22] Hachem J., Karamchandani N., Diggavi S. Multi-level coded caching. Proceedings IEEE ISIT; 2014.

[23] Karamchandani N., Niesen U., Maddah-Ali M.A., Diggavi S.N. 'Hierarchical coded caching'. *IEEE Transactions on Information Theory*. 2016;**62**(6):3212–29.

[24] Han W., Liu A., Lau V.K.N. 'Degrees of freedom in cached MIMO relay networks'. *IEEE Transactions on Signal Processing*. 2015;**63**(15):3986–97.

[25] Han W., Liu A., V K N L. 'Improving the degrees of freedom in MIMO interference network via PHY caching'. Proceedings of 2015 IEEE Global Communications Conference (GLOBECOM); 2015. pp. 1–6.

[26] Naderializadeh N., Maddah-Ali M.A., Avestimehr A.S. 'Fundamental limits of cache-aided interference management'. *arXiv preprint arXiv*. 2016;**160204207**.

[27] Zhang J., Elia P. 'Fundamental limits of cache-aided wireless bc: interplay of coded-caching and CSIT feedback'. *arXiv preprint arXiv*. 2016;**151103961**.

[28] Liu A., VKN L. 'How much cache is needed to achieve linear capacity scaling in backhaul-limited dense wireless networks'. *IEEE/ACM Transactions on Networking*. 2016;**99**:1–10.

[29] Shokrollahi A. 'Raptor codes'. *IEEE Transactions on Information Theory*. 2006;**52**(6):2551–67.

[30] Ozgur A., Leveque O. 'Throughput-Delay tradeoff for hierarchical cooperation in AD *hoc* wireless networks'. *IEEE Transactions on Information Theory*. 2010;**56**(3):1369–77.

[31] Ghaderi J., Xie L.L., Shen X. 'Hierarchical cooperation in ad hoc networks: optimal clustering and achievable throughput'. *IEEE Transactions on Information Theory*. 2009;**55**(8):3425–36.

[32] Niesen U., Shah D., Wornell G.W. 'Caching in wireless networks'. *IEEE Transactions on Information Theory*. 2012;**58**(10):6524–40.

[33] Ji M., Caire G., Molisch A.F. 'Fundamental limits of caching in wireless D2D networks'. *IEEE Transactions on Information Theory*. 2016;**62**(2):849–69.

[34] Yu Q., Maddah-Ali M.A., Avestimehr A.S. 'The exact rate-memory tradeoff for caching with uncoded prefetching'. 2017 IEEE International Symposium on Information Theory (ISIT); 2017. pp. 1613–7.

[35] Jeon S.-W., Hong S.-N., Ji M., Caire G., Molisch A.F. 'Wireless multihop device-to-device caching networks'. *IEEE Transactions on Information Theory*. 2017;**63**(3):1662–76.

[36] Lee S.H., Chung S.Y. 'Capacity scaling of wireless ad hoc networks: Shannon meets Maxwell'. *IEEE Transactions on Information Theory*. 2012;**58**(3):1702–15.

[37] Avestimehr A.S., Diggavi S.N., Tse D.N.C, . 'Wireless network information flow: a deterministic approach'. *IEEE Transactions on Information Theory*. 2011;**57**(4):1872–905.

Chapter 12

Wireless device-to-device caching networks with distributed MIMO and hierarchical cooperations

Jiajia Guo[1,2], Jinhong Yuan[1], and J Andrew Zhang[3]

This chapter first categorizes and summarizes wireless caching on its capability of raising the wireless network throughput in centralized and decentralized network topology. More precisely, what is the throughput scaling law of the wireless caching networks, i.e., how does the network throughput scale with the caching size, the library size, and the number of nodes in different wireless networks? Then, a novel caching scheme for device-to-device (D2D) networks is presented and the throughput scaling law is derived, which shows that the average aggregate throughput of the network scales almost linearly with the number of nodes. Both analytical and numerical results show that the proposed scheme outperforms existing ones when the local cache size is limited.

12.1 Introduction

With the wireless data traffic predicted to increase dramatically in the next few years, tremendous efforts of academia and industry have been made to satisfy this explosive demand while minimizing the bandwidth and energy expenditure. Among the many proposals trying to solve this issue, wireless caching has been proposed as a cost-effective way to handle the high traffic requirement [1–5]. Although the caching technique has been widely used in computer networks, it is until recently that the wireless communication society realized its usefulness after exploiting two basic facts: (1) the content delivered in the networks is "cacheable," which means that the same content is requested by different users, though the requests usually occur at different times; (2) caching shortens the distance between the content and the requester, which naturally alleviates the network burden and reduces the resource

[1]University of New South Wales, Australia
[2]Nurlink Technology Cooperation, United States
[3]University of Technology Sydney, Australia

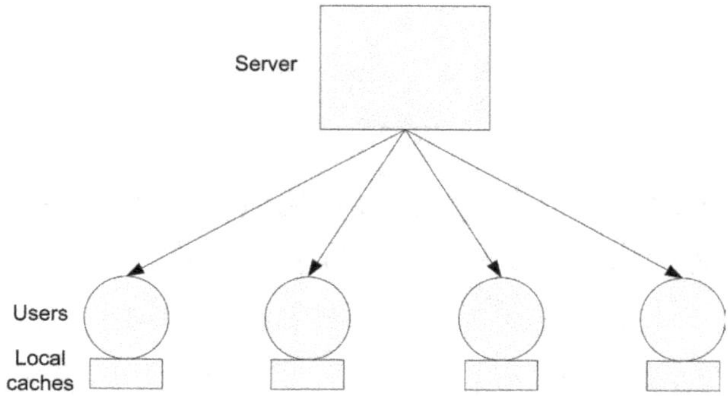

Figure 12.1 A centralized network model

consumption. These properties endow the wireless caching with a promising future for the next-generation cellular networks [6–8].

This chapter is organized as follows. In Section 12.2, we introduce the concept of scaling law, and summarize existing works based on their utilized network topology: centralized and decentralized. Then, we introduce the system model and formulate the problem in Section 12.3. We state the main results of this chapter in Section 12.4. Section 12.5 presents an achievable scheme consisting of both the caching placement strategy and the delivery policy. For the proposed scheme, we then derive the throughput and the outage probability, obtain the scaling law and optimize the throughput in Section 12.6. Section 12.7 provides some numerical results. In the end, Section 12.8 concludes the chapter.

12.2 Wireless caching networks

12.2.1 Caching in centralized networks

Consider a centralized network with one server (or macro-cell base station (MBS)) and several users (or small-cell base stations (SBS)), as depicted in Figure 12.1. Each user is equipped with a local cache of size M_c, which is filled with content files from the core networks through a backhaul link [9–11]. The backhaul link from the server to users is limited, either of low-capacity or expensive.

For this model, an important information theoretical work by M.A. Maddah-Ali and U. Niesen [12] focused on minimizing the backhaul requirement for any random request (from a set of known files) of users. They formulated the problem through the concept of an achievable memory-rate pair (M_c, R). Denote the smallest rate that (M_c,R) is achievable by $R^*(M_c)$, which describes the *memory-rate tradeoff* of the caching problem. We have

$$R^*(M_c) \triangleq \inf\{R : (M_c, R) \text{ is achievable}\}. \tag{12.1}$$

In this scheme, the files in the library are divided into packets and users cache coded subsets of such packets. In the content delivery phase, multicasting opportunities are

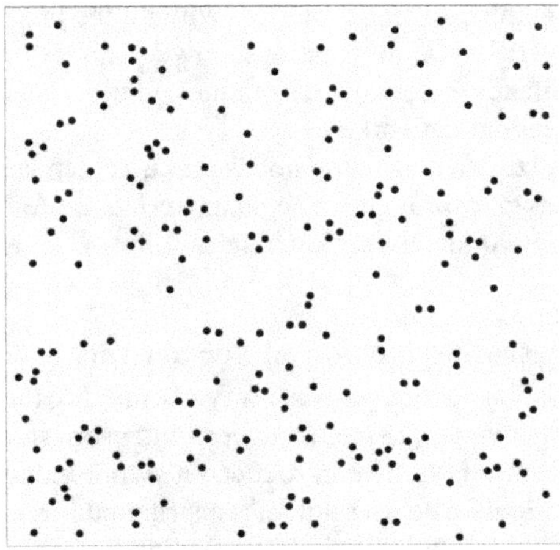

Figure 12.2 A decentralized network model

created through coding across the required packets. Each user will be able to resolve its wanted file by combining this multicasted message and its locally cached coded packets.

Then, assuming that the library size is m and the number of users is K, the authors provided an achievable rate $R^C(M_c)$ as an upper bound of $R^*(M_c)$ as follows;

$$R^C(M_c) = K\left(1 - \frac{M_c}{m}\right) \cdot \frac{1}{1+\frac{KM_c}{m}}. \tag{12.2}$$

Above result suggests that with this scheme, the number of users K and the size of local caches M_c can almost linearly contribute to the achievable rate. This scheme is of zero outage and all the users can be served by one common message from the server. However, its required code length is combinatorial with the number of users, which is infeasible for networks with many users.[a] Besides, changing even a single file in the library requires a significant reconfiguration of the user caches, which makes the cache update very difficult.

12.2.2 Caching in decentralized networks

Consider a network with n wireless nodes, which are uniformly and independently distributed in a unit square as depicted in Figure 12.2. Each node has a transmit power constraint and a local cache size of M_c files. The size of caches in this chapter

[a]Note that given the caching policy and the user demands, the problem of minimizing the load over the shared link is essentially an index coding problem, which is not covered in this chapter.

is represented by the number of standard files. Without the presence of a centralized server or base station (BS), communications between nodes are considered in this network. Each node intends to download its requested file from a library with m files through this cached random network.

Unlike the centralized network, the nodes and user demand here are both random. How can the network throughput be characterized? We introduce the tool in conventional ad hoc networks without caching as follows.

12.2.2.1 Capacity scaling law for ad hoc networks

The capacity scaling law is recognized as a way of studying the fundamental trade-offs in large wireless networks [13] and has been intensely studied since the initial work by Gupta and Kumar [14]. The introduced network model in [14], which has been widely used, considers n nodes randomly distributed in a unit area and arbitrarily grouped into source-destination pairs. The scaling law characterizes the behavior of the capacity of such networks when the number of nodes n becomes large. The works in [14, 15] show that, for such networks, a multihop strategy is capable of achieving an aggregate throughput scaling of $\Theta\left(\sqrt{n}\right)$. This important scaling result indicates that the capacity can increase rather than saturate, with the number of users n. Note that this result is achieved based on a *protocol model*, where nodes within a certain distance are able to communicate and only one node is allowed to transmit at a time within a radius [14]. The protocol model assumes that the signals received from nodes other than the source are interference and regarded as noise, which degrades the link performance. Under this assumption, the optimal strategy is to maximize the number of simultaneous transmissions (spatial reuse) [16].

On the other hand, it has been widely accepted that broadcast and superposition are the two main distinguishing features of wireless medium [17]. Beyond the protocol model, a *physical model* was introduced in [16] to take into consideration wireless channel properties such as pathloss, fading, additive white Gaussian noise (AWGN) noise and interference. Based on this model, when the network is dense (fixed area with node density growing), a hierarchical cooperation scheme proposed in [16, 18, 19] achieves a near-linear throughput scaling of $\Theta\left(n^{1-\epsilon}\right)$, where ϵ can be arbitrarily small. Furthermore, applying the treating-interference-as-noise (TIN) condition in [20], Hong and Caire [21] characterized the achievable average per-link rate (not just the scaling law) for both hierarchical cooperation and multihop schemes. They showed that the signal-to-noise ratio and the number of hierarchical stages in [16, 18, 19] can be optimized, and the optimized hierarchical cooperation schemes yield significant rate gains over multihop strategy, even under realistic conditions.

12.2.2.2 Caching networks

A decentralized D2D caching network was first studied by Ji, Caire, and Molisch [22–24]. The considered caching network consists of n nodes randomly distributed in a unit area, which is similar to the conventional ad hoc model. The difference is that each node in a caching network is equipped with a local cache and requests files from a library according to a priori popularity distribution.

In [22–24], Ji, Caire, and Molisch formulated the problem by using a throughput-outage trade-off as follows. For a given network and request probability mass function, an outage-throughput pair (p, t) is achievable if there exists a cache placement scheme and an admission control and transmission scheduling policy with outage probability $p_0 \leq p$ and minimum per-user average throughput $T_{\min} \geq t$. The largest per-user average throughput $T^*(p)$ that (p, t) is achievable describes the *throughput-outage tradeoff* of the caching problem. It is given by

$$T^*(p) \triangleq \sup\{t : (p, t) \text{ is achievable}\}. \tag{12.3}$$

Then the authors try to solve the following optimization problem:

$$\max_{\text{placement/delivery policy}} \quad T_{\min}, \text{ subject to } p_0 \leq p. \tag{12.4}$$

The goal of this caching network is to get each node's request served through D2D links. Following the protocol model in [14], the network is divided into clusters of equal size, which is independent of the users' request and the cache placement. A user can look for the requested file only inside its own cluster.

They first maximized the probability that any user finds its requested file inside its own cluster by optimizing the caching distribution for given request probability mass function. The given solution suggested an interesting water-filling-like caching placement strategy. Then, with this optimized caching distribution, the corresponding $T^*(p)$ is calculated, which describes the throughput-outage trade-off of D2D caching schemes.

They showed that for a single-hop cache network, with an optimized caching placement strategy and cluster size, the network throughput scales as $\Theta\left(\frac{nM_c}{m}\right)$, growing linearly with the size of local caches M. This demonstrates that equipping caches is a cost-effective solution since the cost of providing caches is much lower than providing bandwidth. This nice scaling is achieved by exploiting spatial reuse of concurrent multiple short range D2D transmissions allowed by local caches at each node [22]. With the size of local caches growing, the requested content can be served by a shorter range D2D transmission and therefore the number of simultaneous active links increases, improving the network throughput. They later improved this scaling to $\Theta\left(n\sqrt{\frac{M_c}{m}}\right)$ in [25, 26] by adopting the multihop strategy, where nodes get served by multi-relaying through the network. This even better scaling is achieved by reducing the number of transmissions within each hopping cell when increasing the size of local cache. With a larger local cache, the requested content can be served by a shorter range D2D transmission, which reduces the number of data paths within each hopping cell. Assuming a constant aggregate achievable rate for one hopping cell, the achievable rate of each data path is increased, leading to an improved network throughput.

The work in [3] improved the scaling law of the ad hoc networks from $\Theta\left(\frac{1}{n}\right)$ to $\Theta\left(1\right)$ by optimizing the content distribution so that homogeneous opportunistic coordinated multi-point (CoMP) transmissions can be supported. However, this opportunistic CoMP is available by caching the same content on different helpers, which actually reduces the cache size for placement. The trade-off between the opportunistic CoMP gain and the cache size reduction is not mentioned by the authors there.

12.2.3 Motivations

Based on the reviews of existing works, this chapter mainly focuses on the decentralized approach, since it largely counts on the cooperation of end users where an potential improvement from a proper relaying strategy can be expected. Especially, the key of designing a decentralized scheme is to optimize the content distribution and transmission scheduling so that (1) the probability that any user finds its requested file in its accessible neighbors is maximized and (2) the interference of different transmitters is minimized.

Current studies in [22–26] are based on the protocol model in [14]. With the success of employing the physical model for conventional ad hoc networks, we are curious about how the caching network capacity would scale if the physical model is used and more wireless channel properties are exploited. We notice that the work in [7] uses the physical model and enlarges the network throughput by employing maximum distance separable code (MDS) for the cached content to create cache-induced opportunistic CoMP. However, we are further interested in whether hierarchical cooperations in caching networks can achieve a better throughput scaling than the multihop scheme as in conventional ad hoc networks. To address these questions, we investigate the caching network based on the physical model as well as considering hierarchical cooperations.

Another motivation of this work is the concern about the limited cache size in practice. We notice that compared with the vast library that users may request, the local cache size at each node is rather limited in realistic networks. Especially, in a D2D network, the cache size of a single device cannot be sufficiently large in many cases. For example, consider the on-demand video streaming case, the file library could be up to 1 000 TB while the available cache size of a user device is usually less than 1 TB. In addition, users usually are willing to contribute only a small fraction of their local caches. This conscious motivates us to design a caching scheme for the small cache case (small $\frac{M_c}{m}$), where the network throughput will be poor according to the existing works [24–26]. Thus, in this work, our primary focus is on the small cache case when designing the caching scheme.

12.3 System model and problem formulation

Consider a network with n wireless nodes, which are uniformly and independently distributed in a unit square, as depicted in Figure 12.2. Each node u,

$u \in \mathcal{U} = \{1, \cdots n\}$, has an average transmit power of P watts and a local cache size of M_c files. The size of caches in this chapter is represented by the number of standard files (F bits per file). Without the presence of a server or a BS, D2D communications between nodes are considered in this network. A library with m files is denoted by $\mathcal{F}$, $\mathcal{F} = \{f_1, \cdots, f_m\}$. Each node intends to download its requested file from the library $\mathcal{F}$ through this cached D2D network.

A caching scheme is performed in two phases: caching placement phase and caching delivery phase. The library $\mathcal{F}$ is generated once and kept fixed during the two phases. The cached messages are stored at each node in the caching placement phase and remain fixed during the delivery phase. Each node requests its own intended file from the library and gets served through wireless transmissions in the delivery phase. Multiple requests made by one node can be handled consecutively. In this chapter, we will work on maximizing the network throughput by jointly designing both phases. To this end, we introduce some useful definitions as follows.

12.3.1 Definition 1 (Caching placement strategy)

A cache placement strategy Π_c is a rule to assign files from the library $\mathcal{F}$ to a node's local cache in the caching placement phase. Let $\mathcal{M}_u$ represent the local storage of node u. Then, a particular cache placement G_u at node u can be viewed as a mapping from the library $\mathcal{F}$ to the local memory $\mathcal{M}_u$. One realization of caching placement at all nodes is denoted by $G \triangleq \{G_u, u \in \mathcal{U}\}$. Note that the caching placement is done without a priori knowledge of the nodes' actual requests.

12.3.2 Definition 2 (Users' request)

At each request time, each node makes a request to a file f_i, $f_i \in \mathcal{F}$, randomly and independently. Denote the index of node u's requested file by i_u. Then, one realization of all nodes' requests, denoted by R, can be represented by

$$R = \{i_1, \cdots, i_n\}\,. \tag{12.5}$$

In this chapter, we consider that for any node in the network, its requested file index i_u is uniformly distributed over $\{1, \cdots, m\}$. This stands as the worst-case scenario where users' requests are quite scattered rather than gathered. This assumption is also adopted in [12] and fits the spirit in [22–24] considering a "heavy tail" Zipf request distribution.

12.3.3 Definition 3 (Transmission policy)

A transmission policy Π_t is a rule of designing and scheduling the D2D transmissions in the delivery phase. Generally, it includes two parts: one is the transmission protocol $\mathcal{P}$ describing the feasible D2D links in the network, such that these D2D links can provide reliable communications, considering the physical constraints such as power and interference; and the other is the transmission scheduling $\mathcal{S}$ describing the activated D2D links at one time. Denote by y a set of simultaneous transmission links.

12.3.4 Definition 4 (Average aggregate throughput)

For a given caching placement G, node requests R, and a set of transmission links y, define $t_u(G,R,y)$ as the number of useful received information bits per time-slot by node u during one delivery phase. Adding $t_u(G,R,y)$ from all nodes together, we have the aggregate throughput of the network as

$$\widetilde{T}_n(G,R,y) = \sum_{u\in\mathcal{U}} t_u(G,R,y). \tag{12.6}$$

Let T_n denote the *average aggregate throughput*, and

$$T_n = E_R\left[E_S\left[\widetilde{T}_u(G,R,y)\right]\right] = \sum_R\sum_y \widetilde{T}_u(G,R,y)Pr\{y|R,G\}\,Pr\{R\} \tag{12.7}$$

where E_R means averaging over the randomness of nodes' actual requests and E_S means averaging over the randomness caused by scheduling during all delivery phases. We see that T_n is a function of the caching placement strategy Π_c and the transmission policy Π_t, i.e.,

$$T_n \sim f(\Pi_c,\Pi_t). \tag{12.8}$$

12.3.5 Definition 5 (Outage probability)

For a given caching placement G, given node requests R and a given set of transmission links y, if a node's request cannot be satisfied through the delivery phase, we say that this node is in outage. Denote the number of nodes in outage by $N_o(G,R,y)$ in one delivery phase, i.e.,

$$N_o(G,R,y) = \sum_u \mathbf{1}\left(t_u(G,R,y) = 0\right). \tag{12.9}$$

Define the *outage probability* of the network as

$$p_{out} = \frac{E_R[E_S[N_o(G,R,y)]]}{n}, \tag{12.10}$$

where E_R and E_S are described under (12.7). Similarly, p_{out} is a function of the caching placement strategy Π_c and the transmission policy Π_t.

In this chapter, we focus on the scaling of the average aggregate throughput T_n with an increasing number of n, while the power constraint of each node, the allocated bandwidth and the occupied spatial area of all nodes remain constant. In addition, we will assume that the size of the library m and the size of local cache M_c increase with n. Otherwise, if m and M_c remain constant when $n \to \infty$, a node can find its intended file within its neighbors with probability one regardless of any Π_c and Π_t. Thus, throughout the chapter, we assume that

$$m = n^{\alpha} \text{ and } M_c = n^{\beta}, \tag{12.11}$$

where $\alpha > 0$ and $\beta > 0$. We further assume that $0 < \alpha - \beta < 1$. This is because if $\alpha - \beta < 0$, each node is able to store the entire library in its local memory, which is not the focus of this chapter. On the other hand, if $\alpha - \beta > 1$, the total memory size of the network nM_c is smaller than the library size m. This will cause that the outage probability goes essentially large [22], making the system not useful for practical deployment.

The main goal of this chapter is to study the following two optimization problems. Let $T_n = \Theta\left(n^{\varphi}\right), \varphi \geq 0$. The first goal is, in the study of scaling laws, to jointly optimize the caching strategy Π_c and the transmission policy Π_t so that the exponent φ is maximized while the outage probability is vanishing, i.e.,

$$\begin{aligned} \mathscr{P}_0 : & \max_{\Pi_c, \Pi_t} \varphi \\ s.t. & \lim_{n \to \infty} p_{out} = 0. \end{aligned} \tag{12.12}$$

Beyond the scaling law, it is desirable to jointly optimize Π_c and Π_t so that the throughput T_n is maximized with a target outage probability ϵ when designing a wireless caching network. This can be represented by problem $\mathscr{P}_1$ as follows;

$$\begin{aligned} \mathscr{P}_1 : & \max_{\Pi_c, \Pi_t} T_n \\ \text{s.t.} & \ p_{out} \leq \epsilon. \end{aligned} \tag{12.13}$$

With the open and general definitions mentioned above, we see that optimizing Π_c and Π_t over all possible candidates have a prohibitive complexity. To address above problems, in this chapter, we focus on designing a transmission policy Π_t exhibiting a good scaling law of the average aggregate throughput T_n. A caching placement strategy Π_c is designed to serve the transmissions.

12.4 Achievable scaling law and throughput

This section states the main results of this chapter. We first introduce an achievable scaling law when the number of nodes n is sufficiently large as follows [27, 28].

Theorem 1: For the considered caching wireless network with a library of m files and local caches of M_c files, if nM_c, as $n \to \infty$, the following scaling law is achievable with high probability:

$$T_n = \Theta\left(n^{\frac{t}{t+1}}\right), \tag{12.14}$$

where $t, t \geq 1$, is an integer constant independent of n.

Proof. An achievable scheme is presented in Section 12.5 and the proof is given in Section 12.6. The choice of the parameter t will be discussed in Section 12.6.

Remark 1: From (12.14), with t increases, $\frac{t}{t+1} \to 1$. This indicates that with the help of caching, the average throughput of the network can increase almost linearly with the number of nodes n. Note that the scaling $\Theta\left(n^{\frac{t}{t+1}}\right)$ in this chapter is different from the scaling $\Theta\left(\frac{nM_c}{m}\right)$ in [12, 22] or $\Theta\left(n\sqrt{\frac{M_c}{m}}\right)$ in [25, 26].

Figure 12.3 compares the throughput scaling laws of the caching network in [12, 22, 25, 26] and Theorem 1. Recall that the most interesting region is that $nM_c > m$, i.e., $0 < \alpha - \beta < 1$. We see that $1 \leq \frac{nM_c}{m} < n$, $\sqrt{n} \leq n\sqrt{\frac{M_c}{m}}$, and in our result $\sqrt{n} \leq n^{\frac{t}{t+1}}$

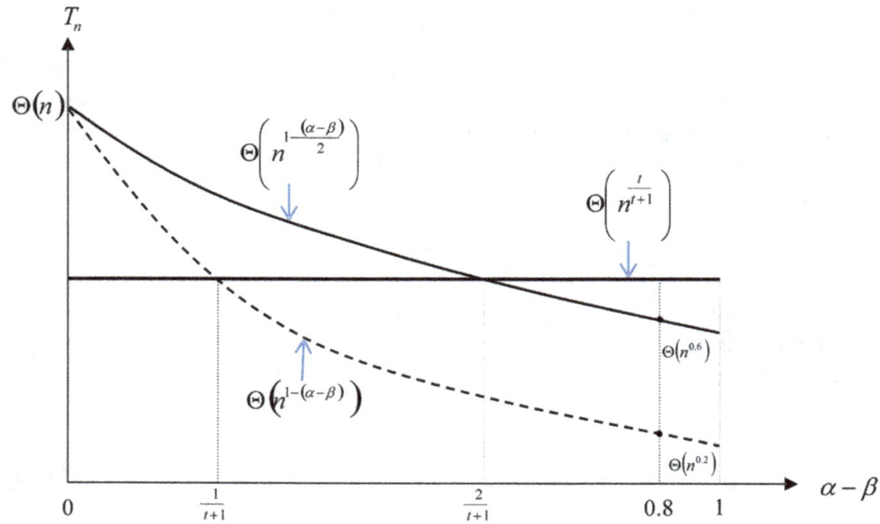

Figure 12.3 *Achievable throughput scaling laws of $\Theta(nM_c/m)$ in [12, 22], $\Theta(n\sqrt{M_c/m})$ in [25, 26], and $\Theta(n^{t/(t+1)})$ in (12.14), respectively. The proposed scheme outperforms current schemes when $\alpha - \beta > 1/(t+1)$ or $2/(t+1)$, while is inferior when $\alpha - \beta < 1/(t+1)$ or $2/(t+1)$.*

(since $t \geq 1$). This is shown in the figure where these three scalings exhibit different throughput ranges in the region of $0 < \alpha - \beta < 1$. It is also immediate to notice that the scalings in [12, 22] and [25, 26] are both functions of $\alpha - \beta$, the relative exponent of n between the library size m and the cache size M_c; however, the scaling in Theorem 1 is irrelevant with $\alpha - \beta$. We will discuss the reason in Section 12.6.

Furthermore, it can be observed that our scaling performs better under a cache-size limited situation (with $\alpha - \beta$ large). For example, consider a caching network with $\alpha = 1$ and $\beta = 0.2$. This setup corresponds to a disadvantageous caching case where there are many files in the library, however the local cache at each node is relatively small. For example, when $n = 10^8$, there are 10^8 files in the library while each node only has a cache size of 40 files. The aggregate average throughput in [12, 22] and [25, 26] will be $\Theta\left(n^{0.2}\right)$ and $\Theta\left(n^{0.6}\right)$, respectively, while the proposed scheme achieves $\Theta\left(n^{0.75}\right)$ when a typical value $t = 3$ is applied according to (12.14). It is also worth mentioning that since this scheme is specifically designed for the small cache case, it does not perform as good as current schemes when $\alpha - \beta < \frac{1}{t+1}$ or $\frac{2}{t+1}$.

The main difference between this proposed scheme and existing ones is the use of the spatial network resources. In fact, no matter one-hop [24], multihop [26], or hierarchical cooperations [29], the nice scaling with the size of caches is mainly achieved by exploiting the spatial reuse in the network. The growth of the cache size increases the chance of spatial reuse, leading to a growing network throughput.

However, since this work aims at the small cache case, we do not focus on the spatial reuse in designing the caching scheme. Instead, we try to exploit the spatial degree of freedom and increase the cache-hit probability by allowing one node to access more other nodes, to better facilitate the design for the small cache situation.

In addition to the scaling law, for the proposed achievable scheme, we also derive an explicit expression of the throughput T_n and then solve the optimization problem of maximizing T_n when achieving a given outage probability ϵ. The result is shown in the following theorem.

Theorem 2: For the considered caching wireless network with n nodes, a library of m files and local caches of M_c files where $nM_c \geq m$, to achieve a target outage probability ϵ, the following average aggregate throughput is achievable when n is sufficiently large:

$$T_n = \begin{cases} \frac{1}{t+1}\left(\frac{n}{C}\right)^{\frac{t}{t+1}}, & \text{if } B_U \geq \left(\frac{n}{C}\right)^{\frac{t}{t+1}} \\ \frac{nB_U}{n+tCB_U^{\frac{t+1}{t}}}, & \text{if } B_U < \left(\frac{n}{C}\right)^{\frac{t}{t+1}}, \end{cases} \tag{12.15}$$

where C is a system-dependent constant and B_U is an upper bound for the number of packets of a file, depending on the target outage probability ϵ, the library size m and the cache size M_c.

Proof. An achievable scheme is presented in Section 12.5 and the proof is given in Section 12.6. Details of parameters C and B_U are defined later in (12.53) and (12.38), respectively.

Remark 2: From Theorem 2, we see that the optimal throughput can be in two regimes, depending on system parameters such as the target outage probability ϵ, the library size m, and the cache size M_c. For example, if M_c and ϵ are large, the optimal throughput T_n will probably fall in the first line of (12.15). This is different from the pure scaling in Theorem 1, which is irrelevant with these parameters.

In the rest of this chapter, we will present an achievable scheme with the scaling law in (12.14) in Section 12.5, and then prove these two theorems in Section 12.6.

12.5 An achievable scheme

We note that the throughput of a dense network is fundamentally limited by the interference between simultaneous transmissions. The following scheme is proposed to deal with the interference by integrating distributed MIMO, hierarchical cooperations, and caching.

12.5.1 Caching placement phase

Now, we propose an uncoded, distributed, and randomized placement strategy Π_c.

Before the cache placement, each file in the library is partitioned into B packets of equal size. Here, we assume that the size of each file is sufficiently large to

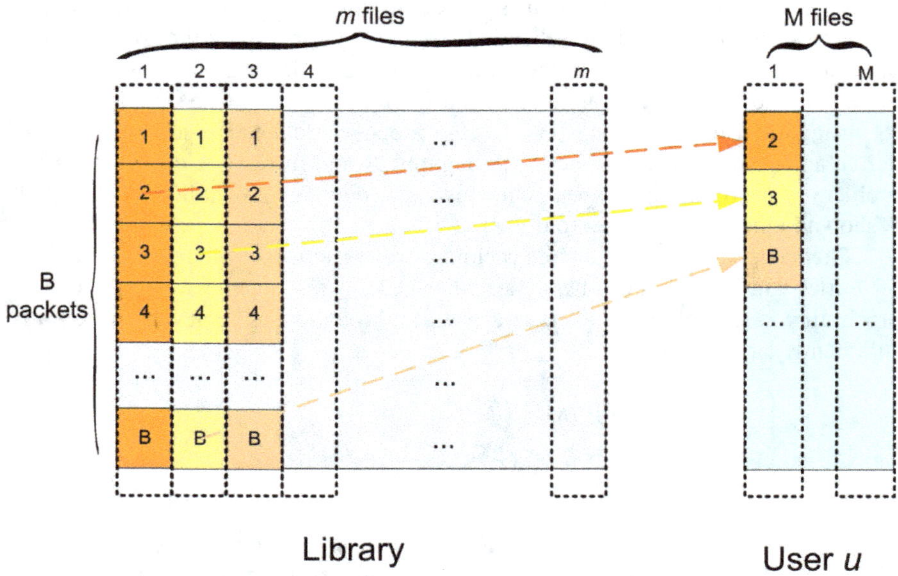

Figure 12.4 The proposed caching placement strategy

support the required number of packet B as determined later.[b] More specifically, we partition a file $f_i \in \mathscr{F}$ into packets of equal size $b_{i,j}$, $j \in \{1, \ldots, B\}$, i.e.

$$f_i = \bigcup_{j \in \{1,\cdots,B\}} b_{i,j}. \tag{12.16}$$

Thus, there are mB packets in the library. Note that B is an important parameter to be determined.

During the cache placement, each node fills in its local cache in an i.i.d. manner according to the same placement strategy Π_c. Each node randomly chooses $M_C B$ packets from the mB packets of the library and stores them in its own cache, as depicted in Figure 12.4. Denote the lth packet in node u's cache $\mathscr{M}_u$ by c_u^l and we have

$$\mathscr{M}_u = \bigcup_{l \in [1, M_c B]} c_u^l. \tag{12.17}$$

In order to enable the multiplexing gain in a MIMO transmission in the delivery phase, it is required that different packets of a file are simultaneously transmitted from different nodes. To achieve this, the caching placement must guarantee that each node caches packets from different files. We describe this process in details as follows.

[b]Although the file size F, when it is small, can limit the performance of caching networks, it will not become a bounding factor in this work. This is because for the proposed scheme, is always much smaller than the file size in practical scenarios.

First, each node u decides which files to choose from, by randomly generating M_cB file indexes $\{i_u^l\}$, $l \in \{1, \cdots, M_cB\}$, $i_u^l \in \{1, \cdots, m\}$. We use the following rule to determine the chosen file indexes for node u.

1. If the number of packets stored in each node M_cB is less than the number of files m (i.e., $M_cB \leq m$), the packets for each node cache should be chosen from different files. That is, $\forall l_1, l_2 \in [1, M_cB]$, $i_u^{l_1} \neq i_u^{l_2}$. This can be recognized as a similar process of blindly drawing M_cB balls from a box of m different balls at a time.
2. If M_cB, the file indexes are generated as follows. First, all file indexes $1, \cdots, m$ are chosen for $W = \left\lfloor \frac{M_cB}{m} \right\rfloor$ times. That is, for $w \in [1, W]$, $\forall l \in [(w-1)m + 1, wm]$, $i_u^l = l - (w-1)m$. Then, for the rest of $(M_cB - Wm)$ packets, their file indexes must be different. That is, $\forall l_1, l_2 \in [Wm + 1, M_cB]$, $i_u^{l_1} \neq i_u^{l_2}$.

After the file indexes are determined, node u randomly picks up one of the packets from file $f_{i_u^l}$ with probability $\frac{1}{B}$ and fills it in the local cache $\mathcal{M}_u$.

Denote by j_u^l the lth packet chosen by node u in the i_u^l file, we have that

$$c_u^l = b_{i_u^l, j_u^l}, \text{ and } \mathcal{M}_u = \bigcup_{l \in [1, M_cB]} b_{i_u^l, j_u^l}. \tag{12.18}$$

These two steps ensure that the packets stored at one node come from different files to the most extent, which facilitates the delivery phase to be discussed next.

12.5.2 Caching delivery phase

12.5.2.1 Preliminaries

12.5.2.1.1 Channel model

We use the line-of-sight *physical model* in [16, 18, 19]. Denote the channel coefficient from node k to node u by h_{uk}, then

$$h_{uk} = \sqrt{G_A} (r_{uk})^{-\frac{\gamma}{2}} \exp(j\theta_{uk}), \tag{12.19}$$

where r_{uk} is the distance between node k to node u, θ_{uk} is the random phase, uniformly distributed in $[0, 2\pi)$, G_A is the antenna gain, and γ is the pathloss exponent of the environment. We have $\gamma \geq 2$ for the far-field assumption.

12.5.2.1.2 User clustering

Divide the entire network into square cells of area A_c, as shown in Figure 12.5. We call each square cell a *cluster* in this chapter. Let A_c be a function of n and recall that the entire network is a unit square, then we have

$$A_c = n^{-\eta}, \; \eta > 0. \tag{12.20}$$

We see that η is an important parameter determining the cluster size, which will be chosen later.

The number of nodes in one cluster is denoted by N_c. From literatures [14, 25], we know that

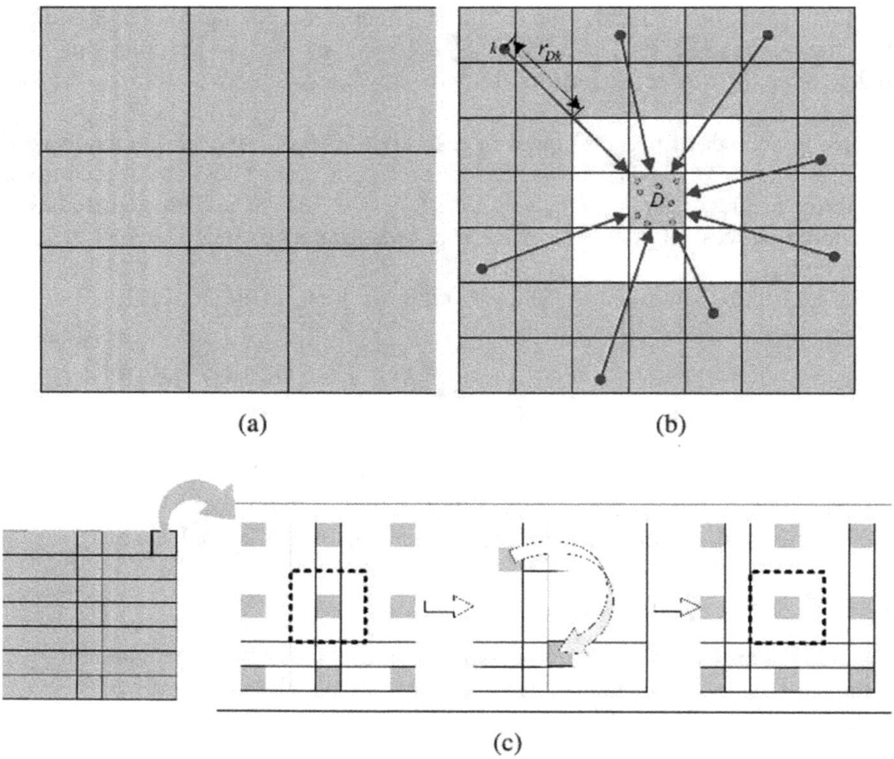

Figure 12.5 *Caching delivery phase of the proposed scheme: (a) Clustering: Divide the entire network into square cells of area A_c where $A_c = n^{-\eta}$, $\eta > 0$. η is a parameter determining the cluster size. The bigger η is, the smaller each cluster is. (b) Stage I: Distributed MIMO. The middle grey cluster represents the current destination cluster D. The white area denotes its neighbouring clusters $\mathcal{N}$. The surrounding grey area contains all candidates of source nodes. The distance from transmitter k to the destination cluster D is denoted by r_{Dk}. (c) Stage II: Hierarchical Cooperations. All clusters work simultaneously. Within one cluster, a three-phase hierarchical cooperation is employed including clustering, distributed MIMO and quantize-and-forward. A time-division protocol is used for the first and third phases. For example, in this figure, only one cluster is allowed to transmit among 9 clusters (L = 3).*

$$N_c = \Theta\left(n^{1-\eta}\right). \tag{12.21}$$

We denote by D the cluster that the destination node belongs to and by $\mathcal{N}$ all the neighboring clusters of D. We then design the caching delivery phase in two stages as follows.

12.5.2.2 Stage I: Distributed MIMO

In this stage, we design the transmission policy Π_t utilizing the fact that each node caches different packets of different files from previous caching placement (we will show that this is achieved with high probability in Section 12.6). We first consider the transmission policy for serving a single node and then extend it to serving all nodes in the network.

Recall that node u requires one file f_{i_u} with B packets. In order to collect all B packets of file f_{i_u}, we need to determine B source nodes that store the B packets in their caches, respectively. Then, to serve node u, a virtual multiplexing MIMO transmission can be formed from these B source nodes to the destination cluster D. For any requested file f_{i_u} of node u, B source nodes, $u_1^S, \cdots, u_B^S \in \mathcal{U}$, are selected according to the following criteria.

1. Each of the source nodes has one different packet of the requested file f_{i_u} in its local cache, i.e.,

$$\exists l_1, \cdots, l_B \in \left[1, M_c B\right], \; c_{u_1^S}^{l_1} \cup c_{u_2^S}^{l_2} \cup \cdots \cup c_{u_B^S}^{l_B} = f_{i_u}. \tag{12.22}$$

2. All the source nodes cannot lie in the neighboring clusters of node u, i.e.,

$$u_1^S, \cdots, u_B^S \notin \mathcal{N}, \tag{12.23}$$

which will be justified in Appendix 12.9.1.

Note that if there are multiple candidates of source nodes, we can randomly pick up any B candidates that meet the criteria above. The criteria allow a destination node to choose source nodes from the entire network except for the neighboring clusters $\mathcal{N}$.

In the following, we first focus on the case where all source nodes are not in the destination cluster D. We will then show that it is equivalent to the case where some of the source nodes are in cluster D in terms of achieving the same aggregate throughput.

After selecting the source nodes $u_1^S, \cdots, u_B^S$ as transmitters, we choose all the N_c nodes in the destination cluster D as receivers. Each packet of file f_{i_u} is simultaneously sent from the B transmitters to the N_c receivers in a multiplexing way, as shown in Figure 12.5. An MIMO transmission will be formed as long as the observations at all receivers can be jointly processed. We will discuss how to realize this joint processing in Stage II. Note that to achieve multiplexing transmission, it is required that all the source nodes are synchronized for their transmissions.

So far, we have described the transmission policy Π_t for serving a single node. For serving all nodes in the network, the same Π_t is applied in a time-division manner.

12.5.2.3 Stage II: Hierarchical cooperations

The goal of this stage is to collect and jointly process the MIMO observations from Stage I at the destination node u. To this end, we apply hierarchical operations similar to that in [13, 21] as follows.

Within a destination cluster, each node quantizes the received signal and sends it to the destination node. The destination node then jointly processes the N_c copies of superimposed signals received from previous multiplexing transmissions nodes. Thus an MIMO transmission from the source nodes to the destination node is formed through the two-stage cooperations. Note that the quantization does not change the linear scaling (of the number of independent transmitting streams) of MIMO capacity, which is proven in [16].

We notice that each node in the cluster wants to send independent messages to all other nodes in the same cluster. This communication problem is referred to as a *network multiple access problem* in [13, 21]. In this problem, each of the n nodes in the network has $n - 1$ independent information messages, one for other $n - 1$ node (all messages have the same size in bits). For this problem, a hierarchical three-phase cooperative transmission was proposed employing a similar idea of clustering, distributed MIMO, and quantize-and-forward [13]. A hierarchical scheme with an improved scheduling, referred to as Method 4 in [21], achieves the best scaling and transmission rate by far to our best knowledge. We adopt this design for the network multiple access problem in Stage II to achieve the scaling.

12.6 Performance of the proposed scheme

In this section, we first derive an explicit expression of the aggregate throughput for the proposed scheme. Then, we analyze the outage probability and derive an asymptotic expression of its upper bound. Using these results, we then prove Theorem 1 and Theorem 2.

12.6.1 Aggregate throughput of the proposed scheme

We calculate the aggregate throughput for the proposed scheme as follows.

12.6.1.1 Required time-slots for Stage I

The distributed MIMO transmissions in Stage I can be explicitly described in the following. Let $\mathbf{x}$ be a $B \times 1$ vector denoting the signals transmitted from the source nodes, and let $\mathbf{y}$ be an $N_c \times 1$ vector denoting the signals received at the destination cluster D. Then, we have that

$$\mathbf{y} = \mathbf{H}\mathbf{x} + \mathbf{z}, \tag{12.24}$$

where is the $N_C \times B$ channel matrix where the $\{u,k\}$-th h_{uk} element is defined in (12.19), and $\mathbf{z}$ is an i.i.d. Gaussian noise vector. In order to guarantee reliable transmissions from the source nodes to the destination cluster, it is required that the number of independent transmitting streams B should be no more than the number of receiving antennas N_C, i.e.,

$$B \leq N_c. \tag{12.25}$$

Let the geographical distance between transmitter k and destination cluster D be r_{Dk} as shown in Figure 12.5. We use the following transmit power control mechanism such that

$$\mathbf{E}\left[\left|x_k\right|^2\right] = \frac{P}{B}\left(r_{Dk}\right)^{\gamma}. \tag{12.26}$$

We see that the transmit power control mechanism is similar to that in [13, 16]. We then have the following lemma.

Lemma 1: Under the channel model described in (12.19), applying the transmit power control in (12.26), a long-distance distributed MIMO transmission from the source nodes to the destination cluster D achieves an aggregate rate scaling linearly with the number of transmitters B.

The proof is given in Appendix 12.9.1.

We see that the transmit power constraint at each node is satisfied in our design. This is because when serving one node, each node only needs to transmit $\frac{P}{B}\left(r_{Dk}\right)^{\gamma}$ watts and only a fraction of $\frac{B}{n}$ nodes are transmitting. So in order to serve all n nodes in the network, on average, each node needs a total transmit power

$$\frac{P}{B}\left(r_{Dk}\right)^{\gamma} \cdot \frac{B}{n} \cdot n = P\left(r_{Dk}\right)^{\gamma} \leq P, \tag{12.27}$$

which satisfies the transmit power constraint.

Based on the designs of Stage I, we calculate the required time-slots for serving all nodes in Stage I as follows. We know that each packet has $\frac{F}{B}$ bits and all the B packets of a file are simultaneously transmitted in a multiplexing way. Using Lemma 1, for any requested file of one node, a reliable transmission from source nodes to the destination cluster will take $\frac{F}{B}$ time-slots. Therefore, in total, $\frac{F}{B}n$ time-slots are needed for serving all n nodes in Stage I.

12.6.1.2 Required time-slots for Stage II

In this stage, we use hierarchical cooperations within one cluster. Recalling the results in [21], the time to complete the network multi-access transmission for a network of n nodes with t hierarchical stages and Q-bit quantization is given by

$$t_n = tL^2 Q^{\frac{t-1}{2}} n^{\frac{t+1}{t}}, \tag{12.28}$$

where

$$L = \left\lceil 2^{\frac{3+\frac{\gamma}{\ln 2}}{\gamma}} + 1 \right\rceil \tag{12.29}$$

is the reuse factor, meaning that each cluster has one transmission opportunity every L^2 time-slots. By substituting the number of nodes n by N_c in our problem, we obtain the time t_{N_c} for completing transmissions for a cluster of n nodes as

$$t_{N_c} = \frac{F}{B} t L^2 Q^{\frac{t-1}{2}} N_c^{\frac{t+1}{t}}. \tag{12.30}$$

The optimal number of hierarchical stages t for maximizing the throughput can be found by an exhaustive search for a given number of nodes N_c according to [21].

Here, we discuss the case where some of the source nodes are in the destination cluster D. If some packets of the requested file f_{i_u}, say N_D packets, can be found within cluster D, in Stage I, the number of transmitted packets will be $(N_c - N_D)$. Thus, we can reduce the number of receivers to $(N_c - N_D)$ and the consumed time-slots for MIMO transmissions in Stage I remain unchanged. In Stage II, the number of transmitted packets to node u will remain as $(N_c - 1)$. This is because node u will only need to receive MIMO observations from $(N_c - N_D - 1)$ nodes, in addition to receiving the N_D packets from its neighbors. It is convenient to arrange that the neighbors with N_D packets are not scheduled as receivers in MIMO transmissions. Thus, Stage II can still be seen as a network multiple access problem with N_c nodes and it will also occupy the same time-slots. As a result, we see that in terms of the aggregate throughput, the case where some packets can be found within cluster D is equivalent to the case where no source nodes are in cluster D.

12.6.1.3 Aggregate throughput

From previous designs for both stages, we know that nF bits in total are delivered to their destinations in $\frac{F}{B}n + t_{N_c}$ time-slots. Thus, the aggregate throughput is

$$T_n = \frac{\mathrm{nF}}{\frac{F}{B}n + t_{N_c}} = \frac{nB}{n + tL^2 Q^{\frac{t-1}{2}} N_c^{\frac{t+1}{t}}} \tag{12.31}$$

It is clear that in order to maximize T_n, the number of nodes in one cell N_c should be as small as possible. Since $B \le N_c$, the optimal N_c^* should be

$$N_c^* = B. \tag{12.32}$$

Remark 3: From (12.32), we see that the optimal number of nodes in one cluster N_c and the number of packets of a file B are coupled. On one hand, if choosing a large cluster size (a large N_c), the number of packets of a file B should also be large. This is essentially attributed to the design in Stage I: with large clusters (a large N_c), the number of independent transmitting streams B should also be large in order to utilize the multiplexing gain to the most extent; on the other hand, with a large B, N_c must be large enough to support reliable multiplexing transmissions in Stage I.

Then, with (12.32), we get that

$$T_n = \frac{nB}{n + tL^2 Q^{\frac{t-1}{2}} B^{\frac{t+1}{t}}}. \tag{12.33}$$

We see that the number of packets of a file B is a critical design parameter in both the caching placement and delivery phase. By optimizing over B, the caching strategy Π_c and the transmission policy Π_t are jointly optimized. Therefore, in the proposed scheme, the throughput T_n no longer needs to be optimized over Π_c and Π_t as in problem $\mathscr{P}_0$ and $\mathscr{P}_1$. Instead, we only need to optimize T_n over the number of packets B.

12.6.2 Outage probability of the proposed scheme

In this section, we examine the outage probability of the proposed scheme and derive an asymptotic expression of the upper bound of the outage probability.

Theorem 3: As $n \to \infty$, the outage probability in the proposed scheme is upper bounded by

$$p_{out} \leq 1 - e^{-mB \cdot e^{-\frac{\left(n-\left(L^2-1\right)N_c\right)M_c}{m}}}. \tag{12.34}$$

The proof is given in Appendix 12.9.2.

According to (12.34), we see that p_{out} increases with the number of nodes in one cluster N_c. Recall that $N_c \geq B$. Therefore, to achieve a small outage probability p_{out}, it is required that

$$N_c^* = B. \tag{12.35}$$

This agrees with the requirement for maximizing T_n in (12.32). Thus, we have

$$p_{out} \leq p_{out}^U \triangleq 1 - e^{-mB \cdot e^{-\frac{\left(n-\left(L^2-1\right)B\right)M}{m}}}. \tag{12.36}$$

where p_{out}^U is referred to as the outage probability upper bound.

Remark 4: In order to achieve a target outage probability ϵ, it is suffice to let

$$p_{out}^U \leq \epsilon. \tag{12.37}$$

Let $B = B_U$ denote the solution to the equation

$$p_{out}^U = \epsilon. \tag{12.38}$$

Here, ϵ represents the maximum number of packets of a file, allowed by a given library size m, local cache size M_c, and the target outage probability ε. We see that decreases with B_U, and increases with M_c and ε. Especially, increasing the local cache size M_c would allow an exponentially larger number of packets B. To achieve the outage probability, it is suffice to let

$$B \leq B_U. \tag{12.39}$$

12.6.3 Throughput scaling law of the proposed scheme

Based on (12.33), the scaling law of the throughput can be easily obtained as

$$T_n = \frac{nB}{n+tL^2Q^{\frac{t-1}{2}}B^{\frac{t+1}{t}}} = \frac{nB}{n+\Theta\left(B^{\frac{t+1}{t}}\right)}. \tag{12.40}$$

By maximizing T_n over the number of packets B, we obtain

$$T_n^* = \Theta\left(n^{\frac{t}{t+1}}\right) \tag{12.41}$$

when

$$B^* = \Theta\left(n^{\frac{t}{t+1}}\right) \tag{12.42}$$

It is interesting to see that the optimal throughput scaling T_n^* and the corresponding B^* are on the same order, i.e.,

$$T_n^* = \Theta\left(B^*\right). \tag{12.43}$$

This is because the improvement of the throughput from $\Theta(1)$ to $\Theta\left(n^{\frac{t}{t+1}}\right)$ is mainly attributed to the multiplexing gain achieved by distributed MIMO. For instance, a smaller B means that one file is divided into fewer packets, and the number of simultaneous transmissions is reduced, which decreases the throughput T_n.

However, it is also important to note that T_n^* does not always increase as B increases. In fact, with B increasing, on one hand, the number of simultaneous transmissions in Stage I increases, leading to an increasing number of total transmitted useful bits nB. On the other hand, the size of the receiving cluster also increases, consuming more time-slots for completing the traffic within one cluster in Stage II. Therefore, the optimal scaling T_n is not achieved when B is maximized, but when the trade-off between the total traffic and the consumed time-slots in Stage II is optimized. Although from Section 12.6 we can learn that B_U, the upper bound of B, increases exponentially with the size of local caches M_c, the throughput scaling T_n cannot increases exponentially with M_c. Thus, for the proposed scheme, the optimal scaling T_n is independent of M_c and m, as long as $nM_c \geq m$ satisfied.

Remark 5: We further explain this relationship in the context of the distributed MIMO channel. The proposed two-stage caching delivery can be regarded as an equivalent distributed MIMO channel with finite backhaul capacity. The transmitting and receiving antennas respectively correspond to the source nodes and the receiving nodes in the scheme. In order to achieve the capacity of this channel, the backhaul capacity should increase with the number of antennas (nodes). Otherwise, when the number of antennas (nodes) goes large, the channel capacity will be limited by the backhaul capacity and it will not linearly increase with the number of antennas (nodes). We call this the *multiplexing-backhaul tradeoff* in the proposed scheme. Therefore, as the size of cache increases, the number of transmit antennas (source nodes) grows and then the network throughput in the first stage increases due to a larger spatial degree of freedom gain. However, to guarantee the linear scaling of MIMO transmissions, the required number of receive antennas (the number of nodes within the receiving cluster) also increases. This increases the traffic within the receiving cluster and decreases the network throughput in the second stage. This explains why the achievable throughput scaling is not a function of the size of the cache and the library, but is fundamentally determined by the design of the caching network.

Furthermore, using the above results, we can also determine the cluster size in our design as follows. We see that to achieve the optimal throughput, $N_c^* = B$ and $B_n^* = \Theta\left(n^{\frac{t}{t+1}}\right)$ must be satisfied at the same time. Recall (12.21), η must satisfy that $\Theta\left(n^{\frac{t}{t+1}}\right) = \Theta\left(n^{1-\eta}\right)$, i.e.,

$$\eta = \frac{1}{t+1}. \tag{12.44}$$

We now prove that the throughput $\Theta\left(n^{\frac{t}{t+1}}\right)$ can be achieved with a vanishing outage probability as $n \to \infty$. To see more clearly about the scaling of p_{out} with n, we further extend (12.36) as follows. It is easy to verify that as $n \to \infty$, $\frac{n-(L^2-1)B}{n} \to 1$. Therefore, $n - (L^2 - 1)\, B$can be approximated by n when n goes sufficiently large. Recall that $B^* = \Theta\left(n^{\frac{t}{t+1}}\right)$, then we have

$$p_{out} \leq 1 - e^{-n^{\varepsilon} e^{-n^{\tau}}}, \tag{12.45}$$

where

$$\varepsilon = \alpha + \frac{t}{t+1}, \text{ and } \tau = 1 - (\alpha - \beta). \tag{12.46}$$

Applying L'Hospital's Rule, we have that

$$\lim_{n\to\infty} n^{\varepsilon} e^{-n^{\tau}} = \lim_{n\to\infty} \frac{\varepsilon n^{\varepsilon-\tau+1}}{\tau e^{n^{\tau}}} \overset{(a)}{=} 0, \tag{12.47}$$

where the equality (a) is obtained by keeping differentiating the numerator and the denominator until the power exponent of the numerator is negative. Thus, we have

$$\lim_{n\to\infty} p_{out} = 0. \tag{12.48}$$

meaning that the scaling $T_n = \Theta\left(n^{\frac{t}{t+1}}\right)$ is achieved with high probability.

This finishes the proof of Theorem 1.

12.6.4 Optimized throughput under a target outage Probability

Based on the throughput and outage probability analysis in Sections 12.5 and 12.6, we see that problem $\mathscr{P}_1$ in (12.13) now becomes

$$\mathscr{P}_2 : \max_{B} \frac{nB}{n + tL^2 Q^{\frac{t-1}{2}} B^{\frac{t+1}{t}}} \text{s.t. } \mathrm{B} \leq \mathrm{B_U}. \tag{12.49}$$

Since the constraint is linear and the objective function is concave, which can be verified by taking the second derivative in B, $\mathscr{P}_2$ is a convex optimization problem. Now introducing a Lagrange multiplier λ, $\lambda \geq 0$, we form the Lagrangian

$$L(B, \lambda) = -\frac{nB}{n+tL^2 Q^{\frac{t-1}{2}} B^{\frac{t+1}{t}}} + \lambda\left(B - B_U\right). \tag{12.50}$$

The solution to $\mathscr{P}_2$ can be found by applying the Karush-Kuhn-Tucker (KKT) condition to (12.50) and we obtain

$$\begin{cases} \left(-\frac{nB}{n+tL^2Q^{\frac{t-1}{2}}B^{\frac{t+1}{t}}}\right)' + \lambda = 0 \\ \lambda\left(B - B_U\right) = 0 \end{cases} . \tag{12.51}$$

Then we get

$$B^* = \min\left\{B^{opt}, B_U\right\}. \tag{12.52}$$

where $B^{opt} \triangleq \left(\frac{n}{L^2Q^{\frac{t-1}{2}}}\right)^{\frac{t}{t+1}}$. Define a system-dependent constant

$$C \triangleq L^2Q^{\frac{t-1}{2}}. \tag{12.53}$$

Thus, the corresponding optimal throughput can be written as

$$T_n = \begin{cases} \frac{1}{t+1}\left(\frac{n}{C}\right)^{\frac{t}{t+1}}, & \text{if } B_U \geq \left(\frac{n}{C}\right)^{\frac{t}{t+1}} \\ \frac{nB_U}{n+tCB_U^{\frac{t+1}{t}}}, & \text{if } B_U < \left(\frac{n}{C}\right)^{\frac{t}{t+1}}, \end{cases} \tag{12.54}$$

This establishes Theorem 2.

Remark 6: According to (12.34), p_{out} increases with the number of packets B. In order to achieve a low outage probability, the corresponding B should be small. However, a smaller B also decreases the throughput T_n according to the second line in (12.54) and vice versa. We refer to this as the *throughput-outage tradeoff*, similar to that in [22].

The throughput in (12.54) is illustrated in Figure 12.6. Recall that B_U is the maximum number of packets of a file determined by a given library size m, local cache size M_c, and the target outage probability ϵ. For given values of m, M_c, and ϵ, if $B_U \geq B^{opt}$, as shown in the position of long dash line in Figure 12.6, the optimal T_n^* can be achieved when $B^* = B^{opt}$. However, if B_U^{opt}, as shown in the position of solid line, the throughput will be limited by U_B and cannot achieve the optimal T_n. This reveals that in the caching network with distributed MIMO and hierarchical cooperations, the limits of the throughput are twofold. One is the multiplexing-backhaul trade-off between the total traffic and the consumed time-slots in Stage II, which corresponds to the first line in (12.54). The other one is the throughput-outage

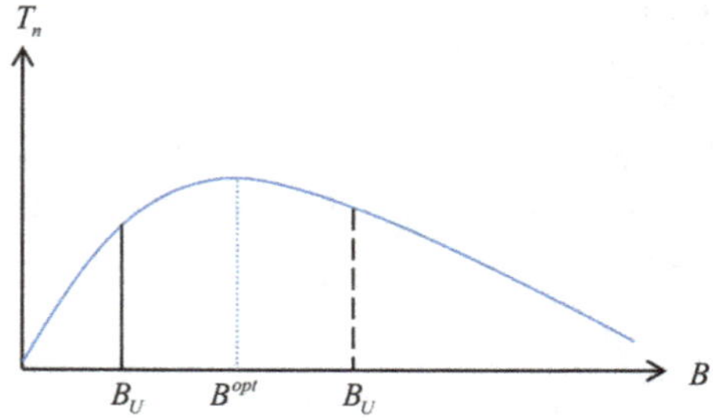

Figure 12.6 Achievable throughput as the function ofB

trade-off as shown in the second line in (12.54). Especially, a low target outage probability ϵ would result in a rather small U_B, decreasing the network throughput. In other words, in this situation the network throughput will be limited by the throughput-outage trade-off. Recall that U_B increases with the local cache size M_c. Therefore, in this regime, the local cache size M_c is able to contribute to the throughput, since a large M_c would result in a large U_B and an increased throughput.

12.7 Numerical results

We use examples to demonstrate the performance of the proposed scheme and compare it with the multihop caching scheme in [25, 26]. For a network of area $A = 1\,\text{km}^2$, we consider two networks operating at microWave and mmWave, respectively. One network operates at the carrier frequency of 3 GHz ($\lambda = 0.1$ m) with the number of users $n \leq 10^4$. The other one operates at a carrier frequency of 28 GHz ($\lambda \approx 0.01$ m) with the number of users $n \leq 10^5$. This meets the spatial degree of freedom condition in [18], which allows distributed MIMO and hierarchical cooperations. Also, this is a realistic setup that can be found in many scenarios. The work in [21] shows that for such a network, the optimal number of hierarchical stages $t = 3$ and the optimal quantization bit $Q = 1$. We set the target outage probability $\epsilon = 10^{-4}$ and let $\gamma = 3$ for the microWave network, and let $\epsilon = 10^{-5}$, $\gamma = 7$ for the mmWave network, respectively. We fix $\alpha = 1$ for simplicity, which corresponds to the case where the number of files in the library equals to the number of nodes. Applying (12.54), we get the optimized throughput of the network as

$$T_n = \begin{cases} 0.02n^{\frac{3}{4}}, & \text{if } B_U \geq 0.09n^{\frac{3}{4}} \\ \frac{nB_U}{n+75(B_U)^{\frac{4}{3}}}, & \text{if } B_U < 0.09n^{\frac{3}{4}} \end{cases} \tag{12.55}$$

We can further derive that for the multihop caching scheme [21, 26]

$$T_n \geq \frac{n}{2L^2}\sqrt{\frac{M_c}{m}}, \tag{12.56}$$

where L is determined through (12.29).

Figure 12.7 compare the throughput in [25, 26] and in (12.54) for these two networks respectively. Both figures show that the proposed scheme exhibits a near-linear behavior as n grows. This means that the network is mainly limited by the *multiplexing-backhaul tradeoff* (shown in the first line in (12.54)) under realistic parameters. We observe that in both figures the proposed scheme outperforms the multihop caching scheme when the local cache size is limited (small β).

For example, in Figure 12.7, the throughput of the proposed scheme with $\beta = 0.4$ is better than the multihop caching scheme with $\beta = 0.56$. This corresponds to a magnificent difference of the local size caches considering that the number of nodes n is large. For example, when $n = 10^4$, $\beta = 0.4$ means that the local cache has 40 files while $\beta = 0.56$ represents a local cache with 174 files. In Figure 12.7, the throughput of a mmWave network demonstrates a similar

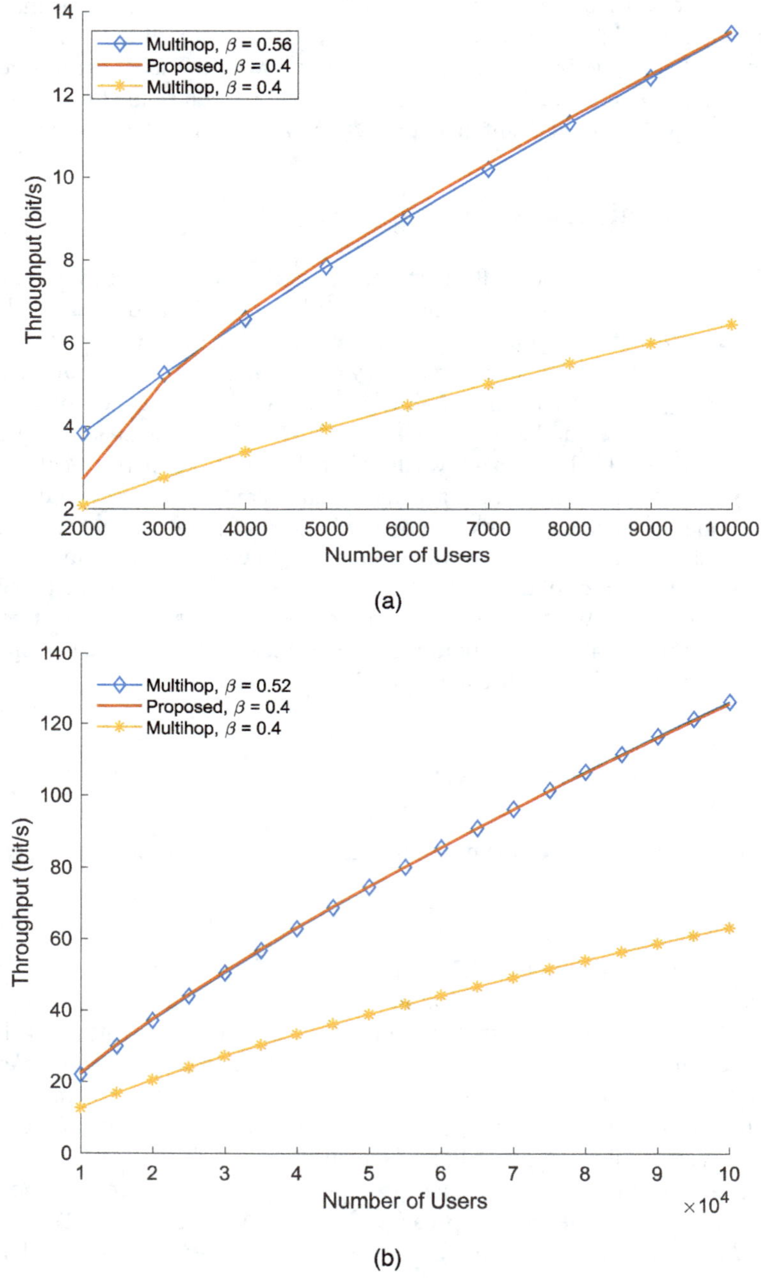

Figure 12.7 Achievable throughput of a D2D caching network as a function of the number of users n

behavior. This is because the distributed MIMO technology increases the number of simultaneous transmissions and improves the throughput of the network. If the local cache size is abundant ($\beta > 0.6$), the multihop caching scheme outperforms the proposed scheme as can be seen from both figures. This is because in this case the proposed scheme is limited by the multiplexing-backhaul trade-off, while the throughput of the multihop caching scheme grows unboundedly with the size of local caches. However, $\beta = 0.6$ is a fairly high requirement for the local cache size. For example, when $n = 10^5$, $\beta = 0.6$ requires a local cache of 1 000 files. This means the proposed scheme is more "cache-economic" than the multihop caching scheme.

Moreover, we infer from Theorem 1 that the scaling of $\Theta(n)^{\frac{t}{t+1}}$ outperforms multihop scaling of $\Theta(n)^{1-\frac{\alpha-\beta}{2}}$ when $\alpha - \beta > \frac{2}{t+1}$. With $\alpha = 1$ and $t = 3$ in this example, the scheme will outperform multihop schemes when $\beta < 0.5$. This conjecture is confirmed by the observation in both figures that the proposed scheme outperforms multihop schemes roughly when $\beta < 0.5$.

12.8 Conclusions

This chapter introduced and investigated the throughput scaling problem in a wireless D2D network where each node is equipped with a local cache and would like to download files from a pre-fixed library. The proposed two-stage caching delivery can be regarded as an equivalent distributed MIMO channel with finite backhaul capacity. First, distributed MIMO is applied between source nodes and the neighbors of the destination node to increase the number of simultaneous transmissions in the network. Then, hierarchical cooperations are used within one cluster based on existing works in [13, 21] to provide a high backhaul capacity. An uncoded random caching placement strategy is established to serve this cooperative caching delivery. Analytical results showed that the average aggregate throughput of the proposed scheme scales almost linearly with n, with a vanishing outage probability. Numerical results showed that the proposed scheme outperforms typical existing schemes when the local cache size is limited.

Future works may consider but not limited to the following directions. First, current work used the classic point-to-point channel model. The topology of multi-point wireless channels such as two-way relay channel and Y-channel can be utilized in the context of caching, to create a more efficient caching placement algorithm and delivery scheme [30, 31]. Second, advanced channel coding, especially the lattice code, has the potential to further improve the caching delivery in wireless networks [32, 33]. Last but not least, caching can be more beneficial in hybrid D2D-cellular networks [34]. Compared with the pure D2D network in this chapter, a direct link from BS is expected to largely increase the cache-hit probability.

12.9 Appendix

12.9.1 Proof of Lemma 1

Recall that the distance between transmitter k and receive cluster D is r_{Dk}. Observe that for any node $i \in D$,

$$r_{Dk} \leq r_{ik} \leq r_{Dk} + 2\sqrt{2A_c}. \tag{12.57}$$

When node i is not in the neighboring cluster of D, we also have $r_{Dk} \geq \sqrt{A_c}$. Then,

$$1 \leq \frac{r_{ik}}{r_{Dk}} \leq 1 + 2\sqrt{2}. \tag{12.58}$$

This means that

$$\left(r_{ik}\right)^{-\gamma} = \rho_{ik}\left(r_{Dk}\right)^{-\gamma}, \tag{12.59}$$

where $\rho_{kj} \in \left[b, 1\right]$ and b is independent of B or n. Note that a boundary from the cluster D to the source nodes such as $r_{kD} \geq \sqrt{A_c}$ must be satisfied to reach (12.59). This implies that the source nodes cannot be chosen from the neighboring clusters.

Then, (12.24) can be rewritten as

$$\mathbf{y} = \mathbf{Hx} + \mathbf{z} = \sqrt{\frac{GP}{B}}\mathbf{F}\widetilde{\mathbf{x}}+\mathbf{z}. \tag{12.60}$$

where $f_{ik} = \rho_{ik} \exp\left(j\theta_{ik}\right)$ is a scaled version of h_{ik}, $\widetilde{\mathbf{x}}=\left[\widetilde{x}_1, \cdots, \widetilde{x}_B\right]$ and $\mathbf{E}\left(\left|\widetilde{x}_k\right|^2\right) = 1$.

The mutual information is

$$I\left(\mathbf{x}; \mathbf{y}, \mathbf{F}\right) = E\left[\log\det\left(I + \frac{G_A P}{N_0 WB}\mathbf{FF^*}\right)\right]. \tag{12.61}$$

Applying a similar proof with Appendix I in [16], we can get that for any v,

$$I\left(\mathbf{x}; \mathbf{y}, \mathbf{F}\right) \geq B\log\left(1 + SNRv\right)\frac{\left(b^2-v\right)^2}{2}. \tag{12.62}$$

As v is independent of n, we see that $I\left(\mathbf{x}; \mathbf{y}, \mathbf{F}\right)$ grows at least linearly with B. This implies the linear scaling for the distributed MIMO transmission. This establishes Lemma 1.

12.9.2 Proof of Theorem 3

We first decompose the outage probability p_{out} and obtain that $p_{out} \leq 1 - p_{hit}$, where p_{hit} is defined in Definition 6. We then derive an asymptotic expression of p_{hit} and prove Theorem 3.

12.9.2.1 Decomposition of the outage probability p_{out}

Rewrite the outage probability p_{out} defined in Section 12.3 as

$$\begin{aligned} p_{\text{out}} &= \frac{\sum\limits_R\sum\limits_y\sum\limits_u \mathbf{1}\left\{t_u(G,R,y)=0\right\}\Pr\{y \mid R, G\}\Pr\{R\}}{n} \\ &\overset{(a)}{\leq} \sum_R\sum_y \mathbf{1}\left\{t_u(G,R,y)=0\right\}\Pr\{y \mid R, G\}\Pr\{R\}, \end{aligned} \tag{12.63}$$

where the inequality (a) comes from the following inequality

$$\sum_u \mathbf{1}\left\{t_u\left(G,R,y\right)=0\right\} \le n\cdot\mathbf{1}\left\{t_u\left(G,R,y\right)=0\right\},$$

which is derived from the fact that for different destination nodes, their corresponding $\mathbf{1}\left\{t_u\left(G,R,y\right)=0\right\}$ are not independent. This is because the potential source nodes for different destination nodes may overlap, and also different destination nodes may request the same file.

We then decompose p_{out} into two parts by considering the causes of outage. Generally, a node's request cannot be satisfied due to two reasons: one is that this node cannot find its intended file within its communication range (referred to as *no cache-hit*), and the other is that the communication fails in actual delivery phases. We have that

$$\begin{aligned}&\textstyle\sum_y \mathbf{1}\{t_u(G,R,y)=0\}\Pr\{y\mid R,G\}\\&=p_{\text{out}}^{\text{delivery}}+\Pr\{\text{No cache-hit for a given request } R \text{ of one user}\}\end{aligned} \tag{12.64}$$

As such, (12.63) can be further derived as

$$\begin{aligned}p_{\text{out}} &\le \sum_R \left(p_{\text{out}}^{\text{delivery}}+\Pr\{\text{No cache-hit for a given request } R \text{ of one user}\}\right)\Pr\{R\}\\&=\sum_R p_{\text{out}}^{\text{delivery}}\Pr\{R\}+\Pr\{\text{No cache-hit for any request of one user}\}\end{aligned}$$

Recall that in the transmission policy Π_t, we consider reliable transmissions in each delivery stage, and hence we have $p_{out}^{delivery}=0$. Thus, for the proposed scheme,

$$\begin{aligned}p_{\text{out}} &\le \Pr\{\text{No cache-hit for any request of one user}\}\\&=1-\Pr\{\text{Cache-hit for any request of one user}\}\end{aligned} \tag{12.65}$$

We will focus on the cache-hit probability in the following analysis.

Definition 6 (Cache-hit probability): Define $\mathscr{A}_u$ as the cache-hit event for user u, where for any requested file of node u, its B packets can be found in B different nodes among the entire networks *except for its neighboring clusters*, i.e.,

$$\mathscr{A}_u \triangleq \left\{\begin{array}{c}\forall f_u\in\{1,\cdots m\},\ \exists u_1^s,\cdots,u_B^s\in\{1,\cdots n\}\backslash\mathscr{N},\\ \exists l_1,\cdots,l_B\in\{1,\cdots M_cB\},\ c_{u_1^s}^{l_1}\cap c_{u_2^s}^{l_2}\cap\cdots\cap c_{u_B^s}^{l_B}=f_u.\end{array}\right\} \tag{12.66}$$

The probability of $\mathscr{A}_u$, denoted by p_{hit}, is referred to as the *cache-hit probability* in the proposed scheme, i.e.,

$$p_{hit}\triangleq\Pr\{\mathscr{A}_u\}. \tag{12.67}$$

Then,

$$p_{out}\le 1-p_{hit}. \tag{12.68}$$

12.9.2.2 Analysis of the cache-hit probability p_{hit}

It is difficult to directly calculate this cache-hit probability due to the complexity of the event $\mathscr{A}_u$ and the proposed caching strategy Π_c. To tackle this problem, we derive p_{hit} in the following way.

For a destination node u, the set of total source nodes is $\mathscr{U}\backslash\mathscr{N}$. The total caches of all potential source nodes are denoted as

$$\mathscr{M}_u^S = \bigcup_{u_S\in\mathscr{U}\backslash\mathscr{N}} \mathscr{M}_{u_S}. \tag{12.69}$$

To proceed, we need the following lemma.

Lemma 2: Based on the proposed caching strategy Π_c, for node u, the cache-hit event $\mathscr{A}_u$ happens as long as every packet of its requested file is placed in the total source cache $\mathscr{M}_u^S$, i.e.,

$$\Pr\{\mathscr{A}_u\} = \Pr\left\{\forall f_{i_u} \in \mathscr{F}, \forall j \in \{1,\cdots,B\}, b_{i_u,j} \in \mathscr{M}_u^S\right\} \tag{12.70}$$

Lemma 2 can be proved by contradiction based on the proposed Π_c in Section 12.4, and the proof is omitted due to page limit. Lemma 2 indicates that the B packets of a file can be found at different nodes with probability one, as long as all the B packets are placed in the total source cache $\mathscr{M}_u^S$ (no matter at which node). To ensure $\mathscr{A}_u$, it is required that all packets of all files in the library have been placed in the total source cache $\mathscr{M}_u^S$. Since the goal here is to calculate the probability of collecting all mB packets in total caches, we do not need to distinguish which file a collected packet belongs to, when calculating this probability. Therefore, we transform the problem of calculating $\Pr\{\mathscr{A}_u\}$ into the following question Ψ:

Ψ : According to the placement strategy Π_c, given mB packets in total, how many source nodes (denote this number by V) are required before having placed all mB packets in the total source cache $\mathscr{M}_u^s$? (12.71)

For node u, the cache-hit event $\mathscr{A}_u$ happens when the required number of source nodes V is smaller than the actual number of source nodes. Recall that for node u, there are a total of $n-\left(L^2-1\right)N_c$ potential source nodes. Thus,

$$p_{hit} = \Pr\{\mathscr{A}_u\} = \Pr\left(V \le n-\left(L^2-1\right)N_c\right). \tag{12.72}$$

The problem Ψ is a special type of the Coupon Collector's problem in probability theory: coupons in groups of constant size. In the literature, there is an expression of the expected number of groups to complete the collection. However, this expression does not present an explicit analytical result. Hence, we need to develop a different analytical expression for our problem. Let p_0 be the probability of any packet in the library being chosen at each node, we have the following lemma.

Lemma 3: According to the proposed caching strategy Π_c, the probability of any packet in the library being picked at each node is $\frac{M}{m}$, i.e.,

$$p_0 = \frac{M}{m}. \tag{12.73}$$

Proof. We omit the details of the proof due to the page limit. The approach is to first consider that each node randomly chooses M_cB files from m files with probability $\frac{M_cB}{m}$, and then picks up one of the packets of the chosen files with probability $\frac{1}{B}$.

The caching placement with the proposed Π_c can be regarded as picking one packet at a time, with a probability p_0 determined in Lemma 3. Denote a packet in the library by k, $k \in \{1, 2, \cdots, mB\}$. Similar to the Coupon Collector's problem, we can then obtain the probability of getting a new packet k as

$$p_k = \frac{mB-(k-1)}{mB} \cdot M_cB. \tag{12.74}$$

12.9.2.3 Asymptotic expression of the cache-hit probability

Based on the analysis above, we now treat the whole caching placement process as selecting one packet at a time with probability p_0.

Lemma 4: When the number of nodes n is sufficiently large, the probability of the required number of source nodes V less than the number of source nodes $n - (L^2 - 1) N_c$ is

$$\Pr\left(V < n - \left(L^2 - 1\right) N_c\right) = \Pr\left(V < \frac{mB}{M_cB}\log mB + a\frac{mB}{M_cB}\right) \to e^{-e^{-a}} \tag{12.75}$$

where $a = \left(n - \left(L^2 - 1\right) N_c\right) \frac{M_cB}{mB} - \log mB$.

Proof. The proof is similar to that in [35], only with a different drawing probability $p_0 = \frac{M}{m}$. We omit the details of the proof due to the space limit.

Then,

$$p_{hit} = \Pr\left(V - KN_c\right) \to e^{-e^{-a}} = e^{-mB \cdot e^{-\frac{\left(n-\left(L^2-1\right)N_c\right)M}{m}}}. \tag{12.76}$$

Since $0 < \alpha - \beta < 1$, as $n \to \infty$, we have $n^{1-(\alpha-\beta)} \to \infty$, $e^{-\frac{nM}{m}} \to 0$ and $p_{hit} \to 1$. This indicates that all mB packets can be collected from $n - \left(L^2 - 1\right) N_c$ source node candidates with probability one as n becomes very large. The larger the size of local memory M_c is, the more quickly that $p_{hit} \to 1$ as $n \to \infty$.

This proves the outage probability results in Theorem 3.

References

[1] Poularakis K., Iosifidis G., Sourlas V., Tassiulas L. 'Exploiting caching and multicast for 5G wireless networks'. *IEEE Transactions on Wireless Communications*. 2016;**15**(4):2995–3007.

[2] Cui Y., Jiang D., Wu Y. 'Analysis and optimization of caching and multicasting in large-scale cache-enabled wireless networks'. *IEEE Transactions on Wireless Communications*. 2016;**15**(7):5101–5112.

[3] Golrezaei N., Shanmugam K., Dimakis A.F. 'FemtoCaching: wireless video content delivery through distributed caching helpers'. proceedings IEEE INFOCOM; Orlando; 2012. pp. 1107–15.

[4] Tao M., Chen E., Zhou H., Yu W. 'Content-centric sparse multicast beamforming for cache-enabled cloud ran'. *IEEE Transactions on Wireless Communications*. 2016;**15**(9):6118–31.

[5] Zhuo H., Tao M., Chen E. 'Content-centric multicast beamforming in cache-enabled cloud radio access networks'. Proceedings IEEE Global Communications Conference; San Diego; 2015. pp. 1–6.

[6] Dai B., Yu W. 'Joint user association and content placement for cache-enabled wireless access networks'. Proceedings IEEE International Conference on Acoustics, Speech and Signal Processing (ICASSP); Shanghai; 2016. pp. 3521–5.

[7] Liu A., Lau V.K.N. 'Asymptotic scaling laws of wireless ad hoc network with physical layer caching'. *IEEE Transactions on Wireless Communications*. 2016;**15**(3):1657–64.

[8] Zhang L., Xiao M., Wu G., Li S. 'Efficient scheduling and power allocation for D2D-assisted wireless caching networks'. *IEEE Transactions on Communications*. 2016;**64**(6):2438–52.

[9] Rao J., Feng H., Yang C., Chen Z., Xia B. 'Optimal caching placement for D2D assisted wireless caching networks'. Proceedings IEEE International Conference on Communications (ICC); Kuala Lumpur; 2016. pp. 1–6.

[10] Wu H., Wang L., Svensson T., Han Z. 'Resource allocation for wireless caching in socially-enabled D2D communications'. Proceedings IEEE International Conference on Communications (ICC); Kuala lumpur; 2016. pp. 1–6.

[11] Gregori M., Gómez-Vilardebó J., Matamoros J. 'Efficient scheduling and power allocation for D2D-assisted wireless caching networks'. *IEEE Journal on Selected Areas in Communications*. 2016;**34**(5):1222–34.

[12] Maddah-Ali M.A., Niesen U. 'Fundamental limits of caching'. *IEEE Transactions on Information Theory*. 2014;**60**(5):2856–67.

[13] Ozgur A., Leveque O. 'Throughput-delay trade-off for hierarchical cooperation in ad hoc wireless networks'. International Conference on Telecommunications; St. Petersburg; 2008. pp. 1–5.

[14] Gupta P., Kumar P.R. 'The capacity of wireless networks'. *IEEE Transactions on Information Theory*. 2000;**46**(2):388–404.

[15] Franceschetti M., Dousse O., Tse D.N.C., Thiran P. 'Closing the gap in the capacity of wireless networks via percolation theory'. *IEEE Transactions on Information Theory*. 2007;**53**(3):1009–18.

[16] Ozgur A., Leveque O., Tse D.N.C. 'Hierarchical cooperation achieves optimal capacity scaling in ad hoc networks'. *IEEE Transactions on Information Theory*. 2007;**53**(10):3549–72.

[17] Avestimehr A.S., Diggavi S.N., Tse D.N.C. 'Wireless network information flow: a deterministic approach'. *IEEE Transactions on Information Theory*. 2011;**57**(4):1872–905.

[18] Ozgur A., Leveque O., Tse D. 'Spatial degrees of freedom of large distributed MIMO systems and wireless ad hoc networks'. *IEEE Journal on Selected Areas in Communications*. 2013;**31**(2):202–14.

[19] Ozgur A., Johari R., DNC T. *Operating Regimes of Large Wireless Networks*. Now Publishers, Inc; 2011.

[20] Geng C., Naderializadeh N., Avestimehr A.S., Jafar S.A. 'On the optimality of treating interference as noise'. *IEEE Transactions on Information Theory*. 2015;**61**(4):1753–67.

[21] Hong S.-N., Caire G. 'Beyond scaling laws: on the rate performance of dense device-to-device wireless networks'. *IEEE Transactions on Information Theory*. 2015;**61**(9):4735–50.

[22] Ji M., Caire G., Molisch A.F. 'Optimal throughput-outage trade-off in wireless one-hop caching networks'. Proceedings IEEE International Symposium on Information Theory Proceedings (ISIT); Istanbul; 2013. pp. 1461–5.

[23] Ji M., Caire G., Molisch A.F. 'Wireless device-to-device caching networks: basic principles and system performance'. *IEEE Journal on Selected Areas in Communications*. 2016;**34**(1):176–89.

[24] Ji M., Caire G., Molisch A.F. 'Fundamental limits of caching in wireless D2D networks'. *IEEE Transactions on Information Theory*. 2016;**62**(2):849–69.

[25] Jeon S.W., Hong S.N., Ji M., Caire G. 'Caching in wireless multihop device-to-device networks'. Proceedings IEEE International Conference on Communications (ICC); London; 2015. pp. 6732–7.

[26] Jeon S.-W., Hong S.-N., Ji M., Caire G., Molisch A.F. 'Wireless multihop device-to-device caching networks'. *IEEE Transactions on Information Theory*. 2017;**63**(3):1662–76.

[27] Guo J., Yuan J., Zhang J. 'Wireless device-to-device caching networks with distributed MIMO and hierarchical cooperations'. Proceedings IEEE Global Communication Conference; Singapore; 2017. pp. 1–6.

[28] Guo J., Yuan J., Zhang J. 'An achievable throughput scaling law of wireless device-to-device caching networks with distributed MIMO and hierarchical cooperations'. *IEEE Transactions on Wireless Communications*. 2018;**17**(1):492–505.

[29] Liu A., Lau V.K.N., Caire G, Lau V G.C. 'Cache-induced hierarchical cooperation in wireless device-to-device Caching networks'. *IEEE Transactions on Information Theory*. 2018;**64**(6):4629–52.

[30] Guo J., Yang T., Yuan J., Zhang J.A. 'Linear vector physical-layer network coding for MIMO two-way relay channels: design and performance analysis'. *IEEE Transactions on Communications*. 2015;**63**(7):2591–604.

[31] Guo J., Yang T., Yuan J., Zhang J.A. 'A novel linear physical-layer network coding scheme for Y-channel without transmitter CSI'. *IEEE Transactions on Vehicular Technology*. 2018;**67**(10):10049–53.

[32] Qiu M., Huang Y.-C., Shieh S.-L., Yuan J. 'A lattice-partition framework of downlink non-orthogonal multiple access without SiC'. *IEEE Transactions on Communications*. 2018;**66**(6):2532–46.

[33] Qiu M., Yang L., Xie Y., Yuan J. 'On the design of multi-dimensional irregular repeat-accumulate lattice codes'. *IEEE Transactions on Communications*. 2018;**66**(2):478–92.

[34] Guo J., Yu W., Yuan J. 'Enhancing cellular performance through device-to-device distributed MIMO'. *IEEE Transactions on Communications*. 2018;**66**(12):6096–109.

[35] Erdös P. 'On a classical problem of probability theory'. *Citeseer*. 1961;**60**(5):215–20.

Chapter 13

Randomised geographic caching and its applications in wireless networks

Anastasios Giovanidis[1] *and Bartłomiej Błaszczyszyn*[2]

The randomised (or probabilistic) geographic caching is a proactive content placement strategy that has attracted a lot of attention, because it can simplify a great deal cache-management problems at the wireless edge. It diversifies content placement over caches and applies to scenarios where a request can be possibly served by multiple cache memories. Its simplicity and strength is due to randomisation. It allows one to formulate continuous optimisation problems for content placement over large homogeneous geographic areas. These can be solved to optimality by standard convex methods and can even provide closed-form solutions for specific cases. This way the algorithmic obstacles from NP-hardness are avoided and optimal solutions can be derived with low computational cost. Randomised caching has a large spectrum of applications in real-world wireless problems, including FemtoCaching, multi-tier networks, device-to-device (D2D) communications, mobility, mm-wave, security, UAVs and more. In this chapter, we will formally present the main policy with its applications in various wireless scenarios. We will further introduce some very useful extensions related to unequal file sizes and content placement with neighbourhood dependence.

13.1 Introduction

The idea to bring multimedia content closer to the user and offload cellular base stations has gained momentum during the research and development phase of 5th generation (5G) cellular networks [1], as well as in the context of Internet-of-Things (IoT). In the seminal paper by Shanmugam *et al.* [2], the authors introduced a novel architecture where so-called *helper stations* can be deployed as an additional support to cellular communications. These 'helpers' are low complexity stations with weak connection to the backhaul, which are equipped with storage space to cache popular multimedia content. They are seen as small local data-centres that can

[1]Sorbonne University, CNRS, LIP6, France
[2]Inria-ENS, France

offload traffic, which otherwise would be served over the Internet through the backhaul. This way, congestion can be avoided because each multimedia content needs to be fetched once and stored at the local helper. Users asking for this content can be served directly by the locally stored copy for so long as this is made available. Furthermore, the density of the helpers can allow for high-quality video reception, thus improving the user Quality-of-Experience (QoE). The original idea of installing helpers was gradually extended to include cache-memories on multiple-tiers of the heterogeneous cellular network [3] including the base station. For example, cache storage space can be installed at the Baseband Unit (BBU) or the BBU pool for Cloud-RAN architectures. Furthermore, the mobile devices [4, 5] can also make use of their own storage space to cache content of interest for the whole network, which can be transmitted to other users via direct D2D links bypassing the cellular infrastructure. These ideas extend in practice the concept of Content Centric Network architectures [6] to include the wireless-edge.

In all these scenarios, it is very important to decide what content to cache where, and for how long. There are two main types of cache-management policies, the *proactive* [1] and the *online* [7]. The former suggests that the whole memory inventory should be determined in advance based on measurements of local content popularity and be updated infrequently over long-time windows. The latter favours a frequent update of the cache inventory every time a new request arrives, with the aim to follow closely the evolution of present requests. It includes the time-to-live (TTL) and the least-recently-used (LRU) [8, 9] replacement principle.

13.1.1 The FemtoCaching content placement

In the FemtoCaching paper [2], the authors proposed a proactive content placement policy resulting from a binary optimisation problem. They assume all user and station positions are fixed and known. In a dense node deployment with frequency reuse, since one user can be covered by several stations or helpers, the user-station association forms a bipartite graph. Assuming further that a user can be served by any covering station, the content placement variables $x_{i,j} \in \{0, 1\}$ are the unknowns, which should decide whether to place content j on the cache of station i or not. The aim of the placement is to minimise the expected download time for files. In the special case of unit delay ([2], eq. 3) the problem simplifies to *maximising sum hit probability*

$$\begin{array}{ll} \max_x & \sum_{u=1}^{U} \left[1 - \sum_{j=1}^{F} a_j \prod_{i\in\mathcal{H}(u)}(1 - x_{ij})\right] \quad \texttt{[BinCache]} \\ \text{subject to} & \sum_{j=1}^{F} x_{ij} \leq K, \quad i = 1, \cdots, N \quad \mathbf{x} \in \{0, 1\}^{N\times F} \end{array} \tag{13.1}$$

In the above, the stations are indexed by $i = 1, \ldots, N$, the users by $u = 1, \ldots, U$ and the content by $j = 1, \ldots, F$. A user asks for content j with probability a_j. Also, $\mathcal{H}(u)$ is the set of stations covering user u, which can eventually serve the request, if this is available on-cache. The objective function is the expected hit rate summed over all users. To see this, let us focus on user u covered by the station set $\mathcal{H}(u)$. The user can ask for content j with some probability a_j. Then his request will be served by the local helpers (we call this a *hit*), if content j is stored in at least one cache among the set $\mathcal{H}(u)$. This is written as $1 - \prod_{i\in\mathcal{H}(u)} \left(1 - x_{i,j}\right) \in \{0,1\}$. This expression is 0 only if $x_{i,j} = 0$ for all covering stations. By taking expectation over the user request probability, we obtain $\sum_{j=1}^{F} a_j \left(1 - \prod_{i\in\mathcal{H}(u)} \left(1 - x_{i,j}\right)\right)$ and note that $\sum_{j=1}^{F} a_j = 1$. Finally, summing up over all users U we get the objective.

The constraint set limits the number of content placed in each cache memory $i = 1, \ldots, N$. The authors assume that all files are of equal size – or that these can be broken into chunks of equal size. Then, all cache-memories have a capacity equal to $K \geq 1$ files, K being a natural number. Since the decision variables $(x_{i,1}, \ldots, x_{i,F})$ for station i are all binary, their sum should not exceed the capacity K.

We call this problem BinCache because it is a binary cache placement problem. Its solution will place a different set of K files on each station, so that higher hit rate is achieved through content diversity. The above is shown to be NP-hard, and the authors propose approximation algorithms for a *sub-optimal* solution with some guaranteed distance from optimality. Their algorithm is based on a continuous relaxation of the unknowns $\tilde{x}_{i,j} \in \left[0,1\right]$ and subsequent pipage rounding. It has complexity of the order $O((U+N)^{7/2}F^{7/2})$, which is growing too high for realistic numbers of users, stations and files. For the case when every user is connected to at most 4 helpers (i.e. $card(\mathcal{H}(u)) \leq 4$), their approximation ratio is $1 - (1 - 1/4)^4 \approx 68\%$ from the optimal. Extensions of the BinCache include the joint caching and association problem [10], and the consideration of memory leasing [11], which have similar inherited issues.

13.1.2 The randomised approach

The exploding algorithmic complexity as the number of nodes and the catalog size increases, the sub-optimality of the solution and the requirement for known fixed positions of a given user set, render BinCache impractical. To overcome all these drawbacks, we have suggested in [12] a continuous reformulation of the cache placement problem. The problem and the new policy resulting from its solution are known in the literature as *randomised (or probabilistic) geographic caching*. We refer to the specific problem with cache helpers formulated in [12] and its solution as RandCache. The formulation of the simplified problem RandCache is based on two principal ideas. (i) The first one is the randomness in user and station positions, assuming homogeneity over the whole area. (ii) The second is the randomness (actually independence) of content placement. These two ideas are of course very much related.

(i) We consider a *typical user* rather than a specific given set of users. Assuming that users take positions homogeneously on the plane with some density per unit area, this representative user can be picked at random among them, and thus has a statistical viewpoint of the network. Let us denote this sampled user by u_0. Since the user was picked at random, the cardinality of stations covering him $card(\mathcal{H}(u_0))$ is a random variable with some distribution. To empirically find this distribution, we calculate the percentage p_0 of users present in the system that are not covered by any station, i.e. those users with $card(\mathcal{H}(u)) = 0$, then the percentage p_1 of users covered by a single station with $card(\mathcal{H}(u)) = 1$ and so on. Hence, the coverage number of the typical user is drawn from the empirical distribution $\{p_m\}$, $m = 0, 1, 2, \ldots$. Note that, this formulation does not need to know the exact coverage set $\mathcal{H}(u)$ per user, just the statistics of the coverage number.

Another way to estimate $\{p_m\}$ is by a measurement campaign as follows: we split the whole planar area with a grid. At each vertex of the grid we measure how many stations cover this point. Then, similarly as above, we can find that approximately p_0 percentage of the points in the grid are not covered by any station, p_1 are covered by exactly one and so on. The denser the grid, the finer our estimation over the coverage number mass function. The reason we can do so is that we assume that the requesting user can be selected at random over the plane and can take uniformly any position. We thus presume an homogeneous distribution of users on the plane.

(ii) Since only the number of covering stations is important and not in detail which station covers which user, it is only natural to consider content placement also probabilistically. A randomised placement ensures that content j needs to be present in a *percentage* of all stations equal to $0 \leq b_j \leq 1$, without specifying which stations in particular. In other words, if we pick a station at random we expect that this station will hold content j in its cache with probability b_j. It is important to note that content should be placed *independently* of the user positions and *independently* of station positions, in order to measure the performance of the typical user.

With the above, the problem BinCache becomes continuous and takes the following form, called RandCache,

$$\begin{aligned} \max_{\mathbf{b}} \quad & \mathbb{E}\left[1 - \textstyle\sum_{j=1}^{F} a_j \left(1 - b_j\right)^{card(\mathcal{H})}\right] \quad \texttt{[RandCache]} \\ \text{subject to} \quad & \textstyle\sum_{j=1}^{F} b_j \leq K, \\ & \mathbf{b} \in \left[0, 1\right]^F. \end{aligned} \tag{13.2}$$

Observe here, that the new objective is in expectation, and should be compared with the objective of BinCache after dividing the latter by the number of users U to obtain an average expression of hit probability over the given set of users U. The expectation in RandCache is taken over the random variable $card(\mathcal{H})$, i.e. the cardinality of the covering set of base stations, which has probability mass function $\{p_m\}$, for $m = 1, \ldots, M$. Here, M is the maximum number of covering stations and we can allow $M \to \infty$ to include all possible scenarios. Then the objective can be equivalently written as

$$f_{cell}(\mathbf{b}) = 1 - \sum_{j=1}^{F} a_j \sum_{m=1}^{M} p_m \left(1 - b_j\right)^m . \tag{13.3}$$

Observe that in the randomised version the product $\left(1 - b_j\right)^m$ replaces the product $\prod_{i \in \mathcal{H}(u)} (1 - x_{i,j})$ with $card(\mathcal{H}(u)) = m$ in the binary one. The product $\left(1 - b_j\right)^m$ is due to *independence* of content placement among stations, i.e. if $i = 1, \ldots, m$ station covers the user $\mathbb{P}\left(\{x_{1,j} = 0\} \cap \ldots \cap \{x_{m,j} = 0\}\right) = \mathbb{P}\left(x_{1,j} = 0\right)^m = (1 - b_j)^m$. The probability b_j substitutes every element of the decision vector $(x_{1,j}, \ldots, x_{N,j})$. We could interpret the probability b_j as the percentage of present copies for content j when the number of stations grows large,

$$\lim_{N\to\infty} \frac{1}{N}\sum_{i=1}^{N} x_{i,j} = b_j \quad \in [0,1]. \tag{13.4}$$

A direct benefit from the independent randomisation is that the RandCache problem has F continuous unknowns compared to the $N \times F$ binary unknowns of the BinCache problem, because it treats all stations in a homogeneous way, thus removing the index $i = 1, \ldots, N$. It has, thus, much smaller dimension.

Furthermore, it can be shown that the new objective is concave, and the problem is convex and separable in the variables b_j, so we can find very efficient algorithms to solve it. After having obtained the optimal solution $(b_1^*, \ldots, b_F^*)$, however, it is not obvious how to place content on stations, while respecting these probabilities (frequencies). As we will see, there is a particular way to place batches of K contents on each station, which is very different from the binary policy. We need to emphasise that the RandCache is not just a linear relaxation of the BinCache problem.

The randomised content placement from [12] found application in the years that followed in many interesting edge-caching scenarios, including:

- *Heterogeneous (multi-tier) networks* with Nakagami fading [13], two-tier [14], and multi-tier placement [3, 15, 16], area spectral efficiency [17], cooperative transmission [18]
- *mm-wave communications* [19], with multiple antennas [20]
- *Device-to-device (D2D) communications* [21] and variations with spatial content correlation [22]
- *Unmanned Aerial Vehicle (UAV) communications* [23, 24]
- *Opportunistic spectrum access* [25]
- *Time-varying content popularity* [26]
- *Effect of retransmissions in throughput* [27]
- *File secrecy and physical-layer security* [28]
- *Joint content caching and recommendations* [29]
- *Content classes of unequal file size* [30]
- *Networks of caches* [31]

A distributed solution for the RandCache problem and its variations is proposed in [32] using arguments from game theory and simulated annealing. The authors in [31] have proposed the randomised vector placement approach for optimal content placement in a network of caches, arising in ICNs, CDNs, and P2P systems. Randomised placement has been recently applied for online recommendations [33].

In what follows, we will present in detail the randomised policy in Section 13.2. The ingredients of the RandCache problem are given in Section 13.3 and its solution steps in Section 13.4. In Section 13.5 we extend the randomised policy, to treat unequal file sizes. Furthermore, we present two additional applications, one for D2D networks with mobility in Section 13.6, and one with competition among several content providers for cache space in Section 13.7. Finally, in Section 13.8 we depart from the assumption of independent random placement among caches to introduce policies of content placement with cache dependence.

13.2 Random independent placement policy

We have shown in the introduction how the randomisation of content placement allowed us to move from the `BinCache` problem to the `RandCache` problem. The random independent placement policy is in fact much more general than the specific problem instance in `RandCache`. Furthermore, this policy has an interesting way to be implemented.

More specifically, in the generic caching problem we assume that there is a set of stations $i = 1, \ldots, N$, and each has a memory of size $K \geq 1$ contents installed. The catalog of possible contents is of size F and contents are indexed by $j = 1, \ldots, F$ and all have *unit size*. The memory inventory of station i contains a subset of the content catalog with a number of elements not larger than K. To keep a general notation that will include many cases, suppose that the decision to place content j in cache i is a binary random variable (r.v.)

$$X_{i,j} = \begin{cases} 1, & \text{with probability } b_{i,j} \in [0, 1] \\ 0, & \text{with probability } 1 - b_{i,j} \end{cases}. \tag{13.5}$$

Note that each of these probabilities refers to a specific station i and can differ from one station to the other. These probabilities generally take values within the interval $b_{i,j} \in [0, 1]$.

- *Deterministic policies:* To avoid mixed time-sharing policies, a way to proceed is to enforce $b_{i,j} \in \{0, 1\}$. This way we can define a *deterministic* policy, where $X_{i,j} = x_{i,j} \in \{0, 1\}$ with probability 1. To have a feasible policy, we need to ensure that no more than K items are installed per cache. The set of feasible deterministic placements for our problem with N stations is then defined as

$$\mathcal{F}_{determ} = \left\{ \mathbf{x} \mid \mathbf{x} \in \{0, 1\}^{N \times F}, \sum_{j=1}^{F} x_{i,j} \leq K, \; i = 1, \ldots, N \right\}. \tag{13.6}$$

The way the set is defined guarantees that at most one item j is placed in each cache i. The above set is the constraint set of `BinCache`. Depending on the objective function of the optimisation problem, its solution will provide an optimal vector $\mathbf{x}^*$ among the feasible ones of size $N \times F$. Once this vector is determined, each cache memory i is filled with the contents given by the solution $x^*_{i,j} = 1$.

- *Randomised homogeneous policies:* A different category of policies results by enforcing the same probabilistic placement for all stations. This would read $b_{i,j} = b_j$ for all stations $i = 1, \ldots, N$ and we refer to it as randomised homogeneous. Each station should place a copy of content j *independently* with probability b_j. We write

$$b_j = \mathbb{P}\left(X_{i,j} = 1\right), \quad i = 1, \ldots, N. \tag{13.7}$$

This would mean that a copy of content j should be present in a b_j percentage of the N stations. Of course, this interpretation makes sense for N large enough.

For a randomised policy to be feasible, each realisation should guarantee that at most K contents are placed in each station, as in the deterministic case. This constraint reads,

$$\sum_{j=1}^{F} X_{i,j} \leq K, \tag{13.8}$$

where the left hand-side is a random sum. All is left is to find a policy which places content with probability (13.7) in each cache, while satisfying the constraint (13.8). Of course, any such policy will necessarily satisfy in expectation,

$$\mathbb{E}\left[\sum_{j=1}^{F} X_{i,j}\right] = \sum_{j=1}^{F} \mathbb{P}\left(X_{i,j} = 1\right) \stackrel{1.7}{=} \sum_{j=1}^{F} b_j \leq K. \tag{13.9}$$

This is not a trivial task. Let us proceed naively and decide to place content j in each cache i by a random coin toss (`Bernoulli` variable) with probability b_j. If we repeat this process independently for all contents $j = 1, \ldots, F$ although we do respect (13.7), however we cannot guarantee constraint (13.8) that there will be less or equal to K different files per cache. To see this, let us use an example with size $K = 2$ memory slots, catalog $F = 4$ contents and a placement vector for the F contents $(0.5, 0.5, 0.5, 0.5)$, such that $\sum_{j=1}^{4} b_j = 2$. If we perform an independent coin toss, the probability to select all 4 contents and overflow the memory space is $(0.5)^4 = 0.0625$ (larger than zero). The independent coin toss (Bernoulli trial) will not guarantee the constraint (13.8). So, we need to propose a different way to place content appropriately.

The idea is as follows: Instead of deciding independently whether to place or not content j in some cache, let us better choose an entire vector of K contents. The K-sized vectors should be drawn independently one from the other but should respect that the frequency of appearance of each content j is b_j as in (13.7).

With the above method, the probability of appearance of content j in a memory of size K is exactly equal to b_j.

In a sense, this policy is a *random vector generator of size* $K > 1$, which generalises the discrete random generator (for $K = 1$). Since we have found a way to place content while respecting both (13.7) and (13.8), the feasible set for any randomised homogeneous placement problem becomes,

$$\mathcal{F}_{random} = \left\{ \mathbf{b} \mid \mathbf{b} \in [0, 1]^F, \ \sum_{j=1}^{F} b_j \leq K \right\}. \tag{13.10}$$

Random placement policy

We are given a vector of placement probabilities $\mathbf{b} = (b_1, \ldots, b_F)$ for the F contents of the catalog, satisfying (13.7) and (13.9), and a cache memory of size $K \geq 1$. We split the cache into K continuous memory intervals of unit length and place them one under the other, as shown in Figure 13.1. The F contents of the catalog are picked one after the other without replacement and in any order. Their values b_j fill in the memory intervals without leaving a gap. When a unit interval is full and cannot contain the entire b_j, the remainder of b_j fills in the next unit memory interval below. The content probabilities b_j are continuously placed one next to the other until all contents have been selected. Due to the constraint (13.9) the process does not exceed the memory space K. To simplify, suppose that the memory is full up to K (although this is not necessary). In order to randomly choose a set of exactly K contents, with a joint distribution that follows (13.7) and (13.9), we pick uniformly at random a number within the interval [0, 1] and draw a vertical line intersecting the memory space covered by exactly K distinct contents. The contents are distinct because $b_j \in [0, 1]$.

The above is just the constraint set for RandCache. We can use any objective function that treats all caches homogeneously to derive the optimal feasible solution $\mathbf{b}^*$. Once the optimal probability vector is known, the K-sized memories of all stations can be filled in by randomly drawing vectors using the random placement policy.

In the example of Figure 13.1 we see that the following batches of $K = 3$ items can be drawn from our method: $\mathcal{C}_1 = \{‘1’, ‘2’, ‘3’\}$, $\mathcal{C}_2 = \{‘1’, ‘2’, ‘4’\}$, $\mathcal{C}_3 = \{‘1’, ‘2’, ‘5’\}$, $\mathcal{C}_4 = \{‘1’, ‘3’, ‘5’\}$, $\mathcal{C}_5 = \{‘2’, ‘3’, ‘5’\}$ and $\mathcal{C}_6 = \{‘2’, ‘3’, ‘6’\}$. What is interesting is that the frequency of each batch depends on the width occupied on the unitary interval. We can observe, e.g. that the most frequent batch from the six possible is the $\mathcal{C}_2 = \{‘1’, ‘2’, ‘4’\}$, because it occupies a wider interval. If we measure the frequency of each batch $(\rho_1, \ldots, \rho_6)$ we can reproduce the values of b_j, $j = 1, \ldots, 6$. To see this, take content ‘1’. This content is present inside the batches $\mathcal{C}_1, \mathcal{C}_2, \mathcal{C}_3, \mathcal{C}_4$. So, to shift from batch frequencies to item frequencies we just sum-up $\rho_1 + \rho_2 + \rho_3 + \rho_4 = b_1$. In general,

$$b_j = \sum_k \rho_k \mathbf{1}_{\{j \in \mathcal{C}_k\}}, \tag{13.11}$$

where as explained above ρ_k is the frequency of the k-th possible batch, and the indicator function $\mathbf{1}_{\{j \in \mathcal{C}_k\}} = 1$ if content j is inside the k-th batch, otherwise 0.

The random placement policy selects independently, at random a vector of K contents per station, thus it is not a solution tailored for some specific position, contrary to the deterministic policy. It only tries to statistically guarantee some performance of the system in expectation. This loss in detail is what brings the benefits of continuity and reduction in the number of unknowns. Furthermore, since the users in real scenarios are mobile, the optimal deterministic solution is bound to change the whole time. Hence, it might be preferable for scenarios with mobility to apply the randomised rather than the deterministic solution. The same holds for scenarios with D2D communications, where nodes are mobile and so they change constantly position relative to each other. In both cases, the statistics of coverage number are stable, but the actual instance of the user-station coverage bipartite graph changes constantly. On the other hand, the random vector generator presented above will not work well in cases where the stations are positioned inhomogeneously, i.e. more dense at some areas and less at others. In such scenarios, the random placement policy needs to be appropriately adapted (see also Section 13.8).

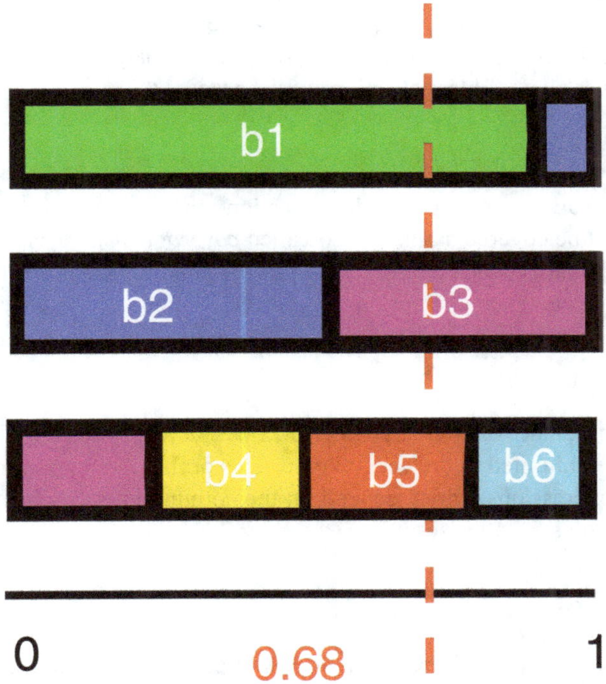

Figure 13.1 *A realisation of the probabilistic placement policy for* F = 6 *contents and* K = 3 *memory slots. We first draw uniformly a random number within the interval* $[0, 1]$*, say we find* 0.68*. The vertical line at this point intersects with each of the three memory slots at a specific content. We conclude from the figure that contents* {'1', '3', '5'} *will be cached.*

13.3 The typical user

We consider a wireless network, where stations are positioned on the two-dimensional (2D) plane. The performance of the network is evaluated at the Cartesian origin $(0,0)$, which we denote as the *typical user* u_o, following our discussion in the introduction. The typical user is characterised by two distributions: one for the content requests and one for the number of covering stations.

In order for the typical user performance to be representative, the statistics of the network should be unchanged whatever point we choose for the evaluation. For the traffic, this can be achieved if requests network-wide are drawn independently from the same probability distribution. For the coverage number, it depends on how stations are positioned relatively to the users. Assuming that stations are placed following a homogeneous Poisson Point Process (PPP) in 2D with constant density $\lambda_s > 0$, then such process has the properties of stationarity and isotropy [34] and the typical user approach is valid [35]. We denote the PPP of the stations by Φ_s and a realisation of this process is $\phi_s = \{\chi_i\}$, where χ_i are the deterministic coordinates of the i-th station in this specific realisation. We will evaluate the performance at the typical user over all possible realisations of the process Φ_s.

13.3.1 Coverage number

In cellular networks a user at a random location may be covered by one or more BSs or may not be covered at all. The so-called *coverage number* is a random variable (r.v.) [36], that depends on the features of the communications scheme and the network parameters (such as transmission power). We denote the coverage number by $\mathcal{M}$,

$$p_m := \mathbb{P}\left[\mathcal{M} = m\right], m = 0, 1, \dots. \tag{13.12}$$

The maximum number of covering BSs is $M \in \mathbb{N}^+ \cup \{\infty\}$, where we let the coverage number be unlimited in the general case. This probability mass function p_m can take specific expressions, which depend on the communication scheme evaluated. Obviously, it holds for all cases

$$\sum_{m=0}^{M} p_m = 1. \tag{13.13}$$

In what follows we will present special cases of coverage, depending on the communications scenario. An example of multi-coverage is illustrated in Figure 13.2.

13.3.1.1 SINR model

Here we consider the downlink of a cellular network and study the signal reception at the typical user. In the most general case all stations operate on the entire available frequency spectrum and the user experiences interference of unrelated stations upon signal reception. The quality of coverage at the origin is described by the SINR_o

(from now on SINR). SINR(χ_i) is the SINR at the reception, when user u_o is served by BS χ_i and is defined as

$$\text{SINR}(\chi_i) := \frac{S_i/\ell(r_i)}{\sigma_N^2 W + I - S_i/\ell(r_i)}. \quad (13.14)$$

The constant σ_N^2 is the noise power per Hertz, W is the bandwidth, $I = \sum_{\chi_i \in \Phi} S_i/\ell(r_i)$ is the total received power from the network, $|\chi_i| = r_i$ is the distance of χ_i from o, and $\ell(r) = (Br)^\beta$ is the path-loss function, with constants $B > 0$ and $\beta > 2$. We say that the typical user is covered when SINR(χ_i) $> T$, where T is a predefined positive threshold.

The coverage number $\mathcal{M}(T)$ as a function of the threshold indicates how many stations cover the typical user simultaneously and is the r.v.

$$\mathcal{M}(T) = \sum_{\chi_i \in \Phi} \mathbf{1}[\text{SINR}(\chi_i) \geq T]. \quad (13.15)$$

For the coverage of a user who can choose to be served by different BSs in some realisation of Φ, we make use of a basic result from ([34], Proposition 6.2), see also ([37], Lemma 5.1.2) a special case of which is given in [(38], Lemma 1). It is shown that if m stations cover a user at SINR level T, then the following inequality holds

$$m < 1 + 1/T. \quad (13.16)$$

For example, when $T \geq 1$ then necessarily $m < 1 + 1/T < 2$, which implies that $m \in \{0, 1\}$. Similarly, when $1 > T \geq 1/2$, $m \in \{0, 1, 2\}$, etc. The authors in [36] (see

Figure 13.2 Example of a cellular network with caching on BSs. A user is covered by three different BSs. Each BS has a cache memory of size two, where content can be temporarily saved. Hence, the user can choose among the content in a total of six memory slots.

also [37], (5.26) and (5.35)) have given explicit expressions for the probability that the typical user is covered by exactly m BSs, given T and a PPP for the BS positions. For general shadowing, they have calculated the probability

$$p_m^{\mathrm{SINR}} := \mathbf{P}\left[\mathcal{M}(T) = m\right]. \tag{13.17}$$

The numerical values for p_m^{SINR} can be calculated by the software developed for MATLAB® in [39].

13.3.1.2 Boolean model

For the noise-limited case, where the interference is small compared to noise, we can use the Boolean model to calculate the probability of coverage by m BSs. This is a seed-grain model, where the atoms of the PPP are the seeds. Centred on each atom is a 2D sphere $\mathcal{B}\left(\chi_i, R_b\right)$ that describes the area of coverage. R_b is a fixed radius that can be expressed by communication quantities. Specifically, if we only consider signal attenuation and no fading, the received signal at the boundary should be larger than the threshold, in order to guarantee coverage, i.e. $(\tilde{B}R_b)^{-\beta} \geq T \Rightarrow R_b = T^{-1/\beta}\tilde{B}^{-1}$.

It is shown in ([34], Lemma 3.1) that the number of BSs covering the typical user follows a Poisson distribution with parameter $\nu = \lambda\pi\left(T^{-1/\beta}\tilde{B}^{-1}\right)^2$

$$p_m^{Bool} = \frac{\nu^m}{m!}e^{-\nu}. \tag{13.18}$$

13.3.1.3 Overlaid 2-network Model

Very often in practice, it occurs that two (or more) networks of the same provider operate in parallel over an area, using different infrastructure (nodes) and orthogonal resources (bandwidth). It is typical, e.g. for operators to have one network of base stations for 3G/4G technology and numerous WiFi hotspots within a city. Given that a user may choose between the two to connect to the Internet with his/her cellphone (and assuming for simplicity mathematically independent models of these two networks), the coverage number at the typical user is distributed as the convolution of the coverage probability vectors of the two individual networks $\mathbf{p}^{(1)} = \left[p_0^{(1)}, \ldots, p_M^{(1)}\right]$ and $\mathbf{p}^{(2)} = \left[p_0^{(2)}, \ldots, p_M^{(2)}\right]$, that is (with $p_m^{(\cdot)} := 0$ for $m < 0$),

$$\mathbf{p}^{2NET} = \mathbf{p}^{(1)} * \mathbf{p}^{(2)} \Rightarrow p_m^{2NET} = \sum_{n=0}^{M} p_n^{(1)} p_{m-n}^{(2)}. \tag{13.19}$$

13.3.1.4 Coverage number from measurements

For general placement of wireless access nodes, the probability vector $\mathbf{p}$ can be estimated by measurements over a whole area. For example, we can imagine that a measurement scenario samples the coverage number at several planar points and estimates the probability p_0 as the ratio of points not covered at all divided by the total number of samples. Similarly, p_1 is estimated as the ratio of points covered by exactly one station divided by the total number of sample points and so on.

13.3.2 Content catalog and popularity

Each user – and consequently the typical user at the Cartesian origin – has a request for a specific content (say video file) that he/she wants to receive. In our model, we consider the independent reference model (IRM) for the traffic [40]. This model considers a fixed content library with cardinality F files. The set is denoted by $\mathcal{C} := \{`1', \ldots, `F'\}$, where each element is an entire file, and indexing follows the file-names $j = 1, \ldots, F$. We consider for now that all content has the same size, normalised to 1. To apply this approach for unequal sized files, we can assume that each file can be divided into chunks of equal size. (The general case of unequal file sizes will be treated in Section 13.5.) Furthermore, each content is related to its popularity, which we assume known as a priori. We order the content by popularity: '1' is the most popular content, '2' the second most popular and so on. The popularity follows a distribution $\{a_j\}$. To be consistent with the above ordering, $a_1 \geq a_2 \geq \ldots \geq a_F$. In the IRM model, every request is independent of every other, drawn from the same distribution. Hence, if we focus on the typical user, he/she requests for content j with probability a_j.

Although the type of distribution can be left open in practice data from measurements indicates that content popularity is Zipf distributed [41, 42]. Zipf distribution considers the order from the most popular file with index $j = 1$ to the least popular with index $j = F$. The probability that a user requests content j is equal to

$$a_j = C^{-1} \cdot j^{-\gamma}. \tag{13.20}$$

The Zipf exponent γ is chosen often within the range $\gamma \in (0, 2]$ to limit the difference between the two most popular files, i.e. $a_1/a_2 \leq 4$, but it can vary depending on the scenario. The normalisation factor C is found by summing the mass to unity

$$\sum_{j=1}^{F} a_j = 1. \tag{13.21}$$

A value γ close to zero results in a uniform popularity distribution, whereas larger values of γ produce distributions with increasingly lighter tails. So in Zipf, γ controls the skewness of the popularity.

13.4 Optimal randomised caching in cellular networks

In this section, we will present the original problem of caching in cellular networks with multi-coverage and present the optimal (independently) randomised policy published in [12]. The solution will be described in more detail here. As mentioned in the introduction, when searching for the optimal random placement policy, the problem obtains the RandCache formulation. The objective (13.3) aims to maximise the probability that the typical user will find the content he/she is asking for in one of the BSs he/she is covered from. This is one minus the probability that the user does not find its content anywhere among the covering stations. More precisely, this happens when the user is covered by $m = 0$ stations (i.e. no coverage), or by some number $m > 0$, but the content has not been saved in the cache memory space of any of these. We repeat at this point the optimisation problem [RandCache] (13.2),

$$\begin{array}{ll} \max_{\mathbf{b}} & f_{cell}(\mathbf{b}) := 1 - \sum_{j=1}^{F} a_j \sum_{m=0}^{\infty} p_m (1 - b_j)^m \; \texttt{[RandCache]} \\ \text{s.t} & \mathbf{b} \in \mathcal{F}_{random} \end{array} \tag{13.22}$$

where the feasible set $\mathcal{F}_{random}$ is defined in (13.10). We can control the objective hit probability, by varying the content availability b_j, $j = 1, \dots, F$ within the feasibility region. The size of the cache memory K plays an important role in the constraints. We further note that, in the case that for some content j the optimal placement is $b_j = 1$, i.e. the content j is cached almost surely in all stations, then $\sum_{m=0}^{\infty} p_m(1 - b_j)^m |_{b_j=1} = p_0$, i.e. the non-coverage probability.

Remark 1: For the objective in (13.22) (same as (13.3)) we can make the following observation. Let $z_j := 1 - b_j$. Then we have

$$\psi(z_j) := \mathbb{E}\left[z_j^{\mathcal{M}}\right] = \sum_{m=0}^{\infty} p_m z_j^m. \tag{13.23}$$

The expression is the *probability-generating function* (or $\mathcal{Z}$*-transform* when using z^{-1} instead of z) of the random variable $\mathcal{M}$ defined in (13.12) as the number of stations covering the typical user. By replacing this in (13.3) we get

$$f_{cell}(\mathbf{b}) := 1 - \sum_{j=1}^{J} a_j \psi(1 - b_j). \tag{13.24}$$

Before we move on to the general solution of [RandCache], let us first formulate and solve a toy version of the problem, with $F = 2$ different types of content, $K = 1$ memory slot per BS and $M = 2$, meaning that the typical user can only be in 3-coverage states, i.e. 0-coverage, 1- and 2-coverage, with $m \in \{0, 1, 2\}$. Its solution will provide intuition on how to solve the most general one.

$$\begin{array}{ll} \max_{(b_1,b_2)} & \sum_{j=1}^{2} a_j(1 - p_0 - p_1(1 - b_j)^1 - p_2(1 - b_j)^2) \texttt{[RandK1F2]} \\ \text{s.t.} & b_1 + b_2 \leq 1 \\ & b_1, b_2 \in [0, 1] \end{array}$$

13.4.1 *Solution for [*`RandK1F2`*]*

The toy problem can be further simplified by making use of the equalities $p_0 + p_1 + p_2 = 1$ and $a_1 + a_2 = 1$. We also observe that the objective function is *monotone increasing* in the unknowns b_1 and b_2 for their range of values in the feasible set. As a consequence, the optimal solution should satisfy the constraint with equality $b_1 + b_2 = 1$. Using these three equalities, the problem can be expressed only with the parameter set (a_1, p_1, p_2) and has a single unknown b_1

$$\begin{array}{lll} \max & -p_2 b_1^2 + b_1\left(2a_1\left(p_1+p_2\right)-p_1\right)+\left(1-a_1\right)\left(p_1+p_2\right) & [\texttt{RandK1F2}-\texttt{r}] \\ \text{s.t} & b_1 \in \left[0,1\right] & \end{array}$$

We denote the objective function of $[\texttt{RandK1F2}-\texttt{r}]$ by $f_{toy}(b_1)$. Now, the only constraint is that $b_1 \in \left[0, 1\right]$. Let us take the first derivative of the objective function,

$$\frac{df_{toy}}{db_1} = -2p_2 b_1 + 2a_1\left(p_1+p_2\right) - p_1.$$

When the derivative is positive the function is increasing, otherwise decreasing.

$$\frac{df_{toy}}{db_1} \geq 0 \bigcap b_1 \in [0,1] \Rightarrow 0 \leq b_1 \leq \min\left\{1, \frac{2a_1\left(p_1+p_2\right)-p_1}{2p_2}\right\},$$

where in the above we limit the b_1 inside the interval [0, 1]. We observe that the objective function is increasing in b_1 until some upper value b_1^*. If this upper value is larger than 1, then the objective is increasing in the entire interval [0, 1]. Otherwise, it is increasing up to b_1^* and then decreasing in the interval $[b_1^*, 1]$. Then the solution of the problem is simply equal to b_1^*. This is because, as long as b_1 moves from 0 to larger, the objective is also increasing (positive derivative). This is true until b_1^* with objective $f_{toy}(b_1^*)$, after which the objective value decreases. In fact, it can be shown from the second derivative that the function is concave

$$\frac{d^2 f_{toy}}{db_1^2} = -2p_2 \leq 0, \tag{13.25}$$

so there is a unique maximum at b_1^*. To sum up,

Proposition 1: The solution for the $[\texttt{RandK1F2}-\texttt{r}]$ problem is given by the placement probability for content type '1' equal to

$$b_1^* = \begin{cases} \frac{2a_1(p_1+p_2)-p_1}{2p_2} & , if\, 0.5 \leq a_1 < 1 - \frac{p_1}{2(p_1+p_2)} \\ 1 & , \text{otherwise} \end{cases} \tag{13.26}$$

The condition for $b_1^* < 1$ is that the popularity of file '1' is within the interval $0.5 \leq a_1 < 1 - p_1/(2p_1 + 2p_2) < 1$. This results from the feasibility constraint $0 \leq b_1^* < 1$, by replacing $b_1^* = \frac{2a_1(p_1+p_2)-p_1}{2p_2}$. The right-hand side is then found directly. The left-hand side inequality gives $a_1 \geq p_1/(2p_1 + 2p_2)$, which is always true because $p_1/(2p_1 + 2p_2) \geq 0.5$ and the files are ordered in popularity, so for $F = 2$ necessarily $a_1 \geq 0.5$, being the most popular one.

Interestingly, as (13.26) shows, the optimal solution b_1^* is just an increasing affine function of the popularity a_1. The more popular the file '1' the more often this file is placed. The optimal value ranges from $b_1^* = 0.5$ when $a_1 = 0.5$, to $b_1^* = 1$ when $a_1 = 1 - \frac{p_1}{2(p_1+p_2)}$. Obviously $b_2^* = 1 - b_1^*$.

To evaluate the solution, we first plot of the objective function $f_{toy}(b_1)$ over b_1, and for various values of p_1 and p_2. To facilitate comparison, we keep $a_1 = 0.6$ for the popularity of the first content and the sum $p_1 + p_2 = 0.95$, so that $p_0 = 0.05$. For

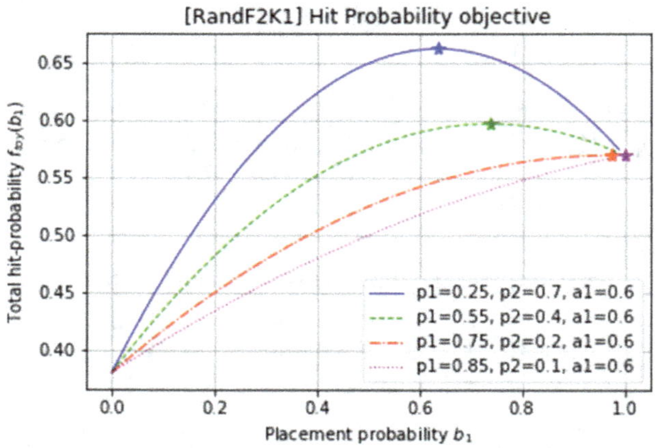

Figure 13.3 *Four curves of the hit probability (objective of [RandK1F2-r]), for different values of* p_1 *and* p_2*, but for constant sum* $p_1 + p_2 = 0.95$ *and popularity* $a_1 = 0.6$*. The star marks the maximum.*

each pair (p_1, p_2) we find the maximum b_1^*. We denote the maximal value of the objective function by $f_{toy}^* = f_{toy}(b_1^*)$. Our findings are summarised in the table below:

p_1	p_2	b_1^*	f_{toy}^*
0.25	0.7	0.6357	0.6629
0.55	0.4	0.7375	0.5976
0.75	0.2	0.9750	0.5701
0.85	0.1	1	0.5700

Figure 13.3 that follows illustrates the related plots. To explain better we note that

- All curves are concave.
- For $b_1 = 0$: $f_{toy}(0) = (1 - a_1)(p_1 + p_2) = a_2(1 - p_0)$. In our case $(1 - 0.6) * 0.95 = 0.38$ for all curves.
- For $b_1 = 1$: $f_{toy}(1) = a_1(p_1 + p_2)$. In our case $0.6 * 0.95 = 0{,}57$ for all curves.
- For $b_1 = b_1^*$ there are two cases:
  - If $b_1^* = 1$ then $f_{toy}(b_1^*) = a_1(p_1 + p_2)$.
  - If $b_1^* = \frac{2a_1(p_1+p_2)-p_1}{2p_2} < 1$ then the maximal value is equal to

$$f_{toy}(b_1^*) = f_2(0) + (b_1^*)^2 p_2. \tag{3.27}$$

Even in this simple toy example, the policy of placing the most popular content '1' everywhere ($b_1 = 1$) is not necessarily optimal. When the probability of coverage by 2 BSs is considerable, a portion of the slots should be occupied by the second most popular content '2'. This depends also on the content popularity ratio a_2/a_1. To understand this, we plot in Figure 13.4 (left) the optimal caching policy as a function of p_1/p_2. We see that on the left part of the x-axis, where the probability of coverage by 2 stations is much larger than by a single one, the placement frequency is split among file '1' and '2', whereas at the right part of the x-axis, where most users are covered by a single station only file '1' is placed. In Figure 13.4 (right) the total hit is plotted, where it is clearly seen at the right part of the x-axis, that placing both items with the appropriate frequency has important performance improvements. The hit probability with $b_1 = 1$ everywhere at the right part of the x-axis is simply equal to $a_1(1 - p_0)$, i.e. the probability to be covered by at least 1 station and ask for content '1'.

Conclusions: This toy example shows that the solution can be found from a convex program with equality constraint. The optimal policy is characterised by diversity of content placement, in rich multi-coverage environments. Always caching the most popular contents is sub-optimal in general.

13.4.2 Problem solution [RandCache]

We move now to the solution of the general problem. The analysis and optimisation is very much facilitated by identifying certain properties of the problem at hand.

Lemma 1: The objective function of [RandCache] in (13.22) has the following two properties:

- **P.1:** It is separable with respect to the optimisation variables $b_1, \ldots, b_F$ and can be rewritten as

$$f_{cell}(\mathbf{b}) = \sum_{j=1}^{J} a_j \left(1 - \sum_{m=0}^{\infty} p_m (1 - b_j)^m\right) = \sum_{j=1}^{F} a_j g(b_j). \tag{13.28}$$

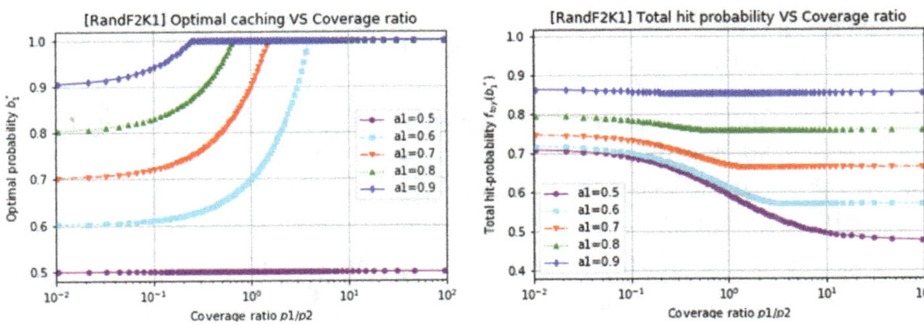

(a) (b)

Figure 13.4 Variation of (left) the optimal caching policy b_1^*, and (right) the optimal hit-probability objective, with respect to the coverage ratio p_1/p_2. The evaluation takes five different values of a_1 into consideration.

- **P.2:** It is an increasing and concave function of b_j, $\forall j$. Consequently it is a concave function of the vector $(b_1, \ldots, b_F)$.

Proof. Property 1 results by rewriting the objective function after replacing $1 = \sum_{j=1}^{J} a_j$. The proof of Property 2 becomes trivial due to the separability property. We can see that $\forall j$:

$$\frac{dg}{db_j} = \sum_{m=1}^{\infty} p_m m(1 - b_j)^{m-1} \geq 0,$$

$$\frac{d^2 g}{db_j^2} = -\sum_{m=2}^{\infty} p_m m(m-1)(1 - b_j)^{m-2} \leq 0.$$

Hence the objective function is increasing (non-decreasing) and concave on b_j, $\forall j$. Since by **P.1** it is also separable, the Hessian is a diagonal matrix, with negative entries on the diagonal equal to $a_j \frac{d^2 g}{db_j^2}$ and hence negative semi-definite. We conclude that the objective function is a concave function of the vector $(b_1, \ldots, b_F)$.

Since the objective function is concave by **P.2** and the constraint set $\mathcal{F}_{random}$ is linear, the optimisation problem can be solved as a *convex program*. We will make use of the Lagrangian relaxation and the subgradient method (see [43] and [44]).

13.4.2.1 Lagrangian relaxation

We first relate the dual variable $\mu \geq 0$ to the sum constraint inequality (13.9). This is called a Lagrange multiplier. We write down the Lagrangian function of the optimisation problem

$$L(b_1, \ldots, b_F, \mu) = \sum_{j=1}^{F} a_j \left(1 - \sum_{m=0}^{\infty} p_m (1 - b_j)^m\right) + \mu \left(K - \sum_{j=1}^{F} b_j\right). \quad (13.29)$$

The remaining constraint set is

$$\mathcal{F}_2 := \{b_j \in [0, 1], \forall j = 1, \ldots, J\}. \quad (13.30)$$

We can systematically find the optimal primal and dual variables by solving a min-max problem. Additionally, in our case where we deal with a convex program, the optimal value of the min-max problem is equal to the optimal value of the original problem [RandCache] with objective function

$$\begin{aligned} f^*_{cell} := \max_{\mathcal{F}_{random}} f_{cell}(b_1, \ldots, b_F) &= \min_{\mu \geq 0} \max_{\mathcal{F}_2} L(b_1, \ldots, b_F, \mu) \\ &= f_{cell}(b^*_1, \ldots, b^*_F). \end{aligned} \quad (13.31)$$

We then say that the duality gap between the original [RandCache] problem and the min-max problem is zero. We should further mention that the optimal primal and dual variables b^*_j, $\forall j$ and μ^* are related by the *Complementary Slackness Condition* (CSC), which basically forces the dual variables to be zero when the constraint is active.

$$\mu^* \left(K - \sum_{j=1}^{J} b_j^* \right) = 0 \Rightarrow \begin{cases} \mu^* = 0, & \text{if } K - \sum_{j=1}^{J} b_j^* > 0. \\ \mu^* \geq 0, & \text{if } K - \sum_{j=1}^{J} b_j^* = 0. \end{cases} \tag{13.32}$$

The following Lemma characterises the solution:

Lemma 2: At the optimal solution, the sum constraint inequality (13.9) is inactive, i.e. the optimal solution satisfies

$$b_1^* + \ldots + b_F^* = K. \tag{13.33}$$

Proof. Suppose that the inequality is inactive, i.e. strictly $< K$ for the optimal solution. But then, for any $\ell \leq F$, we can increase $b_\ell^* \rightarrow b_\ell^* + \epsilon$, so that the constraint is satisfied with equality. Substituting in the objective function $b_\ell^* + \epsilon$ instead of b_ℓ^* the value of the function will increase, because f_{cell} is increasing over b_ℓ (by property **P.2**). Hence the primal optimal solution cannot leave the constraint inactive and (13.33) is true.

13.4.2.2 Relaxed primary problem decomposition

Taking into account Lemma 2 and the CSC in (13.32), we first consider a fixed dual price $\mu \geq 0$ and solve the *relaxed primary problem* of (13.31), which we rewrite as

$$q(\mu) = \max_{\mathcal{F}_2} L(b_1, \ldots, b_F, \mu). \tag{13.34}$$

From Lemma 1, we can see that the above problem is separable in the variables b_j, i.e. given μ we should solve F separate problems, one for each content

$$\begin{aligned} \textbf{max} \quad & -\mu b_j + a_j \left(1 - p_0 - \textstyle\sum_{m=1}^{\infty} p_m (1 - b_j)^m\right) \text{ [SubRand} - \text{j]} \\ s.t. \quad & 0 \leq b_j \leq 1 \end{aligned}$$

Lemma 3: Given fixed $\mu \geq 0$, the solution of [subRand − j] is

$$b_j(\mu) = \begin{cases} 1, & \text{if } a_j p_1 > \mu \\ \omega(\mu), & \text{if } a_j p_1 \leq \mu \leq a_j \mathbb{E}[\mathcal{M}] \\ 0, & \text{if } a_j \mathbb{E}[\mathcal{M}] < \mu \end{cases}, \tag{13.35}$$

where $\mathbb{E}[\mathcal{M}] = \sum_{m=1}^{\infty} m p_m$ and $\omega(\mu)$ is the solution over b_j of the equation

$$a_j \sum_{m=1}^{\infty} p_m m (1 - b_j)^{m-1} = \mu \tag{13.36}$$

We take the first derivative of the objective function of [subRand − j]

$$-\mu + \frac{dg}{db_j} = -\mu + a_j \sum_{m=1}^{\infty} m p_m (1 - b_j)^{m-1}. \tag{13.37}$$

We distinguish between three cases:

- If $-\mu + \frac{dg}{db_j} > 0$, in the entire domain of b_j, the objective function is increasing and the optimal solution is $b_j = 1$. Since from **(P.2)** it holds $d^2g/db_j^2 \leq 0$, the function $-\mu + \frac{dg}{db_j}$ is decreasing in b_j, taking its minimum value for $b_j = 1$. Hence, it suffices to verify that $-\mu + \frac{dg}{db_j}|_{b_j=1} > 0$, which from the expression in (13.37) gives the condition $a_j p_1 > \mu$.

- If $-\mu + \frac{dg}{db_j} < 0$, in the entire domain of b_j, the objective function is decreasing and the optimal solution is $b_j = 0$. With the same argument about concavity, the maximum value of $-\mu + \frac{dg}{db_j}$ is obtained for $b_j = 0$. Hence, it suffices to verify $-\mu + \frac{dg}{db_j}|_{b_j=0} < 0$, which from (13.37) gives the condition $\mu_j \sum_{m=1}^{\infty} mp_m$.
- In the intermediate case, there exists a point $b_j = \omega \in (0, 1)$, such that $-\mu + \frac{dg}{db_j} = 0$. The exact value ω is found by solving (13.36).

Having written the primal variables as functions of the dual $b_j(\mu)$ in (13.35) we only need to solve the dual problem over the dual variable μ, see (13.31). Using the notation in (13.34), the dual problem is

$$\min_{\mu \geq 0} \; q(\mu). \tag{13.38}$$

13.4.2.3 Subgradient method for the dual

One of the most standard ways to solve (13.38) is by the *subgradient method*, which can find the optimal μ^* in a finite number of steps, even in the case where the function $q(\mu)$ is non-differentiable. The method involves a sequence of updates for the value of $\mu^{(k)}$, $k = 1, \ldots$. For each $\mu^{(k)}$ the relaxed primary problem is solved and the solution over $b_j(\mu^{(k)})$ is used for the next update that gives $\mu^{(k+1)}$. The method converges when $|\mu^{(k+1)} - \mu^{(k)}| < \epsilon$ or ideally $|\mu^{(k+1)} - \mu^{(k)}| = 0$. The update steps are described by the equation

$$\mu^{(k+1)} = \left[\mu^{(k)} + s^{(k)}\left(K - \sum_{j=1}^{F} b_j\left(\mu^{(k)}\right)\right)\right]^{+}, \quad k = 1, 2, \ldots. \tag{13.39}$$

The term in the parenthesis is the *subgradient* of our problem (practically it is the subgradient of the Lagrangian in (13.29) taking $\frac{dL}{d\mu}$), whereas the operation $[\ldots]^+$ sets to zero the case of negative value inside the brackets, to guarantee $\mu^{(k)} \geq 0$. Finally, $s^{(k)}$ is the step size of the iteration, which is determined a priori. For decreasing $s^{(k)}$, and more precisely for a sequence that is *non-negative, non-summable and diminishing*, i.e $s^{(k)} \geq 0$, $\lim_{k\to\infty} s^{(k)} = 0$ and $\sum_{k=1}^{\infty} s^{(k)} = \infty$ [45], the method is guaranteed to converge to the optimal solution, i.e. $q(\mu^{(k)}) \overset{k\to\infty}{\to} q(\mu^*) = f^*_{cell}$. Following the analysis in ([45], par. 3.3) the optimal choice in the case of our problem is $s^{(k)} = K/\sqrt{k}$.

13.4.2.4 A tailored method

Our problem [RandCache] has a special structure that allows for a tailored method for the dual problem. This may be simpler to implement and faster to converge than the subgradient one. Specifically, from Lemma 2 we know that the optimal solution satisfies the sum constraint with equality. We will prove the following Theorem, that allows to build our own special algorithm

Theorem 1: As μ decreases from $\mu^{high} = a_1\mathbb{E}\left[\mathcal{M}\right]$ (or increases from $\mu^{low} = a_K p_1$) the sum $\sum_{j=1}^{F} b_j(\mu)$ crosses the horizontal line K at a **unique** μ^*, such that

$$b_1\left(\mu^*\right) + \ldots + b_F\left(\mu^*\right) = K, \tag{13.40}$$

where $b_j\left(\mu\right)$ is given in (13.35). The value μ^* is the optimal one for the dual variable. Furthermore, the optimal primal variables are found by replacing $\mu = \mu^*$ in Lemma 3, i.e. $b_j^* = b_j\left(\mu^*\right), \forall j$.

We will prove this using Lemmas 2 and 3, to first derive Lemmas 4 and 5.

Lemma 4: The solution $b_j\left(\mu\right),\ j = 1, \ldots, F$ of the relaxed primary problem, given in (13.35), is non-increasing in μ.

Proof. To see this, let μ_1 and μ_2 be two distinct values of the dual variables. Without loss of generality, assume $\mu_1 > \mu_2$. We want to show that

$$\mu_1 > \mu_2 \Rightarrow b_j\left(\mu_1\right) \leq b_j\left(\mu_2\right),\ \forall j. \tag{13.41}$$

We write the expression of $b_j\left(\mu\right)$ as a combination of all cases in (13.35) using indicator functions

$$b_j\left(\mu\right) = \mathbf{1}_{\left\{a_j p_1 > \mu\right\}} + \omega\left(\mu\right)\mathbf{1}_{\left\{a_j p_1 \leq \mu \leq a_j \mathbb{E}\left[\mathcal{M}\right]\right\}} + 1 - \mathbf{1}_{\left\{a_j \mathbb{E}\left[\mathcal{M}\right] < \mu\right\}}. \tag{13.42}$$

Then for $\mu_1 > \mu_2$ we have

$$\begin{aligned} &\mathbf{1}_{\left\{a_j p_1 > \mu_1\right\}} \leq \mathbf{1}_{\left\{a_j p_1 > \mu_2\right\}} \\ &1 - \mathbf{1}_{\left\{a_j \mathbb{E}\left[\mathcal{M}\right] < \mu_1\right\}} \leq 1 - \mathbf{1}_{\left\{a_j \mathbb{E}\left[\mathcal{M}\right] < \mu_2\right\}}. \end{aligned}$$

For the remaining term, we observe from (13.36) that $\omega(a_j p_1) = 1$, whereas $\omega(a_j\mathbb{E}\left[\mathcal{M}\right]) = 0$. Also, the $\omega\left(\mu\right)$ is decreasing in μ. To see this,

$$\begin{aligned} \mu_1 > \mu_2 &\Rightarrow a_j \sum_{m=1}^{\infty} p_m m (1 - b_j^{(1)})^{m-1} > a_j \sum_{m=1}^{\infty} p_m m (1 - b_j^{(2)})^{m-1} \\ &\Rightarrow \sum_{m=1}^{\infty} p_m m \left[(1 - b_j^{(1)})^{m-1} - (1 - b_j^{(2)})^{m-1}\right] > 0. \end{aligned} \tag{13.43}$$

The sign of $(1 - b_j^{(1)})^{m-1} - (1 - b_j^{(2)})^{m-1}$ is the same for any $m > 1$, so it should definitely hold $(1 - b_j^{(1)})^{m-1} - (1 - b_j^{(2)})^{m-1} > 0$, for all m, hence also for $m = 2$. This gives the relation $b_j^{(2)} > b_j^{(1)}$. This combined with the other two observations and the expression in (13.42) proves that (13.41). The inequality is strict $\forall j$ only when $a_1 p_1 < \mu_2 < \mu_{1F}\mathbb{E}\left[\mathcal{M}\right]$ (worst case).

Lemma 5: For $K < F$, the sum function $\sum_{j=1}^{F} b_j(\mu)$ is strictly decreasing for $\mu \in [a_K p_1, a_1 \mathbb{E}[\mathcal{M}]]$. It is equal to 0 when $\mu > a_1 \mathbb{E}[\mathcal{M}]$ and it is larger than K when $\mu < a_K p_1$.

Proof. To see this, we use the previous Lemma and the expression in (13.42) for the $b_j(\mu)$. The sum of the b_js is necessarily non-increasing. To show that it is strictly decreasing in the claimed interval, we observe that the sum will decrease as long as at least one b_j decreases. The domain of μ where no b_j changes is either for $\mu < \min_j a_j p_1 = a_F p_1$ and all $b_j = 1$ or for $\mu > \max_j a_j \mathbb{E}[\mathcal{M}] = a_1 \mathbb{E}[\mathcal{M}]$ and all $b_j = 0$. The complement of this domain is the claimed interval.

But since there are only $K < F$ slots available in the memory, there cannot be more than K most popular contents with $b_j = 1$, hence we replace the lower limit of the interval by $a_K p_1$.

Since the sum is monotone, then the solution μ^* is unique and we have also identified the interval where the solution lies, i.e. $\mu^* \in [a_K p_1, a_1 \mathbb{E}[\mathcal{M}]]$. Using this, we can propose an algorithmic solution.

Algorithm: The proposed numerical algorithm to find the solution $\mathbf{b}^*$ with some given precision is based on the **bisection method**.

- Step $\ell = 0$: Initialise the interval that contains the solution,

$$[\mu^{(0,low)}, \mu^{(0,high)}] = [\, a_K p_1, a_1 \mathbb{E}[\mathcal{M}]\,].$$

Also, choose a desired precision $\epsilon > 0$.

- Iterate for $\ell = 0, 1, \ldots$:
  - Find the mid-point of the interval

$$\mu^{(\ell+1)} = \mu^{(\ell,low)} + \frac{1}{2}\left(\mu^{(\ell,high)} - \mu^{(\ell,low)}\right)$$

  - Evaluate the sum

$$S^{(\ell+1)} = \sum_{j=1}^{F} b_j\left(\mu^{(\ell+1)}\right)$$

  - If $S^{(\ell+1)}$ then $\mu^{(\ell+1,high)} = \mu^{(\ell+1)}$. Else $\mu^{(\ell+1,low)} = \mu^{(\ell+1)}$.

- Repeat the iteration until

$$\left\|\mu^{(\ell)} - \mu^{(\ell+1)}\right\| < \epsilon.$$

The method converges *linearly* and every iteration halves the distance of $\mu^{(\ell)}$ from the optimal value μ^*. In fact, one needs at most $\log_2\left(a_1 \mathbb{E}[\mathcal{M}] - a_K p_1\right) - \log_2(\epsilon)$ iterations, to achieve ϵ-precision.

A difficulty encountered when applying the algorithm is to find the values for $b_j\left(\mu^{(\ell+1)}\right)$. These satisfy the polynomial equalities of the form (13.36), which do not have a closed form solution when M is large. We can also solve these equalities over the individual b_j's by use of the *bisection method.*

13.4.2.5 Verification for the [*RandK*1*F*2]

We can now verify the solution of [RandK1F2] in (13.4.1) using Theorem 1. Because $m \in \{0, 1, 2\}$, the $\omega\left(\mu\right)$ obtains an explicit expression and (13.35) now becomes

$$b_1\left(\mu\right) = \begin{cases} 1, & \text{if } a_1p_1 > \mu \\ 1 + \frac{a_1p_1-\mu}{2a_1p_2}, & \text{if } a_1p_1 \leq \mu \leq a_1(p_1+2p_2) \\ 0, & \text{if } a_1(p_1+2p_2) < \mu \end{cases}, \tag{13.44}$$

and similarly for $b_2(\mu)$.

Since $K = 1$, the case $b_1^* = 1$, $b_2^* = 0$ happens only when $a_1p_1 > \mu > a_2(p_1+2p_2)$. The last inequality, with $a_1 + a_2 = 1$ and solving for a_1 is rewritten as $a_1 > 1 - \frac{p_1}{2(p_1+p_2)}$. The case where $b_1^* = 0$, $b_2^* = 1$ holds when $a_2p_1 > \mu_1(p_1+2p_2)$ and the inequality is rewritten as $a_1 < \frac{p_1}{2(p_1+p_2)}$.

In the case where $b_1^* \in \left(0,1\right)$ and $b_2^* \in \left(0,1\right)$ we solve the equation $b_1(\mu) + b_2(\mu) = 1$ to find μ^*, i.e.

$$\begin{aligned} 1 &= 1 + \tfrac{a_1p_1-\mu^*}{2a_1p_2} + 1 + \tfrac{a_2p_1-\mu^*}{2a_2p_2} \Rightarrow \mu^* \\ &= 2a_1(1-a_1)(p_1+p_2), \quad \text{with } a_2 = 1 - a_1. \end{aligned} \tag{13.45}$$

We see from the middle condition in (13.44) that

$$\begin{aligned} a_1p_1 \leq \mu^* &= 2a_1(1-a_1)(p_1+p_2)_1(p_1+2p_2) \\ &\Rightarrow 1 - \tfrac{p_1}{2(p_1+p_2)} \geq a_1 > \tfrac{p_1}{2(p_1+p_2)}. \end{aligned} \tag{13.46}$$

Given the value of μ^* in (13.45) the optimal primal variables are $b_1^* = b_1(2a_1a_2(p_1+p_2)) = 1 + \frac{a_1p_1-2a_1a_2(p_1+p_2)}{2a_1p_2} = \ldots = \frac{2a_1(p_1+p_2)-p_1}{2p_2}$. Hence, it results in the same solution as before (see (13.26)).

13.4.2.6 Evaluation of [RandCache]

We can now evaluate the user hit probability when the optimal randomised placement policy is implemented. In this specific problem instance, the coverage number probability in Section 13.3.1.2 is Poisson distributed (13.18) with $M = \infty$ and parameter $\nu = \pi(T^{-1/\beta})^2$. With fading exponent β=4, we choose a range of threshold values $T \in \left[10^{-2}, 10^3\right]$, $F = 25$ files in the catalog and a cache memory equal to $K = 5$. We plot in Figure 13.5 the optimal hit probability by solving the [RandCache] problem. Note that the popularity is Zipf, with three possible exponents $\gamma \in \{0.56,\ 0.9,\ 1.6\}$.

We observe that for all popularity distributions, the total hit probability (objective function) tends to 1 when the threshold gets smaller. A smaller threshold implies a larger ν and hence a larger expected value $\mathbb{E}[\mathcal{M}] = \nu$. So, the optimal cache placement can take full advantage of a rich multi-coverage environment and offer diversity of files to the user, so that the hit probability reaches 1. Of course, the hit probability tends to 0 as ν gets smaller (larger threshold), because in such cases p_0 tends to 1.

We can compare the optimal randomised caching scheme with a scheme where only the K most popular files are always placed. In such case, denoted by Pop in the figure, the hit probability is always

$$f_{pop} = (1 - p_0)\sum_{j=1}^{K} a_j = (1 - e^{-\nu})\sum_{j=1}^{K} a_j. \tag{13.47}$$

Such policy that does not offer diversity among cache memories is strongly sub-optimal in case of multi-coverage, as the figure illustrates.

13.5 Treating unequal file sizes

The randomised content placement policy and the solution of RandCache have been based on the strong assumption that all files have equal size. Although one can argue that files can be split into chunks of equal size, this is not true in general. If we want to analyse more realistic cache systems, we will associate a tuple (a_j, z_j) per catalog item j, where z_j is the known file size in *bytes*, e.g. a video can have size of 0.83 [*Gbyte*]. Consequently, the cache should have size K_b, in contrast to the K used until this point. Then, we want to place the content to caches, respecting the available memory size; the feasibility constraint (13.8) now takes the form

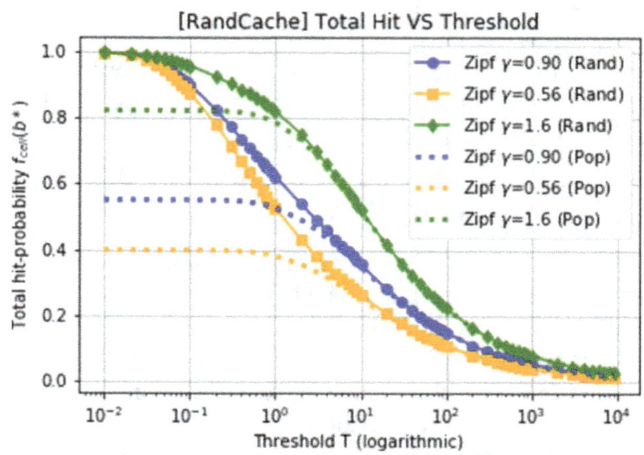

Figure 13.5 Evaluation of the optimal [RandCache] policy and comparison with the most popular policy [Pop] for the Boolean coverage model and various file popularity distributions

$$\sum_{j=1}^{F} X_{i,j} z_j \leq K_b. \tag{13.48}$$

This is a *binary Knapsack inequality*, and every randomised placement policy satisfies – similarly to (13.9) – the inequality

$$\mathbb{E}\left[\sum_{j=1}^{F} X_{i,j} z_j\right] = \sum_{j=1}^{F} b_j z_j \leq K_b, \tag{13.49}$$

where $b_j = \mathbb{P}(X_{i,j} = 1)$ is the placement probability (or frequency) of content j. We cannot use in a straightforward way the construction in Figure 13.1 to sample batches, because these batches are not guaranteed to satisfy the Knapsack constraint!

The brute-force way would be to find the exhaustive set of feasible content batches

$$\mathcal{C} = \left\{\mathbf{x} \mid \mathbf{x} \in \{0, 1\}^F, \ \mathbf{x} \cdot \mathbf{b} \leq K\right\}. \tag{13.50}$$

From the above set we should further remove the feasible vectors $\tilde{\mathbf{x}} \in \mathcal{C}$, for whom there exists $\mathbf{x}' \in \mathcal{C}$, such that $\tilde{\mathbf{x}} \leq \mathbf{x}'$. The inequality is element-wise, meaning that $\tilde{x}_j \leq x'_j$, for all j. Then, if both $\mathcal{C}_1 = \{\text{"1"}\}$ and $\mathcal{C}_2 = \{\text{"1"}, \text{"4"}\}$ are feasible, we keep only the second batch, because otherwise memory space is left unused. We denote this more fine set by $\mathcal{C}^*$, and the feasible batches by $\mathcal{C}_\ell \in \mathcal{C}^*$, where the index runs over $\ell = 1, \ldots, L$. Here $L = card(\mathcal{C}^*)$.

We can now associate a frequency ρ_ℓ per batch, $\ell = 1, \ldots, L$, which satisfies

$$\sum_{\ell=1}^{L} \rho_\ell = 1. \tag{13.51}$$

The randomised policy would be the following: select a number within the interval $(0, 1)$ uniformly at random, and pick the batch $k + 1$, $k \in \{0, \ldots, L - 1\}$, if the random number is within the interval $\left(\sum_{\ell=1}^{k} \rho_\ell, \sum_{\ell=1}^{k+1} \rho_\ell\right]$, with the convention $\sum_{\ell=1}^{0} \rho_\ell = 0$.

But one important detail remains open: how to determine the ρ_ℓs from the given (or optimal) probabilities b_js. We can use the fact that the frequency of object j is the sum of frequencies of batches where the b_j appears, i.e.

$$b_j = \sum_{\ell=1}^{L} \rho_\ell \mathbf{1}_{\{j \in \mathcal{C}_\ell\}}, \quad j = 1, \ldots, F. \tag{13.52}$$

There are F such equalities and L unknowns. If we could solve the system $\mathbf{A}\rho = \mathbf{b}$, then we could determine the batch frequencies. Often, however $L > F$ or $L < F$ so the system is under- or over-determined. For such cases, we could use the Moore-Penrose pseudo-inverse, $\rho = \left(\mathbf{A}^T\mathbf{A}\right)^{-1} \mathbf{A}^T\mathbf{b}$, but this does not guarantee that the system actually has a solution.

In any case, an interesting observation from the brute-force method is that there is empty space in the cache with average size equal to

$$E = \sum_{\ell=1}^{L} \rho_\ell \left(K_b - \sum_{j=1}^{F} x_j^{(\ell)} z_j \right) \geq 0. \tag{13.53}$$

This observation actually motivates the following much simpler method.

13.5.1 Method: overflow constraint

A more practical method is to replace the constraint (13.49) by the tighter constraint

$$\sum_{j=1}^{F} b_j z_j \leq K_b - \delta. \tag{13.54}$$

We saw previously that most feasible batches do not fill in the K_b memory completely. The above-tightened inequality guarantees that the Knapsack constraint $\sum_{j=1}^{F} X_{i,j} z_j \leq K_b$ is violated only with some small probability ϵ. To see this,

$$\begin{aligned} \mathbb{P}\left(\sum_{j=1}^{F} X_{i,j} z_j > K_b\right) &\overset{Markov}{\leq} \frac{\mathbb{E}\left[\sum_{j=1}^{F} X_{i,j} z_j\right]}{K_b} = \frac{\sum_{j=1}^{F} b_j z_j}{K_b} \\ &\overset{(1.54)}{\leq} 1 - \frac{\delta}{K_b} = \epsilon(\delta). \end{aligned} \tag{13.55}$$

We can also choose a desired violation bound ϵ, to determine the appropriate δ

$$\delta(\epsilon) = (1 - \epsilon) K_b. \tag{13.56}$$

For example, suppose we require a violation probability of 10%. Then $\epsilon = 0.1$ and $\delta(0.1) = 0.9K_b$, so that the right-hand side of (13.54) is $0.1K_b$.

Now, we can use a similar construction as in Figure 13.1, but without predefining the number of unit memories (it was K before). We place the b_js one next to the other, say starting from $j = 1$ and on. If their sum at some point is larger than 1 then we split and move to the next line, until all b_js are placed. Then, similarly as in the random placement policy, we choose a number between 0 and 1 uniformly at random, draw a vertical line intersecting the above construction and we pick this batch of distinct contents. Note here, that now the number of contents can vary per batch. This method guarantees that the batches are drawn so that $\mathbb{P}(X_{i,j} = 1) = b_j$. Remember that the b_js were chosen such that the batches violate the memory constraint with probability ϵ.

13.5.2 [RandCache] with cache overflow constraint

We can now formulate and solve the RandCache problem, including the unequal file sizes. We use the overflow constraint in (13.54), where $\delta(\epsilon)$ is chosen based on the desired level of Knapsack violation, as in (13.56). The new feasible set of the random placement policy is

$$\mathcal{F}_{random}(\mathbf{z}, \delta) = \left\{ \mathbf{b} \mid \mathbf{b} \in [0,1]^F, \sum_{j=1}^{F} b_j z_j \leq K_b - \delta \right\}. \tag{13.57}$$

The problem takes the form

$$\begin{aligned} \max_{\mathbf{b}} \quad & f_{cell}(\mathbf{b}) := 1 - \sum_{j=1}^{F} a_j \sum_{m=0}^{\infty} p_m \left(1 - b_j\right)^m \quad [Rand - z]. \\ s.t \quad & \mathbf{b} \in \mathcal{F}_{random}(\mathbf{z}, \delta) \end{aligned} \tag{13.58}$$

In the above, the objective is the same as in the RandCache case, i.e. it is the average cache hit probability. The file sizes affect only the constraints.

This variation can be solved following the standard procedure with the tailored method we have presented for the RandCache problem. For completion, we show here the solution for Rand − z. Following the Lagrange relaxation, the primal variables are solved, similarly to Lemma 3. For fixed dual $\mu \geq 0$

$$b_j(\mu; z_j) = \left\{ \begin{array}{cc} 1, & if \ \frac{a_j}{z_j} p_1 > \mu \\ \omega(\mu; z_j), & if \ \frac{a_j}{z_j} p_1 \leq \mu \leq \frac{a_j}{z_j} \mathbb{E}\left[\mathcal{M}\right] \\ 0, & if \ \frac{a_j}{z_j} \mathbb{E}\left[\mathcal{M}\right] < \mu \end{array} \right\} \tag{13.59}$$

where, now $\omega(\mu; z_j)$ is the solution over b_j of the equation

$$\frac{a_j}{z_j} \sum_{m=1}^{\infty} p_m m (1 - b_j)^{m-1} = \mu. \tag{13.60}$$

Similarly, μ^* is the unique μ such that

$$b_1(\mu^*) z_1 + \ldots + b_F(\mu^*) z_F = K_b - \delta. \tag{13.61}$$

Structure of the solution: The difference compared to the equal (unit) size solution RandCache is subtle, and reveals interesting design insights for the optimal placement policy of Rand − z. When the file sizes **z** are taken into account, the primal solution takes the form (13.59). Compared to (13.35), it has the same structure, but one detail: the popularities a_j are replaced everywhere by the ratio a_j/z_j. Previously, we had ordered the files in decreasing order of popularity and optimally cached them with decreasing placement probability $b_j(\mu)$ following that content order. Now we should order the files by the ratio a_j/z_j and place them with decreasing probability $b_j(\mu; z_j)$ following the new order. So, this means that the size plays a major role in the placement decision. Even if a file is moderately popular, when its size is very large, it will be ordered low, and hence will have low placement probability. This is because, we want to avoid overflowing the cache memory. Instead, a very small-sized file with moderate popularity will be ordered high, and thus will be placed very frequently, because it is not so dangerous to overflow the cache.

13.6 Node mobility

A very common characteristic in cellular networks is node mobility. This is manifested as users moving from one cell to the other. In D2D or vehicle-to-vehicle (V2V) communications, mobility plays a major role. User mobility results in loss of user association with a specific station, and a shift of service towards a different station nearby. Mobility requires an appropriate design of content placement policies. We follow the analysis in [5].

Suppose that a user requests content j at some time $t = 0$. This user is covered by a number of stations – say m. Till now, we considered the hit probability, i.e. the probability that the content is cached in at least one covering station. But, the objective function in RandCache assumes silently that the typical user will not change position (and stick to its station association), until the file is completely downloaded, something not necessarily true. Suppose, instead, that the user does find the content cached in one of the covering stations, say station i, but at time $t = \tau$ she/he moves and the link is lost.

If during the interval $(0, \tau]$ the signal-to-noise ratio (SNR) with the associated station i is strong enough, then the whole file can be downloaded, i.e.,

$$\tau W \log_2 \left(1 + SNR(R_i)\right) \geq z_j. \tag{13.62}$$

The above uses the Shannon formula with bandwidth W. The $SNR(R_i) = SR_i^{-\beta}/W\sigma_N^2$, see also (13.14). If the inequality (13.62) is not fulfilled, then the file is not completely transmitted, so there is no hit. Hence,

$$\Psi_{i,j} = X_{i,j} \cdot \mathbf{1}_{\left\{\tau W \log_2(1+SNR(R_i)) \geq z_j\right\}} \in \{0, 1\} \tag{13.63}$$

is the binary random variable for hit or miss, that takes into account not only content availability but also file size, user mobility and distance from the user. Let us consider station positions that follow a 2D PPP of density $\lambda \geq 0$, so there is an infinity of stations positioned on the plane, indexed by $i = 1, 2, \ldots$. The hit probability that accounts for file download completion is called here *service success probability*. It is the probability that at least one station i' satisfies $\Psi_{i',j} = 1$, so the objective now is

$$f_{d2d}(\mathbf{b}) = \sum_{j=1}^{F} a_j \sum_{i=1}^{\infty} \mathbb{P}\left[\Psi_{i,j} \geq 1\right] = \sum_{j=1}^{F} a_j \left(1 - e^{-b_j \lambda h(\tau, z_j)}\right), \tag{13.64}$$

where $h(\tau, z_j) = \pi \Gamma\left(\frac{3}{2}\right) \sqrt{S/W} \sigma_N^{-1} / \sqrt{2^{z_j/(W\tau)} - 1}$. The expression is taken from ([5], Cor. 1) and refers to the case of Rayleigh fading and error exponent $\beta = 4$. The proof can also be found in the same reference.

The expression (13.64) satisfies the two properties from Lemma 1. It is separable in b_j. From the first and second partial derivatives, it is monotone increasing and concave in b_j. So the problem

$$\begin{aligned} \max_{\mathbf{b}} \quad & f_{d2d}\left(\mathbf{b}\right) := \textstyle\sum_{j=1}^{F} a_j \left(1 - e^{-b_j \lambda h(\tau, z_j)}\right) \quad [\texttt{Rand} - \texttt{D2D}]. \\ \text{s.t} \quad & \mathbf{b} \in \mathcal{F}_{random}(\mathbf{z}, \delta) \end{aligned} \tag{13.65}$$

can be solved again using the method we have described for RandCache. We just mention that the derivative of the sub-problem subD2D – j is equal to

$$-\mu z_j + \tfrac{dg}{db_j} = -\mu z_j + a_j \lambda h(\tau, z_j) e^{-b_j \lambda h(\tau, z_j)} \tag{13.66}$$

13.7 Competition among content providers

An interesting extension of the work, presented in ([46], Chapter 6), is when $S > 1$ content providers compete to share the same cache space and place their catalog. The providers are indexed by $s = 1, \ldots, S$ and each catalog has no overlap with any other, it is thus a disjoint set of contents. We index the contents of catalog s by $j_s = 1, \ldots, F_s$. The total memory space K_b per cache, will be split among the S providers, and each will get a share k_s. The following inequality holds

$$\sum_{s=1}^{S} k_s \leq K_b. \tag{13.67}$$

Suppose that once k_s space is given to a provider, the catalog is placed following the random placement policy with placement variables $\mathbf{b}_s = (b_{s,1}, \ldots, b_{s,F_s})$. We get the following feasible set

$$\mathcal{F}_{rand,CP}(\mathbf{z}_1, \ldots, \mathbf{z}_S, \delta) = \left\{ \begin{array}{c} (\mathbf{b}_1, \ldots, \mathbf{b}_S) \mid \mathbf{b}_s \in [0, 1]^{F_s}, \mathbf{k} \in [0, K_b]^S, \\ \sum_{s=1}^{S} k_s \leq K_b, \\ \sum_{j=1}^{F_S} b_{s,j} z_{s,j} \leq k_s - \delta, \ s = 1, \ldots, S \end{array} \right\}. \tag{13.68}$$

This set is characterised by $S + 1$ inequalities, that is 1 inequality for the cache-space allocation and S inequalities for the content placement per allocated space k_s. This type of problem keeps the structure of the original `RandCache` problem, and its solution can be easily derived, using similar methods.

13.8 Beyond independent caches

The random independent content placement and the related `RandCache` formulation, although it has many advantages, it can still result in hit probabilities much lower than those derived by detailed binary placement, both for a given realisation of user-station positions, e.g. from `BinCache`, as well as on average. The reason is that the caching decisions are done independently for different stations, so that two or more stations jointly covering a location may propose a complementary content only *by chance*. A possible way of improving upon this situation would consist in some content placement with cooperative decisions made between caches to guarantee content diversity among neighbouring nodes (see related works [47] and [48]).

The randomised policies can also be tuned in this direction by relaxing the assumption of independent placement. This was done in [49] where the dependence between caching decisions of different stations is described by a Gibbs distribution. Coming back to the setting of Section 13.1.1, with a given, finite configuration of stations, let us define the joint probability of placing the content at different stations according to the variables $\mathbf{x} \in \{0, 1\}^{N \times F}$, (with $\sum_{j=1}^{F} x_{i,j} \leq K (i = 1, ..., N)$) by

$$\pi_\beta(\mathbf{x}) := \frac{e^{\beta h(\mathbf{x})}}{Z_\beta}, \tag{13.69}$$

where

$$h(\mathbf{x}) = \int_{\mathcal{A}} \left[1 - \sum_{j=1}^{F} a_j \prod_{i \in \mathcal{H}(u)} (1 - x_{i,j}) \right] \mathrm{d}u$$

is the total hit probability in the deployment region $\mathcal{A}$ replacing the discrete sum over discrete user locations considered in (13.1), β is called the 'inverse temperature' (motivated by literature from statistical physics) and $Z_\beta = \sum_{\mathbf{x} \in \{0,1\}^{N \times F}} e^{\beta h(\mathbf{x})}$ is the normalising constant called the 'partition function'.

Note that, $\lim_{\beta \to \infty} \sum_{\mathbf{x} \in \arg\max_{\mathbf{x}' \in \{0,1\}^{N \times F}} h(\mathbf{x}')} \frac{e^{\beta h(\mathbf{x})}}{Z_\beta} = 1$. Hence, if we choose a configuration $\mathbf{x}$ with probability $\pi_\beta(\mathbf{x})$, then, for sufficiently large β, the chosen configuration will be optimal (for the given configuration of stations, on average for different homogeneous user location in $\mathcal{A}$) with probability close to 1.

In most cases, there is no closed form expression for $\pi_\beta(\mathbf{x})$. Note, however, that we do not need to know probabilities $\pi_\beta(\mathbf{x})$. The only thing we need is to be able to sample from this distribution (so as to place the content according to the sampled values of $\mathbf{x}$ at different stations). And there are good ways of doing this. Indeed, there are several Gibbs sampling algorithms that simulate a discrete-time Markov chain on the space $\{0, 1\}^{N \times F}$ having $\pi_\beta(\cdot)$ as the stationary probability distribution. In the particular case of our hit function $h(\mathbf{x})$ (which is called the Hamiltonian of the Gibbs distribution) these algorithms can be even made decentralised, in the sense that different stations can run them by updating their content knowing only the current choice of the content of the neighbouring stations – i.e. those with overlapping coverage regions. Moreover, the popularity can be learned during this process, and the parameter β can be appropriately increased to approach the optimal solution, a technique known as Simulated Annealing. We refer the reader to [49] for more details.

Observe that the above solution is proposed for a finite collection of stations. Gibbs models usually do not scale very well with the number of nodes (here stations) going to infinity. Mathematically challenging problems arise in this context, related to the long range propagation of the dependence between caching decisions of different stations. This problem was recently studied in [50] in a general mathematical context going beyond Gibbs distributions.

References

[1] Bastug E., Bennis M., Debbah M. 'Living on the edge: the role of proactive caching in 5G wireless networks'. *IEEE Communications Magazine*. 2014;**52**(8):82–9.

[2] Shanmugam K., Golrezaei N., Dimakis A.G., Molisch A.F., Caire G. 'FemtoCaching: wireless content delivery through distributed caching helpers'. *IEEE Transactions on Information Theory*. 2013;**59**(12):8402–13.

[3] Wen J., Huang K., Yang S., Li V.O.K. 'Cache-enabled heterogeneous cellular networks: optimal tier-level content placement'. *IEEE Transactions on Wireless Communications*. 2017;**16**(9):5939–52.

[4] Jeon S.-W., Hong S.-N., Ji M., Caire G., Molisch A.F. 'Wireless multihop device-to-device caching networks'. *IEEE Transactions on Information Theory*. 2017;**63**(3):1662–76.

[5] Jarray C., Giovanidis A. 'Successful file transmission in mobile D2D networks with caches'. *Computer Networks*. 2018;**147**(6):162–79.

[6] Kurose J. 'Information-centric networking: the evolution from circuits to packets to content'. *Computer Networks*. 2014;**66**(7):112–20.

[7] Garetto M., Leonardi E., Martina V. 'A unified approach to the performance analysis of caching systems'. *ACM Transactions on Modeling and Performance Evaluation of Computing Systems*. 2016;**1**(3):1–28.

[8] Jiang B., Nain P., Towsley D. 'On the convergence of the TTL approximation for an LRU cache under independent stationary request processes'. *ACM Transactions on Modeling and Performance Evaluation of Computing Systems*. 2018;**3**(4):1–31.

[9] Giovanidis A., Avranas A. 'Spatial multi-LRU caching for wireless networks with coverage overlaps'. *ACM SIGMETRICS Performance Evaluation Review*. 2016;**44**(1):403–5.

[10] Poularakis K., Iosifidis G., Tassiulas L. 'Approximation algorithms for mobile data caching in small cell networks'. *IEEE Transactions on Communications*. 2014;**62**(10):3665–77.

[11] Krolikowski J., Giovanidis A., Di Renzo M. 'A decomposition framework for optimal edge-cache leasing'. *IEEE Journal on Selected Areas in Communications*. 2018;**36**(6):1345–59.

[12] Blaszczyszyn B., Giovanidis A. 'Optimal geographic caching in cellular networks'. 2015 IEEE International Conference on Communications (ICC); 2015. pp. 3358–63.

[13] Chae S.H., Choi W. 'Caching placement in stochastic wireless caching helper networks: channel selection diversity via caching'. *IEEE Transactions on Wireless Communications*. 2016;**15**(10):6626–37.

[14] Cui Y., Jiang D. 'Analysis and optimization of caching and multicasting in large-scale cache-enabled heterogeneous wireless networks'. *IEEE Transactions on Wireless Communications*. 2017;**16**(1):250–64.

[15] Li K., Yang C., Chen Z., Tao M. 'Optimization and analysis of probabilistic Caching in N -Tier heterogeneous networks'. *IEEE Transactions on Wireless Communications*. 2018;**17**(2):1283–97.

[16] Serbetci B., Goseling J. 'On optimal geographical caching in heterogeneous cellular network'. 2017 IEEE Wireless Communications and Networking Conference (WCNC); 2017. pp. 1–6.

[17] Liu D., Yang C. 'Caching policy toward maximal success probability and area spectral efficiency of cache-enabled HetNets'. *IEEE Transactions on Communications*. 2017;**65**(6):2699–714.

[18] Wen W., Cui Y., Zheng F.-C., Jin S., Jiang Y. 'Random caching based cooperative transmission in heterogeneous wireless networks'. *IEEE Transactions on Communications*. 2018;**66**(7):2809–25.

[19] Zhao J., Zhao S., Qu H., Ren G., Shi Y. 'Analysis and optimization of probabilistic caching in micro/millimeter wave hybrid networks with dual connectivity'. *IEEE Access*. 2018;**6**:72372–80.

[20] Zhu Y., Zheng G., Wang L., Wong K.-K., Zhao L. 'Content placement in cache-enabled sub-6 GHz and millimeter-wave multi-antenna dense small cell networks'. *IEEE Transactions on Wireless Communications*. 2018;**17**(5):2843–56.

[21] Chen Z., Pappas N., Kountouris M. 'Probabilistic caching in wireless D2D networks: cache hit optimal versus throughput optimal'. *IEEE Communications Letters*. 2017;**21**(3):584–7.

[22] Malak D., Al-Shalash M., Andrews J.G. 'Spatially correlated content caching for device-to-device communications'. *IEEE Transactions on Wireless Communications*. 2018;**17**(1):56–70.

[23] Lin X., Xia J., Wang Z. 'Probabilistic caching placement in UAV-assisted heterogeneous wireless networks'. *Physical Communication*. 2019;**33**(12):54–61.

[24] Song F., Li J., Ding M., *et al.* 'Probabilistic caching for small-cell networks with terrestrial and aerial users'. *IEEE Transactions on Vehicular Technology*. 2019;**68**(9):9162–77.

[25] Emara M., Elsawy H., Sorour S., Al-Ghadhban S., Alouini M.-S., Al-Naffouri T.Y. 'Optimal caching in 5G networks with opportunistic spectrum access'. *IEEE Transactions on Wireless Communications*. 2018;**17**(7):4447–61.

[26] Gao J., Zhang S., Zhao L., Shen X. 'The design of dynamic probabilistic caching with time-varying content popularity'. *IEEE Transactions on Mobile Computing*. 2020;**20**(4):1672–84.

[27] Krishnan S., Afshang M., Dhillon H.S. 'Effect of Retransmissions on optimal caching in cache-enabled small cell networks'. *IEEE Transactions on Vehicular Technology*. 2017;**66**(12):11383–7.

[28] Yang Q., Wang H.-M., Zheng T.-X. 'Delivery-secrecy tradeoff for cache-enabled stochastic networks: content placement optimization'. *IEEE Transactions on Vehicular Technology*. 2018;**67**(11):11309–13.

[29] Liu D., Yang C. 'A learning-based approach to joint content caching and recommendation at base stations'. *2018 IEEE Global Communications Conference*. 2018:1–7.

[30] Choi M., Kim J., Moon J. 'Wireless video caching and dynamic streaming under differentiated quality requirements'. *IEEE Journal on Selected Areas in Communications*. 2018;**36**(6):1245–57.

[31] Ioannidis S., Yeh E. 'Adaptive Caching Networks with Optimality Guarantees'. Proceedings of the 2016 ACM SIGMETRICS International Conference on Measurement and Modeling of Computer Science. SIGMETRICS '16; New York, NY, USA, Association for Computing Machinery; 2016. pp. 113–24.

[32] Avrachenkov K., Goseling J., Serbetci B. 'A low-complexity approach to distributed cooperative caching with geographic constraints'. Proceedings of the ACM on Measurement and Analysis of Computing Systems; 2017. pp. 1–25.

[33] Giannakas T., Giovanidis A., Spyropoulos T. 'Soba: session optimal MDP-based network friendly recommendations'. IEEE International Conference on Computer Communications. INFOCOM; 2021.

[34] Baccelli F., Błaszczyszyn B. 'Stochastic geometry and wireless networks: volume I theory'. *Foundations and Trends® in Networking*. 2009;**3**(3-4):249–449.

[35] Andrews J.G., Baccelli F., Ganti R.K. 'A tractable approach to coverage and rate in cellular networks'. *IEEE Transactions on Communications*. 2011;**59**(11):3122–34.

[36] Keeler H.P., Błaszczyszyn B., Karray M.K. 'SINR-based k-coverage probability in cellular networks with arbitrary shadowing'. 2013 IEEE International Symposium on Information Theory; 2013. pp. 1167–71.

[37] Błaszczyszyn B., Haenggi M., Keeler P. *Stochastic Geometry Analysis of Cellular Networks*. Cambridge: Cambridge University Press; 2018.

[38] Dhillon H.S., Ganti R.K., Baccelli F., Andrews J.G. 'Modeling and analysis of K-tier downlink heterogeneous cellular networks'. *IEEE Journal on Selected Areas in Communications*. 2012;**30**(3):550–60.

[39] Keeler H.P. *SINR-based k-coverage probability in cellular networks. MATLAB Central File Exchange [online]*. 2020. Available from https://www.mathworks.com/matlabcentral/fileexchange/40087-sinr-based-k-coverage-probability-in-cellular-networks [Accessed 24 Sep 2021].

[40] Dan A., Towsley D. 'An approximate analysis of the LRU and FIFO buffer replacement schemes'. Proceedings of the 1990 ACM SIGMETRICS Conference on Measurement and Modeling of Computer Systems. SIGMETRICS '90; New York, NY, USA, Association for Computing Machinery; 1990. pp. 143–52.

[41] Newman M.E.J. 'Power laws, Pareto distributions and Zipf's law'. *Contemporary Physics*. 2005;**46**(5):323–51.

[42] Breslau L., Pei Cao L.F. 'Web caching and Zipf-like distributions: evidence and implications'. IEEE INFOCOM '99. Conference on Computer Communications. Proceedings. Eighteenth Annual Joint Conference of the IEEE Computer and Communications Societies. The Future Is Now; 1999. pp. 126–34.

[43] Boyd S., Vandenberghe L. *Convex Optimization*. Cambridge: Cambridge University Press; 2004.

[44] Bertsekas D.P. 'Network optimization: continuous and discrete models'. Athena Scientific; 1998.

[45] Boyd S. *Notes for EE364b Subgradient methods*. Stanford University; 2014.

[46] Krolikowski J. Optimal content management and dimensioning in wireless networks.[PhD Thesis Networking and Internet Architecture [csNI], NNT: 2018SACLS452, tel-02124294]; 2018.

[47] Malak D., Al-Shalash M., Andrews J.G. 'Spatially correlated content caching for device-to-device communications'. *IEEE Transactions on Wireless Communications*. 2018;**17**(1):56–70.

[48] Malak D., Médard M., Yeh E.M. 'Spatial soft-core caching'. IEEE International Symposium on Information Theory (ISIT); 2019. pp. 2009–13.

[49] Chattopadhyay A., Błaszczyszyn B., Keeler H.P. 'Gibbsian on-line distributed content caching strategy for cellular networks'. *IEEE Transactions on Wireless Communications*. 2017;**17**(2):969–81.
[50] Błaszczyszyn B., Hirsch C. 'Optimal stationary markings'. *arXiv*. 2020;**200108074**.

Chapter 14

Pricing-aided proactive caching for mobile services

John Tadrous[1], *Atilla Eryilmaz*[2], *and Hesham El-Gamal*[2]

14.1 Summary

In this chapter, we study the potential of harnessing "proactive caching" at end-users, supported with pricing incentives, to maximize the profit of wireless content providers (CPs). CP consistently encounters time varying, yet partially predictable, demand characteristics. The disparate demand levels throughout the course of the day yield excessive service cost in the peak hours that substantially hurt the reaped profit. With the CP's ability to track and statistically predict future requests of its users, we propose to enable *proactive caching* of the peak hour demand during the off-peak times. Thus network traffic will be smoothed out, while end-users' activity patterns are undisturbed. In addition, the CP is able to assign personalized pricing policies that strike a best balance between enhancing certainty about the future demand for optimal proactive caching, and maximizing the revenue collected from end-users. These proactive caching decisions accompanied with smart data pricing offer a win-win situation to both the CP and the end-users. In this chapter, we propose a joint proactive caching and content pricing scheme that archives such a win-win situation. Comparing the proposed system's performance with the baseline scenario of the existing practice of no-proactive caching, we show that the CP attains profit gain that grows with the number of users, at least, as the first derivative of the cost function. Moreover, end-users that receive proactive caching services make strictly positive savings. Thus, we essentially demonstrate the win-win situation to be reaped by exploiting consistent users' activity. In addition, we reveal significant (potentially unbounded) gains and low-complexity means of the smart pricing paradigm for proactive caching.

We begin this chapter with an introduction to the problem and a discussion of relevant solutions in Section 14.2. We mathematically describe the proposed system setup and formulate the content-based pricing problem together with proactive content caching in Section 14.3. We then study the single-user case of the problem in Section 14.4. In particular, we derive necessary and sufficient condition on the

[1]Department of Electrical and Computer Engineering, Gonzaga University, USA
[2]Department of Electrical and Computer Engineering, Ohio State University, USA

activity of the end-users and their preferences so that smart pricing with proactive caching can always yield a positive profit and savings gain for the CP and end-user relative to the flat pricing policy without proactive service. We develop an algorithm that attains these gains and prove its convergence. Afterwards, we address the multi-user case of the problem in Section 14.5, where we tackle non-convexity of the price allocation problem. Successive approximations of convex problems are proposed, and centralized and distributed algorithms are developed to implement the joint allocation of pricing and proactive caching. In the same section, we also show that the CP can reap profit gains that grow unboundedly with the number of users through proactive services. Moreover, all users who receive such services make positive savings gain. Thus we highlight the potential for the win-win situation offered by leveraging the predictability of human behavior. Finally, we conclude the work in Section 14.6.

14.2 Introduction

The significant transformation of the data traffic from simple low-throughput communication to high data rate multimedia applications (e.g., HD video streaming) has raised major concerns on the ability of data networks to support reliable data delivery at reasonable costs for content providers (CPs) (e.g., YouTube, CNN, and Netflix). On the other hand, network resources consistently experience an under-utilization issue during the off-peak time because of the lack of user access.

There has been an urgent need for more advanced and intelligent techniques to boost the network resource management. Pricing of network access resources and data content has been a key factor to consider while alleviating the network resource management problem. One of the proposed remedies is the dynamic spectrum access (DSA) technique which is tied to the wireless resource utilization. DSA has been introduced to allow out-of-band (cognitive or secondary) users to access the under-utilized bands of a primary network at lower prices [1]. DSA, however, offers only a partial solution to the problem as most users already remain idle during the off-peak time leaving the network resources under-utilized. While DSA implicitly relies on pricing incentives to enhance the spectrum utilization, it does not offer an effective solution to the excessive demands during the peak time of the day.

More research directions have been opened that rely on service pricing. Load-level-based pricing strategies have been developed in [2–4] based on the load-level of the system with prices potentially increasing when the load rises and decreasing otherwise. The adopted models, however, fall short of capturing the inherent variability of the system loads not only due to pricing but also due to users' activity patterns. In [5–7], the variable user activity patterns have been taken into account through a joint pricing and scheduling framework that incentivizes users to delay their demand to the off-peak time through optimized discounts. The potentially reaped gains of these approaches rely on the trade-off between the economic responsiveness of the users and the preferred time of access. Clearly, users will have to change their regular activity patterns in order to retrieve low-priced service, for example, a user who wants to access the network while in public transit, may have to defer demand to later time when network prices are less, and thus get pricing discount at the expense of inconvenient access pattern. Hence, the CP gains are not surprising.

In a parallel direction, WiFi infrastructure has been proposed to off-load the cellular traffic for wireless users with WiFi connection capabilities [8, 9]. The idea is to re-route the users demand through WiFi access points covering the users' connections through possible pricing incentives [10–12]. While the emerging results reveal a potential of promising gains, the approach does not effectively tackle the excessive costs and low quality-of-service (QoS) in places with no WiFi coverage (e.g., public transit, rallies, stadiums, etc.). Moreover, it is questionable that large amounts of traffic will be efficiently off-loaded through WiFi since users by default connect to WiFi as a cheaper and more reliable source of data communication, and therefore, the fraction of users who direct their demand to the cellular provider while WiFi covered is expected to be minor.

Proactive caching, on the other hand, is a promising paradigm to alleviate the network resource management problem without relying directly on pricing incentives [13–16]. The technique aims to exploit the *predictable* human demand together with the large storage offered by smart wireless devices in smoothing out the wireless network traffic over time by *proactively serving peak-hour requests during the off-peak time*. Thus, users need not change their regular activity patterns, yet they will be promised better QoS. By moving caching to end-users, CPs can attain higher performance gains as compared to the use of caching at the small base stations only since optimized individual proactive services can further utilize available resources. In addition, it enables personalized incentives for enhanced certainty about future demand. Figure 14.1 illustrates the idea of proactive caching.

We have demonstrated the potential gains of proactive resource allocation, under large-timescale optimization, for unicast and multicast networks as well as networks with heterogeneous QoS requirements under perfectly predictable demand. Further, authors in [15, 17] show that delay reduction gains from proactive caching at end-users grow exponentially in the proactive service window.

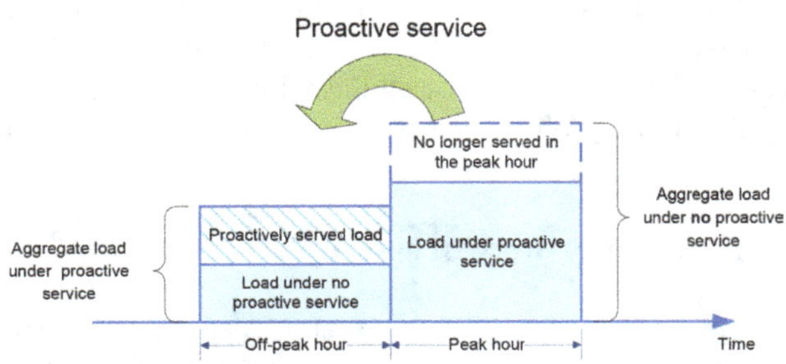

Figure 14.1 Network load through proactive caching at end-users. Peak-hour demand is proactively cached at end-users during the off-peak hour to minimize the large peak-to-average demand ratio.

Proactive caching at end-users, however, depends on the certainty in individual users' preferences and demand profiles. The higher the certainty, the more accurate the proactive caching. Incentives can hence be offered to end-users to shape their preferences and render their demand more predictable. In [14], we have introduced the notion of *demand shaping* where modified content valuations are proposed in place of the original user-perceived valuations of data items. The modified valuations are optimized to strike a best balance between maximizing the certainty about the user preferences and minimizing the distance to the original user-perceived valuations. To that end, we have adopted an *entropy* of the original user preferences as a measure of how flexible the user is to accept modified valuations. Undetermined users exhibit more uniform preferences over data item, hence higher entropy and higher flexibility to follow modified valuations since they are undetermined. More opinionated users exhibit concentrated preferences towards specific data items, hence less entropy and less flexibility to follow modified valuations. Nevertheless, there is no concern about those users' inflexibility to follow modified valuations since their demand profiles are sufficiently predictable to successfully employ proactive caching.

In this chapter, we consider pricing as an alternative incentive to modified content valuations. Pricing discounts can be offered to render the users' demand profiles more predictable while ensuring that the users do not have to change their demand activities over time. The idea is to price data content based on the user preferences so as to enhance the certainty about future demand, enabling the CP to attain efficient proactive download performance while preserving the statistical user activity pattern unchanged over time. This approach is different from the commonly known techniques, e.g. [6], that utilize time-dependent pricing to statistically shift the end-users' demand over time to the off-peak hour, which potentially conflict with the regular end-users' activities. In [18], we have shown in a single-user scenario that not only does the proposed joint proactive service and pricing schemes improve the CP's profit, but also the user pays less without changing the activity pattern over time.

In this chapter, we generalize our results to the multi-user system served by a content source where the impact of the coupling between different users' demand is pronounced. We establish asymptotic bounds on the extra profit the CP can make through proactive caching. In addition, we develop a distributed algorithm where users can determine proactive caching with the promise of less payment while the CP makes more profit, a win-win situation for all involved parties.

14.3 Content provider network setup

We consider a system comprising a CP and a set of N users. We consider the CP to be a content source such as YouTube, Netflix, CNN, Facebook, etc. In a time-slotted fashion, users generate random requests targeting content from one of the possible M data items available at the CP. The CP in turn responds to such users' requests on a timely basis. The content of a data item could be a movie (as on YouTube or

Netflix), a soundtrack (as on Pandora), a news update (as on CNN or Fox News), a social network update (as on Facebook or Twitter), etc.

We assume that such contents are updated consistently for every time slot, where the slot duration represents the time over which a user consumes the content of requested data item. This can span a few minutes to possibly hours. Thus, in the slot duration, the CP completes delivery of requested data content, which has been partially served over the previous slot.

14.3.1 Users' demand profiles

At time slot t, user n may decide to consume a total of $S_{n,t}$ amount of data, where $\check{S} \leq S_{n,t} \leq \hat{S}$, $\forall n, t$, and $0 < \check{S} \leq \hat{S} < \infty$. This consumption is from one of the M data items, per slot. The user decides to consume content from item m at time t with probability $P_{n,t}(m)$, $m = 1, \cdots, M$. We denote the demand profile for user n at time t by the vector $\mathbf{p}_{n,t} = \{P_{n,t}(m)\}_{m=1}^{M}$.

Each user exhibits a different willingness-to-pay value for every data item, which potentially captures the value of such item to the user. The willingness-to-pay value for item m as recognized by user n is denoted by $v_n(m)$. We assume the CP can measure these values through different methods (see, e.g., [19] and the references therein), and through the significant advances in machine learning and collaborative filtering for personalized recommendation techniques [20, 21]. While the content of data items may be updated for every time slot, we assume willingness-to-pay value to vary slowly and over a larger time horizon, e.g., weeks or months. This can be thought of as the data item being a Youtube channel, or a category of Netflix movies toward which the user exhibits a constant interest despite the particular content being changing. The CP assigns a price $y_{n,t}(m) \leq v_n(m)$ to the data item m when it is requested by user n at time t. The impact of such price on the probability of requesting item m is captured through the mapping $P_{n,t}(m) = \phi_{m,n,t}(y_{n,t}(m))$, which is non-negative and non-increasing in $y_{n,t}(m)$. The price $y_{n,t}(m)$ is not necessarily restricted to *currency*. Instead, it can be a proxy to virtual tokens, points, or quota assigned to the users based on their subscriptions.

The random activity of the users' demand is modeled as follows:

- The demand of user n to item m at slot t is captured by a random variable $\mathbb{I}_{n,t}(m)$ where

$$\mathbb{I}_{n,t}(m) = \begin{cases} 1, & \text{with probability } P_{n,t}(m), \\ 0, & \text{with probability } 1 - P_{n,t}(m). \end{cases} \tag{14.1}$$

- Any two different users n, k have $\mathbb{I}_{n,t}(m)$ independent of $\mathbb{I}_{k,t}(j)$ for all m, j.
- Each user n can consume content from at most one data item per time slot. Hence $\sum_{m=1}^{M} \mathbb{I}_{n,t}(m) \leq 1$.
- Any user n is idle at slot t with probability $q_{n,t} := 1 - \sum_{m=1}^{M} P_{n,t}(m)$.

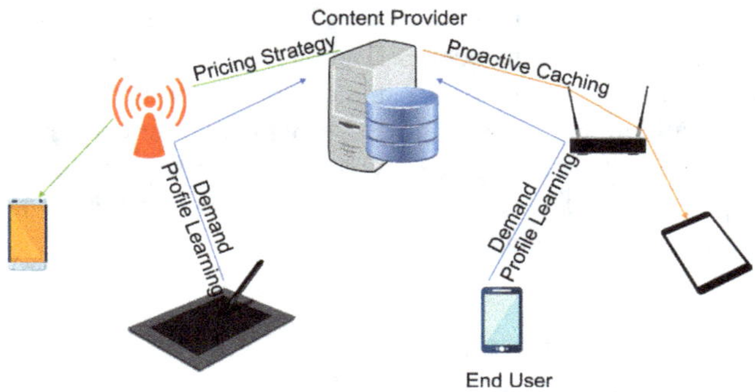

Figure 14.2 *CP network setup: Users generate consistent demand that can be learned over time. CP responds to demand with proactive caching and optimized prices.*

Further, motivated by the consistent user activity patterns (e.g., in [22]), and the reported results on human-behavioral patterns in [23], we assume that the demand profile of each user follows a *cyclostationary* pattern that repeats consistently in a period of T time slots. The T-slot period can be interpreted as a single day through which the activity of each user varies each hour but occurs with the same statistics consistently each day. That is, $S_{n,t} = S_{n,t+kT}$ and $q_{n,t} = q_{n,t+kT}$ for any non-negative integer k, and t=0, 1, ….

14.3.2 Proactive caching decisions

To balance its aggregate load over time, the CP assigns one-slot-ahead proactive caching values from the potential demand of the next slot to be served in the current. In particular, the CP sets $x_{n,t+1}(m)$ as the proactive download from the content of item m to be served to user n with the demand of time t. Figure 14.2 shows CP network components and the corresponding interactions while Figure 14.3 depicts the time-domain activities of the end-users and the CP. The assumption of only one-slot ahead proactive service is to account for content freshness guarantees so that the cached content is relevant to the user at the time of consumption. In addition, it also serves as the worst-case scenario where proactive services applied with longer lookahead intervals can always achieve higher performance gains than those analyzed in the sequel.

14.3.3 Incurred cost

The CP incurs costs proportional to the aggregate load at each slot. In particular, if L is the total CP load at a specific slot, then the cost incurred in such slot is $C(L)$, for some smooth, strictly convex, and monotonically increasing cost function $C : \mathbb{R}_+ \rightarrow \mathbb{R}_+$. Denoting by $\mathbf{x}$ and $\mathbf{y}$, respectively, the collections of proactive caching and service prices for all users, time slots, and data items, the CP's time average expected cost, thus, can be written as

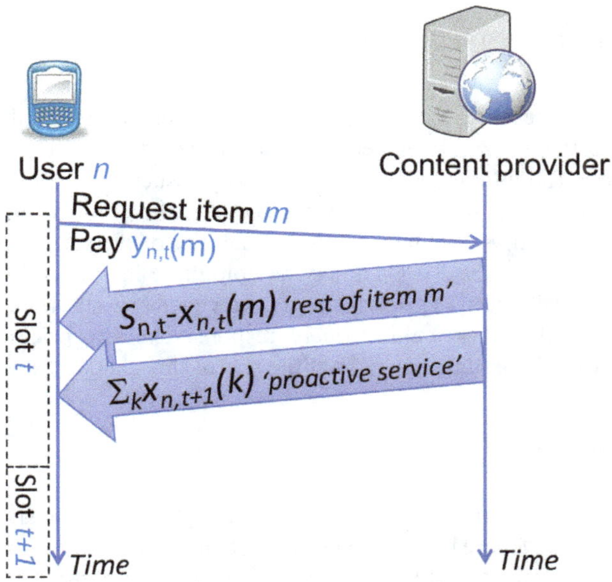

Figure 14.3 Time-domain diagram of end-user and CP interactions

$$\eta(\mathbf{x},\mathbf{y}) := \frac{1}{T}\sum_{t=0}^{T-1}\mathbb{E}\left[C\left(\sum_{m=1}^{M}\sum_{n=1}^{N}(S_{n,t}-x_{n,t}(m))\mathbb{I}_{n,t}(m)+x_{n,t+1}(m)\right)\right]. \tag{14.2}$$

Note the contribution of the proactive download $x_{n,t+1}(m)$ to the total load at time t, which translates to a cost that the CP incurs.

14.3.4 Revenue

The CP also reaps revenue due to the users' payments for the requested content. Such revenue essentially depends on the users' demand profiles which are shaped through the pricing policy used. The time average revenue (or user payments) received by the CPs is given by

$$\mu(\mathbf{y}) := \frac{1}{T}\sum_{t=0}^{T-1}\sum_{m=1}^{M}\sum_{n=1}^{N}y_{n,t}(m)\phi_{n,m,t}(y_{n,t}(m)). \tag{14.3}$$

The CP's net profit is then given by $R(\mathbf{x},\mathbf{y}) := \mu(\mathbf{y}) - \eta(\mathbf{x},\mathbf{y})$, and its ultimate objective is to find the optimal allocation of $\mathbf{x}$ and $\mathbf{y}$ that maximizes such a profit. The time average profit maximization to be solved by the CP is formulated as:

$$R^* := \max_{\mathbf{x},\mathbf{y}} \quad R(\mathbf{x},\mathbf{y}) \tag{14.4}$$

$$\text{subject to,} \quad x_{n,T}(m) = x_{n,0}(m), \forall m, n, \tag{14.5}$$

$$0 \le x_{n,t}(m) \le S_{n,t}, \quad \forall m, n, t, \tag{14.6}$$

$$0 \le y_{n,t}(m) \le v_n(m), \quad \forall m, n, \tag{14.7}$$

$$\sum_{m=1}^{M} \phi_{m,n,t}(y_{n,t}(m)) = 1 - q_{n,t}, \quad \forall n, t. \tag{14.8}$$

Here, the activity of each user over time is kept unchanged through the constraints $\sum_{m=1}^{M} \phi_{m,n,t}(y_{n,t}(m)) = 1 - q_{n,t}$, which ensures same access probability.

In the given formulation, there are convexity issues with the following. (i) The objective function is not concave in $(\mathbf{x}, \mathbf{y})$ due to the product form of the cost as a function of proactive caching and the probability of demand as a function of price. (ii) The functions $\phi_{m,n,t}(y_{n,t}(m))$ may be non-affine in $y_{n,t}(m)$, hence yielding non-convex constraint set [24]. Interestingly, however, the studies carried out in [6] have revealed that users exhibit *linear* responsiveness to pricing incentives, which is what we will assume throughout the rest of this chapter. In particular, we will adopt

$$\phi_{m,n,t}(y_{n,t}(m)) := \tfrac{1-q_{n,t}}{D_{n,t}}(v_n(m) - y_{n,t}(m)), \quad \forall m, n, t, \tag{14.9}$$

where $D_{n,t}$ is a normalizing constant. Further, we will provide remarks along with the main results to cover the case when $\phi_{m,n,t}$ is not an affine function. With the given form for $\phi_{m,n,t}$, constraints (14.8), reduce to an average price constraint

$$\tfrac{1}{M}\sum_{m=1}^{M} y_{n,t}(m) = \bar{y}_{n,t}, \quad \forall n, t, \quad \text{where } \bar{y}_{n,t} := \sum_{m} v_n(m) - D_{n,t}. \tag{14.10}$$

Our main focus in the following sections is to tackle the inherent non-convexity issue of the objective function, and show the system gains under suboptimal, yet efficient, algorithms. We note that, in (14.4), we assume the CP has *only* statistical knowledge about the users activity, which is justified since most CPs consistently log and track user interactions with the network over time and can construct the needed demand profiles through the aid of machine learning tools, such as collaborative filtering. Hence, in the sequel, we focus on the design of off-line algorithms which take place as a one shot that determines pricing and proactive caching for every slot. In the next section, we begin with the single-user scenario for its simplicity, then generalize the design to multi-user case in the subsequent section.

14.4 The single-user scenario: decoupled users' demand

In this section, we consider $N = 1$ and omit the subscript n from all notations. While the single-user scenario facilitates an ease of exposition, it also fits the setup whereby the contribution of the users' requests to the aggregate cost is not coupled by the convex cost function. In this case, the proactive download and price allocation problem (14.4) reduces to

$$R^* := \max_{\{\mathbf{x},\mathbf{y}\}_{t=0}^{T-1}} \frac{1}{T}\sum_{t=0}^{T-1}\frac{1-q_t}{D_t}\left[\sum_{m=1}^{M}(v(m)-y_t(m))\left(y_t(m)\right.\right.$$
$$\left.\left.-C\left(S_t - x_t(m) + \sum_{j=1}^{M} x_{t+1}(j)\right)\right) - q_t C\left(\sum_{j=1}^{M} x_{t+1}(j)\right)\right] \tag{14.11}$$

subject to constraints (14.5–14.7), and (14.10), with $N = 1$.

It can be noted from (14.11) that the non-convexity of the objective function is still restraining the achievability of the globally optimal solution. However, an efficient solution to it can still be attained as described in the following section.

14.4.1 Single-user iterative pricing and proactive download

Consider a suboptimal algorithm that takes an iterative sequential form of price allocation followed by proactive caching steps. In particular, we start the algorithm with some initial point $(\mathbf{x}^{(0)}, \mathbf{y}^{(0)})$, then, at every iteration k, the algorithm goes through the following two steps:

- **Step 1: Profit maximization**

$$R^{(k)} := \max_{y \in \mathscr{Y}} \; R(\mathbf{X}^{(K-1)}, y) \tag{14.12}$$

where $\mathbf{x}^{(k-1)}$ is the optimal solution of **Step 2** at iteration $k-1$,

$$\mathscr{Y} := \left\{ \mathbf{y} : 0 \leq y_t(m) \leq v(m), \quad \frac{1}{M}\sum_{m=1}^{M} y_t(m) = \bar{y}_t, \forall m, t \right\}.$$

- **Step 2: Cost minimization**

$$\eta^{(k)} := \min_{\mathbf{x} \in \mathcal{X}} \; \eta(\mathbf{x}, \mathbf{y}^k) \tag{14.13}$$

where $\mathbf{y}^{(k)}$ is the optimal solution of **Step 1** at iteration k,

$$\mathscr{X} := \{ \mathbf{x} : x_T(m) = x_0(m), \quad 0 \leq x_t(m) \leq S_t, \quad \forall m, t \}.$$

The algorithm utilizes the observation that the profit maximization problem when given a proactive download control $\mathbf{x}$ is a convex function in $\mathbf{y}$. Moreover, given a price allocation strategy $\mathbf{y}$, the profit function is convex in $\mathbf{x}$, and its maximization is equivalent to minimizing the expected cost.

The following theorem establishes the convergence of the proposed suboptimal algorithm.

Theorem 1: The sequence of CP profits $\{R^{(k)}\}_k$ generated through the iterative solutions of (14.12), (14.13) is increasing and convergent.

Proof. We have for any iteration k,

$$R^{(k)} = \max_{\mathbf{y}\in\mathscr{Y}} R(\mathbf{x}^{(k-1)}, \mathbf{y}) \geq R(\mathbf{x}^{(k-1)}, \mathbf{y}^{(k-1)}) \geq R(\mathbf{x}^{(k-2)}, \mathbf{y}^{(k-1)}) = R^{(k-1)} \tag{14.14}$$

where the last inequality follows since $x^{(k-1)}$ $.:= \arg min_{\mathbf{x}}\ \eta(\mathbf{x}, \mathbf{y}^{(k-1)}) = \arg min_{\mathbf{x}\in\mathscr{Y}},R(\mathbf{x}, \mathbf{y}^{(k-1)})$

Thus, $R^{(k)} \geq R^{(k-1)}$, $\forall k \geq 0$. Now, since $R^{(k)}$ is bounded above by $R^{(k)} \leq \sum_{m=1}^{M} v(m) < \infty$, then it converges to a finite limit $R^* := R(\mathbf{x}^*, \mathbf{y}^*)$ with $\mathbf{x}^*, \mathbf{y}^*$ are the limit points of $\{\mathbf{x}^{(k)}\}_k$, $\{\mathbf{y}^{(k)}\}_k$, respectively.

Thus, starting from any initial point $(\mathbf{x}^{(0)}, \mathbf{y}^{(0)})$, the algorithm converges to a point $(\mathbf{x}^*, \mathbf{y}^*)$ for which $R(\mathbf{x}^*, \mathbf{y}^*) \geq R(\mathbf{x}^{(0)}, \mathbf{y}^{(y)})$. In Section 14.4.2, we investigate the performance of the algorithm in terms of CP profit and end-user payments.

14.4.2 Performance analysis

For performance evaluation, we compare the results of the developed algorithm in the previous subsection 14.4.1 with a baseline profit and user expected payment in which no-proactive caching downloads are scheduled. In other words, the proactive download control of the baseline scenario is $\hat{\mathbf{x}} := \mathbf{0}$, and the associated pricing policy is $\hat{\mathbf{y}} := \arg\max_{\mathbf{y}\in\mathcal{Y}} R(\mathbf{y}, \mathbf{0})$. The expected profit, cost, and user payments under the baseline scenario are denoted by $\hat{R}$, $\hat{\eta}$, and $\hat{\mu}$, respectively.

In our analysis, we focus on the performance of the proposed algorithm when the initial point is of the baseline scenario, i.e., $(\mathbf{x}^{(0)}, \mathbf{y}^{(0)}) = (\mathbf{0}, \hat{\mathbf{y}})$. The idea behind such a choice is to highlight the potential gains to be reaped by such system when proactive caching is employed. In fact, under the baseline scenario, the CP can assign time-dependent and data-item-dependent prices to maximize its profit. However, with the additional degree of freedom of proactive service, such prices can be optimized to realize more gains.

Letting $(\mathbf{x}^*, \mathbf{y}^*)$ denote the limit point of the proposed algorithm when the initial point is $(\mathbf{0}, \hat{\mathbf{y}})$, we define $R^* := R(\mathbf{x}^*, \mathbf{y}^*)$, $\eta^* = \eta(\mathbf{x}^*, \mathbf{y}^*)$, and $\mu^* = \mu(\mathbf{y}^*)$. We compare the profit and savings gain of the proposed algorithm relative to the baseline through the quantities: profit gain $\Delta R := R^* - \hat{R}$, savings gain $\Delta\mu := \hat{\mu} - \mu^*$, and cost reduction gain $\Delta\eta := \hat{\eta} - \eta^*$. Clearly, we have $\Delta\eta = \Delta R + \Delta\mu$. In other words, the cost reduction achieved through proactive caching is divided over the profit gain and savings gain.

In the following theorem, we establish a lower bound on the profit gain.

Theorem 2: Let $m_t^* := \arg\max_{m\in\{1,\cdots,M\}} \phi_{m,t}(\hat{y}_t(m))$, and $\mathscr{T} := \left\{t : \phi_{m_t^*,t}(\hat{y}_t(m_t^*))C'(S_t) > (1 - q_{t-1})C'(S_{t-1})\right\}$, $\forall t$. Then,

$$\Delta R \geq \frac{1}{T}\sum_{t\in\mathcal{T}} \chi_t \left(\phi_{m_t^*,t}(\hat{y}_t(m_t^*))C'(S_t - \chi_t) - \mathbb{E}\left[C'\left(\sum_{j=1}^{M} S_{t-1}\hat{\mathbb{I}}_{t-1}(j) + \chi_t\right)\right]\right), \tag{14.15}$$

where C' is the first derivative of C,

$$\chi_t := \arg\max_{0\le x\le S_t} x\left(\phi_{m_t^*,t}(\hat{y}_t(m_t^*))C'(S_t - x) - \mathbb{E}\left[C'\left(\sum_{j=1}^{M} S_{t-1}\hat{\mathbb{I}}_{t-1}(j) + x\right)\right]\right), \tag{14.16}$$

and $\hat{\mathbb{I}}_t(m)$ is the random variable capturing the event that the user requests item m at time t under pricing policy $\hat{\mathbf{y}}$.

Proof. Please refer to Section 14.7.1.

The established bound in Theorem 2 essentially relies on the existence of a sufficient difference between the marginal costs of two consecutive slots, clearly, if $\mathscr{T}$ is non-empty, then the lower bound is strictly positive. For instance, some users may be busy at work until the end of the business day, when they decide to follow up the news and request data content. The CP can therefore partially supply such data before the end of the business day so as to regulate the traffic over time. We also note that, if $\mathscr{T} = \{\}$, then $\Delta R = 0$, which potentially captures the event of a user with balanced demand over time, for which the CP does not gain by shifting the load in time.

Now, we study the effect of proactive caching on the savings gain, $\Delta\mu$. We first establish the statement that proactive caching cannot yield a larger expected payment than the baseline scenario. Subsequently, we characterize the necessary and sufficient conditions for strictly positive savings gain.

Lemma 1: The expected payment by the user under the proposed iterative algorithm satisfies $\mu(\mathbf{y}^*) \le \mu(\hat{\mathbf{y}})$.

Proof. Since $R(\mathbf{0},\hat{\mathbf{y}}) = \max_{\mathbf{y}\in\mathcal{Y}} \mu(\mathbf{0},\mathbf{y}) - \eta(\mathbf{0},\mathbf{y})$ with (14.17)

$$\eta(\mathbf{0},\mathbf{y}) = \sum_{t,m} C(S_t)\phi_{m,t}(y_t(m)), \tag{14.18}$$

the constraint that $\sum_m \phi_{m,t}(y_t(m)) = 1 - q_t$ implies $\eta(\mathbf{0},\mathbf{y}) = \sum_t C(S_t)(1-q_t)$ is independent of $\mathbf{y}$. In this case, profit maximization will reduce to maximizing the expected payment. Therefore, $\mu(\mathbf{0},\hat{\mathbf{y}})$ is the maximum expected payment by the user.

The user will be charged the maximum possible payment when the CP cannot modify its load dynamics over time. With the proactive caching, however, the smoothed-out load characteristics will never yield extra payments on the user. The following theorem characterizes an even stronger result.

Theorem 3: Suppose that $\max_m v(m) > \min_m v(m)$, then the savings gain satisfies $\Delta\mu > 0$ if and only if

$$\phi_{m_0,t_0}(\hat{y}_{t_0}(m_0))C'(S_{t_0}) > (1 - q_{t_0-1})C'(S_{t_0-1}), \tag{14.19}$$

for some time slot t_0, and data item m_0.

Proof. Please refer to Section 14.7.2.

Remark 1: In Theorem 3, suppose that $\min_m v(m) \ge |2\bar{y}_{t_0} - \bar{v}|$, where $\bar{v} = \frac{1}{M}\sum_m v(m)$. Then Condition (14.19) reduces to

$$\frac{1}{M} + \frac{1}{2}\cdot\frac{\max_m v(m) - \bar{v}}{D_{t_0}} > \frac{(1-q_{t_0-1})C(S_{t_0-1})}{(1-q_{t_0})C(S_{t_0})}. \tag{14.20}$$

The requirement on $\min_m v(m)$ facilitates a full characterization of $\hat{\mathbf{y}}$ as a function of $(v(m))_m$ and D_t.

The LHS of (14.20) is the probability of requesting the service with maximum willingness-to-pay value, given that the user is going to request a service at time t_0. Here, $\hat{y}_{t_0}(m) = \frac{v_m+\bar{v}}{2} - \frac{D_{t_0}}{M}$. Thus, it is required to have this probability greater than the ratio between user activity at the previous slot $t_0 - 1$ over the activity at t_0.

Remark 2: In the proofs of Theorems 1–3, we used a general form for $\phi_{m,t}$ which is not necessarily linear in $y_t(m)$. This therefore extends the results of those theorems to any algorithm that yields a locally optimal solution to (14.11) under general economic responsiveness.

14.4.3 Distributed implementation

As we have seen that the end-user has a potential to achieve positive savings gain through proactive caching, the CP can involve the user to carryout proactive download decisions of the proposed iterative algorithm in response to the user at assigned prices. In particular, the CP can run **Step 1** of the algorithm and assign a certain pricing policy that maximizes the profit given the adopted proactive download strategy by the end-user.

The end-user reacts to such pricing policy by executing **Step 2** of the algorithm so as to balance its load over time, reduce the cost at the CP's end, and essentially attain less payments. The two parties will iterate their respective steps until convergence. From a game-theoretic perspective, the point $(\mathbf{x}^*, \mathbf{y}^*)$ to which the algorithm converges is a Nash equilibrium point of the dynamic coordinated game incorporating the CP and end-user as its players, with the common objective of maximizing the profit gain [25].

14.4.4 Numerical results for the single-user scenario

We consider a setup of $T = 5$ time slots with the user inactivity captured by $(q_t)_t = (0.03, 0.9, 0.02, 0.01, 0.9)$. The number of data items is $M = 3$, the data item size is $S_t = 100$, $\forall t$, the willingness-to-pay values are $(2.99, 2.24, 2.10)$ for the three items, and the average price per item $\bar{y}_t = 2$ for all slots. We consider a quadratic cost function $C(L) = BL^2$ with $B = 2 \times 10^{-4}$.

We run the single-user iterative algorithm starting from the no-proactive caching initial point $(\mathbf{0}, \hat{\mathbf{y}})$. Convergence results of the expected profit as a function are plotted versus the iteration number.

Figure 14.4 depicts the convergence results of the system profit and user payment under the proposed single-user iterative algorithm. Clearly, convergence points imply positive profit gain associated with positive savings gain, i.e., a win-win situation.

To further elaborate on the effect of pricing strategy on the certainty of the user demand, we evaluate the entropy of the normalized profile $\psi_{m,t} = \phi_{m,t}/(1 - q_t)$ and plot it versus the iteration number [26]. The value of $\psi_{m,t}$ captures the probability of requesting item m at time t conditioned on the event that the user requests at least

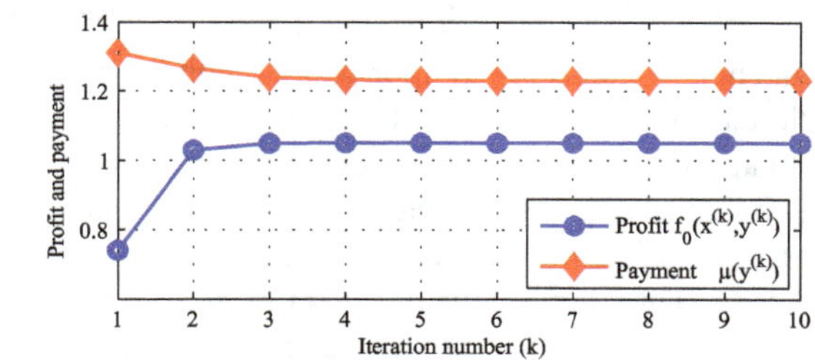

Figure 14.4 Convergence of profit and user payment in the single-user scenario

one item at time t. Figure 14.5 plots the entropy of the normalized demand profile for all slots $t = 0, \ldots, 4$ with the iteration number. The peak-hour slots 0 and 2 witness reduced entropy values to improve the proactive caching performance. Note that, no-proactive caching is applied for slot 3, since it is preceded by a higher activity slot. The off-peak slots 1 and 4 are not assigned proactive caching values either.

14.5 The multi-user scenario: coupled demand

The win-win situation established in the single-user scenario whereby the CP achieves higher profit while the end-user attains lower payment, compared with the no-proactive caching scenario, will inspire the development of an iterative algorithm that essentially yields a similar win-win situation to CPs and end-users who employ proactive content caching. In the iterative single-user algorithm, we have observed that the separation of the price allocation and proactive download decisions leads to two convex problems which can be solved iteratively until convergence. In the multi-user case, the price allocation step suffers from a product of prices for many users that need to be jointly optimized, thus rendering the price allocation problem non-convex. In this section, we tackle such an additional difficulty through tools from non-convex optimization [27] and establish performance bounds as in the previous section.

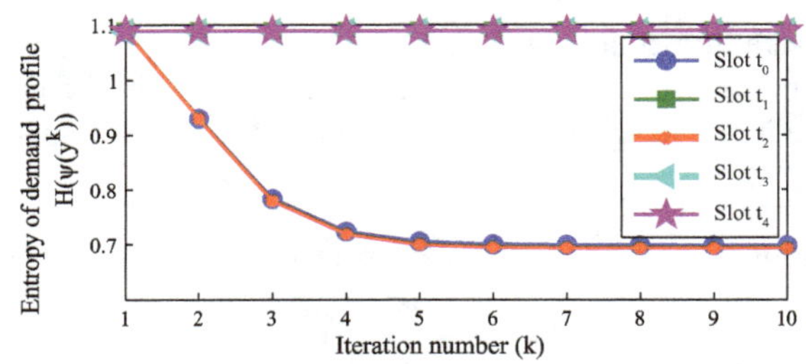

Figure 14.5 Entropy of the normalized demand profiles

14.5.1 Multi-user iterative pricing and proactive download

We consider the sequence of solutions to approximate convex problems of (14.4). In particular, in each iteration k, we replace the objective function $R(\mathbf{x},\mathbf{y})$, which is non-concave, with an approximate concave function $\tilde{R}^{(k)}(\mathbf{x},\mathbf{y})$, and we use the solution to the resulting convex problem to generate a new approximate function $\tilde{R}^{(k+1)}$, and so on. The series of solutions to these approximate convex problems converges to a point that satisfies the KKT conditions of (14.4).

Lemma 2: Let $\tilde{R}^{(k)}$ be a concave function in $(\mathbf{x},\mathbf{y})$ that replaces the objective function R of (14.4) at iteration k. Denote by $(\mathbf{x}^{(k-1)},\mathbf{y}^{(k-1)})$ the optimal solution to the resulting convex optimization problem at the $(k-1)^{st}$ iteration, $k = 1, 2, \ldots$. If

1. $\tilde{R}^{(k)}(\mathbf{x},\mathbf{y}) \leq R(\mathbf{x},\mathbf{y})$for all feasible $(\mathbf{x},\mathbf{y})$,
2. $\nabla\tilde{R}^{(k)}(\mathbf{x}^{(k-1)},\mathbf{y}^{(k-1)}) = \nabla R(\mathbf{x}^{(k-1)},\mathbf{y}^{(k-1)})$,
3. $\tilde{R}^{(k)}(\mathbf{x}^{(k-1)},\mathbf{y}^{(k-1)}) = R(\mathbf{x}^{(k-1)},\mathbf{y}^{(k-1)})$,

then $R(\mathbf{x}^{(k-1)},\mathbf{y}^{(k-1)}) < R(\mathbf{x}^{(k)},\mathbf{y}^{(k)}), \forall k$, and the sequence $\{(\mathbf{x}^{(k)},\mathbf{y}^{(k)})\}_k$ converges to a point $(\mathbf{x}^*,\mathbf{y}^*)$ which is a locally optimal solution to (14.4).

The given lemma is a special case of Theorem 1 in [27], which aims to provide local optimal solutions to non-convex problems. Thus, the resulting local optimal solution will essentially lead to a better profit performance than that of the baseline scenario when no-proactive caching is carried out.

In the following theorem, we provide a particular approximation to R of (14.4) at each new iteration k that satisfies the requirements of Lemma 2.

Theorem 4: For R being the objective function of (14.4), the approximate function

$$\tilde{R}^{(k)}(\mathbf{x},\mathbf{y}) = \mu(\mathbf{y}) - \eta(\mathbf{x},\mathbf{y}^{(k-1)}) - \frac{1}{T}\sum_{t=0}^{T-1}\sum_{m,n} h_{n,t}^{(k)}(m) y_{n,t}(m), \tag{14.21}$$

where

$$h_{n,t}^{(k)}(m) = \left.\frac{\partial\eta(\mathbf{x}^{(k-1)},\mathbf{y})}{\partial y_{n,t}(m)}\right|_{(\mathbf{x}^{(k-1)},\mathbf{y}^{(k-1)})}, \qquad \forall m,n,t, \tag{14.22}$$

at iteration $k \geq 1$ is concave in $(\mathbf{x},\mathbf{y})$. Further, the sequence of solutions to the problem resulting from replacing R with $\{\tilde{R}^{(k)}\}_k$ converges to a locally optimal solution of (14.4).

Proof. Please refer to Section 14.7.3.

With the construction of the approximate functions $\tilde{R}^{(k)}$ in the given theorem, we now present the iterative price and proactive download algorithm.

- Start with some initial point $(\mathbf{x}^{(0)},\mathbf{y}^{(0)})$.
- At each iteration $k \geq 1$, compute $\tilde{R}^{(k)}$ as in (14.21), and solve the convex problem

$$\max_{\{\mathbf{x},\mathbf{y}\}_{m,n,t=0}^{T-1}} \tilde{R}^{(k)}(\mathbf{x},\mathbf{y}) \tag{14.23}$$

subject to, constraints (14.4)–(14.6), (14.8). (14.24)

In the following subsection, we focus on the performance analysis of the proposed multi-user iterative algorithm.

14.5.2 Performance analysis

To quantify the performance of the multi-user iterative algorithm, we utilize the same metrics of the single-user counterpart. In particular, we take the initial point of the algorithm as the baseline no-proactive caching scenario. That is, $(\mathbf{x}^{(0)}, \mathbf{y}^{(0)}) = (0, \hat{y})$, where $\hat{\mathbf{y}} = \arg\max_{\mathbf{y}\in\mathscr{Y}} R(\mathbf{0}, \mathbf{y})$, and here we take the set $\mathscr{Y}$ as:

$$\mathscr{Y} := \left\{\mathbf{y} : 0 \le y_{n,t}(m) \le v_n(m), \quad \sum_{m=1}^{M} \phi_{m,n,t}(y_{n,t}(m)) = 1 - q_{n,t}, \forall m, n, t\right\}. \quad (14.25)$$

We use the notation $(\mathbf{x}^*, \mathbf{y}^*)$ to capture the point to which the sequence of solutions $\{(\mathbf{x}^{(k)}), \mathbf{y}^{(k)}\}_k$ generated according to the multi-user iterative algorithm converges. Further we use ΔR, $\Delta\mu$, and $\Delta\eta$ to denote the profit, savings, and cost reduction gains, respectively, where $\Delta R = R(\mathbf{x}^*, \mathbf{y}^*) - R(\mathbf{0}, \hat{\mathbf{y}})$, $\Delta\mu = \mu(\hat{\mathbf{y}}) - \mu(\mathbf{y}^*)$, and $\Delta\eta = \eta(\mathbf{0}, \hat{\mathbf{y}}) - \eta(\mathbf{x}^*, \mathbf{y}^*)$.

To characterize a lower bound on the profit gain, we introduce the following notion of active users:

Definition 1 (Active users): For each data item m and time slot t, we define a set $\mathscr{B}_t(m)$ of active users as:

$$\mathscr{B}_t(m) : \; = \left\{n : \mathbb{E}\left[\hat{\mathbb{I}}_{n,t}(m) C'(\hat{L}_t) - C'(\hat{L}_{t-1})\right] > 0\right\}, \qquad (14.26)$$
$$t = 0, \cdots, T-1, \quad m = 1, \cdots, M,$$

where $\hat{L}_t := \sum_{m,n} S_{n,t} \hat{\mathbb{I}}_{n,t}(m)$. We also use $B_t(m) := |\mathscr{B}_t(m)|$ is the cardinality of set $\mathscr{B}_t(m)$.

The active users of slot t w.r.t. item m are those users who, in the expected sense, create higher marginal cost in slot t by requesting item m than the marginal cost of the previous slot. These active users have a high potential to improve the cost reduction through a proactive service of their demand. We will show in the sequel that existence of such active users is necessary and sufficient for positive system gains. In the given definition, the expectation $\mathbb{E}\left[\hat{\mathbb{I}}_{n,t}(m) C'(\hat{L}_t) - C'(\hat{L}_{t-1})\right]$ captures the *marginal* contribution of user n to the cost of time slot t, when requests item m, over the cost of the previous time slot $t-1$. This, for instance, may attribute to the disparate activity levels of the users between times $t-1$ and t.

Now, we establish the following lower bound on the profit gain.

Theorem 5 (Lower bound on profit gain): For the sets of active users $\mathscr{B}_t(m)$ as defined,

$$\begin{aligned}\Delta R &\geq \frac{1}{T}\sum_{t=0}^{T-1}\tilde{x}_t\sum_{m=1}^{M}\sum_{n\in\mathcal{B}_t(m)}\mathbb{E}\left[\hat{\mathbb{I}}_{n,t}(m)C'\left(\hat{L}_t-\sum_{j=1}^{M}\sum_{k\in\mathcal{B}_t(j)}\tilde{x}_t\hat{\mathbb{I}}_{k,t}(j)\right)\right.\\ &\left.-C'\left(\hat{L}_{t-1}+\sum_{j=1}^{M}\sum_{k\in\mathcal{B}_t(j)}\tilde{x}_t\right)\right]>0,\end{aligned} \tag{14.27}$$

where

$$\begin{aligned}\tilde{x}_t &= \arg\max_{0\leq\tilde{x}\leq\tilde{S}}\tilde{x}\sum_{m=1}^{M}\sum_{n\in\mathcal{B}_t(m)}\mathbb{E}\left[\hat{\mathbb{I}}_{n,t}(m)C'\left(\hat{L}_t-\sum_{j=1}^{M}\sum_{k\in\mathcal{B}_t(j)}\tilde{x}\hat{\mathbb{I}}_{k,t}(j)\right)\right.\\ &\left.-C'\left(\hat{L}_{t-1}+\sum_{j=1}^{M}\sum_{k\in\mathcal{B}_t(j)}\tilde{x}\right)\right]\end{aligned} \tag{14.28}$$

Proof. Please refer to Section 14.7.4.

The given lower bound is achieved by allocating equal proactive caching values to all active users of a certain time slot and keeping the pricing policy unchanged. While such a scheme is strictly suboptimal, it reveals insights on the users, data items, and time slots that can essentially benefit the system, through proactive content caching, beyond the baseline.

We further consider the asymptotic behavior of the lower bound (14.27) when the number of users N grows to infinity. The idea is to study the impact of the number of users (and active users) on the potential profit gain when the user base expands significantly as in large rallies, stadiums, supermarkets, public transit, etc.

To that end, we start our investigations with the assumption that $q_{n,t} < 1-\epsilon$ for all n, t and some $\epsilon > 0$. That is, each user can request a data item at any time slot with a positive probability, and this probability will not vanish as the number of users grows to infinity. Hence $\hat{L}_t \to \infty$ almost surely as $N \to \infty$, $\forall t$.

We also capture the scaling of the number of active users with N through two main assumptions. In the first, we consider the case when such a number grows to infinity with N, whereas, in the second, we consider the case when the number of active users is finite and does not scale with N.

Assumption 1 (Scaling of the number of active users): Assume that there exists some non-decreasing function $g : \mathbb{N} \to \mathbb{N}$ such that $g(N) \to \infty$ as $N \to \infty$, and for every time slot t and data item m, the limit

$$\beta_t(m) := \lim_{N\to\infty}\frac{B_t(m)}{g(N)} \tag{14.29}$$

exists, and $\beta_t(m) < \infty$, $\forall m = 1, \cdots, M, t = 0, \cdots, T-1$.

Thus, the function $g(N)$ captures the maximum possible scaling of the active users of any item m, and time slot t with the total number of users N.

We define $\check{P}_{m,t} := \inf_{n\in\mathcal{B}_t}\{\phi_{m,n,t}(\hat{y}_{n,t}(m))\}$.

Theorem 6 (Asymptotic profit gain for infinite number of active users): Under Assumption 1, suppose that the cost function C satisfies

$$\lim_{L\to\infty} \frac{L^{\delta}}{C'(L)} = 0, \quad \text{for some } \delta > 0. \tag{14.30}$$

Also, suppose that $\beta_t(m) > 0$ for some data item m and time slot t, with

$$\liminf_{N\to\infty}\{\check{P}_{m,t}\} > \limsup_{N\to\infty} \frac{\mathbb{E}[L_{t-1}]}{\mathbb{E}[L_t]}. \tag{14.31}$$

Then, $\Delta R(N) = \Omega(g(N)C'(\gamma N))$, for some $\gamma > 0$. In other words, the profit gain grows with N at least as $g(N)C'(\gamma N)$.

Proof. Please refer to Section 14.7.5.

Note that, the assumption on $\beta_t(m) > 0$ for some m, t ensures that the number of active users grows to infinity as $g(N)$ with N. Also, Condition (14.31) requires the least possible contribution of an active user to the system not to vanish as the number of users grows to infinity. It can be noted from Theorem 6 that the achieved profit gain depends on the growth of the *active users* of the system with the total number of users. These are the users that essentially allow for more gains by receiving proactive service to smooth out the network load over time and reduce the incurred costs. The first derivative of the cost function captures the rate of increase in the incurred cost as the number of users grows. Yet, it also factors-in in the lower bound demonstrating that proactive service enables superlinearly increasing profit gain with the number of users.

Remark 3 (Finite number of active users): If $0 < B_t(m) \leq B$ for all m, t, and some positive integer B, then the reaped profit gain grows with N as $\Omega(C'(\gamma N))$. That is, even under a finite number of active users, the CP is able to reap a profit grain that grows unboundedly with the number of users as $C'(\gamma N)$.

Now, we shed light on the potential savings gain $\Delta\mu$ for the users. To consider the savings of a particular user, we denote the expected payment for user n by $\mu_n(\mathbf{y})$, where

$$\mu_n(\mathbf{y}) = \frac{1}{T}\sum_{t=0}^{T-1}\sum_{m=1}^{M} y_{n,t}(m)\phi_{m,n,t}(y_{n,t}(m)) \tag{14.32}$$

and we consider the savings gain for user n as $\Delta\mu_n := \mu_n(\hat{\mathbf{y}}) - \mu_n(\mathbf{y})$.

Lemma 3: Under the proposed multi-user algorithm, the savings gain for each user n satisfies $\Delta\mu_n \geq 0$.

Proof. Starting from the initial point $(\mathbf{0}, \hat{\mathbf{y}})$, we have

$$R(\mathbf{0}, \hat{\mathbf{y}}) = \max_{\mathbf{y}\in\mathscr{Y}\,\mu(\mathbf{y})} -\frac{1}{T}\sum_{t=0}^{T-1}\mathbb{E}\left[C\left(\sum_{n=1}^{N} S_{n,t}\sum_{m=1}^{M}\mathbb{I}_{n,t}(m)\right)\right] \tag{14.33}$$

But $\sum_{m=1}^{M}\mathbb{I}_{n,t}(m)$ is a Bernoulli random variable with parameter $1 - q_{n,t}$, i.e., independent of $\{y_{n,t}(m)\}_m$, for all n, t. Thus, $\hat{\mathbf{y}} := \arg\max_{\mathbf{y}\in\mathscr{Y}} \mu(\mathbf{y})$. Now, since $\mu(\mathbf{y})$ can be decoupled to the superposition of per-user payments $\mu(\mathbf{y}) = \sum_{n=1}^{N}\mu_n(\mathbf{y})$, it turns out that $\mu_n(\hat{\mathbf{y}}) \geq \mu_n(\mathbf{y})$ for all $\mathbf{y} \in \mathscr{Y}$.

Hence, users will not pay more than their original subscription fees under no-proactive caching. However, unlike in the single-user scenario, where we characterized in Theorem 3 a necessary and sufficient condition that depends only on the initial prices and willingness-to-pay values to have a positive savings gain, in the multi-user case, the dependencies of proactive caching decisions of one user on the pricing policies for the others render the derivations of similar conditions for the multi-user case analytically intractable.

Nevertheless, we can still draw some insights on the users with positive savings gain.

Theorem 7: Suppose that $\max_m v_{n_0}(m) > \min_m v_{n_0}(m)$, for some user n_0. Then $\Delta\mu_{n_0} > 0$ if and only if $x^*_{n_0,t_0}(m_0) > 0$ for some data item m_0 and time slot t_0.

Proof. Please refer to Section 14.7.6.

Thus, the users whose proactive caching converges to positive values must receive a positive savings gain, whereas the users with zero proactive caching make neither gain nor loss. However, it can be noted that the profit gain reaped by the CP must be met with savings gain reaped by at least one user. This is formalized in the following Theorem.

Theorem 8: Under the proposed multi-user iterative algorithm, $\Delta R > 0$ and $\Delta\mu > 0$ if and only if the set of active users $\mathscr{B}_t(m)$ is non-empty for some time slot and data item.

Proof. Please refer to Section 14.7.7.

Theorem 8 shows that the existence of active users in the system is necessary and sufficient for profit and savings gain. The attained win-win situation attributes to the judicious utilization of the predictable nature of the user behavior. In particular, the CP trades pricing discounts for enhanced certainty which improves the proactive caching performance and thus more profit. Users who receive proactive caching also receive pricing discounts so as to access proactively served content with higher probability. Thus, both parties win.

Remark 4 (Generalizing $\phi_{m,n,t}$): In Theorems 6–8, we have developed the results given that the proposed algorithm converges to a locally optimal solution of the problem with $\Delta R > 0$, which is attained under the affine form of $\phi_{m,n,t}(y_{n,t}(m))$. However, the same results also hold for any locally optimal solution for the problem with $\Delta R > 0$ and $\phi_{m,n,t}$ taking a more general form of non-negative and non-decreasing function in the employed pricing policy.

Remark 5 (Multiple-slot ahead proactive service): In the given analysis, we have focused on the one-slot ahead proactive service for fresh content guarantees. Yet, for data items with more static content, proactive service can be applied up to ρ slots ahead, $\rho \geq 1$. The profit gain can then be lower bounded through an extended definition of active users in which user n is active with respect to item m and time t if there exists a time slot $t' \in \{t-\rho, \cdots, t\}$ such that $\mathbb{E}\left[\hat{\mathbb{I}}_{n,t}(m)C'(\hat{L}_t) - C'(\hat{L}_{t'})\right] > 0$. Thus, this shows more opportunities for existence of active users in the system, and promises of larger scaling orders of the function $g(N)$ of Theorem 6.

Having gained some insights on the performance of the proposed algorithm, and the incentives for users to save money, we consider the distributed implementation of it in Section 14.5.3.

14.5.3 Distributed implementation

The need for distributed implementation of the algorithm is to reduce the centralized complexity at the CP and enable system scalability. Inspired by the single-user scenario where the distributed implementation took place through the sequential solution of price allocation and corresponding optimal proactive caching, here we also aim to leverage such dynamics to control the algorithm. In particular, we observe that the optimization (14.23) can be decomposed to price allocation and proactive caching control. Hence, we propose to run the distributed version of the multi-user algorithm through the iteration of the two consecutive steps of price and proactive caching assignment.

Users will determine the optimal proactive caching corresponding to the prices set by the CP, then the CP will respond with optimized pricing. The allocation of optimal proactive download in a distributed fashion is essentially more difficult than in the single-user case, since users need to iterate on the proactive caching within the proactive caching allocation step itself until they converge to the minimizer of the average cost under the current pricing. Thus, there will be an inner loop within the proactive caching step.

Also, it can be noted from the previous subsection 14.5.2 that not all users may necessarily attain a positive savings gain through the proposed algorithm. Hence, before the proactive caching step of each iteration k, the CP can compute the new sets of active users defined as:

$$\begin{aligned}\mathscr{B}_t^{(k)}(m):\ &= \left\{n : \mathbb{E}\left[\mathbb{I}_{n,t}^{(k-1)}(m) C'(L_t^{(k-1)}) - C'(L_{t-1}^{(k-1)})\right] > 0\right\} \\ &, t = 0, \cdots, T-1,\ \ m = 1, \cdots, M,\end{aligned} \tag{14.34}$$

where $L_t^{(k-1)} := \sum_{n=1}^{N} S_{n,t} \sum_{m=1} \mathbb{I}_{n,t}^{(k-1)}(m)$, and $\mathbb{I}_{n,t}^{(k-1)}(m)$ is the indicator that user n requests item m at time t under the pricing policy $\mathbf{y}^{(k-1)}$.

Then, only the active users will be called to participate in such an iteration. The following lemma reveals a nice property of the distributed implementation.

Lemma 4: Starting from the baseline initial point $(\boldsymbol{x}^{(0)}, \boldsymbol{y}^{(0)}) = (\mathbf{0}, \hat{\boldsymbol{y}})$, the updated sets of active users according to (14.34) satisfy $\mathscr{B}_t^{(k)}(m) \subseteq \mathscr{B}_t^{(k-1)}(m)$, for all m, t, k.

Proof. By induction. We see from the proof of Theorem 8 that if $n \notin \mathscr{B}_t^{(0)}(m)\ \forall m, t$, i.e., n is not an active user under $\hat{\mathbf{y}}$, then $y_{n,t}^{(1)}(m) = \hat{y}_{n,t}(m), \forall m, t$. Now,

$$\mathbb{E}\left[\mathbb{I}_{n,t}^{(1)}(m) C'\left(L_t^{(1)}\right)\right] - \mathbb{E}\left[C'\left(L_{t-1}^{(1)}\right)\right] = \mathbb{E}\left[\hat{\mathbb{I}}_{n,t}(m) C'\left(\hat{L}_t\right)\right] - \mathbb{E}\left[C'\left(\hat{L}_{t-1}\right)\right], \tag{14.35}$$

since $\mathbb{E}[C'(L_{t-1})]$ is independent of the pricing policy. Thus, $n \notin \mathscr{B}_t^{(1)}(m), \forall m, t$. We can also apply the same procedure with all subsequent iterations to show that $n \notin \mathscr{B}_t^{(k)}(m), \forall m, t, k$.

Now, suppose that $\mathscr{B}_t^{(k-1)}(m) \subseteq \mathscr{B}_t^{(k-2)}(m)$, and pick a user $n \in \mathscr{B}_t^{(k-2)}(m)$ but $n \notin \mathscr{B}_t^{(k-1)}(m)$, for all m, t. Then, clearly, $x_{n,t}^{(k-1)}(m) = 0, \forall m, t$, which in turn implies $\mathbf{y}^{(k)} = \hat{\mathbf{y}}$, and hence $n \notin \mathscr{B}_t^{(k)}(m)$ for the reason that $\mathbb{E}[C'(L_{t-1})]$ is independent of the pricing policy.

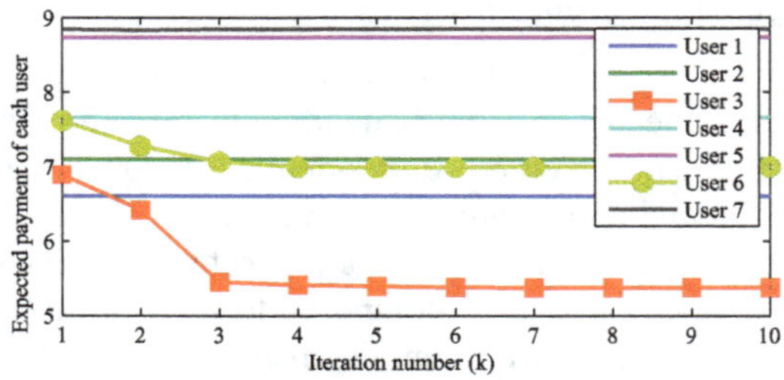

Figure 14.6 Convergence of the expected per-user payment

Thus, users that have been found *not-active* at a given iteration can be safely excluded from participation in the distributed algorithm for the rest of iterations. Consequently, the system will efficiently avoid unnecessary computations. We present our proposed distributed version of the iterative multi-user algorithm in Algorithm 1.

We can see from the algorithm that after each price allocation decision, selected users iterate on optimizing their proactive caching, with the coordination of the CP, which supplies them with the needed information encapsulated in the personalized cost functions $\eta_n(\mathbf{x}_n, \mathbf{x}_{-n}^{(k,l-1)})$ to minimize, until convergence to the best proactive caching policy corresponding to the set prices. Thus users need not exchange their own information with each other. We also note that the iterations on the proactive caching (i.e., the inner loop) always converge to $\mathbf{x}^{(k)}$, the optimal proactive caching of (14.23) since the problem is convex in $\mathbf{x}$.

14.5.4 Numerical results for the multi-user scenario

For the multi-user case, we show the convergence results in Figures 14.6 and 14.7. In particular, we consider a system with $N = 7$ users and $M = 5$ data items. We assume that users exhibit the same activity probabilities $q_{n,t}$ as in the single-user case, that is, $(q_{n,t})_t = (0.03, 0.9, 0.02, 0.01, 0.9)$, $\forall n$. We also use the quadratic cost function, but with B chosen such that the baseline profit is zero, i.e., $R(\mathbf{0}, \hat{\mathbf{y}}) = 0$.

We have randomly generated the system parameters as

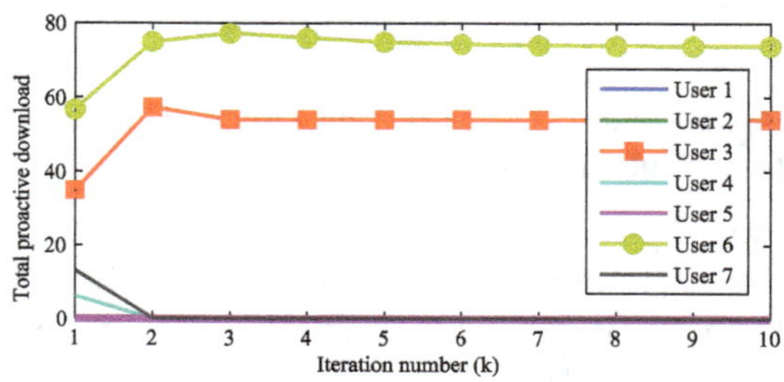

Figure 14.7 Convergence of aggregate proactive caching per user

$$
\left[S_{n,t}\right]_{n,t} = \begin{bmatrix} 8.69 & 9.06 & 39.18 & 53.07 & 0.23 \\ 64.26 & 4.71 & 21.12 & 39.53 & 8.44 \\ 5.81 & 2.16 & 54.11 & 60.39 & 8.10 \\ 9.91 & 4.39 & 13.73 & 22.18 & 0.57 \\ 18.48 & 1.63 & 28.28 & 47.73 & 9.87 \\ 71.12 & 8.56 & 23.13 & 78.42 & 6.89 \\ 1.28 & 8.03 & 45.56 & 41.29 & 9.69 \end{bmatrix}, \tag{14.36}
$$

Algorithm 1 Distributed Multi-user Iterative Pricing and Proactive Caching

Initialization: choose the initial point $(\mathbf{x}^{(0)}, \mathbf{y}^{(0)})$, minimum convergence accuracy $\varepsilon > 0$, and set the iteration index $k = 1$.
while $\|(\mathbf{x}^{(k)}, \mathbf{y}^{(k)}) - (\mathbf{x}^{(k-1)}, \mathbf{y}^{(k-1)})\| > \varepsilon$ *outer loop* **do**
    Content provider: Compute the sets $\mathscr{B}_t^{(k)}(m)$ as in (15.23).
    Content provider: Solve

$$
\begin{aligned}
\mathbf{y}^{(k)} = \arg\max_{\mathbf{y}} \mu_{\mathbf{y}} - \frac{1}{T}\sum_{t=0}^{T-1}\sum_{m=1}^{M}\sum_{n=1}^{N} h_{n,t}^{(k)}(m) y_{n,t}(m)\\
\text{s.t.,}\quad 0 \le y_{n,t}(m) \le v_n(m),\\
\frac{1}{M}\sum_{m=1}^{M} y_{n,t}(m) = \bar{y}_{n,t} \forall m, n, t.
\end{aligned}
$$

    Content provider: Set an iteration index $l = 1$ for the inner loop, and let $\mathbf{x}_n^{(k,0)} := (x_{n,t}^{(k)}(m))_{m,t}, \forall n \in \cup_{m,t}\mathscr{B}_t^{(k)}(m)$.
    while $\|\mathbf{x}^{(k,l)} - \mathbf{x}^{(k,l-1)}\| > \varepsilon$ *inner loop* **do**
        Content provider: To each user $n \in \cup_{m,t}\mathscr{B}_t^{(k)}(m)$ compute $\eta_n(\mathbf{x}_n, \mathbf{x}_{-n}^{(k,l-1)})$.
        for $n \in \cup_{m,t}\mathscr{B}_t^{(k)}(m)$ **do**
            User n solve $\mathbf{x}_n^{(k,l)} := \arg\min_{\mathbf{x}_n} \eta_n(\mathbf{x}_n, \mathbf{x}_{-n}^{(k,l-1)})$

$$
\begin{aligned}
\text{s.t.,}\quad x_{n,0}(m) = x_{n,T}(m),\\
0 \le x_{n,t}(m) \le S_{n,t}, \quad \forall m, t,
\end{aligned}
$$

            where $\mathbf{x}_n = (x_{n,t}(m))_{m,t}$ is the proactive caching decision vector for user n, and $\eta_n(\mathbf{x}_n, \mathbf{x}_{-n})$ is the expected CP's cost as a function of $\mathbf{x}_n$, with the proactive caching for the rest of users $\mathbf{x}_{-n} = (\mathbf{x}_k)_{k\neq n}$ given.
        end for
        $l = l + 1$.
    end while
    $k = k + 1$.
end while

$$\left[v_n(m)\right]_{n,m} = \begin{bmatrix} 3.40 & 3.59 & 6.16 & 3.1 & 5.25 \\ 2.26 & 6.93 & 3.87 & 2.90 & 5.01 \\ 2.22 & 3.40 & 2.43 & 4.80 & 2.19 \\ 3.24 & 4.69 & 10.43 & 3.30 & 6.03 \\ 5.13 & 2.85 & 12.38 & 2.60 & 10.15 \\ 3.81 & 2.11 & 4.37 & 2.24 & 7.54 \\ 5.31 & 6.37 & 2.04 & 7.39 & 15.17 \end{bmatrix}, \tag{14.37}$$

and $D_{n,t} = D_{n,\tau}$, for all t, τ, with $(D_{n,1})_n$= (11.47, 10.98, 5.04, 17.70, 23.09, 10.01, 26.28).

In Figure 14.6, the evolution of the user payments is plotted throughout the algorithm iterations. It is obvious that users 3 and 6 (who are active according to Definition 1) can achieve strictly positive savings gain, while the rest of the users remain at their original payments. The disparate differences between consumption levels from slot 1 to slot 2 in case of user 3 and from slot 4 to slot 0 in case of user 6 render them perfect candidates to receive proactive caching. In addition, based on their willingness-to-pay values, the initial normalized entropy of their profiles is least amongst all users, thus CP is more certain about their future demand.

To confirm Theorem 7, we plot the total amount of proactive caching optimized for each user throughout the iterative algorithm evolution. Clearly, users 3 and 6 who achieved positive savings gain are receiving positive proactive caching service.

We also considered the scaling of system gains with the number of users. We run the simulation for $T = 4$ slots, with $q_{n,t} = q_t$, $\forall n$, and $(q_t)_t = (0.32, 0.9, 0.01, 0.95)$, and we consider $M = 4$ items. We take $B = 10^{-3}$, and generate $v_n(m)$ as follows $v_n(m) = 2 + e_n(m)$, where $e_n(m)$ is a realization of an exponential distribution with mean m. We also set $D_{n,t} = \sum_{m=1}^{M} e_n(m)$, $\forall t$. The amount of consumption for user n and time t is generated based on $1 - q_t$ as follows. If $q_t < 0.5$ then $S_{n,t}$ is uniformly distributed on [50, 70], and $S_{n,t}$ is uniformly distributed on [0, 10] if $q_t \geq 0.5$. The idea behind such a choice is to increase the likelihood of active users scaling linearly with N. We averaged the results over a multitude of simulation runs and plotted the profit and savings gain in Figures 14.8 and 14.9. Further, we also plot the results of the *finite-horizon* scenario, when cyclostationarity of demand profiles does not hold possibly because of fast varying willingness-to-pay values, thus the CP needs to optimize each cycle of T-slots independently from the previous one.

We can see from Figure 14.8 that the profit gain satisfies the established lower bound in Theorem 6 as it grows at least as N^2, where N is the number of users. Scaling of savings gain is shown in Figure 14.9 with an illustrating linear curve of $0.8N$ to give an insight on the growth rate of the aggregate savings. Obviously, there is a linear increase of such gains, which essentially promises constant saving shares per user. In addition, finite-horizon optimization attains considerable gains despite the limitation of independent resource allocation over cycles. In this result, the average savings gain per user are 30.2% for cyclostationary case, and 20.6% for finite-horizon case. Further, the percentage of average excess load due to proactive caching inaccuracies are 9.4% and 3.4% for the two respective cases. That is, the battery loss associated with proactive caching is proportional to ~ 9.4%, ~ 3.4% of additional load consumption.

In Figure 14.10, we study the impact of number of data items M when $N = 12$ users. As M grows, CP suffers more uncertainty and loses profit gain. However, savings gain is not significantly impacted since CP still needs to maintain a balance between high-pricing incentives, to combat against uncertainty, and service costs.

14.6 Conclusion

In this chapter, we have demonstrated the promise of achieving a win-win situation between CPs and end-users through proactive content caching supported with personalized data pricing. Through proactive caching, a CP can regulate its service load over time to minimize the peak-average by proactively caching future peak-hour demand during the off-peak hour. Smart data pricing assigns service prices to data content based on the individual end-users' preferences to render their demand more certain, hence improve the quality of caching performance. We have thus addressed the profit maximization problem for a CP that experiences time varying, yet predictable, demand which assumes high peak-to-average ratio. The CP has been allowed to employ joint smart pricing and proactive caching so as to balance its load-level over time, without enforcing users to physically change their regular demand activities. We have analyzed such a system under single- and multi-user setups and developed efficient algorithms that yield strictly improved profit gain for the CP and savings gain for the end-user as compared with the baseline scenario of no-proactive caching, i.e., a win-win situation. We have established bounds on the asymptotic scaling performance of the profit gain and showed that all users that receive proactive data services must attain positive savings gain.

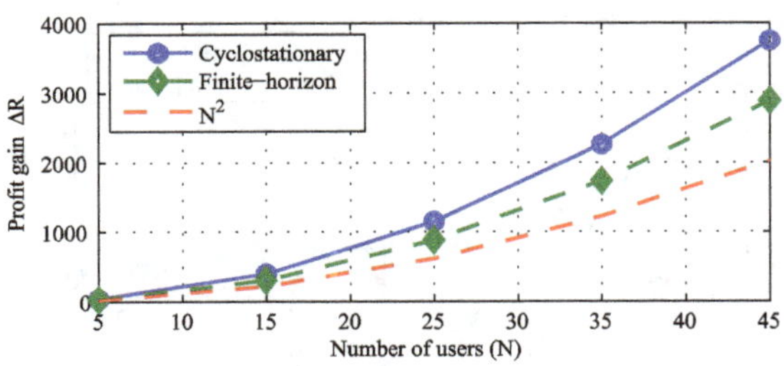

Figure 14.8 Expected profit gain grows with the number of users as N^2

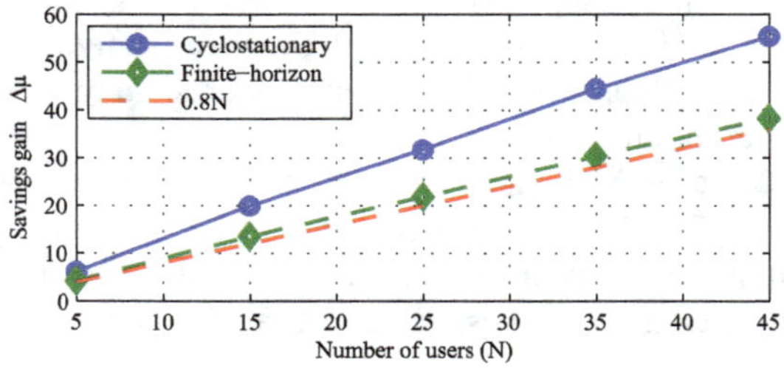

Figure 14.9 Expected savings gain vs. the number of users N

14.7 Proofs of main results

14.7.1 Proof of Theorem 2

By the monotonicity of $\{R^{(k)}\}_k$, we have $\Delta R \geq R(x^{(1)}, \hat{\mathbf{y}}) - R(\mathbf{0}, \hat{\mathbf{y}})$, i.e., $\Delta R \geq \eta(\mathbf{0}, \hat{\mathbf{y}}) - \eta(\mathbf{x}^{(1)}, \hat{\mathbf{y}})$. Further, we use a suboptimal proactive download strategy rather than $\mathbf{x}^{(1)}$. In such a strategy, we set $x_t(m) = \chi_t$ if $m = m_t^*$, and $t \in \mathscr{T}$, and $x_t(m) = 0$, otherwise. Hence,

$$\begin{aligned}
&\Delta R \geq \eta(\mathbf{0}, \hat{\mathbf{y}}) - \eta(\mathbf{x}, \hat{\mathbf{y}}) \\
&= \tfrac{1}{T}\sum_{t=0}^{T-1} \mathbb{E}\left[C\left(\sum_{m=1}^{M} S_t \hat{\mathbb{I}}_t(m)\right)\right] - \mathbb{E}\left[C\left(\sum_{m=1}^{M}(S_t - x_t(m))\hat{\mathbb{I}}_t(m) + x_{t+1}(m)\right)\right] \\
&\overset{(a)}{\geq} \tfrac{1}{T}\sum_{t=0}^{T-1} \mathbb{E}\left[C'\left(\sum_{m=1}^{M}(S_t - x_t(m))\hat{\mathbb{I}}_t(m) + x_{t+1}(m)\right)\left(\sum_{m=1}^{M} x_t(m)\hat{\mathbb{I}}_t(m) - x_{t+1}(m)\right)\right] \\
&\overset{(b)}{=} \tfrac{1}{T}\sum_{t=0}^{T-1}\sum_{m=1}^{M} x_t(m)\mathbb{E}\left[\hat{\mathbb{I}}_t(m)C'\left(\sum_{j=1}^{M}(S_t - x_t(j))\hat{\mathbb{I}}_t(j) + x_{t+1}(j)\right)\right. \\
&\left. - C'\left(\sum_{j=1}^{M}(S_{t-1} - x_{t-1}(j))\hat{\mathbb{I}}_{t-1}(j) + x_t(j)\right)\right] \\
&\overset{(c)}{\geq} \tfrac{1}{T}\sum_{t\in\mathscr{T}} \chi_t \left(\phi_{m_t^*,t}(\hat{y}_t(m_t^*))C'(S_t - \chi_t) - E\left[C'\left(\sum_{j=1}^{M} S_{t-1}\mathbb{I}_{t-1}(j) + \chi_t\right)\right]\right),
\end{aligned} \tag{14.38}$$

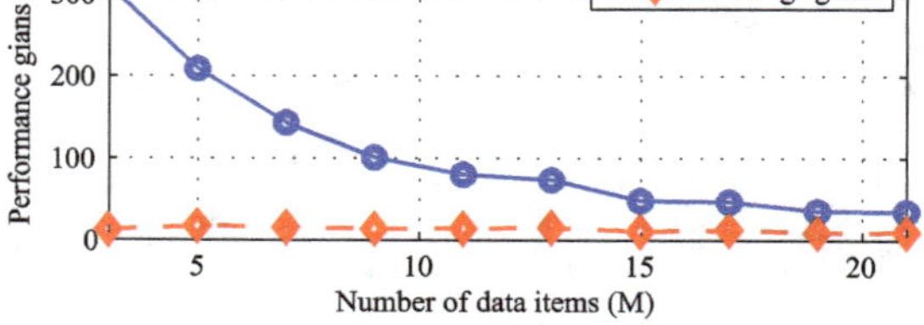

Figure 14.10 Impact of number of data items M *on performance gains*

Inequality (a) follows by the mean value theorem for random variables [28] while noting that C' is monotonically increasing and non-negative. Equality (b) follows by rearranging the terms. Inequality (c) follows since we omit the non-negative values of $x_{t+1}(j)$ and $x_{t-1}(j)$, $\forall j$ which can only reduce the RHS of Equality (b).

14.7.2 *Proof of Theorem 3*

Since $\max_m v(m) > \min_m v(m)$, then $\hat{y}_{t_0}(m_1) \neq \hat{y}_{t_0}(m_2)$ for some two items m_1, m_2. Hence, if $x^*_{t_0}(m_0) > 0$, we get $\max_m x^*_{t_0}(m) > \min_m x^*_{t_0}(m)$, which eventually yields $\mu(\mathbf{y}^*) < \mu(\mathbf{0}, \hat{\mathbf{y}})$. Now, we show that (14.19) is necessary and sufficient to have $x^*_{t_0}(m_0) > 0$.

($\Rightarrow$) Suppose that (14.19) holds. We set $x_t(m) = 0$ for all $(m, t) \neq (m_0, t_0)$. Consider the difference between the expected costs,

$$\sum_{t=0}^{T-1}\sum_{m=1}^{M}\left(C(S_t) - C\left(S_t - x_t(m) + \sum_j x_{t+1}(j)\right)\right)\phi_{m,t}(\hat{y}_t(m))$$
$$+ C(0) - C\left(\sum_j x_{t+1}(j)\right) q_t$$
$$= (C(S_{t_0}) - C(S_{t_0} - x_{t_0}(m_0)))\phi_{m_0,t_0}(\hat{y}_{t_0}(m))$$
$$+ \left(C(S_{t_0-1}) - C(S_{t_0-1} + x_{t_0}(m_0))\right)(1 - q_{t_0-1}) + (C(0) - C(x_{t_0}(m_0)))q_{t_0-1} \geq x_{t_0}(m_0)G(x_{t_0}(m_0)), \tag{14.39}$$

where

$$\begin{aligned} G(x_{t_0}(m_0)) :&= C'(S_{t_0} - x_{t_0}(m_0))\phi_{m_0,t_0}(\hat{y}_{t_0}(m_0)) \\ &-C'(x_{t_0}(m_0))q_{t_0-1} - C'(S_{t_0-1} + x_{t_0}(m_0))(1 - q_{t_0-1}). \end{aligned} \tag{14.40}$$

The last inequality follows by the mean value theorem.

Since G is decreasing in $x_{t_0}(m_0)$, and $G(0) = C'(S_{t_0})\phi_{m_0,t_0}(\hat{y}_{t_0}(m_0)) - C'(S_{t_0-1})(1 - q_{t_0-1}) > 0$ by hypothesis. By the continuity of C', there exists $x > 0$ for which $xG(x) > 0$. That is, $x_{t_0}(m_0) > 0$ will yield a strictly reduced cost as compared to the no-proactive download scenario.

($\Leftarrow$) Suppose that $\mathbf{x}^* > 0$. Further, suppose toward contradiction (14.19) does not hold. Since $\mathbf{x}^* > 0$, the cost reduction due to proactive service can only be upper bounded by a positive value. That is,

$$\sum_{t=0}^{T-1}\sum_{m=1}^{M}\left(C(S_t) - C\left(S_t - \hat{x}_t(m) + \sum_j \hat{x}_{t+1}(j)\right)\right)\phi_{m,t}(\hat{y}_t(m)) - C\left(\sum_j x^*_{t+1}(j)\right) q_t$$
$$\leq \sum_{t=0}^{T-1}\sum_{m=1}^{M} C'(S_t)\left(x^*_t(m) - \sum_j x^*_{t+1}(j)\right)\phi_{m,t}(\hat{y}_t(m))$$
$$= \sum_{t=0}^{T-1}\sum_{m=1}^{M} x^*_t(m)(\phi_{m,t}(\hat{y}_t(m))C'(S_t) - (1 - q_{t-1})C'(S_{t-1})), \tag{14.41}$$

hence if (14.19) does not hold for any m or t, the RHS of the last expression will be non-positive, which contradicts the hypothesis that $\mathbf{x}^* > 0$, with positive cost reduction. Therefore (14.19) must hold.

14.7.3 Proof of Theorem 4

First, we note that $\tilde{R}^{(k)}$ is concave in $(\mathbf{x}, \mathbf{y})$ since $\mu(\mathbf{y})$ is a quadratic form of $\mathbf{y}$ with a negative definite Hessian, and $-\frac{1}{T}\sum_{t=0}^{T-1}\sum_{m,n} h^{(k)}_{n,t}(m)y_n, t(m)$ is an affine function of $\mathbf{y}$. Finally, $\eta(\mathbf{x}, \mathbf{y}^{(k-1)})$ is a convex function of $\mathbf{x}$ by the definition of the cost function C.

Second, we consider the three conditions specified in Lemma 2. Since R is continuous in $(\mathbf{x}, \mathbf{y})$ and is defined over a bounded and closed feasible set, then it has a global maximum value $U > 0$. Such a value can be subtracted from $\tilde{R}^{(k)}$ as defined to keep Condition (1) of Lemma 2 satisfied. However, subtracting a constant from the objective function does not affect the solution, which is the main point of interest. Therefore, Condition (1) of Lemma 2 is not necessary in this case. For Condition (2) of Lemma 2, we have

$$\left.\frac{\partial \tilde{R}^{(k)}(\mathbf{x},\mathbf{y})}{\partial x_{n,t}(m)}\right|_{(\mathbf{x}^{(k-1)},\mathbf{y}^{(k-1)})} \begin{aligned} &= \left.\frac{\partial \eta(\mathbf{x},\mathbf{y}^{(k-1)})}{\partial x_{n,t}(m)}\right|_{(\mathbf{x}^{(k-1)},\mathbf{y}^{(k-1)})} \\ &= \left.\frac{\partial R(\mathbf{x},\mathbf{y})}{\partial x_{n,t}(m)}\right|_{(\mathbf{x}^{(k-1)},\mathbf{y}^{(k-1)})}, \quad \forall m,n,t. \end{aligned} \tag{14.42}$$

Likewise,

$$\left.\frac{\partial \tilde{R}^{(k)}(\mathbf{x},\mathbf{y})}{\partial y_{n,t}(m)}\right|_{(\mathbf{x}^{(k-1)},\mathbf{y}^{(k-1)})} \begin{aligned} &= \frac{\partial \mu(\mathbf{y})}{\partial y_{n,t}(m)} - \left.\frac{\partial \eta(\mathbf{x}^{(k-1)},\mathbf{y})}{\partial y_{n,t}(m)}\right|_{(\mathbf{x}^{(k-1)},\mathbf{y}^{(k-1)})} \\ &= \left.\frac{\partial R(\mathbf{x},\mathbf{y})}{\partial y_{n,t}(m)}\right|_{(\mathbf{x}^{(k-1)},\mathbf{y}^{(k-1)})}, \quad \forall m,n,t. \end{aligned} \tag{14.43}$$

Thus Condition (2) of Lemma 2 is satisfied.

Finally, Condition (3) of the same lemma need not be satisfied since it is mainly stated in Theorem 1 [27] for a non-convex constraint function that has to be replaced by a convex approximate. Condition (3) mainly implies the satisfaction of the complementary slackness conditions by both the approximate and the original constraint functions. Since we are interested only in the objective function, Condition (3) of Lemma 2 is not necessary for convergence to a KKT point.

14.7.4 Proof of Theorem 5

We have $\Delta R \geq R(\mathbf{x}^{(1)}, \hat{\mathbf{y}}) - R(\mathbf{0}, \hat{\mathbf{y}})$, where $\mathbf{x}^{(1)}$ is the proactive caching solution to the first iteration of the proposed multi-user algorithm when the starting point is $(\mathbf{0}, \hat{\mathbf{y}})$. Thus, $\mathbf{x}^{(1)}$ is also the minimizer of $\eta(\mathbf{x}, \hat{\mathbf{y}})$ under the constraints of (14.23).

Hence, we have $\Delta R \geq \eta(\mathbf{0}, \hat{\mathbf{y}}) - \eta(\mathbf{x}^{(1)}, \hat{\mathbf{x}})$. Further, suppose we use the sub-optimal proactive caching policy $\tilde{\mathbf{x}}$ rather than $\mathbf{x}^{(1)}$ at the first iteration, whereby $\tilde{x}_{n,t}(m) := \tilde{x}_t, n \in \mathcal{B}_t(m)$, and $\tilde{x}_{n,t}(m) = 0$ otherwise. Hence, the profit gain satisfies

$$\begin{aligned}
\Delta R &\geq \frac{1}{T}\sum_{t=0}^{T-1}\mathbb{E}\left[C(\hat{L}_t) - C\left(\hat{L}_t - \sum_{m=1}^{M}\sum_{n\in\mathcal{B}_t(m)}\tilde{x}_t\hat{\mathbb{I}}_{n,t}(m) + \sum_{m=1}^{M}\sum_{n\in\mathcal{B}_{t+1}(m)}\tilde{x}_{t+1}\right)\right]\\
&\overset{(a)}{\geq} \frac{1}{T}\sum_{t=0}^{T-1}\mathbb{E}\left[C'\left(\hat{L}_t - \sum_{m=1}^{M}\sum_{n\in\mathcal{B}_t(m)}\tilde{x}_t\hat{\mathbb{I}}_{n,t}(m) + \sum_{m=1}^{M}\sum_{n\in\mathcal{B}_{t+1}(m)}\tilde{x}_{t+1}\right)\right.\\
&\quad\left.\times\left(\sum_{m=1}^{M}\sum_{n\in\mathcal{B}_t(m)}\tilde{x}_t\hat{\mathbb{I}}_{n,t}(m) - \sum_{m=1}^{M}\sum_{n\in\mathcal{B}_{t+1}(m)}\tilde{x}_{t+1}\right)\right]\\
&\overset{(b)}{\geq} \frac{1}{T}\sum_{t=0}^{T-1}\sum_{m=1}^{M}\tilde{x}_t\sum_{n\in\mathcal{B}_t(m)}\mathbb{E}\left[\hat{\mathbb{I}}_{n,t}(m)C'\left(\hat{L}_t - \sum_{j=1}^{M}\sum_{k\in\mathcal{B}_t(k)}\tilde{x}_t\hat{\mathbb{I}}_{k,t}(j)\right)\right.\\
&\quad\left.-C'\left(\hat{L}_{t-1} + \sum_{j=1}^{M}\sum_{k\in\mathcal{B}_t(j)}\tilde{x}_t\right)\right] > 0.
\end{aligned} \tag{14.44}$$

Inequality (a) holds by the first order condition on the convexity of the cost function C. Inequality (b) follows by rearranging the terms of the RHS of Inequality (a) and replacing the terms $\sum_{j=1}^{M}\sum_{k\in\mathcal{B}_{t+1}(j)}\tilde{x}_{t+1}, -\sum_{j=1}^{M}\sum_{k\in\mathcal{B}_{t-1}(k)}\tilde{x}_{t-1}\mathbb{I}_{k,t-1}(j)$ with zeros while noting that C' is monotonically increasing on its domain. Finally, the last strict inequality holds by the definition of $\tilde{x}_t$ in Theorem 5 and the definition of active users.

14.7.5 Proof of Theorem 6

We have $\Delta R \geq R(\mathbf{x}^{(1)}, \hat{\mathbf{y}}) - R(\mathbf{0}, \hat{\mathbf{y}})$, this reduces to $\Delta R(N) \geq \eta(\mathbf{0}, \hat{\mathbf{y}}) - \eta(\mathbf{x}^{(1)}, \hat{\mathbf{y}})$, the cost reduction after the first iteration of the multi-user iterative algorithm. To show that $\eta(\mathbf{0}, \hat{\mathbf{y}}) - \eta(\mathbf{x}^{(1)}, \hat{\mathbf{y}}) = \Omega(g(N)C'(\gamma N))$, we consider two steps. In the first step, we show that if $\liminf_{N\to\infty}\tilde{x}_t > 0$ then

$$\liminf_{N\to\infty}\frac{1}{g(N)C'(\gamma\cdot N)}\sum_{n\in\mathcal{B}_t(m)}\mathbb{E}\left[\hat{\mathbb{I}}_{n,t}(m)C'\left(\hat{L}_t - \sum_{j=1}^{M}\sum_{k\in\mathcal{B}_t(j)}\tilde{x}_t\hat{\mathbb{I}}_{k,t}(j)\right)\right] > 0, \tag{14.45}$$

for some $\gamma > 0$. In the other step, we prove that $\liminf_{N\to\infty}\tilde{x}_t > 0$, and

$$\liminf_{N\to\infty}\frac{\mathbb{E}\left[\sum_{n\in\mathcal{B}_t(m)}\hat{\mathbb{I}}_{n,t}(m)C'\left(\hat{L}_t - \sum_{j=1}^{M}\sum_{k\in\mathcal{B}_t(j)}\tilde{x}_t\hat{\mathbb{I}}_{k,t}(j)\right)\right]}{\sum_{n\in\mathcal{B}_t(m)}\mathbb{E}\left[C'\left(\hat{L}_{t-1} + \sum_{j=1}^{M}\sum_{k\in\mathcal{B}_t(j)}\tilde{x}_t\right)\right]} > 1. \tag{14.46}$$

Step 1: Suppose that $\liminf_{N\to\infty} \tilde{x}_t > 0$. By Fubini's theorem, we can move the summation inside the expectation, since all the summands are non-negative. Also, by Fatou's lemma, we have

$$\begin{aligned}
&\liminf_{N\to\infty}\frac{1}{g(N)C'(\gamma\cdot N)} \sum_{n\in\mathscr{B}_t(m)} \mathbb{E}\left[\hat{\mathbb{I}}_{n,t}(m)C'\left(\hat{L}_t - \sum_{j=1}^{M}\sum_{k\in\mathscr{B}_t(j)} \tilde{x}_t\hat{\mathbb{I}}_{k,t}(j)\right)\right] \\
&\geq \mathbb{E}\left[\liminf_{N\to\infty}\frac{\sum_{n\in\mathscr{B}_t(m)}\hat{\mathbb{I}}_{n,t}(m)}{g(N)} \times \liminf_{N\to\infty}\frac{C'\left(\sum_{j=1}^{M}\sum_{k=1,k\neq n}^{N}(S_{k,t}-\tilde{x}_t)\hat{\mathbb{I}}_{k,t}(j)\right)}{C'(\gamma N)}\right] \\
&\overset{(a)}{=} \liminf_{N\to\infty}\frac{\sum_{n\in\mathscr{B}_t(m)}P_{n,t}(m)}{g(N)} \times\mathbb{E}\left[\liminf_{N\to\infty}\frac{C'\left(N\cdot\sum_{j=1}^{M}\frac{\sum_{k=1,k\neq n}^{N}(S_{k,t}-\tilde{x}_t)\hat{\mathbb{I}}_{k,t}(j)}{N}\right)}{C'(\gamma N)}\right].
\end{aligned} \tag{14.47}$$

Note that, on the LHS of Equality (a), we have removed the contribution of user n in the argument of C' which can only reduce its value, thus yielding $\hat{\mathbb{I}}_{n,t}(m)$ independent of the argument of C'. Hence, on the RHS, we split the expectation over the product. But $1 - q_{n,t} > \varepsilon$ for all n,t, for any $n \in \mathscr{B}_t(m)$, and $\beta_t(m) > 0$ by hypothesis.

$$\liminf_{N\to\infty}\sum_{n\in\mathscr{B}_t(m)}\frac{P_{n,t}(m)}{g(N)} = \beta_t(m)\liminf_{B_t(m)\to\infty}\sum_{n\in\mathscr{B}_t(m)}\frac{P_{n,t}(m)}{B_t(m)} > 0 \tag{14.48}$$

On the other hand, Kolmogorov's strong law of large numbers implies

$$\begin{aligned}
\gamma :&= \lim_{N\to\infty}\sum_{j=1}^{M}\frac{\sum_{k=1,k\neq n}^{N}(S_{k,t}-\tilde{x}_t)\hat{\mathbb{I}}_{k,t}(j)}{N} \\
&\overset{a.s.}{=} \lim_{N\to\infty}\sum_{j=1}^{M}(S_{k,t}-\tilde{x}_t)\frac{\sum_{k\neq n}P_{k,t}(j)}{N} > 0,
\end{aligned} \tag{14.49}$$

since $q_{n,t} < 1, \forall n, t$. Hence, we have

$$\begin{aligned}
&\mathbb{E}\left[\liminf_{N\to\infty}\frac{\sum_{n\in\mathscr{B}(m)}P_{n,t}(m)}{g(N)} \times \liminf_{N\to\infty}\frac{C'\left(N.\sum_{j=1}^{M}\frac{\sum_{k=1,k\neq n}^{N}(S_{k,t}-\tilde{x}_t)\mathbb{I}_{k,t}(j)}{N}\right)}{C'(\gamma N)}\right] \\
&\geq \beta_t(m).\liminf_{B_t(m)\to\infty}\sum_{n\in\mathscr{B}_t(m)}\frac{P_{n,t}(m)}{B_t(m)} > 0.
\end{aligned} \tag{14.50}$$

Step 2: In this step, we prove that there exists $\chi > 0$, independent of N, for which if $\tilde{x}_t = \chi$, then

$$\liminf_{N\to\infty} \frac{\sum\limits_{n\in\mathscr{B}_t(m)} \mathbb{E}\left[\hat{\mathbb{I}}_{n,t}(m) C'\left(\hat{L}_t - \sum\limits_{j=1}^{M}\sum\limits_{k\in\mathscr{B}_t(j)} \chi \hat{\mathbb{I}}_{n,t}(j)\right)\right]}{\sum\limits_{n\in\mathscr{B}_t(m)} \mathbb{E}\left[C'\left(\hat{L}_{t-1} + \sum\limits_{j=1}^{M}\sum\limits_{k\in\mathscr{B}_t(j)} \chi\right)\right]} > 1. \tag{14.51}$$

We set $\tilde{x}_t = \chi$, independent of N, and we will prove that $0 < \chi < \check{S}$. By Fubini's theorem and Fatou's lemma, as in ***Step 1***, it suffices to prove that

$$\liminf_{N\to\infty} \frac{\sum\limits_{n\in\mathscr{B}_t(m)} \mathbb{E}\left[\hat{\mathbb{I}}_{n,t}(m) C'\left(\hat{L}_t - \sum\limits_{j=1}^{M}\sum\limits_{k\in\mathscr{B}_t(j)} \chi \hat{\mathbb{I}}_{n,t}(j)\right)\right]}{\sum\limits_{n\in\mathscr{B}_t(m)} \mathbb{E}\left[C'\left(\hat{L}_{t-1} + \sum\limits_{j=1}^{M}\sum\limits_{k\in\mathscr{B}_t(j)} \chi\right)\right]} > 1. \tag{14.52}$$

for some $0 < \chi < \check{S}$.

By Condition (14.30), we have

$$\begin{aligned}
&\liminf_{N\to\infty} \frac{\mathbb{E}\left[\sum\limits_{n\in\mathscr{B}_t(m)} \hat{\mathbb{I}}_{n,t}(m) C'\left(\hat{L}_t - \sum\limits_{j=1}^{M}\sum\limits_{k\in\mathscr{B}_t(j)} \chi \hat{\mathbb{I}}_{k,t}(j)\right)\right]}{\mathbb{E}\left[\sum\limits_{n\in\mathscr{B}_t(m)} C'\left(\hat{L}_{t-1} + \sum\limits_{j=1}^{M}\sum\limits_{k\in\mathscr{B}_t(j)} \chi\right)\right]} \\
&= \liminf_{N\to\infty} \frac{\mathbb{E}\left[\sum\limits_{n\in\mathscr{B}_t(m)} \hat{\mathbb{I}}_{n,t}(m) \left(\hat{L}_t - \sum\limits_{j=1}^{M}\sum\limits_{k\in\mathscr{B}_t(j)} \chi \hat{\mathbb{I}}_{k,t}(j)\right)^{\delta}\right]}{\mathbb{E}\left[\sum\limits_{n\in\mathscr{B}_t(m)} \left(\hat{L}_{t-1} + \sum\limits_{j=1}^{M}\sum\limits_{k\in\mathscr{B}_t(j)} \chi\right)^{\delta}\right]} \\
&\overset{(a)}{\geq} \liminf_{N\to\infty} \frac{\mathbb{E}\left[\sum\limits_{n\in\mathscr{B}_t(m)} \hat{\mathbb{I}}_{n,t}(m) \left(\sum\limits_{j=1}^{M}\sum\limits_{k=1}^{N} (S_{k,t} - \chi) \hat{\mathbb{I}}_{k,t}(j)\right)^{\delta}\right]}{\mathbb{E}\left[\sum\limits_{n\in\mathscr{B}_t(m)} \left(\hat{L}_{t-1} + \sum\limits_{j=1}^{M}\sum\limits_{k=1}^{N} \chi\right)^{\delta}\right]} \\
&\overset{(b)}{=} \liminf_{N\to\infty} \frac{\mathbb{E}\left[\sum\limits_{n\in\mathscr{B}_t(m)} \frac{\hat{\mathbb{I}}_{n,t}(m)}{B_t(m)} \left(\sum\limits_{j=1}^{M}\sum\limits_{k=1}^{N} \frac{(S_{k,t} - \chi) \hat{\mathbb{I}}_{k,t}(j)}{N}\right)^{\delta}\right]}{\mathbb{E}\left[\left(\sum_{j=1}^{M}\sum_{k=1}^{N} \frac{S_{k,t-1}\hat{\mathbb{I}}_{k,t-1}(j)}{N} + M\chi\right)^{\delta}\right]},
\end{aligned} \tag{14.53}$$

where Inequality (a) follows since we extend the negative sum $-\sum_{j=1}^{M}\sum_{k\in\mathscr{B}_t(j)} \chi\hat{\mathbb{I}}_{n,t}(j)$ to include the terms from outside set $\mathscr{B}_t(j), j = 1, \dots M$, while noting that C' is monotonically increasing. Further, we increase denominator by extending the sum to

$\sum_{j,k}\chi$. Equality (b) follows through multiplying both the numerator and denominator by $N^{\delta}B_t(m)$.

Now, by Kolmogorov's strong law of large numbers, we define the quantities:

$$\begin{aligned} c_1(m) &:= \liminf_{N\to\infty}\sum_{n\in\mathscr{B}_t(m)}\frac{\hat{\mathbb{I}}_{n,t}(m)}{B_t(m)} \overset{a.s.}{=} \liminf_{N\to\infty}\sum_{n\in\mathscr{B}_t(m)}\frac{P_{n,t}(m)}{B_t(m)}, \\ c_2(j) &:= \limsup_{N\to\infty}\sum_{k=1}^{N}\frac{\hat{\mathbb{I}}_{k,t-1}(j)}{N} \overset{a.s.}{=} \limsup_{N\to\infty}\sum_{k=1}^{N}\frac{P_{k,t-1}(j)}{N}, \\ c_3(j) &:= \liminf_{N\to\infty}\sum_{k=1}^{N}\frac{\hat{\mathbb{I}}_{k,t}(j)}{N} \overset{a.s.}{=} \liminf_{N\to\infty}\sum_{k=1}^{N}\frac{P_{k,t}(j)}{N}. \end{aligned} \tag{14.54}$$

Since $1-q_{n,\tau}<\epsilon,\ \forall n,\tau,\ \beta_t(m)>0$, and $\liminf_{N\to\infty}\check{P}_t(m),\ \forall k\in\mathscr{B}_t(m)$, we conclude $c_1(m)\overset{a.s.}{>}0,\ c_2(j)\overset{a.s.}{>}0,\ j=1,\cdots,M$, and $c_3(j)\overset{a.s.}{>}0 j=1,\cdots,M$. Hence,

$$\begin{aligned} &\frac{\mathbb{E}\left[\liminf_{N\to\infty}\sum_{n\in\mathscr{B}_t(m)}\frac{\hat{\mathbb{I}}_{n,t}(m)}{B_t(m)}\left(\sum_{j=1}^{M}\sum_{k=1}^{N}\frac{(S_{k,t}-\chi)\hat{\mathbb{I}}_{k,t}(j)}{N}\right)^{\delta}\right]}{\mathbb{E}\left[\limsup_{N\to\infty}\left(\sum_{j=1}^{M}\sum_{k=1}^{N}\frac{S_{k,t-1}\hat{\mathbb{I}}_{k,t-1}(j)}{N}+M\chi\right)^{\delta}\right]} \\ &\overset{(c)}{=} \frac{c_1(m)\left(\sum_{j=1}^{M}(S_3-\chi)c_3(j)\right)^{\delta}}{\left(M\chi+\sum_{j=1}^{M}S_2c_2(j)\right)^{\delta}}, \end{aligned} \tag{14.55}$$

where $S_3 := \liminf_{N\to\infty}\sum_{k=1}^{N}S_{k,t}/N$, and $S_2 := \limsup_{N\to\infty}\sum_{k=1}^{N}S_{k,t-1}/N$. The RHS of Equality (c) is strictly greater than 1 if and only if

$$\chi < \frac{\left(c_1(m)^{\frac{1}{\delta}}\sum_{j=1}^{M}c_3(j)S_3\right)\%-\sum_{j=1}^{M}c_2(j))S_2}{\left(M+\sum_{j=1}^{M}c_3(j)\right)}. \tag{14.56}$$

Now, to show that $\chi>0$, it suffices to prove that

$$\left(c_1(m)^{\frac{1}{\delta}}\sum_{j=1}^{M}c_3(j)S_3\right)-\sum_{j=1}^{M}c_2(j)S_2>0. \tag{14.57}$$

We have by the definition of set $\mathscr{B}_t(m)$ that

$$\sum_{n\in\mathscr{B}_t(m)}\mathbb{E}\left[\hat{\mathbb{I}}_{n,t}(m)C'(\hat{L}_t)-C'(\hat{L}_{t-1})\right]>0. \tag{14.58}$$

By Condition (14.30), consider $n\in\mathscr{B}_t(m)$, for sufficiently large N, we obtain

$$P_{n,t}(m) > \mathbb{E}\left[\left(\hat{L}_{t-1}\right)^{\delta}\right] / \mathbb{E}\left[\left(\hat{S}+\hat{L}_t\right)^{\delta}\right] \tag{14.59}$$

which implies $\sum_{n\in\mathscr{B}} \frac{P_{n,t}(m)}{B_t(m)} > \frac{\mathbb{E}\left[(\hat{L}_{t-1})^{\delta}\right]}{\mathbb{E}\left[(\hat{S}+\hat{L}_t)^{\delta}\right]}$

$$\liminf_{N\to\infty} \sum_{n\in\mathscr{B}} \frac{P_{n,t}(m)}{B_t(m)} > \left(\frac{\left(\sum_{j=1}^{M} c_2(j)S_2\right)}{\left(\sum_{j=1}^{M} c_3(j)S_3\right)}\right)^{\delta}, \tag{14.60}$$

by (14.31). Hence, $\left(c_1(m)\right)^{\frac{1}{\delta}} \frac{\left(\sum_{j=1}^{M} c_3(j)S(j)\right)}{\left(\sum_{j=1}^{M} c_2(j)S(j)\right)} > 1$.

14.7.6 Proof of Theorem 7

The condition $\max_m v_{n_0}(m) > \min_m v_{n_0}(m)$ ensures $\{y^*_{n_0,t}(m)\}$ are not identical, hence $\{x^*_{n_0,t_0}(m)\}_m$ are also non-identical if $x^*_{n_0,t_0}(m_0) > 0$.

($\Rightarrow$)Suppose that $x^*_{n_0,t_0}(m_0) > 0$. Under the multi-user iterative algorithm, since $(\mathbf{x}^*, \mathbf{y}^*)$ is the limit point of the sequence of solutions generated by the algorithm, then we can write

$$\begin{aligned}
\mathbf{y}^* &= \arg\max_{\mathbf{y}\in\mathscr{Y}} \mu(\mathbf{y}) - \frac{1}{T}\sum_{t=0}^{T-1}\sum_{n=1}^{N}\sum_{m=1}^{M} h_{n,t}(m)y_{n,t}(m) \\
&= \arg\max_{\mathbf{y}\in\mathscr{Y}} \sum_{n=1}^{N} \mu_n(\mathbf{y}) + \frac{1}{T}\sum_{t=0}^{T-1}\sum_{m=1}^{M} y_{n,t}(m)\mathbb{E}\Bigg[C\Bigg(S_{n,t} - x^*_{n,t}(m) \\
&\quad + \sum_{k\neq n}\sum_{j=1}^{M}(S_{k,t} - x^*_{k,t}(j))\mathbb{I}^*_{k,t}(j) + \sum_{k,j} x^*_{k,t+1}(j)\Bigg)\Bigg].
\end{aligned} \tag{14.61}$$

Now, that $x^*_{n_0,t_0}(m_0) > 0$, the superposition

$$\frac{1}{T}\sum_{t=0}^{T-1}\sum_{m=1}^{M} y_{n_0,t}(m)\mathbb{E}\left[C\left(S_{n_0,t} - x^*_{n_0,t}(m) + \sum_{k\neq n}\sum_{j=1}^{M}(S_{k,t} - x^*_{k,t}(j))\mathbb{I}^*_{k,t}(j) + \sum_{k,j} x^*_{k,t+1}(j)\right)\right] \tag{14.62}$$

is dependent on $\{y_{n_0,t_0}(m)\}_m$, thus, yielding the pricing policy $y^*_{n_0,t_0}(m) \neq \hat{y}_{n_0,t_0}(m)$, $\forall_m$. But $\mu_n(\mathbf{y})$ is strictly concave in $\mathbf{y}$, and $\hat{\mathbf{y}}$ is its unique maximizer. Thus, $\mu_n(\mathbf{y}^*) < \mu_n(\hat{\mathbf{y}})$.

($\Leftarrow$)If $x^*_{n_0,t}(m) = 0$, $\forall m$, then the superposition (14.61) is independent of $\{y_{n_0,t}(m)\}$, $\forall m, t$. Hence, $y^*_{n_0,t}(m) = \hat{y}_{n_0,t}(m)$, and $\Delta\mu_n = 0$.

14.7.7 Proof of Theorem 8

In Theorem 5, we have proved the sufficiency part. Here prove the necessity part, that is $\Delta R > 0$ and $\Delta\mu > 0$ imply that $\mathscr{B}_t(m)$ is non-empty for some time slot and data item.

($\Leftarrow$)Suppose that $\mathscr{B}_t(m)$ is empty for all m, t. We have the proposed algorithm $R(\mathbf{x}^{(1)}, \hat{\mathbf{y}}) \geq R(\mathbf{0}, \hat{\mathbf{y}})$, with equality if and only if $\mathbf{x}^{(1)} = \mathbf{0}$. So, suppose toward contradiction $\mathbf{x}^{(1)} \neq 0$. Hence, $\eta(\mathbf{0}, \hat{\mathbf{y}}) - \eta(\mathbf{x}^{(1)}, \hat{\mathbf{y}})$. By the mean value theorem for random variables, we have

$$\begin{aligned}\eta(\mathbf{0}, \hat{\mathbf{y}}) \quad & -\eta(\mathbf{x}^{(1)}, \hat{\mathbf{y}}) \leq \frac{1}{T}\sum_{t=0}^{T-1} \mathbb{E}\left[C'\left(\hat{L}_t\right)\left(\sum_{m=1}^{M}\sum_{n=1}^{N} x_{n,t}^{(1)}(m)\hat{\mathbb{I}}_{n,t}(m)\right.\right. \\ & \left.\left. - \sum_{m=1}^{M}\sum_{n=1}^{N} x_{n,t+1}^{(1)}(m)\right)\right] \\ & \overset{(a)}{=} \frac{1}{T}\sum_{t=0}^{T-1} \mathbb{E}\left[\sum_{m=1}^{M}\sum_{n=1}^{N} x_{n,t}^{(1)}(m)(\hat{\mathbb{I}}_{n,t}(m)C'(\hat{L}_t) - C'(\hat{L}_{t-1}))\right] \\ & = \frac{1}{T}\sum_{t=0}^{T-1}\sum_{m=1}^{M}\sum_{n=1}^{N} x_{n,t}^{(1)}(m)\mathbb{E}\left[\mathbb{I}_{n,t}(m)C'(\hat{L}_t) - C'(\hat{L}_{t-1})\right] \\ & \overset{(b)}{\leq} \frac{1}{T}\sum_{t=0}^{T-1}\sum_{m=1}^{M}\sum_{n\in\mathscr{B}_t(m)}^{N} S_{n,t}\mathbb{E}\left[\hat{\mathbb{I}}_{n,t}(m)C'(\hat{L}_t) - C'(\hat{L}_{t-1})\right].\end{aligned} \tag{14.63}$$

Equality (a) follows by rearranging the terms, and Inequality (b) follows by summing only over the elements of $\mathscr{B}_t(m)$ which can only increase the RHS. Now, since $\mathscr{B}_t(m)$ is empty for all m, t, it turns out that the RHS of (b) is zero, which is a contradiction. Finally, with $\mathbf{x}^{(1)} = \mathbf{0}$, we have $\mathbf{y}^{(2)} = \hat{\mathbf{y}}$, yielding $\mathbf{x}^{(2)} = \mathbf{0}$, and so on, leading to zero profit and savings gains.

References

[1] Niyato D., Hossain E. 'Competitive pricing for spectrum sharing in cognitive radio networks: dynamic game, inefficiency of NASH equilibrium, and collusion'. *IEEE Journal on Selected Areas in Communications*. 2008 Jan;**26**(1):192–202.

[2] Paschalidis I.C., Tsitsiklis J.N. 'Congestion-dependent pricing of network services'. *IEEE/ACM Transactions on Networking*. 2000 April;**8**(2):171–84.

[3] Viterbo E., Chiasserini C.F. 'Dynamic pricing for connection-oriented services in wireless networks'. *12th IEEE International Symposium on Personal, Indoor and Mobile Radio Communications. PIMRC 2001. Proceedings (Cat. No.01TH8598)*. **1**; 2001.

[4] Heidarpour B., Dziong Z., Lau W.C., Vakilinia S. 'Pricing the volume-based data services in cellular wireless markets'. *IEEE Transactions on Network and Service Management*. 2018 March;**15**(1):260–74.

[5] Tao X., Wu J., Lu J. 'Dynamic pricing strategy for delay tolerant service aggregation multicast in wireless networks'. 2013 IEEE Wireless Communications and Networking Conference Workshops (WCNCW); 2013. pp. 178–82.

[6] Ha S., Sen S., Joe-Wong C. 'TUBE: time-dependent pricing for mobile data'. Proceedings of the ACM SIGCOMM 2012 Conference on Applications, Technologies, Architectures, and Protocols for Computer Communication. SIGCOMM '12; New York, NY, USA: ACM; 2012. pp. 247–58.

[7] Sen S., Joe-Wong C., Ha S., Chiang M. 'Time-dependent pricing for multimedia data traffic: analysis, systems, and trials'. *IEEE Journal on Selected Areas in Communications*. 2019 July;**37**(7):1504–17.

[8] Lee J., Yi Y., Chong S. 'Economics of WiFi offloading: trading delay for cellular capacity'. 2013 IEEE Conference on Computer CommunicationsWorkshops (INFOCOM WKSHPS); 2013. pp. 357–62.

[9] Gao L., Iosifidis G., Huang J., *et al.* 'Economics of mobile data offloading'. 2013 Proceedings IEEE INFOCOM; 2013. pp. 3303–8.

[10] Wang K., Lau F.C.M., Chen L., Schober R. 'Pricing mobile data offloading: a distributed market framework'. *IEEE Transactions on Wireless Communications*. 2016 Feb;**15**(2):913–27.

[11] Kostopoulos A., Papafili I., Fortetsanakis G., *et al.* 'Pricing wireless access services: the effect of offloading and users' bounded rationality'. 2016 12th Annual Conference on Wireless On-demand Network Systems and Services (WONS); 2016. pp. 1–4.

[12] Poularakis K., Iosifidis G., Tassiulas L. 'Joint deployment and pricing of next-generation WiFi networks'. *IEEE Transactions on Communications*. 2019 Sep;**67**(9):6193–205.

[13] Tadrous J., Eryilmaz A., Gamal H.E. 'Proactive resource allocation: harnessing the diversity and multicast gains'. *IEEE Transactions on Information Theory*. 2013 Aug;**59**(8):4833–54.

[14] Tadrous J., Eryilmaz A., El Gamal H. 'Proactive content download and user demand shaping for data networks'. *IEEE/ACM Transactions on Networking*. 2015 Dec;**23**(6):1917–30.

[15] Huang L., Zhang S., Chen M., Liu X. 'When backpressure meets predictive scheduling'. *IEEE/ACM Transactions on Networking*. 2016 Aug;**24**(4):2237–50.

[16] Tadrous J., Eryilmaz A. 'On optimal proactive caching for mobile networks with demand uncertainties'. *IEEE/ACM Transactions on Networking*. 2016 October;**24**(5):2715–27.

[17] Zhang S., Huang L., Chen M., Liu X. 'Proactive serving decreases user delay exponentially: the light-tailed service time case'. *IEEE/ACM Transactions on Networking*. 2017;**25**(2):708–23.

[18] Tadrous J., Eryilmaz A., Gamal H.E. 'Joint pricing and proactive caching for data services: global and user-centric approaches'. 2014 IEEE Conference on Computer Communications Workshops (INFOCOM WKSHPS); 2014. pp. 616–21.

[19] Breidert C., Hahsler M., Reutterer T. 'A review of methods for measuring willingness-to-pay'. *Innovative Marketing*. 2015;**1**.

[20] Guo Y., Liu Q. 'E-commerce personalized recommendation system based on multi-agent'. *2010 Seventh International Conference on Fuzzy Systems and Knowledge Discovery*. **4**; 2010. pp. 1999–2003.

[21] Netflix Prize [online]. Available from https://www.netflixprize.com/ [Accessed 18 Aug 2021].

[22] RRDtool Gallery [online]. Available from http://oss.oetiker.ch/rrdtool/gallery/index.en.html.

[23] Song C., Qu Z., Blumm N., Barabási A.-L. 'Limits of predictability in human mobility'. *Science*. 2010;**327**(5968):1018–21.

[24] Boyd S., Vandenberghe L. *Convex Optimization*. New York, NY, USA: Cambridge University Press; 2004.

[25] MacKenzie A.B., DaSilva L.A. *Game Theory for Wireless Engineers (Synthesis Lectures on Communications)*. Morgan & Claypool Publishers; 2006.

[26] Cover T.M., Thomas J.A. *Elements of Information Theory (Wiley Series in Telecommunications and Signal Processing)*. New York, NY, USA: Wiley-Interscience; 2006.

[27] Marks B.R., Wright G.P. 'Technical note—a general inner approximation algorithm for nonconvex mathematical programs'. *Operations Research*. 1978;**26**(4):681–3.

[28] Crescenzo A.D. 'A probabilistic analogue of the mean value theorem and its applications to reliability theory'. 1999;**36**(3):706–19.

Chapter 15
Mobile edge caching: an optimal auction approach

Xuanyu Cao[1], Junshan Zhang[2], and H. Vincent Poor[3]

With an explosive growth of wireless data, the sheer volume of mobile traffic is challenging the capacity of current wireless systems. To tackle this challenge, mobile edge caching has emerged as a promising paradigm, in which service providers (SPs) prefetch some popular content items in advance and cache them locally at the network edge. When requested, those locally cached content items can be directly delivered to users with low latency, thus, alleviating the traffic load over backhaul channels during peak hours and enhancing the quality-of-experience (QoE) of users simultaneously. Owing to the limited available cache space, it makes sense for the SP to cache the most profitable content items. Nevertheless, users' true valuations of content items are their private knowledge, which is unknown to the SP in general. This information asymmetry poses a significant challenge for effective caching at the SP side. Further, the cached content items can be delivered with different levels of quality, which needs to be chosen judiciously to balance delivery costs and user satisfaction. To tackle these difficulties, we propose an optimal auction mechanism from the perspective of the SP. In the auction, the SP determines the cache space allocation over content items and user payments based on the users' (possibly untruthful) reports of their valuations so that the SP's expected revenue is maximized. The advocated mechanism is designed to elicit true valuations from the users (incentive compatibility) and to incentivize user participation (individual rationality). In addition, we devise a computationally efficient method for calculating the optimal cache space allocation and user payments. We further examine the optimal choice of the content delivery quality for the case with a large number of users and derive a closed-form solution to compute the optimal delivery quality. Finally, extensive simulations are implemented to evaluate the performance of the proposed optimal auction mechanism, and the impact of various model parameters is highlighted to obtain engineering insights into the content caching problem.

[1]Coordinated Science Lab, University of Illinois at Urbana-Champaign, USA
[2]School of Electrical, Computer and Energy Engineering, Arizona State University, USA
[3]Department of Electrical Engineering, Princeton University, USA

15.1 Introduction

The last decades have witnessed a dramatic proliferation of wireless data. It was projected that the volume of mobile traffic in 2020 will become 1 000 times of that in 2010 [1]. Compounding the issue of congested wireless networks is the rapid growth in social network traffic, and indeed many people view content on their mobile devices. In particular, a content item can become viral, in the sense that it has a rapid increase in popularity in a short time. This has resulted in tremendous pressure on wireless service providers (SPs), because moving a large volume of data into and out of the cloud wirelessly requires substantial spectrum resources, and meanwhile may incur annoying latency and jitter.

A key observation is that many content requests by users are highly repetitive, which would lead to numerous redundant transmissions. With this insight, mobile edge caching is emerging as a new paradigm to alleviate the unprecedented demand on network traffic. Residing on the network edge, mobile edge caching can offer storage resources to mobile users through low-latency wireless connections, facilitating a number of mobile services. It is forecast that, by caching contents at the network edge, up to 35% traffic on the backhaul can be reduced. Specifically, SPs can prefetch popular content items and cache them locally at the network edge during off-peak hours when the cellular systems have plenty of resources. As such, the traffic burden of the costly backhaul transmissions is significantly mitigated during peak hours while the quality-of-experience (QoE) of users is also enhanced due to the low transmission delay between local caches and users.

Owing to its promising prospect, content caching has attracted extensive research interests in the past few years. Many works have been devoted to various resource allocation issues in caching systems to optimize content placement, caching, routing, and delivery. Different parties in caching systems, such as content providers (CPs), SPs, and users, are often non-cooperative organizations or individuals seeking to maximize their own benefits instead of the overall social welfare, which are amenable to mechanism design [2] and game theory [3]. This has spurred a recent tide in designing incentive mechanisms and game-theoretic solutions to content caching networks [4–10].

In most existing works on caching systems, user demands for content items are assumed to be either deterministically given or known with fixed distributions. However, in practice, rational users often make strategic content access decisions in response to the content caching and pricing schemes defined by the SPs or CPs. The goal of the users is to maximize their own payoffs selfishly based on their valuations of the content items. Moreover, users' true valuations of content items are usually private knowledge unknown to the SPs/CPs. This information asymmetry makes user demands hardly predictable and may compromise the SPs' profits in content caching systems.

In this chapter, we consider the design of incentive mechanisms for strategic users requesting content items from an SP. To alleviate the traffic load during peak hours, the SP may download some content items from CPs during off-peak hours in advance and cache them at the network edge such as in small-cell base stations (SBSs). The prefetched content items can enhance the QoE of users due to the reduction of transmission latency and the users, in turn, pay the SP for QoE improvement. Owing to the limited cache/storage capacity, the SP will only prefetch those most profitable content items from their respective CPs. Nevertheless, the SP is not aware of the users' true valuations or willingness-to-pay, which hinders the SP from choosing profitable content items to prefetch. In this chapter, we design an optimal auction mechanism [11] to maximize the SP's profit. The users report their valuations (not necessarily truthfully) to the SP, who decides the amount of content to be cached and the payment of users based on the reports. The main contributions of this chapter are summarized as follows.

- An optimal auction mechanism is designed for the SPs to prefetch content items and charge users, in a way that maximizes their expected revenue while ensuring truthful reports and user participation. Different from existing optimal auction formulations [11–14], in this chapter, the auctioneer (SP) has multiple objects (content items) to sell and the transfer of objects (content delivery) incurs costs, which can be controlled by judicious design of delivery quality. The same content can be shared by multiple users and this introduces coupling.
- A systematic method of finding the optimal cache space allocation over content items is presented under certain regularity conditions. In addition, based on the special structure of the optimal cache space allocation scheme, we devise a computationally efficient method for calculating the optimal user payments.
- Based on the proposed optimal auction mechanism for content caching, we further examine the optimal choice of the content delivery quality for the case with a large number of users. A closed-form solution is presented to determine an optimal delivery quality so as to maximize the expected revenue of the SP.
- Extensive numerical experiments are conducted to evaluate the performance of the proposed optimal auction mechanism. The impact of various model parameters is studied empirically to obtain engineering insights into the content caching problem.

15.2 System model and problem formulation

Consider a model with one SP, m CPs, and n users, as illustrated in Figure 15.1. The SP may be in charge of one or several base stations serving users within some region. The CPs provide (sell) contents to the SP, who transmits them to the interested mobile users. To mitigate the traffic burden during peak hours, the SP may prefetch (download) some content items and store them in its local cache at the base stations during off-peak hours in advance. When requested (typically during peak hours), the prefetched content items at the cache can be delivered to users with low latency, leading to better QoE for users [15]. In turn, the users pay the SP, who thus

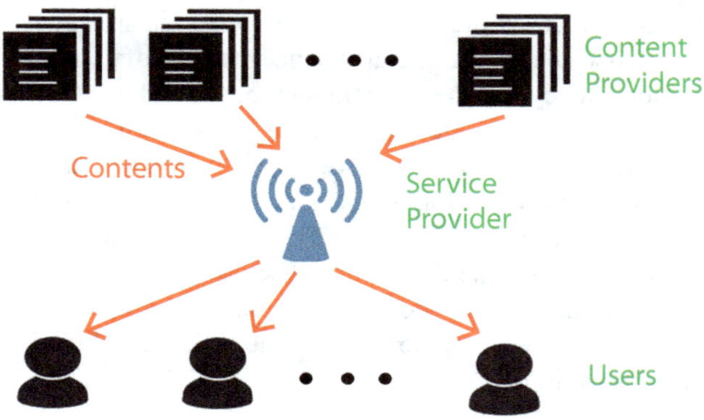

Figure 15.1 An illustration of the caching system under study

gains additional profit. In what follows, we detail the model of the content caching system in Figure 15.1 and formulate the mechanism design problem.

15.2.1 System model

Henceforth, we call the content item from CP i content item i, $i = 1, ..., m$. Denote the set of content items that interest user j as $\mathcal{S}_j \subset \{1, ..., m\}$, $j = 1, ..., n$. For instance, a Netflix client may be interested in certain categories of movies. On the other hand, denote the set of users interested in content item i as $\Omega_i \subset \{1, ..., n\}$. Thus, $i \in \mathcal{S}_j$ is equivalent to $j \in \Omega_i$. Each user j has a valuation (or willingness-to-pay) t_j of those content items that interest him/her, i.e., content items in $\mathcal{S}_j$. The valuation t_j is called the type of user j and depends on factors such as the user's demand for content items and economic conditions (richer users tend to have higher valuations or willingness-to-pay). The valuation t_j is only privately known by user j and is unknown to the SP and other users. To model this uncertainty to others, we assume that t_j is a random variable (r.v.) over the interval $\mathbb{T}_j = [\underline{a}_j, \overline{a}_j]$ with probability density function (PDF) $f_j(t_j)$ and cumulative distribution function (CDF) $F_j(t_j)$. The realization of the r.v. t_j is only revealed to user j while the distribution functions are common knowledge to all users and the SP. The valuations $\boldsymbol{t} = [t_1, ..., t_n]$ are assumed to be independent across users. The joint type space of all users is denoted as $\mathbb{T} = \mathbb{T}_1 \times \cdots \times \mathbb{T}_n$ while the joint PDF is denoted as $f(\boldsymbol{t}) = \Pi_{j=1}^{n} f_j(t_j)$.

To download content i, the SP needs to pay CP i with unit price r_i, which is set by CP i and considered fixed hereafter. After acquiring content items, the SP delivers them to users as per requests with certain content delivery/transmission quality $\theta > 0$, e.g., the video quality of video contents. To deliver content items to k users with quality θ, the SP incurs a unit cost of $kh(\theta)$, where $h : \mathbb{R}_+ \mapsto \mathbb{R}_+$ is the cost function of content delivery. The quality parameter θ is controlled by the SP. If user j receives content items in $\mathcal{S}_j$ with quality θ, his/her unit satisfaction level will be θt_j, which depends on both his/her valuation t_j and the content delivery quality θ.

15.2.2 Problem formulation for auction mechanism design

Given the limited available cache size at the base stations, the SP aims at downloading the most profitable content items, i.e., those items that interest many users with high valuations. The main challenge faced in such a case by the SP is that he/she does not know the private valuations of different users. What the SP knows is only the distribution of the user valuations. Thus, an incentive mechanism is imperative for the SP to elicit truthful valuation information from users and to extract maximal profit. According to the direct revelation principle [16], each user j first reports his/her type (possibly untruthfully) as $\tau_j \in \mathbb{T}_j$, which may deviate from his/her true type t_j. After collecting all the reports $\boldsymbol{\tau} = [\tau_1, ..., \tau_n] \in \mathbb{T}$ from all users, the SP allocates the cache space over content items and charges users according to a mechanism specified by the tuple $\langle \boldsymbol{p}(\cdot), \boldsymbol{x}(\cdot) \rangle$. Here, we define $\boldsymbol{p}(\cdot) = [p_1(\cdot), ..., p_m(\cdot)]^{\mathrm{T}}$ with $p_i : \mathbb{T} \mapsto [0, 1]$ prescribing the fraction of caching space used to store content item i, i.e., $p_i(\boldsymbol{\tau})$ is the fraction of caching space for content item i given the reports $\boldsymbol{\tau}$. Furthermore, we define $\boldsymbol{x}(\cdot) = [x_1(\cdot), ..., x_n(\cdot)]^{\mathrm{T}}$ with $x_j : \mathbb{T} \mapsto \mathbb{R}$ prescribing the payment of user j, i.e., $x_j(\boldsymbol{\tau})$ is the payment of user j when the reports are $\boldsymbol{\tau}$. If all the users report their true types, the expected revenue (ER) of the SP is:

$$\mathrm{ER} \triangleq \int_{\mathbb{T}} \left[\sum_{j=1}^{n} x_j(\boldsymbol{t}) - \sum_{i=1}^{m} p_i(\boldsymbol{t}) r_i - \sum_{i=1}^{m} p_i(\boldsymbol{t}) |\mathfrak{R}_i| h(\theta) \right] f(\boldsymbol{t}) d\boldsymbol{t}, \tag{15.1}$$

where $|\mathfrak{R}_i|$ is the cardinality of the set $\mathfrak{R}_i$ and the integrands correspond to the user payments, content acquisition costs, and content delivery costs, respectively. Given users' reports $\tau \in \mathbb{T}$ and their true types $\boldsymbol{t} \in \mathbb{T}$, the ex-post utility of user j under the mechanism is:

$$u_j(\boldsymbol{\tau}, \boldsymbol{t}) \triangleq \left[\sum_{i \in \mathfrak{S}_j} p_i(\boldsymbol{\tau}) \right] \theta t_j - x_j(\boldsymbol{\tau}). \tag{15.2}$$

Denote $\mathbb{T}_{-j} = \mathbb{T}_1 \times \cdots \times \mathbb{T}_{j-1} \times \mathbb{T}_{j+1} \times \cdots \times \mathbb{T}_n$ as the type space excluding user j and $f_{-j}(\boldsymbol{t}_{-j}) = \Pi^{n}_{j'=1, j' \neq j} f_{j'}(t_{j'})$ as the joint PDF of types excluding user j. Thus, if user j reports τ_j while his/her true type is t_j and other users are truth-telling, his/her interim (expected) utility is:

$$v_j(\tau_j, t_j) \triangleq \int_{\mathbb{T}_{-j}} u_j(\tau_j, \boldsymbol{t}_{-j}, t_j, \boldsymbol{t}_{-j}) f_{-j}(\boldsymbol{t}_{-j}) d\boldsymbol{t}_{-j}. \tag{15.3}$$

Define the truth-telling interim utility of user j as $\widetilde{v}_j(t_j) \triangleq v_j(t_j, t_j)$. To incentivize truthful reports of user types, a key step is to guarantee that truth-telling is a Bayesian Nash equilibrium [3], i.e.,

$$\widetilde{v}_j(t_j) \geq v_j(\tau_j, t_j), \ \forall \tau_j, t_j \in \mathbb{T}_j, \ \forall j = 1, ..., n, \tag{15.4}$$

which is called the *incentive compatibility* (*IC*) constraint. In addition, to incentivize users' participation of the auction, the mechanism should satisfy the following (interim) *individual rationality* (*IR*) constraint:

$$\widetilde{v}_j(t_j) \geq 0, \ \forall t_j \in \mathbb{T}_j, \ \forall j = 1, ..., n, \tag{15.5}$$

which means non-negative truth-telling interim utility. Further, the mechanism obviously needs to satisfy the following *fraction feasibility* (*FF*) constraint:

$$p_i(\boldsymbol{t}) \geq 0, \ i = 1, ..., m, \ \sum_{i=1}^{m} p_i(\boldsymbol{t}) \leq 1, \ \forall \boldsymbol{t} \in \mathbb{T}. \tag{15.6}$$

Overall, the optimization problem faced by the SP is to design an optimal auction mechanism $\langle \boldsymbol{p}(\cdot), \boldsymbol{x}(\cdot) \rangle$ and optimal delivery quality θ such that the IC, IR, and FF constraints are satisfied and its ER is maximized, i.e.,

$$\begin{aligned} &\text{Maximize}_{\boldsymbol{p},\boldsymbol{x},\theta} \ \text{ER in (15.1)} \\ &\text{subject to} \begin{cases} \text{FF in (15.6)}, \\ \text{IC in (15.4)}, \\ \text{IR in (15.5)}, \\ \theta > 0. \end{cases} \end{aligned} \tag{15.7}$$

There are several differences between the optimal auction problem (15.7) in this chapter and the classical Myerson's optimal auction [11] as well as other applications of optimal auctions in various resource allocation problems in [12–14]. First, unlike traditional optimal auction problems, the auctioneer (**SP**) has multiple objects (m contents from different CPs) to sell. Different objects (content items) have different popularity (captured by Ω_i) among users and the acquisition of different objects incurs different costs (captured by r_i). The same content item can be shared by multiple users and hence brings forth coupling, i.e., different users may be interested in the same content item so that caching it by the SP can benefit multiple users. In fact, exploiting this redundancy in content usage is the keystone of information-centric networks and caching systems so that the frequently requested content items need not be transmitted repetitively through the costly backhaul channels. Second, different from the existing optimal auction frameworks, in this chapter, content delivery incurs costs that are controlled by the auctioneer through the delivery quality parameter $\boldsymbol{\theta}$. Increasing $\boldsymbol{\theta}$ boosts users' utilities and thus incentivizes them to pay more. From the perspective of the SP, this enhances its profit extraction from users at the expense of higher content delivery costs. Thereby, a judiciously chosen $\boldsymbol{\theta}$ is important to achieve the optimal profit-cost trade-off.

15.3 Optimal auction design

In this section, we solve the optimal auction problem in (15.7). We first fix the content delivery quality $\theta > 0$ and solve for the optimal auction mechanism $\langle \boldsymbol{p}(\cdot), \boldsymbol{x}(\cdot) \rangle$. To this end, we transform the IC, IR constraints, and the objective function into more tractable forms, and find the optimal mechanism after imposing some regularity conditions on the type distributions (Theorem 1).

We next show how to compute the optimal payment efficiently (Proposition 3) and demonstrate a property of the optimal mechanism (Proposition 4). Once obtaining the optimal mechanism $\langle \boldsymbol{p}^*(\cdot), \boldsymbol{x}^*(\cdot)\rangle$, we further endeavor to find the optimal content delivery quality θ under the assumption that the number of users n is very large (Theorem 2), which is well justified by the prevalence of various content services and mobile networks.

15.3.1 Optimal auction with fixed content delivery quality

In this section, we fix θ and solve for the optimal $\langle \boldsymbol{p}(\cdot), \boldsymbol{x}(\cdot)\rangle$. In other words, for fixed $\theta > 0$ we solve the following problem:

$$\text{Maximize}_{\boldsymbol{p},\boldsymbol{x}} \ \text{ER in (15.1)}$$

$$\text{subject to} \begin{cases} \text{FF in (15.6)}, \\ \text{IC in (15.4)}, \\ \text{IR in (15.5)}. \end{cases} \tag{15.8}$$

From (15.2) and (15.3), we write

$$v_j(\tau_j, t_j) = \theta t_j \widetilde{p}_j(\tau_j) - \widetilde{x}_j(\tau_j), \ \forall \tau_j, t_j, \ \forall j = 1, \ldots, n, \tag{15.9}$$

where we have defined

$$\widetilde{p}_j(\tau_j) \stackrel{\Delta}{=} \int_{\mathbb{T}_{-j}} \left[\sum_{i \in \mathcal{S}_j} p_i(\tau_j, \boldsymbol{t}_{-j}) \right] f_{-j}(\boldsymbol{t}_{-j}) d\boldsymbol{t}_{-j}, \tag{15.10}$$

$$\widetilde{x}_j(\tau_j) \stackrel{\Delta}{=} \int_{\mathbb{T}_{-j}} x_j(\tau_j, \boldsymbol{t}_{-j}) f_{-j}(\boldsymbol{t}_{-j}) d\boldsymbol{t}_{-j}. \tag{15.11}$$

We can transform the IC and IR constraints into equivalent tractable forms according to the following lemma, the proof of which is omitted.

Lemma 1. Suppose the FF constraint in (15.6) holds. Then, a necessary and sufficient condition for IC in (15.4) and IR in (15.5) is:

i. $\widetilde{p}_j(\cdot)$ *is an increasing function for each* $j = 1, \ldots, n$.
ii. *For any* $t_j \in \mathbb{T}_j$, $j = 1, \ldots, n$: $\widetilde{v}_j(t_j) = \widetilde{v}_j(\underline{a}_j) + \theta \int_{\underline{a}_j}^{t_j} \widetilde{p}_j(\tau_j) d\tau_j$.
iii. *For each* $j = 1, \ldots, n$: $\widetilde{v}_j(\underline{a}_j) \geq 0$.

Based on Lemma 1, we can simplify the expected revenue, i.e., the objective function of problem (15.8), as follows.

Lemma 2. For any feasible mechanism of problem (15.8), i.e., $\langle \boldsymbol{p}(\cdot), \boldsymbol{x}(\cdot)\rangle$ satisfying the FF, IC, and IR constraints, the expected revenue becomes

$$\begin{aligned} \text{ER} = & -\textstyle\sum_{j=1}^{n} \widetilde{v}_j(\underline{a}_j) + \theta \int_{\mathbb{T}} \left\{ \sum_{j=1}^{n} \left[t_j - \frac{1 - F_j(t_j)}{f_j(t_j)} - \frac{h(\theta)}{\theta} \right] \left(\sum_{i \in \mathcal{S}_j} p_i(\boldsymbol{t}) \right) \right\} f(\boldsymbol{t}) d\boldsymbol{t} \\ & - \int_{\mathbb{T}} \left[\textstyle\sum_{i=1}^{m} p_i(\boldsymbol{t}) r_i \right] f(\boldsymbol{t}) d\boldsymbol{t}. \end{aligned} \tag{15.12}$$

Proof. The proof is relegated to the Appendix.

Combining Lemmas 1 and 2, we see that problem (15.8) is equivalent to

$$\text{Maximize}_{\boldsymbol{p},\boldsymbol{x}} \text{ ER in (15.12)}$$
$$\text{subject to} \begin{cases} \text{FF in (15.6),} \\ \text{(i), (ii), (iii) in Lemma 1.} \end{cases} \tag{15.13}$$

According to the definitions, $\widetilde{v}_j(\cdot)$ is related to both $\boldsymbol{p}(\cdot)$ and $\boldsymbol{x}(\cdot)$ while $\widetilde{p}_j(\cdot)$ only depends on $\boldsymbol{p}(\cdot)$ (independent of $\boldsymbol{x}(\cdot)$). We note that, for ER in (15.12), only the first term $-\sum_{j=1}^{n}\widetilde{v}_j(\underline{a}_j)$ depends on $\boldsymbol{x}(\cdot)$ and the remaining terms are independent of $\boldsymbol{x}(\cdot)$. Furthermore, constraints FF and (i) in Lemma 1 depend on $\boldsymbol{p}(\cdot)$ only and are independent of $\boldsymbol{x}(\cdot)$. Therefore, if we fix some $\boldsymbol{p}(\cdot)$ satisfying constraints FF and (i) of Lemma 1 and optimize over $\boldsymbol{x}(\cdot)$, problem (15.13) becomes

$$\text{Maximize}_{\boldsymbol{x}} \; -\sum_{j=1}^{n}\widetilde{v}_j(\underline{a}_j)$$
$$\text{subject to (ii), (iii) in Lemma 1.} \tag{15.14}$$

The solution to problem (15.14) is given as follows.

Proposition 1. For fixed $\boldsymbol{p}(\cdot)$ satisfying constraints FF and (i) of Lemma 1, the optimal solution to problem (15.14) is

$$x_j^*(\boldsymbol{t}) \triangleq \theta t_j \sum_{i\in\mathcal{S}_j} p_i(\boldsymbol{t}) - \theta\int_{\underline{a}_j}^{t_j}\sum_{i\in\mathcal{S}_j} p_i(\tau_j,\boldsymbol{t}_{-j})d\tau_j,\ \forall j=1,...,n,\ \forall \boldsymbol{t}\in\mathbb{T}, \tag{15.15}$$

which achieves the optimal value of 0.

Proof. According to (iii) in Lemma 1, we see that the optimal value of (15.14) is no greater than 0. From (15.15), we know that, for any $\boldsymbol{t}_{-j}\in\mathbb{T}_{-j}$, $x_j^*(\underline{a}_j,\boldsymbol{t}_{-j}) = \theta\underline{a}_j\sum_{i\in\mathcal{S}_j}p_i(\underline{a}_j,\boldsymbol{t}_{-j})$. So, (iii) of Lemma 1 holds with equality for $\boldsymbol{x}^*(\cdot)$, i.e., $\widetilde{v}_j(\underline{a}_j)=0$ for each j. Substituting (15.15) in the left-hand side of (ii) in Lemma 1, interchanging the order of integrals and noting the definition of $\widetilde{p}_j(\cdot)$, we can show that (ii) also holds for $\boldsymbol{x}^*(\cdot)$. So, $\boldsymbol{x}^*(\cdot)$ is feasible for problem (15.14). Meanwhile, it also achieves an objective function value of 0. Thus, it is optimal for problem (15.14).

Next, we optimize over $\boldsymbol{p}(\cdot)$. According to Proposition 1, we only need to solve the following problem:

$$\text{Maximize}_{\mathrm{p}} \int_{\pi}\left\{\sum_{j=1}^{n}\left[\theta t_j - \frac{\theta(1-F_j(t_j))}{f_j(t_j)} - h(\theta)\right]\left[\sum_{i\in S_j}p_i(t)\right] - \sum_{i=1}^{m}p_i(t)r_i\right\}f(t)dt$$
$$\text{subject to :} \begin{cases} \text{FF in (15.6)} \\ \text{(i) in Lemma} \end{cases} \tag{15.16}$$

Define $c_j(t_j) \triangleq t_j - \frac{1-F_j(t_j)}{f_j(t_j)}$, for any $j = 1, ..., n$, $t_j \in \mathbb{T}_j$. Note that $c_j(\cdot)$ is a function depending only on the distribution of the true type of user j. To facilitate solving problem (15.16), we make the following regularity assumption, which is common in the mechanism design literature [11–13].

Definition 1. [Regularity] The optimal auction problem (15.8) is *regular* if $c_j(\cdot)$ is an increasing function on $\mathbb{T}_j$ for each $j = 1, ..., n$.

We remark that this regularity condition holds for many familiar distributions, e.g., uniform distributions and exponential distributions. One may refer to [17] for a partial list. Under the regularity assumption, we can compute the optimal solution to problem (15.16) as follows.

Proposition 2. Suppose the optimal auction problem (15.8) is regular. Then, the optimal solution $\boldsymbol{p}^*(\cdot)$ to problem (15.16) can be constructed as follows. For each $\boldsymbol{t} \in \mathbb{T}$, define $k = \arg\max_{i=1,...,m} \left\{\sum_{j \in \Omega_i} [\theta c_j(t_j) - h(\theta)] - r_i\right\}$, whose value depends on $\boldsymbol{t}$. Then, the optimal $\boldsymbol{p}^*(\boldsymbol{t})$ is given as

$$\begin{cases} p_k^*(\boldsymbol{t}) = 1, \ p_i^*(\boldsymbol{t}) = 0, \ \forall i \neq k, \ \text{if } \max_{i=1,...,m} \left\{\sum_{j \in \Omega_i} [\theta c_j(t_j) - h(\theta)] - r_i\right\} > 0, \\ p_i^*(\boldsymbol{t}) = 0, \ \forall i = 1, ..., m, \ \text{otherwise}. \end{cases} \tag{15.17}$$

Proof. We first omit the constraint (i) of Lemma 1 and study the following relaxed version:

$$\begin{aligned} &\text{Maximize}_{\boldsymbol{p}} \ \text{The objective function of (15.16)} \\ &\text{subject to: FF in (15.16).} \end{aligned} \tag{15.18}$$

Later, we will show that the optimal solution to this relaxed problem happens to satisfy (i) as well, i.e., it is also optimal for the original problem (15.16). We note that both the objective and the constraint of problem (15.18) are decoupled across types. So, problem (15.18) can be solved for each given type $\boldsymbol{t} \in \mathbb{T}$. For fixed $\boldsymbol{t} \in \mathbb{T}$, by making use of the definition of $c_j(\cdot)$ and rearranging terms in the objective function, we can rewrite the corresponding problem as:

$$\begin{aligned} &\text{Maximize}_{p(\boldsymbol{t})} \ \sum_{i=1}^{m} \left\{ \sum_{j \in \Omega_i} [\theta c_j(t_j) - h(\theta)] - r_i \right\} p_i(\boldsymbol{t}) \\ &\text{subject to:} \begin{cases} p_i(\boldsymbol{t}) \geq 0, \ i = 1, ..., m, \\ \sum_{i=1}^{m} p_i(\boldsymbol{t}) \leq 1. \end{cases} \end{aligned} \tag{15.19}$$

Problem (15.19) is a simple linear program (LP), whose optimal solution is clearly $\boldsymbol{p}^*(\boldsymbol{t})$ given in (15.17). It now remains to show that $\boldsymbol{p}^*(\cdot)$ satisfies statement (i) in Lemma 1, i.e., $\int_{\mathbb{T}_{-j}} \left[\sum_{i \in \mathbb{S}_j} p_i^*(\boldsymbol{t})\right] f_{-j}(\boldsymbol{t}_{-j}) d\boldsymbol{t}_{-j}$ is an increasing function of $t_j \in \mathbb{T}_j$ for every $j \in \{1, ..., n\}$, so that the aforementioned relaxation is exact.

To this end, for any $j \in \{1, ..., n\}$, consider arbitrary $\tau_j, t_j \in \mathbb{T}_j$ with $\tau_j \leq t_j$. By the regularity assumption, we have $c_j(\tau_j) \leq c_j(t_j)$. From the construction of $\boldsymbol{p}^*(\cdot)$ in (15.17), we know that $\sum_{i \in \mathcal{S}_j} p_i^*(\tau_j, \boldsymbol{t}_{-j})$ is either 0 or 1, for any $\boldsymbol{t}_{-j} \in \mathbb{T}_{-j}$. Next, for arbitrarily fixed $\boldsymbol{t}_{-j} \in \mathbb{T}_{-j}$, we distinguish two cases.

Case 1: $\sum_{i \in \mathcal{S}_j} p_i^*(\tau_j, \boldsymbol{t}_{-j}) = 1$. In such a case, there exists some $l \in \mathcal{S}_j$ such that $p_l^*(\tau_j, \boldsymbol{t}_{-j}) = 1$. Hence, according to the construction of $\boldsymbol{p}^*(\cdot)$, we have

$$\theta c_j(\tau_j) - h(\theta) + \sum_{s \in \Omega_l, s \neq j} [\theta c_s(t_s) - h(\theta)] - r_l \tag{15.20}$$

$$= \max \left\{ \max_{i \in \mathcal{S}_j} \left\{ \theta c_j(\tau_j) - h(\theta) + \sum_{s \in \Omega_i, s \neq j} [\theta c_s(t_s) - h(\theta)] - r_i \right\}, \max_{i \notin \mathcal{S}_j} \left\{ \sum_{s \in \Omega_i} [\theta c_s(t_s) - h(\theta)] - r_i \right\} \right\} \tag{15.21}$$

$$> 0. \tag{15.22}$$

Thus, for any $i \in \mathcal{S}_j : \theta c_j(\tau_j) - h(\theta) + \sum_{s \in \Omega_l, s \neq j} [\theta c_s(t_s) - h(\theta)] - r_1 \geq \theta c_j(\tau_j) - h(\theta) + \sum_{s \in \Omega_i, s \neq j} [\theta c_s(t_s) - h(\theta)] - r$, adding $\theta c_j(t_j) - \theta c_j(\tau_j)$ on both sides, we obtain

$$\sum_{s \in \Omega_l} [\theta c_s(t_s) - h(\theta)] - r_l \geq \sum_{s \in \Omega_i} [\theta c_s(t_s) - h(\theta)] - r_i, \ \forall i \in \mathcal{S}_j. \tag{15.23}$$

On the other hand, for any $i \notin \mathcal{S}_j$,

$$\sum_{s \in \Omega_i} [\theta c_s(t_s) - h(\theta)] - r_i \tag{15.24}$$

$$\leq \theta c_j(\tau_j) - h(\theta) + \sum_{s \in \Omega_l, s \neq j} [\theta c_s(t_s) - h(\theta)] - r_l \tag{15.25}$$

$$\stackrel{(a)}{\leq} \sum_{s \in \Omega_l} [\theta c_s(t_s) - h(\theta)] - r_l, \tag{15.26}$$

where (a) follows from the fact that $c_j(\tau_j) \leq c_j(t_j)$. Combining inequalities (15.23) and (15.26), we thus conclude $l = \arg\max_{i=1,\ldots,m} \left\{ \sum_{r \in \Omega_i} [\theta c_s(t_s) - h(\theta)] - r_i \right\}$. Moreover, we note that $\max_{i=1,\ldots,m} \left\{ \sum_{s \in \Omega_i} [\theta c_s(t_s) - h(\theta)] - r_i \right\} = \sum_{s \in \Omega_l} [\theta c_s(t_s) - h(\theta)] - r_l \geq \theta c_j(\tau_j) - h(\theta) + \sum_{s \in \Omega_l, s \neq j} [\theta c_s(t_s) - h(\theta)] - r_l > 0$,where the second last step results from $c_j(\tau_j) \leq c_j(t_j)$ and the last step follows from (15.22). Hence, according to the construction of $\boldsymbol{p}^*(\cdot)$, we have $p_l^*(\boldsymbol{t}) = 1, \ p_i^*(\boldsymbol{t}) = 0, \ \forall i \neq l$. So, $\sum_{i \in \mathcal{S}_j} p_i^*(\boldsymbol{t}) = 1 = \sum_{i \in \mathcal{S}_j} p_i^*(\tau_j, \boldsymbol{t}_{-j})$.

Case 2: $\sum_{i \in \mathcal{S}_j} p_i^*(\tau_j, \boldsymbol{t}_{-j}) = 0$. In such a case, since $\sum_{i \in \mathcal{S}_j} p_i^*(\boldsymbol{t})$ is either 0 or 1, we evidently have $\sum_{i \in \mathcal{S}_j} p_i^*(\tau_j, \boldsymbol{t}_{-j}) \leq \sum_{i \in \mathcal{S}_j} p_i^*(\boldsymbol{t})$.

Combining cases 1 and 2, we always have $\sum_{i \in \mathcal{S}_j} p_i^*(\tau_j, \boldsymbol{t}_{-j}) \leq \sum_{i \in \mathcal{S}_j} p_i^*(\boldsymbol{t})$, for any $\boldsymbol{t}_{-j} \in \mathbb{T}_{-j}$. Thus,

$\int_{\mathbb{T}_{-j}}\left[\sum_{i\in\mathbb{S}_j}p_i^*(\tau_j,\boldsymbol{t}_{-j})\right]f_{-j}(\boldsymbol{t}_{-j})d\boldsymbol{t}_{-j}\leq\int_{\mathbb{T}_{-j}}\left[\sum_{i\in\mathbb{S}_j}p_i^*(\boldsymbol{t})\right]f_{-j}(\boldsymbol{t}_{-j})d\boldsymbol{t}_{-j}$. Hence, $\boldsymbol{p}^*(\cdot)$ satisfies statement (i) in Lemma 1 and it is optimal for problem (15.16).

Remark 1. The optimal cache space allocation mechanism $\boldsymbol{p}^*(\cdot)$ can be explained as follows. Suppose all users report their true types $\boldsymbol{t}$, as guaranteed by the IC constraint. If the SP allocates all cache space to content i, i.e., $p_i(\boldsymbol{t})=0$, $p_l(\boldsymbol{t})=0$, for any $l\neq i$, then the social welfare comprised the SP's profit and all users' utilities is $\sum_{j\in\Omega_i}(\theta t_j-h(\theta))-r_i$. If we replace the true valuation t_j with a *virtual* valuation $c_j(t_j)$, then the virtual social welfare becomes $\sum_{j\in\Omega_i}(\theta c_j(t_j)-h(\theta))-r_i$. Thus, according to Proposition 2, the optimal allocation mechanism $\boldsymbol{p}^*(\boldsymbol{t})$ is to allocate all the cache space to the content i that yields the greatest virtual social welfare if this greatest virtual social welfare is positive. Otherwise, if the greatest virtual social welfare is negative (i.e., the virtual social welfare of every content is negative), then the SP will not cache any content. Therefore, the optimal allocation mechanism is to maximize the virtual social welfare (including zero in which case the SP caches nothing). Note that this assertion can only facilitate our comprehension of the mechanism. It cannot be directly implemented since the virtual social welfare is an ex-post quantity, i.e., it depends on the realizations of types $\boldsymbol{t}$ which are private information of users. In fact, the reason that we use the virtual valuation $c_j(t_j)$ instead of the true valuation t_j is just to incentivize users to report their private types truthfully.

Propositions 1 and 2 together specify the optimal auction mechanism $\langle\boldsymbol{p}^*(\cdot),\boldsymbol{x}^*(\cdot)\rangle$ for problem (15.8), as summarized in the following theorem.

Theorem 1. Suppose the optimal auction problem (15.8) is regular. The optimal solution $\langle\boldsymbol{p}^*(\cdot),\boldsymbol{x}^*(\cdot)\rangle$ to problem (15.8) is given as follows. For each fixed $\boldsymbol{t}\in\mathbb{T}$, define $k=\arg\max_{i=1,\ldots,m}\left\{\sum_{j\in\Omega_i}[\theta c_j(t_j)-h(\theta)]-r_i\right\}$. Then, the optimal cache space allocation mechanism $\boldsymbol{p}^*(\boldsymbol{t})$ at this particular $\boldsymbol{t}$ is prescribed by equation (15.17). Furthermore, the optimal payment mechanism is given by

$$x_j^*(\boldsymbol{t})=\theta t_j\sum_{i\in\mathbb{S}_j}p_i^*(\boldsymbol{t})-\theta\int_{a_j}^{t_j}\sum_{i\in\mathbb{S}_j}p_i^*(\tau_j,\boldsymbol{t}_{-j})d\tau_j,\ \forall j\in\{1,\ldots,n\},\ \forall\boldsymbol{t}\in\mathbb{T}. \tag{15.27}$$

According to Theorem 1, to compute the optimal payment $x_j^*(\boldsymbol{t})$, we need to calculate the integral $\int_{a_j}^{t_j}\sum_{i\in\mathbb{S}_j}p_i^*(\tau_j,\boldsymbol{t}_{-j})d\tau_j$ for arbitrarily given $\boldsymbol{t}$, which is hard to implement directly. Thanks to the special structure of the optimal allocation mechanism $\boldsymbol{p}^*(\cdot)$, we can compute this integral (and thus the payment $x_j^*(\boldsymbol{t})$) efficiently for any given $\boldsymbol{t},j$, as stated in the following.

Proposition 3. Suppose the optimal auction problem (15.8) is regular. For each given $\boldsymbol{t}\in\mathbb{T}$ and $j\in\{1,\ldots,n\}$, we define a number β_j and a function $\phi_j:\mathbb{T}_j\mapsto\mathbb{R}$ as follows (their dependence on $\boldsymbol{t}$ is suppressed):

$$\beta_j \triangleq \max\left\{0, \max_{i \notin \mathbb{S}_j}\left\{\sum_{s\in\Omega_i}[\theta c_s(t_s) - h(\theta)] - r_i\right\}\right\}, \tag{15.28}$$

$$\phi_j(\tau_j) \triangleq \theta c_j(\tau_j) - h(\theta) + \max_{i\in \mathbb{S}_j}\left\{\sum\nolimits_{s\in\Omega_i, s\neq j}[\theta c_s(t_s) - h(\theta)] - r_i\right\}, \quad \forall \tau_j \in \mathbb{T}_j. \tag{15.29}$$

Then, we have

$$\int_{\underline{a}_j}^{t_j} \sum_{i\in \mathbb{S}_j} p_i^*(\tau_j, \boldsymbol{t}_{-j}) d\tau_j = \begin{cases} t_j - \underline{a}_j, & \text{if } \phi_j(\underline{a}_j) \geq \beta_j, \\ 0, & \text{if } \phi_j(t_j) < \beta_j, \\ t_j - \xi_j, & \text{if } \phi_j(\underline{a}_j) < \beta_j \leq \phi_j(t_j). \end{cases} \tag{15.30}$$

In the last case of (15.30), we have defined

$$\xi_j \triangleq c_j^{-1}\left(\frac{1}{\theta}\left[\beta_j + h(\theta) - \max_{i\in \mathbb{S}_j}\left\{\sum_{s\in\Omega_i, s\neq j}[\theta c_s(t_s) - h(\theta)] - r_i\right\}\right]\right), \tag{15.31}$$

where $c_j^{-1}(\cdot)$ denotes the inverse function of $c_j(\cdot)$[a].

Proof. Fix arbitrary $\boldsymbol{t} \in \mathbb{T}$ and $j \in \{1, ..., n\}$. Define

$$l \triangleq \arg\max_{i\in \mathbb{S}_j}\left\{\sum_{s\in\Omega_i}[\theta c_s(t_s) - h(\theta)] - r_i\right\} = \arg\max_{i\in \mathbb{S}_j}\left\{\sum_{s\in\Omega_i, s\neq j}[\theta c_s(t_s) - h(\theta)] - r_i\right\}, \tag{15.32}$$

in which we suppress the dependence of l on $\boldsymbol{t}$ and j to avoid cluttered notation. Suppose l is the unique maximum for (15.32), which happens with probability 1. Thus, for any $\tau_j \in \mathbb{T}_j$, $i \in \mathbb{S}_j$, $i \neq l$, we have

$$\phi_j(\tau_j) = \theta c_j(\tau_j) - h(\theta) + \sum_{s\in\Omega_l, s\neq j}[\theta c_s(t_s) - h(\theta)] - r_l \tag{15.33}$$

$$> \theta c_j(\tau_j) - h(\theta) + \sum_{s\in\Omega_i, s\neq j}[\theta c_s(t_s) - h(\theta)] - r_i. \tag{15.34}$$

So, from the construction of $\boldsymbol{p}^*(\cdot)$ in Proposition 2, we know that $p_i^*(\tau_j, \boldsymbol{t}_{-j}) = 0$. Hence, we have

[a]A sufficient condition for the existence of the inverse function $c_j^{-1}(\cdot)$ is that $c_j(\cdot)$ is strictly increasing and continuous. Otherwise, one can replace $c_j^{-1}(z_j)$ with $\inf\{t_j' \in \mathbb{T}_j | c_j(t_j') \geq z_j\}$ since the regularity assumption has already guaranteed that $c_j(\cdot)$ is (weakly) increasing.

$$\int_{\underline{a}_j}^{t_j} \sum_{i \in \mathbb{S}_j} p_i^*(\tau_j, \boldsymbol{t}_{-j}) d\tau_j = \int_{\underline{a}_j}^{t_j} p_l^*(\tau_j, \boldsymbol{t}_{-j}) d\tau_j. \tag{15.35}$$

Note that $\phi_j(\cdot)$ is an increasing function on $\mathbb{T}_j$. We thus distinguish three cases.

Case 1: $\phi_j(\underline{a}_j) \geq \beta_j$. In such a case, $\phi_j(\tau_j) \geq \beta_j$, for any $\tau_j \in [\underline{a}_j, t_j]$. From the definition of β_j in (15.28), (15.33), (15.34), and the construction rule of $\boldsymbol{p}^*(\cdot)$, one can easily see that $p_l^*(\tau_j, \boldsymbol{t}_{-j}) = 1$, for any $\tau_j \in [\underline{a}_j, t_j]$. Thus, $\int_{\underline{a}_j}^{t_j} p_l^*(\tau_j, \boldsymbol{t}_{-j}) d\tau_j = t_j - \underline{a}_j$.

Case2: $\phi_j(t_j) < \beta_j$. In such a case, we have $\phi_j(\tau_j) < \beta_j$, for any $\tau_j \in [\underline{a}_j, t_j]$. Analogously, we can assert that $p_l^*(\tau_j, \boldsymbol{t}_{-j}) = 0$, for any $\tau_j \in [\underline{a}_j, t_j]$. So, $\int_{\underline{a}_j}^{t_j} p_l^*(\tau_j, \boldsymbol{t}_{-j}) d\tau_j = 0$.

Case 3: $\phi_j(\underline{a}_j) < \beta_j \leq \phi_j(t_j)$. In such a case, from the definition of ξ_j in (15.31), we know that it is the unique solution of the equation $\phi_j(\xi_j) = \beta_j$ over the interval $[\underline{a}_j, t_j]$. Hence, for $\tau_j \in [\underline{a}_j, \xi_j)$, we have $\phi_j(\tau_j) < \beta_j$ and thus $p_l^*(\tau_j, \boldsymbol{t}_{-j}) = 0$. On the other hand, for $\tau_j \in [\xi_j, t_j]$, we have $\phi_j(\tau_j) \geq \beta_j$ and thus $p_l^*(\tau_j, \boldsymbol{t}_{-j}) = 1$. Combining these two situations, we derive $\int_{\underline{a}_j}^{t_j} p_l^*(\tau_j, \boldsymbol{t}_{-j}) d\tau_j = t_j - \xi_j$.

Combining the results in all three cases and noting relation (15.35), we conclude the proposition.

We note that the right-hand side of (15.30) can be calculated easily for arbitrarily given $\boldsymbol{t} \in \mathbb{T}$ and $j \in \{1, ..., n\}$. Therefore, Proposition 3 provides us a simple way of computing the optimal payment $x_j^*(\boldsymbol{t})$. Moreover, making use of Proposition 3, we can show the following intuitively reasonable proposition of the optimal mechanism $\langle \boldsymbol{p}^*(\cdot), \boldsymbol{x}^*(\cdot) \rangle$.

Proposition 4. Suppose the optimal auction problem (15.8) is regular. For any $\boldsymbol{t} \in \mathbb{T}$ and any $j \in \{1, ..., n\}$, if $\sum_{i \in \mathbb{S}_j} p_i^*(\boldsymbol{t}) = 0$, then $x_j^*(\boldsymbol{t}) = 0$.

Proof. Suppose $\sum_{i \in \mathbb{S}_j} p_i^*(\boldsymbol{t}) = 0$. We will show by contradiction that $\phi_j(t_j) < \beta_j$. Otherwise, if $\phi_j(t_j) \geq \beta_j$, then we have

$$\theta c_j(t_j) - h(\theta) + \sum_{s \in \Omega_l, s \neq j} [\theta c_s(t_s) - h(\theta)] - r_l \geq \beta_j, \tag{15.36}$$

where we define $l \triangleq \arg\max_{i \in \mathbb{S}_j} \left\{ \sum_{s \in \Omega_i, s \neq j} [\theta c_s(t_s) - h(\theta)] - r_i \right\}$. From the definition of β_j in (15.28), we have for any $i \notin \mathbb{S}_j$:

$$\sum_{s \in \Omega_l} [\theta c_s(t_s) - h(\theta)] - r_l \geq \sum_{s \in \Omega_i} [\theta c_s(t_s) - h(\theta)] - r_i. \tag{15.37}$$

From the definition of l, we can see that (15.37) holds for $i \in \mathbb{S}_j$ as well. Hence, we have $l = \arg\max_{i=1,...,m} \left\{ \sum_{s \in \Omega_i} [\theta c_s(t_s) - h(\theta)] - r_i \right\}$. From (15.36) and $\beta_j \geq 0$, we further know that $\max_{i=1,...,m} \left\{ \sum_{s \in \Omega_i} [\theta c_s(t_s) - h(\theta)] - r_i \right\} \geq 0$. According to the construction of $\boldsymbol{p}^*(\boldsymbol{t})$, we thus have $p_l^*(\boldsymbol{t}) = 1$ and $p_i^*(\boldsymbol{t}) = 0$ for any $i \neq l$. Noting that $l \in \mathbb{S}_j$, we get $\sum_{i \in \mathbb{S}_j} p_i^*(\boldsymbol{t}) = 1$, which is a contradiction. So, we must have $\phi_j(t_j) < \beta_j$, which implies $\int_{\underline{a}_j}^{t_j} \sum_{i \in \mathbb{S}_j} p_i^*(\tau_j, \boldsymbol{t}_{-j}) d\tau_j = 0$ according to Proposition 3. Hence, according to the construction of $\boldsymbol{x}^*(\cdot)$ in Theorem 1, we obtain $x_j^*(\boldsymbol{t}) = 0$.

Remark 2. Proposition 4 asserts that, in the optimal mechanism, a user does not pay anything if no content item of his/her interest is cached by the SP, as expected.

In such a case, his/her ex-post utility is also zero, i.e., $u(\boldsymbol{t},\boldsymbol{t})=0$ (c.f. Equation (15.2)). Furthermore, we note that for each $\boldsymbol{t}\in\mathbb{T}$, $\boldsymbol{p}^*(\boldsymbol{t})$ is either $\mathbf{0}$ or $\boldsymbol{e}_k$ for some $k\in\{1,\ldots,m\}$, where $\boldsymbol{e}_k=[0,\ldots,0,1,0,\ldots,0]^{\mathsf{T}}$ (the sole 1 takes place at the kth entry). Thus, for the condition in Proposition 4 to hold, we distinguish the following two cases for each $\boldsymbol{t}\in\mathbb{T}$. If there exists k such that $\boldsymbol{p}^*(\boldsymbol{t})=\boldsymbol{e}_k$, then $\sum_{i\in\mathbb{S}_j}p_i^*(\boldsymbol{t})=0$, for any $j\notin\Omega_k$. If $\boldsymbol{p}^*(\boldsymbol{t})=\mathbf{0}$, then obviously $\sum_{i\in\mathbb{S}_j}p_i^*(\boldsymbol{t})=0$ for any $j=1,\ldots,n$. The assertion in Proposition 4 will be confirmed empirically through numerical experiments in Section 15.4.

15.3.2 Optimal determination of content delivery quality θ

The optimization problem (15.8) is for fixed delivery quality $\theta>0$. In this section, we let θ vary and determine the optimal θ to further maximize the ER, i.e., solving the optimization problem (15.7). By using the optimal auction mechanism $\langle\boldsymbol{p}^*(\cdot),\boldsymbol{x}^*(\cdot)\rangle$ specified in Theorem 1, the corresponding optimal value of the ER becomes a function of θ as follows:

$$\mathrm{ER}^*(\theta)=\int_{\mathbb{T}}\left\{\sum_{i=1}^{m}\left[\sum_{j\in\Omega_i}(\theta c_j(t_j)-h(\theta))-r_i\right]p_i^*(\boldsymbol{t})\right\}f(\boldsymbol{t})\mathrm{d}\boldsymbol{t}, \tag{15.38}$$

which is difficult to evaluate in general, based on the construction of $\boldsymbol{p}^*(\cdot)$ in Proposition 2. For analytical tractability, we next focus on the case meeting the following assumptions.

Assumption 1. The number of users n is large.

Assumption 2. The type distributions of all users are the same, i.e., $f_j(\cdot)=f(\cdot)$, $F_j(\cdot)=F(\cdot)$, $\underline{a}_j=\underline{a}>0$, $\overline{a}_j=\overline{a}$ for any $j\in\{1,\ldots,n\}$. Moreover, this common distribution satisfies the regularity condition in Definition 1.

Assumption 3. For each content item $i\in\{1,\ldots,m\}$, each user $j\in\{1,\ldots,n\}$ is included in Ω_i, i.e., user j is interested in content item i, with probability $q_i\in[0,1]$ independently.

Assumption 4. The cost function $h(\cdot)$ satisfies the following properties: $h(0)=0$; h is convex; $h'(0)=0$; and $\lim_{\theta\to+\infty}h'(\theta)=\infty$.

Assumption 1 is applicable to many practical scenarios since the user density has increased dramatically with the advancement of mobile networks and devices. Assumption 2 is a homogeneity hypothesis commonly used in the analysis of large-scale systems. Assumption 3 is widely adopted in the popularity modeling of content-centric networks, where q_i characterizes the popularity of content item i, e.g., the Zipf distribution of content popularity [18–20]. Assumption 4 collects the common properties of cost functions in resource allocation [21]. Under these assumptions, the optimal θ for problem (15.7), or equivalently the θ that maximizes $\mathrm{ER}^*(\theta)$ in (15.38), can be computed as in the following theorem.

Theorem 2. Suppose that Assumptions 1-4 hold. Then, the optimal θ for problem (15.7), maximizing $\mathrm{ER}^*(\theta)$ in (15.38), is given by $\theta^*=(h')^{-1}(\underline{a})$, where $(h')^{-1}$ denotes the inverse function of h', the derivative of $h(\cdot)$.

Proof. From Assumption 2, we note that $c_j(\cdot) = c(\cdot), \forall j = 1, ..., n$. For any $\boldsymbol{t} \in \mathbb{T}$ and any $i \in \{1, ..., m\}$, we have

$$\frac{1}{n}\left\{\sum_{j\in\Omega_i}[\theta c(t_j) - h(\theta)] - r_i\right\} = \frac{|\Omega_i|}{n}\left[\frac{\theta}{|\Omega_i|}\left(\sum_{j\in\Omega_i} c(t_j)\right) - h(\theta)\right] - \frac{r_i}{n}. \quad (15.39)$$

By the strong law of large numbers (SLLN) [22], we know that

$$\frac{|\Omega_i|}{n} = \frac{1}{n}\sum_{j=1}^{n} \mathbb{1}(j \in \Omega_i) \xrightarrow{\text{a.s.}} \mathbb{E}[\mathbb{1}(1 \in \Omega_i)] = q_i, \quad (15.40)$$

where $\mathbb{1}(\cdot)$ is the indicator function. Moreover, by SLLN, we have

$$\frac{1}{|\Omega_i|}\sum_{j\in\Omega_i} c(t_j) \xrightarrow{\text{a.s.}} \mathbb{E}[c(t_1)] = \int_{\mathbb{T}_1}\left[t_1 - \frac{1 - F(t_1)}{f(t_1)}\right]f(t_1)dt_1 = \underline{a}, \quad (15.41)$$

where we use integration by parts in the last step. Substituting these limits in (15.39), we obtain

$$\frac{1}{n}\left\{\sum_{j\in\Omega_i}[\theta c(t_j) - h(\theta)] - r_i\right\} \xrightarrow{\text{a.s.}} q_i[\theta\underline{a} - h(\theta)]. \quad (15.42)$$

Denote $k = \arg\max_{i=1,...,m} q_i$. According to the construction of $\boldsymbol{p}^*(\cdot)$ in Proposition 2, we distinguish two cases. If $\theta\underline{a} - h(\theta) > 0$, then $p_k^*(\boldsymbol{t}) = 1$, $p_i^*(\boldsymbol{t}) = 0, \forall i \neq k, \forall \boldsymbol{t} \in \mathbb{T}$. In such a case, for any $\boldsymbol{t} \in \mathbb{T}$,

$$\begin{aligned}
&\frac{1}{n}\sum_{i=1}^{m}\left\{\sum_{j\in\Omega_i}[\theta c_j(t_j) - h(\theta)] - r_i\right\}p_i^*(\boldsymbol{t}) \\
&= \frac{1}{n}\left\{\sum_{j\in\Omega_k}[\theta c_j(t_j) - h(\theta)] - r_k\right\} \\
&\xrightarrow{\text{a.s.}} q_k[\theta\underline{a} - h(\theta)]
\end{aligned} \quad (15.43)$$

and $\text{ER}^*(\theta) \approx nq_k[\theta\underline{a} - h(\theta)]$. Otherwise, if $\theta\underline{a} - h(\theta) \leq 0$, then $p_i^*(\boldsymbol{t}) = 0$, $\forall i = 1, .., m, \forall \boldsymbol{t} \in \mathbb{T}$ and $\text{ER}^*(\theta) = 0$. According to Assumption 4, we know that

$$[\theta\underline{a} - h(\theta)]\big|_{\theta=0} = 0, \quad \frac{d}{d\theta}[\theta\underline{a} - h(\theta)]\big|_{\theta=0} = \underline{a} > 0. \quad (15.44)$$

Thus, $\theta\underline{a} - h(\theta) > 0$ for small enough $\theta > 0$. As such, the optimal θ^* that maximizes $\text{ER}^*(\theta)$ is

$$\theta^* = \arg\max_{\theta>0}\{\theta\underline{a} - h(\theta)\} = (h')^{-1}(\underline{a}), \quad (15.45)$$

where we make use of the convexity of $h(\cdot)$ in Assumption 4.

15.4 Numerical results

In this section, extensive numerical experiments are carried out to evaluate the performance of the proposed optimal auction mechanism for content caching. In particular, the impact of various model parameters is studied empirically to get engineering insights into the content caching problem. All results involving expectations are average over 10^4 independent trials.

Consider a content caching system (c.f. Figure 15.1) with m=3 CPs and n=10 users. Each content item interests 6 randomly chosen users as follows: $\mathfrak{N}_1 = \{1,3,4,5,6,10\}$, $\mathfrak{N}_2 = \{1,3,5,7,8,9\}$, and $\mathfrak{N}_3 = \{1,2,3,5,9,10\}$. The content delivery cost function is $h(\theta) = \alpha\theta^2$, where $\alpha = 0.1$ for now. We set the content delivery quality to be $\theta = 1$ temporarily. The acquisition costs of the three content items are set to be $r_1 = 4.2036$, $r_2 = 1.2714$, and $r_3 = 4.0714$, which are chosen randomly. In the following simulations, we consider two user type distributions, namely a uniform distribution and an exponential distribution, both of which satisfy the regularity condition in Definition 1. For uniformly distributed user types, we set the lower and upper bounds of user type supports to be $\underline{a}_j = 1 + 0.1(j-1)$ and $\overline{a}_j = 4 + 0.1(j-1)$, $j = 1, \ldots, n$, respectively. In addition, for exponentially distributed user types, i.e., $f_j(t_j) = \lambda_j e^{-\lambda_j t_j}$ for $t_j \geq 0$, we set the distribution parameters to be $\lambda_j = \frac{1}{10+0.4(j-1)}$, $j = 1, \ldots, n$, so that the expected types of users form an arithmetic progression. For either uniformly distributed types or exponentially distributed types, the expected types of users increase with user index.

We first study the expected payment $\mathbb{E}[x_j^*(\boldsymbol{t})]$, the expected utility $\mathbb{E}[u_j(\boldsymbol{t},\boldsymbol{t})] = \mathbb{E}\left[\left(\sum_{i\in\mathfrak{S}_j} p_i^*(\boldsymbol{t})\right)\theta t_j\right] - \mathbb{E}[x_j^*(\boldsymbol{t})]$, and the expected fraction of content items $\mathbb{E}\left[\sum_{i\in\mathfrak{S}_j} p_i^*(\boldsymbol{t})\right]$ of each user j. The results for uniformly distributed user types and exponentially distributed user types are shown in Figure 15.2(a) and 15.2(b), respectively. From Figure 15.2(a), we observe that the users with higher expected fractions of content items generally tend to have higher expected utilities and payments. Furthermore, among users with similar expected fractions of content items (e.g., users 1, 3, and 5), those with higher expected types pay more, i.e., the SP extracts more profit from those users with higher valuations than those with lower valuations. These phenomena highlight the fairness of the proposed optimal auction mechanism. Analogous observations can be drawn from Figure 15.2(b) for exponentially distributed user types. In addition, the expected cache space allocations $\mathbb{E}[\boldsymbol{p}^*(\boldsymbol{t})]$ are $[0.17, 0.657, 0.146]^\mathsf{T}$ and $[0.183, 0.259, 0.175]^\mathsf{T}$ for uniformly distributed types and exponentially distributed types, respectively. We remark that the expected idle cache space $\mathbb{E}[1 - \mathbf{1}^\mathsf{T}\boldsymbol{p}^*(\boldsymbol{t})]$ at the SP for exponentially distributed types is greater than that for the uniformly distributed types. The reason is that the exponential distribution often realizes low types (the PDF is a decreasing function), which may lead to the decision of no caching at the SP. Besides, we note that the expected caching space allocated to content item 2 is the largest among the three content items, in accordance with the large valuations of users in $\mathfrak{N}_2$ (the user indices in $\mathfrak{N}_2$ is the largest among the three content items). Moreover, we study the payment, utility, and fraction of content items for users in one typical realization of types $\boldsymbol{t}$. The results are shown in Figure 15.2(c) and 15.2(d) for uniformly distributed types and exponentially distributed types, respectively. In either case, the cache space allocation is $\boldsymbol{p}^*(\boldsymbol{t}) = [0, 1, 0]^\mathsf{T}$, i.e., the entire cache space is allocated to

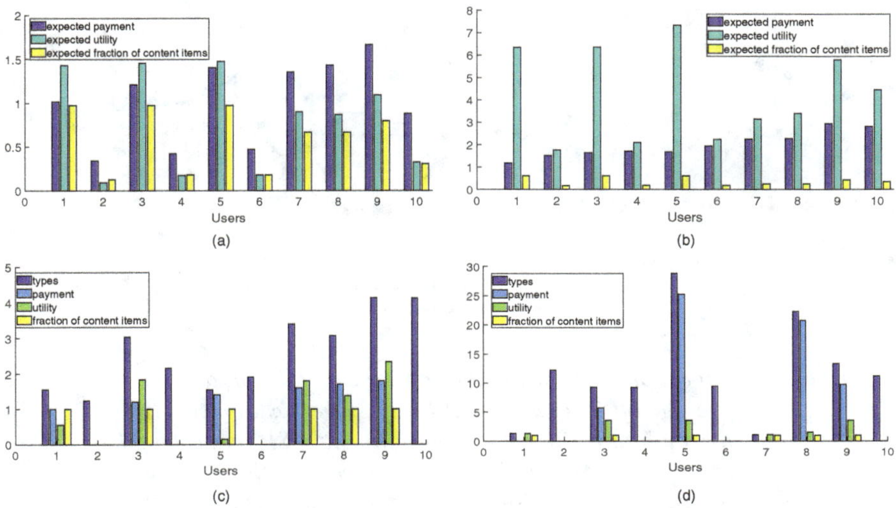

Figure 15.2 *The payment, utility, and fraction of content items of each user. (a) Uniform distribution: expected results. (b) Exponential distribution: expected results. (c) Uniform distribution: one realization. and (d) Exponential distribution: one realization.*

content item 2. We observe that payments are non-zero only for those users whose fraction of content items is positive, i.e., those in $\mathfrak{N}_2$ in this case, as guaranteed by Proposition 4. In addition, for users in $\mathfrak{N}_2$ with similar type realizations (e.g., users 3 and 8 in Figure 15.2(c)), those with higher *expected* types pay more. The reason is that to incentivize users with high-expected types to report truthfully (as requested by the IC constraint), the SP charges them relatively high payments even if their reports are low. This is to deter these users from reporting low types while possessing high true types, which are very probable since their expected types are high.

Next, we investigate the impact of type distributions on the ER of the SP and the expected average utility of users. Suppose the user types are i.i.d. uniformly distributed with $\underline{a}_j = \underline{a}$, $\overline{a}_j = \overline{a}$ for any j. First, we keep the length of the support of the type distributions, i.e., $\overline{a} - \underline{a}$, to be fixed at 3 and let the expected type, i.e., $\frac{1}{2}(\overline{a} + \underline{a})$, vary from 2 to 6.5. In other words, we set $(\underline{a}, \overline{a})$ to be $(0.5, 3.5), (1, 4), (1.5, 4.5), ..., (5, 8)$. In such a way, the level of uncertainty of types remains unchanged while the expected types increase. The ER of the SP and the expected average utility of users are shown in Figure 15.3(a) and 15.3(d), respectively. We observe that the ER of the SP increases with the expected valuation of users. This is reasonable because users with higher valuations tend to pay more to the SP. In addition, as the expected valuation of users increases, the expected average utility of users first increases and then saturates. This suggests that, when users' valuations are high enough, the SP may reap all the excess valuations from users in the optimal auction mechanism, whose goal is to maximize the SP's ER. Second, we keep the expected type of users

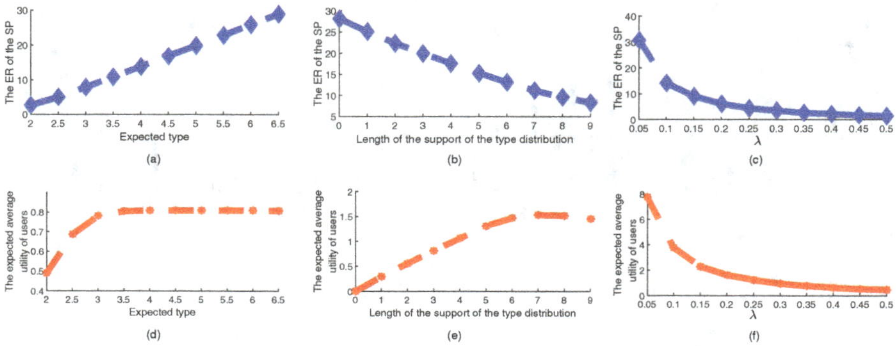

Figure 15.3 *Impact of type distributions on the ER of the SP and the expected average utility of users. (a) ER of SP versus expected type for uniformly distributed types. (b) ER of SP versus length of support for uniformly distributed types. (c) ER of SP versus λ for exponentially distributed types. (d) Expected average utility of users versus expected type for uniformly distributed types. (e) Expected average utility of users versus length of support for uniformly distributed types. (f) Expected average utility of users versus λ for exponentially distributed types.*

$\frac{1}{2}(\overline{a}+\underline{a})$ fixed at 5 and let the length of the support $\overline{a}-\underline{a}$ vary from 0 to 9. In such a case, the level of uncertainty in user types increases while the expected types remain unaltered. The corresponding ER of the SP and the expected average utility of users are shown in Figure 15.3(b) and 15.3(e), respectively. Unsurprisingly, the ER of the SP decreases with the length of the support of type distributions, since the uncertainty of user valuations hinders the SP's profit extraction. Moreover, as the length of support increases, the expected average utility of users first increases because the uncertainty in users' private valuations prevents the SP from charging too much in payments. In particular, the expected average utility of users vanishes when the length of the support is zero. In such a case, the user types become deterministic so that the SP can reap all possible profit from the users. Interestingly, the expected average utility of users decreases slightly after the length of the support is large enough (c.f. Figure 15.3(e) when the length of support is larger than 7). The reason may be that, when user types are very uncertain, the SP cannot extract enough profit to compensate for the content delivery and acquisition costs so that he/she often chooses not to cache anything. This in turn hurts the expected utility of users. Third, we consider i.i.d. exponentially distributed user types with distribution parameters $\lambda_j = \lambda$ for all j. We let λ increase so that the expected types $(1/\lambda)$ decrease. The corresponding ER of the SP and the expected average utility of users are shown in Figure 15.3(c) and 15.3(f), respectively. We observe that, as the expected user valuation decreases (i.e., λ increases), both the ER of the SP and the expected average

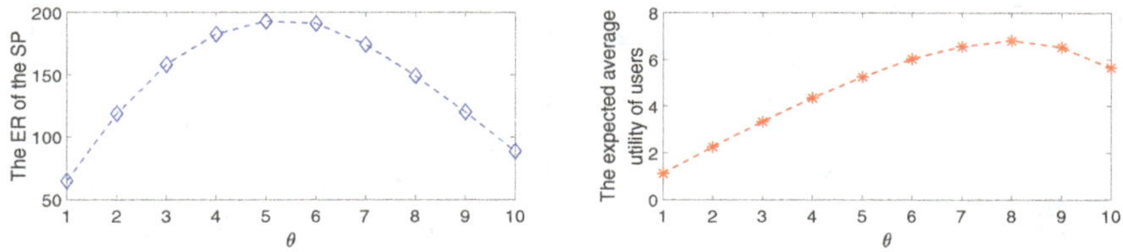

(a) Impact of θ on the ER of the SP (b) Impact of θ on the expected average utility of users

Figure 15.4 Impact of θ on the ER of the SP and the expected average utility of users

utility of users decrease. This is reasonable since high valuations of users benefit both the SP and the users themselves.

Furthermore, we examine the impact of the content delivery quality θ on the SP and the users. Suppose user types are i.i.d. uniformly di stributed on $[\underline{a}, \overline{a}]$, where we set $\underline{a} = 1$ and $\overline{a} = 4$. Suppose we have a (relatively) large number of users $n=100$ and each Ω_i is formed according to the procedure in Assumption 3 with $q_1 = 0.7$, $q_2 = 0.5$, and $q_3 = 0.4$. It can be easily verified that the quadratic content delivery cost function $h(\theta) = \alpha\theta^2$ satisfies Assumption 4, where α is set to be 0.1. Thus, all Assumptions 1–4 hold true. According to Theorem 2, the optimal θ^* that maximizes the ER of the SP is $\theta^* = \frac{a}{2\alpha} = 5$. To confirm this empirically, we report the ER of the SP and the expected average utility of users with varying θ in Figure 15.4(a) and 15.4(b), respectively. From Figure 15.4(a), we observe that, in accordance with our theoretical result, the optimal θ^* to maximize the ER of the SP is 5. From Figure 15.4(b), we further see that the value of θ that maximizes the expected average utility of users is 8, which is larger than 5, the optimal θ^* for the SP. This is reasonable since improving the delivery quality θ directly benefits users and only indirectly benefits the SP through users' payments. As such, the users tend to prefer higher delivery quality than the SP does.

Next, we study the impact of content popularity, which is captured by $|\Omega_i|$. We return to the case of $n = 10$ users whose types are uniformly distributed with parameters $\underline{a}_j = 1 + 0.1(j - 1)$ and $\overline{a}_j = 4 + 0.1(j - 1)$ for all j. For simplification, we let each content item interest the same number of users k, i.e., $|\Omega|_i = k$ for all i, and each Ω_i consists of k randomly chosen users. The ER of the SP and the expected average utility of users with varying k, i.e., varying number of users interested in each content item, are illustrated in Figure 15.5(a) and 15.5(b), respectively. We observe that both the ER of the SP and the expected average utility of users increase with content popularity. As content items become more popular, the SP can reap profits from more users. On the other hand, each user is charged a lower price and her utility can be boosted.

We also examine the impact of the number of users n on the ER of the SP and the expected average utility of users in Figure 15.6(a) and 15.6(b), respectively. We set each $|\Omega_i|$ to be $0.6n$, i.e., each content item interests a constant proportion of

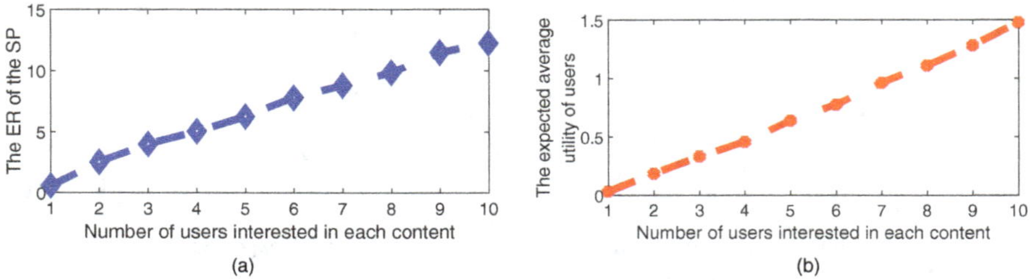

Figure 15.5 *Impact of the content popularity (captured by* $|\Omega_i|$*). (a) Impact of the content popularity on the ER of the SP. (b) Impact of the content popularity on the expected average utility of users.*

users. We observe that the ER of the SP increases with n because the SP can collect more profit from more users. The expected average utility of users increases at first and gradually saturates for large n. This suggests that, in the optimal auction mechanism, the SP will not further discount its content prices if the number of users is already large enough.

In addition, we investigate the impact of the content delivery cost function, which is controlled by the coefficient α in the quadratic cost function used here. The greater the α is, the higher the content delivery cost for the same delivery quality θ, which is set to be 0.1 here. We report the ER of the SP and the expected average utility of users in Figure 15.7(a) and 15.7(b), respectively. We observe that, as α increases, both the ER of the SP and the expected average utility of users decrease while the former decreases faster. This is reasonable since increasing delivery cost directly hurts the SP and only reduces users' utility indirectly through increasing payments.

15.5 Conclusion

In this chapter, we have designed an optimal auction mechanism for content caching to maximize the expected revenue of the SP. This mechanism takes into

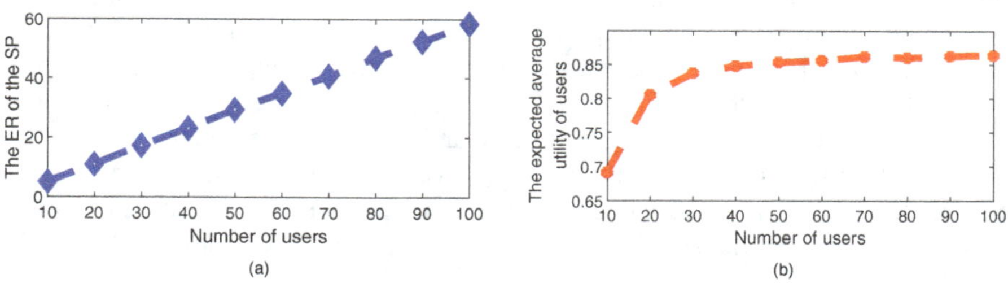

Figure 15.6 *Impact of the number of users* n. *(a) Impact of the number of users on the ER of the SP. (b) Impact of the number of users on the expected average utility of users.*

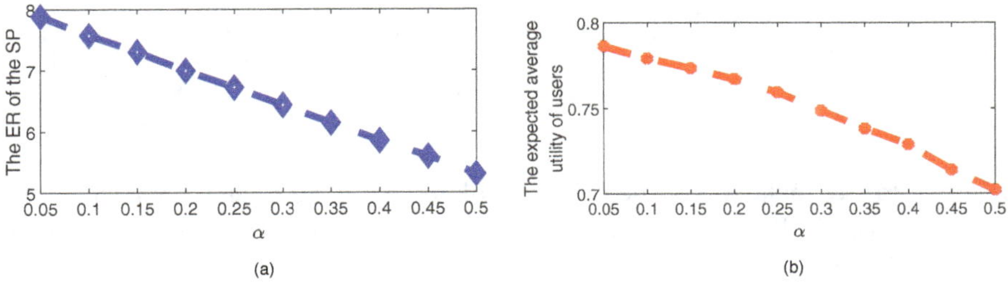

Figure 15.7 Impact of the content delivery cost function (captured by α). (a) Impact of the content delivery cost function on the ER of the SP. (b) Impact of the content delivery cost function on the expected average utility of users.

consideration both the content acquisition costs and the content delivery costs. It incentivizes truthful reports of the private user types (incentive compatibility) and user participation of the mechanism (individual rationality). Moreover, computationally efficient methods of calculating the optimal cache space allocation and user payments have been presented, and it has been shown that a user does not need to pay anything if no content of his/her interest is cached by the SP. We further examine the optimal choice of the content delivery quality under the reasonable hypothesis of a large number of users and have derived a simple formula to compute the optimal delivery quality. Finally, extensive simulations have been carried out to evaluate the performance of the proposed mechanism and the impact of various model parameters has been highlighted.

Appendix: Proof of Lemma 2

For a feasible mechanism $\langle \boldsymbol{p}(\cdot), \boldsymbol{x}(\cdot)\rangle$, according to Lemma 1, we know that statements (i), (ii), and (iii) in Lemma 1 hold. From the definition of $\widetilde{v}_j(t_j)$, we get, for any $t_j \in \mathbb{T}_j$

$$\widetilde{v}_j(t_j) = \int_{\mathbb{T}_{-j}} \left[\left(\sum_{i\in \mathbb{S}_j} p_i(\boldsymbol{t})\right)\theta t_j - x_j(\boldsymbol{t})\right] f_{-j}(\boldsymbol{t}_{-j})d\boldsymbol{t}_{-j}. \tag{15.46}$$

Along with statement (ii) of Lemma 1, we obtain

$$\int_{\mathbb{T}_{-j}} x_j(\boldsymbol{t})f_{-j}(\boldsymbol{t}_{-j})d\boldsymbol{t}_{-j} = \int_{\mathbb{T}_{-j}} \left[\sum_{i\in\mathbb{S}_j} p_i(\boldsymbol{t})\right]\theta t_j f_{-j}(\boldsymbol{t}_{-j})d\boldsymbol{t}_{-j} - \widetilde{v}_j(\underline{a}_j) - \theta\int_{\underline{a}_j}^{t_j}\widetilde{p}_j(\tau_j)d\tau_j \tag{15.47}$$

$$\overset{(a)}{=} -\widetilde{v}_j(\underline{a}_j) + \theta \int_{\mathbb{T}_{-j}} \left\{ \left[\sum_{i \in \mathcal{S}_j} p_i(\boldsymbol{t}) \right] t_j - \int_{\underline{a}_j}^{t_j} \sum_{i \in \mathcal{S}_j} p_i(\tau_j, \boldsymbol{t}_{-j}) d\tau_j \right\} f_{-j}(\boldsymbol{t}_{-j}) d\boldsymbol{t}_{-j}, \tag{15.48}$$

where in (a) we make use of the definition of $\widetilde{p}_j(\tau_j)$ in (15.10) and interchange the order of integrals. Performing the operation $\widetilde{p}_j(\tau_j)$ on both sides of the above equation, we obtain

$$\begin{aligned} \int_{\mathbb{T}} x_j(\boldsymbol{t}) f(\boldsymbol{t}) d\boldsymbol{t} = & -\widetilde{v}_j(\underline{a}_j) + \theta \int_{\mathbb{T}} \left[\sum_{i \in \mathcal{S}_j} p_i(\boldsymbol{t}) \right] t_j f(\boldsymbol{t}) d\boldsymbol{t} \\ & -\theta \int_{\mathbb{T}_{-j}} \left\{ \int_{\mathbb{T}_j} \int_{\underline{a}_j}^{t_j} \left[\sum_{i \in \mathcal{S}_j} p_i(\tau_j, \boldsymbol{t}_{-j}) \right] f_j(t_j) d\tau_j dt_j \right\} f_{-j}(\boldsymbol{t}_{-j}) d\boldsymbol{t}_{-j} \end{aligned} \tag{15.49}$$

$$\begin{aligned} &\overset{(a)}{=} -\widetilde{v}_j(\underline{a}_j) + \theta \int_{\mathbb{T}} \left[\sum_{i \in \mathcal{S}_j} p_i(\boldsymbol{t}) \right] t_j f(\boldsymbol{t}) d\boldsymbol{t} \\ &-\theta \int_{\mathbb{T}_{-j}} \int_{\mathbb{T}_j} [1 - F_j(t_j)] \left[\sum_{i \in \mathcal{S}_j} p_i(\boldsymbol{t}) \right] dt_j f_{-j}(\boldsymbol{t}_{-j}) d\boldsymbol{t}_{-j} \end{aligned} \tag{15.50}$$

$$= -\widetilde{v}_j(\underline{a}_j) + \theta \int_{\mathbb{T}} \left[t_j - \frac{1 - F_j(t_j)}{f_j(t_j)} \right] \left[\sum_{i \in \mathcal{S}_j} p_i(\boldsymbol{t}) \right] f(\boldsymbol{t}) d\boldsymbol{t}, \tag{15.51}$$

where in (a) we interchange order of integrals:

$$\begin{aligned} &\int_{\mathbb{T}_j} \int_{\underline{a}_j}^{t_j} \left[\sum_{i \in \mathcal{S}_j} p_i(\tau_j, \boldsymbol{t}_{-j}) \right] f_j(t_j) d\tau_j dt_j = \int_{\mathbb{T}_j} \int_{\tau_j}^{\bar{a}_j} \left[\sum_{i \in \mathcal{S}_j} p_i(\tau_j, \boldsymbol{t}_{-j}) \right] f_j(t_j) dt_j d\tau_j \\ &= \int_{\mathbb{T}_j} [1 - F_j(t_j)] \left[\sum_{i \in \mathcal{S}_j} p_i(\boldsymbol{t}) \right] dt_j \end{aligned} \tag{15.52}$$

Plugging (15.51) into (15.1), we get

$$\begin{aligned} ER = & -\sum_{j=1}^{n} \widetilde{v}_j(\underline{a}_j) + \theta \int_{\mathbb{T}} \left\{ \sum_{j=1}^{n} \left[t_j - \frac{1-F_j(t_j)}{f_j(t_j)} \right] \left[\sum_{i \in \mathcal{S}_j} p_i(\boldsymbol{t}) \right] \right\} f(\boldsymbol{t}) d\boldsymbol{t} \\ & - \int_{\mathbb{T}} \left\{ \sum_{i=1}^{m} p_i(\boldsymbol{t}) \left[r_i + |\Omega_i| h(\theta) \right] \right\} f(\boldsymbol{t}) d\boldsymbol{t} \end{aligned} \tag{15.53}$$

$$\begin{aligned} &\overset{(a)}{=} -\sum_{j=1}^{n} \widetilde{v}_j(\underline{a}_j) + \theta \int_{\mathbb{T}} \left\{ \sum_{j=1}^{n} \left[t_j - \frac{1-F_j(t_j)}{f_j(t_j)} - \frac{h(\theta)}{\theta} \right] \left[\sum_{i \in \mathcal{S}_j} p_i(\boldsymbol{t}) \right] \right\} f(\boldsymbol{t}) d\boldsymbol{t} \\ &- \int_{\mathbb{T}} \left[\sum_{i=1}^{m} p_i(\boldsymbol{t}) r_i \right] f(\boldsymbol{t}) d\boldsymbol{t}, \end{aligned} \tag{15.54}$$

where in (a) we make use of $\sum_{j=1}^{n} \sum_{i \in \mathcal{S}_j} p_i(\boldsymbol{t}) = \sum_{i=1}^{m} p_i(\boldsymbol{t}) |\Omega_i|$.

References

[1] Osseiran A., Braun V., Hidekazu T. 'The foundation of the mobile and wireless communications system for 2020 and beyond: challenges, enablers and technology solutions'. Proceedings of the IEEE 77th Vehicular Technology Conference; 2013. pp. 1–5.

[2] Nisan N., Ronen A. 'Algorithmic mechanism design'. Proceedings of the Thirty-first Annual ACM Symposium on Theory of Computing; 1999. pp. 129–40.

[3] Osborne M.J., Rubinstein A. *A Course in Game Theory*. MIT Press; 1994.

[4] Dai J., Liu F., Li B., Li B., Liu J. 'Collaborative caching in wireless video streaming through resource auctions'. *IEEE Journal on Selected Areas in Communications*. 2012;**30**(2):458–66.

[5] Li J., Chen H., Chen Y., Lin Z., Vucetic B., Hanzo L. 'Pricing and resource allocation via game theory for a small-cell video caching system'. *IEEE Journal on Selected Areas in Communications*. 2016;**34**(8):2115–29.

[6] Liu T., Li J., Shu F., Tao M., Chen W., Han Z. 'Design of contract-based trading mechanism for a small-cell caching system'. *IEEE Transactions on Wireless Communications*. 2017;**16**(10):6602–17.

[7] Pacifici V., Dán G. 'Distributed caching algorithms for interconnected operator CDNs'. *IEEE Journal on Selected Areas in Communications*. 2017;**35**(2):380–91.

[8] Pacifici V., Dán G. 'Coordinated selfish distributed caching for peering content-centric networks'. *IEEE/ACM Transactions on Networking*. 2016;**24**(6):3690–701.

[9] Hajimirsadeghi M., Mandayam N.B., Reznik A. 'Joint caching and pricing strategies for popular content in information centric networks'. *IEEE Journal on Selected Areas in Communications*. 2017;**35**(3):654–67.

[10] Wang L., Tyson G., Kangasharju J., Crowcroft J. 'Milking the cache cow with fairness in mind'. *IEEE/ACM Transactions on Networking*. 2017;**25**(5):2686–700.

[11] Myerson R.B. 'Optimal auction design'. *Mathematics of Operations Research*. 1981;**6**(1):58–73.

[12] Nadendla V.S.S., Brahma S.K., Varshney P.K. 'Optimal spectrum auction design with 2-D truthful revelations under uncertain spectrum availability'. *IEEE/ACM Transactions on Networking*. 2017;**25**(1):420–33.

[13] Cao N., Brahma S., Varshney P.K. 'Target tracking via crowdsourcing: a mechanism design approach'. *IEEE Transactions on Signal Processing*. 2015;**63**(6):1464–76.

[14] Chun S.H., La R.J. 'Secondary spectrum trading—auction-based framework for spectrum allocation and profit sharing'. *IEEE/ACM Transactions on Networking*. 2013;**21**(1):176–89.

[15] Wang C., He Y., Yu F.R., Chen Q., Tang L. 'Integration of networking, caching, and computing in wireless systems: a survey, some research issues, and challenges'. *IEEE Communications Surveys & Tutorials*. 2018;**20**(1):7–38.

[16] Myerson R.B. 'Incentive compatibility and the bargaining problem'. *Econometrica*. 1979;**47**(1):61–73.

[17] Baron D.P., Besanko D., Regulation B.D. 'Regulation, asymmetric information, and auditing'. *The RAND Journal of Economics*. 1984;**15**(4):447–70.

[18] Golrezaei N., Dimakis A.G., Molisch A.F. 'Scaling behavior for device-to-device communications with distributed caching'. *IEEE Transactions on Information Theory*. 2014;**60**(7):4286–98.

[19] Ji M., Caire G., Molisch A.F. 'The throughput-outage tradeoff of wireless one-hop caching networks'. *IEEE Transactions on Information Theory*. 2015;**61**(12):6833–59.

[20] Jeon S.-W., Hong S.-N., Ji M., Caire G., Molisch A.F. 'Wireless multihop device-to-device caching networks'. *IEEE Transactions on Information Theory*. 2017;**63**(3):1662–76.

[21] Georgiadis L., Neely M.J., Tassiulas L., *et al.* 'Resource allocation and cross-layer control in wireless networks'. *Foundations and Trends® in Networking*. 2006;**1**(1):1–144.

[22] Gallager R.G. *Discrete Stochastic Processes*. **321**. Springer Science & Business Media; 2012.

Chapter 16

Contract theory-based approach for caching

Tingting Liu[1,2], Jun Li[3], Feng Shu[4], and Zhu Han[5]

16.1 Introduction

With the dramatically growing demands on video entertainment and social connections, wireless multimedia communications have played an important role in future cellular networks. Unfortunately, the capacity of current networks has not been able to keep a similar pace with the tremendous growth of cellular traffic [1–3]. At the same time, it is pointed out that the next generation of mobile communication systems will have to meet the following requirements: high throughput, high access amount, high data rate, low power consumption and low latency [4, 5]. In order to fulfill these requirements, researchers have begun conceiving 5G cellular networks [6, 7] and are focusing on key techniques [8], such as millimeter wave technology [9], massive multiple-input multiple-output [10] and super density heterogeneous networks [11]. However, these techniques may need to intensively change the hardware equipment and network protocols, imposing high cost and complexity to current communication systems. It is urgent to find a solution which is more economical and convenient.

At the same time, by observing network traffic, it can be found that there are numerous repetitive requests of popular contents, e.g., video clips, raised by mobile users (MU), leading to heavy traffic pressures on the backhaul channels [12]. The redundancy of transmission in the current networks can be effectively mitigated by storing popular contents into local caches of network nodes located at the edge of wireless networks, known as wireless caching [13–15]. Wireless caching technology, which is economical and easy to realize, generally consists of two stages: a data placement stage [16] and a data delivery stage [17]. In the first stage, files such as popular contents are pre-cached to the local storage during off-peak time. While

[1]The School of Information and Communication Engineering, Nanjing Institute of Technology, China
[2]The School of Electronic Science and Engineering, Nanjing University, China
[3]The School of Electronic and Optical Engineering, Nanjing University of Science and Technology, China
[4]The School of Information and Communication Engineering, Hainan University, China
[5]Department of Electrical and Computer Engineering, University of Houston, USA

in the second stage, MUs may acquire data directly from local facilities which are adjacent to them. As such, wireless caching mechanism can release pressures on backhaul channels and bring contents closer to the MUs.

Most of the existing works on wireless caching focus on the optimization of the aforementioned two stages in order to improve the caching efficiency. Meanwhile, some researchers begin to investigate a small-cell caching system from the commercial perspective. Particularly, the authors in [18] research on a commercial video-caching system using the Stackelberg game by treating the SBSs as a specific type of resource within the scenario of one network service provider (NSP) and multiple content providers (CPs). Furthermore, the authors extend their researches to the scenario with one CP and multiple NSPs [19].

However, in the aforementioned two works, the utilities of both involved sides are assumed to be known to each other, which is the information symmetric environment [20]. When trading is modeled in an information asymmetry scenario, i.e., one side has certain private information, which cannot be observed by the other side, contract theory is provided as an effective way to solve this kind of problem [21]. Most of the existing works on contract theory can be classified into two categories: one is adverse selection referring to the problems of hidden information [22–28], and the other one is moral hazard referring to the problems of hidden actions [29, 30].

16.1.1 Related works

The related works on contract theory-based approach for caching can be classified into two camps, i.e., the adverse selection and moral hazard. We will discuss them from two aspects in the following subsections.

16.1.1.1 Adverse selection

Adverse selection-type contract theory-based approach for caching [31] designs an incentive contract between a mobile network operator (MNO) and multiple CPs in a fixed network. The proposed model exploits the strategic interdependence between the CPs, and the optimization problem is defined to maximize the utility of the CPs while the MNO is ensured to be budget balanced. In [32], the authors design a contract-theoretic inventive mechanism for CPs which are classified into different types based on two-dimensional factors. The base stations offer the optimal contracts for CPs to motivate them to help offload traffic. In [33], the authors propose a caching strategy and utilize the contract theory to solve joint resource allocation problem in a wireless network virtualization environment. In [34], the authors study the incentive mechanism problem based on contract theory between the infrastructure provider (InP) and service providers (SPs). An InP designs the optimal contracts to motivate SPs to rent caching spaces and maximize its utility. In [35], the authors study hierarchical game problems, including a cache allocation problem and a user association problem, in cloud radio access networks. Particularly, the cache allocation problem between the network operator and CPs is solved by contract theory. In [36], the authors consider a two-layer network, where each layer provides a different quality of service (QoS) to MUs. The optimal contracts for CPs are designed by taking account of the service differences of the two layers. In [37], the authors investigate the optimal contract design problem in ultra-dense cellular networks where the SBSs are divided into multiple tiers in a dynamic on-off working mode.

16.1.1.2 Moral hazard

In moral hazard-type contract theory-based approach for caching, in [38], the authors investigate cooperative caching based on multi-hop communications, and the incentive mechanism developed for a moral hazard scenario is presented to encourage the cooperative users caching and transmitting contents. In [39], the authors consider a problem of revenue sharing contracts between CPs to a common service provider (CSP). Moral hazard framework is employed to determine the optimal contract in order to maximize the CPs' profits. In [40], the authors investigate the moral hazard contract framework in the wireless network virtualization scenario which is an enable technology used in enhancing the wireless caching performance. In the concerned scenario, based on the optimal contract, the SPs can purchase some reused physical resources and enhance the spectrum efficiency.

16.1.2 Small-cell caching issues

In a small-cell caching system, there exists an asymmetric information scenario, where the NSP, as a network facility monopolist, intends to lease the SBSs to the CPs for maximizing its profits, while it does not know the specific type of each CP. Accordingly, we are inspired to solve the problem of asymmetric information within the framework of contract theory. Here, the SBSs owned by the NSP are divided into different *qualities*, i.e., various fractions of the SBSs. Moreover, the CPs can be classified into different *types* according to their popularity among the MUs. We model the trading in a monopoly market, where the monopolist NSP sets the contract entries with combinations of quality and price for its goods, and each buyer CP with a specific type chooses an appropriate entry to sign. The proposed scheme, together with the derived solutions, can offer proper economical incentives to the NSP. All the proofs in the chapter can be found in [41].

The remainder of this chapter is organized as follows: The system model is presented in Section 16.2. The contract-based service model is elaborated in Section 16.3. The optimal contract design and solutions are developed in Section 16.4. Other resource allocation and pricing strategies are discussed in Section 16.5. Numerical results are presented in Section 16.6, and conclusions are drawn in Section 16.7.

16.2 System model

We focus on a commercial small-cell caching system with one NSP, N CPs and multiple MUs. In this system, by viewing the SBSs as a kind of resource of the NSP, each CP rents a certain fraction of these SBSs from the NSP for placing its popular contents. An MU affiliated with a CP may directly download data from its nearby SBSs that have been rented by this CP. Otherwise, the MU has to acquire

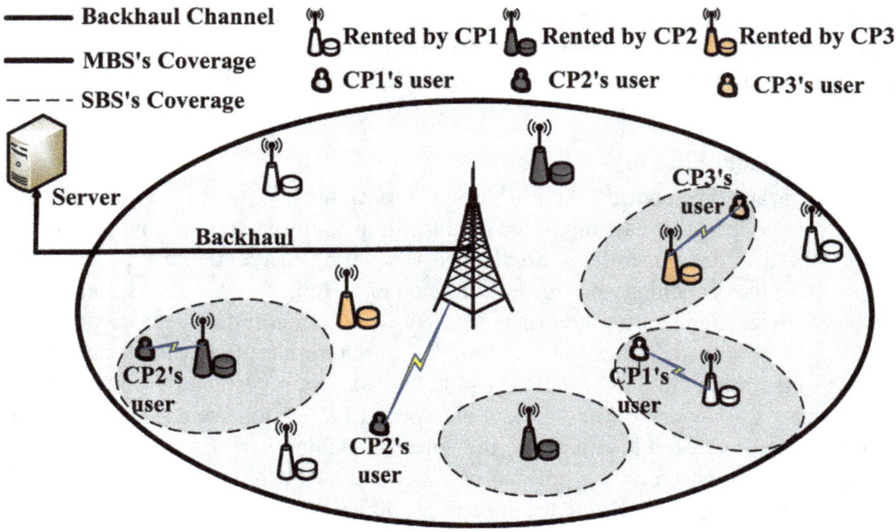

Figure 16.1 *An example model with one NSP and three CPs, where CP1 rents four SBSs, CP2 rents three SBSs, and CP3 rents two SBSs*

data from the MBS. An example system is depicted in Figure 16.1, where three CPs rent a number of SBSs owned by the NSP. In this section, network model is first presented in Section 16.2.1. Then, caching and procedures are given in Section 16.2.2. Caching procedures are elaborated in Section 16.2.3. A table of notations used throughout this chapter is given in Table 16.1.

16.2.1 Network model

Let us consider a small cell network consisting of multiple SBSs owned by the monopolist NSP. We denote the set of N CPs by $\mathscr{V} = \{\mathscr{V}_1, \cdots, \mathscr{V}_v, \cdots, \mathscr{V}_N\}$ and the NSP by $\mathcal{L}$. We assume that SBSs are equipped with an uniform transmission power of P and the same caching size of Q files. Also, the spatial distributions of the SBS and the MUs are modeled as two independent homogeneous Poisson point processes (PPP) Φ and Ψ, with densities λ and ζ, respectively. SBSs transmit on the orthogonal channels to the MBS, and hence we do not consider interferences induced by the MBSs.

We consider a typical MU located at the origin and an SBS located at x. The path-loss attenuation between an SBS and the typical MU is denoted by $||x||^{-\alpha}$, where α is the path-loss exponent. The channel between an SBS and the typical MU is denoted by h_x, where we have $h_x \sim \exp(1)$. The noise is modeled as the additive white Gaussian noise (AWGN) with mean zero and variance σ^2.

Table 16.1 *Table of Notations*

Notations	Meanings
$\mathscr{V}_v$	The vth CP
$\mathscr{L}$	The NSP
$\mathscr{F}_i$	The th file
N	The number of CPs
F	The number of files
Q	The maximum file number stored in each SBS
γ	The CPs' popularity distribution parameter
β	The files' popularity distribution parameter
$\mathscr{E}_{v,i}$	The event that an MU of $\mathscr{V}_v$ requests a file $\mathscr{F}_i$ from the SBS
τ_v	The fraction of the SBS assigned to $\mathscr{V}_v$
π_v	The corresponding price of τ_v
p_i	The popularity of $\mathscr{F}_i$
θ_v	The popularity of $\mathscr{V}_v$
λ	The density of SBSs within a unit area (/UA)
ζ	The density of MUs within a unit area (/UA)
K	Average file downloading demands from each MU within a unit period (/UP)
s^{bh}	Average backhaul cost for a file transmission
s^{ld}	Average local downloading saving for a file transmission
κ	The scale parameter between s^{bh} and s^{ld}

A saturate network is considered in this model, where all SBSs are powered on and keep transmitting to their subscribers. Hence, the signal-to-interference-and-noise-ratio (SINR) at the typical MU from an SBS, whose location is x, can be expressed as

$$\rho(x) = \frac{Ph_x||x||^{-\alpha}}{\sum_{x' \in \Phi \backslash x} Ph_{x'}||x'||^{-\alpha} + \sigma^2}. \tag{16.1}$$

Note that if $\rho(x)$ is no less than a predefined threshold δ, this typical MU can be covered by the SBS located at x.

16.2.2 Popularity and preferences

Let us denote the file set by $\mathscr{F} = \{\mathscr{F}_1, \mathscr{F}_2, \cdots, \mathscr{F}_F\}$ consisting of F files. The requested probabilities by the MUs, i.e., popularity distribution, of the files are denoted by $\mathbf{p} = \{p_1, \cdots, p_i, \cdots, p_F\}$, which has been shown to follow the Zipf distribution [42–44] as

$$p_j = \frac{1/j^\beta}{\sum_{f=1}^{F} 1/f^\beta}, \quad j = 1, \cdots, F, \tag{16.2}$$

where the positive value β represents the files' popularity. Note that each SBS can store at most Q files, and we assume $Q < F$. In this chapter, we assume that all the CPs generate the same file set $\mathscr{F}$. Note that different CPs may have different file sets in a practical system. However, this does not affect the design procedure for the optimal contract.

Usually, MUs have different preferences to the N CPs, combining with many factors including personal favor, quality of service and the charging standards, and so on. We again utilize the Zipf distribution to present MUs' preferences to $\mathcal{V}$, denoted by $\boldsymbol{\Theta} = \{\theta_1, \cdots, \theta_v, \cdots, \theta_N\}$,

$$\theta_v = \frac{1/v^{\gamma}}{\sum_{j=1}^{N} 1/j^{\gamma}}, \quad v = 1, \cdots, N, \tag{16.3}$$

where γ is a positive value and determines the distribution of CPs' popularity among the MUs [42]. The parameter γ can be determined by questionnaires conducted by a consulting firm or through learning of the historical data flow by the NSP. Moreover, the distribution of MUs' preferences can be any function other than Zipf, since it does not affect the design of the optimal contract.

16.2.3 Caching procedures

In this subsection, we describe the content caching procedure in a small-cell caching system. There are three stages in our system. In the first stage, each CP rents a certain fraction of the SBSs from $\mathscr{L}$ for placing its files. We denote the fraction vector by $\boldsymbol{\Gamma} = \{\tau_1, \tau_2, \cdots, \tau_N\}$, in which τ_v represents the fraction of the SBSs assigned to $\mathscr{V}_v$. In general, we have $\tau_v = 0$, if a CP $\mathscr{V}_v$ does not rent any fraction from $\mathscr{L}$. Obviously, fractions cannot be negative or infinity, and thus we have $\sum_{v=1}^{N} \tau_v = 1$ and $\tau_v \geq 0$. We assume that the SBSs rented by $\mathscr{V}_v$ are uniformly and randomly distributed and can be modeled as a thinned homogeneous PPP Φ_v of density $\tau_v \lambda$.

In the second stage, the data placement commences during off-peak time after each CP obtains an access to the SBSs. Owing to the limited caching volume of each SBS, the Q most popular files are placed for increasing the caching efficiency.

In the third stage, an MU of $\mathscr{V}_v$ requests a file $\mathscr{F}_i \in \mathscr{F}$. It first searches the SBSs in Φ_v and connects to the nearest SBS that caches the requested file and covers it. If such an SBS exists, the subscriber obtains this file directly from the caching of this SBS. This event is defined by $\mathscr{E}_{v,i}$. Otherwise, the MU triggers the transmission via backhaul channels of the NSP for remotely dispatching the requested file, leading to an extra cost on $\mathscr{L}$.

Followed by the same derivation steps in [45], the probability $\Pr(\mathscr{E}_{v,i})$ of the event $\mathscr{E}_{v,i}$ can be derived as

$$\Pr(\mathscr{E}_{v,i}) = \frac{\tau_v}{\tau_v A(\delta, \alpha) + (1 - \tau_v) C(\delta, \alpha) + \tau_v}, \quad i = 1, 2, \cdots, Q, \tag{16.4}$$

where $A(\delta, \alpha) = \frac{2\delta}{\alpha - 2} {}_2F_1(1, 1 - \frac{2}{\alpha}; 2 - \frac{2}{\alpha}; -\delta)$ and $C(\delta, \alpha) = \frac{2}{\alpha} \delta^{\frac{2}{\alpha}} B(\frac{2}{\alpha}, 1 - \frac{2}{\alpha})$. Furthermore, ${}_2F_1(.)$ in the function $A(\delta, \alpha)$ is the hypergeometric function and

$B(.,.)$ is the beta function in $C(\delta,\alpha)$. From (16.4), we can see that the probability of $\Pr(\mathscr{E}_{v,i})$ is independent of the transmission power P and the intensity λ of the SBSs.

16.3 Contract-based service model

We now focus our attention on modeling the contract-based trading on the SBSs in our small-cell caching system as shown in Figure 16.1. Since the NSP $\mathscr{L}$ owns the network facilities and dominates the trading process, we consider a monopoly market, where $\mathscr{L}$ is the monopolist and designs different contract entries for the CPs in $\mathscr{V}$ for incentivizing them to participate in the trading market. On the other hand, each CP is free to accept or decline any contract offered by $\mathscr{L}$ based on its own goodness. Contract theory is proposed to solve the profit maximization problem for a monopoly market, where there is only one monopolist controlling the resources. Considering the competitive sellers' scenario, other game theory schemes such as matching theory, auction theory or Stackelberg game theory might be applicable. Incorporating the caching and data transmission processes, we elaborate on the contract issues in our small-cell caching system in the following.

16.3.1 NSP model

We assume that $\mathscr{L}$ sets the contract entries $\{\boldsymbol{\Gamma},\boldsymbol{\Pi}\}$, where the set of prices is denoted by $\boldsymbol{\Pi}=\{\pi_1,\pi_2,\cdots,\pi_N\}$, and each quality τ_v in $\boldsymbol{\Gamma}$ corresponds to a price π_v. The SBSs are divided into different fractions with each fraction being regarded as a specific *quality*. The CPs are free to decide whether or which quality to purchase. The given discussions are the contract-related construction and commitment taking place before caching procedures.

We assume that there are averagely K file downloading demands from each MU within a unit period (*/UP*), e.g., */month*, and the average backhaul cost for a file transmission is s^{bh}. The profits of $\mathscr{L}$ in terms of per unit area times unit period (*/UAP*), e.g., */month* $\cdot$ *km*^{2}, come from leasing the SBSs plus the saved transmission costs over backhaul channels. If a CP rents the SBSs with quality τ_v, the first Q files $\mathscr{F}_i, i\leq Q$ will be cached in each local SBSs, since they are the most Q popular files among the users. This is the most efficient way to place files. Please be noted that each CP rents multiple SBSs and each CP caches the same set of files, i.e., the most Q popular files, to each SBSs it rents. Based on (16.4), the saved backhaul costs (*/UAP*) of $\mathscr{L}$ can be expressed by

$$S_v^{\text{BH}}=\sum_{f=1}^{F}p_f\theta_v\zeta Ks^{\text{bh}}g(\tau_v),\quad \forall v, \tag{16.5}$$

where

$$g(\tau_v) = \begin{cases} \Pr(\mathcal{E}_{v,f}), & f \le Q; \\ 0, & f > Q. \end{cases} \tag{16.6}$$

The CP_v's backhaul saving of a file $\mathscr{F}_f$ can be calculated as $p_f\theta_v\zeta K s^{\text{bh}} g(\tau_v)$, in which $\theta_v\zeta$ represents the MUs' density affiliate to CP and $p_f K$ means the downloading demand for the file $\mathscr{F}_f$ from each MU. Thus, $\theta_v\zeta p_f K$ is the downloading demand of $\mathscr{F}_f$ of the MUs affiliated to CP_v. $g(\tau_v)$ that represents the probability of the event that an MU can successfully download $\mathscr{F}_f$ from the local SBS. Therefore, $\theta_v\zeta p_f K g(\tau_v)$ is the number of successful downloading of the file $\mathscr{F}_f$ requested by the MUs affiliated to CP_v from the local SBSs instead of the backhaul channels. Thus, $\theta_v\zeta p_f K g(\tau_v) s^{\text{bh}}$ is saved backhaul cost of the NSP from the successful local downloading of the file $\mathscr{F}_f$ requested by the MUs affiliated with CP_v. Then the utility of $\mathscr{L}$ can be formulated as

$$U_{\text{NSP}}(\mathbf{\Theta}, \mathbf{\Gamma}, \mathbf{\Pi}) = \sum_{v=1}^{N} (\pi_v + S_v^{\text{bh}}). \tag{16.7}$$

Note that there is no cost presented in (16.7). Usually, costs of $\mathscr{L}$ come from two aspects: the power cost for transmitting cached files requested and the storage consumption for caching. These costs can be considered to be included in the original price firstly and then be nulled by subtracting them. Hence, in this chapter we focus on the remaining part of the price π_v, and we represent the net profit of $\mathscr{L}$ as in (16.7).

16.3.2 CP model

If $\mathscr{V}_v$ signs a contract of $\{\tau_v, \pi_v\}$, its MUs who locally trigger the event $\mathscr{E}_{v,i}$, $\forall i \le Q$, will save a long-distance transmission for $\mathscr{V}_v$, referred to as local downloading savings (LDS). The LDS for a file transmission is denoted by s^{ld}. Based on (16.4), we can calculate the overall savings S_v^{ld} (*/UAP*) of $\mathscr{V}_v$ as

$$S_v^{\text{ld}} = \sum_{f=1}^{Q} p_f\theta_v\zeta K s^{\text{ld}} g(\tau_v), \quad \forall v. \tag{16.8}$$

According to the popularity distribution of the CPs, we classify these CPs into N different types. The definition of CPs' types is given in Definition 1.

Definition 1: CPs' types: We use popularity defined in (16.3) to represent the types of the CPs in $\mathscr{V}$, which are sorted in a descending order as $\theta_1 > \cdots > \theta_v > \cdots > \theta_N$.

A higher type implies greater popularity, and hence more data traffic can be potentially incurred on the backhaul channels. Please note that it is the information asymmetric environment, where the exact values of the CP types are private information. The NSP does not know the specific type of each CP and is only aware of the distribution information about the CP types, i.e., the Zipf parameter γ.

To simplify the notations, we define $M^{\text{ld}} \triangleq \sum_{f=1}^{Q} p_f\zeta K s^{\text{ld}}$ and $M^{\text{bh}} \triangleq \sum_{f=1}^{Q} p_f\zeta K s^{\text{bh}}$. Thus, we have $S_v^{\text{bh}} = M^{\text{bh}}\theta_v g(\tau_v)$. For the clearance of discussion, we further define the valuation of the quality τ_v by the corresponding LDS S_v^{LD} in (16.8), i.e.,

$$V(\theta_v, \tau_v) \triangleq S_v^{\text{LD}} = M^{\text{ld}}\theta_v g(\tau_v), \quad \forall v, \tag{16.9}$$

which is increasing with the type θ_v and is also a strictly increasing and concave function of τ_v, i.e., $\partial V/\partial \tau_v > 0$ and $\partial^2 V/\partial \tau_v^2 < 0$. Simply speaking, $V(\theta_v, \tau_v)$ can be viewed as the benefits of type θ_v received from employing quality τ_v. The utility of $\mathcal{V}_v$ can be formulated as

$$U_v(\theta_v, \tau_v, \pi_v) = V(\theta_v, \tau_v) - \pi_v. \tag{16.10}$$

A rational CP will not accept a negative utility U_v, and thus we have $V(\theta_v, \tau_v) \geq \pi_v$.

16.4 Optimal contract design and solution

In our contract-based caching system, the NSP $\mathscr{L}$ designs the contract and each CP in $\mathscr{V}$ is free to choose an appropriate entry. The goal of $\mathscr{L}$ is to maximize U_{NSP} by offering the optimal contract entries $\{\tau_v^*, \pi_v^*\}, v = 1, 2, \cdots, N$. In this section, the optimal contract problem is first formulated in Section 16.4.1. Then, constraints are reduced in Section 16.4.2, and solution to the optimal contract problem is discussed in Section 16.4.3.

16.4.1 Contract formulation

A feasible contract, which is capable of attracting the CPs to buy certain fractions of SBSs for placing their contents, must comply with the feasibility constraint on the individual rationality (IR) and incentive compatibility (IC) [46] for all N types. First, the definition of IR constraint is given in Definition 2.

Definition 2: IR constraint: Given that each CP is rational, it will not accept a contract when it receives a negative utility for its type. Therefore, the IR constraint can be expressed as

$$V(\theta_v, \tau_v) - \pi_v \geq 0, \quad \forall v. \tag{16.11}$$

In other words, to stimulate $\mathscr{V}_v$ renting the SBSs, the rewards obtained must compensate its cost. In the case of $U_v < 0$, $\mathscr{V}_v$ will purchase nothing. The definition of IC constraint is given in Definition 3.

Definition 3: IC constraint: The incentive compatibility constraint means that $\mathscr{V}_v$ cannot gain more utility by accepting a contract entry which is not designed for its type. That is,

$$V(\theta_v, \tau_v) - \pi_v \geq V(\theta_v, \tau_{\tilde{v}}) - \pi_{\tilde{v}}, \ \forall \tilde{v} \in \{1, \cdots, N\}, \ \tilde{v} \neq v, \tag{16.12}$$

where the contract $\{\tau_{\tilde{v}}, \pi_{\tilde{v}}\}$ is designed for type $\theta_{\tilde{v}}$. In other words, $\mathscr{V}_v$ with type θ_v obtains the maximum utility if and only if it chooses the contract $\{\tau_v, \pi_v\}$ designed for its type.

A feasible contract must guarantee the IC and IR constraints, and we focus on designing the optimal contract that maximizes the utility of $\mathscr{L}$ subject to these two basic constraints. We formulate the optimization of the contract problem as:

$$\begin{aligned} &\{\tau_v^*, \pi_v^*\} = \arg\max \ U_{\text{NSP}}, \\ &\text{s.t. IR(16.10), IC(16.11)}, \ \sum_{\mu} \tau_\mu = 1, \ \tau_\mu \geq 0. \end{aligned} \tag{16.13}$$

16.4.2 Constraints reduction

The problem in (16.13) is not straightforward to solve, since there are N IR constraints and $N(N-1)$ IC constrains. We simplify these constraints followed by a standard method before solving the optimization problem [46]. Firstly, we simplify the IR constraints by Lemma 1.

Lemma 1: (to simplify the IR constraints): For an optimal solution, given that the IC constraints are satisfied, the IR constraint for the lowest type θ_N is binding, i.e.,

$$V(\theta_N, \tau_N) - \pi_N = 0. \tag{16.14}$$

Proof: Please refer to [41].

The IR constrains can be reduced by Lemma 1, which indicates that the price π_N for the lowest type should be equal to the valuation of quality τ_N, i.e., the lowest type θ_N gains zero profit, while other CPs' profits are larger than that of the binding one.

Next, we will prove that the IC constraints can be simplified by the following lemmas. We begin by introducing the necessary conditions, i.e., Lemmas 2 and 3, for IC constrains.

Lemma 2: If the contract satisfies the IC constraints, the following condition holds true: given the quality $\tau_v > \tau_{\tilde{v}}$, if and only if the price satisfies $\pi_v > \pi_{\tilde{v}}$.

Proof: Please refer to [41].

Lemma 2 presents an important property for a feasible contract, i.e., a higher quality corresponds to a higher price and vice versa.

Lemma 3: If the contract satisfies the IC constraints, the quality τ_v of the SBSs monotonically increases with $\mathcal{V}_v$'s type θ_v, i.e., if $\theta_v > \theta_{\tilde{v}}$, then we have $\tau_v > \tau_{\tilde{v}}$.

Proof: Please refer to [41].

Lemma 3 implies that the quality τ_v is monotonically increasing with the type θ_v when the contract satisfies the IC constraints. This means that the quality assigned to a higher type must be larger than that to a lower one. Recall that type θ_v indicates the requesting proportion of the MUs, and the quality τ_v is the fraction of the SBSs assigned to θ_v. In a feasible contract, $\mathcal{V}_v$ with a higher type θ_v is allocated with a larger fraction τ_v of the SBSs, and hence achieves a higher coverage and downloading probability.

From Lemmas 2 and 3, we find that the necessary conditions for IC constraints, given the types $\theta_1 > \theta_2 > \ldots > \theta_N$ are $\tau_1 > \tau_2 > \cdots > \tau_N$ and $\pi_1 > \pi_2 > \cdots > \pi_N$.

In addition, the corresponding sufficient conditions for IC constraints are shown in the following lemma. Before we come to the sufficient conditions for IC constraints, two definitions of local downward incentive constraint (LDIC) and local upward incentive constraint (LUIC) are introduced in Definitions 4 and 5, respectively.

Definition 4: Local Downward Incentive Constraint (LDIC): The IC constraint between two types θ_v and θ_{v-1} satisfies $V(\theta_v, \tau_v) - \pi_v \geq V(\theta_v, \tau_{v-1}) - \pi_{v-1}$.

Definition 5: Local Upward Incentive Constraint (LUIC): The IC constraint between types θ_v and θ_{v+1} satisfies $V(\theta_v, \tau_v) - \pi_v \geq V(\theta_v, \tau_{v+1}) - \pi_{v+1}$.

Then, the sufficient conditions for IC constraints are presented in Lemma 4.

Lemma 4: If utility functions satisfy the LDIC and LUIC simultaneously, the IC constraints will also be satisfied.

Proof: Please refer to [41].

Lemma 4 attests that the LDIC and LUIC are sufficient conditions for the IC constraints. Thus, the IC constraints can be replaced by the LDIC and LUIC constraints. Furthermore, we employ Lemma 5 to reduce the LDIC and LUIC constraints.

Lemma 5: (to reduce the IC constraints): For maximizing the NSP's profits, the IC constraints can be replaced by

$$V(\theta_v, \tau_v) - \pi_v = V(\theta_v, \tau_{v+1}) - \pi_{v+1}. \tag{16.15}$$

Proof: Please refer to [41].

The proof of Lemma 5 testifies that the LDIC can be derived from LUIC. Thus, we can reduce the IC constraints to LUIC and further to (16.14).

16.4.3 Solution to optimal contract

In this section, we will work on the solution to problem in (16.13) to obtain an optimal contract. By using the lemmas in the previous section, the optimization problem in (16.13) can be rewritten as

$$\begin{aligned} &\{\tau_v^*, \pi_v^*\} = \arg\max\ U_{\text{NSP}}, \\ &\text{s.t.}\ \ \text{IR}(16.13),\ \ \text{IC}(16.14),\ \ \sum_{\mu} \tau_\mu = 1,\ \ \tau_\mu \geq 0. \end{aligned} \tag{16.16}$$

We iterate the IR and IC constraints in (16.16), and obtain

$$\begin{aligned} &\{\tau_v^*\} = \text{argmax} \sum_{\mu=1}^{N} \left(V(\theta_N, \tau_N) +) \sum_{u=\mu}^{N} w_u + S_v^{bh} \right), \\ &\text{s.t.} \sum_{\mu} \tau_\mu = 1, \tau_\mu \geq 0, \end{aligned} \tag{16.17}$$

where

$$w_u = \begin{cases} 0, & u = N; \\ V(\theta_u, \tau_u) - V(\theta_u, \tau_{u+1}), & u = 1, \cdots, N-1. \end{cases} \tag{16.18}$$

Generally, the LDS s^{ld} obtained by $\mathcal{V}_v$ has a certain relationship with the s^{bh} induced by a file transmission over backhaul channels. For instance, the benefit of a CP from LDS may be partially supported by the NSP for saving its backhaul transmissions. Generally speaking, a higher value of the saved backhaul cost will contribute to a greater benefit from local downloading. Therefore, we assume that there is a linear relationship between s^{bh} and s^{ld}, i.e., $s^{\text{bh}} = \kappa s^{\text{ld}}$, where $\kappa \geq 0$ represents the relationship between s^{bh} and s^{ld}. Substituting (16.18) and $s^{\text{bh}} = \kappa s^{\text{ld}}$ in (16.17), we rewrite the optimization problem in (16.15) as:

$$\{\tau_v^*\} = \operatorname{argmax} \sum_{\mu=1}^{N} R_\mu, \\ \text{s.t.} \sum_{\mu} \tau_\mu = 1, \tau_\mu \geq 0, \tag{16.19}$$

where $R_1 \triangleq (1+\kappa)V(\theta_1, \tau_1)$, and

$$R_v \triangleq (v+\kappa)V(\theta_v, \tau_v) - (v-1)V(\theta_{v-1}, \tau_v), \quad \forall v \geq 2. \tag{16.20}$$

We find that the optimal R_v is only associated with quality τ_v and independent of other qualities $\tau_{\tilde{v}}, \tilde{v} \neq v$. Obviously,

$$\frac{\partial R_1}{\partial \tau_1} = \frac{(1+\kappa)M^{\mathrm{ld}}\theta_1 C(\delta,\alpha)}{\left(\tau_1 A(\delta,\alpha) + (1-\tau_1)C(\delta,\alpha) + \tau_1\right)^2} > 0, \tag{16.21}$$

and

$$\frac{\partial^2 R_1}{\partial \tau_1^2} = \frac{-2(1+\kappa)M^{\mathrm{ld}}\theta_1 C(\delta,\alpha)\left(A(\delta,\alpha) - C(\delta,\alpha) + 1\right)}{\left(\tau_1 A(\delta,\alpha) + (1-\tau_1)C(\delta,\alpha) + \tau_1\right)^3} < 0. \tag{16.22}$$

Thus, R_1 is an increasing and concave function of τ_1.

For $R_v, v \geq 2$, we have the first-order derivation with regard to τ_v as

$$\frac{\partial R_v}{\partial \tau_v} = \frac{[(v+\kappa)\theta_v - (v-1)\theta_{v-1}]M^{\mathrm{ld}}C(\delta,\alpha)}{(\tau_v A(\delta,\alpha) + (1-\tau_v)C(\delta,\alpha) + \tau_v)^2}, \quad \forall v \geq 2. \tag{16.23}$$

The second-order derivation of R_v on τ_v is calculated as

$$\frac{\partial^2 R_v}{\partial \tau_v^2} = \frac{-2[(v+\kappa)\theta_v - (v-1)\theta_{v-1}]M^{\mathrm{ld}}C(\delta,\alpha)(A(\delta,\alpha) - C(\delta,\alpha) + 1)}{(\tau_v A(\delta,\alpha) + (1-\tau_v)C(\delta,\alpha) + \tau_v)^3}, \quad \forall v \geq 2. \tag{16.24}$$

Substituting $\theta_v = \frac{1/v^\gamma}{\sum_{i=1}^N 1/i^\gamma}$ and $\theta_{v-1} = \frac{1/(v-1)^\gamma}{\sum_{i=1}^N 1/i^\gamma}$ in (16.23) and (16.24), we get (16.25)

$$\frac{\partial R_v}{\partial \tau_v} = \frac{\left[(v+\kappa)\frac{1}{v^\gamma} - (v-1)\frac{1}{(v-1)^\gamma}\right]M^{\mathrm{ld}}C(\delta,\alpha)}{\left(\sum_{i=1}^N 1/i^\gamma\right)\left(\tau_v A(\delta,\alpha) + (1-\tau_v)C(\delta,\alpha) + \tau_v\right)^2}, \quad \forall v \geq 2, \tag{16.25}$$

and

$$\frac{\partial^2 R_v}{\partial \tau_v^2} = \frac{-2\left[(v+\kappa)\frac{1}{v^\gamma} - (v-1)\frac{1}{(v-1)^\gamma}\right]M^{\mathrm{ld}}C(\delta,\alpha)(A(\delta,\alpha) - C(\delta,\alpha) + 1)}{\left(\sum_{i=1}^N 1/i^\gamma\right)\left(\tau_v A(\delta,\alpha) + (1-\tau_v)C(\delta,\alpha) + \tau_v\right)^3}, \quad \forall v \geq 2. \tag{16.26}$$

Observing (16.25) and (16.26), we define the common part of the numerators of (16.25) and (16.26) as

$$f_v(\gamma) \triangleq (v+\kappa)\frac{1}{v^\gamma} - (v-1)\frac{1}{(v-1)^\gamma}, \quad \forall v \geq 2. \tag{16.27}$$

The property of R_v with regard to τ_v is determined by $f_v(\gamma)$. What is more, we substitute (16.9) and (16.3) in (16.20), and find that R_v can be represented by $f_v(\gamma)$ as:

$$R_v = \frac{f_v(\gamma)}{\sum_{j=1}^{N} 1/j^\gamma} M^{\text{ld}} g(\tau_v), \quad \forall v \geq 2. \tag{16.28}$$

Thus, we come to Remark 1.

Remark 1: R_v, $\forall v \geq 2$ has the same properties of positive and negative as $f_v(\gamma)$.

What is more, from (16.25) and (16.26), the concavity and convexity of R_v, $\forall v \geq 2$, is associated with the value of $f_v(\gamma)$. Before we determine the concavity and convexity of R_v, $\forall v \geq 2$, the variation of $f_v(\gamma)$ versus γ is shown in Theorem 1.

Theorem 1: There is a unique root $\hat{\gamma}(v)$ such that $f_v\left(\hat{\gamma}(v)\right) = 0$, given v, which can be calculated as

$$\hat{\gamma}(v) = \frac{\ln(v+\kappa) - \ln(v-1)}{\ln v - \ln(v-1)}, \quad v \geq 2. \tag{16.29}$$

In addition, we have $f_v(\gamma) > 0$ when $0 < \gamma < \hat{\gamma}(v)$, and $f_v(\gamma) < 0$, otherwise.

Proof: Please refer to [41].

Based on Theorem 1, we have the following remark.

Remark 2: Therefore, according to Theorem 1, when $\gamma > \hat{\gamma}(v)$, we have $f_v(\gamma) < 0$, thus R_v is monotonically decreasing and convex to τ_v. When $0 < \gamma < \hat{\gamma}(v)$, we have $f_v(\gamma) > 0$ and thus R_v is monotonically increasing and concave to τ_v.

Thereby, we have Corollary 1.

Corollary 1: The value of $\hat{\gamma}(v)$, $\forall v \geq 2$, increases with the growth of v.

Proof: Please refer to [41].

From Corollary 1, we can see that a larger index v leads to a larger value of $\hat{\gamma}(v)$, $\forall v \geq 2$. Figure 16.2 plots the curves of $f_v(\gamma)$ versus γ in case of $\kappa = 1$. Theorem 16.4.3 and Corollary 1 can be verified using this figure. In specific, from Figure 16.2, we have $\hat{\gamma}(2) = 1.585$, $\hat{\gamma}(3) = 1.710$, $\hat{\gamma}(4) = 1.776$, and $\hat{\gamma}(5) = 1.817$. Thus, as shown by Corollary 1, $\hat{\gamma}(v)$ grows monotonously with the value of v. In addition, from this figure, given v, there is only one cross point between $f_v(\gamma)$ and the x-axis, thereby validating Remark 2.

Furthermore, we define

$$\hat{\gamma}_{min} \triangleq \min\{\hat{\gamma}(v), v = 1, \cdots, N\} = \hat{\gamma}(2), \tag{16.30}$$

and

$$\hat{\gamma}_{max} \triangleq \max\{\hat{\gamma}(v), v = 1, \cdots, N\} = \hat{\gamma}(N). \tag{16.31}$$

We have three cases according to the value of Zipf parameter γ, i.e., $\gamma \geq \hat{\gamma}_{max}$, $0 \leq \gamma \leq \hat{\gamma}_{min}$, and $\hat{\gamma}_{min} < \gamma < \hat{\gamma}_{max}$. In the following sections, we will conduct the optimization on the problem in (16.19) for these three cases.

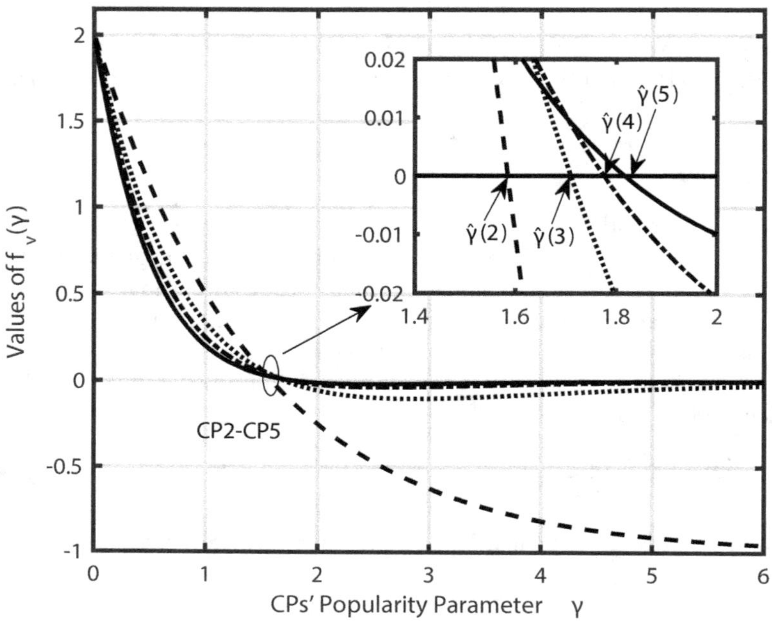

Figure 16.2 Values of $f_v(\gamma)$ in (16.24), when $N = 5$, $\kappa = 1$, γ varies from 0.01 to 6

16.4.3.1 Case 1 $\gamma \geq \hat{\gamma}_{\max}$

In this case, according to Remark 2, R_v is a decreasing and convex function of τ_v, $\forall v \geq 2$. Therefore, the optimal solution of $\arg\max R_v, \forall v \geq 2$, is $\tau_v = 0$. Also, R_1 is always an increasing function of τ_1. We consider the constraint $\sum_{\mu=1}^{N} \tau_\mu = 1, \tau_\mu \geq 0$, as well as the monotonicity of τ_v in Lemma 3, i.e., $\tau_1 > \tau_2 > \cdots > \tau_N$, the rational solution in this case is to set $\tau_1 = 1$ and $\tau_v = 0, \forall v \geq 2$. Thus, it is optimal to assign type θ_1 with the entire fraction of SBSs as long as $\gamma \geq \hat{\gamma}_{max}$. The optimal fraction τ_v^* for *Case 1* can be expressed by

$$\tau_v^* = \begin{cases} 0, & v > 1, \\ 1, & v = 1. \end{cases} \tag{16.32}$$

Substituting (16.32) into (16.14) in Lemma 1 and (16.15) in Lemma 5, we have

$$\pi_v^* = \begin{cases} 0, & v > 1, \\ V(\theta_1, \tau_1^*), & v = 1. \end{cases} \tag{16.33}$$

The utility of $\mathscr{L}$ can be thereby calculated as $U_{\text{NSP}}^* = (1 + \kappa)V(\theta_1, \tau_1^*)$.

16.4.3.2 Case 2 $0 < \gamma \leq \hat{\gamma}_{min}$

In this case, according to Remark 2, R_v is an increasing and concave function of $\tau_v, v = 1, \cdots, N$. We resort to the optimization method of Lagrangian multipliers [47] and obtain

$$\tilde{\tau}_1^* = \max\left\{\frac{\sqrt{\frac{(1+\kappa)\theta_1 M^{\mathrm{d}} C(\delta,\alpha)}{\xi}} - C(\delta,\alpha)}{A(\delta,\alpha) - C(\delta,\alpha) + 1}, 0\right\}, \tag{16.34}$$

and

$$\tilde{\tau}_v^* = \max\left\{\frac{\sqrt{\frac{((v+\kappa)\theta_v - (v-1)\theta_{v-1}) M^{\mathrm{d}} C(\delta,\alpha)}{\xi}} - C(\delta,\alpha)}{A(\delta,\alpha) - C(\delta,\alpha) + 1}, 0\right\}, \forall v \geq 2, \tag{16.35}$$

where $\xi = \left(\frac{\sqrt{(1+\kappa)\theta_1 M^{\mathrm{d}} C(\delta,\alpha)} + \sum_{i=2}^{N^*} \sqrt{((i+\kappa)\theta_i - (i-1)\theta_{i-1}) M^{\mathrm{d}} C(\delta,\alpha)}}{A(\delta,\alpha) - (N^*-1)C(\delta,\alpha) + 1}\right)^2$, and N^* satisfies the constraint that $\tau_v \geq 0$. Meanwhile, we need to check whether the given $\tilde{\tau}_v^*$, $\forall v$, satisfies the monotonicity constraint in Lemma 3. If the constraint is satisfied, i.e., there is $\tilde{\tau}_1^* \geq \tilde{\tau}_2^* \geq \cdots \geq \tilde{\tau}_N^*$, the expressions of $\tilde{\tau}_v^*$ in (16.34) and (16.35) are the optimal solutions.

In case the constraint is not satisfied for certain range of v, e.g., $[\underline{v}, \bar{v}]$ with $\tilde{\tau}_{\underline{v}}^* < \cdots < \tilde{\tau}_{\bar{v}}^*$, bunching-and-ironing method [46] will be adopted for adjusting the values of $\tilde{\tau}_v^*$, $\forall v \in [\underline{v}, \bar{v}]$ to ensure $\tilde{\tau}_{\underline{v}}^* = \cdots = \tilde{\tau}_{\bar{v}}^*$. In this case, the optimization problem is reformulated as

$$\begin{aligned} &\{\tau_v^*\} = \operatorname{argmax} \sum_{\mu=1}^{N} R_\mu, \\ &\text{s.t.} \sum_{\mu=1}^{N} \tau_\mu = 1, \tau_\mu \geq 0, \ \tau_{\dot{v}} = \tau_{\underline{v}} = \tau_{\bar{v}}, \ \forall \dot{v} \in [\underline{v}, \bar{v}]. \end{aligned} \tag{16.36}$$

Obviously, Lagrangian multipliers can be applied again to the given problem in order to obtain an optimal solution, with the monotonicity constraint in Lemma 3 being satisfied.

After obtaining an optimal quality τ_v^*, the optimal price can thus be derived recursively by Lemmas 1 and 5 as

$$\pi_v^* = V(\theta_N, \tau_N^*) + \sum_{i=v}^{N} w_i^*, \tag{16.37}$$

where

$$w_v^* = \begin{cases} 0, & v = N, \\ V(\theta_v, \tau_v^*) - V(\theta_v, \tau_{v+1}^*), & v = 1, \cdots, N-1. \end{cases} \tag{16.38}$$

Consequently, the utility of $\mathcal{L}$ can be calculated as

$$U_{\text{NSP}}^* = (1+\kappa)V(\theta_1, \tau_1^*) + \sum_{v=2}^{N} \left((v+\kappa)V(\theta_v, \tau_v^*) - (v-1)V(\theta_{v-1}, \tau_v^*)\right). \tag{16.39}$$

16.4.3.3 Case 3 $\hat{\gamma}_{min} < \gamma < \hat{\gamma}_{max}$

According to Theorem 1 and Corollary 1, given γ, there exists $v_{th} \in \{3, \cdots, N\}$ such that $f_{v_1}(\gamma) < 0$, $\forall 2 \leq v_1 \leq v_{th} - 1$, and $f_{v_2}(\gamma) > 0$, $\forall v_{th} \leq v_2 \leq N$. According to Remark 2, in case of $2 \leq v_1 \leq v_{th} - 1$, R_{v_1} is decreasing and convex to τ_{v_1} due to $f_{v_1}(\gamma) < 0$. Therefore, we need to set $\tau_{v_1}^* = 0$ for maximizing R_{v_1}. At the same time, to comply with the monotonicity in Lemma 3, i.e., $\tau_{v_1}^* \geq \tau_{v_2}^*$, we have $\tau_{v_2}^* = 0$, $\forall v_{th} \leq v_2 \leq N$. Consequently, the rational and feasible solution in this case is $\tau_1 = 1, \tau_v = 0, \forall v \geq 2$.

16.5 Discussions on other schemes

16.5.1 Uniform quality scheme

In this section, we discuss the uniform quality scheme, where the NSP deliberately imposes the same quality (fraction) $\tau_v = \frac{1}{N}$ on different CPs. Then by substituting $\tau_v^* = \tau_{v+1}^* = \frac{1}{N}$ in (16.37) and (16.38), the optimal price of the uniform quality scheme can be calculated as

$$\pi_v^* = V\left(\theta_N, \frac{1}{N}\right), \quad \forall v. \tag{16.40}$$

From this optimal price, we can see that it is uniform for all types of CPs, i.e., it is equal to the valuation of quality $\frac{1}{N}$ by the lowest type θ_N. In this case, the NSP can simply provide one contract option to all the CPs instead of diverse options. Furthermore, the utility of the NSP can be calculated as

$$U_{\text{NSP}}^{\text{UQC}} = \sum_{v=1}^{N} \left(\pi_v^* + \kappa V\left(\theta_v, \frac{1}{N}\right)\right). \tag{16.41}$$

16.5.2 Stackelberg game scheme

In this section, we formulate the interaction between the NSP $\mathscr{L}$ and the CPs $\mathscr{V}$ within the framework of a Stackelberg game, which has been well investigated in [45]. In the Stackelberg game scheme, $\mathscr{L}$ acts as a leader who moves first and $\mathscr{V}$ are the followers who move subsequently. We assume that $\mathscr{L}$ offers a price vector $\boldsymbol{s} \triangleq \{s_1, \cdots, s_v, \cdots, s_N\}$, where s_v is defined as the price of each SBS rented by $\mathscr{V}_v$. According to the price vector $\boldsymbol{s}$, each CP then determines the optimal fraction of the resources for maximizing its profit. The fraction vector is denoted by $\boldsymbol{\mu} \triangleq \{\mu_1, \cdots, \mu_v, \cdots, \mu_N\}$, where μ_v is the fraction rented by $\mathcal{V}_v$.

The utility functions of $\mathcal{L}$ and $\boldsymbol{\mathcal{V}}$ can be expressed as

$$U_{\text{NSP}}^{\text{SGS}} = \sum_{v=1}^{N} \left(s_v \mu_v + M^{\text{bh}} \theta_v g(\mu_v) \right), \tag{16.42}$$

and

$$U_v^{\text{SGS}} = M^{\text{ld}} \theta_v g(\mu_v) - s_v \mu_v, \tag{16.43}$$

respectively, where $M^{\text{bh}} \theta_v g(\mu_v)$ is the saved backhaul cost, $M^{\text{ld}} \theta_v g(\mu_v)$ denotes the savings of $\mathcal{V}_v$ for employing the caching mechanism, and $s_v \mu_v$ is the fees for leasing/renting a fraction of μ_v of the SBSs. Note that the profits of the NSP and each CP in the Stackelberg game scheme in (16.43) correspond to (16.7) and (16.10) in the contract-based scheme, respectively.

The Stackelberg game is formulated as follows,

1. Optimization formulation of the NSP:

$$\max_{s^*} \; U_{\text{NSP}}^{\text{SGS}}(\boldsymbol{s}, \boldsymbol{\mu}), \text{ s.t. } \sum_{v=1}^{N} \mu_v = 1, \; \mu_v > 0. \tag{16.44}$$

2. Optimization formulation of the CPs:

$$\max_{\mu^*} \; U_v^{\text{SGS}}. \tag{16.45}$$

The goal of the Stackelberg game is to find point (s^*, μ^*) forming an equilibrium, where both sides of the game will not gain more utilities if they deviate their strategies.

According to [45], we write the optimal solutions as follows:

$$s_v^* = \frac{C(\delta, \alpha) s^{\text{ld}} (\sum_{j=1}^{N} \sqrt[3]{\Gamma_j})^2 \sqrt[3]{\Gamma_v}}{(A(\delta, \alpha) + (N-1) C(\delta, \alpha) + 1)^2}, \tag{16.46}$$

and

$$\mu_v^* = \left(\sqrt{\frac{\Gamma_v s^{\text{ld}} C(\delta, \alpha)}{(A(\delta, \alpha) - C(\delta, \alpha) + 1)^2}} \sqrt{\frac{1}{s_v}} - \frac{C(\delta, \alpha)}{A(\delta, \alpha) - C(\delta, \alpha) + 1} \right)^+. \tag{16.47}$$

where $\Gamma_v \triangleq \sum_{f=1}^{Q} p_f \zeta K \theta_v$. The utility of $\mathcal{L}$ can thus be calculated as

$$U_{\text{NSP}}^{\text{SGS}} = \sum_{v=1}^{N} \frac{\xi_v \left(-C(\delta, \alpha) s_v + \Gamma_v s^{\text{ld}} \right)}{A(\delta, \alpha) - C(\delta, \alpha) + 1}, \tag{16.48}$$

where ξ_v is an indicator, with $\xi_v = 1$ if $s_v < \frac{\Gamma_v s^{\text{ld}}}{C(\delta,\alpha)}$, and $\xi_v = 0$ otherwise.

16.5.3 Discussions on the Stackelberg game scheme and the contract scheme

Generally, the Stackelberg game is based on information symmetric scenarios, where the CPs know the price charged on the resources, and the NSP $\mathcal{L}$ is aware of the quantity of the resources to be purchased by each CP. Even though $\mathcal{L}$ is the leader in this game, its strategy is designed based on the actions the followers employ, thereby allowing the CPs to purchase an optimal quota of the resources for maximizing their own profits. In other words, the Stackelberg game offers both sides with autonomies to maximize their own profits.

By contrast, in the contract-based scheme, $\mathcal{L}$ acts as a monopolist in the market, and designs the contract entries for various CPs so as to achieve maximum profit. The NSP $\mathcal{L}$ will not consider the profits of the CPs, though it needs to consider the IR and IC constraints when designing the contract, for the purpose of attracting each CP to accept the offered contract. However, compared with the Stackelberg game, these IR and IC constraints are quite weak in terms of maximizing CPs' profits. Thus, in the contract-based scheme, the monopolist NSP will gain a higher profit than in a scheme with the Stackelberg game, while the CPs will suffer from obtaining lower profits relative to those gained from the Stackelberg game.

16.5.4 Discussions on auction scheme

Auction is a branch of game theory [48]. It is also widely used in the asymmetric information environment where there are usually not enough commodities for all the potential buyers and the buyers' valuations of the commodities are unknown [49]. There are three roles in auction game: the bidder who is the buyer, the seller who wants to sell commodities, and the auctioneer who conducts the auction processes, usually, the seller can be an auctioneer itself. The final prices are determined by the competition among the potential buyers after bidding. Comparing to the auction approach, contract scheme is more efficient, since the seller has determined and posted the final prices at the beginning of the game. The buyer needs to determine whether to join the game and which contract to sign [50]. This has proven that the two approaches are equivalent in extracting profit under ideal conditions.

16.5.5 Discussions on a heterogeneous system

In a heterogeneous network, the macro-cell base station, small-cell base station, and femto access point with different transmission powers coexist in the system. They can be divided into multiple layers according to their transmission powers. As a piece of possible future work, the fraction as well as the transmission power can be regarded as trading resources, based on which the NSP designs an optimal contract.

16.6 Numerical results

In this section, we conduct numerical analysis to evaluate the performance of our contract-based caching system. The main parameters are listed in Table 16.2.

Table 16.2 Simulation Parameters

Parameter	**Value**
Path-loss parameter	$\alpha = 4$
Transmit power	$P = 10$ W
SINR threshold	$\delta = 0.02$
Density of SBS	$\lambda = 20/\text{Km}^2$
Density of MUs	$\zeta = 80/\text{Km}^2$
Files' popularity parameter	$\beta = 0.2$
File number	$F = 100$
Storage size	$Q = 40$
Number of CPs	$N = 5$
Coefficient κ	$\kappa = 1$
Local downloading savings	s^{ld}=1
Files demand from each MU within a unit period	$K = 50$/month

16.6.1 Discussions on R_v and $\hat{\gamma}(v)$

First, we make an investigation of the convexity and concavity of R_v in Remark 2 by numerical results under different values of γ and v. Note that from Figure 16.2, we have $\hat{\gamma}_{min}$=1.585 and $\hat{\gamma}_{max}$=1.817. As we have derived, in the case of $\gamma \in [\hat{\gamma}_{max}, +\infty)$, R_v is decreasing and convex to τ_v, while in the case of $\gamma \in [0, \hat{\gamma}_{min}]$, R_v is increasing and concave to τ_v. When $\gamma \in (\hat{\gamma}_{min}, \hat{\gamma}_{max})$, the convexity or concavity of function R_v depends on the values of v.

Figure 16.3 plots the curve of R_v versus τ_v with three values of γ. In particular, Figure 16.3(a) considers the case of $\gamma = 1 \in [0, \hat{\gamma}_{min}]$. From this figure, it is evident that the curves of all the R_v from R_1 to R_5 are concave and increasing to τ_v. Increasing the value of γ up to $1.68 \in (\hat{\gamma}_{min}, \hat{\gamma}_{max})$, we demonstrate the corresponding curves of R_v versus τ_v in Figure 16.3(b). Observing this figure, R_2 becomes convex to τ_2. However, the remaining curves of R_v still remain concave. Also, Figure 16.3(c) illustrates the corresponding curves of R_v versus τ_v at $\gamma = 2 \in [\hat{\gamma}_{max}, +\infty)$. From this figure, only R_1 remains concave, while other curves from R_2 to R_5 change to be convex. These results are consistent with the theoretical analysis in Section 16.4.3.

16.6.2 Impact of γ on optimal price and quality

We now investigate the impact of the popularity parameter γ on the optimal contract design. Figure 16.4 demonstrates the bar chart of the CPs' type (i.e., the CPs' popularity, which is inversely proportional to order), the optimal price and quality versus the order of the CPs given $\gamma = 0.4$ under our proposed scheme. From Figure 16.4(a), we can see that the type of CPs, which varies from 0.2856 to 0.15, decreases as their order increases. In Figure 16.4(b), when the type of CP θ_v is large, the optimal price charged by $\mathscr{L}$ is correspondingly high. Moreover, the optimal price declines as the type of CP decreases. Figure 16.4(c) shows that there are same trends as those in Figure 16.4(a) and 16.4(b) in terms of the optimal quality. Thus, a CP with a larger

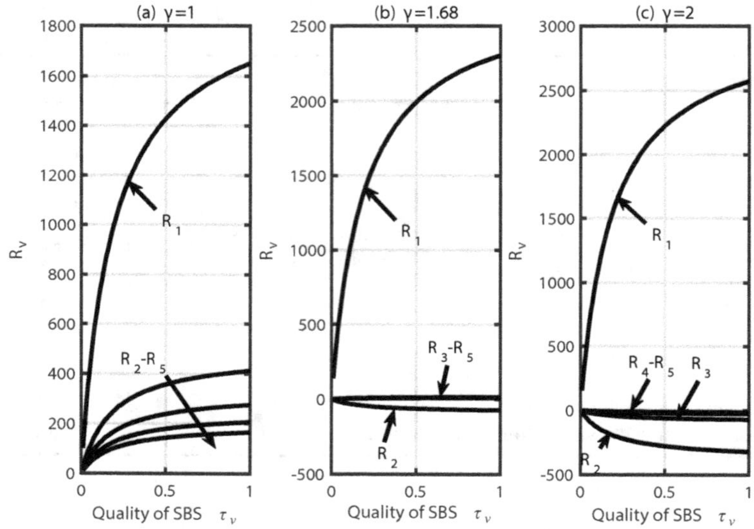

Figure 16.3 *Verification of convexity and concavity of function* R_v *over* τ_v *(*$\gamma = 1$*, 1.68 and 2)*

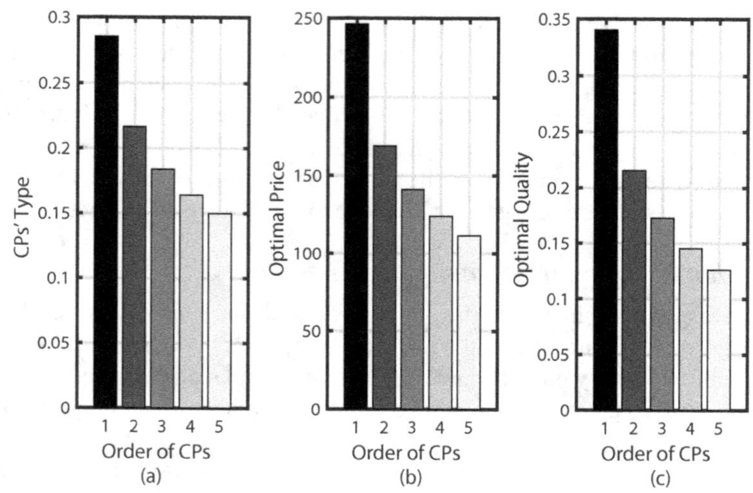

Figure 16.4 *Type, optimal price, and quality of CPs utilizing the proposed scheme (*$\gamma = 0.4$*)*

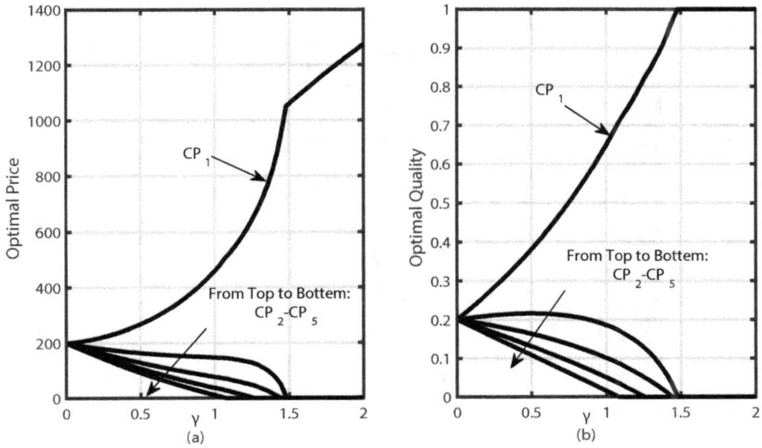

Figure 16.5 *(a) Optimal price design for each CP utilizing the proposed scheme. (b) Optimal quality design for each CP utilizing the proposed scheme.*

type tends to be provided with a higher quality of resources (please be aware that higher quality here means more fractions). These results are consistent with Lemmas 2 and 3 in Section 16.4.2.

In Figure 16.5(a), we change the value of CPs' popularity parameter γ to see how it affects the optimal price. In the small-value region of γ, for example $\gamma = 0.1$, the prices of all CPs are approximately identical. As the value of γ increases from 0.1 to 2, the distinction among the prices of all CPs becomes significant. Hence, the value of γ plays an important role in pricing.

Figure 16.5(b) illustrates the curves of optimal quality versus γ. Observing this figure, we find the following results: For small values of γ, the approximate same quality is allocated to all CPs while the quality difference among all CPs becomes large with an increase in the value of γ. For instance, $\gamma = 1.2$ and $\gamma = 2$, the quality of CP1 is far larger than those of the remaining CPs. In other words, CP1 is dominant compared to other CPs.

16.6.3 Comparisons of NSPs' and CPs' profits achieved by the three different schemes

Figure 16.6 plots the bar chart of the NSPs' and CPs' profits among the proposed, the uniform quality and the Stackelberg game schemes when γ is set to 1.2. From the figure, we can see that the NSP can gain more profits by using the proposed scheme relative to using the other two schemes. This is due to the fact that the proposed contract-based scheme is conducted by monopolists, whose aim is to maximize their own profit. As we discussed in Section 16.5.2, in the contract design process, the NSP squeezes more profits from CPs by using the proposed scheme relative to the

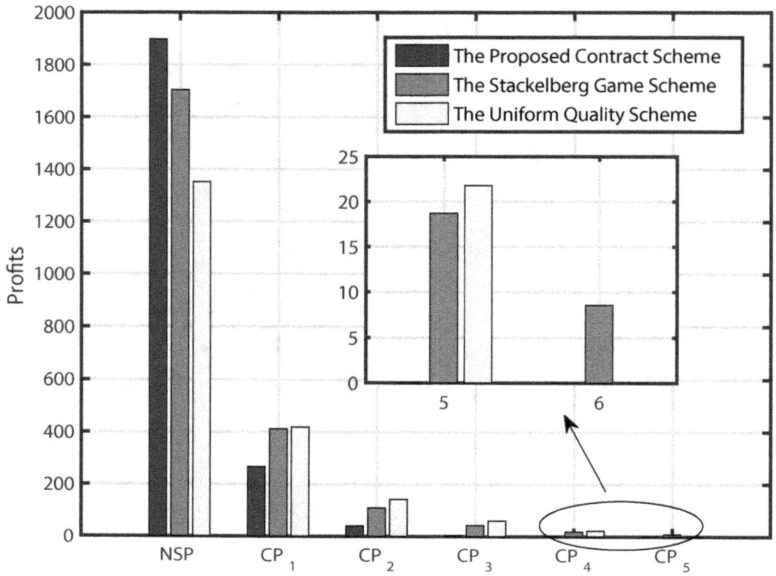

Figure 16.6 Comparisons among the profits of the proposed, the uniform quality, and the Stackelberg game schemes (γ = 1.2)

Stackelberg game scheme. Also, we observe that when the uniform quality scheme is adopted, the profits achieved by NSP are less than that achieved by the proposed scheme. This can be explained by the following facts. As per the discussion provided in Section 16.5.1, the uniform quality scheme, which charges a uniform price on all CPs, is a special case of the proposed scheme. The uniform quality scheme also has to guarantee the IR constraint, especially for the lowest type of CPs. As such, the uniform price in (16.40) cannot be set too high. Otherwise, the uniform quality scheme cannot guarantee the IR constraint for the lowest type of CPs. This leads to the fact that the prices charged on other high types of CPs are relatively low, which implies that the profits achieved by the NSP in the uniform quality scheme is lower than that achieved in the proposed scheme.

In addition, we compare the profits gained by the CPs in the aforementioned three different schemes in Figure 16.6. We first observe that the profits achieved by the CPs decrease as their indices increase. Focusing on one particular CP, we find that each CP gains minimum profit in the proposed scheme. Furthermore, in this figure we find that the profits of CP5 are all zeros in the proposed and the uniform quality schemes, while they are positive in the Stackelberg game scheme. The lowest type of CP, i.e., the type of CP5, is designed to gain zero profit in the proposed scheme as shown in Lemma 1, while in the Stackelberg game model, the CP who gains zero profit will opt out for renting any resource from the NSP.

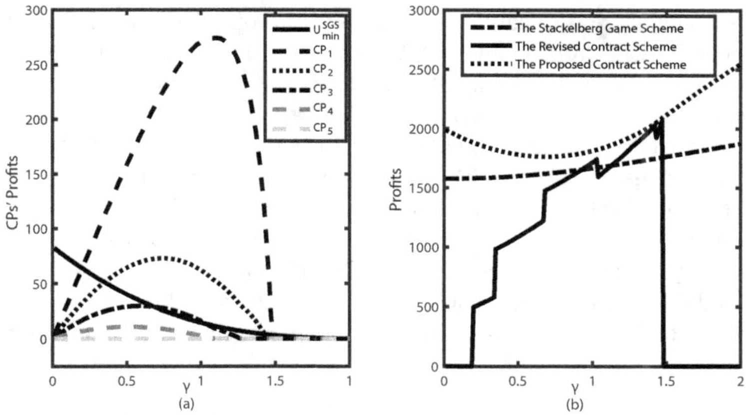

Figure 16.7 *(a) Profit of each CP when the revised contract scheme is employed, γ varies from 0.01 to 2. (b) Total profit of the NSP when the Stackelberg game scheme and the revised contract scheme are employed.*

In order to make a comprehensive comparison between the Stackelberg game scheme and the contract scheme, we investigate a novel contract scheme with the revised IR constraint, which is henceforth referred to as "revised contract scheme". We denote by $U^{SGS}_{v,\min}$ the achieved utility of the CP with the lowest popularity in the Stackelberg scheme. Then, a CP in the revised contract scheme will accept the offered contract if this CP's utility is no less than $U^{SGS}_{v,\min}$ and reject the contract otherwise.

In Figure 16.7(a), we vary the CPs' popularity parameter γ from 0.01 to 2 to clarify the relationship between the minimum profit $U^{SGS}_{\min}$ in the Stackelberg game scheme and CPs' profits in the revised contract scheme. It can be seen in Figure 16.7(a) that, at first, the profits of CP_1 to CP_5 are all smaller than $U^{SGS}_{\min}$ until $\gamma \approx 0.2$. Then, the profit of CP_1 becomes larger than $U^{SGS}_{\min}$, and the profits of CP_2 and CP_3 become larger than $U^{SGS}_{\min}$ at $\gamma \approx 0.35$ and $\gamma \approx 0.68$, subsequently. However, when $\gamma \approx 1.04$, the profit of CP_3 drops below $U^{SGS}_{\min}$ at first, followed by CP_2 at $\gamma = 1.43$ and CP_1 at $\gamma = 1.48$. It can be inferred from Figure 16.7(a) that CP_1 accepts the offered contract in the interval $\gamma \in [0.2, 1.48]$, CP_2 accepts the offered contract in the interval $\gamma \in [0.35, 1.43]$ and CP_3 accepts the offered contract in the interval $\gamma \in [0.68, 1.04]$.

In Figure 16.7(b), we compare the NSP's total profit between the Stackelberg game scheme and the revised contract scheme when γ varies from 0.01 to 2. We also plot the NSP's profit utilizing the proposed contract scheme as a benchmark. The total profit of the NSP using the revised contract scheme is the summations of CP_1, CP_2, and CP_3, while CP_4 and CP_5 always reject the offered contracts. It can be seen that the overall profit gained by the NSP in the Stackelberg game scheme is not always larger than that in the revised contract scheme. When $\gamma = 1.48$ in the revised contract scheme, there is a sharp drop in the NSP's profit, due to the reason that the profit gained by CP_1 drops below $U_{v,\min}^{\text{SGS}}$ as shown in Figure 16.7(a). According to the criteria, we set at the beginning of this section, CP_1 will definitely not accept the offered contract. Thus, the NSP's profit drops to zero.

The proceeding simulations highlight that even under such a strict IR constraint, the Stackelberg game scheme does not always outperform the revised contract scheme regarding NSP's revenue. Generally, the contract scheme only focuses on maximizing the NSP's profit under the IR and IC constraints. On the other hand, in the Stackelberg game, both the NSP and the CPs try to maximize their own profits to achieve the Nash equilibrium. In this sense, the NSP in the contract scheme acts as a monopolist in the market and thus will gain more profits than that in the Stackelberg game.

16.6.4 Performance of social welfare among three schemes

In what follows, we will evaluate the performance of the proposed scheme from the social welfare perspective. Social welfare is the sum of the utilities of NSP $\mathscr{L}$ and all CPs in $\mathscr{V}$. The expression for social welfare is defined by

$$U_{\text{SW}} = U_{\text{NSP}} + \sum_{v=1}^{N} U_v = \sum_{v=1}^{N} S_v^{\text{BH}} + \pi_v + S_v^{\text{LD}} - \pi_v = \sum_{v=1}^{N} S_v^{\text{BH}} + S_v^{\text{LD}}, \tag{16.49}$$

which is actually the sum of the saved backhaul cost from $\mathcal{L}$ and the LDS gained from CPs. As per (16.49), we find that the price π_v does not affect the social welfare. This price π_v is an internal transfer between $\mathcal{L}$ and $\mathcal{V}_v$. It can be canceled out when the social welfare is considered. Based on (16.5), (16.8), and the simulation setup, i.e., $s^{\text{ld}} = 1$ and $\kappa = 1$ in Table 16.2, we rewrite (16.40) as $U_{\text{SW}} = \sum_{v=1}^{N} 2M^{\text{ld}}\theta_v g(\tau_v)$.

Likewise, the social welfare expressions for the uniform quality and the Stackelberg game schemes are

$$U_{\text{SW}}^{\text{UQC}} = \sum_{v=1}^{N} 2V\left(\theta_v, \frac{1}{N}\right) = \sum_{v=1}^{N} 2M^{\text{ld}}\theta_v g\left(\frac{1}{N}\right) \tag{16.50}$$

and

$$U_{\text{SW}}^{\text{SGS}} = U_{\text{NSP}}^{\text{SGS}} + \sum_{v=1}^{N} U_v^{\text{SGS}} = \sum_{v=1}^{N} 2M^{\text{ld}}\theta_v g(\mu_v), \tag{16.51}$$

respectively. Comparing (16.50) with (16.51), we find that the difference between the uniform quality and the Stackelberg game schemes lies in the fractions $\frac{1}{N}$ and μ_v assigned to the CPs.

In Figure 16.8, we compare the social welfare among these three schemes. We vary the popularity parameter γ to examine its impact on social welfare. From Figure 16.8, when γ is small, e.g., $\gamma = 0.1$, the social welfare of the three schemes is almost the same. It is due to the fact that when γ is relatively small, the popularity of different CPs is similar. In such a scenario, the optimal qualities determined by

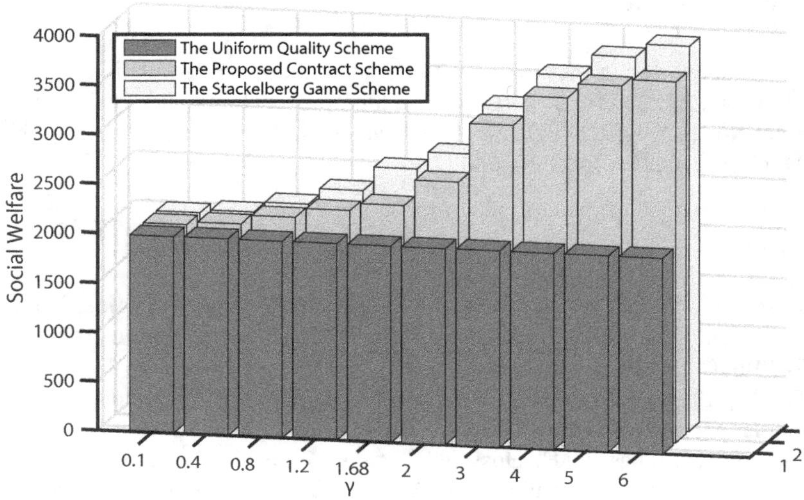

Figure 16.8 Comparisons among the social welfare of the proposed, the uniform quality, and the Stackelberg game schemes

the three schemes are nearly equal. As we increase γ from 0.1 to a larger value, the social welfare of the proposed and the Stackelberg game schemes increases. In addition, we find that the social welfare for the uniform quality scheme in Figure 16.8 is a constant, regardless of the values of γ. This is clear, since we can rewrite (16.50) as $\sum_{v=1}^{N} 2M^{\mathrm{ld}}\theta_v g\left(\frac{1}{N}\right) = 2M^{\mathrm{ld}} g\left(\frac{1}{N}\right)$. Furthermore, the Stackelberg game scheme achieves a larger social welfare relative to the proposed scheme. This is because the Stackelberg game scheme is to optimize the profits of both sides, while the proposed contract scheme is to maximize the monopolist $\mathcal{L}$'s profits.

16.7 Conclusions

In this chapter, we study a contract-based trading scheme for a commercialized small-cell caching system consisting of one NSP, several CPs and multiple MUs. We model this system as a monopoly market, where the NSP owns the network facilities, and offers contract entries to different CPs in maximizing its own profits. First, we formulate the utility functions of the NSP and CPs by modeling the SBSs and the MUs as two independent PPP. Then the optimal contract problem is developed in an information asymmetric scenario, where the NSP only has the information about the distribution of CPs' popularity. Next, the feasibility of the proposed scheme is derived. Afterwards, the optimal contract solutions are provided when the CPs' popularity parameter γ takes different values. Simulation results verify the effectiveness of the proposed scheme by showing its superiority to the benchmarks in maximizing the NSP's profit.

Acknowledgements

This work is supported in part by the national natural science foundation of China under grant No. 61702258, in part by the China postdoctoral science foundation under grant No. 2021M691540, in part by US MURI, US NSF CNS-1717454, CNS-1731424, CNS-1702850, CNS-1646607.

References

[1] Bastug E., Bennis M., Debbah M. 'Living on the edge: the role of proactive caching in 5G wireless networks'. *IEEE Communications Magazine*. 2014;**52**(8):82–9.

[2] Ge X., Cheng H., Guizani M., Han T. '5G wireless backhaul networks: challenges and research advances'. *IEEE Network*. 2014;**28**(6):6–11.

[3] Siddique U., Tabassum H., Hossain E., *et al.* 'Wireless backhauling of 5G small cells: challenges and solution approaches'. *IEEE Wireless Communications*. 2015;**22**(5):22–31.

[4] Andrews J.G., Buzzi S., Choi W., *et al.* 'What will 5G be?' *IEEE Journal on Selected Areas in Communications*. 2014;**32**(6):1065–82.

[5] Osseiran A., Boccardi F., Braun V., *et al.* 'Scenarios for 5G mobile and wireless communications: the vision of the METIS project'. *IEEE Communications Magazine*. 2014;**52**(5):26–35.

[6] Wang C.X., Haider F., Gao X., *et al.* 'Cellular architecture and key technologies for 5G wireless communication networks'. *IEEE Communications Magazine*. 2014;**52**(2):122–30.

[7] Bangerter B., Talwar S., Arefi R., *et al.* 'Networks and devices for the 5G era'. *IEEE communications magazine*. 2014;**52**(2):90–6.

[8] Boccardi F., Heath R.W., Lozano A., *et al.* 'Five disruptive technology directions for 5G'. *IEEE Communications Magazine*. 2014;**52**(2):74–80.

[9] Rappaport T.S., Sun S., Mayzus R., *et al.* 'Millimeter wave mobile communications for 5G cellular: it will work!' *IEEE Access*. 2013;**1**:335–49.

[10] Lu L., GY L., Swindlehurst A.L., *et al.* 'An overview of massive MIMO: benefits and challenges'. *IEEE Journal of Selected Topics in Signal Processing*. 2014;**8**(5):742–58.

[11] Niu Z., Wu Y., Gong J., *et al.* 'Cell zooming for cost-efficient green cellular networks'. *IEEE Communications Magazine*. 2010;**48**(11):74–9.

[12] Wang X., Chen M., Taleb T., *et al.* 'Cache in the air: exploiting content caching and delivery techniques for 5G systems'. *IEEE Communications Magazine*. 2014;**52**(2):131–9.

[13] Liu D., Chen B., Yang C., *et al.* 'Caching at the wireless edge: design aspects, challenges, and future directions'. *IEEE communications magazine*. 2016;**54**(9):22–8.

[14] Paschos G., Bastug E., Land I., *et al.* 'Wireless caching: technical misconceptions and business barriers'. *IEEE Communications Magazine*. 2016;**54**(8):16–22.

[15] Bastug E., Bennis M., Zeydan E., *et al.* 'Big data meets telcos: a proactive caching perspective'. *Journal of Communications and Networks*. 2015;**17**(6):549–57.

[16] Golrezaei N., Shanmugam K., Dimakis A.G. 'FemtoCaching: Wireless video content delivery through distributed caching helpers'. IEEE International Conference on Computer Communications (INFOCOM); Orlando, FL, USA; 2012. pp. 1107–15.

[17] Liu H., Chen Z., Tian X., *et al.* 'On content-centric wireless delivery networks'. *IEEE Wireless Communications*. 2014;**21**(6):118–25.

[18] Li J., Chen W., Xiao M., *et al.* 'Efficient video pricing and caching in heterogeneous networks'. *IEEE Transactions on Vehicular Technology*. 2016;**65**(10):8744–51.

[19] Li J., Sun J., Qian Y., *et al.* 'A commercial video-caching system for small-cell cellular networks using game theory'. *IEEE Access*. 2016;**4**:7519–31.

[20] Liu T., Li J., Shu F. 'D2D-enabled wireless caching using Stackelberg game'. International Conference on Wireless Communications Signal Processing (WCSP); Yangzhou China; 2016. pp. 1–5.

[21] Zhang Y., Gu Y., Pan M. 'Financing contract with adverse selection and moral hazard for spectrum trading in cognitive radio networks'. IEEE China Summit and International Conference on Signal and Information Processing (ChinaSIP); Chengdu China; 2015. pp. 601–5.

[22] Gao L., Wang X., Xu Y., *et al.* 'Spectrum trading in cognitive radio networks: a contract-theoretic modeling approach'. *IEEE Journal on Selected Areas in Communications*. 2011;**29**(4):843–55.

[23] Luo Y., Gao L., Huang J. 'Spectrum reservation contract design in TV white space networks'. *IEEE Transactions on Cognitive Communications and Networking*. 2015;**1**(2):147–60.

[24] Kordali A.V., Cottis P.G. 'A contract-based spectrum trading scheme for cognitive radio networks enabling hybrid access'. *IEEE Access*. 2015;**3**:1531–40.

[25] Sheng S.P., Liu M. 'Profit incentive in trading nonexclusive access on a secondary spectrum market through contract design'. *IEEE/ACM Transactions on Networking*. 2014;**22**(4):1190–203.

[26] Li Y., Zhang J., Gan X., *et al.* 'A contract-based incentive mechanism for delayed traffic offloading in cellular networks'. *IEEE Transactions on Wireless Communications*. 2016;**15**(8):5314–27.

[27] Hasan Z., Bhargava V.K. 'Relay selection for OFDM wireless systems under asymmetric information: a contract-theory based approach'. *IEEE Transactions on Wireless Communications*. 2013;**12**(8):3824–37.

[28] Zhang Y., Song L., Saad W., *et al.* 'Contract-based incentive mechanisms for device-to-device communications in cellular networks'. *IEEE Journal on Selected Areas in Communications*. 2015;**33**(10):2144–55.

[29] Wang W., Chen L., Shin K.G., *et al.* 'Thwarting intelligent malicious behaviors in cooperative spectrum sensing'. *IEEE Transactions on Mobile Computing*. 2015;**14**(11):2392–405.

[30] Zhang Y., Tran N.H., Niyato D. 'Multi-dimensional payment plan in fog computing with moral hazard'. IEEE International Conference on Communication Systems (ICCS); Shenzhen China; 2016. pp. 1–6.

[31] Hamidouche K., Saad W., Debbah M. 'Breaking the economic barrier of caching in cellular networks: incentives and contracts'. IEEE Global Communications Conference (GLOBECOM); Washington DC; 2016. pp. 1–6.

[32] Ma R., Wang L., Ding Y. 'Adverse selection based incentive mechanism for cooperative caching in distributed storage system'. 9th International Conference on Wireless Communications and Signal Processing (WCSP); Nanjing China; 2017. pp. 1–6.

[33] Yang Z., Zhang B., Wang L. 'Contract for joint communication and computing in wireless caching systems'. 11th International Conference on Wireless Communications and Signal Processing (WCSP); Xian China; 2019. pp. 1–6.

[34] Le THT T.N.H., PL V. 'Contract-based cache partitioning and pricing mechanism in wireless network slicing'. IEEE Global Communications Conference (GLOBECOM); Singapore; 2017. pp. 1–6.

[35] Le THT T.N.H., PL V. 'Joint cache allocation with incentive and user association in cloud radio access networks using hierarchical game'. *IEEE Access: Practical Innovations, Open Solutions*. 2019:20773–88.

[36] Liu T., Li J., Shu F., *et al.* 'Incentive mechanism design for two-layer wireless edge caching networks using contract theory'. *IEEE Transactions on Services Computing*. 2018:1.

[37] Li J., Chu S., Shu F., *et al.* 'Contract-based small-cell caching for data disseminations in ultra-dense cellular networks'. *IEEE Transactions on Mobile Computing*. 2019;**18**(5):1042–53.

[38] Ding Y., Wang L., Yang Z. 'Moral hazard based incentive mechanism for cooperative caching in multi-hop communications'. IEEE Globecom Workshops(GC Wkshps); Singapore; 2017. pp. 1–6.

[39] Malik F., Hanawal M.K. 'Revenue sharing between service providers in the internet: a moral hazard approach'. 2019 11th International Conference on Communication Systems Networks (COMSNETS); 2019. pp. 420–3.

[40] Guo C., Zhang Y., Wang L. 'Efficient resource allocation in wireless network virtualization: a joint design of adverse selection and moral hazard'. 2017 IEEE/CIC International Conference on Communications in China (ICCC); 2017. pp. 1–6.

[41] Liu T., Li J., Shu F., *et al.* 'Design of contract-based trading mechanism for a small-cell caching system'. *IEEE Transactions on Wireless Communications*. 2017;**16**(10):6602–17.

[42] Cha M., Kwak H., Rodriguez P., *et al.* 'I tube, you tube, everybody tubes: analyzing the world's largest user generated content video system'. ACM SIGCOMM Conference on Internet Measurement; 2007. pp. 1–13.

[43] Breslau L., Cao P., Fan L. 'Web caching and Zipf-like distributions: evidence and implications'. Eighteenth Annual Joint Conference of the IEEE International Conference on Computer and Communications (INFOCOM), vol. 1; New York, USA; 1999. pp. 126–34.

[44] Hefeeda M., Saleh O. 'Traffic modeling and proportional partial caching for peer-to-peer systems'. *IEEE/ACM Transactions on Networking*. 2008;**16**(6):1447–60.

[45] Li J., Chen H., Chen Y., *et al.* 'Pricing and resource allocation via game theory for a small-cell video caching system'. *IEEE Journal on Selected Areas in Communications*. 2016;**34**(8):2115–29.

[46] Bolton P., Dewatripont M. *Contract Theory*. The MIT Press; 2005.

[47] Boyd S., Vandenberghe L. *Convex Optimization*. Cambridge University Press; 2013.

[48] Krishna V. *Auction Theory*. Academic Press; 2002.

[49] Zhang Y., Lee C., Niyato D., Wang P., Zhang Y., Lee C. 'Auction approaches for resource allocation in wireless systems: a survey'. *IEEE Communications Surveys & Tutorials*. 2013;**15**(3):1020–41.

[50] Wang X., Montgomery A., Srinivasan K. 'When auction meets fixed price: a theoretical and empirical examination of buy-it-now auctions'. *Quantitative Marketing and Economics*. 2008;**6**(4):339–70.

Chapter 17
Mean-field game-theoretic edge caching

Hyesung Kim[1,2], *Jihong Park*[3], *Mehdi Bennis*[4], *Seong-Lyun Kim*[1], *and Mérouane Debbah*[5,6]

17.1 Introduction

Mobile networks are envisaged to be extremely densified in 5G and beyond to cope with the ever-growing user demand [1–4]. Edge caching is a key enabler of such an ultra-dense network (UDN), through which popular content is prefetched at each small base station (SBS) and downloaded with low latency [5, 6] while alleviating the significant backhaul congestion between a data server and a large number of SBSs [7]. Focusing on this, in this chapter we study the content caching strategy of an *ultra-dense edge caching network* (*UDCN*). Optimizing the content caching of a UDCN is a crucial yet challenging problem. Owing to the sheer amount of SBSs, even a small misprediction of user demand may result in a large amount of useless data cached in capacity-limited storages. Furthermore, the variance of interference is high due to short inter-SBS distances [8], making it difficult to evaluate cached data downloading rates, which is essential in optimizing the caching file sizes. To resolve these problems, we first present a spatio-temporal user demand model in continuous time, in which the long-term and short-term content popularity variations at a specific location are modeled using the Chinese restaurant process (CRP) and the Ornstein-Uhlenbeck (OU) process, respectively. Based on this, we aim to develop a scalable and distributed edge caching algorithm by leveraging the mean-field game (MFG) theory [9, 10].

To this end, at first the problem of optimizing distributed edge caching strategies in continuous time is cast as a non-cooperative stochastic differential game (SDG). As a game player, each SBS decides how much portion of each content file is prefetched by minimizing its long run average (LRA) cost that depends on the

[1]School of Electrical and Electronic Engineering, Yonsei University, South Korea
[2]Samsung Research, South Korea
[3]School of Information Technology, Deakin University, Australia
[4]Centre for Wireless Communications, University of Oulu, Finland
[5]CNRS, CentraleSupélec, Universit'e Paris-Saclay, France
[6]Lagrange Mathematical and Computing ResearchCenter, France

prefetching overhead, cached file downloading rates under inter-SBS interference, and overlapping cached files among neighboring SBSs, i.e., content overlap. This minimization problem is tantamount to solving a partial differential equation (PDE) called the Hamilton-Jacobi-Bellman equation (HJB) [11]. The major difficulty is that the HJB solution of an SBS is intertwined with the HJB solutions of other SBSs, as they interact with each other through the inter-SBS interference and content overlap. The complexity of this problem thus increases exponentially with the number of SBSs, which is unfit for a UDCN. Alternatively, exploiting MFG, we decouple the SBS interactions in a way that each SBS interacts only with a virtual agent whose behavior follows the state distribution of the entire SBS population, known as mean-field (MF) distribution. For a given SBS population, the MF distribution is uniquely and derived by locally solving a PDE, called the Fokker-Planck-Kolmogorov (FPK) equation. Consequently, the optimal caching problem at each SBS boils down to solving a single pair of HJB and FPK, regardless of the number of SBSs. Such an MF approximation guarantees achieving the epsilon Nash equilibrium (NE) [9, 12], when the number of SBSs is very large (theoretically approaching infinity) while their aggregate interactions are bounded. Both conditions are satisfied in a UDCN [13, 14], mandating the use of MFG.

To describe the MFG-theoretic caching framework and show its effectiveness for a UDCN, this chapter is structured as follows. Related works on UDCN analysis and MFG-theoretic approaches are briefly reviewed in Section 17.2. The network, channel, and caching models as well as the spatio-temporal dynamics of user demand, caching, and interference are described in Section 17.3. For the SBS supporting a reference user, its optimal caching problem is formulated and solved using MFG in Section 17.4. The performance of the MFG-theoretic edge caching is numerically evaluated in terms of LRA and the content overlap amount in Section 17.5, followed by concluding remarks in Section 17.6.

17.2 Related works

Edge caching in cellular networks has received significant attention in 5G and beyond [6, 7, 15]. In the context of MFG-theoretic edge caching in a UDN, we briefly review its preceding user demand models, interference analysis, and MFG-based applications as follows.

User Demand Model and Interference Analysis. The effectiveness of edge caching is affected significantly by user demand according to content popularity. The user demand model in most of the works on edge caching relies commonly on the Zipf's law. Under this model, the content popularity in the entire network region is static and follows a truncated power law [16], which is too coarse to reflect spatio-temporal content popularity dynamics in a UDCN. A time-varying user demand model has been considered in [17, 18] while ignoring spatial characteristics, which motivated us to seek for a more detailed user demand model reflecting spatio-temporal content popularity variations.

The spatial characteristics of interference dynamics have been analyzed in [19, 20] using stochastic geometry. These works, however, rely on a globally static user demand model and thus, ignore the temporal and local dynamics of interference [21]. By contrast, in this chapter we consider the spatio-temporal content popularity dynamics, and analyze their impact on interference.

The impacts of SBS densification on interference in a UDN have been investigated in [8, 22–27], in which the interference dynamics are governed by the spatial dynamics of user demand, i.e., locations [8]. Interestingly, these works neglect temporal user demand variations. It is worth noting that a recent study [13] has considered spatio-temporal user demand fluctuations. However, it does not take into account the temporal content popularity correlation. The gap has been filled by its follow-up work [28] that models the correlated content popularity using the CRP, which is addressed in this chapter.

MFG Applications. The MFG theory is built upon an asymptotically large number of agents in a non-cooperative game. This fits naturally with a UDN within which assuming an infinite number of SBSs becomes reasonable [1, 2, 8, 29]. In this respect, SBS transmit power control and the user mobility management in a UDN have been studied in [22, 23]. For a massive number of drones, their rate-maximizing altitude control and collision-avoid path planning have been investigated in [30] and [31–34], respectively. In a similar vein, in this chapter we utilize the MFG theory to simplify the spatio-temporal analysis on interference and content overlap in a UDCN. One major limitation of MFG-based methods is that solving a pair of HJB-FPK PDEs may still be challenging when the agent's state dimension is large. In fact, existing PDE solvers rely mostly on the Euler discretization method such that the derivatives in a PDE are approximated using finite differences. To guarantee the convergence of a numerical PDE solution, the larger state dimension is considered, the finer discretization step size is required under the Courant-Friedrichs-Lewy (CFL) condition [35], increasing the computing complexity. To avoid this problem, in this chapter we describe the state of each SBS separately for each content file, reducing the dimension of each PDE. Alternatively, machine learning methods have been applied in recent works [32, 33, 36] through which solving a PDE is recast as a regression learning problem. By leveraging this method, incorporating large-sized edge caching states (e.g., joint optimization of transmit power, caching strategy, and mobility management) could be an interesting topic for future research.

17.3 System model

17.3.1 Network, channel, and caching model

In this section, we describe a downlink UDN under study, followed by its communication channel and caching models.

Network Model. SBSs and their users are independently and uniformly distributed in a two-dimensional plane $\mathbb{R}^2$ with finite densities, forming two independent Poisson point processes (PPPs) [8, 37]. Following [8], the network is assumed to be a UDN such that SBS density λ_b is much higher than user density λ_u, i.e., $\lambda_b \gg \lambda_u$.

In this UDN, the i-th user is located at coordinates $y_i \in \mathbb{R}^2$, and receives signals from multiple SBSs located within a reception ball $b(y_i, R)$ centered at y_i with radius $R > 0$, as depicted in Figure 17.1. The radius R can be determined based on the noise floor so that the average received signal power should be larger than the noise floor. When $R \rightarrow \infty$, the reception ball model becomes identical to a conventional PPP-based network model in stochastic geometric analysis [37, 39].

Channel and Antenna Pattern Models. The transmitted signals from SBSs experience path-loss attenuation and multi-path fading. Specifically, the path loss $l_{k,i}$ from the k-th SBS located at $z_k \in \mathbb{R}^2$ to the i-th user at $y_i \in \mathbb{R}^2$ is given as $l_{k,i} = \min(1, \|z_k - y_i\|^{-\alpha})$, where $\alpha > 2$ is the path-loss exponent. The transmitted signals experience an independent and identically distributed fading with the coefficient $g_{k,i}(t)$. We assume that the fading coefficient is not temporally correlated. Consequently, the received signal power at the i-th user is given as $S(t) = P|h_{k,i}(t)|^2$, where P denotes the transmit power of every SBS, and $h_{k,i}$ is the channel gain determined by $|h_{k,i}(t)|^2 = l_{k,i}|g_{k,i}(t)|^2$. Next, the transmission of each SBS is directional using N_a antennas. Following [40], the beam pattern follows a sectored uniform linear array model, in which the center of the main lobe beam points at the receiving user. The main lobe gain is given as N_a with the beam width $\theta_{N_a} = 2\pi/\sqrt{N_a}$ while ignoring side lobes.

Caching Model. Consider a set $\mathcal{M}$ of M content files in total, each of which is encoded using the maximum distance separable dateless code [41]. At time t, a fraction $p_{k,j}(t) \in [0, 1]$ of the j-th content file with size L_j is prefetched to the k-th SBS from a remote server through a capacity-limited backhaul link, as shown in Figure 17.1. The SBS is equipped with a data storage of size $C_{k,j}$ assigned for the content file j, and therefore we have $p_{k,j}(t)L_j \leq C_{k,j}$. Each user in the network requests the j-th content file with probability x_j. Within the user's content request range $R_c > 0$, there exists a set $\mathcal{N}$ of N SBSs [8]. If multiple SBSs cached the requested file (i.e., cache hitting), then the user downloads the file from a randomly selected SBS. If there is no SBS caching for the requested file (i.e., cache missing), then the file is downloaded to a randomly selected SBS from the remote server via the backhaul, which is then delivered to the user from the SBS. At time t, the goal of the k-th SBS is to determine its file caching fraction vector $p_k(t) = \{p_{k,1}(t), \cdots, p_{k,j}(t), \cdots p_{k,M}(t)\}$.

17.3.2 Spatio-temporal dynamics of demand, caching, and interference

The effectiveness of caching strategies is affected by spatio-temporal dynamics of content popularity among users, backhaul and storage capacities in SBSs, and interference across SBSs, as we elaborate next.

User Demand Dynamics. The user demand on content files is often modeled as a Zipf distribution [16]. Such a long-term user demand pattern in a wide area is too coarse to capture the spatial demand and its temporal variations [21], calling for a detailed spatio-temporal user demand model for a UDCN. To this end, we consider that each SBS regularly probes the content popularity within the distance R_s, and the content popularity for each SBS follows an independent stochastic process. For the

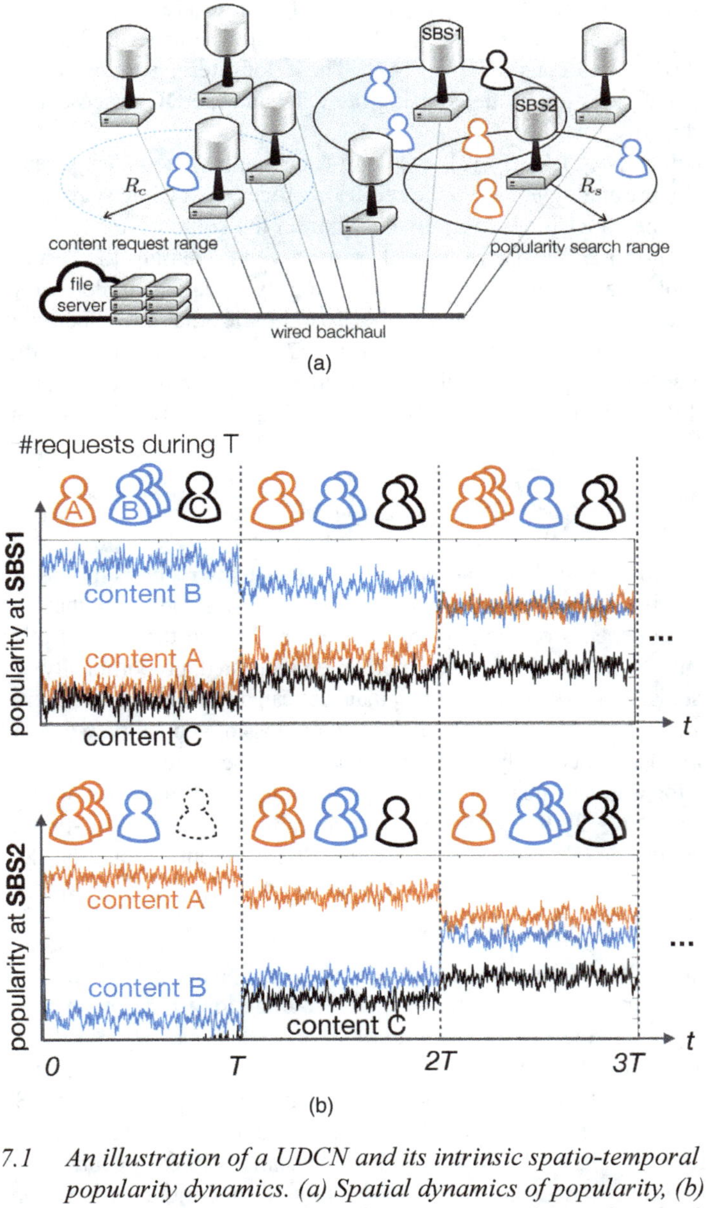

Figure 17.1 An illustration of a UDCN and its intrinsic spatio-temporal popularity dynamics. (a) Spatial dynamics of popularity, (b) Temporal dynamics where the content popularity changes for long-term and short-term durations. The long-term dynamics is captured by the Chinese restaurant process, which determines the mean popularity for a certain time period T*. During this period, the instantaneous popularity is captured by the mean reversion model following the OU process [38].*

content popularity of each SBS, its temporal dynamics are described by the long-term fluctuations across time $t = T, 2T, \cdots, \kappa T$ and short-term fluctuations over $t \in [(\kappa - 1)T, \kappa T]$ as considered in [42]. These long-term and short-term content popularity dynamics are modeled using the CRP and the OU process, respectively, as detailed next.

Following the CRP [5, 43], the long-term content popularity variations are described by the analogy of the table selection process in a Chinese restaurant. Here, UDCN becomes the Chinese restaurant, wherein the content files and users are the tables and customers in the restaurant, respectively. Treating the restaurant table seating problem as a long-term content popularity updating model, we categorize content files into two groups: the set $U_k^r(\kappa T)$ of the files that have been requested by $N_k(\kappa T)$ users at least once at SBS k until time κT; and the set $U_k^u(\kappa T)$ of the files that have not yet been requested until then. For these two groups, the mean popularity $\mu_{k,j}(\kappa T)$ of the j-th content file at SBS k during $t \in [(\kappa - 1)T, \kappa T]$ is given by:

$$\mu_{k,j}(\kappa T) = \begin{cases} \frac{n_{k,j}(\kappa T) - \nu}{N_k(\kappa T) + \theta} & \text{for } j \in U_k^r(\kappa T) \\ \frac{\nu |U_k^r(\kappa T)| + \theta}{N_k(\kappa T) + \theta} & \text{for } j \in U_k^u(\kappa T), \end{cases} \tag{17.1}$$

where $n_{k,j}(\kappa T)$ is the number of accumulated downloading requests for the j-th content file at SBS k until time κT, and θ and ν are positive constants. Note that each content file popularity depends on the popularity of other files and the number of other files. Consequently, more popular files are more often requested, proportionally to the previous request history $n_{k,j}(\kappa T)$, while unpopular files can also be requested with a probability proportional to θ and ν. For simplicity without loss of generality, we omit the index κT of $\mu_{k,j}(\kappa T)$ and focus only on the case when $\kappa = 1$ hereafter.

Next, for a given mean content popularity $\mu_{k,j}$, at time t during a short-term period $t \in [0 \leq t \leq T]$, the content request probability $x_{k,j}(t)$ of the j-th file at SBS k is described by the OU process [42], a stochastic differential equation (SDE) given as follows:

$$\mathrm{d}x_{k,j}(t) = r(\mu_{k,j} - x_{k,j}(t))\mathrm{d}t + \eta \mathrm{d}W_{k,j}(t), \tag{17.2}$$

where $W_{k,j}(t)$ is the Wiener process, and r and η are positive constants. It describes that the short-term content popularity drifts from the long-term mean content popularity $\mu_{k,j}$ by $x_{k,j}(t)$ and fluctuates randomly by $W_{k,j}(t)$.

Figure 17.1 illustrates the user demand pattern generated from the aforementioned long-term and short-term content popularity dynamics. As observed by SBS 1 and SBS 2, for the same content files A, B, and C, these two spatially separated SBSs have different popularity dynamics, while at the same SBS each content popularity is updated according to a given temporal correlation. Furthermore, as shown in SBS 2 at around $t = T$, the previously unpopular file C can emerge as an up-do-date popular file.

Caching Dynamics. The remaining storage capacity varies according to the instantaneous caching strategy. Let us assume that SBSs have a finite storage size and discard content files at a rate of $e_{k,j}$ from the storage unit in order to make space for caching other contents. Considering the discarding rate, we model the evolution law of the storage unit at SBS k as follows:

$$\mathrm{d}Q_{k,j}(t) = (e_{k,j} - L_j p_{k,j}(t))\mathrm{d}t, \tag{17.3}$$

where $Q_{k,j}(t)$ denotes the remaining storage size dedicated to content j of SBS k at time t, and L_j is data size of content j. Note that $L_j p_{k,j}(t)$ represents the data size of content j downloaded by SBS k at time t. Since each user can download the requested file from one of the multiple SBSs within its reception ball, for the given limited storage size, it is crucial to minimize the overlapping content caching while maximizing the cache hitting rates, by determining the file caching fraction $p_{k,j}(t)$ at SBS k. This problem is intertwined with other SBSs' caching decisions, and the difficulty is aggravated under ultra-dense SBS deployment, seeking a novel solution with low-complexity using MFG to be elaborated in Section 17.4.

Interference Dynamics. In a UDN, there is a considerable number of SBSs with no associated user within its coverage. These SBSs become idle and do not transmit any signal according to the definition of UDN ($\lambda_b \gg \lambda_u$) [8]. Hence, this dormant SBS does not cause interference to neighbor SBSs. This leads to a spatially dynamic distribution of interference characterized by users' locations. We assume that active SBSs always have data to transmit. Let us denote the probability of active SBSs by p_a. An aggregate interference is imposed by the active SBSs with probability p_a. Assuming that p_a is homogeneous over SBSs yields $p_a \approx 1 - [1 + \lambda_u/(3.5\lambda_b)]^{-3.5}$ [44]. It provides that the density of interfering SBSs is equal to $p_a\lambda_b$. Then, for a typical user selected uniformly at random, the signal-to-interference-plus-noise ratio (SINR) with N_a number of transmit antennas is given as:

$$\mathrm{SINR}(t) = \frac{N_a P|h(t)|^2}{\sigma^2 + \frac{\theta_{N_a}}{2\pi} N_a I^f(t)}. \tag{17.4}$$

where the aggregate interference $I^f(t)$ depends on the set $\Phi_R(p_a\lambda_b)$ of active SBS coordinates within the reception ball of radius R, given by $I^f(t) = \sum_k^{|\Phi_R(p_a\lambda_b)|} P|h_{k,i}(t)|^2$. The term $\frac{\theta_{N_a}}{2\pi}N_a$ in (17.4) is given by the directional beam pattern. Given uniformly distributed (i.e., isotropically distributed) users, an SBS becomes the interferer with probability $\theta_{N_a}/(2\pi)$ with the main lobe gain N_a and the beam width $\theta_{N_a} = 2\pi/\sqrt{N_a}$. Note that the interference term $I^f(t)$ depends on the spatial locations of SBSs and users (through p_a). This becomes a major bottleneck in analyzing UDCN, calling for a tractable way of handling $I^f(t)$ using MFG to be discussed in Section 17.4.

17.4 Game-theoretic formulation for edge caching

We utilize the framework of non-cooperative games to devise a fully distributed algorithm. The goal of each SBS k is to determine its own caching amount $p_{k,j}^*(t)$ for content j in order to minimize the LRA cost. This cost is determined by spatio-temporally varying content request probability, network dynamics, content overlap, and aggregate inter-SBS interference. As the SBSs' states and content popularity evolve, the caching strategies of the SBSs must adapt accordingly. Minimizing the

LRA cost under the spatio-temporal dynamics can be modeled as a dynamic SDG [45]. In the following subsection, we specify the impact of other SBSs' caching strategies and inter-SBS interference in the SDG by defining the LRA cost.

17.4.1 Cost functions

An instantaneous cost function $J_{k,j}(t)$ defines the LRA cost. It is affected by backhaul capacity, remaining storage size, average rate per unit bandwidth, and overlapping contents among SBSs. SBS k cannot download more than $B_{k,j}(t)$, defined as the allocated backhaul capacity for downloading content j at time t. In the proposed LRA cost, the download rate $L_j p_{k,j}(t)$ is prevented from exceeding the backhaul capacity constraint $B_{k,j}(t)$ by the backhaul cost function $\phi_{k,j}$ as $\phi_{k,j}(p_{k,j}(t)) = -\log(B_{k,j}(t) - L_j p_{k,j}(t))$. If $L_j p_{k,j}(t) \geq B_{k,j}(t)$, the value of the cost function $\phi_{k,j}$ goes to infinity. This form of cost function is widely used to model barrier or constraint of available resources as in [18]. As cached content files occupy the storage, it causes processing latency [46] or delay to search requested files by users. This overhead cost is proportional to the cached data size in the storage unit. To incorporate this, a storage cost function is proposed based on the occupation ratio of the storage unit normalized by the storage size as follows:

$$\psi_{k,j}(Q_{k,j}(t)) = \gamma(C_{k,j} - Q_{k,j}(t))/C_{k,j}, \tag{17.5}$$

where $Q_{k,j}(t)$ is the storage cost function at time t, and γ is a constant storage cost parameter. Then, the global instantaneous cost is given by:

$$J_{k,j}(p_{k,j}(t), p_{-k,j}(t)) = \frac{\phi_{k,j}(p_{k,j}(t))(1 + I^{r}_{k,j}(p_{-k,j}(t)))}{\mathcal{R}_k(t, \hat{I}^{f}(t))x_j(t)} + \psi_{k,j}(Q_{k,j}(t)), \tag{17.6}$$

where $I^{r}_{k,j}(p_{-k,j}(t))$ denotes the expected amount of overlapping content per unit storage size, $C_{k,j}$, $p_{-k,j}(t)$ is a vector of caching control variable of all the other SBSs except SBS k, $\hat{I}^{f}(t)$ denotes the normalized aggregate interference from other SBSs with respect to the SBS density and the number of antennas, and $\mathscr{R}_k(t)$ is the average downlink rate per unit bandwidth. The cost increases with the amount of overlapping contents and aggregate interference, which are described in the next Section 17.4.2. From the global cost function (17.6), the LRA caching cost is given by:

$$\mathscr{J}_{k,j} = \mathbb{E}\left[\int_t^T J_{k,j}(p_{k,j}(t), p_{-k,j}(t))\,\mathrm{d}t\right]. \tag{17.7}$$

17.4.2 Interactions through content overlap and interference

The caching strategy of an SBS inherently makes an impact on the caching control strategies of other SBSs. These *interactions* can be defined and quantified by the amount of overlapping contents and interference. These represent major bottlenecks for optimizing distributed caching for two reasons: first, they undergo changes with respect to the before-mentioned spatio-temporal dynamics, and it is hard to acquire the knowledge of other SBSs' caching strategies directly. In this context, our purpose is to estimate these interactions in a distributed fashion without full knowledge of other SBSs' states or actions.

Content Overlap. As shown in Figure 17.1a, in UDNs, there may be overlapping contents downloaded by multiple SBSs located within radius R_c from the randomly selected typical user. For example, let us consider that these neighboring SBSs cache the most popular contents with the intention of increasing the local caching gain (i.e., cache hit). Since only one of the SBS candidates is associated with the user to transmit the cached content file, caching the identical content of other SBSs becomes a waste of storage and backhaul usage. In this context, overlapping contents increase redundant cost due to inefficient resource utilization [47]. The amount of overlapping contents is determined by other SBSs' caching strategies. We define the content overlap function $I^r_{k,j}(p_{-k,j}(t))$ as the expected amount of overlapping content per unit storage size $C_{k,j}$, which is given by:

$$I^r_{k,j}(p_{-k,j}(t)) = \frac{1}{C_{k,j}N_{r(j)}}\sum_{i\neq k}^{|\mathcal{N}|} p_{i,j}(t), \tag{17.8}$$

where $N_{r(j)}$ denotes the number of contents whose request probability is asymptotically equal to x_j. It can be defined as: cardinality of the following set: $\{m|m \in M \text{ s.t. } |x_m - x_j| \leq \epsilon\}$. When the value of ϵ is sufficiently small, $N_{r(j)}$ becomes the number of contents whose request probability is equal to that of content j. If there is a large number of contents with equal request probabilities, a given content is selected randomly and cached. Hence, the occurrence probability of content overlap decreases with a higher diversity of content caching.

Inter-SBS Interference. In a UDN, user location determines the location of interferers, or the density of the user determines the density of interfering SBSs. It is because there are SBSs that have no users in their own coverage and become dormant without imposing interference to their neighboring SBSs. These spatial dynamics of interference in UDN is a bottleneck for optimizing distributed caching such that an SBS in a high interference environment cannot deliver the cached content to its own users. To incorporate this spatial interaction, following the interference analysis in UDNs [22], the interference normalized by SBS density and the number of antennas is given by:

$$\hat{I}^f(t) = (\lambda_u \pi R)^2 N_a^{-\frac{1}{2}} \lambda_b^{-\frac{\alpha}{2}} \left(1 + \frac{1 - R^{2-\alpha}}{\alpha - 2}\right) P\mathsf{E}_g[|g(t)|^2], \tag{17.9}$$

where $\hat{I}^f(t)$ denotes the normalized interference with respect to SBS density and the number of antennas. It gives us an average downlink rate per unit bandwidth $\mathscr{R}_k(t)$ and its upper bound in UDN as follows:

$$\mathscr{R}_k(t) = \mathsf{E}_{S,I^f}\left[\log(1 + \mathsf{SINR}(t))\right] \tag{17.10}$$

$$\leq \mathsf{E}_S \log\left(1 + \frac{S_k(t)}{\frac{\sigma^2}{N_a \lambda_b^{\alpha/2}} + \mathsf{E}_{I^f}[\hat{I}^f(t)]}\right), \tag{17.11}$$

where σ^2 is the noise power. Note that (17.11) shows the effect of interference on the upper bound of an average SE. It is because we consider that only the SBSs within the pre-defined reception ball cause interference to a typical user. Hence, the equality in (17.11) holds, when the size of reception ball R goes to infinity, including all the SBSs in the networks as interferers.

17.4.3 Stochastic differential game for edge caching

As the SBSs' states and content popularity evolve according to the dynamics (17.2) and (17.3), an individual SBS's caching strategy must adapt accordingly. Hence, minimizing the LRA cost under the spatio-temporal dynamics can be modeled as a dynamic SDG, where the goal of each SBS k is to determine its own caching amount $p^*_{k,j}(t)$ for content j to minimize the LRA cost $\mathscr{J}_{k,j}(t)$ (17.7):

$$\textbf{(P1)} \quad v_{k,j}(t) = \inf_{p_{k,j}(t)} \mathscr{J}_{k,j}(t) \tag{17.12}$$

$$\text{subject to} \quad \mathrm{d}x_j(t) = r(\mu - x_j(t))\mathrm{d}t + \eta \mathrm{d}W_j(t), \tag{17.13}$$

$$\mathrm{d}Q_{k,j}(t) = (e_{k,j} - L_j p_{k,j}(t))\mathrm{d}t. \tag{17.14}$$

In problem **P1**, the state of SBS k and content j at time t is defined as $s_{k,j}(t) = \{x_j(t), \mathscr{R}_k(t), Q_{k,j}(t)\}$, $\forall k \in \mathscr{N}, \forall j \in \mathscr{M}$. The SDG for edge caching is defined by $(\mathscr{N}, \mathscr{S}_{k,j}, \mathscr{A}_{k,j}, \mathscr{J}_{k,j})$ where $\mathscr{S}_{k,j}$ is the state space of SBS k and content j, $\mathscr{A}_k$ is the set of all caching controls $\{p_{k,j}(t), 0 \leq t \leq T\}$ admissible for the state dynamics.

To solve problem **P1**, the long-term average of content request probability μ is necessary for the dynamics of content request probability (17.13). To determine the value of μ, the mean value $m_k(t)$ of the cardinality of the set $U^r_k(t)$ needs to be obtained. Although the period $\{0 \leq t \leq T\}$ is not infinite, we assume that the inter-arrival time of the content request is sufficiently smaller than T and that numerous content requests arrive during that period. Then, the long-term average of content request probability μ becomes an asymptotic mean value $(t \to \infty)$. Noting that $\sum_j n_{k,j}(t) = N_k(t)$, the mean value of $m_k(t)$ is asymptotically given by [48] as follows:

$$\langle |U^r_k(t)| \rangle \simeq \begin{cases} \frac{\Gamma(\theta+1)}{\alpha\Gamma(\theta+\alpha)} N_k(t)^\alpha & \text{for } \alpha > 0 \\ \theta \log(N_k(t) + \theta) & \text{for } \alpha = 0 \end{cases} \tag{17.15}$$

where expression $\langle . \rangle$ is the average value, and $\Gamma(.)$ is the Gamma function.

Problem **P1** can be solved by using a backward induction method where the minimized LRA cost $v_{k,j}(t)$ is determined in advance through solving the following N-coupled HJB equations.

$$\begin{aligned} 0 = \partial_t v_{k,j}(t) + \inf_{p_{k,j}(t)} \Bigg[& J_{k,j}(p_{k,j}(t), p_{-k,j}(t)) + \frac{\eta^2}{2} \partial^2_{xx} v_{k,j}(t) \\ & + \underbrace{(e_{k,j} - L_j p_{k,j}(t))}_{(A)} \partial_{Q_k} v_{k,j}(t) + \underbrace{r(\mu - x_j(t))}_{(B)} \partial_x v_{k,j}(t) \Bigg] \end{aligned} \tag{17.16}$$

The HJB equations (17.16) for $k = 1, ..., N$ have a unique joint solution if the drift functions defining temporal dynamics (A) and (B) and the cost function (17.6) are smooth [11]. Since this smoothness is satisfied, we can assure that a unique solution of equation (17.16) exists. The optimal joint solution of HJB equations achieves NE as problem **P1** is a non-cooperative game wherein players do not share their state or strategy [11, 12]. The unique minimized cost $v^*_{k,j}(t)$ of problem **P1** and its corresponding NE can be defined as follows:

Definition 1: The set of SBSs' caching strategies $\mathbf{p}^* = \{p^*_{1,j}(t), ..., p^*_{N,j}(t)\}$, where $p^*_{k,j}(t) \in \mathscr{A}_{k,j}$ for all $k \in \mathscr{N}$, is a Nash equilibrium, if for all SBS k and for all admissible caching strategy set $\{p_{1,j}(t), ..., p_{N,j}(t)\}$, where $p_{k,j}(t) \in \mathcal{A}_{k,j}$ for all $k \in \mathcal{N}$, it is satisfied that

$$\mathscr{J}_{k,j}(p^*_{k,j}(t), p^*_{-k,j}(t)) \leq \mathscr{J}_{k,j}(p_{k,j}(t), p^*_{-k,j}(t)), \tag{17.17}$$

under the temporal dynamics (17.13) and (17.14) for common initial states $x_j(0)$ and $Q_{k,j}(0)$.

Unfortunately, this approach is accompanied with high computational complexity in achieving the NE (17.17), when N is larger than two because an individual SBS should take into account other SBSs' caching strategies $p_{-k,j}(t)$ to solve the inter-weaved system of N HJB equation (17.16). Furthermore, it requires collecting the control information of all other SBSs including their own states, which brings about a huge amount of information exchange among SBSs. This is not feasible and impractical for UDNs. For a sufficiently large number of SBSs, this problem can be transformed to a MFG, which can achieve the ϵ-Nash equilibrium [45].

17.4.4 Mean-field game for edge caching

To reduce the aforementioned complexity in solving the SDG **P1**, the following features are utilized. When the number of SBSs becomes large, the influence of every individual SBS can be modeled with the effect of the collective (aggregate) behavior of the SBSs. MFG theory enables us to transform these multiple interactions into a single aggregate interaction, called MF interaction, via MF approximation. According to [38], this approximation holds under the following conditions: (i) a large number of players, (ii) the exchangeability of players under the caching control strategy, and (iii) finite MF interaction. If these conditions are satisfied, the MF approximation can provide the optimal solution which the original SDG achieves.

Condition (i) corresponds to the definition of UDNs. For condition (ii), players (i.e., SBSs) in the formulated SDG are said to be exchangeable or indistinguishable under the control $p_{k,j}(t)$ and the states of players and contents if the player's control is invariant by their indices and decided by only their own states. In other words, permuting players' indices cannot change their own control strategies. Under this exchangeability, it is sufficient to investigate and re-formulate the problem for a generic SBS by dropping its index k.

The MF interactions (17.8) and (17.9) should asymptotically converge to a finite value under the given conditions. The content overlap (17.8) in MF regime, called MF overlap, goes to zero when the number of contents per SBS is extremely large, i.e., $M \gg N$. Such a condition implies that the cardinality of the set consisting of asymptotically equal content popularity goes to infinity. In other words, $N_{r(j)}$ tends to infinity yielding that the expected amount of overlapping content per unit storage size $I^r_{k,j}(\boldsymbol{p}_{-k,j}(t))$ becomes zero. In terms of interference, the MF interference converges as the ratio of SBS density to user density tends to infinity, i.e., $N_a \lambda_b^\alpha / (\lambda_u R)^4 \to \infty$ [22]. This condition corresponds to the notion of UDN [8] or massive MIMO ($N_a \to \infty$). Thus, the MF approximation can be utilized as the conditions inherently hold for UDCNs.

To approximate the interactions from other SBSs, we need a state distribution of SBSs and contents at time t, called MF distribution $m_t(x(t), Q(t))$. The MF distribution is derived from the following empirical distribution.

$$M_t^{(N\times M)}(x(t), Q(t)) = \frac{1}{NM} \sum_{j=1}^{M} \sum_{k=1}^{N} \delta_{\{x_j(t), Q_k(t)\}} \tag{17.18}$$

When the number of SBSs increases, the empirical distribution $M_t^{(N\times M)}(x_j(t), Q(t))$ converges to $m_t(x_j(t), Q(t))$, which is the density of contents and SBSs in state $(x_j(t), Q(t))$. Note that we omit the SE $\mathscr{R}(t)$ from the density measure to only consider temporally correlated state without loss of generality.

To this end, we derive a FPK equation [45] that is a partial differential equation capturing the time evolution of the MF distribution $m_t(x_j(t), Q(t))$ under dynamics of the popularity $x_j(t)$ and the available storage size $Q(t)$. The FPK equations for $m_t(x_j(t), Q(t))$ subject to the temporal dynamics (17.2) and (17.3) are given as follows:

$$\begin{aligned} 0 &= \partial_t m_t(x_j(t), Q(t)) + r(\mu_j - x_j(t))\partial_x m_t(x_j(t), Q(t)) \\ &+ (e_j - L_j p_j(t))\partial_Q m_t(x_j(t), Q(t)) - \frac{\eta^2}{2}\partial_{xx}^2 m_t(x_j(t), Q(t)). \end{aligned} \tag{17.19}$$

Let us denote the solution of the FPK equation (17.19) as $m_t^*(x_j(t), Q(t))$. Exchangeability and existence of the MF distribution allow us to approximate the interaction $I^r_{k,j}(p_{-k,j}(t))$ as a function of $m_t^*(x_j(t), Q(t))$ as follows:

$$I_j^r(t, m_t^*(x_j(t), Q(t))) = \int_Q \int_x \frac{m_t^*(x_j, Q) p_j(t, x(t), Q(t))}{C_{k,j} N_{r(j)}} \mathrm{d}x \mathrm{d}Q. \tag{17.20}$$

This interaction from (17.20) can be estimated without observing other SBSs' caching strategies. Thus, it is not necessary for an SBS to have full knowledge of the states or the caching control policies of other SBSs. An SBS needs to solve only a pair of equations, namely the FPK equation (17.19) and the following modified HJB is obtained by applying the MF approximation (17.20) to (17.16):

$$\begin{aligned} 0 = \partial_t v_j(t) + \inf_{p_j(t)} \Bigg[& J_j(p_j(t), I_j(t, m_t^*(x_j(t), Q(t))) + \frac{\eta^2}{2}\partial_{xx}^2 v_j(t) \\ & + (e_j - L_j p_j(t))\partial_Q v_j(t) + r(\mu - x_j(t))\partial_x v_j(t) \Bigg]. \end{aligned} \tag{17.21}$$

The FPK equation (17.19) and the HJB equation (17.21) are intertwined with each other for the MF distribution and the optimal caching amount, which depends on the optimal trajectory of the LRA cost $v_j^*(t)$. The optimal LRA cost $v_j^*(t)$ is found by applying backward induction to the single HJB equation (17.21). Also, its corresponding MF distribution (state distribution) $m_t^*(x_j(t), Q(t))$ is obtained by forward solving the FPK equation (17.19). These solutions of the HJB and FPK equations $[m_t^*(x_j(t), Q(t)), v_j^*(t)]$ define the mean-field equilibrium (MFE), defined as follows:

Definition 2: The generic caching strategies $p_j^*(t)$ achieve an MFE if for all admissible caching strategies set $\{p_{1,j}(t), \ldots, p_{N,j}(t)\}$ where $p_{k,j}(t) \in \mathscr{A}_{k,j}$ for all $k \in \mathscr{N}$ it is satisfied that

$$\mathscr{J}_j(p_j^*(t), m_t^*(x_j(t), Q(t)) \leq \mathscr{J}_j(p_j(t), m_t^*(x_j(t), Q(t)), \tag{17.22}$$

under the temporal dynamics (17.13) and (17.14) for an initial MF distribution m_0. The MFE corresponds to the ε-Nash equilibrium:

$$\mathscr{J}_{k,j}(p_{k,j}^*(t), p_{-k,j}^*(t)) \leq \mathscr{J}_j(p_j^*(t), m_t^*(x_j(t), Q(t)) - \epsilon, \tag{17.23}$$

where ε asymptotically becomes zero for a sufficiently large number of SBSs.

Let us define $p_j^*(t)$ as an optimal caching control strategy, which achieves the MFE yielded by the optimal caching cost trajectory $v_j^*(t)$ and MF distribution $m_t^*(x_j(t), Q(t))$. The solution $p_j^*(t)$ is given by the following proposition.

Proposition 1. *The optimal caching amount is given by*

$$p_j^*(t) = \frac{1}{L_j}\left[B_j(t) - \frac{1 + I_j^r(t, m_t^*(x_j(t), Q(t)))}{\mathscr{R}(t, I^f(t))x_j(t)\partial_Q v_j^*}\right]^+, \tag{17.24}$$

where $m_t^*(x(t), Q(t))$ and $v_j^*(t)$ are the solutions of (17.19) and (17.21), respectively.

Proof: The optimal control of the differential game with HJB equations is the argument of the infimum term (17.21) [11].

$$\begin{aligned} p_j^*(t) = \arg \inf_{p_j(t)} \Big[& J_j(p_j(t), I_j(t, m_t^*(x_j(t), Q(t))) + \frac{\eta^2}{2}\partial_{xx}^2 v_j(t) \\ & + (e_j - L_j p_j(t))\partial_Q v_j(t) + r(\mu - x_j(t))\partial_x v_j(t)\Big] \end{aligned} \tag{17.25}$$

The infimum term (17.25) is a convex function of $p_j(t)$ for all time t, since its first and second-order derivatives are lower than zero. Hence, we can apply Karush-Khun-Tucker (KKT) conditions and get a sufficient condition for the unique optimal control $p_j^*(t)$ by finding a critical point given by:

$$\frac{\partial}{\partial p_j(t)}\left[J_j(p_j(t), I_j(t, m_t^*(x_j(t), Q(t))) + (e_j - L_j p_j(t))\partial_Q v_j(t)\right] = 0. \tag{17.26}$$

Owing to the convexity, the solution of equation (17.26) is the unique optimal solution described as follows:

$$p_j^*(t) = \frac{1}{L_j}\left[B_j(t) - \frac{1 + I_j^r(t, m_t^*(x_j(t), Q(t)))}{\mathscr{R}(t, I^f(t))x_j(t)\partial_Q v_j^*}\right]^+. \tag{17.27}$$

Remark that $p_j^*(t)$ is a function of $m_t^*(x_j(t), Q(t))$ and v_j^*, which are the solutions of equations (17.19) and (17.21), respectively. The expression of $p_j^*(t)$ (17.26) provides the final versions of the HJB and FPK equations as follows:

$$
\begin{aligned}
0 = {} & \partial_t v_j(t) - \frac{\log\left(B_j(t) - \left[B_j(t) - \frac{1+I_j^r(t, m_t^*(x_j(t)Q(t)))}{\mathscr{R}(t, I^f(t))x_j(t)\partial_Q v_j}\right]^+\right)}{\mathscr{R}(t, I^f(t))x_j(t)} \\
& \times(1+I_j^r(t, m_t^*(x_j(t), Q(t)))) + \frac{\alpha(C-Q(t))}{C} + r(\mu_j - x_j(t))\partial_x v_j(t) + \\
& \left(e_j - \left[B_j(t) - \frac{1+I_j^r(t, m_t^*(x,Q))}{\mathscr{R}(t, I^f(t))x_j(t)\partial_Q v_j}\right]^+\right)\partial_Q v_j(t) + \frac{\eta^2}{2}\partial_{xx}^2 v_j(t), \\
0 = {} & \partial_t m_t(x_j(t), Q(t)) + r(\mu_j - x_j(t))\partial_x m_t(x_j(t), Q(t)) - \\
& \frac{\eta^2}{2}\partial_{xx}^2 m_t(x_j(t), Q(t)) \\
& + \left(e_j - \left[B_j(t) - \frac{1+I_j^r(t, m_t(x_j(t), Q(t)))}{\mathscr{R}(t, I^f(t))x_j(t)\partial_Q v_j^*}\right]^+\right)\partial_Q m_t(x_j(t), Q(t)).
\end{aligned}
\tag{17.28}
$$

From these equations, we can find the values of $v_j^*(t)$ and $m_t^*(x(t), Q(t))$. Note that the smoothness of the drift functions in the dynamic equation and the cost function (17.6) assure the uniqueness of the solution [11].

Proposition 1 provides the optimal caching amount of $p_j^*(t)$ in a water-filling fashion of which the water level is determined by backhaul capacity $B_j(t)$. Noting that the average rate per unit bandwidth $\mathscr{R}(t)$ increases with the number of antenne N_a and SBS density λ_b, SBSs cache more contents from the server when they can deliver content to users with high wireless capacity. Also, SBSs diminish the caching amount of content j, when the estimated amount of content overlap $I_j^r(t, m_t^*(x_j(t), Q(t)))$ is large.

Remark that the existence and uniqueness of the optimal caching control strategy are guaranteed. The optimal caching algorithm converges to a unique MFE, when the initial conditions m_0, $x_j(0)$, and $Q(0)$ are given. The specific procedure of this MF caching algorithm is described in Algorithm 1.

Algorithm 1 Mean-Field Caching Control

Require: $x_j(t)$, m_0, $B(t)$ and $Q(0)$

1: Find the optimal trajectory of caching cost and state distribution $[v_j^*(t), m_t^*(x_j(t), Q(t))]$ by solving HJB (17.21) and FPK (17.19) equations

2: Calculate $I_j^r(t, m_t^*(x_j(t), Q(t)))$, $I^f(t)$ and $\partial_Q v_j^*$

3: Compute the instantaneous caching amount $p_j^*(t)$

$$: \quad p_j^*(t) = \frac{1}{L_j}\left[B_j(t) - \frac{1+I_j^r(t, m_t^*(x_j(t), Q(t)))}{\mathscr{R}(t, I^f(t))x_j(t)\partial_Q v_j^*}\right]^+$$

4: Get values of $[x_j(t), Q(t)]$ according to the dynamics

5: Go line 2

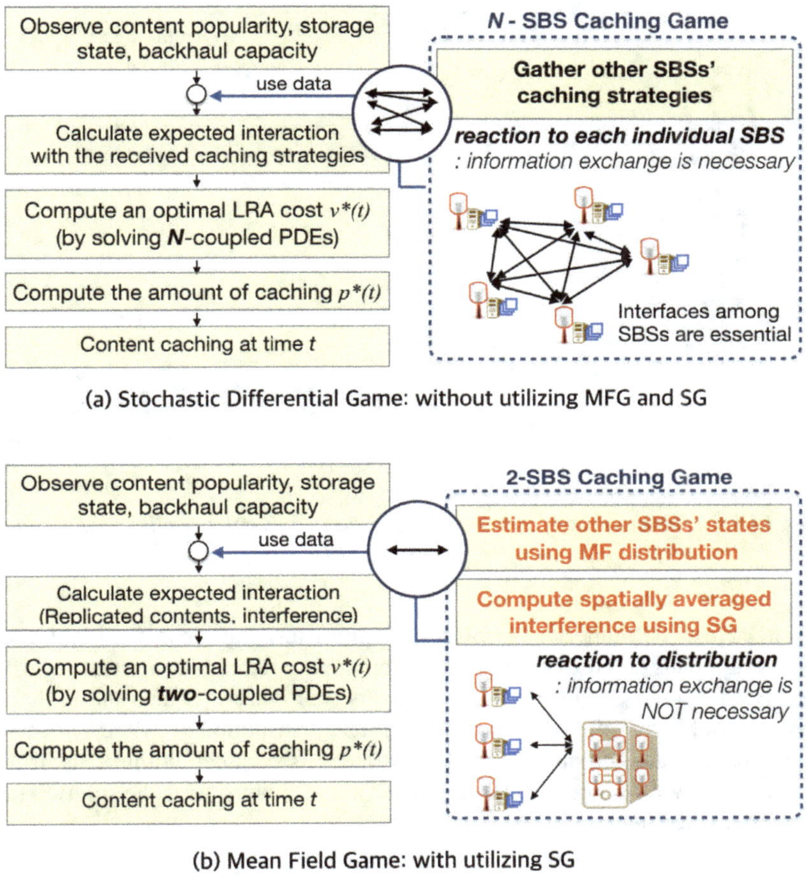

Figure 17.2 *Ultra-dense edge caching flow charts according to the approaches of SDG and MFG, respectively. (a) In the framework of SDG, we solve the game of N SBSs (players) interacting with each individual SBS. (b) By incorporating MFG theory and SG into the framework, we can estimate the collective interaction of other SBSs. This relaxes the N-SBS caching game to a two-SBS caching game.*

The respective processes of solving **P1** in ways of SDG and MFG are depicted in Figure 17.2. Remark that the solution of the MFG becomes equivalent to that of the N-player SDG **P1** as N increases. The complexity of the proposed method is much lower as compared to solving the original N-player SDG **P1**. The number of PDEs to solve for one content is reduced to two from the number of SBSs N. Thus, the complexity is consistent even though the number of players N becomes large. This feature is verified via simulations, as shown in Figure 17.3, which represents the number of iterations required to solve the HJB-FPK equations (17.21) and (17.19) as a function for different SBS densities λ_b. Here, it is observed that the caching

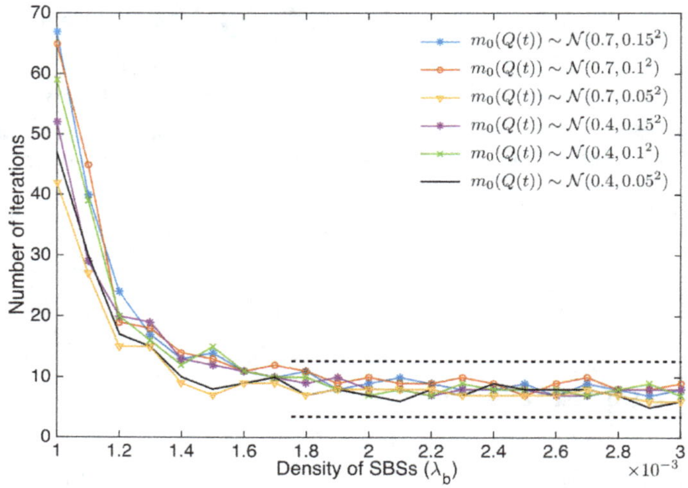

Figure 17.3 *The number of iterations required to solve the coupled HJB and FPK equations for different densities of SBSs*

problem **P1** can be numerically solved in only a few iterations for highly dense networks. It means that the computational complexity remains consistent regardless of the SBS density λ_b, or the number of players N. Figure 17.3 also shows that this consistency holds for different initial storage state distribution of SBSs. The number of iterations to reach the optimal caching strategy is bounded within tens of iterations even for low SBS density. The proposed algorithm provides the solution faster for more densified networks.

17.5 Numerical results

Numerical results are provided for evaluating the proposed algorithm under spatio-temporal content popularity and network dynamics illustrated in Figure 17.1. Let us assume that the initial distribution of the SBSs m_0 is given as normal distribution and that the storage size $Q(t)$ belongs to a set [0, 1] for all time t. Considering Rayleigh fading with mean one, the parameters are configured, as shown in Table 17.1. To solve the coupled PDEs (the first step of Algorithm 1) using a finite element method, we used the MATLAB PDE solver.

17.5.1 Mean-field equilibrium achieved by the proposed MF caching algorithm

To demonstrate that the proposed MF caching algorithm achieves the MFE, it is assumed that SBSs have full knowledge of contents request probability, which implies that perfect popularity information is available at SBSs. The trajectory of the proposed caching algorithm and MF distribution is analyzed numerically when

Table 17.1 Key simulation parameters

Parameter	Value
SBS density λ_b	0.005, 0.02, 0.035, 0.05 (SBSs/m^2)
User density λ_u	10^{-4}, 2.5×10^{-4} (users/m^2)
Transmit power P	23 dBm
Noise floor	−70 dBm
Number of contents	20
CRP parameters θ,v	$\theta = 1$, $v = 0.5$
Reception ball radius R	$10/\sqrt{\pi}$ km
Network size	20 km × 20 km
File discarding rate e_j	0.1

the content request probability is static. In this case, the caching control strategies do not depend on the evolution law of the content popularity. Specifically, in HJB (17.21) and FPK (17.19) equations, the derivative terms with respect to content request probability x become zero.

Figure 17.4 shows the evolution of the optimal caching amount $p^*(t)$ with respect to the storage state and time. The value of $p^*(t)$ is maintained lower than the content request probability to reduce the content overlap and prevent redundant backhaul and storage usage.

Figure 17.5 shows heat maps representing the instantaneous density of SBSs having the remaining storage size $Q(t)$ in terms of the MF distribution $m_t^*(Q(t))$ for a content during the period $\{0 \leq t \leq T\}$, where $T = 1$. A bright-colored point means there are many SBSs with unoccupied storage size corresponding to the point. It is observed that the unoccupied storage space of SBSs does not diverge from each other as the proposed algorithm brings SBSs' state in the MFE. At this equilibrium, the amount of cached content file decreases when the content popularity x becomes low. This tendency corresponds to the trajectory of the optimal caching probability in Figure 17.4. Almost every SBS has cached the content over time, but not used its entire storage. The remaining storage saturates even though the content popularity is equal to 0.9. This implies that SBSs adjust the caching amount of popular content in consideration of the content overlap expected to possibly increase the cost.

17.5.2 Performance evaluation in terms of long run average caching cost

This section evaluates the performance of the proposed MF caching algorithm under the spatio-temporal content popularity dynamics. In addition, we evaluate the robustness of our scheme to imperfect popularity information in terms of the LRA caching cost. To this end, we compare the performance of the proposed MF caching algorithm with the following caching algorithms.

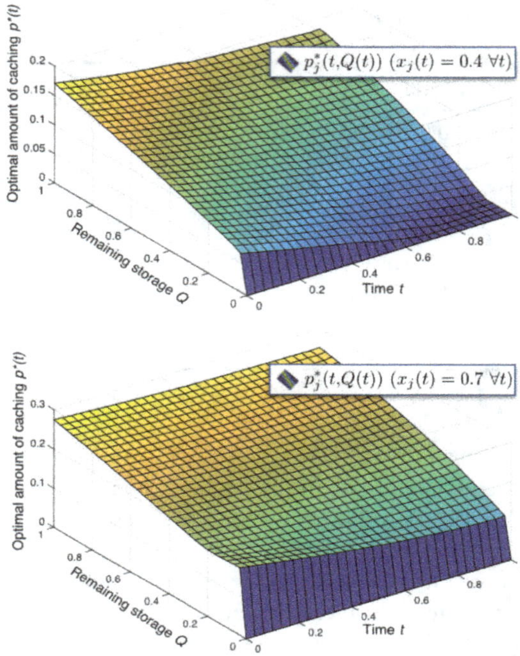

Figure 17.4 *The optimal caching amount $p^*(t)$ at the MF equilibrium under two different content popularities 0.4 and 0.7, assuming that the content popularity is static. The initial MF distribution $m_0(Q(0))$ is given as $\mathcal{N}(0.7, 0.05^2)$.*

- *Baseline caching algorithm* that does not consider the amount of content overlap but determines the instantaneous caching amount $\hat{p}_j(t)$ proportionally to the instantaneous request probability $x_j(t)$ subject to current backhaul, storage state, and interference described as follows: $\hat{p}_j(t) = \frac{1}{L_j}\left[B_j(t) - \frac{1}{1+\mathcal{R}(t, I^j(t))x_j(t)}\right]^+$.
- *Uniformly random caching* that randomly determines the caching amount following the uniform distribution.

LRA Cost Comparison. Figure 17.6 shows the LRA cost evaluation of the proposed MF caching algorithm, uniformly random caching, and the baseline caching algorithm, which disregard the content overlap among neighboring SBSs. The LRA costs over time for different user density λ_u are evaluated numerically. The proposed caching control algorithm reduces about 24% of the LRA cost as compared to the caching algorithm without considering the content overlap. This performance gain is due to avoiding redundant content overlap and having an SBS under lower interference environment to cache more contents. As the user density λ_u becomes higher for a fixed SBS density λ_b, the final values of the LRA cost increase for all the three

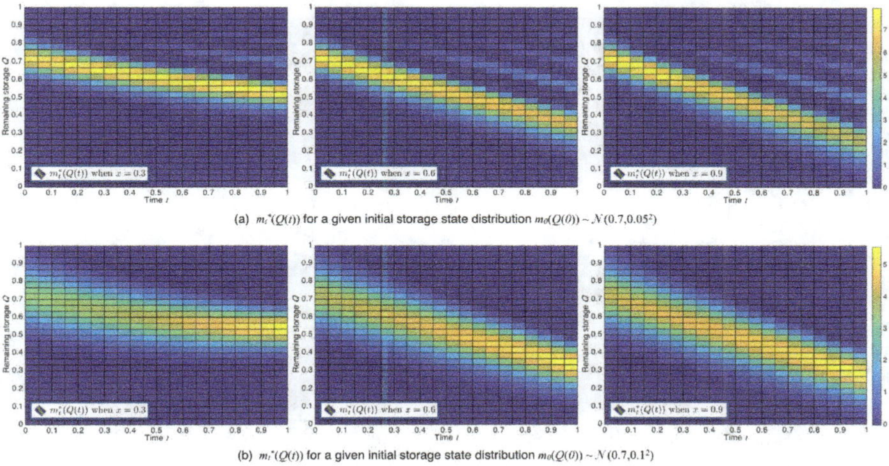

Figure 17.5 *A heat map illustration of the MF distribution $m_t^*(Q(t))$ that represents the instantaneous density of SBSs having the remaining storage space $Q(t)$ for an arbitrary content during a long-term period $\{0 \leq t \leq T\}$, when the proposed MF caching algorithm is applied. A bright-colored point means there are many SBSs with unoccupied storage size corresponding to the point. It shows the temporal evolution of the density of SBSs with respect to different content popularity x_j, and initial distribution $m_0(Q(0))(B(t) = 1, N_{r(j)} = 20, \lambda_u = 0.001, \lambda_b = 0.03)$.*

caching schemes. When UDCNs are populated by numerous users, the fluctuation of spatial dynamics of popularity increases and the number of SBSs having associated users increases. Hence, both the aggregate interference imposed by the SBSs and the content popularity severely change over the spatial domain. In this environment, the advantage of the proposed algorithm compared to the popularity-based algorithm becomes larger, yielding a higher gap between the final values of the produced LRA cost.

Demand Misprediction Impact. Accurate content popularity information may not be available at SBSs due to misprediction or estimation error of content popularity. It is thus assessed how the proposed algorithm is robust against imperfect popularity information (IPI) given as follows:

$$\hat{x}(t) = x(t) + \Delta(t), \tag{17.29}$$

where $\hat{x}(t)$ denotes a content request probability estimated by an SBS, and $\Delta(t)$ represents an observation error for the request probability $x(t)$ at time t. An SBS has perfect popularity information (PPI) if $\Delta(t)$ is equal to zero for all t (i.e., $\hat{x}(t) = x(t)$). The magnitude of Δ determines the accuracy of the popularity.

For numerical evaluations, an observation error Δ is assumed to follow a normal distribution $\mathcal{N}(0.2, 0.001^2)$. SBSs respectively determine their own caching

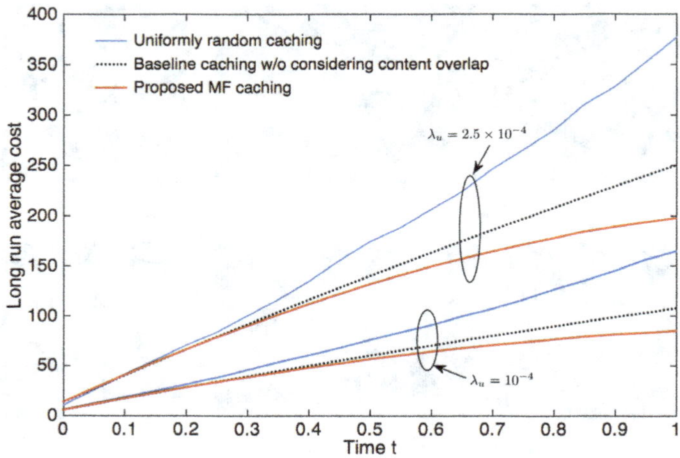

Figure 17.6 Long-run average costs of the caching strategies with respect to different user density λ_u. ($Q(0) = 0.7, x(0) = 0.3, \eta = 0.1$).

control strategies based on imperfect content request probability $\hat{x}(t)$ (17.29) instead of PPI $x(t)$. With this IPI, the LRA caching cost over time is evaluated as shown in Figure 17.7. The impact of IPI increases with the number of SBSs because redundant caching occurs at several SBSs. Also, the LRA increment due to IPI is evaluated for our MF caching algorithm and the popularity-based one for different SBS density, i.e., the number of neighboring SBSs, as shown in Figure 17.8. The numerical

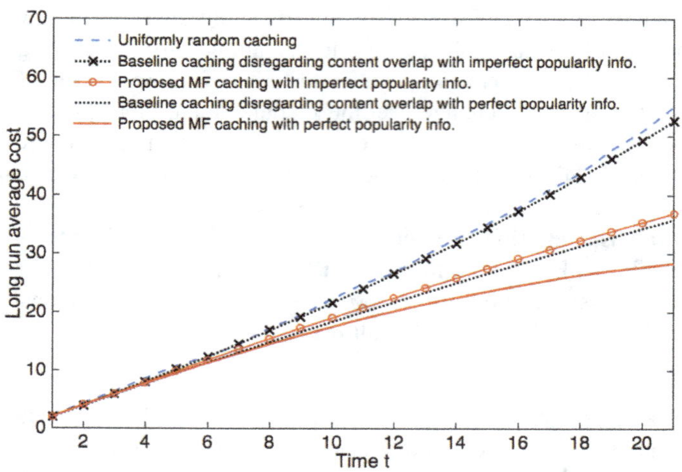

Figure 17.7 Long-run average costs of different caching strategies with perfect and imperfect popularity information. ($Q(0) = 0.7, x(0) = 0.3, \eta = 0.1$).

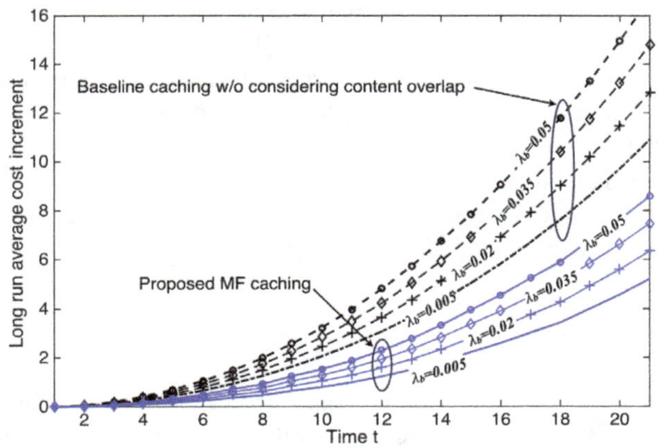

Figure 17.8 *LRA cost increment due to imperfect popularity information. For different SBS density λ_b, the proposed MF caching and the baseline caching without considering the content overlap are compared ($Q(0) = 0.7, x(0) = 0.3, \eta = 0.1$).*

results corroborate that the proposed algorithm is more robust against imperfect information of content popularity in comparison with the popularity-based benchmark scheme. In particular, our caching strategy reduces about 50% of the LRA cost increment as compared to the popularity-based baseline method.

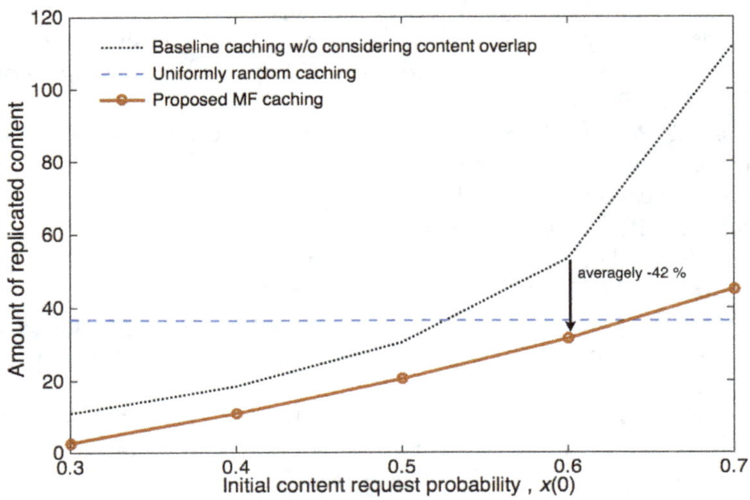

Figure 17.9 *The amount of overlapping contents per storage usage ($Q(0) = 0.7, \eta = 0.1$)*

Figure 17.9 shows the amount of overlapping contents per storage usage as a function of the initial content probability $x(0)$. The proposed MF caching algorithm reduces caching content overlap by an average of 42% as compared to popularity-based caching. However, MF caching algorithm yields a higher amount of content overlap than random caching does when the content request probability becomes high. The reason is that the random policy downloads the contents regardless of their popularity, so the amount of content overlap remains steady. On the other hand, MF caching increases the downloaded volume of popular content.

17.6 Conclusion

In this chapter, scalable and distributed edge caching in a UDCN has been investigated. To accurately reflect time-varying local content popularity, spatio-temporal content popularity modeling, and interference analysis have been applied in optimizing the edge caching strategy. Finally, by leveraging MFG, the computing complexity of optimizing the caching strategy has been reduced to a constant overhead from the cost exponentially increasing with the number of SBSs in conventional methods. Numerical simulations corroborate that the proposed MFG-theoretic edge caching yields lower LRA costs while achieving more robustness against imperfect content popularity information, compared to several benchmark schemes ignoring content popularity fluctuations or cached content overlap among neighboring SBSs.

References

[1] Kamel M., Hamouda W., Youssef A. 'Ultra-dense networks: A'. *IEEE Communications Surveys & Tutorials*. 2016;**18**(4):2522–45.

[2] Zander J. 'Beyond the ultra-dense barrier: Paradigm shifts on the road beyond 1000x wireless capacity'. *IEEE Wireless Communications*. 2017;**24**(3):96–102.

[3] Shim T., Park J., Ko S.-W., Kim S.-L., Lee B., Choi J. 'Traffic convexity aware cellular networks: a vehicular heavy user perspective'. *IEEE Wireless Communications*. 2016;**23**(1):88–94.

[4] Popovski P., Nielsen J.J., Stefanovic C., *et al.* 'Wireless access for ultra-reliable low-latency communication: principles and building blocks'. *IEEE Network*. 2018;**32**(2):16–23.

[5] Baştuğ E., Bennis M., Debbah M. 'Living on the edge: the role of proactive caching in 5g wireless networks'. *IEEE Communications Magazine*. 2014;**52**(8):82–9.

[6] Tamoor-ul-Hassan S., Bennis M., Nardelli P.H.J., Latva-aho M. 'Caching in wireless small cell networks: a storage-bandwidth tradeoff'. *IEEE Communications Letters*. 2016;**20**(6):1175–8.

[7] Wang X., Chen M., Taleb T., Ksentini A., Leung V. 'Cache in the air: exploiting content caching and delivery techniques for 5G systems'. *IEEE Communications Magazine*. 2014;**52**(2):131–9.

[8] Park J., Kim S.-L., Zander J. 'Tractable resource management with uplink decoupled millimeter-wave overlay in ultra-dense cellular networks'. *IEEE Transactions on Wireless Communications*. 2016;**15**(6):4362–79.

[9] Caines P.E., Games M.F. *Encyclopedia of systems and control*. London: Springer; 2014. pp. 1–6.

[10] Şen N., Caines P.E. 'Mean field games with partial observation'. *SIAM Journal on Control and Optimization*. 2019;**57**(3):2064–91.

[11] Oksendal B. *Stochastic differential equations*. Springer; 2003.

[12] Couillet R., Perlaza S.M., Tembine H., Debbah M. 'Electrical vehicles in the smart grid: a mean field game analysis'. *IEEE Journal on Selected Areas in Communications*. 2012;**30**(6):1086–96.

[13] Kim H., Park J., Bennis M., Kim S.-L., Debbah M. 'Ultra-dense edge cachingunder spatio-temporal demand and network dynamics'. Proceedings of IEEE International Conference on Communications (ICC); Paris, France, May; 2017.

[14] Kim H., Park J., Bennis M., Kim S.-L., Debbah M. 'Mean-field game theoretic edge caching in ultra-dense networks'. *IEEE Transactions on Vehicular Technology*. 2020;**69**(1):935–47.

[15] Elbamby M.S., Perfecto C., Liu C.-F., *et al.* 'Wireless edge computing with latency and reliability guarantees'. *Proceedings of the IEEE*. 2019;**107**(8):1717–37.

[16] Breslau L., Cao P., Fan L., Phillips G., Shenker S. 'Web caching and zipf-like distribution: evidence and implications'. Proceedings of IEEE International Conference on Computer Communications; New York, USA; 1999 Mar.

[17] Gregori M., Gómez-Vilardebó J., Matamoros J., Gündüz D. 'Wireless content caching for small cell and D2D networks'. *IEEE Journal on Selected Areas in Communications*. 2016;**34**(5):1222–34.

[18] Hamidouche K., Saad W., Debbah M., Poor H.V. 'Mean-field games for distributed caching in ultra-dense small cell networks'. Proceedings of 2016 American Control Conference (ACC); Boston, MA, USA; July 2016.

[19] Perabathini B., Baştuğ E., Kountouris M., Debbah M., Conte A. 'Caching at the edge: A green perspective for 5g networks'. Proceedings of IEEE International Conference on Communications (ICC); London, UK; 2015 Jun.

[20] Malak D., Al-Shalash M., Andrews J.G. 'Spatially correlated content caching for device-to-device communications'. *IEEE Transactions on Wireless Communications*. 2018;**17**(1):56–70.

[21] Zink M., Suh K., Gu Y., Kurose J. 'Watch global, cache local: Youtube network traffic at a campus network: Measurements and implications'. Proceedings of the SPIE/ACM Multimedia Computing and Networking Conference; 2008.

[22] Park J., Bennis M., Kim S.-L., Debbah M. 'Spatio-temporal network dynamics framework for energy-efficient ultra-dense cellular networks'. Proceedings of IEEE Global Communications Conference (GLOBECOM); Washington, DC; Dec 2016.

[23] Park J., Jung S.Y., Kim S.-L., Bennis M., Debbah M. 'User-centric mobility management in ultra-dense cellular networks under spatio-temporal

dynamics'. Proceedings of IEEE Global Communications Conference (GLOBECOM); Washington, DC; Dec 2016.

[24] Hamidouche M., Baştug E., Park J. 'Downlink performance of dense antenna deployment: to distribute or concentrate?'. 2017 IEEE 28th Annual International Symposium on Personal, Indoor, and Mobile Radio Communications; 2017. pp. 1–6.

[25] Park J., Kim D.M., Popovski P., Kim S-LKim S-L. 'Revisiting frequency reuse towards supporting ultra-reliable ubiquitous-rate communication'. Proceedings of IEEE WiOpt Workshop on Spatial Stochastic Models for Wireless Networks (SpaSWiN); Paris, France; May 2017.

[26] Kim J., Park J., Kim S., Kim S.-L., Sung K.W., Kim K.S. 'Millimeter-wave interference avoidance via building-aware associations'. *IEEE Access*. 2018;**6**:10618–34.

[27] Nemati M., Park J., Choi J. 'RIS-assisted coverage enhancement in millimeter-wave cellular networks'. *IEEE Access*. 2020;**8**:188171–85.

[28] Kim H., Park J., Bennis M., Kim S.-L., Debbah M. 'Mean-field game theoretic edge caching in ultra-dense networks'. *IEEE Transactions on Vehicular Technology*. 2020;**69**(1):935–47.

[29] Park J., Samarakoon S., Bennis M., Debbah M. 'Wireless network intelligence at the edge'. *Proceedings of the IEEE*. 2019;**107**(11):2204–39.

[30] Zhang Z., Li L., Liang W., *et al.* 'Downlink interference management in dense drone small cells networks using mean-field game theory'. Proceedings of International Conference on Wireless Communications and Signal Processing; Hangzhou, China; Oct 2018.

[31] Kim H., Park J., Bennis M., Kim S.-L. 'Massive uav-to-ground communication and its stable movement control: A mean-field approach'. Proceedings of IEEE Workshop on Signal Processing Advances in Wireless Communications; Kalamata, Greece; Jun 2018.

[32] Shiri H., Park J., Bennis M. 'Massive autonomous uav path planning: A neural network based mean-field game theoretic approach'. Proceedings of IEEE Global Communications Conference (GLOBECOM); Waikoloa, HI, USA; Dec 2019.

[33] Shiri H., Park J., Bennis M. 'Communication-efficient massive UAV online path control: federated learning meets mean-field game theory'. *IEEE Transactions on Communications*. 2020;**68**(11):6840–57.

[34] Park J., Samarakoon S., Shiri H. Extreme URLLC: vision, challenges, and key enablers. 2020. ArXiv preprint, arXiv:2001.09683 [Accessed 12 Aug 2021].

[35] Courant R., Friedrichs K., Lewy H. 'On the partial difference equations of mathematical physics'. *IBM Journal of Research and Development*. 1967;**11**(2):215–34.

[36] Shiri H., Park J., Bennis M. 'Remote UAV online path planning via neural network-based opportunistic control'. *IEEE Wireless Communications Letters*. 2020;**9**(6):861–5.

[37] Haenggi M. *Stochastic geometry for wireless networks*. Cambridge University Press; 2012.

[38] Mériaux F., Lasaulce S., Tembine H. 'Stochastic differential games and energy-efficient power control'. *Dynamic Games and Applications*. 2013;**3**(1):3–23.

[39] Salbaroli E., Zanella A. 'Interference analysis in a poisson field of nodes of finite area'. *IEEE Transactions on Vehicular Technology*. 2008;**58**(4):1776–83.

[40] Venugopal K., Valenti M.C., Heath R.W. 'Interference in finite-sized highly dense millimeter wave networks'. Proceedings of Information Theory and Applications Workshop (ITA); Feb 2015.

[41] Bioglio V., Gabry F., Land I. 'Optimizing MDS codes for caching at the edge'. Proceedings of IEEE Global Communications Conference (GLOBECOM); San Diego, CA, USA; Dec 2015.

[42] Qiu T., Ge Z., Lee S., Wang J., Zhao Q., Xu J. 'Modeling channel popularity dynamics in a large IPTV system'. Proceedings of ACM joint conference on Measurement and modeling of computer systems (SIGMETRICS); New York, NY, USA; 2009. pp. 276–86.

[43] Griffiths T.L., Ghahramani Z. 'The Indian buffet process: an introduction and review'. *Journal of Machine Learning Research*. 2011;**12**:1185–224.

[44] Yu S.M., Kim S.-L. 'Downlink capacity and base station density in cellular networks'. Proceedings of IEEE WiOpt Workshop on Spatial Stochastic Models for Wireless Networks (SpaSWiN 2013); May 2013.

[45] Guéant O., Lasry J.-M., Lions P.-L. 'Mean field games and applications'. *Paris Princeton Lectures on Mathematical Finance 2010*. Berlin Heidelberg: Springer; 2011. pp. 205–66.

[46] Cao Y., Tao M., Xu F., Liu K. 'Fundamental Storage-Latency tradeoff in Cache-Aided MIMO interference networks'. *IEEE Transactions on Wireless Communications*. 2017;**16**(8):5061–76.

[47] Chun B.-G., Chaudhuri K., Wee H., Barreno M., Papadimitriou C.H., Kubiatowicz J. 'Selfish caching in distributed systems: A game-theoretic analysis'. Proceedings of the Twenty-Third Annual ACM Symposium on Principles of Distributed Computing; Jul 2004. pp. 21–30.

[48] Bassetti B., Zarei M., Cosentino Lagomarsino M., Bianconi G. 'Statistical mechanics of the "chinese restaurant process": Lack of self-averaging, anomalous finite-size effects, and condensation'. *Physical Review E*. 2009;**80**(6):066118.

Chapter 18

Caching with time-varying popularity profiles

B N Bharath[1], K G Nagananda[2], D Gündüz[3], and H V Poor[4]

This chapter considers the problem of content caching at the small-cell base stations (sBSs) in a heterogeneous wireless network. A cost function called the "offloading loss," which measures the fraction of the requested files that are not available in the sBS caches, is used as a metric for evaluating the performance of the proposed caching scheme. In contrast to the previous approaches that consider time-invariant and perfectly known popularity profiles, caching with non-stationary and statistically dependent popularity profiles (assumed unknown, and hence, estimated) is studied from a learning-theoretic perspective. A high probability bound on the offloading loss difference referred to as a probably approximately correct (PAC) result is derived. Here, the offloading loss difference refers to the error between the estimated and the optimal offloading loss. The difference is shown to be a function of the Rademacher complexity, the β-mixing coefficient, the number of time slots, and a measure of discrepancy between the estimated and true popularity profiles. Using insights from this bound, a practical cache update algorithm is proposed. Simulation results are presented to show its superiority over periodic updates. This chapter also presents caching performance analyses for Bernoulli and Poisson request models.

18.1 Introduction

The unprecedented growth in wireless data traffic requires a major change in the design of next generation wireless infrastructures [1]. The first major step taken toward handling this explosion in data demand is to deploy small-cell base stations (sBSs) to offload part of the wireless data traffic from a macro base station (BS) [2, 3]. Although, this alleviates the problem of uplink/downlink data bottleneck, high backhaul link-capacity requirement during peak hours still persists. There have

[1]Indian Institute of Technology, India
[2]Lehigh University, USA
[3]Imperial College London, UK
[4]Princeton University, USA

been several other proposals to improve the wireless architecture, including the fog radio access network concept with edge computing, integrating WiFi access points to share the load, to name a few. These new design approaches provide the flexibility of proactively caching popular contents at the edge of the wireless network. This not only reduces the delivery latency by bringing the contents closer to the end users but also the peak backhaul load [4].

Several aspects of caching have been addressed ranging from energy-efficient caching of files in a device-to-device (D2D) network [5–8] to proactive caching strategy [8] to coded caching [9–11]. For example, [8] shows that proactive caching increases the energy efficiency of D2D communications, while the advantages of caching on mobile social networks is reported in [12]. An information-centric architecture and caching-based content-centric networking for energy-efficient content distribution are proposed in [13] and [14], respectively. In [15–17], results on caching video files are presented and their benefits are highlighted. Joint benefits of coded storage at the sBSs and coded caching at the edge devices are studied in [18]. A low complexity solution to coded caching with beamforming is considered in [19]. Centralized coded caching for a radio access combination networks is studied in [20]. Caching at SBSs with multicast transmission to users is considered in [21–23]. In [24, 25], it is shown that in caching, a fraction of the files along with multicast transmission will further increase the efficiency of the system. Benefits of coded caching over noisy broadcast and interference channels are considered in [26–29]. Please see [30] for an overview of cache aided interference management techniques.

Most of the aforementioned caching mechanisms assume either *a priori* knowledge of the popularity profile of the cached contents or the demands are stationary; and hence, the popularity can be estimated efficiently using simple estimators. An unknown yet stationary popularity distribution was first considered in [31, 32]. The works of [33–37] concern learning-based approaches to estimate the popularity profile with theoretical guarantees, which highlight the implications of learning the popularity profile and user preferences on the performance. However, these solutions do not generalize well when the demands are changing in a non-stationary fashion. In practical world, however, in most cases the popularity profile of contents is a function of time (e.g., video on demand) [38–41]. Motivated by these applications and the growing significance of caching in improving the quality of service for end-users during peak traffic periods, we analyze the performance of a random caching strategy for a *non-stationary* popularity profile with statistical dependence across time.

In this chapter, we consider a heterogeneous network with users, BSs, and sBSs; all distributed according to independent PPPs. Each user is assumed to request files from a finite library of popular contents. The sBSs are assumed to be equipped with limited cache memories to store contents partially or in full, forming a distributed content network with communication constraints. A protocol model for communication is proposed, and a cost function, which captures the backhaul link overhead called the "offloading loss," is considered. For the sake of simplicity, the sBSs are assumed to employ random caching strategies. The offloading loss at time t, which depends on the popularity profile, is denoted by $\mathscr{I}(t)$. The goal is to obtain risk bounds on this offloading loss under unknown non-stationary demands (popularity profile), possibly correlated across time. Under a Poisson request model (see Assumption1), the BS observes the requests during the first t time slots, and computes an estimate of the popularity profile. It then chooses the caching probabilities $\pi \triangleq (\pi_1, \pi_2, \ldots, \pi_N)$, where N is the number of popular content items that can be cached, which minimizes the estimated offloading loss $\hat{\mathscr{I}}(t)$. The BS distributes this caching strategy to various sBSs in the coverage area to facilitate them to store content items in their caches. Owing to the varying popularity profile, it is necessary to refresh the caches frequently, say after every T time slots, albeit at an additional cost. Thus, it is important to investigate the minimum periodicity T of cache updates that guarantees the desired offloading loss.

We will derive probably approximately correct (PAC) type guarantees on the *offloading loss difference* $\Delta_{\mathscr{I}}(t, T)$, which is defined as the difference between the offloading loss incurred by using an outdated caching policy obtained by optimizing $\hat{\mathscr{I}}(t)$ at time $t + T$, and the optimal offloading loss at time $t + T$. We show that $\Delta_{\mathscr{I}}(t, T) < \varepsilon$ with a probability of at least $1 - \delta$ for any $\delta > \zeta$ and $\varepsilon > 0$, where ζ is a function of the β-mixing coefficient, the number of content items N, and the user density. The β-mixing coefficient is a measure of statistical dependency of the time-varying popularity profiles. If the popularity profile process is "sufficiently" mixing, i.e., if the process becomes almost independent after a sufficiently long time, and if the user density is very high, then the desired ε can be achieved for negligibly small $\delta > 0$. In particular, to achieve a fixed probability $\delta > \zeta$, we require the error ϵ to be a function of N, the rate of change of the popularity profile, and the Rademacher complexity, which is a measure of the difficulty in estimating the offloading loss.

The following are the main findings [42, 43]: (1) the error ϵ increases with N; (2) higher user density helps in achieving the desired smaller error ϵ with higher probability (i.e., ζ becomes smaller), thereby improving the caching performance—this is because higher user density results in more requests, allowing a better estimate of the popularity profile; (3) the higher the correlation of the popularity profile across time (defined in terms of the β-mixing coefficient), the longer the waiting time t to achieve a target error level ε with probability $1 - \delta$; (4) the error ε is a function of the rate of change of the popularity profile, and hence, the cache refresh period T. Thus, outdated cache contents lead to a larger error for a given δ, and a rapidly varying popularity profile requires more frequent updates to achieve the desired error performance; (5) a higher Rademacher complexity results in poorer error performance; and (6) in general, when the user requests are independent and identically distributed (i.i.d.), the error performance is better compared to non-stationary and statistically dependent requests. For stationary popularity profiles and large t, frequent cache updates are not necessary to achieve the desired performance.

Finally, motivated by our theoretical bounds, we present a novel caching algorithm that updates the cache contents only if the discrepancy that captures the rate at which the popularity profile is changing is large. We demonstrate the benefits of this cache update policy compared to periodic cache updates through simulations.

To the best of our knowledge, this is the first time that random caching is studied with non-stationary, statistically dependent, and unknown popularity profiles from a learning theory perspective.

The remainder of the chapter is organized as follows. In Section 18.2, we present the system model and introduce the notation. The problem statement is introduced in Section 18.3, while the main results are presented in Section 18.4. Performance analyses for Bernoulli and Poisson request models are analyzed in Section 18.5. Numerical results are presented in Section 18.6. Concluding remarks are provided in Section 18.7.

18.2 User request model

Let the stochastic process $X_v(\tau) \in \{1, 2, \ldots, N\}$ denote the index of the requested file by user $v \in \Phi_u$ at time $\tau \in \mathbb{R}$. For example, each user can maintain an independent local Poisson clock and makes a request whenever the local clock ticks. For any two users $v, w \in \Phi_u$, the request processes $X_v(\tau)$ and $X_w(\tau)$ are independent. For the ease of analysis, let us divide the time into slots of size $\Delta > 0$ each. Further, for each $v \in \Phi_u$, $\{X_v(\tau), \tau \in \mathbb{R}\}$ is a non-stationary and statistically dependent stochastic process across time slots, but the process $v \in \Phi_u$ within each time slot (i.e., $\tau \in [i\Delta, (i+1)\Delta)$, $i = 1, 2, \ldots$) is assumed to be stationary. Further, we assume that there is a "typical" BS at the origin with a coverage radius of $R > 0$. The BS estimates the popularity of the content items based on the requests it receives. Essentially, at a given time slot t, the BS collects requests (for t time slots) from all the users in the BS's coverage area to estimate the popularity profile of the requested files. Let $n_u \sim \text{Poiss}(\pi\lambda_u R^2)$ denote the number of users in its coverage area. The random arrival instants of the requests from different users are supposed to satisfy the following assumption.

Assumption 1: There exist constants $0 \leq \alpha_{\min} \leq \alpha_{\max} \leq 1$ such that for any random $n_u = n \geq 1$ users in the coverage area of the BS, the number of requests in $n_u = n \geq 1$ time slots, denoted by $r_a \in \mathbb{N}$, satisfies $\Pr\{\alpha_{\min} na \leq r_a \leq \alpha_{\max} na \mid n_u = n\} > \zeta_{a,n}$ for some $\zeta_{a,n} > 0$.

It turns out that the results based on this assumption can be used to derive performance guarantees when the arrival process is a homogenous Poisson point process (see Section 18.4). Further, we assume that the request instants and the number of requests within a time slot are independent of the files requested. The set of request instants at which the requests from all users in the coverage area of the BS arrive within the ith time slot is denoted by $\mathscr{R}_i$. Let $X(\tau) \triangleq \bigcup_{v \in \Phi_u \cap \|v\|_2 \leq R} \{X_v(\tau)\}$ denote the set of requests from all the users in the coverage area of the BS at time $\tau \in \mathbb{R}$. Note that if two or more users request for the same file at time $\tau \in \mathbb{R}$, then it is counted as the same index due to the union in the definition of $X(\tau)$. However, this event does not occur almost surely. The set of requests from all users in time slots t_1 to t_2 is denoted by $X_{t_1,t_2} \triangleq \{X(\tau) : \tau \in \mathscr{R}_{t_1,t_2}\}$, where $\mathscr{R}_{t_1,t_2} \triangleq \bigcup_{i=t_1}^{t_2} \mathscr{R}_i$ (see Figure 18.1). After receiving requests $X_{1,t}$ within first t time slots, the BS computes the empirical estimate of the popularity profile, i.e., the probability of the ith file being requested is estimated as follows:

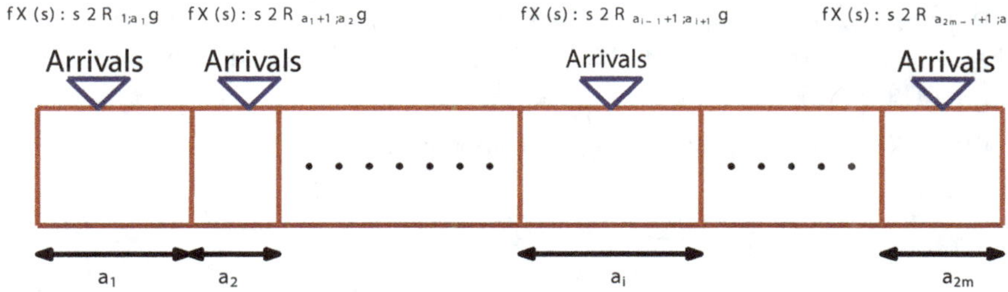

Figure 18.1 *A time period consisting of t time slots, each of duration* Δ*, is divided into 2m blocks, where the* i^{th} *block consists of* a_i *slots, and* $t = \sum_{i=1}^{2m} a_i$.

$$\hat{p}_{i,t} = \frac{1}{r_t} \sum_{s \in \mathscr{R}_{1,t}} \mathbb{1}\{X(s) = i\}, \; i = 1, \ldots, N, \tag{18.1}$$

where $r_t \triangleq |\mathscr{R}_{1,t}|$ is the total number of requests in the first t slots, and the indicator function $\mathbb{1}\{X(s) = i\}$ is one when the event $\{X(s) = i\}$ occurs, and zero otherwise. The accuracy of the estimate $\hat{\mathscr{P}}^{(t)} \triangleq \{\hat{p}_{i,t} : i = 1, 2, \ldots, N\}$ depends on: (i) the number of available samples, which in turn is related to the number of users in the coverage area of the BS, (ii) the number of requests per user, and (iii) the behavior of the process $X(s)$. The estimate in (18.1) is valid only when there is a positive number of user requests, which is guaranteed by the given Assumption 1. In the next section, we present the performance measure for the given model and state the main problem addressed in this work.

18.3 Problem statement

We consider a typical user located at the origin denoted by $o \in \Phi_u$. At time slot $t \in \mathbb{N}$, the "offloading loss" is defined as

$$\mathscr{I}(\Pi^{(t)}, \mathscr{P}^{(t)}, X_{1,t-1}) \triangleq \frac{B}{R_0} \Pr\{f_o \notin \mathcal{N}_u \mid X_{1,t-1}\}, \tag{18.2}$$

where $\Pi^{(t)}$ denotes the caching policy, $\mathscr{P}^{(t)} \triangleq \{p_1(t), p_2(t), \ldots, p_N(t)\}$ is the popularity profile in slot t, R_0 and $\frac{B}{R_0}$ denote the rate supported by the BS and the time overhead incurred in transmitting the file from the BS to the user, respectively, and f_o denotes the file requested by the typical user in the tth slot. In (18.2), with a slight abuse of notation, $f_0 \notin \mathcal{N}_u$ denotes the event that the requested file f_0 is not present in the caches of the neighboring sBSs. The offloading loss is the scaled probability of the content requested by user o not being cached by any of the sBSs within its communication range conditioned on the requests received by the BS until the beginning of time slot t, i.e., $X_{1,t-1}$. We employ the following random caching strategy, which enables us to derive a closed form expression for the offloading loss at time t.

Random caching strategy: At time t (determined by the BS), each sBS $s \in \Phi_s$ caches content items in an i.i.d. fashion by generating M indices distributed according to $\Pi^{(t)} \triangleq \left\{\pi_i(t) : \sum_{i=1}^{N} \pi_i(t) = 1,\right\}$ (see [44]).

We seek to solve the following optimization problem:

$$\min_{\Pi^{(\tau)} \in \mathscr{P}_\pi : \tau \in \mathbb{N}} \limsup_{t \to \infty} \frac{1}{t} \sum_{\tau=1}^{t} \mathscr{I}(\Pi^{(\tau)}, \mathscr{P}^{(\tau)}, X_{1,\tau-1}), \tag{18.3}$$

where $\mathscr{P}_\pi$ denotes the N-dimensional probability simplex. An expression for $\mathscr{I}(\Pi^{(t)}, \mathscr{P}^{(t)}, X_{1,t-1})$ is given in the following theorem, whose proof can be obtained by replacing p_i by $p_{X,i}(t)$ in the proof of Theorem 1 found in [35, Appendix A].

Theorem 1.

The average offloading loss at time t for the random caching strategy $\Pi^{(t)}$ is given by

$$\mathscr{I}(\Pi^{(t)}, \mathscr{P}^{(t)}, X_{1,t-1}) = \sum_{i=1}^{N} g(\pi_i(t)) p_{X,i}(t), \tag{18.4}$$

where

$p_{X,i}(t) \triangleq \Pr\{f_i \text{ requested by } o \text{ in slot } t | X_{1,t-1}\}$,

and

$g(\pi_i(t)) \triangleq \frac{B}{R_0} \exp\{-\lambda_u \pi \gamma^2 [1 - (1 - \pi_i(t))^M]\}$.

Even assuming that the conditional probabilities $p_{X,i}(t)$ are perfectly known, the complexity involved in solving (18.3) can be high owing to the fact that the caching policy at time t depends on $X_{1,t}$, which grows with t. In practice, the conditional probability $\Pr\{f_i \text{ requested } | X_{1,t-1}\}$ is unknown and has to be estimated. Also, the BS may not have enough samples to compute a reasonably good estimate of this conditional probability. Hence, it is reasonable to consider unconditional probability in the definition of the offloading loss. Thus, one can minimize the offloading loss $\mathscr{I}(\Pi^{(t)}, \mathscr{P}^{(t)}) \triangleq \left[\sum_{i=1}^{N} g(\pi_i(t)) p_i(t)\right]$, where $p_i(t)$ is the probability of the ith file being requested at time t. However, the values of $p_i(t)$ are unknown; and hence, an estimate of the popularity profile needs to be used in place of $\mathscr{P}^{(t)}$. More precisely, at time t, let $\hat{\Pi}_t^*$ denote the caching policy obtained using an estimate $\hat{\mathscr{P}}^{(t)}$; that is,

$$\hat{\Pi}_t^* = \arg \min_{\Pi^{(t)} \in \mathscr{R}\pi} \mathscr{T}(\Pi^{(t)}, \hat{\mathscr{P}}^{(t)}). \tag{18.5}$$

Suppose that the cache contents chosen by the optimal caching policy at time t are used to satisfy user demands over the period $(t, t+T]$. Let us consider the offloading loss in using $\hat{\Pi}_t^*$ at a later time, say at time $t+T$. The offloading loss at time $t+T$ is given by $\hat{\mathscr{I}}^*(t+T) \triangleq \mathscr{I}(\hat{\Pi}_t^*, \mathscr{P}^{(t+T)})$. Further, let Π_{t+T}^* denote the optimal caching policy at time $t+T$ using perfect knowledge of the popularity profile $\mathscr{P}^{(t+T)}$; that is,

$$\Pi_{t+T}^* = \arg \min_{\Pi^{(t+T)} \in \mathscr{P}\pi} \mathscr{I}(\Pi^{(t+T)}, \mathscr{P}^{(t+T)}), \tag{18.6}$$

with the corresponding offloading loss $\mathscr{I}^*(t+T) \triangleq \mathscr{I}(\Pi^*_{t+T}, \mathscr{P}^{(t+T)})$. Similar to [35], the central theme of this work is the analysis of the *offloading loss gap*, $\Delta_{\mathscr{I}}(t, T) \triangleq \hat{\mathscr{I}}^*(t+T) - \mathscr{I}^*(t+T)$. For example, if $\Delta_{\mathscr{I}}(t, T)$ is small, then each term in (18.3) is small, which results in a small average offloading loss. This approach is central to the analyses of prediction problems involving non-stationary stochastic processes [45].

The number of requests in any given slot and the requested file index are independent. For example, if the arrivals are Poisson, then the number of requests in any two disjoint intervals is independent. However, the files requested across time are correlated. This assumption is reasonable when the popularity depends on, for example, the files that are trending due to their popularity elsewhere, while a user's decision to browse is independent of the popularity. The unconditional probability does not lead to the independence of the files requested in any slot t from the files requested in future slots. Moreover, an estimate of the popularity profile at time slot t depends on the past requests. However, for future work we aim to investigate generalization bounds retaining the conditioning on the past requests, which makes the offloading loss $\mathscr{I}(\Pi^{(t)}, \mathscr{P}^{(t)}, X_{1,t-1}) \triangleq \frac{B}{R_0} \Pr\{f_o \notin \mathscr{N}_u \mid X_{1,t-1}\}$ at any given slot t random.

18.4 Main results

We study risk bounds on the offloading loss difference, $\Delta_{\mathscr{I}}(t, T)$, when the popularity profile is non-stationary. Essentially, for any $\varepsilon > 0$, we seek to identify a risk bound $\delta > 0$, such that

$$\Pr\left\{\hat{\mathscr{I}}^*(t+T) - \mathscr{I}^*(t+T) > \varepsilon\right\} < \delta. \tag{18.7}$$

First, we relate (18.7) to an expression in terms of the estimation error in the following theorem.

Theorem 2.

For the estimate of the popularity profile in (18.1), the following bound holds:

$$\Pr\left\{\hat{\mathscr{I}}^*(t+T) - \mathscr{I}^*(t+T) > \varepsilon\right\} \le 2\Pr\{\mathscr{A}_T(X_{1,t}) > \varepsilon\}, \tag{18.8}$$

where $\mathscr{A}_T(X_{1,t}) \triangleq \sup_{\Pi \in \mathscr{P}_\pi} \left|\sum_{i=1}^N g(\pi_i)(\hat{p}_{i,t} - p_{i,t+T})\right|$, and $g(\pi_i)$ is defined in Theorem 1.

Proof. See [43, Appendix A].

The term $\Pr\{A_T(X_{1,t}) > \varepsilon\}$ can be bounded as follows:

$$\begin{aligned}
\Pr\{\mathscr{A}_T(X_{1,t}) > \varepsilon\} &= \sum_{j=0}^{\infty} \Pr\{\mathscr{A}_T(X_{1,t}) > \varepsilon \mid n_u = j\} \Pr\{n_u = j\} \\
&\le \Pr\{n_u = 0\} \\
&\quad + \textstyle\sum_{j=1}^{\infty} \Pr\{\mathscr{A}_T(X_{1,t}) > \varepsilon \mid n_u = j\} \Pr\{n_u = j\} \\
&= \exp\{-\lambda_u \pi R^2\} \\
&\quad + \textstyle\sum_{j=1}^{\infty} \Pr\{\mathscr{A}_T(X_{1,t}) > \varepsilon \mid n_u = j\} \Pr\{n_u = j\}
\end{aligned} \tag{18.9}$$

Next, we derive an upper bound on $\Pr\{\mathscr{A}_T(X_{1,t}) > \varepsilon | n_u = j\}$. The term $\mathscr{A}_T(X_{1,t})$ depends on $\hat{p}_{i,t}$, which involves the sum of non-stationary random variables that are possibly correlated across time. In order to apply the standard large deviation bounds, we must convert the sum of non-stationary dependent random variables to a sum of blocks of independent random vectors through a coupling argument, which is explained next.

For a given stochastic process $X_{1,\infty}$, and $s \in \mathbb{N}$, let $\mathbb{P}_{\tau,\tau+s}(\star)$ and $\mathbb{P}_{1\to\tau}(\star) \otimes \mathbb{P}_{\tau+s\to\infty}(\star)$ denote the joint and product distributions of the stochastic processes $X_{1,\tau}$ and $X_{\tau+s,\infty}$, respectively. If $X_{1,\tau}$ and $1 \le \tau \le \infty$ are independent, then $\|\mathbb{P}_{\tau,\tau+s}(\star) - \mathbb{P}_{1\to\tau}(\star) \otimes \mathbb{P}_{\tau+s\to\infty}(\star)\|_{\mathrm{TV}} = 0$, where $\| \star \|_{\mathrm{TV}}$ denotes the total variational norm. Thus, for a given s, this difference, maximized over all $1 \le \tau \le \infty$, is a natural measure of the dependency between $X_{1,\tau}$ and $X_{\tau+s,\infty}$. This is commonly referred to as the β-mixing coefficient, and for $s \in \mathbb{N}$, it is given by

$$\beta(s) \triangleq \sup_{1\le\tau\le\infty} \|\mathbb{P}_{\tau,\tau+s}(\star) - \mathbb{P}_{1\to\tau}(\star) \otimes \mathbb{P}_{\tau+s\to\infty}(\star)\|_{TV}. \tag{18.10}$$

A stochastic process is said to be β-mixing if $\beta(s) \to 0$ as $s \to \infty$. For a given stochastic process that is β-mixing, two well-separated sequences of the process are approximately independent, where the approximation error is given by $\beta(s)$. Thus, we assume that the request process $X(t)$ is a β-mixing stochastic process, i.e., $\beta(s) \to 0$ as $s \to \infty$.

We now provide the details of the coupling argument, through which the dependent stochastic process is replaced by independent blocks of random variables. This will facilitate the use of a concentration inequality; in particular, McDiarmid's inequality. Fix $m \in \mathbb{N}$, and consider $2m$ consecutive blocks, where the block i, $i \in \{1, 2, \ldots, 2m\}$ consists of a_i time slots and $t \triangleq \sum_{j=1}^{2m} a_j$ is the total number of time slots (see Figure 18.1). Let $a_0 \triangleq 0$. Consider the time instants at which the requests arrive corresponding to odd and even blocks defined as $\mathbb{T}_o^{(t)} \triangleq \bigcup_{j:j=0,2,4,\ldots,2(m-1)} \mathscr{R}_{a_j+1,a_{j+1}}$ and $\mathbb{T}_e^{(t)} \triangleq \bigcup_{j:j=1,3,5,\ldots,2m-1} \mathscr{R}_{a_j+1,a_{j+1}}$, respectively. Thus, the requests corresponding to the odd and even blocks are given by $X_{1,t}^e \triangleq \{X(s) : s \in \mathbb{T}_e^{(t)}\}$ and $X_{1,t}^o \triangleq \{X(s) : s \in \mathbb{T}_o^{(t)}\}$, respectively. In order to use a coupling argument, define new stochastic process $\tilde{X}(\tau)$, $\tau \in \mathbb{R}$, such that for a fixed $\mathscr{R}_{a_{i-1}+1,a_i}$, $\{\tilde{X}(\tau) : \tau \in \mathscr{R}_{a_{i-1}+1,a_i}\}$ and $\{X(\tau) : \tau \in \mathscr{R}_{a_{i-1}+1,a_i}\}$ have the same distribution, $i = 1, 2, \ldots, 2m$. Now, consider $\tilde{X}_{1,t}^h \triangleq \{\tilde{X}(s) : s \in \mathbb{T}_h^{(t)}\}$, $h \in \{e, o\}$, such that the requests in the even (and odd) blocks of $\tilde{X}_{1,t}$ are independent. However, within each block, the random variables can be correlated arbitrarily. We can always construct such a stochastic process, and the pair $(X(s), \tilde{X}(s))$ is called a *coupling* (see Figure 18.1). We define $\tilde{X}_{1,t}^e$ and $\tilde{X}_{1,t}^o$ similarly to $X_{1,t}^e$ and $X_{1,t}^o$, respectively.

The following theorem provides a bound on the performance guarantees in terms of the β-mixing coefficient.

Theorem 3.

For the given model, and the popularity estimate in (18.1), with a probability of at least $\delta' > 0$, the following holds

$$\hat{\mathscr{I}}^*(t+T) - \mathscr{I}^*(t+T) < \min\{\mathbb{E}[\mathscr{A}_T(\tilde{X}^e_{1,t})], \mathbb{E}[\mathscr{A}_T(\tilde{X}^o_{1,t})]\} + \frac{N\alpha_{\max}Ba_{\max}}{\alpha_{\min}R_0a_{\min}}\sqrt{\frac{\log\left(\frac{2}{\delta'}\right)}{2m}}, \tag{18.11}$$

where

$$\delta' \triangleq \delta/2 - \exp\{-\lambda_u\pi R^2\} - \sum_{i=2}^{2m-1}\beta(a_i) - e^{-\lambda_u\pi R^2}\sum_{j=1}^{\infty}\sum_{i=1}^{2m}(1-\zeta_{a_i,j})\frac{(\lambda_u\pi R^2)^j}{j!} > 0. \tag{18.12}$$

Further,

$$\mathscr{A}_T(\hat{X}^{(h)}_{1,t}) \triangleq \sup_{\Pi\in\mathscr{P}_\pi}\left|\sum_{i=1}^{N} g(\pi_i)\left(\hat{P}^h_{i,t} - P_{i,t} + T\right)\right|, \tag{18.13}$$

where

$$\hat{p}^h_{i,t} \triangleq \frac{1}{\left|\mathbb{T}^{(t)}_h\right|}\sum_{s\in\mathbb{T}^{(t)}_h}\mathbb{1}\{\tilde{X}(s) = i\}, h \in \{e, o\}. \tag{18.14}$$

Proof. See [43, Appendix B].

Note that $\delta' > 0$ implies a bound on δ. Next, we bound $\min\{\mathbb{E}[\mathscr{A}_T(\tilde{X}^e_{1,t})], \mathbb{E}[\mathscr{A}_T(\tilde{X}^o_{1,t})]\}$ to get the desired result. This bound depends on the Rademacher complexity and the non-stationarity of the stochastic process. We begin with the following definition.

Definition 1: (Rademacher complexity): The Rademacher complexity of $\mathscr{P}_\pi$ is defined as [46, chapter 3]

$$\mathscr{R}^{(t)}_h \triangleq \mathbb{E}_{\tilde{X},\sigma}\frac{1}{\left|\mathbb{T}^{(t)}_h\right|}\sup_{\Pi\in\mathscr{P}_\pi}\sum_{i=1}^{N} g(\pi_i)\left|\sum_{s\in\mathbb{T}^{(t)}_h}\sigma_{i,s}\mathbb{1}\{\tilde{X}(s) = i\}\right|, \tag{18.15}$$

where the Rademacher random variables $\sigma_{i,s} \in \{-1, 1\}$, $i = 1, 2, \ldots, N$ for $s \in \mathbb{T}^{(t)}_h$ are i.i.d. with probability 1/2, $\boldsymbol{\sigma} \triangleq \{\sigma_{i,s} \in \{-1, 1\} : i = 1, 2, \ldots, N, s \in \mathbb{T}^{(t)}_h\}$, and $h \in \{e, o\}$.

Next, we provide one of the main results of this work.

Theorem 4.

For the given model, and the popularity estimate in (18.1), with a probability of at least $1 - \delta$, the following holds:

$$\hat{\mathscr{I}}^*(t+T) < \mathscr{I}^*(t+T) + 2\max\{\mathscr{R}^{(t)}_e, \mathscr{R}^{(t)}_o\} + \max\{\Delta^{(e)}_{t,T}, \Delta^{(o)}_{t,T}\} + \frac{N\alpha_{\max}Ba_{\max}}{R_0a_{\min}\alpha_{\min}}\sqrt{\frac{a_{\max}\log\left(\frac{2}{\delta'}\right)}{t}}, \tag{18.16}$$

where $\mathscr{R}_h^{(t)}$ is the Rademacher complexity, $\Delta_{t,T}^{(h)} \triangleq \sup_{\Pi \in \mathscr{P}_\pi} \sum_{i=1}^{N} g(\pi_i) d_i^{(h)}(t,T)$, $a_{\max} \triangleq \max_{1 \le i \le 2m} a_i$, $d_i^{(h)}(t,T) \triangleq \frac{1}{\left|\mathbb{T}_h^{(t)}\right|} \sum_{s \in \mathbb{T}_h^{(t)}} \left|p_{i,s} - p_{i,t+T}\right|$, $h \in \{e,o\}$, and $\delta' > 0$ is as defined in Theorem 3 with $m = \lceil \frac{t}{a_{\max}} \rceil$.

Proof. See [43, Appendix C].

Remarks

1. The error ε increases linearly with N. To compensate for larger values of λ_u, the waiting time t should be of the order of N^2; a similar observation was also made in [35]. As λ_u increases, a lower value of δ can be achieved. In general, as $\lambda_u \to \infty$, $\delta = 0$ cannot be achieved due to the dependence of the stochastic process across time, i.e., $\beta(a) > 0$, $a > 0$.
2. The error ε decreases as ϵ increases. When the requests are i.i.d., $a_{\max} = 1$, and hence, ϵ is small. Thus, when the requests are correlated we incur a penalty of a_{max}, since the error decreases as $a_{\max} = 1$ compared to $\sqrt{1/t}$ for i.i.d. requests. The error can be reduced by choosing $a_{\max} = 1$, i.e., $i = 1, \ldots, 2m$, $i = 1, \ldots, 2m$. Since $\beta(x)$ is a monotonically decreasing function of x, the probability of achieving a lower error is very small, indicating a trade-off between the error and the probability with which the bound in (18.9) holds. Also, the lower values of δ' result in higher error. This requires the value of m to be small. However, m scales as $t/a_{\max}$, which indicates that if $a_{\max} = \mathcal{O}(\sqrt{t})$, then the last term in the error goes down as $1/t^{1/4}$ instead of $1/\sqrt{t}$. On the other hand, for larger values of m, the value of δ' is small provided the β-mixing coefficient reduces at a smaller rate compared to $1/\sqrt{t}$; this indicates that one should have sufficiently fast decaying β-mixing for better performance. The last term in the expression for δ' depends on $\zeta_{a_{i,j}}$, whose effect is studied by looking at specific examples, such as the Bernoulli and Poisson models for user requests, as detailed in the next section.
3. The error ϵ increases with $\frac{\alpha_{\max}}{\alpha_{\min}}$. The higher this ratio, the larger the variation in the number of requests. On the other hand, the lower this ratio, the smaller the error which indicates a greater number of requests. The non-stationarity of the process is captured through $\Delta_{t,T}^{(h)}$, $h \in \{e,o\}$. For a stationary process $\Delta_{t,T}^{(h)} = 0$, $h \in \{e,o\}$.
4. When the user requests are i.i.d., the error does not vanish as $t \to \infty$, because the Rademacher complexity will not tend to zero as $t \to \infty$. This indicates the difficulty in estimating the offloading loss, or equivalently the popularity profile, for a given caching policy.
5. The only term that depends on T is $\max\{\Delta_{t,T}^{(e)}, \Delta_{t,T}^{(o)}\}$. The frequency with which the cache update should be done depends on $\Delta_{t,T}^{(h)}$, $h \in \{e,o\}$. For instance, if $\Delta_{t,T}^{(h)}$, $h \in \{e,o\}$ is high, then the updates should be more frequent.
6. The error is directly proportional to the number of bits per file and inversely proportional to the rate at which the file is transmitted from the SBS to the users.

18.5 Bernoulli and Poisson requests

In this section, we consider Bernoulli and Poisson request models and analyze the implications on the results derived so far.

18.5.1 Bernoulli request model

Let $X_u^k \in \{0, 1\}$, $u \in \Phi_u$ denote the request made by user $X_u^k \in \{0, 1\}$ for a cached file, in the kth slot. In the Bernoulli model, it is assumed that $X_u^k \in \{0, 1\}$ is i.i.d. across users and slots. Further, a user makes a request with probability p in each time slot, independent of the file it requests, i.e., $\Pr\{X_u^k = 1\} = p$. The slot width $\Delta > 0$ is chosen such that at most one file is requested. Conditioned on the event that a set of requests are made from several users, the files requested follow a non-stationary dependent random process. This simplified assumption makes the analysis of the offloading loss guarantees tractable. To provide theoretical guarantees for this model, from the general result in Theorem 4, it suffices to prove an upper bound on the probability of the event $\{r_{a_i} < \alpha_{\min} n a_i\} \bigcup \{r_{a_i} > \alpha_{\max} n a_i\}$ in the ith block of size a_i, conditioned on the presence of n users, i.e.,

$$\begin{aligned}\Pr\left\{r_{a_i} < \alpha_{\min} n a_i \bigcup r_{a_i} > \alpha_{\max} n a_i \mid n_u = n\right\} \le \\ \Pr\left\{r_{a_i} < \alpha_{\min} n a_i \mid n_u = n\right\} + \Pr\left\{r_{a_i} > \alpha_{\max} n a_i \mid n_u = n\right\},\end{aligned} \tag{18.17}$$

where r_{a_i} is the total number of requests in a_i slots, which is the sum of $a_i n$ independent Bernoulli random variables, leading to $\mathbb{E}[r_{a_i} \mid n_u = n] = a_i np$. Towards this end, we use the following result:

Theorem 5. Let $X_1, X_2, \ldots, X_k$ be independent Bernoulli random variable with $\Pr\{X_i = 1\} = p$ and $\Pr\{X_i = 0\} = 1 - p$. Then, for $X \triangleq \sum_{i=1}^n X_i$, and $\lambda > 0$, we have

$$\Pr\{X \le \mathbb{E}[X] - \lambda\} \le \exp\{-\lambda^2/2np\}, \tag{18.18}$$

and

$$\Pr\{X \ge \mathbb{E}[X] + \lambda\} \le \exp\left\{-\frac{\lambda^2}{2(np + \lambda/3)}\right\}. \tag{18.19}$$

Using Theorem 5 conditioned on the event $\{n_u = n\}$, we have the following theorem.

Theorem 6. For the Bernoulli model with $0 < p < \alpha_{\min} < \alpha_{\max}$, we have

$$\Pr\left\{r_{a_i} < \alpha_{\min} n a_i \bigcup r_{a_i} > \alpha_{\max} n a_i \mid n_u = n\right\} \le 2\exp\left\{-\frac{\psi_p a_{\min} n}{2p}\right\}, \tag{18.20}$$

for $i = 1, 2, \ldots, 2m$, and $n \ge 1$. Here, $\psi_p \triangleq \min\left\{\frac{a_{\min}(p - \alpha_{\max})^2}{1 + \frac{a_{\max}(\alpha_{\min} - p)}{3}}, (p - \alpha_{\min})^2\right\}$.

By using (18.17), we have

$$\Pr\left\{\alpha_{\min} n a_i < r_{a_i} < \alpha_{\max} n a_i \mid n_u = n\right\} \ge 1 - 2\exp\left\{-\frac{\psi_p a_{\min} n}{2p}\right\} \triangleq \zeta_{a,n}. \tag{18.21}$$

Using this in the expression for δ' in Theorem 4, and after some algebraic manipulation, we obtain the following result.

Theorem 7. For the Bernoulli request model with $0 < \alpha_{\min} < \alpha_{\max}$, and the popularity estimate in (18.1), the following holds with a probability of at least $1-\delta$

$$\hat{\mathscr{T}}^*(t+T) \leq \mathscr{T}^*(t+T) + 2\max\{\mathscr{R}_e^{(t)}, \mathscr{R}_o^{(t)}\} + \max\{\Delta_{t,T}^{(e)}, \Delta_{t,T}^{(o)}\} + \frac{NBa_{\max}\alpha_{\max}}{a_{\min}R_0\alpha_{\min}}\sqrt{\frac{a_{\max}\log\left(\frac{2}{\delta'}\right)}{t}}, \tag{18.22}$$

where $\Delta_{t,T}^{(h)} \triangleq \sup_{\Pi\in\mathcal{P}} \sum_{i=1}^{N} g(\pi_i) d_i^{(h)}(t, T)$. Further,

$$\delta' = \frac{\delta}{2} - \left(\exp\{-\lambda_u \pi R^2\} + \sum_{i=2}^{2m-1} \beta(a_i) + 4m\left[e^{-\lambda_u\pi R^2}(e^{-\lambda_u\pi R^2 e^{-\phi_P}} - 1)\right]\right) > 0, \tag{18.23}$$

where $\phi_p \triangleq \frac{a_{\min}\psi_P}{2p}$, and ψ_p is as defined in Theorem 6.

From this theorem, the following observations can be made. First, assuming that $a_{\min}$ and $a_{\max}$ grow as $\mathcal{O}(\sqrt{t})$ (which implies that $m = \mathcal{O}(\sqrt{t})$), the last term in the error in (7) goes to zero as $1/t^{1/4}$, while the other terms are not affected by this choice. For $m = \mathcal{O}(\sqrt{t})$, the second term in the expression for δ' tends to zero as $t \to \infty$, provided that $\beta(\sqrt{t}) \to 0$ as $t \to \infty$. This demands a faster decay rate of β-mixing. The last term in the expression for δ' tends to $-\infty$ as $t \to \infty$, resulting in a larger value of δ', and hence, reducing the error. As a result of this, asymptotically in t, any value of $\delta > 0$ is a valid choice. Thus, by choosing δ sufficiently close to 0, a high probability result on the performance can be obtained.

18.5.2 Poisson request model

We assume that the requests follow a Poisson model as defined in this section.

Assumption 2 :The number of requests across users in any interval follows an independent homogenous Poisson process with arrival rate λ_r. Conditioned on the number of requests, the requested files follow a non-stationary, possibly dependent stochastic process.

As in the previous section, we first provide a bound on $\zeta_{a_i,n}$ for each i.

Theorem 8. For the Poisson request model, with $\alpha_{\min} = \frac{\Delta\lambda_r}{e^2}$ and $\alpha_{\max} = \Delta\lambda_r e$, the following bound holds

$$Pr\left\{r_{a_i} < \alpha_{min} n\alpha_i \cup r_{a_i} > \alpha_{max} n\alpha_i \mid n_u = n\right\} \leq 2\, exp\{-n\alpha_{min}\lambda_r\Delta\} \tag{18.24}$$

Theorem 9. For the Poisson request model with the popularity estimate in (18.1), with a probability of at least $1 - \delta$, the following holds

$$\hat{\mathscr{I}}^*(t+T) \leq \mathscr{I}^*(t+T) + 2\max\{\mathscr{R}_e^{(t)}, \mathscr{R}_o^{(t)}\} + \max\{\Delta_{t,T}^{(e)}, \Delta_{t,T}^{(o)}\} + \frac{NBa_{\max}e}{a_{\min}R_0}\sqrt{\frac{a_{\max}\log\left(\frac{2}{\delta'}\right)}{t}}, \tag{18.25}$$

where $\Delta_{t,T}^{(h)} \triangleq \mathbb{E}\left[\sup_{\Pi\in\mathscr{P}} \sum_{i=1}^{N} g(\pi_i) d_i^{(h)}(t, T)\right]$. Further,

$$\delta' = \frac{\delta}{2} - \left(\exp\{-\lambda_u \pi R^2\} + \sum_{i=2}^{2m-1} \beta(a_i) + 4m \left[e^{-\lambda_u \pi R^2} (e^{-\lambda_u \pi R^2 e^{-a_{min} \lambda r \Delta}} - 1) \right] \right) > 0,$$

As in the Bernoulli case, a better performance can be achieved by choosing $m = \mathcal{O}(\sqrt{t})$ and $a_i = \mathcal{O}(\sqrt{t})$ for all i. It can also be seen that as λ_r (and Δ) increases, a smaller value of δ is possible leading to a better performance. However, unlike the Bernoulli model, the bound is independent of $\alpha_{\min}$ and $\alpha_{\max}$. The results presented for the models considered here lead to a simple yet effective algorithm for updating the cache when the popularity profile is varying across time. Next, we provide the details of this algorithm along with numerical simulations.

18.6 Cache update algorithm and numerical results

In this section, we present a cache update algorithm by following Theorem 4, and the corresponding simulation results. Theorem 4 suggests that the sBSs should update their caches at the time instants at which the error becomes large. The only relevant term is

$$\max\{\Delta_{t,T}^{(e)}, \Delta_{t,T}^{(o)}\} \leq \Delta_{t,T} \triangleq \frac{1}{\left|\mathbb{T}_e^{(t)} \cup \mathbb{T}_o^{(t)}\right|} \sup_{\Pi \in \mathscr{P}_\pi} \sum_{i=1}^{N} \sum_{s \in \mathbb{T}_o^{(t)} \cup \mathbb{T}_e^{(t)}} g(\pi_i) \left| p_{i,s} - p_{i,t+T} \right|. \tag{18.26}$$

The following cache update mechanism is employed:

1. Initialize $t = 0$ and $T = 0$. Update the caches randomly.
2. If $\hat{\Delta}_{t,T} >$ threshold, then update the caches using the caching probability obtained by solving $\hat{\Pi}_{t+T}^{*} = \arg\min_{\Pi^{(t+T)} \in \mathscr{P}_\pi} \mathscr{T}(\Pi^{(t+T)}, \hat{\mathscr{P}}^{(t+T-1)})$, where $\hat{\mathscr{P}}^{(t+T-1)}$ is the estimate obtained using (18.1), and set $T = t$. Here, $\hat{\Delta}_{t,T}$ denotes an estimate of $\lambda_B = 0.00001$, and threshold > 0 determines the error achieved.
3. Set $t \leftarrow t + 1$ and go to step 2.

We define the fetching cost as the average number of files downloaded at each cache update. The simulation setup consists of sBSs and users distributed according to PPPs with densities $\lambda_B = 0.00001$ and $\lambda_u = 0.0001$, respectively. The number of files is $N = 100$, and the coverage of the BS and sBSs are 1 000 m and 500 m, respectively. We let $\gamma = 500$. The deterministic arrival rate corresponds to a deterministic variation in the distribution of the popularity profile once every 150 slots; while the random change corresponds to a random change in the popularity profile which occurs once in every 100 slots on an average. In the deterministic variation scenario, a random set of 3 pairs of files are chosen and are permuted in a uniformly random fashion.

In the random variation scenario, two pairs of indices are randomly and uniformly permuted at random times. The requests follow a Poisson arrival model with rates $\lambda_r = 0.09$ and 0.01 for the scenarios corresponding to random and deterministic changes, respectively. Requests for the files are generated using a Zipf distribution with parameter $\theta = 0.8$. Thus, the arrival is non-stationary but independent across

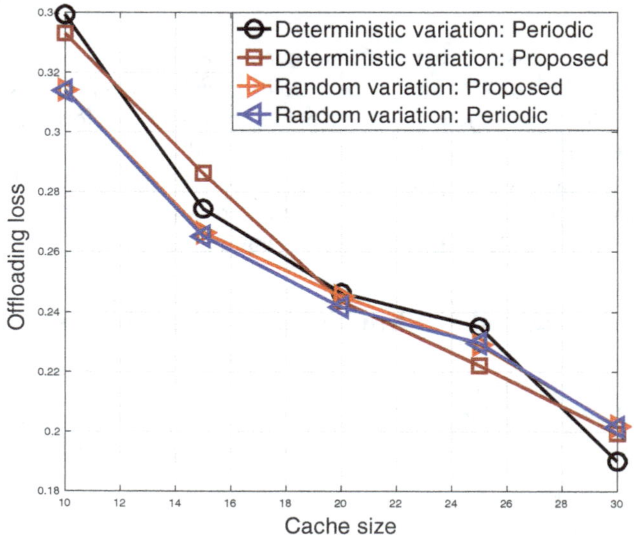

Figure 18.2 Offloading loss as a function of the cache size.

time. This non-stationarity results in oscillations in the curves. The requests from a typical user at the origin are used to evaluate the offloading loss. Figure 18.2 shows the offloading loss with $B = R_0$ as a function of the cache size for the two scenarios as mentioned earlier. The periodic updates are carried out every 5 time slots. It is clear from the figure that, for the random variation scenario, the performance of the proposed scheme and the periodic scheme are almost the same. However, we observe in Figure 18.3 that the fetching cost of the proposed scheme is lower, as the periodic update scheme requires far too many updates. This confirms that by appropriately choosing the threshold values, the proposed scheme outperforms the periodic cache update scheme for specific scenarios. The variation in the fetching cost for the proposed (deterministic) scheme is an artifact of choosing the threshold. For the deterministic variation case, it can be seen in Figure 18.3 that for certain cache sizes (10, 20, and 25), the offloading loss of the proposed scheme outperforms periodic caching, while it performs poorly for other cache sizes. However, the fetching cost is lower than that of the periodic update scheme for all the cache sizes. This shows that in order to achieve a smaller offloading loss, it is better to update more frequently; while in other scenarios (cache size = 15), it is possible to achieve both a lower offloading loss and a lower fetching cost. A smaller offloading loss can be achieved by lowering the threshold value at the expense of the fetching cost. The gain of the proposed scheme depends on how frequently the popularity profile changes. For example, when the popularity profile changes slowly, the gain is small; but the frequency of updates will also be less in the proposed scheme.

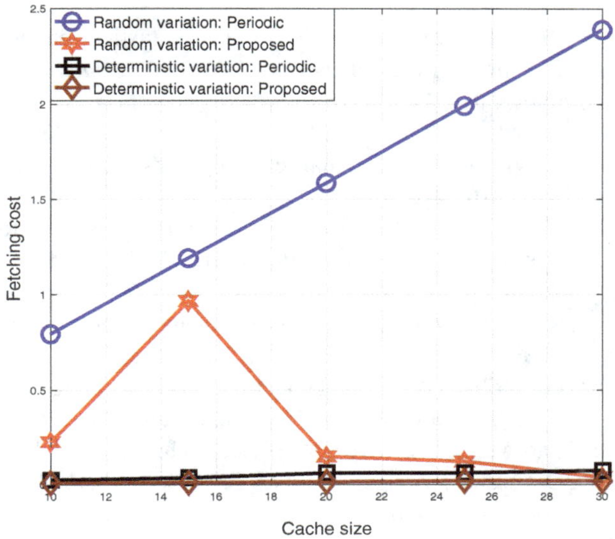

Figure 18.3 *Fetching cost versus cache size for two different scenarios of arrival processes*

18.7 Concluding remarks

A learning-theoretic analysis of content caching in heterogenous networks with non-stationary, statistically dependent, and unknown popularity profiles has been carried out. A PAC result on the offloading loss is presented in Theorem 4, based on the following caching algorithm: At every slot t, the BS computes an estimate of the Rademacher complexity and the discrepancy based on the available requests. The optimal caching policy is employed at the BS based on these estimates, and the cache content items at the sBSs are updated only if the discrepancy in the popularity profile is larger than a pre-specified threshold (to be determined based on the error tolerance). A detailed analysis of this algorithm can be studied as a potential future extension of this work. We have also presented performance analyses for the Bernoulli and Poisson request models. The work summarized in this chapter can be considered as an initial step toward understanding the fundamental aspects of learning-aided proactive content caching and delivery in wireless edge networks with non-stationary user and content dynamics.

References

[1] Furuskar A., Charles J., Frodigh M. 'Refined statistical analysis of evolution approaches for wireless networks'. *IEEE Transactions on Wireless Communications*. 2015;**5**:2700–10.

[2] Bennis M., Simsek M., Czylwik A., Saad W., Valentin S., Debbah M. 'When cellular meets WiFi in wireless small cell networks'. *IEEE Communications Magazine*. 2013;**51**(6):44–50.

[3] Chou S.F., Chiu T.C., YJ Y. 'Mobile small cell deployment for next generation cellular networks'. Proceedings of IEEE Global Communications Conference; 2014. pp. 4852–7.

[4] Niesen U., Shah D., Wornell G.W. 'Caching in wireless networks'. *IEEE Transactions on Information Theory*. 2012;**58**(10):6524–40.

[5] Ji M., Caire G., Molisch A.F. 'Fundamental limits of caching in wireless D2D networks'. *IEEE Transactions on Information Theory*. 2016;**62**(2):849–69.

[6] Zhang L., Xiao M., Wu G., Li S. 'Efficient scheduling and power allocation for D2D-assisted wireless caching networks'. *IEEE Transactions on Communications*. 2016;**64**(6):2438–52.

[7] Chen B., Yang C. 'Caching policy for cache-enabled D2D communications by learning user preference'. *IEEE Transactions on Communications*. 2018;**66**(12):6586–601.

[8] Gregori M., Gómez-Vilardebó J., Matamoros J. 'Wireless content caching for small cell and D2D networks'. *IEEE Journal on Selected Areas in Communications*. 2016;**5**:1222–34.

[9] Maddah-Ali M.A., Niesen U. 'Fundamental limits of caching'. *IEEE Transactions on Information Theory*. 2014;**60**(5):2856–67.

[10] Golrezaei N., Dimakis A.G., Molisch A.F. 'Wireless device-to-device communication with distributed caching'. Proceedings of IEEE International Symposium Information Theory; 2012. pp. 2781–5.

[11] Yang Q., Gunduz D. 'Coded caching and content delivery with heterogeneous distortion requirements'. *IEEE Transactions on Information Theory*. 2018;**64**(6):4347–64.

[12] Wu Y., Yao S., Yang Y. 'Semigradient-based cooperative caching algorithm for mobile social networks'. Proceedings of IEEE Global Communications Conference; Washington, DC, USA; 2016. pp. 1–6.

[13] Wang X., Chen M., Taleb T., Ksentini A., Leung V. 'Cache in the air: exploiting content caching and delivery techniques for 5G systems'. *IEEE Transactions on Wireless Communications*. 2014;**52**:131–9.

[14] Fang C., Yu F., Huang T., Liu J., Liu Y. 'A survey of energy-efficient caching in information-centric networking'. *IEEE Communications Magazine*. 2014;**52**(11):122–9.

[15] Pedarsani R., Maddah-Ali M.A., Niesen U. 'Online coded caching'. Proceedings of IEEE International Conference on Communications; 2014. pp. 1878–83.

[16] Golrezaei N., Mansourifard P., Molisch A.F. 'Base-station assisted device-to-device communications for high-throughput wireless video networks'. *IEEE Transactions on Wireless Communications*. 2014;**7**:3665–76.

[17] Li J., Xiao M., Chen W. 'Efficient video pricing and caching in heterogeneous networks'. *IEEE Transactions on Vehicular Technology*. 2016;**10**:8744–51.

[18] Mital N., Gunduz D., Ling C. 'Coded caching in a multi-server system with random topology'. *IEEE Transactions on Communications*. 2020;**68**(8):4620–31.

[19] Zhao J., Amiri M.M., Gündüz D. 'Multi-antenna coded content delivery with caching: a low-complexity solution'. *IEEE Transactions on Wireless Communications*. 2020;**19**(11):7484–97.

[20] Roushdy A., Motahari A., Nafie M., Gunduz D. 'Cache-aided combination networks with interference'. *IEEE Transactions on Wireless Communications*. 2020;**19**(1):148–61.

[21] Dai B., Liu Y.F., Yu W. 'Optimized base-station cache allocation for cloud radio access network with multicast backhaul'. *IEEE Journal on Selected Areas Communications*. 2018;**36**(8):1737–50.

[22] Tao M., Chen E., Zhou H., Yu W. 'Content-centric sparse multicast beamforming for cache-enabled cloud RAN'. *IEEE Transactions on Wireless Communications*. 2016;**15**(9):6118–31.

[23] Ugur Y., Awan Z.H., Sezgin A. 'Cloud radio access networks with coded caching'. Proceedings of ITG Workshop on Smart Antennasa; 2016. pp. 1–5.

[24] Park S.H., Simeone O., Shitz S.S. 'Joint optimization of cloud and edge processing for fog radio access networks'. *IEEE Transactions on Wireless Communications*. 2016;**15**(11):7621–32.

[25] Cui Y., Jiang D., Wu Y. 'Analysis and optimization of caching and multicasting in large-scale cache-enabled wireless networks'. *IEEE Transactions on Wireless Communications*. 2016;**15**(7):5101–12.

[26] Mohammadi Amiri M., Gunduz D. 'Cache-aided content delivery over erasure broadcast channels'. *IEEE Transactions on Communications*. 2018;**66**(1):370–81.

[27] Naderializadeh N., Maddah-Ali M.A., Avestimehr A.S. 'Fundamental limits of cache-aided interference management'. *IEEE Transactions on Information Theory*. 2017;**63**(5):3092–107.

[28] Sengupta A., Tandon R., Simeone O. 'Fog-aAided wireless networks for content delivery: fundamental latency tradeoffs'. *IEEE Transactions on Information Theory*. 2017;**63**(10):6650–78.

[29] Pujol Roig J.S., Tosato F., Gunduz D. 'Interference networks with caches at both ends'. Proceedings of IEEE International Conference on Communications; 2017. pp. 1–6.

[30] Tao M., Gunduz D., Xu F., Roig J.S.P. 'Content caching and delivery in wireless radio access networks'. *IEEE Transactions on Communications*. 2019;**7**:4724–49.

[31] Blasco P., Gündüz D. 'Learning-based optimization of cache content in a small cell base station'. Proceedings of IEEE International Conference on Communications. Sydney; NSW, Australia; 2014. pp. 1897–903.

[32] Blasco P., Gündüz D. 'Multi-armed bandit optimization of cache content in wireless infostation networks'. Proceedings of IEEE International Symposium Information Theory; 2014. pp. 51–5.

[33] Golrezaei N., Shanmugam K., Dimakis A. 'Femto caching: wireless video content delivery through distributed caching helpers'. *IEEE Transactions on Information Theory*. 2013;**12**:8402–13.

[34] Tatar A., de Amorim M.D., Fdida S., Antoniadis P. 'A survey on predicting the popularity of web content'. *Journal of Internet Services and Applications*. 2014;**5**(1):1–20.

[35] Bharath B.N., Nagananda K.G., Poor H.V. 'A learning-based approach to caching in heterogenous small cell networks'. *IEEE Transactions on Communications*. 2016;**64**(4):1674–86.

[36] Song J., Sheng M., Quek T.Q.S., Xu C., Wang X. 'Learning based content aching and sharing for wireless networks'. *IEEE Transactions on Communications*. 2017:1.

[37] Chen B., Yang C. 'Caching policy for cache-enabled D2D communications by learning user preference'. Proceedings of IEEE Vechicular Technology Conference Spring; Nanjing, China; 2016. pp. 1–6.

[38] Cha M., Kwak H., Rodriguez P. 'Analyzing the video popularity characteristics of large-scale user generated content systems'. *IEEE/ACM Transactions on Networking*. 2009;**5**:1357–70.

[39] Szabo G., Huberman B.A. 'Predicting the popularity of online content'. *Communications ACM*. 2010;**53**(8):80–8.

[40] Kim H., Park J., Bennis M. 'Ultra-dense edge caching under spatio-temporal demand and network dynamics'. Proceedings of IEEE International Conference on Communications; Paris, France; 2017. pp. 1–7.

[41] Somuyiwa S.O., Gyorgy A., Gunduz D. 'A reinforcement-learning approach to proactive caching in wireless networks'. *IEEE Journal on Selected Areas in Communications*. 2018;**36**(6):1331–44.

[42] Bharath B.N., Nagananda K.G., Gündüz D. 'Learning-based content caching with time-varying popularity profiles'. Proceedings of IEEE Global Communications Conference; Singapore; 2017. pp. 1–6.

[43] Bharath B.N., Nagananda K.G., Gündüz D., Poor H.V. 'Caching with time-varying popularity profiles: a learning-theoretic perspective'. *IEEE Transactions on Communications*. 2018;**66**(9):3837–47.

[44] Ji M., Caire G., Molisch A.F. 'Optimal throughput-outage trade-off in wireless one-hop caching networks'. Proceedings of IEEE International Symposium Information Theory; 2013. pp. 1461–5.

[45] Kuznetsov V., Mohri M., Zilles S. 'Generalization bounds for time series prediction with non-stationary processes' in Auer P., Clark A., Zeugmann T. (eds.). *Algorithmic Learning Theory*. Springer; 2014. pp. 260–74.

[46] Mohri M., Rostamizadeh A., Talwalkar A. *Foundations of machine learning*. MIT Press; 2012.

Chapter 19
Reinforcement learning for caching with space-time popularity dynamics

Alireza Sadeghi[1], Georgios B Giannakis[1], Gang Wang[1], and Fatemeh Sheikholeslami[2]

With the tremendous growth of data traffic over wired and wireless networks along with the increasing number of rich-media applications, caching is envisioned to play a critical role in next-generation networks. To intelligently prefetch and store contents, a cache node should be able to learn what and when to cache. Considering the geographical and temporal content popularity dynamics, the limited available storage at cache nodes, as well as the interactive influence of caching decisions in networked caching settings, developing effective caching policies is practically challenging. In response to these challenges, this chapter presents a versatile reinforcement learning-based approach for near-optimal caching policy design, in both single-node and network caching settings under dynamic space-time popularities. The policies presented here are complemented using a set of numerical tests, which showcase the merits of the presented approach relative to several standard caching policies.

19.1 Introduction

The term cache was initially introduced in computer systems in 1970s to describe a small capacity memory with practically fast access [1–3]. This idea was later applied to the Internet, where content requests were typically routed to just a few central servers [4, 5]. Mimicking the computer caches, an Internet cache is to be deployed at the edge of the network, closer to end-users, and thus to serve requests locally. Recently, the exponential growth of mobile video traffic, in conjunction with advances in machine learning and artificial intelligence has boosted caching to the front line of research in wired and wireless networks. In this regard, caching has been investigated from different perspectives.

[1]Department of Electrical and Computer Engineering, and Digital Technology Center, University of Minnesota, USA
[2]Bosch Center for Artificial Intelligence, USA

Several works have considered *static* content popularities, which yield tractable caching models at the price of accuracy. For instance, a multi-armed bandit approach that accounts for the demand history and unknown popularities can be found in [6]. Coded, convexified, and distributed extensions of this problem were studied in [7], context and trend-aware learning approaches in [8, 9], and coordinated-distributed extensions in [10]. From a learning perspective, the trade-off between the "accuracy" of learning a static popularity, and the corresponding learning "speed" was investigated [11] and [12].

Nonetheless, popularities in practice are *dynamic*, which means they fluctuate over a time horizon. For instance, half of the top 25 requested Wikipedia articles change on a daily basis [2, 13], which motivates well adaptive caching strategies that can account for popularity dynamics [14–19]. To approximate the evolution of popularities, a Poisson shot noise model was adopted in [15], for which an age-based caching solution was proposed in [16]. In addition, postulating a Markovian evolution for popularities, reinforcement learning-based approaches were recently studied in [20–24]. Albeit reasonable for discrete states, these approaches cannot deal with large continuous state-action spaces. To cope with such spaces, deep reinforcement learning approaches have been considered for content caching in [25, 26].

The aforementioned works have focused on devising caching policies for a single entity. A more common setting in next-generation networks, however, involves a network of interconnected caching nodes. For example, today's content delivery networks, such as Akamai [27], have tree network structures. On the other hand, it has been shown that optimizing a network of connected caches jointly can further improve performance [28].

This chapter aspires to glean some of the recent advances in caching with space-time popularities, through a suite of reinforcement learning tools. The collection here is by no means exhaustive. Upon modeling the space-time popularity dynamics in a cellular network as a Markovian process, a reinforcement learning formulation of the caching task is laid out. After reviewing the classical reinforcement learning algorithms, Q-learning in particular, a scalable approach is derived. Next, a networked caching scenario is considered where a parent node is connected to several leaf nodes to serve end-user file requests. Inspired by the tree network structure in Akamai, a two-timescale formulation of the caching problem is developed. To model the interaction between caching decisions of parent and leaf nodes along with the space-time evolution of file requests, a scalable deep reinforcement learning approach based on hyper deep Q-networks (DQNs) is developed.

Notation. Lower- (upper-) case boldface letters denote column vectors (matrices), whose (i,j)-th entry is denoted by $[\,.\,]_{i,j}$. Calligraphic symbols are reserved for sets, while $\top$ stands for transposition. The operator $\mathbb{E}$ denotes the expectation, and vector $\mathbf{1}$ represents the all-one vector.

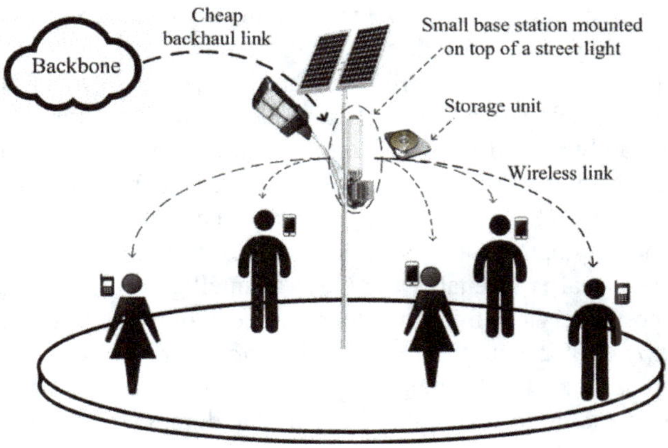

Figure 19.1 Local section of a HetNet

19.2 Single-node caching: Modeling and problem statement

Consider a subsection of a cellular network, which is simply a single small base station (SB) connected to the backbone network through a low-bandwidth, high-delay backhaul link. The SB is equipped with M units to store unit-size content (files); see Figure 19.1. We suppose that caching is to be carried out in a slotted fashion over $t = 1\,2, \ldots,$ where, at the end of each slot, the cache control unit (CCU)-enabled SB selects M files "intelligently" from a total of $F \gg M$ available ones, and prefetches them for possible reuse in the subsequent slot. The structure of a slot is depicted in Figure 19.2. At the beginning of a time slot, the user file requests are revealed, and the so-called content delivery phase takes place. The second phase, pertains to "information exchange," where the SBs transmit their locally observed popularity profiles to the network operator, and in return receive the estimated network-wide global popularity profile. Finally, "cache placement" is to happen where a selection of files is stored for the next time slot. These slots may not be of equal length, as the starting time may be set a priori, for example, at 3 AM, 11 AM, or 4 PM, when the network load is relatively low; or, slot intervals may be dictated to the CCU by the network operator "on the fly." Generally, a slot begins when the network is at an

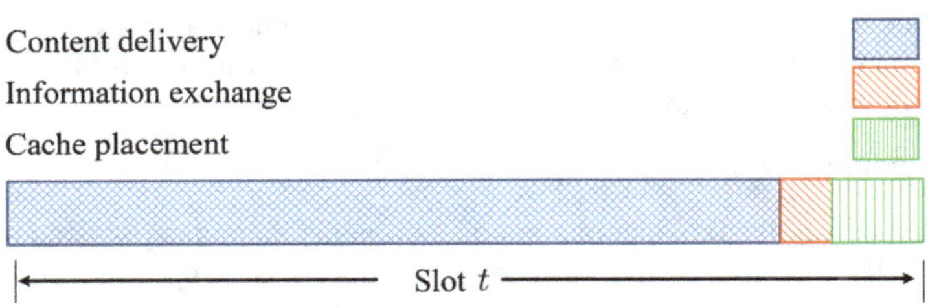

Figure 19.2 The slot structure

off-peak period, and its duration coincides with the peak traffic time when the pertinent costs of serving users are high.

Users request a subset of files from the set $\mathscr{F} := \{1, 2, \ldots, F\}$ of available ones, during the content delivery of slot t. If a requested file is stored in the cache, it will be immediately served locally, without incurring any cost. Conversely, if it is not available in the cache, then the SB must fetch it from the cloud through a congested backhaul link, thus incurring a certain cost, due to possible electricity price surges, processing cost, or a sizable delay resulting in low QoE and user dissatisfaction. The goal is to enable the CCU to intelligently select the cache contents, so that costly services from the cloud can be avoided as often as possible. To this aim, let $\mathbf{a}(t) \in \mathscr{A}$ denote the $F \times 1$ binary *caching action vector* at slot t, with $\mathscr{A} := \{\mathbf{a} | \mathbf{a} \in \{0, 1\}^F, \mathbf{a}^\top \mathbf{1} = M\}$ being the set of all feasible actions; where having $[\mathbf{a}(t)]_f = 1$ is tantamount to saving the file f for the duration of slot t; and $[\mathbf{a}(t)]_f = 0$ otherwise.

During the content delivery phase, the CCU computes an $F \times 1$-vector of the *local popularity profile* $\mathbf{p}_L(t)$ per slot t, whose f-th entry captures the expected local demand for file f, defined as

$$\left[\mathbf{p}_{\mathrm{L}}(t)\right]_f := \frac{\text{Number of local requests for } f \text{ at slot } t}{\text{Number of all local requests at slot } t}. \tag{19.1}$$

Similarly, the network operator estimates an $F \times 1$ *global popularity profile* vector $\mathbf{p}_G(t)$ and transmits it to all CCUs. Having the estimated local and global file popularities by the end of the information exchange phase of slot t, we define the overall system state as

$$\mathbf{s}(t) := \left[\mathbf{p}_G^\top(t), \mathbf{p}_L^\top(t), \mathbf{a}^\top(t)\right]^\top. \tag{19.2}$$

While at slot $t-1$, our *objective* is to leverage historical observations of states $\{\mathbf{s}(\tau)\}_{\tau=0}^{t-1}$, and corresponding costs, to learn an optimal action for the next slot, that is, $\mathbf{a}^*(t)$. The ensuing sections formulate this objective from a reinforcement learning vantage point.

19.2.1 Costs and caching strategies

Serving file requests is costly, where the overall cost comprises the superposition of three terms. The first term $c_{1,t}$ corresponds to the cost of refreshing the cache contents locally. In its general form, $c_{1,t}(\cdot)$ is a function of the upcoming action $\mathbf{a}(t)$, and available contents at the cache according to current caching action $\mathbf{a}(t-1)$, where the subscript t captures the possibility of incurring a time-varying cost for refreshing the cache. A reasonable choice of $c_{1,t}(\cdot)$ is

$$c_{1,t}(\mathbf{a}(t), \mathbf{a}(t-1)) := \lambda_{1,t}\mathbf{a}^\top(t)\left[\mathbf{1} - \mathbf{a}(t-1)\right]. \tag{19.3a}$$

The second cost $c_{2,t}(\mathbf{s}(t))$ is incurred during the operational phase of slot t to satisfy local user file request. A prudent choice for this cost must: (i) penalize requests for files already cached much less than requests for files not stored and (ii) be a non-decreasing function of popularities $[\mathbf{p}_L]_f$. For simplicity, we choose

$$c_{2,t}(\mathbf{s}(t)) := \lambda_{2,t}\left[\mathbf{1} - \mathbf{a}(t)\right]^\top \mathbf{p}_L(t) \tag{19.3b}$$

which solely penalizes the non-cached files in a descending order of their local popularities.

The third type of cost captures the "mismatch" between caching action $\mathbf{a}(t)$ and the global popularity profile $\mathbf{p}_G(t)$. Indeed, it is reasonable to consider the global popularity of files as a surrogate of what the local profiles must look like in the near future; thus, keeping the caching action close to $\mathbf{p}_G(t)$ may reduce future possible costs. Note that having a relatively small number of local requests may just give a rough estimate of local demands, while the global popularity profile can serve as side information in tracking the evolution of content popularities across the network. To account for this issue, we introduce the third type of cost as

$$c_{3,t}(\mathbf{s}(t)) := \lambda_{3,t}\left[\mathbf{1} - \mathbf{a}(t)\right]^\top \mathbf{p}_G(t) \tag{19.3c}$$

penalizing the files not cached according to the global popularity profile $\mathbf{p}_G(\cdot)$ provided by the network operator, thus promoting adaptation of caching policies close to global demand trends.

Upon taking action $\mathbf{a}(t)$, the *aggregate cost conditioned* on the popularity vectors revealed, can be expressed as follows:

$$\begin{aligned} C_t\left(\mathbf{S}(t-1), \mathbf{a}(t)|\mathbf{p}_G(t), \mathbf{p}_L(t)\right) : &= c_{1,t}\left(\mathbf{a}(t), \mathbf{a}(t-1)\right) + c_{2,t}(\mathbf{s}(t)) + c_{3,t}(\mathbf{s}(t)) \\ &= \lambda_{1,t}\mathbf{a}^\top(t)\left(\mathbf{1} - \mathbf{a}(t-1)\right) + \lambda_{2,t}(\mathbf{1} - \mathbf{a}(t))^\top \mathbf{p}_L(t) \\ &+ \lambda_{3,t}(\mathbf{1} - \mathbf{a}(t))^\top \mathbf{p}_G(t) \end{aligned} \tag{19.4}$$

where weights $\lambda_{1,t}$, $\lambda_{2,t}$, and $\lambda_{3,t}$ control the relative significance of the corresponding terms. As asserted earlier, the cache-refreshing cost is considered to be less than that of fetching, which justifies $\lambda_{1,t} \ll \lambda_{2,t}$. In addition, setting $\lambda_{3,t} \ll \lambda_{2,t}$ is of interest when the local popularity profiles are of acceptable accuracy, or, if tracking local popularities are of higher importance. In particular, setting $\lambda_{3,t} = 0$ corresponds to the special case where the caching cost is decoupled from the global popularity profile evolution. On the other hand, setting $\lambda_{2,t} \ll \lambda_{3,t}$ is desirable in networks where globally popular files are of high significance, for instance, when users have high mobility and may change SBs rapidly, or, when a few local requests prevent the SB from estimating the local popularity profiles accurately. Figure 19.3 depicts the evolution of popularities and action vectors along with the aggregate conditional costs across slots.

19.2.2 Markovian popularity profile evolution

We consider Markovian evolution for user requests (popularities) both globally and locally as depicted in Figure 19.4. Global popularity profiles are assumed to be generated through an underlying Markov chain with $|\mathscr{P}_G|$ states collected in the set $\mathscr{P}_G := \left\{\mathbf{p}_G^1, \ldots, \mathbf{p}_G^{|\mathscr{P}_G|}\right\}$; and likewise, for the local popularity profiles having states

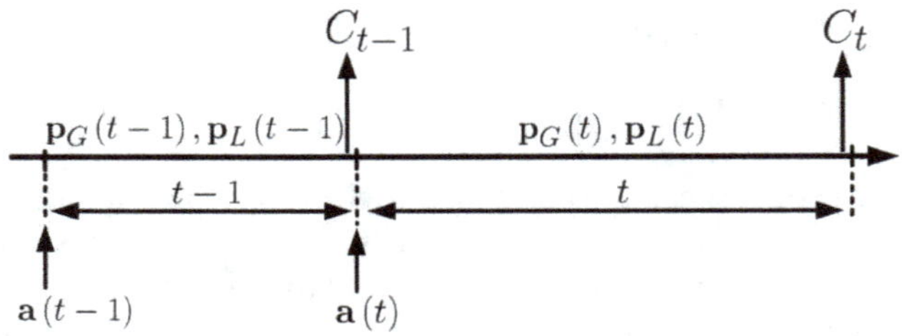

Figure 19.3 A schematic depicting the evolution of key quantities across time slots. Duration of slots can be unequal.

$\mathscr{P}_L := \left\{\mathbf{p}_L^1, \ldots, \mathbf{p}_L^{|\mathscr{P}_L|}\right\}$. Although $\mathscr{P}_G$ and $\mathscr{P}_L$ are known, the underlying transition probabilities of the two Markov chains are unknown in practice.

Given $\mathscr{P}_G$ and $\mathscr{P}_L$ as well as feasible caching decisions from the set $\mathscr{A}$, the overall set of states in the network is

$$\mathscr{S} := \left\{\mathbf{s} | \mathbf{s} = \left[\mathbf{p}_G^\top, \mathbf{p}_L^\top, \mathbf{a}^\top\right]^\top, \mathbf{p}_G \in \mathscr{P}_G, \mathbf{p}_L \in \mathscr{P}_L, \mathbf{a} \in \mathscr{A}\right\}. \tag{19.5}$$

In the proposed reinforcement learning-based caching, the underlying transition probabilities for global and local popularity profiles are *unknown*. Therefore, the learner should find an optimal policy interactively by making sequential decisions, and observing the corresponding costs. The following section formulates the optimal

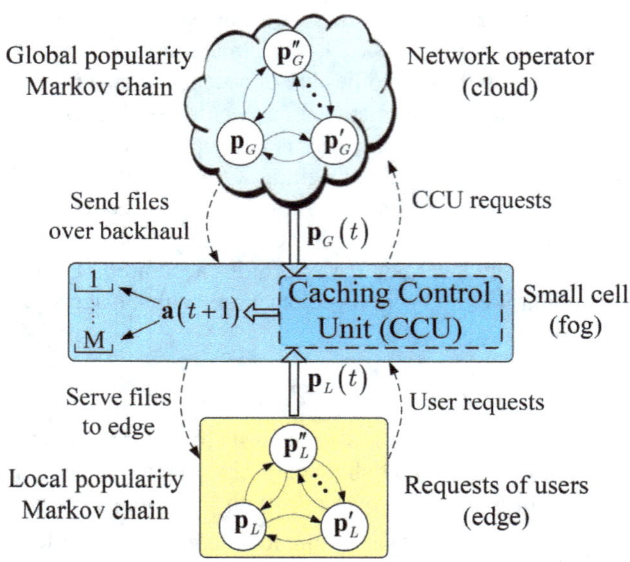

Figure 19.4 Schematic of network structure with required communications between a SB with the network operator

caching problem and provides an efficient solver to handle the "curse of dimensionality" typically emerging with reinforcement learning [29].

19.2.3 Reinforcement learning formulation

Due to the random nature of user requests, the cost (19.4) is random with mean

$$\begin{aligned}\overline{C}_t(\mathbf{s}(t-1), \mathbf{a}(t): &= \mathbb{E}_{\mathbf{p}_G(t),\mathbf{p}_L(t)}\left[C_t\left(\mathbf{s}(t-1), \mathbf{a}(t) \middle| \mathbf{p}_G(t), \mathbf{p}_L(t)\right)\right] \\ &= \lambda_1 \mathbf{a}^\top(t)\left[\mathbf{1} - \mathbf{a}(t-1)\right] + \lambda_2 \mathbb{E}\left[(\mathbf{1} - \mathbf{a}(t))^\top \mathbf{p}_L(t)\right] \\ &+ \lambda_3 \mathbb{E}\left[(\mathbf{1} - \mathbf{a}(t))^\top \mathbf{p}_G(t)\right]\end{aligned} \tag{19.6}$$

where the expectation is taken with respect to (wrt) $\mathbf{p}_L(t)$ and $\mathbf{p}_G(t)$, while the weights are $\lambda_{1,t} = \lambda_1$, $\lambda_{2,t} = \lambda_2$, and $\lambda_{3,t} = \lambda_3$. Let $\pi : \mathcal{S} \to \mathcal{A}$ denote the policy function, which maps state $\mathbf{s} \in \mathcal{S}$ to the action set, where for a given $\pi(\cdot)$, at the state $\mathbf{s}(t)$, caching is carried out via $\mathbf{a}(t+1) = \pi(\mathbf{s}(t))$ dictating what files to store for the next slot $(t+1)$. The caching performance is measured through the so-termed state value function

$$V_\pi\left(\mathbf{s}(t)\right) := \lim_{T\to\infty} \mathbb{E}\left[\sum_{\tau=t}^{T} \gamma^{\tau-t}\overline{C}\left(\mathbf{s}\left[\tau\right], \pi\left(\mathbf{s}\left[\tau\right]\right)\right) \middle| \mathbf{s}(t)\right] \tag{19.7}$$

which is the expected discounted cost incurred over an infinite time horizon, with future terms discounted by a factor $\gamma \in [0, 1)$. Furthermore, tuning γ trades-off current for future costs, and it also accounts for modeling uncertainties or dynamics.

The objective here is to seek an optimal policy π^* such that the expected discounted cost is minimized for any initial state $\mathbf{s}$

$$\pi^* = \arg\min_{\pi\in\Pi} V_\pi\left(\mathbf{s}\right), \quad \forall \mathbf{s} \in \mathcal{S} \tag{19.8}$$

where Π is the set of all feasible policies. The optimization problem in (19.8) is a sequential decision-making problem, which can be solved using the Bellman equations, as elaborated in next section.

19.2.4 Optimality conditions

The Bellman equations provide necessary conditions for optimality of a policy in a sequential decision-making problem. Let $[\boldsymbol{P}^a]_{\mathbf{ss}'}$ denote the transition probability of going from the current state $\mathbf{s}$ to the next one $\mathbf{s}'$ while action $\mathbf{a}$ is taken; that is,

$$[\boldsymbol{P}^{\mathbf{a}}]_{\mathbf{ss}'} := \Pr\left\{\mathbf{s}(t) = \mathbf{s}' \middle| \mathbf{s}(t-1) = \mathbf{s}, \pi(\mathbf{s}(t-1)) = \mathbf{a}\right\}. \tag{19.9}$$

Bellman equations express the state value function by (19.7) in a recursive fashion as follows [29, p. 47]:

$$V_\pi\left(\mathbf{s}\right) = \overline{C}\left(\mathbf{s}, \pi(\mathbf{s})\right) + \gamma \sum_{\mathbf{s}'\in\mathcal{S}} [\boldsymbol{P}^{\pi(\mathbf{s})}]_{\mathbf{ss}'} V_\pi\left(\mathbf{s}'\right), \forall \mathbf{s}, \mathbf{s}' \tag{19.10}$$

which amounts to the superposition of $\overline{C}$ plus a discounted version of future state value functions under a given policy π. Specifically, $\overline{C}$ in (19.6) can be written as

$$\overline{C}\left(\mathbf{s}, \pi(\mathbf{s})\right) = \sum_{\mathbf{s}' := [\mathbf{p}'_G, \mathbf{p}'_L, \mathbf{a}'] \in \mathcal{S}} [\boldsymbol{P}^{\pi(\mathbf{s})}]_{\mathbf{ss}'} C\left(\mathbf{s}, \pi(\mathbf{s}) \middle| \mathbf{p}'_G, \mathbf{p}'_L\right) \tag{19.11}$$

where $C(\mathbf{s}, \pi(\mathbf{s})|\mathbf{p}'_G, \mathbf{p}'_L)$ is defined in (19.4). Given $[\boldsymbol{P}^{\mathbf{a}}]_{\mathbf{ss}'}$, one can readily obtain $\{V_\pi(\mathbf{s}), \forall \mathbf{s}\}$ by solving (19.10), and eventually the optimal policy π^* in (19.13) using the so-termed policy iteration algorithm [29, pg. 79]. In practice however, $\boldsymbol{P}^{\mathbf{a}}$ is unknown. This motivates the use of adaptive dynamic programming (ADP) and model-free reinforcement learning algorithms, such as Q-learning. The ADP estimates $[\boldsymbol{P}^{\mathbf{a}}]_{\mathbf{ss}'}$ for all $\mathbf{s}, \mathbf{s}' \in \mathcal{S}$, and $\mathbf{a} \in \mathcal{A}$, as iterations proceed [30, pg. 834]. Unfortunately, ADP algorithms are often very slow and impractical, as they must estimate $|\mathcal{S}|^2 \times |\mathcal{A}|$ probabilities. In contrast, the Q-learning algorithm to be elaborated next finds the optimal π^* as well as $V_\pi(\mathbf{s})$, while circumventing the need for estimating $[P^a]_{\mathbf{ss}'}, \forall \mathbf{s}, \mathbf{s}'$ [29, pg. 140].

To describe how this algorithm works in our context, we introduce the so-termed state-action value function for a given policy π[29, pg. 62]

$$Q_\pi\left(\mathbf{s}, \mathbf{a}'\right) := \overline{C}\left(\mathbf{s}, \mathbf{a}'\right) + \gamma \sum_{\mathbf{s}' \in \mathcal{S}} [\boldsymbol{P}^{\mathbf{a}'}]_{\mathbf{ss}'} V_\pi\left(\mathbf{s}'\right) \tag{19.12}$$

which is commonly referred to as the "Q-function." The Q-factor $Q_\pi(\mathbf{s}, \mathbf{a})$ basically captures the expected current cost of taking a particular action $\mathbf{a}$ when the system is in state $\mathbf{s}$, followed by the discounted value of the future states, provided that the future actions are taken by following the policy π. The next section uses the Q-function to find an optimal policy.

19.2.5 Optimal caching via Q-learning

Q-learning is a reinforcement learning scheme to infer the optimal policy π^* by estimating the optimal state-action value function $Q^*(\mathbf{s}, \mathbf{a}') := Q_{\pi^*}(\mathbf{s}, \mathbf{a}'),\ \forall \mathbf{s}, \mathbf{a}'$ "on the fly." Upon obtaining $Q^*(\mathbf{s}, \mathbf{a}')$, the optimal policy can be easily extracted [30, pg. 67] as:

$$\pi^*(\mathbf{s}) = \arg\min_{\boldsymbol{\alpha}} Q^*(\mathbf{s}, \boldsymbol{\alpha}), \quad \forall \mathbf{s} \in \mathcal{S}. \tag{19.13}$$

Furthermore, the Q-function and $V(\cdot)$ under π^* are related

$$V^*(\mathbf{s}) := V_{\pi^*}(\mathbf{s}) = \min_{\boldsymbol{\alpha}} Q^*(\mathbf{s}, \boldsymbol{\alpha}) \tag{19.14}$$

which in turn yields

$$Q^*\left(\mathbf{s}, \mathbf{a}'\right) = \overline{C}\left(\mathbf{s}, \mathbf{a}'\right) + \gamma \sum_{\mathbf{s}' \in \mathcal{S}} [\boldsymbol{P}^{\mathbf{a}}]_{\mathbf{ss}'} \min_{\boldsymbol{\alpha} \in \mathcal{A}} Q^*\left(\mathbf{s}', \boldsymbol{\alpha}\right). \tag{19.15}$$

The well-known Q-learning algorithm relies on (19.15) to approximate $Q^*(\mathbf{s}, \mathbf{a}')$. This algorithm is tabulated in Algorithm 1. In this algorithm, the agent updates its estimated $\hat{Q}(\mathbf{s}(t-1), \mathbf{a}(t))$ as $C\left(\mathbf{s}(t-1), \mathbf{a}(t) \middle| \mathbf{p}_G(t), \mathbf{p}_L(t)\right)$ is observed. That is, given

$\mathbf{s}(t-1)$, Q-learning takes action $\mathbf{a}(t)$, and upon observing $\mathbf{s}(t)$, it incurs cost and forms the instantaneous error

$$\varepsilon\left(\mathbf{s}(t-1), \mathbf{a}(t)\right) := \frac{1}{2}\left(C\left(\mathbf{s}(t-1), \mathbf{a}(t)\right) + \gamma \min_{\boldsymbol{\alpha}} \widehat{Q}\left(\mathbf{s}(t), \boldsymbol{\alpha}\right) - \widehat{Q}\left(\mathbf{s}(t-1), \mathbf{a}(t)\right)\right)^2 \tag{19.16}$$

and correspondingly updates the Q-function through a stochastic gradient descent step to minimize error. As a result, one can easily write the update rule in short as

$$\begin{aligned}\hat{Q}_t\left(\mathbf{s}(t-1), \mathbf{a}(t)\right) = {} & (1-\beta_t)\hat{Q}_{t-1}\left(\mathbf{s}(t-1), \mathbf{a}(t)\right) \\ & +\beta_t\left[C\left(\mathbf{s}(t-1), \mathbf{a}(t)\middle|\mathbf{p}_{\mathrm{G}}(t), \mathbf{p}_{\mathrm{L}}(t)\right) + \gamma \min_{\boldsymbol{\alpha}} \hat{Q}_{t-1}\left(\mathbf{s}(t), \boldsymbol{\alpha}\right)\right]\end{aligned} \tag{19.17}$$

while keeping the rest of the entries in $\hat{Q}_t(\cdot,\cdot)$ unchanged.

Algorithm 1: Caching via Q-learning at CCU

Initialize $\mathbf{s}(0)$ randomly and $\hat{Q}_0(\mathbf{s},\mathbf{a}) = 0\ \forall \mathbf{s},\mathbf{a}$

for $t = 1,2,\ldots$ **do**

  Take action $\mathbf{a}(t)$ chosen probabilistically by

$$\mathbf{a}(t) = \begin{cases} \arg\min_{\mathbf{a}} \hat{Q}_{t-1}\left(\mathbf{s}(t-1),\mathbf{a}\right) & \text{w.p. } 1-\varepsilon_t \\ \text{random } \mathbf{a} \in \mathscr{A} & \text{w.p. } \varepsilon_t \end{cases}$$

  $\mathbf{p}_L(t)$ and $\mathbf{p}_G(t)$ are revealed based on user requests

  Set $\mathbf{s}(t) = \left[\mathbf{p}_G^\top(t), \mathbf{p}_L^\top(t), \mathbf{a}(t)^\top\right]^\top$

  Incur cost $C\left(\mathbf{s}(t-1), \mathbf{a}(t)\middle|\mathbf{p}_{\mathrm{G}}(t), \mathbf{p}_{\mathrm{L}}(t)\right)$

  Update

$$\begin{aligned}\hat{Q}_t\left(\mathbf{s}(t-1),\mathbf{a}(t)\right) = {} & (1-\beta_t)\hat{Q}_{t-1}\left(\mathbf{s}(t-1),\mathbf{a}(t)\right) \\ & + \beta_t\left[C\left(\mathbf{s}(t-1),\mathbf{a}(t)\middle|\mathbf{p}_{\mathrm{G}}(t),\mathbf{p}_{\mathrm{L}}(t)\right) + \gamma \min_{\boldsymbol{\alpha}} \hat{Q}_{t-1}\left(\mathbf{s}(t),\boldsymbol{\alpha}\right)\right]\end{aligned}$$

end

Regarding convergence of the Q-learning algorithm, a necessary condition ensuring $\hat{Q}_t(\cdot,\cdot) \to Q^*(\cdot,\cdot)$ is that all state-action pairs must be updated [31] continuously. Under this and the usual stochastic approximation conditions that will be specified later, $\hat{Q}_t(\cdot,\cdot)$ converges to $Q^*(\cdot,\cdot)$ with probability 1 [32], for a detailed description. Finite-time error bounds of the Q-learning algorithm with function approximation can be found in [33, 34]. To guarantee visiting all state-action pairs, various exploration-exploitation algorithms have been proposed. In this section, we have adopted an ε-greedy action selection algorithm. During initial iterations, or when the CCU observes considerable shifts in content popularities, setting high ε_t promotes exploration in order to learn the underlying dynamics. On the other hand, in stationary settings and once "enough" observations are made, small values of ε_t are desirable as they enable agent actions to approach the optimal policy.

The main limitation of the Q-learning algorithm is its slow convergence, which is due to the independent updates of the Q-values. Fortunately, Q-values are related, and leveraging these relationships leads to multiple updates per observation, hence, accelerating convergence. In the ensuing section, the structure of the problem at hand is exploited to develop a linear function approximation of the Q-function, which in turn offers scalability.

19.2.6 Scalable caching

Despite the simplicity of the updates as well as the optimality guarantees of the Q-learning algorithm, it is not scalable due to the possibly large state-action spaces present in real networks. For instance, the Q-table is of size $|\mathscr{P}_G||\mathscr{P}_L||\mathscr{A}|^2$, where $|\mathscr{A}| = \binom{F}{M}$ encompasses all possible selections of M out of F files. Hence, the Q-table size grows prohibitively with F, rendering the convergence of the table entries, as well as the policy iterates unacceptably slow. Furthermore, the action selection in $\min_{\boldsymbol{\alpha}\in\mathscr{A}} Q(\mathbf{s},\mathbf{a})$ entails an expensive exhaustive search over the entire feasible action set $\mathscr{A}$ whose cardinality can be huge. Function approximation schemes are appealing as they can endow Q-learning with scalability to handle real-world problems [35–39]. In our problem, a delicate linear approximation for $Q(\mathbf{s},\mathbf{a})$ is inspired by the additive form of the instantaneous costs in (19.4). Specifically, we propose to approximate $Q(\mathbf{s},\mathbf{a}')$ as follows

$$Q(\mathbf{s},\mathbf{a}') \simeq Q_G(\mathbf{s},\mathbf{a}') + Q_L(\mathbf{s},\mathbf{a}') + Q_R(\mathbf{s},\mathbf{a}') \tag{19.18}$$

where Q_G, Q_L, and Q_L correspond to the global and local popularity mismatches and the cache-refreshing cost, respectively. Recall that the state vector $\mathbf{s}$ consists of three subvectors as $\mathbf{s} := [\mathbf{p}_G^\top, \mathbf{p}_L^\top, \mathbf{a}^\top]^\top$. Corresponding to the global popularity subvector, we model the first term of the approximation in (19.18) as

$$Q_G(\mathbf{s},\mathbf{a}') := \sum_{i=1}^{|\mathscr{P}_G|}\sum_{f=1}^{F} \theta_{i,f}^G \mathbb{1}_{\{\mathbf{p}_G=\mathbf{p}_G^i\}} \mathbb{1}_{\{[\mathbf{a}']_f=0\}} \tag{19.19}$$

where the sums are over possible global popularity profiles as well as files, and where the indicator function $1_{\{\cdot\}}$ is 1 if its argument holds true; and 0 otherwise; while $\theta_{i,f}^G$ captures the average "overall" cost if the system is in global state $\mathbf{p}_G^i$, and the CCU decides not to cache the fth content. By defining the $|\mathscr{P}_G| \times |\mathscr{F}|$ matrix with (i,f)-th entry $[\boldsymbol{\Theta}^G]_{i,f} := \theta_{i,f}^G$, one can rewrite (19.18) as:

$$Q_G(\mathbf{s},\mathbf{a}') = \boldsymbol{\delta}_G^\top(\mathbf{p}_G)\boldsymbol{\Theta}^G(\mathbf{1}-\mathbf{a}') \tag{19.20}$$

where

$$\boldsymbol{\delta}_G(\mathbf{p}_G) := \left[\delta\left(\mathbf{p}_G - \mathbf{p}_G^1\right), \cdots, \delta\left(\mathbf{p}_G - \mathbf{p}_G^{|\mathscr{P}_G|}\right)\right]^\top. \tag{19.21}$$

Similarly, we can approximate the second summand (19.18) as

$$\begin{aligned} Q_L(\mathbf{s},\mathbf{a}') : &= \sum_{i=1}^{|\mathscr{P}_L|}\sum_{f=1}^{F}\theta_{i,f}^{L}\mathbb{1}_{\{\mathbf{p}_L=\mathbf{p}_L^i\}}\mathbb{1}_{\{[\mathbf{a}']_f=0\}} \\ &= \boldsymbol{\delta}_L^{\top}(\mathbf{p}_L)\boldsymbol{\Theta}^L(\mathbf{1}-\mathbf{a}') \end{aligned} \tag{19.22}$$

where $[\boldsymbol{\Theta}^L]_{i,f} := \theta_{i,f}^L$, and

$$\boldsymbol{\delta}_L(\mathbf{p}_L) := \left[\delta(\mathbf{p}_L-\mathbf{p}_L^1),\ldots,\delta(\mathbf{p}_L-\mathbf{p}_L^{|\mathscr{P}_L|})\right]^{\top} \tag{19.23}$$

with $\theta_{i,f}^L$ modeling the average overall cost for not caching file f when the local popularity is in state $\mathbf{p}_L^i$.

Finally, we model the third summand in (19.18) corresponding to the cache refreshing cost, as follows

$$\begin{aligned} Q_R(\mathbf{s},\mathbf{a}') : &= \sum_{f=1}^{F}\theta^R\mathbf{1}_{\{[\mathbf{a}']_f=1\}}\mathbf{1}_{\{[\mathbf{a}]_f=0\}} \\ &= \theta^R\mathbf{a}'^{\top}(1-\mathbf{a}) \\ &= \theta^R\left[\mathbf{a}'^{\top}(1-\mathbf{a})+\mathbf{a}^{\top}\mathbf{1}-\mathbf{a}'^{\top}\mathbf{1}\right] \\ &= \theta^R\mathbf{a}^{\top}(\mathbf{1}-\mathbf{a}') \end{aligned} \tag{19.24}$$

where θ^R is the long-time-averaged cache-refreshing cost per content. The constraint $\mathbf{a}^{\top}\mathbf{1}=\mathbf{a}'^{\top}\mathbf{1}=M$ is to factor out the term $\mathbf{1}-\mathbf{a}'$, which will become useful shortly.

Upon collectively denoting all parameters by $\Lambda := \{\boldsymbol{\Theta}^G,\boldsymbol{\Theta}^L,\theta^a\}$, the Q-function is readily approximated as (cf. (19.18))

$$\widehat{Q}_{\Lambda}(\mathbf{s},\mathbf{a}') := \underbrace{\left(\boldsymbol{\delta}_G^{\top}(\mathbf{p}_G)\boldsymbol{\Theta}^G+\boldsymbol{\delta}_L^{\top}(\mathbf{p}_L)\boldsymbol{\Theta}^L+\theta^R\mathbf{a}^{\top}\right)}_{\boldsymbol{\psi}(\mathbf{s}):=}(\mathbf{1}-\mathbf{a}'). \tag{19.25}$$

The original task of learning $|\mathscr{P}_G||\mathscr{P}_L||\mathscr{A}|^2$ parameters in Algorithm 1 is now reduced to that of learning Λ containing $\left(|\mathscr{P}_G|+|\mathscr{P}_L|\right)|\mathscr{F}|+1$ parameters.

19.2.7 Learning Λ

Given the current parameter estimates $\{\widehat{\boldsymbol{\Theta}}_{t-1}^G,\widehat{\boldsymbol{\Theta}}_{t-1}^L,\hat{\theta}_{t-1}^R\}$ at the end of the information exchange phase of slot t, the so-called temporal difference error is given by

$$\hat{e}\left(\mathbf{s}(t-1),\mathbf{a}(t)\right) := C\left(\mathbf{s}(t-1),\mathbf{a}(t)\right)+\gamma\, min_{\mathbf{a}'}\hat{Q}_{\Lambda_{t-1}}\left(\mathbf{s}(t),\mathbf{a}'\right)-\hat{Q}_{\Lambda_{t-1}}\left(\mathbf{s}(t-1),\mathbf{a}(t)\right). \tag{19.26}$$

Using the definition

$$\widehat{\varepsilon}\left(\mathbf{s}(t-1),\mathbf{a}(t)\right) := \frac{1}{2}\left(\widehat{e}\left(\mathbf{s}(t-1),\mathbf{a}(t)\right)\right)^2, \tag{19.27}$$

the parameter update rules are obtained following stochastic gradient descent to minimize the loss in (19.27) as follows [30, p. 847]:

$$\begin{aligned} \hat{\boldsymbol{\Theta}}_t^G &= \hat{\boldsymbol{\Theta}}_{t-1}^G-\alpha_G\nabla_{\boldsymbol{\Theta}^G}\widehat{\varepsilon}\left(\mathbf{s}(t-1),\mathbf{a}(t)\right) \\ &= \hat{\boldsymbol{\Theta}}_{t-1}^G+\alpha_G\,\widehat{e}\left(\mathbf{s}(t-1),\mathbf{a}(t)\right)\ \nabla_{\boldsymbol{\Theta}^G}\widehat{Q}_{\Lambda_{t-1}}(\mathbf{s}(t-1),\mathbf{a}(t)) \\ &= \hat{\boldsymbol{\Theta}}_{t-1}^G+\alpha_G\,\widehat{e}\left(\mathbf{s}(t-1),\mathbf{a}(t)\right)\ \boldsymbol{\delta}_G(\mathbf{p}_G(t-1))(\mathbf{1}-\mathbf{a}(t))^{\top} \end{aligned} \tag{19.28}$$

and

$$
\begin{aligned}
\hat{\boldsymbol{\Theta}}_t^L &= \hat{\boldsymbol{\Theta}}_{t-1}^L - \alpha_L \nabla_{\boldsymbol{\Theta}^L} \hat{\varepsilon}\left(\mathbf{s}(t-1), \mathbf{a}(t)\right) \\
&= \hat{\boldsymbol{\Theta}}_{t-1}^L + \alpha_L \, \hat{e}\left(\mathbf{s}(t-1), \mathbf{a}(t)\right) \; \nabla_{\boldsymbol{\Theta}^L} \widehat{Q}_{\Lambda_{t-1}}(\mathbf{s}(t-1), \mathbf{a}(t)) \\
&= \hat{\boldsymbol{\Theta}}_{t-1}^L + \alpha_L \, \hat{e}\left(\mathbf{s}(t-1), \mathbf{a}(t)\right) \; \boldsymbol{\delta}_L(\mathbf{p}_L(t-1))(\mathbf{1} - \mathbf{a}(t))^\top
\end{aligned}
\tag{19.29}
$$

along with

$$
\begin{aligned}
\hat{\theta}_t^R &= \hat{\theta}_{t-1}^R - \alpha_R \nabla_{\theta^R} \hat{\varepsilon}\left(\mathbf{s}(t-1), \mathbf{a}(t)\right) \\
&= \hat{\theta}_{t-1}^R + \alpha_R \, \hat{e}\left(\mathbf{s}(t-1), \mathbf{a}(t)\right) \; \nabla_{\theta^R} \widehat{Q}_{\Lambda_{t-1}}(\mathbf{s}(t-1), \mathbf{a}(t)) \\
&= \hat{\theta}_{t-1}^R + \alpha_R \, \hat{e}\left(\mathbf{s}(t-1), \mathbf{a}(t)\right) \; \mathbf{a}^\top(t-1)(\mathbf{1} - \mathbf{a}(t)).
\end{aligned}
\tag{19.30}
$$

Algorithm 2: Scalable Q-learning

Initialize $\mathbf{s}(0)$ randomly, $\widehat{\boldsymbol{\Theta}}_0^G = \mathbf{0}$, $\widehat{\boldsymbol{\Theta}}_0^L = \mathbf{0}$, $\hat{\theta}_0^R = 0$, and thus $\widehat{\boldsymbol{\psi}}(\mathbf{s}) = \mathbf{0}$

for $t = 1, 2, \ldots$ **do**

Take action $\mathbf{a}(t)$ chosen probabilistically by

$$
\mathbf{a}(t) = \begin{cases} M \text{ best files via } \widehat{\boldsymbol{\psi}}(\mathbf{s}(t-1)) & \text{w.p. } 1 - \varepsilon_t \\ \text{random } \mathbf{a} \in \mathcal{A} & \text{w.p. } \varepsilon_t \end{cases}
$$

where $\hat{\boldsymbol{\psi}}(\mathbf{s}) := \boldsymbol{\delta}_G^\top(\mathbf{p}_G)\widehat{\boldsymbol{\Theta}}^G + \boldsymbol{\delta}_L^\top(\mathbf{p}_L)\widehat{\boldsymbol{\Theta}}^L + \widehat{\theta}^R \mathbf{a}^\top$

$\mathbf{p}_G(t)$ and $\mathbf{p}_L(t)$ are revealed based on user requests

Set $\quad \mathbf{s}(t) = \left[\mathbf{p}_G^\top(t), \mathbf{p}_L^\top(t), \mathbf{a}(t)^\top\right]^\top$

Incur cost $\quad C\left(\mathbf{s}(t-1), \mathbf{a}(t) \middle| \mathbf{p}_{\mathrm{G}}(t), \mathbf{p}_{\mathrm{L}}(t)\right)$

Find $\quad \hat{\varepsilon}\left(\mathbf{s}(t-1), \mathbf{a}(t)\right)$

Update $\quad \widehat{\boldsymbol{\Theta}}_t^G$, $\widehat{\boldsymbol{\Theta}}_t^L$ and $\hat{\theta}_t^R$ based on (1.22)-(1.24)

end

The pseudocode for this scalable approximation of the Q-learning scheme is tabulated in Algorithm 2.

The advantage of this scalable Q-learning is threefold.

- The large state-action space in the Q-learning algorithm is handled by reducing the number of parameters from $|\mathcal{P}_G||\mathcal{P}_L||\mathcal{A}|^2$ to $\left(|\mathcal{P}_G| + |\mathcal{P}_L|\right)|\mathcal{F}| + 1$.
- In contrast to single-entry updates in the exact Q-learning Algorithm 1, $(F - M)$ entries in $\widehat{\boldsymbol{\Theta}}^G$ and $\widehat{\boldsymbol{\Theta}}^L$ as well as θ^R, are updated per observation using (19.22) to (19.24), which leads to a much faster convergence.
- The exhaustive search in $\min_{\mathbf{a} \in \mathcal{A}} Q\left(\mathbf{s}, \mathbf{a}\right)$ required in exploitation and also in the error evaluation (19.17), is circumvented. Specifically, it holds that (cf. (19.20))

$$
\min_{\mathbf{a}' \in \mathcal{A}} Q(\mathbf{s}, \mathbf{a}') \approx \min_{\mathbf{a}' \in \mathcal{A}} \boldsymbol{\psi}^\top(\mathbf{s}) \left(\mathbf{1} - \mathbf{a}'\right) = \max_{\mathbf{a}' \in \mathcal{A}} \boldsymbol{\psi}^\top(\mathbf{s}) \; \mathbf{a}'
\tag{19.31}
$$

where $\boldsymbol{\psi}(\mathbf{s}) := \boldsymbol{\delta}_G^\top(\mathbf{p}_G)\boldsymbol{\Theta}^G + \boldsymbol{\delta}_L^\top(\mathbf{p}_L)\boldsymbol{\Theta}^L + \theta^R\mathbf{a}^\top$. The solution of (19.25) is readily given by $[\mathbf{a}]_{\nu_i} = 1$ for $i = 1, \ldots, M$, and $[\mathbf{a}]_{\nu_i} = 0$ for i, where $[\boldsymbol{\psi}(\mathbf{s})]_{\nu_F} \leq \cdots \leq [\boldsymbol{\psi}(\mathbf{s})]_{\nu_1}$ are sorted entries of $\boldsymbol{\psi}(\mathbf{s})$.

19.2.8 Numerical tests

This section tests the performance of the proposed Q-learning algorithm and its scalable approximation. To compare the proposed algorithms with an optimal caching policy, which is the best policy computed under *known* transition probabilities for global and local popularity Markov chains, we first simulated a small network with $F = 10$ contents, and caching capacity $M = 2$ at the local SB. Global popularity profile is modeled by a two-state Markov chain with states $\mathbf{p}_G^1$ and $\mathbf{p}_G^2$, that are drawn from Zipf distributions having parameters $\eta_1^G = 1$ and $\eta_2^G = 1.5$, respectively [40]. That is, for state $i \in \{1, 2\}$, the F contents are assigned a random ordering of popularities, and then sorted accordingly in a descending order. Given this ordering and the Zipf distribution parameter η_i^G, the popularity of the f-th content is set to

$$[\mathbf{p}_G^i]_f = \frac{1}{f^{\eta_i^G} \sum_{l=1}^{F} 1 / l^{\eta_i^G}}, \qquad i = 1, 2 \tag{19.32}$$

where the summation normalizes the components to follow a valid probability mass function, while $\eta_i^G \geq 0$ controls the skewness of popularities.

Furthermore, state transition probabilities of the global popularity Markov chain were drawn randomly. Similarly, local popularities were modeled by a two-state Markov chain, with states $\mathbf{p}_L^1$ and $\mathbf{p}_L^2$, whose entries were drawn from Zipf distributions with parameters $\eta_1^L = 0.7$ and $\eta_2^L = 2.5$, respectively. The transition probabilities of the local popularity Markov chain were generated randomly.

Caching performance was assessed under three cost-parameter settings: (s1) $\lambda_1 = 10, \lambda_2 = 600, \lambda_3 = 1\,000$; (s2) $\lambda_1 = 600, \lambda_2 = 10, \lambda_3 = 1\,000$, and (s3) $\lambda_1 = 10, \lambda_2 = 10, \lambda_3 = 1\,000$. In all numerical tests, the optimal caching policy was found by utilizing the policy iteration algorithm with known transition probabilities. In addition, Q-learning in Algorithm 1 and its scalable approximation in Algorithm 2 were run with $\beta_t = 0.8$, $\alpha_G = \alpha_L = \alpha_R = 0.005$, and $\epsilon_t = 0.05$.

Figure 19.5 depicts the observed cost versus iteration (time) index averaged over 1 000 realizations. It is seen that the caching cost via $|\mathcal{A}| = 45$-learning, as well as through its scalable approximation converges to that of the optimal policy. As anticipated, even for this small-size network, with $|\mathscr{P}_G| = |\mathscr{P}_L| = 2$ and $|\mathscr{A}| = 45$, the Q-learning algorithm converges slowly to the optimal policy, especially under s1, while its scalable approximation exhibits faster convergence. The reason for the slower convergence under (s1) is that the corresponding cost parameters of local and global popularity mismatches are set high, and thus convergence of the $F = 1{,}000$ -learning algorithm as well as that of the caching policy essentially relies on learning both the global and local popularity Markov chains. In contrast, under (s2), λ_2 corresponding to the local popularity mismatch is low; hence, the impact of local

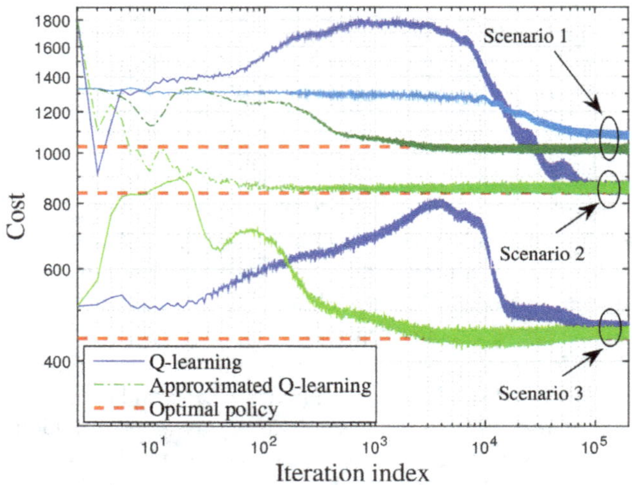

Figure 19.5 Performance of the proposed algorithms

popularity Markov chain on the optimal policy is reduced, giving rise to a simpler policy, thereby a faster convergence.

Having demonstrated the accuracy and efficiency of the proposed algorithms, we next simulated a larger network with $F = 1\,000$ available files, and a cache capacity of $M = 10$, offering a total of $\binom{1\,000}{10} \simeq 2 \times 10^{23}$ feasible caching actions. In addition, we set the local and global popularity Markov chains to have $|\mathscr{P}_L| = 40$ and $|\mathscr{P}_G| = 50$ states, for which the underlying state transition probabilities were drawn randomly, and Zipf parameters were drawn uniformly over the interval $(2, 4)$.

Figure 19.6 plots the performance of Algorithm 2 under (s4) $\lambda_1 = 100$, $\lambda_2 = 20$, $\lambda_3 = 20$, (s5) $\lambda_1 = 0$, $\lambda_2 = 0$, $\lambda_3 = 1\,000$, and (s6) $\lambda_2 = 0$, $\lambda_2 = 1\,000$, $\lambda_3 = 600$. The exploration-exploitation parameter was set to $\varepsilon_t = 1$ for $t = 1, 2, \ldots, 7 \times 10^5$, in order to greedily explore the entire state-action space in initial iterations, and $\varepsilon_t = 1\,/\,(\text{iteration index})$ for $t > 7 \times 10^5$.

Finding an optimal policy in (s5) and (s6) requires a prohibitively sizable memory as well as it incurs an extremely high computational complexity, which is impractical for this network. However, having large cache-refreshing costs with $\lambda_1 \gg \lambda_2, \lambda_3$ in (s4) forces the optimal caching policy to freeze its cache contents, making the optimal caching policy predictable in this setting. Despite the very limited storage capacity of $10/1\,000 = 0.01$ of available files, reinforcement learning-enabled caching offers considerable cost reduction, while the proposed approximated Q-learning endows the approach with scalability and light-weight updates.

The proposed Algorithms 1 and 2 deal with only a single caching entity. However, in more common settings, a cache node resides within a network of interconnected caches. For example, most of today's content delivery networks, such as Akamai, have a tree network structure [27]. Therefore, the caching policy design for a network of interconnected caches has become a common practice in

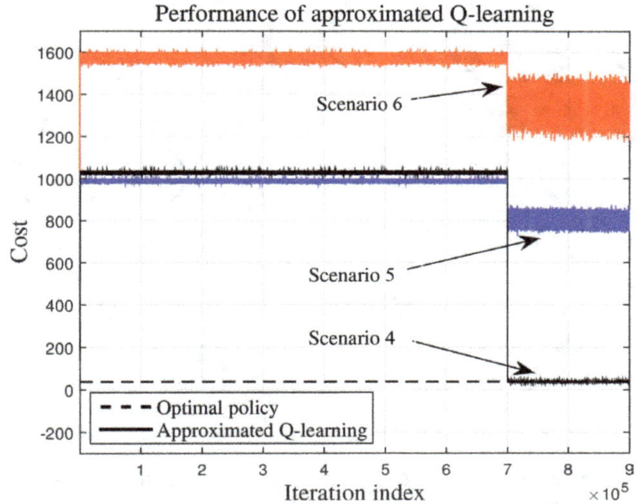

Figure 19.6 *Performance in large state-action space scenario*

recent contributions; see [25, 41]. Designing practical caching policies for such networks involve the following challenges: (i) topology-adaptive caching: caching decisions of a node in a network of caches, influence decisions of all other nodes, and thus a desired caching policy must adapt to the network topology as well as policies of the neighboring nodes; (ii) complex dynamics: content popularities are random, and exhibit unknown space-time, heterogeneous, and often non-stationary dynamics over the entire network; and (iii) large and continuous state-space: because of the sheer size of the available content, caching nodes, and possible realizations of requests, the decision space is huge. These considerations prompted us to address the network caching problem from a reinforcement learning perspective.

19.3 Network caching with space-time popularity dynamics

Consider a two-level network caching, where a parent node is connected to multiple leaf nodes to serve end-user file requests. Indeed, this two-level network constitutes a building block of the popular tree hierarchical cache networks in [27] (see also Figure 19.7). To model the interactions between caching decisions of parent and leaf nodes along with the space-time evolution of file requests, we develop a scalable deep reinforcement learning approach based on a hyper DQN implementation developed in our prior works [42, 43]. The objective is to endow the caching policy of the parent node with the capability of adapting itself to local policies of leaf nodes and space-time evolution of file requests.

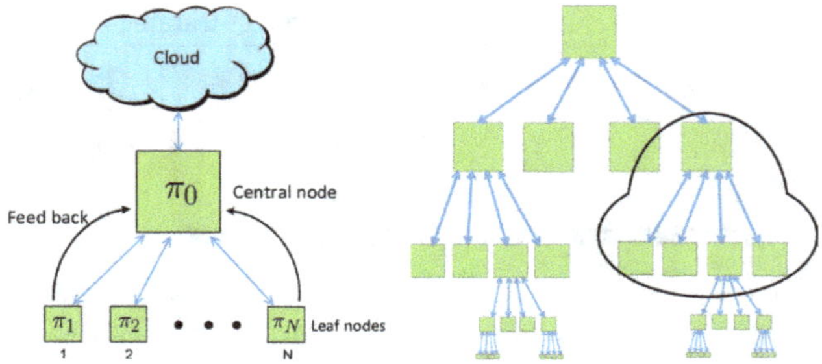

Figure 19.7 The considered network of caching nodes (left), and a hierarchical tree network of caches (right)

19.3.1 Modeling and problem statement

Consider the two-level network of interconnected caching nodes in Figure 19.7 (left), where a parent node is connected to several leaf nodes indexed by $n \in \mathcal{N} := \{1, \ldots, N\}$. The parent node is also connected to the cloud through a congested backhaul link. One could consider this network as part of a larger hierarchical caching system, where the parent node is connected to a higher level caching node instead of the cloud, as depicted in Figure 19.7 (right); see also [44, 45].

All nodes in this network store files to serve file requests. Every leaf node serves its locally connected end-users, by providing their requested files. If a requested content is available locally at a leaf node, the content will be served immediately at no cost. If it is not available locally due to limited caching capacity, the content will be fetched from its parent node at a certain cost. Similarly, if the file is available at the parent node, it will be served to the leaf at no cost; otherwise, the file must be fetched from the cloud at a higher cost.

Every leaf node follows its local caching policy to store popular files locally, while the parent node should globally store the popular files that are typically not stored by the leaf nodes. Leaf nodes are closer to end-users; therefore, they frequently receive user file requests that exhibit rapid temporal evolution at a *fast* timescale. In contrast, the parent node observes an aggregate of requests over a large number of users that are not served by the N leaf nodes, which naturally exhibit smaller fluctuations and thus evolve at a *slow* timescale.

Building on this observation, we consider a two-timescale approach to managing such a tree network of caching nodes. To that end, let $\tau = 1, 2, \ldots$ denote the slow time intervals, each of which is further divided into T fast time slots indexed by $t = 1, \ldots, T$; see Figure 19.8 for a depiction. We assume that the network state remains unchanged during each fast time slot t, but can change from t to $t+1$. Suppose that the F files in the cloud are collected in the set $\mathcal{F} = \{1, \ldots, F\}$. At the beginning of each slot $\boldsymbol{a}_n(t+1, \tau) = \pi_n(\boldsymbol{s}_n(t, \tau))$, every leaf node n selects a subset of files in $\mathcal{F}$ to prefetch and store for possible use in this slot. To determine which files to store, every leaf node relies on a local caching policy function π_n, to take action $\boldsymbol{a}_n(t+1, \tau) = \pi_n(\boldsymbol{s}_n(t, \tau))$ at the beginning of slot $t+1$, based on its *state* vector $\boldsymbol{s}_n$ at

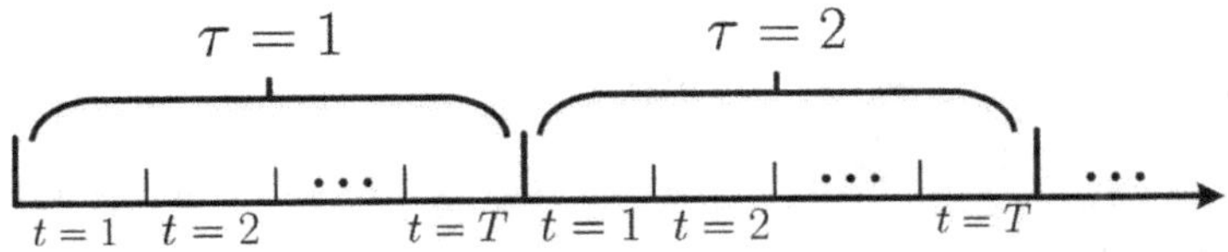

Figure 19.8 Slow and fast time slots

the end of slot t. We define the state vector $\boldsymbol{s}_n(t,\tau) := \boldsymbol{r}_n(t,\tau) := [r_n^1(t,\tau) \cdots r_n^F(t,\tau)]^\top$ to collect the number of requests received at leaf node n for individual files over the duration of slot t on interval τ. Likewise, to serve file requests that have not been served by leaf nodes, the parent node takes action $\boldsymbol{a}_0(\tau)$ to store files at the beginning of every interval τ, according to a certain policy π_0.

19.3.2 Two-timescale problem formulation

File transmission over links consumes resources, including energy, time, and bandwidth. Among possible choices, we consider the following cost for node $n \in \mathcal{N}$, at slot $t + 1$ of interval τ to serve a request

$$\begin{aligned} \boldsymbol{c}_n(\pi_n(\boldsymbol{s}_n(t,\tau)), \boldsymbol{r}_n(t+1,\tau), \boldsymbol{a}_0(\tau)) &:= \boldsymbol{r}_n(t+1,\tau) \odot (\mathbf{1} - \boldsymbol{a}_0(\tau)) \\ &\odot (\mathbf{1} - \boldsymbol{a}_n(t+1,\tau)) + \boldsymbol{r}_n(t+1,\tau) \odot (\mathbf{1} - \boldsymbol{a}_n(t+1,\tau)) \end{aligned} \tag{19.33}$$

where $\boldsymbol{c}_n(\cdot) := [c_n^1(\cdot) \; \cdots \; c_n^F(\cdot)]^\top$ concatenates the cost for serving individual files per node n; symbol $\odot$ denotes entry-wise vector multiplication; entries of $\boldsymbol{a}_0$ and $\boldsymbol{a}_n$ are either 1 (cache, hence no need to fetch), or 0 (no-cache, hence fetch); and $\mathbf{1}$ stands for the all-one vector. Specifically, the second term in (19.27) is the cost of the leaf node fetching files for end-users, while the first term is that of the parent fetching from the cloud.

We again model file requests as Markov processes with unknown transition probabilities. During any interval τ, a reasonable caching policy for leaf node $n \in \mathcal{N}$ minimizes the expected cumulative cost; that is,

$$\pi_{n,\tau}^* := \arg\min_{\pi_n \in \Pi_n} \mathbb{E}\Big[\sum_{t=1}^{T} \mathbf{1}^\top \boldsymbol{c}_n(\pi_n(\boldsymbol{s}_n(t,\tau)), \boldsymbol{r}_n(t+1,\tau), \boldsymbol{a}_0(\tau))\Big] \tag{19.34}$$

where n represents the feasible policies for node n. Although solving (19.28) is in general challenging, efficient near-optimal solutions have been introduced [23, 24, 46]. The remainder will thus focus only on designing the caching policy π_0 for the parent node.

To have a caching policy that accounts for both leaf node policies and file popularity dynamics, the parent collects relevant local information from all leaf nodes. At the end of interval τ, each leaf node $n \in \mathcal{N}$ first obtains the time-average state vector $\bar{\boldsymbol{s}}_n(\tau) := (1/T)\sum_{t=1}^{T} \boldsymbol{s}_n(t,\tau)$ and, at the same time, forms and forwards the per-node vector $\bar{\boldsymbol{s}}_n(\tau) \odot (\mathbf{1} - \pi_n(\bar{\boldsymbol{s}}_n(\tau)))$ to its parent node. This vector has average number of file requests received during interval τ, and zero entries if π_n stores the corresponding files. Using the latter, the parent node forms its weighted state vector

$$\mathbf{s}_0(\tau) := \sum_{n=1}^{N} w_n \bar{\mathbf{s}}_n(\tau) \odot (\mathbf{1} - \pi_n(\bar{\mathbf{s}}_n(\tau))) \tag{19.35}$$

with the weights $\{w_n \geq 0\}$. Similarly, having received at the end of interval $\tau + 1$ the slot-averaged costs

$$\begin{aligned} &\bar{c}_n(\pi_0(\mathbf{s}_0(\tau)); \{\mathbf{a}_n(t, \tau+1), \mathbf{r}_n(t, \tau+1)\}_{t=1}^{T}) \\ &= \frac{1}{T} \sum_{t=1}^{T} c_n(\mathbf{a}_n(t, \tau+1), \mathbf{r}_n(t, \tau+1), \mathbf{a}_0(\tau+1)) \end{aligned} \tag{19.36}$$

from all leaf nodes, the parent node computes the cost

$$c_0(\mathbf{s}_0(\tau), \pi_0(\mathbf{s}_0(\tau))) = \sum_{n=1}^{N} w_n \bar{c}_n(\boldsymbol{\pi}_0(\mathbf{s}_0(\tau)); \{\mathbf{a}_n(t, \tau+1), \mathbf{r}_n(t, \tau+1)\}_{t=1}^{T})). \tag{19.37}$$

Having observed $\{\mathbf{s}_0(\tau'), \mathbf{a}_0(\tau'), c_0(\mathbf{s}_0(\tau'-1), \mathbf{a}_0(\tau')\}_{\tau'=1}^{\tau}$, the goal is to find an optimal policy function π_0^* to take a caching action $\mathbf{a}_0(\tau+1) = \pi_0^*(\mathbf{s}_0(\tau))$.

Since requests $\{\mathbf{r}_n(t, \tau)\}_{n,t}$ are Markovian and the present actions $\{\mathbf{a}_n(t, \tau)\}_{n,t}$ affect future costs, the optimal reinforcement learning policy for the parent node minimizes the expected discounted cost overall leaf nodes in the long run, namely

$$\pi_0^* := \arg\min_{\pi_0 \in \Pi_0} \mathbb{E}\Big[\sum_{\tau=1}^{\infty} \gamma^{\tau-1} \mathbf{1}^\top c_0(\mathbf{s}_0(\tau), \pi_0(\mathbf{s}_0(\tau)))\Big] \tag{19.38}$$

where Π_0 is a set of feasible policies.

Clearly, the decision at a given state $\mathbf{s}_0(\tau+1)$, that is, $\mathbf{a}_0(\tau+1) = \pi(\mathbf{s}_0(\tau))$, influences the next state $\mathbf{s}_0(\tau+1)$ through $\pi_{n,\tau}(\cdot)$ in (19.29), as well as the cost $c_0(\cdot)$ in (19.31). Therefore, problem (19.32) is a discounted infinite time horizon Markov decision process (MDP). The ensuing section develops a deep reinforcement learning approach to solve this problem.

19.3.3 Deep reinforcement learning-based caching

Reinforcement learning deals with action-taking policy function estimation in an environment with dynamically evolving states, so as to minimize a long-term cumulative cost. By interacting with the environment (through successive actions and observed states and costs), reinforcement learning seeks a policy function (of states) to draw actions from, in order to minimize the average cumulative cost as in (19.32) [47]. At this point, it is instructive to recall basic notions of MDPs [47, p. 310].

To solve (19.32), let us again define the *value function* to indicate the quality of policy π_0, starting from initial state $c_0(\cdot)$ as

$$V_{\pi_0}(\mathbf{s}_0(0)) := \mathbb{E}\Big[\sum_{\tau=1}^{\infty} \gamma^{\tau-1} \mathbf{1}^\top c_0(\mathbf{s}_0(\tau), \pi_0(\mathbf{s}_0(\tau))) \,|\mathbf{s}_0(0)\Big]. \tag{19.39}$$

For brevity, the time index and the subscript 0 referring to the parent node will be dropped whenever it is clear from the context. To find $V_\pi(\cdot)$, one can rely on the Bellman equation, which basically relates the value of a policy at one state to the values of the remaining states [48, p. 46]. Leveraging (19.33), the Bellman equation for value function $V_\pi(\cdot)$ is given by

$$V_\pi(\mathbf{s}) = \mathbb{E}\left[\mathbf{1}^\top \mathbf{c}(\mathbf{s}, \pi(\mathbf{s})) + \gamma \sum_{\mathbf{s}'} P_{\mathbf{ss}'}^{\pi(\mathbf{s})} V_\pi(\mathbf{s}')\right] \tag{19.40}$$

where the average immediate cost can be found as

$$\mathbb{E}\left[\mathbf{1}^\top \mathbf{c}(\mathbf{s}, \pi(\mathbf{s}))\right] = \sum_{\mathbf{s}'} P_{\mathbf{ss}'}^{\pi(\mathbf{s})} \mathbf{1}^\top \mathbf{c}\left(\mathbf{s}, \pi(\mathbf{s}) | \mathbf{s}'\right) \tag{19.41}$$

and $P_{\mathbf{ss}'}^{\pi(\mathbf{s})}$ is the underlying transition probability of going from current state $\mathbf{s}$ to the next state $\mathbf{s}'$ upon taking action $\pi(\mathbf{s})$.

Using the value function, we can further define the Q-function for policy π as

$$Q_\pi(\mathbf{s}, \mathbf{a}) := \mathbb{E}\left[\mathbf{1}^\top \mathbf{c}(\mathbf{s}, \mathbf{a})\right] + \gamma \sum_{\mathbf{s}'} P_{\mathbf{ss}'}^{\mathbf{a}} V_\pi(\mathbf{s}'). \tag{19.42}$$

As discussed earlier, finding an optimal policy amounts to estimating $Q_{\pi^*}(\mathbf{s}, \mathbf{a})$, which entails estimating a function defined over state and action spaces. In the problem at hand however, the state vector variables are continuous. In this case, (re-) visiting every state-action pair is clearly impossible. To overcome this challenge, we advocate the recently popular methods of deep learning to judiciously generalize from only a few observed pairs. Deep learning approaches [48, 49] have recently demonstrated remarkable performance in diverse applications, such as object detection, speech recognition, and language translation, to just name a few. DNNs are capable of extracting compact low-dimensional features from high-dimensional data. Wedding DNNs with reinforcement learning, we put forward a deep reinforcement learning approach for the network caching problem, which can effectively deal with the "curse of dimensionality."

To this aim, we pursue a parametric model to estimating $Q(\mathbf{s}, \mathbf{a})$ with a DNN. Specifically, we consider a deep feedforward NN having L fully connected layers, with input of $F \times 1$ state vector $\mathbf{s}(\tau)$ as in (19.29), and $F \times 1$ output cost vector $\mathbf{o}(\tau + 1)$ [50]. Each hidden layer $l \in \{2, \ldots, L-1\}$ comprises n_l neurons with ReLU activation functions $h(z) := \max(0, z)$ for $z \in \mathbb{R}$ ([51], for properties of ReLU networks); neurons of the L-th output layer use a softmax non-linearity to yield the f-th entry of $\mathbf{o}$, an estimated long-term cost o_f, if file f is not stored. Having a limited cache capacity at the parent node M ($< F$), the M largest entries of the DQN output $\mathbf{o}(\tau + 1)$ are to be chosen judiciously by the decision module in Figure 19.9 to obtain the action vector $\mathbf{a}(\tau + 1)$. The DNN spits out $\mathbf{o}(\tau + 1)$ as a predicted cost vector for interval $\tau + 1$, based on which the caching decision $\mathbf{a}(\tau + 1)$ is found.

The objective is to train the DQN to find a set of weights collected in the vector $\boldsymbol{\theta}$ that parameterizes the input–output relationship $\mathbf{o}(\tau + 1) = Q(\mathbf{s}(\tau); \boldsymbol{\theta}_\tau)$. To recursively update $\boldsymbol{\theta}_\tau$ to $\boldsymbol{\theta}_{\tau+1}$, consider two successive intervals along with corresponding states $\mathbf{s}(\tau)$ and $\mathbf{s}(\tau + 1)$; the action $\mathbf{a}(\tau + 1)$ taken at the beginning of interval $\tau + 1$; and, the cost $\mathbf{c}(\tau + 1)$ revealed at the end of interval $\tau + 1$. The instantaneous approximation of the optimal cost-to-go from interval $\tau + 1$ is given by

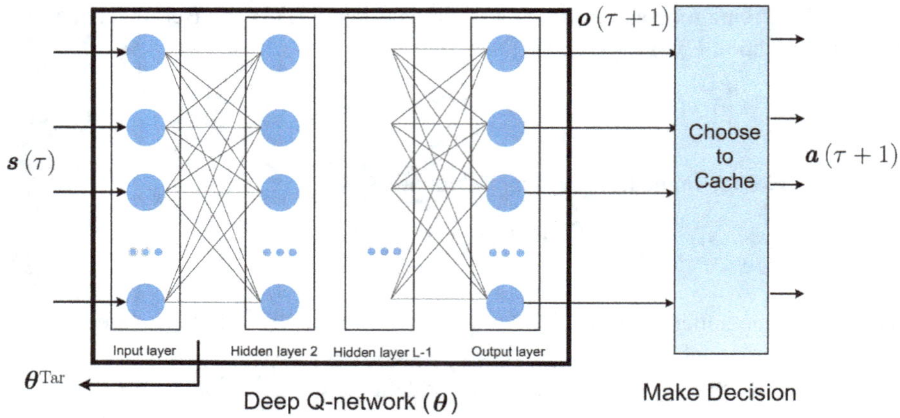

Figure 19.9 *Deep* Q-*network*

$\boldsymbol{c}(\tau+1)+\gamma Q(\boldsymbol{s}(\tau+1);\boldsymbol{\theta}_\tau)$, where $\boldsymbol{c}(\tau+1)$ is the immediate cost, and $Q(\boldsymbol{s}(\tau+1);\boldsymbol{\theta}_\tau)$ represents the predicted cost-to-go from interval $\tau+2$ that is provided by our DQN with $\boldsymbol{\theta}_\tau$, and discounted by γ. Since our DQN offers $Q(\boldsymbol{s}(\tau);\boldsymbol{\theta}_\tau)$ as the predicted cost for interval $\tau+1$, the prediction error of this cost as a function of $\boldsymbol{\theta}_\tau$ is given by

$$\boldsymbol{\delta}(\boldsymbol{\theta}_\tau) := [\overbrace{\boldsymbol{c}(\tau+1)+\gamma Q(\boldsymbol{s}(\tau+1);\boldsymbol{\theta}_\tau)}^{\text{target cost-to-go from interval } \tau+1} - Q(\boldsymbol{s}(\tau);\boldsymbol{\theta}_\tau)] \odot (\mathbf{1}-\boldsymbol{a}(\tau+1)) \tag{19.43}$$

and has non-zero entries for files not stored at interval $\tau+1$. Define for future use the so-termed experience $\boldsymbol{E}_{\tau+1} := [\boldsymbol{s}(\tau), \boldsymbol{a}(\tau+1), \boldsymbol{c}(\tau+1), \boldsymbol{s}(\tau+1)]$; and adopt as loss function the ℓ_2-norm of $\boldsymbol{\delta}(\boldsymbol{\theta}_\tau)$, namely

$$\mathscr{L}(\boldsymbol{\theta}_\tau) = \|\boldsymbol{\delta}(\boldsymbol{\theta}_\tau)\|_2^2 \tag{19.44}$$

that DNN weights will be selected to minimize. As the cardinality of $\boldsymbol{\theta}$ is much smaller than $|\mathscr{S}||\mathscr{A}|$, the DQN is effectively trained with few experiences, and generalizes to unseen state vectors. Unfortunately, the DQN model inaccuracy can propagate in the cost prediction error in (19.37) that can in turn cause instability. This is due to (i) correlated experiences used to update the DQN parameters $\mathrm{mod}(\tau, \mathrm{C}) = 0$; and, (ii) the influence of any changes in policy on subsequent experiences. To handle these challenges, remedies include the so-called *experience replay* and *target network* to update the DQN weights [50]. In experience replay, the parent node stores all past experiences $\boldsymbol{E}_\tau$ in $\mathscr{E} := \{\boldsymbol{E}_1, \ldots, \boldsymbol{E}_\tau\}$, and utilizes a batch of B uniformly sampled experiences from this data set, namely $\{\boldsymbol{E}_{i_\tau}\}_{i=1}^B \sim U(\mathscr{E})$. By sampling and replaying previously observed experiences, experience replay overcomes the two challenges. Moreover, to have target values in (19.37), a second NN (called target network) with structure identical to the DQN is invoked with parameter vector $\boldsymbol{\theta}^{\mathrm{Tar}}$. The $\boldsymbol{\theta}^{\mathrm{Tar}}$ will be periodically replaced by $\boldsymbol{\theta}_\tau$ every C training iterations of the DQN. With a randomly sampled experience $\boldsymbol{E}_{i_\tau} \in \mathscr{E}$, the prediction error with the target cost-to-go estimated using the target network is

Algorithm 3: Deep reinforcement learning for adaptive caching.

Initialize: $\boldsymbol{s}(0)$, $\boldsymbol{s}_n(t,\tau), \forall n$, $\boldsymbol{\theta}_\tau$, and $\boldsymbol{\theta}^{\text{Tar}}$
for $\tau = 1, 2, \ldots$ **do**
    Take action $\boldsymbol{a}(\tau)$ via exploration-exploitation
    $\boldsymbol{a}(\tau) = \begin{cases} \text{Best files via } Q(\boldsymbol{s}(\tau-1); \boldsymbol{\theta}_\tau) & \text{w.p. } 1-\varepsilon_\tau \\ \text{random } \boldsymbol{a} \in \mathscr{A} & \text{w.p. } \varepsilon_\tau \end{cases}$
    for $t = 1, \ldots, T$ **do**
        for $n = 1, \ldots, N$ **do**
            Take action $\boldsymbol{a}_n$ using local policy
            $\boldsymbol{a}_n(t,\tau) = \begin{cases} \pi_n(\boldsymbol{s}(t-1,\tau)) & \text{if } t \neq 1 \\ \pi_n(\boldsymbol{s}(T,\tau-1)) & \text{if } t = 1 \end{cases}$
            Requests $\boldsymbol{r}_n(t,\tau)$ are revealed
            Set $\boldsymbol{s}_n(t,\tau) = \boldsymbol{r}_n(t,\tau)$
            Incur $\boldsymbol{c}_n(\cdot)$, cf. (19.26)
        end
    end
    Leaf nodes
    Set $\bar{\boldsymbol{s}}_n(\tau) := (1/T)\sum_{t=1}^{T} \boldsymbol{s}_n(t,\tau)$
    Send $\bar{\boldsymbol{s}}_n(\tau) \odot (\mathbf{1} - \pi_n(\bar{\boldsymbol{s}}_n))$ to parent node
    Send $\bar{\boldsymbol{c}}_n(\cdot)$ cf. (19.29), to parent node
    Parent node
    Set $\boldsymbol{s}(\tau) := \sum_{n=1}^{N} w_n \bar{\boldsymbol{s}}_n(\tau) \odot (\mathbf{1} - \pi_n(\bar{\boldsymbol{s}}_n))$
    Find $c(s(\tau-1), \boldsymbol{a}(\tau))$
    Save $(\boldsymbol{s}(\tau-1), \boldsymbol{a}(\tau), \boldsymbol{c}(s(\tau-1), \boldsymbol{a}(\tau)), \boldsymbol{s}(\tau))$ in $\mathscr{E}$
    Uniformly sample B experiences from $\mathscr{E}$
    Find $\nabla \mathscr{L}^{\text{Tar}}(\boldsymbol{\theta})$ for these samples, using (19.39)
    Update $\boldsymbol{\theta}_{\tau+1} = \boldsymbol{\theta}_\tau - \beta_\tau \nabla \mathscr{L}^{\text{Tar}}(\boldsymbol{\theta})$
    If $\text{mod}(\tau, \text{C}) = 0$, then update $\boldsymbol{\theta}^{\text{Tar}} = \boldsymbol{\theta}_\tau$
end

$$\boldsymbol{\delta}^{\text{Tar}}(\boldsymbol{\theta}; \boldsymbol{E}_{i_\tau}) := [c(i_\tau+1) + \gamma Q(\boldsymbol{s}(i_\tau+1); \boldsymbol{\theta}^{\text{Tar}}) - Q(\boldsymbol{s}(i_\tau); \boldsymbol{\theta})] \odot (\mathbf{1} - \boldsymbol{a}(i_\tau+1)). \tag{19.45}$$

Different from (19.37), target values here are found through the target network with weights $\boldsymbol{\theta}^{\text{Tar}}$. In addition, the experience used here is randomly drawn from past experiences in $\mathscr{E}$. As a result, a reasonable loss function to be optimized is

$$\mathscr{L}^{\text{Tar}}(\boldsymbol{\theta}) = \mathbb{E} \|\boldsymbol{\delta}^{\text{Tar}}(\boldsymbol{\theta}; \boldsymbol{E})\|_2^2 \tag{19.46}$$

where the expectation is taken with respect to the uniformly sampled experience $\boldsymbol{E}$. In practice however, only a batch of B samples is available to update $\boldsymbol{\theta}_\tau$, so the expectation will be replaced by the sample average. Finally, upon minimizing the loss in (19.39) the DQN weights will be updated. Incorporating these remedies, Algorithm 3 tabulates our deep reinforcement learning-based adaptive caching scheme.

19.3.4 Numerical tests

Here, we present several numerical tests to assess the performance of the proposed deep reinforcement learning-based caching scheme. We consider a parent node with $N = 5$ leaf nodes, where every leaf node implements a local caching policy π_n with

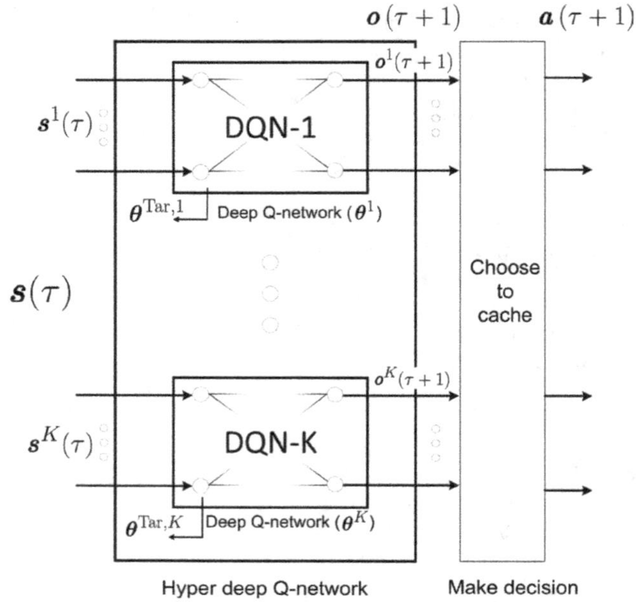

Figure 19.10 Hyper deep Q-network for scalable caching

capacity to store $M_n = 5$ files. Every leaf node receives end-user requests per fast time slot, and each slow-timescale interval contains $T = 2$ slots. Content popularity exhibits different Markovian dynamics locally at leaf nodes. Having no access to local policies $\{\pi_n\}$, the parent node not only should learn file popularities along with their temporal evolutions, but also learn the caching policies of leaf nodes. To endow our approach with scalability to handle $F \gg 1$, we advocate hyper Q-network implementation, where files are first split into K smaller groups of sizes $F_1, \ldots, F_K$ with $F_k \ll F$. This yields the representation $\boldsymbol{s}^\top(\tau) := [\boldsymbol{s}^{1\top}(\tau), \ldots, \boldsymbol{s}^{K\top}(\tau)]$, where $\boldsymbol{s}^k \in \mathbb{R}^{F_k}$. By running K DQNs in parallel, every DQN-k now spits out the associated predicted costs of input files through $\boldsymbol{o}^k(\tau) \in \mathbb{R}^{F^k}$. Concatenating all these outputs, one obtains the predicted output cost vector of all files as $\boldsymbol{o}^\top(\tau+1) := [\boldsymbol{o}^{1\top}(\tau+1), \ldots, \boldsymbol{o}^{K\top}(\tau+1)]$; see Figure 19.10.

The numerical tests consider $F = 1\,000$ files with $K = 25$ and $\{F_k = 20\}$, where only $M_0 = 50$ can be stored. To further assess the performance, we adopt a *non-causal* optimal policy as a benchmark, which assumes having knowledge of future requests and stores the most frequently requested files. In fact, this is the best policy that one may ideally implement. Further, practically used cache schemes including the LRU, LFU, and FIFO [52] are also simulated. A difference between LRU, LFU, FIFO, and our proposed approach is that they refresh the cache whenever a request is received, while our scheme refreshes the cache only at the end of every time slot. By drawing 100 samples randomly from all 2 000 time intervals, the instantaneous reduced cost and the empirical cumulative distribution function (CDF) obtained over these 100 random samples are depicted in Figure 19.11.

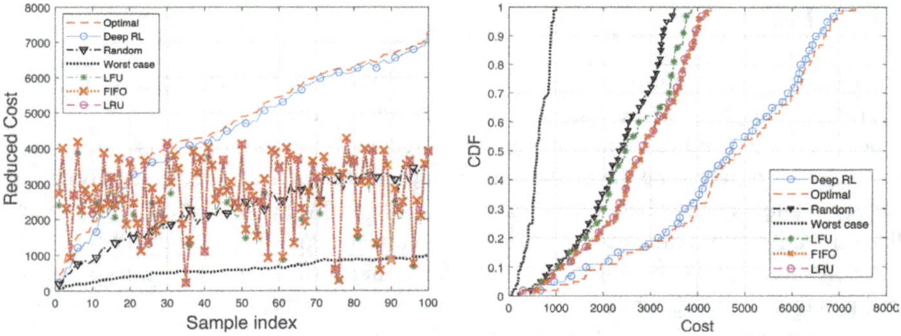

Figure 19.11 *Instantaneous reduced cost for different policies (left); CDF of reduced cost for different policies (right)*

These plots further illustrate how the deep reinforcement learning policy performs relative to the alternatives, and in particular, it approximates closely to the optimal policy. LRU, LFU, and FIFO make caching decisions based on instantaneous observations, and can refresh the cache many times within each slot. Yet, our proposed policy as well as the optimal one here learns from all historical observations to cache, and refreshes the cache only once per slot. Because of this difference, the former policies outperform the latter at the very beginning of Figure 19.11 (left), but they do not adapt to the underlying popularity evolution and are outperformed by our learning-based approach after a number of slots. The merit of our approach is further illustrated by the CDF of the reduced cost depicted in Figure 19.11 (right).

In the last test, the number of leaf nodes is increased from 10 in the previous experiment to $N = 1\,000$. Figure 19.12 (left) and (right) showcases that the deep reinforcement learning performance approaches that of the optimal one as the number of nodes increases. This is likely because the more the leaf nodes are, the smoother the popularity fluctuations become, and the easier it is for deep reinforcement learning to learn the optimal policy.

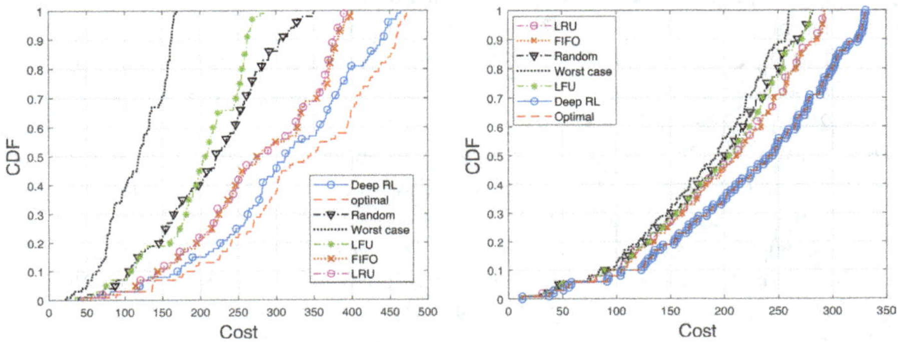

Figure 19.12 *Performance of all policies for* N = *10 nodes (left) and* N = *1 000 leaf nodes (right)*

19.4 Discussion

This chapter has reviewed some recent reinforcement learning-enabled caching approaches for both single- and multi-node settings. After carefully accounting for space-time popularities, conventional reinforcement learning-based policies were introduced. To overcome the challenges arising due to the high dimensionality in the state- or action-spaces, function approximation schemes were developed. In particular, linear function approximation as well as deep neural network-based models were advocated. Although, numerical tests highlighted the merits of the novel approaches, several practical challenges remain in our future research agenda. For instance, bringing computation and communication resources closer to the end-users along with caching will further enhance the network performance. However, jointly optimizing the available resources is a formidable challenge. In this context, innovative resource allocation schemes are needed, where toolkits, e.g., graph-based representations, non-convex optimization schemes, state-of-the-art machine learning methods, statistical signal processing techniques, sparse and low-rank models, only to name a few, can play critical roles to realize the vision of smart networking.

References

[1] Paschos G.S., Iosifidis G., Caire G. 'Cache optimization models and algorithms'. *ArXiv*. 2019:1–127.

[2] Paschos G.S., Iosifidis G., Tao M., Towsley D., Caire G. 'The role of caching in future communication systems and networks'. *IEEE Journal on Selected Areas in Communications*. 2018;**36**(6):1111–25.

[3] Bastug E., Bennis M., Debbah M. 'Living on the edge: the role of proactive caching in 5G wireless networks'. *IEEE Communications Magazine*. 2014;**52**(8):82–9.

[4] Wang X., Han Y., Wang C., Zhao Q., Chen X., Chen M. 'In-edge AI: intelligentizing mobile edge computing, caching and communication by Federated learning'. *IEEE Network*. 2019;**33**(5):156–65.

[5] Liu D., Chen B., Yang C., Molisch A.F. 'Caching at the wireless edge: design aspects, challenges, and future directions'. *IEEE Communications Magazine*. 2016;**54**(9):22–8.

[6] Blasco P., Gündüz D. 'Learning-based optimization of cache content in a small cell base station'. Proceedings of International Conference on Communications; 2014. pp. 1897–903.

[7] Sengupta A., Amuru S., Tandon R. 'Learning distributed caching strategies in small cell networks'. Proceedings of the International Symposium on Wireless Communication Systems; 2014. pp. 917–21.

[8] Müller S., Atan O., van der Schaar M., Klein A. 'Context-aware proactive content caching with service differentiation in wireless networks'. *IEEE Transactions on Wireless Communications*. 2017;**16**(2):1024–36.

[9] Li S., Xu J., van der Schaar M., Li W. 'Trend-aware video caching through online learning'. *IEEE Transactions on Multimedia*. 2016;**18**(12):2503–16.

[10] Leonardi E., Neglia G. 'Implicit coordination of caches in small cell networks under unknown popularity profiles'. *IEEE Journal on Selected Areas in Communications*. 2018;**36**(6):1276–85.

[11] Li J., Shakkottai S., Lui J.C.S., Subramanian V. 'Accurate learning or fast mixing? Dynamic adaptability of caching algorithms'. *IEEE Journal on Selected Areas in Communications*. 2018;**36**(6):1314–30.

[12] Bharath B.N., Nagananda K.G., Poor H.V. 'A learning-based approach to caching in heterogenous small cell networks'. *IEEE Transactions on Communications*. 2016;**64**(4):1674–86.

[13] Hasslinger G., Ntougias K., Hasslinger F., *et al.* 'Performance evaluation for new web caching strategies combining LRU with score based object selection'. *Computer Networks*. 2017;**125**(7):172–86.

[14] Kwak J., Paschos G., Iosifidis G. 'Dynamic cache rental and content caching in elastic wireless CDNs'. International Symposium on Modeling and Optimization in Mobile, Ad Hoc, and Wireless Networks (WiOpt); 2018. pp. 1–8.

[15] Traverso S., Ahmed M., Garetto M., *et al.* 'Temporal locality in today's content caching: Why it matters and how to model it'. *ACM SIGCOMM Comput Commun Rev*. 2013;**43**(5):5–12.

[16] Leconte M., Paschos G., Gkatzikis L. 'Placing dynamic content in caches with small population'. International Conference on Computer Communications, 10-15; 2016. pp. 1–9.

[17] Pedarsani R., Maddah-Ali M.A., Niesen U. 'Online coded caching'. *IEEE/ACM Transactions on Networking*. 2016;**24**(2):836–45.

[18] Sadeghi A., Sheikholeslami F., Marques A.G., Giannakis G.B. 'Reinforcement learning for adaptive caching with dynamic storage pricing'. *IEEE Journal on Selected Areas in Communications*. 2019;**37**(10):2267–81.

[19] Azimi S.M., Simeone O., Sengupta A., Tandon R. 'Online edge caching and wireless delivery in fog-aided networks with dynamic content popularity'. *IEEE Journal on Selected Areas in Communications*. 2018;**36**(6):1189–202.

[20] Sreedevi A.G., Rama Rao T. 'Reinforcement learning algorithm for 5G indoor device-to-device communications'. *Transactions on Emerging Telecommunications Technologies*. 2019;**30**(9):e3670.

[21] Lu L., Jiang Y., Bennis M. Distributed edge caching via reinforcement learning in fog radio access networks. 2019 Vehicular Technology Conference; 2019. pp. 1–6.

[22] Guo B., Zhang X., Sheng Q., Yang H. 'Dueling deep-Q-network based delay-aware cache update policy for mobile users in FOG radio access networks'. *IEEE Access*. 2020;**8**:7131–41.

[23] Sadeghi A., Sheikholeslami F., Giannakis G.B. 'Optimal and scalable caching for 5G using reinforcement learning of space-time popularities'. *IEEE Journal of Selected Topics in Signal Processing*. 2018;**12**(1):180–90.

[24] Somuyiwa S.O., György A., Gunduz D. 'A reinforcement-learning approach to proactive caching in wireless networks'. *IEEE Journal on Selected Areas in Communications*. 2018;**36**(6):1331–44.

[25] Sadeghi A., Wang G., Giannakis G.B. 'Deep reinforcement learning for adaptive caching in hierarchical content delivery networks'. *IEEE Transactions on Cognitive Communications and Networking*. 2019;**5**(4):1024–33.

[26] Zhong C., Gursoy M.C., Velipasalar S. 'A deep reinforcement learning-based framework for content caching'. Conference on Information Sciences and Systems; 2018. pp. 1–6.

[27] Nygren E., Sitaraman R.K., Sun J. 'The Akamai network: a platform for high-performance Internet applications'. *ACM SIGOPS Operating Systems Review*. 2010;**44**(3):2–19.

[28] Maddah-Ali M.A., Niesen U. 'Fundamental limits of caching'. *IEEE Transactions on Information Theory*. 2014;**60**(5):2856–67.

[29] Sutton R.S., Barto A.G. *Reinforcement Learning: An Introduction*. Cambridge, MA, USA: MIT Press; 2016.

[30] Russell S., Norvig P. *Artificial Intelligence: A Modern Approach*. Prentice-Hall, Upper Saddle River, NJ, USA; 2010.

[31] Watkins C.J.C.H., Dayan P. 'Q-learning'. *Machine Learning*. 1992;**8**(3-4):279–92.

[32] Papadimitriou C.H., Tsitsiklis J.N. 'The complexity of Markov decision processes'. *Mathematics of Operations Research*. 1987;**12**(3):441–50.

[33] Wang G., Li B., Giannakis G.B. 'A multistep Lyapunov approach for finite-time analysis of biased stochastic approximation'. *ArXiv*. 2019:190904299.

[34] Sun J., Wang G., Giannakis G.B. 'Finite-sample analysis of decentralized temporal-difference learning with linear function approximation'. Proceedings of 23rd International. Conference on Artificial Intelligence and Statistics; Palermo, Italy; 2020.

[35] Zhang N., Zheng K., Tao M. 'Using grouped linear prediction and accelerated reinforcement learning for online content caching'. 2018 IEEE International Conference on Communications; 2018. pp. 1–6.

[36] Geramifard A. 'A tutorial on linear function approximators for dynamic programming and reinforcement learning'. *Foundations and Trends® in Machine Learning*. 2013;**6**(4):375–451.

[37] Mahadevan S. 'Learning representation and control in Markov decision processes: new frontiers'. *Foundations and Trends® in Machine Learning*. 2008;**1**(4):403–565.

[38] Haj-Ali A., Ahmed N.K., Willke T., *et al.* 'Deep reinforcement learning in system optimization'. *arXiv preprint arXiv*. 2019:190801275.

[39] Hu Y., Jiang Y., Bennis M. 'Distributed edge caching in ultra-dense fog radio access networks: a mean field approach'. 2018 IEEE Vehicular Technology Conference; 2018. pp. 1–6.

[40] Breslau L., Cao P., Fan L. 'Web caching and Zipf-like distributions: evidence and implications'. International Conference on Computer Communications; 1999. pp. 126–34.

[41] Dai J., Hu Z., Li B. 'Collaborative hierarchical caching with dynamic request routing for massive content distribution'. International Conference on Computer Communications; Orlando, FL, USA; 2012. pp. 2444–52.

[42] Sadeghi A., Wang G., Giannakis G.B. 'Deep reinforcement learning for adaptive caching in hierarchical content delivery networks'. *IEEE Transactions on Cognitive Communications and Networking*. 2019;**5**(4):1024–33.

[43] Yang Q., Wang G., Sadeghi A., Giannakis G.B., Sun J. 'Two-timescale voltage control in distribution grids using deep reinforcement learning'. *IEEE Transactions on Smart Grid*. 2019;**11**(3):2313–23.

[44] Ramadan E., Babaie P., Zhang Z.L. 'A framework for evaluating caching policies in a hierarchical network of caches'. IFIP Networking Conference and Workshops, IEEE; 2019. pp. 1–9.

[45] Babaie P., Ramadan E., Zhang Z.L. 'Cache network management using big cache abstraction'. IEEE INFOCOM: IEEE International Conference on Computer Communications IEEE; 2019. pp. 226–34.

[46] Blasco P., Gündüz D. 'Learning-based optimization of cache content in a small cell base station'. IEEE International Conference on Communications; Sydney, Australia; 2014. pp. 1897–903.

[47] Sutton R.S., Barto A.G. *Reinforcement Learning: An Introduction*. Cambridge, MA: MIT Press; 2018.

[48] Sun R. 'Optimization for deep learning: theory and algorithms. arXiv'. 2019;**191208957**.

[49] Zhang L., Wang G., Giannakis G.B. 'Real-time power system state estimation and forecasting via deep unrolled neural networks'. *IEEE Transactions on Signal Processing*. 2019;**67**(15):4069–77.

[50] Mnih V., Kavukcuoglu K., Silver D., *et al.* 'Human-level control through deep reinforcement learning'. *Nature*. 2015;**518**(7540):529–33.

[51] Wang G., Giannakis G.B., Chen J. 'Learning ReLU networks on linearly separable data: algorithm, optimality, and generalization'. *IEEE Transactions on Signal Processing*. 2019;**67**(9):2357–70.

[52] Dan A., Towsley D. 'An approximate analysis of the LRU and FIFO buffer replacement schemes'. *ACM*. 1990;**18**.

Chapter 20

Push-based wireless converged networks with user caching

Zhiyong Chen[1], Kongtao Wang[2], and Hui Liu[3]

With rapid development of the emerging information and communication technologies, e.g., the mobile Internet, the Internet of Things, and social networks, mobile applications continue to be enriched, and the demand for mobile data is showing an explosive growth trend. It is widely expected that the data traffic of mobile devices in 2030 will be 1,000 times the traffic in 2020. Nowadays, the mobile data is mainly delivered through the mobile cellular network, e.g., 4G/5G. However, recent studies have shown that the fast growing traffics would soon become a significant burden on the cellular infrastructure and cause severe congestion in the near future [1].

Although the development of wireless network never seems to be enough to cope with the exponential growth of mobile data, it has been observed that unlike voice and message services, the multimedia service has its unique property of diversity, i.e., a few numbers of popular videos account for a majority of user requests. In fact, there always exist some contents that are much more popular than others. Therefore, among the huge quantity of multimedia contents, only a small portion is accessed frequently by majority of users. Meanwhile, requests for hot videos, although massive, arrive at different times. As a result, wireless broadcasting becomes an efficient way to deliver such contents to unlimited number of wireless users. By contrast, the cellular network, which is optimized for unicasting services, is more suitable for delivering unique information to individual users. A converged network, which combines broadcasting with cellular system, is capable of not only delivering the contents of high interest to all the users simultaneously but also accommodating on-demand requests through unicasting, and thus significantly improves the network capacity [2–4].

Meanwhile, today's smart mobile devices possess a tremendous amount of computing power and storage space. Taking the latest generation of iPhone (iPhone XS), for example, iPhone XS also offers 512 GB in-built storage, similar to the storage size of a computer in 2010. Thus, a converged network can take full advantage

[1]Cooperative Medianet Innovation Center, Shanghai Jiao Tong University, China
[2]Shenzhen Stock Exchange, China
[3]University of Washington, USA

of the user devices' caching resource, where the most popular contents are pushed through broadcasting and then can be stored in the user device [5]. When a particular content is requested by the users, then the users can simply retrieve it from the storage. As a result, no unicast is needed, and the traffics is offloaded from the cellular network [6, 7].

In this chapter, we design a framework of push-based and caching-enabled content delivery in a converged broadcasting and cellular network with user caching, where the most popular contents are pushed through broadcasting to user terminals, while cellular networks accommodate the unique requests. The scheme would offload the multimedia transmission from the cellular network and improve the network capacity. To achieve the capacity of the converged network, we first build a comprehensive model, including the network structure, the content characteristics, and the user requests. With this model, we analyze the transition of the multimedia set and the offloading effect using queuing theory and finally figure out the network capacity expression. We further present an analysis on several important aspects of the converged network: the average downloading waiting time, the system energy consumption, and the user request block rate. The theoretical results imply that the push-based scheme can not only improve the network capacity and decrease the system energy consumption but also provide better user experience (QoS) through eliminating the downloading waiting time. To verify the theoretical analysis, we also build up a simulation platform to imitate the behavior of the converged network. The simulation results validate the theoretical analysis. Both of them reveal that the converged network is superior to cellular-only networks in terms of the network capacity, energy consumption, and user experience.

20.1 Modeling of the converged network

In order to analyze the converged network theoretically, we first perform mathematical modeling on it. The model contains several important aspects of the converged network.

20.1.1 System model

Figure 20.1 shows the system model of the converged network, including a wireless broadcasting system and a cellular system. We assume that mobile users can receive signals from these cellular and broadcasting systems. For the sake of simplicity, we consider statistically identical cells in the network, e.g., the transmission rate of each cellular sub-channel. Besides, we assume that all users within the coverage area can receive the broadcasting service. Therefore, we can focus on the analysis of one cell and the results obtained can be extended to all other cells to measure the system performance.

As shown in Figure 20.1, there are N sub-channels in the cellular network to transmit information for users, and all users in the cell can receive broadcasting data from the broadcasting channel. Let B_b (MHz) and B_c (MHz) be the bandwidth of the wireless broadcasting channel and the cellular sub-channel, respectively. Let C_b(bps/Hz) and C_c (bps/Hz) be the average bandwidth efficiency for the broadcasting channel and the cellular sub-channel, respectively. Thus, we can obtain the transmission capacity of the

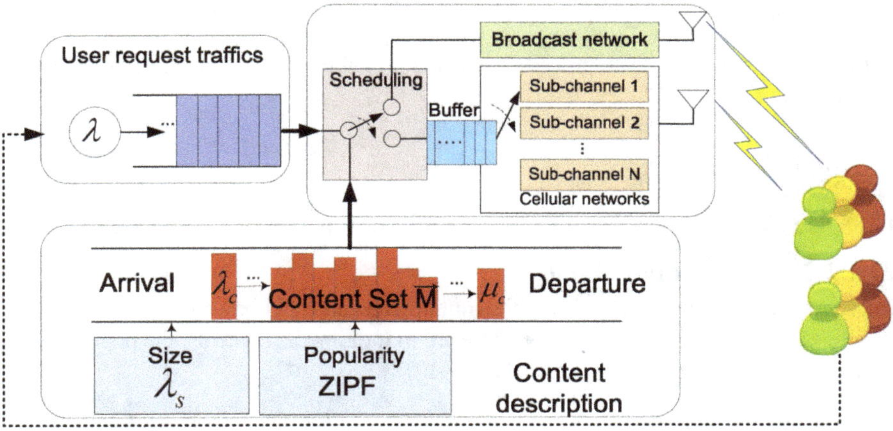

Figure 20.1 Pushing contents through broadcasting in converged networks

broadcasting channel and the cellular sub-channel, namely as $B_b C_b$ and $B_c C_c$ (Mbps), respectively. In addition, the cellular network has B_n buffers to cache the user requests.

As a result, the multimedia contents can be delivered to users in two ways in the converged network: (i) pushing via. the broadcasting system, all users can receive the pushed contents and (ii) transmitting via. the sub-channels of the cellular system. Let $\mathrm{M} = \{\mathrm{M}_1, \ldots, \mathrm{M}_m\}$ be the set of multimedia contents. Here, m denotes the dimension of the content set. Let $\mathrm{M}' = \{\mathrm{M}_{i_1},\ldots,\mathrm{M}_{i_k}\}$, $\{\mathrm{i}_1,\ldots,\mathrm{i}_k\} \subset \{i, \ldots, m\}$ be the multimedia contents that have been delivered by the broadcasting system and cached in the users. When a multimedia content in M′ is requested by a user, the user only needs to obtain this content from the local storage. Therefore, the cellular network does not need to transmit this multimedia content to the user, which is equivalent to offload the multimedia traffic from the cellular network. However, if a content is not delivered completely before a user's request, then the cellular network needs to serve the request by allocating a sub-channel. There are many bases for scheduling multimedia content to be pushed via. the broadcasting system. One of the most common solutions is to push the most popular multimedia contents. In this chapter, the scheduler always updates the contents in M′, so that they are the most popular among M.

20.1.2 Characterization of a multimedia content

First, we introduce three important attributes {life, popularity, size}, which affect the network performance in characterizing a multimedia content.

20.1.2.1 Content life span

We model content creation as a birth-death process with birth rate λ_c and death rate μ_c. As shown in Figure 20.1, we use an exponential random variable to characterize the life span of a content. Thus, we have the probability density function (PDF) as following

$$f(x) = \mu_c e^{-\mu_c x}, x \geq 0. \tag{20.1}$$

For the content as birth-death process, the stable probability of the process staying in the mth state is

$$P_c(m) = \frac{(\lambda_c/\mu_c)^m e^{-\lambda_c/\mu_c}}{m!}. \tag{20.2}$$

20.1.2.2 Content popularity

In general, the popularity statistics of the multimedia contents obey the Zipf distribution, which is a long tail style curve. When there are m multimedia contents, the popularity of the *i*th ranked content can be expressed as [8–10]

$$f_m^i = \frac{1/i^v}{\sum_{j=1}^{m} 1/j^v}, v \geq 0 \tag{20.3}$$

20.1.2.3 Content size

Content size is another important attribute. Considering that the size of the most popular multimedia content is limited, only a small part has a large amount of data, we use a long tail style distribution to approximate the content size distribution. With the broadcasting rate $B_C C_C$, the cumulative distribution function (CDF) of the transmission time for a content using the broadcasting channel is:

$$P\left(\frac{X_s}{B_b C_b} < y\right) = 1 - e^{-\lambda_s B_b C_b y} \tag{20.4}$$

20.1.3 Traffic offloading effect in the converged network

Next, we describe the traffic offloading effect in a converged network. In this chapter, we are interested in how many users can be supported in the converged network. To quantify the effect, we model the user request arrival process in a single cell as a Poisson process with rate λ. We further assume that each user request is associated with only one multimedia content. When the desired content is one of the pushed contents in M', the corresponding user request is offloaded, relieving the transmission burden of cellular networks. Otherwise, the network operator needs to assign an available cellular sub-channel to transmit the desired multimedia content to the particular user.

20.2 Characterization of traffic offloading

20.2.1 Size of M'

Let us define the content life span as L_c, and the time interval between the content arrival time and the broadcast finish time as L_b. If a content is scheduled, we obtain

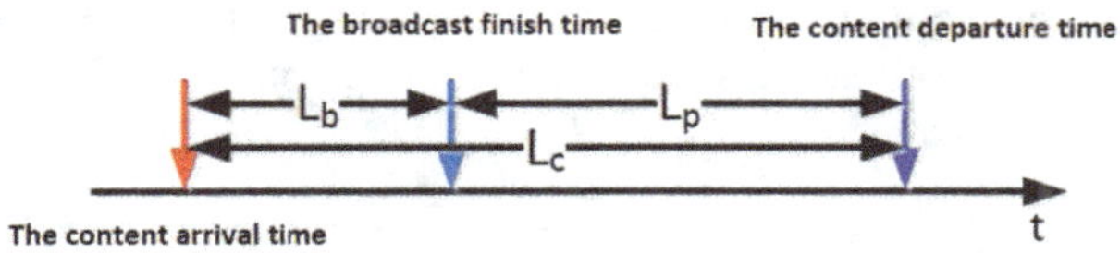

Figure 20.2 Content life span

the residual live time after it is delivered via. the broadcasting system denoted by $L_p = L_c - L_b$, as shown in Figure 20.2.

We have Proposition 1 to show an important property of the residual alive time L_p as following.

Proposition 1: *L_p is a μ_c exponential random variable.*

It can be proved that the probability that a content departs from the content set M before the broadcast finishes is $\theta = \mu_c / (\mu_c + \lambda_s B_b C_b)$, as $\lambda_s B_b C_b >> \mu_c, \theta \to 0$. The effective broadcasting rate becomes $(1 - \theta) B_b C_b = B'_b C_b$. In other words, the broadcasting capacity is not used fully. However, the difference is trivial in the cases where $\theta \to 0$. According to Proposition 1, we conclude that the output of the broadcasting system M′ can also be described by a birth-death process. Specifically, the birth rate is $\lambda_s B'_b C_b$, while the death rate is μ_c, and the maximum state is the size of the content set M: m. Accordingly, the stable probability that the size of M′ is k can be expressed as

$$P_d(k|m) = \frac{(\lambda_s B'_b C_b / \mu_c)^k}{k!} \frac{1}{\sum_{j=0}^{m} \frac{(\lambda_s B'_b C_b / \mu_c)^j}{j!}}. \tag{20.5}$$

20.2.2 Traffic offloading effects

In order to analyze the traffic offloading effect, let us define the following notations.

Notation 1: *$\vec{s} = (m, k)$ represents the situation where the size of M is m and the size of M′ is k.*

Notation 2: *S denotes the set of $\vec{s} = (m, k)$ where $0 \leq k \leq m, 0 \leq m \leq \infty$. Therefore, $\vec{s} \in S$.*

Notation 3: *$P_s(\vec{s})$ is the probability of $\vec{s}$.*

Notation 4: *$P_{off}(\vec{s})$ denotes the probability that the user request is offloaded in the situation of $\vec{s}$.*

As shown in the previous sections, M and M′ can all be described by the birth-death process. Consider the general situation of $\vec{s} = (m, k)$. The probability $P_s(\vec{s})$ is the product of the probability that the size of M is m and the size of M′ is k. Therefore, it can be expressed as

$$P_s(\vec{s}) = P_c(m) P_d(k|m) = \frac{(\lambda_c / \mu_c)^m e^{-\lambda_c/\mu_c}}{m!} \frac{\frac{(\lambda_s B'_b C_b / \mu_c)^k}{k!}}{\sum_{j=0}^{m} \frac{(\lambda_s B'_b C_b / \mu_c)^j}{j!}}. \tag{20.6}$$

In this situation, as the user request arrives and requires a content, the probability that the user request is offloaded is the sum of Zipf's popularity of the k pushed contents, and it can be written as:

$$P_{off}\left(\vec{s}\right) = \sum_{i=1}^{k} f_m^i = \sum_{i=1}^{k} \frac{1/i^v}{\sum_{j=1}^{m} 1/j^v}. \tag{20.7}$$

In cases of k or m is equal to *0*, we have $P_{off}\left(\vec{s}\right) = 0$, although in general the probability of these scenarios is negligible. Therefore,

$$P_{off}(\vec{s}) = \begin{cases} \sum_{i=1}^{k} \frac{1/i^v}{\sum_{j=1}^{m} 1/j^v}, & for\ m > 0, k > 0 \\ 0, & otherwise \end{cases} . \tag{20.8}$$

Based on the results regarding $P_s\left(\vec{s}\right)$ and $P_{off}\left(\vec{s}\right)$, we can now articulate the offloaded traffic characteristics. Consider the time interval $T\left(T \to \infty\right)$, during which we have a set of request arrival moments $\{t_i\}, i = 1, ..., a + b$, of the user request process. Thus, there are $N_r = a + b\ \left(a > 0, b > 0\right)$ user requests. Correspondingly, let us denote $\left(\vec{s}_{t_1}, \vec{s}_{t_2}, ..., \vec{s}_{t_{a+b}}\right)$ at moment $\left(t_1, t_2, ..., t_{a+b}\right)$, and the offloading probabilities are $\left(P_{off}\left(\vec{s}_{t_1}\right), P_{off}\left(\vec{s}_{t_2}\right), ..., P_{off}\left(\vec{s}_{t_{a+b}}\right)\right)$. Denote the number of requests offloaded as N_b, and the number of requests that enter into the cellular network as N_c. To obtain further insight, we provide the probability of $N_b = a$ and $N_c = b$ in the following theorem.

Theorem 1: The probability that $N_b = a$ and $N_c = b$ in time interval T is

$$P\left(N_b = a, N_c = b\right) = \frac{e^{-\lambda O_b T}\left(\lambda O_b T\right)^a}{a!} \frac{e^{-\lambda O_c T}\left(\lambda O_c T\right)^b}{b!} \tag{20.9}$$

where $N_b = a$ and $O_c = \sum_{\vec{s} \in \mathbf{S}} P_s\left(\vec{s}\right)\left(1 - P_{off}\left(\vec{s}\right)\right)$.

Theorem 1 implies that the offloaded traffics is Poisson process with parameter $O_b\lambda$, while the traffic entering into the cellular network is Poisson process with parameter $O_c\lambda = \left(1 - O_b\right)\lambda$.

20.3 Performance analysis of the converged network

20.3.1 The capacity improvement of the converged network

The network capacity is defined as the number of users supported subject to a given block rate P_b, for both the converged network and the cellular-only network. The block rate is generally defined as

$$P_b = \frac{R_b}{R_t} \tag{20.10}$$

where R_b and R_t denote the number of blocked requests and total requests, respectively. Given a pre-defined block rate P_b, we can calculate the capacity improvement of the converged network in the following proposition.
Proposition 2: The network capacity of the converged network is I_c times that of the cellular-only network,

$$I_c = \begin{cases} \infty, \text{ for } \mathrm{P_b} > \mathrm{O_c} \\ I_b/O_c, \text{ otherwise} \end{cases}. \tag{20.11}$$

Here, I_b can be found through the following steps:

1. Determine the traffic intensity λ_1 at which the block rate is P_b in the cellular-only network;
2. Improve P_b by $1/O_c$ times to be P_b/O_c;
3. Determine the traffic intensity λ_2 at which the block rate is P_b/O_c in the cellular-only network;
4. $I_b = \lambda_2/\lambda_1$.

The more the traffics are offloaded, the smaller the O_c becomes and the greater the network capacity improvement is. Notice that since the offloaded traffic remains to be Poisson process and the service time of the cellular sub-channel is of exponential distribution, it is straightforward to use the Erlang C (finite capacity) formula to get the block rate, even in the converged network.

20.3.2 The average downloading time $\mathbb{E}[T_w]$

In the push-based converged network, the waiting time of the user request associated with a pushed content is eliminated since no transmission from the multimedia server through the cellular sub-channels is needed. Only those contents that are not pushed require unicast downloading through the cellular system. Considering the downloading time elimination effect, we define the request waiting time for a content M in the converged network T_w as

$$T_w(M) = \begin{cases} 0, \text{ for } \mathrm{M} \in \mathrm{M'} \\ T_c(M), \text{ for } \mathrm{M} \notin \mathrm{M'} \end{cases} \tag{20.12}$$

where $T_c(M)$ denotes the transmission time in the cellular network. The average downloading time is then given by

$$\mathbb{E}[T_w] = P(M \in M') \times 0 + P(M \notin M') \times \mathbb{E}[X_s/B_cC_c] = P(M \notin M')\,\mathbb{E}[T_c]. \tag{20.13}$$

In this analysis, we have shown that the probability that the user request cannot be offloaded by the pushed contents is O_c. Therefore, we have $P(M \notin M') = O_c$. It can be proved that

$$\mathbb{E}[T_c] = \frac{1}{\lambda_s B_c C_c} + \pi_0 \frac{\rho(N\rho)^N}{\lambda(1-\rho)^2} N! \tag{20.14}$$

$$\pi_0 = \left(\sum_{k=0}^{N} \frac{E^k}{k!} + \frac{N^N}{N!} \sum_{k=N+1}^{N+B_n} \frac{E^k}{N^k} \right)^{-1} \tag{20.15}$$

where $\rho = \frac{\lambda}{N\lambda_s B_c C_c}$, $E = \frac{\lambda}{\lambda_s B_c C_c}$, N is the number of cellular sub-channels, and B_n is the number of buffers. Therefore, the average downloading time in the converged network is $\mathbb{E}\left[T_w\right] = O_c \left(\frac{1}{\lambda_s B_c C_c} + \pi_0 \frac{\rho (N\rho)^N}{\lambda (1-\rho)^2 N!} \right)$.

The expression reflects the downloading time reduction effect in the converged network. The QoS is improved by reducing the average service waiting time $\mathbb{E}\left[T_w\right]$. As the majority of the requests can be accommodated immediately, the user experience is enhanced significantly.

20.3.3 System energy consumption

To evaluate the energy consumption tradeoffs, we consider the large scale path-loss and fast fading model for both the broadcasting and unicasting: $P_r = P_t |h|^2 / r^n$, where P_r and P_t denote the received power density at the user terminals and the transmission power density of the station, respectively, r is the distance between one transmitter and the users, n and $|h|^2$ reflect the large scale path-loss effect and the fast fading effect, respectively. Under this assumption, the bandwidth efficiency C can be expressed as

$$C = \mathbb{E}_h \left\{ log_2 \left(1 + \frac{P_t |h|^2}{r^n N_0} \right) \right\}, \tag{20.16}$$

where N_0 is the power density of additive white Gaussian noise (AWGN).

We consider that the broadcasting system covers a large service area with radius r_b km. For unicasting, cells in the cellular network with radius of r_b km are deployed to fill up the broadcasting area. In other words, the broadcasting system has a coverage area of around r_b^2 / r_c^2 cells. In the broadcasting system, rather than adjusting the transmission power for a single user, the transmitter aims at covering users even at the edge of its service area.

As for the cells, we assume that the users are distributed with PDF $f_r(r) = 2r/r_c^2$, where $0 \leq r \leq r_c$. The base station can adjust its transmission power for users to achieve the requested data rate.

In the following, we denote the transmission power of the broadcasting as P_{broad}, and the power of a cellular sub-channel as P_{sub}. P_{broad} is chosen so that the broadcasting bandwidth efficiency is C_b bps/Hz within the coverage area. The cellular transmitter consumes the power of P_{sub} to deliver C_c bps/Hz at all distance r. The average power of a sub-channel within a cell is considered as the power consumption of a cellular sub-channel. Since there is no explicit expression for P_{broad} and P_{sub}, they are determined through extensive numerical simulations.

Under these assumptions, the average power consumption for the cellular-only network and the converged network in a long period T is given by

$$\bar{P}_{cell} = \frac{E_{cell}(T)}{T}, \bar{P}_{con} = \frac{E_{con}(T)}{T}, \tag{20.17}$$

where $E_{cell}(T)$ is the energy consumption of the cellular-only network in the period of T, it consists of the energy consumption of the overall r_b^2/r_c^2 base stations; $E_{con}(T)$ is the energy consumption of the converged network, which is made of the energy consumption of the broadcasting tower and the overall r_b^2/r_c^2 base stations. The following propositions give the average power consumption of the cellular-only network and the converged networks.

Proposition 3: The average power consumption of the cellular-only network:

$$\bar{P}_{cell} = \frac{r_b^2}{r_c^2} \frac{P_{sub}\lambda\left(1 - P_b\left(\lambda\right)\right)}{\lambda_s B_c C_c}, \tag{20.18}$$

where $P_b(\lambda)$ is the request block rate at λ.

Proposition 4: The average power consumption of the converged network:

$$\bar{P}_{con} = P_{broad} + \frac{r_b^2}{r_c^2} \frac{\%P_{sub}O_c\lambda\left(1 - P_b\left(O_c\lambda\right)\right)}{\lambda_s B_c C_c}, \tag{20.19}$$

where $P_b(O_c\lambda)$ is the request block rate at $O_c\lambda$.

Accordingly, the average power consumption of both networks is a function of the user request arrival rate λ. Equation in Proposition 4 provides a straightforward way to understand the power consumption in the converged network. We observe that $\bar{P}_{con}$ is made up of two parts: the power of the broadcasting system P_{broad} and the latter cellular part. Despite the extra power consumption in broadcasting, the pushing process relieves the transmission burden of the cellular sub-channels, which in turn saves energy. Therefore, the power consumption in the converged network is not always greater than the cellular-only network. We will show with numerical examples that in the scenario of quick request arrivals, the converged network consumes much less energy than the cellular-only network.

20.3.4 Extreme case analysis for the converged network

Several special cases are now investigated to provide further insight into the converged network behaviors, including the traffic offloading effect, the downloading time reduction effect and the total system power consumption.

1. $B_b \to 0$. In the scenario where the broadcasting bandwidth $B_b \to 0$, or the content pushing capability of the broadcasting system is extremely low, it can be proved that

$$P_d\left(k|m\right) = \begin{cases} 1, \textit{ for } k = 0, \\ 0, \textit{ otherwise.} \end{cases} \tag{20.20}$$

This means that no content can be pushed. The expression of \$O_{b}\$ can be simplified to

$$O_b = \sum_{m,k=0} P_c(m)P_{off}(m, 0) = 0. \tag{20.21}$$

In the scenario, the converged network degrades into the cellular-only network, thus all the user requests would be carried through the cellular sub-channels. Thus, $O_c = 1 - O_b = 1$, then

$$\mathbb{E}\left(T_w\right) = \frac{1}{\lambda_s B_c C_c} + \pi_0 \frac{\rho\left(N\rho\right)^N}{\lambda\left(1-\rho\right)^2 N!}, \; \bar{P}_{con} = \bar{P}_{cell}. \tag{20.22}$$

2. $B_b \to \infty$. It is a scenario where the broadcasting system possesses great content push capability. Based on the assumption, we can prove that when $m \geq 1$

$$P_d(k|m) = \begin{cases} 0, & for\ 0 \leq k \leq m-1 \\ 1, & for\ k=m \end{cases} \tag{20.23}$$

The simplified equation implies that as the broadcasting bandwidth B_b becomes large enough, all the multimedia contents can be pushed through broadcasting. Further, O_b is rewritten as

$$O_b = \sum_{m,k=m} P_c(m) P_{off}(m, m) = 1. \tag{20.24}$$

We can observe that all the user traffics are offloaded, i.e., $O_b = 1$. Therefore, in the scenario $B_b \to \infty$, the converged network turns into a pure content push broadcasting system. It can be shown that

$$\mathbb{E}\left(T_w\right) = 0, \; \bar{P}_{con} = \frac{P_{broad}\lambda_c}{\lambda_s C_b}. \tag{20.25}$$

This case reveals that great pushing capability is able to satisfy all the user requests at the cost of enormous bandwidth occupation, which is not feasible in practice. Given a limited bandwidth, the better strategy is to push the most popular contents through broadcasting while the cellular system accommodates the rest by delivering the remaining parts.

20.4 Numerical results

In this section, we evaluate the performance of the proposed converged network by simulations. The default parameter settings are listed in Table 20.1.

We first evaluate the accuracy of the analytical expressions. We develop a simulation platform according to the system model on Matlab to emulate the behavior of the converged network. Based on the platform, intensive simulations are performed, from which we obtain statistics about the network capacity improvement. As shown in Figure 20.3, the theoretical results coincide exactly with the simulation results. The impact of the content characteristics, the network resource and the user request traffics on the system performance is then discussed as following.

20.4.1 Impact of the network resources

From Figure 20.3, we can observe that the converged network can significantly support more users than that of the cellular-only network. Besides, Figure 20.3 demonstrates the number of users supported in the converged network increases

Table 20.1 Default parameter settings

Parameters	Values
Broadcasting channel bandwidth: B_b	1 (MHz)
Broadcasting radius: r_b	25 (km)
Cellular sub-channel bandwidth: B_c	1 (MHz)
Number of buffers: B_n	20
Number of cellular subchannels: N	20
Cell radius: r_c	0.5 (km)
Broadcast bandwidth efficiency: C_b	1 (bps/Hz)
Cellular bandwidth efficiency: C_c	1 (bps/Hz)
Path-loss exponent: n	
Content birth rate: λ_c	2 500/(3 600*12) (content/s)
Content death rate: μ_c	1/(3 600*12) (1/s)
Content size parameter: λ_s	1/120 (content/Mbits)
Zipf constant: v	0.5, 1
Tolerable block rate	2%
Average user active time: E_u	60 (min)

in an exponential fashion with the broadcasting bandwidth B_b. As shown in Figures 20.4 and 20.5, we show the impact of the network resources, such as the broadcasting bandwidth B_b and the cellular sub-channel bandwidth B_c, on the network performance. We can see from Figure 20.4 that the average downloading time decreases with the increasement of B_b.

20.4.2 Impact of the content characteristics

The content characteristics, defined by three attributes $\{L_c, f^i_m, X_s\}$, are essential parts of the system model. Here, the impact of the content attributes on the network performance is evaluated individually. By varying the values of the content update parameters λ_c, the size parameter λ_s, the effect of the content characteristics on the network capacity is quantified, as shown in Figure 20.6.

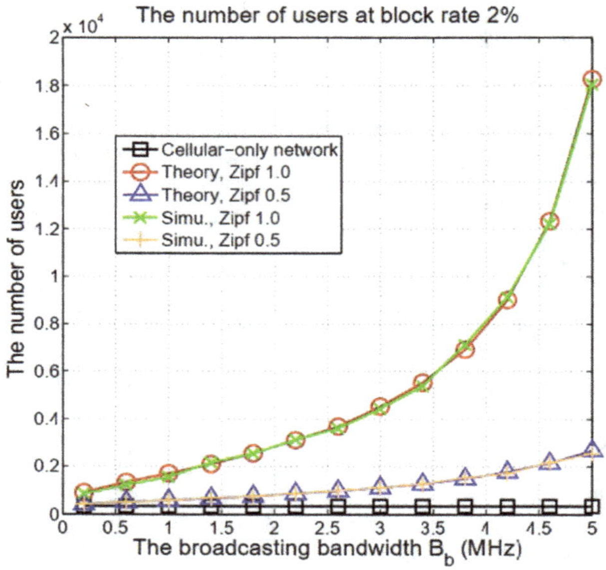

Figure 20.3 *The number of users supported in a cell of the converged network and the cellular-only network*

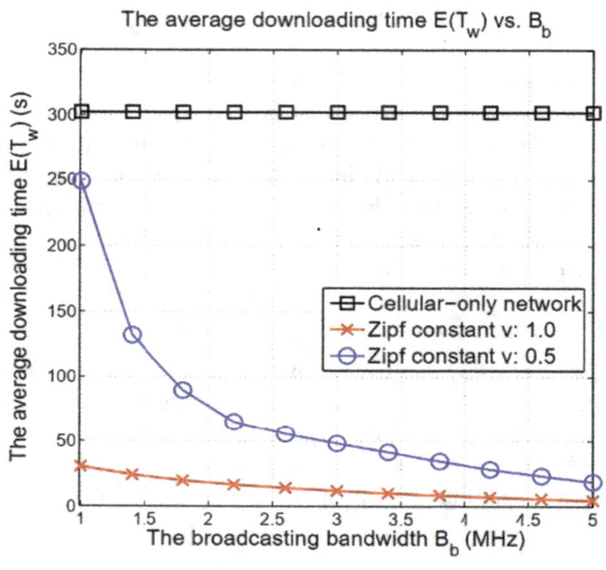

Figure 20.4 *The impact of the broadcasting bandwidth* B_b *on the average downloading time* $\mathbb{E}\left[T_w\right]$ *when* $B_c = 1MH_z$

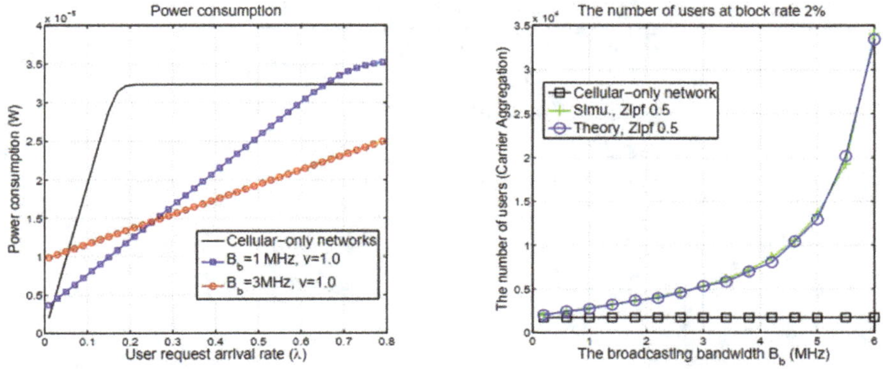

Figure 20.5 *Power consumption comparison for different* B_b *settings and the number of users supported using LTE-A carrier aggregation*

20.4.3 Impact of the user request traffics

To obtain further insight, we compare the block rate performance for the converged network and the cellular-only network by varying the user request traffic parameter λ. The result is shown in Figure 20.7.

20.4.4 User request classification

In the simulation, we consider two cases for the user requests. One is the VIP class and the other is the normal class. We consider that 10% requests are in the VIP class, while the rest are in the normal class. The requests are served separately in the cellular network. We assign 15% bandwidth (divided into 2 channels) and 15% buffer for the VIP service. The remaining 85% bandwidth (divided into 17 channels) and

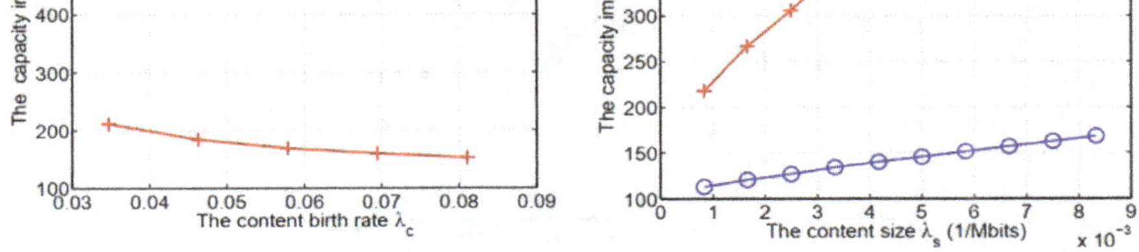

Figure 20.6 *The impact of* λ_c *and* λ_s *on the capacity improvement of the converged network when* B_b = 1 MHz

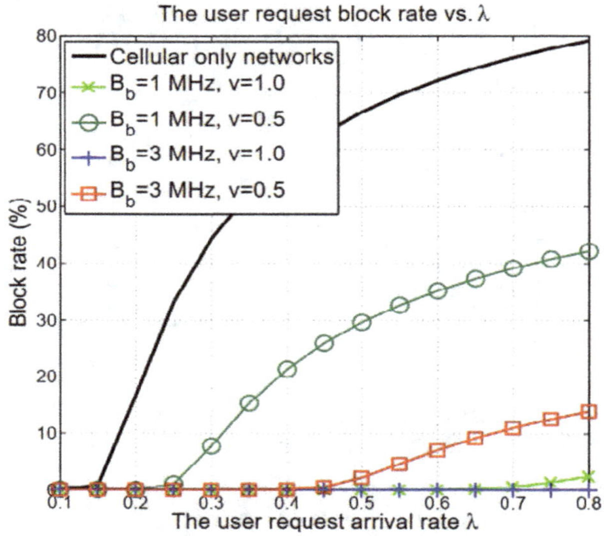

Figure 20.7 *Block rate comparison for the converged network and the cellular-only network when NB_c = 20 MHz*

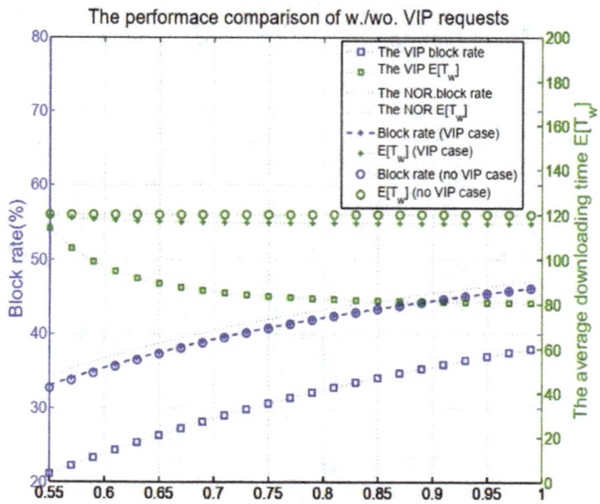

Figure 20.8 *The performance comparison of w/ and w/o VIP requests when* B_b = *3 MHz and* v = 0.5

85% buffers are assigned for the normal request. We evaluate the converged network by using the block rate and the average downloading time, compared with the no-VIP request case. Figure 20.8 shows the simulation results. We can observe that the VIP users obtain better service than the normal users in the case with VIP requests.

20.5 Conclusions

In this chapter, we have conducted theoretical derivation and performance analysis of the converged network. First, we established a mathematical model of the converged network. The model includes multimedia birth and death process, user request arrival, multimedia broadcast scheduling, and so on. Based on the model, in this chapter we have done theoretical analysis and derivation, and obtained the network capacity of the converged network, and then measured the user congestion rate, average download time, and system power consumption. In addition, we built a simulation platform to verify the correctness of the theoretical analysis.

References

[1] Liu H., Chen Z., Qian L. 'The three primary colors of mobile systems'. *IEEE Communications Magazine*. 2016;**54**(9):15–21.

[2] Wang K., Chen Z., Liu H. 'Push-based wireless converged networks for massive multimedia content delivery'. *IEEE transactions on wireless communications*. 2014;**13**(5):2894–905.

[3] Feng H., Chen Z., Liu H. 'Performance analysis of push-based converged networks with limited storage'. *IEEE Transactions on Wireless Communications*. 2016;**15**(12):8154–68.

[4] Sung D.K., Baek S.Y., Hong Y.J. 'Adaptive transmission scheme for mixed multicast and unicast traffic in cellular systems'. *IEEE Transactions on Vehicular Technology*. 2009;**58**(6):2899–908.

[5] Yang C., Yao Y., Chen Z., Xia B. 'Analysis on cache-enabled wireless heterogeneous networks'. *IEEE Transactions on Wireless Communications*. 2016;**15**(1):131–45.

[6] Lykourgiotis A., Birkos K., Dagiuklas T. 'Hybrid broadcast and broadband networks convergence for immersive TV applications'. *IEEE Wireless Communications*. 2014;**21**(3):62–9.

[7] Zhou S., Gong J., Zhou Z., Chen W., Niu Z. 'GreenDelivery: proactive content caching and push with energy-harvesting-based small cells'. *IEEE Communications Magazine*. 2015;**53**(4):142–9.

[8] Maddah-Ali M.A., Niesen U. 'Fundamental limits of caching'. *IEEE Transactions on Information Theory*. 2014;**60**(1):2856–67.

[9] Chen W., Poor H.V. 'Content pushing with request delay information'. *IEEE Transactions on Communications*. 2017;**65**(3):1146–61.

[10] Huang W., Chen W., Poor H.V. 'Request delay-based pricing for proactive caching: a Stackelberg game approach'. *IEEE Transactions on Wireless Communications*. 2017;**18**(6):2903–18.

Chapter 21

On the joint optimization of content caching and recommendations

Livia Elena Chatzieleftheriou[1], Merkourios Karaliopoulos[1], and Iordanis Koutsopoulos[1]

Content caching has been experiencing revived interest within the context of current and next generation wireless networks. It brings content closer to users, decreasing the *aggregate* delivery costs and service delays for network providers. Recommender systems have become integral components of content provision sites. They offer personalized recommendations, increasing the *individual* user satisfaction and engagement with the content provider's platform. Traditionally, caching and recommendation decisions are taken separately. However, there is a recent persistent trend where both network and content providers tend to deploy their own content delivery solutions. In light of this, we explore how the phenomenally conflicting objectives of content caching and recommendation can be jointly addressed.

In this chapter we approach recommender systems as network traffic engineering tools that actively shape content demand to serve both user- and network-centric performance objectives. We introduce a model that captures the coupling between caching decisions and issued recommendations. Based on experimental evidence, we describe the impact of recommendations on user content requests and present a systematic way of engineering the user recommendations. Our viewpoint to recommender systems raises some concerns, not least ethical ones. Hence, we introduce a measure called *preference distortion tolerance* to quantify how much the engineered recommendations distort the original user content preferences. We describe the "recommendation window" as a way to controllably bound the distortion that recommendations will undergo and discuss specific properties that the recommender system should have in order to ensure that the "Quality of Recommendations (QoR)" is kept high by the issued recommendations. We formulate a joint optimization problem for the content to cache and recommend to each user, aiming to maximize the cache hit ratio. The preference distortion tolerance is embedded as a constraint to this problem formulation and marks an equally important dimension for assessing possible solutions. We prove that the problem lacks those properties that would guarantee its approximability, and devise a low-complexity practical

[1]Athens University of Economics and Business, Athens, Greece

algorithm that solves it efficiently. The algorithm is essentially a form of lightweight control over user recommendations, so that the recommended content is both appealing to the end-user and more friendly to the caching system and the network resources. We thoroughly evaluate it, by establishing its main properties and analytical performance bounds. We conduct an extensive sensitivity analysis over various system parameters, both analytically and through simulations with real and synthetic dataset. Next we extend our model and approach for a system in which users can access content from multiple caches, and analyse the intermediate joint caching and user association problem. We then discuss open directions and provide a detailed taxonomy of the related work. Due to space limitations, we do not provide proofs for our results. However, the interested reader may refer to [1–3] for more propositions and their detailed proofs.

21.1 Introduction

Content caching has been experiencing revived interest in recent years within the context of current and next generation wireless cellular networks. The soaring demand for mobile video services pushes caching functionality towards the wireless network edge. Storing popular content at caches close to the user results in enhanced Quality of Experience (QoE) for the end-users and smaller footprints of the bandwidth-demanding mobile video traffic within the wireless network.

On the other hand, the technological advances have totally reshaped the nature of multimedia services. The plethora of content made available to users creates them discomfort: the users either find quickly something of interest, or they abandon the service [4]. Nevertheless, the great variety of immediately available content should turn to users' advantage and recommender systems play a key role to this end.

Recommender systems have become integral components of content provision sites. Their mission is to make personalized recommendations for movies, video clips, music songs or other content items that best match the interests and preferences of individual users. This improves the users' satisfaction and boosts their experience, increasing the number of content downloads and keeping them engaged with the platform. For instance, the Related Video recommendations generate about 30% of the overall views on YouTube [5], whereas the recommender system used by Netflix is considered responsible for about 80% of the hours streamed at Netflix. Moreover, the combined effect of personalization and recommendations saves Netflix more than $1B per year [4].

Typically, the recommendation engine and the caches at the wireless network are owned and managed by different entities. Recommender systems are controlled by content providers through apps that interact with users, whereas the caching infrastructure is typically possessed and controlled by the wireless network operator. Content providers often insert servers storing content within other networks through either their own or third-party Content Delivery Networks (CDNs). In mobile cellular networks, in particular, these servers tend to be placed at their egress nodes rather than at their edge.

However, players with originally distinct roles in the business value chain, such as access network operators and content providers, tend for some time now to deploy their own content delivery solutions, albeit for different reasons. Content providers seek to acquire better control of the network access infrastructure so as to improve the QoE delivered to their subscribers. Netflix Open Connect[a] and Google Global Cache[b] are two widely known examples of CDN solutions owned by content providers. Access network operators, on the other hand, deploy CDNs (telco CDNs) to minimize costs related to the delivery of video traffic through external networks. At the same time, by having storage servers closer to the end-users, telco CDNs can provide their subscribers with faster content access and can gain an advantage over their competitors in the pursuit of customers.

In the case of wireless networks, these trends motivate novel content provisioning scenarios, whereby the coordination of (at least) three different mechanisms can enhance user- and network-centric performance measures. First, *content replication* at the caches of different (small) cells can be used to maximize the locally served demand and optimize the access delays experienced by users, hence their QoE. At the same time, caching reduces the traffic at the backhaul links and maximizes the cache hit ratio. Secondly, *routing user content requests* to different cell caches, it can balance the request load across caches, again improving user QoE but also the network resource usage. Finally, *recommender systems* can be carefully used to *nudge* users towards more network-friendly content request patterns, i.e., they can *shape* content demand for the benefit of the caching mechanism.

Motivation: Recommendations appear, thus, to have interesting repercussions for the design of caching algorithms. Consider, e.g., a cell cache serving the users who are associated with the cell. The rough idea is that the recommender system does not necessarily issue recommendations for content that ranks as the topmost relevant according to the recommendation algorithm; instead, it could recommend to individual users cached content that still matches *adequately* with their preferences and, at the same time, attracts strong demand from many other users. Hence, anticipating that its recommendations affect the content access patterns of users, *the recommender system seeks to gently blunt some of the heterogeneity in users' demands, thus aiming at higher caching efficiency and better users' QoE* (Figure 21.1).

The possibility to route content requests through different (small) cells within reach of the user adds further degrees of freedom to the problem. Besides *which* content to store, the caching decisions concern *where* to store it. Hence, a joint caching and recommendation algorithm could cache and recommend to a user content items that rank second, third, or lower in her preferences, as long as these items are in great demand in at least one of the cells that lie within the coverage of the user. It would instead avoid caching and recommending an item that, say,

[a] openconnect.netflix.com/en/
[b] peering.google.com/

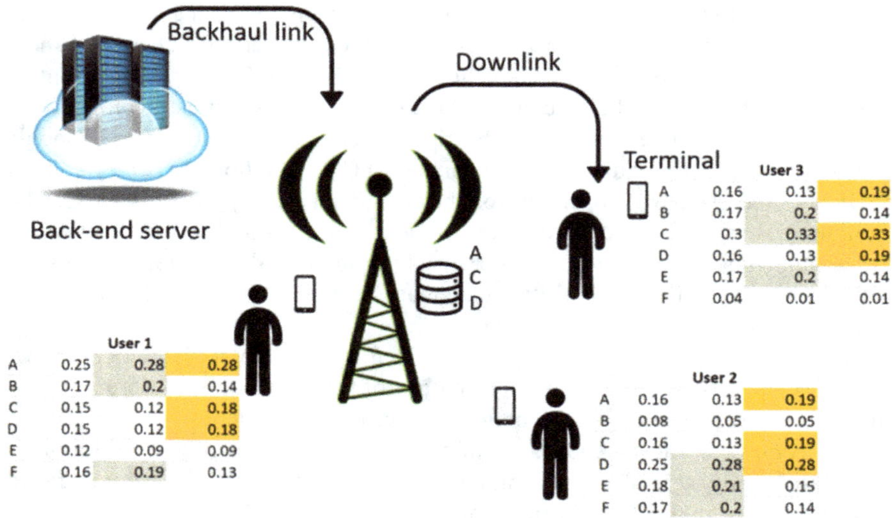

Figure 21.1 *Illustration of the system model and toy example. Caches are co-located with small cells and serve users with content of their preference. The example shows three users with their content preference distributions (2nd column) over six items A–F (1st column). The three items (A, C, D) that attract the highest aggregate preference from all users are locally cached, yielding an expected cache hit ratio of 0.58. Consider a recommender system issuing content recommendations to users and a naive model for their impact: When an item is recommended to a user the preference (demand) for that item is boosted by 0.03. On the contrary, when an item is not recommended to a user, the preference of that user for this item decreases by 0.03. When recommendations are issued for the top-3 items in users' preferences (3rd column), the cache hit ratio drops to 0.55. When recommendations are issued for items that are both cached and of adequate interest to users, yet not necessarily within the top-3 set (4th column), the cache hit ratio increases to 0.67.*

ranks first in the user preferences but is of no interest to other users in the cells the user can associate with.

21.2 System model

We consider a wireless network with a set of small cells and a macrocell that work in conjunction, forming a two-layer heterogeneous network. In current communication networks, macro- and micro-cells are outdoor cells with a typical radius up to a few kilometer and a few hundred meters, respectively.

Table 21.1 Notation table

Symbol	Context	Symbol	Context
S_c	Storage capacity of cache c	M	Number of thematic categories
L_i	Normalized length of item	a_{ui}	Similarity of user u and item
$\mathbf{f}^i$	Feature vector values of content item over the M thematic categories	$\mathbf{f}_u$	Feature vector value of user u over the M thematic categories
p_u^{pref}	Inherent content preference distribution of user u	p_u^{rec}	Probability distribution due to Recommendation
p_u^{req}	Content item request probability distribution of user u (recommended items)	$\widetilde{p}_u^{req}$	Content item request probability distribution of user u (non-recommended items)
R	Number of recommended items	$\mathcal{W}_u$	Recommendation window of user u
K_u	Length of $\mathcal{W}_u$	r_d	Preference distortion tolerance
H_s	Cache hit ratio under scheme S	$\mathcal{P}$	Cache placement
$\mathcal{RC}_u^{in}$	Provisional set of recommended items to u	$\mathcal{RC}_u^{f}$	Final set of recommended items to u
ZD	Zero-distortion scheme	UD	Unbounded-distortion scheme
$CawR$	Caching-aware recommendations Scheme		

21.2.1 Caches, content, users

Our model involves a set of caches $\mathcal{C}$, a catalog of content items (video clips), $\mathcal{I}$, and a set of users, $\mathcal{U}$ (Figure 21.1) (Table 21.1).

Caches: Caches are co-located with wireless network microcells. Each cache $c \in \mathcal{C}$ has limited storage capacity, S_c, which is measured in normalized file size units. At any point in time, it stores a finite set of files, referred to as the cache placement $\mathcal{P}_c$. An additional cache is installed on the macrocell or the cloud, operating as a backend server. This "cache" is assumed to have enough capacity to store copies of the entire catalog.

Content: Content items are relevant to one or more of a set of M thematic categories. The detail and resolution of this categorical separation may vary. Such information may be stored in content metadata in the form of (hierarchical) tags. For example, "soccer" may form a distinct category, but it may also be further split into "English/French/Spanish soccer."

The M thematic categories serve as a feature set that describes items. Namely, each item $i \in \mathcal{I}$ is represented by a feature vector $\mathbf{f}^i$, whose j^{th} element $f^i(j), j \in [1, .., M]$ denotes the score of item i in feature j, i.e., how relevant is item i to thematic category j. These relevance scores assume values in [0,1] and are normalized so that $\sum_{j=1}^{M} f^i(j) = 1, \ \forall i \in \mathcal{I}$.

The content catalog includes content items of different sizes, e.g.,entire movies and their trailers. We denote by $L_i \geq 1$ the normalized size of content item i, i.e., the ratio of its actual size over the smallest item size of the catalog. Indicatively, consider the simple case where files come with only one resolution quality. Then, $l_{min} = 100s$ and $l_{max} = 100min = 6000s$ are typical values for the trailers' and movies' duration so that $L_i = \frac{l_{max}}{l_{min}}$ is in the order of 60[c]. Replicas of each content item may be stored in any subset of the small cell caches, besides the backend cache, depending on the actual caching decisions.

Example 1: Let the feature set be $\mathcal{A}$ = {Biography, Crime, Drama, Comedy, Romance}, and the set of contents $\mathcal{I}$ = {The Godfather, The Irishman, Amélie}. The "Godfather" falls under the Crime and Drama themes, the "Irishman" below Biography, Crime and Drama, and "Amélie" falls under the Comedy and Romance themes. Then, the feature vectors are $\mathbf{f}^{Godfather} = [0, 0.5, 0.5, 0, 0]$, $\mathbf{f}^{Irishman} = [0.33, 0.33, 0.34, 0, 0]$ and $\mathbf{f}^{Amélie} = [0, 0, 0, 0.5, 0.5]$. Alternatively, the feature vectors could be binary, so that a feature assumes the value 1 if a tag is relevant for the movie, and 0, otherwise. For example, the binary feature vector for "Amélie" would be $\mathbf{f}^{Amélie} = [0, 0, 0, 1, 1]$.

Users: At any point in time, each user $u \in \mathcal{U}$ is located within range of a different subset of the network (micro)cells. The user might access different content items through different caches, each time dynamically changing her association depending on the requested content. In section 21.3, we assume that users do not change their association point in the network dynamically in response to content requests, while in section 21.7 we consider user association as an additional (indirect) road for nudging users' preferences on which we have control.

Users are described by similar feature vectors $\mathbf{f}_u \in [0, 1]^M$ as the content items. Each vector element $f_u(j), j \in [1, .., M]$ expresses how much user u is interested in content classified under thematic category j. We normalize these values as well, *i.e.*, $\sum_{j=1}^{M} f_u(j) = 1,\ \forall u \in \mathcal{U}$. Practically, content provision sites draw on the history of users' content downloads, and more broadly their interactions with the site such as possible ratings of content items, to infer these vectors. We elaborate on this in section 21.6.1.

Example 2: Consider the feature set of Example 1 and the set of users $\mathcal{U} = \{u_1, u_2, u_3\}$. A way to quantify the relevance of features for the users is to consider the percentage of tags in the movies they rated. For example: $\mathbf{f}_{u_1} = [0.6, 0.4, 0, 0, 0]$, $\mathbf{f}_{u_2} = [0.2, 0.2, 0.2, 0.2, 0.2]$ and $\mathbf{f}_{u_3} = [0, 0, 0, 0, 1]$.

21.2.2 *Content preferences of users*

The demand for content items is time-varying. It grows for some finite time after the content item first becomes available for download, and then it gradually fades

[c]http://www.stephenfollows.com/long-average-movie-trailer, last accessed on Feb 20, 2019.

out [6, 7]. Our work concerns time scales over which the demand for each item can be considered "fixed," in the order of a few hours within a day [8]. Namely, content demand predictions and caching decisions are made once every such an interval, and user content request patterns change slowly over that interval.

On the user side, we distinguish between inherent content preferences of users and the eventually issued content requests by them. Hence, each user u can be described by a content preference distribution, $p_u^{pref}(i), i \in \mathcal{I}$, with $\sum_{i \in \mathcal{I}} p_u^{pref}(i) = 1$, which captures her original preferences over all items. The preference of user u for item i, $p_u^{pref}(i)$, can be inferred from the feature vectors $\mathbf{f}_u$ and $\mathbf{f}^i$. We use for this purpose the cosine similarity index, a_{ui}, of vectors $\mathbf{f}_u$ and $\mathbf{f}^i$. Note that other similarity measures could also be applicable [9]. Then, the content preference distribution p_u^{pref} over all items is given by:

$$p_u^{pref}(i) = \frac{a_{ui}}{\sum_{i \in J} a_{ui}}, \quad \text{where } a_{ui} = \frac{\sum_{j=1}^{M} \mathbf{f}_u(j) \cdot \mathbf{f}^i(j)}{\sqrt{\sum_{j=1}^{M} (\mathbf{f}_u(j))^2} \sqrt{\sum_{j=1}^{M} (\mathbf{f}^i(j))^2}}. \tag{21.1}$$

Example 3: Consider the item-feature vector of Example 1 and the user-feature vector of Example 2. Then, the users' inherent preferences for the movies "The Godfather," "The Irishman" and "Amélie" are: $p_1^{pref} = [\,0.38, 0.62, 0]$, $p_2^{pref} = [\,0.33, 0.33, 0.33]$, $p_3^{pref} = [\,0, 0, 1]$.

However, the content that users eventually request also depends on the recommendations issued to them, so that the probability with which user u requests item i, $p_u^{req}(i)$, differs from $p_u^{pref}(i)$. We describe our modeling approach to the recommendations' impact in the next paragraph.

21.2.3 The impact of recommendations on user content requests

The system recommendations affect the relative user demand for all content items. In principle, they boost the demand for the recommended items and at the same time proportionately decrease the demand for remaining items. It is less clear how recommender systems *quantitatively* modulate the a priori preferences of a user p_u^{pref} so as to yield the ultimate content request distribution p_u^{req}. Modeling in literature tends to be both intuition- and evidence-driven. In [6], for example, the recommendations are mapped to a new distribution p_u^{rec} over the content items and $p_u^{req}(i)$ is taken to be equal to $\max\{p_u^{pref}(i), p_u^{rec}(i)\}$. On the other hand, there is strong experimental evidence that both the number of recommended items and their order within the recommendation list have a high impact on the a posteriori distribution of the users' demand. For instance, the "Related Video" recommendation lists of YouTube is shown to be the main source of requests for its content, since items ranked higher up in a list of recommended items attract more user interest than items at lower positions [5]. This finding is more relevant for users that access content through small form-factor devices such as mobile phones, in which case it is less convenient to scroll down the whole recommendation list.

Hence, in modeling the impact of recommendations, we make two assumptions:

Assumption 21.2.1. The impact of a recommendation for item i on the demand that is eventually expressed by a user u for this item is a non-increasing function $f()$ of both the item's i position in the recommendation list, $rnk_u(i) \in [1, R]$, and the number of recommended items, R.

In section 21.6.1 we consider specific instances of p_u^{rec} that satisfy this assumption. For a more detailed discussion please refer to section 6.1 in [1].

Assumption 21.2.2. The content request distribution is a convex combination of the two distributions, p_u^{pref} and p_u^{rec} and is given by

$$p_u^{req}(i) = w_u^r \cdot p_u^{rec}(i) + (1 - w_u^r) \cdot p_u^{pref}(i) \tag{21.2}$$

for each of the R items that are recommended to u, and by

$$\widetilde{p}_u^{req}(i) = (1 - w_u^r) \cdot p_u^{pref}(i) \tag{21.3}$$

for each one of the $(|\mathcal{I}| - R)$ items not recommended to user u. The user-specific recommendation weights w_u^r in (21.2) and (21.3) express the importance user u attaches to recommendations in the r-th position of their recommendation list.

Equations (21.2) and (21.3) capture the way recommendations shape content requests that are ultimately issued by user u. The request probabilities are boosted when compared to the initial ones for recommended items, and they decrease for non-recommended ones, so that the resulting content request distribution remains a probability distribution (see Figure 21.2).

21.2.4 Engineering the user recommendations

This capability of recommender systems to shape user demand for content renders them a powerful tool for content demand shaping. This way, recommender systems could be actively used to optimize network-centric performance objectives, in what marks a departure from their nominal user-oriented mission. Our approach to this is summarized in Figure 21.2 and described in what follows.

Assume that the recommender system seeks to recommend R new items to each user u, where R may range from 1 up to a few (e.g., 5–10) items. Instead of issuing recommendations for the top R items in u's content preference distribution p_u^{pref}, the system selects R items among the ones residing within a *recommendation window* $\mathcal{W}_u$ that is defined by the top K_u items, where $K_u > R$, as shown in Figure 21.2.

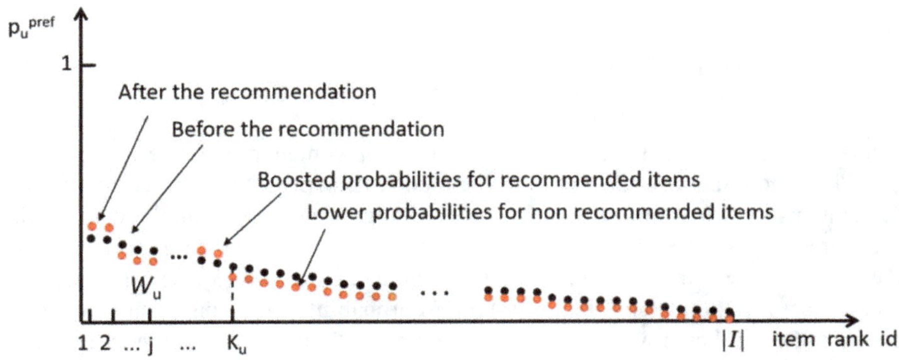

Figure 21.2 *Impact of recommendations on user requests. The content items are ranked in decreasing order of user preference and only items within the recommendation window are recommended. A priori content preference distribution for arbitrary user u, its recommendation window $\mathcal{W}_u$ of size K_u (black), and the resulting content request probability after recommendations (red).*

Namely, the recommender system artificially *inflates* the set of candidate items for recommendation for each user u by a *user-specific* factor K_u/R. When doing so, the system ensures the QoR, preserving two properties addressing the ethical concerns described in section 21.1 regarding the possible manipulation of recommendations:

- It preserves the rank of recommended items in the original user content preferences. If an item i is recommended at higher rank than item j, it holds necessarily that $p_u^{pref}(i) \geq p_u^{pref}(j)$.
- It controllably bounds the distortion that its recommendations introduce to the original user content preferences.

Although the rank of recommended items in the original user content preferences is preserved, it could happen that the absolute rank of file popularities may change after the recommendations. For example, an item that is high in the inherent user's rank and that is not recommended may move lower in her rank than an item that was originally bellow it but has been recommended.

In the worst case, the system will end up recommending items that are ranked in positions $\{K_u - R + 1, K_u - R + 2, \ldots, K_u\}$ in decreasing order of user content preferences (ref. Figure 21.2), instead of the items in top-R positions $\{1, 2, \ldots R\}$. We define the worst-case *user preference distortion* measure, Δ_u to be

$$\Delta_u(K_u, R) = 1 - \frac{\sum\limits_{\substack{j:rnk_u(j)\in\\ [K_u-R+1,K_u]}} p_u^{pref}(j)}{\sum\limits_{j:rnk_u(j)\in[1,R]} p_u^{pref}(j)}. \tag{21.4}$$

where $rnk_u(i), i \in \mathcal{I}$, is the rank of item i in user's u content preferences.

The denominator in (21.4) equals the total request probability of user u for the top R items, which are the ones a typical recommender system would recommend. The numerator, on the other hand, equals to the total request probability of user u for the bottom R items in the recommendation window. Hence, $\Delta_u(K_u, R)$ expresses the worst-case deviation from initial user request probabilities that may result from the choices of our scheme when compared to a typical "honest" recommender system. As such, this metric is admittedly conservative and it denotes an upper bound on the possible distortion of original user request probabilities.

The size of the recommendation window, K_u, introduces an interesting trade-off. Higher K_u values allow for more flexibility in selecting items to recommend to users and shaping their demand in favor of caching efficiency, as it will be seen in the sequel. But at the same time, a higher K_u value may result in higher distortion of user preferences, as can be readily seen in (21.4).

21.3 The joint caching and recommendations problem

In this section we assume that the user associations with network (small) cells depend on the quality of radio signals and the load of the radio network, without taking into account the availability of content at the co-located caches. A direct consequence of this assumption, which is in line with user association practices in current mobile cellular networks, is that the caching decisions can be made independently in each cache-enabled small cell. Any given cache placement $\{P_c : c \in \mathcal{C}\}$ may satisfy a portion of the total demand in the cell as measured by the cache hit ratio[d]

$$H = \frac{\sum\limits_{u\in\mathcal{U}}\sum\limits_{i\in\mathcal{P}} p_u^{req}(i)}{\sum\limits_{u\in\mathcal{U}}\sum\limits_{i\in\mathcal{I}} p_u^{req}(i)}. \tag{21.5}$$

On the user side, the requirement is to maximize the portion of her content requests that can be satisfied by the cell cache (cache hits). This results in lower content access delays and higher user QoE. At the same time, since fewer requests have to be satisfied by the back-end server, the utilization of backhaul links is lower. In other words, the maximization of the cache hit ratio serves both user- and network-oriented objectives.

Formally, let $\{y_i\}$ and $\{x_{ui}\}$, $i \in \mathcal{I}, u \in \mathcal{U}$ be two sets of binary decision variables with $y_i = 1$ when item i is cached and $y_i = 0$, otherwise; and $x_{ui} = 1$ when item

[d]Since in this section the caching decisions are made independently for each cache, we drop the index c from subsequent references to cache placements, i.e., we write $\mathcal{P}$ instead of $\mathcal{P}_c$.

i is recommended to user u and $x_{ui} = 0$ when it is not. The objective of the joint caching and recommendation decisions is then to find the cache and recommendation policy that maximizes the hit ratio, as defined below:

$$\max_{\mathbf{y},\mathbf{x}} \sum_{u \in \mathcal{U}} \sum_{i \in \mathcal{W}_u} y_i (x_{ui} p_u^{req}(i) + (1 - x_{ui}) \widetilde{p}_u^{req}(i)) \tag{21.6}$$

$$s.t. \sum_{i \in \mathcal{I}} y_i L_i \leq S \tag{21.7}$$

$$\sum_{i \in \mathcal{W}_u} x_{ui} = R, \quad \forall u \in \mathcal{U} \tag{21.8}$$

$$y_i, x_{ui} \in \{0, 1\} u \in \mathcal{U}, i \in \mathcal{W}_u. \tag{21.9}$$

In (21.6) and (21.8), $\mathcal{W}_u$ denotes items within the recommendation window of user u. The cardinality of this set is

$$K_u = \max\{k | \Delta_u(k, R) \leq r_d(u)\}, \tag{21.10}$$

where $r_d(u) \in [0, 1)$ denotes the user-specific *preference distortion tolerance*, an upper bound on user preference distortion in (21.4) that should not be exceeded for any user. Our formulation implies that the system could provide users with the opportunity to determine themselves how much distortion tolerance they are willing to tolerate. Inequality (21.7) reflects the cache storage capacity constraint, whereas equalities (21.8) ensure that exactly R items are recommended to every user.

21.3.1 Problem complexity and approximability properties

We refer to the problem (21.6)–(21.9) as the Joint Caching and Recommendations Problem (JCRP). In [1] we prove that:

Proposition 1. The Joint Caching and Recommendations Problem is NP-complete.

As a first step towards an efficient approximation algorithm, we investigate the monotonicity and submodularity of the *JCRP* (*JCRDP*) objective function. We recall the relevant definitions in what follows. Let S be a finite set of elements (interchangeably called universe or ground set) and X, Y any two subsets of S satisfying $X \subseteq Y \subseteq S$. A set function $f: 2^S \rightarrow R$ is called *monotone* if $f(X) \leq f(Y), \forall X, Y \subseteq S$. Moreover, f is called *submodular*, if for any element $e \in S \setminus Y$, it holds that $f_X(e) \geq f_Y(e)$, where $f_A(e) = f(A \cup e) - f(A)$. Namely, the extra marginal benefit from adding an item to a set decreases as this set grows larger. Similarly, f is called supermodular, if for any element $e \in S \setminus Y$, it holds that $f_X(e) \leq f_Y(e)$. In [1] we prove that:

Proposition 2. The objective function of *JCRP* is non monotone.

Proposition 3. The objective function of *JCRP* is neither submodular nor supermodular.

Hence, more general techniques that yield approximability guarantees for optimization problems with (monotone) submodular objective functions, and have been often applied to caching problems, are not applicable to the *JCRP*.

In the following section, we propose an efficient heuristic algorithm for solving the JCRP. Although the algorithm does not lend to rigorous approximability analysis (section 21.4.2), it is computationally simple (section 21.4.1) and exhibits excellent performance (section 21.6).

21.4 An algorithm for the joint caching and recommendation problem

21.4.1 Description of the algorithm

Our Caching-aware Recommendations (CawR) algorithm (see Algorithm 1) proceeds in three steps, as shown in Figure 21.3.

In the first step, a provisional set of recommended items $\mathcal{RC}_u^{in}$ is derived for each user. Input to this step are the content preference probability distributions of users. Recommendations are made for the top-R items in the user preferences but, contrary to what would happen with a typical recommender system, these recommendations are not communicated to the user; they are only relevant as intermediate result of the algorithm's operation.

The second step is the *content placement* step, where we determine which content should be cached. To this end, we compute the content request probabilities according to (21.2), (21.3). All content items are assigned utilities that equal the aggregate request probability they attract:

$$v(i) = \sum_{u \in \mathcal{U}} p_u^{req}(i) \quad i \in \mathcal{I}. \tag{21.11}$$

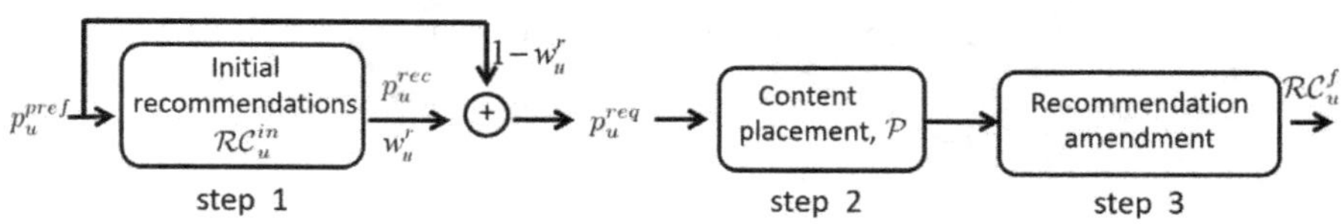

Figure 21.3 Schematic outline of the three-step algorithm for the JCR problem

The optimal placement is then an instance of the 0-1 Knapsack Problem (KP). We use the Dynamic Programming (DP) Fully Polynomial Time Approximation Scheme (FPTAS) algorithm in ([10], §8.2) to obtain an $(1-\varepsilon), \varepsilon > 0$ approximation of the optimal solution; let $\mathcal{P}$ denote this placement.

Finally, in the *recommendation amendment* step, the original recommendations to users are amended so as to maximize the utility (i.e., expected attracted requests) of the cached content. To this end, we first identify for each user u the set of K_u items in her recommendation window from (21.10). Then we compare the item sets $\mathcal{P}$ and $\mathcal{RC}_u^{in}$. Two possibilities exist:

- If $\mathcal{RC}_u^{in} \subseteq \mathcal{P}$, then the original recommendations derived in the initialization step remain intact (and the resulting user preference distortion is zero).
- If $|\mathcal{RC}_u^{in} \bigcap \mathcal{P}| = F_1 \in [0, R-1]$, then the F_1 items that appear in both sets are retained in the final recommendation list; $F = \min\{R - F_1, F_2\}$, $F_2 = |(\mathcal{W}_u \setminus RC_u^{in}) \bigcap \mathcal{P}|$, most preferred cached items appearing in the recommendation window of u (but not in the recommendation set derived in the first step), are added to the recommendation list for u, replacing the bottom-F items in $\mathcal{RC}_u^{in}$; and, if there is still space ($F_1 + F < R$), the remaining recommendations are made for the $(R - F_1 - F)$ least popular items out of the $(R - F_1)$ remaining (non-cached) items in $\mathcal{RC}_u^{in}$.

The final set, $\mathcal{RC}_u^f$, of R items that are recommended to user u are in general different from the equal-size provisional set $\mathcal{RC}_u^{in}$ derived in the first step. Since the values of p_u^{req} after the recommendation amendment step are different, one might think that the algorithm returns to the content placement step and runs another round of steps 2 and 3. However, in [1] we prove that this algorithm terminates in a single round:

> **Proposition 4.** CawR terminates after a single execution of the recommendation amendment step.

Hence, the algorithm essentially determines the cache placement on the basis of the original recommendations to users (with zero user preference distortion). Then, it selectively changes recommendations to nudge individual user preferences towards content that attracts demand from the overall user population. This way, the utility of the cached content, i.e., the demand it attracts, grows and the expected cache hit ratio increases.

In the toy example of Figure 21.1, CawR first sets the provisional recommendation list RC_u^{in} to the R most preferred items for every user, as highlighted in the 2nd column. The resulting demand distributions are shown in the 3rd column. These recommendations are not shown to the end-users. In the second step, CawR computes

Algorithm 1 The CawR algorithm

Input: Probabilities p_u^{pref}, weights w_u^r, recommendation windows $\mathcal{W}_u$, $\forall u \in \mathcal{U}$, number of recommendations R

Output: Content placement $\mathcal{P}$ and recommended item sets $\mathcal{RC}_u^f, \forall u \in \mathcal{U}$.

Step 1:

Set the initial recommendation list $\mathcal{RC}_u^{in}$ to the R most preferred items of the user and p_{rec} to $f(rnk_u(i), R)$ (ref. Assumption 21.2.1)

for every user $u \in \mathcal{U}$ and item $i \in \mathcal{I}$ **do**

  Compute the content request probabilities $p_u^{req}(i)$ from (21.2) and (21.3).

end for

Step 2:

for every item $i \in \mathcal{I}$ **do**

  Compute its utility from (21.11).

end for

Use the DP $(1-\varepsilon)$-approximation algorithm, $\varepsilon > 0$, to solve the 0-1 KSP and derive the content placement $\mathcal{P}$.

Step 3:

for every user u **do**

  Add items in $\mathcal{RC}_u^{in} \bigcap \mathcal{P}$ to the final recommendation list $\mathcal{RC}_u^f$ for u.

  if $\mathcal{RC}_u^f$ is not full **then**

    Add to set $\mathcal{RC}_u^f$ items that are both cached and within $\mathcal{W}_u$ in decreasing order of preference probability till it is filled up.

  end if

  if $\mathcal{RC}_u^f$ is still not full **then**

    Add to the list items out of the remaining ones in $\mathcal{RC}_u^{in}$ in decreasing order of preference probability till the set is filled with R recommended items.

  end if

end for

the utility of each item from (21.11) and caches items A,C,D. In the final step, CawR compares the recommendation window W_u of each user u with $\{A, C, D\}$ and determines the final lists of recommendations $\mathcal{RC}_u^f$, which are issued to the users. In our case, $\mathcal{W}_{u1} = \{A, B, C, D\}$, $\mathcal{W}_{u2} = \{A, C, D, E, F\}$, and $\mathcal{W}_{u3} = \{A, B, C, D, E\}$, respectively. The sets $\mathcal{RC}_u^f$ are those highlighted in the 4th column in Figure 21.1. The cache hit ratio without recommendations is 0.58. When recommendations are issued for the top-3 items in users' preferences (3rd column), the cache hit ratio drops to 0.55, because non-cached items are recommended. On the contrary, when recommendations are issued for items that are both cached and of adequate interest to users, yet not necessarily within the top-3 set (4th column), the cache hit ratio increases to 0.67.

21.4.2 Complexity of the CawR algorithm

In the first step, our algorithm sorts the list of the items and finds the most preferred ones for each user at time $O(|\mathcal{U}| \cdot |\mathcal{I}| \cdot \log |\mathcal{I}| + |\mathcal{U}| \cdot |\mathcal{I}|) = O(|\mathcal{U}| \cdot |\mathcal{I}| \cdot \log |\mathcal{I}|)$. In the second step the algorithm computes a utility for every item and then uses the DP FPTAS algorithm for the 0-1 KSP. This implies a complexity of $O(|\mathcal{I}| + |\mathcal{I}| \cdot |\mathcal{P}|) = O(|\mathcal{I}|^2)$ since the cache capacity is upper bounded by the total catalog size. In the third step, the algorithm compares the items within the recommendation window of each user against the cache placement to define the final recommendations, leading to $O(|\mathcal{U}| \cdot \max_u(|\mathcal{W}_u|) \cdot |\mathcal{P}|)$. Since the size of the recommendation window is naturally bounded by the catalog size, the total computational complexity of CawR is bounded by $O(|\mathcal{U}| \cdot |\mathcal{I}|^2)$.

In the two sections that follow, we draw on both analysis (21.5) and simulations with real and synthetic datasets (21.6) to gain further insight to the properties of the proposed algorithm.

21.5 Properties and performance bounds of our algorithm

In this section, we compare the performance of our algorithm to benchmark recommender schemes and analyze its sensitivity to the maximum distortion tolerance parameters, $\{r_d(u)\}$ and the user content preference distributions, $\{p_u^{pref}\}$.

21.5.1 Benchmark recommender schemes

We consider three alternative schemes for determining which content to cache and which to recommend to each user. They serve as plausible comparison references for our algorithm and help set bounds for its performance.

Zero-distortion (ZD) scheme. The scheme recommends to each user the top-R items in her preferences and caches the C items attracting the highest aggregate demand, after accounting for the impact of recommendations. Hence, the recommendations follow precisely the user preferences, in line with what recommender systems nominally do, and the cache placement adapts to them.

Unbounded-distortion (UD) scheme. This scheme ranks items in order of decreasing aggregate demand over all users, after taking into account the factors $(1 - w_u^r)$ that weigh users' original preferences. It then caches the top-C of those and recommends to all users the *same* top-R items. Contrary to the ZD scheme, it is now the cache placement that determines the individual recommendations, catering for no bounds on the resulting distortion of the inherent user preferences. In [1] we prove that:

Proposition 5. When recommendations follow assumptions 21.2.1 and 21.2.2 and the user preference distortion tolerances are relaxed, $W_u \equiv \mathcal{I},\ \forall u \in \mathcal{U}$ in (21.6) and (21.8), the *UD* scheme is the optimal solution to the JCRP.

Least Frequently Used (LFU) caching algorithm. The algorithm caches items that attract the maximum aggregate demand over all users, but it does not recommend anything to them. It is known that LFU maximizes the cache hit ratio for a single cache in the absence of recommendations.

Denoting the expected cache hit ratio each scheme achieves by $H_s, s \in$ {ZD, UD, CawR}, we can show that

Proposition 6. The expected cache hit ratios achieved by the three schemes that issue recommendations satisfy:

$$H_{ZD} \leq H_{CawR} \leq H_{UD}. \tag{21.12}$$

Thus, the performance of the unbounded- and zero-distortion schemes set an upper and a lower bound, respectively, for what is achievable with CawR. Notably, due to Prop. 5, H_{UD} sets an upper bound to the cache hit ratio under the optimal algorithm, H_{OPT}, for the JCRP (when the distortion constraints are not relaxed), and, eventually, for the cache hit ratio achieved by the CawR algorithm. Namely

$$H_{CawR} \leq H_{OPT} \leq H_{UD}. \tag{21.13}$$

It is less intuitive how LFU compares with the three schemes issuing recommendations since they weigh the inherent user content preferences with factors $(1 - w_u^r)$ to determine the cache placement. We explore these comparative relationships as well as the tightness of the bounds in (21.13) with numerical simulations in section 21.6.

21.5.2 Sensitivity analysis to parameters of CawR

21.5.2.1 Monotonicity and submodularity of H_{CawR} with respect to the distortion tolerance parameters

The parameters $\{r_d(u)\}, u \in \mathcal{U}$ in CawR control the amount of distortion that is tolerated with respect to the original user content preferences. Higher $r_d(u)$ values imply larger W_u sizes, as can be seen from (21.4), and higher chances to find cached items in them. This is interpreted into higher cache hit ratio when at least one of the currently issued recommendations is for a non-cached item. In that case, CawR can replace its recommendation(s) for one (or more) of these items with recommendations for one (or more) of the cached items that are included in the enlarged W_u.

Proposition 7. H_{CawR} is a monotonically increasing function of the distortion tolerance parameters $\{r_d(u), u \in \mathcal{U}\}$.

Less strict claims can be made about the submodularity H_{CawR} of with the $\{r_d(u)\}$ parameters [1].

21.5.2.2 Sensitivity of H_{CawR} to the individual user content preferences

Although the exact values of H_{CawR} and H_{ZD} may vary widely depending on the user content preference distributions, $\{p_u^{pref}, u \in \mathcal{U}\}$, we can state that:

> **Proposition 8.** As the individual content preference distributions become more skewed, H_{ZD} and H_{CawR} tend to converge with each other.

The key remark is that as the distributions $\{p_u^{pref}\}$ become more skewed, the recommendation windows become smaller. Therefore, the additional flexibility of CawR in selecting items to recommend tends to vanish and its recommendations to the users coincide with those of the ZD scheme. In fact, the recommendations of the two schemes exactly coincide when $K_u = R$.

21.6 Experimental evaluation of CawR

21.6.1 Datasets and default parameter settings

We use both synthetic and real datasets [11] to derive the user and item feature vectors, $\mathbf{f}_u$ $u \in \mathcal{U}$ and $\mathbf{f}^i$ $i \in \mathcal{I}$, respectively, and then infer the user content preference distributions p_u^{pref} (see section 21.2). The purpose of using real datasets is three-fold: (a) to show how our model of user preferences and content items can be informed by real data; (b) to drive a more "realistic" evaluation of the cache hit ratio CawR achieves; and (while doing so) (c) to validate the theoretical claims made in section 21.5. With synthetic datasets, on the other hand, we can control the experimentation settings and analyze the sensitivity of our algorithm to important variables such as the (dis)similarity of content preferences across users. Experiments were performed with the MovieLens dataset as well as with synthetic datasets.

MovieLens dataset [11]: We analyze different samples of 700 users and 10.000 items of the catalog. Each of these movies is described by $M = 19$ thematic tags, i.e. the feature set contains 19 themes. In populating our model, we draw on the fact that a user *did* rate a specific item, rather than the actual rating she assigned to it. More specifically, the item tags are directly used to generate the item-feature vectors, as described in Example 1. Then, if $\mathcal{I}_r(u)$ is the set of items that user u has rated, we estimate the element $\mathbf{f}_u(j)$, $j \in [1, M]$, of user's u feature vector as in Example 2.

Synthetic datasets: In this set of experiments, the elements of the two vectors are populated with values drawn from random probability distributions, by default from the standard uniform distribution. The default parameter values for simulations with synthetic datasets are $|\mathcal{U}|$=150 users, $|\mathcal{I}|$=1000 items and $M = 8$ thematic areas.

For simulations with both types of datasets, the normalized item size L_i is sampled from a discrete uniform distribution $U(1, L_{max})$, with default $L_{max} = 4$. Our baseline assumption is that recommendations provide all R items in the list with

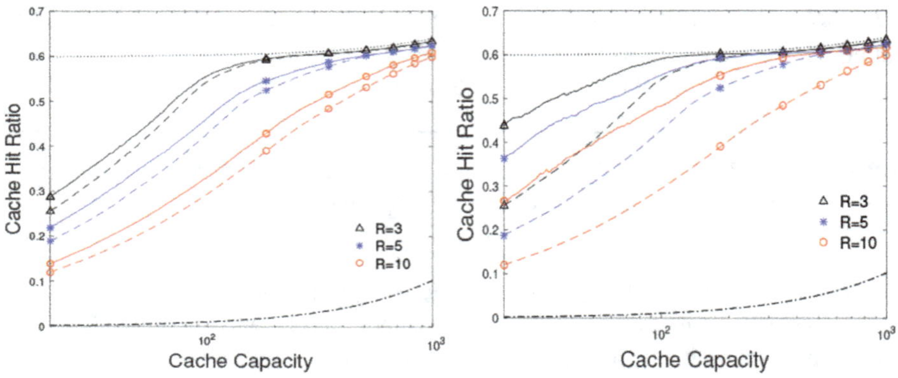

(a) $r_d(u)$=0.01, $\forall u \in \mathcal{U}$ (b) $r_d(u)$=0.1, $\forall u \in \mathcal{U}$

Figure 21.4 Experiments with MovieLens traces. *Cache hit ratio, H, capacity for different number of recommended items, R. UD (dotted line) and LFU (dash-dot line) are not affected by R, solid lines correspond to CawR and dashed ones to ZD.*

an equal boost $p_u^{rec}(i) = 1/R$, which fades out with R. The intuition is that the fewer the recommendations are, the less cognitive load they demand from users to process them, and the more significant their impact is on the eventual user content requests. This assumption is more realistic for users accessing content from large-display devices, where it is more comfortable to scroll down the recommendations' list.

The choices of user recommendation weights, w_u^r, are aligned with experimental evidence in [5], according to which YouTube users request one of the top 10 recommended items with a probability that varies in [0.5, 0.7]. Thus, in our experiments the user recommendation weights are sampled from $U(0.5, 0.7)$.

21.6.2 Trace-Driven simulations

Figure 21.4 and 21.5 validate the performance bounds we found analytically for the cache hit ratio under CawR in section 21.5. The extent to which these bounds are tight depends on the number of issued recommendations per user, R, and the preference distortion tolerance parameter, r_d.

The impact of the number of recommended items on cache hit ratio: On the one hand, R does not affect the performance of UD and LFU schemes. On the other hand, the achievable cache hit ratio under CawR and ZD deteriorates with higher R values. Intuitively, as the recommendation effect is spread across more items, some of it is wasted because it gets harder to find R cached items within the users' recommendation window (in the case of CawR) and among the top-R items in their preferences (in the case of ZD). Hence, the upper bound becomes looser as the number of recommendations grows. Interestingly, so does the lower bound when the relative gain of CawR over the zero-distortion scheme grows with R. This is the case as far as the size of the recommendation window size K_u is higher than R so that CawR has higher chances to find a cached item to recommend. For $R = 10$

and r_d=0.01, ZD performs up to 16.6% worse than CawR in terms of cache hit ratio, especially for really small instances of cache capacity (Figure 21.4a). For r_d=0.1, this gap grows to 121% (Figure 21.4b). At the same time, CawR reaches the 96% and 97% of the performance of the UD scheme, for r_d=0.01 and r_d=0.1 respectively, and realistic cache capacities (Figure 21.4a-21.4b). An alternative way to quantify the benefits of our algorithm is by looking into cache capacity requirements. For r_d=0.01, ZD needs up to 35% more cache capacity than CawR to reach the maximum achievable cache hit ratio when R=10 (Figure 21.4a). For r_d=0.1, this value climbs to 147% (Figure 21.4b). In all cases, the LFU scheme is significantly outperformed by schemes that use recommendations.

The impact of the preference distortion tolerance, r_d, on cache hit ratio: This enhanced flexibility of CawR is also the reason why it approaches the upper bound faster (i.e., for smaller cache) when r_d=0.1 (Figure 21.4b), increasing its advantage over the ZD scheme that is insensitive to r_d.

The only scheme that is affected by the preference distortion tolerance parameter is CawR. Higher values of r_d provide the scheme with more flexibility in recommending items that are simultaneously parts of the cache placement and the user recommendation windows. Hence, as can be seen in Figure 21.5(a-b), but also in Figure 21.4(a-b), the CawR performance increases monotonically with r_d moving away from its lower bound (ZD scheme) towards its upper bound (UD scheme). Indicatively, for $R = 3$ and $r_d = 0.1$, we evidence with CawR cache hit ratios up to 71% higher than those under the ZD scheme at very small caches (Figure 21.5a). The respective performance gain increases to 121% for $R = 10$ (Figure 21.5b).

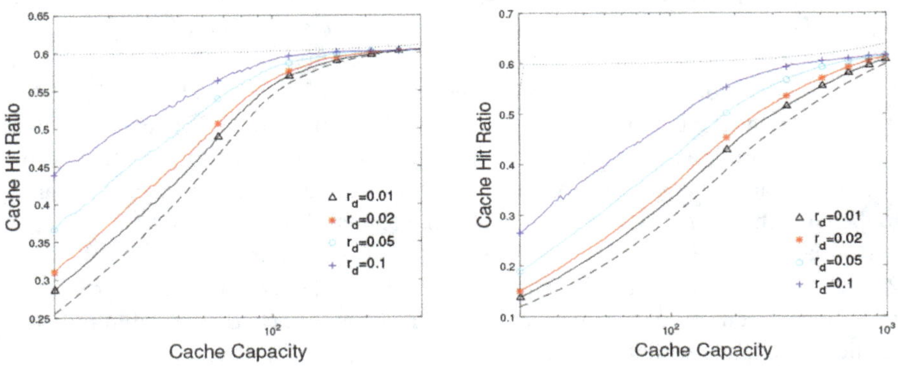

(a) $R = 3$ (b) $R = 10$

Figure 21.5 Experiments with MovieLens traces. *Cache hit ratio, H, capacity as a function of the preference distortion tolerance,* r_d *(identical for all users). Dotted and dashed lines correspond to the UD and ZD scheme, respectively. The four intermediate curves correspond to CawR.*

Regarding cache capacity requirements, even for distortion tolerance values smaller than $r_d = 0.01$ and $R = 3$, CawR needs up to 35% less storage capacity to converge to the upper bound confronted to the ZD scheme. This gain escalates up to 65% when $r_d = 0.1$ (Figure 21.5a).

The impact of cache capacity, catalog size, and number of users on cache hit ratio: A final set of experiments with the MovieLens traces addresses basic scaling properties of the algorithm, i.e., how the cache hit ratio varies as a function of the content catalog size, cache size, and population of users. They report the positive impact of ratio $\alpha = \frac{\text{cache capacity}}{\text{catalog size}} = \frac{C}{|\mathcal{I}|}$ on the cache hit ratio. We further notice that the CawR and the ZD schemes approach the hit ratio of the UD scheme slower when compared to a system with fewer items. Related figures can be found in [1].

21.6.3 Simulations with synthetic datasets

In this subsection, we explore how the performance of our algorithm is affected by the sensitivity of users to recommendations and the heterogeneity in their content preferences.

User recommendation weights: In the experiments of this subsection we consider two scenarios for the range of the user recommendation weights w_u^r in (21.3), namely $w_u^r \in [0, 0.2]$ and $w_u^r \in [0.5, 1]$. The achieved cache hit ratio for each scenario is shown in Figure 21.6.

As is expected, for small recommendation weights, the cache hit ratio under all recommendation schemes is small and comparable to that achieved under the recommendation-agnostic LFU scheme. However, CawR remains attractive when cache space is a concern. With storage spaces smaller than the 5% of the total catalog size, CawR scores similarly to the ZD scheme, needing up to 35% less cache capacity. Compared to the UD scheme, CawR reaches the 89% of the performance of the UD scheme, performing always more than 14% better than ZD.

On the contrary, for larger recommendation weight values, UD, ZD and CawR schemes differ more clearly from each other. The cache hit ratio under CawR is 33% higher than under ZD, for modest cache capacities in the order of 2% of the catalog size. In terms of cache capacity, CawR equals ZD needing up to 32% less cache capacity even for capacities lower than the 2% of the total catalog size. Compared to the UD scheme, for high recommendation weights, CawR reaches the 97% of the performance of the UD scheme, performing always more than 6% better than the ZD scheme. Overall, Figure 21.6 indicates that the more users assign importance to recommendations made to them, the higher the advantage of our algorithm over the "honest" recommendation scheme.

Heterogeneity in user content preferences: Due to space limitations, this section is briefly presented. For a more detailed presentation of the results regarding the heterogeneity in user content preferences, the interested reader may see [1].

The heterogeneity in user content preferences can manifest itself in multiple ways. To control heterogeneity, we generate the user content preference distribution as a convex combination of two distributions: a user-specific component p_u^{ego}, which is modeled as described earlier in section 21.2.2; and a second user-agnostic

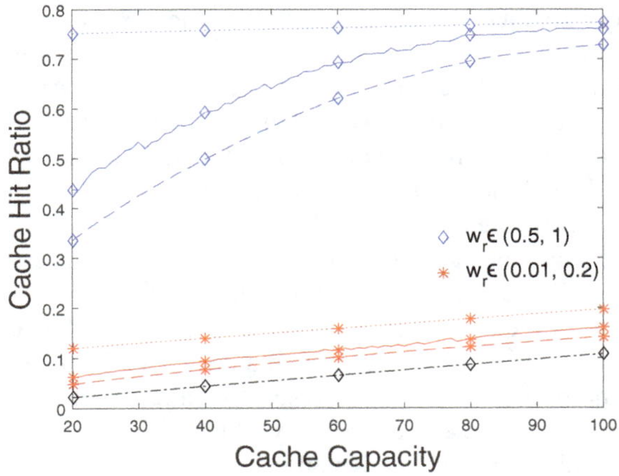

Figure 21.6 Experiments with Synthetic datasets. *Cache hit ratio capacity as a function of user recommendation weights, R=10,* r_d*=0.01,* $|\mathcal{I}|$ = 1.000 *items. Dotted lines correspond to the UD scheme, dashed lines to ZD, dash-dot lines to LFU, and solid lines to the CawR.*

probability distribution, p^{ext}, which is modeled after a Zipf distribution, in line with experimental evidence [12].

We have experimented with different values for w_e^u, concluding that for $w_u^e = 0$ the three schemes that issue recommendations collapse to one. Moreover, the total achieved cache hit ratio is higher when the content preference distributions are homogeneous ($w_u^e = 0\ \forall u \in \mathcal{U}$). Moreover, we experimented with the impact of the content preference distribution. Namely, we let various Zipf and uniform distributions serve as the user content preference distributions, permuting the order of items for each one of them. We confirm that the performance of CawR, much as that of other caching schemes, is quite sensitive to the shape of the original user content preferences, which set hard bounds on what is achievable. Our results are inline with the previously proposed Propositions.

21.7 User associations as additional control parameter: extending the model

Until now, our analysis has assumed that the associations of users to small cells are driven by physical proximity and radio signal quality considerations, in line with current practices in mobile cellular networks. As a result, the caching and recommendation decisions are taken independently in each cell, as detailed in section 21.4.

However, both in current generation cellular systems and, even more, in the ultra dense radio network architectures envisaged for 5G and beyond systems [13], users have the choice to associate with multiple different cells. Controlling then which users are associated with each cell and which content is stored at each cell-colocated cache may offer additional degrees of freedom when trying to satisfy locally the user demand for content and alleviate the load on the backhaul links. While recommendations shape the individual user demand, the user associations intervene on its spatial distribution across the radio network.

In principle, the two mechanisms may be exercised over different time scales. For instance, with online caches the cache placement may be reiterated upon each and every content item request by a user. Alternatively, the cached content may be determined periodically (e.g., every day or half a day) out of estimates for the local content demand and remain fixed for the respective interval of time. On the other hand, a new user association to a small cell may let its cache intact or trigger changes that budget for her individual content demand distribution.

In general, the more aggressively the network carries out these control functions, the better the performance it can achieve (smaller content access delay on the user side, offloading of mobile backhaul) at the expense of computational resources. More recently, in light of the disruptive network capacity and latency targets advertised by 5G cellular networks [14], the wireless community has been investigating scenarios of joint content caching and user association, where the stored content across the network small caches and the associations of users to small cells are simultaneously determined [15–17]. Putting this into the context of real mobile networks, it essentially implies that the cache placements and user associations are revisited every time a new user joins the network associating with a cell. We expand on this thread adding the recommendation dimension to the problem and explore how these three control mechanisms, i.e., content caching, recommendations and user associations can be jointly tuned to enhance the network performance.

21.7.1 Model extension

We consider, as before, a two-layer heterogeneous network with cache-enabled small cells: multiple Small Base Stations (SBSs) and a Macro Base Station (MBS). A user can be associated either to *one* SBS c, *or* to the MBS M. The association of a user to an SBS is the preferred alternative since it can exploit the capacity and caching capability of the SBSs more efficiently, preserving the radio resources of the macro cell. Nevertheless, a user can always associate with the MBS, as a fallback solution, when the association to an SBS is not feasible. At any given time slot, each user $u \in \mathcal{U}$ is located within the range of a different subset of SBSs. We define as $\mathcal{N}(u)$ the "neighborhood" of user u, the set of cells user u can be associated with. Each SBS can serve a set of mobile users within its range, while the MBS can serve the users within range of any SBS. Each cell c has a limited service rate, B_c, split among all user devices associated with it. A minimum downlink data rate is guaranteed to each user u, incurring an association cost b_{cu} for her association to cache c.

This cost is strongly related to the physical distance between the user and the cell. In fact, a more distant user generates higher association cost in terms of transmit power for the cell, network interference or even power consumption (for her own device). Thus, the number of users that an SBS can serve is limited by the aggregate *association cost*, which should not be higher than B_c, for all cache-enabled SBSs $c \in \mathcal{C}$.

21.7.2 The Joint Caching, Recommendations and Association problem

The objective of the Joint Caching, Recommendation, and Association (JCRA) problem is to maximize the portion of total demand that can be satisfied by all caches, assuming also control over the user association process. Let $\{x_{ui}\}$, $\{y_{ic}\}$ and $\{z_{cu}\}$ be three sets of binary decision variables, with $x_{ui} = 1$ if item i is recommended to user u and $x_{ui} = 0$ otherwise; $y_{ic} = 1$ if item i is cached in cache c and $y_{ic} = 0$ otherwise; $z_{cu} = 1$ if user u is associated to cache c and $z_{cu} = 0$ otherwise. Then, reusing the notation in section 21.3, the JCRA problem is formulated as:

$$\max_{\mathbf{y},\mathbf{x},\mathbf{z}} \sum_{u \in \mathcal{U}} \sum_{c \in \mathcal{C}} \sum_{i \in \mathcal{I}} y_{ic} z_{cu} (x_{ui} p_u^{req}(i) + (1 - x_{ui}) \widetilde{p}_u^{req}(i)) \tag{21.14}$$

$$s.t. \sum_{i \in \mathcal{I}} y_{ic} L_i \leq S_c, \qquad \forall c \in \mathcal{C} \tag{21.15}$$

$$\sum_{u \in \mathcal{U}} b_{cu} z_{cu} \leq B_c, \qquad \forall c \in \mathcal{C} \tag{21.16}$$

$$\sum_{c \in \mathcal{N}(u)} z_{cu} \leq 1, \qquad \forall u \in \mathcal{U}, \tag{21.17}$$

$$x_{ui} = 0, \qquad \forall i \notin \mathcal{W}_u, u \in \mathcal{U}, \tag{21.18}$$

$$\sum_{i \in \mathcal{I}} x_{ui} = R, \qquad \forall u \in \mathcal{U} \tag{21.19}$$

$$y_{ic}, x_{ui}, z_{cu} \in \{0, 1\}, \; u \in \mathcal{U}, i \in \mathcal{I}, c \in \mathcal{C}, \tag{21.20}$$

Constraints (21.15) and (21.16) reflect the cache storage and service cost constraints for each cache, respectively. Constraints (21.17) capture the unique association of users either to cells in their neighbourhood or to the MBS. Constraints (21.18) and (21.19) ensure that the content items recommended to each user are always within her recommendation window and that exactly R items will be recommended to her, respectively.

Controlling the user-to-cache association creates an additional optimization potential. However, it makes the (already difficult) optimization problem even more difficult. In [3], we use generalization arguments to prove that:

Proposition 9. The Joint Caching, Recommendations and Association problem is NP-hard.

21.7.3 Heuristic algorithms for the JCRA problem

We present two heuristic solutions to the JCRA problem. The core of both algorithms is the same, i.e., an iterative process for jointly determining the user associations to the small cells and the content to be stored at the small cell caches (see section 21.7.3.1). What changes between the two algorithms is the way recommendations are introduced.

The first heuristic proceeds in three phases:

1. *Initialization phase:*
   (a) At each SBS, assume that all users within its range are associated with it.
   (b) Determine the cache placements at the SBSs caches by solving independent instances of the 0-1 KP, where the utilities of content items equal the aggregate user demand they satisfy if honest recommendations are issued to users (ref. section 21.4, $r_d = 0$).
2. *Iterative phase: Until the cache hit ratio in (21.14) is no further improved do:*
   (a) Given the cache placements determined in (1b), update the user associations at each SBS solving an instance of Generalized Assignment Problem (GAP). Users correspond to items and cells to bins, the cell-specific user costs are the association costs $\{b_{cu}\}$ and the cell-specific user profits are the user demands covered by the current cell cache placement assuming honest recommendations.
   (b) With the new user associations computed in (2a), recompute the content item utility values and solve anew the C0-1 KP instances for each SBS assuming honest recommendations.
3. *Recommendation amendment phase:* Adjust recommendations to be both within each user's recommendation window W_u and the cache placement at the cell of user's attachment. The user associations and the cell cache placements are the outputs of the last iteration in (2a) and (2b), respectively.

This heuristic mimics the rationale of the CawR scheme in Algorithm 1. Namely, it uses honest recommendations, corresponding exactly to the user content preferences, as an intermediate step to determine the cache placements (here, also, the user associations). It then amends the recommendations in line with the cache placements to boost the portion of demand that can be locally served by the small cell caches.

An apparent alternative would be to not account for (honest) recommendations at all in the first and second phase of this heuristic and only derive them once as part of the third phase, after the iterations for the determination of cached content and user associations terminate. With this second heuristic, recommendations are *caching-aware* and still boost the part of user content demand that can be satisfied locally. Yet, they do not have an impact on what is eventually cached at the small cells and who are the users served by each of them.

Both heuristics demand an algorithmic solution for the intermediate joint content caching and user association problem. We turn to this problem next.

21.7.3.1 The Intermediate Joint Caching and Association problem

The intermediate Joint Caching and Association (JCAP) problem results from the JCRA problem when omitting the recommendation-related variables and constraints.

$$\max_{\mathbf{y},\mathbf{z}} \sum_{u\in\mathcal{U}} \sum_{c\in\mathcal{C}} \sum_{i\in\mathcal{I}} y_{ic} z_{cu} p_u^{pref}(i)) \tag{21.21}$$

$$s.t.(21.15), (21.16), (21.17) \tag{21.22}$$

$$y_{ic}, z_{cu} \in \{0,1\},\ u \in \mathcal{U}, i \in \mathcal{I}, c \in \mathcal{C}, \tag{21.23}$$

JCAP is an instance of bilinear programming, a special class of non-convex quadratic programming. To show that it is NP-hard, it suffices to remark that when we fix variables $\{y_{ic}\}$, we get the generalized-assignment type problem first treated in [18]. This problem is equivalent to the maximum GAP[e], where SBSs correspond to agents (knapsacks) and users to jobs (items) with agent-specific requirements $\{b_{uc}\}$ and profits $\{\sum_{i\in\mathcal{I}} y_{ic} p_u^{pref}(i)\}$, $c \in \mathcal{N}_u$. Since the maximum GAP is an NP-hard problem, the generalization (JCAP) of its equivalent problem in [18] is NP-hard as well. We tackle this problem in [20] and [21], but due to space limitation we will restrict our presentation to the first.

In [20], we propose an iterative algorithm for solving JCAP that corresponds to the first two phases of the heuristic for the JCRA problem (see section 21.7.3), if recommendations are let aside. We first prove its correctness and discuss its computational complexity. Then we proceed in a two-phase evaluation of the algorithm. In the first phase, we consider small problem instances and compare it against the optimal solution, as this can emerge from linearization techniques and use of standard ILP solvers. Table 21.2 gathers statistics on ΔH, the empirical gap between the cache hit rates achieved by the two solutions and $G = \frac{\Delta H}{HR_{opt}}$x100%, the percentage ratio of ΔH to the optimal value. The median of G over all experiments is practically 0% and the two solutions coincide in more than half of the experiments under two scenarios for the content demand distribution across the network: uniform at random and with strong spatial locality properties.

Table 21.2 Accuracy of the iterative heuristic

Comparison Scenario	var Users random demand	var Users Clustered demand	var Items random demand
median Δ H(G)	0.001(0.2%)	0(0%)	0.004(0.81%)
95th perc. ΔH(G)	0.097(17.7%)	0.1(14.2%)	0.081(13.3%)
max ΔH(G)	0.161(29.2%)	0.21(25%)	0.189(35.2%)

[e]The inequalities in (21.17) are replaced by equalities in the maximum GAP. The equivalence of the two problems is shown in e.g., [19], pp. 190-191

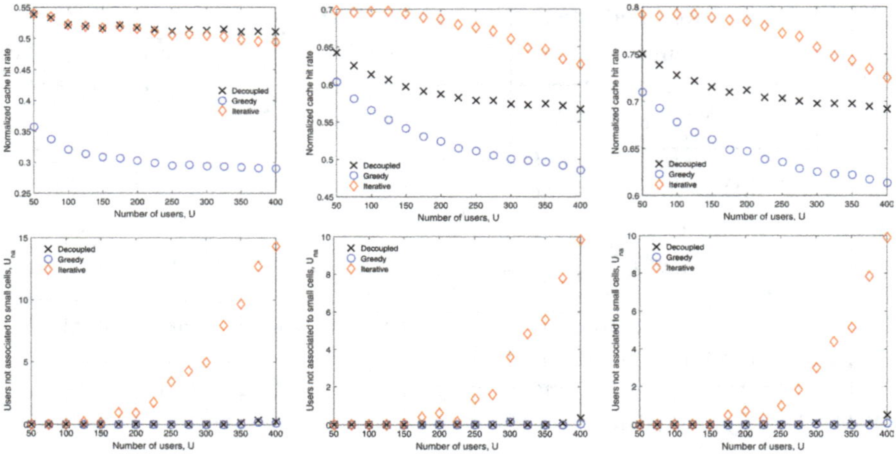

a. No spatial locality, L_a = 0:3 b. Spatial locality N_{cl} = 10, L_a = 0:1 c. Spatial locality N_{cl} = 10, L_a = 0:2

Figure 21.7 Comparison of the three heuristics as a function of the number of users: I=1000, C=20, B_c=200, l_{max}=12, b_{max}=20

In a second phase, we consider more realistic problem instances and compare the heuristic against simpler algorithms for tackling the user association and content caching tasks: a greedy-like algorithm and another heuristic, called *Decoupled*, which decouples the user association decisions from the content caching ones and tackles user associations first.

Of particular interest are the experiments that look into the sensitivity of the algorithms to the existence or not of spatial locality in the content demand. Under no spatial locality, the decoupled heuristic competes with our iterative heuristic, and even slightly outperforms it for high numbers of users, as shown in Figure 21.7a. At those user levels, the decoupled approach can match all users to SBSs, whereas our iterative heuristic is forced to associate a few with the MBS. The non-satisfied demand of these users corresponds to the marginal performance gain of the decoupled heuristic. The greedy heuristics cannot compete with either of the two alternatives.

When there is spatial locality, the performance of all three heuristics improves (Figure 21.7b). In this scenario, it matters more *which* users will be grouped in a cell rather than *how many*. Factoring the content preferences of users in its decisions, the iterative heuristic clearly outperforms the decoupled one. This performance gain fades out as the number of users grows since the latter manages to squeeze more users in small cells. For more details on the performance characteristics of the iterative heuristic, the interested reader is referred to [20].

21.8 Open directions

Understanding the precise way recommendations shape individual content preferences is a research thread per se, mainly pursued in the context of marketing and recommender systems. An open research direction consists in translating these largely experimental findings into solid models and validating the presented findings with them as starting points. This should address the human behavior heterogeneity, possibly leveraging data-driven machine learning techniques and principles from behavioral science.

In this chapter, we considered one round of interaction between the users and the system. An interesting extension would be to consider how the dynamics of this interaction evolve in the long-term under such recommendations, addressing aspects like the trust users build in the system and including possible system penalization (e.g., churn effects). In a fully transparent system, where users are aware of the nudging recommendation practices, users might even choose themselves and trade the distortion their recommendations will undergo.

Another direction is to consider that the different mechanisms (content caching, recommendations and user association) are controlled by different entities with opposing interests and objectives. For example, the recommendation is controlled by the content provider with the aim to maximize the user engagement and relevance of content recommended; the caching is controlled by the infrastructure provider with the aim to maximize hit ratio; and the association is controlled by the user to minimize the delivery delay or the energy consumption. Game-theoretical tools, either cooperative or not, could be used to describe such kind of scenarios, with each player holding her own objective.

21.9 Related work

Related works about the interplay between caching and Recommender Systems (RS) are divided in two main categories: RS are considered either as predicting tools in order to decide a better caching placement (e.g., [6, 22]), or as traffic engineering mechanisms, used as demand-shaping tools. In both categories, RS are used as proxies for inferring the content popularity. However, the last are distinctly different regarding the way they approach RS: not just as alternative predictors of content demand but also as demand-shaping tools that can actively be used to trade-off user- and network-centric performance objectives. In fact, their approach to increasing the utility of cache content could be seen as dual to the first: Rather than struggling for accurate predictions of the users' demand for content, they nudge the users' demand towards content items that are common in their preferences.

Due to space limitations, we will provide a taxonomy only for the second category (Figure 21.8), in which our work has made significant contribution. We identify five key factors that differentiate existing works that use the RS as demand-shaping tools: Do they jointly optimize RS and caching decisions? How are recommendations accepted by users? Which is the RS mechanism they use to issue recommendations? Do they take into consideration the general Quality of Service (QoS) that users will receive? Last but not least, are they concerned about the Quality of the issued Recommendations (QoR) and the ethics around "nudged" recommendations?

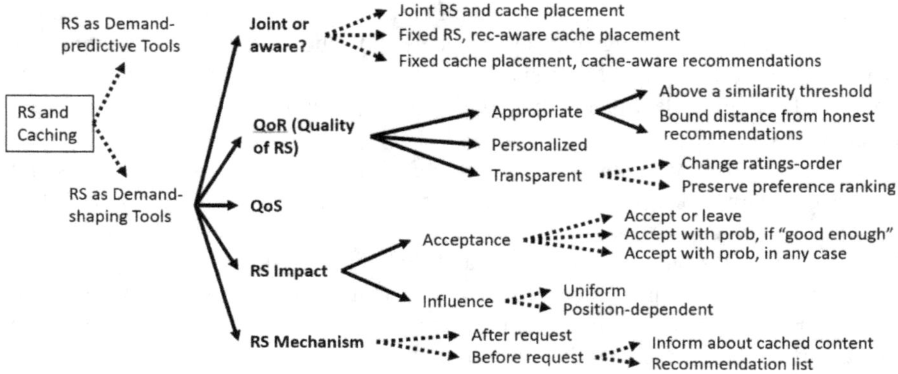

Figure 21.8 Taxonomy of works considering the interplay between RS and wireless caching. Solid lines under a category imply that all subcategories can be chosen (the various criteria that characterize works), while dotted lines imply that a work can belong only to one subcategory (either it considers uniform or position-dependent recommendation influence)

We briefly discuss these factors and their main subcategories, presenting some representative works.

Joint or aware? Works are distinguished based on the relation of caching and recommendation decisions. The "soft cache hits" in [23] considers systems where users requesting uncached content are informed about cached content which is similar to their request, and systems where alternative content is directly delivered to them. For a given cache placement, the goal is to find a set of recommendations that will maximize the cache hit ratio. On the contrary, a cache placement is a priori given and only recommendations are decided for sequential recommendations in [24, 25] and list reordering [5] or rating shaping [8] are proposed, aiming at the same goal. The same in [26], aiming to the service cost minimization in P2P networks. We are actually different from this thread of works in *jointly taking both recommendations and caching decisions*.

Influenced by [5], in [2] we introduced a mathematical model that captures the coupling between caching decisions and personalized recommendations. Then [1, 27–30], studied how to jointly take these decisions under different assumptions. In [3] the user-to-AP association is an additional decision taken jointly with the others, capturing the case of Telefonica, the giant network provider that also holds the huge Latin-American content provider Movistar. Works [31, 32] jointly consider recommendation and content pushing decisions, while [33] considers coded caching and recommendations.

RS Impact. We consider two factors that characterize the impact of RS on the users' final requests: The probability of acceptance of the recommendation and, if accepted, the distribution of the influence across various positions in the recommendations list. Regarding the acceptance, in [26] it is assumed that users either accept the recommendation, or leave the system. Other works consider users that accept a recommendation with a user-specific [1–3] or universal [24, 25, 27, 30, 34] probability if it is "good enough," while in [23, 33] they accept it with a probability in any case, and in [35] this probability is learned. Regarding the impact after acceptance, in [2] we assumed uniform influence across items in the recommendation list, while in [1] we introduced a model according to which the demand attracted by recommendations in a non-increasing function of both the item's position in the recommendation list and the number of recommended items, inline with experimental evidence [5] and providing insights about characteristics of the problem. Similar models are considered also in [3, 25, 29].

RS mechanism. Most of the works provide recommendations before the users' requests, either as a recommendation list as described above, or by informing users about the cached content [35]. In [23] the authors recommend alternative cached content to users *after* their request for uncached content.

Quality of Service (QoS). The QoS is taken into consideration in [3] by considering that users receive content from a BS if the received SINR is greater than threshold, ensuring it as an embedded constraint in the optimization problem. In [29] the authors consider a probability of users offloading content from caches, based on their distances, while in [34] users are directly asked through questionnaires.

Quality of Recommendations (QoR) and ethics. We consider three factors that characterize the QoR: Appropriateness, personalization and transparency. *Appropriateness* is about being *good enough* and matching adequately the users' preferences. *Personalization* is about taking into consideration the specific inherent preferences of the *individual users* when issuing recommendations, in contrast to issuing generic recommendations to all users in the system. These are not equivalent: A recommendation can be personalized while not being adequate, e.g., by violating the user's preferences over an accepted threshold. It can be appropriate for many users but not personalized, e.g., when issuing recommendations about the most popular items to all users. In that case, the recommendations will be adequate to the majority of users, albeit not personalized. *Transparency* is about being transparent with the user, in anything that could violate their preferences. In both of the previous cases, recommendations can be non-transparent if users are not informed about their preferences' violation.

In [26] and [8] these aspects are not taken into consideration: The authors in [26] propose to recommend content cached by peers and those in [8] to modify the ratings shown to users, without personalization or adequateness constraints. In [33], although the personalized model of [2] is adopted, the interplay with coded caching does not allow to consider the distortion of the users' preferences in the issued recommendations. Some other works consider adequate but not personalized recommendations. In [29], recommendations that are above a certain (global) threshold of quality, same for all uses in the system. Works [5, 23, 24, 27] and [26], indirectly consider a kind of adequateness in recommendations, in the sense that they recommend cached content similar to that requested by users, or reorder the recommendation lists of YouTube putting the cached on top of them [5]. However, this item-item similarity-based approach does not necessarily lead to personalized recommendations. Moreover, it could end up recommending items from the same, very restricted, pool of cached items.

Our viewpoint to RS raises some concerns, not least ethical ones. To this end, in [2] we introduced a measure called *preference distortion tolerance*, to quantify and strictly bound how much the engineered recommendations distort the original user content preferences. The algorithm we proposed is essentially a form of light-weight control over user recommendations so that the recommended content is both appealing to the end-user and more friendly to the caching system and the network resources. An important consideration of the conservative [1, 2] or more general [3] quality metrics we presented is that recommendations should be *both personalized and adequate*, following their original reason of existence: to provide users with *personalized, appealing* content, in order to increase their overall satisfaction and engagement with the platform. Additionally, to the best of our understanding, discussion about *ethics and transparency* of network-related recommendations is only made in [1, 2] and [3], where both a bounded distortion from the users' inherent preferences is guaranteed and a preferences rank-preserving recommendation list is issued. By *carefully* nudging the individual user demand towards content that attracts preference from many users, the RS can result in higher cache hit ratios and enhanced QoE for users. Our results suggest that the proposed approach could yield significant gains for the network performance and the users' satisfaction without disrespecting their individual preferences.

21.10 Conclusions

This thread of works is motivated by the trend that wants both content providers and network operators also assuming roles in content delivery by owning and managing content delivery networks. We have looked into the possible benefits that can arise for the end-users and the network when there is some coordination between recommender systems and caching decisions. This coordination, at least in these works, implies that recommender systems actively engineer the recommendations issued to users in ways that enhance the caching performance. Practically, this engineering consists in recommending content that may not necessarily rank top in the inferred user content preferences but still score high in them. By carefully nudging the individual user demand towards content that attracts preference from many users, the recommender system can result in higher cache hit ratios and enhanced QoE for end-users.

Practitioners in areas like e-commerce are more familiar with this demand-shaping (more broadly: behavior-shaping) dimension of recommendations. There is strong evidence that the willingness to consume can be affected by online recommendations. Their manipulation there aims at nudging consumers to spend more on products and services. We rather advocate their "manipulation" for "good" purpose, as an additional network traffic engineering tool that can be used to jointly optimize or balance user- and network-oriented performance objectives.

We have attempted to show this potential in the context of wireless networks with small cells. At the same time, we tried to explicitly and systematically address ethical concerns that are raised by this approach. Simulation results show that the proposed caching-aware recommender systems bring significant caching performance gains that persist over a broad range of parameters for the diversity in users' preferences, the capacity of caches, the number of recommended items and the content catalog size.

This chapter includes material from the following paper: "Jointly Optimizing Content Caching and Recommendations in Small Cell Networks," IEEE Transactions on Mobile Computing, 2018. Reproduced by permission of IEEE.

Funding

This work was supported by the CHIST-ERA grant CHIST-ERA-18-SDCDN-004, through the General Secretariat for Research and Innovation (GSRI), grant number T11EPA4-00056. This work was supported by the Hellenic Foundation for Research and Innovation (H.F.R.I.) under theFirst Call for H.F.R.I. Research Projects to support Faculty members and Researchers and the procurementof high- cost research equipment grant (Project Number: HFRI-FM17-352).

References

[1] Chatzieleftheriou L.E., Karaliopoulos M., Koutsopoulos I. 'Jointly optimizing content caching and recommendations in small cell networks'. *IEEE Transactions on Mobile Computing*. 2019;**18**(1):125–38.

[2] Chatzieleftheriou L.E., Karaliopoulos M., Koutsopoulos I. Caching-aware recommendations: Nudging user preferences towards better caching performance. *In: Proc. IEEE INFOCOM*; 2017. pp. 1–9.

[3] Chatzieleftheriou L.E., Darzanos G., Karaliopoulos M., Koutsopoulos I. 'Joint user association, content caching and recommendations in wireless edge networks'. *ACM SIGMETRICS Performance Evaluation Review*. 2018;**46**(3):12–17.

[4] Gomez-Uribe C.A., Hunt N. 'The netflix recommender system: algorithms, business value, and innovation'. *ACM Trans on Management Information Systems*. 2016;**6**(4):13:1–13:19.

[5] Krishnappa D.K., Zink M., Griwodz C., Halvorsen P. 'Cache centric video recommendation: an approach to improve the efficiency of YouTube caches'. *ACM Transactions on Multimedia Computing, Communications, and Applications (TOMM)Tomm*. 2015;**11**(4):1–20.

[6] Verhoeyen M. Optimizing for video storage networking with recommender systems'. *Bell Labs Technical Journal*. 2012;**16**(4):97–113.

[7] Lee M.-C., Molisch A.F., Sastry N., Raman A. 'Individual preference probability modeling and parameterization for video content in wireless caching networks'. *IEEE/ACM Transactions on Networking*. 2019 Apr;**27**(2):676–90.

[8] Tadrous J., Eryilmaz A., El Gamal H., *et al.* 'Proactive content download and user demand shaping for data networks'. *IEEE/ACM Transactions on Networking*. 2015;**23**(6):1917–30.
[9] Vegelius J., Janson S., Johansson F. 'Measures of similarity between distributions'. *Quality and Quantity*. 1986;**20**(4):437–41.
[10] Vazirani V.V. *Approximation algorithms*. Springer; 2001.
[11] Harper F., Konstan J. 'The movielens datasets: history and context'. *ACM Transactions on Interactive Intelligent Systems (TiiS)*. 2015;**5**(4):1–19.
[12] Paschos G., Iosifidis G., Caire G. 'Cache optimization models and algorithms'. *Foundations and Trends in Communications and Information Theory*. 2020;**16**(3 - 4):156–345.
[13] Andreev S., Petrov V., Dohler M., Yanikomeroglu H. 'Future of ultra-dense networks beyond 5G: harnessing heterogeneous moving cells'. *IEEE Communications Magazine*. 2019;**57**(6):86–92.
[14] Andrews J.G., Buzzi S., Choi W., *et al.* 'What will 5G be?' *IEEE Journal on Selected Areas in Communications*. 2014 June;**32**(6):1065–82.
[15] ElBamby M.S., Bennis M., Saad W., *et al.* Content-aware user clustering and caching in wireless small cell networks. *In: International Symposium on Wireless Communications Systems (ISWCS)*; 2014. pp. 945–9.
[16] Pantisano F., Bennis M., Saad W., *et al.* Cache-aware user association in backhaul-constrained small cell networks. *In: International Symposium on Modeling and Optimization in Mobile, Ad Hoc, and Wireless Networks (WiOpt). IEEE*; 2014. pp. 37–42.
[17] Wang Y., Tao X., Zhang X., Mao G. 'Joint caching placement and user association for minimizing user download delay'. *IEEE Access*. 2016;**4**:8625–33.
[18] Chalmet L., Gelders L. Lagrangian relaxations for a generalized assignment-type problem. *In: Proc. Second European Congress on Operations Research*; 1976. pp. 103–9.
[19] Martello S., Toth P. 'Knapsack Problems'. *Algorithms and Computer Implementations*. New York, NY, USA: John Wiley & Sons, Inc; 1990.
[20] Karaliopoulos M., Chatzieleftheriou L.E., Darzanos G., Koutsopoulos I . On the joint content caching and user association problem in small cell networks. *In: IEEE International Conference on Communications Workshops (ICC Workshops)*; 2020. pp. 1–6.
[21] Darzanos G., Chatzieleftheriou L.E., Karaliopoulos M., Koutsopoulos I. Content preference-aware user association and caching in cellular networks. *In: WiOPT Workshop on Content Caching and Delivery in Wireless Networks (CCDWN)*; 2020. pp. 1–8.
[22] Kaafar M.A., Berkovsky S., Donnet B., *et al.* 'On the potential of recommendation technologies for efficient content delivery networks'. *ACM SIGCOMM Computer Communication Review*. 2013;**43**(3):74–7.
[23] Sermpezis P., Giannakas T., Spyropoulos T., Vigneri L. 'Soft cache hits: improving performance through recommendation and delivery of related content'. *IEEE Journal on Selected Areas in Communications*. 2018;**36**(6):1300–13.

[24] Giannakas T., Sermpezis P., Spyropoulos T. Show me the Cache: Optimizing Cache-Friendly Recommendations for Sequential Content Access. *In: IEEE WoWMoM*; 2018. pp. 1–9.

[25] Giannakas T., Spyropoulos T., Sermpezis P. The Order of Things: Position-Aware Network-friendly Recommendations in Long Viewing Sessions. *In: WiOpt*; 2019. pp. 1–6.

[26] Munaro D., Delgado C., Menasché D.S. Content recommendation and service costs in swarming systems. *In: ICC*; 2015. pp. 1–6.

[27] Kastanakis S., Sermpezis P., Kotronis V., Dimitropoulos X. CABaRet: Leveraging Recommendation Systems for Mobile Edge Caching. *In: ACM SIGCOMM workshops: MECOMM*; 2018. pp. 1–7.

[28] Liu D., Yang C. A Learning-based approach to joint content caching and recommendation at base stations. *In: IEEE GLOBECOM*; 2018. pp. 1–6.

[29] Qi K., Chen B., Yang C., *et al.* Optimizing caching and recommendation towards user satisfaction. *In: 2018 10th International Conference on Wireless Communications and Signal Processing (WCSP)*; 2018. pp. 1–7.

[30] Guo K., Yang C. 'Temporal-spatial recommendation for caching at base stations via deep reinforcement learning'. *IEEE Access*. 2019;7:58519–32.

[31] Lin Z., Chen W. Joint pushing and recommendation for susceptible users with time-varying connectivity. *In: IEEE GLOBECOM*; 2018. pp. 1–6.

[32] Liu D., Yang C. 'A deep reinforcement learning approach to proactive content pushing and recommendation for mobile users'. *IEEE Access*. 2019;7:83120–36.

[33] Zhu B., Chen W. Coded caching with joint content recommendation and user grouping. *In: IEEE GLOBECOM*; 2018. pp. 1–6.

[34] Sermpezis P., Kastanakis S., Pinheiro J.I., *et al*. 'Towards QoS-Aware recommendations'. *In: arXiv: 1907.06392*. 2019:1–6.

[35] Guo K., Yang C., Liu T. Caching in base station with recommendation via q-learning. *In: 2017 IEEE Wireless Communications and Networking Conference (WCNC)*; 2017. pp. 1–6.

Chapter 22

Caching with recommendation and reinforcement learning

Dong Liu[1] *and Chenyang Yang*[2]

The benefit of proactive caching at mobile devices hinges on the accurate prediction of user demands and mobility, which, however, is hard to achieve due to the uncertain user behavior. In this chapter, we show how to leverage personalized content recommendation to reduce the uncertainty in user demands for increasing the caching gain at mobile devices. Specifically, a joint content pushing and recommendation problem is formulated for maximizing the net profit of the mobile network operator. To cope with the challenges in modeling and learning user behavior, a reinforcement learning (RL) framework is introduced for solving the problem. Since the joint problem encounters the curse of dimensionality due to very large action and state spaces, it is decomposed into two sub-problems where two agents (namely a *recommendation agent* and a *pushing agent*) having different goals cooperate. Then, the sub-problems are solved by using the double deep-Q network with dueling architecture. Our simulation results show that the learned recommendation and pushing policies increase the net profit remarkably when compared with the baseline policies.

22.1 Introduction

Mobile edge caching is a promising way to support the exponentially increasing traffic demands and enhance user experience [1–4]. By caching popular contents at base stations (BSs), the service latency and backhaul load can be reduced. Meanwhile, the throughput and energy efficiency of the network can be increased remarkably [5–7]. By proactively pushing the contents into the cache of mobile devices under favorable channel conditions [3, 8, 9], if the user requests a pushed content under hostile channel condition, the transmission over wireless channel can be further reduced.

Because of the vast amount of contents available on the Internet but relatively limited cache size at individual user device, the caching gain of proactive pushing

[1]Beihang University, China
[2]Beihang University, China

hinges on accurately predicting the user behavior. Although *user preference*, defined as the probability distribution of a user requesting each content in a library [10, 11] or as the rating given by the user for each content [12], can be learned via various machine learning methods such as collaborative filtering [11–13], precisely predicting when and where a user will request which content is still not easy. As a result, the caching gain, such as the cache-hit ratio, at mobile devices will not be satisfactory [3].

On another note, user demands are increasingly driven by recommendation systems, whose original goal is to relieve users from information overload via recommending the best matching contents to the user. This, in turn, improves user stickiness and further boosts the number of content requests [14]. For example, integrated with powerful recommendation algorithms, Douyin (named as TikTok in the US), a short video app developed in China, was the most downloaded iPhone app in the world for the first quarter of 2018 [15]. Considering that people often do not know which content they really like until it is shown to them, recommendation systems can be leveraged as a powerful tool to shape user demand so that the user behavior is less uncertain and hence it has the potential to unleash the caching gain.

So far, wireless edge caching and recommendation have been considered to be operated by different entities, i.e., mobile network operators (MNOs) and content providers, respectively. As a result, caching policy and recommendation policy were designed independently in different communities [1–14, 16–18]. Yet content caching and recommendation are coupled, since recommendation affects which content a user may request, which further determines the caching policy. Recently, there is an increasing trend of integrating recommendation with edge caching [19–23]. In [19], by modifying the rating of contents shown to each user, the uncertainty in user demands is reduced so that the cost of content provider for proactive pushing can be saved. In [20–22], by recommending contents that are both appealing to each user and cached at the BS, the cache-hit ratio of BSs can be increased. In [20], the caching policy was first optimized based on recommending top-N matching contents according to each individual user's preference. Then the recommendation list was adjusted based on the cached contents so that a cached content is more likely to be recommended [21], proposed to recommend related content to the user if the originally requested content is not cached at BSs, and then optimized the caching policy for maximizing a "soft" cache-hit ratio.

However, in [19–21], how user demands are affected by recommendation is assumed known as a priori. Specifically, in [19, 20], the request probability distribution of a user over all the contents after recommendation is assumed known. In [21], the probability of a user accepting a recommended content is assumed known. Those assumptions are not realistic, because the user's reaction to the recommendation is not known unless implementing controlled experiments for measuring the content requests of a user before and after recommendation. In [22], the influence of recommendation on user demands was modeled by introducing a user-specific psychological threshold. Then, an ε-greedy algorithm was proposed to learn the joint recommendation and caching policy via interactions with users. In [23], RL was adopted to learn the caching policy for BSs with recommendation, where the recommended contents are chosen from the cached contents at the BS, which improves the caching gain by making the users aware which contents are cached at the BS. However, the preferences of different user are not distinguished in [23] and hence the recommended contents are identical for each individual user. In reality, user preferences are heterogeneous [11]. Hence, some users may not be satisfied with the recommended contents, leading to performance loss when compared with personalized recommendation as evaluated in [22].

In this chapter, we introduce our recent work [24] that leverages personalized content recommendation to reduce the uncertainty of user demands for increasing the caching gain at mobile devices. Specifically, a joint content pushing and recommendation problem is formulated to maximize the net profit of the MNO. To overcome the challenges in modeling and learning user behavior, we establish an RL framework for solving the problem. Since the straightforward formulation of RL encounters with the curse of dimensionality due to very large action and state spaces, the joint problem is decomposed into two sub-problems. The goal of the first problem (i.e., the recommendation problem) is to recommend a preferred content to the user for increasing the user stickiness, which boosts user requests and hence the revenue of MNO. The goal of the second problem (i.e., the pushing problem) is to reduce the transmission cost via proactive pushing based on the learned recommendation policy. Consequently, we introduce two RL agents with different sub-goals that are trained sequentially, but work together after training to achieve the final goal. By leveraging the ability of recommendation in reducing user demand uncertainty, the action candidates of the pushing problem can be narrowed down. Furthermore, the state and action of the pushing problem are redesigned to enhance the generalization of action values from the experienced states to the unexperienced states. Finally, we employ deep reinforcement learning (DRL) algorithm, specifically double deep Q-network (DDQN) with dueling architecture [25], for solving both sub-problems. Our simulation results show that the net profit can be increased remarkably by the DRL-based recommendation and pushing policies when compared with baseline policies. The performance comparison of DDQN and several other DRL algorithms are also provided.

The rest of this chapter is organized as follows. Section 22.2 introduces the system model. In Section 22.3, we formulate the joint recommendation and pushing problem, decompose the joint problem into two sub-problems, and solve them by DRL. Simulation results are provided in Section 22.4. Conclusions and future works are discussed in Section 22.5.

22.2 System model

In this section, we first introduce the basic terminology and concepts in RL as well as the network architecture to support DRL. Then, we describe operation of content recommendation and pushing in the considered wireless network, as well as the basic idea of joint recommendation and pushing.

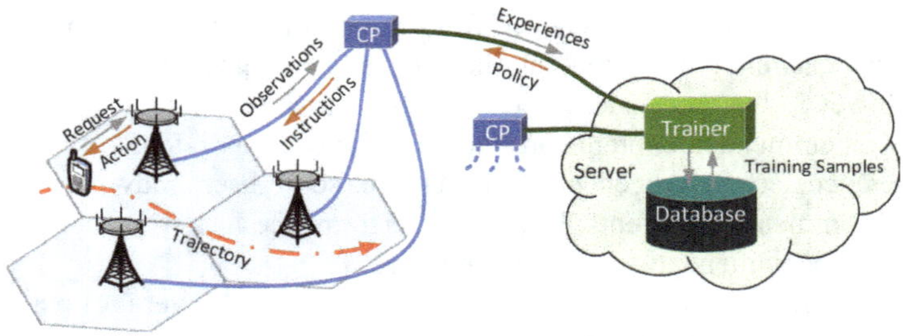

Figure 22.1 *Learning-enabled network architecture. Both the CP and the trainer jointly serve as the RL agent.*

22.2.1 Basic terminology and concepts in RL

A standard RL problem can be formulated as a Markov decision process (MDP), where an *agent* learns how to achieve a goal from its interactions with the *environment* [26]. For an MDP with finite-time horizon, the agent and environment interact in a sequence of discrete time steps $t = 1, 2, \cdots, T$, which constitutes an *episode*. At each time step t, the agent observes the *state* of the environment $\mathbf{s}(t)$ and executes an *action* $\mathbf{a}(t)$. Then, the agent receives a reward $r(t)$ from the environment and transits into a new state $\mathbf{s}(t+1)$. The interaction of the agent with the environment is then captured by a vector $\mathbf{e}(t) \triangleq [\mathbf{s}(t), \mathbf{a}(t), r(t), \mathbf{s}(t+1)]$, termed as the *experience*.

The goal of the agent is to learn a *policy* from its experiences to maximize an expected *return*, which reflects the cumulative reward received by the agent during the T-time-step episode. The policy (denoted by $\boldsymbol{\pi}$) determines which action should be executed in each state. The expected return is defined as $\mathbb{E}[\sum_{i=1}^{T} \gamma^{i-1} r(i)]$, where γ denotes the discount factor.

22.2.2 Learning-enabled network architecture

To enable RL for mobile edge caching and recommendation, we demonstrate a learning-enabled cellular network architecture proposed by [27] in Figure 22.1.

As shown in Figure 22.1, multiple BSs within an area are connected to a central processor (CP) and each CP is connected to a server, which contains a trainer and a database. The CP handles user association and monitors the status of mobile users via the associated BS and sent the collected experiences to the database in the server. A centralized trainer at the server learns the policy based on the stored experiences, and issues the learned policy to the CPs. Each CP implements the learned policy and sends instructions for controlling the associated BS to execute actions.

Both the CP and the trainer jointly act as the so-called agent in RL parlance, but reside in different network entities owing to the following reasons. The CP is located closer to BSs for reducing the delay and signaling overhead. The trainer is centralized at the top of the network for enabling sufficient utilization of computing resources and for efficient learning via experiences gathered throughout the whole network.

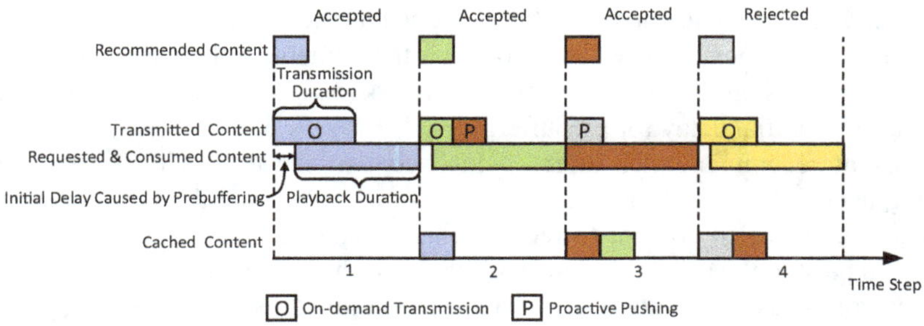

Figure 22.2 *An example session with four requests. The contents are distinguished by the color of each block.*

22.2.3 Recommendation, pushing, and user requests

We consider the scenario where mobile users are traveling across multiple cells covered by a CP when requesting contents. For simplicity, we model the user mobility as a Markov process, where the probability distribution of the next location of a user only depends on its current location and last location.

Each user may request multiple contents from a library stored at the server (or at each BS) sequentially during a period of time, e.g., watching short videos one by one using a mobile app. Such a period of continuously sending requests by a user is termed as a *session* (i.e., an episode in RL parlance), which can be divided into multiple discrete time steps, as shown in Figure 22.2. The user requests and consumes one content in each time step.

At the beginning of each time step, the associated BS of the user recommends a content to the user from a pre-determined recommendation candidate set containing F contents[a] according to the policy stored at the CP. The recommendation is presented to the user by showing the title or thumbnail of the content through the mobile app, which incurs negligible transmission cost compared with sending the whole content. If the user accepts the recommendation, the user will request and consume the recommended content, and then terminate the session with probability q_1. If the user rejects the recommendation, the user will instead request and consume another content according to a specific distribution, and then terminate the session with probability q_2. Because users are more likely to continue requesting contents if their preferred contents are recommended to them, we assume $q_1 < q_2$, which reflects the effect of recommendation in improving user stickiness.

Meanwhile, at each time step, the associated BS of the user can also proactively push one content to the user. If a user requests a content pre-stored in its own cache, the

[a] We assume that the recommendation candidates have been winnowed down to hundreds of contents by machine learning methods, say by deep neural networks [28].

content can be directly fetched from the cache so that the user can enjoy zero latency. Otherwise, the associated BS has to transmit the content to the user. Such an on-demand transmission introduces cost to the network (e.g., energy consumption of BS) and causes an initial delay for prebuffering.

Content recommendation affects which content the user will request, while pushing itself introduces transmission cost. In each time step, the CP should intelligently decide which content to recommend, whether or not to push, and which content to push to the user. Otherwise, if the BS pushes a content under bad channel condition or the user does not request the pushed content during the session, pushing actually increases the total transmission cost instead of saving it.

In Figure 22.2, we provide an example session for illustrating the operation of recommendation and pushing. The session begins when the user opens the mobile app. The BS recommends a content at the beginning of each time step. In time step 1, the user accepts the recommended content, which is then send to the user via on-demand transmission because it has not been cached in the mobile device. Supposing that the channel condition in time step 1 is not good, the BS does not push any content in time step 1. Moreover, supposing that the channel condition becomes better in time steps 2 and 3, the BS pushes the contents to be recommended in time steps 3 and 4, respectively. When the user accepts the recommendation in time step 3, it can directly consume the content without prebuffering because the recommended content is already pushed to the user in time step 2. In time step 4, the user rejects the recommendation and terminates the session after it consumes the 4th requested content.

The preference of a user is reflected by its request history, and a user will accept a recommendation with high probability if it prefers the recommended content. This suggests that the probability of a user accepting a recommendation (termed as *acceptance probability*) depends on its recently requested contents [29]. Let $\mathcal{Q}$ denote the set of recently requested contents of a user and $p_{\mathcal{Q},j}$ denote the acceptance probability of content j; conditioned on that the recently requested content set is $\mathcal{Q}$. For notational simplicity, we consider the simplified case where the acceptance probability only depends on the user's most recently requested content as an illustration. In this case, $p_{\mathcal{Q},j}$ is degenerated into p_{ij}, which denotes the probability that the users who have just requested content i will accept the recommendation of content j. Consequently, the reaction of a user to recommendation can be characterized by a matrix $\mathbf{P} = [p_{ij}]_{F\times F}$.

Since it is easier for a user to simply accept the recommendation of a sufficiently attractive content rather than searching for a preferred content among the vast amount of available content, the probability of the user requesting the recommended content is higher than the probability of requesting the same content without recommendation. In a nutshell, a good recommendation can reduce the uncertainty in user requests.

22.3 Recommendation and pushing by DRL

In this section, we first formulate recommendation and pushing as a joint problem. Then, we show that it is hard to solve the joint problem if we implement RL or DRL in a straightforward manner because of the very large state and action spaces. Finally, we decompose the joint optimization into two sub-problems and resort to DRL to solve them.

22.3.1 Problem formulation

Let $m(t)$ denote the transmission cost of a content at time step t. Considering that the energy consumed at a BS for transmitting a content to a user increases with the duration of transmission, $m(t)$ can be modeled as proportional to the transmission duration. By further considering that the transmission duration is inversely proportional to the average data rate [7], then, $m(t)$ is modeled by

$$m(t) = \frac{\beta}{R(t)} \tag{22.1}$$

where $R(t)$ denotes the average rate of a user and β is a known scaling factor representing the cost for transmitting a content per unit time.

Upon assuming that each user is associated with the BS having the strongest average received power and each BS serves the associated users over orthogonal time-frequency resources, $R(t)$ can be expressed as

$$R(t) = \mathbb{E}_{h_i}\left[W \log_2\left(1 + \frac{P h_0 d_0(t)^{-\alpha}}{\sum_{i \in \Phi \backslash 0} P h_i d_i(t)^{-\alpha} + \sigma^2}\right)\right] \tag{22.2}$$

where W is the transmission bandwidth for the user, P and σ^2 are the transmit power of each BS and the noise power, respectively, α is the path-loss exponent. h_0 and $d_0(t)$ denote, respectively, the small-scale fading channel and the distance between the associated BS to the user, Φ denotes the set of the BSs in the network and $\sum_{i \in \Phi \backslash 0} P h_i d_i(t)^{-\alpha}$ is the total interference power from other BSs. Finally, $\mathbb{E}_{h_i}$ denotes the expectation taken over the small-scale fading.

Then, the total transmission cost in time step t can be expressed as

$$m(t)[I_o(t) + I_p(t)] \tag{22.3}$$

where $I_o(t)$ and $I_p(t)$ are the indicator functions. We define $I_o(t) = 1$ if the requested content is not cached by the user in time step t and hence requires on-demand transmission, and define $I_o(t) = 0$ otherwise. We define $I_p(t) = 1$ if pushing occurs in time step t, and define $I_p(t) = 0$ otherwise.

Let π denote the joint recommendation policy and pushing policy. We can formulate the following problem for maximizing the average accumulated net profit of MNO during each session.[b]

$$\max_{\pi} \mathbb{E}[\sum_{t=1}^{T} \eta - \sum_{t=1}^{T} m(t)[I_o(t) + I_p(t)]] \tag{22.4a}$$

$$s.t. \sum_{f=1}^{F} c_f(t) \leq C, \ \forall t \tag{22.4b}$$

where η is another scaling factor representing the revenue of each request whose value is determined by how much the MNO charges a user, $c_f(t) = 1$ if content f is cached in time step t and $c_f(t) = 0$ otherwise, C reflects the cache capacity. The first and second terms of (22.4a) reflect the accumulated revenue and transmission cost for the session, respectively. The expectation in (22.4a) is taken over all random variables in each time step, including user locations, fading channels, content requests, and whether a session terminates after a content is consumed (which determines the session length T). Equation (22.4b) is the cache size constraint.

The user behaviors, including the mobility patterns, acceptance probability matrix $\mathbf{P}$, the request distribution among contents when rejects a recommendation, and the session end probabilities q_1 and q_2, as well as channel distributions are all not known a priori, which calls for a model-free approach, we resort to RL to solve the optimization problem.

22.3.2 RL framework for joint optimization

To implement RL algorithms, it is important to design the state, action, and reward for the agent. A straightforward formulation of the RL framework for (22.4) is as follows.

State: The recommendation and pushing action in time step t should depend on a user's last consumed content with index denoted as $f_q(t-1)$,[c] the cache status of the user denoted as $\mathbf{c}(t) = [c_1(t), \cdots, c_F(t)]$, as well as the costs to transmit a content in current and future time steps. The transmission cost in a time step depends on the channel condition of the user, which further depends on the user location. According to the Markov property of user mobility, the future transmission cost is determined by the transmission cost in the last two time steps. Therefore, we include the transmission costs $m(t-1)$, $m(t-2)$ into the state.[d] Note that the difference between the costs of successive time steps can reflect the change of channel conditions. For example, if

[b]We can also consider other goals, such as minimizing the transmission delay, by adjusting the reward.

[c]If the acceptance probability depends on multiple recently consumed contents, then $f_q(t-2), f_q(t-3), \ldots,$ should also be included in the state.

[d]To capture more complex user mobility pattern, $m(t-3)$, $m(t-4), \ldots$ can also be included into the state.

$m(t-1) < m(t-2)$, then it implies that the user was moving toward the cell center in time step $(t-1)$. The whole state can be denoted by vector

$$\mathbf{s}(t) = [f_\mathrm{q}(t-1), \mathbf{c}(t), m(t-1), m(t-2)] \tag{22.5}$$

where $f_\mathrm{q}(t-1)$ and $\mathbf{c}(t)$ have $F \times \sum_{i=0}^{C} \binom{F}{i}$ possible discrete combinations, $m(t-1)$ and $m(t-2)$ have continuous values.

Action: Let $f_\mathrm{r}(t)$ and $f_\mathrm{p}(t)$ denote the indexes of the content to be recommended and the content to be pushed in time step t, respectively. In particular, we define $f_\mathrm{p}(t) = 0$ when no content is pushed in time step t. When the cache of the user is full and a content has to be pushed to the user, the least-recently consumed content in the cache will be removed to ensure the cache size constraint (22.4b) because a consumed content is not likely to be requested by the user again.[e] Then, the action can be denoted by

$$\mathbf{a}(t) = [f_\mathrm{r}(t), f_\mathrm{p}(t)] \tag{22.6}$$

which has $F \times (F+1)$ possible combinations.

Reward: Considering that our goal is to maximize the average accumulated net profit of MNO during a session, the reward can be naturally designed as the net profit earned in each time step, i.e.,

$$r(t) = \eta - m(t)\left[I_o(t) + I_p(t)\right] \tag{22.7}$$

The *action-value function* (i.e., the Q-function) of policy π is defined as

$$Q^\pi(\mathbf{s}, \mathbf{a}) \triangleq \mathbb{E}\left[\sum_{i=t}^{T} \gamma^{i-t} r(i) \,\middle|\, \mathbf{s}(t) = \mathbf{s}, \mathbf{a}(t) = \mathbf{a}, \pi\right] \tag{22.8}$$

which is the expected return achieved by taking action $\mathbf{a}$ at state $\mathbf{s}$ and thereafter taking action according to policy π. Note that when the discount factor $\gamma = 1$, the expected return is actually the average accumulated net profit.

The *optimal action-value function* is the action-value function of the optimal policy

$$Q^*(\mathbf{s}, \mathbf{a}) \triangleq \max_\pi Q^\pi(\mathbf{s}, \mathbf{a}) \tag{22.9}$$

With the optimal action-value function $Q^*(\mathbf{s}, \mathbf{a})$, the optimal policy can be readily obtained from

$$\pi^*(\mathbf{s}) \triangleq arg \max_\mathbf{a} Q^*(\mathbf{s}, \mathbf{a}) \tag{22.10}$$

Therefore, the goal of the agent can be achieved by learning $Q^*(\mathbf{s}, \mathbf{a})$ for every single state-action pair.

Q-learning has been widely used to learn $Q^*(\mathbf{s}, \mathbf{a})$ from experience $\mathbf{e}(t) = [\mathbf{s}(t), \mathbf{a}(t), r(t), \mathbf{s}(t+1)]$ by the following iteration

[e]If all the cached contents have not been consumed yet, the least-recently cached content will be removed [24].

$$Q(\mathbf{s}(t), \mathbf{a}(t)) \leftarrow Q(\mathbf{s}(t), \mathbf{a}(t)) + \delta \left[r(t) + \gamma \max_{\mathbf{a}} Q(\mathbf{s}(t+1), \mathbf{a}) - Q(\mathbf{s}(t), \mathbf{a}(t)) \right] \tag{22.11}$$

which has guaranteed convergence to $Q^*(\mathbf{s}, \mathbf{a})$ relying on sufficient experiences and properly chosen learning rate δ [30].

However, the state and action spaces of the straightforward formulation (also termed as the *Non-decomposed* framework in the sequel) are too large for RL algorithms (such as Q-learning) to learn the action values in each possible state. Although DRL uses function approximation for reducing the dimensions, the problem is still very hard to solve considering the combinatorial actions and states involved.

22.3.3 Decomposed RL framework

The large action and state spaces are resulted from the coupling between recommendation and pushing. On one hand, recommendation affects the content request and hence affects which content should be pushed. On the other hand, pushing policy determines the cached content, based on which recommendation policy can be manipulated for guiding the user to request a cached content. For example, instead of recommending the most attractive but non-cached content, the BS can recommend a less attractive but cached content to the user [20].

Nonetheless, the gain from exploiting the impact of pushing on recommendation is much lower than the gain from exploiting the impact of recommendation on pushing. This can be explained by the following facts. When the recommendation policy is designed to recommend a preferred content and the pushing policy is designed to push the recommended content in advance, it is of high probability that user's requested content has been cached. On the contrary, if the pushing policy fails to push the preferred contents, then the cached content is not preferred by the user and hence it is of low probability that the user's requested content is cached even if the BS recommends a cached content. This is because a user is not likely to accept the recommendation of a non-preferred content. In a nutshell, recommendation does not help much in stimulating the user to request a cached but not preferred content.

Moreover, compared with recommendation which is able to directly increase the net profit by boosting the revenue from the growth of user requests [14], pushing helps increase the net profit indirectly be reducing the cost, which only matters when the revenue has already been boosted.

Based on the above analysis, we design the pushing policy based on the recommendation policy, instead of the other way around. Furthermore, benefited from the ability of recommendation in reducing the uncertainty of user requests as previously explained, it is possible to precisely predict the future requested content from a well-learned recommendation policy.

Therefore, we decompose the joint problem into two sub-problems by distinguishing the roles of recommendation and pushing policies explicitly. In particular, we introduce two agents. The first agent learns the recommendation policy, and the second agent learns the pushing policy based on the learned recommendation policy.

22.3.3.1 Recommendation sub-problem

The goal of recommendation is to enhance user stickiness by providing the preferred contents to a user. If the recommended contents match the taste of a user sufficiently well, the user will accept the recommendation and will request more contents during a session. Therefore, we design the recommendation policy for maximizing the average accumulated revenue $\mathbb{E}\left[\sum_{t=1}^{T}\eta\right]$.

The action and reward of the recommendation agent are denoted by $a_1(t) = f_r(t)$ and $r_1(t) = \eta$, respectively. Because which content to recommend should depend on the user's last consumed content, the state in recommendation problem is represented by $s_1(t) = f_q(t-1)$. Then, the index of the content recommended in time step t is

$$f_r(t) = \pi_r(f_q(t-1)) \tag{22.12}$$

where $\pi_r(\cdot)$ denotes the recommendation policy.

The state and action have F possible values, respectively. Compared with the straightforward formulation, $\mathbf{c}(t)$, $m(t-1)$, and $m(t-2)$ are excluded from the state and the $f_p(t)$ is excluded from the action. Hence, the number of possible combinations of the state and action in the decomposed recommendation sub-problem is much smaller.

22.3.3.2 Pushing sub-problem

The goal of pushing is to reduce the cost of the network and improve user experience by pre-downloading contents that the user is likely to request in future time steps. If a requested content was pushed in advance under better channel condition than that when user initiates the request, then the transmission cost can be reduced. Hence, we design pushing policy for minimizing the average accumulated cost $\sum_{t=1}^{T} m(t)[I_o(t) + I_p(t)]$ based on a well-learned recommendation policy.

Although the intention of pushing policy is to push the contents that the user is likely to request in future time steps, in the straightforward RL formulation of the joint problem, the agent does not explicitly predict which contents the user will request in the future. Instead, it directly learns the joint recommendation and pushing policy by maximizing the expected return (i.e., the accumulated net profit), which requires huge number of training samples for trial-and-error due to the very large state and action spaces. By contrast, when we decompose the joint problem into two sub-problems and learn the recommendation policy first, the contents to be requested in multiple time steps ahead can be predicted with high precision because recommendation is capable of reducing the uncertainty in user requests. Therefore, the state and action spaces of the pushing problem can be reduced dramatically, as detailed below.

When the recommendation policy is well-learned, the user will accept the recommendation with high probability. Therefore, it is of high precision to predict that the user will request the recommended content. Let $\hat{f}_q(t)$ us denote predicted content to be requested in time step t. We have $\hat{f}_q(t) = f_r(t)$. Then, the content to be recommended in the next time step can also be predicted by employing (22.12) as

$$\hat{f}_{\rm r}(t+1) = \pi_{\rm r}(\hat{f}_{\rm q}(t)) = \pi_{\rm r}(f_{\rm r}(t)) = \pi_{\rm r}(\pi_{\rm r}(f_{\rm q}(t-1))) \tag{22.13}$$

In a similar fashion, we can predict that the content to be recommended (and hence to be requested by the user) in the next n-step of time step t as

$$\hat{f}_{\rm r}(t+n) = \pi_{\rm r}(\cdots \pi_{\rm r}(f_{\rm q}(t-1))) \triangleq \pi_{\rm r}^{n+1}(f_{\rm q}(t-1)) \tag{22.14}$$

Because the precise prediction of the next n-step recommendation requires that the user accepts the recommended contents in the 1st $\sim$ $(n-1)$th steps, the prediction becomes less precise with the increase of n. Therefore, it is not necessary to push the content to be recommended in too many time steps ahead. Further considering the limited cache size of the user device, we restrict that the pushing agent only pushes the contents that are predicted to be recommended within the next N $(N \leq C)$ steps commencing from time step t (termed *pushing candidate set*), which can be formulated as

$$f_{\rm p}(t) \in \{f|f = \hat{f}_{\rm r}(t+n) \text{ for } n = 1, \cdots, N\} \tag{22.15}$$

This reduces the number of possible actions of the pushing agent from $F+1$ to $N+1$ (including no pushing) for each state.

However, if we simply use the index of the content to be pushed, i.e., $f_{\rm p}(t)$, to represent the action, the possible actions to be executed are distinct for each possible value of $f_{\rm q}(t-1)$ (which is a part of the state vector of the pushing agent as will be introduced later), because the pushing candidate set are, in general, different for different values of $f_{\rm q}(t-1)$. This makes it difficult for the pushing agent to generalize the action values from experienced states to the unexperienced states [24].

Fortunately, we can find a mapping from any given value of $f_{\rm q}(t-1)$ to the index of the content to be pushed in time step t by substituting (22.14) to (22.15) as,

$$f_{\rm p}(t) = \pi_{\rm r}^{n+1}(f_{\rm q}(t-1)) \tag{22.16}$$

With this mapping, the agent only needs to decide how many time steps ahead a content to be recommended should be pushed in the current time step t, i.e., to decide n, instead of directly deciding the specific index of the content to be pushed, i.e., $f_{\rm p}(t)$. Hence, the action can be represented by $a_2(t) = n$. When $n = 1, \cdots, N$, the content to be recommended in the next n-step (i.e., $\hat{f}_{\rm r}(t+n)$) is pushed in time step t. In particular, we define $n = -1$ if no content is pushed. In this way, the possible action set, i.e., $\{-1, 1, 2, \cdots, N\}$, is the same for each possible value of $f_{\rm q}(t-1)$, which makes the agent able to learn the optimal Q-function with less experiences. Then, the index of the content to be pushed at time step t can be expressed as

$$f_{\rm p}(t) = \pi_{\rm r}^{a_2(t)+1}(f_{\rm q}(t-1)) \tag{22.17}$$

In particular, we define $\pi_{\rm r}^{0}(\cdot) \triangleq 0$, i.e., no content is pushed.

Since the agent only selects content from the pushing candidate set to push, it is the cache status of the contents to be recommended (rather than the cache status of all the recommendation candidates) that affects whether a content to push is already in the cache. We define $\tilde{c}_n(t) = 1$ if content $\hat{f}_{\rm r}(t+n)$ is cached and $\tilde{c}_n(t) = 0$ otherwise. Then, the cache status can be represented as $\tilde{\mathbf{c}}(t) \triangleq [\tilde{c}_0(t), \tilde{c}_1(t), \cdots, \tilde{c}_N(t)]$ (instead of $\mathbf{c}(t)$ defined in section III-B) for the pushing agent. Now, the number of possible cache status is reduced from $\sum_{i=0}^{C}\binom{F}{i}$ to 2^N. In practice, $2^N \leq \sum_{i=0}^{C}\binom{F}{i}$ due to the limited cache size at user device, i.e., $N \leq C \ll F$.

Further including $m(t-1)$ and $m(t-2)$, we can represent the state vector of the pushing agent as

$$\mathbf{s}_2(t) = [f_q(t-1),\ \tilde{\mathbf{c}}(t),\ m(t-1),\ m(t-2)] \tag{22.18}$$

The reward of the pushing agent is the minus value of the transmission cost

$$r_2(t) = -m(t)[I_o(t) + I_p(t)] \tag{22.19}$$

By decoupling recommendation from pushing, limiting the number of possible actions, and representing the state and action of the pushing agent in a proper manner, the two sub-problems are now ready for implementing DRL.

DRL algorithms can be generally divided into value-based methods such as deep Q-network (DQN), and policy-based (or more specifically, actor-critic) methods. Among the state-of-the-art policy-based RL algorithms, DDPG [31] is only applicable in continuous action space, while A2C [32] and PPO [33] are on-policy algorithms that are less sample efficient. Moreover, A2C and PPO converge to stochastic policies in general, while a deterministic recommendation policy is more appealing for our problem at hand. This is because a deterministic recommendation policy makes it easier for the agent to accurately predict the contents to be recommended in future time steps, relying on which the pushing candidates set is determined. Therefore, we employ value-based DRL for solving the sub-problems.

22.3.4 DRL with dueling DDQN

We use two Q-networks to approximate the optimal action-value functions for the recommendation and pushing agents, respectively. The Q-networks are implemented based on an improved version of DQN, namely dueling DDQN [25], which modifies the original DQN [34] with dueling network architecture and introduces double Q-learning update [35] for more efficient and stable learning.

Specifically, the Q-networks of the two agents are shown in Figure 22.3, which are stored at CPs. The network parameters $\boldsymbol{\theta}_r, \boldsymbol{\theta}_p$ are learned by the trainer at the server as shown in Figure 22.1 and then sent to CPs to update the Q-networks, based on which the CP selects actions for the BS to execute under each state.

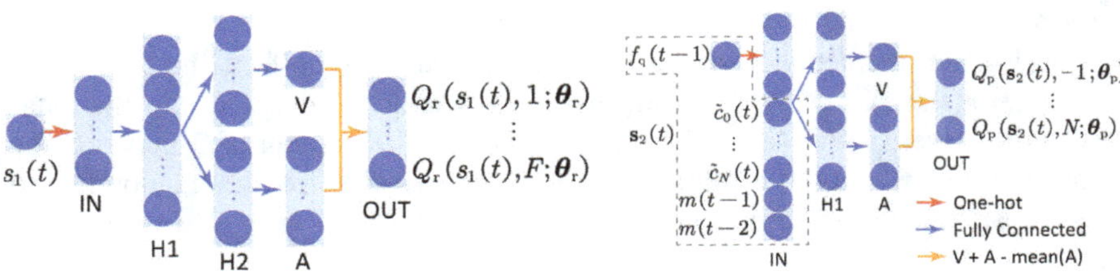

Figure 22.3 Q-networks with dueling architecture

For the recommendation agent, the state $s_1(t) = f_q(t-1)$ is the index of last consumed content. The state is first converted into a F-dimensional one-hot vector and then goes through two fully-connected hidden layers. The second hidden layer H2 is equally split and then flows as two streams, V-stream $V_r(s_1(t); \boldsymbol{\theta}_r)$ and A-stream $A_r(s_1(t), a_1(t); \boldsymbol{\theta}_r)$, forming the dueling architecture [25]. This enables more efficient learning by separately estimating the state value and advantages for each action. The output layer implements the following mapping

$$\begin{aligned} Q_r(s_1(t), a_1(t); \boldsymbol{\theta}_r) &= V_r(s_1(t); \boldsymbol{\theta}_r) + A_r(s_1(t), a_1(t); \boldsymbol{\theta}_r) \\ &\quad - \frac{1}{F}\sum_{a=1}^{F} A_r(s_1(t), a; \boldsymbol{\theta}_r) \end{aligned} \tag{22.20}$$

For the pushing agent, $f_q(t-1)$ as a part of state $\mathbf{s}_2(t)$, is first converted into an F-dimensional one-hot vector, and then concatenated with remaining parts of $\mathbf{s}_2(t)$ as the input layer. After going through a fully connected hidden layer and the dueling architecture, the output layer (similar to (22.20)) yields the approximated action-value $Q_p(\mathbf{s}_2(t), a_2(t); \boldsymbol{\theta}_p)$.

Considering that the training procedures for the two networks are the same, we omit all the subscripts in the following for notational simplicity. The parameter $\boldsymbol{\theta}$ of the Q-network is learned by minimizing the loss function

$$L(\boldsymbol{\theta}) = \mathbb{E}\left[\left(y(t) - Q(\mathbf{s}(t), \mathbf{a}(t); \boldsymbol{\theta})\right)^2\right] \tag{22.21}$$

with

$$y(t) = r(t) + \hat{Q}(\mathbf{s}(t+1), \arg\max_{\mathbf{a}} Q(\mathbf{s}(t+1), \mathbf{a}; \boldsymbol{\theta}); \hat{\boldsymbol{\theta}}) \tag{22.22}$$

where $\hat{Q}(\mathbf{s}, \mathbf{a}; \hat{\boldsymbol{\theta}})$ is the target network, whose architecture is the same as the Q-network but is updated by $\hat{\boldsymbol{\theta}} \leftarrow \tau\boldsymbol{\theta} + (1-\tau)\hat{\boldsymbol{\theta}}$ with very small value of τ for reducing the correlations between the action-values $Q(\mathbf{s}, \mathbf{a}; \boldsymbol{\theta})$ and the target values $y(t)$, which improves the convergence of learning [34].

In (22.22), we employ the double Q-learning update where the optimal action is chosen by the Q-network, i.e.,

$$\mathbf{a}_{\max} \triangleq \arg\max_{\mathbf{a}} Q(s(t+1), \mathbf{a} : \boldsymbol{\theta}) \tag{22.23}$$

Yet the optimal action value is evaluated by the target network, i.e., $\hat{Q}(\mathbf{s}(t+1), \mathbf{a}_{\max}; \hat{\boldsymbol{\theta}})$, which can reduce the overestimation of action value compared with traditional Q-learning update [35].

During the interaction with the environment, each CP collects the experience $\mathbf{e}(t) = [\mathbf{s}(t), \mathbf{a}(t), r(t), \mathbf{s}(t+1)]$ from all the users within its coverage into the database at the server as $\mathscr{D} = \{\mathbf{e}(1), \cdots, \mathbf{e}(t-1)\}$. During the training of Q-network, the trainer iteratively samples a mini-batch of the experiences from $\mathscr{D}$ randomly and updates parameter θ with gradient descent

$$\boldsymbol{\theta} \leftarrow \boldsymbol{\theta} - \frac{\delta}{|\mathscr{B}|}\nabla_{\boldsymbol{\theta}} \sum_{j\in\mathscr{B}} \left(y(j) - Q(\mathbf{s}(j), \mathbf{a}(j); \boldsymbol{\theta})\right)^2 \tag{22.24}$$

in each iteration, where $\mathcal{B}$ denotes the set of indexes of sampled experiences. By resampling the experiences stored in the database to update (22.24) (termed as *experience replay* [34]) the data efficiency can be improved and the correlation among training samples can be reduced for improving the convergence of learning.

As previously explained, effective pushing requires the precise prediction of the contents to be requested in future time steps, which further depends on high acceptance probability of the recommendation. Moreover, recommendation causes negligible transmission cost compared with pushing. To avoid high transmission cost during learning, the recommendation policy should be trained first, while the pushing agent should not push any content until enough experiences are gathered (say after L_1 episodes) for the recommendation agent to learn a recommendation policy with high acceptance probability. In addition, we let the pushing agent learn the pushing policy in an emulated environment while not really pushing any content, until the pushing policy is learned well enough (say after L_2 episodes). Otherwise, the pushing agent may not be able to push the right content at a proper time step, which may make pushing cause more transmission cost than no pushing and hence impairs the online performance.

To be more specific, between the $(L_1 + 1)$th episode and the L_2th episode, the recommendation agent continues to take actions in real environment. Meanwhile, the pushing agent observes the user requests and transmission costs from the real environment but only takes actions in the virtual environment. In the emulated environment, the recommendation agent operates the same as that in the real environment, and the user's reaction to the recommendation is also the same as that in the real environment. Hence, the reward obtained by the recommendation agent is the same as that obtained from the real environment, i.e., $r_1(t)$. Besides, in the emulated environment, the observations related to user mobility is the same as those observed in the real environment (because pushing does not affect user mobility), and hence the cost for transmitting a content at time step t is also the same as that observed in the real environment, i.e., $m(t)$. Yet, the pushing agent operates (i.e., pushes contents) in the emulated environment to learn the pushing policy. Then, based on (22.19), the reward obtained by the pushing agent in the emulated environment can be computed by

$$r'_2(t) = -m(t)[I'_{\mathrm{o}}(t) + I'_{\mathrm{p}}(t)] \tag{22.25}$$

where $I'_{\mathrm{o}}(t) = 1$ if the requested content is not cached by the user in the emulated environment, and $I'_{\mathrm{o}}(t) = 0$ otherwise, $I'_{\mathrm{p}}(t) = 1$ if a content is pushed in the emulated environment, and $I'_{\mathrm{p}}(t) = 0$ otherwise. Based on what content is pushed in the emulated environment, the cache status in the next time step (denoted by $\mathbf{c}'(t+1)$) can be obtained. Because pushing only affects the cache status in the state vector defined in (22.5), the next state of the emulated environment can be obtained as $\mathbf{s}'(t+1) = [f_{\mathrm{q}}(t), \mathbf{c}'(t+1), m(t), m(t-1)]$, where the only difference with the next state in the real environment is the cache status. The experiences obtained in the emulated environment are also put into the database as training samples.

Algorithm 1 Content recommendation and pushing learning

Initialize replay memory, i.e., the database $\mathscr{D}$.
Initialize Q-network Q_r, Q_p with random weights $\boldsymbol{\theta}_r$, $\boldsymbol{\theta}_p$.
Initialize target network $\hat{Q}_r, \hat{Q}_p$ with weights $\hat{\boldsymbol{\theta}}_r = \boldsymbol{\theta}_r$, $\hat{\boldsymbol{\theta}}_p = \boldsymbol{\theta}_p$.
for episode $= 1, 2, \cdots$ **do**
  for time step $t = 1, \cdots, T$ **do**
    With probability ε_r select a random recommendation $a_1(t)$, otherwise $a_1(t) = \pi_r(s_1(t)) \triangleq \arg\max_a Q_r(s_1(t), a; \boldsymbol{\theta}_r)$.
    if episode $\leq L_1$ **then**
      Do not push any content, $a_2(t) = -1$.
    else
      Update the predicted next 1- to N-step recommendations by
      $\hat{f}_r(t+1) = \pi_r^2(f_q(t-1)), \cdots$
      $\hat{f}_r(t+N) = \pi_r^{N+1}(f_q(t-1))$.
      With probability ε_p randomly select $a_2(t) \in \{-1, 1, \cdots, N\}$,
      otherwise select $a_2(t) = \arg\max_n Q_p(\mathbf{s}_2(t), n; \boldsymbol{\theta}_r)$.
    end if
    if $L_1 <$ episode $\leq L_2$ **then**
      if $t = 1$ **then**
        Initialize the emulated environment, $\mathbf{s}'(t) = \mathbf{s}(t)$.
      end if
      Execute action $\mathbf{a}(t) = [a_1(t), 0]$, observe reward $r_1(t)$, and record transmission cost $m(t)$.
      Execute action $\mathbf{a}'(t) = [a_1(t), \pi_r^{a_2(t)+1}(f_q(t-1))]$ in the emulated environment, obtain the reward for pushing agent by substituting the recorded $m(t)$ into (1.25), and observe next state $\mathbf{s}'(t+1)$.
      Store experience $[\mathbf{s}'(t), \mathbf{a}'(t), r_1(t), r_2'(t), \mathbf{s}'(t+1)]$ in $\mathscr{D}$.
    else
      Execute action $\mathbf{a}(t) = [a_1(t), \pi_r^{a_2(t)+1}(f_q(t-1))]$,
      observe reward $r_1(t)$, $r_2(t)$ and next state $\mathbf{s}(t+1)$
      Store experience $[\mathbf{s}(t), \mathbf{a}(t), r_1(t), r_2(t), \mathbf{s}(t+1)]$ in $\mathscr{D}$.
    end if
    Randomly sample a mini-batch of experiences from $\mathscr{D}$ as
    $\mathscr{B} = \{[\mathbf{s}(j), \mathbf{a}(j), r_1(j), r_2(j), \mathbf{s}(j+1)]\}$
    Get $s_1(t)$ from $\mathbf{s}(t)$
    Get $a_{\max} = \arg\max_a Q_r(s_1(j+1), a; \boldsymbol{\theta}_r)$ and set
    $$y_1(j) = \begin{cases} r_1(j), \text{ if episode terminates at step } j+1 \\ r_1(j) + \gamma \hat{Q}_r(s_1(j+1), a_{\max}; \hat{\boldsymbol{\theta}}_r), \text{ otherwise} \end{cases}$$
    Perform a gradient descent step minimizing
    $(y_1(j) - Q_r(s_1(t), a_1(t); \boldsymbol{\theta}_r))^2$ with respect to $\boldsymbol{\theta}_r$
    if episode $> L_1$ **then**
      Get state representation $\mathbf{s}_2(j)$ from $\mathbf{s}(j)$ based on π_r
      Get $n_{\max} = \arg\max_n Q_p(\mathbf{s}_2(j+1), n; \boldsymbol{\theta}_p)$ and set
      $$y_2(j) = \begin{cases} r_2(j), \text{ if episode terminates at step } j+1 \\ r_2(j) + \gamma \hat{Q}_p(\mathbf{s}_2(j+1), n_{\max}; \hat{\boldsymbol{\theta}}_p), \text{ otherwise} \end{cases}$$
      Perform a gradient descent step minimizing
      $(y_2(j) - Q_p(\mathbf{s}_2(t), a_2(t); \boldsymbol{\theta}_p))^2$ with respect to $\boldsymbol{\theta}_p$
    end if
    Update the target networks:
    $\hat{\boldsymbol{\theta}}_r \leftarrow \tau \boldsymbol{\theta}_r + (1-\tau)\hat{\boldsymbol{\theta}}_r$, $\hat{\boldsymbol{\theta}}_p \leftarrow \tau \boldsymbol{\theta}_p + (1-\tau)\hat{\boldsymbol{\theta}}_p$
  end for
end for

In each time step of the first L_1 episodes, the recommendation agent either randomly selects a content to recommend with probability ε_r or selects the optimal action $a_1(t) = \arg\max_a Q_r(s_1(t), a; \boldsymbol{\theta}_r)$ based on the stored Q-network at the CP with probability $1 - \varepsilon_r$. The ε-greedy method is able to balance the trade-off between finding a better action (i.e., exploration) and maximizing the return based on the current Q-network (i.e., exploitation).

In each time step between the $(L_1 + 1)$th and the L_2th episodes, the recommendation agent continues to act ε-greedily, and the pushing agent starts to act ε-greedily in the emulated environment to learn the pushing policy.

After the L_2th episode, both agents act ε-greedily in real environment to learn the recommendation policy and pushing policy simultaneously. The overall learning procedure is given in Algorithm 1.

22.4 Simulation results

In this section, we illustrate how the DRL-based policy behaves and compares the performance of the DRL-based policy with the baseline policies via simulation. We also compare the adopted DRL algorithm with several state-of-the-art policy-based DRL algorithms.

22.4.1 Simulation environment setup

We consider a scenario where users are moving along a straight road across multiple cells, as shown in Figure 22.4. The transmit power of each BS is 46 dBm, and the distance between adjacent BSs is 500 m. The pathloss is modeled by $35.3 + 37.6 \log_{10}(d_i)$ in dB, and the small-scale channel is modeled by Rayleigh fading. The duration of each time step is set to 10 s, which is common for short videos, such as in Douyin app [15]. In each time step, each user is with probability 0.8 to move forward with a random speed randomly selected from (5 ± 0.5) m/s, and is with probability 0.2 to stop (due to traffic light or traffic jam).

Since the net profit actually depends on the ratio between η and β, we normalize the revenue coefficient as $\eta = 1$ without loss of optimality. Considering that the transmission cost may exceed the revenue for a user located in cell-edge area, as shown in Figure 22.4, we set the scaling factor for the transmission cost to $\beta = 3$. By

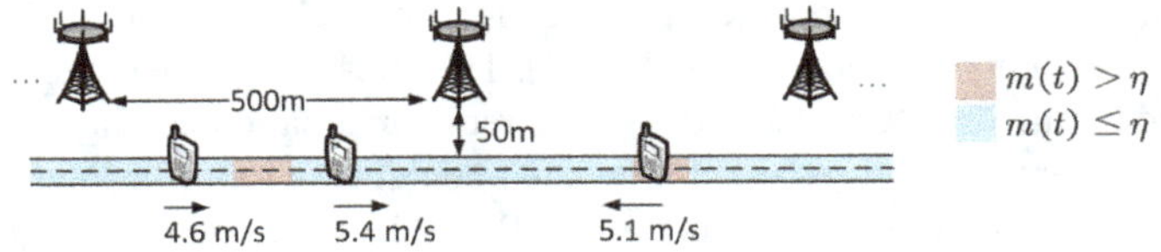

Figure 22.4 Simulation scenario

this setting, the transmission cost will be higher than the revenue when the distance between user and BS is greater than 220 m.

The cache size of each mobile device is $C = 5$ while the recommendation candidate set consists of $F = 100$ contents. The size of the pushing candidate set is $N = 5$. The request probability for content f when rejecting the recommendation (or when no content is recommended) is modeled by Zipf distribution $p_f = f^{-\zeta} / \sum_{j=1}^{N_f} j^{-\zeta}$ with skewness parameter $\zeta = 0.6$ [36].

To evaluate the performance of a recommendation policy, one has to know whether a user accepts the recommended content or not. Existing datasets, such as the MovieLens dataset [37] and the Netflix dataset [14], only provide the rating of contents or the request records, which cannot reflect whether or not a user accepts a recommendation. Therefore, the performance of a recommendation policy is not able to be evaluated with existing datasets. To evaluate the performance of a recommendation policy, one has to conduct real experiments, such as A/B test [38], which however is rather labor intensive and time consuming. In this chapter, to evaluate the policy learned by DRL, we provide a simulated environment based on well-acknowledged intuitions and facts revealed from real data sets analysis.

Specifically, we set the acceptance probability matrix $\mathbf{P} = [p_{ij}]_{F \times F}$ for each user in the following form,

$$\mathbf{P} = \begin{bmatrix} ☹ & ☻ & \cdots & ☻ & ☹ & ☹ & \cdots & ☹ \\ ☹ & ☹ & ☻ & \cdots & ☻ & ☹ & \cdots & ☹ \\ & & & & \cdots & & & \\ ☻ & \cdots & ☻ & ☹ & ☹ & ☹ & \cdots & ☹ \end{bmatrix} \tag{22.26}$$

where "☹" represents a value randomly chosen from $[0, 0.1]$, and "☻" represents a value randomly chosen from $[0.9, 1]$. If $p_{ij} \in [0, 0.1]$, the users who just have consumed content i are likely to reject the recommendation of content j. Otherwise, if $p_{ij} \in [0.9, 1]$, the users who just have consumed content i are likely to accept the recommendation of content j. Compared with the user's content request distribution without recommendation (such as the Zipf distribution with $\zeta = 0.6$, where the most likely requested content has a request probability of $1^{-0.6} / \sum_{j=1}^{100} j^{-0.6} = 0.07$), the uncertainty in user request can be reduced remarkably when recommending a content with acceptance probability greater than 0.9 (i.e., the content with "☻"). We note that although the users share the same structure of **P**, but the specific values of p_{ij} differ among users.

In each row of **P**, 5% of the elements are randomly selected from $[0.9, 1]$ and the rest of 95% elements are chosen from $[0, 0.1]$. This means that the contents likely to be accepted by the users are 5% of the whole recommendation candidate set. Those 5% of the contents can be regarded as related contents, e.g., the contents with the similar genre. To show a clear structure of **P**, we have re-ordered the indexes of contents so that similar contents lie together with each other.

Because a user is less likely to accept the recommendation of a content that has just been consumed, the diagonal elements of **P** is randomly selected from $[0, 0.1]$.

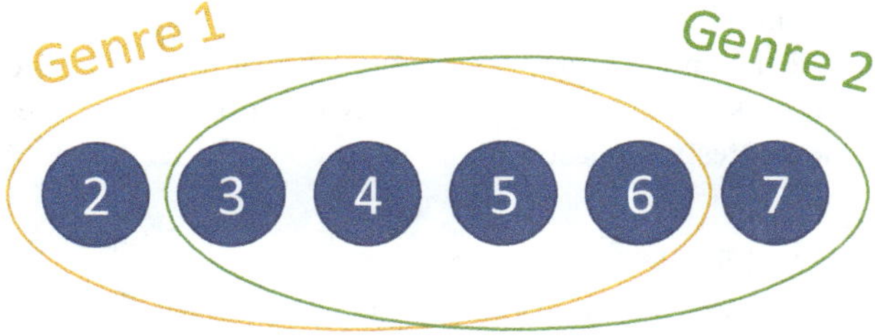

Figure 22.5 *Genre overlapping*

Considering that the requests of a user are also not likely to bounce between two contents, e.g., a user is not likely to request contents in the sequence such as $1 \rightarrow 2 \rightarrow 1 \rightarrow 2$ during a session, p_{ij} and p_{ji} should not lie in [0.9, 1] simultaneously. One possible way to satisfy this constraint is to set the ith row as the circulating shift of the $(i-1)$th row. This circulating-shift structure can also reflect the fact that a content may have multiple genres. Take the first two rows of **P**, for example, as shown in Figure 22.5, contents $2 \sim 6$ have the same genre (say genre 1, *Romance*), and contents $3 \sim 7$ have the same genre (say genre 2, *Comedy*). Then, the contents $3 \sim 6$ belong to both genres 1 and 2.

Despite that the structure of **P** given by (22.26) may not capture more complex user reaction towards recommendation, it agrees with intuition and can reflect some basic facts revealed by real datasets [10, 37], such as the non-repeated requests, no bouncing requests between two contents, and contents having multiple genres. One advantage of using simulated environment is that the reactions of users towards the recommendation can be flexibly controlled by adjusting the parameters of matrix **P**, which enables us to analyze the impact of user behavior on recommendation and pushing. Moreover, using such a circulating-shift structure of **P**, it is not hard to find that the optimal recommendation policy maximizing the average accumulated revenue, i.e., $\mathbb{E}[\sum_{t=1}^{T} \eta]$, is actually to recommend a content with "☻" based on the user's last consumed content. For instance, when the last consumed content of a user is content 1, we look up the 1st row of **P** given in (22.26) and find that contents $2 \sim 6$ is with "☻." Then, any one of contents $2 \sim 6$ can be recommended. It is worthy to note that although we consider a certain structure of **P** in simulation, the proposed RL framework does not use any prior knowledge concerning such a structure.

To capture the effect of recommendation in enhancing user stickiness, the session ending probabilities for each user varies within $q_1 = 0.05 \pm 0.005$ if the user accepts the recommendation or varies within $q_2 = 0.5 \pm 0.05$ otherwise.

The simulations are conducted on a work station with one Intel Core i7-8700K CPU and one Nvidia Geforce GTX 1080Ti GPU. The DRL algorithms are implemented using TensorFlow 1.8.0 [39] with Python 3.6 on Windows 10.

22.4.2 Fine-tuned hyper-parameters of Algorithm 1

The hidden layers H1 and H2 consist of 800 and 600 nodes, respectively, and we use the rectified linear unit (ReLU) as the activation function. Both V and A layers have no activation function. The output layers of Q_r and Q_f have F and $N+1$ nodes, respectively, where each node returns an action value of the input state.

The whole process of simulation contains two consecutive phases, namely the training phase and the testing phase.

22.4.2.1 Training phase

The training phase of the recommendation agent starts from the first episode. The exploration probability is set to $\varepsilon_r = 1$ for the first 5×10^3 episodes and then decreases linearly to 0.01 within the next 3.5×10^4 episodes. The training phase of the pushing agent starts from the $L_1 = 4 \times 10^4$th episode, during which the pushing agent first interacts with the emulated environment within the following 6.5×10^3 episodes, and then starts to interact with real environment from the $L_2 = 4.65 \times 10^4$th episode. The exploration probability is set to $\varepsilon_p = 1$ at the 4×10^4th episode and then decreases linearly to 0.01 within the next 10^4 episodes. The Adam method [40] is used to adjust the learning rate during training, and the initial learning rate is set to $\delta = 10^{-4}$. The update rate for the target network is $\tau = 0.0025$ and the mini-batch size for gradient descent is $|\mathscr{B}| = 32$. The replay memory size is $|\mathscr{D}| = 10^5$ and the discount factor is set as $\gamma = 1$.

22.4.2.2 Testing phase

The testing phase commences from the 5×10^4th episode, during which the exploration probability remains zero and the neural network weights $\boldsymbol{\theta}_r$, $\hat{\boldsymbol{\theta}}_r$, $\boldsymbol{\theta}_p$, $\hat{\boldsymbol{\theta}}_p$ are frozen and no longer updated.

22.4.3 Comparing with policy-based DRL algorithms

We first compare the adopted dueling DDQN (value-based method) with several policy-based DRL algorithms, which are PPO, A2C, and the DDPG with the k-nearest neighbor (KNN) search [41] (termed as "DDPG + KNN" for short). We take the recommendation sub-problem as an example for illustration. The details of these policy-based DRL algorithms are given in the appendix of [24]. The returns achieved by different DRL algorithms are provided in Figure 22.6.

We can see from Figure 22.6 that dueling DDQN converges faster than all the policy-based algorithms. "DDPG + KNN" converges faster than PPO and A2C, but can only achieve 50% of the return achieved by dueling DDQN. Both PPO and A2C can achieve a similar return to dueling DDQN after convergence. Because both PPO and A2C are on-policy algorithms, they are less sample-efficient than off-policy algorithm that uses experience replay such as DQN and hence converge slower. Therefore, we employ dueling DDQN for our DRL framework in the following experiments.

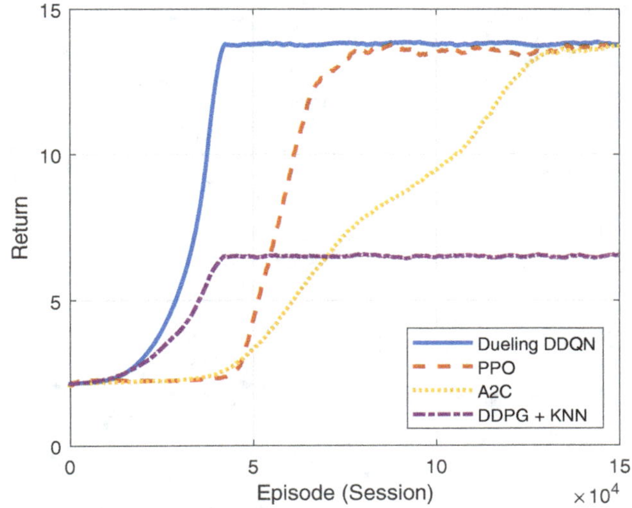

Figure 22.6 *Comparison of dueling DDQN with policy-based DRL algorithms in the recommendation sub-problem. All experiments are run for 50 different random seeds each. The curves are smoothed by averaging over a window of 50 episodes.*

22.4.4 Comparing with baseline policies

To show the gain respectively from optimizing recommendation and optimizing pushing by decomposing the joint optimization problem, the following baselines are compared with the policy learned by DRL:

1. *"Optimal Rec. & No pushing"*: In each time step, the BS recommends a content according to the optimal recommendation policy that maximizes the average accumulated revenue $\mathbb{E}[\sum_{t=1}^{T} \eta]$ and does not push any content. The optimal recommendation policy is obtained from the known structure of **P** as we explained in Section 22.4.1, which recommends a content with "☻" based on the last consumed content. This policy can be regarded as the performance upper-bound of the state-of-the-art recommendation policies without pushing.
2. *"No Rec. & Optimal pushing"*: The BS does not recommend any contents to each user and pushes the most preferred contents of each user. This policy provides a performance upper-bound of the state-of-the-art pushing policies in the absence of recommendation.
3. *"Non-decomposed"*: This policy is learned by directly applying dueling DDQN to the straightforward RL framework for joint pushing and recommendation optimization in Section 22.3.2. More details of the implementation of this method can be found in [24].

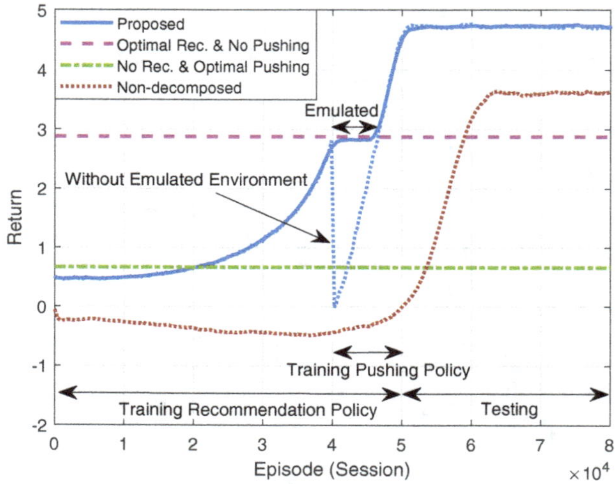

Figure 22.7 *Learning curve. All experiments are run for 50 different random seeds each. The curves are smoothed by averaging over a window of 50 episodes.*

In Figure 22.7, we show the learning curve of proposed RL framework. For the first 4×10^4 episodes, the recommendation policy is learned without pushing any contents to the user and finally achieves 5x of the return (i.e., accumulated net profit) over "*No Rec. & Optimal pushing.*" From the 4×10^4th to the 4.65×10^4th episode, the pushing agent learns the pushing policy in the emulated environment while does not push any content in the real environment. As a result, the return is almost the same as "*Optimal Rec. & No pushing.*" After that, the pushing agent starts to interact with the real environment. The return finally achieves 8x of the return over "*No Rec. & Optimal pushing.*" We also show the performance if pushing agent directly interacts with real environment (i.e., without the emulated environment) at the 4×10^4th episode in the figure. The return first drops because pushing policy has not been learned well enough, which causes higher transmission cost compared with no pushing. We can see that "*Non-decomposed*" is inferior to the proposed decomposed framework, which is resulted from the "delayed reward" as explained in the sequel. Simulation results reveal that "*Non-decomposed*" tends to not push anything to the user. Without pushing, "*Non-decomposed*" actually minimizes the immediate cost in the current time step (because pushing incurs transmission cost) and hence maximizes the immediate reward but does not necessarily maximize the accumulated reward in the long run. In other word, the benefit of pushing may only be observed several time steps later (e.g., when the user experiences worse channel condition and requests the pushed content).

In Figure 22.8, we compare the return breakdown for understanding where the gain of the net profit comes from. Compared with "*No Rec. & Optimal pushing*"

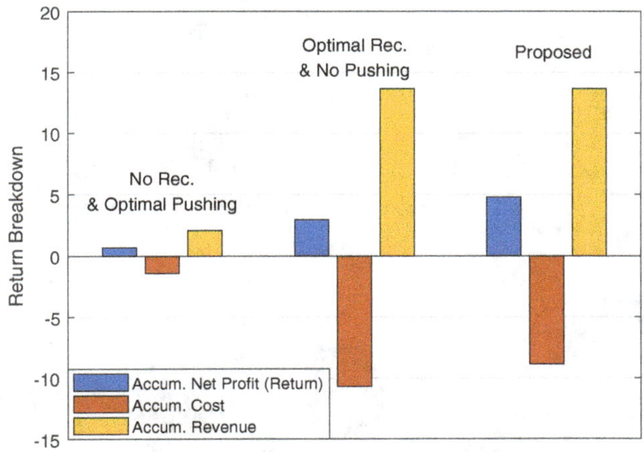

Figure 22.8 Return breakdown after Algorithm 1 converges

the accumulated revenue achieved by "*Optimal Rec. & No pushing*" is boosted 6x because a user will request more contents during a session with recommendation. However, with the increase of content requests, the accumulated transmission cost also increases significantly. Our proposed framework can achieve the same accumulated revenue as "*Optimal Rec. & No pushing*", which suggests that the learned recommendation policy is optimal. By learning the optimal pushing policy, the accumulated transmission cost of our proposed framework can be reduced and hence the accumulated net profit increases significantly compared with baselines.

22.4.5 Behavior of the pushing policy

In Figure 22.9, we show the average achievable rate (which reflects the channel condition), which content is pushed to a user (if any), and whether a requested content is cached in each time step during a simulated session to illustrate how the learned pushing policy behaves. The result is obtained from one snapshot of a session during the testing phase.

From Figure 22.9(a), we can see that the contents pushed in time steps $6 \sim 9$ are the contents to be recommended in time steps $10 \sim 13$, respectively. The user accepts the recommendation in time steps $10 \sim 13$, leading to cache hit events shown by the blue bars. Because the requested contents have already been cached, on-demand transmission is no longer required in time steps $10 \sim 13$, where the average achievable rate is low (i.e., transmission cost is high). These pushing actions validates that the learned pushing policy can intelligently adapt to the mobility pattern of the users and the propagation environment of the network, so that the contents to be requested under bad channel condition are proactively pushed under good channel condition. In addition, we can see that only the content to be recommended in four-time steps ahead

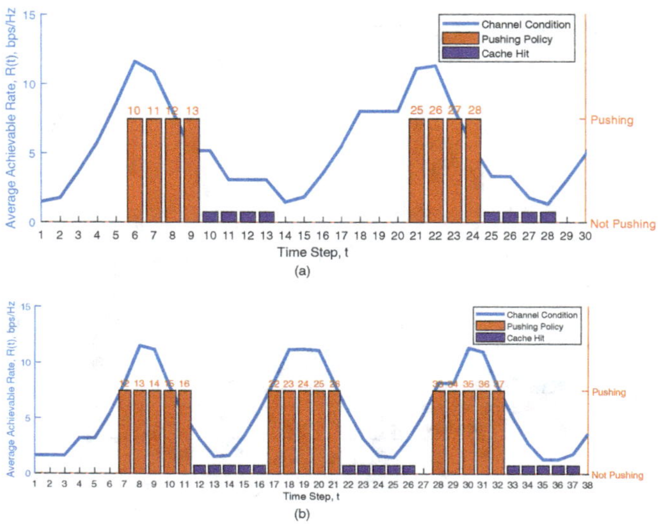

Figure 22.9 *Illustration of the learned pushing policy in a simulated session. The number above each orange bar denotes the time step where currently pushed content will be recommended to the user. For example, in (a), the content pushed in time step 6 is the content to be recommended in time step 10.*

(i.e., $n = 4$) is pushed to the user, although the size of pushing candidate set is five. This is because the prediction of the next n-step recommendation (and hence user request) becomes less precise with the increase of n, as we explained in Section 22.3.3.2.

In Figure 22.9b, we show the impact of prediction precision of the next-n step recommendation by adjusting the acceptance probability. Specifically, the elements with "☻" in **P** are set to one, which means that users will definitely accept the recommendation of a sufficiently attractive content. This is the most optimistic case where the prediction of next-n step recommendation is perfect as long as the session does not terminate. We can see that, with more precise prediction of the next-n step recommendation, the content pushed to the user in the current time step is the content to be recommended in five-time steps ahead. Compared with Figure 22.9(a), more contents are pushed under good channel conditions and hence the cache hit probability under bad channel conditions increases, which reduces the overall transmission cost. Moreover, since the acceptance probability is higher, a user is more likely to continue requesting contents, which results in longer session duration, e.g., $T = 38$ as shown in Figure 22.9(b). Consequently, the accumulated revenue during a session increases.

22.5 Conclusion and future works

In this chapter, we considered joint content pushing and recommendation for mobile users to increase the net profit of MNO without comprising user experience. To cope with the challenge caused by unknown user behaviors, we resorted to RL for solving the problem. Considering that the straightforward RL formulation of the joint pushing and recommendation problem has very large state and action spaces, which makes the problem hard to solve even using DRL algorithms, we decoupled the joint problem into a recommendation sub-problem and a pushing sub-problem by distinguishing the roles of recommendation and pushing. By first training the recommendation agent to learn the recommendation policy, the pushing agent is capable of explicitly predicting when a user will request which content. This can further reduce the state and action spaces and can also reduce the transmission cost during interactions with the environment. The cost can be further cut down by learning the pushing policy via interactions in an emulated environment. Simulation results showed that the revenue of MNO can be boosted by enhancing user stickiness by recommendation. The transmission cost of the network can be further reduced by integrating proactive pushing with content recommendation. As a final result, the net profit of MNO can be increased remarkably.

Although the result of this chapter is still preliminary, it reveals the potential of joint recommendation and caching by DRL to relieve the negative impact of user behavior uncertainty on proactive caching in wireless edge. Due to the lack of experiments and models for users' interaction with content recommendation, the evaluation of the proposed framework in real world scenarios are left for future works. It is worthy to note that content recommendation alone is a hot research topic in machine learning field. Readers can refer to [42–44] for recent advances in DRL-based content recommendation. As a proof-of-concept framework of integrating caching with recommendation, this chapter assumes that the recommendation candidates have been readily winnowed down to hundreds of items so that a content can be represented as an one-hot vector avoiding huge dimension. An alternative while promising approach to tackle huge number of contents is to represent the content by a low-dimensional continuous vector via embedding [45–48]. Therefore, how to integrate embedding into the joint recommendation and caching framework, as well as the corresponding performance, deserve further investigation.

References

[1] Golrezaei N., Molisch A.F., Dimakis A.G., Caire G. 'Femtocaching and device-to-device collaboration: a new architecture for wireless video distribution'. *IEEE Communications Magazine*. 2013 April;**51**(4):142–9.

[2] Zeydan E., Bastug E., Bennis M., *et al.* 'Big data caching for networking: moving from cloud to edge'. *IEEE Communications Magazine*. 2016;**54**(9):36–42.

[3] Liu D., Chen B., Yang C., Molisch A.F. 'Caching at the wireless edge: design aspects, challenges, and future directions'. *IEEE Communications Magazine*. 2016;**54**(9):22–8.

[4] Li L., Zhao G., Blum R.S. 'A survey of caching techniques in cellular networks: research issues and challenges in content placement and delivery strategies'. *IEEE Communications Surveys & Tutorials*. 2018;**20**(3):1710–32.

[5] Li X., Wang X., Li K., Han Z., Leung V.C.M. 'Collaborative multi-tier caching in heterogeneous networks: modeling, analysis, and design'. *IEEE Transactions on Wireless Communications*. 2017;**16**(10):6926–39.

[6] Liu D., Yang C. 'Energy efficiency of downlink networks with caching at base stations'. *IEEE Journal on Selected Areas in Communications*. 2016;**34**(4):907–22.

[7] Liu J., Bai B., Zhang J., Letaief K.B. 'Cache placement in Fog-RANs: from centralized to distributed algorithms'. *IEEE Transactions on Wireless Communications*. 2017;**16**(11):7039–51.

[8] Wang K., Chen Z., Liu H. 'Push-based wireless converged networks for massive multimedia content delivery'. *IEEE Transactions on Wireless Communications*. 2014;**13**(5):2894–905.

[9] Chen W., Poor H.V. 'Content pushing with request delay information'. *IEEE Transactions on Communications*. 2017;**65**(3):1146–61.

[10] Lee M.C., Molisch A.F., Sastry N. 'Individual preference probability modeling for video content in wireless caching networks'. Proceedings of the IEEE GLOBECOM; 2017.

[11] Chen B., Yang C. 'Caching policy for cache-enabled D2D communications by learning user preference'. *IEEE Transactions on Communications*. 2018;**66**(12):6586–601.

[12] Ekstrand M.D. 'Collaborative filtering recommender systems'. *Foundations and Trends® in Human–Computer Interaction*. 2010;**4**(2):81–173.

[13] Hofmann T. 'Latent semantic models for collaborative filtering'. *ACM Transactions on Information Systems*. 2004;**22**(1):89–115.

[14] Gomez-Uribe C.A., Hunt N. 'The netflix recommender system: algorithms, business value, and innovation'. *ACM Transactions on Management Information Systems*. 2016;**6**(4):13.

[15] Wittenberg M., Douyin I. *China's incredibly sticky short video app* [online]. 2018Posted on medium. Available from https://mondaynote.com/introducing-douyin-chinas-ridiculously-sticky-short-video-app-ab005727d89e [Accessed Sep 3, 2018].

[16] Singhal A., Sinha P., Pant R. 'Use of deep learning in modern recommendation system: a summary of recent works'. *International Journal of Computer Applications*. 2017;**180**(7):17–22.

[17] Liu D., Yang C. 'Caching policy toward maximal success probability and area spectral efficiency of cache-enabled HetNets'. *IEEE Transactions on Communications*. 2017;**65**(6):2699–714.

[18] Chen Z., Lee J., Quek T.Q.S., Kountouris M. 'Cooperative caching and transmission design in cluster-centric small cell networks'. *IEEE Transactions on Wireless Communications*. 2017;**16**(5):3401–15.

[19] Tadrous J., Eryilmaz A., El Gamal H. 'Proactive content download and user demand shaping for data networks'. *IEEE/ACM Transactions on Networking*. 2015;**23**(6):1917–30.

[20] Chatzieleftheriou L.E., Karaliopoulos M., Koutsopoulos I. 'Caching-aware recommendations: nudging user preferences towards better caching performance'. Proceedings of the IEEE INFOCOM; Atlanta, GA, USA; 2017.

[21] Sermpezis P., Giannakas T., Spyropoulos T., Vigneri L. 'Soft cache hits: improving performance through recommendation and delivery of related content'. *IEEE Journal on Selected Areas in Communications*. 2018;**36**(6):1300–13.

[22] Liu D., Yang C. 'A learning-based approach to joint content Caching and recommendation at base stations'. Proceedings of the IEEE Globecom; Abu Dhabi, UAE; 2018.

[23] Guo K., Yang C., Liu T. 'Caching in base station with recommendation via Q-learning'. Proceedings of the IEEE WCNC; San Francisco, CA, USA; 2017.

[24] Liu D., Yang C. 'A deep reinforcement learning approach to proactive content pushing and recommendation for mobile users'. *IEEE Access*. 2019;**7**:83120–36.

[25] Wang Z., Schaul T., Hessel M. 'Dueling network architectures for deep reinforcement learning'. Proceedings of the ICML; New York City, NY, USA; 2016.

[26] Sutton R.S., Barto A.G. *Reinforcement Learning: An Introduction*. Cambridge: MIT Press; 1998.

[27] Calabrese F., Wang L., Ghadimi E., *et al.* Learning radio resource management in RANs: framework, opportunities and challenges. arXiv preprint: 161110253. Available from http://arxiv.org/abs/1611.10253.

[28] Covington P., Adams J., Sargin E. 'Deep neural networks for YouTube recommendations'. Proceedings of the ACM RecSys; Boston, MA, USA; 2016.

[29] Shani G., Heckerman D., Brafman R.I. 'An MDP-based recommender system'. *Journal of Machine Learning Research*. 2005;**6**(9):1265–95.

[30] Watkins C.J.C.H., Dayan P. 'Q-learning'. *Machine Learning*. 1992;**8**(3-4):279–92.

[31] Lillicrap T.P., Hunt J.J., Pritzel A. 'Continuous control with deep reinforcement learning'. Proceedings of the ICLR; San Juan, Puerto Rico; 2016.

[32] Mnih V., Badia A.P., Mirza M. 'Asynchronous methods for deep reinforcement learning'. Proceedings of the ICML; New York City, NY, USA; 2016.

[33] Schulman J., Wolski F., Dhariwal P. Proximal policy optimization algorithms. *arXiv preprint arXiv:170706347*; 2017.

[34] Mnih V., Kavukcuoglu K., Silver D., *et al.* 'Human-level control through deep reinforcement learning'. *Nature*. 2015;**518**(7540):529–33.

[35] Van Hasselt H., Guez A., Silver D. 'Deep reinforcement learning with double Q-learning'. Proceedings of the AAAI; Phoenix, Arizona, USA; 2016.

[36] Gill P., Arlitt M., Li Z. 'Youtube traffic characterization: a view from the edge'. Proceedings of the ACM SIGCOMM IMC; San Diego, CA, USA; 2007.

[37] Harper F.M., Konstan J.A. 'The movielens datasets: history and context'. *ACM Transactions on Interactive Intelligent Systems*. 2015;**5**(4):19:1–19:19.

[38] Kohavi R., Longbotham R., Sommerfield D., Henne R.M. 'Controlled experiments on the web: survey and practical guide'. *Data Mining and Knowledge Discovery*. 2009;**18**(1):140–81.

[39] Abadi M., Agarwal A., Barham P., *et al. TensorFlow: Large-scale machine learning on heterogeneous systems* [online]. 2015. Available from http://tensorflow.org/ [Accessed Sep 3 2018].

[40] Kingma D.P., Ba J. 'Adam: a method for stochastic optimization'. Proceedings of the ICLR; Banff, AB, Canada; 2014.

[41] Dulac-Arnold G., Evans R., van Hasselt H. Deep reinforcement learning in large discrete action spaces. *arXiv preprint arXiv:151207679*; 2015.

[42] Zhao X., Zhang L., Ding Z. 'Recommendations with negative feedback via pairwise deep reinforcement learning'. Proceedings of the ACM SIGKDD; London, UK; 2018.

[43] Zou L., Xia L., Ding Z. 'Reinforcement learning to optimize long-term user engagement in recommender systems'. Proceedings of the ACM SIGKDD; Anchorage, AK, USA; 2019.

[44] Chen H., Dai X., Cai H., *et al.* 'Large-scale interactive recommendation with tree-structured policy gradient'. *Proceedings of the AAAI Conference on Artificial Intelligence*. 2019;**33**:3312–20.

[45] Barkan O., Koenigstein N. 'Item2vec: neural item embedding for collaborative filtering'. 2016 IEEE MLSP; Salerno, Italy; 2016.

[46] Liang D., Altosaar J., Charlin L. 'Factorization meets the item embedding: Regularizing matrix factorization with item co-occurrence'. Proceedings of the ACM Recsys; Boston, MA, USA; 2016.

[47] Shi C., Hu B., Zhao W.X., Yu P.S. 'Heterogeneous information network embedding for recommendation'. *IEEE Transactions on Knowledge and Data Engineering*. 2018;**2**:357–70.

[48] Tang J., Wang K. 'Personalized top-N sequential recommendation via convolutional sequence embedding'. Proceedings of the ACM WSDM; Melbourne, Australia; 2018.

Chapter 23

Social networking-driven caching

Kuo Chun Tsai[1], Li Wang[2], Huaqing Wu[3], Lianming Xu[2], and Zhu Han[1]

23.1 Introduction

As a promising technology in future communication systems and networks, wireless caching is envisioned to revolutionize the wireless resource management in 5G and B5G. By storing content files closer to the user end, caching can effectively alleviate the load on the content server and reduce the content retrieving delay. In particular, distributed storage/caching system (DSS/DCS) has gradually attracted the attention of academia and industry by storing content on multiple cache servers in a distributed manner to improve system storage efficiency, information security, and scalability. To further enhance system robustness and reliability, the social networks should also be taken into consideration to evaluate the relationship among caching entities for comprehensive caching scheme design. In this section, we will first introduce the basic concept of caching, and then the significance of considering social networks in caching will be presented.

23.1.1 Concepts of caching

The concept of caching is first introduced in computer systems to store data in a relatively small memory for quick access. By taking advantage of the memory access patterns, which generally exhibit temporal locality and spatial locality, small caches can achieve significant system performance improvement. Later the idea of caching is applied to the Internet. Rather than retrieving web pages from a central server, it is advantageous to replicate popular web pages in small servers (i.e., caches) around the world to alleviate server congestion, decrease network bandwidth consumption, and reduce content retrieving delay. With the explosion of Internet traffic in the late 1990s, the cache management became more intricate, leading to the prosperity of Content Delivery Networks (CDNs). The main research of CDN in the past

[1]University of Houston, USA
[2]Beijing University of Posts and Telecommunications, China
[3]University of Waterloo, Canada

decades includes: the location to deploy servers (server placement); the storage capacity allocated to each server (cache annotation); selection of files cached on servers (content placement); the routing of content from cache to requesting users (routing policy), etc. However, new challenges arise nowadays in CDN to support multifarious services with more rigorous requirements.

Recently, the application of caching techniques in wireless networks has attracted broad interests to facilitate content delivery. To enhance network capacity, increasing the physical layer transmission rate or densely deploying base stations (BSs) is cost-ineffective and outpaced by the fast-growing mobile data traffic. As a promising technique to address this issue, wireless caching provides potential solutions including: (1) deep caching in the core networks to reduce content retrieving delay; (2) caching in BSs to mitigate the pressure on backhaul links with limited throughput; (3) caching on mobile devices to take advantage of device-to-device communications; and (4) encoded caching to facilitate broadcast transmission.

According to different caching entities, wireless caching can be classified into three categories: caching in core network and BSs, caching in small base stations (SBSs), and caching in user equipments (UEs).

Caching in the core network and BSs: The rapid growth of wireless services has led to an exponential increase in cellular network traffic. In order to reduce the traffic load on the backhaul links of cellular networks, it is effective to implement traffic localization to cache requested data in BSs with large storage capacity. The basic idea of caching in the core network and BSs is described as follows. In the caching process, the core network first caches popular/frequently requested content to the BSs. When a user requests the content from a BS, the BS first checks whether it has cached the content. If the BS has cached the content, the BS transmits the content directly to the user. If the BS has not cached the content, the BS can obtain content from nearby BSs, and then transmit the content to the user. Therefore, the traffic load of the core network and the backhaul links will be greatly reduced.

Caching in SBSs: Due to the small size, low power consumption, and flexible deployment of SBSs, caching in SBSs helps shorten the communication distance, reduce the required transmission power, improve system capacity, and enhance spectrum utilization efficiency. By caching content in SBSs, users' content requests can be directly served by the SBS, reducing the backhaul link occupation and the congestion of the macroBS especially during the peak traffic hours. However, the interference among ultra-dense SBSs and the energy consumption of cellular networks are critical problems and require further investigation.

Caching in UEs: With the proliferation of smart phones and the continuous improvement of hardware technology, the storage capacity of UEs is constantly increasing, thus caching in UEs is widely studied and discussed. Due to the application of smart phones in social areas, it is very feasible to use smart phones to achieve short-distance D2D communications. Caching in UEs means that UEs can cache popular content during the off-peak traffic hours. When nearby UEs request the cached content files, the content can be transmitted via D2D communications without requesting the content from the BS. The main advantages of caching in UEs with D2D communications include:

1. Content transmission between short-distance users through D2D communication can effectively reduce transmission latency;
2. By utilizing the social information of the user equipment, the distribution rule of content requests can be provided, thereby facilitating a suitable content caching scheme design;
3. By using the cooperative caching method of D2D communication, the user equipments can cache frequently requested/popular content files to effectively reduce the traffic load of the BS and avoid the excessive resource consumption caused by repeated content delivery;
4. By caching popular files to mobile users, the storage capacity of the system can be greatly increased.

23.1.2 Socially aware caching networks

Despite the benefits brought by wireless caching with D2D communications, it brings new challenges for caching scenarios, such as the unreliable D2D links and unwillingness for content sharing due to user selfishness. Traditional wireless networks mainly focus on the wave propagation characteristics and the physical locations of users, instead of their social characteristics. In D2D-enabled wireless caching, however, wireless communication technologies, interest similarities, and social ties should be jointly considered to develop a comprehensive solution since they are highly related to the design of D2D-based caching schemes [1]. *Interest similarity* describes the phenomenon that users in social networks have similar interests in movies, TV shows, books, games, or other social content. *Social ties* are related to the interactions among users, including the existing social structures, shared activities, and personal attributes.

In mobile networks with user mobility, it is not practical to assume that D2D links are always stable as assumed in most existing works [2]. Therefore, in addition to the physical link condition, the social information is also a decisive factor to ensure the stability of D2D links. In general, mobile users with relatively strong social ties have a high probability to share content with each other and form relatively stable D2D links. For example, family members, colleagues, and club members generally have common interests and are willing to share content with each other. Therefore, the probability of mobile user cooperation can be assessed to achieve more efficient resource sharing when taking social characteristics into consideration. The effectiveness of wireless caching schemes via D2D links can be further enhanced with proper resource allocation, which can leverage both physical channel state information and social interactions.

With content caching in BSs, SBSs, and UEs, D2D and device-to-infrastructure (D2I) communications can effectively and efficiently increase the system social welfare as the number of end users in the network increases. We will refer to D2D and D2I communications as D2X communications for simplicity. Although applying the D2X communication technology to the mobile social network seems to be a huge step across the limitation of the traditional wireless network framework, the D2X communication technology is unable to show the outstanding benefit if the system randomly caches data into arbitrary cache entities. To overcome this problem, the caching strategies should be carefully designed, by utilizing approaches including deep learning (e.g., long short-term memory (LSTM) neural network) and game theory (e.g., the Irving's stable fixture (ISF) algorithm) methods.

23.1.3 Organization

The rest of the chapter is organized as follows. In Section 23.2, we discuss about fundamental design principles for caching with D2D communications and hypergraph theory [3] for multi-dimension information. A socially aware D2D caching scheme is also discussed. Next, the D2I-based caching scheme is studied in mobile social networks in Section 23.3, where matching theory and deep learning-based solutions are proposed. Finally, we conclude this chapter in Section 23.4.

23.2 Exploit socially enabled D2D communications in caching

In this section, we will investigate socially aware wireless caching with D2D communications. The fundamental design principles will first be investigated to leverage the interplay among communication link conditions, social ties, and interest similarities. We then introduce a hypergraph-based model to facilitate the exploitation of the multi-dimensional information. After that, we investigate the social interaction-based resource allocation to jointly optimize spectrum efficiency and energy efficiency in D2D-based wireless caching networks, while guaranteeing the quality of service (QoS) requirements of users [4].

23.2.1 Social characteristics consideration

The social relationships contain face-to-face social relationships and online social relationships [5]. Face-to-face social relationships can be established by localized interactions, and online social relationships rely on social media, such as Facebook, Instagram, Linkedin, Twitter, etc. These two types of social relationships can be modeled from the following aspects.

Social ties: In order to realize the direct communication, which cannot be guaranteed because of the mobility of nodes, we need to consider the social interaction. Generally, two main factors are used to characterize the social interaction features among users: contact duration and contact frequency [6]. Contact duration evaluates the time interval during which mobile users are within communication range for information exchange. Specifically, if the transmission process is accomplished within the contact duration, the data are deemed to be successfully transmitted. Contact frequency means the times for those nodes' encountering. We can formulate and derive social ties between mobile users by analyzing their contact duration and contact frequency. With the analysis and prediction on human–social relationships, the accuracy of human mobility analysis can be enhanced, thus facilitating the resource management and improving users' quality of experience (QoE).

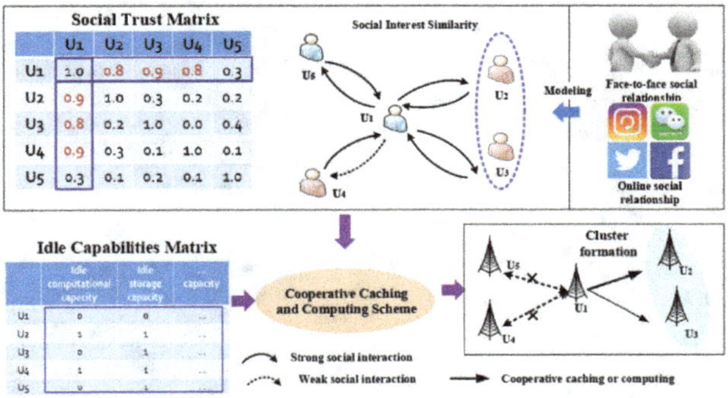

Figure 23.1 An illustration of social trust in social relationships

Social trust, which is widely accepted as a major component of human social relationships [7], is another evolution of social ties and plays a vital role in determining the level of altruism among users [8]. Social trust can be leveraged in D2D communications by selecting the most trustworthy device to cooperate as a content server, a friendly jammer, or a reliable relay [9]. Notice that social trust can affect users' willingness to participate in cooperative caching or content sharing, due to the potential user selfishness. Therefore, incentive algorithms are required to stimulate the cooperation among users in proactive caching and content sharing. In specific, social trust can be quantified by a non-negative value $s \in [0, 1]$, where $s = 1$ means unconditional availability, while $s = 0$ means no availability. Social trust between each pair of users can be quantified by a matrix, as shown in Figure 23.1 [10].

Interest similarity: Within a certain period, users in a certain area are more likely to have the same interests. Social interest similarity is a metric to quantify how similar the interests of two users are. Social interest similarity can be used for partitioning the network into user groups to improve the performance of content sharing and reduce the complexity of algorithms. For instance, people with common interests are likely to request for the same content, and thus, they can be formed as a group [11] to facilitate content distribution. Furthermore, considering that users usually are only willing to communicate with their neighbors with tight social relationships for privacy purposes, it can be efficient to leverage the interest similarity to perform proactive content sharing.

Physical Link: D2D communications allow neighboring devices to communicate with each other directly for content sharing. Thus, the quality and reliability of physical links (D2D communication links) must be taken into consideration. Basically, D2D communication links among devices can eventually establish a physically connected graph, which can be also called the physical graph. A physical graph is generally built by considering both physical distance and communication link conditions (e.g., pathloss, shadowing, fading.).

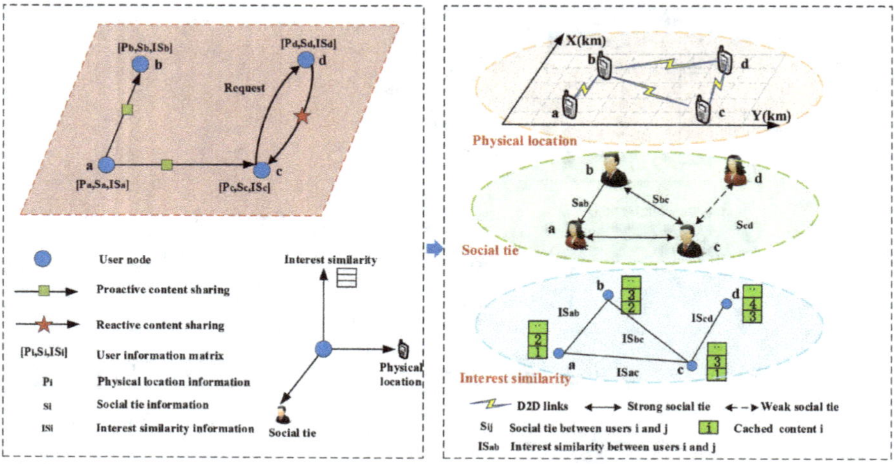

Figure 23.2 Socially aware D2D-enabled caching framework

23.2.2 Socially enabled D2D caching with hypergraph

Motivated by the above, social ties, common interests, and physical link conditions should be jointly considered in D2D caching scenarios, as shown in Figure 23.2. We will then introduce a hypergraph-based model to represent and integrate the multi-dimensional information. In this part, the definition and the application of hypergraphs in socially enabled D2D caching systems will be introduced.

23.2.2.1 Hypergraph definition and construction

In traditional graphs, there exist vertices and edges, where two vertices can be connected through an edge. Incidence matrices or adjacency matrices are used to represent the graphs. In a traditional graph with V vertices and E edges, an incidence matrix has V rows and E columns. If the i-th vertex is incident to the j-th edge, entry (i,j) in the incidence matrix is set to 1, otherwise it is set to 0. The adjacency matrix has V rows and V columns, and the entry (i,j) in the matrix is 1 when the i-th and j-th vertices are adjacent to each other, otherwise it is 0. For weighted graphs, the adjacency matrices can also be weighted, where the entry (i,j) is the weight of edge connecting vertices i and j. An example of a traditional graph and its corresponding incidence matrix is given in Figure 23.3a.

Hypergraph is a generalized graph, in which an edge can connect any subset of the given set of vertices instead of exactly two vertices as defined in a traditional graph [12]. Let $V = \{v_1, v_2, \ldots, v_n\}$ be a finite set of vertices, and then a hypergraph H is a family $H = (E_1, E_2, \ldots, E_m)$, where each hyperedge E_i is a subset of V such that $E_i \neq \varnothing$ and $\bigcup_{i=1}^{m} E_i = V$. Unlike traditional graphs, there is no one-to-one correspondence between a hypergraph and its adjacency matrix. Therefore, a hypergraph is generally represented by its incidence matrix. We define the number of all the edges containing vertex x as the degree of vertex x, denoted by $D(x)$. Figure 23.3b

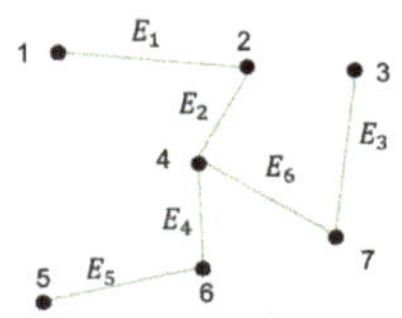

	E_1	E_2	E_3	E_4	E_5	E_6
1	1	0	0	0	0	0
2	1	1	0	0	0	0
3	0	0	1	0	0	0
4	0	1	0	1	0	1
5	0	0	0	0	1	0
6	0	0	0	1	1	0
7	0	0	1	0	0	1

(a) An example of a traditional graph with its incidence matrix

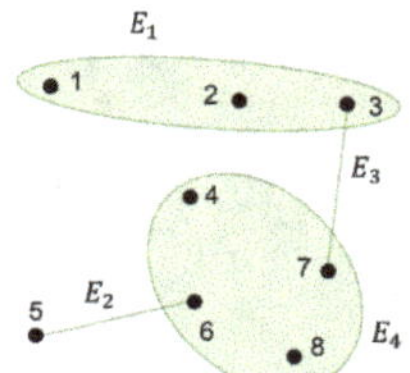

	E_1	E_2	E_3	E_4
1	1	0	0	0
2	1	0	0	0
3	1	0	1	0
4	0	0	0	1
5	0	1	0	0
6	0	1	0	1
7	0	0	1	1
8	0	0	0	1

(b) An example of a hypergraph with its incidence matrix

Figure 23.3 Illustrations of traditional graphs and hypergraphs

gives an example of a hypergraph, where the left part is a hypergraph with four hyperedges and the right part is the corresponding incidence matrix. For example, the hyperedge E_4 contains v_4, v_6, v_7, and v_8, and thus in incident matrix, elements (4,4), (6,4), (7,4), and (8,4) are set to 1.

23.2.2.2 Hypergraph formulation

To formulate the socially aware D2D caching, it is imperative to propose a unified modeling and analysis framework based on hypergraph.

Considering the scenario where D2D communications share the spectrum with cellular users (CUEs), the major concerns in physical layer are spectrum management and interference mitigation among D2D users and CUEs. In the social ties layer, the reliability and the duration of a D2D link can be evaluated by analyzing the social thrust, which is a key premise for D2D link establishment. Furthermore, in the common interests layer, multiple D2D devices may request for the same content, which can be leveraged for content delivery scheme design to achieve a high content dissemination efficiency. It is difficult to model the relationships among these factors for socially aware D2D-enabled caching system design. Therefore, we propose a hypergraph-based model to formulate this system.

Denote by N^d the set of D2D users in a socially aware D2D-enabled caching system, where $n_i^{\mathrm{d}} \in N^{\mathrm{d}}$ is the i-th D2D user. Let N^c denote the set of content files, where $n_k^{\mathrm{c}} \in N^{\mathrm{c}}$ is the k-th content, and let N^s denote the set of cellular subchannels in the system, where $n_m^{\mathrm{S}} \in N^{\mathrm{S}}$ is the m-th subchannel. Then the set of vertices in the hypergraph is $N = N^{\mathrm{d}} \cup N^{\mathrm{c}} \cup N^{\mathrm{S}}$. The set of hyperedges is denoted by E, where the j-th hyperedge E_j is a subset of N. Notice that a hyperedge exists if and only if:

1. For two D2D users in N^{d}, one of them requests for content n_k^{c}, and the other one has cached this content;
2. These two D2D users are within communication range, have a strong social tie, and are willing to share content via a D2D link;
3. For a D2D link sharing a CUE's resource on subchannel n_m^{S}, the target transmission rate R can be guaranteed;
4. The interference between the D2D link and the CUE sharing the same resource is smaller than a threshold.

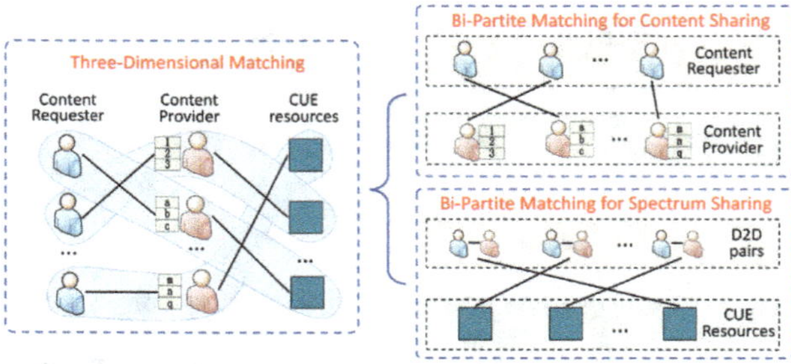

Figure 23.4 Decomposition of three-dimensional matching into hierarchical bi-partite matchings

When multiple D2D links are allowed to share the same spectrum resource, the hyperedge can include all the co-channel D2D devices in $\mathcal{N}^{\mathrm{d}}$.

The graph constructed based on the above principles is a three-partite hypergraph, which can be leveraged for the content caching scheme design with a three-dimensional matching among the D2D transmitters (content providers), D2D receivers (content requesters), and the CUE spectrum resources, as shown in the left part of Figure 23.4. Considering that the hypergraph-based three-dimensional matching problem is NP-hard and difficult to solve, we can divide it into hierarchical bi-partite matching problems: bi-partite matching for content sharing and bi-partite matching for spectrum sharing, as shown in the right part of Figure 23.4. The content sharing bi-partite matching determines the optimal connections between the content providers and content requesters to form D2D pairs. Then, based on the formed D2D pairs, the spectrum sharing bi-partite matching optimizes the allocation of the CUE spectrum resources to each D2D pair. This hierarchical bi-partite matching framework can facilitate the socially aware D2D-based content sharing scheme design with a low time complexity.

23.2.3 Socially aware caching scheme

23.2.3.1 Resource allocation problem in wireless caching

We consider that there are N CUEs and Jcontent requesters (CRs) in a wireless cellular system. CRs require to download content from adjacent content servers (CSs) via D2D links by reusing the uplink (UL) resource of the CUEs or turn to the BS via the cellular downlink. The bandwidth is the same for each UL and DL channel, denoted by B. CUEs and CRs are randomly distributed in the cell. CSs are randomly distributed within a disc area with radius r centered at the reference CR.

Denote by d_{max} the maximum allowable distance for D2D communication. We define $\mathcal{M}_j$ as the initial candidate set of CSs for the J-th CR. If there are more than one CS within d_{max} from the J-th CR, the best M CSs will be chosen as the initial candidate set. However, if the number of achievable CSs is fewer than M, all the available CSs will be kept in $\mathcal{M}_j$, i.e., $|\mathcal{M}_j| \leq M$.

Given that D2D transmission may fail because of the mobility and maximum communication distance constraint, social interaction is also involved to evaluate the D2D link stability and transmission success probability. We consider time-sensitive services, i.e., the content is deemed to be successfully delivered only when the data transmission is accomplished during one single encounter time. Therefore, the probability that the j-th CR can receive content f successfully from the m-th CS by reusing the UL spectrum resource of CUE i in cellular D2D underlay, can be derived by considering both physical condition and social interactions:

$$P^i_{j,m,f} = \int_0^{\infty} \frac{\frac{P^d_{m,j} g_{m,j}}{\sigma_N^2}}{\frac{P^d_{m,j} g_{m,j}}{\sigma_N^2} + \frac{P^c_i h_{i,j}}{\sigma_N^2}\left(2^{\frac{Z_f/B}{t \cdot q_{m,j}}} - 1\right)} \exp\left(-\frac{2^{\frac{Z_f/B}{t \cdot q_{m,j}}} - 1}{\frac{P^d_{m,j} g_{m,j}}{\sigma_N^2}} - t\right) dt, \tag{23.1}$$

where $P^d_{m,j}$ and P^c_i are transmit powers of CS m and CUE i, $g_{m,j}$ and $h_{i,j}$ represent the channel power gains from CS m to CR j and from CUE i to CR j, σ_N^2 is the background noise power, $q_{m,j}$ is the average contact time duration between CR j and CS m, and Z_f is the size of content f in bits.

Notice that CR j can successfully receive content f from the adjacent CSs only when there is at least one CS with content f within d_{max} and the chosen CS can transmit content f successfully to CR j. Therefore, the probability that CR j can successfully receive content f from adjacent CSs by reusing the UL spectrum resource of CUE i can be expressed as

$$P^i_{j,f} = p^s_f \cdot \sum_{m \in \mathcal{M}_j} (\rho_{m,j} \cdot P^i_{j,m,f}), \tag{23.2}$$

where p^s_f is the probability that a CR can find at least one CS with desired content f. $\rho_{m,j}$ is defined to indicate the association between CR j and CS m, i.e.,

$$\rho_{m,j} = \begin{cases} 1, & \text{CS } m \text{ is selected to deliver content to CR } j, \\ 0, & \text{otherwise.} \end{cases} \tag{23.3}$$

$\rho_{m,j}$ must satisfy $\sum_{m \in \mathcal{M}_j} \rho_{m,j} \leq 1$ to ensure that there is at most one CS to help CR j for each request.

Then the achievable data rate of CR j to get content f from adjacent CS m by sharing the UL spectrum of CUE i is

$$R^d_j = B \log\left(1 + \frac{P^d_{m,j} g_{m,j}}{P^c_i h_{i,j} + \sigma_N^2}\right). \tag{23.4}$$

In this case, we can express the achievable data rate of CUE i sharing its resource with CR j as

$$R^c_{i,j} = B \log\left(1 + \frac{P^c_i g_{i,B}}{P^d_{m,j} h_{m,B} + \sigma_N^2}\right), \tag{23.5}$$

where $g_{i,B}$ and $h_{m,B}$ represent the channel power gains from CUE i to BS and from CS m to BS, respectively.

Moreover, we denote $v_{i,j}$ as an indicator representing the resource sharing between CR j and CUE i, i.e.,

$$v_{i,j} = \begin{cases} 1, & \text{CR } j \text{ reuses the resource of CUE } i, \\ 0, & \text{otherwise.} \end{cases} \tag{23.6}$$

Here we assume that the resource of CUE i can be reused by at most one D2D link at the same time, and one D2D link can reuse the resource of at most one CUE at one time. Thus, $v_{i,j}$ must satisfy $\sum_{j=1}^{J} v_{i,j} \leq 1$ and $\sum_{i=1}^{N} v_{i,j} \leq 1$. There are totally $\sum_{j\in\mathscr{R}} \sum_{i\in\mathscr{C}} v_{i,j}$ D2D links that request to reuse the resource of CUEs for content sharing.

However, if the D2D data transmission fails, the power is still consumed. The bandwidth resource reused by CR in this case is equal to B. Combining (23.2)–(23.5), the average total data rate is

$$R_{D2D} = \sum_{i\in\mathscr{C}} \sum_{j\in\mathscr{R}} \left[v_{i,j} \cdot p_f^s \cdot \sum_{m\in\mathscr{M}_j} \left(\rho_{m,j} p_{j,m,f}^i\right) \left(R_j^d + R_{i,j}^c\right) \right], \tag{23.7}$$

the average bandwidth consumption is

$$B_{D2D} = B \cdot \sum_{j\in\mathscr{R}} \sum_{i\in\mathscr{C}} v_{i,j}, \tag{23.8}$$

and the average power consumption of D2D links and CUEs is

$$P_{D2D} = \sum_{j\in\mathscr{R}} \sum_{i\in\mathscr{C}} \left[v_{i,j} \left(P_{m,j}^d + P_i^c \right) \right]. \tag{23.9}$$

When CR j fails to get desired content from adjacent CSs, the CR will turn to the BS in the cellular mode for help. In this case, each CR occupies a dedicated DL channel, and the CUEs occupy orthogonal UL channels without spectrum sharing. Therefore, the achievable data rate of CR j receiving content from the BS can be expressed as,

$$R_j^B = B \log \left(1 + \frac{P_B g_{B,j}}{\sigma_N^2} \right), \tag{23.10}$$

where P_B is the fixed transmission power at the BS and $g_{B,j}$ is the channel gain from BS to CR j. Besides, the data rate of CUE i without sharing its resource with any D2D links is

$$R_i^c = B \log \left(1 + \frac{P_i^c g_{i,B}}{\sigma_N^2} \right). \tag{23.11}$$

For users using orthogonal spectrum, which include CUEs without spectrum sharing and the CRs downloading the desired content from the BS, the average total data rate is

$$\begin{aligned} R_{BS} &= \sum_{j\in\mathcal{R}}\left[\left(1-\sum_{i\in\mathcal{C}}\left(v_{i,j}\cdot p_f^s\cdot\sum_{m\in\mathcal{M}_j}(\rho_{m,j}p_{j,m,f}^i)\right)\right)R_j^B\right] \\ &+\sum_{i\in\mathcal{C}}\left[\left(1-\sum_{j\in\mathcal{R}}\left(v_{i,j}\cdot p_f^s\cdot\sum_{m\in\mathcal{M}_j}(\rho_{m,j}p_{j,m,f}^i)\right)\right)R_i^c\right], \end{aligned} \tag{23.12}$$

where $\left(1-\sum_{i\in\mathcal{C}}\left(v_{i,j}\cdot p_f^s\cdot\sum_{m\in\mathcal{M}_j}(\rho_{m,j}p_{j,m,f}^i)\right)\right)$ is the probability that CR j cannot get content via D2D links, $\left(1-\sum_{j\in\mathcal{R}}\left(v_{i,j}\cdot p_f^s\cdot\sum_{m\in\mathcal{M}_j}(\rho_{m,j}p_{j,m,f}^i)\right)\right)$ is the probability that CUE i does not share resources with D2D links. The average bandwidth consumption is defined as

$$\begin{aligned} B_{BS} &= \sum_{j\in\mathcal{R}}\left[\left(1-\sum_{i\in\mathcal{C}}\left(v_{i,j}\cdot p_f^s\cdot\sum_{m\in\mathcal{M}_j}(\rho_{m,j}p_{j,m,f}^i)\right)\right)B\right] \\ &+\sum_{i\in\mathcal{C}}\left[\left(1-\sum_{j\in\mathcal{R}}\left(v_{i,j}\cdot p_f^s\cdot\sum_{m\in\mathcal{M}_j}(\rho_{m,j}p_{j,m,f}^i)\right)\right)B\right], \end{aligned} \tag{23.13}$$

and the average power consumed by these users is

$$\begin{aligned} P_{BS} &= \sum_{j\in\mathcal{R}}\left[\left(1-\sum_{i\in\mathcal{C}}\left(v_{i,j}\cdot p_f^s\cdot\sum_{m\in\mathcal{M}_j}(\rho_{m,j}p_{j,m,f}^i)\right)\right)P_B\right] \\ &+\sum_{i\in\mathcal{C}}\left[\left(1-\sum_{j\in\mathcal{R}}\left(v_{i,j}\cdot p_f^s\cdot\sum_{m\in\mathcal{M}_j}(\rho_{m,j}p_{j,m,f}^i)\right)\right)P_i^c\right]. \end{aligned} \tag{23.14}$$

Based on the above discussion, the optimization problem is to solve a three-dimensional optimization problem that can perform CS selection, spectrum resource allocation, and power allocation jointly. The goal is to maximize the system efficiency by jointly considering spectrum efficiency and energy efficiency, leveraging both physical link condition and social interactions. The mixed integer non-linear optimization problem can be formulated as

$$\max_{P_i^c,\,P_{m,j}^d,\,\rho_{m,j},\,v_{i,j}} \sum_{f=1}^{F}\frac{R_{D2D}+R_{BS}}{(B_{D2D}+B_{BS})(P_{D2D}+P_{BS})}. \tag{23.15}$$

$$s.t.\quad 0\le P_i^c\le P_{\max}^c,\, 0\le P_{m,j}^d\le P_{\max}^d,\, 0\le P_i^c+P_{m,j}^d\le P_{\max}, \tag{23.16a}$$

$$\frac{P_i^c g_{i,B}}{P_{m,j}^d h_{m,B}+\sigma_N^2}\ge \mathrm{SINR}_c,\ \frac{P_{m,j}^d g_{m,j}}{P_i^c h_{i,j}+\sigma_N^2}\ge \mathrm{SINR}_d \tag{23.16b}$$

$$\sum_{m\in\mathcal{M}_j}\rho_{m,j}\le 1,\ \rho_{m,j}\in\{0,1\}, \tag{23.16c}$$

$$\sum_{j=1}^{J}v_{i,j}\le 1,\ \sum_{i=1}^{N}v_{i,j}\le 1,\ v_{i,j}\in\{0,1\}. \tag{23.16d}$$

Constraint (23.16a) demonstrates the individual power constraint and the joint power constraint. Constraint (23.16b) guarantees the communication quality of the involved users by satisfying the minimum signal-to-interference-plus-noise ratio (SINR) requirements. Constraint (23.16c) regulates that a CR can receive a certain content from at most one CS at one time, and constraint (23.16d) represents that the resource of a CUE can be accessed by at most one D2D link and one D2D link can reuse the spectrum of at most one CUE at one time.

23.2.3.2 Social-based problem optimization

Notice that it is difficult to find the global optimal solution of the optimization problem in (23.16) directly, because it is an NP-hard problem which involves both continuous variables and binary variables. Therefore, we first determine the initial candidate CS set for each CR and then propose a heuristic algorithm to decouple the optimization problem into several sub-problems to achieve a sub-optimal solution.

Recall that if there are less than M achievable CSs, we keep all the achievable CSs in its initial candidate set. If a CR can find more than M CSs with the desired content within the distance of d_{max}, we choose the best M CSs as the initial set, and the selection of the best M candidates will provide a higher success probability when transmitting content to CR. Referring to the success probability expressed in (23.1), we can find that with better physical link condition (i.e., larger $\frac{p^d_{m,j} g_{m,j}}{\sigma_N^2}$) and stronger social tie (i.e., larger $q_{m,j}$), a higher transmission success probability can be achieved when the other parameters are fixed. Therefore, we evaluate the social information $q_{m,j}$ and the physical information g_{mj} when we determine the candidate CS set. For simplicity, we define the following parameter

$$\theta_{m,j} = \varepsilon \frac{g_{m,j}}{g_{j,\max}} + (1-\varepsilon)\frac{q_{m,j}}{q_{j,\max}}, \quad 0 \le \varepsilon \le 1, \tag{23.17}$$

where $g_{j,\max}$ and $q_{j,\max}$ are the maximum channel gain and maximum contact time duration between CR j and its adjacent CSs with desired content. Basically, a larger $\theta_{m,j}$ indicates a higher transmission success probability. Therefore, when selecting the candidate CSs in $\mathcal{M}_j$, we first find CSs with desired content within the communication distance. If the number of candidate CSs is larger than M, we calculate $\theta_{m,j}$ for every CS m and choose m CSs with the highest $\theta_{m,j}$.

After determining the candidate CS set, we propose a heuristic algorithm to split the integer non-linear optimization problem in (23.15) into the following steps:

1. Allocate initial spectrum resource for each CR which requires content delivery, and initialize transmit power for data transmission;
2. Perform CS selection for each CR by optimizing the combinatorial variable $\rho_{m,j}$;
3. Implement the spectrum resource allocation for the paired D2D links;
4. Manage power optimization for CUEs and CSs based on the result of D2D link formation and spectrum resource allocation;
5. Repeat 2 to 4 until converged.

Notice that steps 2 and 3 involve only binary variables, $\rho_{m,j}$ and $v_{i,j}$, respectively. Matching algorithms such as the Kuhn-Munkres (KM) algorithm [13] can be adopted to solve this type of problems. However, the KM algorithm incurs high-computational complexity, especially with a large number of CSs, CRs, and CUEs. To reduce the computational complexity without severely degrading the matching performance, a low-complexity inverse popularity pairing order (IPPO) algorithm was proposed in [14]. Taking the CS-CR matching problem in step 2 as an example, to increase the probability of overall matching success without iteration, the IPPO sorts the CRs in ascending order of the available edges, and starts with the CR with the fewest edges. After finding the best pairing partner for the least popular CR, the matching process continues to deal with the second CR in the list with the next fewest number of edges, and so forth. The IPPO can effectively guarantee user fairness and incur low time complexity with near-optimal performance.

23.2.3.3 Simulation

In this part, we evaluate the performance of the proposed heuristic scheme which maximizes the system efficiency by proper CS selection and wireless resource allocation. In the following simulation results, KM refers to the case where we use the KM algorithm to optimize CS selection and spectrum allocation in our heuristic-based mixed integer nonlinear optimization problem. IPPO refers to the case where we use the IPPO algorithm to optimize CS selection and spectrum allocation. In addition to the KM and the IPPO algorithms, we also take the Random algorithm which selects cooperative nodes randomly as a benchmark for performance comparison.

Figure 23.5 shows the impact of the number of candidate CSs for each CR and the number of CUEs on the achievable system efficiency. Figure 23.6 shows the corresponding average simulation time consumption of using different matching algorithms in the procedure of CS selection and spectrum allocation. We can see that a larger number of CUEs and a larger number of candidate CSs for each CR both lead to better system efficiency, but at the cost of higher-computational time consumption. The reason is that more CUEs lead to a higher probability that better spectrum resource allocation can be found. More candidate CSs accessible for each CR also indicates a higher probability that a better transmitter can be found for data transmission. Besides, the computational time increases with more user nodes as we can expect.

To sum up, the achievable system efficiency of the KM algorithm outperforms that of the IPPO and Random algorithms; however, the performance gap between the KM and the IPPO is insignificant comparing with the performance gap between the KM and the Random algorithm. On the other hand, IPPO costs a little more time than the Random algorithm but is significantly less time-consuming than KM. Thus, IPPO provides a very good performance-complexity trade-off.

23.3 D2X-based caching with deep learning

In the previous section, we have demonstrated that caching content in user equipments can effectively improve the system performance. However, what we

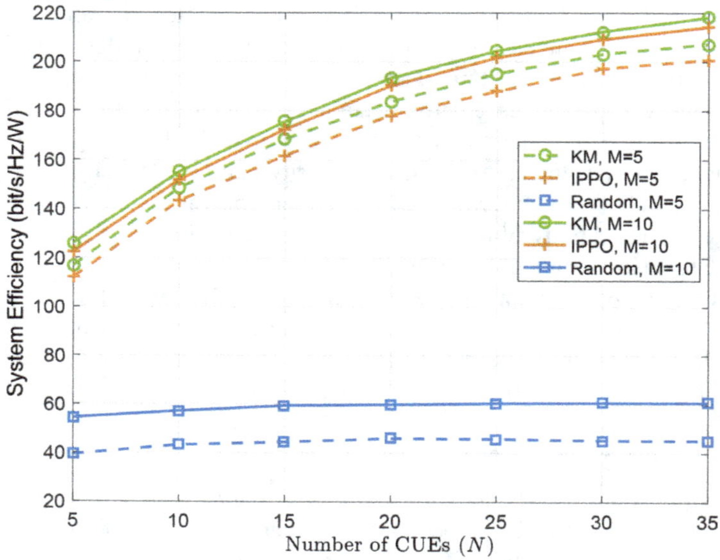

Figure 23.5 System efficiency

discussed in the previous section only applies in the wireless social network that covers a small region of area. The D2D communication technique is just one small part when we talk about the entire social network. With the growth of mobile social media, the amount of data grows exponentially [15–17] and the social media network has not only spread nationwide but also internationally. However, with the limitation of data transmission rate and the capability of the hardware, the

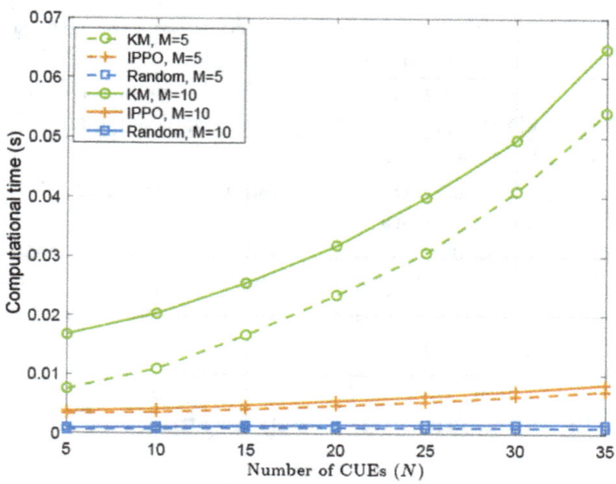

Figure 23.6 Simulation time

exponential data growth of mobile social media network causes significant burden to communication networks. As mentioned before, D2D transmission only works for short-distance communications, which is no longer sufficient when we consider the entire social network. Hence, in order to overcome this problem, the use of BSs and SBSs is indispensable.

Since most of the users rely on their mobile devices to access social networks, the data transmission performance among the wireless networks is extremely important for users [18]. With the increasing density of SBSs, the backhaul capacity becomes the bottleneck to effectively utilizing the heterogeneous small cell networks (HSCNs) [19]. In particular, this approach requires high-speed backhaul to connect the BSs or the core network to SBSs, which is one of the main drawbacks of this approach. Currently, there are researches considering the scenarios where the user density will be comparable to the density of SBSs [20]. In this regime, deploying high-speed fiber backhaul to improve the D2I transmission rate is not a good option due to the extremely high implementation and operation cost. In order to overcome this problem, the use of the caching capabilities of BSs, SBSs, and UEs is necessary. By caching the mobile social media data closer to the users who have a high interest on that data, the users can fetch the data directly from another user devices without reaching the servers via a BS or SBS. This strategy is called D2D communication [21–24], which can not only utilize the HSCNs more efficiently but also provide high QoS experience due to the reduction of service delay [25]. Moreover, caching data in SBSs and data transmission via D2D communications are envisioned to be in the next-generation mobile networks in order to satisfy the reliable service at anytime and anywhere for mobile social media network users [26, 27]. Some relative works can be found in [1, 21, 26, 28–30]. In [1], the authors show the framework that improves the caching-based socially aware D2D communication. In [21], the authors discuss the matching process for the D2D and D2I communications by using the well-labeled data and the calculation of the information value for each data. The challenges of caching data in SBSs are discussed in [26] and [30]. Authors of [28] present a data dissemination architecture of a distributed caching scheme. In [29], the authors develop a caching-and-delivery scheme which enables data to be prefetched. From these works, we can see how the D2D and the D2I communications can effectively and efficiently increase the system social welfare as the number of the end users in the network increases.

Although applying the D2X communication technology to the mobile social network seems to be a huge step across the limitation of the traditional wireless networks framework, it is unable to show the outstanding benefits if the system randomly caches data into arbitrary cache entities. To overcome this problem, in this section, we adopt the ISF algorithm [21] and the deep learning LSTM neural network into the caching system design [31]. First, the ISF algorithm can build up a stable many-to-many matching framework for cache entities among the network. Furthermore, with LSTM neural network, each cache entity has a better idea on the value of caching a data to itself in order to make correct caching decisions. In this section, we will show the details of applying the ISF algorithm and the LSTM neural network in the caching scheme design.

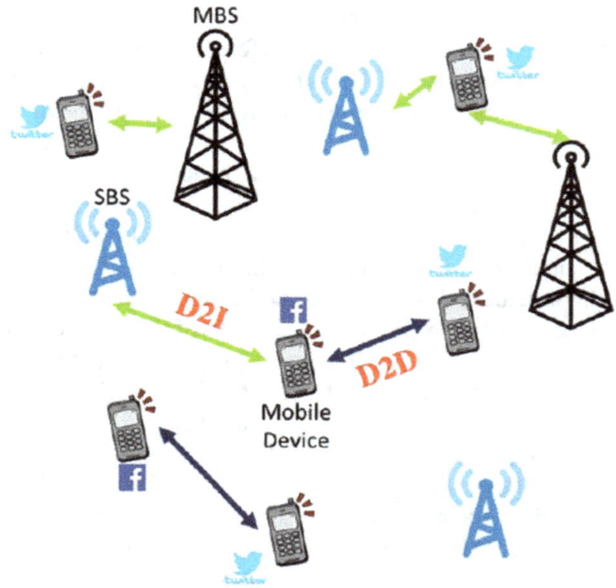

Figure 23.7 *System model: content sharing via D2X communications*

23.3.1 System model

In this simulated system environment, we used the social media network from Twitter as the background where the posts are mainly related to 2016 U.S. presidential election. In the system model, we assume there are m_{ud} users and each user has its own mobile device which is in the set of $C^{ud} = \{c_1^{ud}, ..., c_{m_{ud}}^{ud}\}$. In our work, we consider BSs as one type of cache entities and they are directly connected to the servers. We assume there are m_b BSs cache entities in the set of $C^b = \{c_1^b, ..., c_{m_b}^b\}$ and m_s SBSs cache entities in the set of $C^s = \{c_1^s, ..., c_{ms}^s\}$. In addition to BSs and SBSs, since user mobile devices also have the ability to store data, we can treat each user mobile device as a cache entity. In our system model, cache entities are randomly located in the system as shown in Figure 23.7. In order to provide a unified study on the data caching allocation solution [32], without loss of generality, we assume that all cache entities of the same type have identical properties. Thus, cache entities in C^{ud} have an identical cache storage capacity of $cap_{ud} \in \mathbb{R}^K$ where K is the number of data types which will be discussed later on. Similarly, cache entities of BSs or SBSs have identical cache storage capacities of $cap_b \in \mathbb{R}^K$ and $cap_s \in \mathbb{R}^K$, 23.3.4.1 Preference list respectively. For better understanding and simplicity, in the rest of this section, the set of $\mathcal{N} = \{n_1, ..., n_{m_{ud}+m_b+m_s}\}$ denotes the joint set of C^{ud}, C^b, and C^s cache entities. We also assume that each cache entity n_i initially caches a certain amount of data in various categories such as sports, weather, political, education, etc. We then represent the quantity of data of each data type cached by cache entity n_i to be $Q_i = \{q_1^i, ..., q_K^i\}$. After the initial data caching, content exchanging can happen when any two cache entities are linked via D2X communications to exchange a certain amount of the cached data.

Depending on users' request demands and interests, different data types and the contents of the data have different priorities (weights) for each cache entity. For example, assume there is a piece data $\mathfrak{D}$ that is categorized as a political type of data and its content is about Donald Trump. There are two cache entities, $\mathcal{X}$ and $\mathcal{Y}$, where both of the cache entities are interested in political type of data. Hence, the priority of the data type for both $\mathcal{X}$ and $\mathcal{Y}$ is the same. However, $\mathcal{X}$ is interested in the data content which involves the keywords of "President Trump" and $\mathcal{Y}$ is interested in "Hillary Clinton." Hence, cache entity $\mathcal{X}$ is more likely to cache data $\mathfrak{D}$ than cache entity $\mathcal{Y}$. Thus, we can construct a positive semidefinite matrix for the weights for each incoming data as $WT_i = \{wt_1, ..., wt_K\}$. The total information $val_{i,j}$ that cache entity n_i receives from another cache entity n_j can be represented as

$$val_{i,j} = \sum_{j=1}^{K} wt_j \cdot q_j^i, \tag{23.18}$$

where $i \neq j$ and $\cdot$ is the dot product operation. Notice that BSs and SBSs can also exchange the cached data with the user mobile device cache entities since it is a two-side data transmission process. The main differences between user mobile devices, BSs, and SBSs are the cache storage capacity and service delay. In addition, the old data cached in each cache entity will get replaced while the cache entity is performing data exchanging process.

By exchanging data via D2X communications with another cache entity, the cache entities can effectively improve their own knowledge of the current network, increase the cache hit rate, and reduce service latency. By constructing a D2X link matrix of all cache entities, we can quantify the exchanged data value in the entire network. Define the transmission matrix ρ as

$$\rho = \{\rho_{i,j} | n_i, n_j \in \mathcal{N}\}, \tag{23.19}$$

where

$$\rho_{i,j} = \begin{cases} 1, & \text{if } n_i \text{ and } n_j \text{ can be connected via a D2X link,} \\ 0, & \text{otherwise.} \end{cases} \tag{23.20}$$

In addition, the capacity of the user mobile devices, BSs, and SBSs determines the number of the radio interfaces equipped on each cache entities.

23.3.2 *Problem formulation*

While maximizing the information value during data transmission, we also need to consider the quality of the data transmission when performing data exchanging process between two cache entities. The quality of the data transmission can be quantified by transmission rate $r_{i,j}$ from cache entity n_i to cache entity n_j. The transmission rate is defined as

$$r_{i,j} = l_{i,j} \log\left(1 + \frac{p_{i,j} g_{i,j}}{\sigma_{i,j}^2}\right), \tag{23.21}$$

where $l_{i,j}$ is the sub-band assigned for transmission from cache entity n_i to n_j and the noise of the band is denoted by $\sigma_{i,j}$. $p_{i,j}$ and $g_{i,j}$ are the transmission power and the channel propagation gain from cache entity n_i to n_j. Channel propagation gain $g_{i,j}$ is determined by a constant system parameter k_p, the fast fading gain $\beta_{i,j}$, the slow fading gain $\zeta_{i,j}$, and the distance $d_{i,j}$ between cache entities n_i and n_j, i.e.,

$$g_{i,j} = k_p \beta_{i,j} \zeta_{i,j} d_{i,j}^{-\alpha}, \tag{23.22}$$

where α is the path loss exponent.

Each cache entity n_i can communicate with multiple cache entities via D2X communications as long as the capacity of the cache entity n_i can handle the data. The utility u_i, i.e., the information value of a cache entity is defined as the summation of the data value it gains from each data exchange process via D2X communications during a period of time T. With the matrix ρ defined in Section 23.3.1, the utility u_i of each mobile social media cache entity n_i is calculated by

$$u_i = \sum_{n_j \in \mathbb{C}_i} T \cdot \rho_{i,j} \cdot val_{i,j} \cdot r_{i,j}, \tag{23.23}$$

where $\mathbb{C}_i$ denotes the set of cache entities n_j that exchanges data with n_i.

The more data a cache entity exchanges with other cache entities around it, the higher interest degree the cached data holds. In particular, data with a higher interest degree means that more users are attracted by the cached data compared to others. This leads to higher cache hit rate when mobile social media users request for self-interests data. Therefore, the objective of this chapter is to maximize the social welfare by maximizing the total received information value for each cache entity during a certain communication period T. Based on the objective and the constraints in the system model we discussed above, the problem can be formulated as follows:

$$\max_{\rho_{i,j}} \sum_{ud_i \in UD} \sum_{n_j \in N} T \cdot \rho_{i,j} \cdot val_{i,j} \cdot r_{i,j}, \tag{23.24}$$

$$\text{s.t.} \begin{cases} C1: \sum_{n_j \in \mathcal{N}} \rho_{i,j} \le cap_i, & \forall n_i \in \mathcal{N}, \\ C2: \gamma_{j,i} = \rho_{i,j} \frac{p_{i,j} g_{i,j}}{\sigma_{i,j}^2} \ge \gamma_{\min}, & \\ C3: \rho_{i,j} = 0, & \forall i = j. \end{cases} \tag{23.25}$$

(23.21) defines the system objective that maximizes the exchanged data value between cache entities during a certain communication period T. In other words, the system is trying to maximize the utility for all users' mobile social media devices. The capacity requirement of each cache entity is indicated by the system constraint $C1$. In addition, there is a minimum requirement on the signal to noise ratio (SNR) for each cache entity which is represented by the system constraint $C2$. Finally, since there is no point for one cache entity to connect to itself and exchange data, there will have no D2D link nor D2I link for a cache entity to assign to itself as shown in constraint $C3$.

The formulated problem is a mix integer linear programming (MILP) optimization problem. In general, the MILP problem is NP-hard according to [33]. To obtain a centralized solution, the global information of the system is usually required. When the cache entities increase, the system complexity increases significantly. Since the D2X communication networks rely on the wireless networks, it will be difficult to gather the global information from all cache entities due to the high mobility and the various channel conditions of each cache entity. Thus, this problem motivates us to find a distributive approach where each cache entity can make its own decisions without knowing the global information from the other cache entities.

23.3.3 Mobile social media natural language processing using deep learning

Mobile social media users often request and upload a significant amount of data from their mobile devices when they are using mobile social media applications such as Twitter, Facebook, etc. In order to reduce the data transmission latency, caching the right data to increase the cache hit rate is necessary. To achieve this goal, applying the natural language processing on the data using deep learning can accurately classify the data to the right category and the cache entities can have a better determination on whether to cache the data or not.

It is a challenge for machines to perform natural language processing (NLP) especially with a small amount of labeled data available. NLP can break down into two major parts: semantics and sentiment. Compared to the sentiment of a sentence, the semantic of a sentence is relatively easy to analyze due to the solid development of grammar in natural language. On the other hand, the sentiment of a sentence usually depends not only on one of the words in the sentence but also on the words before and after the current word and sometimes depends on the sentences before and after. The most common way to perform sentence sentiment analysis is studying the statistic from a huge amount of well-labeled data, by applying convolutional neural network (CNN) [34] or LSTM [35].

In this section, we will first discuss the importance of data preprocessing where some information in the data should be removed in order to increase the accuracy of the deep learning model. We then show the benefits of the LSTM and the detail of the LSTM architecture which includes the choice of the hyperparameters, regularization process, etc.

23.3.3.1 Data preprocessing

23.3.3.1.1 Data preprocessing

Here, we consider a post as a piece of data. A mobile social media post is the collection of sentences. Mobile social media enables users to share their thoughts along with additional information such as tags of other users, URLs, etc. When performing sentence analysis, the additional information will be removed from the post since it usually contains useless information.

23.3.3.1.2 Sentence preprocessing

Each sentence is the collection of words where each word contains a certain amount of information of the data (Figure 23.8). In another word, the combination the

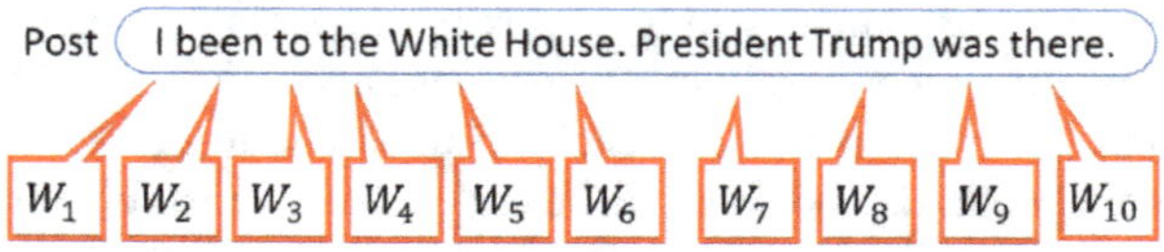

Figure 23.8 Post tokenization

features of all the words in the data can be used to build up the statement of the data, which is the semantic and sentiment of the data.

It is very straightforward to think that a long sentence contains more useful information than a short sentence. Unfortunately, this statement is not necessarily true. In the case of having grammar and spelling errors, some features of the sentence might end up being noise to the sentence when applying the sentence analysis, which might decrease the analysis accuracy. To overcome grammar and spelling errors, we will perform the stemming and lemmatizing process on each word. Keep in mind that word modification process will only modify the words in the sentence but the length of the sentence will remain the same.

With respect to the computing complexity, the length of the sentence is the major factor that affects the complexity. The longer the sentence, the higher complexity of the calculation is. We can reduce the length of the sentence by filtering out some less significant words which are known as the stop words. Stop words usually refer to the words that appear most common in a language and have less impact on the sentence feature when removing them from the sentence. After filtering out the unnecessary words from the sentence for all sentences in the dataset, we adjust the number of features in the data to size H which is a hyperparameter that will be discussed in Section 23.3.3.2. In other words, if the data have more than H features, this data will be truncated to size of H. However, if the data have less features than H, we zero-pad for the features to H to make the computation consistent.

Although stemming, lemmatizing, and removing stop words can reduce computing complexity, the processed sentence contains less information than the original sentence which might end up reducing accuracy when performing sentence analysis. This is the trade-off between complexity and accuracy [36]. Reducing complexity usually has a higher priority than increasing accuracy due to the small effect on accuracy from sentence preprocessing.

23.3.3.2 Sentence sentiment analysis using deep learning

Our deep learning model uses the LSTM as the architecture of the model. LSTM is a type of recurrent neural network (RNN) where the neurons in the neural network will remember the certain degrees of the previous state of the neuron. LSTM allows forward and backward connections between neurons. Although in most of the cases the training process of the CNN is relatively faster than that of the LSTM due to the calculation parallelism, the neurons in CNN are unable to have memory about the previous state. As we mentioned before, the sentiment of a sentence highly depends not only on each individual words in the sentence but also on the sequence of the words. Therefore, LSTM is more suitable for the sentence sentiment analysis task.

In the traditional RNN, the vanishing and exploding gradients problem is the major obstacle to the neural network performance since RNN highly depends on the magnitude of the weights in the transition matrix. If the weights in the transition matrix are small, it can lead to vanishing gradients due to the multiplication process from each timestamp and the learning process will be either very slow or even stopped [37]. On the other hand, if the weights in the transition matrix are large, eventually the gradients will explode and diverge the learning process.

LSTM is an implementation of the traditional RNN where LSTM has a new structure called a memory cell. Each neuron in LSTM is formed by four main components: input gate, forget gate, output gate, and memory cell [37]. Each neuron can decide which input data to be accepted from another cell, what data to be stored in the memory cell, and where the output data to be written to depending on the degrees of importance of the data [35, 38, 39]. This is kind of similar to the digital storage for the computer hard drives except that the gates are analog where the activation function for input and output gates is often to be the hyperbolic tangent (tanh) and the forget gate often uses the sigmoid activation function. Since the LSTM adopts the concept of the back-propagation from RNN, the differentiable analog signal is suitable for the LSTM [40].

23.3.3.2.1 Dataset

Natural Language Toolkit (NLTK) [41] is a free toolkit made for Python programming to deal with natural language processing. In order to train and test classifiers used in this work, we gather the dataset provided from multiple database [41–45]. Some data are well labeled while some only include information about the time, location, gender, length, etc. The most common and important label for all the data is the sentence sentiment which is either positive or negative. In addition, the dataset we got from [41–45] contains multiple categories including political opinions, Twitter tweets, movie reviews, restaurant reviews, etc. With varying categories of sentences, we can reduce the overfitting during training phase of our LSTM. By applying data preprocessing to data x_i and storing along with the associated sentiment label, we get $d_i = (x_i^*, s_i)$ where x_i^* is the data x_i after the data preprocessing process and $s_i \in \{0, 1\}$ represents the sentence sentiment, i.e.,

$$s_i = \begin{cases} 1, & \text{if } x_i \text{ has positive sentiment,} \\ 0, & \text{if } x_i \text{ has negative sentiment.} \end{cases} \tag{23.26}$$

23.3.3.2.2 Hyperparameters

In our LSTM model, the hyperparameters used are as follows: The maximum number of features is the dictionary size which is 20000 purified words from our database, the maximum length of a data varies from 25 to 200, each piece of data will be transformed into a vector of size 128, the sigmoid function will be used as the activation function for the neurons, epoch is set to 50 which is the number of times to iterate over the training data arrays, batch size is 32 which is the number of samples per gradient update, and the dropout rate is the rate that each feature will be dropped out during the training process which is set to

0.5. Some of hyperparameters are chosen via grid search from our dataset and some are suggested by [46–48]. In Section 23.3.5, we will show tuning on the hyperparameters and the numerical results of the tuned model.

Since our dataset does not come with a standard development set, we randomly select 10% of our training set as our development set. We adapt the Adadelta update rule [49] for the training process which calculates the stochastic gradient descent over shuffled batches.

23.3.3.2.3 Classification

Once finishing training the LSTM neural network, all the parameters in the network have been set to optimize the accuracy for the input data. When we input a data into our LSTM, the features of the data will be extracted out and proceeded to each neuron. The output wt of the LSTM neural network is the weight wt of the data we discussed in (23.18). wt is a scalar value where $wt \in (0, 1)$ since the last layer of the LSTM is the sigmoid unit. The predict class $CLASS_{predict}$ for the data with output wt greater or equal to 0.5 will be set to positive (i.e., with positive sentiment). On the other hand, the data with output less than 0.5 will be assigned to negative sentiment data, i.e.,

$$CLASS_{predict} = \begin{cases} \text{positive}, & \text{if } wt \geq 0.5, \\ \text{negative}, & \text{otherwise.} \end{cases} \tag{23.27}$$

23.3.4 Proposed algorithm

As we can see in the MILP optimization problem in (23.24), the computational complexity increases exponentially with the increasing of the size of the mobile social media network. Therefore, it is necessary to find a distributed way to solve this problem. In this section, we propose a distributed solution with lower computational complexity but similar performance when compared with the college admission-based content caching algorithm proposed in [50].

In the contest-aware data caching problem in [50], data are able to request to be cached in cache entities and the cache entities can determine whether to accept the piece of data or not. A piece of data ranks the cache entities by its preferences, e.g., the data transmission delay. On the other hand, cache entity will also rank the data that request for caching by the data popularity and size. Intuitively, cache entity can only consider the data with the size less or equal to its capacity before its capacity is full. Generally, in the matching theory, although the matching-based content caching algorithms mentioned in some articles (e.g., [29, 50]) can solve such a stable matching problem in a post-condition manner, lots of the matching algorithms for data caching are either one-to-one or one-to-many matching which is not practical in the real world since one piece of data may be cached in multiple cache entities and one cache entity can also cache multiple data. Another problem with some matching algorithms, such as the college admission algorithm proposed in [50], is that the preference of each cache entity to the data is only based on some physical restrictions such as distance, transmission rate, hardware efficacy, etc. Most of the preference lists for the cache entities over the data in the matching algorithms either neglect the content of the data or fail to quantify the importance of the data based on the content of the data.

The solution to above problems is to combine the idea of the ISF algorithm [21] with our proposed deep learning data analysis to determine the meaning of the content of each piece of data. With the data analysis via the LSTM model in Section 23.3.3, the cache entities will have a better understanding of the content of the data when determining whether to cache the data or not. Therefore, the cache entities will have more confidence when generating the preference list. In the following sections, we will discuss how the cache entity generates its preference list and the ISF matching algorithm that will be used for the D2X communications.

23.3.4.1 Preference list

To model as the ISF game, each cache entity n_i in the model is considered as an agent in ISF. Based on ISF, each cache entity will first search for an acceptable set of cache entities $\mathbb{S}_i$, where $\mathbb{S}_i \subseteq \mathcal{N}$, to itself who satisfy the SNR requirement, i.e., the *C*2 condition in (23.25). Once the searching process is done, the cache entity n_i will start to rank each cache entity n_j in the acceptable set $\mathbb{S}_i$ by the scalar value of the product of the transmission rate and the total information value. We will call the scalar value the "score," i.e., $score_{i,j}$.

$$score_{i,j} = r_{j,i} \cdot val_{j,i}. \tag{23.28}$$

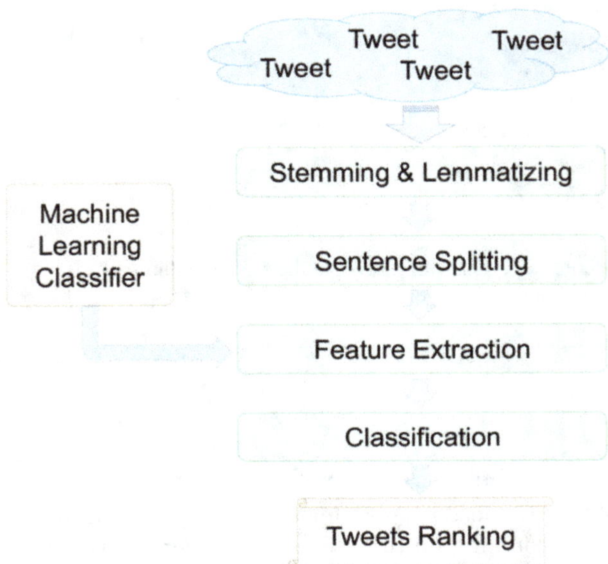

Figure 23.9 Tweets ranking

After the calculation, each cache entity in the acceptable set $\mathbb{S}_i$ will be assigned an associated score. Finally, we sort the list based on the scores in decreasing order and produce the preference list PL_i for cache entity n_i, i.e.,

$$PL_i = \text{sort}\{n_j \in \mathbb{S}_i\} \text{ in decending order of } score_{i,j}. \tag{23.29}$$

We name the algorithm as the LSTM caching preference algorithm in this work, as shown in Figure 23.9.

Algorithm 1 LSTM Caching Preference Algorithm

Definition:
$count$: an arbitrary variable for counting the index of a set;
$\mathcal{N}$: the set of all cache entities; PL_i: the preference list for cache entity n_i;
Q_j: the data set of n_j; $\gamma_{j,i}$: the signal to noise ratio (SNR) from n_j to n_i;
$\gamma_{\min}$: the minimum SNR requirement; $\mathbb{S}_i$: the set of acceptable cache entities for n_i;
$\mathcal{T}_i$: the temporary matrix for n_i where the first column contains the cache entities and the second column contains the score with respect to the cache entities in the first column;
$val_{i,j}$: total information value that n_i receives from n_j;
WT_j: the weight matrix for Q_j determined by n_i;
Initialization:
$count = 0$; $\mathbb{S}_i = \varnothing$; $\mathcal{T}_i = \varnothing$;
Input: n_i
Output: PL_i
for $n_j \in \mathcal{N}$ **do**
    if $\gamma_{j,i} \geq \gamma_{\min}$ and $i \neq j$ **then**
        Insert n_j into $\mathbb{S}_i$;
    end if
end for
for $n_j \in \mathcal{S}_i$ **do**
    $\mathcal{T}_i$[count][1]= n_j;
    for $q_{m,n} \in Q_j$ **do**
        WT_j[m][n]=confident level of $LSTM(q_{m,n})$;
    end for
    $val_{i,j} = \sum_{v=1}^{K} wt_v \cdot q_v^j$; $\mathcal{T}$[count][2]= $val_{i,j}$; $count += 1$;
end for
Sort $\mathcal{T}$ in decreasing order with respect to the second column of $\mathcal{T}$;
Replace PL_i with the first column of $\mathcal{T}$;
return PL_i

23.3.4.2 Two-sided matching game

The purpose of the two-sided many-to-many matching game is to match both sides of cache entities according to their preference to others in the opposite side [51]. Applying machine learning and matching theory to the mobile social media networks draws many interests.

In the two-sided matching game, we use the ISF algorithm suggested in [21]. There are two phases in the ISF algorithm. In the first phase, cache entities will generate their own preference lists. Here we modify the algorithm by using the deep learning neural network to generate the preference lists. Based on cache entity's self-interest on the data from its acceptable set of cache entities, we can produce the preference list for the cache entity according to the strategy in Algorithm 1. The second phase of the ISF algorithm adopts the matching algorithm suggested in [52] where the algorithm uses the preference lists generated in phase one to determine the stable matches between the cache entities. In the stable fixture game, a stable matching means that there is no player who has the incentive to change their partner(s). According to [21], once the cache entity n_i gets its own preference list, the ISF algorithm will match as many cache entities on the preference list as long as the total data from all matched cache entities does not exceed its capacity. The ISF algorithm can be more understandable using a chess tournament for explanation. Each player in the tournament will be matched with multiple players based on their skill level. Each chess game happens individually which means if player 1 plays against player 2 in the first game, then against player 3 in the second game, this does not require player 2 to play against player 3 and neither does it mean player 2 cannot play against other players 4, 5, or 6.

23.3.5 Performance evaluation

23.3.5.1 Simulation setup

In the simulation, we assume there are $N = [0, 200]$ mobile social media devices and eight BSs and SBSs. The capacity of each mobile social media device is set to 4. Also, the positions of cache entities are uniformly and randomly assigned in a square area within 2 50 000 square meters centered at Houston, TX. We focus on the Twitter users in our simulation. The data to simulate our system are collected from July 1, 2016 to October 1, 2016, which is the period of [50] presidential election. Thus, the main interest topic for users in this system is set to "Supporting rate for Donald Trump and Hillary Clinton?" Since our data are the tweets in the simulation, we will use the word "tweet" as the data in the later part of this section. During the simulation, although the number of tweets will increase along with the number of user devices, we will not compare the amount of tweets in the system with the number of user devices due to simplicity.

We apply sentence sentiment analysis through LSTM, which is the implementation of the RNN, in finding the interests of different sets of tweets to each user. As comparison, we also use other classifiers in the simulation which include CNN with and without drop rate of 0.5, Naïve Bayes Classifier, Linear Support Vector Classifier, Bernoulli Naïve Bayes Classifier, and Logistic Regression Classifier. We also adjust the maximum length of the tweets from 25 words to 200 words to see the difference between accuracy of the LSTM model.

As a remark, due to the lack of well-labeled tweets, the dataset used in the simulation for training and testing our classifiers is from database [41–45]. We build up our training set by randomly picking 80,000 sample data. As we mentioned in

Table 23.1 Classifiers' accuracy

Classifier	Accuracy %
LSTM_ml200	86.724
LSTM_ml100	83.024
CNN_0.50	76.6833
CNN_0.00	76.1726
Naïve Bayes	72.4398
Logistic Regression	71.6867
Linear SVC	68.5241
MLP	64.4578
Bernoulli Naïve Bayes	56.7996

Section 23.3.3.2, 10% of our training set will be used as development set. Notice that the data samples from [41–45] are very general, which means it can be used for most sentence sentiment analysis but might result in lower accuracy when classifying sentences into a particular category. In our case, our category is "politics" since most Twitter users focused on [50] presidential election.

For performance comparison, in addition to the IFS matching algorithm that is mentioned in Section 23.3.4.2, we also propose another two heuristic mechanisms which are the Without_D2D and the ISF_1-to-1. For the Without_D2D mechanism, all user mobile social media devices can only connect to the BSs and the SBSs near themselves. On the other hand, the ISF_1-to-1 mechanism is similar as the IFS model. The only difference is that the capacity of each cache entity is set to one. This means the matching algorithm will be a one-to-one matching algorithm instead of many-to-many matching.

23.3.5.2 Numeric result

We observe the accuracy for each of the classifiers in Table 23.1. With the Naïve Bayes classifier as a benchmark, we can see that the LSTM models not only work better than non-deep learning classifiers but also perform better than the CNN classifiers.

To show the interest of one of the cache entities, we take the subset of our tweets set to simulate. The tweets from the subset are all posted at neared area. We divide the tweets into two sets, one is about Donald Trump and the other is about Hillary Clinton. In Figure 23.10, we can see that most tweets that mentioned about Donald Trump are more of anti-Trump. On the other hand, tweets mentioned Hillary Clinton seem to be more neutral than those mentioning Donald Trump. Based on the histogram, we see the data content distribution among the system.

We then compare the ISF matching algorithm with the heuristic mechanisms, i.e., the Without_D2D and the ISF_1-to-1. For all three mechanisms, the number of user mobile social media devices varies from 20 to 160 devices with step size of 20. From Figure 23.11, we can see that all three mechanisms improve the throughput of the system as the user devices increase especially for the ISF where the throughput is

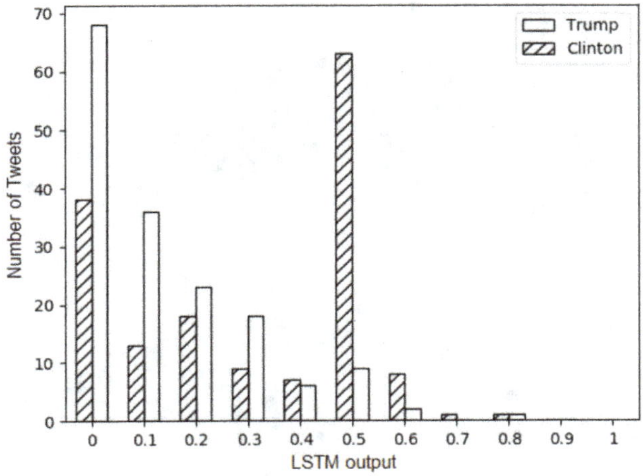

Figure 23.10 Support rate—Trump vs. Hillary

the system total exchanged data rate [21]. This observation is within the expectation since more devices joining the network means more data fed into the network which leads to higher data exchange rate among the system and improves the system social welfare. The curves in Figure 23.11 show the average user device performance on the data exchange process. Once again, the ISF curve shows the highest performance throughout the increasing of the number of user devices. Since the ISF mechanism uses D2X many-to-many communication, the user devices can exchange data not only with another user device but also with the BS and the SBS. On the other hand, the curves of the Without_D2D and the ISF_1-to-1 show the dropping performance. The reason for the Without_D2D having the worst performance is due to the high traffic in the network while each user device can only be connected to one of the BSs or the SBSs via D2I communications. Since the BSs and the SBSs have their own capacity limits, not all user devices are able to connect to a BS or an SBS at the same

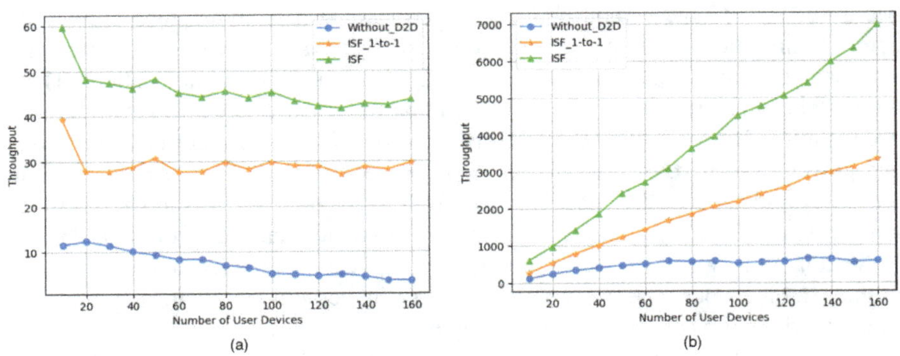

Figure 23.11 System social welfare (a) and average device performance (b)

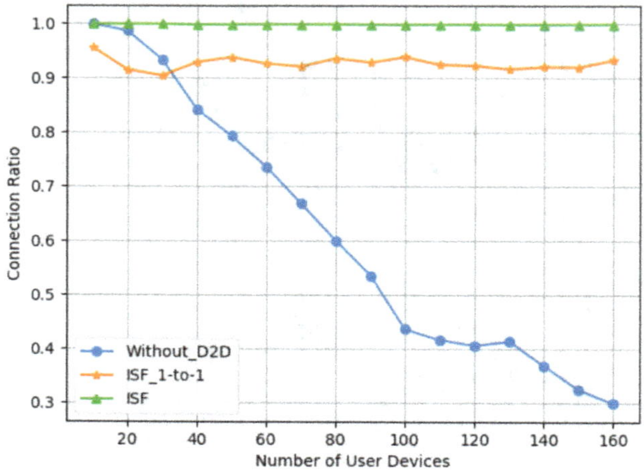

Figure 23.12 Devices connection ratio

time. Therefore, some user devices might idle for certain periods of time before connecting to a BS or an SBS. Thus, the performance of the Without_D2D mechanism is the worst among the three curves. Figure 23.12 provides the evidence supporting the discussion for Figure 23.11. We can see that the Without_D2D connection ratio drops significantly when the user devices increase. On the other hand, the connection ratio of the ISF curve remains around 100% and the connection ratio of ISF_1-to-1 curve is just slightly lower than that of the ISF curve. From Figure 23.11a, b and Figure 23.12, we can tell the importance of combining the many-to-many matching game with D2X communications in ISF algorithm.

23.4 Conclusion

In this chapter, we have first presented related research on social network-driven caching, including background of caching, social in caching and application methods. In Section 23.2, we have introduced D2D scenario and utilized hypergraph to abstract compact multiple-layer information in terms of physical location, common interest, and social ties for a caching system via socially enabled D2D communications. Then the system efficiency enhancement in D2D-based wireless caching networks has been studied. By optimizing the establishment of D2D links, spectrum allocation, and power allocation, we can effectively improve the jointly considered spectrum efficiency and energy efficiency. We have further proposed a heuristic-based mixed integer non-linear programming algorithm to solve the joint optimization problem. In Section 23.3, we have further studied the application of D2X communications in the heterogeneous wireless mobile social media networks. A context-aware data caching problem has been investigated by using the deep learning data analysis to evaluate the data interest degrees for the cache entities. Combining the data interest degrees, the locality and the capacity of each cache entity, and the transmission rate, we are able to formulate the data caching optimization problem. To solve the optimization problem, we have proposed an algorithm based on the Ivring's stable fixture algorithm and LSTM deep learning model, which can significantly increase the system social welfare and enrich the cached information in the cache entities. Simulation results have shown the effectiveness of our proposed ISF algorithm with LSTM deep learning model.

References

[1] Bai B., Wang L., Han Z., Chen W., Svensson T. 'Caching based socially-aware D2D communications in wireless content delivery networks: a hypergraph framework'. *IEEE Wireless Communications*. 2016;**23**(4):74–81.

[2] Zhang Y., Pan E., Song L., Saad W., Dawy Z., Han Z. 'Social network aware device-to-device communication in wireless networks'. *IEEE Transactions on Wireless Communications*. 2015;**14**(1):177–90.

[3] Berge C. *Graphs and Hypergraphs*. North-Holland; 1973.

[4] Wu H., Wang L., Svensson T. Resource allocation for wireless caching in socially-enabled D2D communications. IEEE International Conference on Communications; 2016. pp. 1–6.

[5] Zhao H., Zhou H., Yuan C., Huang Y., Chen J. 'Social discovery: exploring the correlation among three-dimensional social relationships'. *IEEE Transactions on Computational Social Systems*. 2015;**2**(3):77–87.

[6] Wang L., Wu H., Stüber G.L. 'Cooperative jamming-aided secrecy enhancement in P2P communications with social interaction constraints'. *IEEE Transactions on Vehicular Technology*. 2017;**66**(2):1144–58.

[7] Liu G., Yang Q., Wang H. Assessment of multi-hop interpersonal trust in social networks by three-valued subjective logic. IEEE INFOCOM 2014 - IEEE Conference on Computer Communications; 2014. pp. 1698–706.

[8] Wang L., Tang H., Čierny M. 'Device-to-device link admission policy based on social interaction information'. *IEEE Transactions on Vehicular Technology*. 2015;**64**(9):4180–6.

[9] Wang L., Wu H., Wang W., Chen K.-C. 'Socially enabled wireless networks: resource allocation via bipartite graph matching'. *IEEE Communications Magazine*. 2015;**53**(10):128–35.

[10] Ai Y., Wang L., Han Z., Zhang P., Hanzo L. 'Social networking and caching aided collaborative computing for the internet of things'. *IEEE Communications Magazine*. 2018;**56**(12):149–55.

[11] Wu Z., Wang L., Araniti G. 'Exploiting social-interest interactions on user clustering and content dissemination in device-to-device communications'. IEEE/CIC International Conference on Communications; China; 2015. pp. 1–6.

[12] Li Q., Negi R. 'Maximal scheduling in wireless ad hoc networks with Hypergraph interference models'. *IEEE Transactions on Vehicular Technology*. 2012;**61**(1):297–310.

[13] West D.B. *Introduction to Graph Theory*. **Vol. 2**. Upper Saddle River, NJ: Prentice Hall; 1996.

[14] Wang L., Wu H. 'Fast pairing of device-to-device link underlay for spectrum sharing with cellular users'. *IEEE Communications Letters*. 2014;**18**(10):1803–6.

[15] Han Z., Hong M., Wang D. *Signal Processing and Networking for Big Data Applications*. Cambridge: Cambridge University Press; 2017.

[16] Chen Z., Xu J., Tang J., Kwiat K.A., Kamhoua C.A., Wang C. 'GPU-accelerated high-throughput online stream data processing'. *IEEE Transactions on Big Data*. 2016;**4**(2):191–202.

[17] Dinc E., Ozger M., Ates A.F. Crowdsourcing-based mobile network tomography for xG wireless systems. IEEE Symposium on Computers and Communication (ISCC). IEEE; 2016. pp. 346–51.

[18] Yilmaz T., Akan O.B. 'State-of-the-art and research challenges for consumer wireless communications at 60 GHz'. *IEEE Transactions on Consumer Electronics*. 2016;**62**(3):216–25.

[19] Andrews J.G., Buzzi S., Choi W., *et al.* What will 5G be?' *IEEE Journal on Selected Areas in Communications*. 2014;**32**(6):1065–82.

[20] Quek T.Q.S., de La Roche G., Güvenç I. *Small Cell Networks: Deployment, PHY Techniques, and Resource Management*. Cambridge: Cambridge University Press; 2013.

[21] Gu Y., Cai L.X., Pan M. Exploiting the stable fixture matching game for content sharing in D2D-based LTE-V2X communications. IEEE Global Communications Conference (GLOBECOM). IEEE; 2016. pp. 1–6.

[22] Li Y., Gursoy M.C., Velipasalar S. Joint mode selection and resource allocation for D2D communications via vertex coloring. GLOBECOM 2017-2017 IEEE global communications conference. IEEE; 2017. pp. 1–6.

[23] Gao C., Tang J., Sheng X., Zhang W., Zou S., Guizani M. 'Enabling green wireless networking with device-to-device links: a joint optimization approach'. *IEEE Transactions on Wireless Communications*. 2015;**15**(4): 2770–9.

[24] Meshgi H., Zhao D., Zheng R. 'Optimal resource allocation in multicast device-to-device communications underlaying LTE networks'. *IEEE Transactions on Vehicular Technology*. 2017;**66**(9):8357–71.

[25] Calin D., Claussen H., Uzunalioglu H. 'On femto deployment architectures and macrocell offloading benefits in joint macro-femto deployments'. *IEEE Communications Magazine*. 2010;**48**(1):26–32.

[26] Bastug E., Bennis M., Debbah M. 'Living on the edge: the role of proactive caching in 5G wireless networks'. *IEEE Communications Magazine*. 2014;**52**(8):82–9.

[27] Khalid N., Akan O.B. 'Experimental throughput analysis of low-THz MIMO communication channel in 5G wireless networks'. *IEEE Wireless Communications Letters*. 2016;**5**(6):616–9.

[28] Li J., Chen Y., Lin Z., Chen W., Vucetic B., Hanzo L. 'Distributed caching for data dissemination in the downlink of heterogeneous networks'. *IEEE Transactions on Communications*. 2015;**63**(10):3553–68.

[29] Hachem J., Karamchandani N., Diggavi S. Content caching and delivery over heterogeneous wireless networks. 2015 IEEE conference on computer communications (INFOCOM). IEEE; 2015. pp. 756–64.

[30] Zhou S., Gong J., Zhou Z., Chen W., Niu Z. 'GreenDelivery: proactive content caching and push with energy-harvesting-based small cells'. *IEEE Communications Magazine*. 2015;**53**(4):142–9.

[31] Wang J., Tang J., Xu Z. Spatiotemporal modeling and prediction in cellular networks: a big data enabled deep learning approach. IEEE INFOCOM 2017-IEEE Conference on Computer Communications. IEEE; 2017. pp. 1–9.

[32] Liu N., Li Z., Xu J. A hierarchical framework of cloud resource allocation and power management using deep reinforcement learning. IEEE 37th International Conference on Distributed Computing Systems (ICDCS). IEEE; 2017. pp. 372–82.

[33] Pochet Y., Wolsey L.A. *Production Planning by Mixed Integer Programming*. Berlin: Springer Science & Business Media; 2006.

[34] Wu H., Liu B., Li Y. Tube caching: an effective caching scheme in Content-Centric networking. IEEE/ACM 24th International Symposium on Quality of Service (IWQoS). IEEE; 2016. pp. 1–2.

[35] Hochreiter S., Schmidhuber J. 'Long short-term memory'. *Neural Computation*. 1997;**9**(8):1735–80.

[36] Xu Q., Zheng R. When data acquisition meets data analytics: a distributed active learning framework for optimal budgeted mobile crowdsensing. IEEE INFOCOM 2017-IEEE Conference on Computer Communications. IEEE; 2017. pp. 1–9.

[37] Nicholson C., Gibson A., Patterson J. *A beginner's guide to recurrent networks and lstms* [online]. 2017. Available from https://deeplearning4j.org/lstm.html [Accessed 1 May 2017].

[38] Gers F.A., Schmidhuber J., Cummins F. 'Learning to forget: continual prediction with LSTM'. *IET Conference Proceedings*. 1999:850–5.

[39] Bengio Y., Simard P., Frasconi P. 'Learning long-term dependencies with gradient descent is difficult'. *IEEE Transactions on Neural Networks*. 1994;**5**(2):157–66.

[40] Hochreiter S., Bengio Y., Frasconi P. *Gradient flow in recurrent nets: the difficulty of learning long-term dependencies*. Wiley-IEEE Press; 2001.

[41] Pang B., Lee L. 'A sentimental education: sentiment analysis using subjectivity summarization based on minimum cuts'. Proceedings of the 42nd Annual Meeting on Association for Computational Linguistics; Spain: Association for Computational Linguistics Barcelona; 2004.

[42] Pechenizkiy M., Tromp E., Demirtas E., *et al. Databases and hypermedia group, TU/e* [online]. Available from http://www.win.tue.nl/~mpechen/projects/smm/ [Accessed Sep 2021].

[43] Sanders N. *Twitter sentiment corpus* [online]. Available from http://www.sananalytics.com/lab/twitter-sentiment/ [Accessed May 2018].

[44] Naji L. *Twitter sentiment analysis training corpus (dataset)* [online]. 2012. Available from http://thinknook.com/twitter-sentiment-analysis-training-corpus-dataset-2012-09-22/ [Accessed Sep 2021].

[45] Kotzias D., Denil M., De Freitas N. From group to individual labels using deep features. Proceedings of the 21th ACM SIGKDD International Conference on Knowledge Discovery and Data Mining. ACM; 2015. pp. 597–606.

[46] Hinton G.E., Srivastava N., Krizhevsky A., *et al.* Improving neural networks by preventing co-adaptation of feature detectors'. *ArXiv*. 2012:arXiv:12070580.

[47] Britz D. *Implementing a CNN for text classification in TensorFlow* [online]. 2016. Available from http://www.wildml.com/2015/12/implementing-a-cnn-for-text-classification-in-tensorflow/ [Accessed Sep 2021].

[48] iamtrask. *Anyone can learn to code an LSTM-RNN in python (*Part 1: RNN) [online]. 2015. Available from http://iamtrask.github.io/2015/11/15/anyone-can-code-lstm/ [Accessed Sep 2021].

[49] Zeiler M.D. 'ADADELTA: an adaptive learning rate method'. *ArXiv*. 2012:arXiv:12125701.

[50] Chang Z., Gu Y., Han Z. 'Context-aware data caching for 5G heterogeneous small cells networks'. 2016 IEEE International Conference on Communications (ICC). IEEE; 2016. pp. 1–6.

[51] Abdulkadiroglu A., Sönmez T. 'Matching markets: theory and practice'. *Advances in Economics and Econometrics*. 2013;**1**:3–47.

[52] Irving R.W., Scott S. 'The stable fixtures problem—a many-to-many extension of stable roommates'. *Discrete Applied Mathematics*. 2007;**155**(16):2118–29.

Chapter 24

Caching for secure social networks

F. Richard Yu[1] *and Tiffany Ying He*[2]

Social networks have continuously been expanding and trying to be innovative. The recent advances of edge caching, as well as computing and communications (3C), can have significant impacts on mobile social networks (MSNs). MSNs can leverage these new paradigms to provide a new mechanism for users to share resources (e.g., information, computation-based services). The intrinsic nature of social networks, i.e., the trust formed through social relationships among users, can be exploited to enable users to share resources securely. Specifically, mobile edge computing (MEC), in-network caching, and device-to-device (D2D) communications are applicable. When considering the trust-based MSNs with MEC, caching, and D2D, a novel deep reinforcement learning approach can be used to automatically make a decision for optimally allocating the network resources. The decision is made purely through observing the network's states, rather than any handcrafted or explicit control rules, which makes it adaptive to variable network conditions. Google TensorFlow can be used to implement the proposed approach, and to show the effectiveness of the proposed secure scheme.

24.1 Social networks with caching, computing, and communications

MSNs have been developing rapidly, which can provide a variety of social services and applications to mobile users [1]. Millions of mobile users interact with each other in MSNs, and MSNs will become one of the most important networking paradigms in future wireless mobile networks [2]. MSNs have always been trying to be innovative and leverage new technologies. The recently proposed integrated framework of computing, caching, and communication (3C) can have positive impacts on MSNs. The integration of the 3C framework and MSNs can help create a new mechanism to share resources among users. The available resources that can be shared extend beyond information, for instance, the storage devices and computing

[1]Department of Systems and Computer Eng, Carleton University, Canada
[2]College of Computer Science and Software Engineering, Shenzhen University, China

capability contributed by users can also be shared. Additionally, the MSNs have an inherent advantage, i.e., the trust of mobile users is naturally formed through the social relationships and interactions. This makes the resource sharing between users more reliable. In this chapter, the specific 3C framework with *in-network caching, mobile edge computing* (MEC), and *device-to-device* (D2D) communications is considered.

24.1.1 In-network caching

In-network caching, as a promising technology, can effectively reduce the duplicate content transmission in the network. Recent studies of applying the in-network caching technique in MSNs show that traffic loads and latency can be significantly reduced by caching contents in MSNs [2].

When caching technology is integrated into mobile networks, where to cache, what to cache, and how to cache are the major factors that make a significant influence on the system performance [3]. In [4], the authors investigate the proactive storage allocation problem over BSs in cellular networks, which is proved to be NP-hard. To get a low-complexity solution, a heuristic method is utilized and the convergence is proved by strict theoretic deduction. A caching-enabled D2D communication scheme is designed in [5], where the caching strategy is optimized by jointly considering users' social relationships and common interests with the constraint of hit ratio, delay, and caching capacity. In fact, contents with the most popularity should be cached within limited caching capacity. Since the content popularity is varying with time, learning-based caching policies are proposed in [6]. The authors formulate the distributed caching over small BSs (SBSs) as a reinforcement learning problem, and with the help of coded caching, the complicated caching problem is solved by being reduced to a convex optimization problem. In [7], Qiao *et al.* consider the mobility pattern of mobile users and formulate the mobility-aware caching problem as an optimization problem to maximize the caching utility. The problem is solved by using a polynominal-time heuristic solution. He *et al.* [8] exploit deep reinforcement learning to address the resource allocation problems in cache-enabled interference alignment networks [8, 9].

24.1.2 Mobile edge computing

With MEC, computation resources are placed at the edge of wireless mobile networks in physical proximity to mobile users [10]. Compared to traditional mobile cloud computing, MEC can provide faster interactive response by low-latency connections. Therefore, MEC has been envisioned as a promising technique to offer agile and ubiquitous computation augmenting mobile services and applications, including social services and applications [11, 12].

With the increasing popularity of compute-intensive applications, such as augmented reality, interactive video games, etc., MEC is becoming a very promising paradigm that enables the possibility of offloading the computation tasks from resource-limited mobile devices to more powerful edge servers. MEC can bring various benefits, where minimizing energy consumption and minimizing latency are two major optimization objectives [3]. In the literature, various resource allocation approaches have been proposed to solve the formulated optimization problems. For example, Zhang *et al.* [13] propose an energy-efficient computation offloading scheme that jointly optimizes radio resource allocation and offloading to minimize the energy consumption of the offloading system with the latency constraint. In their scheme, the mobile devices are first classified into three types and then wireless channels are allocated to mobile devices based on their priority iteratively. By doing this, the optimization problem can be solved in polynomial complexity. Liu *et al.* [14] design an optimal computation task scheduling policy for MEC systems. They first analyze the average delay of each task and average power consumption at the mobile device side under the proposed scheduling policy using Markov chain theory. Then they formulate a delay minimization problem with the power constraint. An efficient one-dimensional search algorithm is used to derive the optimal offloading policy. In [15], the authors jointly optimize the radio resources and computational resources to minimize the total energy expenditure, where an iterative algorithm based on successive convex approximation technique is adopted to solve the formulated non-convex optimization problem. In [16], a distributed game theoretic approach is proposed to solve the NP-hard efficient computation offloading problem. Chen *et al.* [17] choose a model-free RL technique to solve the optimal traffic offloading strategy for heterogeneous cellular networks with dynamic traffic.

24.1.3 D2D communications

Moreover, with D2D communications, users in close proximity can directly communicate with each other via D2D links, instead of accessing BSs exclusively. As a promising approach to offload traffic from BSs, D2D communications can enable the sharing of radio connectivity and direct information delivery between two close users [18, 19]. By exploiting D2D communications, the content-centric social networks can benefit from the large number of mobile devices involved. That is because the integration of D2D into content-centric social networks can enable the content caching not only in the air but also in mobile devices [20].

24.1.4 Integration of in-network caching, MEC, and communications

The comprehensive resource allocation schemes for efficient integration of computing, caching, and communication resources to achieve the optimal performance have been developed, even though not yet been widely investigated. In [21], Zhou *et al.* design a novel information-centric heterogeneous network framework, and with the virtualization technology, communication, computing, and caching resources are shared among users. They formulate the virtual resource allocation strategy as a joint optimization problem, which is a non-convex NP-hard problem. An alternating direction method of multiplier (ADMM) approach is used to solve the problem by simplifying and relaxing the non-convex problem into a convex one. ADMM method is also used in [12] to achieve the optimal resource allocation strategy in wireless networks with MEC and in-network caching being jointly considered. He *et al.*

propose an integrated framework that can enable dynamic orchestration of networking, caching, and computing resources, where the dynamics are modeled as finite-state Markov chains [22, 23]. The authors use the deep Q-leaning approach to solve the complex problems with large number of system states and actions. Research on this topic of resource allocation strategies for integrated systems of computing, caching, and communication will continue with the evolving of each of these three technologies.

Although some works have been done on applying recent advances of MEC, in-network caching, and D2D to improve the performance of MSNs, the *knowledge of social relationships* among users in MSNs is largely ignored in these new paradigms to improve the reliability and efficiency of resource sharing and delivery in MSNs. In fact, social relationships have been investigated to enhance wireless networking. For example, Zhang *et al.* [24] exploit social ties to achieve a reliable D2D communication, which can improve packet transmission and reduce the workload on the network infrastructures. Pan *et al.* [25] utilize social information as an essential element for designing forwarding algorithms in ad hoc networks. Social relationships also play an important role for routing in wireless environments [26]. In the next section, a social trust scheme will be introduced.

24.2 Social trust scheme with uncertain reasoning

Security is always an important aspect in wireless applications and services [27, 28]. Trust-based security schemes are important detection-based approaches in MSNs. The definition of trust in MSNs is similar to that in sociology, where trust is interpreted as degrees of the belief that an entity will carry out tasks as expected [29].

In this section, how to obtain the trust values of mobile users will be derived. We evaluate the trustworthiness of a mobile user by a real number Tr ranging from 0 to 1. The trust value Tr is jointly determined based on direct observations and indirect observations. The direct observation trust of a mobile user is defined as the estimated degree of trustworthiness from its directly connected mobile users based on their past experiences. However, the subjective evaluation from direct connections may be prejudiced, and therefore in order to be more objective and impartial, we also consider the rating of trust from other indirectly connected mobile users. Here, the trust value from direct observations is denoted as Tr^D and the trust value from indirect observations as Tr^{InD}. By combining these two trust values, a more accurately estimated trust value of a mobile user can be obtained as

$$Tr = \omega Tr^D + (1 - \omega)Tr^{InD} \tag{24.1}$$

where ω is the weight coefficient to adjust the weightiness between direct and indirect observations and $0 \leq \omega \leq 1$.

The trust evaluation procedure can be visually explained in Figure 24.1. In the following, how to obtain the trust evaluation from direct observations and indirect observations by using the Bayesian inference approach and Dempster-Shafer theory [30] will be discussed, respectively.

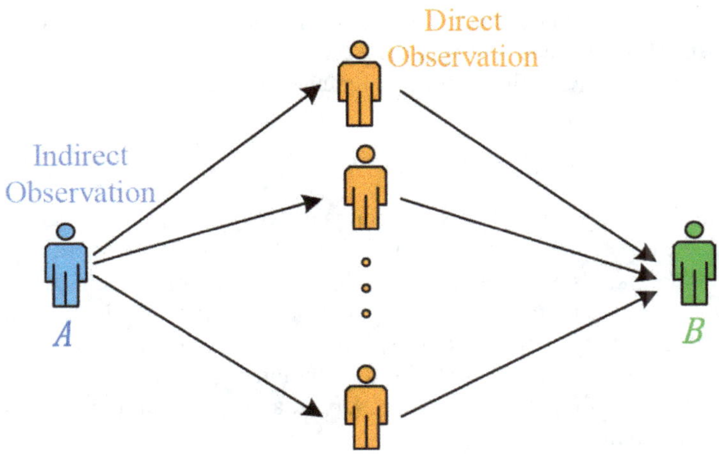

Figure 24.1 *Social trust evaluation with both direct observation and indirect observation*

24.2.1 Trust evaluation from direct observations

In the direct observations, consider that an observing mobile user can overhear the data forwarded by the observed mobile user, and identify the observed mobile user's malicious behaviors, such as discarding or modifying some of the original data.

Through multiple observations of the observed mobile user's behavior, the observing mobile user can evaluate the trust value by exploiting Bayesian inference [31], which is a method of statistical inference using the Bayes' theorem to update the probability for a hypothesis with more evidence becomes available.

Under the Bayesian framework, we model the trust of a mobile user as a continuous random variable, denoted as Φ, where ϕ takes values from 0 to 1. We assume that Φ follows a beta distribution [32], i.e., $\Phi \sim Beta(a, b)$, which is defined as follows with parameters a and b

$$Beta(a,b) = \frac{\phi^{a-1}(1-\phi)^{b-1}}{\int_0^1 \phi^{a-1}(1-\phi)^{b-1}\mathrm{d}\phi}, \tag{24.2}$$

for $0 \leq \phi \leq 1$. Here, since Φ is assumed to obey a beta distribution, the trust value can be represented by the two parameters a and b.

We summarize our belief of the trust Φ in a probability distribution iteratively as more observations are available. Assume that the prior probability density function (pdf) at the $(t-1)$th observation is known. Then, according to the Bayes' theorem, the posterior distribution at the tth observation can be obtained with the pdf as

$$f_t(\phi) = \frac{f_t(x_t|\phi, y_t)f_{t-1}(\phi)}{\int_0^1 f_t(x_t|\phi, y_t)f_{t-1}(\phi)\,\mathrm{d}\phi}, \tag{24.3}$$

where x_t and y_t are the number of data packets that has been forwarded correctly and the number of packets received by the observed mobile user at the tth observation, respectively; $f_t(x_t|\phi, y_t)$ is the likelihood function, which follows a binomial distribution as

$$f_t(x_t|\phi, y_t) = \binom{y_t}{x_t}\phi^{x_t}(1-\phi)^{y_t-x_t}. \tag{24.4}$$

In Bayesian inference, the beta distribution is the conjugate prior probability distribution for the binomial distribution [31, 32]. Since the likelihood fuction $f_t(x_t|\phi, y_t)$ follows a binomial distribution, the priori distribution $f_{t-1}(\phi)$ is definitely assumed to follow a beta distribution, which reflects what is already known about the distribution of Φ at the $(t-1)$th observation. Given that the priori distribution $f_{t-1}(\phi)$ follows a beta distribution, the posterior distribution $f_t(\phi)$ also follows a beta distribution. Particularly, if $f_{t-1}(\phi) \sim Beta(a_{t-1}, b_{t-1})$ and x_t, y_t from the tth observation are also given, then

$$f_t(\phi) \sim Beta(a_{t-1} + x_t, b_{t-1} + y_t - x_t), t \geq 1. \tag{24.5}$$

At the very beginning of the observation, there is no evidence about the distribution of Φ, thus we assume that it follows a uniform distribution, i.e., $f_0(\phi) \sim Beta(1, 1)$. Summarily, we can say that $f_t(\phi)$ follows $Beta(a_t, b_t)$ with parameters

$$\begin{aligned} a_t &= a_{t-1} + x_t, \quad & b_t &= b_{t-1} + y_t - x_t, \\ a_0 &= 1, & b_0 &= 1. \end{aligned} \tag{24.6}$$

The trust value can be represented with the mathematical expectation of beta distribution as

$$\mathbb{E}_t[\Phi] = \frac{a_t}{a_t + b_t}. \tag{24.7}$$

Intuitively, the trust value of a mobile user is 0.5 at the beginning and updated continuously by using the follow-up observations.

From the above discussion, we can see that past experiences play an important role in the Bayesian inference. In fact, recent behaviors of a mobile user should weight more in the trust evaluation. Here, we introduce a punishment factor for reputation fading, which gives more weights on misbehaviors in the Bayesian inference. The trust evaluation formula in (24.7) is revised as follows:

$$\mathbb{E}_t[\Phi] = \frac{a_t}{a_t + \tau b_t}, \tag{24.8}$$

where τ is the punishment factor and $\tau \geq 1$. With the increment of τ, the trust value declines quickly.

The punishment factor makes the trust evaluation more realistic and reliable. First, if a mobile user once behaves maliciously, compared with those who have no bad records, its trust value will be lowered much more. Second, the trust value will not recover quickly even if he behaves well recently because of the constraint of the punishment factor. This helps distinguish the malicious mobile users quickly and prevent them disrupting others' trust evaluation. Based on the above deduction, the trust value from direct observation, Tr^D, is defined as:

$$Tr^D = \mathbb{E}_t[\Phi]. \tag{24.9}$$

24.2.2 Trust evaluation from indirect observations

Apart from the direct observations, indirect observations from other mobile users are also very important in assessing the trustworthiness of an observed mobile user. Considering the indirect observations helps mitigate the situation that an observed mobile user is loyal to one mobile user but cheating on others. Assume that an observing mobile user collects observations from several other mobile users (also called subsidiary observing mobile user) and combines the collected evidence into a decision about the observed mobile user's trust value. However, these subsidiary observing mobile users may be untrustworthy or the evidence offered by them is unreliable.

The Dempster-Shafer theory can be used as an effective way to handle the uncertainty issue and combine the evidence from multiple subsidiary observers [33]. The core of this theory is based on two ideas: the degrees of belief about a proposition can be obtained from multiple subjective probabilities of a related theme, and these degrees of belief can be combined together under the condition that they are from independent evidence [30]. In the indirect observation, we assume that there are more than one subsidiary observing mobile users and the evidence provided by them is mutually independent.

24.2.2.1 Belief function

As an introduction to the belief function, suppose that each subsidiary observing mobile user has two states, i.e., trustworthy and untrustworthy with probabilities p_1 and $1 - p_1$, respectively. Assume that mobile user 1 claims that the observed mobile user B is trustworthy. If mobile user 1 is trustworthy, its statement is regarded as true; however, if mobile user 1 is untrustworthy, its statement is not necessarily untrue. We think that mobile user 1 provides p_1 degree of belief in the observed mobile user's trustworthiness (hypothesis H), 0 degree of belief in the untrustworthiness (hypothesis $\overline{H}$), and $1 - p_1$ degree in the uncertainty, i.e., either trustworthy or untrustworthy (hypothesis U). Each hypothesis is assigned with a basic probability value $m(H)$, $m(\overline{H})$, $m(U)$ taking values from the interval [0, 1], respectively. In our scheme, we assign the basic probability value with the direct observation trust value. For example, if mobile user 1 believes that mobile user B is trustworthy, then the basic probability value for each possible hypothesis is:

$$\begin{aligned} m_1(H) &= T_{A1}^D, \\ m_1(\overline{H}) &= 0, \\ m_1(U) &= 1 - T_{A1}^D, \end{aligned} \tag{24.10}$$

where T_{A1}^D represents the trust value of mobile user 1 from the direct observation of mobile user B.

Oppositely, if mobile user 1 thinks that user B is untrustworthy, the basic probability value will be:

$$
\begin{aligned}
&m_1(H) = 0,\\
&m_1(\overline{H}) = T_{A1}^{D},\\
&m_1(U) = 1 - T_{A1}^{D}.
\end{aligned}
\tag{24.11}
$$

Concerning the formal definition of the belief function, let Ω be the universe of discourse, i.e., the set representing all possible states of the considered system, $\Omega = \{trustworthy, untrustworthy\}$. The power set 2^{Ω} refers to the set of all subsets of Ω, here $2^{\Omega} = \{\varnothing, \{trustworthy\}, \{untrustworthy\}, \Omega\}$. Any hypothesis A_i refers to a subset of 2^{Ω} and is mapped into a basic probability $m(A_i)$ that reflects the proportion of total belief assigned to hypothesis A_i. These two conditions should be satisfied: $m(\varnothing) = 0$ and $\sum_{A_i \subseteq \Omega} m(A_i) = 1$. For any hypothesis B, the belief function is defined as

$$
beli(B) = \sum_{A_i \subseteq B} m(A_i),
\tag{24.12}
$$

which represents the strength of the evidence that supports hypothesis B's provability.

24.2.2.2 Dempster's rule of combining belief functions

Based on the definition of belief function, the Dempster-Shafer theory combines multiple users' belief. Assuming that $beli_1(B)$ and $beli_2(B)$ are two belief functions over the same universe of discourse, Ω. The orthogonal sum of $beli_1(B)$ and $beli_2(B)$ can be defined as

$$
beli(B) = beli_1(B) \oplus beli_2(B) = \frac{\sum_{i,j,A_i \cap A_j = B} m_1(A_i)m_2(A_j)}{\sum_{i,j,A_i \cap A_j \neq \varnothing} m_1(A_i)m_2(A_j)},
\tag{24.13}
$$

where $A_i, A_j \subseteq \Omega$.

In our scenario, the combined degree of belief from mobile user 1 and mobile user 2 can be calculated as follows [30]:

$$
\begin{aligned}
m_1(H) \oplus m_2(H) &= \frac{1}{K}[m_1(H)m_2(H) + m_1(H)m_2(U) + m_1(U)m_2(H)],\\
m_1(\overline{H}) \oplus m_2(\overline{H}) &= \frac{1}{K}[m_1(\overline{H})m_2(\overline{H}) + m_1(\overline{H})m_2(U) + m_1(U)m_2(\overline{H})],\\
m_1(U) \oplus m_2(U) &= \frac{1}{K}m_1(U)m_2(U),
\end{aligned}
\tag{24.14}
$$

where

$$
\begin{aligned}
K &= m_1(H)m_2(H) + m_1(H)m_2(U) + m_1(U)m_2(U) + m_1(U)m_2(H)\\
&\quad + m_1(U)m_2(\overline{H}) + m_1(\overline{H})m_2(\overline{H}) + m_1(\overline{H})m_2(U).
\end{aligned}
\tag{24.15}
$$

Following the rule of combination of belief, we can combine more degree of belief from other mobile users. Based on the Dempster-Shafer theory, T_{AB}^{N} is defined as:

$$
Tr_{AB}^{InD} = m_1(H) \oplus m_2(H) \ldots \oplus m_n(H),
\tag{24.16}
$$

24.3 System model with 3C framework

24.3.1 Network model

We consider the scenario with K BSs and M mobile users that are considered as transmitters in the potential D2D communications. The network is operated by a central controller. Assume that there are L subscribed mobile users issuing video requests to the network operator, which is responsible for assigning a proper content provider to each requester, i.e., associating with one BS or setting up a one-to-one D2D communication. The sets are denoted as $\mathcal{C}_K = \{1, 2, ..., K\}$, $\mathcal{C}_M = \{1, 2, ..., M\}$, and $\mathcal{L} = \{1, 2, ..., L\}$, respectively. All the K BSs, the M D2D transmitters and the L mobile users are equipped with both caching and computing capabilities. The unicast D2D communication is available only when the mobile user's requested video is stored in the transmitter's cache. In addition, the transmitters can also provision mobile edge computing if they have spare computing capacity left at the considered moment.

In our system model, we assume that the lth mobile user, denoted as s_l, issues a video request. First, the network controller checks all the video providers, i.e., all the BSs and D2D transmitters. We denote $\mathcal{C} = \mathcal{C}_K \bigcup \mathcal{C}_M$ be the set of video providers, and let $c_i(1 \leq i \leq K)$ be the ith BS and $c_j(K+1 \leq j \leq K+M)$ be the jth transmitter. Each video provider's cache is labeled by a content indicator which indicates whether or not the requested video is being stored.

If the requested video is being stored at any of the video provider's caches, and the versions match up as well, the network controller will build up a communication between mobile user s_l and the optimal video provider. However, if all the stored videos' versions mismatch with the requested one, video transcoding should be performed at either mobile user s_l itself or at the video provider's side. On the other hand, if the requested video is not being stored at any of the caches, it has to associate with a proper BS.

Note that in this work, we use video version to represent video specification (e.g., H.263, H.264, MPEG2, or MPEG4). With the rapid growth of mobile services, increasing volumes of videos are played by mobile devices. Consequently, service providers often need to transcode the video contents into different specifications (e.g., bit rate, resolution, quality) with different QoS (e.g., delay) for heterogeneous mobile devices, networks, and user preferences.

24.3.2 Communication model

Denote $h_{s_l}^{c_p}$ as the channel gain between mobile user s_l and content provider $c_p, \forall l \in \mathcal{L}, \forall p \in \mathcal{C}$. According to the Shannon theorem, the achievable rate of mobile user s_l when associating with provider c_p can be expressed as

$$r_{s_l}^{c_p} = B \log_2 \left(1 + \frac{p_T h_{s_l}^{c_p}}{N_o B} \right), \tag{24.17}$$

where B denotes the bandwidth allocated to each mobile user, p_T is the equally transmitted power, and N_o is the noise spectral density.

Here, we consider the realistic wireless channels, and actually $h_{s_l}^{c_p}$ is a continuous random variable. Such assumption is intractable for analysis [34], and therefore we model the wireless channels as the finite-state Markov channels (FSMC), which may achieve performance improvement compared with the traditional assumption of static channels. In our model, the channel gain $h_{s_l}^{c_p}$ is discretized and quantized into H levels: $\mathcal{H}_0$, if $h_0^* \leqslant h_{s_l}^{c_p} < h_1^*$; $\mathcal{H}_1$, if $h_1^* \leqslant h_{s_l}^{c_p} < h_2^*$; ...; $\mathcal{H}_{H-1}$, if $h_{sl}^{cp} \geq h_{H-1}^*$. The boundary values for a particular environment under certain criterion should be optimized for better performance [35]. However, for simplicity, uniformly setting the boundary values of FSMC is widely used in the literature (e.g., [35]). In this chapter, all the boundary values from h_1^* to h_{H-1}^* are increasing in the same distance with each other. Each level corresponds to a state of the Markov chain, and therefore a H-element state space is formed. Due to the relationship between $h_{s_l}^{c_p}$ and $r_{s_l}^{c_p}$, for convenience we use $r_{s_l}^{c_p}$ to describe a state of the wireless channel. We consider the dynamic process, and the channel state realization of $r_{s_l}^{c_p}$ at time instant t is denoted as $\Upsilon_{s_l}^{c_p}(t)$. Based on a certain transitional probability, $\Upsilon_{s_l}^{c_p}(t)$ varies from one state to another as one time slot is gone. The transitional probability of $\Upsilon_{s_l}^{c_p}(t)$ from one state j_s to another state k_s is denoted as $\psi_{j_s k_s}(t)$. The $H \times H$ channel state transitional probability matrix between mobile user s_l and provider c_p is shown as:

$$\Psi_{s_l}^{c_p}(t) = \left[\psi_{j_s k_s}(t)\right]_{H\times H}, \tag{24.18}$$

where $\psi_{j_s k_s}(t) = \Pr\left(\Upsilon_{s_l}^{c_p}(t+1) = k_s \mid \Upsilon_{s_l}^{c_p}(t) = j_s\right)$.

24.3.3 Cache model

Assume that there are totally I video contents in the network that mobile users can request for. The set of the contents is denoted as $\mathcal{I} = \{1, 2, ..., I\}$, which is ranked by popularity. Here, we consider finite cache capacity, i.e., each provider caches some of the I video contents and refreshes the contents periodically. We consider that mobile user s_l has a video request $v_i, i \in \mathcal{I}$. After receiving the request message, the network controller checks every provider's content indicator for video v_i, i.e., $x_{c_p}^{v_i}, \forall p \in \mathcal{C}$. The content distribution indicator $x_{c_p}^{v_i} = 1$ means that video v_i is being stored in the cache of provider p; otherwise $x_{c_p}^{v_i} = 0$. Here, $x_{c_p}^{v_i}$ is viewed as a random variable, representing the state of the cache and is modeled using a two-state (i.e., states 0 and 1) Markov chain [36]. The transitional probability matrix of the state $x_{c_p}^{v_i}$ is defined as:

$$\Gamma_{c_p}^{v_i}(t) = [\delta_{a_s b_s}(t)]_{2\times 2}, \tag{24.19}$$

where $\delta_{a_s b_s}(t) = \Pr\left(x_{c_p}^{v_i}(t+1) = b_s \mid x_{c_p}^{v_i}(t) = a_s\right)$ and $a_s, b_s \in \{0, 1\}$.

When the cache capacity is assumed to be finite, the transitional probability matrix of the cache state can be derived based on different cache refreshment strategies. An important one is the least recently used (LRU) cache refreshment policy. The transitional probability matrix can be obtained using the following Markov chain flow matrix [36],

$$\Lambda_i = \begin{bmatrix} -\gamma_i & 0 & \cdots & 0 & 0 & \gamma_i \\ \zeta_i + \mu_i & -\beta - \mu_i & \cdots & 0 & 0 & \gamma_i \\ \vdots & \vdots & \vdots & \vdots & \vdots & \vdots \\ \mu_i & 0 & \vdots & \zeta_i & -\beta - \mu_i & \gamma_i \\ \mu_i & 0 & \vdots & 0 & \zeta_i & -\zeta_i - \mu_i \end{bmatrix}. \tag{24.20}$$

Here $\zeta_i = \beta - \gamma_i$, and γ_i represents the average request rate for the ith popular video v_i that can be denoted as

$$\gamma_i(t) = \frac{\beta}{\rho i^{\alpha}}, \tag{24.21}$$

where the request for v_i is assumed to arrive as a Poisson process with rate β [36], and the probability follows a Zipf-like distribution. Therefore, the probability of requesting content v_i is $1/\rho i^{\alpha}$, where $\rho = \sum_{i=1}^{I} 1/i^{\alpha}$ and α is the Zipf slope with $0 < \alpha < 1$. The video content is refreshed periodically, and the lifetime of any video content v_i follows an exponential distribution with mean $1/\mu_i$.

24.3.4 Computing model

On the condition that the versions of the requested video at all the providers' caches do not match with the requested version, the associated provider has to extract the current video content and construct a computation task based on the input data information, as well as the number of CPU cycles. In the computation model, we construct the computation task as $Q_{s_l}^{v_i} = \{o_{s_l}^{v_i}, q_{s_l}^{v_i}\}$. The first parameter $o_{s_l}^{v_i}$ represents the size of the requested-version video, and the second parameter $q_{s_l}^{v_i}$ indicates the required number of CPU cycles that is needed to finish the computation task.

The computation capability of provider c_p assigned to mobile user s_l is denoted as $f_{s_l}^{c_p}$, which can be measured by the number of CPU cycles per second [16, 21]. In our network, multiple mobile users may associate with the same provider and share the computing device simultaneously, which would result in the fact that the computation capability for provider c_p is not exactly known at the next time instant. Therefore, the computation capability $f_{s_l}^{c_p}$ is modeled as a random variable, and equally divided into N discrete levels denoted by $\mathcal{E} = \{\mathcal{E}_0, \mathcal{E}_1, \ldots, \mathcal{E}_{N-1}\}$. The realization of the random variable $f_{s_l}^{c_p}$ is denoted as $F_{s_l}^{c_p}$ at time slot t. We model the transition of the computation capability level of a provider as a Markov process. The transitional probability matrix of $F_{s_l}^{c_p}$ can be expressed as:

$$\Theta_{s_l}^{c_p}(t) = \left[\iota_{x_s y_s}(t)\right]_{N \times N}, \tag{24.22}$$

where $\iota_{x_s y_s}(t) = \Pr\left(F_{s_l}^{c_p}(t+1) = y_s \mid F_{s_l}^{c_p}(t) = x_s\right)$, and $x_s, y_s \in \mathcal{E}$.

The consumed time for executing the task $Q_{s_l}^{v_i}$ at provider c_p can be obtained as $t_{s_l,c_p} = \frac{q_{s_l}^{v_i}}{F_{s_l}^{c_p}(t)}$. Furthermore, the rate of computation, i.e., the number of bits computed per second, can be given

$$r_{s_l,c_p}^{\text{comp}}(t) = a_{s_l,c_p}^{\text{comp}}(t)\frac{o_{s_l}^{v_i}}{t_{s_l,c_p}} = a_{s_l,c_p}^{\text{comp}}(t)\frac{f_{s_l}^{c_p}(t)o_{s_l}^{v_i}}{q_{s_l}^{c_p}}, \tag{24.23}$$

where $a_{s_l,c_p}^{\text{comp}}(t) \in \{0, 1\}$ represents whether or not the computation task is decided to be performed at provider c_p.

24.4 Formulation as a reinforcement learning problem

In this section, an optimization problem of the integrated trust-based social network with the 3C framework is formulated as a deep reinforcement learning problem. Assume that a mobile user requests for a video content to the integrated network. For the network operator, it should decide which BS or a D2D transmitter is assigned to serve the requesting user, whether or not the video transcoding (i.e., MEC) should be performed, and whether or not newly emerged contents should be cached. The network operator needs a comprehensive consideration of many factors, including: the wireless channel conditions, whether or not the requested content is stored at the local cache, whether or not the content version is matched up, the computational capacity, the trustworthiness of a D2D transmitter. Here, dynamic scenarios are considered, i.e., the available network conditions are varying with time. Deep *Q*-learning algorithm is used to solve the formulated optimization problem.

24.4.1 Deep Q-learning

Reinforcement learning is a category of machine learning, where an agent learns to take actions on the environment and tries to obtain the most reward from the environment even though it faces with much uncertainty about the environment. The agent has to make a trade-off between the exploration and exploitation, and adjusts its actions based on the delayed rewards. Usually, a reinforcement learning problem can be described by using a Markov decision process (MDP). More advantageously, reinforcement learning can handle the complex situations that the state, explicit transitional probability, and immediate reward are not known [37], which is called model-free reinforcement learning.

Note: In order not to make confusion for readers from different backgrounds, here we briefly explain some related terms with reinforcement learning. The term dynamic programming (DP) refers to a collection of algorithms that can be utilized to compute optimal policies with the knowledge of a perfect model of the environment. However, classical DP algorithms are of limited applicability due to the assumption of a perfect model and the curse of dimensionality, which means the computational requirements grow exponentially with the number of system states [38]. To solve these problems, an alternative strategy that uses iterative algorithms to help estimate the value function is invented. This general strategy has been referred to as forward dynamic programming, iterative dynamic programming, adaptive dynamic programming, reinforcement learning, and neuro-dynamic programming [39]. The term that is being increasingly adopted is approximate dynamic programming (ADP). However, artificial intelligence community continues to use "reinforcement learning," and in control theory community, the term "neuro-dynamic programming" is popularly used [40, 41].

Q-learning is one of the most widely used model-free reinforcement learning. Q-function is defined as the state-action function, and it can be obtained in a recursive manner using the available state, action, and reward information. In Q-learning, in order to obtain the best policy, it is necessarily required to estimate the Q-function. Sometimes the Q-function is implemented by a non-linear approximator, such as a neural network. However, a neural network achieves flexibility at the cost of stability [42]. Recently, a more advanced deep Q-learning is proposed, which approximate the Q-function with a deep neural network, and it is proven to be more stable and advantageous [42]. Compared with the traditional Q-learning, deep Q-learning has several outstanding improvements.

First of all, it goes from an one-layer fully connected ordinary neural network to a multiple-layer sparsely connected convolutional neural network. Convolutional neural networks, also called convolutional networks, play an important role in the evolution of deep learning, and they are some of the first deep models that can perform well [43]. Convolutional networks exploit convolution in substitute of matrix multiplication in at least one of the layers. For the convolution layer, the convolution operation is performed on the two arguments: *the input* and *the kernel*. The kernel is the representation for extracting a specific kind of feature (such as an edge, a contour, sharpening, an object identity, etc.), and it is also obtained through learning. The kernel is smaller than the input, and it slides over the input with a certain stride to produce *a feature map*. Different kernels produce different feature maps. Compared with the input, the size of the feature maps is significantly reduced, and these obtained feature maps are used as the input for the next layer. Convolution also makes it work that the input can be of variable size.

Another operation, called *p*ooling, appears in almost all the convolutional networks. Pooling, also referred as subsampling or downsampling, modifies the outputted feature map by only retaining some specific statistical information. Pooling includes several types, such as *the max pooling*, *the average pooling*, *the sum pooling*, *the L^2 norm pooling*, etc. In the case of max pooling, the largest element within a predefined rectangular neighborhood of the feature map will be taken as the representative output of that neighborhood. Pooling can be used as the next layer of convolution, and it can reduce the dimension of the feature map and make the network invariant to small translations or distortions.

Deep Q-learning also utilizes a novel strategy called experience replay. It stores the past experiences into a replay memory and randomly, samples mini-batches from the pool to train the deep neural network, which refrains the agent from only concentrating on what the network is currently doing.

Additionally, deep Q-learning adopts two separate convolutional networks to generate the target Q values and the estimated Q values. Since deep Q-network is trained toward the target value by minimizing the loss function, using one network for both the estimated Q values and target Q values would lead to falling into feedback loops. For the sake of stabilizing the training process, the target deep Q-network's parameter is fixed and periodically updated.

In this chapter, Minh *et al.*'s deep Q-network architecture [42] is used to train the learning process, but we decompose the Q-function, $Q(x, a)$, into two components, i.e., the value function $V(x)$, and the advantage function $A(x)$. Our deep Q-network separately computes the value and advantage functions, and combines them together into a single Q -function at the final fully connected layer. This helps achieve a more robust estimate of the state value by decoupling it from specific actions.

24.4.2 System state

The system states for a subscriber s_l requesting video v_i at time slot $t \in \{0, 1, \ldots, T-1\}$ mainly includes five components: the realization $\Upsilon_{s_l}^{c_p}(t)$ of the random variables $\gamma_{s_l}^{c_p}$ (channel state), the realization $F_{s_l}^{c_p}(t)$ of the random variables $f_{s_l}^{c_p}$ (computation capability), the realization $X_{v_i}^{c_p}(t)$ of the random variables $x_{v_i}^{c_p}$ (content indicator), the version indicator $y_{v_i}^{c_p}$, and the trust value index $Tr_{s_l}^{c_p}$ for all the video providers, i.e., $\forall p \in \mathcal{C}$. Consequently, the system state vector can be described as follows.

$$\begin{aligned}\chi_{s_l}(t) &= \Big[\Upsilon_{s_l}^{c_1}(t), \Upsilon_{s_l}^{c_2}(t), \ldots, \Upsilon_{s_l}^{c_{K+M}}(t),\\ &F_{s_l}^{c_1}(t), F_{s_l}^{c_2}(t), \ldots, F_{s_l}^{c_{K+M}}(t),\\ &X_{v_i}^{c_1}(t), X_{v_i}^{c_2}(t), \ldots, X_{v_i}^{c_{K+M}}(t),\\ &y_{v_i}(t), Tr_{s_l}^{c_1}(t), Tr_{s_l}^{c_2}(t), \ldots, Tr_{s_l}^{c_{K+M}}(t)\Big].\end{aligned} \tag{24.24}$$

Here, the trust index for all the BSs should be set to 1, i.e., $Tr_{s_l}^{c_p}(t) = 1, \forall c_p \in \mathcal{C}_K$; meanwhile, this index for all the D2D transmitters should be less than 1, i.e., $Tr_{s_l}^{c_p}(t) < 1, \forall c_p \in \mathcal{C}_M$. $y_{v_i}(t)$ denotes the indicator for whether or not the cached video version matches with the requested version. If yes $y_{v_i}(t) = 1$, if no $y_{v_i}(t) = 0$.

24.4.3 System action

In the system, the network controller decides which video provider is assigned to the subscribed user, whether or not the computation offloading (video transcoding) should be performed, and whether or not the video provider should cache the new video. The network controller's composite action for subscriber s_l is given by

$$\mathbf{a}_{s_l}(t) = \{\mathbf{a}_{s_l}^{\text{comm}}(t), \mathbf{a}_{s_l}^{\text{comp}}(t), \mathbf{a}_{s_l}^{\text{cache}}(t)\}, \tag{24.25}$$

where the row vectors $\mathbf{a}_{s_l}^{\text{comm}}(t), \mathbf{a}_{s_l}^{\text{comp}}(t), \mathbf{a}_{s_l}^{\text{cache}}(t)$ are interpreted in more details as follows:

- The first row vector $\mathbf{a}_{s_l}^{\text{comm}}(t)$ is defined as

$$\mathbf{a}_{s_l}^{\text{comm}}(t) = [a_{s_l,c_1}^{\text{comm}}(t), a_{s_l,c_2}^{\text{comm}}(t), \ldots, a_{s_l,c_{K+M}}^{\text{comm}}(t)], \tag{24.26}$$

where $a^{\text{comm}}_{s_l,c_p}(t)$ is the association indicator and $a^{\text{comm}}_{s_l,c_p}(t) \in \{0, 1\}$. If the subscribed user is associated with video provider c_p, $a^{\text{comm}}_{s_l,c_p}(t) = 1$, otherwise $a^{\text{comm}}_{s_l,c_p}(t) = 0$.

- The second row vector $\mathbf{a}^{\text{comp}}_{s_l}(t)$ is defined as

$$\mathbf{a}^{\text{comp}}_{s_l}(t) = [a^{\text{comp}}_{s_l,c_1}(t), a^{\text{comp}}_{s_l,c_2}(t), \ldots, a^{\text{comp}}_{s_l,c_{K+M}}(t)], \tag{24.27}$$

where $a^{\text{comp}}_{s_l,c_p}(t)$ indicates the computation offloading decision and $a^{\text{comp}}_{s_l,c_p}(t) \in \{0, 1\}$. If the network controller has decided that the computation task should be performed at the provider c_p's device, $a^{\text{comp}}_{s_l,c_p}(t) = 1$; on the other hand, if the computation is decided to be executed on subscriber's own mobile device, $a^{\text{comp}}_{s_l,c_p}(t) = 0$.

- The third row vector $\mathbf{a}^{\text{cache}}_{s_l}(t)$ is defined as

$$\mathbf{a}^{\text{cache}}_{s_l}(t) = [a^{\text{cache}}_{s_l,c_1}(t), a^{\text{cache}}_{s_l,c_2}(t), \ldots, a^{\text{cache}}_{s_l,c_{K+M}}(t)], \tag{24.28}$$

where $a^{\text{cache}}_{s_l,c_p}(t)$ means that whether or not the video provider c_p should cache the newly emerged video or the new version, and $a^{\text{cache}}_{s_l,c_p}(t) \in \{0, 1\}$. If the network controlled decides that c_p caches the video, $a^{\text{cache}}_{s_l,c_p}(t) = 1$, otherwise $a^{\text{cache}}_{s_l,c_p}(t) = 0$.

24.4.4 Reward function

The system reward is set to be the total revenue (i.e., utility) of the network operator. The network operator charges the subscribed user s_l for associating with the video provider, which is denoted as λ_{s_l} unit price per bps. On the condition that video transcoding (computation offloading) is decided to be executed on video provider's side, the operator can charge for its computing service, which is defined as ν_{s_l} unit price per bps. In addition, if the network controller decides to let the BSs or D2D transmitters cache the new video or new version, the operator gets a potential revenue on estimated backhaul save, which is denoted as κ_{s_l} unit price per Hz.

On the other hand, the operator has to pay for the rented spectrum and backhaul bandwidth. The unit price for the usage of spectrum is defined as δ_{c_p} per Hz for provider c_p. Moreover, if video transcoding is performed, a certain amount of energy will be consumed for the computing, the network operator is obliged to pay η_{c_p} unit price for per consumed Joule. The cost of caching the video content in the memory of provider c_p is denoted as ς_{c_p} unit price per unit space.

The system reward for serving subscribed user s_l requesting video v_i is defined as the network operator's total revenue, and it is formulated as a function of the system states and actions. The system actions determine whether or not the reward can be optimized. Here, we define the reward function for a specific subscribed user s_l as:

$$\begin{aligned} R_{s_l}^{v_i}(t) &= \sum_{c_p \in C} \left[R_{s_l,c_p}^{\text{comm}}(t) + R_{s_l,c_p}^{\text{comp}}(t) + R_{s_l,c_p}^{\text{cache}}(t) \right] \\ &= \sum_{c_p \in C} a_{s_l,c_p}^{\text{comm}}(t) \left(\lambda_{s_l} Tr_{s_l}^{c_p}(t) \Upsilon_{s_l}^{c_p}(t)(1 - w^{\text{comp}} a_{s_l,c_p}^{\text{comp}}(t)) - \delta_{c_p} B \right. \\ &\quad \left. - (1 - X_{v_i}^{c_p}(t)) \sigma^{c_p} \Upsilon_{s_l}^{c_p}(t) \right) \\ &\quad + \sum_{c_p \in C} (1 - y_{v_i}^{c_p}) a_{s_l,c_p}^{\text{comp}}(t) (\nu_{s_l} Tr_{s_l}^{c_p}(t) \frac{F_{s_l}^{c_p}(t) o_{s_l}^{v_i}}{q_{s_l}^{v_i}} - \eta_{c_p} q_{s_l}^{c_p} e_{c_p}) \\ &\quad + \sum_{c_p \in C} a_{s_l,c_p}^{\text{cache}}(t) (\kappa_{s_l} Tr_{s_l}^{c_p}(t) \Upsilon_{s_l}^{c_p}(t) - \varsigma_{c_p} o_{s_l}^{v_i}). \end{aligned} \tag{24.29}$$

The above reward function is composed of three terms, i.e., the earnings from communications, computing, and caching, respectively. Here, the trust index $Tr_{s_l}^{c_p}(t)$ is added to each incoming revenue term. For the first communication earnings, $\lambda_{s_l} Tr_{s_l}^{c_p}(t) \Upsilon_{s_l}^{c_p}(t)(1 - w^{\text{comp}} a_{s_l,c_p}^{\text{comp}}(t))$ denotes the available earnings for providing the video transmitting service to subscriber s_l. Note that if video transcoding is decided to be performed, the video will be compressed and the corresponding income will be reduced. $\delta_{c_p} B$ is the cost for consuming the radio spectrum bandwidth. $(1 - X_{v_i}^{c_p}(t)) \sigma^{c_p} \Upsilon_{s_l}^{c_p}(t)$ is the cost for possible consumed backhaul bandwidth: if $X_{v_i}^{c_p}(t) = 1$, i.e., the requested video exists in the cache of provider c_p, no backhual cost is needed; otherwise $X_{v_i}^{c_p}(t) = 0$, the video has to be fetched from the Internet and backhaul cost is unavoidable. Note that σ^{c_p} is a very important parameter. It is not realistic to build up a D2D communication when the requested video does not exist in its cache. Thus, σ^{c_p} should be set to be an extremely large number for $c_p, \forall c_p \in C_M$. For the second computing earnings, $(1 - y_{v_i}^{c_p})$ is used to indicate whether or not the version is matching. $\nu_{s_l} Tr_{s_l}(t)^{c_p} \frac{F_{s_l}^{c_p}(t) o_{s_l}^{v_i}}{q_{s_l}^{v_i}}$ represents the gaining for implementing video transcoding, and $\eta_{c_p} q_{s_l}^{c_p} e_{c_p}$ is the expenditure for the energy consumption, where e_{c_p} denotes consumed energy for running one CPU cycle. For the third caching earnings, $\kappa_{s_l} Tr_{s_l}(t)^{c_p} \Upsilon_{s_l}^{c_p}(t)$ is the estimated future income of the saved backhaul bandwidth if the new video or new version is decided to be cached. Here the saved backhaul bandwidth is equal to the wireless communication rate. At last, $\varsigma_{c_p} o_{s_l}^{v_i}$ denotes the cost for caching the content.

$R_{s_l}^{v_i}(t)$ is the system's immediate reward, i.e., the network operator gets $R_{s_l}^{v_i}(t)$ at state $\chi_{s_l}(t)$ when action $\mathbf{a}_{s_l}(t)$ is performed in time slot t. However, a maximum immediate value is not a guarantee for the maximum long-term future rewards. Thus, we should also consider a big picture. A future reward with a discount factor φ is reasonable, which can be denoted as

$$R_{s_l,v_i}^{f} = \max_{\pi} \mathbb{E} \left[\sum_{t=0}^{t=T-1} \varphi^t R_{s_l}^{v_i}(t) \right], \tag{24.30}$$

where π represents a Q-learning policy, i.e., a series of actions where a specific action will be executed given a specific system state, and φ^t approximates to be zero when T is large enough. In fact, a condition for terminating the training process should be set.

The objective of adopting deep Q-learning into our network model is to help find an optimization policy that can maximize the cumulated future rewards for the network operator.

24.5 Simulation results and discussions

In this section, the performance of the proposed scheme is evaluated using computer simulations. TensorFlow [44] is used to implement deep Q-learning. For performance comparison, the following four algorithms are presented: (1) Proposed scheme, which considers MEC, caching, D2D, as well as both direct observation and indirect observation; (2) Proposed scheme w.o. indirect observation, which does not consider indirect observation; (3) An existing scheme w.o. mobile edge computing [45]; and (4) An existing scheme w.o. D2D communications [2].

24.5.1 Simulation settings

In the simulations, an MSN consisting of 5 BSs, and 15 D2D transmitters is considered. The radius of the cell is set to 500 m. Because D2D communications typically perform within short ranges, a clustered-based distribution model [46] is utilized, where multiple users are located within one cluster with a radius of 50 m. Moreover, it is assumed that there are totally ten types of contents distributed in the network, and each content cache state follows the Markov model. We set the cache state transition probability of staying in the same state in the BS as 0.6 (0.3 in the D2D transmitter) and set the transition probability from one state to another as 0.4 (0.7 in the D2D transmitter). We further assume a Zipf popularity distribution in the simulations, with $\theta = 1.5$ [47, 48]. The computation states of MEC servers follow the Markov model. We assume that the computation state of the MEC server can be very low, low, medium, high, and very high.

It is assumed that there are two types of D2D transmitters in the network: normal nodes, which share the content and perform computing normally, and compromised nodes, which modify contents maliciously. The BSs are assumed to be not compromised due to the physical security of BSs. It is also assumed that the number of compromised nodes is much less than the total number of nodes in the network. The attackers are independent. Hence, there is no collusion attack in the network.

In the simulations, a GPU-based server is used, which has four Nvidia GPUs with version GTX TITAN. The CPU is Intel Xeno E5-2683 v3 with 128G memory. The software environment is TensorFlow 0.12.1 with Python 2.7 on Ubuntu 14.04 LTS.

Specifically, the training process of the deep convolutional networks is shown in Figure 24.2. We formulate the 5 BSs and 15 D2D transmitters as the rows and the five levels of computational capacity as the columns in a grid image. Whether or not the required content is in the local cache is characterized by different colors

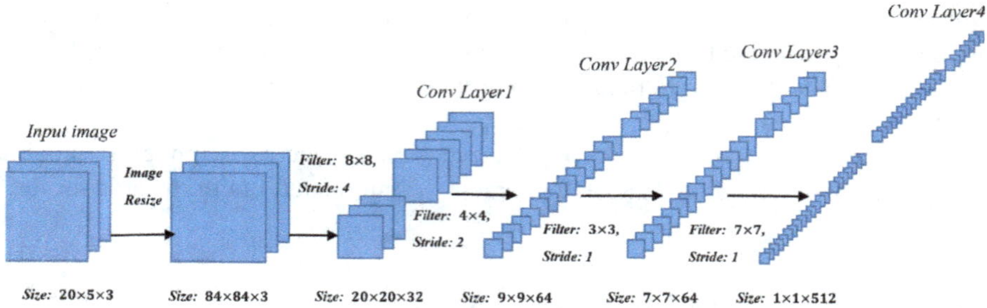

Figure 24.2 The training process of the deep convolution networks

in each small grid. The initial input image of size 20 × 5 × 3 is firstly resized into 84 × 84 × 3. Through four layers of convolutional operations, the output is 512 nodes, and these nodes will be fully connected to be trained in the deep reinforcement learning.

24.5.2 Simulation results

The convergence performance of the proposed scheme using the deep Q-learning algorithm is shown in Figure 24.3. It can be observed from Figure 24.3 that at the beginning of the learning process, the total utility of the proposed scheme is very low. As the number of the episodes increases, the total utility increases until it converges to a relatively stable value, which is around 5 900. It can also be observed that

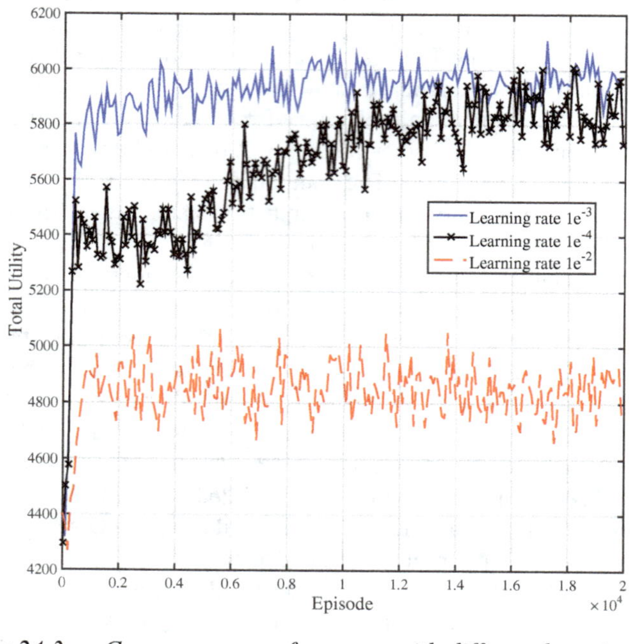

Figure 24.3 Convergence performance with different learning rates

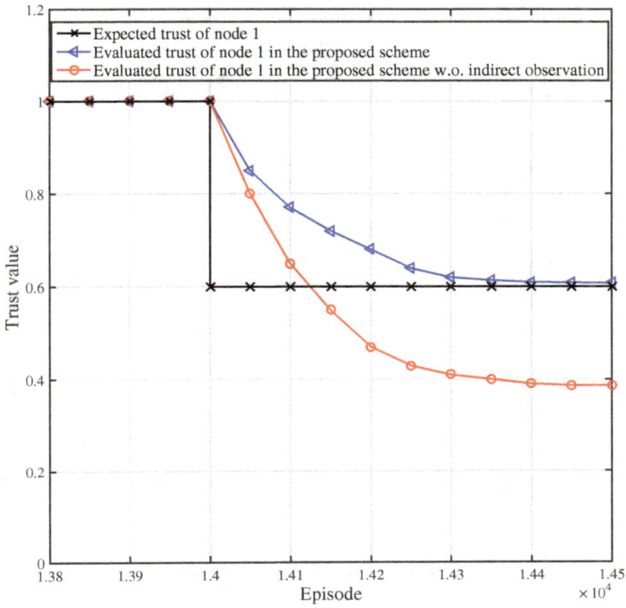

Figure 24.4 The trust value vs. the episodes

the convergence performance of the proposed scheme with different learning rates in the deep Q-learning algorithm. The learning rate has effects on the convergence performance. For example, the convergence of the proposed scheme is faster when the learning rate is 0.001 compared to the case when the learning rate are 0.0001. However, this does not mean that a larger learning rate is always better, because a larger learning rate may result in local optimum in the problem instead of the global optimum. For example, in Figure 24.3, with the learning rate of 0.01, the algorithm converges to the total utility of around 4 900, which is much lower than the converged values achieved by other learning rates. Therefore, it is important to choose an appropriate learning rate for a specific problem.

The performance of the proposed social trust scheme is also studied. Assume that, at episode 14K, D2D transmitter 1 changes its maliciousness to 0.6. Our goal is to observe the accuracy of trust tracking in this scenario where the malicious behavior of a D2D transmitter changes rapidly. Figure 24.4 shows the trust tracking of the system using direct observation with Bayesian inference and the proposed scheme using both direct observation with Bayesian inference and indirect observation with the Dempster-Shafer theory. We can observe from Figure 24.4 that only direct observation can result in inaccurate trust values. By contrast, the proposed scheme can track the trust value accurately with both direct observation and indirect observation.

The number of malicious D2D transmitters in the network also has a significant impact on the performance the network. Here, the system utility is investigated with the number malicious D2D transmitters, from 1 to 5, in a 15-D2D-transmitter

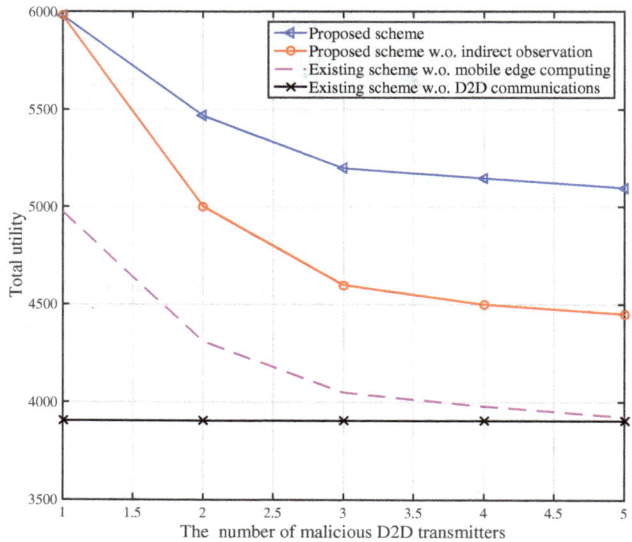

Figure 24.5 The total utility vs. the number of malicious D2D transmitters

environment. Figure 24.5 shows that, as the number of malicious D2D transmitters increases, the utility drops dramatically. It can be concluded that the network is deeply affected by the number of malicious D2D transmitters. Compared with other schemes, the proposed scheme can achieve the best performance with the existence of malicious users.

24.6 Chapter conclusion

MSNs have become one of the most important networking paradigms in future wireless mobile networks. Recent advances of caching, as well as computing and communications, can have significant impacts on the performance of MSNs. Specifically in this chapter, recent advances in mobile edge computing, in-network caching, and D2D communications in MSNs are studied. In addition, the knowledge of social relationships in these new paradigms is investigated to improve the security and efficiency of MSNs, where a social trust scheme is present with both direct observation using Bayesian inference and indirect observation using the Dempster-Shafer theory. Deep Q-learning approach is used to study this complicated system. Simulation results were presented to show the effectiveness of the proposed scheme.

References

[1] Vastardis N., Kun Yang. 'Mobile social networks: architectures, social properties, and key research challenges'. *IEEE Communications Surveys & Tutorials*;15(3):1355–71.

[2] Su Z., Xu Q. 'Content distribution over content centric mobile social networks in 5G'. *IEEE Communications Magazine*. 2015;**53**(6):66–72.

[3] Wang S., Zhang X., Zhang Y., Wang L., Yang J., Wang W. 'A survey on mobile edge networks: convergence of computing, caching and communications'. *IEEE Access*. 2017;**5**:6757–79.

[4] Gu J., Wang W., Huang A. 'Proactive storage at caching-enable base stations in cellular networks'. Proceedings of IEEE International Symposium on Personal, Indoor, and Mobile Radio Communications (PIMRC); London, UK; 2013. pp. 1543–7.

[5] Bai B., Wang L., Han Z., *et al.* 'Caching based socially-aware D2D communications in wireless content delivery networks: a hypergraph framework'. *IEEE Wireless Communications*. 2016;**23**(4):74–81.

[6] Sengupta A., Amuru S., Tandon R. 'Learning distributed caching strategies in small cell networks'. Proceedings of the 11th International Symposium on Wireless Communication Systems; 2014. pp. 917–21.

[7] Guan Y., Xiao Y., Feng H. 'MobiCacher: mobility-aware content caching in small-cell networks'. Proceedings of 2014 IEEE Global Communications Conference; Austin, TX, USA; 2014. pp. 4537–42.

[8] He Y., Liang C., F R Y. 'Optimization of cache-enabled opportunistic interference alignment wireless networks: a big data deep reinforcement learning approach'. Proceedings of IEEE International Conference on Communications; Paris, France; 2017. pp. 1–6.

[9] He Y., Zhang Z., Yu F.R., *et al.* 'Deep-reinforcement-learning-based optimization for cache-enabled opportunistic interference alignment wireless networks'. *IEEE Transactions on Vehicular Technology*. 2017;**66**(11):10433–45.

[10] Patel M., Naughton B., Chan C. 'Mobile-edge computing introductory technical white paper'. White Paper, Mobile-edge Computing (MEC) Industry Initiative; 2014. pp. 854–64.

[11] Kumar N., Zeadally S., Rodrigues J.J.P.C. 'Vehicular delay-tolerant networks for smart grid data management using mobile edge computing'. *IEEE Communications Magazine*. 2016;**54**(10):60–6.

[12] Wang C., Liang C., Yu F.R., Chen Q., Tang L. 'Computation offloading and resource allocation in wireless cellular networks with mobile edge computing'. *IEEE Transactions on Wireless Communications*. 2017;**16**(8):4924–38.

[13] Zhang K., Mao Y., Leng S., *et al.* 'Energy-efficient offloading for mobile edge computing in 5G heterogeneous networks'. *IEEE Access*. 2016;**4**:5896–907.

[14] Liu J., Mao Y., Zhang J. 'Delay-optimal computation task scheduling for mobile-edge computing systems'. Proceedings of IEEE International

Symposium on Information Theory (ISIT); Barcelona, Spain; 2016. pp. 1451–5.

[15] Sardellitti S., Scutari G., Barbarossa S. 'Joint optimization of radio and computational resources for multicell mobile-edge computing'. *IEEE Transactions on Signal and Information Processing over Networks*. 2015;**1**(2):89–103.

[16] Chen X., Jiao L., Li W., *et al.* 'Efficient multi-user computation offloading for mobile-edge cloud computing'. *IEEE/ACM Transactions on Networking*. 2016;**24**(5):2795–808.

[17] Chen X., Wu J., Cai Y., *et al.* 'Energy-efficiency oriented traffic offloading in wireless networks: a brief survey and a learning approach for heterogeneous cellular networks'. *IEEE Journal on Selected Areas in Communications*. 2015;**33**(4):627–40.

[18] Asadi A., Wang Q., Mancuso V. 'A survey on device-to-device communication in cellular networks'. *IEEE Communications Surveys & Tutorials*;16(4):1801–19.

[19] Yang C., Li J., Semasinghe P., Hossain E., Perlaza S.M., Han Z. 'Distributed interference and energy-aware power control for ultra-dense D2D networks: a mean field game'. *IEEE Transactions on Wireless Communications*. 2017;**16**(2):1205–17.

[20] Meng Y., Jiang C., Chen H.-H., *et al.* 'Cooperative device-to-device communications: social networking perspectives'. *IEEE Network*. 2017;**31**(3):38–44.

[21] Zhou Y., Yu F.R., Chen J., *et al.* 'Resource allocation for Information-Centric Virtualized heterogeneous networks with in-network caching and mobile edge computing'. *IEEE Transactions on Vehicular Technology*. 2017;**66**(12):11339–51.

[22] He Y., Zhao N., Yin H., Networking I. 'Integrated networking, caching, and computing for connected vehicles: a deep reinforcement learning approach'. *IEEE Transactions on Vehicular Technology*. 2018;**67**(1):44–55.

[23] He Y., Yu F.R., Zhao N., *et al.* 'Software-defined networks with mobile edge computing and caching for smart cities: a big data deep reinforcement learning approach'. *IEEE Communications Magazine*. 2017;**55**(12):31–7.

[24] Zhang Y., Pan E., Song L., Saad W., Dawy Z., Han Z. 'Social network aware device-to-device communication in wireless networks'. *IEEE Transactions on Wireless Communications*. 2015;**14**(1):177–90.

[25] Hui P., Crowcroft J., Yoneki E. 'Bubble rap: social-based forwarding in delay-tolerant networks'. *IEEE Transactions on Mobile Computing*. 2011;**10**(11):1576–89.

[26] Bulut E., Szymanski B.K. 'Friendship based routing in delay tolerant mobile social networks'. Proceedings of IEEE Global Communications Conference (GLOBECOM); Miami, FL, USA; 2010. pp. 1–5.

[27] Hamedani K., Liu L., Atat R., *et al.* 'Reservoir computing meets smart grids: attack detection using delayed feedback networks'. *IEEE Transactions on Industrial Informatics*. 2018;**14**(2):734–43.

[28] Atat R., Liu L., Chen H., Wu J., Li H., Yi Y. 'Enabling cyber-physical communication in 5G cellular networks: challenges, spatial spectrum sensing,

and cyber-security'. *IET Cyber-Physical Systems: Theory & Applications*. 2017;**2**(1):49–54.

[29] Chen I.-R., Bao F., Guo J. 'Trust-based service management for social internet of things systems'. *IEEE Transactions on Dependable and Secure Computing*. 2016;**13**(6):684–96.

[30] Chen T.M., Venkataramanan V. 'Dempster-Shafer theory for intrusion detection in ad hoc networks'. *IEEE Internet Computing*. 2005;**9**(6):35–41.

[31] Changiz R., Halabian H., FR Y. 'Trust establishment in cooperative wireless networks'. Proceedings - IEEE Military Communications Conference MILCOM; San Jose, CA, USA; 2010. pp. 1074–9.

[32] Zouridaki C., Mark B.L., Hejmo M., *et al.* 'Hermes: a quantitative trust establishment framework for reliable data packet delivery in MANETs'. *Journal of Computer Security*. 2007;**15**(1):3–38.

[33] Bu S., Yu F.R., Liu X.P., *et al.* 'Distributed combined authentication and intrusion detection with data fusion in high-security mobile ad hoc networks'. *IEEE Transactions on Vehicular Technology*. 2011;**60**(3):1025–36.

[34] Parzysz F., Di Renzo M., Verikoukis C. 'Power-availability-aware cell association for energy-harvesting small-cell base stations'. *IEEE Transactions on Wireless Communications*. 2017;**16**(4):2409–22.

[35] Hong Shen Wang., Moayeri N. 'Finite-state Markov channel-a useful model for radio communication channels'. *IEEE Transactions on Vehicular Technology*. 1995;**44**(1):163–71.

[36] Gomaa H., Messier G.G., Williamson C., Davies R., *et al.* 'Estimating instantaneous cache hit ratio using Markov chain analysis'. *IEEE/ACM Transactions on Networking*. 2013;**21**(5):1472–83.

[37] Ong H.Y., Chavez K., Hong A. 'Distributed deep Q-learning'. *arXiv preprint arXiv*. 2015;**150804186**.

[38] Sutton R.S., Barto A.G. *Reinforcement learning: an introduction*. Cambridge, MA: MIT Press; 2011.

[39] Powell W.B. *Approximate dynamic programming: solving the curses of dimensionality*. **703**. Hoboken: John Wiley & Sons; 2007.

[40] Imani M., Braga-Neto U.M. 'Control of gene regulatory networks with noisy measurements and uncertain inputs'. *IEEE Transactions on Control of Network Systems*. 2017;**99**:1.

[41] Al-Tamimi A., Lewis F.L., Abu-Khalaf M. 'Discrete-time nonlinear HJB solution using approximate dynamic programming: convergence proof'. *IEEE Transactions on Systems, Man, and Cybernetics. Part B, Cybernetics: A Publication of the IEEE Systems, Man, and Cybernetics Society*. 2008;**38**(4):943–9.

[42] Mnih V., Kavukcuoglu K., Silver D., *et al.* 'Human-level control through deep reinforcement learning'. *Nature*. 2015;**518**(7540):529–33.

[43] Goodfellow I., Bengio Y., Courville A. *Deep learning*. Cambridge, MA: MIT Press; 2016.

[44] Abadi M., Agarwal A., Barham P., *et al.* 'TensorFlow: large-scale machine learning on heterogeneous systems'. *ArXiv*. 2015;**160304467**.

[45] Zhao Y., Song W., Han Z. 'Social-Aware data dissemination via device-to-device communications: fusing social and mobile networks with incentive constraints'. *IEEE Trans Services Computing*. 2017;**99**:1–1.

[46] Kaufman B., Aazhang B. 'Cellular networks with an overlaid device to device network'. Proceedings of Asilomar Conference on Signals, Systems and Computers; Pacific Grove, CA; 2008. pp. 1537–41.

[47] Lee D.H., Choi K.W., Jeon W.S., *et al.* 'Two-stage semi-distributed resource management for device-to-device communication in cellular networks'. *IEEE Transactions on Wireless Communications*. 2014;**13**(4):1908–20.

[48] Blasco P., Gunduz D. 'Learning-based optimization of cache content in a small cell base station'. Proceedings of IEEE International Conference on Communications (ICC); Sydney, Australia; 2014. pp. 1897–903.

Chapter 25
Caching for CRAN and UAVs

Mingzhe Chen[1,2], Zhaohui Yang[3], Walid Saad[4], Changchuan Yin[5], and Shuguang Cui[2]

25.1 Introduction

In cloud radio access networks (CRANs), mobile users are serviced by a massive number of remote radio heads (RRHs) that are acted as distributed antennas and controlled by cloud-based baseband units (BBUs) via wired or wireless fronthaul links [1]. To improve spectral efficiency, cloud-based cooperative signal processing techniques can be executed centrally at the BBUs [2]. However, despite the ability of CRAN systems to run such complex signal processing functions centrally, their performance remains limited by the capacity of the fronthaul (users to BBU) and backhaul (BBUs to core) links [1]. Indeed, given the massive nature of a CRAN, relying on fiber fronthaul and backhaul links may be infeasible. Consequently, capacity-limited wireless backhaul and fronthaul connections are being studied for CRANs such as in [3]. To overcome these limitations, one can make use of content caching techniques [4, 5] in which users can directly obtain contents stored at the cloud or RRHs. However, deploying caching strategies in a CRAN environment faces many challenges that include optimized cache placement, cached content update, and cached content delivery.

A number of problems related to caching in CRANs have been studied in the existing works such as in [6–10]. However, most existing works assume a known content request distribution (Zipf distribution) that is then used to design an effective caching algorithm and, as such, they do not consider a proactive caching algorithm that can predict the content request distribution of each user. Some recent works such as in [11–15] have studied a number of ideas related to the analysis of users'

[1]Department of Electrical Engineering, Princeton University, USA
[2]Shenzhen Research Institute of Big Data and School of Science and Engineering, The Chinese University of Hong Kong, China
[3]Centre for Telecommunications Research, Department of Informatics, King's College London, UK
[4]Bradley Department of Electrical and Computer Engineering, Virginia Tech, USA
[5]Beijing Key Laboratory of Network System Architecture and Convergence, Beijing University of Posts and Telecommunications, China

data for content caching in CRAN. However, most of these existing caching works [11–15] were typically restricted to static networks without mobility and ultra dense users. Note that, in these contributions [11–15], the content is stored at the terrestrial static base stations. However, caching content at a static base station cannot effectively serve the mobile users as they move outside the coverage range of the base station. In addition, while a mobile user moves to a new cell, its requested content may not be available at the new base station and, consequently, the users cannot be serviced properly. In such case, to serve mobile users, one can make use of the long-term statistics of user mobility to predict the user association, and the efficiency of content caching will be significantly improved. For proactive caching, the users' future locations can also enable seamless handover and content download. Meanwhile, one can track the users' mobility patterns so as to improve the caching efficiency. To this end, unmanned aerial vehicles (UAVs) [16, 17] can be used as flying base stations to dynamically cache the popular contents, track the mobility pattern of the corresponding users and, then, effectively serve them. In this case, due to the high altitude and flexible deployment of the UAVs, they can establish reliable communication links to the users by mitigating the blockage effect.

In this chapter, we first study the application of echo state network to predict the users' mobility and content request distribution for determining optimal cached contents in CRAN. Then, we study the deployment of cache-enabled UAVs to service ground mobile users.

25.2 Echo state networks for proactive caching

In this section, we study how a network can use machine learning frameworks to predict the users' behaviors so as to determine the optimal cached contents. In particular, we develop a novel prediction algorithm based on the echo state networks to predict users' content request distributions and mobility patterns. Based on these predictions, the optimal cached contents can be determined by a sublinear algorithm. Next, we first introduce a basic system model. Then, we present a novel prediction algorithm based on an echo state network to predict the users' content request distributions and mobility patterns. Based on these predictions, the optimal cached contents can be determined. We conclude by evaluating the performance of proposed caching scheme via simulations.

25.2.1 Basic problem

Consider a cache-enabled CRAN that consists of a set $\mathcal{U}$ of U users served by a set $\mathcal{R}$ of R RRHs. The RRHs are connected to the cloud pool of the BBUs via capacity-constrained wired fronthaul links. Let v_F be the maximum fronthaul transmission rate for all users. As shown in Figure 25.1, RRHs which have the same content request distributions are grouped into a virtual cluster with $\mathcal{M} = \mathcal{M}_1 \cup \ldots \cup \mathcal{M}_M$ and M being the number of virtual clusters. We assume that each user will always connect to its nearest RRHs cluster and can request at most one content at each time slot τ. The virtual clusters with their associated users allow the CRAN to use zero-forcing

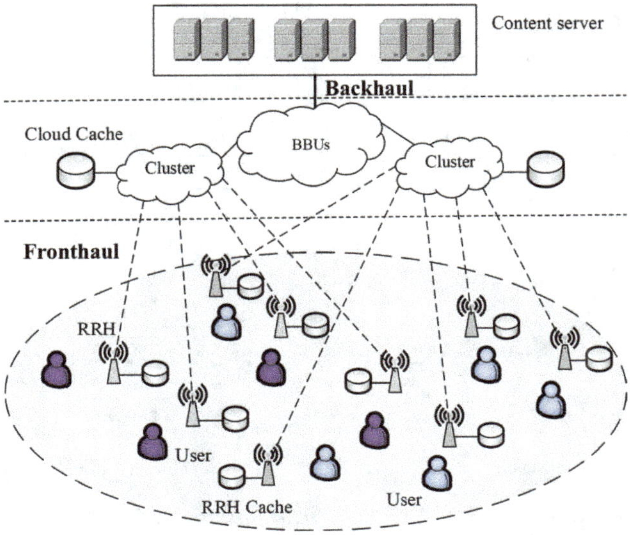

Figure 25.1 A CRAN using clustering and caching

dirty paper coding (ZF-DPC) of multiple-input multiple-output (MIMO) systems to eliminate cluster interference. Virtual clusters are connected to the content servers via capacity-constrained wired backhaul links. The capacity of the backhaul link is limited with v_B being the maximum backhaul transmission rate for all users. Since each RRH may connect to more than one user, the RRH may have more than one type of content request distribution thus belonging to more than one cluster.

We consider daily periodic mobility of users, which is collected once every H time slots. In our model, each user is assumed to be moving from the current location to a target location at a constant speed, and this user will seamlessly switch to the nearest RRH as it moves. We ignore the RRH handover time duration that a user needs to transfer from one RRH to another.

Given each user's periodic mobility, we consider the contents stored, separately, at the RRHs and cloud. Caching at the cloud allows to offload the backhaul traffic and overcome the backhaul capacity limitations. In particular, the cloud can store the popular contents that all users request from the content servers thus alleviating the backhaul traffic. Caching at the RRH, referred to as RRH cache hereinafter, will only store the popular content that the associated users request. The RRH cache can significantly offload the traffic and reduce the transmission delay of both the fronthaul and backhaul. We assume that each content can be transmitted to a given user during time slot τ. In our model, a time slot represents the time duration during which each user has an invariant content request distribution. During each time slot, each user can receive several contents. The RRH cache is updated each time slot τ and the cloud cache is updated during T_τ time slots. We assume that the cached content update of each RRH depends only on the users that are associated with this RRH. We also assume that the content server stores a set $\mathcal{N} = \{1, 2, \ldots, N\}$ of all

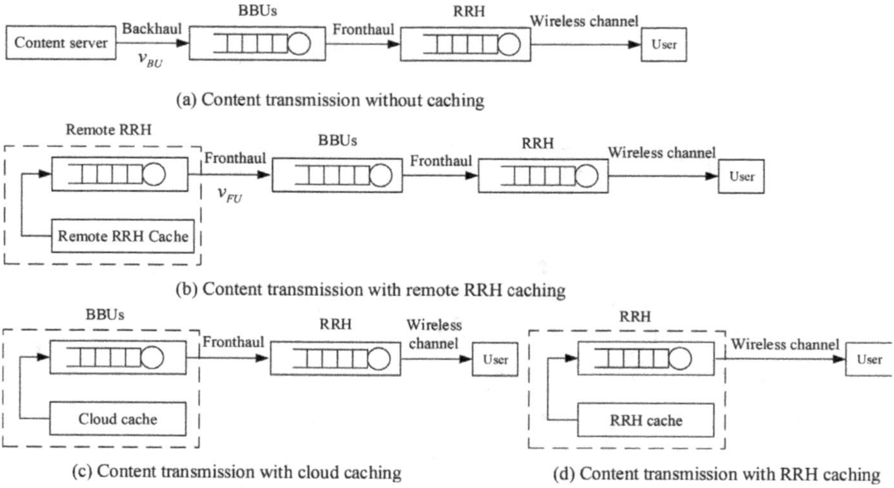

Figure 25.2 Content transmission in CRANs

contents required by all users. All contents are of equal size L. The set of C_c cloud cache storage units is given by $\mathscr{C}_c = \{1, 2, \cdots, C_c\}$, where $C_c \leq N$. The set of C_r RRH cache storage units is given by $\mathscr{C}_r = \{1, 2, \cdots, C_r\}$, where $C_r \leq N$, $r \in \mathscr{R}$.

As shown in Figure 25.2, contents can be sent from: (a) a content server, (b) a remote RRH cache storage unit, (c) a cloud cache storage unit, or (d) an RRH cache storage unit to the user. An *RRH* refers to an RRH that the user is already associated with, while a *remote RRH* refers to other RRHs that store the user's required content but are not associated to this user. We assume that each content can be transmitted independently, and different contents are processed at different queues. The transmission rate of each content, v_{BU}, from the content server to the BBUs is:

$$v_{BU} = \frac{v_B}{N_B}, \tag{25.1}$$

where N_B is the number of the contents that must be transmitted from the backhaul to the BBUs. Since the content transmission rates, from the cloud cache to the BBUs and from the RRH cache to the local RRH, can occur at a rate that is higher than that of the backhaul and fronthaul links, we ignore the delay of these links. After transmitting the content to the BBUs, the content is delivered to the RRHs over fronthaul links. We also assume that the transmission rate from the RRH to the BBUs is the same as the rate from the BBUs to the RRH. Subsequently, the transmission rate, v_{FU}, of each content from the BBUs to the RRHs is $v_{FU} = \frac{v_F}{N_F}$, where N_F is the number of the users that request contents that must be transmitted from the fronthaul to the RRHs. After transmitting the content to the RRHs, the content is transmitted to the users over wireless channels. Therefore, the total transmission link of a specific content consists of one of the following links: (a) content server-BBUs-RRH-user, (b) cloud cache-BBUs-RRH-user, (c) RRH cache-RRH-user, and (d) remote RRH cache-remote RRH-BBUs-RRH-user. To mitigate interference over wireless links, the RRHs can be clustered based on the content requests to leverage MIMO techniques. This, in turn, can also increase the effective capacity for each user, since the RRHs can cooperate and use ZF-DPC to transmit their data to the users. Therefore, the received signal-to-interference-plus-noise ratio (SINR) of user i from the nearest RRH $k \in \mathcal{M}_i$ at time t is:

$$\gamma_{t,ik} = \frac{P d_{t,ik}^{-\beta} \|h_{t,ik}\|^2}{\sum_{j \in \mathcal{M} / \mathcal{M}_i} P d_{t,ij}^{-\beta} \|h_{t,ij}\|^2 + \sigma^2}, \tag{25.2}$$

where $h_{t,ik}$ is the Rayleigh fading parameter and $d_{t,ik}^{-\beta}$ is the path loss at time t, with $d_{t,ik}$ being the distance between RRH k and user i at time t, and β being the path loss exponent. σ^2 is the power of the Gaussian noise, and P is the transmit power of each RRH, assumed to be equal for all RRHs. We also assume that the downlink bandwidth of each user is B. Since the user is moving and the distance between the RRH and user is varying, the channel capacity between RRH k and user i at time t will be $C_{t,ik} = B\log_2(1 + \gamma_{t,ik})$. Since each user is served by the nearest RRH, we use $d_{t,i}$, $h_{t,i}$, $C_{t,i}$, and $\gamma_{t,i}$ to refer to $d_{t,ik}$, $h_{t,ik}C_{t,ik}$, and $\gamma_{t,ik}$, for simplicity.

Since the capacity $C_{t,i}$ does not account for delay, it cannot characterize the quality-of-service (QoS) of requested content. In contrast, we introduce the notion of an effective capacity, as defined in [18], to capture the maximum content transmission rate of a channel with a specific QoS guarantee. First, we introduce the notion of a QoS exponent that allows quantifying the QoS of a requested content and, then, we define the effective capacity. The QoS exponent related to the transmission of a given content n to a user i with a stochastic waiting queue length $Q_{i,n}$ is [18]:

$$\theta_{i,n} = \lim_{q \to \infty} \frac{\log_2 \Pr[Q_{i,n} > q]}{q}, \tag{25.3}$$

where i is the system allowable threshold of queue length. For a large threshold value q_{max}, the buffer violation probability of content n for user i can be approximated by:

$$\Pr[Q_{i,n} > q_{\max}] \approx e^{-\theta_{i,n} q_{\max}}. \tag{25.4}$$

This approximation is obtained from the large deviation theory. Then, the relation between buffer violation probability and delay violation probability for user i with content n is [18]:

$$\Pr[D_{i,n} > D_{\max}] \leq k\sqrt{\Pr[Q_{i,n} > q_{\max}]}, \tag{25.5}$$

where $D_{i,n}$ is the delay of transmitting content n to user i and $D_{\max}$ is the maximum tolerable delay of each content transmission. Here, k is a positive constant and the maximum delay $q_{\max} = cD_{\max}$, with c being the transmission rate over the transmission links. Therefore, $\theta_{i,n}$ can be treated as the QoS exponent of user i transmitting content n which also represents the delay constraint. A smaller $\theta_{i,n}$ reflects a looser QoS requirement, while a larger $\theta_{i,n}$ expresses a more strict QoS requirement.

The QoS exponent pertaining to the transmission of a content n to user i with delay $D_{i,n}$ is [19]:

$$\theta_{i,n} = \lim_{D_{\max}\to\infty} \frac{-\log \Pr\left(D_{i,n} > D_{\max}\right)}{D_{\max} - N_h L/v}, \tag{25.6}$$

where N_h indicates the number of hops of each transmission link and v indicates the rate over the wired fronthaul and backhaul links. Based on (25.3)–(25.6), the cumulative distribution function of delay of user i transmitting content n with a delay threshold $D_{\max}$ is given by:

$$\Pr\left(D_{i,n} > D_{\max}\right) \approx e^{-\theta_{i,n}\left(D_{\max} - N_h/v\right)}. \tag{25.7}$$

The corresponding *QoS exponents* pertaining to the transmission of a content n to a user i can be given as follows: (a) content server-BBUs-RRH-user $\theta_{i,n}^{S}$, (b) cloud cache-BBUs-RRH-user $\theta_{i,n}^{A}$, (c) local RRH cache-RRH-user $\theta_{i,n}^{O}$, and (d) remote RRH cache-remote RRH-BBUs-RRH-user $\theta_{i,n}^{G}$. To achieve the same QoS and delay of transmitting content n over the wired fronthaul and backhaul links, the QoS exponents of the four transmission links of content n with v_{BU} and v_{FU} must satisfy the following conditions:

$$\text{a) } \theta_{i,n}^{S} = \frac{\theta_{i,n}^{O}}{1 - 2L/v_{BU}D_{\max}}, \quad \text{b) } \theta_{i,n}^{A} = \frac{\theta_{i,n}^{O}}{1 - L/v_{FU}D_{\max}},$$
$$\text{c) } \theta_{i,n}^{G} = \frac{\theta_{i,n}^{O}}{1 - 2L/v_{FU}D_{\max}}.$$

Given these basic definitions, the effective capacity of each user is given next. Since the speed of each moving user is constant, the cumulative channel capacity during the time slot τ is given as $C_{\tau,i} = \sum_{t=1,2,\ldots,\tau} C_{t,i} = \mathbb{E}_{d_i,h_i}[C_{t,i}]$, where $\mathbb{E}_{d_i,h_i}[C_{t,i}]$ is the expectation of $C_{t,i}$ with respect to distribution of d_i and h_i. Therefore, the effective capacity of user i receiving content n during time τ is given by [18]:

$$E_{\tau,i}\left(\theta_{i,n_{i\tau},\tau}^{j}\right) = -\frac{1}{\theta_{i,n_{i\tau},\tau}^{j}\tau} \log_2 \mathbb{E}_{d_i,h_i}\left[e^{-\theta_{i,n_{i\tau},\tau}^{j} C_{\tau,i}}\right], \tag{25.8}$$

where $n_{i\tau}$ represents the content that user i requests at time slot τ and $j \in \{O, A, S, G\}$ indicates the link that transmits the content n to user i. Based on (25.8), the sum effective capacity of all moving users during time slot k is:

$$E_k = \sum_{i\in\mathcal{U}} E_{k,i}\left(\theta_{i,n_{ik},k}^{j}\right). \tag{25.9}$$

The sum effective capacity $\bar{E}$ is analyzed during T time slots. Therefore, the long-term effective capacity $\bar{E}$ is given by $\bar{E} = \frac{1}{T}\sum_{k=1}^{T} E_k$. $\bar{E}$ actually captures the delay and QoS of contents that are transmitted from the content server, remote RRHs, and caches to the network users during a period T.

Given this system model, our goal is to develop an effective caching scheme and content RRH clustering approach to reduce the interference and offload the traffic of the backhaul and fronthaul based on the predictions of the users' content request distributions and periodic mobility patterns. To achieve this goal, we formulate a QoS and delay optimization problem whose objective is to maximize the long-term sum effective capacity. This optimization problem of caching involves predicting the content request distribution and periodic location for each user and finding optimal contents to cache at the BBUs and RRHs. This problem can be formulated as follows:

$$\max_{\mathscr{C}_c,\mathscr{C}_r} \bar{E} = \max_{\mathscr{C}_c,\mathscr{C}_r} \frac{1}{T}\sum_{k=1}^{T}\sum_{i\in\mathscr{U}} E_{k,i}\left(\theta^{j}_{i,n_{ik},k}\right), \tag{25.10}$$

$$\text{s. t.} \quad m\bigcap f=\emptyset, m\neq f, m,f\in\mathscr{C}_c \text{ or } m,f\in\mathscr{C}_r, \tag{25.10a}$$

$$\mathrm{j}\in\{O,A,S,G\}, \tag{25.10b}$$

$$\mathscr{C}_c,\mathscr{C}_r,n_{ik},\subseteq\mathscr{N},r\in\mathscr{R}, \tag{25.10c}$$

where $\mathscr{C}_c$ and $\mathscr{C}_r$ are the set of contents that stored in the cloud cache and RRH cache, (25.10a) captures the fact that each cache storage unit in the RRH and cloud stores a single, unique content, (25.10b) represents that the link selection of each content transmission, and (25.10c) indicates that the cached contents come from the content server. Here, we note that, storing contents in the cache can increase the rates v_{BU} and v_{FU} of the backhaul and fronthaul which, in turn, results in the increase of the effective capacity. Meanwhile, the prediction of each user's mobility pattern can be combined with the prediction of the user's content request distribution to determine which content to store in which RRH cache. Such intelligent caching will, in turn, result in the increase of the effective capacity.

25.2.2 Echo state networks for content prediction and mobility

The optimization problem in (25.10) is challenging to solve since the effective capacity depends on the prediction of the content request distribution which determines the optimal cached contents. The effective capacity also depends on the prediction of the user's mobility pattern that will determine the user association thus affecting the RRH caching. In fact, since the RRH and cloud caching need to be aware of the content request distribution of each user in advance, the optimization problem is difficult to solve using conventional optimization algorithms that cannot predict the user's content request distribution. To address this challenge, we propose an echo state network (ESN) [10] based prediction algorithm for predicting content request distribution and mobility pattern for each user. ESNs are an emerging type of recurrent neural networks [10] that can track the state of a network and predict the future information, such as content request distribution and user mobility pattern, over time.

25.2.2.1 Content distribution prediction

In this section, we formulate the ESN-based content request distribution prediction algorithm that consists of four components: (a) agents, (b) input, (c) output, and (d) ESN model. The ESN will allow us to build the content request distribution based on each user's context. The proposed ESN-based prediction approach is thus defined by the following key components:

- *Agents*: The agents in our ESNs are the BBUs. Since each ESN scheme typically performs the prediction for just one user, the BBUs must implement U ESN algorithms at each time slot.
- *Input:* The ESN takes an input vector $\boldsymbol{x}_{t,j} = [x_{tj1}, \cdots, x_{tjK}]^{\mathrm{T}}$ that represents the context of user j at time t including content request time, week, gender, occupation, age, and device type (e.g., tablet or smartphone). The vector $\boldsymbol{x}_{t,j}$ is then used to determine the content request distribution $\boldsymbol{y}_{t,j}$ for user j. For example, the types of videos and TV programs that interest young teenage students will be significantly different from those that interest a much older demographic. Indeed, the various demographics and user information will be critical to determine the content request preferences of various users. Here, K is the number of properties that constitute the context information of user j.
- *Output:* The output of the ESN at time t is a vector of probabilities $\boldsymbol{y}_{t,j} = [p_{tj1}, p_{tj2}, \ldots, p_{tjN}]$ that represents the probability distribution of content request of user j, where p_{tjn} is the probability that user j requests content n at time t.
- *ESN Model:* An ESN model can approximate the function between the input $\boldsymbol{x}_{t,j}$ and output $\boldsymbol{y}_{t,j}$, thus building the relationship between each user's context and the content request distribution. For each user j, an ESN model is essentially a dynamic neural network, known as the dynamic reservoir. Mathematically, the dynamic reservoir consists of the input weight matrix $W_j^{\alpha,in} \in \mathbb{R}^{N_w \times K}$, and the recurrent matrix $W_j^{\alpha} \in \mathbb{R}^{N_w \times N_w}$, where N_w is the number of the dynamic reservoir units that the BBUs use to store the context of user j. The output weight matrix $W_j^{\alpha,out} \in \mathbb{R}^{N \times (N_w + K)}$ is trained to approximate the prediction function. $W_j^{\alpha,out}$ essentially reflects the relationship between context and content request distribution for user j. The dynamic reservoir of user j is therefore given by the pair $(W_j^{\alpha,in}, W_j^{\alpha})$ which is initially generated randomly via a uniform distribution and W_j^{α} is defined as a sparse matrix with a spectral radius less than one [10]. $W_j^{\alpha,out}$ is also initialized randomly via a uniform distribution. By training the output matrix $W_j^{\alpha,out}$, the proposed ESN model can predict the content request distribution based on the input $\boldsymbol{x}_{t,j}$, which will then provide the samples for the sublinear algorithm that effectively determines which content to cache.

Given these basic definitions, we introduce the dynamic reservoir state $v_{t,j}^{\alpha}$ of user j at time t which is used to store the states of user j as follows:

$$v_{t,j}^{\alpha} = f\left(W_j^{\alpha} v_{t-1,j}^{\alpha} + W_j^{\alpha,in} x_{t,j}\right), \tag{25.11}$$

where $f(\cdot)$ is the tanh function. Suppose that each user j has a content request at each time slot. Then, the proposed ESN model will output a vector that captures the content request distribution of user j at time t. The output yields the actual distribution of content request at time t:

$$y_{t,j}\left(x_{t,j}\right) = W_{t,j}^{\alpha,out}\left[v_{t,j}^{\alpha}; x_{t,j}\right], \tag{25.12}$$

where $W_{t,j}^{\alpha,out}$ is output matrix $W_j^{\alpha,out}$ at time t. In other words, (25.12) is used to build the relationship between input $x_{t,j}$ and the output $y_{t,j}$. In order to build this relationship, we need to train $W_{t,j}^{\alpha,out}$. A linear gradient descent approach is used to derive the following update rule,

$$W_{t+1,j}^{\alpha,out} = W_{t,j}^{\alpha,out} + \lambda^{\alpha} \left(e_{t,j} - y_{t,j}\left(x_{t,j}\right)\right) \left[v_{t,j}^{\alpha}; x_{t,j}\right]^{\mathrm{T}}, \tag{25.13}$$

where λ^{α} is the learning rate and $e_{t,j}$ is the real content request distribution of user j at time t. Indeed, (25.13) shows how an ESN can approximate to the function of (25.12).

25.2.2.2 Mobility prediction

In this section, we introduce a minimum complexity ESN algorithm to predict the user trajectory. Unlike the ESN prediction algorithm of the previous section, the ESN prediction algorithm of the user mobility proposed here is based on the minimum complexity dynamic reservoir and adopts an offline method to train the output matrix. The main reason behind this is that the prediction of user mobility can be taken as a time series and needs more data to train the output matrix. Therefore, we use a low complexity ESN to train the output matrix and predict the position of each user. The ESN will help us predict the user's position based on the positions that the user has visited over a given past history, such as the past few weeks, for example. Here, the mobility prediction ESN will also include four components, with the BBUs being the agents, and the other components being:

- *Input:* $m_{t,j}$ represents the current location of user j. This input $m_{t,j}$ combining with the history input data $\left[m_{t-1,j}, \ldots, m_{t-M,j}\right]$ determines the positions $\boldsymbol{s}_{t,j}$ that the user is expected to visit. Here, M denotes the number of the history data that an ESN can record.
- *Output:* $\boldsymbol{s}_{t,j} = \left[s_{tj1}, \cdots, s_{tjN_s}\right]^{\mathrm{T}}$ represents the position that user j may visit for the next steps, where N_s represents the number of positions that user j is expected to visit in the next N_s time duration H.
- *Mobility Prediction ESN Model:* An ESN model builds the relationship between the user's context and positions that the user will visit. For each user j, an ESN model will be combined with the input $m_{t,j}$ to record the position that the user has visited over a given past history. The ESN model consists of the input weight matrix $W_j^{in} \in \mathbb{R}^{W\times 1}$, the recurrent matrix $W_j \in \mathbb{R}^{W\times W}$, where W is the number of units of the dynamic reservoir that the BBUs use to store position records of user j, and the output weight matrix $W_j^{out} \in \mathbb{R}^{N_s\times W}$. The generation of W_j^{in} and W_j^{out} is similar to the content distribution prediction approach. W_j is defined as a full rank matrix defined as follows:

$$W_j = \begin{bmatrix} 0 & 0 & \cdots & w \\ w & 0 & 0 & 0 \\ 0 & \ddots & 0 & 0 \\ 0 & 0 & w & 0 \end{bmatrix}, \tag{25.14}$$

where w can be set as a constant or follows a distribution, such as uniform distribution. Given these basic definitions, we use a linear update method to update the dynamic reservoir state $v_{t,j}$ of user j, which is used to record the positions that user j has visited as follows:

$$v_{t,j} = W_j v_{t-1,j} + W_j^{in} m_{t,j}. \tag{25.15}$$

The position of output $s_{t,j}$ based on $v_{t,j}$ is given by:

$$s_{t,j} = W_j^{out} v_{t,j}. \tag{25.16}$$

In contrast to (25.13), W_j^{out} of user j is trained in an offline manner using ridge regression [10]:

$$W_j^{out} = s_j v_j^{\mathrm{T}} \left(v_j^{\mathrm{T}} v_j + \lambda^2 \mathrm{I} \right)^{-1}, \tag{25.17}$$

where $v_j = \left[v_{1,j}, \ldots, v_{N_{tr},j} \right] \in \mathbb{R}^{W \times N_{tr}}$ is the reservoir states of user j for a period N_{tr}, s_j is the output during a period N_{tr}, and I is the identity matrix.

25.2.3 Sublinear algorithm for clustering and caching

The predictions of the content request distribution and user mobility pattern in Section 25.2.2 must now be leveraged to determine which content to cache at the RRHs, cluster the RRHs at each time slot, and identify which contents to store in cloud cache during a given period. However, it is challenging for the BBUs to scan each content request distribution prediction among the thousands of users' content request distribution predictions resulting from the ESNs' output within a limited time. In addition, in a dense CRAN, the BBUs may not have the entire knowledge of the users' contexts and distributions of content request in a given period, thus making it challenging to determine which contents to cache as per (25.10). To address these challenges, we propose a novel *sublinear approach* for caching [20].

A sublinear algorithm is typically developed based on random sampling theory and probability theory [20] to perform effective big data analytics. In particular, sublinear approaches can obtain the approximation results to the optimal result of an optimization problem by only looking at a subset of the data for the case in which the total amount of data is so massive that even linear processing time is not affordable. For our model, a sublinear approach will enable the BBUs to compute the average of the content request percentage of all users thus determining cached content at the cloud without scanning through the massive volume of data pertaining to the users' content request distributions. Moreover, using a sublinear algorithm enables the BBUs to determine the similarity of two users' content request distributions by only scanning a portion of each content request distribution. Compared to traditional stochastic techniques, a sublinear algorithm can control the trade-off between algorithm processing time or space, and algorithm output quality. Such algorithms can use only a few samples to compute the average content request percentage within the entire content request distributions of all users.

Next, we introduce how to use sublinear algorithms for clustering and determining caching contents. The detailed clustering step is specified as follows:

- The cloud predicts the users' content request distribution and mobility patterns.
- Based on the users' content request distribution and locations, the cloud can estimate the users' RRH association.
- Based on the users' RRH association, the cloud can determine each RRH's content request distribution and then cluster the RRHs into several groups. For any two RRHs, when the difference of their content request distributions is below χ, the cloud will cluster these two RRHs into the same group. Here, we use the sublinear Algorithm 8 in [20] to calculate the difference between two content request distributions.

Based on the RRHs' clustering, we compute the average of the content request percentage of all users. This percentage is used to determine which content to cache in the cloud. Based on the prediction of content request distribution and mobility for each user resulting from the output of the ESN schemes, each RRH must determine the contents to cache according to the ranking of the average content request percentage of its associated users, as given by the computed percentages. For example, denote $p_{r,1}$ and $p_{r,2}$ as the predictions of content request distribution for two users that are associated with RRH r. The average content request percentage is given as $p_r = (p_{r,1} + p_{r,2})/2$. Based on the ranking of the average content request percentage of the associated users, the RRH selects C_r contents to store in the cache as follows:

$$\mathscr{C}_r = \arg\max_{\mathscr{C}_r} \sum_{n \in \mathscr{C}_r} p_{rn}, \tag{25.18}$$

where $p_{rn} = \sum_{i \in \mathscr{U}_r} p_{rin} E_{k,i}(\theta^O_{i,n,k}) / N_r$ is the average weighted percentage of the users that are associated with RRH r requesting content n, $\mathscr{U}_r$ is the set of users that are associated with RRH r, and N_r is the number of users associated with RRH r.

To determine the optimal cached contents at cloud, the cloud needs to update the content request distribution of each user to compute the distribution of the requested content that must be transmitted via fronthaul links based on the associated RRH cache. We define the distribution of the requested content that must be transmitted via fronthaul links using the updated content request distribution, $p'_{r,1} = [p'_{r11}, \ldots, p'_{r1N}]$. The difference between $p_{r,1}$ and $p'_{r,1}$ is that $p_{r,1}$ contains the probability of the requested content that can be transmitted from the RRH cache. For example, we assume that content n is stored at the cache of RRH r, which means that content $n \in \mathscr{C}_r$, consequently, $p'_{r1n} = 0$. Based on the updated content request distribution, the BBUs can compute the average percentage of each content within the entire content request distributions. For example, let $p' = \sum_{\tau=1}^{T} \sum_{i=1}^{U} p'_{\tau,i} E_{k,i}(\theta^A_{i,k}) / TU$ be the average of the updated content request probability during T, where $p'_{\tau,i}$ is the updated content request distribution of user i during time slot τ. Consequently, the BBUs select C_c contents to store at the cloud cache according to the rank of the average updated content request percentage p' which is:

$$\mathscr{C}_c = \arg\max_{\mathscr{C}_c} \sum_{n \in \mathscr{C}_c} p'_n. \tag{25.19}$$

However, within a period T, the BBUs cannot record the updated content request distributions for all users as this will lead to a massive amount of data that is equal to $N \cdot U \cdot T$. The sublinear approach can use only a few updated content request distributions to approximate the actual average updated content request percentage. Moreover, the sublinear approach can control the deviation from the actual average updated content request percentage as well as the approximation error. Since the calculation of the percentage of each content is independent and the method of computing each content is the same, we introduce the percentage calculation of one given content. We define ϵ as the error that captures the deviation from the actual percentage of each content request. Let δ be a confidence interval which denotes the probability that the result of sublinear approach exceeds the allowed error interval. To clarify the idea, we present an illustrative example. For instance, assume that the actual result for the percentage of content n is $\alpha = 70\%$ with $\varepsilon = 0.03$ and $\delta = 0.05$. This means that using a sublinear algorithm to calculate the percentage of content request of type n can obtain a result whose percentage ranges from 67% to 73% with 95% probability. Then, the relationship between the number of the updated content request distributions N_n that a sublinear approach needs to calculate the percentage of content n, ε, and δ can be given by [20]:

$$N_n = -\frac{\ln \delta}{2\varepsilon^2}. \tag{25.20}$$

From (25.20), we can see that a sublinear algorithm can transform a statistical estimation of the expected value into a bound with error deviation ϵ and confidence interval δ. After setting ϵ and δ, the sublinear algorithm can just scan N_n updated content request distributions to calculate the average percentage of each content. Based on the average updated content request percentage, the BBUs store the contents that have the high percentages.

Based on the above formulations, the algorithm based on ESNs and sublinear algorithms is shown in Algorithm 1. Note that, in step 8 of Algorithm 1, a single RRH may belong to more than one cluster since its associated users may have different content request distributions. As an illustrative example, consider a system having two RRHs: an RRH a has two users with content request distributions $p_{a,1}$ and $p_{a,2}$, an RRH b has two users with content request distributions $p_{b,1}$ and $p_{b,3}$, and an RRH c that is serving one user with content request distribution $p_{c,2}$. If $p_{a,1} = p_{b,1}$ and $p_{a,2} = p_{c,2}$, the BBUs will group RRH a and RRH b into one cluster ($p_{a,1} = p_{b,1}$) and RRH a and RRH c into another cluster ($p_{a,2} = p_{c,2}$). In this case, the RRHs that are grouped into one cluster will have the highest probability to request the same contents.

Table 25.1 *System parameters*

Parameters	Values	Parameters	Values
r	1 000 m	P	20 dBm
R	1 000	β	4
B	1 MHz	λ^{α}	0.01
L	10 Mbit	S	25
θ_s^O	0.05	T	300
N_w	1 000	σ^2	−95 dBm
C_c, C_r	6, 3	D_{max}	1
K	7	N_s	10
δ	0.05	ε	0.05
H	3	λ	0.5
T_τ	30	$\mathcal{X}$	0.85

Algorithm 1 Algorithm with ESNs and sublinear algorithms

Require: The set of users' contexts, x_t and m_t;
Ensure: initialize $W_j^{\alpha,in}$, W_j^{α}, $W_j^{\alpha,out}$, W_j^{in}, W_j, W_j^{out}, $y_j = 0$, $s_j = 0$, ε, and δ
for time T_τ **do**
  update the output weight matrix $W_{T_\tau+1,j}^{out}$ based on (25.17)
  obtain prediction $s_{T_\tau+1,j}$ based (25.16)
  for time τ **do**
    obtain prediction $y_{\tau+1,j}$ based on (25.12)
    update the output weight matrix $W_{\tau+1,j}^{\alpha,out}$ based on (25.13)
    determine which content to cache in each RRH based on (25.18)
    cluster the RRHs
  end for
  calculate the content percentage for each content based on (25.20)
  determine which content to cache in cloud based on (25.19)
end for

25.2.4 Simulation results

For simulations, the content request data that the ESN uses to train and predict content request distribution is obtained from *Youku* of *China network video index*. The detailed parameters are listed in Table 25.1. The mobility data is measured from real data generated at the *Beijing University of Posts and Telecommunications*. Note that the content request data and mobility data sets are independent. To map the data, we record the students' locations during each day and map arbitrarily the students' locations to one user's content request activity from Youku. The results are compared to three schemes: (a) optimal caching strategy with complete information, (b) random caching with clustering, and (c) random caching without clustering. All statistical

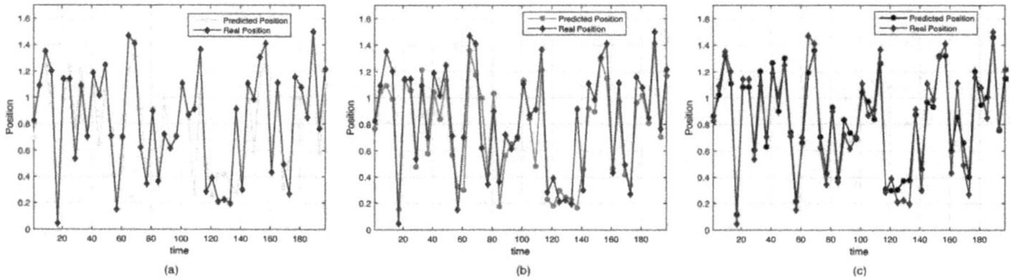

Figure 25.3 The ESNs prediction of the users' mobility as the training dataset N_{tr} varies

results are averaged over 5 000 independent runs. Note that, the benchmark algorithm (a) is based on the assumption that the CRAN already knows the entire content request distribution and mobility pattern. Hereinafter, we use the term "error" to refer to the sum deviation from the estimated distribution of content request to its real distribution.

Figure 25.3 evaluates the accuracy of using ESN for predicting the users' mobility patterns. First, in Figure 25.3, we show how ESN can predict the users' mobility patterns as the size of the training dataset N_{tr} (number of training data to train W^{out}) varies. The considered training data is the user's context during a period. In Figure 25.3, we can see that, as the size of the training dataset increases, the proposed ESN approach achieves more improvement in terms of the prediction accuracy. Therefore, we can conclude that the choice of an appropriate size of the training dataset is an important factor that affects the ESN prediction accuracy of the users' mobility patterns.

Figure 25.4 shows how the sum of the effective capacities of all users in a period changes as the number of the storage units at the cloud cache varies. In Figure 25.4, we can see that, as the number of the storage units increases, the effective capacities

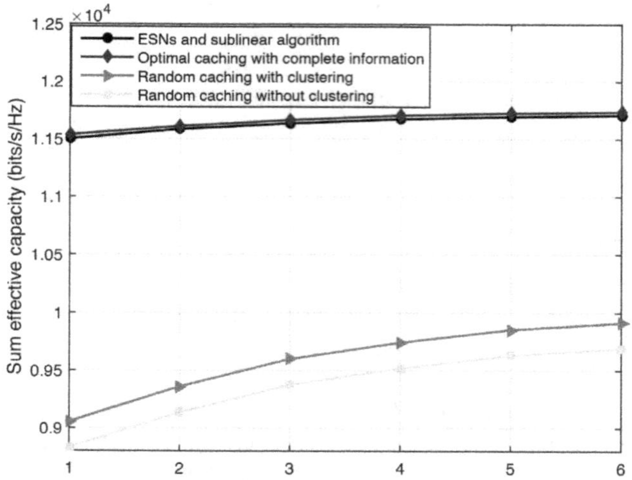

Figure 25.4 Sum effective capacity vs. the number of the storage units at cloud cache

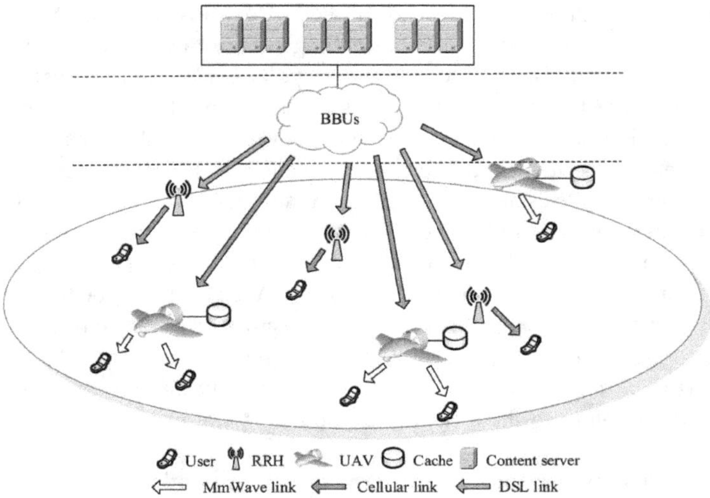

Figure 25.5 A CRAN with cache-enabled UAVs [15]

of all considered algorithms increase since having more storage allows offloading more contents from the content server, which, in turn, will increase the effective capacity for each content. From Figure 25.4, we can see that the proposed algorithm can yield up to 27.8% and 30.7% improvements in terms of the sum effective capacity compared with random caching with clustering and random caching without clustering for the case with one cloud cache storage unit. These gains are due to the fact that the proposed approach can store the contents based on the ranking of the average updated content request percentage of all users as computing by the proposed ESNs and sublinear algorithm.

25.3 Echo state networks for cache-enabled UAVs

In this section, we introduce the use of ESNs for the deployment of cache-enabled UAVs in CRAN. We consider a CRAN system that services a set $\mathcal{U}$ of U mobile users via a set $\mathcal{R}$ of R RRHs, as shown in Figure 25.5. The RRHs are grouped into E clusters using *K-mean clustering* approach [21] so that zero-forcing beamforming (ZFBF) [22] can be used to service the users. In this system, a set $\mathcal{K}$ of K UAVs equipped with cache storage units are deployed to act as flying cache-enabled RRHs to serve the ground users along with the terrestrial RRHs. For the UAV-to-users communication links, since the high altitude of the UAVs can significantly reduce the blocking effect caused by obstacles, we consider air-to-ground UAV transmissions using the millimeter wave (mmWave) frequency spectrum. Meanwhile, the terrestrial RRHs transmit over the cellular band and are connected to the BBUs via capacity-constrained wired fronthaul links. Further, the cloud connects to the content servers via fiber backhaul links. The transmissions from the cloud to the UAVs occurs over wireless fronthaul links using sub-6 GHz band. Consequently, the UAVs' wireless fronthaul links may interfere with the RRHs-users' links.

In our model, the content server stores a set $\mathcal{N}$ of all N contents required by all users. The contents are of equal size L. Caching at the UAVs, referred to as "UAV cache" hereinafter, will be used to store the popular content that the users request. By caching predicted content, the transmission delay from the content server to the UAVs can be significantly reduced as each UAV can directly transmit its stored content to the users. Different from caching at the RRHs or BBUs, caching at UAVs allows servicing mobile users when their QoE requirement cannot be satisfied by the RRHs. We denote the set of C cached contents in the storage units of UAV k by $\mathcal{C}_k$, where $C \leq N$ and $k \in \mathcal{K}$. For simplicity, we assume that each user can request at most one content during each specified time slot τ. We also let Δ_τ be the duration of time slot τ that also represents the maximum transmission duration of each content. We assume that the content stored at the UAV cache will be refreshed every period that consists of T time slots and this caching update is performed at off peak hours when the UAVs return to their cloud-based docking stations for purposes such as battery charge.

In our system, we assume that the users can move continuously. The locations of each user are collected by the BBUs once every H time slots. We assume that each user moves between two collected locations at a constant speed. The mobility pattern for each user will then be used to determine the content that must be cached as well as the optimal location of each UAV thus naturally impacting the QoE of each user.

In this model, the associations of the mobile users with the UAVs or the RRHs can change depending on the QoE requirement. Since the users are moving continuously, the locations of the UAVs must change accordingly so as to serve the users effectively. However, for tractability, we assume that the UAVs will remain static during each content transmission. In essence, the UAVs will update their locations according to the mobility of the users after each content transmission is complete at a current location.

Next, we introduce the models for transmission links between BBUs and UAVs, UAVs and users, and RRHs and users. For ease of exposition, a time slot τ is discretized into F equally spaced time intervals Δ_t, i.e., $\Delta_\tau = F\Delta_t$. The time interval Δ_t is chosen to be sufficiently small so that each user's location can be considered constant during t as in [23].

The mmWave propagation channel of the UAVs-user link is modeled using the standard log-normal shadowing model of [24]. The standard log-normal shadowing model can be used to model the line-of-sight (LoS) and non-line-of-sight (NLoS) links by choosing specific channel parameters. Therefore, the LoS and NLoS path loss of UAV k located at $(x_{\tau,k}, y_{\tau,k}, h_{\tau,k})$ transmitting a content to user i at interval t of time slot τ is [25] (in dB):

$$l_{t,ki}^{\text{LoS}}\left(w_{\tau,t,k}, w_{\tau,t,i}\right) = L_{FS}\left(d_0\right) + 10\mu_{\text{LoS}} \log\left(d_{t,ki}\left(w_{\tau,t,k}, w_{\tau,t,i}\right)\right) + \chi_{\sigma_{\text{LoS}}}, \quad (25.21)$$

$$l_{t,ki}^{\text{NLoS}}\left(w_{\tau,t,k}, w_{\tau,t,i}\right) = L_{FS}\left(d_0\right) + 10\mu_{\text{NLoS}} \log\left(d_{t,ki}\left(w_{\tau,t,k}, w_{\tau,t,i}\right)\right) + \chi_{\sigma_{\text{NLoS}}}, \quad (25.22)$$

where $w_{\tau,t,k} = [x_{\tau,k}, y_{\tau,k}, h_{\tau,k}]$ is the coordinate of UAV k during time slot τ with $h_{\tau,k}$ being the altitude of UAV k at time slot τ. Also, $w_{\tau,t,k} = [x_{t,i}, y_{t,i}]$ is the time-varying coordinate of user i at interval t is the free space path loss given by 20 log $(d_0 f_c 4\pi / c)$ with d_0 being the free-space reference distance, f_c being the carrier frequency and c being the speed of light. μ_{Los} and μ_{NLos} are the path loss exponents for LoS and NLoS links. $\chi_{\sigma_{\mathrm{LoS}}}$ and $\chi_{\sigma_{\mathrm{NLoS}}}$ are the shadowing random variables which are, respectively, represented as the Gaussian random variables in dB with zero mean and σ_{LoS}, σ_{NLoS} dB standard deviations. $d_{t,ki}(w_{\tau,t,k}, w_{\tau,t,i}) = \sqrt{(x_{t,i} - x_{\tau,k})^2 + (y_{t,i} - y_{\tau,k})^2 + h_{\tau,k}^2}$ is the distance between user i and UAV k.

In our model, the probability of LoS connection depends on the environment, density and height of buildings, the locations of the user and the UAV, and the elevation angle between the user and UAV. The LoS probability is given by:

$$\Pr\left(l_{t,ki}^{\mathrm{LoS}}\right) = \left(1 + X\exp\left(-Y[\phi_t - X]\right)\right)^{-1}, \tag{25.23}$$

where X and Y are constants which depend on the environment (rural, urban, dense urban, or others) and $\phi_t = \sin^{-1}\left(h_{\tau,k}/d_{t,ki}\left(w_{\tau,t,k}, w_{\tau,t,i}\right)\right)$ is the elevation angle. Clearly, the average path loss from the UAV k to user i at interval t is [26]:

$$\bar{l}_{t,ki}\left(w_{\tau,t,k}, w_{\tau,t,i}\right) = \Pr\left(l_{t,ki}^{\mathrm{LoS}}\right) \times l_{t,ki}^{\mathrm{LoS}} + \Pr\left(l_{t,ki}^{\mathrm{NLoS}}\right) \times l_{t,ki}^{\mathrm{NLoS}}, \tag{25.24}$$

where $\Pr\left(l_{t,ki}^{\mathrm{NLoS}}\right) = 1 - \Pr\left(l_{t,ki}^{\mathrm{LoS}}\right)$. Based on the path loss, the average signal-to-noise ratio (SNR) of user i located at $w_{\tau,t,i}$ from the associated UAV k at interval t is given by:

$$\gamma_{t,ki}^{\mathrm{V}} = \frac{P_{t,ki}}{10^{\bar{l}_{t,ki}\left(w_{\tau,t,k}, w_{\tau,t,i}\right)/10}\sigma^2}, \tag{25.25}$$

where $P_{t,ki}$ is the transmit power of UAV k at time t, and σ^2 is the variance of the Gaussian noise. We assume that the total bandwidth available for each UAV is B_V which is equally divided among the associated users. The channel capacity between UAV k and user i for each content transmission will be $C_{\tau,ki}^{\mathrm{V}} = \frac{1}{F_{\tau,i}} \sum_{t=1}^{F_{\tau,i}} \frac{B_V}{U_k} \log_2\left(1 + \gamma_{t,ki}^{\mathrm{V}}\right)$, where U_k is the number of the users associated with UAV k and $F_{\tau,i}$ is the number of the intervals that user i uses to receive a content during time slot τ.

For the BBUs-UAVs (ground-to-air) link, we consider probabilistic LoS and NLoS links over the licensed band. Since the distance of the UAVs fronthaul link may be larger compared to the distance of the UAV-user link, the cellular band can provide a more reliable transmission and a smaller path loss compared to the mmWave channel. In such a model, NLoS links experience higher attenuations than LoS links due to the shadowing and diffraction loss. The LoS and NLoS path loss from the BBUs to UAV k at time t of time slot τ can be given by [27]:

$$L_{t,k}^{\mathrm{LoS}} = d_{t,ki}\left(w_{\tau,t,k}, w_{\tau,t,B}\right)^{-\beta}, \tag{25.26}$$

$$L_{t,k}^{\mathrm{NLoS}} = \eta d_{t,ki}\left(w_{\tau,t,k}, w_{\tau,t,B}\right)^{-\beta}, \tag{25.27}$$

where $w_{\tau,t,B} = [x_B, y_B]$ is the location of the BBUs, and β is the path loss exponent. The LoS connection probability and the average SNR of the link between the BBUs and UAV k can be calculated using (25.24)–(25.26).

In our model, RRHs are grouped into E clusters. Then, the RRHs in each cluster use ZFBF to improve the users' rates. The received signals of the users associated with RRHs cluster q at interval t is:

$$b_{t,q} = \sqrt{P_R} H_{t,q} F_{t,q} a_{t,q} + n, \tag{25.28}$$

where $H_{t,q} \in \mathbb{R}^{U_q \times R_q}$ is the path loss matrix with U_q being the number of users associated with RRH cluster q, and R_q is the number of RRHs' antennas. P_R is the transmit power of each RRH which is assumed to be equal for all RRHs. $a_{t,q} \in \mathbb{R}^{U_q \times 1}$ is the transmitted content at interval t and $n_{t,q} \in \mathbb{R}^{U_q \times 1}$ is the noise power. Also, $F_{t,q} = H^{\mathrm{H}}_{t,q} \left(H_{t,q} H^{\mathrm{H}}_{t,q}\right)^{-1} \in \mathbb{R}^{R_q \times U_q}$ is the beamforming matrix [28]. We also assume that the bandwidth of each user associated wth the RRHs is B. Then, the SINR of user i in cluster $\mathcal{M}_q$ at interval t will be:

$$\gamma^{\mathrm{H}}_{t,qi} = \frac{P_R \left\| h_{t,qi} f_{t,qi} \right\|^2}{\underbrace{\sum_{j=1, j\neq}^{E} \sum_{u \in \mathcal{U}_j} P_R \| h_{t,ju} f_{t,ju} \|^2}_{\text{other cluster RRHs interference}} + \underbrace{P_B g_{t,Bi} d_{t,Bi}^{-\beta}}_{\text{wireless fronthaul interference}} + \sigma^2}, \tag{25.29}$$

where $\mathcal{U}_j$ is the set of users associated with the RRHs in group j, $h_{t,qi} \in \mathbb{R}^{1 \times R_q}$ is the channel gain between the RRHs in cluster $\mathcal{M}_q$ and user i with $h_{t,ki} = g_{t,ki} d_{t,ki} \left(x_i, y_i\right)^{-\beta}$, $g_{t,ki}$ is the Rayleigh fading parameter at interval t, and $d_{t,ki} \left(x_i, y_i\right)$ is the distance between RRH k and user i at interval t. $f_{t,qi} \in \mathbb{R}^{R_q \times 1}$is the beamforming vector. Given (25.29), the channel capacity between RRH cluster $\mathcal{M}_q$ and user i for each content transmission is:

$$C^{\mathrm{H}}_{\tau,qi} = \frac{1}{F_{\tau,i}} \sum_{t=1}^{F_{\tau,i}} B \log_2 \left(1 + \gamma^{\mathrm{H}}_{t,qi}\right). \tag{25.30}$$

Next, we present the QoE model for each user. In the considered CRAN system, contents can be transmitted to the users via three types of links: (a) content server-BBUs-RRHs-user, (b) content server-BBUs-UAV-user, and (c) UAV cache-user. Since the backhaul link is assumed to be fiber and, therefore, its delay is neglected. We assume that the capacity of the wired fronthaul links between the BBUs and the RRHs is limited to a maximum rate of v_F for all users. Consequently, the fronthaul rate for each user receiving a content from the RRHs will be $v_{FU} = v_F / N_{FR}$ with N_{FR} being the number of the users receiving contents from the RRHs. Thus, the delay of a user i receiving content n over the three types of links at each time slot τ can be written as:

$$D_{\tau,i,n} = \begin{cases} \frac{L}{v_{FU}} + \frac{L}{C^{\mathrm{H}}_{\tau,qi}}, & \text{link } (a), \\ \frac{L}{C^{F}_{\tau,k}} + \frac{L}{C^{\mathrm{V}}_{\tau,ki}}, & \text{link } (b), \\ \frac{L}{C^{\mathrm{V}}_{\tau,ki}}, & \text{link } (c), \end{cases} \tag{25.31}$$

where $C^{F}_{\tau,k}$ is the rate of content transmission from the BBUs to UAV k which is calculated analogously to (25.3) and (25.31). The lower bound of the delay for each user i receiving content n given by:

$$\min\left\{\frac{L}{v_F}, \frac{L}{C^{\max}_K}\right\} \le D_{\tau,i,n}, \tag{25.32}$$

where $C^{\max}_K = B_V \log_2\left(1 + \frac{P_{\max}}{10^{(L_{FS}(d_0)+10\mu_{\mathrm{LoS}}\log(h_{\min})-4\sigma_{\mathrm{LoS}})/10}\sigma^2}\right)$ with $P_{\max}$ being the maximum transmit power of each UAV, and $h_{\min}$ being the minimum altitude of the UAV. Using (25.32), we can categorize the sensitivity to the delay into five groups using the popular mean opinion score (MOS) model [29] which is often used to measure the QoE of a wireless user. The mapping between delay and MOS model [29] is:

$$\bar{D}_{\tau,i,n} = \frac{\Delta_\tau - D_{\tau,i,n}}{\Delta_\tau - \min\left\{\frac{L}{v_F}, \frac{L}{C^{\max}_K}\right\}}, \tag{25.33}$$

which is shown in Table 25.2.

The screen size of each device type of the user will also affect the QoE perception of the user. In particular, users who own devices that have larger screens (such as tablets) will be more sensitive to QoE compared to those who own smaller devices (such as small smartphones). We capture the impact of the screen size of each user i using a parameter S_i that reflects the diameter length of the user's device. Typically, devices with a larger screen size can display content at a higher resolution thus requiring a higher data rate. We assume that the rate requirement of user i with device S_i receiving a content n at interval t is $\delta_{S_i,n} = S_i\hat{C}_n$, where $\hat{C}_n$ is the rate requirement of each user receiving content n. The mapping from the rate requirement of user device to the MOS model is:

$$V_{t,i} = \begin{cases} 1, & j \ge \delta_{S_i,n}, \\ 0, & j < \delta_{S_i,n}, \end{cases} \tag{25.34}$$

Table 25.2 Mean opinion score model [29]

QoE	Poor	Fair	Good	Very Good	Excellent
Interval scale	0–0.2	0.2–0.4	0.4–0.6	0.6–0.8	0.8–1

where $j \in \{C^{\mathrm{V}}_{t,ki}, C^{\mathrm{H}}_{t,qi}\}$. Here, $V_{t,i} = 1$ indicates that the user's data rate satisfies the requirement of its device type and 0 represents the user's data rate cannot satisfy the requirement. In this case, the rate requirement of a user's device is mapped to the MOS. The QoE of each user i receiving content n at time slot τ can be given by [29]:

$$Q_{\tau,i,n} = \zeta_1 \bar{D}_{\tau,i,n} + \frac{\zeta_2}{F_{\tau,i}} \sum_{t=1}^{F_{\tau,i}} V_{t,i}, \tag{25.35}$$

where ζ_1 and ζ_2 are weighting parameters with $\zeta_1 + \zeta_2 = 1$.

Here, we first find the minimum rate required to meet the QoE requirement of each user associated with the UAVs. Next, we determine the minimum transmit power of each UAV required to meet the QoE threshold of the associated users. Finally, we formulate the minimization problem. From Table 25.2, we can see that, for $0.8 \leq \bar{D}_{\tau,i,n} \leq 1$, the MOS of delay will be "Excellent," which means that the delay is minimized. In this case, $\bar{D}_{\min} = 0.8$ is the minimum value that maximizes the delay component of user i's QoE, during the transmission of a given content n. We define the rate that achieves the optimal delay as the delay requirement and also, define the rate that meets the rate requirement of device as the device rate requirement. Consider the transmission between a UAV k located at $w_{\tau,t,k}$ and a user i located at coordinates $w_{\tau,t,i}$. From (25.31), the delay requirement for UAV k transmitting content n to user i at time slot τ is:

$$C^{R}_{\tau,ki,n} = \begin{cases} \dfrac{L}{\left(\Delta\tau - \bar{D}_{\min}\left(\Delta\tau - \min\left\{\frac{L}{v_F}, \frac{L}{C^{\max}_k}\right\}\right) - \frac{L}{C^{F}_{\tau,k}}\right)}, n \notin \mathscr{C}_k, \\ \dfrac{L}{\left(\Delta\tau - \bar{D}_{\min}\left(\Delta\tau - \min\left\{\frac{L}{v_F}, \frac{L}{C^{\max}_k}\right\}\right)\right)}, n \in \mathscr{C}_k. \end{cases} \tag{25.36}$$

From (25.36), we can see that, by storing content n at UAV k, the delay requirement for minimizing delay decreases.

Let $\delta_{S_i,n}$ be the device rate requirement of user i associated with a UAV. Clearly, the QoE is maximized when $C^{\mathrm{V}}_{t,ki} \geq \max\{C^{R}_{\tau,ki,n}, \delta_{S_i,n}\}$. Hence, the minimum rate required to maximize the user's QoE is $\delta^{R}_{i,n} = \max\{C^{R}_{\tau,ki,n}, \delta_{S_i,n}\}$. Based on (25.25), the minimum transmit power needed to guarantee the QoE requirement of user i receiving content n at interval t is:

$$P^{\min}_{t,ki}\left(w_{\tau,t,k}, \delta^{R}_{i,n}, n\right) = \left(2^{\delta^{R}_{i,n} U_k / B_V} - 1\right) \sigma^2 10^{\bar{l}_{t,ki}\left(w_{\tau,t,k}, w_{\tau,t,i}\right)/10}. \tag{25.37}$$

From (25.37), we can see that the minimum transmit power of UAV k transmitting content n to user i depends on the UAV's location, the rate needed to satisfy the QoE requirement of user i, and the transmitted content n.

Given this system model, our goal is to find an effective deployment of cache-enabled UAVs to enhance the QoE of each user while minimizing the transmit power of the UAVs. This problem involves predicting the content request distribution and periodic locations for each user, finding the optimal contents to cache at the

UAVs, determining the users' associations and adjusting the locations,[a] and transmit power of the UAVs. This problem can be formulated as follows:

$$\min_{\mathscr{C}_k, \mathscr{U}_{\tau,k}, w_{\tau,t,k}} \sum_{\tau=1}^{T} \sum_{k \in \mathscr{K}} \sum_{i \in \mathscr{U}_{\tau,k}} \sum_{t=1}^{F_{\tau,i}} P_{\tau,t,ki}^{\min}\left(w_{\tau,t,k}, \delta_{i,n}^{R}, n_{\tau,i}\right), \tag{25.38}$$

$$\text{s. t. } h_{\min} \le h_{\tau,k}, k \in \mathscr{K}, \tag{25.38a}$$

$$m \ne j, m, j \in \mathscr{C}_k, \mathscr{C}_k \subseteq \mathcal{N}, k \in \mathscr{K}, \tag{25.38b}$$

$$0 < P_{\tau,t,ki}^{\min} \le P_{\max}, i \in \mathscr{U}, k \in \mathscr{K}, \tag{25.38c}$$

where $P_{\tau,t,ki}^{\min}$ is the minimum transmit power of UAV k to user i at interval t during time slot τ is the content that user i requests at time slot τ, $\mathscr{U}_{\tau,k}$ is the set of the users that are associated with UAV k at time slot τ. $h_{\min}$ is the minimum altitude that each UAV can reach at time slot τ. Here, (25.38b) captures the fact that each cache storage unit at the UAV stores a single, unique content, and (25.38c) indicates that the transmit power of the UAVs should be minimized. Since the problem as per (25.38) is to satisfy the rate needed for meeting each user's QoE requirement during the next time period T, the predictions of users' behavior will directly impact the solution. From (25.38), we can see that the prediction of the users' mobility patterns enable the BBUs to find the optimal locations of the UAVs. Moreover, by predicting the users' content request distribution the BBUs can determine the most popular content to cache at the UAVs. In addition, since the content transmission link affects the transmit power that the UAV must use to satisfy the user's QoE requirement, the problem of minimizing the transmit powers of the UAVs in (25.38) inherently incorporates the caching constraints.

25.3.1 Optimal location and content caching for UAVs

In this section, we use the content request distribution and mobility patterns predictions resulting from the proposed conceptor ESN algorithm in Section 25.2.2 to solve the problem in (25.38). In our model, a subset of the users selected by the BBUs are connected to the RRHs. The remaining users are clustered into K clusters and each UAV provides service for one cluster. Based on the associations and predictions, we determine which contents to cache at each UAV and find the optimal location of each UAV. Finally, we analyze the implementation and complexity of the proposed algorithm.

25.3.1.1 Users-RRH association

We find the user-RRH association based on the predicted users' locations at the next time interval. A user is associated with RRHs if the following condition is satisfied: ***Theorem 1****: Given minimum $\bar{D}_{\min}$ and device screen size S_i of each user i, user i will be associated with a cluster k of RRHs if the following rate requirement is satisfied:*

[a]Typically, the speed of a UAV can reach up to 30 m/s while the average speed of each pedestrian ground user is less than 2 m/s. Therefore, in our model, we ignore the time duration that each UAV uses to change its location.

$$C^{\mathrm{H}}_{t,qi} \geq \max\left\{\frac{L}{\left(\Delta_\tau - \bar{D}_{\min}\left(\Delta_\tau - \min\left\{\frac{L}{v_F}, \frac{L}{C_k^{\max}}\right\}\right) - \frac{L}{v_{FU}}\right)}, \delta_{S_i,n}\right\}. \tag{25.39}$$

Based on (25.33) and $\bar{D}_{\min}$, the delay is $D_{\tau,i,n} = \Delta_\tau - \bar{D}_{\min}\left(\Delta_\tau - \min\left\{\frac{L}{v_F}, \frac{L}{C_k^{\max}}\right\}\right)$, and, hence, the delay requirement for RRH cluster q transmitting content n to user i during time slot τ will be:

$$C^R_{\tau,qi} = \frac{L}{\Delta_\tau - \bar{D}_{\min}\left(\Delta_\tau - \min\left\{\frac{L}{v_F}, \frac{L}{C_k^{\max}}\right\}\right) - \frac{L}{v_{FU}}}. \tag{25.40}$$

Therefore, the delay requirement during each interval is equal to $C^R_{\tau,qi}$. Since the device rate requirement is $\delta_{S_i,n}$, the rate of RRH cluster q transmitting content n to user i, $C^{\mathrm{H}}_{\tau,qi}$ must satisfy $C^{\mathrm{H}}_{t,qi} \geq \max\{C^R_{\tau,qi}, \delta_{S_i,n}\}$. This completes the proof.

From Theorem 1, we can see that the user-RRH association depends on the fronthaul rate of each user, the delay requirement, and the device rate requirement. Note that the energy consumption of the RRHs is not considered in our optimization problem as the RRHs can have continuous power supply on the ground while the UAVs are powered by on-board batteries with limited energy. Therefore, it is natural to allow the users to first associate with the RRHs when the RRHs can satisfy the users' QoE requirements.

25.3.1.2 Optimal content caching for UAVs

In our model, the remaining users who are not associated with RRHs will be served by the UAVs. In this case, the user-UAV associations need to be determined. To this end, we use K-mean clustering approach [21] in which the users are clustered into K groups. By implementing the K-mean clustering approach, the users that are close to each other will be grouped into one cluster. Thereby, each UAV services one cluster and the user-UAV association will be determined. Then, based on the UAV association, we find the optimal contents to cache at each UAV. Let $p_{j,i} = [p_{j,i1}, p_{j,i2}, \ldots, p_{j,iN}]$ be the content request distribution of user i during period j that consists of H time slots. The optimal cached contents at each UAV can be determined based on the following theorem.

Theorem 2: *The optimal set of contents $\mathscr{C}_k$ to cache at each UAV k during period T is:*

$$\mathscr{C}_k = \arg\max_{\mathscr{C}_k} \sum_{j=1}^{T/H} \sum_{\tau=1}^{H} \sum_{i \in \mathscr{U}_{\tau,k}} \sum_{n \in \mathscr{C}_k} \left(p_{j,in} \Delta P_{j,\tau,ki,n}\right), \tag{25.41}$$

where $\Delta P_{j,\tau,ki,n} = \begin{cases} P^{\min}_{\tau,ki}\left(C^R_{\tau,ki}\right)_{n \notin \mathscr{C}_k} - P^{\min}_{\tau,ki}\left(C^R_{\tau,ki}\right)_{n \in \mathscr{C}_k}, & C^R_{\tau,ki,n \notin \mathscr{C}_k} \geq \delta_{S_i,n}, \\ P^{\min}_{\tau,ki}\left(\delta_{S_i,n}\right)_{n \notin \mathscr{C}_k} - P^{\min}_{\tau,ki}\left(C^R_{\tau,ki}\right)_{n \in \mathscr{C}_k}, & \delta_{S_i,n} > C^R_{\tau,ki,n \notin \mathscr{C}_k}, \end{cases}$ with $P^{\min}_{\tau,ki}\left(w_{\tau,i,k}, C^R_{\tau,ki}, n\right)$ being simplified to $P^{\min}_{\tau,ki}(C^R_{\tau,ki})$.

Since the delay requirement $C^R_{\tau,ki,n}$ depends on the contents at the UAV cache, it can be written as $\delta^R_{i,n} = \max\left\{C^R_{\tau,ki,n(n\in\mathscr{C}_k)}, C^R_{\tau,ki,n(n\notin\mathscr{C}_k)}, \delta_{S_i,n}\right\}$. Let $P^{\min}_{\tau,ki} = \sum_{t=1}^{F_{\tau,i}} P^{\min}_{j,\tau,t,ki}$. Then the reduction of UAV transmit power by content caching during time slot τ of period j will be:

$$\Delta P_{j,\tau,ki,n} = \begin{cases} P^{\min}_{\tau,ki}\left(C^R_{\tau,ki}\right)_{n\notin\mathscr{C}_k} - P^{\min}_{\tau,ki}\left(C^R_{\tau,ki}\right)_{n\in\mathscr{C}_k}, & C^R_{\tau,ki,n\notin\mathscr{C}_k} \geq \delta_{S_i,n}, \\ P^{\min}_{\tau,ki}\left(\delta_{S_i,n}\right)_{n\notin\mathscr{C}_k} - P^{\min}_{\tau,ki}\left(C^R_{\tau,ki}\right)_{n\in\mathscr{C}_k}, & \delta_{S_i,n} > C^R_{\tau,ki,n\notin\mathscr{C}_k}. \end{cases} \tag{25.42}$$

Considering the fact that the content request distribution changes once every H time slots, the power minimization problem for UAV k during a period that consists of H time slots is:

$$\begin{aligned} \min_{\mathscr{C}_k} \sum_{\tau=1}^{T} \sum_{i\in\mathscr{U}_{\tau,k}} P^{\min}_{\tau,ki} \min_{\mathscr{C}_k} &= \sum_{j=1}^{T/H} \sum_{\tau_j=1}^{H} \sum_{i\in\mathscr{U}_{\tau,k}} P^{\min}_{\tau_j,ki} \\ &= \min_{\mathscr{C}_k} \sum_{j=1}^{T/H} \sum_{\tau=1}^{H} \sum_{i\in\mathscr{U}_{\tau,k}} P^{\min}_{j,\tau,ki} \overset{(a)}{\Leftrightarrow} \max_{\mathscr{C}_k} \sum_{j=1}^{T/H} \sum_{\tau=1}^{H} \sum_{i\in\mathscr{U}_{\tau,k}} \Delta P_{j,\tau,ki,n}, \\ &\overset{(b)}{=} \max_{\mathscr{C}_k} \sum_{j=1}^{T/H} \sum_{\tau=1}^{H} \sum_{i\in\mathscr{U}_{\tau,k}} \left(\sum_{n\in\mathscr{C}_k} \left(p_{j,in}\Delta P_{j,\tau,ki,n}\right) + \sum_{n\notin\mathscr{C}_k} \left(p_{j,in}\Delta P_{j,\tau,ki,n}\right) \right), \\ &= \max_{\mathscr{C}_k} \sum_{j=1}^{T/H} \sum_{\tau=1}^{H} \sum_{i\in\mathscr{U}_{\tau,k}} \sum_{n\in\mathscr{C}_k} \left(p_{j,in}\Delta P_{j,\tau,ki,n}\right), \end{aligned} \tag{25.43}$$

where (a) follows from the fact that minimizing the transmit power of the UAVs is equivalent to maximizing the reduction of the UAVs' transmit power caused by caching, and (b) is obtained by computing the average power reduction using content request probability distribution of each user. This completes the proof. From Theorem 25.3.1, we can see that the content caching depends on the pre-knowledge of users association as well as the content request distribution of each user. Therefore, by predicting the mobility pattern and content request distribution for each user, we can determine the optimal content to cache.

25.3.1.3 Optimal locations of UAVs

Here, we determine the optimal UAVs' locations where the UAVs can serve their associated users using minimum transmit power. Once each UAV selects the suitable contents to cache, the transmission link (BBUs-UAV-user or UAV-user) for each content and the delay requirement $C^R_{\tau,ki,n}$ in (25.37) are determined. In this case, the rate $\delta^R_{i,n}$ which is used to meet the QoE requirment of each user is also determined. Next, we derive a closed-form expression for the optimal location of UAV k during time slot τ in two special cases.

Theorem 3: To minimize the transmit power of UAV k, the optimal locations of UAV k during time slot τ for cases: (a) UAV k positioned at low altitudes compared to the size of its corresponding coverage, $h_{\tau,k}^2 \ll \left(x_{t,i} - x_{\tau,k}\right)^2 + \left(y_{t,i} - y_{\tau,k}\right)^2$ and $\mu_{\mathrm{NLoS}} = 2$, (b) UAV k is placed at high altitudes compared to the size of its corresponding coverage, $h_{\tau,k}^2 \gg \left(x_{t,i} - x_{\tau,k}\right)^2 + \left(y_{t,i} - y_{\tau,k}\right)^2$, is given by:

$$x_{\tau,k} = \frac{\sum\limits_{i \in \mathcal{U}_{\tau,k}} \sum\limits_{t=1}^{F_{\tau,i}} x_{t,i} \psi_{t,ki}}{\sum\limits_{i \in \mathcal{U}_{\tau,k}} \sum\limits_{t=1}^{F_{\tau,i}} \psi_{t,ki}}, y_{\tau,k} = \frac{\sum\limits_{i \in \mathcal{U}_{\tau,k}} \sum\limits_{t=1}^{F_{\tau,i}} y_{t,i} \psi_{t,ki}}{\sum\limits_{i \in \mathcal{U}_{\tau,k}} \sum\limits_{t=1}^{F_{\tau,i}} \psi_{t,ki}}, \tag{25.44}$$

where $\psi_{t,ki} = \left(2^{\delta_{i,n}^R/B} - 1\right)\sigma^2 10^{\left(L_{FS}(d_0) + \chi_\sigma\right)/10}$ with $\sigma = \begin{cases} \sigma_{\mathrm{NLoS}}, & \text{for case } a), \\ \sigma_{\mathrm{LoS}}, & \text{for case } b). \end{cases}$

Using Theorem 25.3.1, we can find the optimal locations of the UAVs given the user association and altitude $h_{\tau,k}$ for the two special cases. For more generic cases, it is highly challenging to find the optimal UAVs' locations using derivation, since the UAV's altitude depends on the x and y coordinates of the UAV. Therefore, we use a learning algorithm given in [30] and [31] to find a sub-optimal solution. The learning algorithm can learn the network state and exploit different actions to adapt the UAV's location according to the network. After the learning step, each UAV will find a sub-optimal location to service the users in a power-efficient way.

25.3.2 Simulation results

For our simulations, the content request data that the ESN uses to train and predict content request distribution is obtained from *Youku* of *China network video index*. Here, one circular CRAN area with a radius $r = 500$ m is considered with $U = 70$ uniformly distributed users and $R = 20$ uniformly distributed RRHs. The

Table 25.3 System parameters [15]

Parameter	Value	Parameter	Value	Parameter	Value
F	1 000	Y	0.13	P_B	30 dBm
X	11.9	N	25	P_R	20 dBm
$\chi_{\sigma LoS}$	5.3	H	10	P_{max}	20 W
N_{tr}	1 000	d_0	5 m	σ^2	−95 dBm
N_s	12	λ	0.01	h_{min}	100 m
N_x	4	β	2	B	1 MHz
μ_{LoS}	2	μ_{NLoS}	2.4	$\delta_{Si,n}$	5 Mbit/s
$\mathcal{X}$	15	ζ_1	0.5	f_c	38 GHz
$\chi_{\sigma NLoS}$	5.27	η	100	B_v	1 GHz
K	5	C	1	L	1 Mbit
T	120	ζ_2	0.5	N_w	1 000

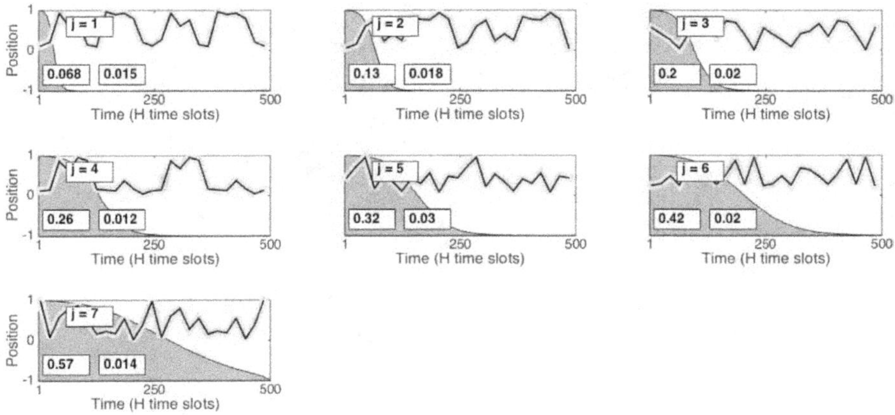

Figure 25.6 Mobility patterns predictions of conceptor ESN algorithm

detailed parameters are listed in Table 25.3. Actual pedestrian mobility data is measured from 100 students at the *Beijing University of Posts and Telecommunications*. We recorded the daily mobility pattern of each student and collected their locations every hour during 9:00 am–12:00 midnight in a period of over two months. For comparison purposes, we investigate: (a) optimal algorithm that has a prior knowledge of the accurate user's mobility patterns and content request distribution, (b) ESN algorithm in [6] to predict the content request distribution and mobility pattern, and (c) random caching with ESN algorithm in [6] to predict content request distribution. All statistical results are averaged over 5 000 independent runs. The accuracy of ESN prediction is measured by normalized root mean square error.

Figure 25.6 shows how the memory of the conceptor ESN reservoir changes as the number of the mobility patterns that were learned by the conceptor ESN varies. Here, one mobility pattern represents the users' trajectory in one day and the colored region represents the memory used by the conceptor ESN. In Figure 25.6, we can see that the memory usage increases as the number of the learned mobility patterns increases. This is due to the fact the conceptor ESN uses a limited memory to learn mobility patterns. From Figure 25.6, we can also see that the conceptor ESN uses less memory for learning mobility pattern 2 compared to pattern 6. In fact, mobility pattern 2 is similar to mobility pattern 1, and, hence, the conceptor ESN requires only a small amount of memory to learn mobility pattern 2. However, the conceptor ESN needs to use more memory to learn mobility pattern 6. Clearly, when a new mobility pattern needs to be learned, the proposed approach needs to learn the difference between the learned mobility patterns and the new one.

Figure 25.7 shows how the total transmit power of the UAVs in a time period changes as the number of the users varies. In Figure 25.7, we can see that the total UAV transmit power of all algorithms increases as the number of the users increases. This is due to the fact that the number of the users associated with the RRHs and the capacity of the wireless fronthaul link of UAVs is limited. Therefore, the UAVs

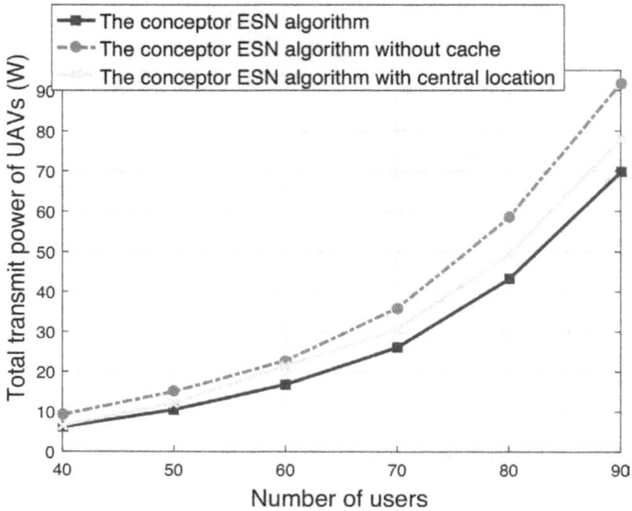

Figure 25.7 Total transmit power as the number of users varies

need to increase their transmit power to satisfy the QoE requirement of each user. Figure 25.7 shows that the proposed algorithm can yield, respectively, 33.3% and 20% gains with respect to reducing the total transmit power compared to the proposed algorithm without cache and the proposed algorithm without optimizing the UAVs' locations for a network with 80 users.

25.4 Summary

Caching is envisioned to be a key technique that offloads the traffic over the fronthaul and backhaul links in CRAN. In this chapter, we introduced two caching applications in CRAN. First, we have developed and analyzed a scenario in which the cache is placed at RRHs and BBU to service users. We have used effective capacity metric to capture the content transmission delay over fronthaul and backhaul links and designed a machine learning-based algorithm for the predictions of users' mobility patterns and content request distributions. We have discussed how to use the predictions of users' mobility patterns and content request distributions to determine the optimal cached contents at RRHs and BBU. Through numerical analysis, we have evaluated the performance of predictions for the learning algorithm and the benefit of using caching for CRANs. Second, we have developed and analyzed a scenario in which the cache-enabled UAVs are used to service the ground mobile users in a CRAN. We have shown that, in this model, a variety of challenges need to be overcome such as modeling users' QoE, developing a air-to-ground mmWave channel model, and designing a learning algorithm for predicting users' mobility patterns and content request distributions. We have discussed how the predictions of users' mobility patterns and content request distributions to determine the optimal locations of UAVs and optimal caching contents of each UAV. Through numerical analysis, we have evaluated the performance of predictions for the learning algorithm and the benefit of the use of cache-enabled UAVs.

This chapter has only scratched the surface of the emerging topic of caching in CRANs. While our focus was on the deployment of caching storage and cache-enabled UAVs in CRAN, many important problems remain open. On the one hand, it is of interest to develop new cooperative cache-enabled UAVs models in which all of the cache-enabled UAVs will cooperate with each other to improve the caching hit ratio so as to improve the users' QoE. On the other hand, it is imperative to analyze the content correlation for caching. In particular, one can use neural networks to investigate the content correlation in time domain and space domain. Based on the content correlation, each BS can store the contents that are the most related to other contents and, hence, the caching efficiency and hit ratio can be improved. Finally, one can study the use of federated learning [32, 33] to analyze and predict users' behaviors and wireless environment to determine the optimal caching contents while reducing the energy consumption of data transmission.

References

[1] Peng M., Sun Y., Li X., Mao Z., Wang C. 'Recent advances in cloud radio access networks: system architectures, key techniques, and open issues'. *IEEE Communications Surveys & Tutorials*. 2016;**18**(3):2282–308.

[2] Tran T.X., Hajisami A., Pompili D. 'Cooperative hierarchical caching in 5G cloud radio access networks'. *IEEE Network*. 2017;**31**(4):35–41.

[3] Stephen R.G., Zhang R. 'Joint millimeter-wave fronthaul and OFDMA resource allocation in ultra-dense CRAN'. *IEEE Transactions on Communications*. 2017;**65**(3):1411–23.

[4] Bastug E., Bennis M., Debbah M. 'Living on the edge: the role of proactive caching in 5G wireless networks'. *IEEE Communications Magazine*. 2014;**52**(8):82–9.

[5] Saad W., Bennis M., Chen M. 'A vision of 6G wireless systems: applications, trends, technologies, and open research problems'. *IEEE network*. 2020;**34**(3):134–42.

[6] Chen M., Saad W., Yin C., Debbah M. 'Echo state networks for proactive caching in cloud-based radio access networks with mobile users'. *IEEE Transactions on Wireless Communications*. 2017;**16**(6):3520–35.

[7] Parvez I., Rahmati A., Guvenc I., Sarwat A.I., Dai H. 'A survey on low latency towards 5G: RAN, core network and caching solutions'. *IEEE Communications Surveys & Tutorials*. 2018;**20**(4):3098–130.

[8] Yang Z., Pan C., Pan Y., *et al.* 'Cache placement in two-tier HetNets with limited storage capacity: cache or buffer?' *IEEE Transactions on Communications*. 2018;**66**(11):5415–29.

[9] Zhao J., Liu Y., Mahmoodi T. 'Resource allocation in cache-enabled cran with non-orthogonal multiple access'. Proceedings of IEEE International Conference on Communications (ICC); Kansas City, MO, USA; 2018.

[10] Chen M., Challita U., Saad W., Yin C., Debbah M. 'Artificial neural networks-based machine learning for wireless networks: a tutorial'. *IEEE Communications Surveys & Tutorials*. 2019;**21**(4):3039–71.

[11] Lau C.P., Alabbasi A., Shihada B. 'An efficient content delivery system for 5G CRAN employing realistic human mobility'. *IEEE Transactions on Mobile Computing*. 2019;**18**(4):742–56.

[12] Dai J., Liu D. 'MAPCaching: A novel mobility aware proactive caching over C-RAN'. Proceedings of IEEE International Symposium on Personal, Indoor, and Mobile Radio Communications (PIMRC); Canada; Montreal, QC; 2017.

[13] Xing H., Cui J., Deng Y. 'Energy-Efficient Proactive Caching for Fog Computing with Correlated Task Arrivals'. Proceedings of IEEE International Workshop on Signal Processing Advances in Wireless Communications (SPAWC); France, Cannes; 2019.

[14] Chen M., Saad W., Yin C. 'Echo-liquid state deep learning for 360° content transmission and Caching in wireless VR networks with cellular-connected UAVs'. *IEEE Transactions on Communications*. 2019;**67**(9):6386–400.

[15] Chen M., Mozaffari M., Saad W., Yin C., Debbah M., Hong C.S. 'Caching in the sky: proactive deployment of cache-enabled unmanned aerial vehicles for optimized quality-of-experience'. *IEEE Journal on Selected Areas in Communications*. 2017;**35**(5):1046–61.

[16] Hu Y., Chen M., Saad W. 'Joint access and backhaul resource management in satellite-drone networks: a competitive market approach'. *IEEE Transactions on Wireless Communications*;19(6):3908–23.

[17] Chen M., Saad W., Yin C. 'Liquid state machine learning for resource and Cache management in LTE-U Unmanned aerial vehicle (UAV) networks'. *IEEE Transactions on Wireless Communications*. 2019;**18**(3):1504–17.

[18] Wu D., Negi R. 'Effective capacity: a wireless link model for support of quality of service'. *IEEE Transactions on Wireless Communications*. 2003;**2**(4):630–43.

[19] Zhao Z., Peng M., Ding Z., Wang W., Poor H.V. 'Cluster content caching: an energy-efficient approach to improve quality of service in cloud radio access networks'. *IEEE Journal on Selected Areas in Communications*. 2016;**34**(5):1207–21.

[20] Wang D., Han Z. *Sublinear algorithms for big data applications*. Springer Berlin Heidelberg; 2015.

[21] Hoppner F., Klawonn F. *Clustering with size constraints*. Springer Berlin Heidelberg; 2008.

[22] Yoo T., Goldsmith A. 'On the optimality of multiantenna broadcast scheduling using zero-forcing beamforming'. *IEEE Journal on Selected Areas in Communications*. 2006;**24**(3):528–41.

[23] Zeng Y., Zhang R., Lim T.J. 'Throughput maximization for UAV-enabled mobile Relaying systems'. *IEEE Transactions on Communications*. 2016;**64**(12):4983–96.

[24] Rappaport T.S. *Wireless communications: principles and practice*. Upper Saddle River, NJ: Prentice-Hall; 2002.

[25] Rappaport T.S., Gutierrez F., Ben-Dor E., Murdock J.N., Qiao Y., Tamir J.I. 'Broadband millimeter-wave propagation measurements and models using adaptive-beam antennas for outdoor urban cellular communications'. *IEEE Transactions on Antennas and Propagation*. 2013;**61**(4):1850–9.

[26] Al-Hourani A., Kandeepan S., Jamalipour A. 'Modeling air-to-ground path loss for low altitude platforms in urban environments'. Proceedings of IEEE Global Communications Conference (GLOBECOM); Austin, TX, USA; 2014.

[27] Mozaffari M., Saad W., Bennis M., Debbah M. 'Unmanned aerial vehicle with underlaid device-to-device communications: performance and tradeoffs'. *IEEE Transactions on Wireless Communications*. 2016;**15**(6):3949–63.

[28] Somekh O., Simeone O., Bar-Ness Y., Haimovich A.M., Shamai S. 'Cooperative multicell zero-forcing beamforming in cellular downlink channels'. *IEEE Transactions on Information Theory*. 2009;**55**(7):3206–19.

[29] Mitra K., Zaslavsky A., Ahlund C., Modelling C.-A.Q. 'Measurement and prediction in mobile computing systems'. *IEEE Transactions on Mobile Computing*. 2015;**14**(5):920–36.

[30] Chen M., Saad W., Yin C. 'Echo state networks for self-organizing resource allocation in LTE-U with uplink–downlink decoupling'. *IEEE Transactions on Wireless Communications*. 2017;**16**(1):3–16.

[31] Chen M., Saad W., Yin C. 'Virtual reality over wireless networks: Quality-of-Service model and learning-based resource management'. *IEEE Transactions on Communications*. 2018;**66**(11):5621–35.

[32] Chen M., Yang Z., Saad W., *et al.* 'A joint learning and communications framework for federated learning over wireless networks'. *arXiv preprint*. 2019.

[33] Chen M., Poor H.V., Saad W., *et al.* 'Convergence time optimization for federated learning over wireless networks'. *arXiv preprint*. 2020.

Chapter 26

Dynamic computation offloading and resource allocation in cache-assisted mobile edge computing systems

Samrat Nath[1] and Jingxian Wu[2]

Mobile edge computing (MEC) is one of the most promising techniques for next-generation wireless communication systems. In this chapter, we study the problem of dynamic caching, computation offloading, and resource allocation in cache-assisted multi-user MEC systems with stochastic task arrivals. There are multiple computationally intensive tasks in the system, and each mobile user (MU) needs to execute a task either locally or remotely in an MEC server by offloading the task data. Popular tasks can be cached in the MEC server to avoid duplicates in offloading. The cached contents can be either obtained through user offloading or fetched from a remote cloud. The objective is to minimize the long-term average of a cost function, which is defined as a weighted sum of energy consumption, delay, and cache contents' fetching costs. The weighting coefficients associated with the different metrics in the objective function can be adjusted to balance the trade-off among them. The optimum design is performed with respect to four decision parameters: whether to cache a given task, whether to offload a given uncached task, how much transmission power should be used during offloading, and how much MEC resources to be allocated for executing a task. We propose to solve the problem by developing a dynamic scheduling policy based on deep reinforcement learning (DRL) with the deep deterministic policy gradient (DDPG) method. Simulation results demonstrate that the proposed algorithm outperforms other existing strategies such as deep Q-network (DQN).

26.1 Introduction

Driven by the ever-increasing popularity of mobile devices and the visions of 5G communications and internet-of-things (IoT), there has been an exponential growth

[1]Walmart Inc, USA
[2]Department of Electrical Engineering, University of Arkansas, USA

of demand in numerous types of mobile applications, such as face and gesture recognition, online interactive gaming, virtual and augmented reality (VR/AR), etc. However, with limited battery life and limited computational capacity of a typical mobile device, it is often hard to meet the stringent requirements and quality of experience (QoE) of these mobile applications, which are usually very computation-intensive, delay-sensitive, and highly energy-consuming. The past decade has seen the mobile cloud computing (MCC) emerging as a way to bridge the gap between these applications and the resource-constrained mobile devices by leveraging the computation power of remote public clouds such as Amazon Web Services and Microsoft Azure. However, one inherent drawback of MCC is the long propagation distances from the end-user mobile devices to the remote cloud servers that cause high latency for mobile applications. Therefore, in recent years, a paradigm shift in mobile computing is happening from the conventional MCC to the mobile edge computing (MEC) [1].

In 5G wireless networks, edge devices, such as small-cell base stations (BSs), wireless access points (APs), etc. are densely deployed close to the MUs. These edge devices each have a computation capacity comparable with that of a high-performance computer server and at every time instant, a large number of devices typically remain idle. Hence, a significant amount of computation and storage resources are harvested at the network edges, which can be leveraged for delivering ubiquitous mobile computing [1]. Unlike MCC, the key aspect of MEC is to push mobile computing to the network edges in order to utilize the resources available at the edges. In the MEC paradigm, MUs can offload computation tasks through the wireless network to the edge devices, e.g. servers connected with BSs. Through the computation offloading of delay-sensitive and computation-intensive tasks, MEC promises to tackle the key challenges for materializing 5G vision by considerably reducing the computation latency and mobile energy consumption, and thus improving the QoE of mobile applications. Therefore, the topic of computation offloading in MEC systems has been rapidly gaining a lot of interests in recent years.

The management of computation, power, and communication resources in an MEC system has a large influence on the efficiency of computation offloading. There have been many studies on computation offloading algorithms in the literature with different types of design objectives and resource allocation schemes [2–11]. Generally, the computation offloading approaches in MEC can be categorized into two classes, namely, binary computation offloading [2–8] and partial computation offloading [9–11]. In binary computation offloading, an MU can either execute its computational task on the local device or offload that task entirely to the MEC server at the edge. On the contrary, partial computation offloading offers more flexibility by allowing the MU to offload fractional parts of the task to the server, while the rest of the task is executed locally. The alternating direction method of multipliers (ADMM) is applied in [3] in order to solve the joint optimization problem of computation offloading and resource allocation. In addition to computation offloading and resource allocation, the study in [8] also considers the problem of channel allocation in a multi-user multi-channel MEC system, with the objective of jointly minimizing the delay, energy consumption, and deadline penalty of all the users. In [11], Lyapunov optimization is adopted in order to develop an online algorithm for jointly managing the radio and computational resources of multi-user MEC systems. An iterative search algorithm is proposed to study the energy-latency trade-off for energy-aware offloading in [4]. However, dynamic channel conditions and/or dynamic task arrivals are not considered in most of the above-mentioned studies. In practice, the MEC systems have stochastic channel conditions and time-varying task arrivals.

Even though the MUs are located close to the densely distributed edge devices/servers in the MEC paradigm, they still suffer from latency and energy consumption due to computation offloading. The concept of task/content caching at the edge has been considered as a promising approach in order to further improve the performance of MEC systems [2, 10, 12–14]. Task caching usually refers to caching task application and related data in the MEC server at the edge. If a task is not cached, it either needs to be executed locally at the MU or needs to be offloaded to the server before execution or it needs to be executed on both local device and server jointly (partial offloading). However, cached tasks can be executed directly in the MEC server. Therefore, caching popular tasks in the MEC server can avoid redundant duplicates in data transmissions and thus further reduce the delay and energy consumption. The problem of dynamic caching with the consideration of time-varying popularity of the tasks is studied in [12–14]. However, in these works, only the storage capacity of the MEC server is considered, while it is assumed that the server has enough computational capacity to execute all the offloaded tasks. This is not a valid assumption since both the computational capacity and storage capacity of the MEC server are limited in practice. Although the works in [2, 10] consider both storage and computation constraints, they provide only offline solutions for static systems. Therefore, it is of vital importance to investigate the joint problem of online task caching, task offloading, and resource allocation policy in dynamic MEC systems.

All the above-mentioned works formulate the complex joint caching, computation offloading, and resource allocation as optimization problems, which are generally non-convex and very hard to solve. A reinforcement learning (RL)-based algorithm can be a promising technique to tackle this challenge since it can provide online solutions to the complex sequential decision-making problems without needing any prior information of the stochastic system [15]. In addition, the explosive growth of interest in the deep neural networks (DNNs) has been driving many researchers in recent years to adopt various deep reinforcement learning (DRL) algorithms such as the DQN method [16], the deep deterministic policy gradient (DDPG) algorithm [17] for solving these problems [5–9]. DRL is considered as an advanced RL technique implemented with DNNs. The useful function approximation property of DNNs allow DRL to solve large-scale problems, where the conventional RL methods become infeasible [18]. The proposed solutions in [5, 6] apply DQN which requires the discretization of the optimization variables, while the solutions in [7–9] deal with the continuous space of optimization variables by adopting DDPG.

Although task caching and computation offloading can considerably reduce the delay and energy consumption of MUs, determining the optimal strategy for caching, offloading, and resource allocation in a dynamic MEC system is a very challenging problem due to the time-varying nature of the system, the heterogeneity of the tasks, and the limited resources at the MEC server. Therefore, the objective of this chapter is to develop an online DRL-based scheme for dynamic caching, computation offloading (binary), and resource allocation in a resource-constrained multi-user MEC system by addressing the following four key questions.

1. Whether a given task should be cached in the MEC server?
2. Whether a given uncached task should be executed locally at an MU or offloaded to the MEC?
3. How much transmission power should be allocated to a given MU for task offloading? and
4. How much computational resources should be allocated by the MEC server for a given task?

In the proposed MEC framework, the time-domain is divided into uniform slots, and the channel conditions, task popularity, and task arrivals are considered to be time-varying and stochastic. At the start of each time slot, the caching decision, offloading decision, power allocation, and computational resource allocation are determined centrally by the BS, and then the results are forwarded to the MU. The MEC server is assumed to have limited storage and computational resource capacity and all the MUs have individual upper bounds on transmission power. Specifically, we formulate an optimization problem to minimize the long-term average of a cost function, which is defined as a weighted sum of energy consumption, delay, and fetching cost of cached content. The problem is formulated under the framework of RL and solved by adopting a DDPG-based method. Simulation results demonstrate the superiority of the proposed DDPG-based solution over other existing algorithms such as DQN.

The rest of this chapter is organized as follows. In Section 26.2, the system model and problem formulation for the dynamic caching, computation offloading, and resource allocation of the MEC system is presented. In Section 26.3, some preliminaries on RL and DRL are introduced followed by the design of the proposed DRL-based algorithm. Simulation results are demonstrated in Section 26.4, while Section 26.5 concludes this chapter.

26.2 System model and problem formulation

Consider a multi-user MEC system that consists of one BS with M antennas, one MEC server, and a set of $N \leq M$ single-antenna mobile users denoted by $\mathcal{N} = \{1, 2, \cdots, N\}$. The BS is connected directly with the MEC server with a cache size of D (in bits) and computing capacity of F (in CPU cycles per second). A discrete-time model is adopted for the MEC system, where the time is divided into slots with equal length T_s (in seconds) and indexed by $\mathcal{T} = \{0, 1, \cdots\}$. The system model is shown in Figure 26.1.

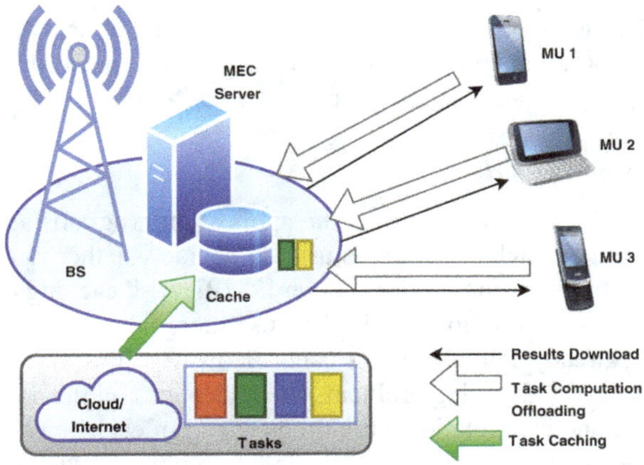

Figure 26.1 System model

26.2.1 Task model

Assume there are K heterogeneous tasks denoted by the set $\mathcal{K} \triangleq \{1, 2, \cdots, K\}$. Each task $k \in \mathcal{K}$ is characterized by two parameters; d_k (in cycles per second) denotes the amount of computing resource required for the task and b_k (in bits) denotes the size of computation input data, such as program codes and input parameters. One example of application scenario for such tasks is virtual reality (VR), where the MUs need to execute various types of computation-intensive VR application tasks such as rendering scenes, recognizing and tracking objects, etc.

The number of MUs is assumed to be more than the number of tasks ($N \geq K$). This is because some computing tasks have higher popularity (e.g., rendering scenes in VR), which are repeatedly requested and executed multiple times [2]. At the beginning of each time slot, each MU requests one task from the set $\mathcal{K}$, where multiple users may simultaneously request a particular task. Denote $k_t^n \in \mathcal{K}$ as the task requested by the nth MU at the start of slot t and $\boldsymbol{k}_t \triangleq [k_t^1, \cdots, k_t^N]^T$ as the task request vector for all MUs.

The popularity of each task $\phi_{k,t}$ is dynamic and follows Zipf distribution [19]. The popularity profile vector is defined as $\boldsymbol{\phi}_t \triangleq [\phi_{1,t}, \cdots, \phi_{K,t}]^T$. Given that the popularity rank of task k during slot t is $z_{k,t} \in \mathcal{K}$, the popularity of the corresponding task can be expressed as

$$\phi_{k,t} = \frac{z_{k,t}^{-\eta}}{\sum_{l=1}^{K} z_{l,t}^{-\eta}}, \tag{26.1}$$

where the Zipf parameter $\eta \geq 0$ controls the skewness of popularity. Specifically, $\eta = 0$ yields a uniform spread of popularity among the tasks, and the popularity difference among the tasks becomes larger with a larger η.

To model the time-varying nature of task popularity, the popularity profile $\boldsymbol{\phi}_t$ is modeled by a V-state Markov chain, represented by V different popularity profiles

$\boldsymbol{\phi}^{(1)}, \cdots, \boldsymbol{\phi}^{(V)}$ [20]. Each profile is modeled by Zipf distributions with parameters η_v. So, at each time slot t, the popularity profile $\boldsymbol{\phi}_t$ will follow one of these V states and each task $k \in \mathcal{K}$ will be assigned popularity ranks $z_{k,t}$ randomly.

26.2.2 Caching model

Based on the definition of task caching, the application programs are all cached in the MEC server, and a caching policy is used to decide whether to cache the input data of the computing task in the MEC server [2, 10]. Task caching can reduce task latency and energy consumption of MU because there is no need for transmitting the data of a task already available in the cache. However, how to assign the limited caching capacity is a challenging problem, which depends on the dynamic popularity ($\phi_{k,t}$), size (b_k), and computational requirement (d_k) of each task.

Denote $c_{k,t} \in \{0, 1\}$ as the binary variable that represents the caching decision for task k in slot t and $\boldsymbol{c}_t = [c_{1,t}, \cdots, c_{K,t}]^T$ denotes the corresponding decision vector for all tasks. If $c_{k,t} = 1$, then the input data for computing task k is cached in the MEC server at the end of slot t and the corresponding data can be utilized in the next time slot $t + 1$ for executing the task entirely at the server. Consequently, the system will not experience any delay and will not incur any energy cost corresponding to the data transmission from MUs to the server for that particular task. Therefore, the users' QoE will be significantly improved. However, all the tasks cannot be cached due to the limited storage size of the MEC server. Therefore, the caching decision variable must meet the following constraint.

$$\sum\nolimits_{k=1}^{K} \mathbf{1}(c_{k,t} = 1)b_k \leq D, \quad \forall t \in \mathcal{T}, \tag{26.2}$$

where $\mathbf{1}(\mathcal{E})$ is the indicator function with $\mathbf{1}(\mathcal{E}) = 1$ if the event $\mathcal{E}$ is true and 0 otherwise.

Moreover, it is assumed that the input data of all the tasks are available in a remote cloud, as shown in Figure 26.1. The BS can download/fetch the data for any task from the cloud via a backhaul link by incurring fetching cost. In addition, the MEC can also cache the data of one or more offloaded tasks in the current slot for future use without incurring the fetching cost. Given the caching decision vector $\boldsymbol{c}_t$ at slot t and $\boldsymbol{c}_{t-1}$ at slot $t - 1$, the overall fetching cost of the MEC system associated with the caching decision at slot t is defined as

$$C_t = \sum\nolimits_{k=1}^{K} \mathbf{1}(c_{k,t-1} = 0, c_{k,t} = 1, k \notin \mathcal{K}_t^{o})g_k, \tag{26.3}$$

where $\mathcal{K}_t^{o}$ is the set of tasks offloaded to the MU at the t-th slot, g_k (in bits) is the fetching cost for downloading the input data for task k from the remote cloud server. Naturally, g_k depends on the size of task data b_k.

Define $\mathcal{K}_t^{c} \subset \mathcal{K}$ as the set of tasks that have been cached during time slot $t - 1$ and are available for utilization at the beginning of slot t, i.e., $\mathcal{K}_t^{c} \triangleq \{k : c_{k,t-1} = 1\}$. If a task requested by the n-th MU is not cached, i.e., $k_t^n \notin \mathcal{K}_t^{c}$, then either the task needs to be executed in the MU locally, or the task data needs to be uploaded to the

MEC server for remote execution. Details regarding task execution and computation are presented in Section 26.2.4.

26.2.3 Communication model

The BS and the MUs form a multi-user multiple-input multiple-output (MIMO) system, with M antennas at the BS serving N single-antenna MUs. The channel conditions between the BS and MUs are described by the $M \times N$ channel matrix $\boldsymbol{H}_t = [\boldsymbol{h}_{1,t}, \cdots, \boldsymbol{h}_{N,t}]$, where $\boldsymbol{h}_{n,t} \in \mathbb{C}^{M\times 1}$ is the channel vector of the nth MU. In order to characterize the temporal channel correlation between consecutive slots, the Gaussian Markov block fading autoregressive model [21] is employed. The channel vector for the nth MU can be expressed as

$$\boldsymbol{h}_{n,t} = \rho_n \boldsymbol{h}_{n,t-1} + \sqrt{1-\rho_n^2}\boldsymbol{e}_t, \tag{26.4}$$

where ρ_n is the normalized channel correlation coefficient for the n-th MU, the error vector $\boldsymbol{e}_t \in \mathbb{C}^{M\times 1}$ is uncorrelated with $\boldsymbol{h}_{n,t}$, and it is complex Gaussian distributed with zero mean and covariance matrix $\sigma_e^2 \boldsymbol{I}_M$, with I_M being a size M identity matrix. Moreover, $\rho_n = J_0(2\pi f_n^d T_s)$ according to Jake's fading spectrum, where f_n^d is the Doppler frequency of the n-th MU, T_s is the slot duration, and $J_0(\cdot)$ is the zero-order Bessel function of the first kind [22].

Denote the transmission power of the n-th MU at the t-th slot as $p_{n,t} \in [0, P_n^{\max}]$, where $P_n^{\max}$ is the maximum transmission power of the n-th MU. The BS manages the uplink transmissions of multiple single-antenna MUs by adopting the linear detection algorithm zero-forcing (ZF) [9, 23]. With the ZF detector at the BS, the signal-to-interference-plus-noise ratio (SINR) for the signal from the n-th MU is

$$\gamma_{n,t} = \frac{p_{n,t}}{\sigma^2[(\sqrt{\boldsymbol{P}_t}^T \boldsymbol{H}_t^H \boldsymbol{H}_t \sqrt{\boldsymbol{P}_t})^{-1}]_{nn}}, \tag{26.5}$$

where $\boldsymbol{P}_t = \mathrm{diag}\{\boldsymbol{p}_t\}$ is a diagonal matrix with $\boldsymbol{p}_t = [p_{1,t}, \cdots, p_{N,t}]^T$ on its main diagonal, σ^2 is the noise power, the operators $\boldsymbol{A}^T$ and $\boldsymbol{A}^H$, respectively, represent the matrix transpose and Hermitian operations, respectively, and $[\boldsymbol{A}]_{mn}$ denotes the (m, n)-th element of the matrix $\boldsymbol{A}$. For those MUs that do not offload in a given slot, their transmission power in that slot will be set to 0. The transmission data rate from the n-th MU to the BS at slot t can be expressed as [9]

$$r_{n,t} = W \log_2(1 + \gamma_{n,t}), \tag{26.6}$$

where W is the system bandwidth.

26.2.4 Computation model

Define $\mathcal{N}_t^c$ as the set of MUs that have requested the tasks at time slot t, with the requested tasks available in the cache of the MEC server, i.e., $\mathcal{N}_t^c \triangleq \{n : k_t^n \in \mathcal{K}_t^c\}$. Similarly, define $\mathcal{N}_t = \mathcal{N} \setminus \mathcal{N}_t^c$ as the set of MUs with requested tasks at time slot t unavailable in the cache. The computation tasks for the MUs belonging to the set $\mathcal{N}_t^c$ are executed in the MEC server during slot t by default. Thus the offloading decision only needs to be performed for $n \in \mathcal{N}_t$.

Denote $x_{n,t} \in \{0, 1\}$ as the computation offloading decision variable of the nth MU at slot t, where $n \in \mathcal{N}_t$. Specifically, if $x_{n,t} = 0$, then the n-th MU decides to execute its task locally, and if $x_{n,t} = 1$, then n-th MU decides to offload the data of its current computation task to the MEC server via the wireless link. The computation offloading decision vector for all MUs is denoted by $\boldsymbol{x}_t \triangleq [x_{n,t}]_{n\in\mathcal{N}_t}$.

26.2.4.1 Local execution

In the local execution approach, the n-th MU executes its computation task k_t^n locally using its own CPU. Denote f_n^l as the computation capability (in CPU cycles per second) of the n-th MU. Computational capabilities may differ across various MUs. The computation time of task k_t^n by local execution can then be expressed as

$$T_{n,t}^l = \frac{d_{k_t^n}}{f_n^l}. \tag{26.7}$$

The corresponding energy consumption is

$$E_{n,t}^l = \zeta_n d_{k_t^n}, \tag{26.8}$$

where the coefficient ζ_n denotes the energy consumption per CPU cycle, which depends on the chip architecture at the MU. In this chapter, we set $\zeta_n = 10^{-27}(f_n^l)^2$ [24].

26.2.4.2 MEC server execution

In this approach, the MEC server executes the computation task on behalf of the MU. This approach consists of three steps. First, the n-th MU uploads its task data of size $b_{k_t^n}$ to the BS through the wireless channel, and the BS forwards that data to the MEC server. Second, the MEC server allocates part of its computational resources to execute the task. Finally, the MEC server returns the execution results of the task to the n-th MU.

In the first step, the transmission delay incurred by task offloading by the n-th MU during slot t can be computed as

$$T_{n,t}^x = \frac{b_{k_t^n}}{r_{n,t}}. \tag{26.9}$$

where $r_{n,t}$ is the uplink data rate of the n-th MU as shown in (26.6). The corresponding energy consumption due to transmission is expressed as

$$E_{n,t}^x = p_{n,t} T_{n,t}^x = \frac{p_{n,t} b_{k_t^n}}{r_{n,t}}. \tag{26.10}$$

In the second step during task execution, the processing delay incurred by the MEC server is computed as

$$T_{n,t}^p = \frac{d_{k_t^n}}{f_{n,t}}, \tag{26.11}$$

where $f_{n,t}$ denotes the computational resource (in CPU cycles per second) allocated by the MEC server to the n-th MU during slot t. During this step, the n-th MU is assumed to be in idle state and the energy consumption of the MU is considered as negligible. The energy consumption of the MEC server is not considered either.

Denote $\boldsymbol{f}_t \triangleq [f_{1,t}, \cdots, f_{N,t}]^T$ as the MEC computational resource allocation vector for all the MUs. The total amount of allocated resource cannot exceed the entire computational resource at the MEC server, i.e.,

$$\sum_{n=1}^{N} \mathbf{1}(x_{n,t} \neq 0 \vee n \in \mathcal{N}_t^c) f_{n,t} \leq F, \quad \forall t \in \mathcal{T}, \tag{26.12}$$

where $\vee$ represents the logical 'or' operation. It is worth noting that the MEC server will not allocate any computational resource to an MU unless it offloads its computation task to the server or the corresponding task data is available in the server cache.

In the final step, the MU downloads the output data from the MEC server. In general, the size of the computation output data is much smaller than that of the computation input data for many applications. Besides, the download data rate is in general much higher compared to the uplink data rate. Hence, similar to many studies such as [25–27], the delay and energy consumption during the final step is not considered.

26.2.5 Problem formulation

Given the computation offloading decision vector $\boldsymbol{x}_t$, the energy consumption and computation delay for the n-th MU can be calculated, respectively, as

$$E_{n,t} = \mathbf{1}(n \notin \mathcal{N}_t^c)[\mathbf{1}(x_{n,t} = 1)E_{n,t}^x + \mathbf{1}(x_{n,t} = 0)E_{n,t}^l], \tag{26.13}$$

$$T_{n,t} = \mathbf{1}(n \notin \mathcal{N}_t^c)[\mathbf{1}(x_{n,t} = 1)(T_{n,t}^x + T_{n,t}^p) + \mathbf{1}(x_{n,t} = 0)T_{n,t}^l] + \mathbf{1}(n \in \mathcal{N}_t^c)T_{n,t}^p. \tag{26.14}$$

The overall cost of all MUs in the MEC system during slot t is defined as

$$J_t = \sum_{n=1}^{N} E_{n,t} + \sum_{n=1}^{N} \omega_n T_{n,t} + \omega_c C_t, \tag{26.15}$$

where ω_n (in W) denotes the weight parameter associated with the delay at the n-th MU, and ω_c (in J/bit) denotes the weight for the fetching cost associated with caching. The weigh parameters ω_n and ω_c control the trade-off among energy, delay, and fetching cost. Different MUs might have different requirements regarding the task execution delay. For example, for MUs prioritizing faster execution, ω_n can be set to a large value.

The objective is to minimize the long-term average cost of the MEC system, which is computed as

$$\bar{J} = \mathbb{E}\left[\lim_{|\mathcal{T}| \to \infty} \frac{1}{|\mathcal{T}|} \sum_{t \in \mathcal{T}} J_t\right], \tag{26.16}$$

where $\mathbb{E}(\cdot)$ denotes the mathematical expectation. The optimization problem is formulated as follows:

$$\mathbf{P1} : \min_{\boldsymbol{c}, \boldsymbol{x}, \boldsymbol{p}, \boldsymbol{f}} \bar{J} \tag{26.17}$$

$$\text{s.t. } (C1)\; c_{k,t} \in \{0,1\},\ \forall k \in \mathcal{K},\ \forall t \in \mathcal{T} \tag{26.18}$$

$$(C2)\; x_{n,t} \in \{0,1\},\ \forall n \in \mathcal{N}_t,\ \forall t \in \mathcal{T} \tag{26.19}$$

$$(C3)\; p_{n,t} \leq P_n^m,\ \forall n \in \{n : x_{n,t} = 1\},\ \forall t \in \mathcal{T} \tag{26.20}$$

$$(C4)\; \sum_{n=1}^{N} \mathbf{1}(x_{n,t} = 1 \vee n \in \mathcal{N}_t^c) f_{n,t} \leq F,\ \forall t \in \mathcal{T} \tag{26.21}$$

$$(C5)\; \sum_{k=1}^{K} \mathbf{1}(c_{k,t} = 1) b_k \leq D,\ \forall t \in \mathcal{T} \tag{26.22}$$

$$(C6)\; T_{n,t} \leq T_s,\ \forall n \in \mathcal{N},\ \forall t \in \mathcal{T} \tag{26.23}$$

Here, $(C3)$ represents the maximum transmission power constraint imposed on each MU that offloads computation, $(C4)$ indicates that the total amount of allocated resources cannot exceed the total computational resource at the MEC server, $(C5)$ specifies that the total amount of cached data cannot exceed the cache size at the MEC server, and $(C6)$ represents the constraint that each MU must execute its task either locally or in the MEC server within one time slot.

The optimization is performed with respect to the binary caching decision vector $\boldsymbol{c} \in \mathcal{B}^K$, the binary offloading decision vector $\boldsymbol{x} \in \mathcal{B}^{|\mathcal{N}_t|}$, the power allocation vector $\boldsymbol{p} \in \mathcal{P}^N$, and the MEC computation resource allocation vector $\boldsymbol{f} \in \mathcal{F}^N$, where $\mathcal{B} = \{0,1\}$, $|\mathcal{N}_t|$ is the cardinality of the set $\mathcal{N}_t$, $\mathcal{P} = \{u \in \mathbb{R} | 0 < u \leq P_n^{\max}\}$ is the set of real numbers, $\mathcal{F} = \{v \in \mathbb{R} | 0 \leq F\}$, and $\mathbb{R}$ is the set of real numbers.

The optimal solution to **P1** requires complete information regarding the mathematical models of the system, such as the statistical distributions of the requests and channel conditions. Such information is in general not available in a practical system. Moreover, **P1** is a mixed-integer nonlinear programming, which is challenging to solve even with all the statistical distributions. One feasible approach to overcome these challenges is to design an online solution that can efficiently make the decisions regarding caching, computation offloading, and resource allocation in real-time through interactions with the system. Therefore, instead of applying conventional optimization methods to solve the NP-hard problem **P1**, we propose a DRL-based method to find the optimal $\boldsymbol{c}, \boldsymbol{x}, \boldsymbol{p}$, and $\boldsymbol{f}$.

26.3 DRL-based solution for dynamic caching, computation offloading, and resource allocation

DRL can be considered as a combination of DNN and RL. In order to solve **P1** with a DRL-based method, we will first reformulate **P1** under the RL framework. Then the details of the proposed DRL-based solution are presented.

26.3.1 RL framework

The RL framework is usually defined based on Markov decision process (MDP) with the underlying Markov property, which states that the evolution of the Markov process in the future depends only on the present state and does not depend on the past history. MDP can be solved by using classical dynamic programming. One of the main differences between dynamic programming and RL is that the latter does not require knowledge of the underlying mathematical model of the MDP, such as the Markovian transition probabilities. The RL can implicitly learn the underlying model by interacting with the environment. Generally, the RL framework is well-suited for providing online solutions to complicated sequential decision-making problems and near-optimal solutions to large-scale MDPs where exact dynamic programming methods become infeasible [15].

The RL framework consists of an agent, an environment, and three key elements: a set of possible states $\mathscr{S}$, a set of available actions $\mathscr{A}$, and a reward function $\mathscr{R} : \mathscr{S} \times \mathscr{A} \rightarrow \mathbb{R}$. The RL agent continually learns and makes decisions through the interactions with the environment in discrete time steps. In each time step t, the agent observes the state of the environment $\boldsymbol{s}_t \in \mathscr{S}$ and takes an action $\boldsymbol{a}_t \in \mathscr{A}$. The agent's behavior is defined by a policy. Here, we consider a deterministic policy μ, which deterministically maps a state to a specific action, i.e., $\mu : \mathscr{S} \rightarrow \mathscr{A}$. After executing the action, the environment returns a scalar reward $r_t = \mathscr{R}(\boldsymbol{s}_t, \boldsymbol{a}_t)$ and makes a transition of state from $\boldsymbol{s}_t$ to $\boldsymbol{s}_{t+1}$.

The infinite-horizon discounted return is defined as the sum of all rewards ever obtained by the agent, but discounted by how far off in the future they are obtained as

$$R(\xi) = \sum\nolimits_{t=0}^{\infty} \gamma^t r_t, \tag{26.24}$$

where $\gamma \in [0, 1]$ is the discount factor, and the trajectory $\xi = (\boldsymbol{s}_0, \boldsymbol{a}_0, \boldsymbol{s}_1, \boldsymbol{a}_1, \cdots)$ is a sequence of states and actions leading to the sequence of rewards $\{r_t\}_{t=0}^{\infty}$. The action-value function $Q^{\mu}(\boldsymbol{s}, \boldsymbol{a})$ (also known as Q-function) represents the expected return under the policy μ with $\boldsymbol{s}$ as the initial state and $\boldsymbol{a}$ as the initial action,

$$Q^{\mu}(\boldsymbol{s}, \boldsymbol{a}) = \mathbb{E}[R(\xi)|\boldsymbol{s}_0 = \boldsymbol{s}, \boldsymbol{a}_0 = \boldsymbol{a}], \tag{26.25}$$

where the expectation is performed with respect to all randomness in the environment.

The goal of the RL agent is to learn the optimal policy μ^* that chooses the optimal action greedily in state $\boldsymbol{s}$ such that

$$\mu^*(\boldsymbol{s}) = \operatorname*{argmax}_{\boldsymbol{a}} Q^*(\boldsymbol{s}, \boldsymbol{a}), \tag{26.26}$$

where $Q^*(\boldsymbol{s}, \boldsymbol{a})$ is the optimal Q-function, i.e., the Q-function obtained by following the optimum policy.

In order to interpret problem **P1** in the RL framework, we define the key elements according to the system model as follows.

26.3.1.1 State

The state of a system is considered as a set of parameters that can be used to describe the system. Based on the system model presented in this chapter, the system state at an arbitrary time slot t is defined as

$$s_t \triangleq \{k_t, H_t, c_{t-1}\}, \tag{26.27}$$

where k_t is the task request vector, H_t is the channel matrix, and c_{t-1} is the caching decision in the previous slot, which in turn represents the status of the MEC server cache at the beginning of current time slot. The randomness in the system is governed by the state variables k_t and H_t. At the start of each slot, k_t is made available to the system, and H_t for the upcoming uplink transmission can be estimated by the channel reciprocity [9]. The dimension of the state vector s_t is $K + (M + 1)N$.

26.3.1.2 Action

Based on the observed system state s_t, the RL agent will choose actions a_t based on the decision variables in **P1**, i.e.,

$$a_t \triangleq \{c_t, x_t, p_t, f_t\}, \tag{26.28}$$

where c_t is the caching decision, x_t is the computation offloading decision, p_t is the MU transmission power allocation decision, and f_t is MEC server computational resource allocation decision. The dimension of the action vector a_t is $K + 3N$.

26.3.1.3 Reward

Given a particular state s_t and an action a_t at time slot t, it is evident that the overall system cost in (26.15) can be expressed as a function of the state-action pair, i.e.,

$$J_t = \mathscr{J}(s_t, a_t), \tag{26.29}$$

where the function $\mathscr{J}$ maps the state-action pair to the system cost. We define the reward function of the state-action pair (s_t, a_t) in the RL framework as

$$r_t \triangleq -\mathscr{J}(s_t, a_t). \tag{26.30}$$

It is noteworthy to mention that although RL algorithms maximize the infinite-horizon expected discounted return, these algorithms can also be used to approximate the true expected infinite-horizon non-discounted return, when the discount factor $\gamma \to 1$ [28]. Therefore, the average system cost in (26.16) will be minimized by applying the policy learned via RL agent.

It is hard to obtain the exact solution of the RL problem with high-dimensional state and action spaces by directly maximizing the Q-function. We propose to tackle this issue by obtaining an approximate solution of the RL problem using DRL with DDPG. Details are provided in the next subsection.

26.3.2 DRL-based solution with DDPG

Many approaches in RL utilize the recursive relationship of the state-action function known as the Bellman equation. Under the optimal policy μ^*, the Bellman optimality equation for the state-action function can be written as

$$Q^*(s, a) = \mathbb{E}\left[\mathscr{R}(s, a) + \gamma \max_{a'} Q^*(s', a')\right], \tag{26.31}$$

where $\boldsymbol{s}'$ denotes the next state being transitioned from state $\boldsymbol{s}$ under action $\boldsymbol{a}$. A feasible method to solve the RL problem is the well-known Q-learning algorithm [29]. Q-learning comes from a class of model-free RL algorithms known as temporal-difference (TD) learning, which combines the concepts of Monte Carlo methods and dynamic programming methods. With Bellman optimality equation as the core of the algorithm, Q-learning solves Q-function through a value iteration updating approach as

$$Q(\boldsymbol{s},\boldsymbol{a}) \leftarrow Q(\boldsymbol{s},\boldsymbol{a}) + \alpha\left[\mathscr{R}(\boldsymbol{s},\boldsymbol{a}) + \gamma \max_{\boldsymbol{a}'} Q(\boldsymbol{s}',\boldsymbol{a}') - Q(\boldsymbol{s},\boldsymbol{a})\right], \tag{26.32}$$

where $\delta = \mathscr{R}(\boldsymbol{s},\boldsymbol{a}) + \gamma \max_{\boldsymbol{a}'} Q(\boldsymbol{s}',\boldsymbol{a}') - Q(\boldsymbol{s},\boldsymbol{a})$ is the TD error and α is the learning rate. It is proven that the Q-learning algorithm converges with probability one [15].

As the dimensions of the state space and action space increase, the complexity in solving (26.32) grows exponentially, which is known as the curse of dimensionality. In order to address this issue, a DRL algorithm can be an efficient alternative. This is because the powerful function approximation properties of DNNs allow DRL algorithms to learn low-dimensional representations for RL problems. The DQN method [16] exploits the architecture of DNN to approximate the Q function with a finite number of parameters and thus to facilitate solving (26.32).

Even though DQN can successfully solve problems in high-dimensional state spaces, it can handle only discrete and low-dimensional action spaces. Specifically, when there are a finite number of discrete actions, finding optimal policy according to (26.26) is relatively simple. However, for problems with continuous action spaces such as **P1**, the action space has to be discretized before applying DQN. The discretization of the action space results in loss of precision. Moreover, as the number of discretization levels increases, the computational complexity of DQN grows exponentially.

We propose to address this challenge by applying DDPG [17], which extends DRL algorithms to continuous action space and continuous state space. In DDPG, an actor-critic approach is adopted by using two separate DNNs, where the critic network $Q(\boldsymbol{s},\boldsymbol{a}|\boldsymbol{\theta}^Q)$ approximates the Q-function, and the actor network $\mu(\boldsymbol{s}|\boldsymbol{\theta}^\mu)$ approximates the policy function μ. Here, $\boldsymbol{\theta}^Q$ and $\boldsymbol{\theta}^\mu$ are the neural network weights of the critic and actor networks, respectively. Therefore, unlike DQN, instead of running an expensive optimization subroutine for computing $\max_{\boldsymbol{a}} Q(\boldsymbol{s},\boldsymbol{a})$ each time, the DDPG algorithm approximates it with $\max_{\boldsymbol{a}} Q(\boldsymbol{s},\boldsymbol{a}) \approx Q\big(\boldsymbol{s},\mu(\boldsymbol{s}|\boldsymbol{\theta}^\mu)|\boldsymbol{\theta}^Q\big)$.

Details of the proposed solution are described in Algorithm 1. In the proposed DDPG framework, four-layer fully connected neural networks with two hidden layers are used for both the actor and critic networks. The number of neurons in the two hidden layers are $8N$ and $6N$, respectively. The neural networks use the rectified linear unit (ReLU) as the activation function for all hidden layers, while the final output layer of the actor network uses a sigmoid layer to bound the actions. Ornstein-Uhlenbeck process [30] is used to provide temporally correlated noise for action exploration, while the adaptive moment estimation (Adam) method [31] is adopted for updating the neural network parameters.

Algorithm 1 Proposed Solution using DDPG

Input: System model parameters, number of episodes $K_{\max}$, number of time steps in each episode $T_{\max}$, replay buffer size $|\mathscr{R}_B|$, mini-batch size B, learning rates for critic network α^Q and actor network α^μ, and update rates ζ^Q and ζ^μ for the target critic network and target actor network, respectively.

Initialization:
  Randomly initialize the critic network $Q(\boldsymbol{s},\boldsymbol{a}|\boldsymbol{\theta}^Q)$ and actor network $\mu(\boldsymbol{s}|\boldsymbol{\theta}^\mu)$ with weights $\boldsymbol{\theta}^Q$ and $\boldsymbol{\theta}^\mu$, respectively from a uniform distribution $[-3\times 10^{-3}, 3\times 10^{-3}]$.
  Initialize target networks Q' and μ' with weights $\boldsymbol{\theta}^{\mu'} \leftarrow \boldsymbol{\theta}^\mu, \boldsymbol{\theta}^{Q'} \leftarrow \boldsymbol{\theta}^Q$.
  Initialize the experience replay buffer $\mathscr{R}_B$ as an empty array.
for each episode $k = 1, 2, \cdots, K_{\max}$ **do**
  Randomly generate an initial state $\boldsymbol{s}_1$
  for each episode $t = 1, 2, \cdots, T_{\max}$ **do**
    Determine the decision vectors by selecting an action $\boldsymbol{a}_t = \mu(\boldsymbol{s}_t|\boldsymbol{\theta}^\mu) + \Delta\mu$ using the current policy μ and exploration noise $\Delta\mu$, which is generated by following the Ornstein-Uhlenbeck process [30].
    Execute action $\boldsymbol{a}_t$ and observe the reward $r_t = -\mathscr{J}(\boldsymbol{s}_t, \boldsymbol{a}_t)$ and the new state $\boldsymbol{s}_{t+1}$ from the simulation environment.
    Save the transition $(\boldsymbol{s}_t, \boldsymbol{a}_t, r_t, \boldsymbol{s}_{t+1})$ into the replay buffer $\mathscr{R}_B$. If $\mathscr{R}_B$ is full, discard the oldest samples.
    Randomly sample a mini-batch of B transitions $\{(\boldsymbol{s}_i^{(b)}, \boldsymbol{a}_i^{(b)}, r_i^{(b)}, \boldsymbol{s}_{i+1}^{(b)})\}_{b=1}^B$ from $\mathscr{R}_B$.
    for each transition b in the mini-batch **do**
      Compute the target value $y^{(b)}$ as

$$y^{(b)} = r_i^{(b)} + \gamma Q'(\boldsymbol{s}_{i+1}^{(b)}, \mu'(\boldsymbol{s}_{i+1}^{(b)}|\boldsymbol{\theta}^{\mu'})|\boldsymbol{\theta}^{Q'}) \tag{1.26}$$

    end for
    Update the critic network $Q(\boldsymbol{s},\boldsymbol{a}|\boldsymbol{\theta}^Q)$ by one-step gradient descent as $\boldsymbol{\theta}^Q \leftarrow \boldsymbol{\theta}^Q - \alpha^Q \nabla_{\boldsymbol{\theta}^Q} L^Q$, where the loss L^Q is

$$L^Q = \frac{1}{B}\sum_{b=1}^{B} \left[y^{(b)} - Q(\boldsymbol{s}_i^{(b)}, \boldsymbol{a}_i^{(b)}|\boldsymbol{\theta}^Q)\right]^2. \tag{1.27}$$

    Update the actor network $\mu(\boldsymbol{s},\boldsymbol{a}|\boldsymbol{\theta}^Q)$ by using one-step sampled policy gradient ascent as $\boldsymbol{\theta}^\mu \leftarrow \boldsymbol{\theta}^\mu - \alpha^\mu \nabla_{\boldsymbol{\theta}^\mu} J^\mu$, where $J^\mu \triangleq \mathbb{E}_{\boldsymbol{s},\boldsymbol{a}}[Q^\mu(\boldsymbol{s},\boldsymbol{a})]$, and

$$\nabla_{\boldsymbol{\theta}^\mu} J^\mu \approx \frac{1}{B}\sum_{b=1}^{B} \nabla_{\boldsymbol{a}} Q(\boldsymbol{s}_i^{(b)}, \boldsymbol{a}|\boldsymbol{\theta}^Q)\,\Big|_{\boldsymbol{a}=\boldsymbol{a}_i^{(b)}} \times \nabla_{\boldsymbol{\theta}^\mu} \mu(\boldsymbol{s}_i^{(b)}|\boldsymbol{\theta}^\mu). \tag{1.28}$$

    Update the target networks:

$$\boldsymbol{\theta}^{Q'} \leftarrow \zeta\boldsymbol{\theta}^Q + (1-\zeta)\boldsymbol{\theta}^{Q'} \text{ and } \boldsymbol{\theta}^{\mu'} \leftarrow \zeta\boldsymbol{\theta}^\mu + (1-\zeta)\boldsymbol{\theta}^{\mu'}$$

  end for
end for

Output: Optimal policy μ^*.

26.4 Simulation results

Simulation results are presented in this section to illustrate the performance of the proposed algorithm with DDPG. Unless specified otherwise, the default settings of the MEC system are set as follows: the number of MUs is $N = 5$, the number of antennas BS is $M = 6$, the coverage radius of the small cell BS is $d_m = 50$ m, the cache size of the MEC server is $D = 200$ Megabytes (MB), the computational resource of the MEC server $F = 5$ GHz, the CPU frequency of each MU is $f_n^l = 1$ GHz, the channel bandwidth is $W = 20$ MHz, and the duration of a time slot is $T_s = 1$ second.

At the beginning of every episode, the channel vector of each MU is initialized as $\boldsymbol{h}_{n,0} \sim \mathcal{CN}(\boldsymbol{0}, h_0(d_0/d_n)^{\beta}\boldsymbol{I}_M)$, where $h_0 = -30$ dB, $d_0 = 1$ m, the path-loss exponent is $\beta = 3$ [9], and d_n (in meters) denotes the distance from the BS to the n th MU. In each episode, the locations of MUs are randomly set such that they are uniformly scattered throughout the coverage region, and the locations are independent in different episodes. The channel vectors, $\boldsymbol{h}_{n,t}$, $\forall n \in \mathcal{N}$, are updated according to (26.4), where the channel correlation coefficient is $\rho_n = 0.95$ and the error vector is $\boldsymbol{e}_t \sim \mathcal{CN}(\boldsymbol{0}, h_0(d_0/d_m)^{\beta}\boldsymbol{I}_M)$. The MU's maximum allowed transmission power is $P_n^{\max} = 2$ W, $\forall n \in \mathcal{N}$, and the background noise power is $\sigma^2 = 10^{-9}$W [9]. The energy-delay trade-off parameters are $\omega_n = 1$ W for all MUs, and the weight for fetching cost is $\omega_c = 10^{-8}$ J/bit.

There are $K = 4$ computation tasks. The number of CPU cycles required to complete the tasks d_k is uniformly distributed between $[\bar{d} - 0.05, \bar{d} + 0.05]$ Gigacycles with $\bar{d}$ being the average computations per task. The data sizes of the computation tasks b_k is uniformly distributed between $[\bar{b} - 5, \bar{b} + 5]$ MB with $\bar{b}$ being the average data size. Unless specified otherwise, the default values are $\bar{d} = 0.5$ Gigacycles and $\bar{b} = 75$ Mb. The fetching cost of each task is assumed to be the same as the data size of corresponding task, i.e., $g_k = b_k$. Moreover, the popularity profile $\boldsymbol{\phi}_t$ is modeled by a three-state Markov chain, represented by three different popularity profiles $\boldsymbol{\phi}^{(1)}, \boldsymbol{\phi}^{(2)}$, and $\boldsymbol{\phi}^{(3)}$ [20]. These profiles are modeled by Zipf distributions with parameters $\eta_1 = 1, \eta_2 = 1.2$, and $\eta_3 = 1.5$, respectively. So, at each time slot t, the popularity profile $\boldsymbol{\phi}_t$ will follow one of these three states and each task $k \in \mathcal{K}$ will be assigned popularity ranks $z_{k,t}$ randomly. The Markov transition probabilities are given by the transition matrix

$$\boldsymbol{\tau} \triangleq \begin{bmatrix} \tau_{1,1} & \tau_{1,2} & \tau_{1,3} \\ \tau_{2,1} & \tau_{2,2} & \tau_{2,3} \\ \tau_{3,1} & \tau_{3,2} & \tau_{3,3} \end{bmatrix} = \begin{bmatrix} 0.5 & 0.3 & 0.2 \\ 0.1 & 0.6 & 0.3 \\ 0.25 & 0.35 & 0.4 \end{bmatrix}, \tag{26.33}$$

where $\tau_{i,j}$ represents the transition probability from state i to state j, for $i, j \in \{1, 2, 3\}$. Please note that these states are different from the system states defined in our problem formulation.

The hyper-parameters for training the neural networks in the proposed DDPG-based algorithm are described in Table 26.1. Figure 26.2 illustrates the performance of the system's average cost ($\bar{J}$) at different learning rates, α^Q, of the critic network

Table 26.1 Hyper-parameters for training neural networks

Parameter name	Value
Number of training episodes (K_{max})	2 000
Number of steps in each episodes (T_{max})	100
Experience reply buffer size ($\lvert \mathscr{R}_B \rvert$)	50 000
Mini-batch size (B)	128
Learning rate for critic network (α^Q)	10^{-4}
Learning rate for actor network (α^μ)	10^{-3}
Soft update rate for target networks (ζ)	10^{-3}

during the training stage. The curves are smoothed using a moving average of 50 episodes. Learning rate controls how quickly or slowly a neural network model learns a problem, i.e., updates the weights θ^Q or θ^μ. Generally, a large learning rate allows the model to learn faster, at the cost of possibly arriving on a sub-optimal solution. A smaller learning rate may yield a better or even globally optimal solution but may take significantly longer to train. Based on the results, the learning rate of 10^{-4} for the critic network yields the best performance in terms of $\bar{J}$ under the current system configuration. The performance becomes stable after about 1 400 episodes. Using a similar analysis, we found that the best learning rate for the actor network is 10^{-3}.

To evaluate the performance of policy μ^* learned by the proposed DDPG-based solution, testing results are averaged over 1 000 episodes, with each episode consisting of 100 steps. Results obtained from the proposed algorithm are compared to four baseline strategies described as follows.

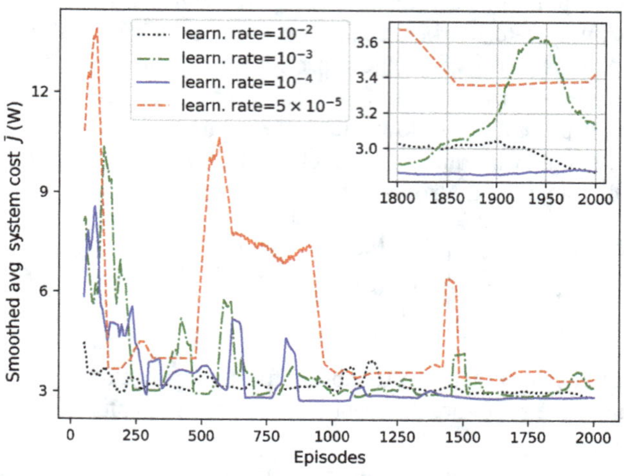

Figure 26.2 Convergence performances under different learning rates for critic network

Popularity-based caching and local execution (*PCL*): The MEC server caches the data of the computing tasks with the maximum number of requests, till reaching the caching capacity. Computational resource F is distributed equally to each MU that requests a cached task. The MUs that request uncached tasks execute their tasks by local CPU.

Popularity-based caching and full offloading (*PCO*): The MEC server caches the data of the computing tasks with the maximum number of requests, till reaching the caching capacity. All tasks are executed at the MEC server. That means, all MUs who requested uncached tasks will offload their tasks to the MEC server and transmit the corresponding data with the maximum power available. The computational resource F is distributed uniformly across N MUs.

Randomized caching, offloading, and resource allocation (*RCOR*): The MEC server caches the data of computing tasks randomly, till reaching the caching capacity. For MUs requesting uncached tasks, all the offloading and resource allocation decisions are also taken randomly.

DQN-based solution: DQN [16] can only be implemented on systems with discrete state and action spaces. The support spaces for $\boldsymbol{p}$ and $\boldsymbol{f}$ are both discretized uniformly into finite L levels each. Therefore, the size of the action space becomes $2^K(2L^2)^N$ for K tasks and N MUs. We arbitrarily set $L = 3$, train the DQN with the same values of hyper-parameters as mentioned in Table 26.1, and maintain the same neural network architecture as mentioned in Section 26.3.2. Moreover, ε-greedy exploration method is adopted for exploring the actions during network training with $\varepsilon = 0.01$.

Figure 26.3 shows the average system cost ($\bar{J}$) as a function of the computational resource capacity of the MEC server (F) with various algorithms. Due to the extra computational resources from the MEC, the performances of all algorithms improve as F increases, but with different slopes. The PCL approach has the smallest

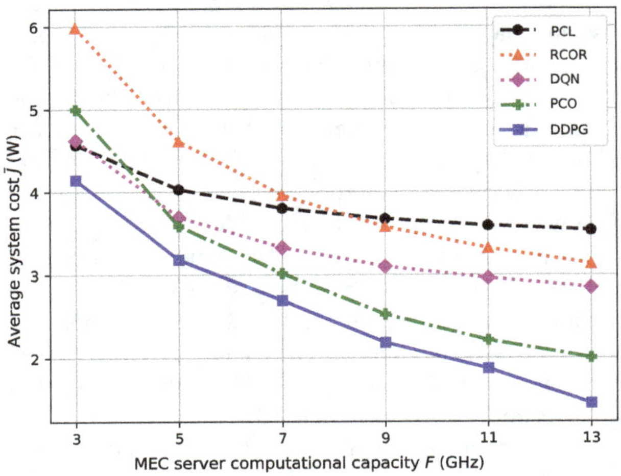

Figure 26.3 Average system cost vs. computational capacity of the MEC server

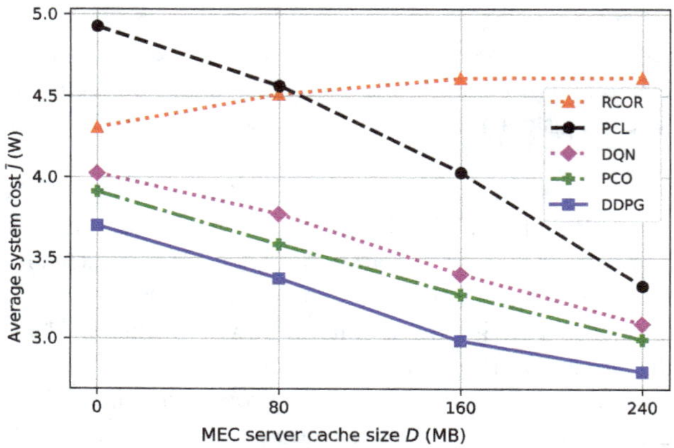

Figure 26.4 Average system cost vs. cache size of the MEC server

absolute slope because the average number of MUs using the MEC resources in each time slot is the least among all approaches. The proposed DDPG-based algorithm achieves the best performance. The performance gap between the DDPG-based algorithm and DQN algorithm becomes larger as F increases. Specifically, the performance gap increases from 10.3% when $F = 3$ to 49% when $F = 13$. This means that the DDPG-based algorithm can better utilize the MEC resources. Even though with a larger number of discrete levels L, it may be possible to get better results by the DQN approach, it will yield a very high-dimensional action space with prohibitively high computational complexity.

Figure 26.4 shows $\bar{J}$ as a function of the cache size D of the MEC server. Here, the results obtained at $D = 0$ represents the case without caching. The average system cost decreases as D becomes larger for all approaches except the RCOR strategy. A larger cache size allows the MEC server to cache more tasks, thus reducing the transmission delay and the corresponding energy consumption incurred by the MUs. However, the RCOR strategy demonstrates a different behavior because caching decision in the RCOR strategy is changed frequently in a random manner, which leads to much larger average fetching cost than other approaches. The larger fetching cost of RCOR negates the advantage of lower delay and energy cost. For the cache-assisted MEC system, the proposed DDPG-based algorithm has the best performance, followed by PCO, DQN, PCL, and RCOR, respectively.

Figure 26.5 shows the effects of different computation offloading and resource allocation approaches on $\bar{J}$, and the results are shown as functions of the average computation resources required for each task ($\bar{d}$). In all the approaches, the policy for determining the decision variables $\boldsymbol{x}, \boldsymbol{p}$, and $\boldsymbol{f}$ differs, while the caching decision variable $\boldsymbol{c}$ is obtained according to the proposed DDPG-based solution. Here, LE refers to the local execution approach as defined in the PCL strategy, FO refers to the full offload approach as defined in the PCO strategy, and ROR refers to the randomized offloading and resource allocation strategy. In all the approaches, $\bar{J}$ gets larger as $\bar{d}$ increases, since more computation-intensive tasks incur more processing delay

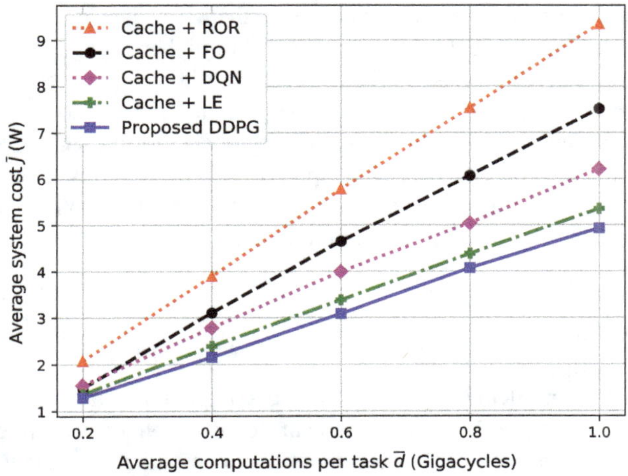

Figure 26.5 *Effect of computation offloading and resource allocation*

and energy consumption. The proposed DDPG-based solution achieves the best performance, followed by LE, DQN, FO, and ROR, respectively. The performance gaps between the proposed algorithm and all the other approaches become larger as $\bar{d}$ increases, which means the proposed solution can achieve better MEC resource allocation, especially when more computations resources are needed.

Figure 26.6 demonstrates the trade-off relationship between average energy consumption and average delay of the system. Various trade-off points are obtained by changing the values of ω_n for all $n \in \mathcal{N}$ MUs. Please note that here we consider the

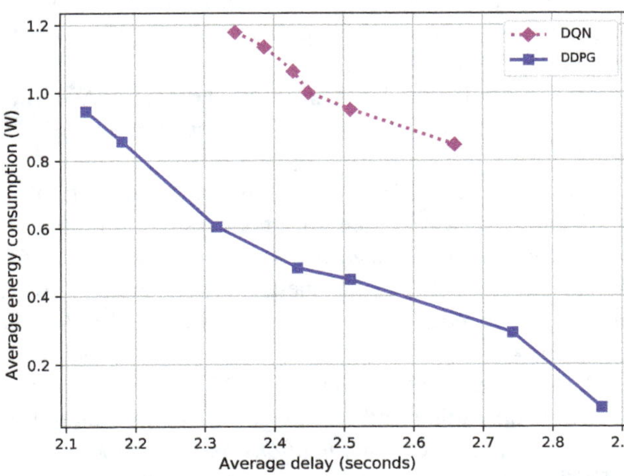

Figure 26.6 *Trade-off between average energy consumption and average delay with trade-off parameter* $\omega_n = \omega \in (0.01, 100)$ W $\forall n \in \mathcal{N}$

time-average not the ensemble average across all the MUs. The average delay experienced by all the MUS can be decreased at the cost of higher energy consumption, and vice versa. Moreover, the proposed DDPG-based algorithm shows better trade-off performance compared to the DQN-based solution, i.e., the MUs governed by the DDPG-based policy experience comparatively less average delay for the same average energy consumption. Since all the other baseline approaches have fixed policies, ω_n does not have any impact on the decision variables. Therefore, no trade-off relationships are observed in the PCL, PCO, and RCOR approaches.

26.5 Conclusion

We have studied the problem of dynamic caching, computation offloading, and resource allocation in cache-assisted MEC systems with stochastic wireless channel conditions. In the MEC system, multiple MUs execute random computation-intensive tasks either locally or remotely in an MEC server. Popular tasks can be fetched from a remote cloud and cached in the MEC server to avoid unnecessary duplicates in transmissions. We have formulated the problem under the MDP framework, which aims at minimizing the long-term average system cost that includes a weighted sum energy consumption, delay, and cache fetching cost, under the constraint of limited storage and computational resource at the MEC server. A DDPG-based algorithm has been developed to solve the problem. Simulation results have shown that the proposed algorithm outperforms other existing approaches such as DQN.

References

[1] Mao Y., You C., Zhang J., Huang K., Letaief K.B. 'A survey on mobile edge computing: the communication perspective'. *IEEE Communications Surveys & Tutorials*. 2017;**19**(4):2322–58.

[2] Hao Y., Chen M., Hu L., Hossain M.S., Ghoneim A. 'Energy efficient task caching and offloading for mobile edge computing'. *IEEE Access*. 2018;**6**:11365–73.

[3] Wang C., Liang C., Yu F.R., Chen Q., Tang L. 'Computation offloading and resource allocation in wireless cellular networks with mobile edge computing'. *IEEE Transactions on Wireless Communications*. 2017;**16**(8):4924–38.

[4] Zhang J., Hu X., Ning Z., *et al.* 'Energy-latency tradeoff for energy-aware offloading in mobile edge computing networks'. *IEEE Internet of Things Journal*. 2017;**5**(4):2633–45.

[5] Huang L., Feng X., Zhang C., Qian L., Wu Y. 'Deep reinforcement learning-based joint task offloading and bandwidth allocation for multi-user mobile edge computing'. *Digital Communications and Networks*. 2019;**5**(1):10–17.

[6] Wang J., Zhao L., Liu J., Kato N. 'Smart resource allocation for mobile edge computing: a deep reinforcement learning approach'. *IEEE Transactions on Emerging Topics in Computing*. 2019:1.

[7] Nath S., Wu J. 'Dynamic computation offloading and resource allocation for multi-user mobile edge computing'. *IEEE Global Communications Conference (GLOBECOM)*; Taipei, Taiwan; 2020. pp. 1–6.

[8] Nath S., Li Y., Wu J. 'Multi-user multi-channel computation offloading and resource allocation for mobile edge computing'. Proceedings of the IEEE International Conference on Communications (ICC). Dublin, Ireland; 2020. pp. 1–6.

[9] Chen Z., Wang X. 'Decentralized computation offloading for multi-user mobile edge computing: a deep reinforcement learning approach'. *EURASIP Journal on Wireless Communications and Networking*. 2020;**2020**(1):1–21.

[10] Liu P., Xu G., Yang K., Wang K., Meng X. 'Jointly optimized energy-minimal resource allocation in cache-enhanced mobile edge computing systems'. *IEEE Access*. 2018;**7**:3336–47.

[11] Mao Y., Zhang J., Song S.H., Letaief K.B. 'Stochastic joint radio and computational resource management for multi-user mobile-edge computing systems'. *IEEE Transactions on Wireless Communications*. 2017;**16**(9):5994–6009.

[12] Chunlin L., Zhang J. 'Dynamic cooperative caching strategy for delay-sensitive applications in edge computing environment'. *The Journal of Supercomputing*. 2020;**76**(10):7594–618.

[13] Xu J., Chen L., Zhou P. 'Joint service caching and task offloading for mobile edge computing in dense networks'. Proceedings IEEE Conference on Computer Communications (INFOCOM); Honolulu, HI, USA; 2018. pp. 207–15.

[14] Yang P., Zhang N., Zhang S. 'Dynamic mobile edge caching with location differentiation'. Proceedings of the IEEE Global Communications Conference (GLOBECOM); Singapore; 2017. pp. 1–6.

[15] Sutton R.S., Barto A.G. *Reinforcement Learning: An Introduction*. Cambridge, MA, USA: MIT Press; 2018.

[16] Mnih V., Kavukcuoglu K., Silver D., *et al.* 'Playing atari with deep reinforcement learning'. *arXiv*. 2013;**13125602**.

[17] Lillicrap T.P., Hunt J.J., Pritzel A., *et al.* 'Continuous control with deep reinforcement learning'. *arXiv*. 2015;**150902971**

[18] Arulkumaran K., Deisenroth M.P., Brundage M., Bharath A.A. 'Deep reinforcement learning: a brief survey'. *IEEE Signal Processing Magazine*. 2017;**34**(6):26–38.

[19] Shanmugam K., Golrezaei N., Dimakis A.G., Molisch A.F., Caire G. 'FemtoCaching: wireless content delivery through distributed caching helpers'. *IEEE Transactions on Information Theory*. 2013;**59**(12):8402–13.

[20] Sadeghi A., Sheikholeslami F., Giannakis G.B. 'Optimal dynamic proactive Caching via reinforcement learning'. Proceedings of the 19th International Workshop on Signal Processing Advances in Wireless Communications (SPAWC); Kalamata, Greece: IEEE; 2018. pp. 1–5.

[21] Suraweera H.A., Tsiftsis T.A., Karagiannidis G.K., Nallanathan A. 'Effect of feedback delay on amplify-and-forward relay networks with beamforming'. *IEEE Transactions on Vehicular Technology*. 2011;**60**(3):1265–71.

[22] Abramowitz M., Stegun I.A. *Handbook of Mathematical Functions with Formulas, Graphs, and Mathematical Tables*. **vol. 55**. New York: Dover; 1964.

[23] Ngo H.Q., Larsson E.G., Marzetta T.L. 'Energy and spectral efficiency of very large multiuser MIMO systems'. *IEEE Transactions on Communications*. 2013;**61**(4):1436–49.

[24] Wen Y., Zhang W., Luo H. 'Energy-optimal mobile application execution: taming resource-poor mobile devices with cloud clones'. Proceedings of the IEEE Conference on Computer Communications (INFOCOM); Orlando, FL, USA; 2012. pp. 2716–20.

[25] Chen X., Jiao L., Li W., Fu X. 'Efficient multi user computation offloading for mobile-edge cloud computing'. *IEEE/ACM Transactions on Networking*. 2016;**24**(5):2795–808.

[26] Zhang K., Mao Y., Leng S., *et al.* 'Energy-efficient offloading for mobile edge computing in 5G heterogeneous networks'. *IEEE Access*. 2016;**4**:5896–907.

[27] Li J., Gao H., Lv T. 'Deep reinforcement learning based computation offloading and resource allocation for MEC'. Proceedings of the Wireless Communications and Networking Conference (WCNC); Barcelona, Spain; 2018. pp. 1–6.

[28] Adelman D., Mersereau A.J. 'Relaxations of weakly coupled stochastic dynamic programs'. *Operations Research*. 2008;**56**(3):712–27.

[29] Watkins C.J.C.H., Dayan P. 'Q-learning'. *Machine Learning*. 1992;**8**(3):279–92.

[30] Uhlenbeck G.E., Ornstein L.S. 'On the theory of the Brownian motion'. *Physical Review*. 1930;**36**(5):823–41.

[31] Kingma D.P., Ba J. 'Adam: a method for stochastic optimization'. *arXiv*. 2014;**14126980**.

Index

achievable rate-memory pair 55–8
achievable scaling law 312–14
achievable scheme 52–5
achievable throughput-outage tradeoff 229, 231, 237
active users 419–22
adaptive caching 557
additive white Gaussian noise (AWGN) 344, 466, 572
ad hoc networks, scaling laws for 223
adverse selection 464–5
aggregate proactive caching 424
aggregate throughput 354–7
alternating direction method of multiplier (ADMM) approach 677
approximate dynamic programming (ADP) 686
asymmetric fading statistics 28
asynchronous content reuse 221, 318
auction scheme 480
auction theory 8
average aggregate throughput 348
average delivery rate derivation 36–8
AWGN: *see* additive white Gaussian noise (AWGN)

baseband unit (BBU) 372
baseline caching algorithm 511
baseline schemes 312–14
Bayesian inference approach 678
Bayesian Nash equilibrium 443
belief function 681–2
Bellman equations 543
Bernoulli request model 529–30
binary caching 288
binary Knapsack inequality 395
BinCache 373–6, 399
bisection method 392
β-mixing coefficient 519, 521, 526
Boolean model 382
broadcast bit pipes 107
broadcast channels (BCs)
  erasure broadcast channels
    achievable rate-memory pair 55–8
    achievable scheme 52–5
    preliminaries 51–2
    problem formulation 49–51
    upper bound 58–9
  error-free broadcast channels 44–7, 47–9
  Gaussian broadcast channels
    centralized cache-aided broadcasting 61–6
    decentralized cache-aided broadcasting 66–8
    lower bound 68–70
    preliminaries 60–1
    problem formulation 59–60
  Maddah-Ali and Niesen (MAN) scheme 44–7, 47–9
  numerical experiments 70–1
  overview 43

cache-aided coded multicast (CACM) 120
cache-aided error-free broadcast channel 106
cache-aided interference networks 101–3

cache allocation policy 137–8
cache-assisted multihopping gain 285
cache-assisted multihop scheme 308–10
cache content placement optimization 286
  low-complexity algorithm 303–5, 326
  problem formulation 302–3
cache-enabled unmanned aerial vehicles
  echo state networks for 713–9
    optimal location and content caching 719–22
    simulation results 722–24
cache encoder 129, 206, 212
cache-hit probability 365–7
cache-induced hierarchical cooperation 286
  cache content placement 297–8
  content delivery, capacitated graph for 298–302
  key components 296–7
  overview 321–4
  preliminaries 290–6
  scaling laws of 311–14
  throughput performance 306, 329–33
    comparison with cache-assisted multihop scheme 308–10
    under Zipf popularity distribution 307–8, 326–9
cache-induced MIMO cooperation gain 285
cache-induced opportunistic MIMO cooperation 284
cache placement codebook generation 206, 212
cache update algorithm 531–3
caching
  in base stations 644
  contents 440
  in core network 644
  delivery phase 351–4
  dynamics 498–9
  fundamental limits 2–4
  in large mobile networks 4–7
  mobile network applications 12–13
  overview 1
  over wireless channels 3–4
  placement phase 349–51
  placement strategy 345
  recommendation systems 10–12
  resource allocation 7–9
  scheme 288
  in small base stations 644
  with time-varying popularity 10–12
  in user equipments 644
caching-aware recommendations (CawR) algorithm
  complexity 594
  description 592–94
  parameters, sensitivity analysis to 596–7
caching broadcast channel with a wire and cache tapping adversary of type II (CBC-WCT II) 191
caching networks with correlated sources
  average rate-memory function 130–1
  dynamic case 141–2
    dynamic correlation-aware scheme 142–7
    greedy group coloring 147
  peak rate-memory function 130–1
  problem formulation 128–30
  static case
    Gray-Wyner multiple-request (GW-MR) 138–41
    Gray-Wyner multiple-request CACM 133–5
    Gray-Wyner network 132–3
    multiple-request cache-aided coded multicast 135–38
CACM: *see* cache-aided coded multicast (CACM)

capacity scaling law
for ad hoc networks 344
in D2D caching networks 283
in wireless ad hoc networks 282–3
capacity scaling laws 286
CawR algorithm: *see* caching-aware recommendations (CawR) algorithm
CC-CM scheme: *see* cooperative cache-aided coded multicast (CC-CM) scheme
CDF: *see* cumulative distribution function (CDF)
CDNs: *see* content delivery networks (CDNs)
cellular networks, randomised geographic caching in 383–94
cellular users (CUEs) 649–52
centralized cache-aided broadcasting 61–6
centralized caching scenario 45–6, 160
centralized networks
caching in 342–3
model 342
central processor (CP) 620
CFL condition: *see* Courant-Friedrichs-Lewy (CFL) condition
Chinese restaurant process (CRP) 493, 497
cloud-based baseband units (BBUs) 699
cloud radio access networks (CRANs) 13
with cache-enabled unmanned aerial vehicles 713
content transmission 702
overview 699–100
using clustering and caching 701
clustering scheme 225
CNN: *see* convolutional neural network (CNN)
coded caching 99, 189–90, 223
broadcast channel 199–17
combination network with security requirements 197–199
delivery phase 192–3
for error-free channels 2–3
multi-cell network (*see* Multi-cell network)
overview 17–18
placement phase 192
private 194–6
with secure delivery 191–4
single-cell network
content delivery model 18–19
equivalent content delivery rate 20
extensions 25–7
physical-layer channel 19–20
scalability 20
scalable content delivery 20–5
coded-caching-enabled multicasting 318–19
coded MapReduce 96
common service provider (CSP) 465
communication model 683–4
computing model 685–6
content caching
benchmark recommender schemes 595–6
caching-aware recommendations (CawR)
algorithm 594
parameters, sensitivity analysis to 596–7
experimental evaluation 597–8
joint caching and recommendations problem 590–2
algorithm 592–600
Joint Caching, Recommendation, and Association (JCRA) problem 603–4
overview 581–2
quality of experience 582–84
recommender systems (RS) 607
system model 584–90
user associations 601
model extension 602–3
content delivery model 18–19
content delivery networks (CDNs) 582, 583, 643
content distribution prediction 697–99

content/information-centric networking (CCN/ICN) 99
cache-aided interference networks 101–3
performance limits 103–6
separation architecture 106–13
information-theoretic outer bounds 112–13
low-SNR regime 113
network layer 109
physical layer 110–12
content life span 469–70
content popularity 255–63, 470
content provider (CP) 408, 409, 618
model 470–1
optimal price and quality 481, 482
optimization formulation 479
profits 483–86
network setup 408–09
incurred cost 410–11
proactive caching decisions 410
revenue 411–412
users' demand profiles 409–10
content replication 583
contract scheme 480
contract theory 8–9
contract theory-based approach
adverse selection 464–5
auction scheme 480
contract-based service model 469
content providers model 468–9
network service provider model 469–70
contract scheme 480
heterogeneous system 480
moral hazard 465
numerical analysis 480–7
optimal contract design and solution
constraints reduction 472–3
contract formulation 471
solution 473–8
overview 463–4
small-cell caching issues 465
Stackelberg game scheme 478–80
system model 465–6
caching procedures 468–9
network model 466–7
popularity 467–8
preferences 467–8
uniform quality scheme 478
converged networks
average downloading time 571–2
capacity improvement 570–1
extreme case analysis for 573–4
multimedia content, characterization of 567–8
numerical results 574–9
overview 565–6
performance analysis 570–74
system energy consumption 572–3
system model 566–7
traffic offloading characterization 568–72
traffic offloading effect 568
convex nonlinear program 265
convolutional neural network (CNN) 661
cooperation layer 299–300
cooperative cache-aided coded multicast (CC-CM) scheme 127, 128
coordinated multi-point (CoMP) 284, 346
correlated video content delivery
over cache-aided broadcast networks 122–3
correlation-aware cache-aided coded multicasting 123–6
distortion-memory trade-offs 126–28
correlation-aware cache-aided coded multicasting 123–6
correlation-aware cache encoder 142–4
correlation-aware CACM schemes 123
correlation-aware group coloring 145–7

correlation-aware grouping 144–5
correlation-aware multicast encoder 143–7
cost minimization 413
Courant-Friedrichs-Lewy (CFL) condition 495
coverage number 380–2
CRANs: *see* cloud radio access networks (CRANs)
CRP: *see* Chinese restaurant process (CRP)
CSP: *see* common service provider (CSP)
cumulative distribution function (CDF) 21–2, 34–6, 470, 558

data compression 4
data preprocessing 661
DDPG: *see* deep deterministic policy gradient (DDPG)
decentralized cache-aided broadcasting 66–8
decentralized caching scenario 46–7, 160
decentralized networks
  caching in 343–6
  model 343
decodability 83–4
deep convolution networks 691, 692
deep deterministic policy gradient (DDPG) 729
  algorithm 731
  DRL-based solution with 738–9
  solution 738
Deep learning
  D2X-based caching with 655–8
    mobile social media natural language processing 661–4
    performance evaluation 667–8
    problem formulation 659–61
    proposed algorithm 664–5
    system model 658–9
deep learning 555
deep neural networks (DNNs) 731, 739
deep Q-learning 686–7
deep Q-network (DQN) 555, 556, 729, 741, 745, 748
deep reinforcement learning (DRL) algorithm 617, 729
  for adaptive caching 557
  comparing policy-based 634–5
  comparing with baseline policies 635–7
  decomposed framework 624–7
  with dueling double deep Q-network 627–31
  fine-tuned hyper-parameters 633–4
  framework for joint optimization 622–23
  problem formulation 621–2
  pushing policy behavior 637–8
  simulation environment setup 631–3
deep reinforcement learning-based caching 554–7
degree of freedom (DoF) 3
degrees of freedom (DoF) 102-7, 109, 111, 284
delivery codebook generation 206–7, 212
demand shaping 408
Dempster-Shafer theory 681
Dempster's rule 682
dense network scales 282
deterministic approach 225
deterministic policy 376
device-to-device (D2D) caching networks 5
  complexity 247–49
  improved schemes 241–2
  in mobile networks 246–7
  multihop coded D2D caching networks 242–5
  multihop networks 234–9
  under realistic channel model 245–6
  scaling laws without caching 223

single-hop-coded D2D caching 239–41
single-hop networks 227
advantages 232
scaling law results 228–32
system model
network model 225
popularity model 224
scaling analysis, goal of 225
video distribution 221–2
discrete memoryless source (DMS) 159
distortion-memory trade-offs 123, 126–8
distributed caching algorithm 275–7
DNNs 555: *see* deep neural networks (DNNs)
double deep Q-network (DDQN) 617
doubly symmetric binary source (DSBS) 166, 167
downlink training phase 90
DRL algorithms: *see* deep reinforcement learning (DRL) algorithm
dueling double deep Q-network (DDQN) 627–31
D2X-based caching
with deep learning 655–7
mobile social media natural language processing 661–4
performance evaluation 667–70
problem formulation 659–61
proposed algorithm 664–7
system model 658–9
dynamic correlated content 125–6
dynamic correlation-aware cache-aided coded multicast scheme (DCA-CACM scheme) 142–4, 147, 149
dynamic correlation-aware scheme 142–7
dynamic pricing 7–8
dynamic programming (DP) 686
dynamic spectrum access (DSA) technique 406
echo state networks (ESN)
for cache-enabled unmanned aerial vehicles 713–9
optimal location and content caching 719–20
simulation results 722–24
clustering and caching, sublinear algorithm 708–11
content distribution prediction 705–7
mobility prediction 707–8
for proactive caching 700–5
simulation results 711–13
edge caching 120
mean-field game (MFG) theory 503–9
antenna pattern models 496
applications 495
caching dynamics 498–9
caching model 496
channel model 496
control 507
cost functions 500
interactions 500–1
interference analysis 494–5
interference dynamics 499
network model 495–6
numerical results 507–14
overview 493–4
stochastic differential game 502–3
user demand dynamics 496–8
user demand model 494–5
mobile
content delivery quality θ 452–3
with fixed content delivery quality 445–52, 460–1
numerical results 454–59
optimal auction design 444–45
overview 439–40
problem formulation 441, 443–4
system model 441–2
edge computing, mobile 675–6
edge nodes (ENs) 28–31
entropy 408

equivalent content delivery rate 20
erasure broadcast channels
  achievable rate-memory pair 55–8
  achievable scheme 52–5
  preliminaries 51–2
  problem formulation 49–51
  upper bound 58–9
error-free bit pipes 107
error-free broadcast channels 44–9
error-free broadcast network 110
error-free channels 2–3
ESN: *see* echo state networks (ESN)
Euclidean minimum spanning tree (EMST)-based routing 242
Evolved Multimedia Broadcast Multicast Service (eMBMS) standard 17
extended multi-cell networks 28
extended network scales 282

fair scheduling 28
Fatou's lemma 432, 433
feedback bottleneck 75, 84–5
FemtoCaching content placement 372–3
fixed content delivery quality 445–52, 460–1
Fokker-Planck-Kolmogorov (FPK) 494, 504–9
fractional caching 288
fraction feasibility (FF) constraint 444, 445
f-separable distortion 161
Fubini's theorem 432, 433
fundamental limit 477

Gács-Körner CI 170–1
Gaussian broadcast channels
  centralized cache-aided broadcasting 61–6
  decentralized cache-aided broadcasting 66–8
  lower bound 68–70
  preliminaries 60–1
  problem formulation 59–60
generalized assignment problem (GAP) 604, 605
genie-aided lower bound 170–1, 174–5, 175–7
global caching gain 104, 189, 244
Gray-Wyner coding scheme 4
Gray-Wyner decoder 133
Gray-Wyner encoder 133
Gray-Wyner multiple-request (GW-MR) 132, 138–41
  optimality 140–1
  rate-memory functions 133–5
  for two files and K receivers 138–41
Gray-Wyner multiple-request CACM 133–5
Gray-Wyner network 132–3
Gray-Wyner source coding system 164–5
Greedy group coloring 147
GW-MR: *see* Gray-Wyner multiple-request (GW-MR)

Hamilton-Jacobi-Bellman (HJB) 494, 506
Hamming distortion 155, 161, 166
harmonic broadcasting 229–30
helper stations 371
heterogeneous small cell networks (HSCNs) 657
hierarchical cache content placement 296–8, 317–18
hierarchical cooperation 282
HJB: *see* Hamilton-Jacobi-Bellman (HJB)
HSCNs: *see* heterogeneous small cell networks (HSCNs)
hyper deep Q-network 558

imperfect popularity information (IPI) 511

incentive compatibility (IC) constraint 443–5, 471
independent delivery mechanism 318
independent random caching 228
independent reference model (IRM) 383
individual rationality (IR) constraint 443–5, 471
information-theoretic outer bounds 112–13
infrastructure provider (InP) 464
initialization phase 604
in-network caching 676, 677–8
interference-alignment gain 104
interference analysis 494–5
interference avoidance scheme 225
interference dynamics 499
internet traffic 120
inter-SBS interference 501–2
inverse popularity pairing order (IPPO) algorithm 655
IPPO algorithm: *see* inverse popularity pairing order (IPPO) algorithm
Irving's stable fixture (ISF) algorithm 657, 665, 669
iterative phase 604
ITLinQ scheduling approach 249

JCRA problem: *see* Joint Caching, Recommendation, and Association (JCRA) problem
joint caching 11–12, 85–93
Joint Caching and Association (JCAP) problem 605–6
Joint Caching and Recommendations Problem (JCRP) 591
Joint Caching, Recommendation, and Association (JCRA) problem 603–04
  heuristic algorithms 604
  Joint Caching and Association (JCAP) problem 607–6
joint optimization 622–24

Karush-Kuhn-Tucker (KKT) conditions 359, 420, 505
KKT conditions: *see* Karush-Kuhn-Tucker (KKT) conditions
K-mean clustering approach 713
k-nearest neighbor (KNN) search 634
Kolmogorov's strong law 432, 434
Kuhn-Munkres (KM) algorithm 655

Lagrangian multipliers 477
Lagrangian relaxation 388–9
layered approach 182–5
LC-U scheme: *see* local cache-aided unicast (LC-U) scheme
LDIC: *see* local downward incentive constraint (LDIC)
LDS: *see* local downloading savings (LDS)
leader-based coded multicasting scheme 241
learning-enabled network architecture 618
least frequently used (LFU) caching algorithm 595
lifetime-based model 11
linear detection algorithm zero-forcing (ZF) 735
local cache-aided unicast (LC-U) scheme 127, 128
local cache effect 271–2
local caching gain 104, 244
local downloading savings (LDS) 470, 473
local downward incentive constraint (LDIC) 472
local popularity profile 542
local upward incentive constraint (LUIC) 472–3
long run average (LRA) cost 493, 502, 511
long short-term memory (LSTM) 662–4
  caching preference algorithm 666
  classification 664

hyperparameters 663
neural network 657
sentence sentiment analysis 668
low-complexity cache content placement algorithm 303–5
L-parallel channels 22–3
LSTM: *see* long short-term memory (LSTM)
LUIC: *see* local upward incentive constraint (LUIC)

macro base station (MBS) 466, 602
Maddah-Ali and Niesen (MAN) scheme 3, 17–19, 44–6, 47–9
Markov decision process (MDP) 11, 554, 686
Markovian popularity profile evolution 541–3
maximum distance separable (MDS) code 28, 29, 346
maximum distance separable-coded caching 289, 301
Maxwell's equations 282
MBS: *see* macro base station (MBS)
MCC: *see* mobile cloud computing (MCC)
MDP: *see* Markov decision process (MDP)
mean-field caching control 508
mean-field equilibrium (MFE) 505
mean-field game-theoretic edge caching 503–9
antenna pattern models 496
applications 495
caching dynamics 498–9
caching model 496
channel model 496
control 507
cost functions 500
interactions 500–1
interference analysis 494–5
interference dynamics 499
network model 495–496
numerical results 507–14
overview 493–4
stochastic differential game 502–3
user demand dynamics 496–8
user demand model 494–5
mean-field game theory 9
mean opinion score (MOS) model 717
MEC: *see* mobile edge computing (MEC)
memory-rate tradeoff 342
MFE: *see* mean-field equilibrium (MFE)
MISO-BC: *see* multiple-input single-output broadcast channel (MISO-BC)
mixed integer non-linear optimization problem 653
mmWave propagation channel 714
MNO: *see* mobile network operator (MNO)
mobile cloud computing (MCC) 730
mobile edge caching 615
numerical results 458–63
optimal auction design 444–5
content delivery quality θ 452–3
with fixed content delivery quality 445–52, 460–1
overview 440–1
problem formulation 441, 443–4
system model 441–2
mobile edge computing (MEC) 13, 676–7, 730, 731
framework 732
integration 676–7
server execution 736–7
mobile network operator (MNO) 464, 615, 616
mobile networks
caching in 4–7
device-to-device caching in 246–7
mobile social networks (MSNs)
cache model 684–5
communication model 683–4
computing model 685–6
D2D communications 677–8

formulation as reinforcement learning problem
deep Q-learning 686–7
reward function 689–90
system action 688–9
system state 688
in-network caching 676, 677–8
mobile edge computing 676–7, 676–7
network model 683
simulation
results 692–4
settings 691–2
social trust scheme 678–81
trust evaluation
from direct observations 679–81
from indirect observations 681
mobile users (MU) 463–70, 731, 732, 734, 736, 740
mobility prediction 707–8
Monte-Carlo simulation 35
MovieLens dataset 597
MSNs: *see* mobile social networks (MSNs)
multi-antenna cache-aided channels
feedback bottleneck 85
joint caching 85–93
overview 75–6
setting and notation 80–1
subpacketization bottleneck 76–80
extensions 84
impact and insight 81–2
intuition 84
scheme description 82–4
multi-antenna systems
feedback bottleneck of 84–5
single-antenna system *vs.* 82
multicast "bit pipes," 100
multicast decoders 129–30
multicast encoder 129
multi-cell network 28–9
numerical results 38–40
performance analysis 31
average delivery rate derivation 36–8
outage probability derivation 34–6
partial zero-forcing detection 31–4
system model 29–31
multihop coded D2D caching networks 242–5
multihop networks 234–9
multihop transmission gain 244
multimedia content 567–8
multiple-input multiple-output (MIMO)
techniques 6
transmission 246
multiple-input single-output broadcast channel (MISO-BC) 75, 76, 79, 80, 81, 84, 86, 87, 88, 89
multiple-request cache encoder 133
multiple-request CACM 135–38
multiple-request multicast decoder 133
multiple-request multicast encoder 133
multiplexing-backhaul tradeoff 358, 361
multiplicative real-DoF gains 82
MZipf distribution 224, 228, 233

Nash equilibrium (NE) 494
Natural Language Toolkit (NLTK) 663
Netflix 582
network caching
with space-time popularity dynamics 551–2
deep reinforcement learning-based caching 554–7
modeling statement 552–3
numerical tests 557–9
problem statement 552–4
two-timescale problem formulation 553–4
network layer 109–10
network model 683

network multiple access problem 354
network service provider (NSP) 464
  model 469–70
  optimization formulation 479
  profits 483–6
NLTK: *see* Natural Language Toolkit (NLTK)
node mobility 398–9
non-overlapping multicast bit pipes 107
NSP: *see* network service provider (NSP)

offloading loss 523, 532
open directions 606
optimal auction
  for content caching 459
  content delivery quality θ 453–4
  design 444–5
  with fixed content delivery quality 445–2, 460–1
  mechanism 439, 441
optimal caching 256, 257, 263
  allocation scheme 183
  amount 505
  space allocation mechanism 454
  via Q -learning 544–6
optimal content densities 265–7
optimal payment mechanism 454
ordered cache allocation (OCA) algorithm 183
Ornstein-Uhlenbeck (OU) process 493
outage probability 34–6, 346–7, 357, 364–5
overlaid 2-network Model 382

PAC: *see* probably approximately correct (PAC)
partial differential equation (PDE) 494
partial zero-forcing (PZF) detection 31–4
PCPs: *see* Poisson cluster processes (PCPs)
per-node capacity 256–64, 268–74
physical layer (PHY) 110–12, 300–1
  caching 284–5, 312
  model 319–20
point-to-point communication system 282
Poisson cluster processes (PCPs) 245
Poisson point process (PPP) 6, 223, 236, 245, 466, 495
Poisson request model 530–1
popularity-based caching and full offloading (PCO) 745, 746
popularity-based caching and local execution (PCL) 745, 746
popularity dynamics 10–11
popularity skewness parameter 306
PPP: *see* Poisson point process (PPP)
preference distortion tolerance 581, 591, 610
private coded caching 190
proactive caching
  content provider network setup 408–09
    decisions 410
    incurred cost 410–11
    revenue 411–412
    users' demand profiles 409–10
  multi-user scenario 417
    distributed implementation 423–4
    iterative pricing and proactive download 418–9
    numerical results 424–7
    performance analysis 419–22, 428–30
  overview 406–8
  single-user scenario 412–13
    distributed implementation 416
    iterative pricing and proactive download 413–14, 430–7
    numerical results 416–17
    performance analysis 414–16
probability-generating function 384

probability mass function (PMF) 224
probably approximately correct (PAC) 521
profit maximization 413
proportional cache allocation (PCA) algorithm 183
protocol model 282, 344
pushing 619–20
pushing policy 637–8

Q-learning 11, 544–6
quality of recommendations (QoR) 609
quality of service (QoS) 609
quantize-remap-and-forward (QMF) scheme 322
quasi-static fading channels 21–2

Rademacher complexity 527–9
radio access network (RAN) 3, 284
[RandCache] 374, 391
  with cache overflow constraint 397–8
  evaluation 393–4
  general solution 384
  Lagrangian relaxation 388–9
  problem solution 387–8
  relaxed primary problem decomposition 389–90
[RandK1F2]
  solution 384–7
  verification 393
random aggregate popularity caching with greedy constrained coloring (RAP-GCC) 185
random fractional caching with Greedy constrained coloring (RF-GCC) scheme 127
random independent placement policy 376–9
randomised geographic caching 373
  beyond independent caches 399–401
  in cellular networks 383–94
  competition among content providers 399
  FemtoCaching content placement 372–3
  node mobility 398–9
  overview 371–2
  random independent placement policy 376–9
  randomised approach 373–6
  treating unequal file sizes 394–8
  typical user 379–80
    content catalog and popularity 383
    coverage number 380–2
randomised homogeneous policies 376–7
randomized caching, offloading, and resource allocation (RCOR) 745, 746
rate-distortion (RD) theory
  functions 157–8
  K-user systems with uncorrelated sources 179
    cache allocation across users 183–4
    distortions minimizing 184
    layered approach 180–3
    optimal cache allocation scheme 183
    scalable source coding 179–80
    uncoded placement-linear delivery scheme 183
  overview 157–9
  single-user systems 163–9
    genie-aided lower bound 170–1
    Gray-Wyner source coding system 164–5
    super-user lower bound 171–2
    typical properties 172–3
    Wyner common information 171–2
  system model, description 158–63
  two-user systems 173–4

genie-aided lower bound 175–7
and two files 178–79
RCOR: *see* randomized caching, offloading, and resource allocation (RCOR)
recommendation 619–20
recommendation amendment phase 604
recommender systems (RS) 581, 607, 610
rectified linear unit (ReLU) 633
recurrent neural network (RNN) 662
reinforcement learning (RL)
basic terminology 618
concepts 618
framework 738–9
learning-enabled network architecture 618
overview 537–8, 615–7
single-node caching 539–40
caching strategies 540–1
costs 540–1
formulation 543
Markovian popularity profile evolution 541–3
numerical tests 549–51
optimal caching via Q -learning 544–6
optimality conditions 543–4
scalable caching 546–7
relaxed primary problem decomposition 389–90
remote radio heads (RRHs) 13, 699–704
request delay information (RDI) 10
RF-GCC scheme: *see* random fractional caching with Greedy constrained coloring (RF-GCC) scheme
RNN: *see* recurrent neural network (RNN)
routing layer 299
routing user content requests 583
RRHs: *see* remote radio heads (RRHs)
Ruzsa-Szeméredi graph-based design 248

sBSs: *see* small-cell base stations (sBSs)
scalable caching 546–7, 558
scalable content delivery 20–5
scalable Q-learning 549
scalable source coding 179–81
scalable video coding (SVC) 122
scaling laws
for ad hoc networks 223
of cache-induced hierarchical cooperation 311–14
in extended networks
converse proof 314–17, 333–5
main converse results 314
system scaling regime 310–11
extension
to coded-caching-enabled multicasting 318–19
to more general PHY model 319–20
for n≠M 317–18
limitation 320
notations 227
order notation 283
single-hop networks 228–32
secrecy preservation 4
security analysis 207–8, 212
semi-static wireless D2D caching network 301
sentence preprocessing 661–2
sentence sentiment analysis 662–3, 667
separable distortion functions 161
separation architecture 106–13
information-theoretic outer bounds 112–13
low-SNR regime 113
network layer 109–10
physical layer 109–12
separation-based architecture 100
service providers (SPs) 464

signal-to-interference-and-noise-ratio (SINR) 26, 380–2, 467, 499
single-antenna system
  vs. multi-antenna system 82
single-cell network
  content delivery model 18–19
  equivalent content delivery rate 20
  extensions 25–7
  physical-layer channel 19–20
  scalability 20
  scalable content delivery 20–5
single-hop-coded D2D caching 239–41
single-hop networks 227
  advantages 232
  scaling law results 228–32
single-node caching 539–40
  reinforcement learning
    caching strategies 540–1
    costs 540–2
    formulation 543
    Markovian popularity profile evolution 541–3
    numerical tests 549–51
    optimal caching via Q -learning 544–6
    optimality conditions 543–4
    scalable caching 546–7
single-source rate-distortion function 157
single-user rate-distortion systems 163–9
  rate-distortion (RD) theory
    genie-aided lower bound 170–1
    Gray-Wyner source coding system 164–5
    super-user lower bound 171–2
    typical properties 172–3
    Wyner common information 171–2
SLLN: *see* strong law of large numbers (SLLN)
small base stations (SBSs) 602
small-cell base stations (sBSs) 464–70
  Bernoulli request model 529–30
  cache update algorithm 531–3
  main results 525–28
  numerical results 531–3
  overview 519–20
  Poisson request model 530–1
  problem statement 523
  random caching strategy 524–5
  user request model 522–3
small-cell caching system 465
small-scale Rayleigh fading 31
social networking-driven caching
  characteristics consideration 646–8
  concepts 643–5
  organization 646
  overview 643
  socially aware caching networks 645–6
  socially aware caching scheme 650–55
  socially enabled D2D caching 648–50
social trust scheme 678–82
source determination layer 299
space-time popularity dynamics
  network caching with 551–3
    deep reinforcement learning-based caching 554–7
    modeling statement 552–4
    numerical tests 557–9
    problem statement 552–4
    two-timescale problem formulation 553–4
squared-error distortion 155
Stackelberg game scheme 478–80
stochastic approach 225
stochastic differential game 502–3
strong law of large numbers (SLLN) 453
subgradient method 388
sublinear algorithm 708–11
subpacketization bottleneck 75, 76, 79–81
  extensions 84

impact and insight 81–2
intuition 84
scheme description 88–92
super-user lower bound 171–2
SVC: *see* scalable video coding (SVC)
synthetic datasets 600
system energy consumption 572–3
system reward 689–91

tailored method 391–3
task caching 734
TDMA: *see* time division multiple access (TDMA)
telco Content Delivery Networks (telco CDNs) 583
temporal difference error 547
temporal-difference (TD) learning 741
TensorFlow 700
three-phase cooperative transmission scheme 321–3
throughput analysis 286
throughput-outage tradeoff 238, 343, 360
throughput scaling law 357–8
time division multiple access (TDMA) 78
approach 111
scheme 235, 295
time-varying popularity 10–12
Trace-Driven simulations 598–600
traffic offloading
characterization 568–70
effect 568
traffic pattern 290
transmission distance 267
transmission policy 345
transmit power analysis
centralized cache-aided broadcasting 65–6
decentralized cache-aided broadcasting 68
transmitted vectors design 89–90
treating-interference-as-noise (TIN) 321, 342
tree-graph-based content delivery 296, 318
Tweets ranking 665, 667
two-sided matching game 666–7
two-timescale problem formulation 553–4
two-user rate-distortion systems 173–4
genie-aided lower bound 175–7
and two files 178–80

UAVs: *see* unmanned aerial vehicles (UAVs)
ultra-dense edge caching network (UDCN) 493, 494, 497, 511
ultra-dense network (UDN) 9, 493, 499, 501, 502
unbounded-distortion (UD) scheme 595
uncoded placement-linear delivery scheme 183
unicast bit pipes 107
uniformly random caching 510
uniform permutation traffic 291
uniform quality scheme 478
unmanned aerial vehicles (UAVs) 13
cache-enabled 713–9
content caching 719–22
optimal location 719–22
overview 699
simulation results 722–24
uplink training phase 88
user clustering 351, 353
user demand dynamics 496–8
user demand model 494–5
user preference 617
user request 619–20
classification 577–9
model 522–3
traffics, impact of 577
users' demand profiles 409–10
users-RRH association 719–20

WiFi 409
wireless ad hoc networks 282–3
wireless caching technology 463
wireless channel model 290
wireless communication network 100
wireless device-to-device caching networks
  achievable scaling law and throughput 347–9
  capacity scaling law 283
    in wireless ad hoc networks 282–3
  in centralized networks 340–1
  in decentralized networks 341–4
  delivery phase 351–4
  motivations 344
  numerical results 361–3
  organization 287
  overview 281–2
  performance
    aggregate throughput 354–7
    optimized throughput 359–61
    outage probability 357
    throughput scaling law 357–9
  physical layer caching 284–5
  placement phase 349–51
  problem formulation 344–7
  results 285–6
  system model 287–90, 344–7
  time line 288
wireless network capacity 257–8
  caching 255–6
  caching algorithms 275–8
  content access 258–9
  network model 258
  overview 255–7
  under uniform content popularity
    lower bound 261–2
    numerical results 262–3
    upper bound 260–1
  under Zipf content popularity 263
    different parameters, influence of 269–71
    local cache effects 271–2
    minimum transmission distance 267
    numerical results 272–5
    optimal content densities 265–7
    per-node capacity 268–9
    problem formulation 263–5
  Zipf distribution 258–9
wiretap channel (WTC) II 190–1, 199
worst-case RDM function 162, 166, 168
worst-case user preference distortion 589
Wyner common information 171–2

zero-distortion (ZD) scheme 595
zero-forcing beamforming (ZFBF) 713
zero-forcing dirty paper coding (ZF-DPC) 701–3
zero-one linear programming (ZOLP) 305
Zipf content popularity
  wireless network capacity under 263
    different parameters, influence of 269–71
    local cache effects 271–2
    minimum transmission distance 267
    numerical results 272–5
    optimal content densities 265–7
    per-node capacity 268–9
    problem formulation 263–5
Zipf distribution 5, 6, 258–9
Zipf popularity distribution 283, 286

www.ingramcontent.com/pod-product-compliance
Lightning Source LLC
LaVergne TN
LVHW011535200926
848484LV00001B/269